Mathematical Methods and Theory in Games, Programming, and Economics

VOLUME I:
MATRIX GAMES, PROGRAMMING, AND MATHEMATICAL ECONOMICS

VOLUME II:
THE THEORY OF INFINITE GAMES

SAMUEL KARLIN
Department of Mathematics, Stanford University

TWO VOLUMES BOUND AS ONE

DOVER PUBLICATIONS, INC.
New York

Published in Canada by General Publishing Company, Ltd., 30 Lesmill Road, Don Mills, Toronto, Ontario.
Published in the United Kingdom by Constable and Company, Ltd., 3 The Lanchesters, 162–164 Fulham Palace Road, London W6 9ER.

This Dover edition, first published in 1992, is an unabridged and unaltered republication of the work first published by the Addison-Wesley Publishing Company, Inc., Reading, Massachusetts, in 1959.

Manufactured in the United States of America
Dover Publications, Inc., 31 East 2nd Street, Mineola, N.Y. 11501

Library of Congress Cataloging-in-Publication Data

Karlin, Samuel, 1923–
Mathematical methods and theory in games, programming, and economics / Samuel Karlin.
p. cm.
Reprint. Originally published: Reading, Mass. : Addison-Wesley Pub. Co., 1959.
Includes bibliographical references and indexes.
ISBN 0-486-67020-1 (pbk.)
1. Game theory. 2. Programming (Mathematics) 3. Economics, Mathematical. I. Title.
QA269.K37 1992
519.3—dc20 92-3905
CIP

VOLUME I:
Matrix Games, Programming, and Mathematical Economics

PREFACE

The mathematics involved in decision making is playing an increasingly important role in the analysis of management problems, in economic studies, in military tactics, and in operations research. These areas of application have led to challenging new types of mathematical structures. The development of game theory, linear and nonlinear programming, and mathematical economics marks one significant aspect of this kind of mathematics.

Essentially, these volumes are an attempt at a preliminary synthesis of the concepts of game theory and programming theory, together with the concepts and techniques of mathematical economics, into a single systematic theory. Secondarily, I hope they will prove useful as textbooks and reference books in these subjects, and as a basis for further study and research.

The coverage of these volumes is in no sense exhaustive; I have stressed most what appeals to me most. At the same time, I have tried to supply the necessary routine and basic information on which a proper understanding of game theory and programming depends. The presentation of each subject is designed to be self-contained and completely rigorous, but every effort has been made to bring out the conceptual depth and formal elegance of the over-all theory. In both volumes, the principles of game theory and programming are applied to a variety of simplified problems based on economic models, business decision, and military tactics in an attempt to clarify the key mathematical concepts involved and suggest their applicability to other similar problems.

Each chapter includes a certain amount of advanced exposition, which is generally incorporated in starred (*) sections. Elementary background material is presented in the appendixes, and more advanced background material is incorporated directly in the text.

Each chapter contains problems of various degrees of difficulty, many of them leading to extensions of the theory. Solutions to most of the problems and hints for solving others are given at the end of each Part.

The three Parts (Parts I and II in Volume I, and Volume II, which comprises Part III) are independent of each other except for the material in starred sections, and may be studied independently in conjunction with the relevant appendixes. For the reader's convenience, a diagram showing the interrelationships of the various nonadvanced portions of Parts I and II of Volume I is presented on the following page.

Each chapter ends with a section of historical remarks, general comments on technical details, and citations of books and articles that may profitably be consulted for further information. All references cited are listed in the bibliography at the close of each book. In some instances the historical remarks consist in assigning priorities to notable discoveries; any inaccuracies or omissions along these lines are wholly unintentional and deeply regretted.

My indebtedness extends to many, and not least to Stanford University, California Institute of Technology, the RAND Corporation, and the Office of Naval Research, all of which provided facilities, intellectual stimulation, and encouragement without which these books might never have been written.

Among my academic colleagues, I am grateful to Professors Bowker, Lieberman, Madow, Parzen, and Scarf of Stanford for their constant encouragement; to Professor H. F. Bohnenblust of California Institute of Technology, who introduced me to this field of study and who taught me more than I know how to acknowledge; to my students Rodriguez Restrepo and R. V. Miller, who wrote up my initial lectures on which the material of these books is based; to my students William Pruitt and Charles Stone, each of whom made valuable suggestions for organizing and improving the final manuscript; to Professors Hirofumi Uzawa and K. J. Arrow, who taught me most of what I know about mathematical economics; to my friends Melvin Dresher, Ray Fulkerson, and Harvey Wagner for their helpful comments on the first six chapters; and to Irving Glicksberg for assistance in developing the appendixes.

Finally, I am indebted to J. G. Bell, Jr., for helping to straighten out my syntax; to Frieda Gennerich, Sally Morgan, and Babette Doyle for their superb technical typing; and above all to my wife, who has offered warm encouragement, unbounded patience and kind consideration during these trying years of writing.

SAMUEL KARLIN

Stanford, California
August 1959

LOGICAL INTERDEPENDENCE OF THE VARIOUS CHAPTERS

(*applies only to unstarred sections of each chapter*)

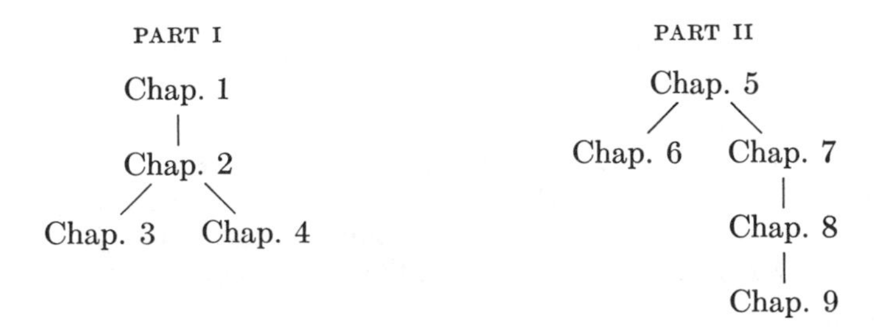

CONTENTS

INTRODUCTION

THE NATURE OF THE MATHEMATICAL THEORY OF DECISION PROCESSES

Section 1. The background. The art of making optimal judgments according to various criteria is as old as mankind; it is the essence of every field of endeavor from volleyball to logistics. The science of making such judgments, as opposed to the mere art, is a newer development, and the mathematical theory of decision-making is newer still—only a little over twenty years old. It has grown fast. Particularly in this last decade, we have seen a rapid spread of the use of scientific disciplines in economic planning, military strategy, business management, and systems engineering. There has been a steady interplay between the more careful formulation of business and logistics problems and the development of the mathematical tools needed for their solution. Along with and partly as a result of advances in game theory, linear programming, and queueing theory, there have grown up new practical disciplines whose very names were unknown forty years ago: operations research, management science, systems analysis, and a score of others. All these disciplines are concerned with the same fundamental task, that of scientifically analyzing various possible courses of action with a view to determining which course is "best" under the circumstances.

What a decision problem is must be clearly defined. Specifically, the possible courses of action, their consequences, the objectives of the various participants, and the nature of the random effects must be ascertained. From here on, the theory tries to evaluate and make comparisons of the various alternatives of action in terms of the objectives. A decision problem typically has four parts: (1) a model, expressing a set of assumed empirical relations among the set of variables; (2) a specified subset of decision variables, whose values are to be chosen by the firm or other decision-making entity; (3) an objective function of the variables, formulated in such a way that the higher its value, the better the situation is from the viewpoint of the firm; and (4) procedures for analyzing the effect on the objective function of alternative values of the decision variables.

The main component of a decision problem is the model. The following definition may help explain the significance and scope embodied in this concept. A model is a suitable abstraction of reality preserving the essential structure of the problem in such a way that its analysis affords insight into both the original concrete situation and other situations which have the same formal structure.

Model building has become increasingly widespread in these past years, particularly in the social sciences. Modern psychologists approach the study of learning processes in terms of a model involving stochastic variables. Econometricians, drawing on game theory and programming theory, have developed global equilibrium models in an attempt to understand the relationship of consumption, production, and price units in a competitive economy. Some political scientists are using the concepts of n-person game theory in an effort to analyze the power relationships that govern the decisions of legislative bodies. More and more business firms are applying rigorous mathematical thinking to management problems, inventory problems, production problems. Engineers are engaged in intensive studies in the new field of systems analysis, whose main technique consists of simulating idealized models of a given operation. The basic mathematics used in these studies is a mixture of the mathematical concepts and thinking processes of statistics, probability, and general decision theory.

The past fifty years have been years of fruitful fermentation in the disciplines of probability, stochastic processes, and statistics on the one hand, and have witnessed the development of game theory and linear programming on the other. These mathematical disciplines are the natural tools for the analytical investigation of problems arising in the social and managerial sciences. In this respect it is interesting to compare briefly the usefulness of probability, statistics, and decision theory with that of the other types of pure and applied mathematics.

The development of classical mathematics was principally inspired by problems based on phenomena in the fields of physics and engineering. In the usual classical physics problem, the variables of the system are seen to satisfy a system of well-defined deterministic relations. All parameters of the problem can be assumed known or accurately measurable, and one seeks to determine the relations between them in order to predict exactly what will happen in similar or analogous circumstances in the future. Until recently scientists were primarily concerned with just such deterministic problems, problems that could be solved, by and large, by the methods of partial differential equations and related mathematical techniques.

Clearly, however, these deterministic concepts and techniques are not altogether suitable for the social sciences, which typically present problems involving uncertainties and variability. Whenever human beings play a crucial role in the formulation of a model, we must allow for the unknown and sometimes random character of human actions, and for the randomness inherent in the environment in general, if we are to arrive at meaningful results. It is therefore inevitable that students of human behavior should be concerned prominently with the theories of probability, statistics, and decision analysis of the kind embodied in game theory and linear programming.

The reverse is also true: mathematicians and statisticians are increasingly concerned with the analytical problems originating from the analysis of human behavior. It is now no longer exclusively the physical sciences and engineering, but also the social and management sciences, that signal promising new directions for mathematical research.

There are two aspects to this research: the construction of the model, in which probability and stochastic processes play a fundamental role, and the normative essence of the model (optimization), the province of statistics, game theory, decision theory, and operations research. In practice these two aspects are inseparable. Obviously, to understand how to optimize the objective function, it is necessary to understand the structure and mechanism of the model; and to construct a model properly, it is necessary to understand what use is to be made of it.

In order to properly delimit the scope of these researches, we also mention the descriptive significance in the study of mathematical model building. This is in contrast to the normative essence of the model which we have emphasized. From the standpoint of mathematical models in psychology and sociology, the bulk of current literature is concerned with the orientation of the model toward empirical adequacy. On the other hand, the model based on situations in management, economics, and statistics is more concerned with the kind of normative questions of rational behavior theory (game and decision theory) to which we devote our main attention.

Inevitably the techniques of probability, statistics, and decision theory interact, and developments in each lead to developments in the others. As a result, the determination of optimum procedures for decision problems must be considered not in terms of half a dozen, or three, or two more or less separate sets of techniques, but in terms of a single over-all analytical methodology.

It is this methodology, in the context of game theory, programming, and mathematical economics, that we seek to delineate in this book. We may summarize our aims as an attempt to unify the mathematical techniques of the field and to help crystallize the concepts underlying these kinds of decision problems. It is not our purpose to propose a formal structure which will encompass all problems in the areas covered. We do not concentrate on definitions and abstractions *per se*. Rather, we attempt to develop in detail the principal mathematical methods that are applicable, putting emphasis wherever possible on mathematical prototypes and their analyses.

Section 2. The classification of the mathematics of decision problems. Our purpose in this section is to describe various useful mathematical classifications of decision problems that are also applicable to the theory of games, programming, and mathematical economics.

I. Decision problems can be divided into two contrasting types—static problems and dynamic problems. A static decision problem is one in which the variables do not involve time explicitly. A dynamic model is one in which time plays a very decisive role.

The distinction between static and dynamic problems is not always clear. Often a static situation is a stationary state of a dynamic situation, i.e., an equilibrium phase achieved by a dynamic process that has been in operation over a long period of time. Or an ostensibly dynamic process can be regarded as static, as when the same variables are introduced at the successive time points as new variables. Nevertheless there is an intrinsic difference in concept and technique between the two kinds of problems which will be manifest throughout the detailed analysis.

The principal technical difference between dynamic (multistage) models and static models is the degree of complexity involved in describing a given strategy or procedure. In a static situation a strategy is selected once and for all to be carried out directly, whereas the strategies available in the dynamic situation are usually complicated functions of information received and actions undertaken in the preceding stages. Theoretically, by defining a strategy suitably, we can reduce any dynamic model to a static model. In practice, however, this operation so complicates the nature of the strategy that it is often impossible to analyze the problem in the resulting terms. The essence of the analysis of a dynamic problem seems to be the exploitation of the recursive, time-dependent nature of the problem and the symmetries made available by the infinite horizon of time.

The significance of dynamic decision problems is obvious, since almost all human activity takes place in time. Events and learning processes are sequential. Most business decisions involve a time element. Economists try to take account of dynamic fluctuations in time. Nonetheless, the recognition and mathematical analysis of dynamic decision problems are of recent origin. A notable innovator in this regard was Wald, whose Sequential Analysis of Statistical Data [244]* contained the essence of the notion of sequential decision-making for the accumulation of information. Other more special contributors to the sequential decision problem include Bellman [19], who saw the generality of the sequential idea and used it in the formulation of several multistage management decision models, and Arrow, Harris, and Marschak [8], who cast a type of inventory decision problem in dynamic form. The theory of multistage games is an extension of the usual theory of games to a dynamic situation. A recent volume [72] shows the importance of this development.

The study of static models serves two purposes. First, a static model may be the best way of describing a static problem whose solution is inter-

---

* References are to the Bibliography.

esting in itself and affords insight into the workings and structure of a given operation. One common static problem of this sort is that of optimizing a given criterion among different procedures where the variables range over a prescribed set in which the time variable is not a special variable. Second, from a technical point of view, many dynamic models must be approached through the study of modified static problems. For example, it is often convenient to reduce the infinite version of the dynamic problem to the case of a finite number of periods N. The solution of the resulting problem is obtained by a simple extension of the static model, after which a passage to the limit on N produces the solution of the full dynamic problem.

II. Decision problems can also be classified in terms of the number of antagonistic participants involved. We note three examples.

(1) If there are at least two irreconcilable participants, we are dealing with problems in the theory of games. Where there are just two such opponents, their objectives are clear: each strives to achieve as great a return as possible, knowing that his opponent will strive for the same objective with all his ability and intelligence. Where there are more than two, their objectives are not so clear. The conceptual framework of this model has not yet been well established, and therefore it is difficult to develop appropriate mathematical methodology for it.

(2) Where there is only one participant involved and his objectives are clearly defined, the basic decision problem reduces technically to a strict maximization of the objective functions subject to the natural constraints of the model. Generally, this situation leads to variational problems, with the solution usually occurring on the boundaries of the domain of the variables. To solve such extreme constrained problems entails developing new techniques. The fields of linear and dynamic programming abound with models of this kind.

(3) Where there is one participant and uncertainties are present, a solution can usually be obtained by combining the methods of statistics with variational techniques.

III. A third breakdown of the general decision model is into deterministic and stochastic cases. Naturally the deterministic models are conceptually and technically much simpler and more frequently amenable to thorough analytical treatment. However, the majority of practical problems involve uncertainties, variability, and all kinds of randomness. To solve problems of this sort requires a thorough acquaintance with the theory of stochastic processes and the corresponding theory of inference in stochastic processes.

IV. Another classification can be made in terms of the complexity of the available strategies (the strategy space). Strategy spaces can be divided into two categories: finite-dimensional and infinite-dimensional. In many of our examples, the infinite-dimensional space is identifiable with a subset of some classical function space. When this is so, the specific structure of this function space may be employed in seeking a solution. This method is used to solve so-called games played over the unit square; it is also useful in the more general finite-action statistical decision problems.

V. Finally, a classification can be made in terms of subject matter: i.e., into game theory, linear programming, multistage decision problems, statistical decision theory, etc. Although many of these subject areas overlap both conceptually and technically, each has certain predominant features which can be used to solve problems that are not amenable to solution by the others.

It is impossible to treat all kinds of decision problems adequately in a single work. We have accordingly devoted these two volumes to problems of game theory, programming, and mathematical economics, leaving some aspects of statistical decision theory and a discussion of dynamic stochastic decision problems to other books now in the making.

Broadly speaking, then, the present two volumes deal primarily with static decision problems in which uncertainties and randomness in the structure of the model are not central. We devote Part I (four chapters) to finite game theory, and Part II (five chapters) to linear programming, nonlinear programming, and mathematical economics. The affinity of mathematical economics to programming methods, concepts, and applications justifies this grouping. In the main these two parts treat static, finite decision problems. The concept of mixed strategy for finite games (implying an element of randomness) comes into the solution of game problems but does not appear in the original formulation of game theory.

Volume II (nine chapters) is concerned with infinite static game theory, in the sense that the strategy spaces are infinite-dimensional and the game has the basic form of involving one play.

Section 3. The main disciplines. In this section, by way of indicating how game theory and programming theory fit into the larger picture, we present a cursory description of the four main disciplines of decision theory. It should be understood, of course, that proper insight into any of these disciplines can be gained only from a detailed study of specific problems and solutions.

1. *Theory of games.* Throughout these volumes we are concerned primarily with two-person, zero-sum games. The mathematical investigations done on the theory of games involving more than two participants is of an introductory nature, primarily because of the conceptual difficulties inherent to the formulation of the model. There has been some elegant and sophisticated work on "solutions" of n-person game models [215, 169], but the techniques and concepts of the subject are relatively undeveloped

and we have therefore decided to omit them from consideration in this book. The challenge of n-person game theory is being taken up, however, and we may hope that the conceptual obstacles will soon be overcome.

In two-person, zero-sum games, each player has available a space of strategies, and there is a known payoff function which represents the yield to Player I as a function of the specific strategies employed by both players. Of course, the yield to Player II is the negative of Player I's yield.

Player I seeks to maximize his return, but affects the pay-off only through control of his own strategy; Player II desires to minimize Player I's yield, since the interests of the players are in strict conflict. This leads to the min-max principle. It is remarkable that one can define a rational mode of behavior for each of two players with irreconcilable opposing interests. The proof of this fact was the fundamental contribution of John von Neumann.

The strict conflict of interests is a realistic assumption for parlor games and military operations, and for a few special situations in business, notably duopoly. Outside of this rather limited domain, one is hard pressed to come up with a truly realistic game problem. This is of little importance, however; the fundamental contribution of game theory is not its solution of practical problems, but its clarification of the basic methods of decision theory. Advances and refinements in game theory have fundamentally influenced the development of statistics, linear and nonlinear programming, mathematical economics, and model building in general.

2. *The theory of programming.* The theory of linear and nonlinear programming is simply the mathematics of maximizing a function of several variables which are subject to constraints. As the adjective "linear" suggests, the constraints and the function maximized are often linear functions of the variables. Unfortunately, the ordinary calculus is usually not applicable to programming problems, since their solutions necessarily fall on the boundaries of the domain over which the free variables range. It is accordingly the task of the theory to devise analytical methods of determining solutions wherever possible, and in the absence of such methods to furnish efficient computational means of approximating a solution. Moderate success has been achieved in both directions.

On the theoretical side, the theory of linear programming is equivalent to the theory of discrete two-person, zero-sum games. The interaction between these two theories has benefited and accelerated the growth of both. An extension of the theory to deal with nonlinear functions has also been partly accomplished.

The applications of linear programming are numerous, especially in the field of activity analysis, which is concerned with planning production operations. A typical linear programming model is the assignment problem, which calls for the assignment of n persons of various talents to n jobs of varying importance in such a way as to maximize the total value of the work done in those jobs. So general a model obviously has any number of practical applications, and the same is true of several other key programming models. The oil industry has particularly benefited from the use of linear programming to determine efficient production procedures.

3. *Multistage decision problems.* Multistage decision problems, sometimes called dynamic programming, are problems involving a number (usually an infinite number) of decisions made in the course of time. Each decision depends on all past decisions and in turn affects all future decisions. A typical example is furnished by the inventory problem. To what level should a merchant raise his available stock in order to maximize his gain? The factors influencing his decision are ordering costs, profits, holding costs, and penalties (e.g., loss of trade, loss of good will) for not having goods available to satisfy demand. A further complication could arise from a lag in delivery of the order he places. This ordering problem comes up periodically, which makes the problem dynamic. Of course what happened in preceding periods will inevitably affect the merchant's decisions.

The same general considerations apply to planning production schedules, determining replacement policies for old equipment, releasing hydroelectric energy, and the like. Basic dynamic models are found also in the fields of sociology and psychology, notably in the study of learning processes and in latent structure analysis. Queueing models furnish almost an inexhaustible source of dynamic decision processes, with important applications to cargo loading, transportation systems, doctor's schedules, etc.

Finally, statistical sequential analysis obviously belongs to the category of multistage procedures. In fact, this is the richest subject of all from the point of view of mathematical content, subtlety, and depth, since it draws on some of the finer points of the theory of stochastic processes, statistical decision theory, game theory, and general decision methodology.

4. *Statistical decision theory.* Statistical decision theory as conceived by Wald is essentially statistics considered from a broader point of view. Almost all types of statistical analysis—testing hypotheses, estimating unknown parameters, prediction, discrimination—can be cast into this general framework, and their relevance and effectiveness can be evaluated in its terms. The spirit and mathematical procedures of statistical decision theory and the theory of games are very much the same, but their objectives differ. Explicitly, the theory of games deals primarily with two conflicting interests, whereas statistical decision theory is concerned with determining optimal procedures, with allowances for uncertainties in nature. In a sense the statistician and nature represent two players engaged in a game, but the relationship between them is not the irreconcilable antagonism upon which the theory of games is predicated.

The general outlook of statistical decision theory has helped increase our understanding of many of the standard principles utilized in statistical practice: for example, the principle of unbiasedness, type-A regions, invariance concepts, and min-max theory, which all represent criteria by which we single out a given mode of action. The classical Neyman-Pearson philosophy of testing hypotheses and estimating unknown parameters by means of confidence intervals is encompassed within the boundaries of statistical decision theory. The scope of this theory is clearly tremendous, and its future influence on the development of specific statistical theory and practice should be very great.

NOTATION

Vectors

A vector $\mathbf{x}$ with n components is denoted by

$$\mathbf{x} = \langle x_1, x_2, \ldots, x_n \rangle.$$

$\mathbf{x} \geq \mathbf{0}$ shall mean $x_i \geq 0$ $(i = 1, 2, \ldots, n)$.

$\mathbf{x} > \mathbf{0}$ shall signify $x_i \geq 0$ and at least one component of $\mathbf{x}$ strictly positive.

$\mathbf{x} \gg \mathbf{0}$ denotes that $x_i > 0$ $(i = 1, 2, \ldots, n)$ (all components are positive).

The inner product of two real vectors $\mathbf{x}$ and $\mathbf{y}$ in E^n is denoted by

$$(\mathbf{x}, \mathbf{y}) = \sum_{i=1}^{n} x_i y_i.$$

In the complex case

$$(\mathbf{x}, \mathbf{y}) = \sum_{i=1}^{n} x_i \bar{y}_i,$$

where $\bar{y}_i$ denotes conjugate complex.

The distance between two vectors $\mathbf{x}$ and $\mathbf{y}$ is denoted by $|x - y|$.

Matrices

A matrix $\mathbf{A}$ in terms of its components is denoted by $\|a_{ij}\|$.

A matrix $\mathbf{A}$ applied to vector $\mathbf{x}$ in the manner $\mathbf{xA}$ gives the vector

$$\langle \sum x_i a_{i1}, \sum x_i a_{i2}, \ldots, \sum x_i a_{in} \rangle.$$

Similarly,

$$\mathbf{Ay} = \langle \sum a_{1j} y_j, \sum a_{2j} y_j, \ldots, \sum a_{nj} y_j \rangle \cdot$$

$(\mathbf{Ax})_j$ denotes the jth component of the vector $\mathbf{Ax}$.

The transpose of $\mathbf{A}$ is denoted by $\mathbf{A}'$.

The determinant of $\mathbf{A}$ is usually designated by $|\mathbf{A}|$ and alternatively by det $\mathbf{A}$.

Convex Sets

The convex set spanned by a set S is denoted alternatively by Co $(S) = [S]$.

The convex cone, spanned by S, is denoted by P_S or $\mathcal{P}_S$.

Distributions

The symbols

$$\mathbf{x}_{t_0}, \mathbf{y}_{t_0} \quad \text{and} \quad \mathbf{I}_{t_0} \qquad (0 \leq t_0 \leq 1)$$

are used interchangeably to represent a distribution defined on the unit interval which concentrates its full mass at the point t_0, that is,

$$\mathbf{x}_{t_0}(\xi) = \begin{cases} 0 & \xi < t_0, \\ 1 & \xi \geq t_0. \end{cases}$$

The symbol

$$\sum_{i=1}^{k} \lambda_i I_{\xi_i}$$

represents the probability distribution function with jumps λ_i located at ξ_i.

Part I

CHAPTERS 1–4

THE THEORY OF MATRIX GAMES

CHAPTER 1

THE DEFINITION OF A GAME AND THE MIN-MAX THEOREM

1.1 Introduction. Games in normal form. The mathematical theory of games of strategy deals with situations involving two or more participants with conflicting interests. The outcome of such games is usually controlled partly by one side and partly by the opposing side or sides; it depends to some extent on chance, but primarily on the intelligence and skill employed by the participants. Aside from games proper, such as poker and chess, there are many conflicting situations to which the theory of games can be applied, notably in certain areas of operations research, economics, politics, and military science.

We shall first consider only two-person zero-sum games, i.e., games with only two participants (competing persons, teams, firms, nations) in which one participant wins what the other loses. It should be noted that the terms of this situation exclude the possibility of any bargaining between the participants. We shall also be concerned with two-person constant-sum games, in which the two players compete irreconcilably for the greatest possible share of the kitty. By suitable renormalization such a game may be converted into a zero-sum game.

A fundamental concept in game theory is that of *strategy*. A strategy for Player I is a complete enumeration of all actions Player I will take for every contingency that might arise, whether the contingency be one of chance or one created by a move of the opposing player. What a strategy is should not be interpreted too naively. It might seem that once Player I has chosen a strategy, every move he makes at any stage is determined, in the sense that he has in mind at the beginning of the game a sequence of moves which he will carry out no matter what his opponent does. However, what we mean by "strategy" is a rule which in determining Player I's ith moves takes into account everything that has happened before his ith turn. The reason a player does not change a strategy during a game is not that the strategy has committed him to a sequence of moves he must make no matter what his opponent does, but that it gives him a move to make in any circumstances that may arise.

It is usual, in describing a game, to regard all possible procedures, good or bad, as possible strategies. Even in simple games the number of possible strategies is often forbidding. Consider the game of ticktacktoe.* Suppose Player I makes the first move. There are nine possible positions for his first cross. Player II then has eight possible moves he can make, and what Player I does at his second opportunity to move will depend on the preceding move of Player II. In this situation there are seven possible moves he may make. Player I's third move will, of course, depend on all the preceding moves of both players; and so on.

A strategy might start off as follows: Player I's first cross is to be made in the upper right-hand square. If Player II marks either the square below or the square to the left, Player I makes his second cross in the center square; if Player II marks the center square, Player I makes his second cross in the lower left-hand square; if Player II marks any of the other five squares, Player I makes his second cross in the square immediately below the first; and thus the description goes on. Player I might even embody in his strategy the possibility of randomizing, according to a fixed probability distribution, among alternatives at a given move.

Clearly, an enormous number of possible strategies present themselves even for such a simple game as ticktacktoe. Although many of them intuitively seem to be poor strategies, we are obliged to include all the possibilities in order to give a complete description of the game. In the course of this book mathematical tools for manipulating and analyzing these large sets of strategies will be developed.

A second fundamental concept in game theory is that of the *pay-off*. The pay-off is the connecting link between the set of strategies open to Player I and the set open to Player II. Specifically, it is a rule that tells how much Player I may be expected to win from Player II if Player I chooses any particular strategy from his set of strategies and Player II chooses any particular strategy from his set. The pay-off function is always evaluated in terms of the appropriate utility units (see the notes to this section on p. 34).

We are now ready for a formal definition of a game.

A game is defined to be a *triplet* $\{X, Y, K\}$, where X denotes the space of strategies for Player I, Y signifies the space of strategies of Player II, and K is a real-valued function of X and Y. Player I chooses a strategy $\mathbf{x}$ from X and Player II chooses a strategy $\mathbf{y}$ from Y. For the pair $\{\mathbf{x}, \mathbf{y}\}$ the pay-off to Player I is $K(\mathbf{x}, \mathbf{y})$ and the pay-off to Player II is $-K(\mathbf{x}, \mathbf{y})$. We shall call K the pay-off kernel.

In the absence of a statement to the contrary, the following conditions are assumed to be satisfied throughout this chapter:

---

* Ticktacktoe is played on a 3×3 matrix grid. Players move alternately and on each turn are allowed to capture one of the remaining free squares. The first player who takes possession of three squares which are on a single horizontal, vertical, or diagonal line wins.

(a) X is a convex, closed, bounded set in Euclidean n-space E^n.

(b) Y is a convex, closed, bounded set in Euclidean m-space E^m.

(c) The pay-off kernel K is a convex linear function of each variable separately. Explicitly,

$$K[\lambda\mathbf{x}_1 + (1 - \lambda)\mathbf{x}_2, \mathbf{y}] = \lambda K(\mathbf{x}_1, \mathbf{y}) + (1 - \lambda)K(\mathbf{x}_2, \mathbf{y})$$

and

$$K[\mathbf{x}, \lambda\mathbf{y}_1 + (1 - \lambda)\mathbf{y}_2] = \lambda K(\mathbf{x}, \mathbf{y}_1) + (1 - \lambda)K(\mathbf{x}, \mathbf{y}_2),$$

where λ is a real number satisfying $0 \leq \lambda \leq 1$.

Several of these limitations will be relaxed in later chapters.

The property that is essential in Chapters 1–4 is that X and Y are convex sets and have the character of finite dimensionality. (A representation of a game as a triplet involving strategy spaces which are finite-dimensional is necessarily a restriction; its justification rests on the fact that numerous actual games are of this kind.) The identification of strategies with points in Euclidean n-space is a convenience that simplifies the mathematical analysis.

An important special class of games is obtained where X is taken as the simplex S^n in E^n, defined as the set of all $\mathbf{x} = \langle x_1, x_2, \ldots, x_n\rangle$ where $x_i \geq 0$ and

$$\sum_{i=1}^{n} x_i = 1,$$

and the space Y is the corresponding simplex T^m in E^m. The pay-off kernel then takes the form

$$K(\mathbf{x}, \mathbf{y}) = \sum_{j=1}^{m} \sum_{i=1}^{n} x_i a_{ij} y_j = (\mathbf{x}, \mathbf{Ay}),$$

where $\mathbf{A}$ is the matrix $\|a_{ij}\|$. In the case of such matrix games we shall often denote the pay-off corresponding to strategies $\mathbf{x}$ and $\mathbf{y}$ as $A(\mathbf{x}, \mathbf{y})$ in place of $K(\mathbf{x}, \mathbf{y})$ to suggest that these games are matrix games, i.e., games in which $K(\mathbf{x}, \mathbf{y}) = (\mathbf{x}, \mathbf{Ay})$.

Certain special strategies consisting of vertex points of X are denoted by $\boldsymbol{\alpha}_i = \langle 0, \ldots, 0, 1, 0, \ldots, 0\rangle$ $(i = 1, \ldots, n)$, where the 1 occurs in the ith component. These are Player I's *pure strategies*. Similarly, the strategies $\boldsymbol{\beta}_j = \langle 0, \ldots, 0, 1, 0, \ldots, 0\rangle$ of Y $(j = 1, \ldots, m)$ are referred to as Player II's pure strategies. Since $A(\boldsymbol{\alpha}_i, \boldsymbol{\beta}_j) = K(\boldsymbol{\alpha}_i, \boldsymbol{\beta}_j) = a_{ij}$, we see that the i, j element of the matrix array A expresses the yield to Player I when Player I uses the pure strategy $\boldsymbol{\alpha}_i$ and Player II employs the pure strategy $\boldsymbol{\beta}_j$.

A strategy $\mathbf{x} = \langle x_1, x_2, \ldots, x_n \rangle$ with no component equal to 1 is called a *mixed strategy*. In view of the relationship

$$\sum_{i=1}^{n} x_i K(\boldsymbol{\alpha}_i, \mathbf{y}) = K(\mathbf{x}, \mathbf{y}), \tag{1.1.1}^*$$

the mixed strategy $\mathbf{x}$ can be effected as follows. An experiment is conducted with n possible outcomes such that the probability of the ith outcome is x_i. The ith pure strategy is used by Player I if and only if the ith outcome has resulted, and the yield to Player I becomes the expected yield from this experiment, namely

$$\sum_{i=1}^{n} x_i K(\boldsymbol{\alpha}_i, \mathbf{y}).$$

This amounts to playing each pure strategy with a specified probability. We can therefore interpret a mixed strategy as a probability distribution defined on the space of pure strategies, and conversely. Similar interpretations apply to the strategy space T^m of Player II.

The usual introductory approach to game theory is to enumerate the spaces of pure strategies for both players and to specify the pay-off matrix **A** corresponding to these pure strategies. The concept of mixed strategy is introduced subsequently, and the pay-off function is then replaced in a natural way by the expected pay-off function. In this book we have started with the more general formulation of a game, in terms of a triplet specifying the complete strategy spaces for both players and the pay-off kernel. In this formulation, the distinction between pure and mixed strategies does not exist. Nevertheless, in many special classes of games it is natural to single out the pure strategies, which span convexly the space of all strategies. This will be done wherever it is advantageous, e.g., whenever pure strategies occur in a natural manner and possess special significance. Broadly speaking, however, we lose no generality by starting directly with two-person zero-sum games in normal form, where the strategy spaces are finite-dimensional.

Again, it has been common in the literature of game theory to make a distinction between games in extensive form and games in normal form and to take the first as a point of departure. A game formulated in extensive form is developed in terms of more primitive concepts such as "play," "chance move," "personal move," and "information structure." A strategy

* Although we have postulated this formula, it is possible to derive it by appealing to a suitable axiom system satisfied by a preference pattern which selects among alternative probability distributions over the space of outcomes resulting from the choice of a pure strategy (see the notes to this section).

is then defined within the framework of these notions, and the analysis of optimal strategies proceeds from this point. Finally, a theorem is proved to the effect that any game in extensive form may be in fact reduced to an equivalent game in normal form.

In contrast, our definition of a game as a triplet begins immediately with the concepts of strategy and pay-off, and is flexible and general enough to encompass all forms of finite game theory, including in particular the structure of games in extensive form. By carefully defining strategies and specifying completely X, Y and $K(\mathbf{x}, \mathbf{y})$ we are able to handle all forms of information patterns that arise. This will become clear as we study specific games.

It may happen that in special instances one can construct two apparent strategy spaces which are in fact equivalent in terms of pay-off. Whenever necessary we shall demonstrate this equivalence. For purposes of mathematical consistency, however, whenever any two given games differ in a specific component, the two games are taken to be distinct. For example, if we enlarge the X strategy space while the components Y and K remain unchanged, we create a new game. This is so even if the added strategies are obviously inferior and cannot affect either player's ultimate choice of an optimal strategy. In practice, as will be seen, the strategy spaces exhibited constitute an exhaustive class of procedures, i.e., a class of procedures that take into account all the fine structure of the model.

1.2 Examples. In dealing with finite matrix games it is sufficient to specify the pure strategies for both players and the corresponding pay-off matrix $\|a_{ij}\|$. The pay-off kernel for arbitrary mixed strategies $\mathbf{x}$ and $\mathbf{y}$ is given by the expression

$$K(\mathbf{x}, \mathbf{y}) = \sum_{j=1}^{m} \sum_{i=1}^{n} x_i a_{ij} y_j. \tag{1.2.1}$$

Example 1. Matching pennies. Players I and II each display simultaneously a single penny. If I matches II, i.e., if both are heads or both are tails, I takes II's penny. Otherwise, II takes I's penny. The pay-off kernel is represented in matrix form as follows:

		Player II	
		H	T
Player I	H	1	−1
	T	−1	1

The first pure strategy for I would be to display heads and the second tails. One possible mixed strategy would be to randomize equally between showing heads and tails (i.e., $\mathbf{x} = \langle \frac{1}{2}, \frac{1}{2} \rangle$).

Example 2. Two-finger Morra. Each player displays either one or two fingers and simultaneously guesses how many the opposing player will show. If both players guess correctly or both guess incorrectly, the game is a draw. If only one guesses correctly, he wins an amount equal to the total number of fingers shown by both players.

In this case each pure strategy will have two components: (a) the number of fingers to show, and (b) the number of fingers to guess. Thus, each strategy can be represented by a pair $\langle a, b \rangle$, where a denotes the first component and b the second. For example, the strategy $\langle 2, 1 \rangle$ for Player I is to show two fingers and guess one. There will be four such pure strategies for each player: $\langle 1, 1 \rangle$, $\langle 1, 2 \rangle$, $\langle 2, 1 \rangle$, and $\langle 2, 2 \rangle$. The pay-off matrix is shown below.

		Player II			
		⟨1, 1⟩	⟨1, 2⟩	⟨2, 1⟩	⟨2, 2⟩
Player I	⟨1, 1⟩	0	2	−3	0
	⟨1, 2⟩	−2	0	0	3
	⟨2, 1⟩	3	0	0	−4
	⟨2, 2⟩	0	−3	4	0

Example 3. Poker model. In this poker model there are only three possible hands, as compared with $\binom{52}{5}$ in regular poker, and all three are considered equally likely to occur. One hand is dealt to each player. There is a preference ordering among the hands: Hand 1 wins over hands 2 and 3, and hand 2 wins over hand 3. The ante is a units. Player I can either pass or bet b units. If Player I bets then Player II can either fold or call. If Player I passes and Player II bets, then Player I again has the option of folding or calling. The three possible courses of play can be diagrammed as follows:

Player I		Player II		Player I
pass	—	pass		
pass	—	bet	—	{fold / call}
bet	—	{fold / call}		

If a pass follows a bet, the bet wins. If both contestants pass or if one contestant calls, the hands are compared and the player with the better hand wins the pot.

A strategy for Player I must provide a set of actions to be taken for each of the three possible hands that he might draw and for all possible contingencies of the play of a hand. We can denote such a strategy by a triplet of pairs $[\{\alpha, \beta\}_1, \{\alpha, \beta\}_2, \{\alpha, \beta\}_3]$. The component α of $\{\alpha, \beta\}_k$ represents for Player I the probability of betting on the first round if he draws hand k. The component β represents the probability of his betting on the second round if he holds hand k. The second component is relevant only if there is a second round, but since a strategy must take all contingencies into account, a strategy in this game must include the values of β.

Similarly, Player II's strategies can be represented by a triplet of pairs $[\{\lambda, \mu\}_1, \{\lambda, \mu\}_2, \{\lambda, \mu\}_3]$. The component λ of $\{\lambda, \mu\}_k$ is the probability of Player II's contesting a bet by Player I if Player II has drawn hand k. The component μ represents the probability of Player II's betting if Player I has previously passed.

In this example the mixed strategies have been introduced directly. In view of the fact that the components of such a strategy are formulated in terms of the behavior at each move of both players, strategies of this sort are often called "behavior strategies." This specific example is fully discussed in Section 4.3. For other examples see Problems 1 through 4 at the end of this chapter.

1.3 Choice of strategies. Suppose that Player II is compelled to announce to Player I what strategy he is going to use, and that he announces $\mathbf{y}_0$. Then Player I, seeking to maximize his pay-off or yield, naturally chooses his strategy $\mathbf{x}_0$ so that $K(\mathbf{x}_0, \mathbf{y}_0) = \max_x K(\mathbf{x}, \mathbf{y}_0)$. The best thing for Player II to do under these circumstances would be to announce $\mathbf{y}_0$ such that

$$\max_x K(\mathbf{x}, \mathbf{y}_0) = \min_y \max_x K(\mathbf{x}, \mathbf{y}) = \bar{v},$$

where $\bar{v}$ (the upper value) can be interpreted as the most that Player I can achieve if Player II employs the strategy $\mathbf{y}_0$.

Suppose the tables are turned and Player I has to announce his strategy $\mathbf{x}_0$. Since Player II is sure to choose $\mathbf{y}_0$ such that

$$K(\mathbf{x}_0, \mathbf{y}_0) = \min_y K(\mathbf{x}_0, \mathbf{y}),$$

Player I can best protect himself by choosing $\mathbf{x}_0$ such that

$$\min_y K(\mathbf{x}_0, \mathbf{y}) = \max_x \min_y K(\mathbf{x}, \mathbf{y}) = \underline{v},$$

where $\underline{v}$ (the lower value) can be interpreted as the most that Player I can guarantee himself independent of Player II's choice of strategy. In order to continue this line of reasoning, we first establish the following simple result.

▶ LEMMA 1.3.1. (a) Let $f(\mathbf{x}, \mathbf{y})$ denote a real-valued function defined on $X \times Y$; then

$$\inf_y \sup_x f(\mathbf{x}, \mathbf{y}) \geq \sup_x \inf_y f(\mathbf{x}, \mathbf{y}) \qquad (\pm\infty \text{ are possible values}). \tag{1.3.1}$$

(b) If X and Y are compact and f is continuous, then sup and inf in (1.3.1) may be replaced by max and min, respectively.

Proof. By definition, for any $\mathbf{y}$, $\sup_x f(\mathbf{x}, \mathbf{y}) \geq f(\mathbf{x}, \mathbf{y})$. Hence $\inf_y \sup_x f(\mathbf{x}, \mathbf{y}) \geq \inf_y f(\mathbf{x}, \mathbf{y})$, and (1.3.1) follows immediately. The simple proof of (b) is omitted.

Lemma 1.3.1 for $f = K$ gives

$$\min_{y \in Y} \max_{x \in X} K(\mathbf{x}, \mathbf{y}) \geq \max_{x \in X} \min_{y \in Y} K(\mathbf{x}, \mathbf{y}).$$

Expressed in terms of the upper and lower values of a game, we obtain

$$\bar{v} \geq \underline{v}.$$

Our principal aim in this chapter is to establish that for a matrix game $\bar{v} = \underline{v} = v$. By an appropriate choice of strategy Player I can guarantee himself the value $\underline{v} = v$, and by judicious play Player II can prevent Player I from achieving more than $\bar{v} = \underline{v}$. Thus, unless Player I has additional information about Player II's mode of behavior—for example, expects him to choose a strategy in a wild manner—Player I should play so as to achieve v. If he departs from the course of action that assures him the value v, his ultimate yield might be less than v. Thus, it is reasonable to call this common value v the value of the game to Player I. Of course, $-v$ is the value to Player II.

All strategies $\mathbf{x}_0$ and $\mathbf{y}_0$ such that $K(\mathbf{x}_0, \mathbf{y}) \geq v$ for all $\mathbf{y}$ and $K(\mathbf{x}, \mathbf{y}_0) \leq v$ for all $\mathbf{x}$ will be referred to as optimal strategies for Players I and II, respectively.

The following simple criterion is often very useful in determining when $\bar{v} = \underline{v}$; moreover, it points up the connection between the existence of optimal strategies and the equality of the lower and upper values.

▶ THEOREM 1.3.1. If there exist $\mathbf{x}_0 \in X$, $\mathbf{y}_0 \in Y$ and a real number v such that

$$K(\mathbf{x}_0, \mathbf{y}) \geq v \qquad \text{for all} \quad \mathbf{y} \in Y$$

and

$$K(\mathbf{x}, \mathbf{y}_0) \leq v \qquad \text{for all} \quad \mathbf{x} \in X,$$

then

$$\bar{v} = \min_y \max_x K(\mathbf{x}, \mathbf{y}) = v = \max_x \min_y K(\mathbf{x}, \mathbf{y}) = \underline{v},$$

and conversely.

Proof. (a) Since

$$K(\mathbf{x}_0, \mathbf{y}) \geq v \qquad \text{for all} \quad \mathbf{y},$$

it follows that

$$\min_y K(\mathbf{x}_0, \mathbf{y}) \geq v \qquad \text{and} \qquad \max_x \min_y K(\mathbf{x}, \mathbf{y}) \geq v.$$

Similarly,

$$\max_x K(\mathbf{x}, \mathbf{y}_0) \leq v;$$

hence

$$\min_y \max_x K(\mathbf{x}, \mathbf{y}) \leq v.$$

This gives

$$\max_x \min_y K(\mathbf{x}, \mathbf{y}) \geq \min_y \max_x K(\mathbf{x}, \mathbf{y}).$$

But

$$\min_y \max_x K(\mathbf{x}, \mathbf{y}) \geq \max_x \min_y K(\mathbf{x}, \mathbf{y})$$

by Lemma 1.3.1. Therefore

$$\min_y \max_x K(\mathbf{x}, \mathbf{y}) = \max_x \min_y K(\mathbf{x}, \mathbf{y}).$$

(b) Conversely, choose $\mathbf{y}_0$ such that

$$\max_x K(\mathbf{x}, \mathbf{y}_0) = \min_y \max_x K(\mathbf{x}, \mathbf{y}) = v,$$

and $\mathbf{x}_0$ such that

$$\min_y K(\mathbf{x}_0, \mathbf{y}) = \max_x \min_y K(\mathbf{x}, \mathbf{y}) = v.$$

Then $K(\mathbf{x}, \mathbf{y}_0) \leq v$ and $K(\mathbf{x}_0, \mathbf{y}) \geq v$. This completes the proof.

▶Corollary 1.3.1. A necessary and sufficient condition that

$$\min_y \max_x K(\mathbf{x}, \mathbf{y}) = \max_x \min_y K(\mathbf{x}, \mathbf{y})$$

is that there exist strategies $\{\mathbf{x}_0, \mathbf{y}_0\}$ such that for all $\mathbf{y}$ in Y and all $\mathbf{x}$ in X

$$K(\mathbf{x}_0, \mathbf{y}) \geq K(\mathbf{x}_0, \mathbf{y}_0) \geq K(\mathbf{x}, \mathbf{y}_0).$$

Obviously, $v = K(\mathbf{x}_0, \mathbf{y}_0)$.

1.4 The min-max theorem for finite matrix games. Throughout this section we shall be concerned exclusively with matrix games. For this special case the strategy spaces X and Y of Players I and II are identifiable with the simplexes S^n and T^m, respectively, and thus each set of all strategies is spanned convexly by the pure strategies. Our objective is to establish the fundamental min-max theorem for these matrix games.

The number of different proofs of this theorem is very large. Von Neumann constructed two proofs: one based on the Brouwer fixed-point theorem (see Section C.2 of the Appendix), the other on the proposition that nonoverlapping convex sets lying in a Euclidean space can be separated by hyperplanes. Other proofs have been given which involve algebraic inequalities and the properties of convex functions. Three of these proofs are reviewed in this section and the next. We have singled out those proofs that are of use in obtaining further information concerning optimal strategies (see Remark 1.4 below).

The first proof presented is based on the idea of separating convex sets by supporting hyperplanes. Specific geometric representations of optimal strategies are developed in the course of the argument. Similar representations will be used later in analyzing the nature of the sets of optimal strategies.

▶LEMMA 1.4.1. If S^n denotes the collection of all vectors in E^n satisfying $x_i \geq 0$ $(i = 1, \ldots, n)$, $\sum_{i=1}^n x_i = 1$, and $\mathbf{b} = \langle b_1, \ldots, b_n \rangle$ is a fixed vector, then

$$\max_{x \in S^n} \sum_{i=1}^n x_i b_i = \max_i b_i. \tag{1.4.1}$$

Proof. Clearly, $\max_i b_i = \sum_{j=1}^n (x_j \max_i b_i) \geq \sum_{j=1}^n x_j b_j$. On the other hand, $\max_i b_i = b_{i_0} \leq \max_{x \in S^n} \sum_{i=1}^n x_i b_i$.

We next develop a geometrical interpretation for strategies and a schematic representation for the value of a game. To this end, we consider the pay-off matrix

$$\mathbf{A} = \left\| \begin{matrix} a_{11} & \cdots & a_{1m} \\ \vdots & & \vdots \\ a_{n1} & \cdots & a_{nm} \end{matrix} \right\| = \|\mathbf{a}_1, \ldots, \mathbf{a}_m\|,$$

where $\mathbf{a}_i$ stands for the column vector $\langle a_{1i}, \ldots, a_{ni} \rangle$. The $\mathbf{a}_i$'s can be viewed as m points in n-dimensional space.

Let Γ denote the convex set in E^n-space spanned by the $\mathbf{a}_i$. Equivalently, Γ is the intersection of all convex sets containing all $\mathbf{a}_i$. Formally,

$$\Gamma = \left\{ \sum_{j=1}^m y_j \mathbf{a}_j \,\middle|\, \sum_{j=1}^m y_j = 1, y_j \geq 0 \right\}.$$

Admissible values. We call a real number λ admissible if there exists a vector $\mathbf{p} \in \Gamma$ such that $\mathbf{p} \leq \boldsymbol{\lambda}$. (Recall that $\mathbf{p} \leq \boldsymbol{\lambda}$ means that every component p_i is smaller than or equal to λ; see p. 10.)

If Ω denotes the set of admissible numbers, then (i) Ω is nonvoid since it contains N, provided $N > a_{ij}$ for all i, j, and (ii) Ω is bounded from below by $-N$, for N sufficiently large.

Let

$$\lambda_0 = \inf_{\lambda \in \Omega} \lambda.$$

In order to show that λ_0 is admissible—i.e., that $\lambda_0 \in \Omega$—we select a sequence $\{\lambda_n\}$ in Ω such that $\lambda_n \to \lambda_0$.

To every λ_n there corresponds at least one vector $\mathbf{q}_n$ in Γ such that $\mathbf{q}_n \leq \boldsymbol{\lambda}_n$. Since Γ is compact and the sequence $\{\mathbf{q}_n\}$ must have at least one limit point $\mathbf{q}_0$, it follows readily that $\mathbf{q}_0 \leq \boldsymbol{\lambda}_0$, and thus λ_0 is admissible.

Consider the translation of the negative orthant O_{λ_0} consisting of all vectors having no component greater than λ_0. The set O_{λ_0} touches Γ with no overlap. In fact, $O_{\lambda_0} \cap \Gamma \supset \mathbf{q}_0$, but if the interior of O_{λ_0} were to contain points of Γ, the meaning of λ_0 would be contradicted.

Supporting planes to O_{λ_0}. We shall now characterize the form of all supporting planes to O_{λ_0} that are oriented so that their normals are directed into the half-space not containing O_{λ_0} (see also Lemma B.1.3 in Appendix B).

Let us consider a hyperplane defined by

$$\sum_{i=1}^{n} \mu_i \xi_i + \mu_0 = 0, \qquad \sum_{i=1}^{n} |\mu_i| \neq 0 \tag{1.4.2}$$

whose normal is directed away from O_{λ_0}. This provides a supporting plane to O_{λ_0} if and only if for every $\boldsymbol{\xi}$ in O_{λ_0}

$$\sum_{i=1}^{n} \mu_i \xi_i + \mu_0 \leq 0, \tag{1.4.3}$$

and there exists a $\boldsymbol{\xi}^0$ in O_{λ_0} for which equality holds. We now show that μ_i can be characterized as follows:

$$\mu_i \geq 0 \quad (i = 1, \ldots, n), \qquad \sum_{i=1}^{n} \mu_i > 0, \qquad \mu_0 = -\lambda_0 \sum_{i=1}^{n} \mu_i. \tag{1.4.4}$$

That μ_i $(1 \leq i \leq n)$ is nonnegative is seen by applying (1.4.3) to the vector $\boldsymbol{\xi} = \langle \xi_1, \ldots, \xi_n \rangle$ with components $\xi_j = \lambda_0$ for $j \neq i$, $\xi_i = -N$, and N sufficiently large. Since (1.4.2) defines a hyperplane, it follows that

$$\sum_{i=1}^{n} \mu_i > 0.$$

Finally, any supporting plane to O_{λ_0} clearly must contain the vertex point $\boldsymbol{\xi} = \langle \lambda_0, \lambda_0, \ldots, \lambda_0 \rangle$. Therefore,

$$0 = \left(\sum_{i=1}^{n} \mu_i\right) \lambda_0 + \mu_0$$

as asserted. Conversely, any set of μ_i satisfying the conditions of (1.4.4) defines a supporting hyperplane to O_{λ_0} whose normal is directed into the half-space not containing O_{λ_0}.

If we change notation from μ_i to x_i, supporting planes to O_{λ_0} whose normal is directed away from O_{λ_0} can be characterized by the relation

$$\sum_{i=1}^{n} x_i \xi_i - \lambda_0 = 0,$$

where $\sum_{i=1}^{n} x_i = 1$ and $x_i \geq 0$.

Geometrical representation of strategies. Strategies $\mathbf{x} = \langle x_1, \ldots, x_n \rangle$ in S^n, i.e.,

$$\sum_{i=1}^{n} x_i = 1, \qquad x_i \geq 0,$$

generate supporting planes to O_{λ_0} that are defined by the equation

$$\sum_{i=1}^{n} x_i \xi_i - \lambda_0 = 0,$$

and conversely.

Strategies $\mathbf{y} = \langle y_1, \ldots, y_m \rangle$ in T^m for Player II correspond to points in Γ. To $\mathbf{y}$ we associate the point $\mathbf{p}$ with coordinates

$$p_i = \sum_{j=1}^{m} a_{ij} y_j.$$

Unfortunately, the correspondence of Y to Γ is not one to one, since subsets of Y map into points of Γ if and only if $\mathbf{A}$ has rank less than m.

Finally, the expected pay-off $A(\mathbf{x}, \mathbf{y})$ can be interpreted as the inner product of the normal to the plane determined by $\mathbf{x}$, and the point of Γ determined by $\mathbf{y}$. Explicitly,

$$(\mathbf{x}, \mathbf{Ay}) = \sum_{i=1}^{n} \sum_{j=1}^{m} x_i a_{ij} y_j.$$

▶THEOREM 1.4.1 (*Min-Max Theorem*). If $\mathbf{x}$ and $\mathbf{y}$ range over S^n and T^m, respectively, then

$$\min_y \max_x A(\mathbf{x}, \mathbf{y}) = \max_x \min_y A(\mathbf{x}, \mathbf{y}) = v.$$

Proof. We construct O_{λ_0} and Γ as described above.

Since O_{λ_0} and Γ are nonoverlapping convex sets by Lemma B.1.2 in Appendix B, there exists a separating plane. Since O_{λ_0} and Γ touch, the separating plane must also be a supporting plane, which may be oriented so that its normal is directed away from O_{λ_0}.

Let the direction numbers of its normal be x_i^0. Then, since

$$\sum_{i=1}^{n} x_i^0 = 1$$

and $x_i^0 \geq 0$, as required by (1.4.4),

$$\sum_{i=1}^{n} x_i^0 \xi_i - \lambda_0 \leq 0 \qquad \text{for all} \quad \boldsymbol{\xi} \in O_{\lambda_0},$$

and

$$\sum_{i=1}^{n} x_i^0 \xi_i - \lambda_0 \geq 0 \qquad \text{for all} \quad \boldsymbol{\xi} \in \Gamma.$$

If $\mathbf{y}$ is any given strategy, then the vector $\boldsymbol{\xi}$, where

$$\xi_i = \sum_{j=1}^{m} a_{ij} y_j,$$

is in Γ, and thus

$$A(\mathbf{x}^0, \mathbf{y}) = \sum_{i=1}^{n} x_i^0 \sum_{j=1}^{m} a_{ij} y_j \geq \lambda_0 \tag{1.4.5}$$

for any strategy $\mathbf{y}$.

If $\mathbf{y}^0$ denotes a strategy corresponding to a point Γ which also belongs to O_{λ_0}, then

$$\sum_{j=1}^{m} a_{ij} y_j^0 \leq \lambda_0 \qquad \text{for all} \quad i, \tag{1.4.6}$$

and hence $A(\mathbf{x}, \mathbf{y}^0) \leq \lambda_0$ for all $\mathbf{x}$ strategies (by Lemma 1.4.1).

The proof of the theorem is complete by virtue of Theorem 1.3.1.

Remark 1.4. We have observed that strategies for I can be viewed as supporting planes, and strategies for II as points in Γ. We further see that optimal strategies for I correspond to separating planes between O_{λ_0} and Γ, and that optimal strategies for II correspond to those points of Γ which have contact with O_{λ_0}. This geometric characterization of optimal strategies will be exploited in Chapter 3.

The interpretations used in the arguments of Theorem 1.4.1 are unsymmetrical, in that the $\mathbf{x}$ strategies correspond to hyperplanes and the $\mathbf{y}$

strategies to points of Euclidean n-space. However, a completely parallel analysis can be developed which associates $\mathbf{y}$ strategies with hyperplanes and $\mathbf{x}$ strategies with points. In this analysis the analog of Γ is the convex set spanned by the row vectors of $\mathbf{A}$, and so on.

It is often of interest to know when there exist optimal pure strategies. This will be the case when there exist pure strategies $\{\boldsymbol{\alpha}_{i_0}, \boldsymbol{\beta}_{j_0}\}$ such that

$$\begin{aligned} a_{i_0,j} &\geq a_{i_0,j_0} \qquad \text{for all} \quad j, \\ a_{i_0,j_0} &\geq a_{i,j_0} \qquad \text{for all} \quad i. \end{aligned} \tag{1.4.7}$$

The first inequality implies that for any mixed strategy $\mathbf{y}$,

$$\sum_{j=1}^{m} a_{i_0,j}y_j \geq a_{i_0,j_0} = v,$$

or $A(\boldsymbol{\alpha}_{i_0}, \mathbf{y}) \geq v$. Similarly, we obtain $v \geq A(\mathbf{x}, \boldsymbol{\beta}_{j_0})$, and thus $\boldsymbol{\alpha}_{i_0}$ and $\boldsymbol{\beta}_{j_0}$ are optimal strategies for Players I and II, respectively.

Where pure optimal strategies exist, the matrix $\mathbf{A}$ is said to have a *saddle point* at (i_0, j_0). A saddle point is an element of a matrix which is both a minimum of its row and a maximum of its column.

***1.5 General min-max theorem.** In this section two alternative proofs of the min-max theorem for general finite games are given. The first proof is based on the Kakutani fixed-point theorem. The second proof relies only on the properties of convex functions.

▶THEOREM 1.5.1. Let $f(\mathbf{x}, \mathbf{y})$ be a real-valued function of two variables $\mathbf{x}$ and $\mathbf{y}$ which traverse C and D, respectively, where both C and D are closed, bounded, convex sets. If f is continuous, convex in $\mathbf{y}$ for each $\mathbf{x}$, and concave in $\mathbf{x}$ for each $\mathbf{y}$, then $\min_y \max_x f(\mathbf{x}, \mathbf{y}) = \max_x \min_y f(\mathbf{x}, \mathbf{y})$.

Proof. For each $\mathbf{x}$, define $\phi_x(\mathbf{y}) = f(\mathbf{x}, \mathbf{y})$. Let

$$B_x = \{\mathbf{y} | \phi_x(\mathbf{y}) = \min_{y \in D} \phi_x(\mathbf{y})\}.$$

It is easily verified that B_x is nonvoid, closed, and convex. The fact that B_x is convex is a consequence of the convexity of the function f in the variable $\mathbf{y}$. For each $\mathbf{y}$, define $\psi_y(\mathbf{x}) = f(\mathbf{x}, \mathbf{y})$. Let

$$A_y = \{\mathbf{x} | \psi_y(\mathbf{x}) = \max_{x \in C} \psi_y(\mathbf{x})\}.$$

Again, A_y is nonvoid, closed, and convex.

---

* Starred sections are advanced discussions that may be omitted at first reading without loss of continuity.

Let the set E be the direct product, $E = C \times D$, of the convex sets C and D, and recall that the direct product of two convex sets is itself convex.

Let g be the mapping which maps the point $\{\mathbf{x}, \mathbf{y}\} \in E$ into the set $(A_y \times B_x)$. Since A_y and B_x are nonvoid, closed, convex subsets of C and D, respectively, $(A_y \times B_x)$ is a nonvoid, closed, convex subset of E. Hence, condition (i) of the Kakutani theorem is fulfilled (see Section C.2 in the Appendix). It is also easy to verify that the continuity requirement of the Kakutani theorem is satisfied. Therefore, invoking the Kakutani theorem, there exists an $\{\mathbf{x}^0, \mathbf{y}^0\} \in (A_{y^0} \times B_{x^0})$, that is,

$$f(\mathbf{x}^0, \mathbf{y}^0) \geq f(\mathbf{x}, \mathbf{y}^0) \qquad \text{for all} \quad \mathbf{x} \in C,$$

$$f(\mathbf{x}^0, \mathbf{y}^0) \leq f(\mathbf{x}^0, \mathbf{y}) \qquad \text{for all} \quad \mathbf{y} \in D.$$

It then follows from Corollary 1.3.1 that

$$\min_{y \in D} \max_{x \in C} f(\mathbf{x}, \mathbf{y}) = \max_{x \in C} \min_{y \in D} f(\mathbf{x}, \mathbf{y}). \tag{1.5.1}$$

Whenever (1.5.1) is satisfied, we say that the game defined by (f, C, D) possesses a value or, alternatively, that the game is determined.

The other proof proceeds as follows. Assume first that $f(\mathbf{x}, \mathbf{y})$ is a strictly convex function of $\mathbf{y}$, for fixed $\mathbf{x}$, and $f(\mathbf{x}, \mathbf{y})$ is a strictly concave function of $\mathbf{x}$, for fixed $\mathbf{y}$. By virtue of the strict convexity, there exists a unique $\mathbf{y}(\mathbf{x})$ for which

$$f[\mathbf{x}, \mathbf{y}(\mathbf{x})] = \min_{y} f(\mathbf{x}, \mathbf{y}) = m(\mathbf{x}).$$

From the uniform continuity of f and the uniqueness of $\mathbf{y}(\mathbf{x})$ it follows readily that $m(\mathbf{x})$ and $\mathbf{y}(\mathbf{x})$ are continuous. Also, since the minimum of a family of concave functions remains concave, $m(\mathbf{x})$ is concave. Let $\mathbf{x}^*$ be a point where

$$m(\mathbf{x}^*) = \max_{x} m(\mathbf{x}) = \max_{x} \min_{y} f(\mathbf{x}, \mathbf{y}).$$

For any $\mathbf{x}$ in C and any t in $0 < t < 1$, we have by the hypothesis that

$$\begin{aligned} f[(1 - t)\mathbf{x}^* + t\mathbf{x}, \mathbf{y}] &> (1 - t)f(\mathbf{x}^*, \mathbf{y}) + tf(\mathbf{x}, \mathbf{y}) \\ &\geq (1 - t)m(\mathbf{x}^*) + tf(\mathbf{x}, \mathbf{y}). \end{aligned}$$

Put $\mathbf{y} = \mathbf{y}[(1 - t)\mathbf{x}^* + t\mathbf{x}] = \tilde{\mathbf{y}}$, so that

$$m[(1 - t)\mathbf{x}^* + t\mathbf{x}] \geq (1 - t)m(\mathbf{x}^*) + tf(\mathbf{x}, \tilde{\mathbf{y}}).$$

But since $m(\mathbf{x}^*) \geq m(\mathbf{x})$ for all $\mathbf{x}$ in C, it follows that

$$m(\mathbf{x}^*) \geq m[(1 - t)\mathbf{x}^* + t\mathbf{x}],$$

and hence

$$f(\mathbf{x}, \tilde{\mathbf{y}}) \leq m(\mathbf{x}^*) = f[\mathbf{x}^*, \mathbf{y}(\mathbf{x}^*)].$$

Now let $t \to 0$, so that $(1 - t)\mathbf{x}^* + t\mathbf{x} \to \mathbf{x}^*$ and $\tilde{\mathbf{y}} \to \mathbf{y}(\mathbf{x}^*)$; from this we deduce that

$$f[\mathbf{x}, \mathbf{y}(\mathbf{x}^*)] \leq f[\mathbf{x}^*, \mathbf{y}(\mathbf{x}^*)] \tag{1.5.2}$$

for any $\mathbf{x}$. If we let $\mathbf{y}(\mathbf{x}^*) = \mathbf{y}^*$ and $v = f(\mathbf{x}^*, \mathbf{y}^*)$, then it is clear from (1.5.2) and the construction of $\mathbf{y}(\mathbf{x})$ that

$$f(\mathbf{x}, \mathbf{y}^*) \leq v \leq f(\mathbf{x}^*, \mathbf{y}).$$

This is equivalent to the min-max theorem. It remains only to relax the conditions of strictness imposed above. To this end, define

$$f_\epsilon(\mathbf{x}, \mathbf{y}) = f(\mathbf{x}, \mathbf{y}) - \epsilon f(\mathbf{x}) + \epsilon g(\mathbf{y}),$$

where

$$f(\mathbf{x}) = \sum_{i=1}^{n} x_i^2, \qquad g(\mathbf{y}) = \sum_{j=1}^{m} y_j^2.$$

Each f_ϵ satisfies the strict convexity-concavity requirements used in the previous argument, and consequently

$$f(\mathbf{x}, \mathbf{y}_\epsilon) - \epsilon f(\mathbf{x}) \leq f_\epsilon(\mathbf{x}, \mathbf{y}_\epsilon) \leq v_\epsilon \leq f_\epsilon(\mathbf{x}_\epsilon, \mathbf{y}) \leq f(\mathbf{x}_\epsilon, \mathbf{y}) + \epsilon g(\mathbf{y})$$

for all $\{\mathbf{x}, \mathbf{y}\}$ in $C \times D$. Take a sequence of ϵ's approaching zero and consider a subsequence $\mathbf{x}_\epsilon \to \mathbf{x}^*$, $\mathbf{y}_\epsilon \to \mathbf{y}^*$, $v_\epsilon \to v^*$. We obtain in the limit

$$f(\mathbf{x}, \mathbf{y}^*) \leq v^* \leq f(\mathbf{x}^*, \mathbf{y}),$$

and the proof is complete.

1.6 Problems.

1. "Stone, Paper, and Scissors." Both players must simultaneously name one of these objects; "paper" defeats "stone," "stone" defeats "scissors," and "scissors" defeats "paper." The player who chooses the winning object wins one unit; if both players choose the same object, the game is a draw. Set up the matrix for this game and describe the optimal strategies.

2. Two cards marked Hi and Lo are placed in a hat. Player I draws a card at random and inspects it. He may pass immediately, in which case he pays Player II the amount $a > 0$; or he may bet, in which case Player II may either pass, paying amount a to Player I, or call. If he calls, he

receives from or pays to Player I the amount $b > a$, according as Player I holds Lo or Hi. Set up the game and solve by the methods of Theorem 1.4.1.

3. Player I draws a card Hi or Lo from a hat. He then has the option of either betting or passing. If he bets, Player II may either fold, in which case he pays Player I an amount a, or call, in which case he wins or loses $b > a$ according as Player I holds Hi or Lo. If Player I passes initially, Player II draws Hi or Lo from a new hat. Player II then has the symmetrical choice of actions, to which Player I must respond with the corresponding pay-off. If Player II passes on the second round, the pay-off is zero. Describe the game and the strategy spaces.

4. Each player must choose a number between 1 and 9. If a player's choice is one unit higher than his opponent's, he loses two dollars; if it is at least two units higher, he wins one dollar; if both players choose the same number, the game is a draw. Set up the matrix for this game.

5. Show that the $n \times n$ matrix $\|a_{ij}\|$ with $a_{ij} = i - j$ has a saddle point, and describe the corresponding optimal strategies.

6. Use Theorem 1.4.1 to solve the game with matrix

$$\begin{Vmatrix} 3 & -1 & 3 & 7 \\ -1 & 9 & 3 & 0 \end{Vmatrix}.$$

7. Two games have matrices $\mathbf{A} = \|a_{ij}\|$ and $\mathbf{B} = \|b_{ij}\|$, respectively; the value of the first game is v, and $b_{ij} = a_{ij} + a$. Show that the value of $\mathbf{B}$ is $v + a$, and that the optimal strategies are the same for both games.

8. Show that each of the two matrices

$$\mathbf{A}_1 = \begin{Vmatrix} 0 & x \\ 1 & 2 \end{Vmatrix}, \qquad \mathbf{A}_2 = \begin{Vmatrix} 2 & 1 \\ x & 0 \end{Vmatrix}$$

has a saddle point, and find values x_1 and x_2 such that

$$v(\mathbf{A}_1 + \mathbf{A}_2) < v(\mathbf{A}_1) + v(\mathbf{A}_2)$$

and

$$v(\mathbf{A}_1 + \mathbf{A}_2) > v(\mathbf{A}_1) + v(\mathbf{A}_2),$$

respectively.

9. Give an example to show that

$$(\mathbf{x}^*, \mathbf{A}\mathbf{y}^*) = \min_{y \in Y} \max_{x \in X} (\mathbf{x}, \mathbf{A}\mathbf{y}) = \max_{x \in X} \min_{y \in Y} (\mathbf{x}, \mathbf{A}\mathbf{y}),$$

is not sufficient proof that $\mathbf{x}^*$ and $\mathbf{y}^*$ are optimal.

10. Let $\mathbf{A} = \|a_{ij}\|$ be an $n \times m$ matrix. Show that either there exists

$$\mathbf{u} = \langle u_1, \ldots, u_n \rangle \qquad \left(\text{with } u_i \geq 0, \sum_{i=1}^{n} u_i = 1\right)$$

such that

$$\sum_{i=1}^{n} u_i a_{ij} \geq 0 \qquad \text{for all} \quad j,$$

or there exists

$$\mathbf{w} = \langle w_1, \ldots, w_m \rangle \qquad \left(\text{with } w_j \geq 0, \sum_{j=1}^{m} w_j = 1\right)$$

such that

$$\sum_{j=1}^{m} a_{ij} w_j < 0 \qquad \text{for all} \quad i.$$

11. Let $\mathbf{B} = \|b_{ij}\|$ be a fixed $n \times m$ game, and let $f(t)$ be a convex increasing function of t. Let Γ be the game with strategies

$$\mathbf{x} = \langle x_1, \ldots, x_n \rangle, \qquad \mathbf{y} = \langle y_1, \ldots, y_m \rangle$$

and pay-off

$$\sum_{i=1}^{n} x_i f\left(\sum_{j=1}^{m} b_{ij} y_j\right).$$

Show that the optimal strategies for $\mathbf{B}$ are also optimal for the new game.

12. Show that a game $\sum x_i a_{ij} y_j$ played over closed sets of strategies S and T has a value if and only if the game played over the convex hulls of S and T has two optimal strategies $\mathbf{x}^* \in S$ and $\mathbf{y}^* \in T$.

13. Let $\mathbf{A}$, $\mathbf{B}$, and $\mathbf{D}$ be matrix games of orders $n_1 \times m_1$, $n_1 \times m_2$, $n_2 \times m_2$ and values a, b, d, respectively. Let $\{\mathbf{x}^0, \mathbf{y}^0\}$ and $\{\mathbf{x}^*, \mathbf{y}^*\}$ be optimal for $\mathbf{A}$ and $\mathbf{D}$, respectively, and let $\langle \xi, 1 - \xi \rangle$ and $\langle \eta, 1 - \eta \rangle$ be optimal for the game

$$\begin{Vmatrix} a & b \\ 0 & d \end{Vmatrix}.$$

Show that if $a < 0 < d$, the strategies

$$\mathbf{x} = \langle \xi x_1^0, \ldots, \xi x_{n_1}^0, (1 - \xi) x_{n_1+1}^*, \ldots, (1 - \xi) x_{n_1+n_2}^* \rangle,$$

$$\mathbf{y} = \langle \eta y_1^0, \ldots, \eta y_{m_1}^0, (1 - \eta) y_{m_1+1}^*, \ldots, (1 - \eta) y_{m_1+m_2}^* \rangle$$

are optimal for the game

$$\begin{Vmatrix} \mathbf{A} & \mathbf{B} \\ 0 & \mathbf{D} \end{Vmatrix}.$$

14. Show that the value of a matrix game is a nondecreasing continuous function of the components of the matrix.

15. Let X and Y be convex compact sets, and let $f(\mathbf{x}, \mathbf{y})$ be concave in $\mathbf{x}$ and convex in $\mathbf{y}$. Show that for any finite set $\mathbf{x}_1, \ldots, \mathbf{x}_n, \mathbf{y}_1, \ldots, \mathbf{y}_m$ there exists a pair $\{\mathbf{x}_0, \mathbf{y}_0\}$ such that

$$\max_{1 \leq i \leq n} f(\mathbf{x}_i, \mathbf{y}_0) \leq \min_{1 \leq j \leq m} f(\mathbf{x}_0, \mathbf{y}_j).$$

16. Let X and Y be convex sets, with Y also compact. Let $f(\mathbf{x}, \mathbf{y})$ be a function defined on $X \times Y$ with the property that f is lower-semicontinuous in $\mathbf{y}$, concave in $\mathbf{x}$, and convex in $\mathbf{y}$. Show that for each real α, either there exists $\mathbf{y}_0$ in Y such that $f(\mathbf{x}, \mathbf{y}_0) \leq \alpha$ for all $\mathbf{x}$ in X, or there exists $\mathbf{x}_0$ in X such that $f(\mathbf{x}_0, \mathbf{y}) > \alpha$ for all $\mathbf{y}$ in Y.

17. Show that if every 2×2 submatrix of a matrix $\mathbf{A}$ has a saddle point, then $\mathbf{A}$ has a saddle point.

18. Let $\mathbf{A}$ be a nonsingular $n \times m$ matrix with $m \geq 3$. Show that if every $n \times m - 1$ submatrix has a strict saddle point, then $\mathbf{A}$ has a saddle point.

19. Let X and Y denote polyhedral convex sets in E^n and E^m space, respectively. Let $\mathbf{A}$ be an $n \times m$ matrix. Define

$$v_2 = \inf_{y \in Y} \sup_{x \in X} (\mathbf{x}, \mathbf{A}\mathbf{y}), \qquad v_1 = \sup_{x \in X} \inf_{y \in Y} (\mathbf{x}, \mathbf{A}\mathbf{y}).$$

Prove that if $v_1 \neq v_2$, then $v_1 = -\infty$ and $v_2 = +\infty$. Construct examples satisfying $v_1 = -\infty$; $v_2 = +\infty$; $v_1 = v_2 = +\infty$; $v_1 = v_2 = -\infty$. (A polyhedral convex set is defined as the intersection of a finite number of closed half spaces.)

Notes and References to Chapter 1

§ **1.1.** Several popular and entertaining discussions of discrete game theory and its applications are available for the nonmathematically-oriented reader, notably Williams [254] and McDonald [176]. A comprehensive critical survey of the concepts of game theory, utility theory, and related subjects is contained in Luce and Raiffa [169].

Historically, the modern mathematical approach to game theory is attributed to von Neumann in his papers of 1928 [238] and 1937 [239], although some game problems were dealt with earlier by Borel [39]. The primary stimulation for the present vigorous research in game theory was the publication in 1944 of the fundamental book on games and economic behavior by von Neumann and Morgenstern [242]. For introductory mathematical expositions in game theory, the reader is directed to McKinsey [179], Gale [96], and Vajda [236].

To the more advanced student we recommend the volumes edited by Kuhn and Tucker [161, 162]. Another volume of studies in game theory has been compiled by Dresher, Tucker, and Wolfe [72]. To the student interested primarily in n-person game theory we call attention to Shapley [216] and a volume edited by Luce and Tucker [170].

Several texts on related subjects incorporate some aspects of game theory. These include Blackwell and Girshick [30], Churchman, Ackoff, and Arnoff [50], McCloskey and Trefethen [175], and Allen [2].

Luce and Raiffa, in addition to their discussions of the underlying concepts of games and decisions, provide in eight appendixes a moderate survey of several of the techniques of the subject.

A handbook of important results in game theory has been compiled by Dresher [66].

Although a knowledge of utility theory is not essential to the analytical study of the structure of games and their solutions, a few words on the subject may be helpful.

Von Neumann and Morgenstern in their classic book introduced an axiom system satisfied by a preference pattern $\succ$ defined on a finite set of outcomes and the space $\mathcal{P}$ of probability distributions of outcomes. That is to say, of any two probability distributions P_1 and P_2 on the space of outcomes, the preference relation selects one as preferable to the other. (In symbols, either $P_1 \succ P_2$ or $P_2 \succ P_1$.)

The axioms, stated loosely, are as follows:

I. If $P_1 \succ P_2$ and $Q_1 \succ Q_2$, then any probability mixture of P_1 and Q_1 is preferred to the same probability mixture of P_2 and Q_2.

II. (Continuity Axiom.) There is no probability distribution on the space of outcomes that is infinitely desirable or infinitely undesirable. Formally, we may express this axiom as follows: If $P_1 \prec P_2 \prec P_3$, there are numbers $\lambda < 1$ and $\mu > 0$ such that

$$\lambda P_1 + (1 - \lambda)P_3 \prec P_2, \qquad \mu P_1 + (1 - \mu)P_3 \succ P_2.$$

When the preference relation obeys these axioms, it can be expressed as a numerical utility function U defined for probability distributions on the set of outcomes. Furthermore, U is linearly convex with respect to the operation of forming convex combinations of elements of $\mathcal{P}$. It is also shown by these authors that U is uniquely determined up to a positive linear affine transformation. That is, if U is a utility function, then $aU + b$ (with $a > 0$ and b real) is the only other possible utility function. The proof is given in Chapter i of [242].

Blackwell and Girshick [30] have extended this result to the case in which the set of outcomes may contain a countable number of elements. Luce and Raiffa [169] subject the von Neumann and Morgenstern axioms to a penetrating review. We recommend their critical discussion to the reader interested in the foundations of game theory.

In the classical approach to the theory of games, it is necessary when defining a strategy to specify decisions for both players which appropriately take account of the complete present state of information. Because of chance moves which may occur in the course of the play the outcomes of a pair of strategies may take the form of a random variable with a known probability distribution. Generally, in choosing among strategies, it is necessary to specify a preference relation which chooses among probability distributions over the set of outcomes. If the von Neumann and Morgenstern axioms are valid for this preference relation, then the preference associated with any pair of pure strategies can be expressed by a number obtained by applying utility notions to distributions over outcomes. In this way the pay-off function can be so constructed as to be consistent with the preference pattern. Moreover, the pay-off operates additively (i.e., is linearly convex) when computing expected values with respect to probability distributions on the set of pure strategies. Hence the formula (1.1.1) obtains.

§ **1.2.** The poker example of this section was first introduced and studied by Kuhn [156]. A general technique for the analysis of parlor games is developed in Volume II Chapter 9 of the present work.

§ **1.3.** The concept of the upper and lower values was proposed by von Neumann, who refers to corresponding situations as the majorant game and minorant game, respectively.

§ **1.4.** Proofs of the fundamental min-max theorems have been given by von Neumann [239], Ville [237], Loomis [168], Karlin [133], Fan [77], Glicksberg [108], and others. All proofs except Loomis' utilize in essence some form of convexity and compactness of the strategy spaces.

§ **1.5.** The first proof in this section is due to Kakutani [130], who obtained, in the course of establishing the min-max theorem, a generalization of the classical Brouwer fixed-point theorem. Kakutani's method depends on certain convexity properties of the mapping function considered. Eilenberg and Montgomery [73] and Begle [18] have shown that this generalized fixed-point theorem is valid when the requirement of simple connectedness for certain sets is substituted for convexity. Further extensions along these lines have been given by Debreu [61]. The second proof in this section was proposed by Shiffman [221].

§ **1.6.** Problems 2 and 3 are based on Borel [39]. The statement of Problem 12 was first observed by I. Glicksberg. The conclusions of Problems 15 and 16 were used by Fan [77]. Problems 17 and 18 were suggested by L. Shapley. Problem 19 is due to Wolfe [255].

CHAPTER 2

THE NATURE OF OPTIMAL STRATEGIES FOR MATRIX GAMES

In this chapter we develop some of the fundamental working tools needed in the analysis of matrix games. Particular attention is directed to characterizing certain basic extreme-point optimal strategies. Since the extreme-point optimal strategies span the set of all optimal strategies, knowing the former is essentially equivalent to knowing the latter; i.e., every optimal strategy is a convex combination of extreme-point optimal strategies. The characterization of the extreme-point optimal strategies can be used to set up a computational scheme for obtaining the full set of optimal strategies.

The concept of dominance among strategies, which is introduced in Section 2.2, enables us to eliminate obviously poor strategies from consideration in the early stages of the consideration of the game. The dominance principle is especially useful in solving games explicitly, since many seemingly large-scale games can be reduced to small matrix games by excluding all dominated strategies. In this chapter four types of dominance are delimited, and the techniques associated with them are developed.

An important subclass of games called completely mixed games is also introduced in this chapter. These are games in which all pure strategies have relevance—i.e., have positive probability—in every optimal strategy. Several examples of such games are considered.

At the close of the chapter, symmetric games are discussed. A symmetric game is the equivalent of a "fair game," one in which the opportunities and possible pay-offs are equal for both players.

2.1 Properties of optimal strategies.

In this section we establish two basic properties of optimal strategies for matrix games. These properties will be used in subsequent sections.

▶ LEMMA 2.1.1. The set of optimal strategies for each player is a convex closed set.

Proof. We shall give the proof for Player I only. If v is the value of the game and $\mathbf{x}_0$ and $\mathbf{x}_0'$ are optimal strategies for Player I, then $A(\mathbf{x}_0, \mathbf{y}) \geq v$ and $A(\mathbf{x}_0', \mathbf{y}) \geq v$ for all $\mathbf{y}$. Thus

$$A[\lambda\mathbf{x}_0 + (1 - \lambda)\mathbf{x}_0', \mathbf{y}] = \lambda A(\mathbf{x}_0, \mathbf{y}) + (1 - \lambda)A(\mathbf{x}_0', \mathbf{y}) \geq v$$

for all $\mathbf{y}$, and therefore $\lambda\mathbf{x}_0 + (1 - \lambda)\mathbf{x}'_0$ is contained in the set X^0 of optimal strategies.

That X^0 is closed follows from the continuity of $A(\mathbf{x}, \mathbf{y})$ as a function of $\mathbf{x}$.

▶LEMMA 2.1.2. If $\mathbf{y}^0$ is an optimal mixed strategy for Player II and $y^0_{j_0} > 0$, then every optimal $\mathbf{x}$ strategy for Player I must have the property that

$$A(\mathbf{x}, j_0) = \sum_{i=1}^{n} x_i a_{ij_0} = v.$$

The corresponding result is true for the other player.

Proof. Let $\mathbf{x}$ be an optimal strategy.

Since $\langle 0, \ldots, 1, 0, \ldots, 0 \rangle$, where 1 occurs only in the rth component, is a possible $\mathbf{y}$ strategy, it follows that

$$\sum_{i=1}^{n} x_i a_{ir} \geq v \qquad \text{for all} \quad r. \tag{2.1.1}$$

Suppose that strict inequality holds for $r = j_0$. Then

$$y^0_{j_0} \sum_{i=1}^{n} x_i a_{ij_0} > v y^0_{j_0},$$

since $y^0_{j_0} > 0$, while for $r \neq j_0$,

$$y^0_r \sum_{i=1}^{n} x_i a_{ir} \geq v y^0_r.$$

Hence

$$\sum_{r=1}^{m} y^0_r \sum_{i=1}^{n} x_i a_{ir} > v \sum_{r=1}^{m} y^0_r = v.$$

This contradicts the assumption that $\mathbf{y}^0$ is optimal. Consequently,

$$\sum_{i=1}^{n} x_i a_{ij_0} = v \qquad \text{for all optimal } \mathbf{x}.$$

For later reference we introduce here some terminology. If x_i, the ith component of a strategy $\mathbf{x}$, is positive, the corresponding pure strategy i is said to be relevant to, essential to, or employed with positive probability in the strategy $\mathbf{x}$. In referring to a pure strategy, we use the terms "ith row" and "ith pure strategy" interchangeably.

A row i such that

$$\sum_{j=1}^{n} a_{ij} y_j = v$$

is said to be an equalizer with respect to the strategy $\mathbf{y}$. In this context Lemma 2.1.2 may be restated as follows: If i is relevant to an optimal strategy $\mathbf{x}^0$, then i is an equalizer with respect to every optimal $\mathbf{y}$ strategy.

The corresponding terminology will be employed with reference to $\mathbf{y}$ strategies and columns. An optimal strategy will also be referred to as a solution to the game.

Finally, we shall frequently use the notation $\mathbf{Ay} = \mathbf{v}$, where $\mathbf{A}$ is a matrix, $\mathbf{y}$ is a vector, and $\mathbf{v}$ is an ordinary scalar which in this context is to be interpreted as the vector $\langle v, v, v, \ldots, v \rangle$ with the appropriate number of components. We shall also use the analogous expressions $\mathbf{Ay} \leq \mathbf{v}$ and $\mathbf{xA} \geq \mathbf{v}$.

2.2 Types of strict dominance. The matrix associated with a game can often be reduced in size by the use of the dominance principle. When this principle is used, the strategies which are optimal with respect to the smaller matrix are also optimal with respect to the larger. Thus the concept of dominance can be used to simplify the problem of obtaining optimal strategies. Specific applications of this principle are given in Chapter 4.

The dominance principle states that if one strategy open to a player is better than another regardless of which strategy his opponent uses, then the inferior strategy may be disregarded in the search for optimal strategies. This principle and its applications are set forth explicitly in the following four theorems. These theorems are stated and proved for row dominance only; the analogous results for column dominance are stated without proof.

▶ DEFINITION 2.2.1. An n-tuple $\boldsymbol{\alpha} = \langle \alpha_1, \alpha_2, \ldots, \alpha_n \rangle$ strictly dominates a second n-tuple $\boldsymbol{\beta} = \langle \beta_1, \ldots, \beta_n \rangle$ if $\alpha_i > \beta_i$ for $i = 1, 2, \ldots, n$.

▶ THEOREM 2.2.1. If the jth row of the pay-off matrix $\mathbf{A}$ strictly dominates the ith row, the ith row can be deleted from the matrix without changing the set of optimal strategies for Player I.

Proof. For optimal $\mathbf{y}^0$

$$v \geq \sum_k a_{jk} y_k^0 > \sum_k a_{ik} y_k^0,$$

and by Lemma 2.1.2 no optimal $\mathbf{x}^0$ can have $x_i^0 > 0$.

▶ THEOREM 2.2.1a. If the jth column strictly dominates the ith column, the jth column can be deleted without changing the set of optimal strategies for Player II.

▶ THEOREM 2.2.2. If the ith row is strictly dominated by a convex combination of the other rows, the ith row can be deleted.

Proof. The hypothesis of the theorem means that there exists a strategy $\mathbf{x} = \langle x_1, x_2, \ldots, x_n \rangle$ with $x_i = 0$ such that

$$a_{ij} < \sum_{k=1}^{n} x_k a_{kj} \qquad \text{for} \quad j = 1, \ldots, m. \tag{2.2.1}$$

If $\mathbf{y}^0$ is optimal for Player II, it follows that

$$\sum_{j=1}^{m} a_{ij} y_j^0 < (\mathbf{x}, \mathbf{A}\mathbf{y}^0) \leq v.$$

The last inequality, in conjunction with Lemma 2.1.2, implies that the ith row cannot be relevant to any optimal $\mathbf{x}$ strategy, and hence the conclusion of the theorem is established.

The condition "without changing the set of optimal strategies" is of course implied in this and the following theorems.

▶Theorem 2.2.2a. If the ith column strictly dominates some convex combination of the other columns, the ith column can be deleted.

▶Theorem 2.2.3. If a convex combination of rows in a set R_1 strictly dominates a convex combination of rows in a set R_2, there exists a row in R_2 that can be deleted.

Proof. Let $\mathbf{x}^1$ be the convex combination of rows in R_1 that strictly dominates the convex combination $\mathbf{x}^2$ of rows in R_2; that is, $\mathbf{x}^1\mathbf{A} \gg \mathbf{x}^2\mathbf{A}$.

Let us assume to the contrary that every row in R_2 is employed with positive probability in some optimal strategy. Then we know from Lemma 2.1.2 that for $\mathbf{y}^0$ optimal, each of the components in R_2 of $\mathbf{A}\mathbf{y}^0$ equals v. Thus we have $v \geq (\mathbf{x}^1, \mathbf{A}\mathbf{y}^0) > (\mathbf{x}^2, \mathbf{A}\mathbf{y}^0) = v$, which is impossible. Hence some row of R_2 may be deleted; i.e., there is a row such that the component of any optimal $\mathbf{x}$ for that row is 0.

▶Theorem 2.2.3a. If a convex combination of columns in a set C_1 strictly dominates a convex combination of columns in a set C_2, there exists a column in C_1 that may be deleted.

▶Theorem 2.2.4. Let

$$\mathbf{A} = \begin{Vmatrix} \mathbf{A}_1 & \mathbf{A}_2 \\ \mathbf{A}_3 & \mathbf{A}_4 \end{Vmatrix}.$$

If every column of $\mathbf{A}_2$ strictly dominates some convex combination of the columns of $\mathbf{A}_1$, and if every row of $\mathbf{A}_3$ is strictly dominated by some convex combination of the rows of $\mathbf{A}_1$, the submatrices $\mathbf{A}_2$, $\mathbf{A}_3$, and $\mathbf{A}_4$ may be deleted.

Proof. Suppose that row i of $\mathbf{A}_3$ is employed with probability λ in some optimal strategy $\mathbf{x}^*$. Since some convex combination of the rows of $\mathbf{A}_1$ strictly dominates i, we can spread the probability λ among these rows of $\mathbf{A}_1$ proportionally to the weightings used in forming the convex combination and obtain a new strategy $\mathbf{x}^0$. We perform this construction for every row i of $\mathbf{A}_3$ which is relevant to the optimal strategy $\mathbf{x}^*$. The final strategy obtained is denoted by $\mathbf{x}'$. Clearly, the expected yield when Player I uses $\mathbf{x}'$ and Player II uses a pure strategy corresponding to a column of $\mathbf{A}_1$ is strictly greater than the yield for the same column when Player I uses $\mathbf{x}^*$. In symbols,

$$(\mathbf{x}', \mathbf{A}\boldsymbol{\beta}_j) > (\mathbf{x}^*, \mathbf{A}\boldsymbol{\beta}_j) \geq v \tag{2.2.2}$$

for any pure strategy $\boldsymbol{\beta}_j$ with j in $\mathbf{A}_1$.

For any column j of $\mathbf{A}_2$ there exists by hypothesis a convex combination $\mathbf{y}^0$ of columns in $\mathbf{A}_1$ which is strictly dominated by j, where we extend $\mathbf{y}^0$ to be a strategy by using 0's for the columns of $\mathbf{A}_2$. It follows that

$$(\mathbf{x}', \mathbf{A}\boldsymbol{\beta}_j) > (\mathbf{x}', \mathbf{A}\mathbf{y}^0) = (\mathbf{x}'\mathbf{A}, \mathbf{y}^0) \geq v \text{ for } j \text{ in } \mathbf{A}_2, \tag{2.2.3}$$

where the last inequality is a consequence of (2.2.2). The inequalities (2.2.2) and (2.2.3) together contradict the meaning of v. Thus no row of $\mathbf{A}_3$ can be employed with positive probability in an optimal strategy.

An analogous argument shows that the columns of $\mathbf{A}_2$ are not relevant.

Although all the dominance criteria are stated above in terms of strict inequalities, it is possible to apply these same theorems to nonstrict dominance. This again leads to a reduced matrix whose optimal strategies are also optimal for the original game, although in this case the sets of optimal strategies for the reduced game and the original game will not necessarily coincide. Once the value of the game and a pair of optimal strategies are ascertained, however, it is easy to check the relevance to any optimal strategy of the rows and columns that were eliminated.

It is possible to apply several of the above theorems successively to an initial matrix and the ensuing reduced matrices. Consider the following example:

$$\mathbf{A} = \begin{Vmatrix} 0 & -3 & -3 & 5 & -3 \\ 1 & 3 & -2 & 2 & 0 \\ 1 & -2 & 1 & 1 & 0 \\ 0 & 0 & -1 & -4 & 1 \\ 3 & -2 & -1 & 6 & -2 \end{Vmatrix}$$

The first row is dominated termwise by the fifth row. In the resulting reduced matrix, the first column dominates the convex combination of the last two columns weighted equally. Deleting the first column, we obtain

$$\begin{Vmatrix} 3 & -2 & 2 & 0 \\ -2 & 1 & 1 & 0 \\ 0 & -1 & -4 & 1 \\ -2 & -1 & 6 & -2 \end{Vmatrix} = \begin{Vmatrix} \mathbf{A}_1 & \mathbf{A}_2 \\ \mathbf{A}_3 & \mathbf{A}_4 \end{Vmatrix},$$

where each of the $\mathbf{A}_i$ represents the corresponding 2×2 submatrix. Applying Theorem 2.2.4, we obtain the 2×2 game

$$\begin{Vmatrix} 3 & -2 \\ -2 & 1 \end{Vmatrix},$$

whose solution for both players is the strategy $\langle \frac{3}{8}, \frac{5}{8} \rangle$. The value of the game is $-\frac{1}{8}$. It is readily verified that the only equalizing rows of $\mathbf{A}$ with respect to the optimal strategy $\langle 0, \frac{3}{8}, \frac{5}{8}, 0, 0 \rangle$ are the second and third rows, and that the same is true for the columns. We therefore conclude that this single strategy $\langle 0, \frac{3}{8}, \frac{5}{8}, 0, 0 \rangle$ is the unique optimal strategy for each player.

2.3 Construction of optimal strategies. It was proved in Chapter 1 that matrix games have a value and that each player has an optimal strategy. However, these proofs present no effective method for finding the value of the game or determining explicit solutions. Before tackling the problem of characterizing solutions in general, let us consider some simple constructive methods of solving certain particularly important games. The large number of $2 \times m$ and $n \times 2$ games that arise in practice justifies consideration of the following methodology.

Example 1. $2 \times m$ game. Consider the game in which Player I has 2 possible pure strategies, Player II has m possible pure strategies, and the pay-off matrix is

$$\begin{Vmatrix} a_{11} & a_{12} & \cdots & a_{1m} \\ a_{21} & a_{22} & \cdots & a_{2m} \end{Vmatrix}.$$

A strategy for Player I can be characterized by an x $(0 \leq x \leq 1)$, where x represents the probability of Player I's choosing strategy 1. If Player II chooses a pure strategy j, the pay-off corresponding to the strategies x can be represented by a line of the form $l_j(x) = xa_{1j} + (1 - x)a_{2j}$.

Let

$$\phi(x) = \text{envelope from below of the set of lines } l_j(x)$$
$$= \inf_j l_j(x) = \min_j l_j(x).$$

Let x_0 be such that $\phi(x_0) = \max_x \phi(x)$. Then $\phi(x_0)$ is the maximum pay-off that Player I can assure himself by choosing x_0 against any pure strategy that Player II might use.

Since a mixed strategy for Player II necessarily consists of a convex combination of the lines $l_j(x)$, it is clear that by choosing x_0 Player I in all circumstances assures himself of at least the amount $v^* = \phi(x_0)$. Hence $v \geq v^*$, where v is the value of the game.

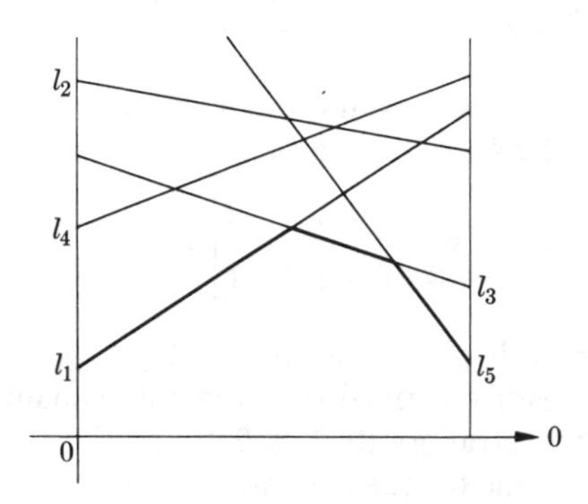

FIGURE 2.1

Now let us see to what amount Player II can restrict Player I. Let us assume that $\phi(x)$ achieves a maximum at a single point x_0 in the interior of the unit interval, as indicated in Fig. 2.1. If there exists a line l_{j_0} passing through $\{x_0, \phi(x_0)\}$ with zero slope, then let the pure strategy j_0 be II's strategy. If there exists no such line through $\{x_0, \phi(x_0)\}$, choose two lines, l_{j_0} and $l_{j_0^*}$, through the point $\{x_0, \phi(x_0)\}$, one with a positive slope and one with a negative slope. (The restriction that the minimum is attained at a single interior point obviously implies the existence of two such lines.) Also choose a weight λ_0 $(0 \leq \lambda_0 \leq 1)$ such that the line $l^* = \lambda_0 l_{j_0}(x) + (1 - \lambda_0) l_{j_0^*}(x)$ has zero slope, i.e.,

$$\lambda_0 l_{j_0}(x) + (1 - \lambda_0) l_{j_0^*}(x) \equiv v^* = \phi(x_0).$$

Let II's strategy be the mixed strategy

$$\langle 0, 0, \ldots, \lambda_0, \ldots, 1 - \lambda_0, \ldots, 0 \rangle,$$

where λ_0 occurs in the j_0 coordinate and $1 - \lambda_0$ in the j_0^* coordinate. If $\phi(x)$ achieves its maximum at more than one point in the interior of the unit interval, it follows that there must exist a line l_{j_0} of the form $l_{j_0}(x) \equiv v^*$. Let II's strategy be the pure strategy j_0 in this case. If the point $x_0 = 0$ is a maximum point of $\phi(x)$, let II's strategy be the pure strategy corresponding to the line passing through $\{0, \phi(0)\}$ whose nonpositive slope has the largest absolute value; and if $x_0 = 1$ is a maximum point, let II's strategy be the pure strategy corresponding to the line passing through $\{1, \phi(1)\}$ with the largest nonnegative slope.

In all these cases, if II uses the designated strategy, I is limited to the amount v^* regardless of what strategy he plays. Therefore the value of the game must be less than or equal to v^*. But we found previously that $v \geq v^*$. Hence $v = v^*$, and the optimal strategies for I and II are those that were suggested as possible optimal strategies.

If the maximum point of $\phi(x)$ occurs at one of the end points, both players have optimal pure strategies. If the function $\phi(x)$ has a plateau, i.e., more than one maximum point, then II has an optimal pure strategy and I has an infinite number of optimal mixed strategies.

For an $n \times 2$ game there are analogous procedures for finding the value and the optimal strategies. The lines $l_i(y) = ya_{i1} + (1 - y)a_{i2}$, where y varies over the unit interval, give the pay-off if I uses the pure strategy i and II the mixed strategy y. The minimum of the function $\psi(y) = \sup_i l_i(y)$ gives the value of the game.

Example 2. $3 \times m$ game. A strategy for Player I in a $3 \times m$ game is designated by a pair $\langle x_1, x_2 \rangle$, where $x_1 \geq 0$, $x_2 \geq 0$, and $x_1 + x_2 \leq 1$. The set of possible strategies can be represented as the shaded triangle in Fig. 2.2.

Another way to represent a strategy for Player I is by a plane triangle (Fig. 2.3). The three vertices correspond to the three pure strategies. Any other point $\mathbf{x}$ in the triangle is a mixed strategy and is denoted by a triple $\langle x_1, x_2, x_3 \rangle$, where x_i is that part of a unit mass that would have to be located at end point i for the center of gravity of the system to be the point $\mathbf{x}$. Thus $\langle x_1, x_2, x_3 \rangle$ designates $\mathbf{x}$ in barycentric coordinates. The important feature of this use of barycentric coordinates to represent the set of mixed strategies is that it is independent of any coordinate system.

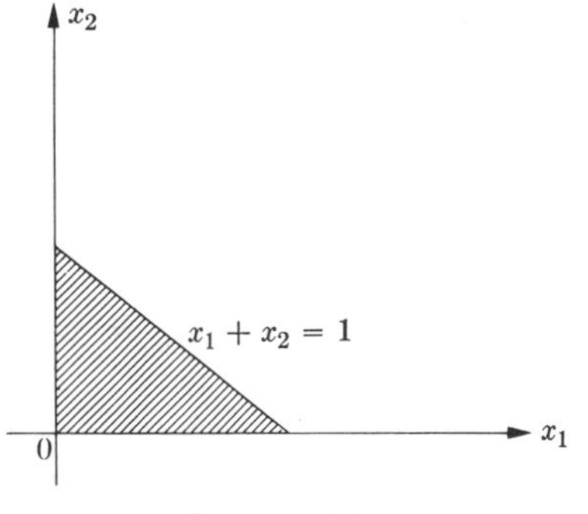

FIGURE 2.2

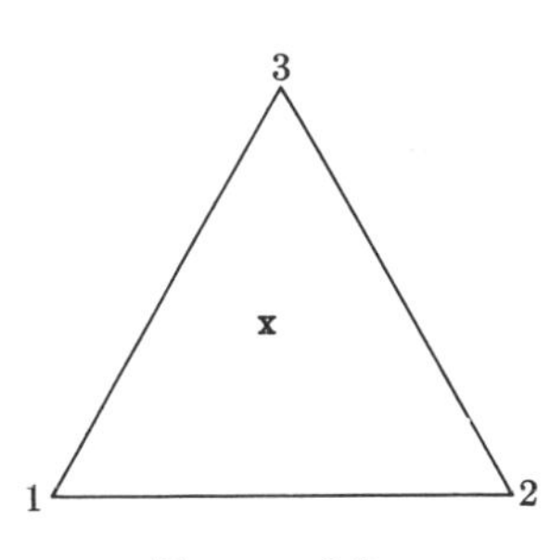

FIGURE 2.3

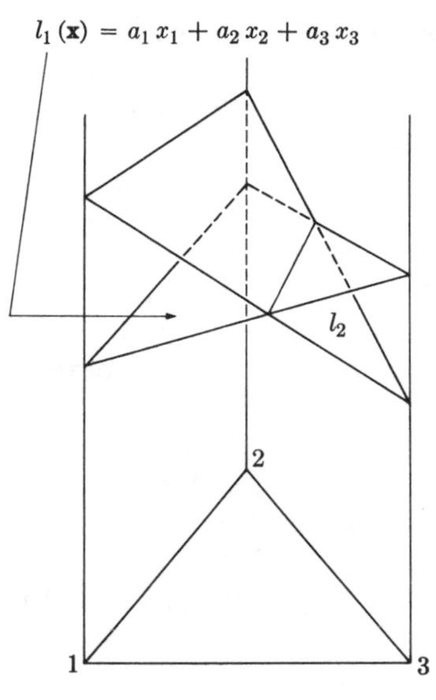

FIGURE 2.4

Over this plane triangle we can construct a set of planes, each plane giving the pay-off if Player II uses a particular pure strategy. This has been done in Fig. 2.4. As before, we form the envelope from below and consider its set of maximum points; this can be a single point, a line, or (in general) a polyhedron. The value of the function $\phi(\mathbf{x}) = \inf_i l_i(\mathbf{x})$ at a maximum point is the value of the game. Points in the projection of the maximum polyhedron on the plane triangle correspond to optimal strategies for Player I. Optimal strategies for Player II are obtained by a correct weighting of the pure strategy planes in a manner analogous to the procedures used in the $2 \times m$ game; i.e., the pure strategy planes are so weighted that the mixed strategy is a plane parallel to the basic triangle.

For $n \times m$ games with both n and m greater than 3, this method becomes impractical.

2.4 Characterization of extreme-point optimal strategies. A strategy $\mathbf{x}^0$ is called an extreme-point optimal strategy if $\mathbf{x}^0$ is an extreme point of the convex set X^0 of optimal strategies (see Section B.2). Extreme-point optimal $\mathbf{y}$ strategies are similarly defined. By way of introduction to the methods used to characterize extreme-point optimal strategies, we shall first analyze a special case of Theorem 2.4.1.

Suppose that $\mathbf{A} = \|a_{ij}\|$ is $n \times n$ and nonsingular. Let $\{\mathbf{x}^0, \mathbf{y}^0\}$ be a pair of optimal strategies and assume that $x_i^0 > 0$ for all i and $y_j^0 > 0$ for all j. Let $\mathbf{e} = \langle 1, 1, \ldots, 1\rangle$ represent a vector which consists of n components each of value 1.

By Lemma 2.1.2

$$\mathbf{A}\mathbf{y}^0 \equiv \mathbf{v}, \qquad \mathbf{x}^0\mathbf{A} \equiv \mathbf{v}, \tag{2.4.1}$$

where v, the value of the game, is to be interpreted either as a scalar or as a vector of suitable size all of whose components are v, depending on the context. It follows that

$$\mathbf{y}^0 = \mathbf{A}^{-1}\mathbf{v}, \qquad \mathbf{x}^0 = \mathbf{v}\mathbf{A}^{-1}$$

and

$$1 = \sum_{j=1}^{n} y_j^0 = (\mathbf{e}, \mathbf{A}^{-1}v\mathbf{e}) = v(\mathbf{e}, \mathbf{A}^{-1}\mathbf{e}).$$

This last relation implies $v \neq 0$. Thus for this case the solution $\mathbf{x}^0$, $\mathbf{y}^0$ and the value v can be explicitly computed by means of the following formulas:

$$v = \frac{1}{(\mathbf{e}, \mathbf{A}^{-1}\mathbf{e})}, \qquad \mathbf{y}^0 = \frac{\mathbf{A}^{-1}\mathbf{e}}{(\mathbf{e}, \mathbf{A}^{-1}\mathbf{e})}, \qquad \mathbf{x}^0 = \frac{\mathbf{e}\mathbf{A}^{-1}}{(\mathbf{e}, \mathbf{A}^{-1}\mathbf{e})}. \tag{2.4.2}$$

Clearly, a pair of solutions can be represented by the formulas of (2.4.2) whenever (2.4.1) holds and $\mathbf{A}$ is nonsingular. The assumption that (2.4.1) is satisfied is in fact weaker than the hypothesis that $x_i^0 > 0$ for all i and $y_j^0 > 0$ for all j.

This example indicates that under suitable conditions, when $\mathbf{A}$ is nonsingular, a pair of solutions of the game, one for each player, can be calculated explicitly in terms of $\mathbf{A}^{-1}$. To generalize this construction, we shall determine a submatrix $\mathbf{M}$ of $\mathbf{A}$ with properties similar to those of the matrix of (2.4.2). The optimal strategies obtained from such submatrices constitute extreme-point solutions. Our next theorem makes explicit this extension of (2.4.2).

▶ **Theorem 2.4.1.** *If $\mathbf{x}^0, \mathbf{y}^0$ are extreme-point optimal strategies for a game with pay-off matrix $\mathbf{A}$, and if $v \neq 0$, then there exists a nonsingular submatrix $\mathbf{M}$ of $\mathbf{A}$ such that*

$$v = \frac{1}{(\mathbf{e}, \mathbf{M}^{-1}\mathbf{e})} \tag{2.4.3a}$$

$$\mathbf{x}^0 = \frac{\mathbf{e}\mathbf{M}^{-1}}{(\mathbf{e}, \mathbf{M}^{-1}\mathbf{e})} \tag{2.4.3b}$$

$$\mathbf{y}^0 = \frac{\mathbf{M}^{-1}\mathbf{e}}{(\mathbf{e}, \mathbf{M}^{-1}\mathbf{e})}, \tag{2.4.3c}$$

where $\mathbf{e}$ is a vector of the same dimension as $\mathbf{M}$ all of whose components are 1.

(Formulas (2.4.3b) and (2.4.3c) exhibit only such components of $\mathbf{x}^0, \mathbf{y}^0$ as correspond to the rows and columns of $\mathbf{A}$ which appear in $\mathbf{M}$. The remaining components of $\mathbf{x}^0, \mathbf{y}^0$ are identically zero.)

For convenience, the proof of this theorem is divided into a series of lemmas and discussions.

By X^0, Y^0 we denote the spaces of optimal strategies for Players I and II, respectively. We know that X^0, Y^0 are closed convex sets (Lemma 2.1.1). By Lemma B.2.4, we know that X^0, Y^0 have extreme points. Let $\mathbf{x}^0$ and $\mathbf{y}^0$ be such points.

We can relabel the rows and columns of $\mathbf{A} = \|a_{ij}\|$ in such a way that the first r components of $\mathbf{x}^0$ and the first s components of $\mathbf{y}^0$ are positive, while the other components are zero. Thus

$$\mathbf{x}^0 = \langle x_1^0, \ldots, x_r^0, 0, \ldots, 0\rangle \qquad r \leq n,$$

$$\mathbf{y}^0 = \langle y_1^0, \ldots, y_s^0, 0, \ldots, 0\rangle \qquad s \leq m.$$

By Lemma 2.1.2, since $\mathbf{x}^0$ is optimal and $y_j^0 > 0$, the inner product of $\mathbf{x}^0$ with the jth column of $\mathbf{A}$ equals v for all $j \leq s$. For the other columns the inner product is greater than or equal to v.

Similarly, the inner product of $\mathbf{y}^0$ with the ith row equals v for all $i \leq r$. For the other rows, the inner product is less than or equal to v.

Again, we can rearrange the rows and columns of matrix $\mathbf{A}$ (while maintaining the first r rows and s columns) so that (a) the inner product of $\mathbf{y}^\circ$ and $\mathbf{x}^\circ$ with each of the first $\bar{r}$ rows and $\bar{s}$ columns, respectively, is exactly equal to v, where $0 < r \leq \bar{r} \leq n$, $0 < s \leq \bar{s} \leq m$; (b) the inner product of $\mathbf{y}^0$ with each of the last $n - \bar{r}$ rows is less than v; and (c) the inner product of $\mathbf{x}^0$ with each of the last $m - \bar{s}$ columns is greater than v.

With the rows and columns arranged in the order indicated, we decompose $\mathbf{A}$ as

$$\mathbf{A} = \begin{Vmatrix} \mathbf{A}_1 & \mathbf{A}_3 & \mathbf{A}_7 \\ \mathbf{A}_2 & \mathbf{A}_4 & \mathbf{A}_8 \\ \mathbf{A}_5 & \mathbf{A}_6 & \mathbf{A}_9 \end{Vmatrix}$$

where $\mathbf{A}_1$ is of size $r \times s$, $\mathbf{A}_2$ of size $(\bar{r} - r) \times s$, $\mathbf{A}_3$ of size $r \times (\bar{s} - s)$, $\mathbf{A}_4$ of size $(\bar{r} - r) \times (\bar{s} - s)$, etc. This notation is used in the following lemma.

▶ **Lemma** 2.4.1. The matrix

$$\mathbf{C} = \begin{Vmatrix} \mathbf{A}_1 \\ \mathbf{A}_2 \end{Vmatrix}$$

defines a linear mapping of E^s into $E^{\bar{r}}$ with the property that only the zero vector is mapped by $\mathbf{C}$ into zero.

Proof (by contradiction). We shall show that if $\mathbf{Cz} = \mathbf{0}$ for $\mathbf{z} \neq \mathbf{0}$, then $\mathbf{y}^0$ is not an extreme point of Y^0. First, note that $\mathbf{Cz} = \mathbf{0}$ implies

$$0 = (\mathbf{x}^0, \mathbf{Az}) = (\mathbf{x}^0\mathbf{A}, \mathbf{z}) = v \sum_{i=1}^{s} z_i.$$

But, since $v \neq 0$,

$$\sum_{i=1}^{s} z_i = 0.$$

Next extend $\mathbf{z}$ to the m-dimensional vector

$$\tilde{\mathbf{z}} = \langle z_1, z_2, \ldots, z_s, 0, 0, \ldots, 0\rangle$$

and choose ϵ so small that $y_i^0 \pm \epsilon z_i \geq 0$ for all i. Then $\mathbf{y}^0 \pm \epsilon\tilde{\mathbf{z}}$ are strategies, since

$$\sum_{i=1}^{m} (y_i^0 \pm \epsilon\tilde{z}_i) = \sum_{i=1}^{s} y_i^0 \pm \epsilon \sum_{i=1}^{m} \tilde{z}_i = 1.$$

Also,

$$\sum_{j=1}^{s} a_{ij}(y_j^0 \pm \epsilon z_j) = \sum_{j=1}^{s} a_{ij}y_j^0 \pm \epsilon \sum_{j=1}^{s} a_{ij}z_j = \begin{cases} v & i \leq \bar{r}, \\ v - \lambda_i \pm \epsilon K_i & i > \bar{r}, \end{cases}$$

where $\lambda_i > 0$, since

$$\sum_{j=1}^{s} a_{ij}y_j^0 < v$$

for all $i > \bar{r}$. Since we may choose ϵ even smaller if necessary, it follows that $\mathbf{y}^0 \pm \epsilon\tilde{\mathbf{z}}$ are optimal strategies. Finally, the relation

$$\mathbf{y}^0 = \tfrac{1}{2}(\mathbf{y}^0 + \epsilon\tilde{\mathbf{z}}) + \tfrac{1}{2}(\mathbf{y}^0 - \epsilon\tilde{\mathbf{z}})$$

shows that $\mathbf{y}^0$ is not extreme. From this contradiction we infer the truth of the lemma.

Next we define $\tilde{\mathbf{C}}$ to be a minimal submatrix of $\mathbf{C}$ such that (a) $\tilde{\mathbf{C}}$ contains the rows and columns of $\mathbf{A}_1$, and (b) $\tilde{\mathbf{C}}\mathbf{z} = \mathbf{0}$ implies $\mathbf{z} = \mathbf{0}$. We know from Lemma 2.4.1 that $\tilde{\mathbf{C}}$ exists.

The minimality of $\tilde{\mathbf{C}}$ means that if $\mathbf{C}^*$ is a proper submatrix of $\tilde{\mathbf{C}}$ containing $\mathbf{A}_1$, condition (b) is violated. We shall use this fact later.

Analogously, we establish the existence of a minimal submatrix $\tilde{\mathbf{D}}$ of $\mathbf{D} = \|\mathbf{A}_1, \mathbf{A}_3\|$ such that (a′) $\tilde{\mathbf{D}}$ contains the rows and columns of $\mathbf{A}_1$, and (b′) $\mathbf{w}\tilde{\mathbf{D}} = \mathbf{0}$ implies $\mathbf{w} = \mathbf{0}$. We are now prepared to construct the submatrix $\mathbf{M}$ needed in Theorem 2.4.1.

Let $\mathbf{M}$ be the smallest submatrix of $\mathbf{A}$ containing $\tilde{\mathbf{C}}$ and $\tilde{\mathbf{D}}$; that is, $\mathbf{M}$ consists of the rows represented in $\tilde{\mathbf{C}}$ and the columns represented in $\tilde{\mathbf{D}}$. Clearly, $\mathbf{M}$ is a matrix of size $r' \times s'$, where $r \leq r' \leq \bar{r}$ and $s \leq s' \leq \bar{s}$. $\mathbf{M}$ is called a kernel of the pair of extreme-point solutions $\{\mathbf{x}^0, \mathbf{y}^0\}$. Although $\mathbf{M}$ is not uniquely determined, since $\tilde{\mathbf{C}}$ and $\tilde{\mathbf{D}}$ were not uniquely defined, in most practical cases $\mathbf{M}$ is unique for a specified $\{\mathbf{x}^0, \mathbf{y}^0\}$.

▶LEMMA 2.4.2. The matrix $\mathbf{M}$ is square and nonsingular.

Proof. We shall prove that $\mathbf{Mz} = \mathbf{0}$ implies $\mathbf{z} = \mathbf{0}$. A similar argument shows that $\mathbf{wM} = \mathbf{0}$ implies $\mathbf{w} = \mathbf{0}$. These two results imply the validity of the lemma.

Let k be a fixed index such that $s + 1 \leq k \leq s'$. If $s = s'$, the argument of the succeeding paragraph is unnecessary. We wish first to show that $\mathbf{Mz} = \mathbf{0}$ implies $z_k = 0$ (the kth component of $\mathbf{z}$ is zero).

Let $\mathbf{D}^*$ be the submatrix of $\tilde{\mathbf{D}}$ obtained by deleting the kth column. Since $\tilde{\mathbf{D}}$ is minimal, there exists a nonzero vector $\mathbf{w}^0$ of the form $\mathbf{w}^0 = \langle w_1^0, w_2^0, \ldots, w_r^0, 0, 0, \ldots, 0\rangle$ such that $\mathbf{w}^0\mathbf{D}^* = \mathbf{0}$ and $\mathbf{w}^0\tilde{\mathbf{D}}$ is nonzero in the kth coordinate only. Equivalently, the kth coordinate a_k of $\mathbf{w}^0\mathbf{M} = \mathbf{w}^0\tilde{\mathbf{D}}$ is nonzero, while all other coordinates are zero. Hence $a_k z_k = (\mathbf{w}^0\mathbf{M}, \mathbf{z}) = (\mathbf{w}^0, \mathbf{Mz}) = 0$; and since $a_k \neq 0$, we deduce that $z_k = 0$.

To sum up, if $\mathbf{Mz} = \mathbf{0}$, then $\mathbf{z}$ necessarily has the form

$$\mathbf{z} = \langle z_1, z_2, \ldots, z_s, 0, 0, \ldots, 0\rangle.$$

But then $\mathbf{Mz} = \mathbf{0}$ is the same as $\tilde{\mathbf{C}}\mathbf{z} = \mathbf{0}$, and $\mathbf{z}$ must be identically zero by Lemma 2.4.1. The proof is complete.

Proof of Theorem 2.4.1. Let $\mathbf{M}$ denote the kernel associated with the extreme-point optimal strategies $\mathbf{x}^0$ and $\mathbf{y}^0$ as defined above. Since $\mathbf{M}$ is a submatrix of

$$\left\|\begin{matrix} \mathbf{A}_1 & \mathbf{A}_3 \\ \mathbf{A}_2 & \mathbf{A}_4 \end{matrix}\right\|,$$

we have

$$\mathbf{M}\dot{\mathbf{y}}^0 = \mathbf{v}, \qquad \dot{\mathbf{x}}^0\mathbf{M} = \mathbf{v}, \tag{2.4.4}$$

where $\mathbf{v}$ is again a vector of the proper size with each component equal to v, and $\dot{\mathbf{y}}^0$ and $\dot{\mathbf{x}}^0$ are used for the vectors $\mathbf{y}^0$ and $\mathbf{x}^0$ with some of the zero components deleted. But $\mathbf{M}$ is nonsingular by Lemma 2.4.2, and therefore the equations of (2.4.4) become equivalent to formulas (2.4.3a, b, c).

The converse proposition to Theorem 2.4.1 takes the following form.

►THEOREM 2.4.2. If $\mathbf{x}^0, \mathbf{y}^0$ are optimal strategies expressed by formulas (2.4.3a, b, c) and $\mathbf{M}$ is a nonsingular submatrix of $\mathbf{A}$, then $\mathbf{x}^0, \mathbf{y}^0$ are extreme-point optimal strategies.

Proof. Assume

$$\mathbf{x}^0 = \frac{\mathbf{x}' + \mathbf{x}''}{2},$$

with $\mathbf{x}'$ and $\mathbf{x}''$ optimal. Since the components of $\mathbf{x}^0$ which do not correspond to rows of $\mathbf{A}$ appearing in $\mathbf{M}$ are identically zero, the corresponding components of $\mathbf{x}'$ and $\mathbf{x}''$ must also be zero. Thus it makes sense to speak about the vectors $\mathbf{x}'\mathbf{M}$ and $\mathbf{x}''\mathbf{M}$.

Since $\mathbf{x}'$ and $\mathbf{x}''$ are optimal,

$$\mathbf{x}'\mathbf{M} \geq \mathbf{v} \tag{2.4.5}$$

and

$$\mathbf{x}''\mathbf{M} \geq \mathbf{v}. \tag{2.4.6}$$

Adding, we obtain $\mathbf{x}^0\mathbf{M} \geq \mathbf{v}$.

If strict inequality occurs in either (2.4.5) or (2.4.6) for some component, then correspondingly $\mathbf{x}^0\mathbf{M} > \mathbf{v}$ for the same component. This is impossible by (2.4.3b). Therefore $\mathbf{x}'\mathbf{M} = \mathbf{v}$ and $\mathbf{x}''\mathbf{M} = \mathbf{v}$; and thus $(\mathbf{x}' - \mathbf{x}'')\mathbf{M} = \mathbf{0}$, which implies $\mathbf{x}' = \mathbf{x}''$, since $\mathbf{M}$ is nonsingular. But if

$$\mathbf{x}^0 = \frac{\mathbf{x}' + \mathbf{x}''}{2},$$

with $\mathbf{x}'$ and $\mathbf{x}''$ optimal, always requires that $\mathbf{x}' = \mathbf{x}''$, then $\mathbf{x}^0$ by definition is an extreme point of X^0. A similar argument applies to the strategy $\mathbf{y}^0$.

The case $v = 0$. If $v = 0$, we can choose any fixed $a \neq 0$ and form the matrix $\|a_{ij} + a\|$. The value of the game is increased by a while the optimal strategies are unchanged (see Problem 7 of Chapter 1). Formulas (2.4.3a, b, c), are valid for the new game, whose value is now $v = a$. Our object now is to exhibit clearly the functional dependence of these formulas on the constant a. Let $|\mathbf{M}|$ denote the determinant of $\mathbf{M}$. When $v \neq 0$, formulas (2.4.3a, b, c) can be written in terms of the adjoint matrix as follows:

$$v = \frac{|\mathbf{M}|}{(\mathbf{e}, \operatorname{adj}(\mathbf{M})\mathbf{e})}, \tag{2.4.7a}$$

$$\mathbf{x}^0 = \frac{\mathbf{e}\operatorname{adj}(\mathbf{M})}{(\mathbf{e}, \operatorname{adj}(\mathbf{M})\mathbf{e})}, \tag{2.4.7b}$$

$$\mathbf{y}^0 = \frac{\operatorname{adj}(\mathbf{M})\mathbf{e}}{(\mathbf{e}, \operatorname{adj}(\mathbf{M})\mathbf{e})}. \tag{2.4.7c}$$

If $v = 0$, and we consider the matrix $\|a_{ij} + a\|$ with $a > 0$, then every extreme-point solution $\{\mathbf{x}^0, \mathbf{y}^0\}$ admits the representation

$$a + v = \frac{|(\mathbf{M} + \mathbf{a})|}{(\mathbf{e}, \operatorname{adj}(\mathbf{M} + \mathbf{a})\mathbf{e})} = a + \frac{|\mathbf{M}|}{(\mathbf{e}, \operatorname{adj}(\mathbf{M})\mathbf{e})},$$

$$\mathbf{x}^0 = \frac{\mathbf{e} \operatorname{adj}(\mathbf{M} + \mathbf{a})}{(\mathbf{e}, \operatorname{adj}(\mathbf{M} + \mathbf{a})\mathbf{e})} = \frac{\mathbf{e} \operatorname{adj}(\mathbf{M})}{(\mathbf{e}, \operatorname{adj}(\mathbf{M})\mathbf{e})},$$

$$\mathbf{y}^0 = \frac{\operatorname{adj}(\mathbf{M} + \mathbf{a})\mathbf{e}}{(\mathbf{e}, \operatorname{adj}(\mathbf{M} + \mathbf{a})\mathbf{e})} = \frac{\operatorname{adj}(\mathbf{M})\mathbf{e}}{(\mathbf{e}, \operatorname{adj}(\mathbf{M})\mathbf{e})},$$

provided $(\mathbf{e}, \operatorname{adj}(\mathbf{M})\mathbf{e})$ is not zero by virtue of the identities

$$(1)\quad |\mathbf{M} + \mathbf{a}| = |\mathbf{M}| + a(\mathbf{e}, \operatorname{adj}(\mathbf{M})\mathbf{e}),$$

$$(2)\quad \mathbf{e} \operatorname{adj}(\mathbf{M} + \mathbf{a}) = \mathbf{e} \operatorname{adj}(\mathbf{M}),$$

$$(3)\quad \operatorname{adj}(\mathbf{M} + \mathbf{a})\mathbf{e} = \operatorname{adj}(\mathbf{M})\mathbf{e}$$

(see Section A.9). Hence (2.4.7a, b, c) are valid also for the case $v = 0$.

Theorem 2.4.1 and its converse can now be restated in terms of the adjoints of square submatrices instead of in terms of inverses. With this new formulation (Theorem 2.4.3) we need no longer distinguish between the two cases $v = 0$ and $v \neq 0$.

▶THEOREM 2.4.3. A necessary and sufficient condition that the optimal strategies $\mathbf{x}^0$ and $\mathbf{y}^0$ be extreme-point optimal strategies and v the value of the game is that there exist a square submatrix $\mathbf{M}$ of $\mathbf{A}$ such that

$$(1)\quad (\mathbf{e}, \operatorname{adj}(\mathbf{M})\mathbf{e}) \neq 0, \qquad (3)\quad \dot{\mathbf{x}}^0 = \frac{\mathbf{e} \operatorname{adj}(\mathbf{M})}{(\mathbf{e}, \operatorname{adj}(\mathbf{M})\mathbf{e})},$$

$$(2)\quad v = \frac{|\mathbf{M}|}{(\mathbf{e}, \operatorname{adj}(\mathbf{M})\mathbf{e})}, \qquad (4)\quad \dot{\mathbf{y}}^0 = \frac{\operatorname{adj}(\mathbf{M})\mathbf{e}}{(\mathbf{e}, \operatorname{adj}(\mathbf{M})\mathbf{e})}, \tag{2.4.8}$$

where $\dot{\mathbf{x}}^0$ is obtained from $\mathbf{x}^0$ by deleting the components corresponding to the rows deleted from $\mathbf{A}$ to obtain $\mathbf{M}$ (the deleted components equaling zero), and $\dot{\mathbf{y}}^0$ is obtained from $\mathbf{y}^0$ analogously.

Theorem 2.4.1 suggests the following systematic method of obtaining the value and extreme-point optimal strategies of a game. Let us restrict our attention to games with value different from zero. Take all possible submatrices $\mathbf{M}$ of $\mathbf{A}$. For each $\mathbf{M}$ check to see if it has an inverse. If no inverse exists, go on to another $\mathbf{M}$. If one does exist, compute $\dot{\mathbf{x}}^0$, $\dot{\mathbf{y}}^0$, and v by formulas (2.4.3a, b, c). The relations $\sum_i \dot{x}_i^0 = 1$ and $\sum_j \dot{y}_j^0 = 1$ automatically hold by virtue of (2.4.3c), but it might happen that $\mathring{x}_i^0 < 0$ or $\mathring{y}_j^0 < 0$ for some i or j. In this event, reject $\mathbf{M}$. If all the terms are nonnegative, form $\mathbf{x}^0$ and $\mathbf{y}^0$ by adding zeros at the coordinates of the deleted rows and columns, respectively. From (2.4.3a) and (2.4.3b) it follows that the inner product of $\mathbf{x}^0$ with any nondeleted column equals v, but a check must be made to see that the inner product of $\mathbf{x}^0$ with a deleted column is greater than or equal to v. Similarly, the inner product of $\mathbf{y}^0$ with any nondeleted row equals v, but a check must be made to see that the inner product of $\mathbf{y}^0$ with a deleted row is less than or equal to v. If all these inner products correctly satisfy the conditions, $\mathbf{x}^0$ and $\mathbf{y}^0$ are extreme-point optimal strategies. If they do not, this $\mathbf{M}$ must be rejected.

When the entire set of submatrices has been processed, we shall have the sets of extreme-point optimal strategies for Players I and II. The sets of optimal strategies are then obtained by forming the convex hull of these extreme points.

It should be emphasized that the class of all possible submatrices of $\mathbf{A}$ includes $\mathbf{A}$ itself and all single elements.

As an illustration of Theorem 2.4.3, consider the pay-off matrix

$$\begin{array}{c} \text{II} \\ \text{I}\begin{Vmatrix} 1 & 1 & 3 \\ 1 & 1 & 0 \\ 0 & 2 & -5 \end{Vmatrix} = \mathbf{A}. \end{array}$$

The four pairs of extreme-point optimal strategies and their corresponding kernels $\mathbf{M}$ are as follows:

$$\mathbf{x}^0 = \langle 1, 0, 0\rangle \qquad \mathbf{y}^0 = \langle 1, 0, 0\rangle \qquad \mathbf{M} = a_{11} = 1$$

$$\mathbf{x}^0 = \langle 1, 0, 0\rangle \qquad \mathbf{y}^0 = \langle \tfrac{1}{2}, \tfrac{1}{2}, 0\rangle \qquad \mathbf{M} = \begin{Vmatrix} 1 & 1 \\ 0 & 2 \end{Vmatrix}$$

$$\mathbf{x}^0 = \langle \tfrac{1}{3}, \tfrac{2}{3}, 0\rangle \qquad \mathbf{y}^0 = \langle 1, 0, 0\rangle \qquad \mathbf{M} = \begin{Vmatrix} 1 & 3 \\ 1 & 0 \end{Vmatrix}$$

$$\mathbf{x}^0 = \langle \tfrac{1}{3}, \tfrac{2}{3}, 0\rangle \qquad \mathbf{y}^0 = \langle \tfrac{1}{2}, \tfrac{1}{2}, 0\rangle \qquad \mathbf{M} = \mathbf{A}.$$

2.5 Completely mixed matrix games. A matrix game is said to be completely mixed (c.m.) if every optimal $\mathbf{x}, \mathbf{y}$ involves every pure strategy with positive probability: i.e., if $\mathbf{x}^0 = \langle x_1^0, \ldots, x_n^0\rangle$ and $\mathbf{y}^0 = \langle y_1^0, \ldots, y_m^0\rangle$ are both optimal, then $x_i^0 > 0$ for all i and $y_j^0 > 0$ for all j. Completely mixed matrix games constitute an important subclass of games for which the solutions are easily determined.

▶THEOREM 2.5.1. The solutions to a game with matrix **A** c.m. are unique.

The assertion follows from Theorem 2.4.3. If $\mathbf{x}^0, \mathbf{y}^0$ are extreme-point optimal strategies, their kernel must be the full matrix **A**. From (2.4.8) it is clear that all extreme-point optimal strategies coincide, since they share a common kernel. Hence the solutions for each player are unique.

▶COROLLARY 2.5.1. If a game **A** is c.m. and $v \neq 0$, then **A** is nonsingular.

This follows from Theorem 2.4.1, since the kernel of any pair of extreme-point solutions must be **A**.

Example 1. Minkowski-Leontief matrix. This is a matrix which occurs in economic theory in the area of activity analysis. A matrix **A** is called a Minkowski-Leontief matrix if (a) **A** is square; (b) $a_{ii} > q \geq 0$; (c) $a_{ij} \leq q$ for $i \neq j$; and

$$\text{(d)} \qquad \sum_{i=1}^{n} a_{ij} > nq \geq 0.$$

Usually in practice $q = 0$. We shall now show that a Minkowski-Leontief matrix generates a c.m. game.

Let $\mathbf{x} = \langle 1/n, 1/n, \ldots, 1/n \rangle$; then

$$\sum_{i=1}^{n} a_{ij}x_i = \frac{1}{n} \sum_{i=1}^{n} a_{ij} > q.$$

Consequently, the value of the game is $v > q \geq 0$. Suppose that $\mathbf{x} = \langle x_1, \ldots, x_n \rangle$ is an optimal strategy, with $x_{j_0} = 0$. Then

$$(\mathbf{xA})_{j_0} = \sum_{i=1}^{n} x_i a_{ij_0} = \sum_{i \neq j_0} x_i a_{ij_0} \leq \sum_{i=1}^{n} x_i q \leq q < v.$$

Hence **x** cannot be optimal. Thus every optimal strategy for Player I uses every component with positive probability.

Suppose that $\mathbf{y} = \langle y_1, \ldots, y_n \rangle$ is an optimal strategy for Player II and $y_{j_0} = 0$. By Lemma 2.1.2,

$$\sum_{j=1}^{n} a_{ij}y_j = v$$

for all i. However, for $i = j_0$,

$$\sum_{j=1}^{n} a_{j_0 j}y_j = \sum_{j \neq j_0} a_{j_0 j}y_j \leq q < v,$$

a contradiction. Thus, every optimal **y** strategy has every component positive. As a corollary, it follows that every Minkowski-Leontief matrix is nonsingular.

As an immediate consequence of Lemma 2.5.1 below and the above example, we find that if **A** is a square matrix such that (a) $a_{ij} \leq q$ for $i \neq j$, (b) $a_{ii} > q \geq 0$, and

$$\text{(c)} \qquad \sum_{j=1}^{n} a_{ij} > nq,$$

then **A** is c.m.

▶ LEMMA 2.5.1. If **A** is completely mixed, then the matrix $\mathbf{B} = \mathbf{A}'$ (**A** transpose) is completely mixed, and conversely.

Proof. Let $\mathbf{w}^0$, $\mathbf{z}^0$ be the unique solutions of **A** for Players I and II, respectively. Then, by Lemma 2.1.2,

$$\mathbf{w}^0\mathbf{A} = \mathbf{v}, \qquad \mathbf{A}\mathbf{z}^0 = \mathbf{v}.$$

Let $\mathbf{B} = \mathbf{A}'$; then $\mathbf{B}' = (\mathbf{A}')' = \mathbf{A}$ and

$$\mathbf{z}^0\mathbf{B} = \mathbf{A}\mathbf{z}^0 = \mathbf{v}, \qquad \mathbf{B}\mathbf{w}^0 = \mathbf{w}^0\mathbf{A} = \mathbf{v}.$$

Hence $\mathbf{z}^0$ and $\mathbf{w}^0$ comprise optimal strategies for the matrix **B** for Players I and II, respectively, with the same value v. They involve all components with positive probability, and they must be unique or the solution of **A** would not be unique by reversing the reasoning. Hence $\mathbf{A}'$ is completely mixed.

The converse follows from the fact that $(\mathbf{A}')' = \mathbf{A}$.

Example 2. Ordered set game. Let $L_0, L_1, \ldots, L_{n-1}$ be disjoint closed sets on the real axis with $L_i < L_{i+1}$; that is, the set L_i lies strictly to the left of the set L_{i+1}. Let

$$\mathbf{A} = \begin{Vmatrix} L_0 & L_1 & L_2 & L_3 & \cdots & L_{n-1} \\ L_{n-1} & L_0 & L_1 & L_2 & \cdots & \cdots \\ L_{n-2} & L_{n-1} & L_0 & L_1 & \cdots & \cdots \\ \vdots & & & & & \\ L_1 & L_2 & \cdots & \cdots & \cdots & L_0 \end{Vmatrix}$$

A is interpreted as a matrix $\|a_{ij}\|$, where the element a_{ij} is a value contained in the L-set appearing in the $\{i, j\}$ place. Algebraically, $a_{i,j} \in L_{j-i}$ if $j \geq i$, and $a_{ij} \in L_{n-i+j}$ if $j < i$.

We assert that **A** is completely mixed. To show this, we assume that **x** is optimal and that $x_j = 0$ for some j. If $i \neq j$, then $a_{ij} > a_{i,j-1}$ (if $j = 0$, then $j - 1$ is interpreted as $n - 1$). Since $x_j = 0$,

$$\sum_{i=1}^{n} x_i a_{ij} > \sum_{i=1}^{n} x_i a_{i,j-1} \geq v.$$

Therefore, by Lemma 2.1.2, no optimal $\mathbf{y}$ strategy can have $y_j > 0$. Clearly, $a_{j+1,k} < a_{j,k}$ if $j \neq k$ (interpret $j + 1 = 0$ if $j = n - 1$); hence if $\mathbf{y}$ is optimal, we obtain

$$\sum_{k=1}^{n} a_{j+1,k} y_k < \sum_{k=1}^{n} a_{jk} y_k \leq v.$$

Again by Lemma 2.1.2, we see that no optimal $\mathbf{x}$ strategy can have $x_{j+1} > 0$. By iteration, we obtain $\mathbf{x} = \mathbf{0}$, which is impossible; therefore every optimal $\mathbf{x}$ has all components positive. Since starting with $\mathbf{y}$ optimal and $y_j = 0$ leads to the same contradiction, we see that every optimal $\mathbf{y}$ also has all components positive. Hence $\mathbf{A}$ is completely mixed.

If the sets L_k are ordered in the reverse direction, i.e.,

$$L_0 > L_1 > L_2 > \cdots,$$

then the corresponding game is completely mixed.

In particular, if the principal diagonal of a 2×2 matrix dominates or is dominated, then the optimal strategies for that matrix are completely mixed. Formally, the game with matrix

$$\mathbf{A} = \begin{Vmatrix} a & c \\ d & b \end{Vmatrix}$$

will be completely mixed provided

$$\max\{a, b\} < \min\{c, d\} \quad \text{or} \quad \min\{a, b\} > \max\{c, d\}.$$

The converse is also valid in this case, for if the principal diagonal does not dominate, a saddle point exists.

No such complete description of completely mixed games exists for games of higher order than 2×2 (see Problem 9).

Example 3. Cyclic matrix. Let

$$\mathbf{A} = \begin{Vmatrix} a_0 & a_1 & a_2 & \cdots & a_{n-1} \\ a_{n-1} & a_0 & a_1 & \cdots & \cdots \\ \vdots & & & & \\ a_1 & a_2 & a_3 & \cdots & a_0 \end{Vmatrix}$$

where $a_i \neq a_j$. This example is similar to the preceding one, with $L_k = \{a_k\}$ consisting of a single point; however, we do not assume here that $a_i < a_{i+1}$, or, more generally, that any ordering relationship is imposed on the quantities a_i.

A sufficient condition that $\mathbf{A}$ be c.m. is that $\det \mathbf{A} \neq 0$. In order to verify this statement, we first consider the form of $\det \mathbf{A}$.

Note that
$$\lambda_r = \sum_{k=0}^{n-1} a_k w^{kr} \qquad (r = 1, \ldots, n)$$
are distinct characteristic roots of $\mathbf{A}$, where w is a primitive nth root of unity. Thus
$$\det \mathbf{A} = \prod_{r=1}^{n} \lambda_r = \prod_{r=1}^{n} \sum_{k=0}^{n-1} a_k w^{kr}.$$

Next let us verify that $\mathbf{x} = \mathbf{y} = \langle 1/n, \ldots, 1/n \rangle$ are optimal strategies and
$$v = \frac{1}{n} \sum a_i.$$
Explicitly,
$$(\mathbf{xA})_j = \sum_{i=1}^{n} x_i a_{ij} = \frac{1}{n} \sum_{i=1}^{n} a_i = v$$
and similarly $\mathbf{Ay} = \mathbf{v}$. Since, by assumption, $\det \mathbf{A} \neq 0$, it follows from Theorem 2.4.2 that $\mathbf{x}^0 = \mathbf{y}^0 = \langle 1/n, \ldots, 1/n \rangle$ are both extreme-point optimal strategies. Let $\mathbf{y}'$ be any extreme-point optimal strategy different from $\mathbf{y}^0$. The pair of extreme-point solutions $\{\mathbf{x}^0, \mathbf{y}'\}$ must have as its kernel the full matrix, and thus $\mathbf{y}' = \mathbf{y}^0$. In a similar manner we find that $\mathbf{x}^0$ is the only optimal strategy for Player I. Hence $\mathbf{A}$ is c.m.

In particular, for $n = 3$, where a_0, a_1, a_2 are not all equal and $a_0 + a_1 + a_2 \neq 0$, we find that
$$\mathbf{A} = \begin{Vmatrix} a_0 & a_1 & a_2 \\ a_2 & a_0 & a_1 \\ a_1 & a_2 & a_0 \end{Vmatrix}$$
is completely mixed. It is enough to verify that $a_0 = a_1 = a_2$ is implied by either $a_0 + a_1 w + a_2 w^2 = 0$ or $a_0 + a_1 w^2 + a_2 w = 0$, where w is a third root of unity. By examining the imaginary part of either equation, we find that $a_1 = a_2$. It follows from the real part that $a_0 = a_1 = a_2$.

In contrast,
$$\begin{Vmatrix} 0 & 1 & 3 & 2 \\ 2 & 0 & 1 & 3 \\ 3 & 2 & 0 & 1 \\ 1 & 3 & 2 & 0 \end{Vmatrix}$$
has a value of $\frac{3}{2}$, and the set of extreme-point optimal strategies for Player I consists of the two strategies $\langle \frac{1}{2}, 0, \frac{1}{2}, 0 \rangle$ and $\langle 0, \frac{1}{2}, 0, \frac{1}{2} \rangle$. The same is true for Player II.

2.6 Symmetric games. A symmetric game is a game with a skew-symmetric pay-off matrix ($\mathbf{A}' = -\mathbf{A}$). The two players have available the same sets of strategies, and both can achieve the same return by employing a given strategy. Such a game is in all senses the fairest to both players because any opportunity open to either player is also open to the other. In fact, if Player I employs the strategy $\mathbf{x} = \langle x_1, \ldots, x_n \rangle$, his yield is the component of the vectors in $\mathbf{xA}$ that corresponds to the pure strategy used by Player II. Similarly, if Player II uses $\mathbf{y} = \langle y_1, y_2, \ldots, y_m \rangle$, the possible yields to Player I are $\mathbf{Ay}$ depending upon which pure strategy he uses. If $\mathbf{x} = \mathbf{y}$, then

$$-\mathbf{Ay} = \mathbf{xA}. \tag{2.6.1}$$

This relationship points up the distinguishing feature of symmetric games: namely, that any outcome achievable by Player I can also be achieved by Player II by using the same strategy.

Since $(\mathbf{x}, \mathbf{Ay}) = 0$ whenever $\mathbf{x} = \mathbf{y}$, we infer that the value of the game is zero. Furthermore, it is clear from (2.6.1) that the sets of optimal strategies X^0 and Y^0 are identical.

These special properties lend some importance and interest to symmetric games. Their main significance, however, derives from the fact that any game can be associated with a symmetric game by an induced correspondence that relates the sets of optimal strategies of the given game with those of the associated symmetric game. Two methods have been proposed for symmetrizing a matrix game of order $n \times m$. The first involves converting the game to a larger symmetric game of order $n + m + 1$. The second associates with the original game a larger symmetric matrix having $n \times m$ rows. We now describe these two symmetrizing procedures.

►THEOREM 2.6.1. If $\mathbf{A}$ is $n \times m$, then there exists a symmetric game $\mathbf{A}_1$ of order $(m + n + 1) \times (m + n + 1)$ such that every optimal strategy $\langle x_1, \ldots, x_n, y_1, \ldots, y_m, \lambda \rangle$ for the game $\mathbf{A}_1$ has the property $\sum x_i = \sum y_i = a > 0$, and $\mathbf{x}/a, \mathbf{y}/a$ are optimal for the game $\mathbf{A}$ with value λ/a; and conversely.

Proof. We may assume $v > 0$ for the game $\mathbf{A}$. Let

$$\mathbf{A}_1 = \left\| \begin{matrix} 0 & \mathbf{A} & -1 \\ -\mathbf{A}' & 0 & 1 \\ 1 & -1 & 0 \end{matrix} \right\|$$

$\mathbf{A}_1$ is skew-symmetric of order $n + m + 1$. Let $\langle x_1, \ldots, x_n, y_1, \ldots, y_m, \lambda \rangle$ be optimal for Player II in the game $\mathbf{A}_1$. Then, since the value of the game $\mathbf{A}_1$ is zero,

$$\mathbf{Ay} - \boldsymbol{\lambda} \leq \mathbf{0}, \qquad -\mathbf{xA} + \boldsymbol{\lambda} \leq \mathbf{0}, \qquad \sum x_i - \sum y_i \leq 0. \tag{2.6.2}$$

Note that $\lambda > 0$; otherwise we would have $\mathbf{Ay} \leq \mathbf{0}$, $\mathbf{xA} \geq \mathbf{0}$ in (2.6.2), which contradicts our assumption that $v > 0$ for the game $\mathbf{A}$. But $\lambda > 0 \rightarrow \sum x_i = \sum y_i$ by Lemma 2.1.2. Finally, $\lambda < 1$ or the second relation of (2.6.2) could not be satisfied. Hence

$$\sum y_i = \sum x_i = \frac{1-\lambda}{2} > 0.$$

If we set

$$a = \frac{1-\lambda}{2},$$

we obtain from (2.6.2) the inequality

$$\mathbf{A}\left(\frac{\mathbf{y}}{a}\right) \leq \frac{\boldsymbol{\lambda}}{a} \leq \left(\frac{\mathbf{x}}{a}\right)\mathbf{A}.$$

Therefore $\mathbf{x}/a$, $\mathbf{y}/a$ are optimal for $\mathbf{A}$, and the value of the game $\mathbf{A}$ is λ/a.

Conversely, if $\mathbf{x} = \langle x_1, \ldots, x_n \rangle$ and $\mathbf{y} = \langle y_1, \ldots, y_m \rangle$ are optimal for $\mathbf{A}$, then (2.6.2) can be verified for the vector

$$\mathbf{w} = \left\langle \frac{\mathbf{x}}{2+v}, \frac{\mathbf{y}}{2+v}, \frac{v}{2+v} \right\rangle.$$

Thus $\mathbf{w}$ is optimal for the game $\mathbf{A}_1$.

Our second method of symmetrizing a game proceeds as follows: Visualize a player engaged in two games $\mathbf{B}'$ and $\mathbf{B}''$, in which he maximizes and minimizes, respectively. The value of this game is the cumulative expected value obtained by playing $\mathbf{B}'$ and $\mathbf{B}''$. Suppose that the indices for $\mathbf{B}'$ are k', l' and the indices for $\mathbf{B}''$ are k'', l''. Player I controls k', l'' and Player II controls k'', l'; hence a strategy i for Player I is a choice of k' and l'', and a strategy j for Player II is a choice of k'' and l'. The expected pay-off is $b'_{k'l'} + b''_{k''l''}$. Suppose now that $\mathbf{B}'' = -\mathbf{B}$ and $\mathbf{B}' = \mathbf{B}$; then $A_{ij} = b_{k'l'} - b_{k''l''}$ for $i = (k', l'')$, $j = (k'', l')$. This game is symmetric, and hence there exists an optimal y_j with components $y_{k''l'}$ satisfying $y_{k''l'} \geq 0$, $\sum y_{k''l'} = 1$. Put

$$\xi_k = \sum_l y_{kl} \qquad \text{and} \qquad \eta_l = \sum_k y_{kl}.$$

Then $\xi_k \geq 0$, $\eta_l \geq 0$ and $\sum \xi_k = \sum \eta_l = 1$.

It is easily shown that $\boldsymbol{\eta}$ and $\boldsymbol{\xi}$ are optimal strategies for $\mathbf{B}$.

2.7 Problems.

1. Use the graphical method to solve the game with matrix

$$\begin{Vmatrix} 1 & 0 & 4 & -1 \\ -1 & 1 & -2 & 5 \end{Vmatrix}.$$

2. Describe a graphical method for solving $m \times 2$ games.

3. Describe all the extreme-point optimal strategies for the game with matrix

$$\mathbf{A} = \begin{Vmatrix} a_1 & & & & \bigcirc \\ & a_2 & & & \\ & & \cdot & & \\ & & & \cdot & \\ & & & & \cdot \\ \bigcirc & & & & a_n \end{Vmatrix}$$

where $a_i > 0$ for all i.

4. Solve Problem 3 without the restriction $a_i > 0$.

5. Use Theorem 2.4.1 to solve the game with matrix

$$\begin{Vmatrix} 0 & 5 & 6 & -4 \\ 3 & 9 & 9 & -6 \\ 3 & -1 & 0 & 2 \end{Vmatrix} \cdot$$

6. Use Theorem 2.4.3 to solve the game of Two-finger Morra, described in Section 1.2.

7. The game with matrix

$$\begin{Vmatrix} 0 & 1 & 1 & 1 & 0 \\ 1 & 1 & 0 & 0 & 0 \\ 1 & 0 & 1 & 0 & 1 \\ 1 & 0 & 0 & 1 & 0 \\ 0 & 0 & 0 & 1 & 1 \end{Vmatrix}$$

has a pair of optimal strategies whose kernel is the full matrix. Find these strategies.

8. The game described in Problem 4 of Chapter 1 has a pair of optimal strategies that use only the last five rows and the last five columns of the matrix. Find these strategies.

9. Show that the game with matrix

$$\begin{Vmatrix} 4 & -3 & -2 \\ -3 & 4 & -2 \\ 0 & 0 & 1 \end{Vmatrix}$$

is completely mixed, and find the optimal strategies.

10. Find all solutions to the game with matrix

$$\left\|\begin{matrix} 4 & 3 & 3 & 2 & 2 & 6 \\ 6 & 0 & 4 & 2 & 6 & 2 \\ 0 & 7 & 3 & 6 & 2 & 2 \end{matrix}\right\|$$

11. Let $\mathbf{A}$ be an $m \times n$ matrix with

$$a_{i,j} \leq a_{i,j+1} \qquad (i = 1, \ldots, m;\ j = 1, \ldots, n - 1).$$

Show that $\mathbf{A}$ always has a saddle point.

12. A cat and a mouse simultaneously enter this blind maze at the points A and B, respectively; they may turn corners but they cannot turn around or stop. They travel at the same speed, and they have enough time to cover four line segments. If the cat finds the mouse, he wins one unit. Find the optimal strategies.

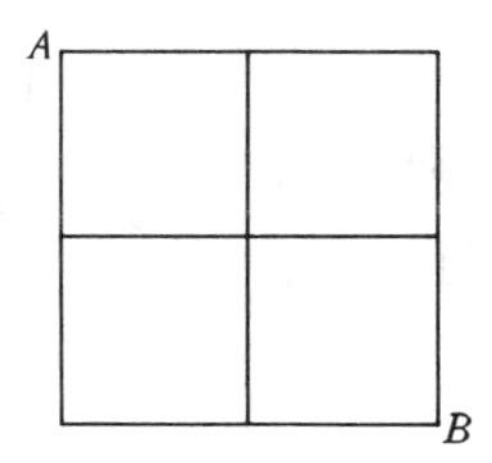

FIGURE 2.5

13. Use the dominance theorems to solve the game

$$\left\|\begin{matrix} 1 & 1 & 0 & 0 & & & & \\ 1 & 1 & 1 & 0 & & \bigcirc & & \\ 0 & 1 & 1 & 1 & & & & \\ \cdot & \cdot & \cdot & \cdot & \cdot & \cdot & \cdot & \cdot \\ \cdot & \cdot & \cdot & \cdot & \cdot & \cdot & \cdot & \cdot \\ & & & & 1 & 1 & 1 & 0 \\ & & \bigcirc & & 0 & 1 & 1 & 1 \\ & & & & 0 & 0 & 1 & 1 \end{matrix}\right\|$$

14. If $\mathbf{A}$ denotes a matrix all of whose elements are nonnegative, and every column of $\mathbf{A}$ has at least one positive component, prove that the value of the game with matrix $\mathbf{A}$ is positive.

15. Prove that the set of optimal strategies in a matrix game constitutes a polyhedral convex set.

16. Let $a_1 < a_2 < a_3 < a_4 < a_5 < a_6$, and

$$\mathbf{A} = \begin{Vmatrix} a_4 & a_3 & a_3 \\ a_1 & a_6 & a_5 \\ a_2 & a_4 & a_3 \end{Vmatrix}$$

Show that any optimal strategy satisfies the conditions

$$x_3 = y_2 = 0, \qquad x_1 > \tfrac{1}{2},$$

$$x_1 > y_1 > x_2, \qquad a_3 < v < a_4.$$

17. Discuss the relationship between the optimal strategies for the game $\mathbf{A}$ and the optimal strategies for the symmetric game

$$\begin{Vmatrix} \mathbf{0} & \mathbf{A} & -\mathbf{1} & \mathbf{1} \\ -\mathbf{A}' & \mathbf{0} & \mathbf{1} & -\mathbf{1} \\ \mathbf{1} & -\mathbf{1} & \mathbf{0} & \mathbf{0} \\ -\mathbf{1} & \mathbf{1} & \mathbf{0} & \mathbf{0} \end{Vmatrix}$$

18. Let G^k $(k = 1, \ldots, K)$ be a game with matrix $\|G^k_{ij}\|$, where $i = 1, \ldots, m$ and $j = 1, \ldots, n$. Let X^k and v^k be the set of optimal maximizing strategies and the value of G^k, respectively. Show that if $\cap X^k \neq 0$, the value v of the game G with matrix

$$\|G_{ij}\| = \left\| \sum_{k=1}^{K} G^k_{ij} \right\|$$

satisfies the inequality

$$v \geq \sum_{k=1}^{K} v^k.$$

19. Consider a game whose matrix $\mathbf{B}$ can be decomposed into MN submatrices $\mathbf{B}^i_j$ with values v^i_j. Show that if for each i there is a common optimal strategy for the games $\mathbf{B}^i_j$, then the value of $\mathbf{B}$ is greater than or equal to the value of the game v whose matrix is $\|v^i_j\|$. Describe the corresponding result in terms of the games $\mathbf{B}^i_j$ (with fixed j), and show that if both conditions hold, the value of $\mathbf{B}$ and its optimal strategies can be determined from the values and optimal strategies of the games $\mathbf{B}^i_j$ and v.

20. Show that there exists no symmetric completely mixed game of even order.

21. Let $a_1, \ldots, a_n$ be positive and $b_1, \ldots, b_{n-1}$ be negative numbers such that $b_{i-1} + a_i + b_i \geq 0$ $(i = 2, \ldots, n-1)$, $a_1 + b_1 \geq 0$, and $a_n + b_{n-1} \geq 0$. Solve the game with matrix

$$\begin{Vmatrix} a_1 & b_1 & & & & \bigcirc \\ b_1 & a_2 & b_2 & & & \\ & b_2 & a_3 & b_3 & & \\ & & \cdot & \cdot & \cdot & \\ & & & \cdot & \cdot & \cdot \\ & & & & \cdot & \cdot \quad \cdot \\ & \bigcirc & & & & b_{n-1} \\ & & & & b_{n-1} & a_n \end{Vmatrix}$$

22. Show that if

$$a_{i,j-1} - 2a_{i,j} + a_{i,j+1} \geq 0$$

for all i and j, then the game with matrix $\mathbf{A}$ always has an optimal strategy in which each player uses at most two components.

Notes and References to Chapter 2

§ **2.1.** The simple properties possessed by the solutions of games described in this section were known and used by von Neumann. They are implicitly contained in [242].

§ **2.2.** Dominance concepts were introduced first in [242]. The two forms of dominance embodied in Theorems 2.2.1 and 2.2.2 are standard equipment in the analysis of game problems. The earliest formulation of the four types of dominance cannot be associated with any individual; in general, it may be ascribed to the Rand Corporation, where many of the more extensive early mathematical developments of game theory were achieved. All four forms of dominance are mentioned in Dresher's handbook [66].

§ **2.3.** The geometric method of this section is well known and has been duly elaborated upon in most elementary treatments of game theory, e.g., [179] and [169]. Borel [39] used these ideas in proving the min-max theorem for 2×2 games.

§ **2.4.** The principal theorem of this section was discovered by Snow and Shapley [219]. One can summarize the main idea of their proof as follows: A point in E^n space can be located as the unique intersection of n particular linearly independent hyperplanes. However, if k hyperplanes through a point are prescribed, some of which may be linearly dependent, then it is necessary to add at least $n - k$ hyperplanes to ensure the point determined. The extra hyperplanes beyond $n - k$ are required by virtue of the dependence relations of the prescribed k hyperplanes.

An elaboration on the geometry underlying this theorem was converted by Motzkin, Raiffa, Thompson, and Thrall [187] into a systematic method for determining both the value and all the solutions of a two-person zero-sum game. Although it is difficult to evaluate the efficiency and practicality of their procedures, the method has an appealing geometric setting.

§ **2.5.** Completely mixed games of order 2×2 were fully investigated by von Neumann [242]. Kaplansky [132] exhibited several special examples of general c.m. games, and extensions of these results were obtained by Bohnenblust, Karlin, and Shapley [36]. Cyclic games constitute a special subclass of games that are invariant with respect to a group of transformations (see also Volume II, Chapter 6 of the present work).

§ **2.6.** The first method of symmetrizing a game was proposed by Gale, Kuhn, and Tucker [98], the second by Brown and von Neumann [42].

§ **2.7.** The result of Problem 18 was noted by M. Woodbury, and that of Problem 19 by Sherman [220]. Problem 20 originated with Kaplansky [132].

CHAPTER 3*

DIMENSION RELATIONS FOR SETS OF OPTIMAL STRATEGIES

In this chapter we investigate the structure of sets of solutions of discrete matrix games. Every such set is known to be a convex set, spanned by only a finite number of extreme-point optimal strategies each of which corresponds to a submatrix with special properties (see Theorem 2.4.3). In the following pages a fundamental relationship is established between the dimensions of the sets of optimal strategies. Particular attention is directed to the class of games whose solutions are unique. For example, it is shown that the class of all games of a given order with unique solutions constitutes an open and dense set within the set of all possible matrix games of the same order. We also solve the problem of constructing a game matrix possessing specified solutions.

3.1 The principal theorems. Before stating the principal theorems we introduce the following notation. Let the vector $\mathbf{a}_i$ denote the ith row of the pay-off matrix $\|a_{ij}\|$, and let the vector $\mathbf{b}_j$ denote the jth column. As previously, $\mathbf{e} = \langle 1, 1, \ldots, 1 \rangle$.

Let X^0 and Y^0 denote the sets of optimal strategies for Players I and II, respectively.

Let I_1 be the set of all row indices i with the property that for some optimal strategy $\mathbf{x}^{(i)}$,

$$x_i^{(i)} > 0.$$

Let I_2 be the set of all row indices i with the property that for every optimal strategy $\mathbf{y}$,

$$\sum_{j=1}^{m} a_{ij} y_j = v.$$

In terms of our previous terminology, the set I_2 consists of all rows which are equalizers with respect to all optimal $\mathbf{y}$ strategies.

Let J_1 and J_2 be the corresponding sets of column indices j satisfying the same properties with respect to the optimal strategies of Player II. More generally, for any strategy $\mathbf{x}$ let $I_1(\mathbf{x})$ be the set of indices i for which

---

* This entire chapter may be omitted at first reading without loss of continuity.

$x_i > 0$, and $I_2(\mathbf{y})$ the set of indices i for which $(\mathbf{a}_i, \mathbf{y}) = v$, and similarly with $J_1(\mathbf{y})$ and $J_2(\mathbf{x})$. It follows that

$$I_1 = \bigcup_{x \in X^0} I_1(\mathbf{x}), \qquad I_2 = \bigcap_{y \in Y^0} I_2(\mathbf{y}),$$

$$J_1 = \bigcup_{y \in Y^0} J_1(\mathbf{y}), \qquad J_2 = \bigcap_{x \in X^0} J_2(\mathbf{x}).$$

Let $\overline{X}$ denote the set of all $\mathbf{x}$ with $I_1(\mathbf{x}) \subseteq I_1$ and $\overline{Y}$ denote the set of all $\mathbf{y}$ with $J_1(\mathbf{y}) \subseteq J_1$. Thus $\overline{X}$ is the smallest face of the fundamental simplex of mixed strategies containing X^0, etc. Our immediate objective is to prove the following two relations:

▶ THEOREM 3.1.1. $I_1 = I_2, J_1 = J_2$.

▶ THEOREM 3.1.2. $\dim \overline{X} - \dim X^0 = \dim \overline{Y} - \dim Y^0$.

Several different types of proofs for Theorems 3.1.1 and 3.1.2 have been given. One proof of Theorem 3.1.2 is based upon representing optimal strategies for Player I as separating hyperplanes and those of Player II as contact points between the two special convex sets, as described in Remark 1.4 of Chapter 1. By carefully counting the number of linearly independent separating planes and noting the dimensions of the contact set, the dimension relations can be established. Another proof of both theorems is based on the proof of the min-max theorem which uses the Kakutani fixed-point theorem; this line of argument will be followed in Chapter 3 of Vol. II, where we consider an extension of these structure theorems to a more general class of games. A third proof, the one given in this chapter, is essentially algebraic in nature, although it has some connections with the geometric proof based on separating convex bodies by hyperplanes.

3.2 Proof of Theorem 3.1.1. The integer-valued function $\omega(\mathbf{x})$ can be used to denote the number of positive components of any strategy $\mathbf{x}$. An optimal strategy $\mathbf{x}$ is said to have full weight if $x_i > 0$ for every i in I_1. Note that if $\mathbf{x}$ has full weight, then $\omega(\mathbf{x}) = n(I_1)$, where $n(I_1)$ is the number of elements of I_1.

▶ LEMMA 3.2.1. $I_1 \subset I_2, J_1 \subset J_2$.

This lemma follows immediately from Lemma 2.1.2.

▶ LEMMA 3.2.2. There exists an optimal strategy $\mathbf{x}$ such that (1) for every i in I_1, $x_i > 0$; (2) the vector $\mathbf{xA}$ equals v for all components corresponding to columns in J_2, and for no others.

Proof. By the definition of I_1, for every $i \in I_1$ there exists $\mathbf{x}^{(i)}$ optimal such that $x_i^{(i)} > 0$. Let $\mathbf{x}^0$ be the average of $\mathbf{x}^{(i)}$. Then $x_i^0 > 0$ for every $i \in I_1$, and $\mathbf{x}^0$ is optimal (Lemma 2.1.1).

Furthermore, for every $j \notin J_2$ there exists $\mathbf{x}^{(j)}$ optimal such that $(\mathbf{x}^{(j)}\mathbf{A})_j > v$. Let $\mathbf{x}^{00}$ be the average of $\mathbf{x}^{(j)}$; $\mathbf{x}^{00}$ is optimal and satisfies condition (2). Finally, we set

$$\mathbf{x} = \frac{\mathbf{x}^0 + \mathbf{x}^{00}}{2}.$$

Then

(1) $\mathbf{x}$ is optimal since X^0 is convex;

(2) $x_i > 0$ for every $i \in I_1$;

(3) $(\mathbf{xA})_j = v$ for $j \in J_2$, (by the definition of J_2);

(4) $(\mathbf{xA})_j > v$ for $j \notin J_2$.

Reduced games. By the reduced game we mean the game whose pay-off matrix $\tilde{\mathbf{A}}$ is the submatrix of $\mathbf{A}$ that keeps only the rows of $\mathbf{A}$ with indices ranging over I_2 and the columns with indices ranging over J_2. The reduced game is also called the essential part of the game.

The sets of optimal strategies for the reduced game will be denoted by X_0^0, Y_0^0.

We observe that $\tilde{\mathbf{A}}$ and $\mathbf{A}$ have the same value v, and that $X^0 \subset X_0^0$, $Y^0 \subset Y_0^0$. To verify these statements, let $\mathbf{x}$, $\mathbf{y}$ be a pair of optimal strategies for the game $\mathbf{A}$. It follows from the definitions of I_1 and J_1 that $\mathbf{x}$ and $\mathbf{y}$ can have positive probability only on those components corresponding to rows and columns of $\tilde{\mathbf{A}}$. Thus they can be considered as strategies for the game $\tilde{\mathbf{A}}$.

Furthermore, if $\mathbf{x}$, $\mathbf{y}$ are optimal for $\mathbf{A}$, then $\mathbf{x}\tilde{\mathbf{A}} = \mathbf{v}$ and $\tilde{\mathbf{A}}\mathbf{y} = \mathbf{v}$ by the definitions of I_2 and J_2. Interpreting these two equations with respect to the game $\tilde{\mathbf{A}}$, we find that $\mathbf{x}, \mathbf{y}$ are optimal for $\tilde{\mathbf{A}}$, and that $v(\tilde{\mathbf{A}}) = v(\mathbf{A}) = v$.

▶ **Lemma 3.2.3.**

$$\dim(X^0) = \dim(X_0^0), \ \dim(Y^0) = \dim(Y_0^0). \tag{3.2.1}$$

Proof. Select $\mathbf{x}$ in X^0 satisfying the conditions of Lemma 3.2.2. Choose $\mathbf{x}'$ in X_0^0 but not in X^0. Then $\mathbf{x}'\mathbf{A} \geq v$ for all columns in J_2, $\mathbf{xA} > v$ for all columns not in J_2, and $\mathbf{xA} = v$ for all columns in J_2. Set $\bar{\mathbf{x}} = \epsilon\mathbf{x}' + (1 - \epsilon)\mathbf{x}$. Then $\bar{\mathbf{x}}\mathbf{A} \geq \mathbf{v}$ for sufficiently small ϵ, so that $\bar{\mathbf{x}}$ is also an optimal strategy of X^0. It follows from the definition of J_2 that $\mathbf{x}'\mathbf{A} = v$ for all columns in J_2.

Since $\mathbf{x}'$ is on the extension of the line segment connecting $\mathbf{x}$ with $\bar{\mathbf{x}}$, we can characterize X_0^0 as the part of the linear extension of X^0 which is contained in the fundamental simplex of strategies S^n for Player I. Therefore dim $(X^0) = \dim (X_0^0)$. The same argument is valid for Y^0.

Our aim is to prove that $I_1 = I_2$ and $J_1 = J_2$. Since the sets I_1, J_1, I_2, J_2 are the same for the reduced game $\tilde{\mathbf{A}}$ as for the original game $\mathbf{A}$, we may without loss of generality restrict ourselves to the reduced game. It is clear from the construction of $\tilde{\mathbf{A}}$ that I_2 and J_2 consist of all $\tilde{\mathbf{A}}$'s row indices and column indices, respectively. In order to complete the proof of Theorem 3.1.1 it is sufficient, by virtue of Lemma 3.2.3, to exhibit a strategy $\mathbf{x}$ in the game $\tilde{\mathbf{A}}$ such that all components of $\mathbf{x}$ are positive, and to find the corresponding optimal strategy $\mathbf{y}$. To this end, we introduce a family of games depending on a parameter defined by matrices

$$\tilde{\mathbf{A}}(\alpha) \qquad (0 \leq \alpha < 1),$$

where the ith row $\mathbf{a}_i(\alpha)$ of $\tilde{\mathbf{A}}(\alpha)$ is given by

$$\begin{aligned} \mathbf{a}_i(\alpha) &= (1 - \alpha)\mathbf{a}_i + \alpha \left(\frac{1}{n} \sum_{i=1}^{n} \mathbf{a}_i\right) \\ &= (1 - \alpha)\mathbf{a}_i + \alpha\bar{\mathbf{a}}. \end{aligned}$$

Let $v(\alpha)$ be the value of the game $\tilde{\mathbf{A}}(\alpha)$. Then $v(\alpha) \leq v$. For let $\mathbf{y}$ be any strategy in Y^0. A direct calculation yields

$$\begin{aligned} \tilde{\mathbf{A}}(\alpha)\mathbf{y} &= \left\langle (1 - \alpha) \sum_j a_{ij}y_j + \alpha \frac{1}{n} \sum_{i=1}^{n} \sum_{j=1}^{m} a_{ij}y_j \right\rangle \\ &\leq (1 - \alpha)\mathbf{v} + \frac{\alpha}{n} n\mathbf{v} = \mathbf{v}. \end{aligned}$$

Hence $v(\alpha) \leq v$.

The key auxiliary result needed to complete the proof of Theorem 3.1.1 is embodied in the following lemma.

►**Lemma 3.2.4.** If there exists an optimal strategy $\mathbf{y}$ in Y_0^0 which is also in the simplex $Y[J_1(\alpha)]$ [the convex hull of the pure strategies with indices in $J_1(\alpha)$] for some α, then $v(\alpha) = v$.

Proof. Construct $\mathbf{y}'$ satisfying the conditions of Lemma 3.2.2 with respect to the $\tilde{\mathbf{A}}(\alpha)$ game. By construction $\mathbf{y}'$ is an interior point of $Y[J_1(\alpha)]$, since it has full weight. Choose $\epsilon > 0$ sufficiently small so that $\mathbf{y}'' = (1 + \epsilon)\mathbf{y}' - \epsilon\mathbf{y}$ is still in $Y[J_1(\alpha)]$. Note first that

$$v(\alpha) \leq \max_i \{((1 - \alpha)\mathbf{a}_i + \alpha\bar{\mathbf{a}}, (1 + \epsilon)\mathbf{y}' - \epsilon\mathbf{y})\}.$$

Since $\mathbf{y}$ is optimal and i is in I_2, and since the indices in I_2 coincide with all the rows of $\tilde{\mathbf{A}}(\alpha)$, we have $(\mathbf{a}_i, \mathbf{y}) = v$ and $(\bar{\mathbf{a}}, \mathbf{y}) = v$ independent of i. Consequently,

$$v(\alpha) \leq \max_i \{(1 + \epsilon)[(1 - \alpha)(\mathbf{a}_i, \mathbf{y}') + \alpha(\bar{\mathbf{a}}, \mathbf{y}')] - \epsilon v\}$$

$$\leq (1 + \epsilon)v(\alpha) - \epsilon v,$$

since $\mathbf{y}'$ is optimal for $\tilde{\mathbf{A}}(\alpha)$. Then $\epsilon v \leq \epsilon v(\alpha)$, or $v \leq v(\alpha)$; and since the reverse inequality always holds, $v = v(\alpha)$. The proof of the lemma is complete.

We shall need the fact that if $\mathbf{y}^\delta \to \mathbf{y}$, then $\tilde{\mathbf{A}}(\alpha_\delta)\mathbf{y}^\delta \to \tilde{\mathbf{A}}\mathbf{y}$ (where $\alpha_\delta \to 0$). This statement is obvious and will not be formally proved.

▶ LEMMA 3.2.5. There exists an $\alpha > 0$ such that $v(\alpha) = v$.

Proof. We must show that the conditions of Lemma 3.2.4 are satisfied.

Let α_δ be a sequence converging to zero. Since $\tilde{\mathbf{A}}$ has a finite number of columns, there exists an infinite number of α_δ's such that the sets $J_1(\alpha_\delta)$ agree. We shall restrict ourselves to this subsequence without changing notation.

Select $\mathbf{y}_\delta$ so that it is optimal for the game $\tilde{\mathbf{A}}(\alpha_\delta)$. Then $\mathbf{y}_\delta$ is in $Y[J_1(\alpha_\delta)]$. Since Y is compact, we may select $\mathbf{y}$ as a limit point for $\mathbf{y}_\delta$. Then $\mathbf{y}$ is in $Y[J_1(\alpha_\delta)]$ for all δ.

It remains to be shown that $\mathbf{y}$ is optimal for the game $\mathbf{A}$. Clearly, $\tilde{\mathbf{A}}(\alpha_\delta)\mathbf{y}_\delta \leq \mathbf{v}(\alpha_\delta) \leq \mathbf{v}$, and therefore $\tilde{\mathbf{A}}\mathbf{y} \leq \mathbf{v}$; hence $\mathbf{y}$ is optimal and in Y_0^0. Applying Lemma 3.2.4, we obtain $v = v(\alpha_\delta)$.

Proof of Theorem 3.1.1. Select α small enough so that $v(\alpha) = v$, and choose $\mathbf{x}$ optimal for $\tilde{\mathbf{A}}(\alpha)$. Form the mixed strategy

$$\mathbf{x}^0 = (1 - \alpha)\mathbf{x} + \frac{\alpha}{n}\mathbf{e},$$

which has full weight (every component is positive). We now show that $\mathbf{x}^0$ is optimal:

$$(\mathbf{x}^0\tilde{\mathbf{A}})_j = \sum_{i=1}^n x_i^0 a_{ij} = (1 - \alpha)\sum_{i=1}^n x_i a_{ij} + \frac{\alpha}{n}\sum_{i=1}^n a_{ij}$$

$$= (1 - \alpha)\sum_{i=1}^n x_i a_{ij} + \frac{\alpha}{n}\left(\sum_{i=1}^n a_{ij}\right)\sum_{i=1}^n x_i$$

$$= \sum_{i=1}^n \left[(1 - \alpha)a_{ij} + \frac{\alpha}{n}\left(\sum_{k=1}^n a_{kj}\right)\right]x_i \geq v(\alpha) = v.$$

This implies that $I_1 = I_2$. The proof of $J_1 = J_2$ is similar.

3.3 Proof of Theorem 3.1.2. In this section we make use of the mapping properties of a matrix. For the general $n \times m$ matrix $\mathbf{B}$, $R(\mathbf{B})$ [or sometimes $\mathcal{R}(\mathbf{B})$] designates the range of the mapping $\mathbf{B}$ viewed as a linear transformation, and $\eta(\mathbf{B})$ the linear space of zeros of $\mathbf{B}$.

The orthogonal complement of a set L in Euclidean space is denoted by $L^{\perp}$.

Finally, if $\mathbf{B}$ is a linear transformation mapping the space E^m into E^n, the following orthogonality relations hold:

(1) $n = \dim R(\mathbf{B}) + \dim R(\mathbf{B})^{\perp}$;

(2) $m = \dim R(\mathbf{B}') + \dim R(\mathbf{B}')^{\perp}$ ($\mathbf{B}'$ is the transpose of $\mathbf{B}$);

(3) if $r(\mathbf{B})$ is the rank of $\mathbf{B}$, then $r(\mathbf{B}') = r(\mathbf{B}) = \dim R(\mathbf{B}) = \dim R(\mathbf{B}')$.

For a fuller discussion of the matrix algebra, we refer the reader to Sections A.3 and A.5 of Appendix A.

Continuing our discussion of dimension relations for sets of optimal strategies, we now establish an important auxiliary theorem. In all that follows, $\widetilde{\mathbf{A}}$ denotes the matrix of the essential part of the game with matrix $\mathbf{A}$.

▶THEOREM 3.3.1. Let X^0 and Y^0 denote the spaces of optimal strategies for Players I and II, respectively. Then (i) if $v \neq 0$,

$$\dim Y^0 = \dim \eta(\widetilde{\mathbf{A}}), \dim X^0 = \dim \eta(\widetilde{\mathbf{A}}');$$

and (ii) if $v = 0$,

$$\dim Y^0 + 1 = \dim \eta(\widetilde{\mathbf{A}}), \dim X^0 + 1 = \dim \eta(\widetilde{\mathbf{A}}').$$

Proof of (*i*). We recall that if $\mathbf{x} \in X^0$, the only nonzero components of $\mathbf{x}$ are those that occur in the rows of $\widetilde{\mathbf{A}}$. Hence we may consider $\mathbf{x}$ as a vector in the Euclidean space of dimension equal to the number of rows of $\widetilde{\mathbf{A}}$. Similarly for $\mathbf{y} \in Y^0$.

Let $\mathbf{y}^0 \in Y^0$ be fixed. Let $\mathbf{y}^1, \ldots, \mathbf{y}^k$ be a maximal set of vectors in Y^0 such that $\mathbf{y}^1 - \mathbf{y}^0, \ldots, \mathbf{y}^k - \mathbf{y}^0$ are linearly independent. It follows from the definition of the essential part of the game that $\widetilde{\mathbf{A}}\mathbf{y}^0 \equiv \mathbf{v}$ and $\widetilde{\mathbf{A}}\mathbf{y}^i \equiv \mathbf{v}$; hence $\widetilde{\mathbf{A}}(\mathbf{y}^i - \mathbf{y}^0) \equiv 0$, and $\{\mathbf{y}^i - \mathbf{y}^0\}$ constitutes a set of linearly independent zeros of $\widetilde{\mathbf{A}}$. Therefore $\dim Y^0 \leq \dim \eta(\widetilde{\mathbf{A}})$.

On the other hand, by Lemma 3.2.2 and Theorem 3.1.1, there exists an optimal strategy $\mathbf{y}^0$ such that $y_j^0 > 0$ for all j (j ranging over the columns of $\widetilde{\mathbf{A}}$). Let $\widetilde{\mathbf{A}}\mathbf{z} = 0$. We argue as in the proof of Lemma 2.4.1: Let $\mathbf{x}^0$ be optimal such that $\mathbf{x}^0\widetilde{\mathbf{A}} = \mathbf{v}$; then $0 = (\mathbf{x}^0, \widetilde{\mathbf{A}}\mathbf{z}) = (\mathbf{x}^0\widetilde{\mathbf{A}}, \mathbf{z}) = v\sum z_i$, and hence $\sum z_i = 0$. Consequently, for ϵ sufficiently small $\mathbf{y}^0 + \epsilon\mathbf{z}$ is an optimal strategy. Hence $\dim \eta(\widetilde{\mathbf{A}}) \leq \dim Y^0$. The proof of the second relation in (i) is similar.

Proof of (ii). If $v = 0$, then a strategy $\mathbf{y}$ is optimal if and only if $\tilde{\mathbf{A}}\mathbf{y} \equiv \mathbf{0}$. That is, any zero of $\tilde{\mathbf{A}}$ which is also a strategy is an optimal strategy. Therefore Y^0 can be obtained from the set $\eta(\tilde{\mathbf{A}})$ by imposing the extra relation $\sum y_i = 1$ intersected with the first orthant. This requirement lowers the dimension by 1. Hence $\dim Y^0 = \dim \eta(\tilde{\mathbf{A}}) - 1$. Similar proofs hold for X^0. Thus the lemma is established.

We are now ready to prove Theorem 3.1.2, which states that

$$\dim \overline{X} - \dim X^0 = \dim \overline{Y} - \dim Y^0.$$

(See page 64 for definitions of $\overline{X}$ and $\overline{Y}$.)

Proof of Theorem 3.1.2. By Lemma 3.2.3, $\dim X^0 = \dim X_0^0$, where X_0^0 denotes the space of optimal strategies for the reduced game; therefore we may restrict ourselves to the reduced game with matrix $\tilde{\mathbf{A}}$.

Since $\mathbf{y} \in \eta(\tilde{\mathbf{A}}) \leftrightarrow \tilde{\mathbf{A}}\mathbf{y} = \mathbf{0} \leftrightarrow (\mathbf{x}, \tilde{\mathbf{A}}\mathbf{y}) = 0 \leftrightarrow (\mathbf{x}\tilde{\mathbf{A}}, \mathbf{y}) = 0$ for all $\mathbf{x}$, we must have

$$\eta(\tilde{\mathbf{A}}) = [R(\tilde{\mathbf{A}}')]^{\perp}; \tag{3.3.1}$$

similarly,

$$\eta(\tilde{\mathbf{A}}') = [R(\tilde{\mathbf{A}})]^{\perp}. \tag{3.3.2}$$

Furthermore, since $\tilde{\mathbf{A}}$ can be viewed as a mapping of the smallest linear space $\mathcal{H}$ containing $\overline{Y}$ into the corresponding linear space $\mathcal{G}$ which contains $\overline{X}$, and since $\tilde{\mathbf{A}}'$ is the corresponding dual transformation of $\mathcal{G}$ into $\mathcal{H}$, it follows from the orthogonality dimension relations (1) and (2) that $\dim \mathcal{G} = \dim \overline{X} + 1 = \dim R(\tilde{\mathbf{A}}) + \dim [R(\tilde{\mathbf{A}})]^{\perp}$ and $\dim \mathcal{H} = \dim \overline{Y} + 1 = \dim R(\tilde{\mathbf{A}}') + \dim [R(\tilde{\mathbf{A}}')]^{\perp}$. Using property (3), equations (3.3.1) and (3.3.2), and Theorem 3.3.1, we get the successive relations

$$\begin{aligned} \dim \overline{Y} &= r(\tilde{\mathbf{A}}') + \dim \eta(\tilde{\mathbf{A}}) - 1, \\ \dim \overline{X} &= r(\tilde{\mathbf{A}}) + \dim \eta(\tilde{\mathbf{A}}') - 1, \\ \dim \overline{Y} - \dim \eta(\tilde{\mathbf{A}}) &= \dim \overline{X} - \dim \eta(\tilde{\mathbf{A}}'), \\ \dim \overline{Y} - \dim Y^0 &= \dim \overline{X} - \dim X^0. \end{aligned} \tag{3.3.3}$$

3.4 The converse of Theorem 3.1.2. If the optimal strategies for both players are unique, say $\mathbf{x} = \langle x_1, \ldots, x_s \rangle$ and $\mathbf{y} = \langle y_1, \ldots, y_r \rangle$, then $\dim X^0 = \dim Y^0 = 0$. By Theorem 3.1.2, $\dim \overline{X} = \dim \overline{Y}$, and hence $r = s$. In this case, both strategies must have the same number of positive components.

Let us consider the converse problem of prescribing single strategies $\mathbf{x}^0$ for Player I and $\mathbf{y}^0$ for Player II and seeking to construct a matrix $\mathbf{A}$ for which $\mathbf{x}^0$ and $\mathbf{y}^0$ are the unique optimal strategies. As indicated, we must require that $\mathbf{x}^0$ and $\mathbf{y}^0$ have the same number of nonzero components.

Letting $\mathbf{x}^0 = \langle x_0, x_1, \ldots, x_k \rangle$ and $\mathbf{y}^0 = \langle y_0, y_1, \ldots, y_k \rangle$ with all x_i and y_j positive, we shall exhibit a matrix of order $(k+1) \times (k+1)$ with $\mathbf{x}^0$ and $\mathbf{y}^0$ the unique optimal strategies. Explicitly, take $v \neq 1/k$ and let

$$\mathbf{A} = \left\| \begin{matrix} \dfrac{v(x_0 + y_0 - 1) + \sum_{i=1}^k x_i y_i}{x_0 y_0}, & \dfrac{v - x_1}{x_0}, & \dfrac{v - x_2}{x_0}, & \ldots, & \dfrac{v - x_k}{x_0} \\ \dfrac{v - y_1}{y_0} & , \quad 1 & , \quad 0 & , \ldots, & 0 \\ \dfrac{v - y_2}{y_0} & , \quad 0 & , \quad 1 & , \ldots, & 0 \\ \vdots & \vdots & \vdots & & \vdots \\ \dfrac{v - y_k}{y_0} & , \quad 0 & , \quad 0 & , \ldots, & 1 \end{matrix} \right\|$$

It is readily verified that $\mathbf{x}^0$ and $\mathbf{y}^0$ are optimal strategies, and that the value of the game is v. It remains to establish the uniqueness.

Let $\mathbf{w} = \langle w_0, w_1, \ldots, w_n \rangle$ be an optimal strategy for Player I. If $j \neq 0$, then by Lemma 2.1.2

$$\frac{v - x_j}{x_0} w_0 + w_j = v,$$

and therefore

$$\sum_{j=1}^k \frac{v - x_j}{x_0} w_0 + \sum_{j=1}^k w_j = kv.$$

But

$$\sum_{j=1}^k x_j = 1 - x_0, \qquad \sum_{j=1}^k w_j = 1 - w_0,$$

so that

$$(vk - 1 + x_0)w_0 + x_0(1 - w_0) = kvx_0$$

or, equivalently,

$$(vk - 1)w_0 = (vk - 1)x_0.$$

Since $vk \neq 1$, it follows that $w_0 = x_0$. Further, since

$$\frac{v - x_j}{x_0} w_0 + w_j = v,$$

$w_j = x_j$ for all j. Thus $\mathbf{x}^0$ is the only optimal strategy for Player I. A similar argument shows that $\mathbf{y}^0$ is the only optimal strategy for Player II.

We now set about extending this result. The more general problem is this: Which strategy sets X^0 and Y^0 can be prescribed in advance such that a matrix **A** can be found for which X^0 and Y^0 are the precise sets of optimal strategies for Players I and II, respectively? A limitation on the choice of X^0 and Y^0 is implied in the dimension relations stated by Theorem 3.1.2. Remarkably enough, however, this is the only restriction. In other words, any sets X^0 and Y^0 satisfying Theorem 3.1.2 for appropriate $\overline{X}$ and $\overline{Y}$ can qualify as the sets of optimal strategies for a matrix **A**. A more precise statement of this result follows.

▶Theorem 3.4.1. Let $\overline{X}$ and $\overline{Y}$ be two fundamental simplexes in E^s and E^r, respectively, described by the relations

$$\sum_{i=1}^{s} x_i = 1, \qquad x_i \geq 0, \qquad \text{and} \qquad \sum_{j=1}^{r} y_j = 1, \qquad y_j \geq 0.$$

Let X^0 and Y^0 be two closed, convex, polyhedral sets contained in $\overline{X}$ and $\overline{Y}$, respectively, and intersecting the interior. If

$$\dim \overline{X} - \dim X^0 = \dim \overline{Y} - \dim Y^0 = q, \tag{3.4.1}$$

then there exists a matrix **A** for which the sets of optimal strategies are X^0 and Y^0 and the value of the game is zero.

Proof. Let X_0^0 and Y_0^0 be the linear extensions of X^0 and Y^0 inside $\overline{X}$ and $\overline{Y}$. We shall first construct a matrix $\tilde{\mathbf{A}}$ with optimal strategies consisting of the sets X_0^0 and Y_0^0.

To this end, let $\mathcal{L}_0$ and $\mathcal{K}_0$ be the smallest linear subspaces containing X_0^0 and Y_0^0, respectively. Since zero belongs to any linear space, we have $\dim \mathcal{L}_0 = \dim X_0^0 + 1 = \dim X^0 + 1$ and $\dim \mathcal{K}_0 = \dim Y_0^0 + 1 = \dim Y^0 + 1$. It follows from (3.4.1) and orthogonality relations (1) and (2) of Section 3.3 that $\dim \mathcal{L}_0^\perp = \dim \mathcal{K}_0^\perp = q$.

Let **P** be the $s \times s$ matrix that represents a projection of E^s onto $\mathcal{L}_0^\perp$. Let **Q** be a matrix of rank q which maps E^r into E^s and which is a one-to-one linear transformation of $\mathcal{K}_0^\perp$ onto $\mathcal{L}_0^\perp$ satisfying $\eta(\mathbf{Q}) = \mathcal{K}_0$. Such a matrix exists by virtue of the fact that $\mathcal{K}_0^\perp$ and $\mathcal{L}_0^\perp$ are of the same dimension. Set $\tilde{\mathbf{A}} = \mathbf{PQ}$. The matrix $\tilde{\mathbf{A}}$ is an $s \times r$ matrix.

The zero manifold of $\tilde{\mathbf{A}}$ is $\mathcal{K}_0$, and that of $\tilde{\mathbf{A}}'$ is $\mathcal{L}_0$. These assertions are established as follows: Since $\mathbf{Q}(\mathcal{K}_0) = \mathbf{0}$ (this means the image of $\mathcal{K}_0$ by the linear mapping **Q**), we deduce that $\eta(\tilde{\mathbf{A}}) \supset \mathcal{K}_0$. On the other hand, since **Q** is a one-to-one linear transformation of $\mathcal{K}_0^\perp$ onto $\mathcal{L}_0^\perp$ and **P** acts as the identity mapping on $\mathcal{L}_0^\perp$, we find that $\tilde{\mathbf{A}}\mathbf{y} = \mathbf{0}$ for **y** in $\mathcal{K}_0^\perp$ implies that $\mathbf{y} = \mathbf{0}$. Since $\mathcal{K}_0$ and $\mathcal{K}_0^\perp$ span E^r as a direct sum, we infer that $\eta(\tilde{\mathbf{A}}) = \mathcal{K}_0$. It follows that $\eta(\tilde{\mathbf{A}}') = R(\tilde{\mathbf{A}})^\perp = (\mathcal{L}_0^\perp)^\perp = \mathcal{L}_0$, and thus our assertions are valid.

Since the value of the game is zero, the optimal strategies for $\tilde{\mathbf{A}}$ are precisely the vectors in X_0^0 and Y_0^0.

We shall now enlarge $\tilde{\mathbf{A}}$ in order to eliminate those strategies of X_0^0 and Y_0^0 which are not in X^0 and Y^0. The following characterization of convex polyhedral sets is needed.

Let F be a closed, convex, polyhedral set of E^n which is also contained in the simplex

$$x_i \geq 0, \qquad \sum_{i=1}^{n} x_i = 1.$$

The dimension of F is at most $n - 1$, and the dimension of any face of F is at most $n - 2$. Let k denote the number of faces of F. By assumption, k is finite.

For any given face of F, there exists at least one $(n - 1)$-dimensional hyperplane B of E^n through the origin which is a supporting plane for F and contains the given face of F. Since the dimension of any face of F is at most $n - 2$, such a hyperplane can clearly be found. Let the direction numbers of the normal to the plane be denoted by $\langle b_1, \ldots, b_n \rangle$. The plane B separates E^n into two half-spaces E_0 and E_1, where E_0 is closed and contains the given face of F, and $E_1 = E_0^c$ (the complement of E_0). Then

$$\mathbf{x} \in B \to \sum_{i=1}^{n} x_i b_i = 0,$$

$$\mathbf{x} \in E_0 \to \sum_{i=1}^{n} x_i b_i \geq 0,$$

$$\mathbf{x} \in E_1 \to \sum_{i=1}^{n} x_i b_i < 0.$$

Let $B^1, \ldots, B^l$ be a set of half-spaces determined by a collection of supporting hyperplanes for the faces of F whose intersection is precisely F. Formally, if $B^j \to \langle b_1^j, \ldots, b_n^j \rangle$, then

$$\mathbf{z} \in F \to \sum_{i=1}^{n} z_i b_i^j \geq 0 = v \qquad \text{for every} \quad j \ (1 \leq j \leq l); \tag{3.4.2}$$

while if $\mathbf{z} \in F^c$, then there exists a value j_0 $(1 \leq j_0 \leq l)$ such that

$$\sum_{i=1}^{n} z_i b_i^{j_0} < 0 = v. \tag{3.4.3}$$

Thus the planes (B^j), together with the conditions (3.4.2) and (3.4.3), completely determine the set F. We remark that although a change in orientation (taking $\{-b_i^j\}$ instead of $\{b_i^j\}$ for *all* i, j) reverses all the inequalities in (3.4.2) and (3.4.3), the set F remains completely characterized.

Construction of **A**. First we construct the set $\{B^j\}$ of planes corresponding to the faces of X^0; we add l new columns to $\tilde{\mathbf{A}}$, the jth column added consisting of the vector $\langle b_1^j, \ldots, b_s^j \rangle$. Next we form the set $\{C^i\}$ of planes corresponding to the faces of Y^0, reversing the inequalities in (3.4.2) and (3.4.3); we add m new rows to $\tilde{\mathbf{A}}$, the ith row consisting of the vector $\langle c_1^i, \ldots, c_r^i \rangle$. Finally, we complete the matrix **A** by filling in zeros for the lower right corner. If $\tilde{\mathbf{A}} = \|a_{ij}\|$, then

$$\mathbf{A} = \begin{Vmatrix} a_{11} & \ldots & a_{1r} & b_1^1 & \ldots & b_1^l \\ \vdots & & \vdots & \vdots & & \vdots \\ a_{s1} & & a_{sr} & b_s^1 & & b_s^l \\ c_1^1 & \ldots & c_r^1 & 0 & \ldots & 0 \\ \vdots & & \vdots & \vdots & & \vdots \\ c_1^m & \ldots & c_r^m & 0 & \ldots & 0 \end{Vmatrix}$$

We shall now verify that X^0, Y^0 are exactly the sets of optimal strategies for **A**. First we note that $\mathbf{x} \in X^0$ and $\mathbf{y} \in Y^0$ implies that

$$\mathbf{x} = \langle x_1, \ldots, x_s, 0, \ldots, 0 \rangle$$

and

$$\mathbf{y} = \langle y_1, \ldots, y_r, 0, \ldots, 0 \rangle,$$

where

$$\mathbf{x} \in X^0 \to \sum_{i=1}^{s} x_i a_{ij} = 0$$

by the construction of $\tilde{\mathbf{A}}$, and

$$\mathbf{x} \in X^0 \to \sum_{i=1}^{s} x_i b_i^j \geq 0 \qquad \text{for all } j \qquad (1 \leq j \leq l)$$

by (3.4.2). Hence $\mathbf{x}$ is optimal for **A**. On the other hand, if $\mathbf{x} \notin X^0$, then by (3.4.3) there exists a j_0 such that $\sum x_i b_i^{j_0} < 0$; hence $\mathbf{x}$ is not optimal. Thus the optimal strategies are exactly the elements of X^0.

The corresponding argument is valid for Y^0 if we remember that $\{C^i\}$ was formed with the inequalities reversed in (3.4.2) and (3.4.3). This completes the proof of Theorem 3.4.1.

The matrix constructed in Theorem 3.4.1 possessing prescribed sets of optimal strategies X^0 and Y^0 is not necessarily the most efficient (i.e., the smallest) such matrix that can be constructed. To obtain the most efficient matrix, we must exercise care in the construction of the additional rows and columns. Specifically, there is no need to construct any plane B^i that duplicates a face of $\overline{X}$. These faces are described by the basic inequalities $x_i \geq 0$ which define the fundamental simplex. By eliminating these extraneous column and row vectors from the matrix **A** we obtain the smallest matrix with the prescribed sets of optimal strategies. The proof of this assertion is left as an exercise for the reader.

Problems 5 and 6 of this chapter also involve the construction procedure of Theorem 3.4.1.

3.5 Uniqueness of optimal strategies. If Players I and II possess unique optimal strategies, the choice of an optimal strategy presents no problem for either player; there is only one to choose. When the sets of optimal strategies are more than zero-dimensional, on the other hand, selecting an optimal strategy involves difficult conceptual questions, the use of additional criteria, and other complications. It is therefore useful and important to know when a game has unique solutions. We shall show in this section that games with unique solutions occur very frequently.

We observe first that for games with unique optimal strategies, $\dim \overline{X} = \dim \overline{Y}$. This is an immediate corollary of Theorem 3.1.2. Equivalently, the maximal weight (number of positive components) of the optimal **x** strategy and the maximal weight of the optimal **y** strategy are the same. Hence the reduced matrix is necessarily square.

Next we introduce the following topological structure on the space of all games. By a neighborhood $U(\mathbf{A})$ of the matrix **A** of order $n \times m$ we mean the set of all possible matrices of the same order whose components are within a prescribed positive distance of the corresponding components of **A**.

Our principal theorem follows.

▶THEOREM 3.5.1. If a matrix **A** has unique optimal strategies **x** and **y**, there exists a neighborhood $U(\mathbf{A})$ such that if **B** is in $U(\mathbf{A})$, then **B** has unique optimal strategies.

The proof depends on the following elementary convergence lemma.

▶LEMMA 3.5.1. Let $\mathbf{A}_n \to \mathbf{A}$. Let X^0 and Y^0 be the sets of optimal strategies for **A**, and let X_n^0 and Y_n^0 be the corresponding sets for the game $\mathbf{A}_n$. If G is an open set containing X^0, then for n sufficiently large, $X_n^0 \subset G$. Moreover, $v_n \to v$.

Proof. The proof of the last statement of the lemma is trivial. Suppose the first part of the lemma is false. Then there exists a subsequence $\{n_i\}$ and a corresponding sequence $\{\mathbf{x}^{n_i}\}$ such that $\mathbf{x}^{n_i}$ is optimal for $\mathbf{A}_{n_i}$ and is in G^c, where G^c is the relative complement of G within S (the simplex of all strategies for Player I).

Since G^c is a closed bounded set, the sequence $\{\mathbf{x}^{n_i}\}$ must have a subsequence converging to a limit $\mathbf{x}^0$ in G^c. Thus, since $\lim v_n = v$, we get $\mathbf{x}^{n_i}\mathbf{A}_{n_i} \geq \mathbf{v}_{n_i}$, and in the limit, $\mathbf{x}^0\mathbf{A} \geq \mathbf{v}$; hence $\mathbf{x}^0$ is in G. But $\mathbf{x}^0$ is also in G^c, which is impossible.

Proof of Theorem 3.5.1. Without loss of generality we may assume $v \neq 0$ (translate all the elements of $\mathbf{A}$ by the same constant, if necessary). Let $\mathbf{A}^n \to \mathbf{A}$, so that $v_n \neq 0$ for n sufficiently large. We define the sets I_1, I_2, I_1^n, I_2^n as in Section 3.1. Explicitly,

$$I_2^n = \{i | \sum_j a_{ij}^n y_j^n = v \quad \text{for all} \quad \mathbf{y}^n \quad \text{optimal for} \quad \mathbf{A}^n\}.$$

Since the total number of row indices i of $\mathbf{A}$ is finite, there can be only a finite number of distinct sets I_2^n. Let I_2^n be any set that occurs infinitely often in $\{I_2^n\}$.

For every $i \subset I_2^n$ we find a subsequence $\{n_i\}$ such that

$$\sum_j a_{ij}^{n_i} y_j^{n_i} = v^{n_i},$$

and in the limit, $\sum_j a_{ij} y_j = v$, where $\mathbf{y}$ is the unique optimal strategy for $\mathbf{A}$. In fact, by Lemma 3.5.1 every limit point of $\mathbf{y}^n$ is the same unique optimal strategy $\mathbf{y}$. Since $\mathbf{y}$ is unique optimal for $\mathbf{A}$, the last equation shows that $i \subset I_2$.

Consequently, $I_2^n \subset I_2$ for all but a finite number of n.

On the other hand, if $x_i > 0$ and $\mathbf{x}^n \to \mathbf{x}$, then for n sufficiently large, $x_i^n > 0$, since $\mathbf{x}$ is the unique optimal strategy for Player I and is thus the only limit point X_n^0 can have (Lemma 3.5.1). Hence $I_1^n \supset I_1$ for n large. Combining, we have $I_2 \supset I_2^n \supset I_1^n \supset I_1$ for n large. But $I_2 = I_1$ by Theorem 3.1.1, and therefore $I_2^n = I_2$ for n large. A similar argument enables us to show that for n sufficiently large, $J_2^n = J_2$.

As observed above, the reduced game $\tilde{\mathbf{A}}$ is a square matrix. Furthermore, since $J_2^n = J_2$ and $I_2^n = I_2$, it follows that the reduced games $\tilde{\mathbf{A}}^n$ are also square matrices, and of the same order as $\tilde{\mathbf{A}}$. But by the hypothesis $\dim X^0 = 0$ and the identity $1 + \dim \overline{X} - \dim X^0 = \text{rank } (\tilde{\mathbf{A}})$ [see (3.3.3)], $\tilde{\mathbf{A}}$ has full rank; and since a small perturbation applied to the components can at worst increase the rank of a matrix, it follows that $\tilde{\mathbf{A}}^n$ must also have full rank for n sufficiently large. Hence, $1 + \dim \overline{X}^n = \text{rank } (\tilde{\mathbf{A}}^n)$. However, $1 + \dim \overline{X}^n - \dim X_n^0 = \text{rank } (\tilde{\mathbf{A}}^n)$. Therefore, the only conclusion is that $\dim X_n^0 = 0$, that is, that the optimal strategy $\mathbf{x}^n$ for $\mathbf{A}^n$ is unique. An analogous argument leads to the same conclusion for the $\mathbf{y}$ optimal strategies, and the proof is complete.

Theorem 3.5.1 states that the set of all matrices with unique solutions is an open set. The next theorem asserts that the complement of this open set is sparse, in the sense that any game with nonunique solutions can be approximated elementwise as closely as desired by games possessing unique solutions. It is gratifying to know that most games have unique solutions; it should be pointed out, however, that in practice we seem constantly to be dealing with games having nonunique solutions.

▶**Theorem** 3.5.2. The set of all matrix games with unique solutions constitutes a dense set in the set of all matrix games.

The proof will be obtained with the aid of the following two lemmas, each possessing independent interest.

▶**Lemma** 3.5.2. Suppose that $\mathbf{B}$ is a matrix with the following properties for all r: (1) the matrix obtained by adjoining a column of 1's to any $(r+1) \times r$ submatrix of $\mathbf{B}$ is nonsingular; (2) the same condition holds for the submatrices of $\mathbf{B}'$; (3) every $r \times r$ submatrix of $\mathbf{B}$ is nonsingular. Then the optimal strategies for $\mathbf{B}$ are unique.

Proof. Let $\mathbf{x}^0$ and $\mathbf{y}^0$ be optimal strategies. Let r be the number of i's such that $x_i^0 > 0$, t the number of j's such that $y_j^0 > 0$, s the number of j's such that $\sum_i a_{ij} x_i^0 = v$, and w the number of i's such that $\sum_j a_{ij} y_j^0 = v$. By Lemma 2.1.2,

$$w \geq r \quad \text{and} \quad s \geq t. \tag{3.5.1}$$

Let $\overline{\mathbf{B}}$ be the $r \times s$ submatrix of $\mathbf{B}$ whose rows correspond to the indices i for which $x_i^0 > 0$, and whose columns correspond to the indices j for which $\sum_i a_{ij} x_i^0 = v$.

Let $\mathbf{C}$ be the matrix of order $s \times (r+1)$ obtained by adjoining a column of 1's to $\overline{\mathbf{B}}'$. Set $\mathbf{x}^0 = \langle x_1^0, \ldots, x_r^0, -v \rangle$.

Then $\mathbf{Cx}^0 = \mathbf{0}$. Suppose that $r + 1 \leq s$. Then we can select any $(r+1) \times (r+1)$ submatrix $\mathbf{D}$ of $\mathbf{C}$. For this $\mathbf{D}$, we have $\mathbf{Dx}^0 \equiv \mathbf{0}$, which is impossible, since $\mathbf{D}$ is nonsingular by assumption (3).

Therefore

$$r + 1 > s. \tag{3.5.2}$$

A similar argument, using the $w \times t$ matrix associated with the vector $\mathbf{y}$, shows that

$$t + 1 > w. \tag{3.5.3}$$

Combining (3.5.1) and (3.5.3) gives us $t + 1 > w \geq r$; hence $t \geq r$. From $s \geq t$ and $t \geq r$ we obtain $s \geq r$, and finally $r + 1 > s$ gives $r \geq s$. It follows readily that $r = s = t = w$.

Since $\mathbf{x}^0$ and $\mathbf{y}^0$ were arbitrary optimal strategies and satisfy the identity $t = s$, we conclude that all optimal $\mathbf{y}$ strategies have exactly the same number of positive components. Therefore the indices of positive components for two $\mathbf{y}$ solutions coincide, for otherwise a convex combination of $\mathbf{y}$ and $\mathbf{y}^0$ would lead to an optimal strategy whose weight exceeds t. Similarly, all optimal $\mathbf{x}$ strategies possess the same $t = s$ positive components.

The equation $r = s$ then indicates that for any two optimal strategies $\mathbf{x}^1$ and $\mathbf{x}^2$,

$$(\mathbf{x}^1 - \mathbf{x}^2)\overline{\mathbf{B}} = \mathbf{0}.$$

Since $\overline{\mathbf{B}}$ is nonsingular by assumption, $\mathbf{x}^1 = \mathbf{x}^2$ and $\dim X^0 = 0$. Similarly, $\dim Y^0 = 0$.

At this point, we exhibit a matrix satisfying the conditions of Lemma 3.5.2. Choose $\alpha_1, \ldots, \alpha_n$ all positive and distinct ($\alpha_i \neq 1$), and take $k_1, \ldots, k_n$ as distinct positive integers. Consider

$$\mathbf{B} = \begin{Vmatrix} \alpha_1^{k_1} & \cdots & \alpha_n^{k_1} \\ \alpha_1^{k_2} & \cdots & \alpha_n^{k_2} \\ \vdots & & \vdots \\ \alpha_1^{k_n} & \cdots & \alpha_n^{k_n} \end{Vmatrix} \tag{3.5.4}$$

We now verify that $\mathbf{B}$ satisfies the conditions of the lemma.

Observe that square submatrices of $\mathbf{B}$ are of the same form as $\mathbf{B}$. If we take an $r \times (r + 1)$ submatrix of $\mathbf{B}$ and adjoin a row of 1's, we again get a matrix of the same form, provided that we assign $k_{n+1} = 0$. Adjoining a column of 1's to an $(r + 1) \times r$ submatrix also provides us with a matrix of the same type where we assign $\alpha_{n+1} = 1$.

Therefore, it suffices to verify that $\det \mathbf{B} \neq 0$ for any choice of distinct $k_i \neq 0$ and α_i. This is the classical Vandermonde type of determinant, which is always nonvanishing.

▶ LEMMA 3.5.3. The set of all matrices satisfying the conditions of Lemma 3.5.2 is dense in the set of all matrices.

Proof. We take $\mathbf{A}$ arbitrary, $\mathbf{B}$ of the form exhibited in (3.5.4). Let $\mathbf{R}$ be any $r \times r$ submatrix of $\mathbf{A} + \epsilon\mathbf{B}$.

The determinant of $\mathbf{R}$ is a polynomial of exact degree r in ϵ. In fact, the coefficient of ϵ^r is the determinant of the corresponding submatrix of $\mathbf{B}$, which is not zero. This polynomial has at most r strictly positive roots.

Similarly, the determinant of every square matrix obtained by adding a row or a column of 1's to an $r \times (r + 1)$ or an $(r + 1) \times r$ submatrix of $\mathbf{A} + \epsilon\mathbf{B}$ is a polynomial of exact degree $r + 1$ in ϵ, and the number of real roots of each such polynomial is at most $r + 1$.

The number of all positive roots ϵ_i obtained from all possible subdeterminants of $\mathbf{A} + \epsilon\mathbf{B}$ is finite.

Hence $\epsilon_0 = \min_i \{\epsilon_i\}$ is strictly positive. If $0 < \epsilon < \epsilon_0$, then $\mathbf{A} + \epsilon\mathbf{B}$ is of type **B** and therefore possesses unique optimal strategies. This completes the proof of the lemma, and the proof of Theorem 3.5.2.

In concluding, it is worth pointing out that Theorem 3.5.1 does not apply to games in which only one player has a unique optimal strategy. For example, let

$$\mathbf{A} = \begin{Vmatrix} 1 & 1 \\ 0 & 1 \end{Vmatrix};$$

then Y^0 is the set $\{\langle y, 1 - y\rangle\}$ $(0 \leq y \leq 1)$, while X^0 is a single strategy $\mathbf{x}^0 = \langle 1, 0\rangle$. Now consider the perturbed game

$$\mathbf{A}_\epsilon = \begin{Vmatrix} 1 + \epsilon & 1 \\ 0 & 1 \end{Vmatrix},$$

with $\epsilon > 0$. In this game $Y^0(\epsilon)$ is a single strategy $\mathbf{y}^0 = \langle 0, 1\rangle$, while $X^0(\epsilon)$ consists of all strategies $\langle x, 1 - x\rangle$, where $1/(1 + \epsilon) \leq x \leq 1$. Hence a small perturbation changes the dimension of X^0 from zero to one.

3.6 Problems.

1. Let **A** be a 3×3 matrix game all of whose optimal strategies use the same components with positive probability. Show that the solution to this game is unique if the number of components used by the two players is the same.

2. Show that the reduced game of a symmetric game is also symmetric.

3. Show that for symmetric games $\dim \overline{X} - \dim X^0$ is even.

4. Show that if $X^0 = Y^0$ and if $\dim \overline{X} - \dim X^0$ is even, there exists a symmetric game with optimal strategies spanned by X^0, Y^0.

5. Construct a matrix game with extreme-point solutions

$$\mathbf{x} = \langle 0, 0, 1\rangle, \qquad \mathbf{y}_1 = \langle \tfrac{1}{3}, \tfrac{2}{3}, 0\rangle, \qquad \mathbf{y}_2 = \langle \tfrac{2}{3}, \tfrac{1}{3}, 0\rangle.$$

6. Construct a matrix game with extreme-point solutions

$$\begin{aligned} \mathbf{x}_1 &= \langle \tfrac{1}{2}, \tfrac{1}{2}, 0\rangle, & \mathbf{y}_1 &= \langle \tfrac{1}{3}, 0, 0, \tfrac{2}{3}\rangle, \\ \mathbf{x}_2 &= \langle \tfrac{1}{3}, \tfrac{1}{3}, \tfrac{1}{3}\rangle; & \mathbf{y}_2 &= \langle 0, \tfrac{1}{2}, 0, \tfrac{1}{2}\rangle, \\ & & \mathbf{y}_3 &= \langle 0, 0, \tfrac{1}{3}, \tfrac{2}{3}\rangle. \end{aligned}$$

7. Show that the infinite matrix game defined by $a_{ij} = i - j$ $(i, j = 1, 2, 3, \ldots)$ has no value.

8. Show that the infinite matrix game defined by

$$a_{ij} = \frac{i - j}{\sqrt{1 + (i - j)^2}} \qquad (i, j = 1, 2, \ldots)$$

has no value.

9. Show that if $c_k > 0$ and $\sum_{k=1}^{\infty} c_k = 1$, the infinite matrix game

$$\begin{Vmatrix} 0 & c_1 - 1 & c_2 & c_3 & \cdots \\ 0 & c_1 & c_2 - 1 & c_3 & \cdots \\ 0 & c_1 & c_2 & c_3 - 1 & \cdots \\ 0 & c_1 & c_2 & c_3 & \cdots \\ 0 & c_1 & c_2 & c_3 & \cdots \\ \vdots & \vdots & \vdots & \vdots & \end{Vmatrix}$$

has a unique solution. Show that this solution does not satisfy the dimension relations of this chapter, and describe the reduced game.

10. Show that $\mathbf{x} = \langle \frac{1}{2}, \frac{1}{2} \rangle$, $\mathbf{y} = \langle 1, 0, 0, \ldots \rangle$ are the only optimal strategies for the infinite game

$$\mathbf{D} = \begin{Vmatrix} d & 2d & \frac{1}{2} & 2d & \frac{1}{4} & \cdots \\ d & 1 & 2d & \frac{1}{3} & 2d & \cdots \end{Vmatrix},$$

where $d \neq 0$. Show that this example violates the dimension relations of this chapter.

11. Show that if $\mathbf{H} = \|h_{ij}\|$ and $\alpha > 0$,

$$\lim_{\alpha \to 0+} \frac{v(\mathbf{A} + \alpha \mathbf{H}) - v(\mathbf{A})}{\alpha} = \max_{x \in X^0(A)} \min_{y \in Y^0(A)} (\mathbf{x}, \mathbf{Hy}).$$

12. Show that

$$\left(\frac{\partial v}{\partial a_{ij}}\right)_+ = \left(\max_{x \in X^0} x_i\right) \left(\min_{y \in Y^0} y_j\right).$$

13. Derive the formula for

$$\left(\frac{\partial v}{\partial a_{ij}}\right)_- .$$

14. Let $\mathbf{A}$ be a fixed 2×2 game, and let $X^0(\mathbf{A})$ and $Y^0(\mathbf{A})$ be the spaces of optimal strategies for $\mathbf{A}$. Show that there exists a neighborhood U of $\mathbf{A}$ such that if $\mathbf{B}$ is in U, then

$$\dim X^0(\mathbf{B}) + \dim Y^0(\mathbf{B}) \leq \dim X^0(\mathbf{A}) + \dim Y^0(\mathbf{A}).$$

15. Consider the game $\mathbf{A}_\sigma = \|e^{-\sigma(i-j)^2}\|$ $(i, j = 1, \ldots, n)$, where σ is a positive parameter and $n \leq 4$. Find the solutions and the value of the game. Prove that for σ large the optimal strategies are unique.

16. Let $\mathbf{A} = \|a_{ij}\|$ be an $m \times n$ matrix in which a_{ij} are independent identically distributed random variables (see Section C.3) whose probability density is $f(\xi)$. Show that the probability that $\mathbf{A}$ has a saddle point

is independent of $f(\xi)$ and equal to

$$\frac{m!n!}{(m+n-1)!}.$$

17. Let **A** and **B** denote matrices consisting of nonnegative components, and suppose that the game $-\mathbf{A}$ and the game **B** have a negative and a positive value, respectively. Prove that there exists a value p_0 $(0 < p_0 < 1)$ and strategies $\mathbf{x}^0$ and $\mathbf{y}^0$ such that $v(\mathbf{C}) = 0, \mathbf{x}^0\mathbf{C} \geq \mathbf{0}, \mathbf{C}\mathbf{y}^0 \leq \mathbf{0}$, and $(\mathbf{x}^0, \mathbf{A}\mathbf{y}^0) > 0$, where

$$\mathbf{C} = (1 - p_0)\mathbf{B} - p_0\mathbf{A}.$$

18. Verify the result of the preceding problem where

$$\mathbf{A} = \begin{Vmatrix} 0 & \cdots & 0 & 1 \\ 0 & \cdots & 1 & 0 \\ \vdots & & \vdots & \vdots \\ 1 & \cdots & 0 & 0 \end{Vmatrix}, \qquad \mathbf{B} = \begin{Vmatrix} 0 & \cdots & 0 & n \\ 0 & \cdots & n-1 & 0 \\ \vdots & & \vdots & \vdots \\ 1 & \cdots & 0 & 0 \end{Vmatrix},$$

by showing that p_0 must be of the form $k/(k+1)$ for some integer $k \leq n$, and find $\mathbf{x}^0$ and $\mathbf{y}^0$ satisfying the desired properties.

19. Let **A** be an $n \times n$ matrix and let X^0 and Y^0 denote the sets of optimal strategies for Players I and II, respectively. Suppose that Y^0 meets the boundary of Y. Show that there exists an optimal $\mathbf{x}^0$ which uses at most $n - 1$ indices.

20. Let W be a bounded, closed, convex polyhedral set lying in E^n which never touches the first orthant except possibly at the origin. Prove that there exists a vector $\mathbf{v} = \langle v_1, v_2, \ldots, v_n \rangle$ such that $v_i > 0$ and $(\mathbf{v}, \mathbf{w}) \leq 0$ for every $\mathbf{w}$ in W.

Notes and References to Chapter 3

§ 3.1. Theorems 3.1.1 and 3.1.2 were first presented by Bohnenblust, Karlin, and Shapley [36]. Gale and Sherman [100] have given an independent proof of Theorem 3.1.2, and Dresher and Karlin [70] have presented proofs of both theorems based on the Kakutani fixed-point theorem. Theorem 3.1.1 also emerges as a consequence of the study of "admissible" points of convex sets by Arrow, Barankin, and Blackwell [5]. Analogous structure theorems for linear programming systems have been obtained by Goldman and Tucker [111].

§ 3.3. Theorem 3.3.1, describing the relationship between sets of optimal strategies and the null space of the linear transformation defined by the matrix of the game, is due to Bohnenblust, Karlin, and Shapley [36].

§ 3.4. The converse of Theorem 3.1.2 is due to Bohnenblust, Karlin, and Shapley [36].

§ 3.5. Theorem 3.5.1 was first presented by Bohnenblust, Karlin, and Shapley [36] (see also Volume II Chapter 3).

§ 3.6. Problems 3 and 4 are due to Gale, Kuhn, and Tucker [98]. Problems 11, 12, and 13 are derived from Mills [181] and Gross [114]. Problem 16 is due to A. Goldman and Problem 17 to Thompson [231] (see also Kemeny, Morgenstern, and Thompson [142]).

CHAPTER 4

SOLUTIONS OF SOME DISCRETE GAMES

Although general numerical methods for solving games exist (see Chapter 6), often the special properties of a particular game can be used to determine its solutions quickly. Intuitive considerations frequently lead to a good initial guess, which can then be modified appropriately to yield the correct solutions.

Our analysis of the five examples in this chapter makes repeated use of the theory previously developed, notably dominance concepts, symmetries, and simple forms of probabilistic arguments. The examples have been selected to emphasize principles of game theory rather than the routine iterative procedures of solving games, and to illustrate methods and types of thinking that are applicable to a wide variety of problems.

The first two examples treated involve elementary matrix games. Both exhibit simple dominance structure and yield to elementary applications of the theory. The third example is the poker model introduced in Section 1.2. In describing the full game for the first two examples, we need only specify for each player the form of the pure strategies and the corresponding pay-off. In the third example it seems easier to deal directly with the full strategy space. The technique employed in this example is typical of a general method of solving poker models.

The fourth example is concerned with the strategies of two opposing political parties in an election campaign—a special case of the general question of how to distribute advertising effort. In the specific model discussed, each of the parties must decide in which of several areas to invest time and money in order to win votes. The explicit solution of the game exhibits some surprising consequences.

The final example involves a more elaborate matrix game. Its background was originally military, but it is presented here as a problem in collective bargaining.

The mathematics in this chapter is mostly elementary. Nevertheless, arguments are displayed in greater detail than is usual, in an effort to convey more completely the methodology and approach upon which the analysis of game problems depends.

4.1 Colonel Blotto game. The Colonel Blotto game is a general title for a class of tactical military games. The game considered in this section is one of the simplest examples.

Two opposing armies are advancing on two posts. The first army, under Colonel Blotto, consists of four regiments; the second, under Captain Kije, consists of three regiments. The army sending the most regiments to either post captures the post and all the regiments sent there by the other side, scoring one point for the captured post and one for each of the captured regiments. Colonel Blotto must decide how to deploy his forces in order to win the most points.

Colonel Blotto has five possible pure strategies. These strategies can be represented as pairs $\{x, y\}$, where x is the number of regiments sent to capture post 1 and y the number of regiments sent to capture post 2; they are $\{4, 0\}$, $\{0, 4\}$, $\{3, 1\}$, $\{1, 3\}$, and $\{2, 2\}$. Similarly, the Captain's four possible strategies can be represented by the ordered pairs $\{3, 0\}$, $\{0, 3\}$, $\{2, 1\}$, and $\{1, 2\}$. The complete pay-off matrix is shown below.

		Kije			
		a $\{3, 0\}$	a $\{0, 3\}$	b $\{2, 1\}$	b $\{1, 2\}$
	$\alpha\{4, 0\}$	4	0	2	1
	$\alpha\{0, 4\}$	0	4	1	2
Blotto	$\beta\{3, 1\}$	1	−1	3	0
	$\beta\{1, 3\}$	−1	1	0	3
	$\gamma\{2, 2\}$	−2	−2	2	2

This game is symmetrical for Colonel Blotto in the strategies $\{4, 0\}$ and $\{0, 4\}$, and also in the strategies $\{3, 1\}$ and $\{1, 3\}$; for Captain Kije strategies $\{3, 0\}$, $\{0, 3\}$, and $\{2, 1\}$, $\{1, 2\}$ are symmetrical. It therefore seems reasonable to search for optimal strategies possessing the analogous symmetry property. Let a denote the probability of Captain Kije's using strategy $\{3, 0\}$ (and $\{0, 3\}$), and b the probability of his using strategy $\{2, 1\}$ (and $\{1, 2\}$). In order to make the mixed strategy $\langle a, a, b, b\rangle$ optimal for Captain Kije, it is necessary and sufficient that

$$4a + 3b \leq v, \qquad 3b \leq v, \qquad -4a + 4b \leq v. \tag{4.1.1}$$

Clearly, the second inequality is implied by the first. Since $2a + 2b = 1$, the inequalities become $4a + 3(\frac{1}{2} - a) \leq v$ or $a + \frac{3}{2} \leq v$, and $-4a + 4(\frac{1}{2} - a) \leq v$ or $-8a + 2 \leq v$.

The values $a^0 = \frac{1}{18}$, $b^0 = \frac{4}{9}$, and $v = \frac{14}{9}$ provide a solution to (4.1.1) with strict inequality in the second relation.

If this does represent an optimal strategy, then Lemma 2.1.2 would require $\beta = 0$, where β is the probability of Colonel Blotto's using $\{3, 1\}$ or $\{1, 3\}$. The mixed strategy $\langle\alpha, \alpha, 0, 0, \gamma\rangle$ will be optimal if and only if

$$4\alpha - 2\gamma \geq v = \tfrac{14}{9}, \qquad 3\alpha + 2\gamma \geq v, \qquad 2\alpha + \gamma = 1.$$

The values $\alpha^0 = \frac{4}{9}$ and $\gamma^0 = \frac{1}{9}$ satisfy these equations. By direct verification, we obtain that $\langle\frac{4}{9}, \frac{4}{9}, 0, 0, \frac{1}{9}\rangle$ is indeed an optimal strategy for Colonel Blotto, that $\langle\frac{1}{18}, \frac{1}{18}, \frac{4}{9}, \frac{4}{9}\rangle$ is an optimal strategy for Captain Kije, and that the value of the game is $\frac{14}{9}$. With the aid of Lemma 2.1.2, it can be immediately established that Colonel Blotto's (Player I's) optimal strategy is unique. However, Captain Kije's set of optimal strategies is not, as can be seen by applying the dimension relations of Theorem 3.1.2 to this example.

4.2 Identification of Friend and Foe (I.F.F. game). Aircraft approaching an air base must signal to the observer at the air base whether they are friendly or hostile. On the basis of the signal the observer must decide to which side, friend or enemy, the incoming aircraft belongs. The pay-off will be

a if the observer identifies a friend as a friend,

b if the observer identifies a friend as an enemy,

c if the observer identifies an enemy as an enemy,

d if the observer identifies an enemy as a friend,

where $a > b, d$ and $c > b, d$.

Suppose that the observer and all friendly aircraft adopt the uncoded signal procedure whereby the friendly aircraft always signal "friend." Let p be the probability that an incoming plane is a friend. This is assumed to be known on the basis of previous experience.

The four possible strategies for the observer are $I_1(F, E) = \{F, E\}$, $I_2(F, E) = \{E, E\}$, $I_3(F, E) = \{E, F\}$, $I_4(F, E) = \{F, F\}$, where, for example, $I_2(F, E) = \{E, E\}$ means that the observer identifies a signal "friend" as an enemy and a signal "enemy" as an enemy.

The enemy has two possible strategies, one to signal "friend" and the other to signal "enemy." These are denoted by F and E, respectively.

If the observer uses I_3 and the enemy strategy is E, the observer will identify a friend as an enemy and receive b with probability p, and he will identify an enemy as a friend and receive d with probability $1 - p$. The total expected pay-off is then $pb + (1 - p)d$. The eight expected pay-offs corresponding to the combinations of pure strategies are summarized in the following matrix:

$$\begin{array}{c} \\ I_1 \\ I_2 \\ I_3 \\ I_4 \end{array} \begin{array}{c} \begin{array}{cc} \quad E \quad & \quad F \quad \end{array} \\ \left\| \begin{array}{cc} pa + (1-p)c & pa + (1-p)d \\ pb + (1-p)c & pb + (1-p)c \\ pb + (1-p)d & pb + (1-p)c \\ pa + (1-p)d & pa + (1-p)d \end{array} \right\| \end{array} .$$

Since $a > b, d$ and $c > b, d$, it is clear that I_1 dominates I_4, I_2 dominates I_3, and in the smaller matrix E dominates F. Hence, the enemy will always signal "friend" and the observer will use I_1 whenever $pa + (1 - p)d > pb + (1 - p)c$, or

$$p > \frac{c - d}{a + c - b - d}.$$

If the inequality is reversed, he will use I_2, and if equality holds, he can use either I_1 or I_2.

The value of the game is $pa + (1 - p)d$ if

$$p > \frac{c - d}{a + c - b - d},$$

and $pb + (1 - p)c$ if

$$p < \frac{c - d}{a + c - b - d}.$$

A more realistic I.F.F. game would involve a scheme of coded signals. Suppose that the observer and friendly aircraft have a system worked out whereby on certain days the friendly aircraft signal "friend" and on others they signal "enemy." If on a particular day the signal "friend" is agreed upon, there are just two sensible strategies for the observer, $I_1(F, E) = \{F, E\}$ and $I_2(F, E) = \{E, E\}$. There are actually two other pure strategies, but these are dominated by I_1 and I_2, so we do not exhibit them. If the signal agreed upon is "enemy," the two sensible pure strategies for the observer are $I_3(F, E) = \{E, E\}$ and $I_4(F, E) = \{E, F\}$.

The expected pay-off matrix is now

$$\begin{array}{c} \\ I_1 \\ I_2 \\ I_3 \\ I_4 \end{array} \begin{array}{c} \begin{array}{cc} \quad E \quad & \quad F \quad \end{array} \\ \left\| \begin{array}{cc} pa + (1-p)c & pa + (1-p)d \\ pb + (1-p)c & pb + (1-p)c \\ pb + (1-p)c & pb + (1-p)c \\ pa + (1-p)d & pa + (1-p)c \end{array} \right\| \end{array} .$$

Let y denote the probability with which the enemy employs strategy E.

Case 1.

$$pb + (1 - p)c < pa + (1 - p)\left(\frac{c + d}{2}\right).$$

The techniques of Section 2.3 apply and yield readily that $y^0 = \frac{1}{2}$ is the unique optimal strategy for the enemy. An optimal strategy for the observer consists of employing I_1 and I_4 each with probability $\frac{1}{2}$.

Case 2.

$$pb + (1 - p)c > pa + (1 - p)\left(\frac{c + d}{2}\right).$$

We obtain a whole set of optimal y strategies for the enemy. Any combination of I_2 and I_3 is an optimal strategy for the observer.

4.3 Poker game. The poker game treated here was introduced previously in Section 1.2. The game is as follows. There are three different hands, 1, 2, and 3, such that $1 < 2 < 3$ ($i < j$ means that j defeats i); each player receives one of the three hands, and antes the amount a; a bet consists of adding an additional amount b. The possible sequences of moves are diagrammed below.

I		II		I
pass	———	{pass, bet}	——— (from bet)	{fold, see}
bet	———	{fold, see}		

If both I and II pass, the hands are compared and the higher hand wins. If a player folds (passes after a bet), the other wins the pot. If a player sees (accepts the other's bet), the hands are compared and the higher wins.

The methods used in solving this game are representative of the analysis of parlor games in general. More important, the same techniques and procedures apply to tactical games of continuous firepower (see Chapters 8 and 9 of Volume II). Because of the importance of the methodology, we shall develop the arguments of the example in some detail, especially in describing the interplay which occurs in determining the maximizing and minimizing optimal strategies.

In this example a strategy for I is a set of three triplets $\langle \alpha_i, \beta_i, \gamma_i \rangle$, where if I gets hand i, α_i is the probability that he bets immediately, β_i is the probability that he passes on the first round and subsequently sees if the occasion arises, and γ_i is the probability that he passes on the first round and subsequently folds if the occasion arises. The quantities α_i, β_i, and γ_i are subject to the following restrictions:

$$0 \leq \alpha_i, \beta_i, \gamma_i \leq 1 \qquad \text{and} \qquad \alpha_i + \beta_i + \gamma_i = 1.$$

A strategy for II is a set of three pairs $\langle x_i, y_i\rangle$, where if II gets hand i, x_i is the probability of his betting after a pass, and y_i is the probability of his seeing after a bet. The quantities x_i and y_i are subject to the restriction $0 \leq x_i, y_i \leq 1$.

Let

$$L(i, j) = \begin{cases} 1 & \text{if} \quad i > j, \\ -1 & \text{if} \quad i < j. \end{cases}$$

Suppose that I receives hand i, thus using strategy $\langle \alpha_i, \beta_i, \gamma_i\rangle$, and suppose that II receives hand j, thus using strategy $\langle x_j, y_j\rangle$. The probability that both I and II will pass is $(\beta_i + \gamma_i)(1 - x_j)$; the pay-off in this case would be $aL(i, j)$. The expected pay-off for a pass by I, a bet by II, and a bet (see) by I is $(a + b)\beta_i x_j L(i, j)$; and the expected pay-off for a pass by I, a bet by II, and a pass (fold) by I is $-a\gamma_i x_j$. The various probabilities and pay-offs if I bets initially are also readily computed. The full expected pay-off function is then seen to be

$$\begin{aligned} 3!\,K(\alpha, \beta, \gamma; x, y) = {} & a \sum_{i \neq j} (\beta_i + \gamma_i)(1 - x_j)L(i, j) \\ & + (a + b) \sum_{i \neq j} \beta_i x_j L(i, j) \\ & - a \sum_{i \neq j} \gamma_i x_j + a \sum_{i \neq j} \alpha_i (1 - y_j) \\ & + (a + b) \sum_{i \neq j} \alpha_i y_j L(i, j). \end{aligned}$$

To obtain a solution to the game we need to find two strategies $\langle x^0, y^0\rangle$ and $\langle \alpha^0, \beta^0, \gamma^0\rangle$ such that

$$K(\alpha^0, \beta^0, \gamma^0; x^0, y^0) \leq K(\alpha^0, \beta^0, \gamma^0; x, y) \tag{4.3.1}$$

for all $\langle x, y\rangle$ and

$$K(\alpha^0, \beta^0, \gamma^0; x^0, y^0) \geq K(\alpha, \beta, \gamma; x^0, y^0) \tag{4.3.2}$$

for all $\langle \alpha, \beta, \gamma\rangle$. In other words, what we want to do is minimize the function $K(\alpha^0, \beta^0, \gamma^0; x, y)$ with respect to $\langle x, y\rangle$, and we hope that the minimum is achieved for x^0, y^0. At the same time we want to maximize $K(\alpha, \beta, \gamma; x^0, y^0)$ in the first set of variables, and we hope that the maximum is achieved for $\langle \alpha^0, \beta^0, \gamma^0\rangle$. This leads to consistency conditions for the optimal strategies $\alpha^0, \beta^0, \gamma^0$ and x^0, y^0.

Suppose that the coefficient of x_i in the pay-off function $K(\alpha^0, \beta^0, \gamma^0; x, y)$ is positive; then, since II wants to minimize, he will require $x_i^0 = 0$. On the other hand, if the coefficient is negative, he will require $x_i^0 = 1$. In the event the coefficient is zero, no immediate constraint is imposed on x_i^0. Similar considerations hold for y_i^0.

The case for I is not quite so simple, because of the added restriction that $\alpha_i + \beta_i + \gamma_i = 1$. What Player I must do in order to maximize $K(\alpha, \beta, \gamma; x^0, y^0)$ is to set that variable (α^0, β^0, or γ^0) equal to 1 which has the largest coefficient. If the two largest coefficients are equal, or all three coefficients are equal, then no restriction on the corresponding α^0, β^0, and γ^0 exists, except that they add up to 1.

These last two paragraphs outline the reasoning with which we shall attempt to solve this poker game.

The variables with their coefficients in the pay-off function are listed below in columns. These expressions were obtained simply by collecting the coefficients from wherever the variable appeared in the function and adding them.

$$
\begin{aligned}
3!\,K(\alpha^0, \beta^0, \gamma^0; x, y) = {} & \\
& y_3[-(\alpha_1^0 + \alpha_2^0)(2a + b)] \\
& y_2[-2a\alpha_1^0 - b(\alpha_1^0 - \alpha_3^0)] \\
& y_1[b(\alpha_2^0 + \alpha_3^0)] \\
& x_3[-b(\beta_1^0 + \beta_2^0)] \\
& x_2[-2a\gamma_3^0 - b(\beta_1^0 - \beta_3^0)] \\
& x_1[-2a(\gamma_2^0 + \gamma_3^0) + b(\beta_2^0 + \beta_3^0)] \\
& + \text{terms independent of } x_i \text{ and } y_j;
\end{aligned}
$$

$$
\begin{aligned}
3!\,K(\alpha, \beta, \gamma; x^0, y^0) = {} & \\
& \alpha_3[2a + b(y_1^0 + y_2^0)] \\
& \beta_3[2a + b(x_1^0 + x_2^0)] \\
& \gamma_3[2a - 2a(x_1^0 + x_2^0)] \\
& \alpha_2[2a + by_1^0 - (2a + b)y_3^0] \\
& \beta_2[b(x_1^0 - x_3^0)] \\
& \gamma_2[-2ax_1^0] \\
& \alpha_1[2a - (2a + b)(y_2^0 + y_3^0)] \\
& \beta_1[-2a - b(x_2^0 + x_3^0)] \\
& \gamma_1[-2a].
\end{aligned}
$$

Since the coefficient of y_3 is very likely to be negative, unless $\alpha_1^0 + \alpha_2^0 = 0$, a good guess for y_3^0 is 1. Similarly, since the coefficient of y_1 is probably positive, a good guess for y_1^0 is 0, and we shall guess that $x_3^0 = 1$, since $-b(\beta_1 + \beta_2)$ is always nonpositive.

The coefficient of γ_3 is most likely to be smaller than the coefficients of α_3 and β_3, so that probably $\gamma_3^0 = 0$. Similarly, by comparing the coefficient of β_1 with that of γ_1, we see that β_1^0 should equal 0.

Our set of variables and coefficients can now be reduced, by eliminating those variables which have been estimated and substituting their values in the remaining coefficients. The resulting coefficients are

$$y_2[-2a\alpha_1^0 - b(\alpha_1^0 - \alpha_3^0)] + x_2[b\beta_3^0] + x_1[-2a\gamma_2^0 + b(\beta_2^0 + \beta_3^0)]$$

and

$$\alpha_3[2a + by_2^0] + \beta_3[2a + b(x_1^0 + x_2^0)] + \alpha_2[-b] + \beta_2[b(x_1^0 - 1)] + \gamma_2[-2ax_1^0] + \alpha_1[-b - (2a + b)y_2^0] + \gamma_1[-2a].$$

No further implications about the coefficients can be deduced without imposing some assumptions on the relative sizes of a and b. The coefficients of α_2, β_2, and γ_2 depend only on the size of x_1, so we shall first attempt to ascertain their weights.

If $2a > b$, then $x_1^0 < b/(2a + b)$ implies that $\beta_2^0 = 0$, $\alpha_2^0 = 0$, $\gamma_2^0 = 1$, which in turn implies that the coefficient of x_1, $[-2a + b\beta_3^0]$, is less than zero. This leads to the conclusion $x_1^0 = 1$, a contradiction.

Thus if $2a > b$, then $x_1^0 \geq b/(2a + b)$. Suppose $x_1^0 > b/(2a + b)$; then $\beta_2^0 = 1$, which implies that the coefficient of x_1^0 is positive and hence that $x_1^0 = 0$, a contradiction. Therefore $x_1^0 = b/(2a + b)$ and $\alpha_2^0 = 0$.

Looking at y_2 and α_1 and supposing $y_2^0 < (2a - b)/(2a + b)$, we obtain

$$y_2^0 < \frac{2a - b}{2a + b} \rightarrow \alpha_1 = 1 \rightarrow \text{the coefficient of } y_2 \text{ is negative} \rightarrow y_2^0 = 1,$$

a contradiction. Furthermore,

$$y_2^0 > \frac{2a - b}{2a + b} \rightarrow \alpha_1 = 0 \rightarrow \text{the coefficient of } y_2 \text{ is nonnegative,}$$

which gives $\alpha_3 = 0$, $\beta_3 = 1$, from which we infer that $x_2^0 = 0$ and

$$x_1^0 = \frac{b}{2a + b} > y_2^0 > \frac{2a - b}{2a + b},$$

so that if we assume $a \geq b$, we have a contradiction.

Case 1. $a > b$. (The case $b > a$ is left as an exercise; see Problems 4–7.)

In this case we have seen that

$$x_1^0 = \frac{b}{2a + b}, \qquad \alpha_2^0 = 0, \qquad \text{and} \qquad y_2^0 = \frac{2a - b}{2a + b}.$$

Now $\beta_3^0 > 0 \rightarrow x_2^0 = 0$. But then $[2a + b(x_1^0 + x_2^0)] < [2a + by_2^0] \rightarrow \beta_3^0 = 0$, a contradiction. Therefore $\beta_3^0 = 0$ and $\alpha_3^0 = 1$. The equations have now been reduced to

$$y_2[-(2a + b)\alpha_1^0 + b] + x_1[-2a\gamma_2^0 + \beta_2^0 b] + x_2 \cdot 0$$

and

$$\beta_2\left[\frac{-2ab}{2a + b}\right] + \gamma_2\left[\frac{-2ab}{2a + b}\right] + \alpha_1[-2a] + \gamma_1[-2a].$$

Since $0 < x_1^0 < 1$, the coefficient of x_1 must be 0; thus $\beta_2^0 = (2a/b)\gamma_2^0$. But $\beta_2 + \gamma_2 = 1$. Hence

$$\beta_2^0 = \frac{2a}{2a+b} \quad \text{and} \quad \gamma_2^0 = \frac{b}{2a+b}.$$

Similarly, since $0 < y_2^0 < 1$, the coefficient of y_2 must be 0. Hence

$$\alpha_1^0 = \frac{b}{2a+b} \quad \text{and} \quad \gamma_1^0 = \frac{2a}{2a+b}.$$

This leaves only x_2 undetermined. Since its coefficient is 0, no restriction emerges directly, but since $\beta_3^0 = 0$, we must have $x_1^0 + x_2^0 \leq y_2^0$. Hence

$$x_2^0 \leq \frac{2(a-b)}{2a+b}.$$

The set of strategies conjectured to be optimal for Case 1 may be summarized as follows:

$$\begin{aligned}
&x_3^0 = 1, && && y_3^0 = 1, \\
&x_2^0 \leq \frac{2(a-b)}{2a+b}, && && y_2^0 = \frac{2a-b}{2a+b}, \\
&x_1^0 = \frac{b}{2a+b}, && && y_1^0 = 0, \\
&\alpha_3^0 = 1, && \beta_3^0 = 0, && \gamma_3^0 = 0, \\
&\alpha_2^0 = 0, && \beta_2^0 = \frac{2a}{2a+b}, && \gamma_2^0 = \frac{b}{2a+b}, \\
&\alpha_1^0 = \frac{b}{2a+b}, && \beta_1^0 = 0, && \gamma_1^0 = \frac{2a}{2a+b}.
\end{aligned} \tag{4.3.3}$$

The functions $K(\alpha, \beta, \gamma; x^0, y^0)$ and $K(\alpha^0, \beta^0, \gamma^0; x, y)$ will now be maximized and minimized, respectively, with the values of (4.3.3) used in the coefficients. If the maximizing and minimizing variables obtained coincide with those above, then the relations of (4.3.1) and (4.3.2) are satisfied and the strategies exhibited are indeed optimal. The analysis is tabulated:

$$y_3\left[\left(\frac{-b}{2a+b}\right)(2a+b)\right] \qquad y_3 = 1 \text{ minimizes}$$

$$y_2\left[-2a\left(\frac{b}{2a+b}\right) - b\left(\frac{-2a}{2a+b}\right)\right] \qquad \text{anything minimizes, in particular } y_2 = \frac{2a-b}{2a+b}$$

$y_1[b]$	$y_1 = 0$ minimizes
$x_3\left[\frac{-2ab}{2a+b}\right]$	$x_3 = 1$ minimizes
$x_2[0]$	anything minimizes, in particular $x_2 \le \frac{2(a-b)}{2a+b}$
$x_1\left[\frac{-2ab}{2a+b} + \frac{2ab}{2a+b}\right]$	anything minimizes, in particular $x_1 = \frac{b}{2a+b}$
$\alpha_3[2a + b(y_2^0)]$	$\alpha_3 = 1$ maximizes
$\beta_3[2a + b(x_1^0 + x_2^0)]$	$\beta_3 = 0$ maximizes
$\gamma_3[2a - 2a(x_1^0 + x_2^0)]$	$\gamma_3 = 0$ maximizes
$\alpha_2[-b]$	$\alpha_2 = 0$ maximizes
$\beta_2\left[-\frac{2ab}{2a+b}\right]$	$\beta_2 + \gamma_2 = 1$ maximizes, in particular
$\gamma_2\left[-\frac{2ab}{2a+b}\right]$	$\beta_2 = \frac{2a}{2a+b}, \gamma_2 = \frac{b}{2a+b}$
$\alpha_1\left[2a - (2a+b)\left(\frac{4a}{2a+b}\right)\right]$	$\alpha_1 + \gamma_1 = 1$ maximizes, in particular
$\gamma_1[-2a]$	$\alpha_1 = \frac{b}{2a+b}, \gamma_1 = \frac{2a}{2a+b}$
$\beta_1[-2a - b(x_2^0 + 1)]$	$\beta_1 = 0$ maximizes

The consistency requirements are satisfied, and thus the strategies are optimal. The value of the game is

$$-\frac{1}{6}\frac{b^2}{2a+b}.$$

A further examination of the reasoning used to derive the form of the solutions of Case 1 shows in fact that we have found them all.

Case 2. $a = b$. The solution for this case is summarized without proof.

$$
\begin{array}{lll}
x_3^0 = 1 & & y_3^0 = 1 \\
x_2^0 = 0 & & y_2^0 = \tfrac{1}{3} \\
x_1^0 = \tfrac{1}{3} & & y_1^0 = 0 \\
\alpha_2^0 = 0 & -3\alpha_1^0 + \alpha_3^0 = 0 & \beta_3^0 + \alpha_3^0 = 1 \\
\beta_1^0 = 0 & -2\gamma_2^0 + \beta_2^0 + \beta_3^0 = 0 & \beta_2^0 + \gamma_2^0 = 1 \\
\gamma_3^0 = 0 & & \alpha_1^0 + \gamma_1^0 = 1
\end{array}
$$

The value of the game is $-a/6.3$.

The two solutions obtained above exhibit several interesting features. For example, Player I never bets the first round with hand 2. We observe in this connection that for $a = b$, Player II also does not bet with hand 2 if Player I has originally passed. This is not necessarily true for $a > b$.

Where $a = b$, Player I sometimes "sandbags," i.e., passes with a high hand on the first round and sees any subsequent bet.

The need for random bluffing comes in as a natural part of the optimal strategies. This is expressed by the fact that x_1^0, α_1^0, and γ_1^0 are all strictly positive. The frequency with which bluffing is desirable depends on the ratio of b to a.

4.4 An advertising example. Two political parties are engaged in an election campaign. There are n localities, labeled 1 through n, which normally vote for the candidate of Party II. In terms of number of voters, these localities rank as follows: $a_1 > a_2 > \cdots > a_n > 0$. Party I has announced that it intends to invade one of these areas in an attempt to win votes. Party II will try to anticipate Party I's campaign with counter-propaganda. Facilities and finances are limited, so that each of the parties can put its effort into only one area. We assume that the campaign is such that if Party I invades area j, the value to him is a_j if the area is not defended by Party II, and pa_j $(0 \leq p < 1)$ if the area is defended. The coefficient p may be considered as a measure of the rate of effectiveness of the party orators, etc.

Each party has n pure strategies. The pay-off matrix is

$$\mathbf{A} = \begin{Vmatrix} pa_1 & a_1 & \cdots & a_1 \\ a_2 & pa_2 & \cdots & a_2 \\ a_3 & a_3 & \cdots & a_3 \\ \vdots & \vdots & & \vdots \\ a_n & a_n & \cdots & pa_n \end{Vmatrix}$$

Intuitively it seems likely that Party I would want to invade the more valuable areas and that Party II would defend the more valuable areas. We try to determine an optimal mixed strategy with the first r rows relevant, keeping r as a parameter to be determined later. Suppose that $\mathbf{x}^0 = \langle x_1^0, x_2^0, \ldots, x_r^0, 0, \ldots, 0\rangle$ denotes an optimal strategy. Assume also, temporarily, that the first r pure strategies belong to an optimal $\mathbf{y}$ strategy. Then, by virtue of Lemma 2.1.2,

$$\sum_{i=1}^{r} a_i x_i^0 - (1 - p)a_k x_k^0 = v \qquad (1 \leq k \leq r). \tag{4.4.1}$$

Hence $a_k x_k^0 = a_1 x_1^0$ for $2 \leq k \leq r$, and by the normalization condition we obtain

$$x_k^0 = \frac{1/a_k}{\sum_{i=1}^r 1/a_i} \qquad \text{and} \qquad v = \frac{r+p-1}{\sum_{i=1}^r 1/a_i}. \tag{4.4.2}$$

It follows easily from (4.4.1) and (4.4.2) that

$$\mathbf{x}^0\mathbf{A} \geq \mathbf{v},$$

with strict inequality for all components j where $j > r$. By Lemma 2.1.2 all optimal $\mathbf{y}$ strategies must concentrate in the first r columns (since we do not know that $\mathbf{x}^0$ is actually optimal for Party I, the analysis should be regarded as a motivated searching for the solution). Assuming that $\mathbf{y}^0 = \langle y_1^0, y_2^0, \ldots, y_r^0, 0, \ldots, 0\rangle$, we determine y_i^0 from

$$\frac{p+r-1}{\sum_{i=1}^r 1/a_i} = pa_k y_k^0 + a_k(1 - y_k^0) \qquad (1 \leq k \leq r)$$

or

$$y_k^0 = \frac{1}{1-p}\left[1 - \frac{(p+r-1)/a_k}{\sum_{i=1}^r 1/a_i}\right] \qquad (1 \leq k \leq r), \tag{4.4.3}$$

which is nonnegative if and only if

$$\frac{1}{r+p-1} \geq \frac{1/a_k}{\sum_{i=1}^r 1/a_i} \qquad (1 \leq k \leq r). \tag{4.4.4}$$

Since a_k strictly decreases, (4.4.4) is equivalent to

$$\frac{1}{r+p-1} \geq \frac{1/a_r}{\sum_{i=1}^r 1/a_i}.$$

The strategy $\mathbf{y}^0$ will satisfy $\mathbf{A}\mathbf{y}^0 \leq \mathbf{v}$ if and only if

$$\frac{1/a_{r+1}}{\sum_{i=1}^r 1/a_i} \geq \frac{1}{r+p-1}. \tag{4.4.5}$$

In order to fulfill the conditions of both (4.4.4) and (4.4.5), we choose the value r as the smallest positive integer less than or equal to n satisfying

$$\frac{1/a_{r+1}}{\sum_{i=1}^r 1/a_i} > \frac{1}{r+p-1} \geq \frac{1/a_r}{\sum_{i=1}^r 1/a_i}. \tag{4.4.6}$$

If no r can be found obeying these inequalities, we define $r = n$. This case can occur only when

$$\frac{1/a_n}{\sum_{i=1}^n 1/a_i} \leq \frac{1}{n+p-1}. \tag{4.4.7}$$

The proof of this assertion is as follows. Note first that if

$$\frac{1}{r+p-1} \geq \frac{1/a_{r+1}}{\sum_{i=1}^{r} 1/a_i},$$

then

$$\frac{1}{r+p} \geq \frac{1/a_{r+1}}{\sum_{i=1}^{r+1} 1/a_i}.$$

(The first inequality is the same as

$$r+p-1 \leq \frac{\sum_{i=1}^{r} 1/a_i}{1/a_{r+1}},$$

and hence

$$r+p \leq \frac{\sum_{i=1}^{r} 1/a_i}{1/a_{r+1}} + \frac{1/a_{r+1}}{1/a_{r+1}},$$

which is the second inequality.)

Next, observe that the right-hand inequality of (4.4.6) is valid for $r = 1$. In view of what has just been shown, we may conclude that (4.4.6) can fail to hold for all r only when (4.4.7) obtains.

The optimal strategies are now explicit. We need only choose r according to (4.4.6) and then define $\mathbf{x}^0$ and $\mathbf{y}^0$ by means of (4.4.2) and (4.4.3), respectively. That these are indeed solutions has been established in the course of the preceding analysis. Except if a_k and p take on special values [for which case the rare equality for the determination of r according to (4.4.6) obtains], the optimal strategies are unique.

We find that both parties concentrate attention on the same r attractive localities, and that they differ solely in having different probabilities of selecting a given locality from among those under consideration.

***4.5 A bargaining example.** The game we shall analyze in this section is given an interpretation of collective bargaining. This interpretation may seem to an extent contrived, since labor-management disputes do not involve a full conflict of interests, but it nonetheless serves to suggest possible applications of game-theory reasoning to industrial situations.

Our example was originally a game of strategy concerned with attack and defense of a hidden object. A defender has an object of great value which he may conceal in any one of many containers; the attacker makes a series of attempts to destroy the object by destroying the containers. By way of illustration, suppose that an A-bomb is to be carried in one of two identical bombers, called P (protected) and F (flank). They fly in forma-

---

*Starred sections are advanced discussions that may be omitted at first reading without loss of continuity.

tion with P behind F, so that a hostile fighter wishing to attack P must pass through F's field of fire and run a risk of being shot down before being able to close on its target. Once it has engaged its target, however, the fighter can destroy it with probability β. The defender (Player II) chooses P or F to carry the bomb. The attacker (Player I) chooses an ordered set of n members $PFFP \ldots P$, where the individual members represent his target choice on each pass provided all previous passes have failed. The integer n represents the number of allowable attacks the fighter can make.

We now revert to the bargaining interpretation, keeping the notation P and F as a reminder of the attack-and-defense implication of the model.

A union and the management of a certain firm are at odds over two issues, F and P. Under all circumstances let β be the probability that the union can settle the issue F, negatively or positively. If F is settled positively (in favor of the union), the game is over; i.e., no further agreements can be reached. If F is settled negatively, let β be the probability of the union's settling P. If F is not settled one way or the other, let $\gamma < \beta$ be the probability of the union's settling P. In other words, if F has not previously been settled, P is harder to settle. (In the attack-and-defense model, it is harder to destroy P when F is intact than when F has been destroyed.)

In any opportunity to bargain, the union must decide which issue to attempt to settle. We shall assume that management *a priori* is inclined to make concessions on only one of the issues, if any at all. Management has two pure strategies; i.e., it must decide which of the two issues it will make concessions on if necessary. (All this in reality is decided ahead of time with a proportion of sureness.) We shall also suppose that the union knows that progress on at most one of the issues is possible. If management is prepared to yield on one of the issues, say P, then we assume that the probability of the union's ascertaining the hopelessness of settling F is still β. Finally, we assume that if the hopelessness of settling F becomes known to the union, the next attempt at negotiation will be directed at issue P. The situation is not symmetrical. If management has decided to yield on F if necessary, the probability of the union's determining the impossibility of negotiating P is γ. The union also has two pure strategies; i.e., it must decide which of the two issues to bargain. Player I is identified with the union and Player II with management.

In the case of only one opportunity for bargaining (opportunity to bargain may correspond to a meeting every three months between union and management representatives), the formal description of the bargaining problem is as follows. Two strategies are available to Player II: (i) to favor yielding on issue P; (ii) to favor yielding on issue F. Player I also has two possible strategies: (i) to bargain P; (ii) to bargain F. The pay-off to Player I is taken as equal to the probability that the issue will be settled positively for the union.

	Yield on P	Yield on F
Bargain P	γ	0
Bargain F	0	β

The value of the game is $v = \beta\gamma/(\beta + \gamma)$, and Player I has the unique optimal mixed strategy

$$\text{play } P \text{ with prob } \frac{\beta}{\beta + \gamma},$$

$$\text{play } F \text{ with prob } \frac{\gamma}{\beta + \gamma}.$$

Player II has the same optimal strategy.

We now generalize the preceding by allowing Player I two opportunities to bargain. Player I now has four pure strategies, and the pay-off kernel becomes

$$\begin{array}{c} \\ P, P \\ P, F \\ F, P \\ F, F \end{array} \begin{array}{c} \begin{array}{cc} \quad P \quad & \quad F \quad \end{array} \\ \left\| \begin{array}{cc} 2\gamma - \gamma^2 & \beta\gamma \\ \gamma & \beta \\ \beta^2 + (1 - \beta)\gamma & \beta \\ \beta^2 & 2\beta - \beta^2 \end{array} \right\| \end{array}$$

The first letter of a pure strategy for Player I indicates the issue to be bargained at the first opportunity, and the second letter the issue to be bargained at the second opportunity if the first session has resulted in no decision or a negative decision. The pay-off kernel is the expected value for Player I or, equivalently, the probability of success according to the strategies chosen by Players I and II. It is understood that if Player I can completely settle either issue, positively or negatively, then that issue may be considered closed.

We now show the derivation of a typical element in the preceding matrix: $\beta^2 + (1 - \beta)\gamma$, appearing in the third row of the first column. Player I's strategy is F, P and Player II's strategy is P. There are two possible outcomes if Player I bargains F: (a) F is settled negatively (with prob β) and the probability now of settling P is β; or (b) F is not settled (with prob $1 - \beta$), and the probability of settling P is γ. The total probability is thus

$$\beta^2 + (1 - \beta)\gamma.$$

We next investigate the solutions of the game. Note that since $\gamma < \beta$, the second row is dominated by the third row. Some elementary manipulations show that the fourth row, although not dominated, is not involved in any optimal strategy for Player I. The solution is as follows:

Case 1. If $\beta^2 - \beta\gamma + \gamma^2 - \gamma \leq 0$, then the optimal strategy for Player I is

$$F,P \text{ with prob } \frac{\gamma(1-\beta) + \gamma(1-\gamma)}{\beta(1-\beta) + \gamma(1-\gamma)},$$

$$P,P \text{ with prob } \frac{\beta(1-\beta) - \gamma(1-\beta)}{\beta(1-\beta) + \gamma(1-\gamma)};$$

the optimal strategy for Player II is

$$\text{favor yielding on } P \text{ with prob } \frac{\beta(1-\gamma)}{\beta(1-\beta) + \gamma(1-\gamma)},$$

$$\text{favor yielding on } F \text{ with prob } \frac{\beta(\gamma-\beta) + \gamma(1-\gamma)}{\beta(1-\beta) + \gamma(1-\gamma)};$$

and the value is

$$v = \beta\gamma \frac{\beta(\gamma-\beta) + 2(1-\gamma)}{\beta(1-\beta) + \gamma(1-\gamma)}.$$

Case 2. If $\beta^2 - \beta\gamma + \gamma^2 - \gamma \geq 0$, then a saddle point exists. Player I chooses F, P, and Player II favors P. The value of the game is

$$v = \beta^2 - \beta\gamma + \gamma.$$

Our final extension of this model is to allow Player I n bargaining opportunities. A strategy is described by an ordered succession of n letters F or P which determines the order of bargaining. We repeat that once an issue is fully settled negatively, the other issue is bargained at the next opportunity.

In studying this general game, we note first that every strategy for Player I in which F is involved r times is dominated by

$$\overbrace{F\ldots F}^{r}\ \overbrace{P\ldots P}^{n-r}.$$

(We have seen an instance of this in the last matrix, where the second row was dominated by the third row.) In other words, if F is to be bargained r times by Player I, then this issue should be bargained first. The argument of the case of two bargaining opportunities shows that $F\ldots FP\ldots$ dominates the similar strategy with the FP interchanged. Iterating this establishes the assertion. Consequently, strategies for Player I are described by an integer r specifying the number of times F is discussed. We now determine the terms in the pay-off matrix $a(r, P)$ and $a(r, F)$, the first of which corresponds to the strategy r for I and P for II.

$$\begin{aligned} a(r, P) = {} & 1 - \text{prob}\,\{\text{I unable to settle } F \text{ in the first } r \text{ bargaining opportunities}\} \cdot \text{prob}\,\{\text{I unsuccessful in bargaining } P | \text{I unable to settle } F\} - \textstyle\sum_{k=1}^{r} \text{prob}\,\{\text{I able to settle } F \text{ at the } k\text{th opportunity}\} \cdot \text{prob}\,\{\text{I unsuccessful in bargaining } P \text{ in the remaining } n-k \text{ trials, given that } F \text{ was settled negatively at the } k\text{th discussion}\} \\ = {} & 1 - (1-\beta)^r(1-\gamma)^{n-r} - r\beta(1-\beta)^{n-1}. \end{aligned}$$

In a similar manner we find

$$a(r, F) = 1 - (1-\beta)^r(1-\gamma)^{n-r} - \gamma(1-\beta)^r \frac{[(1-\gamma)^{n-r} - (1-\beta)^{n-r}]}{\beta - \gamma}.$$

The game is solved by letting r be continuous and making use of the concavity of $a(r, F)$ and $a(r, P)$ in r. Let r_0 be the solution of $da(r, P)/dr = 0$ and let $R = (1-\gamma)/(1-\beta)$. Then

$$r_0 = n - \frac{\ln[\beta/(1-\beta)] - \ln\ln R}{\ln R},$$

provided the right side is positive; otherwise set $r_0 = 0$. If $a(r_0, P) \leq a(r_0, F)$, then I has a pure optimal strategy r_0 and II has a pure optimal strategy P. If $a(r_0, P) \geq a(r_0, F)$, then I has a pure optimal strategy r given by the solution of the equation

$$1 + \frac{r\beta(\beta-\gamma)}{\gamma(1-\beta)} = \left(\frac{1-\gamma}{1-\beta}\right)^{n-r},$$

or approximately $r = n\gamma/(\beta+\gamma)$ when n is large and β is close to γ. II has an optimal strategy which is a mixture of P and F. For the original discrete game, define the integer $r_1 = [r_0]$ and solve the 2×2 matrix game

$$\begin{Vmatrix} a(r_1, P) & a(r_1, F) \\ a(r_1+1, P) & a(r_1+1, F) \end{Vmatrix}.$$

If $a(r_1, P) \leq a(r_1, F)$, we have a pure strategy solution; otherwise the solution is mixed. This technique will be further discussed in a later chapter, in connection with convex games. (See also Section 2.7, Problem 22.)

4.6 Problems.

1. Determine all the solutions of the Colonel Blotto problem of Section 4.1.

2. Let **A** be the $n \times n$ matrix game with

$$a_{ij} = \begin{cases} +1 & \text{if} \quad i \neq j, \\ -1 & \text{if} \quad i = j. \end{cases}$$

Use the symmetry of the game to find the optimal strategies and the value of this game.

3. Consider the I.F.F. coded game with one signal where a signal is an n-tuple whose components consist of the symbols F or E. Using the parameters a, b, c, d and p of Section 4.2, show that the optimal strategy for the friend and observer is to code the signals and select the particular code to be used at random from among all possible codes. Prove that the enemy's optimal strategy is to select a signal at random from among all possible signals. Prove also that if

$$p > \frac{c - d}{c - d + 2^n(a - b)},$$

where n is the number of components to the signal, the observer identifies only the friend's signal as coming from a friend, and that if the opposite inequality holds, the observer identifies every signal as an enemy's. Find the value of this game.

4. Verify that if $a = b$, the poker model of Section 4.3 has the solution described at the end of that section.

5. Verify that if $a < b < 2a$, the poker model of Section 4.3 has the following solution:

$$x_3^0 = 1 \qquad y_3^0 = 1$$

$$x_2^0 = 0 \qquad \frac{2a - b}{2a + b} \leq y_2^0 \leq \frac{b}{2a + b}$$

$$x_1^0 = \frac{b}{2a + b} \qquad y_1^0 = 0$$

$$\alpha_3^0 = 0 \qquad \beta_3^0 = 1 \qquad \gamma_3^0 = 0$$

$$\alpha_2^0 = 0 \qquad \beta_2^0 = \frac{2a - b}{2a + b} \qquad \gamma_2^0 = \frac{2b}{2a + b}$$

$$\alpha_1^0 = 0 \qquad \beta_1^0 = 0 \qquad \gamma_1^0 = 1$$

6. Find the solution of the poker model of Section 4.3, assuming that $2a = b$.

7. Find the solution of the poker model of Section 4.3, assuming that $2a < b$.

8. Player I has three stones marked 1, 2, 3, and Player II has three stones marked -1, -2, -3. The players have a 3×3 checkerboard, on which the squares (1, 3), (2, 2), and (3, 1) are marked 0. To start the game, Player I places one of his stones in any one of the unmarked squares; Player II next places a stone in one of the remaining squares, and so on until the board is full. They then play the matrix game that their moves have set up. Compute the number of pure strategies for each player, and show that Player II has a strategy that prevents Player I from obtaining a positive gain.

9. Show that in Problem 8, Player I can be sure of winning at least 0 units.

10. Player II offers Player I a choice of N boxes, D of which contain a dollar bill each. Player I is to choose C of them, and his pay-off will be the number of dollars they contain. He is allowed to open any number of the boxes before making his choice, provided that he chooses only from among the boxes not yet opened. Player II knows the content of each box and may hand the boxes to Player I in any order. The pay-off to Player II is all the money not acquired by Player I. Determine an optimal strategy for each player.

11. Each of two players writes either 0, 1, or 2 on a piece of paper not visible to his opponent. Player I, knowing his own number, calls a guess of the sum of the two numbers. Next Player II, knowing his number, calls a guess of the sum, but he is not permitted to make the same guess as Player I. Player I receives one unit if his guess is correct; he pays one unit if Player II's guess is correct. If neither guess is correct, the game is a draw and the pay-off is zero. Set up the game and determine optimal strategies for each player.

12. A deck of cards is identified with the integers $1, 2, \ldots, N$. After a thorough shuffling, Players I and II are each dealt a card. Player I may either keep his card or compel Player II to exchange cards with him. Player II may now either keep his card (which may be either the one he was dealt or the one he was compelled to accept), or exchange it for another taken out of the pack at random. When both players have made their choice, their hands are compared and the one with the higher hand receives one unit from his opponent. Solve the game for the case $N = 5$.

13. Each of two poker players places an ante of one unit and is dealt a single card from a standard deck. Player I, knowing what card he holds, has the option of folding immediately or betting any one of the amounts $a_1, \ldots, a_n$. If he bets, then Player II has the option of folding (losing the ante) or seeing the bet. In this case the hands are compared, and the player with the higher hand wins. (Assume that the hands are identified

with the numbers 1, . . . , 52, and that hand i is superior to hand j if $i > j$.) Solve the game when $n = 1$ and $a_1 > 1$.

14. Solve Problem 13 for arbitrary n, assuming that $1 < a_1 < \cdots < a_n$. (See also the general poker models of Volume II Chapter 9.)

15. In an attack-and-defense problem each player has two strategies, and the probability of success for the attack is given by a matrix $\|a_{ij}\|$ with $a_{11} > a_{21}$ and $a_{12} < a_{22}$. In addition, the attacker can (if he chooses) use part of his forces for reconnaissance. If this is done, his probability of success is decreased by a fixed amount r, but the defender's strategy is discovered unless he assigns part of his forces to counter-reconnaissance. This diversion of his forces increases the probability of a successful attack by an amount c. Set up the complete strategy spaces and the pay-off matrix for this game. Use dominance arguments to solve the game.

Notes and References to Chapter 4

§ 4.1. More elaborate Blotto games can be found in Blackett [27].

§ 4.2. The I.F.F. example and its ramifications (Problem 3), as described here, were proposed and solved by the members of the Rand Corporation; see Dresher [66].

§ 4.3. This poker model is due to Kuhn [156]. He did not, however, obtain a complete enumeration of all the optimal strategies of the game, as we do in this section. Kuhn's method was to set up the game initially in terms of the pay-off matrix corresponding to pure strategies and then to reduce the analysis, by dominance arguments, to consideration of a subclass of behavior strategies. Our method, by contrast, is based on a combination of intuition and direct analysis, a common way of dealing with general parlor games; see also Volume II Chapter 9.

§ 4.4. The election campaign model was proposed by Dresher [66].

§ 4.5. This bargaining example was developed at the Rand Corporation as a model of attack and defense; see Dresher [66].

§ 4.6. Problems 4 through 7 are various cases of the poker model of § 4.3. Problem 10 is a special case of a type of game discussed by Gale [95], involving repeated play of games with partial information structure. Problem 12 embodies the essential elements of the classical Le Her game, in which the standard card deck is replaced by a deck consisting of the number sequence 1, 2, . . . , N. The original Le Her game was solved by Dresher [69].

SOLUTIONS TO PROBLEMS OF CHAPTERS 1–4

CHAPTER 1

Problem 2. A strategy for Player I consists of two components, the first giving the probability of betting if he draws Lo, and the second the probability of betting if he draws Hi. A strategy for Player II is given by a single component, the probability of calling if Player I bets. The optimal strategies are

$$\left\{\frac{b-a}{b+a}, 1\right\} \quad \text{and} \quad \left\{\frac{2a}{b+a}\right\},$$

respectively. The value is $a(b-a)/(b+a)$.

Problem 4. Let $\mathbf{A} = \|a_{ij}\|$ $(i, j = 1, 2, \ldots, 9)$ denote the pay-off matrix. Then

$$a_{ij} = \begin{cases} -1 & \text{if } j - i \geq 2, \\ +1 & \text{if } i - j \geq 2, \\ +2 & \text{if } j - i = 1, \\ -2 & \text{if } i - j = 1, \\ 0 & \text{if } i - j = 0. \end{cases}$$

Problem 6. The optimal strategies are $\langle \frac{5}{7}, \frac{2}{7} \rangle$ and $\langle \frac{5}{7}, \frac{2}{7}, 0, 0 \rangle$; the value is $\frac{13}{7}$.

Problem 8. In both $\mathbf{A}_1$ and $\mathbf{A}_2$ the element 1 is a saddle point. Hence $v(\mathbf{A}_1) = v(\mathbf{A}_2) = 1$ and $v(\mathbf{A}_1) + v(\mathbf{A}_2) = 2$. Setting $x = 3$ yields $v(\mathbf{A}_1 + \mathbf{A}_2) = 3 > 2$, and setting $x = -1$ yields $v(\mathbf{A}_1 + \mathbf{A}_2) = 1 < 2$.

Problem 9. Consider the game

$$\begin{Vmatrix} 2 & 0 \\ 0 & 2 \end{Vmatrix}.$$

The value is 1. However, note that $\mathbf{x}^* = \langle \frac{1}{2}, \frac{1}{2} \rangle$, $\mathbf{y}^* = \langle 1, 0 \rangle$ yields $(\mathbf{x}^*, \mathbf{A}\mathbf{y}^*) = 1$ and yet $\mathbf{y}^*$ is not optimal.

Problem 10. Let $\mathbf{u} = \langle u_1, \ldots, u_n \rangle$ be an optimal strategy for Player I and $\mathbf{w} = \langle w_1, \ldots, w_m \rangle$ be an optimal strategy for Player II. Then

$$\sum_{i=1}^{n} u_i a_{ij} \geq v \geq \sum_{j=1}^{m} a_{ij} w_j$$

for all i and j. It follows that if $v \geq 0$ the first alternative holds, while if $v < 0$ the second alternative holds.

Problem 13. In the game with matrix

$$\mathbf{C} = \begin{Vmatrix} a & b \\ 0 & d \end{Vmatrix},$$

the element 0 is a saddle point and $\xi = 0$, $\eta = 1$ are unique optimal strategies.

Thus

$$\mathbf{x} = \langle 0, 0, \ldots, 0, \overset{*}{x}_{n_1+1}, \ldots, \overset{*}{x}_{n_1+n_2} \rangle,$$

$$\mathbf{y} = \langle \overset{0}{y}_1, \ldots, \overset{0}{y}_{m_1}, 0, \ldots, 0 \rangle.$$

If Player II uses any of the first m_1 columns, Player I obtains zero; if Player II uses any of the last m_2 columns, Player I obtains at least $d > 0$. Thus, by taking combinations, $C(\mathbf{x}, \mathfrak{y}) \geq 0$ for any $\mathfrak{y}$. Similarly $C(\mathfrak{x}, \mathbf{y}) \leq 0$ for any $\mathfrak{x}$, and $\{\mathbf{x}, \mathbf{y}\}$ is optimal.

Problem 15. Consider the game with matrix $a_{ij} = f(\mathbf{x}_i, \mathbf{y}_j)$. If $\boldsymbol{\xi} = \langle \xi_1, \ldots, \xi_n \rangle$ and $\boldsymbol{\eta} = \langle \eta_1, \ldots, \eta_m \rangle$ are optimal strategies, then

$$\mathbf{x}^0 = \sum_{i=1}^{n} \xi_i \mathbf{x}_i \qquad \text{and} \qquad \mathbf{y}^0 = \sum_{j=1}^{m} \eta_j \mathbf{y}_j$$

satisfy the inequalities.

Problem 16. If there exists no $\mathbf{x}$ in X for which $f(\mathbf{x}, \mathbf{y}) > \alpha$ for all $\mathbf{y}$ in Y, then the set $B_x = \{\mathbf{y} | f(\mathbf{x}, \mathbf{y}) \leq \alpha\}$ is nonvoid, closed because f is lower-semicontinuous, and convex because f is convex in $\mathbf{y}$. For any finite selection $\mathbf{x}_1, \mathbf{x}_2, \ldots, \mathbf{x}_n$ the set $B_{x_1, x_2, \ldots, x_n} = \{\mathbf{y} | f(\mathbf{x}_i, \mathbf{y}) \leq \alpha \text{ all } i\}$ is also nonvoid, closed, and convex. Otherwise the image of Y in E^n defined by the mapping $\mathbf{y} \to \{f(\mathbf{x}_i, \mathbf{y}) - \alpha\}$ has a convex closure which never touches the negative orthant. A strict separating plane exists with components of the normal ξ_i ($i = 1, 2, \ldots, n$) satisfying $\xi_i \geq 0$ and not all zero. It follows that B_{x_0} is empty where

$$\mathbf{x}_0 = \sum_{i=1}^{n} \xi_i \mathbf{x}_i,$$

a contradiction. Hence the intersection $\bigcap_\gamma B_{x_1, x_2, \ldots, x_n}$ taken over all finite collections of $\mathbf{x}_i$ contains an element $\mathbf{y}_0$, satisfying $f(\mathbf{x}, \mathbf{y}_0) \leq \alpha$ for all $\mathbf{x}$ in X.

Problem 17. Let $\mathbf{A} = \|a_{ij}\|_{n \times m}$, $b_j = \max_{1 \leq i \leq n} a_{ij}$, and i_j be a value of i such that $a_{i_j j} = b_j$. We proceed by induction on m. If $m = 2$, consider the 2×2 submatrix

$$\begin{Vmatrix} a_{i_1 1} & a_{i_1 2} \\ a_{i_2 1} & a_{i_2 2} \end{Vmatrix}.$$

(If $i_1 = i_2$, use this row and any other; if $n = 1$, the theorem is trivial.) If $a_{i_1 1}$ is the saddle point of this 2×2 matrix, then $a_{i_1 1} \leq a_{i_1 2}$ and $a_{i_1 1} = b_1 \geq a_{i1}$ for all i, and hence $a_{i_1 1}$ is a saddle point of the original $n \times 2$ matrix. If $a_{i_2 1}$ is the saddle point of the 2×2 matrix, then $a_{i_2 1} \leq a_{i_2 2}$ and $a_{i_2 1} \geq a_{i_1 1} = b_1 \geq a_{i1}$ for all i, and hence $a_{i_2 1}$ is a saddle point of the original $n \times 2$ matrix. Similar arguments hold for $a_{i_1 2}$ and $a_{i_2 2}$.

Suppose now that the theorem is true for $m - 1$, and let $a_{i_0 j_0}$ be a saddle point for the $n \times (m - 1)$ submatrix consisting of the first $m - 1$ columns of our $n \times m$ matrix. If $a_{i_0 j_0} \leq a_{i_0 m}$, then $a_{i_0 j_0}$ will be a saddle point of the full $n \times m$ matrix. Hence we consider $a_{i_0 j_0} > a_{i_0 m}$. If $a_{i_0 m} = a_{i_m m}$, then $a_{i_0 m}$ is a saddle point of the $n \times m$ matrix. Hence we may also consider only the case $a_{i_0m} < a_{i_mm}$.

Case I: $a_{i_mm} \le a_{i_mj_0}$. Let k be an arbitrary integer between 1 and $n - 1$. Consider the 2×2 submatrix

$$\begin{Vmatrix} a_{i_mk} & a_{i_mm} \\ a_{i_0k} & a_{i_0m} \end{Vmatrix}.$$

Now $a_{i_0m} < a_{i_0j_0} \le a_{i_0k}$, and thus a_{i_0k} cannot be a saddle point of this 2×2 submatrix. Also since $a_{i_0m} < a_{i_mm}$, it follows that a_{i_0m} cannot be a saddle point. If a_{i_mk} is a saddle point, then $a_{i_mk} \ge a_{i_0k} \ge a_{i_0j_0} \ge a_{i_mj_0} \ge a_{i_mm}$. On the other hand, if a_{i_mm} is a saddle point, we also have $a_{i_mk} \ge a_{i_mm}$. Thus a_{i_mm} is a saddle point of the $n \times m$ matrix.

Case II: $a_{i_mm} > a_{i_mj_0}$. We proceed as above; in case a_{i_mk} is the saddle point of the 2×2 submatrix, we have $a_{i_mk} \ge a_{i_mj_0}$ as above. If a_{i_mm} is the saddle point, then $a_{i_mk} \ge a_{i_mm} > a_{i_mj_0}$. Thus $a_{i_mj_0} \le a_{i_mk}$ $(k = 1, 2, \ldots, m)$. Now consider the 2×2 submatrix

$$\begin{Vmatrix} a_{i_mj_0} & a_{i_mm} \\ a_{i_0j_0} & a_{i_0m} \end{Vmatrix}.$$

Clearly $a_{i_0m} < a_{i_0j_0}$, hence $a_{i_0j_0}$ is not a saddle point of this 2×2; $a_{i_0m} < a_{i_mm}$, hence a_{i_0m} is not a saddle point; and $a_{i_mm} > a_{i_mj_0}$, hence a_{i_mm} is not a saddle point. Thus $a_{i_mj_0}$ must be the saddle point of this 2×2 submatrix, and hence $a_{i_mj_0} \ge a_{i_0j_0} \ge a_{ij_0}$ $(i = 1, 2, \ldots, n)$. Thus $a_{i_mj_0}$ is a saddle point of the $n \times m$ matrix.

Problem 18. The example

$$\begin{Vmatrix} 3 & 0 & 4 \\ 2 & 1 & 0 \\ 0 & 1 & 3 \end{Vmatrix}$$

shows that the result is not correct if the strictness assumption is removed.

Problem 19. See Wolfe [255].

CHAPTER 2

Problem 3. The optimal strategy for each player is unique:

$$\mathbf{x}^* = \mathbf{y}^* = \left\langle \frac{1/a_1}{\sum_i 1/a_i}, \frac{1/a_2}{\sum_i 1/a_i}, \ldots, \frac{1/a_n}{\sum_i 1/a_i} \right\rangle; v = \frac{1}{\sum_i 1/a_i}.$$

Problem 8. $\langle 0, 0, 0, 0, \frac{1}{16}, \frac{5}{16}, \frac{4}{16}, \frac{5}{16}, \frac{1}{16} \rangle$.

Problem 10. $v = \frac{10}{3}$; $\mathbf{x}^0 = \langle \frac{1}{3}, \frac{1}{3}, \frac{1}{3} \rangle$ is the unique optimal strategy for Player I. Extreme-point optimal strategies for Player II are:

$$\langle \tfrac{1}{3}, \tfrac{1}{3}, \tfrac{1}{3}, 0, 0, 0 \rangle; \langle \tfrac{1}{2}, \tfrac{1}{3}, 0, \tfrac{1}{6}, 0, 0 \rangle; \langle \tfrac{4}{9}, \tfrac{4}{9}, 0, 0, \tfrac{1}{9}, 0 \rangle;$$
$$\langle \tfrac{1}{3}, 0, 0, \tfrac{1}{2}, 0, \tfrac{1}{6} \rangle; \langle 0, \tfrac{4}{15}, 0, 0, \tfrac{7}{15}, \tfrac{4}{15} \rangle; \langle 0, \tfrac{1}{9}, \tfrac{7}{9}, 0, 0, \tfrac{1}{9} \rangle;$$
$$\langle 0, 0, \tfrac{2}{3}, \tfrac{1}{6}, 0, \tfrac{1}{6} \rangle; \langle 0, 0, 0, \tfrac{1}{3}, \tfrac{1}{3}, \tfrac{1}{3} \rangle.$$

Problem 13. Three cases must be distinguished: $n = 3k$, $n = 3k + 1$, and $n = 3k + 2$. If $n = 3k$,

$$\mathbf{x}^0 = \left\langle 0, \frac{1}{k}, 0, 0, \frac{1}{k}, 0, \dots, 0, \frac{1}{k}, 0 \right\rangle,$$

$$\mathbf{y}^0 = \left\langle \frac{1}{k}, 0, 0, \frac{1}{k}, 0, 0, \frac{1}{k}, 0, \dots, \frac{1}{k}, 0, 0 \right\rangle,$$

$$v = \frac{1}{k}.$$

The other cases are similar. The optimal strategies are not unique.

Problem 16. Column 2 dominates column 3; after deleting column 2, row 1 dominates row 3. It follows that

$$v = \frac{a_4a_5 - a_1a_3}{a_4 + a_5 - a_1 - a_3};$$

$$\mathbf{x}^0 = \left\langle \frac{a_5 - a_1}{a_4 + a_5 - a_1 - a_3}, \frac{a_4 - a_3}{a_4 + a_5 - a_1 - a_3}, 0 \right\rangle;$$

$$\mathbf{y}^0 = \left\langle \frac{a_5 - a_3}{a_4 + a_5 - a_1 - a_3}, 0, \frac{a_4 - a_1}{a_4 + a_5 - a_1 - a_3} \right\rangle.$$

In the original matrix, $A(\mathbf{x}^0, \boldsymbol{\beta}_2) > v$; hence $y_2 = 0$. Similarly, $x_3 = 0$. Since $v = A(\boldsymbol{\alpha}_1, \mathbf{y}^0) = a_4y_1^0 + a_3y_3^0$, and since $y_1^0 > 0$, $y_3^0 > 0$, and $y_1^0 + y_3^0 = 1$, it follows that $a_3 < v < a_4$. By virtue of Lemma 2.1.2, $a_4x_1^0 + a_1x_2^0 = v$ and $a_3x_1^0 + a_5x_2^0 = v$, and similarly for $\mathbf{y}^0$. These relations uniquely determine $\mathbf{x}^0$ and $\mathbf{y}^0$. Now clearly we have $x_1^0 > y_1^0 > x_2^0$, and from $x_1^0 > x_2^0$ and $x_1^0 + x_2^0 = 1$ it follows that $x_1^0 > \frac{1}{2}$.

Problem 18. Let $\mathbf{x}^0 \epsilon \cap X^k$. Then $G^k(\mathbf{x}^0, \mathbf{y}) \geq v^k$ for all $\mathbf{y}$ ($k = 1, \dots, K$). Now if Player II uses $\mathbf{y}$ in the game G, and Player I uses $\mathbf{x}^0$,

$$G(\mathbf{x}^0, \mathbf{y}) = \sum_{i,j} x_i^0 G_{ij} y_j = \sum_{i,j} x_i^0 \sum_k G_{ij}^k y_j = \sum_k \sum_{i,j} x_i^0 G_{ij}^k y_j$$

$$= \sum_k G^k(\mathbf{x}^0, \mathbf{y}) \geq \sum v^k.$$

Hence

$$v \geq \sum_k v^k.$$

Problem 20. Suppose that $\mathbf{A}$ is c.m. symmetric of even order. By Theorem 2.5.1, the kernel must be the entire matrix. Since the value of a symmetric game is zero, $|\mathbf{A}| = 0$ (Theorem 2.4.3). Since $\mathbf{A}$ is singular and skew-symmetric of even order, it has rank at most $n - 2$ (since the rank must be even). But the fact that $\mathbf{A}$ is c.m. implies that $\mathbf{Ay} = \mathbf{0}$ has a solution with all components strictly positive. Since the rank is at most $n - 2$, we may perturb the solution slightly and still have a positive solution with components adding up to 1, thus contradicting the uniqueness.

Problem 22. Proof by induction on the number of columns.

CHAPTER 3

Problem 3. By (3.3.3) $\dim \overline{X} = r(\widetilde{\mathbf{A}}) + \dim \eta(\widetilde{\mathbf{A}}') - 1$. From Theorem 3.3.1, $\dim X^0 + 1 = \dim \eta(\widetilde{\mathbf{A}}')$. Combining these results yields $\dim \overline{X} - \dim X^0 = r(\widetilde{\mathbf{A}})$. By Problem 2, $\widetilde{\mathbf{A}}$ is skew-symmetric and hence has even rank.

Problem 4. See [98].

Problem 6.

$$\begin{Vmatrix} 2 & 1 & 2 & -1 & 1 \\ -2 & -1 & -2 & 1 & 1 \\ 0 & 0 & 0 & 0 & -2 \end{Vmatrix}$$

Problem 8. Let $\mathbf{y}$ and ϵ be given; then choose N so that $y_1 + \cdots + y_N > 1 - \epsilon$. Now choose i so that $i > N/\epsilon$, and let $\mathbf{x}$ assign mass 1 to i. Then $A(\mathbf{x}, \mathbf{y}) = \sum_j a_{ij}y_j \geq (1 - \epsilon)a_{iN} + \epsilon(-1)$. But $a_{iN} \geq 1 - \epsilon$; hence $A(\mathbf{x}, \mathbf{y}) > 1 - 3\epsilon$, and of course $A(\mathbf{x}, \mathbf{y}) \leq 1$. Thus

$$\inf_y \sup_x A(\mathbf{x}, \mathbf{y}) = 1.$$

Similarly,

$$\sup_x \inf_y A(\mathbf{x}, \mathbf{y}) = -1.$$

Problem 10. Use the graphical representation procedure of Section 2.3.

Problem 11. Let $\{\mathbf{x}^0, \mathbf{y}^0\}$ be optimal for $\mathbf{A}$ and $\{\mathbf{x}_\alpha, \mathbf{y}_\alpha\}$ be optimal for $\mathbf{A} + \alpha\mathbf{H}$. Then

$$v(\mathbf{A} + \alpha\mathbf{H}) \leq (\mathbf{x}_\alpha, (\mathbf{A} + \alpha\mathbf{H})\mathbf{y}^0) \leq v(\mathbf{A}) + \alpha(\mathbf{x}_\alpha, \mathbf{H}\mathbf{y}^0)$$

and

$$v(\mathbf{A} + \alpha\mathbf{H}) \geq (\mathbf{x}^0, (\mathbf{A} + \alpha\mathbf{H})\mathbf{y}_\alpha) \geq v(\mathbf{A}) + \alpha(\mathbf{x}^0, \mathbf{H}\mathbf{y}_\alpha).$$

Hence

$$(\mathbf{x}^0, \mathbf{H}\mathbf{y}_\alpha) \leq \frac{v(\mathbf{A} + \alpha\mathbf{H}) - v(\mathbf{A})}{\alpha} \leq (\mathbf{x}_\alpha, \mathbf{H}\mathbf{y}^0).$$

Now let X^* be the set of all possible limit points of sequences $\{\mathbf{x}_{\alpha_k}\}$ as $\alpha_k \to 0$, and let Y^* be defined similarly with respect to the $\mathbf{y}_\alpha$. By Lemma 3.5.1, $X^* \subset X^0(A)$ and $Y^* \subset Y^0(A)$. We clearly have

$$\max_{x \in X^0(A)} \min_{y \in Y^0(A)} (\mathbf{x}, \mathbf{H}\mathbf{y}) \leq \max_{x \in X^0(A)} \min_{y \in Y^*} (\mathbf{x}, \mathbf{H}\mathbf{y}) \leq \lim_{\alpha \to 0+} \frac{v(\mathbf{A} + \alpha\mathbf{H}) - v(\mathbf{A})}{\alpha}$$

$$\leq \max_{x \in X^*} \min_{y \in Y^0(A)} (\mathbf{x}, \mathbf{H}\mathbf{y}) \leq \max_{x \in X^0(A)} \min_{y \in Y^0(A)} (\mathbf{x}, \mathbf{H}\mathbf{y}).$$

Problem 16. First we may confine our attention to the subset of the space of matrices where no two elements are equal. This subset has probability 1. If no two elements of the matrix are equal, there can be only one saddle point. Hence the events $\{a_{ij}$ is a saddle point$\}$ are disjoint events. Thus

$$\Pr\{\mathbf{A} \text{ has a saddle point}\} = \sum_{i,j} \Pr\{a_{ij} \text{ is a saddle point}\}$$

(Pr denotes probability of the event indicated). But any a_{ij} is a saddle point if it is the largest element in its column and the smallest in its row, hence if it is the mth ranking in a collection of $m + n - 1$ arbitrary choices of numbers. The probability of this is

$$\frac{(n-1)!(m-1)!}{(m+n-1)!}.$$

Hence

$$\Pr\{\mathbf{A} \text{ has a saddle point}\} = m \cdot n \frac{(m-1)!(n-1)!}{(m+n-1)!} = \frac{m!n!}{(m+n-1)!}.$$

Problem 17. See Thompson [231].

Problem 19. Suppose that there is no optimal $\mathbf{x}^0$ which uses at most $n - 1$ indices. Then all optimal $\mathbf{x}^0$ use all n indices. But any $\mathbf{x}^0$ using all n indices must have the kernel corresponding to all of $\mathbf{A}$. Hence $\dim X^0 = 0$ and $\dim \overline{X} = n - 1$. It follows that $\mathbf{y}^0$ is unique, and hence that $\dim Y^0 = 0$. Consequently, by Theorem 3.1.2, $\dim \overline{Y} = n - 1$. But this means that $\{\mathbf{y}^0\} = Y^0$ is interior to Y, a contradiction.

Problem 20. Use Theorem 3.5.1. (See also Lemma 5.3.1)

CHAPTER 4

Problem 2. $\mathbf{x}^0 = \mathbf{y}^0 = \langle 1/n, \ldots, 1/n \rangle$; $v = (n-2)/n$.

Problem 3. There are 2^n possible signals. The only reasonable strategies for the observer are (1) to identify all signalers as enemy planes; (2) to identify planes giving the agreed-upon signal as friendly and all others as enemy planes. If (1) is used, the expected pay-off is $pb + (1-p)c$ regardless of the agreed-upon signal. If (2) is adopted, the pay-off is $pa + (1-p)d$ if the enemy uses the same signal, and $pa + (1-p)c$ if the enemy uses any other signal. Hence the matrix

		enemy signals 1	...	2^n
Plan (2)	Code 1	$pa + (1-p)d$	...	$pa + (1-p)c$
	2	$pa + (1-p)c$	...	$pa + (1-p)c$
	$\vdots$	$\vdots$		$\vdots$
	2^n	$pa + (1-p)c$	...	$pa + (1-p)d$
Plan (1): any code		$pb + (1-p)c$	...	$pb + (1-p)c$

Note that if

$$pa + (1-p)\left(\frac{d + (2^n - 1)c}{2^n}\right) > pb + (1-p)c, \tag{a}$$

the last row will be dominated; hence (by symmetry) $\langle 1/2^n, \ldots, 1/2^n \rangle$ is the unique optimal strategy for each player. On the other hand, if the opposite inequality holds, Player I can assure himself of the amount $pb + (1 - p)c$ by using plan (1). If Player I uses any other row, Player II can hold him to the amount of the right side of Eq. (a) by using $\langle 1/2^n, \ldots, 1/2^n \rangle$. Hence, in this case, the observer will identify all signals as enemy signals and the enemy can use any strategy whatsoever. The inequality (a) reduces to that of the problem.

Problem 7.

$$x_3^0 = 1 \qquad y_3^0 = 1$$

$$x_2^0 = 0 \qquad y_2^0 = 0$$

$$x_1^0 = 0 \qquad y_1^0 = 0$$

$$\alpha_2^0 = 0 \qquad \beta_2^0 = 0 \qquad \gamma_2^0 = 1 \qquad \alpha_3^0 + \beta_3^0 + \gamma_3^0 = 1$$

$$\alpha_1^0 = 0 \qquad \beta_1^0 = 0 \qquad \gamma_1^0 = 1 \qquad -2a\gamma_3^0 + b\beta_3^0 \geq 2a$$

Problem 9. One such strategy for Player I is as follows: Place 3 in the square (1, 1). If II doesn't play in (1, 2), place 2 in (1, 2) and in the resulting game play $\langle 1, 0, 0 \rangle$. If II plays in (1, 2), place 2 in (3, 2). If II doesn't play in (3, 3) at his second move, place 1 in (3, 3) and in the resulting game play $\langle 0, 0, 1 \rangle$. If II does play in (3, 3) at his second move, then place 1 in (2, 3). The resulting matrix is

3	$-j_1$	0
$-j_3$	0	1
0	2	$-j_2$

The following strategy (for Player I) now yields zero:

$$\left\langle \frac{j_2 j_3}{3j_2 + 3 + j_2 j_3}, \frac{3j_2}{3j_2 + 3 + j_2 j_3}, \frac{3}{3j_2 + 3 + j_2 j_3} \right\rangle.$$

Problem 11. Optimal strategy for Player I: Pick 0, 1, and 2 each with probability $\frac{1}{3}$ and guess 2.

Optimal strategy for Player II: Pick 0, 1, and 2 each with probability $\frac{1}{3}$. If 0 is chosen and I guesses 2, guess 0. If 1 or 2 is chosen and I guesses 2, guess 3. If I doesn't guess 2, guess 2.

Problem 13. Let α_i = probability I bets on card i and β_j = probability II sees on card j. The pay-off matrix becomes

$$K(\alpha, \beta) = \frac{1}{51.52} \sum_{i \neq j} \{-(1 - \alpha_i) + \alpha_i(1 - \beta_j) + \alpha_i \beta_j L(i, j) a\}.$$

The solution is obtained by employing reasoning analogous to that employed in

Section 4.3. Any optimal strategy has the form

$$\beta_1^0 = \cdots = \beta_{m-1}^0 = 0; \beta_m^0 = m - 52 + \frac{2(51)}{a+1}; \beta_{m+1}^0 = \cdots = \beta_{52}^0 = 1$$

and

$$\alpha_m^0 = \cdots = \alpha_{52}^0 = 1 \quad \text{with} \quad \alpha_1^0, \ldots, \alpha_{m-1}^0 \quad \text{arbitrary}$$

between zero and one except satisfying

$$\sum_{i=1}^{m-1} \alpha_i^0 = \frac{a-1}{a+1}(52 - m) \quad \text{if } \beta_m^0 < 1$$

and satisfying

$$\sum_{i=1}^{m-1} \alpha_i^0 \geq \frac{a-1}{a+1}(52 - m) \quad \text{and} \quad \sum_{i=1}^{m-2} \alpha_i^0 \leq \frac{a-1}{a+1}(53 - m) \quad \text{if } \beta_m^0 = 1.$$

m is equal to the largest integer not exceeding $53 - 2 \cdot 51/(a+1)$.

Problem 15. In addition to the initial strategies labeled 1 and 2, the defender now has two new strategies that may be denoted by $1D$ and $2D$. Here $1D$ represents the strategy where the defender uses his defense 1 against attack and takes counter-measures against reconnaissance. The attacking player, in addition to his initial strategies labeled 1 and 2 has available strategies where he reconnoiters and utilizes the information obtained; for example $1R$ may denote the strategy 1 unless he knows that the enemy is using strategy 2. Similarly, we have the strategy $2R$. There are other possible strategies for player I that ignore or misuse the information obtained by reconnaissance. These are all dominated. The relevant pay-off matrix reduces to

	1	2	$1D$	$2D$
1	a_{11}	a_{12}	$a_{11} + c$	$a_{12} + c$
2	a_{21}	a_{22}	$a_{21} + c$	$a_{22} + c$
$1R$	$a_{11} - r$	$a_{22} - r$	$a_{11} - r + c$	$a_{12} - r + c$
$2R$	$a_{11} - r$	$a_{22} - r$	$a_{21} - r + c$	$a_{22} - r + c$

Part II

CHAPTERS 5–9

LINEAR AND NONLINEAR PROGRAMMING AND MATHEMATICAL ECONOMICS

CHAPTER 5

LINEAR PROGRAMMING

Programming is concerned with determining feasible programs (plans, schedules, allocations) that are optimal with respect to an agreed-upon criterion. Determining an optimal program is usually a matter of selecting from among all feasible programs a program that maximizes (or minimizes) the criterion function.

A feasible program is one that satisfies certain constraints. When these constraints are linear inequalities and the objective (criterion) function is also a linear expression, we have a problem in linear programming. When the objective function to be optimized or the defining constraint relations are nonlinear, we have a problem in nonlinear programming.

The assumptions of linearity, finiteness of the number of variables, and additivity of the gains and losses limit the applications of linear programming. The merits and demerits of these assumptions are not discussed here. In this chapter we are concerned only with the nature of linear programming problems and the properties their solutions must possess.

It might seem that an immediate solution to our problem would be furnished by the calculus, with its standard methods of determining the maxima and minima of functions. Unfortunately, the ordinary calculus can rarely be applied to programming problems, since their solutions fall on the boundaries of the variables domain. Hence we need new techniques for analyzing the qualitative structure of such problems.

Moreover, because of the vast number of practical applications of this theory, it is important to find efficient computational procedures for approximating solutions. This problem will be fully discussed in Chapter 6. The present chapter emphasizes the simple elegance of the theory.

5.1 Formulation of the linear programming problem. The linear programming problem has the following mathematical form:

Given that the variables $x_1, \ldots, x_n$ satisfy the inequalities

$$\begin{aligned} &a_{11}x_1 + a_{12}x_2 + \cdots + a_{1n}x_n \leq b_1, \\ &a_{21}x_1 + a_{22}x_2 + \cdots + a_{2n}x_n \leq b_2, \\ &\vdots \\ &a_{m1}x_1 + a_{m2}x_2 + \cdots + a_{mn}x_n \leq b_m, \\ &x_1 \geq 0, x_2 \geq 0, \ldots, x_n \geq 0, \end{aligned} \tag{5.1.1}$$

find the set of values of $x_1, \ldots, x_n$ which maximizes*

$$(\mathbf{c}, \mathbf{x}) = x_1c_1 + x_2c_2 + \cdots + x_nc_n.$$

Expressed in vector notation, the problem is to find that $\mathbf{x} = \langle x_1, \ldots, x_n \rangle$ which maximizes $(\mathbf{c}, \mathbf{x})$ subject to the constraints $\mathbf{x} \geq \mathbf{0}$ and $\mathbf{Ax} \leq \mathbf{b}$, where $\mathbf{c} = \langle c_1, \ldots, c_n \rangle$, $\mathbf{b} = \langle b_1, \ldots, b_m \rangle$, and $\mathbf{A}$ is the matrix of coefficients of the system of linear inequalities (5.1.1).

A vector $\mathbf{x}$ which satisfies the constraints (5.1.1) is said to be a feasible vector (strategy). The set of all feasible vectors is a polyhedral convex set, since it is the intersection of the $m + n$ half-spaces described by (5.1.1). It may happen that the inequalities are inconsistent, so that there is no feasible vector. Since the contour lines of the function $(\mathbf{c}, \mathbf{x})$ are hyperplanes in n-space, the maximum point of $(\mathbf{c}, \mathbf{x})$ cannot be an interior point of one of the bounding surfaces. Consequently, the maximum is achieved either at a single vertex or at all points on some bounding face. It may happen, of course, that even when the feasible set is nonvoid the problem has no solution, since $(\mathbf{c}, \mathbf{x})$ could become infinite as $\mathbf{x}$ tends to the boundary of the feasible set. The reader may readily construct such examples.

Many important problems in economics and management science can be expressed in the linear programming form without significant distortion or loss of generality. The remainder of this section will be devoted to outlining several such problems.

A. *Diet problem.* There are m different types of food $(i = 1, 2, \ldots, m)$ that supply varying quantities of the n nutrients $(j = 1, 2, \ldots, n)$ that are essential to good health. The n nutrients may be thought of as protein, minerals, vitamins, etc.

Let y_i be the number of units of food i bought, and let b_i be the price per unit of food i. Let c_j be the minimum requirement of nutrient j.

Let a_{ij} be the amount of nutrient content j in one unit of food i.

The question is, how can the requisite nutrients be supplied at minimum cost? That is, subject to the constraint

$$\sum_{i=1}^{m} y_i a_{ij} \geq c_j \qquad (j = 1, 2, \ldots, n),$$

we seek to determine a vector $\mathbf{y}$ which minimizes the total cost

$$\sum_{i=1}^{m} b_i y_i.$$

* Or minimizes. If the problem is one of minimization, the vector inequality $\mathbf{Ax} \leq \mathbf{b}$ obtained from (5.1.1) is usually reversed and takes the form $\mathbf{yA} \geq \mathbf{c}$. In this case, the objective function is changed from $(\mathbf{c}, \mathbf{x})$ into $(\mathbf{y}, \mathbf{b})$. See also the first paragraph of Section 5.2.

Historically, this is one of the earliest linear programs considered. The problem was constructed and solved on the basis of prices for the year 1944, and it was found that approximately $60 a year would finance an adequate diet—adequate, at least, for such rare souls as might take pleasure in an endless round of dried navy beans, cabbage, wheat flour, and the like. Despite this inauspicious beginning, the trial-and-error method used to solve the diet problem proved to be the forerunner of a number of successful linear programming analyses of industrial and management problems. The original analysis is currently used in deciding on the best feed for chickens.

Linear programming problems which closely resemble the diet problem occur in the oil industry—for example, in calculating the cheapest way to blend available quantities and qualities of gasoline into a required assortment of fuels having specified minimum qualities.

B. *Transportation problem.* There are m ports or centers of supply of a certain commodity, and n destinations or markets to which this commodity must be shipped. The ith port possesses an amount s_i of the commodity ($i = 1, 2, \ldots, m$), and the requirements are such that the jth destination is to receive the amount r_j ($j = 1, 2, \ldots, n$).

Let x_{ij} be the quantity of the commodity we intend to ship from port i to destination j.

Let c_{ij} be the cost of shipping one unit from port i to destination j.

Then, subject to $\sum_j x_{ij} \leq s_i$ and $\sum_i x_{ij} \geq r_j$, we seek to minimize the total shipping costs

$$\sum_{i,j} x_{ij} c_{ij}.$$

This transportation model has received more attention than any other linear programming problem; many special computational schemes have been designed exclusively for it. Several linear programming examples arising from apparently different backgrounds—notably the optimal assignment problem and the railway flow problem—are reducible to the transportation problem. The model is investigated in detail in Section 5.9.

C. *Warehouse storage problem.* Suppose that C is the capacity of a warehouse, and let A represent the initial stock size. Now let x_i represent the amount of the commodity bought at the end of the ith period at a unit price of c_i dollars, and y_i represent the amount sold in the ith period at a unit price of b_i dollars. Finally, let n denote the total horizon (number of periods to be considered).

The problem is to determine the $\mathbf{x}$ and $\mathbf{y}$ that maximize the profit function

$$\sum_{i=1}^{n} (-c_i x_i + b_i y_i),$$

subject to the restrictions

$$y_{i+1} \leq A + \sum_{j=1}^{i} (x_j - y_j) \leq C$$

and

$$x_i \geq 0, y_i \geq 0 \qquad (i = 1, 2, \ldots, n).$$

The right-hand inequality in the first expression takes care of the capacity limitations on any buying and selling program. The left-hand inequality states that it is not possible to sell more than is available.

D. *Activity analysis problem.* There are n ways (activities) of producing an output from available stocks of m primary resources.

Let a_{ij} be the amount of the ith material used in operating the jth activity at unit intensity. The intensity of an activity can be related to the number of factories engaged in this one activity, or to the level of labor input, which often is considered as an exogenous resource; any such interpretation will serve.

Let b_i be the available stock of the ith resource.

Let c_j be the value of the output achieved by operating the jth activity at unit intensity.

Let x_j be the level of intensity of the jth activity to be undertaken.

Subject to the restrictions

$$\sum_{j=1}^{n} a_{ij}x_j \leq b_i$$

and $x_j \geq 0$ for all i and j, respectively, we want to choose $\mathbf{x}$ so as to maximize the total value of the output,

$$\sum_{j=1}^{n} c_j x_j.$$

5.2 The linear programming problem and its dual. In the previous section the linear programming problem was defined in terms of finding a vector $\mathbf{x}$ such that $\mathbf{Ax} \leq \mathbf{b}$, $\mathbf{x} \geq \mathbf{0}$, and $(\mathbf{c}, \mathbf{x}) = \sum x_j c_j$ is maximized. This will be referred to as Problem I. A closely related problem, called the dual of Problem I, is to find a vector $\mathbf{y} = \langle y_1, \ldots, y_m \rangle$ that minimizes $(\mathbf{y}, \mathbf{b}) = \sum b_i y_i$ and satisfies the constraints $\mathbf{yA} \geq \mathbf{c}$ and $\mathbf{y} \geq \mathbf{0}$. ($\mathbf{A}$ is an $m \times n$ matrix, $\mathbf{c} = \langle c_1, \ldots, c_n \rangle$, $\mathbf{b} = \langle b_1, \ldots, b_m \rangle$.)

In this dual problem, which we shall call Problem II, minimization has replaced maximization, $\mathbf{b}$ and $\mathbf{c}$ have been interchanged, and the inequality sign in the major constraints has been reversed. Formally, Problems I and II can be characterized as follows:

Problem I:	Problem II:
$\mathbf{x} = \langle x_1, \ldots, x_n \rangle$,	$\mathbf{y} = \langle y_1, \ldots, y_m \rangle$,
$\mathbf{Ax} \leq \mathbf{b}, \mathbf{x} \geq \mathbf{0}$,	$\mathbf{yA} \geq \mathbf{c}, \mathbf{y} \geq \mathbf{0}$,
$\max_x (\mathbf{c}, \mathbf{x})$.	$\min_y (\mathbf{y}, \mathbf{b})$.

If some of the constraining relations in Problem I are equalities, then in Problem II the corresponding components of $\mathbf{y}$ are allowed to vary unrestrictedly (i.e., they may be negative). Moreover, if in Problem I a certain subset of variables x_i vary without restriction, then the corresponding constraining relations of $\mathbf{yA} \geq \mathbf{c}$ are equalities.

In linear programming problems involving economic ideas, the dual problem will usually possess an economic interpretation. In the activity analysis model, for example, the dual variables y_i are usually assigned the meaning of prices; it is common to refer to $\mathbf{y}$ as the "shadow-price" vector, where shadow price may be interpreted as a fictitious price that society gives to a commodity when the economy is operating in a state of perfect competition. (See Chapter 8.) To illustrate these concepts, we return briefly to the activity analysis model and the transportation model of the previous section.

Activity analysis model. Let $\mathbf{y}$ be the external shadow prices that a company should set on its resources in order to reflect their value to society. From the viewpoint of the whole economy, the company's resources would include (besides raw materials and labor) services of management, clerical services, etc. As previously, let c_j be the dollar value of the output achieved by operating activity j at unit intensity. Then, if y_i is the internal price of one unit of the ith resource,

$$\sum_i y_i a_{ij}$$

is the total dollar value (according to these prices) of the output resulting from operating j at unit intensity.

Hence the constraints $\mathbf{yA} \geq \mathbf{c}$ state that internal prices cannot be set to get more value from a product than you put into it. In other words, in a situation of equilibrium the laws of economics for society require no profits. The nature of this assumption and its consequences are fully explored in our general discussion of competitive equilibrium in Chapter 8.

In the dual problem, subject to the constraints $\mathbf{yA} \geq \mathbf{c}$ and $\mathbf{y} \geq \mathbf{0}$, we seek to minimize the value of the total resources, $\sum_i b_i y_i$. As previously, b_i represents the total amount of the ith resource in stock.

Transportation model. The dual constraints of the transportation model are of the form

$$\begin{aligned} c_{ij} &\geq v_j - u_i, \\ u_i &\geq 0 \qquad (i = 1, \ldots, m), \\ v_j &\geq 0 \qquad (j = 1, \ldots, n). \end{aligned} \tag{5.2.1}$$

Here u_i can be interpreted as the shadow price (dollar or value) associated with sending one unit of commodity from port i, and v_j as the shadow price associated with receiving one unit of commodity at destination j. The meaning of equality in (5.2.1) is that under conditions of equilibrium the value to society of one unit of commodity received at destination j is absorbed in the shipping cost c_{ij} and the shadow cost of preparing one unit of commodity for shipment at port i. We shall see later that if the optimal policy calls for a shipment from i to j, then equality must hold in (5.2.1) corresponding to these indices.

The dual to the transportation problem is as follows: Subject to the restrictions of (5.2.1), we wish to select u_i and v_j which maximize

$$\sum_{i=1}^{m} u_i s_i - \sum_{j=1}^{n} v_j r_j,$$

where s_i is the amount available at source i and r_j is the amount demanded at destination j.

The relationships between the linear programming problem and its dual are developed in the following sections. (See also Problem 18, which embodies another meaning of the shadow price.)

5.3 The principal theorems of linear programming (preliminary results). This and the next two sections are primarily concerned with the proof of the two fundamental theorems of linear programming. In this section, we develop some preliminary results concerning convex sets and their application to the linear programming problem. Section 5.4 gives the conditions under which a solution to the problem exists and the relation between the solutions of Problems I and II. Section 5.5 explains the connection between linear programming and game theory.

In seeking a solution to a maximization problem subject to constraints, it is customary to examine the associated Lagrangian form. However, when we have inequality constraints instead of the equalities which are common in the study of variational problems, the analysis of the corresponding Lagrangian form becomes more difficult. Fortunately, the theory of convex sets is exactly what is needed to overcome the difficulties presented by linear inequality constraints. Accordingly, the succeeding Lemma 5.3.1 plays a vital role in the analysis of the Lagrangian form of Problem I. Using the notation of Section 5.2, we may define the Lagrangian form of this problem by

$$\phi(\mathbf{x}, \mathbf{y}) = (\mathbf{c}, \mathbf{x}) + (\mathbf{y}, \mathbf{b} - \mathbf{A}\mathbf{x}) \tag{5.3.1}$$

for $\mathbf{x} = \langle x_1, \ldots, x_n \rangle \geq 0$ and $\mathbf{y} = \langle y_1, \ldots, y_m \rangle \geq 0$.

Problem II is converted into a form of Problem I if the signs of all the parameters are changed, i.e., if $\mathbf{A}$, $\mathbf{b}$, and $\mathbf{c}$ are replaced by $-\mathbf{A}$, $-\mathbf{b}$, and $-\mathbf{c}$. Hence, the Lagrangian form associated with Problem II, consistent with (5.3.1), is defined as

$$\psi(\mathbf{y}, \mathbf{x}) = -(\mathbf{y}, \mathbf{b}) + (\mathbf{yA} - \mathbf{c}, \mathbf{x}) \tag{5.3.2}$$

for $\mathbf{x} \geq \mathbf{0}$ and $\mathbf{y} \geq \mathbf{0}$, which by rearrangement is seen to be $-\phi(x, y)$.

A pair of vectors $\bar{\mathbf{x}}$ and $\bar{\mathbf{y}}$ will be called a *saddle point* of $\phi(\mathbf{x}, \mathbf{y})$ if

$$\bar{\mathbf{x}} \geq \mathbf{0}, \qquad \bar{\mathbf{y}} \geq \mathbf{0}, \tag{5.3.3}$$

and

$$\phi(\mathbf{x}, \bar{\mathbf{y}}) \leq \phi(\bar{\mathbf{x}}, \bar{\mathbf{y}}) \leq \phi(\bar{\mathbf{x}}, \mathbf{y}) \qquad \text{for all} \quad \mathbf{x} \geq \mathbf{0} \quad \text{and} \quad \mathbf{y} \geq \mathbf{0}. \tag{5.3.4}$$

The reader will note parenthetically that this definition of a saddle point agrees with the analogous concept in game theory (Section 1.4). The relation between solutions to Problem I and saddle points of $\phi(\mathbf{x}, \mathbf{y})$ is clarified by Theorem 5.3.1 below.

The following lemma is of basic importance in the proofs of Theorems 5.3.1 and 5.4.1.

▶ LEMMA 5.3.1. Let S be a closed, convex, polyhedral set in Euclidean $(m + n)$-dimensional vector space $(m \geq 1)$ which satisfies the following two properties: (i) if

$$\mathbf{z} = \langle x_1, \ldots, x_n, y_1, \ldots, y_m \rangle = \begin{pmatrix} \mathbf{x} \\ \mathbf{y} \end{pmatrix}$$

is in S and $x_i \geq 0$ $(i = 1, \ldots, n)$, then $\mathbf{y} \not> \mathbf{0}$ [i.e., if $\mathbf{y} \neq \mathbf{0}$, then $\mathbf{y}$ cannot have all components y_j $(1 \leq j \leq m)$ nonnegative]; (ii) S contains at least one point

$$\begin{pmatrix} \mathbf{x}^0 \\ \mathbf{y}^0 \end{pmatrix}$$

for which $\mathbf{x}^0 \geq \mathbf{0}$, $\mathbf{y}^0 \geq \mathbf{0}$ [by hypothesis (i) $\mathbf{y}^0$ is necessarily identically zero]. Then there are vectors $\mathbf{u}$ (of n components) and $\mathbf{v}$ (of m components) such that

(1) $\quad \mathbf{u}^0 \geq \mathbf{0}, \mathbf{v}^0 \gg \mathbf{0};$

(2) $\quad (\mathbf{u}^0, \mathbf{x}) + (\mathbf{v}^0, \mathbf{y}) \leq 0 \qquad \text{for all} \begin{pmatrix} \mathbf{x} \\ \mathbf{y} \end{pmatrix}$ in S.

Remark. The following statement of the geometric version of this lemma may help clarify its meaning. If S contains no point of the positive orthant except on the face $\mathbf{y} = \mathbf{0}$ and does have a point on this face, then S can be separated by a hyperplane from the interior of the positive orthant and strictly from the face $\mathbf{x} = \mathbf{0}$. This lemma can be deduced as an easy consequence of Theorem 3.1.1. However, in order to keep the content of this section independent of all previous advanced discussions, we present the following additional proof, which relies heavily on the duality theory of convex cones (see Section B.3 of the appendix).

Proof. The set S is first enlarged as follows. Define

$$T = \left\{ \begin{pmatrix} \boldsymbol{\xi} \\ \boldsymbol{\eta} \end{pmatrix} \middle| \begin{pmatrix} \boldsymbol{\xi} \\ \boldsymbol{\eta} \end{pmatrix} \leq \begin{pmatrix} \mathbf{x} \\ \mathbf{y} \end{pmatrix} \quad \text{for some} \quad \begin{pmatrix} \mathbf{x} \\ \mathbf{y} \end{pmatrix} \in S \right\}.$$

(The inequality in the definition of T is required to hold for each coordinate.) It is easily verified that T is a closed, convex, polyhedral set which, by assumption (ii), necessarily contains the negative orthant.

Let T^* denote the dual polyhedral cone to T (see Section B.3). Explicitly, we define

$$T^* = \left\{ \begin{pmatrix} \boldsymbol{\alpha} \\ \boldsymbol{\beta} \end{pmatrix} \middle| (\boldsymbol{\alpha}, \boldsymbol{\xi}) + (\boldsymbol{\beta}, \boldsymbol{\eta}) \leq 0 \quad \text{for} \quad \begin{pmatrix} \boldsymbol{\xi} \\ \boldsymbol{\eta} \end{pmatrix} \in T \right\}.$$

The set T^* may then be characterized as

$$T^* = \left\{ \begin{pmatrix} \boldsymbol{\alpha} \\ \boldsymbol{\beta} \end{pmatrix} \middle| \boldsymbol{\alpha} \geq \mathbf{0}, \boldsymbol{\beta} \geq \mathbf{0}, (\boldsymbol{\alpha}, \mathbf{x}) + (\boldsymbol{\beta}, \mathbf{y}) \leq 0 \quad \text{for all} \quad \begin{pmatrix} \mathbf{x} \\ \mathbf{y} \end{pmatrix} \in S \right\}.$$

The conditions $\boldsymbol{\alpha} \geq \mathbf{0}, \boldsymbol{\beta} \geq \mathbf{0}$ are necessary restrictions, since the cone T includes the negative orthant. The second reduction of the defining relations of T^* follows from the definition of T and the fact that $\boldsymbol{\alpha}$ and $\boldsymbol{\beta}$ are required to be nonnegative vectors.

To complete the proof of the lemma it is sufficient to show that for each fixed component j ($1 \leq j \leq m$), T^* must intersect the set

$$H_j = \left\{ \begin{pmatrix} \mathbf{u} \\ \mathbf{v} \end{pmatrix} \middle| \mathbf{u} \geq \mathbf{0}, \ \mathbf{v} \geq \mathbf{0}, \ v_j > 0 \right\}.$$

For in that case an appropriate convex combination of vectors of H_j may be constructed which fits the conclusion of the lemma. Now, to show that T^* intersects H_j, suppose to the contrary that $T^* \subset K_{j_0}$, where

$$K_{j_0} = \left\{ \begin{pmatrix} \mathbf{u} \\ \mathbf{v} \end{pmatrix} \middle| \mathbf{u} \geq \mathbf{0}, \ \mathbf{v} \geq \mathbf{0}, \ v_{j_0} = 0 \right\}.$$

From the duality theory of convex cones, we deduce that

$$T^{**} \supset K_{j_0}^* = \left\{ \begin{pmatrix} \boldsymbol{\xi} \\ \boldsymbol{\eta} \end{pmatrix} \middle| \boldsymbol{\xi} \leq \mathbf{0}, \eta_1 \leq 0, \ldots, \eta_{j_0-1} \leq 0, \right.$$
$$\left. \eta_{j_0+1} \leq 0, \ldots, \eta_m \leq 0 \ \text{ and } \ \eta_{j_0} \text{ arbitrary} \right\}, \tag{5.3.5}$$

where

$$T^{**} = \text{the cone with vertex at the origin spanned by } T, \tag{5.3.6}$$

since T is a closed, convex, polyhedral set which contains the negative orthant. Consider the vector $\mathbf{c} = \begin{pmatrix} \mathbf{0} \\ \delta \end{pmatrix}$ whose first n coordinates are zero and whose last m coordinates are zero except for the j_0th coordinate, which has value $\delta > 0$. Obviously $\mathbf{c}$ belongs to $K^*_{j_0}$. By virtue of the relationship of T and T^{**}, we infer the existence of $\begin{pmatrix} \mathbf{x}^* \\ \mathbf{y}^* \end{pmatrix}$ in S and $\lambda \geq 0$ such that

$$\lambda x_i^* \geq 0 \qquad (i = 1, \ldots, n),$$

$$\lambda y_j^* \geq 0 \qquad (j = 1, \ldots, j_0 - 1, j_0 + 1, \ldots, m),$$

$$\lambda y_{j_0}^* \geq \delta > 0.$$

The last inequality requires $\lambda > 0$; hence $\mathbf{x}^* \geq \mathbf{0}$ and $\mathbf{y}^* > \mathbf{0}$, contrary to the hypothesis of the lemma. The proof of the lemma is complete.

▶**Theorem** 5.3.1. A vector $\bar{\mathbf{x}}$ is a solution to the linear programming Problem I if and only if there exists a vector $\bar{\mathbf{y}}$ such that $\{\bar{\mathbf{x}}, \bar{\mathbf{y}}\}$ is a saddle point of the Lagrangian form $\phi(\mathbf{x}, \mathbf{y})$.

Proof. We shall first prove that if $\bar{\mathbf{x}}$ is a solution to Problem I, then there exists $\bar{\mathbf{y}}$ such that $\{\bar{\mathbf{x}}, \bar{\mathbf{y}}\}$ is a saddle point of $\phi(\mathbf{x}, \mathbf{y})$. Consider in the $(m + 1)$-dimensional vector space a set S defined by

$$S = \left\{ \begin{pmatrix} \mathbf{b} - \mathbf{A}\mathbf{x} \\ (\mathbf{c}, \mathbf{x} - \bar{\mathbf{x}}) \end{pmatrix} \middle| \, \mathbf{x} \geq \mathbf{0} \right\}.$$

A vector of S is an $(m + 1)$-tuple whose first m components constitute the image $\mathbf{b} - \mathbf{A}\mathbf{x}$ of some nonnegative vector $\mathbf{x}$, and whose last component is the real number $(\mathbf{c}, \mathbf{x} - \bar{\mathbf{x}})$ for the same $\mathbf{x}$. Obviously, S is a polyhedral convex set; and $\mathbf{b} - \mathbf{A}\mathbf{x} \geq \mathbf{0}, \mathbf{x} \geq \mathbf{0}$ implies $(\mathbf{c}, \mathbf{x}) - (\mathbf{c}, \bar{\mathbf{x}}) \leq 0$, since $\bar{\mathbf{x}}$ is a solution of Problem I.

Moreover, substituting $\bar{\mathbf{x}}$ for $\mathbf{x}$ in the definition of S produces a nonnegative vector in S. Hence all the requirements of the hypothesis of Lemma 5.3.1 on S are met, and therefore there exists a vector $\bar{\mathbf{y}}$ with m components such that

$$\bar{\mathbf{y}} \geq \mathbf{0} \tag{5.3.7}$$

and

$$(\mathbf{c}, \mathbf{x}) - (\mathbf{c}, \bar{\mathbf{x}}) + (\bar{\mathbf{y}}, \mathbf{b} - \mathbf{A}\mathbf{x}) \leq 0 \qquad \text{for all} \quad \mathbf{x} \geq \mathbf{0}. \tag{5.3.8}$$

Putting $\mathbf{x} = \bar{\mathbf{x}}$ in (5.3.8) and recalling that

$$\mathbf{A}\bar{\mathbf{x}} \leq \mathbf{b}, \tag{5.3.9}$$

we have

$$(\bar{\mathbf{y}}, \mathbf{b} - \mathbf{A}\bar{\mathbf{x}}) = 0 \quad \text{and} \quad (\mathbf{y}, \mathbf{b} - \mathbf{A}\bar{\mathbf{x}}) \geq 0 \qquad \text{for} \quad \mathbf{y} \geq \mathbf{0}. \tag{5.3.10}$$

Equations (5.3.7–10) together show that $\{\bar{\mathbf{x}}, \bar{\mathbf{y}}\}$ is a saddle point of $\phi(\mathbf{x}, \mathbf{y})$.

Next we shall show that if $\{\bar{\mathbf{x}}, \bar{\mathbf{y}}\}$ is a saddle point of $\phi(\mathbf{x}, \mathbf{y})$, then $\bar{\mathbf{x}}$ is a solution to Problem I. The latter part of (5.3.4) can be written as

$$(\bar{\mathbf{y}}, \mathbf{b} - \mathbf{A}\bar{\mathbf{x}}) \leq (\mathbf{y}, \mathbf{b} - \mathbf{A}\bar{\mathbf{x}}) \qquad \text{for all} \quad \mathbf{y} \geq \mathbf{0}. \tag{5.3.11}$$

By choosing the appropriate coordinate of $\mathbf{y}$ sufficiently large, we may deduce that

$$\mathbf{b} - \mathbf{A}\bar{\mathbf{x}} \geq \mathbf{0}, \qquad \bar{\mathbf{x}} \geq \mathbf{0}. \tag{5.3.12}$$

Hence $(\bar{\mathbf{y}}, \mathbf{b} - \mathbf{A}\bar{\mathbf{x}}) \geq 0$. On the other hand, if we insert $\mathbf{y} = \mathbf{0}$ in (5.3.11), it follows that

$$(\bar{\mathbf{y}}, \mathbf{b} - \mathbf{A}\bar{\mathbf{x}}) = 0. \tag{5.3.13}$$

The first part of (5.3.4) is written as

$$(\mathbf{c}, \mathbf{x}) + (\bar{\mathbf{y}}, \mathbf{b} - \mathbf{A}\mathbf{x}) \leq (\mathbf{c}, \bar{\mathbf{x}}) + (\bar{\mathbf{y}}, \mathbf{b} - \mathbf{A}\bar{\mathbf{x}}) \qquad \text{for all} \quad \mathbf{x} \geq \mathbf{0}, \tag{5.3.14}$$

which by (5.3.13) is

$$(\mathbf{c}, \mathbf{x}) + (\bar{\mathbf{y}}, \mathbf{b} - \mathbf{A}\mathbf{x}) \leq (\mathbf{c}, \bar{\mathbf{x}}) \qquad \text{for all} \quad \mathbf{x} \geq \mathbf{0}. \tag{5.3.15}$$

Therefore, for any $\mathbf{x} \geq \mathbf{0}$ such that $\mathbf{b} - \mathbf{A}\mathbf{x} \geq \mathbf{0}$, we have

$$(\mathbf{c}, \mathbf{x}) \leq (\mathbf{c}, \bar{\mathbf{x}});$$

that is, $\bar{\mathbf{x}}$ is a solution to Problem I.

5.4 The principal theorems of linear programming (continued). With the aid of Theorem 5.3.1 we are now prepared to establish, under very general conditions, the principal existence theorem for solutions to Problems I and II. Let us first consider the following duality theorem, which describes the interrelationships between solutions to Problems I and II.

▶THEOREM 5.4.1 (*Duality Theorem*). If Problem I has a solution $\bar{\mathbf{x}}$, then Problem II has a solution $\bar{\mathbf{y}}$ and

$$(\mathbf{c}, \bar{\mathbf{x}}) = (\bar{\mathbf{y}}, \mathbf{b}). \tag{5.4.1}$$

Proof. If $\bar{\mathbf{x}}$ is a solution to Problem I, then by Theorem 5.3.1 there is a vector $\bar{\mathbf{y}}$ such that $\{\bar{\mathbf{x}}, \bar{\mathbf{y}}\}$ is a saddle point of the Lagrangian form $\phi(\mathbf{x}, \mathbf{y})$, where

$$\phi(\mathbf{x}, \mathbf{y}) = (\mathbf{c}, \mathbf{x}) + (\mathbf{y}, \mathbf{b} - \mathbf{A}\mathbf{x}) = -\{(\mathbf{y}, -\mathbf{b}) + (\mathbf{y}\mathbf{A} - \mathbf{c}, \mathbf{x})\}$$
$$= -\psi(\mathbf{y}, \mathbf{x}).$$

Therefore

$$\psi(\mathbf{y}, \bar{\mathbf{x}}) \leq \psi(\bar{\mathbf{y}}, \bar{\mathbf{x}}) \leq \psi(\bar{\mathbf{y}}, \mathbf{x}), \tag{5.4.2}$$

and hence $\{\bar{\mathbf{y}}, \bar{\mathbf{x}}\}$ is a saddle point of $\psi(\mathbf{y}, \mathbf{x}) = (\mathbf{y}, -\mathbf{b}) + (\mathbf{y}\mathbf{A} - \mathbf{c}, \mathbf{x})$. Since $\psi(\mathbf{y}, \mathbf{x})$ is the Lagrangian form of Problem II, $\bar{\mathbf{y}}$ is a solution to Problem II, again by Theorem 5.3.1.

Since

$$(\bar{\mathbf{y}}, \mathbf{b} - \mathbf{A}\bar{\mathbf{x}}) = 0 \quad \text{and} \quad (\bar{\mathbf{y}}\mathbf{A} - \mathbf{c}, \bar{\mathbf{x}}) = 0 \tag{5.4.3}$$

(see 5.3.10), we have

$$(\bar{\mathbf{y}}, \mathbf{b}) = (\bar{\mathbf{y}}, \mathbf{A}\mathbf{x}) = (\bar{\mathbf{y}}\mathbf{A}, \bar{\mathbf{x}}) = (\mathbf{c}, \bar{\mathbf{x}}).$$

▶Theorem 5.4.2 (*Existence Theorem*). If both Problem I and Problem II have feasible vectors, then both have solutions. All pairs of solutions $\bar{\mathbf{x}}$ and $\bar{\mathbf{y}}$ satisfy $(\mathbf{c}, \bar{\mathbf{x}}) = (\bar{\mathbf{y}}, \mathbf{b})$.

Proof. Let $\mathbf{y}^0$ be a feasible vector for Problem II; i.e., $\mathbf{y}^0 \geq \mathbf{0}$ and $\mathbf{y}^0\mathbf{A} \geq \mathbf{c}$. Since the feasible set of Problem I is nonvoid, take any feasible vector $\mathbf{x}$. Then

$$(\mathbf{c}, \mathbf{x}) \leq (\mathbf{y}^0\mathbf{A}, \mathbf{x}) \leq (\mathbf{y}^0, \mathbf{b}). \tag{5.4.4}$$

Therefore, the set $F = \{(\mathbf{c}, \mathbf{x}) | \mathbf{x} \text{ feasible for Problem I}\}$ is bounded above, and since the feasible set F is a closed, polyhedral, convex set, the maximum of $(\mathbf{c}, \mathbf{x})$ is achieved.

The remaining assertions of the theorem follow by applying Theorem 5.4.1.

▶Corollary 5.4.1. If $\mathbf{x}^*$ and $\mathbf{y}^*$ are feasible for Problems I and II, respectively, and $(\mathbf{c}, \mathbf{x}^*) = (\mathbf{y}^*, \mathbf{b})$, then $\mathbf{x}^*, \mathbf{y}^*$ constitute a pair of solutions, and conversely.

Proof. It follows from (5.4.4) that if $\mathbf{x}$ and $\mathbf{y}$ are any pair of feasible vectors for Problems I and II, respectively, then

$$(\mathbf{c}, \mathbf{x}) \leq (\mathbf{y}, \mathbf{b}). \tag{5.4.5}$$

For the special choice of $\mathbf{x}$ and $\mathbf{y}^*$, we find

$$(\mathbf{c}, \mathbf{x}) \leq (\mathbf{y}^*, \mathbf{b}) = (\mathbf{c}, \mathbf{x}^*),$$

and therefore $\mathbf{x}^*$ is a solution to Problem I.

Analogously, $\mathbf{y}^*$ is a solution to Problem II. The converse proposition is the substance of Theorem 5.4.1.

The corollary provides a simple computational criterion for checking whether pairs of feasible vectors are solutions. Moreover, by virtue of (5.4.5) feasible vectors can be used to generate estimates and bounds for Problems I and II. These applications are explored in Chapter 6, where we develop computing methods for solving linear programming problems.

Because of its economic significance, a property of solutions $\{\bar{\mathbf{x}}, \bar{\mathbf{y}}\}$ which is implicitly expressed in (5.4.3) is now stated as a theorem.

▶THEOREM 5.4.3. If $\bar{\mathbf{x}}$ and $\bar{\mathbf{y}}$ are solutions to Problems I and II, respectively, then

(a) $$(\mathbf{A}\bar{\mathbf{x}})_j < b_j \quad \text{implies} \quad \bar{y}_j = 0,$$

(b) $$(\bar{\mathbf{y}}\mathbf{A})_i > c_i \quad \text{implies} \quad \bar{x}_i = 0.$$

These characteristics of solutions are clearly valuable in any computational algorithm, but their basic importance derives from their economic interpretation.

Consider, for example, the activity analysis model and its dual (as formulated in Sections 5.1 and 5.2). In this context part (a) of the theorem states that if $\bar{\mathbf{x}}$ represents an optimal program of effort which does not utilize all of the jth resource, then such a resource is overabundant and has no money value in a competitive market at equilibrium. Part (b) asserts that if the value of the ith activity is less than its operating cost, then that activity is not undertaken (its level is zero). Both conditions make sense economically. The second says in effect that no activity which loses money will be undertaken. The first may be considered a corollary of the law of supply and demand.

***5.5 Connections between linear programming problems and game theory.** In the last two sections we established the existence of solutions to Problems I and II, and described how such solutions are related. In doing so, we introduced a saddle-point or game-type problem (the Lagrangian).

We now establish a connection between solutions of Problems I and II and the solution of the matrix game

$$\begin{array}{cc} & \begin{array}{ccc} n & m & 1 \end{array} \\ \mathbf{B} = \begin{array}{c} n \\ m \\ 1 \end{array} & \left\| \begin{array}{ccc} \mathbf{0} & -\mathbf{A}' & \mathbf{c} \\ \mathbf{A} & \mathbf{0} & -\mathbf{b} \\ -\mathbf{c} & \mathbf{b} & 0 \end{array} \right\| . \end{array}$$

* Starred sections are advanced discussions that may be omitted at first reading without loss of continuity.

The numbers m, n, and 1 outside the matrix denote the number of rows and columns in the various submatrices of B. B is a skew-symmetric matrix; hence the value of the game is zero. Moreover, the sets of strategies for both players are identical (see Section 2.5). A strategy for either player will be denoted by $\langle \mathbf{x}, \mathbf{y}, \lambda \rangle = \langle x_1, \ldots, x_n, y_1, \ldots, y_m, \lambda \rangle$.

▶THEOREM 5.5.1. If the symmetric game **B** has a solution for which $\lambda > 0$, then both linear programming problems have solutions, and conversely.

Proof. Let $\mathbf{z} = \langle x_1, \ldots, x_n, y_1, \ldots, y_m, \lambda \rangle$ be an optimal strategy for the matrix game **B** with $\lambda > 0$. Since the value v of the game is zero, we obtain $\mathbf{Bz} \leq \mathbf{0}$, or

$$-\mathbf{yA} + \mathbf{c}\lambda \leq \mathbf{0}, \qquad \mathbf{Ax} - \mathbf{b}\lambda \leq \mathbf{0}, \qquad -(\mathbf{c}, \mathbf{x}) + (\mathbf{b}, \mathbf{y}) \leq 0.$$

Since $\lambda > 0$, it follows from Lemma 2.1.2 that equality holds in the last relation. Therefore

$$\frac{\mathbf{y}}{\lambda}\mathbf{A} \geq \mathbf{c}, \qquad \mathbf{A}\frac{\mathbf{x}}{\lambda} \leq \mathbf{b}, \qquad \left(\mathbf{c}, \frac{\mathbf{x}}{\lambda}\right) = \left(\mathbf{b}, \frac{\mathbf{y}}{\lambda}\right). \tag{5.5.1}$$

Let $\mathbf{y}' = \mathbf{y}/\lambda$, and $\mathbf{x}' = \mathbf{x}/\lambda$. These vectors are feasible by (5.5.1). Moreover, for any feasible $\mathbf{x}$ in X (the set of all feasible vectors),

$$(\mathbf{c}, \mathbf{x}) \leq (\mathbf{y}'\mathbf{A}, \mathbf{x}) = (\mathbf{y}', \mathbf{Ax}) \leq (\mathbf{y}', \mathbf{b}) = (\mathbf{c}, \mathbf{x}').$$

Hence

$$\max_{x \in X} (\mathbf{c}, \mathbf{x}) = (\mathbf{c}, \mathbf{x}').$$

Similarly,

$$\min_{y \in Y} (\mathbf{y}, \mathbf{b}) = (\mathbf{y}', \mathbf{b}).$$

Consequently, $\mathbf{x}'$ and $\mathbf{y}'$ are solutions to Problems I and II, respectively.

Conversely, let $\mathbf{x} = \langle x_1, \ldots, x_n \rangle$ and $\mathbf{y} = \langle y_1, \ldots, y_m \rangle$ be two solutions to the linear programming Problems I and II, respectively. Set

$$\lambda = \frac{1}{1 + \sum x_i + \sum y_i},$$

and let $\bar{x}_i = \lambda x_i$, $\bar{y}_i = \lambda y_i$, $\mathbf{z} = \langle \bar{x}_1, \ldots, \bar{x}_n, \bar{y}_1, \ldots, \bar{y}_m, \lambda \rangle = \langle \bar{\mathbf{x}}, \bar{\mathbf{y}}, \lambda \rangle$. Clearly, $\sum \bar{z}_i = 1$. We now verify that $\bar{\mathbf{z}}$ is a solution to the symmetric game **B**:

$$\mathbf{yA} \geq \mathbf{c} \rightarrow \lambda\mathbf{yA} \geq \lambda\mathbf{c} \rightarrow -\bar{\mathbf{y}}\mathbf{A} + \lambda\mathbf{c} \leq \mathbf{0},$$

$$\mathbf{Ax} \leq \mathbf{b} \rightarrow \mathbf{A}\lambda\mathbf{x} \leq \lambda\mathbf{b} \rightarrow \mathbf{A}\bar{\mathbf{x}} - \lambda\mathbf{b} \leq \mathbf{0},$$

and

$$(\mathbf{c}, \mathbf{x}) = (\mathbf{y}, \mathbf{b})$$

by Theorem 5.4.1. Combining, we get $\mathbf{B\bar{z}} \leq \mathbf{0}$; hence $\mathbf{\bar{z}}$ is a solution to the game.

By virtue of Theorem 5.5.1, useful ideas from both game theory and linear programming can be applied to problems in either discipline.

***5.6 Extensions of the duality theorem.** Extensions of Problems I and II may be formulated as follows: Let **A**, **B**, **C** denote specified matrices of order $r \times n$, $r \times m$, and $s \times n$, respectively, and let **D** describe an arbitrary matrix of order $s \times m$. The generic vectors **x**, **y**, **u**, and **v** lie respectively in E^n, E^m, E^r, and E^s.

Problem A. From among all matrices **D** which satisfy $\mathbf{Dy} \leq \mathbf{Cx}$, where $\mathbf{Ax} \leq \mathbf{By}$ for some $\mathbf{x} \geq \mathbf{0}$ and $\mathbf{y} \gg \mathbf{0}$, find a maximal matrix **D**.

A matrix **D** is said to be maximal if there exists no $\mathbf{D}'$ which fulfills the conditions given above for some $\mathbf{x} \geq \mathbf{0}$ and $\mathbf{y} \gg \mathbf{0}$ and $\mathbf{D}' > \mathbf{D}$; that is, each component of $\mathbf{D}'$ is at least as large as the corresponding component of **D** and at least one component is larger.

Problem B. Find a minimal matrix **D** satisfying $\mathbf{vD} \geq \mathbf{uB}$ where $\mathbf{uA} \geq \mathbf{vC}$ for some $\mathbf{v} \gg \mathbf{0}$ and $\mathbf{u} \geq \mathbf{0}$.

Remark. If **A** is a general $m \times n$ matrix, $\mathbf{B} = (m \times 1)$, $\mathbf{C} = (1 \times n)$, and $\mathbf{D} = (1 \times 1)$, that is, **D** is a real number, then Problem A is equivalent to Problem I.

There follow extensions of Theorems 5.4.1 and 5.4.2, respectively.

▶THEOREM 5.6.1. If Problem A has a solution **D**, then Problem B has **D** as a solution. Conversely, if Problem B has a solution $\mathbf{\Delta}$, then Problem A has $\mathbf{\Delta}$ as a solution.

▶THEOREM 5.6.2. If there exist vectors $\mathbf{x}_0 \geq \mathbf{0}$, $\mathbf{y}_0 \gg \mathbf{0}$, $\mathbf{u}_0 \geq \mathbf{0}$, and $\mathbf{v}_0 \gg \mathbf{0}$ such that $\mathbf{Ax}_0 \leq \mathbf{By}_0$ and $\mathbf{u}_0\mathbf{A} \geq \mathbf{v}_0\mathbf{C}$, then Problem A and Problem B have identical sets of solutions.

Proof of Theorem 5.6.1. Let **D** be a solution to Problem A. Consider in $(r + s)$-dimensional space a set S defined by

$$S = \left\{ \begin{pmatrix} \mathbf{By} - \mathbf{Ax} \\ \mathbf{Cx} - \mathbf{Dy} \end{pmatrix} \middle| \, \mathbf{x} \geq \mathbf{0}, \mathbf{y} \geq \boldsymbol{\epsilon} \right\},$$

where $\boldsymbol{\epsilon}$ is the vector all of whose components are equal to ϵ (with $\epsilon > 0$

and fixed). The set S is a closed, convex, polyhedral set such that

$$\mathbf{By} - \mathbf{Ax} \geq \mathbf{0},\ \mathbf{x} \geq \mathbf{0},\ \mathbf{y} \geq \boldsymbol{\epsilon} \quad \text{implies} \quad \mathbf{Cx} - \mathbf{Dy} \not> \mathbf{0}. \tag{5.6.1}$$

Otherwise, $\mathbf{Cx} \geq \mathbf{Dy}$ with strict inequality for some component, and one of the elements of $\mathbf{D}$ can be increased without affecting the inequality $\mathbf{Cx} \geq \mathbf{Dy}$; this contradicts the hypothesis that $\mathbf{D}$ is a maximal matrix of Problem A.

Since $\mathbf{D}$ is maximal to Problem A, there are vectors $\mathbf{x}^0 \geq \mathbf{0}$ and $\mathbf{y}^0 \gg \mathbf{0}$ such that

$$\mathbf{Dy}^0 = \mathbf{Cx}^0 \tag{5.6.2}$$

and

$$\mathbf{Ax}^0 \leq \mathbf{By}^0. \tag{5.6.3}$$

We select $\boldsymbol{\epsilon} > 0$ sufficiently small so that $\mathbf{y}^0 \geq \boldsymbol{\epsilon}$ and hence $\mathbf{x}^0$, $\mathbf{y}^0$ qualify in the construction of S. For this $\boldsymbol{\epsilon}$, we see that S contains a vector none of whose coordinates are negative. Thus, conditions (i) and (ii) of Lemma 5.3.1 are satisfied for the set S.

By Lemma 5.3.1, there exist vectors $\mathbf{u}^0$ and $\mathbf{v}^0$ such that $\mathbf{u}^0 \geq \mathbf{0}$, $\mathbf{v}^0 \gg \mathbf{0}$, and

$$(\mathbf{u}^0, \mathbf{By} - \mathbf{Ax}) + (\mathbf{v}^0, \mathbf{Cx} - \mathbf{Dy}) \leq 0 \quad \text{for all} \quad \mathbf{x} \geq \mathbf{0}, \mathbf{y} \gg \boldsymbol{\epsilon}. \tag{5.6.4}$$

Rewriting (5.6.4) gives us

$$(\mathbf{u}^0\mathbf{B} - \mathbf{v}^0\mathbf{D}, \mathbf{y}) + (\mathbf{v}^0\mathbf{C} - \mathbf{u}^0\mathbf{A}, \mathbf{x}) \leq 0 \quad \text{for all} \quad \mathbf{x} \geq \mathbf{0}, \mathbf{y} \gg \boldsymbol{\epsilon}. \tag{5.6.5}$$

For $\mathbf{x} = \mathbf{0}$ in (5.6.5) we get $(\mathbf{u}^0\mathbf{B} - \mathbf{v}^0\mathbf{D}, \mathbf{y}) \leq 0$ for all $\mathbf{y} \gg \boldsymbol{\epsilon}$. Hence

$$\mathbf{u}^0\mathbf{B} \leq \mathbf{v}^0\mathbf{D}. \tag{5.6.6}$$

Since (5.6.5) holds for any $\mathbf{x} \geq \mathbf{0}$, $\mathbf{y} \gg \boldsymbol{\epsilon}$, we have

$$\mathbf{v}^0\mathbf{C} \leq \mathbf{u}^0\mathbf{A}. \tag{5.6.7}$$

Equations (5.6.6) and (5.6.7) show that $\mathbf{D}$ is a feasible matrix for Problem B.

We shall now prove that $\mathbf{D}$ is minimal for Problem B. Assume that $\mathbf{D}$ is not minimal; then there would be a matrix $\boldsymbol{\Delta}$ such that

$$\boldsymbol{\Delta} < \mathbf{D}, \tag{5.6.8}$$

$$\mathbf{v}\boldsymbol{\Delta} \geq \mathbf{uB}, \tag{5.6.9}$$

$$\mathbf{uA} \geq \mathbf{vC}, \tag{5.6.10}$$

for some $\mathbf{u} \geq \mathbf{0}$ and $\mathbf{v} \gg \mathbf{0}$.

It follows that

$$\begin{aligned}(\mathbf{v}, \mathbf{D}\mathbf{y}^0) > (\mathbf{v}, \mathbf{\Delta}\mathbf{y}^0) \geq (\mathbf{u}, \mathbf{B}\mathbf{y}^0) \quad & \text{by (5.6.8) and (5.6.9)} \\ \geq (\mathbf{u}, \mathbf{A}\mathbf{x}^0) \quad & \text{by (5.6.3)} \\ \geq (\mathbf{v}, \mathbf{C}\mathbf{x}^0) \quad & \text{by (5.6.10)} \\ \geq (\mathbf{v}, \mathbf{D}\mathbf{y}^0) \quad & \text{by (5.6.2),}\end{aligned}$$

which is a contradiction. Therefore $\mathbf{D}$ is minimal for Problem B. The second half of the theorem is proved analogously.

Proof of Theorem 5.6.2. Let $\mathbf{B}\mathbf{y}_0 = \mathbf{b}$ and $\mathbf{v}_0\mathbf{C} = \mathbf{c}$, and consider the symmetric game with matrix

$$\left\| \begin{matrix} \mathbf{0} & -\mathbf{A}' & +\mathbf{c} \\ \mathbf{A} & \mathbf{0} & -\mathbf{b} \\ -\mathbf{c} & +\mathbf{b} & 0 \end{matrix} \right\| \cdot$$

Let $\langle \mathbf{x}, \mathbf{u}, \lambda \rangle$ denote an optimal strategy for Player I. Then $(\mathbf{c}, \mathbf{x}) - (\mathbf{b}, \mathbf{u}) \geq 0$. If $\lambda = 0$, then $\mathbf{u}\mathbf{A} \geq \mathbf{0}$ and $\mathbf{A}\mathbf{x} \leq \mathbf{0}$, and consequently

$$0 \leq (\mathbf{u}\mathbf{A}, \mathbf{x}_0) = (\mathbf{u}, \mathbf{A}\mathbf{x}_0) \leq (\mathbf{u}, \mathbf{B}\mathbf{y}_0) = (\mathbf{u}, \mathbf{b}).$$

Similarly, $\mathbf{A}\mathbf{x} \leq \mathbf{0}$ leads to the conclusion that $(\mathbf{c}, \mathbf{x}) \leq 0$. On the other hand, if $\lambda > 0$, then by Lemma 2.1.2 $(\mathbf{c}, \mathbf{x}) - (\mathbf{b}, \mathbf{u}) = 0$. Thus we see that under all circumstances we have $(\mathbf{c}, \mathbf{x}) - (\mathbf{b}, \mathbf{u}) = 0$.

By Theorem 3.1.1, there must exist an optimal strategy with $\lambda > 0$. In any such strategy, $(\mathbf{c}, \mathbf{x}) = (\mathbf{u}, \mathbf{b})$, $\mathbf{u}\mathbf{A} \geq \lambda\mathbf{c}$, and $\lambda\mathbf{b} \geq \mathbf{A}\mathbf{x}$. The vectors $\mathbf{u}$ and $\mathbf{x}$ may presumably be zero. Let $\bar{\mathbf{u}} = \mathbf{u}/\lambda$ and $\bar{\mathbf{x}} = \mathbf{x}/\lambda$. Then $\bar{\mathbf{u}}\mathbf{A} \geq \mathbf{c} = \mathbf{v}_0\mathbf{C}$; $\mathbf{A}\bar{\mathbf{x}} \leq \mathbf{b} = \mathbf{B}\mathbf{y}_0$; and $(\bar{\mathbf{u}}, \mathbf{b}) = (\mathbf{c}, \bar{\mathbf{x}})$, or $(\bar{\mathbf{u}}, \mathbf{B}\mathbf{y}_0) = (\mathbf{v}_0\mathbf{C}, \bar{\mathbf{x}})$, or equivalently $(\bar{\mathbf{u}}\mathbf{B}, \mathbf{y}_0) = (\mathbf{v}_0, \mathbf{C}\bar{\mathbf{x}})$. If $(\bar{\mathbf{u}}\mathbf{B}, \mathbf{y}_0) \neq 0$, define

$$\mathbf{D}\mathbf{y} = \frac{(\bar{\mathbf{u}}\mathbf{B}, \mathbf{y})\mathbf{C}\bar{\mathbf{x}}}{(\bar{\mathbf{u}}\mathbf{B}, \mathbf{y}_0)};$$

otherwise define

$$\mathbf{D}\mathbf{y} = \frac{(\bar{\mathbf{u}}\mathbf{B}, \mathbf{y})\mathbf{1}}{(\mathbf{v}_0, \mathbf{1})} + \frac{(\mathbf{1}, \mathbf{y})\mathbf{C}\bar{\mathbf{x}}}{(\mathbf{1}, \mathbf{y}_0)},$$

where $\mathbf{1}$ is the vector $\mathbf{1} = \langle 1, 1, \ldots, 1 \rangle$. It is readily verified that $\mathbf{D}\mathbf{y}_0 = \mathbf{C}\bar{\mathbf{x}}$ and $\mathbf{v}_0\mathbf{D} = \bar{\mathbf{u}}\mathbf{B}$, and also that the inequalities $\mathbf{A}\bar{\mathbf{x}} \leq \mathbf{B}\mathbf{y}_0$ and $\bar{\mathbf{u}}\mathbf{A} \geq \mathbf{v}_0\mathbf{C}$ are valid. Thus the matrix $\mathbf{D}$ constructed above satisfies the requirements of Problems A and B. It remains only to show that $\mathbf{D}$ is the solution to the problems. Suppose $\mathbf{\Delta} > \mathbf{D}$ has the property that $\mathbf{A}\mathbf{x}' \leq \mathbf{B}\mathbf{y}'$ and $\mathbf{C}\mathbf{x}' \geq \mathbf{\Delta}\mathbf{y}'$ where $\mathbf{y}' \gg \mathbf{0}$ and $\mathbf{x}' \geq \mathbf{0}$. This last fact implies that $\mathbf{C}\mathbf{x}' \geq \mathbf{\Delta}\mathbf{y}' >$

$\mathbf{Dy}'$, with inequality for at least one component. The properties of $\{\bar{\mathbf{u}}, \mathbf{v}_0\}$ imply that

$$\left\|\begin{matrix} \boldsymbol{\xi} \\ \\ \boldsymbol{\eta} \end{matrix}\right\| = \left\|\begin{matrix} -\mathbf{A}' & \mathbf{C}' \\ \\ \mathbf{B}' & -\mathbf{D}' \end{matrix}\right\| \left\|\begin{matrix} \bar{\mathbf{u}} \\ \\ \mathbf{v}_0 \end{matrix}\right\| \leq \mathbf{0}. \tag{5.6.11}$$

Consequently, the inner product of the nonnegative vector $\{\mathbf{x}', \mathbf{y}'\}$ and $\{\boldsymbol{\xi}, \boldsymbol{\eta}\}$ gives a nonpositive result. However,

$$\left\|\begin{matrix} \boldsymbol{\alpha} \\ \\ \boldsymbol{\beta} \end{matrix}\right\| = \left\|\begin{matrix} -\mathbf{A} & \mathbf{B} \\ \\ \mathbf{C} & -\mathbf{D} \end{matrix}\right\| \left\|\begin{matrix} \mathbf{x}' \\ \\ \mathbf{y}' \end{matrix}\right\| \geq \mathbf{0}, \tag{5.6.12}$$

with inequality somewhere in the last set of components corresponding to $\boldsymbol{\beta}$. Thus, since $\mathbf{v}_0 \gg \mathbf{0}$, the inner product of $\begin{pmatrix} \boldsymbol{\alpha} \\ \boldsymbol{\beta} \end{pmatrix}$ and $\{\bar{\mathbf{u}}, \mathbf{v}_0\}$ is strictly positive. But both inner products yield the same formula, one obtained from the other by rearrangement. Hence we obtain a contradiction. The proof of the theorem is complete.

5.7 Warehouse problem. Sections 5.7 through 5.12 are devoted to analytical studies of specific linear programming models.

In this section we examine the warehouse storage problem introduced in Section 5.1. A warehouse of capacity C is used to stock merchandise. The initial amount of merchandise on hand is assumed to be A. It is desired to determine how much stock to buy and sell during n periods of time so as to maximize profit. Buying takes place at the end of each period. Let x_i be the amount of goods bought at the end of the ith period, c_i the cost of the goods at the end of the ith period, y_i the amount of stock sold during the ith period, and b_i the selling price during the ith period. We wish to maximize

$$\pi = \sum_{i=1}^{n} [b_i y_i - c_i x_i].$$

There are two constraints for the problem: (1) no more may be bought than the warehouse can hold, and (2) no more can be sold than is available in stock. Expressed mathematically, these constraints are

$$A + \sum_{j=1}^{i} (x_j - y_j) \leq C \qquad (i = 1, 2, \ldots, n) \tag{5.7.1}$$

and

$$y_{i+1} \leq A + \sum_{j=1}^{i} (x_j - y_j) \qquad (i = 0, 1, \ldots, n-1). \tag{5.7.2}$$

In addition, there are the usual constraints: that is, $x_i \geq 0$ and $y_i \geq 0$ for $i = 1, 2, \ldots, n$.

We first solve the dual problem, after which a solution of the original problem is immediate. The dual problem is to minimize

$$(C - A) \sum_{i=1}^{n} t_i + A \sum_{i=1}^{n} u_i,$$

where $t_i \geq 0$, $u_i \geq 0$, and the corresponding constraints are

$$T_k - U_{k+1} \geq -c_k \; (k = 1, 2, \ldots, n), \quad -T_k + U_k \geq b_k \; (k = 1, 2, \ldots, n);$$

by definition,

$$T_k = \sum_{i=k}^{n} t_i, \qquad U_k = \sum_{i=k}^{n} u_i, \qquad \text{and} \qquad U_{n+1} = 0.$$

The nonnegativity of t_i and u_i are equivalent to the conditions

$$T_{k-1} \geq T_k, \; T_n \geq 0 \qquad \text{and} \qquad U_{k-1} \geq U_k, \; U_n \geq 0. \tag{5.7.3}$$

We seek to minimize $(C - A)T_1 + AU_1$ subject to these constraints.

It is clear from the constraining relations (5.7.3) that the minimum T_k's and U_k's are

$$\begin{aligned} T_n^* &= \max(-c_n, 0), \qquad U_n^* = \max(b_n + T_n^*, 0), \\ T_{n-1}^* &= \max(U_n^* - c_{n-1}, T_n^*), \\ U_{n-1}^* &= \max(b_{n-1} + T_{n-1}^*, U_n), \\ T_{n-2}^* &= \max(U_{n-1}^* - c_{n-2}, T_{n-1}^*), \qquad \text{etc.,} \end{aligned} \tag{5.7.4}$$

and finally the formula for T_1^* and U_1^* obtains. In short, if at any stage a value $U_k(T_k)$ is chosen larger than the value $U_k^*(T_k^*)$, then the constraints imply that T_1 and U_1 exceed T_1^* and U_1^*, respectively. Consequently, T_1^* and U_1^* are the solutions to the dual problem.

It is clear from the formation rules that T_1^* and U_1^* are piecewise linear functions of b_i and c_i. Hence the terms of $(C - A)T_1^* + AU_1^*$ may be rearranged into the form

$$(C - A)T_1^* + AU_1^* = \sum_i [b_i f_i(A, C) - c_i g_i(A, C)], \tag{5.7.5}$$

where $f_i(A, C)$ and $g_i(A, C)$ are functions of A and C only. The fact that f_i and g_i are nonnegative is clear by construction, since in (5.7.4) the coefficients of b_j are always nonnegative and the coefficients of c_j are nonpositive.

We show below that $y_i^0 = f_i(A, C)$ and $x_i^0 = g_i(A, C)$ are feasible vectors. It follows from Corollary 5.4.1 that the pair $\mathbf{x}^0, \mathbf{y}^0$ as defined maximizes profit.

It remains only to verify that x_i^0 and y_i^0 $(i = 1, 2, \ldots, n)$ are feasible. The recurrence relations (5.7.4) imply that one of the following two cases occurs:

Case A.

$$\begin{aligned} U_1^* &= b_1 - c_1 + b_2 - c_2 + \cdots + b_i + T_{i+1}^*, \\ T_1^* &= -c_1 + b_2 - c_2 + \cdots + b_i + T_{i+1}^*, \end{aligned} \tag{5.7.6}$$

for some $1 < i \leq n$, or

$$\begin{aligned} U_1^* &= b_1 + T_2^*, \\ T_1^* &= T_2^*. \end{aligned} \tag{5.7.6a}$$

Case B.

$$\begin{aligned} U_1^* &= b_1 - c_1 + b_2 - c_2 + \cdots - c_j + U_{j+2}^*, \\ T_1^* &= -c_1 + b_2 - c_2 + \cdots - c_j + U_{j+2}^*, \end{aligned} \tag{5.7.7}$$

for some $1 < j \leq n - 1$ (if $j = n - 1$, then assign $U_{j+2}^* = 0$); or

$$\begin{aligned} U_1^* &= U_2^*, \\ T_1^* &= -c_1 + U_2^*. \end{aligned} \tag{5.7.7a}$$

Another application yields

$$T_{i+1}^* = -c_{i+k_1} + b_{i+k_2} - c_{i+k_3} + \cdots,$$

where $1 < k_1 < k_2 < \cdots$ (if $i + k_1 > n$, then $T_{i+1}^* = 0$). Similarly,

$$U_{j+2}^* = +b_{j+2+l_1} - c_{j+2+l_2} + \cdots,$$

where $0 < l_1 < l_2 < \cdots$ (if $j + 2 + l_1 > n - 1$, then $U_{j+2}^* = 0$). Once the form of U_1^* and T_1^* is known according to (5.7.6) or (5.7.7), it is easy to verify directly that the constraints (5.7.1) and (5.7.2) are satisfied.

5.8 Optimal assignment problem. There are m jobs to be done and m workers to do them; one worker is assigned to each job. The value of having worker j do job i is a_{ij} $(a_{ij} > 0)$. The problem is to determine the m assignments that will maximize the total value. An assignment of workers to jobs is denoted by the permutation symbol

$$S = \begin{pmatrix} 1, 2, \ldots, m \\ j_1, j_2, \ldots, j_m \end{pmatrix},$$

where worker j_1 does job 1, worker j_2 does job 2, etc. Hence the problem is to maximize

$$\sum_{i=1}^{m} a_{ij_i}$$

over the $m!$ possible S obtained by making all possible permutations of $1, 2, \ldots, m$, each permutation being an n-tuple $\langle j_1, j_2, \ldots, j_m \rangle$.

An alternative way of describing the procedures is in terms of permutation matrices. A permutation matrix is an $m \times m$ matrix each row and column of which contains one element of value 1 and zeros elsewhere. If $\mathcal{P}$ is the set of all $m \times m$ permutation matrices $\mathbf{P} = \|p_{ij}\|$, then the problem is to find $\mathbf{P}^0$ such that

$$\max_{\mathcal{P}} \sum_{i,j} a_{ij} p_{ij} = \sum_{i,j} a_{ij} p_{ij}^0.$$

Von Neumann treats the assignment problem by reducing it to a $2m \times m^2$ game. In most cases it will be easier to solve the game than to search for the maximum in the set of all permutation matrices. Before demonstrating the equivalence of the assignment problem to a game problem, we consider it as a problem in linear programming.

(1) Consider the set $\mathfrak{X}$ of all $m \times m$ doubly stochastic matrices. That is, $\|x_{ij}\| \in \mathfrak{X}$ implies $\sum_i x_{ij} = 1$ for all j, $\sum_j x_{ij} = 1$ for all i, and $x_{ij} \geq 0$ for all i, j. Clearly, $\mathfrak{X}$ is a convex polyhedral set, since it is the intersection of $2m$ hyperplanes with the positive orthant of m^2-space.

▶ **Lemma 5.8.1.** The matrices in the class $\mathfrak{X}$ form a convex set, and the permutation matrices are its extreme points.

Proof. The convexity of $\mathfrak{X}$ was noted above. Let $\mathbf{X} = \|x_{ij}\|$ be an extreme point.

(i) At most $2m - 1$ elements of $\mathbf{X}$ are strictly positive.

In order to establish this assertion, we consider the following system of equations:

$$\sum_{j=1}^{m} z_{ij} = 0 \qquad (i = 1, 2, \ldots, m),$$
$$\sum_{i=1}^{m} z_{ij} = 0 \qquad (j = 1, 2, \ldots, m). \tag{5.8.1}$$

The system contains $2m$ equations in m^2 variables, but there are at most $2m - 1$ independent equations, since $\sum_i(\sum_j z_{ij}) - \sum_j(\sum_i z_{ij}) = 0$. The matrix $\mathbf{A}$ of the system (5.8.1) is of the form

$$
\mathbf{A} = \left\| \begin{array}{ccccc}
\overbrace{1\,1\cdots 1}^{m} & & & & \\
 & 1\,1\cdots 1 & & & \\
 & & 1\,1\cdots 1 & & \\
 & & & \ddots & \\
 & & & & 1\,1\cdots 1 \\
\begin{matrix}1 & & \\ & \ddots & \\ & & 1\end{matrix} & \begin{matrix}1 & & \\ & \ddots & \\ & & 1\end{matrix} & \begin{matrix}1 & & \\ & \ddots & \\ & & 1\end{matrix} & \cdots & \begin{matrix}1 & & \\ & \ddots & \\ & & 1\end{matrix}
\end{array} \right\| \begin{array}{l} \left.\vphantom{\begin{matrix}1\\1\\1\\1\\1\end{matrix}}\right\} m \\ \left.\vphantom{\begin{matrix}1\\1\\1\end{matrix}}\right\} m \end{array} \tag{5.8.2}
$$

(all entries are 0 except those indicated to be 1). We know that $\mathbf{A}$ is of size $2m \times m^2$ and has rank at most $2m - 1$.

Suppose that $\mathbf{X} = \|x_{ij}\|$ has at least $2m$ nonzero components. Whenever $x_{ij} = 0$, we assign to z_{ij} the value zero in the equations (5.8.1) and view (5.8.1) as a system of equations for the remaining variables. Since there are at least $2m$ undetermined variables and the rank of $\mathbf{A}$ is $2m - 1$ (see Problem 6 of Chapter 6), there exists a nontrivial solution denoted by z_{ij}.

Choose ϵ sufficiently small so that $x_{ij} + \epsilon z_{ij} \geq 0$ and $x_{ij} - \epsilon z_{ij} \geq 0$ for all i, j. Set $\mathbf{X}_1 = \|x_{ij} + \epsilon z_{ij}\|$ and $\mathbf{X}_2 = \|x_{ij} - \epsilon z_{ij}\|$.

It is readily verified that $\mathbf{X}_1$ and $\mathbf{X}_2$ are members of $\mathfrak{X}$.

But $\mathbf{X} = \frac{1}{2}(\mathbf{X}_1 + \mathbf{X}_2)$; hence $\mathbf{X}$ is not an extreme point.

(ii) If $\mathbf{X}$ is an extreme point, then $\mathbf{X}$ is a permutation matrix and conversely.

The proof of this statement is as follows. By (i), if $\mathbf{X}$ is extreme, it can have at most $2m - 1$ nonzero elements. Then there must be at least one row with only one nonzero element in it. Since the sum over this row is 1, the nonzero element must be 1. Moreover, this also makes this element the only nonzero element in its column.

But if $\mathbf{X}$ is an extreme point of the class $\mathfrak{X}$, the matrix obtained by deleting the row and the column with one element is also an extreme point of the class of $(m - 1) \times (m - 1)$ matrices of type $\mathfrak{X}$. Repeating the argument, we see that $\mathbf{X}$ must have only one nonzero element in each row or column, and $\mathbf{X}$ therefore represents a permutation matrix.

Conversely, any matrix that represents a permutation is easily seen to be an extreme point of $\mathfrak{X}$. The proof of the lemma is complete.

Since $\mathcal{P} \subset \mathcal{X}$,

$$\max_{X \in \mathcal{X}} \sum a_{ij}x_{ij} \geq \max_{P \in \mathcal{P}} \sum a_{ij}p_{ij}.$$

But $\|x_{ij}\| \in \mathcal{X}$ is a finite convex combination of extreme points of $\mathcal{X}$, so that (by Lemma 5.8.1)

$$\max_{X \in \mathcal{X}} \sum a_{ij}x_{ij} = \max_{P \in \mathcal{P}} \sum a_{ij}p_{ij}.$$

Thus $\mathbf{X}^0 \in \mathcal{X}$ which maximizes $\sum a_{ij}x_{ij}$ must be either an element of $\mathcal{P}$ or a convex combination of elements of $\mathcal{P}$ all of which maximize $\sum a_{ij}p_{ij}$. Therefore if we know $\mathbf{X}^0$, we can find a $\mathbf{P}^0 \in \mathcal{P}$ which maximizes $\sum a_{ij}p_{ij}$. Consequently, it is sufficient in searching for a maximum to consider $\mathcal{X}$ instead of $\mathcal{P}$.

We further enlarge our domain of consideration to the set $\mathcal{Y}$ of matrices $\|y_{ij}\|$ which satisfy $\sum_i y_{ij} \leq 1$ for all j, $\sum_j y_{ij} \leq 1$ for all i, and $y_{ij} \geq 0$ for all i, j. It follows that

$$\max_{\mathcal{Y}} \sum a_{ij}y_{ij} = \max_{\mathcal{X}} \sum a_{ij}x_{ij},$$

and moreover that the max is achieved only for elements of $\mathcal{X}$.

▶LEMMA 5.8.2. If $\|y^0_{ij}\| \in \mathcal{Y}$ maximizes $\sum_{i,j} a_{ij}y_{ij}$, then $\|y^0_{ij}\| \in \mathcal{X}$.

Proof. Let $\sum_j y^0_{ij} = r_i$ and $\sum_i y^0_{ij} = s_j$; then $\sum_i r_i = \sum_j s_j = K$. If one of the r_i (say r_{i_0}) is less than 1, then one of the s_j (say s_{j_0}) is less than 1, and conversely; and K is less than m.

Let

$$t_{ij} = \frac{(1 - r_i)(1 - s_j)}{m - K};$$

then obviously $t_{i_0 j_0} > 0$. Also $\sum_j t_{ij} = 1 - r_i$ and $\sum_i t_{ij} = 1 - s_j$, so that $t_{ij} + y^0_{ij} \geq 0$ and $\sum_j (t_{ij} + y^0_{ij}) = 1 - r_i + r_i = 1$ for all i, j. Similarly, the sum over i equals 1. Therefore $\|t_{ij} + y^0_{ij}\| \in \mathcal{X} \subset \mathcal{Y}$.

But $\sum_{i,j} a_{ij}(t_{ij} + y^0_{ij}) = \sum_{i,j} a_{ij}y^0_{ij} + c$, where c is a positive number and thus $\|y^0_{ij}\|$ is not a maximum point. This contradicts the hypothesis, hence $\|y^0_{ij}\| \in \mathcal{X}$.

The problem now has a linear programming form. For $\mathbf{y}$ in m^2-space, we wish to maximize $\sum_{i,j} a_{ij}y_{ij}$ subject to the constraints $\mathbf{y} \geq \mathbf{0}$ and $\mathbf{Ay} \leq \mathbf{u}$, where $\mathbf{A}$ is the matrix of Lemma 5.8.1 and $\mathbf{u} = \langle 1, 1, \ldots, 1 \rangle$ with $2m$ components. Thus a solution of this linear programming problem leads to a solution of the optimal assignment problem.

The dual to the linear programming problem is the following. For $\mathbf{z} = \langle t_1, \ldots, t_m, s_1, \ldots, s_m \rangle$, minimize $\sum t_i + \sum s_i$ subject to the constraints $\mathbf{z} \geq \mathbf{0}$ and $\mathbf{zA} \geq \|a_{ij}\|$. Written out in full, the constraints are

$$t_i + s_j \geq a_{ij}, \; t_i \geq 0, \; s_j \geq 0 \qquad \text{for all} \quad i, j.$$

(2) Von Neumann reduced the optimal assignment problem to the two-person zero-sum game with $2m \times m^2$ matrix

$$
\mathbf{B} = \begin{array}{l} m \left\{ \begin{array}{c} \\ \\ \\ \\ \end{array} \right. \\ m \left\{ \begin{array}{c} \\ \\ \\ \\ \end{array} \right. \end{array}
\left\| \begin{array}{cccccccccc}
\overbrace{1/a_{11} \cdots 1/a_{1m}}^{m} & & & & \\
 & 1/a_{21} \cdots 1/a_{2m} & & & \\
 & & \cdot & & \\
 & & & \cdot & \\
 & & & & 1/a_{m1} \cdots 1/a_{mm} \\
1/a_{11} & 1/a_{21} & & \cdots 1/a_{m1} & \\
\cdot & \cdot & & \cdot & \\
\cdot & \cdot & & \cdot & \\
1/a_{1m} & 1/a_{2m} \cdots & & & 1/a_{mm}
\end{array} \right\|
$$

with a zero entered in all places which are not obviously non zero. A solution to the assignment problem can be obtained from an optimal strategy for Player II in this game. Let

$$
\mathbf{z}^0 = \langle z_{11}^0, \ldots, z_{1m}^0, \quad z_{21}^0, \ldots, z_{2m}^0, \ldots, z_{m1}^0, \ldots, z_{mm}^0 \rangle
$$

be an optimal strategy for Player II. Then the inner product of $\mathbf{z}^0$ with the rows of $\mathbf{B}$ must all be less than or equal to v, where v is the value of the game; that is, $\mathbf{B}\mathbf{z}^0 \leq \mathbf{v}$, where $\mathbf{v} = \langle v, v, \ldots, v \rangle$. Since $a_{ij} > 0$, it follows that $v > 0$.

Let

$$
y_{ij}^0 = \frac{z_{ij}^0}{a_{ij} v}.
$$

Clearly, $y_{ij}^0 \geq 0$, and since $\mathbf{z}^0$ is optimal, we obtain $\sum_j y_{ij}^0 \leq 1$ for all i and $\sum_i y_{ij}^0 \leq 1$ for all j. We wish to establish that $\mathbf{y}^0$ maximizes $\sum a_{ij} y_{ij}$. Observe first that $\sum_{ij} a_{ij} y_{ij}^0 = M = 1/v$. Suppose $\bar{\mathbf{y}} \in \mathcal{Y}$ and $\sum_{ij} a_{ij} \bar{y}_{ij} = M' > M$. Define

$$
\bar{z}_{ij} = \frac{a_{ij} \bar{y}_{ij}}{M'}.
$$

Then $\bar{z}_{ij} \geq 0$ and $\sum_{i,j} \bar{z}_{ij} = 1$, so that $\bar{\mathbf{z}}$ is a strategy for Player II in the game with matrix $\mathbf{B}$. The inner product of $\bar{\mathbf{z}}$ with the ith row is

$$
\sum_j \bar{z}_{ij} \frac{1}{a_{ij}} = \frac{1}{M'} \sum_j \bar{y}_{ij} \leq \frac{1}{M'} < v.
$$

For the last m rows, we get

$$
\sum_i \bar{z}_{ij} \frac{1}{a_{ij}} = \frac{1}{M'} \sum_i y_{ij} \leq \frac{1}{M'} < v,
$$

which contradicts the fact that v is the value of the game. Hence $\mathbf{y}^0$ maximizes $\sum a_{ij}y_{ij}$ for $\mathbf{y} \in \mathcal{Y}$.

But previously we found that if $\mathbf{y}^0$ maximizes $\sum a_{ij}y_{ij}$, then $\mathbf{y}^0 \in \mathfrak{X}$; consequently, a solution to the optimal assignment problem can be extracted from a representation of $\mathbf{y}^0$ as a convex combination of extreme points.

An interpretation of the matrix game B as a type of hide-and-seek game can be given. Consider the matrix

$$\begin{Vmatrix} 1/a_{11} & 1/a_{12} & \cdots & 1/a_{1m} \\ 1/a_{21} & 1/a_{22} & \cdots & 1/a_{2m} \\ \vdots & & & \\ 1/a_{m1} & 1/a_{m2} & \cdots & 1/a_{mm} \end{Vmatrix}$$

Player I selects either a row or a column in this matrix and Player II selects a cell. If the cell is located in the row or column that Player I has chosen, Player I gets paid the value of that cell. Otherwise he receives nothing. The reader can verify that the matrix of this game is the matrix $\mathbf{B}$.

5.9 Transportation and flow problem. The transportation model was introduced in Section 5.1; the same notation will be used in this section. Stripped of economic interpretation, the transportation problem reads as follows: Minimize

$$\sum_{j=1}^{n} \sum_{i=1}^{m} x_{ij}c_{ij}$$

subject to the constraints

$$\begin{aligned} &\sum_{i=1}^{m} x_{ij} \geq r_j \qquad (j = 1, \ldots, n), \\ &\sum_{j=1}^{n} x_{ij} \leq s_i \qquad (i = 1, \ldots, m), \qquad x_{ij} \geq 0 \qquad \text{for all} \quad i, j. \end{aligned} \tag{5.9.1}$$

By assumption, $c_{ij} > 0$, $r_j > 0$, and $s_i > 0$. The existence of a solution is obvious.

Throughout the remainder of this section we shall first assume that $\sum_j r_j \leq \sum_i s_i$ (i.e., that the supply is sufficient to satisfy demand).

▶LEMMA 5.9.1. The solution to (5.9.1) is achieved only for vectors $\|x_{ij}\|$ satisfying the relations

$$\sum_i x_{ij} = r_j, \qquad \sum_j x_{ij} \leq s_i, \qquad \text{and} \qquad x_{ij} \geq 0.$$

Proof. Suppose that $\mathbf{x}^0 = \|x_{ij}^0\|$ is a solution to (5.9.1), and that $\sum_i x_{ij_0}^0 = b_{j_0} > r_{j_0}$ for some j_0. Then there exists i_0 such that $x_{i_0 j_0}^0 > 0$. Set $y_{ij} = x_{ij}^0$, except that $y_{i_0 j_0} = x_{i_0 j_0} - \epsilon$; then if $\epsilon > 0$ is chosen sufficiently small, it follows that

$$y_{ij} \geq 0, \qquad \sum_i y_{ij} \geq r_j, \qquad \text{and} \qquad \sum_j y_{ij} \leq s_i.$$

Clearly,

$$\sum_{i,j} y_{ij} c_{ij} < \sum_{i,j} x_{ij}^0 c_{ij},$$

contradicting the fact that $\mathbf{x}^0$ is a solution. Consequently, $\sum_i x_{ij}^0 = r_j$ for every j.

Henceforth, we impose the restriction that $\sum_j r_j = \sum_i s_i$ (i.e., that the supply equals the demand). Then trivially every vector satisfying (5.9.1) necessarily obeys the equalities

$$\sum_i x_{ij} = r_j, \quad \sum_j x_{ij} = s_i \quad \text{and} \quad x_{ij} \geq 0. \tag{5.9.2}$$

Let Γ denote the collection of vectors $\mathbf{x} = \|x_{ij}\|$ satisfying (5.9.2). From here on, in seeking a solution to (5.9.1), we consider only vectors belonging to Γ.

The reader may observe that the optimal assignment problem can be viewed as a special case of the transportation problem, in which $r_i = s_j = 1$ and $m = n$. It is therefore reasonable to expect many of the results valid for the assignment problem to have analogs for the transportation problem. Some of these will be pointed out presently.

The fundamental constraint matrix of the transportation problem is similar to (5.8.2):

$$\mathbf{A} = \begin{array}{c} \\ m\left\{\begin{array}{c}\\ \\ \\ \\ \end{array}\right. \\ n\left\{\begin{array}{c}\\ \\ \\ \\ \end{array}\right. \end{array} \begin{array}{c} x_{11}, \ldots, x_{1n}, \quad x_{21}, \ldots, x_{2n}, \quad \ldots \quad x_{m1}, \ldots, x_{mn} \\ \left\| \begin{array}{ccccccccc} 1 & \cdots & 1 & & & & & & \\ & & & 1 & \cdots & 1 & & & \\ & & & & & & \ddots & & \\ & & & & & & 1 & \cdots & 1 \\ 1 & & & 1 & & & \cdots & 1 & \\ & \ddots & & & \ddots & & & & \ddots \\ & & 1 & & & 1 & \cdots & & 1 \end{array} \right\| \end{array}$$

Let $\boldsymbol{\omega} = \langle s_1, \ldots, s_m, r_1, \ldots, r_n \rangle$. Then the constraint set of (5.9.2) may be expressed by the vector equalities $\mathbf{Ax} = \boldsymbol{\omega}$.

In studying any linear programming problem it is desirable to characterize the extreme points of the constraint set, since the optimum must occur at one of the extreme points. For this purpose the following lemma is useful.

▶ LEMMA 5.9.2. If r_i and s_j not all zero are all integral (i.e., integer-valued), then the extreme points of Γ are contained in the set of all $\mathbf{x} = \|x_{ij}\|$ of Γ with x_{ij} integral. Moreover, there exists at least one row or column with at most one nonzero entry.

The proof is based on the same ideas as that of Lemma 5.8.1. The arguments of Lemma 5.8.1 show that any extreme point $\mathbf{x} \in \Gamma$ is composed of at most $m + n - 1$ nonzero components. Consequently, there exists a row or a column, say a row, with at most one nonzero element, which must be integral (since the r_i's are integral). By removing that row and changing one of the s_j's appropriately, we arrive at an extreme point $\mathbf{x}$ involving $m - 1$ rows and at most n columns, with the new r_i's and s_j's still integral. Repeated applications of this reasoning prove the lemma.

It is an easy step from Lemma 5.9.2 to the following theorem.

▶ THEOREM 5.9.1. If r_i and s_j are integral, then there exists a solution $\mathbf{x}^0$ to (5.9.1) with integral components x_{ij}^0.

Theorem 5.9.1 is very appealing. It says that if supply and demand are expressed in terms of integral, indivisible quantities of a commodity—e.g., 50 refrigerators, 2000 sacks of salt—then the optimal strategy for minimizing costs requires shipping only integral amounts of the commodity from the plants to the various markets, even though *a priori* we permitted strategies involving the transfer of fractional amounts of the commodity.

Network flow problems. The remainder of this section is devoted to the study of network flow problems. This model is closely related, and in certain respects equivalent, to the transportation model. The network problem arises naturally in the study of transportation networks. One is given a network of directed arcs and nodes with two distinguished nodes, one called the source and the other the sink.* A directed arc may be

* The problem in which there are several sources and sinks with flows permitted from any source to any sink is reducible to a single-source, single-sink problem.

The case where flow is allowed in either direction with a restriction on the total capacity of each arc can be reduced to the previous problem with all arcs duplicated once each way. The restriction of the total is no more than the same restriction of flow in each direction, there being no advantage to having flow both ways at once (see also Problem 15 of Chapter 6).

The separate consideration of nodes may similarly be obviated by introducing fictitious arcs from each node to itself (see Problem 14 of Chapter 6).

interpreted as a railway connection between two terminals, and nodes may be thought of as warehouses located at the various intermediate stations. Each directed arc in the network has associated with it a nonnegative integer, its flow capacity; each node also has its capacity, which may be thought of as the capacity of a warehouse. Source arcs (arcs having a terminus at the source) may be assumed to be directed away from the source; sink arcs are always directed toward the sink.

It is desired to find a maximal flow from source to sink, subject to two conditions: (i) the flow in an arc must be in the direction of the arc and may not exceed its capacity, and the flow into a node may not exceed its capacity; (ii) the total flow into any intermediate node must be equal to the flow out of it. The value of any flow is defined to be the sum of the flows in the source arcs. Since any flow is consistent with the capacity restraints, the value of a flow is also equal to the sum of flows in the sink arcs. For example, consider Fig. 5.1, where source and sink are denoted by + and −, respectively, and the capacity limitations are as indicated. The maximal flow for this network is described in Fig. 5.2.

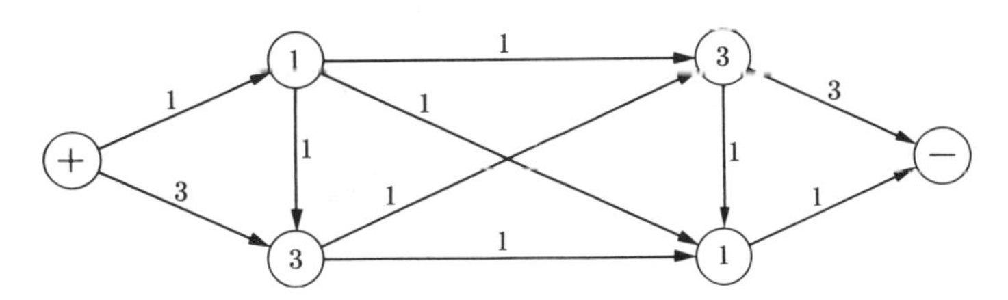

FIGURE 5.1

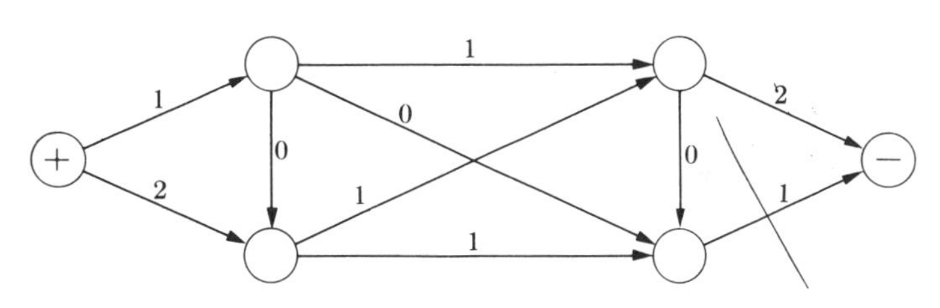

FIGURE 5.2

Let $x_{ij} \geq 0$ ($i \neq j; i, j = 1, \ldots, m$) denote the flow from node i to node j; x_{ii} ($i = 1, \ldots, m$) the total flow through node i; x_{0j} ($j = 1, \ldots, m$) the flow from the source to node j; and x_{i0} ($i = 1, \ldots, m$) the flow from node i to the sink.

For the general network flow problem, we consider the pseudo-transportation array

$$\begin{array}{r} \\ \\ \\ \\ \\ \text{Totals} \end{array}\begin{array}{c} \\ \left[\begin{array}{ccccc} -x_{00} & x_{01} & x_{02} & \cdots & x_{0m} \\ x_{10} & -x_{11} & x_{12} & \cdots & x_{1m} \\ \vdots & & & & \\ x_{m0} & x_{m1} & x_{m2} & \cdots & -x_{mm} \end{array}\right] \\ \begin{array}{ccccc} 0 & \quad 0 & \quad 0 & \cdots & \quad 0 \end{array} \end{array}\begin{array}{c} \text{Totals} \\ 0 \\ 0 \\ \vdots \\ 0 \\ \\ \end{array}$$

schematizing the equations

$$\begin{aligned} -x_{ii} + \sum_{\substack{j \\ j \neq i}} x_{ij} &= 0 \qquad (i = 0, 1, \ldots, m), \\ -x_{jj} + \sum_{\substack{i \\ i \neq j}} x_{ij} &= 0 \qquad (j = 0, 1, \ldots, m). \end{aligned} \tag{5.9.3}$$

Clearly, x_{00} is the total flow, and the problem is to maximize x_{00} subject to (5.9.3) and

$$0 \leq x_{ij} \leq a_{ij} \qquad (i, j) \neq (0, 0), \tag{5.9.4}$$

where the a_{ij} are the given nonnegative constants that describe the capacities of the nodes and arcs. Arcs not present in the network have $a_{ij} = 0$.

This is not a transportation problem in the strict sense because of the minus signs down the diagonal in the array. Notice that the constraining equations (5.9.3) are similar to relations (5.9.2) with all $r_j = s_i = 0$. However, transportation theory (e.g., Theorem 5.9.1) carries over directly to problems of this kind, even with upper bound constraints present. We omit the simple proof of this fact.

***5.10 Maximal-flow, minimal-cut theorem.** A chain is any collection of directed arcs and nodes which leads from source to sink. A "cut" in a directed network is a collection of directed arcs and nodes meeting every directed chain from source to sink. The value of a cut is the sum of the capacities of all its member nodes and arcs. It is clear that the maximal flow cannot exceed the minimal cut value, and conversely. Hence we have the following theorem.

▶THEOREM 5.10.1. The maximal flow value is equal to the minimal cut value.

This theorem is essentially a rephrasing of the duality theorem (Theorem 5.4.1) for the network flow problem. In order to provide a proof, we shall examine the form of the dual problem.

Let

$$
\mathbf{A} = \begin{array}{r}
\\
m+1\left\{\begin{array}{c} \\ \\ \\ \\ \\ \end{array}\right. \\
m+1\left\{\begin{array}{c} \\ \\ \\ \\ \\ \end{array}\right. \\
m^2+2m\left\{\begin{array}{c} \\ \\ \\ \end{array}\right.
\end{array}
\left\|\begin{array}{cccc|cccc|c|cccc}
x_{00}, & x_{01}, & \ldots, & x_{0m}, & x_{10}, & x_{11}, & \ldots, & x_{1m} & \ldots & x_{m0}, & x_{m1}, & \ldots, & x_{mm} \\
-1 & 1 & \cdots & 1 & & & & & & & & & \\
 & & & & 1 & -1 & \cdots & 1 & & & & & \\
 & & & & & & & & \ddots & & & & \\
 & & & & & & & & & 1 & 1 & \cdots & -1 \\
-1 & & & & 1 & & & & \cdots & 1 & & & \\
 & 1 & & & & -1 & & & & & 1 & & \\
 & & \ddots & & & & \ddots & & & & & \ddots & \\
 & & & 1 & & & & 1 & \cdots & & & & -1 \\
\hline
0 & & & & & & & & & & & & \\
\vdots & & & & & & & I & & & & & \\
0 & & & & & & & & & & & &
\end{array}\right\|,
$$

where I stands for the identity matrix of size $m^2 + 2m$.

Let

$$
\mathbf{b} = \langle \overbrace{0, 0, \ldots, 0}^{2m+2}, a_{01}, a_{02}, \ldots, a_{0m}, a_{10}, a_{11}, \ldots, a_{mm} \rangle;
$$

then the matrix of constraints becomes $\mathbf{Ax} \leq \mathbf{b}$, with equality for the first $2m + 2$ components. Since the flow problem seeks to maximize x_{00}, the system of dual constraints in the dual variables u_i $(i = 0, \ldots, m)$ v_j $(j = 0, 1, \ldots, m)$, and ω_{ij} [with $(i, j) \neq (0, 0)$] becomes

$$
\begin{aligned}
u_i + v_j + \omega_{ij} &\geq 0 \qquad (i, j = 0, \ldots, m;\ i \neq j), \\
-u_i - v_i + \omega_{ii} &\geq 0 \qquad (i = 1, \ldots, m), \\
-u_0 - v_0 &\geq 1, \\
\omega_{ij} &\geq 0 \qquad (i, j) \neq (0, 0).
\end{aligned}
\tag{5.10.1}
$$

Let the set of all (u, v, ω) satisfying (5.10.1) be denoted by T. Since the original constraints involved $2m + 2$ equalities, the variables u_i and v_j vary unrestrictedly (see page 117). The optimal $\mathbf{x}$ strategy (except in trivial cases) must have $x_{00} > 0$ by Theorem 5.4.3; hence $-u_0 - v_0 = 1$. Moreover, because one of the equations (5.9.3) is redundant, we may arbitrarily select $u_0 = 0$, which implies that $v_0 = -1$.

At this point the following lemma is needed.

▶ **Lemma** 5.10.1. *If $u_0 = 0$ and $v_0 = -1$, the extreme points of T have u_i equal to 0 or 1 for all i, v_j equal to 0 or -1 for all j, and ω_{ij} equal to 0 or 1 for all i, j.*

Proof. The system (5.10.1) comprises $m^2 + 4m$ variables and $2m^2 + 4m$ inequalities. The extreme points of T are obtained by finding the solutions to all possible sets of $m^2 + 4m$ independent equations in the same number of unknowns. The total set of equations obtainable from (5.10.1) reduces to

$$\begin{aligned} &v_j + \omega_{0j} = 0, \ u_i + \omega_{i0} = 1, \\ &u_i + v_j + \omega_{ij} = 0 \qquad (i, j = 1, \ldots, m; i \neq j), \\ &\omega_{ij} = 0 \qquad (i, j \neq 0, 0), \\ &-u_i - v_i + \omega_{ii} = 0 \qquad (i = 1, \ldots, m). \end{aligned} \tag{5.10.2}$$

Substituting $\omega_{ij} = 0$ wherever this is so, we have a system of equations in the remaining ω_{ij}, u_i, and v_j.

If in the system (5.10.2) the equation $\omega_{i_0j_0} = 0$ is not present, then the associated variable $\omega_{i_0j_0}$ has to be determined by one of the remaining equations which is necessarily of the form

$$\begin{aligned} u_i + v_j + \omega_{ij} &= 0 \qquad (i, j \neq 0, 0 \text{ and } i \neq j), \\ -u_i - v_i + \omega_{ii} &= 0 \qquad (i \neq 0), \\ v_j + \omega_{0j} &= 0 \qquad (j \neq 0), \end{aligned}$$

or

$$u_i + \omega_{i0} = 1 \qquad (i \neq 0).$$

These equations cannot determine u_i and v_j. But we see that u_i and v_j with i, j not zero are determined by $2m$ *independent* equations, each of which has the form

$$u_i + v_j = 0, \qquad v_j = 0, \qquad \text{or} \qquad u_i = 1. \tag{5.10.3}$$

Each u_i $(i = 1, \ldots, m)$ and v_j $(j = 1, \ldots, m)$ is included in some equation of (5.10.3). It is readily seen that a system of *independent* equations of the type (5.10.3) involving all the variables u_i and v_j $(i, j = 1, \ldots, m)$ can only have solutions with u_i equal to 0 or 1 and v_j equal to 0 or -1. On examination of (5.10.2), since $\omega_{ij} \geq 0$, it then follows that ω_{ij} equals 0 or 1 as asserted.

Proof of Theorem 5.10.1. The dual programming problem reduces to the form of minimizing $\sum_{ij} a_{ij}\omega_{ij}$ subject to (5.10.1). Because of the duality theorem (Theorem 5.4.1) and Lemma 5.10.1,

$$\max x_{00} = \sum_{\sigma} a_{ij}, \tag{5.10.4}$$

where σ is the set of arcs i, j and nodes i corresponding to $\omega_{ij} = 1$. By Theorem 5.4.3, $\omega_{ij} = 1$ implies $x_{ij} = a_{ij}$ and $\omega_{ii} = 1$ implies $x_{ii} = a_{ii}$.

It is asserted that σ constitutes a cut. For suppose that all a_{ij} such that $i, j \notin \sigma$ are increased by $\epsilon > 0$. That this increase does not change the solution to the dual problem is evident from (5.10.4) and hence it cannot increase the flow in the network. But if there were some directed chain from source to sink not meeting σ, clearly the flow could be increased. Thus σ is a cut, and since no flow can exceed the value of any cut, σ is a minimal cut.

***5.11 The caterer's problem.** In order to serve meals during the next m days, a caterer will need r_j fresh napkins on the jth day. Normal laundering takes p days at b cents a napkin; rapid laundering takes q days (p and q are integers) at c cents a napkin. Also, new napkins may be purchased at a cents a napkin. How can the caterer arrange matters to meet his needs and minimize his outlays for the m days?

Another interpretation of the model involves military aircraft. "Laundering" corresponds to overhaul of engines, the unit of time is a month rather than a day, etc.

Practical considerations dictate that $r_j \geq 0$ $(j = 1, \ldots, m)$, $a > b$, $a > c$, $c > b$, $p > q$. Let x_j be the number of napkins purchased for use on the jth day, y_j the number sent for normal laundry service, z_j the number sent for rapid laundry service, and s_j the number of soiled napkins on hand on the jth day $(j = 1, \ldots, m)$. Also, let capital letters denote cumulative amounts, e.g.,

$$X_j = \sum_{i=1}^{j} x_i, \qquad \text{etc.}$$

The caterer's problem is as follows.

Problem A. Minimize the total cost

$$a \sum_{j=1}^{m} x_i + b \sum_{j=1}^{m} y_j + c \sum_{j=1}^{m} z_j = aX_m + bY_m + cZ_m \tag{5.11.1}$$

subject to the constraints

$$x_j + y_{j-p} + z_{j-q} \geq r_j, \tag{5.11.2}$$

$$y_j + z_j + s_j - s_{j-1} = r_j, \tag{5.11.3}$$

$$x_j, y_j, z_j, s_j \geq 0, \tag{5.11.4}$$

where $j = 1, \ldots, m$, $s_0 = 0$, and $y_{-k} = z_{-k} = 0$ for $k \geq 0$.

Relation (5.11.2) expresses the need requirement and (5.11.3) states what happens. Implicit in the formulation of these constraints is the reasonable assumption that no napkin will be bought or laundered ahead of time.

It is possible to formulate the caterer's problem as a distribution or transportation problem in which the caterer and each day's hamper of soiled napkins are the origins, and each day's requirement of fresh napkins and the final inventory of soiled napkins are the destinations (see Problem 10 of Chapter 6). Consequently, some of the special computational schemes for the transportation problem can be advantageously adapted to provide a numerical solution to the caterer's problem.

For our purposes here, however, certain features of the caterer's problem are more important than its numerical solution. In the ensuing analysis we follow the methods of Jacobs, who obtained an explicit analytical solution for $p - 1 = q$. Jacobs' approach elegantly displays the technique of reducing a linear programming problem of complex form to a problem of simpler form.

As regards Problem A, two simplifications can be made. First, it is clear that restrictions (5.11.2) may be replaced by

$$x_j + y_{j-p} + z_{j-q} = r_j \qquad (j = 1, \ldots, m). \tag{5.11.2a}$$

Second, napkins are not laundered unless they can be returned by the mth day; so let $y_j = 0$ $(m - p < j \leq m)$ and $z_j = 0$ $(m - q < j \leq m)$. Then the cost function becomes

$$aX_m + bY_{m-p} + cZ_{m-q}.$$

Noticing that $X_m + Y_{m-p} + Z_{m-q} = R_m$, which can be obtained by summing (5.11.2a) over $j = 1, \ldots, m$, we get

$$\begin{aligned} aX_m + bY_{m-p} + cZ_{m-q} &= (a - b)X_m + (c - b)Z_{m-q} + bR_m \\ &= (c - b)\left\{\frac{a - b}{c - b} X_m + Z_{m-q}\right\} + bR_m. \end{aligned}$$

Now Problem A is reduced to

Problem B. Minimize

$$\lambda X_m + Z_{m-q} \tag{5.11.5}$$

subject to the constraints (5.11.2a), (5.11.3), and (5.11.4), where

$$\lambda = \frac{a - b}{c - b} > 0.$$

Now we shall transform the variables in order to eliminate the role of the individual x_j. Accumulating (5.11.2a) up to j, we obtain

$$X_j + Y_{j-p} + Z_{j-q} = R_j; \tag{5.11.6}$$

and accumulating (5.11.3) up to $j - p$, we obtain

$$Y_{j-p} + Z_{j-p} + s_{j-p} = R_{j-p}. \tag{5.11.7}$$

Subtracting (5.11.7) from (5.11.6), we obtain

$$X_j + Z_{j-q} - Z_{j-p} - s_{j-p} = R_j - R_{j-p}. \tag{5.11.8}$$

Let the transformation of (x, y, s, z) into (w, u, v, z) be defined as follows:

$$w = X_m, \quad u_j = X_m - X_j + s_{j-p}, \quad v_j = x_j + y_{j-p}, \quad z_j = z_j.$$

Then (5.11.8), (5.11.2a), and (5.11.4) will be written

$$w + Z_{j-q} - Z_{j-p} - u_j = R_j - R_{j-p}, \tag{5.11.9}$$

$$z_{j-q} + v_j = r_j, \tag{5.11.10}$$

$$w, z_j, u_j, v_j \geq 0, \tag{5.11.11}$$

and the function (5.11.5) to be minimized will be

$$\lambda w + Z_{m-q}. \tag{5.11.12}$$

We shall now prove

▶ LEMMA 5.11.1. Problem B can be reduced to

Problem C. Minimize (5.11.12) subject to the constraints (5.11.9), (5.11.10), and (5.11.11).

Proof. As noted above, to each feasible (x, y, s, z) of Problem B corresponds a feasible (w, u, v, z) of Problem C. The converse is not true. However, every solution of Problem C is the image of a feasible (x, y, s, z) of Problem B.

Let (w, u, v, z) be optimal and define $X_m = w$ and

$$X_j = w - \min_{i \leq j} u_i$$

so that X_j is nondecreasing. The first relation of (5.11.9) implies that $w \geq u_1$; hence the $x_j = X_j - X_{j-1}$ are all nonnegative. Now

$$\min_{1 \leq j \leq m} u_j = 0;$$

for otherwise w and every u_j could be decreased by a fixed amount, contradicting the optimality assumption. Consequently, the two definitions of X_m coincide. Furthermore, $X_m - X_j \leq u_j$; hence $s_{j-p} = u_j - X_m + X_j \geq 0$.

It remains to verify that the $y_j = v_{j+p} - x_{j+p}$ are nonnegative. If $x_k = 0$, where $k = j + p$, then we are finished. So let us assume that $x_k > 0$. Then from the definition of x_k, it follows that $X_k = X_m - u_k$ and $u_i > 0$ for $i < k$. Suppose z_{k-p} positive; then if we reduce z_{k-p} a fixed amount and each of $u_{k-p+q}, u_{k-p+q+1}, \ldots, u_{k-1}$ by the same amount and change v_{k-p+q} appropriately, the constraints of Problem C remain unaffected while the Z_{m-q} are decreased. This contradicts the optimality assumption; hence $z_{k-p} = 0$. Now $s_{k-p} = u_k - X_m + X_k = 0$; therefore

$$y_j = y_{k-p} = r_{k-p} + s_{k-p-1} - s_{k-p} - z_{k-p} = 0 + r_{k-p} + s_{k-p-1} \geq 0.$$

The proof of the lemma is complete.

In view of Lemma 5.11.1, the problem can now be stated as follows.

Problem C′. Minimize $[\lambda w + Z_{m-q}]$ subject to the constraints

$$w + Z_{j-q} - Z_{j-p} \geq R_j - R_{j-p} \qquad (j = 1, 2, \ldots, m), \qquad (5.11.13)$$

$$0 \leq z_{j-q} \leq r_j \qquad (j = 1, 2, \ldots, m), \qquad (5.11.14)$$

$$0 \leq w. \qquad (5.11.15)$$

(The reader may note that we have converted the constraints of Problem C into an equivalent form involving inequalities.)

An algorithm for solutions of the caterer's problem with arbitrary p and q is summarized by Problems 22–26 (see also the notes to this section at the close of the chapter). Here we shall limit our attention to the case $q = p - 1$ and give an explicit form for a solution of Problem C′.

For $q = p - 1$, (5.11.13) reduces to

$$w + z_{j-q} \geq R_j - R_{j-q-1} \qquad (j = 1, \ldots, m). \qquad (5.11.16)$$

From (5.11.15) and (5.11.14), we have

$$w \geq R_j - R_{j-q} \qquad (j = 1, \ldots, m). \qquad (5.11.17)$$

Let

$$\overline{w} = w - \max_{1 \leq i \leq m} \{R_{i-1} - R_{i-p}\},$$

$$H_j = R_j - R_{j-p} - \max_{1 \leq i \leq m} \{R_{i-1} - R_{i-p}\}.$$

Then (5.11.16) and (5.11.17) can be written as

$$\overline{w} + z_{j-q} \geq H_j, \qquad \overline{w} \geq 0.$$

It will now be shown that Problem C′ is reduced to the following form.

Problem C″. Minimize $\lambda\overline{w} + Z_{m-q}$ subject to the constraints

$$\overline{w} + z_{j-q} \geq H_j, \qquad \overline{w} \geq 0, \qquad z_{j-q} \geq 0.$$

Proof. It is clear that a solution to Problem C′ should be feasible for Problem C″.

Let $\overline{w}$, z_j be a solution to Problem C″. Then we shall show that (5.11.14) is indeed satisfied. Since

$$H_j \leq R_j - R_{j-p} - (R_{j-1} - R_{j-p}) = r_j,$$

any $z_{j-q} > r_j$ would not be optimal.

The solution to Problem C″ is implied by the following lemma.

▶LEMMA 5.11.2. Let $a_1 \geq \cdots \geq a_m \geq 0$. A solution to the problem of minimizing $\lambda w + Z_m$ subject to the constraints $w + z_j \geq a_j$, $w \geq 0$, $z_j \geq 0$ $(j = 1, \ldots, m)$ is given by

$$w = a_{[\lambda]+1}, z_j = \max\{0, a_j - a_{[\lambda]+1}\}$$

for $\lambda < m$. If $\lambda \geq m$, the minimum solution is given by

$$w = 0, \qquad z_j = a_j.$$

Proof. For any given $w \geq 0$, Z_m is minimized when

$$z_j = \max\{0, a_j - w\} \qquad (j = 1, \ldots, m).$$

Let $a_k \geq w_k \geq a_{k+1}$, and consider the function

$$\phi(w_k) = \lambda w_k + \sum_{j=1}^{m} \max\{0, a_j - w_k\} = \lambda w_k + \sum_{j=1}^{k} (a_j - w_k)$$

$$= (\lambda - k)w_k + \sum_{j=1}^{k} a_j;$$

then

$$\begin{aligned} \phi(w_k) - \phi(w_{k+1}) &= (\lambda - k)(w_k - w_{k+1}) - (a_{k+1} - w_{k+1}) \\ &\leq 0 \quad \text{if} \quad k \geq \lambda \\ &\geq 0 \quad \text{if} \quad k < \lambda - 1. \end{aligned}$$

Therefore $\phi(w_k)$ will be minimized when

$$w_k = a_k, \qquad k = [\lambda] + 1,$$

as asserted.

Translating the final result to the initial problem, we obtain the solution of Problem A as

$$z_j = \max\{(H_{j+p-1} - \overline{w}), 0\},$$

$$x_j = \max_{i\le j}\{R_i - R_{i-p} - z_{i-p+1}\} - \max_{i\le j-1}\{R_i - R_{i-p} - z_{i-p+1}\},$$

$$y_j = r_{j+p} - x_{j+p} - z_{j+1} \qquad \text{[from (5.11.2a)]},$$

$$s_j = R_j - Y_j - Z_j \qquad \text{[from (5.11.7)]},$$

where

$$H_j = \max\left\{(R_j - R_{j-p} - \max_{1\le i\le m}\ \{R - R_{i-p}\}), 0\right\}$$
$$\qquad\qquad\qquad\qquad i-1$$

and $\overline{w} = H_{[\lambda]+1}$, with $[\lambda]$ the largest integer smaller than or equal to $(a-b)/(c-d)$.

***5.12 Price speculation model.** Various securities can be bought or sold in the course of $m+1$ periods of time. The prices b_k during the kth period are predictable and hence supposedly known. In spite of this unrealistic assumption, the analysis of this deterministic problem is of some interest.

Let X_{k-1} denote the amount of securities on hand at the beginning of the kth period. During that period the change in value is $b_k(X_{k-1} - X_k)$; if $X_k > X_{k-1}$, stock has been purchased, while if $X_k < X_{k-1}$, stock has been sold. The handling costs of the stock during a period are taken to be proportional to the amount of stock on hand; but these costs are not paid until the m periods are over, and thus they do not affect the amount of money available for speculation. The net receipts for the kth period are therefore

$$b_k(X_{k-1} - X_k) - cX_k.$$

Taking account of the interest rate i, the total receipts for all $m+1$ periods are

$$\sum_{k=0}^{m} \alpha^k[b_k(X_{k-1} - X_k) - cX_k], \qquad \text{where} \qquad \alpha = \frac{1}{1+i} \tag{5.12.1}$$

with $X_{-1} = 0$. The amount of money M_k on hand at the end of the kth period satisfies the recursion relation

$$M_k = M_{k-1} + b_k(X_{k-1} - X_k), \tag{5.12.2}$$

with M_0, the initial amount of money available, fixed. It is desired to determine the speculation policy X_k which maximizes the total receipts subject to the constraint $M_k \geq 0$ for all k. In other words, the speculator never owes money, and borrowing is not allowed.

The constraint inequalities can be expressed as

$$0 \leq X_k \leq \frac{M_0 + \sum_{i=0}^{k-1} (b_{i+1} - b_i)X_i}{b_k} \qquad (k = 1, \ldots, m), \qquad (5.12.3)$$

where $b_0 = 0$. Subject to the conditions of (5.12.3) our problem is to determine the policy X which maximizes

$$\sum_{j=0}^{m-1} \alpha^j(\alpha b_{j+1} - b_j - c)X_j + \alpha^m X_m(-b_m - c). \qquad (5.12.4)$$

Obviously, the maximum is achieved for $X_m = 0$ (no stocks should be kept at the end of time m).

The analysis proceeds by determining the optimal X_k backward in time. This is reasonable, since the constraint relation for X_i involves only the preceding X's. For example, if the coefficient $\alpha^{m-1}(\alpha b_m - b_{m-1} - c)$ of X_{m-1} is positive, then the optimal X_{m-1} in (5.12.3) must satisfy the right-hand equality

$$X_{m-1} = \frac{M_0 + \sum_{i=0}^{m-2} (b_{i+1} - b_i)X_i}{b_{m-1}}, \qquad (5.12.5)$$

provided that this expression is positive.

We shall find that the positive values of an optimal policy X_i must correspond to a subset of the indices i with the property $b_{i+1} - b_i > 0$. Consequently, the upper bounds in (5.12.3) are always positive. In short, when prices are dropping, there is no advantage in maintaining stock. However, it still might not pay to keep stock when prices rise too slowly.

We now direct our attention to the explicit determination of the solution. For this purpose some useful implications of the constraint relation (5.12.5) should be noted. Observe that if (5.12.5) holds (that is, $\alpha b_m - b_{m-1} - c > 0$), then the coefficient of X_{m-2} in (5.12.4) reduces to

$$\alpha^{m-2}(\alpha b_{m-1} - b_{m-2} - c) + \frac{b_{m-1} - b_{m-2}}{b_{m-1}} \alpha^{m-1}(\alpha b_m - b_{m-1} - c). \qquad (5.12.6)$$

Hence, if (5.12.6) is positive, then X_{m-2} is to be chosen as large as possible so that [see (5.12.3)]

$$X_{m-2} = \frac{M_0 + \sum_{i=0}^{m-3} (b_{i+1} - b_i)X_i}{b_{m-2}}. \qquad (5.12.7)$$

But upon substituting the value of X_{m-2} in (5.12.7) into (5.12.5), we obtain

$$X_{m-1} = \frac{M_0 + \sum_{i=0}^{m-3} (b_{i+1} - b_i)X_i}{b_{m-2}}.$$

This phenomenon constantly appears: whenever an X_i satisfies its constraint relation with equality in its upper bound in (5.12.3), then all the relations of the higher X_i can be reduced. Furthermore, if (5.12.6) is positive, it follows that $b_{m-1} > b_{m-2}$. If (5.12.6) is negative, then $X_{m-2} = 0$ and (5.12.5) reduces to

$$X_{m-1} = \frac{b_{m-2}}{b_{m-1}} \frac{M_0 + \sum_{i=0}^{m-3} (b_{i+1} - b_i)X_i}{b_{m-2}}.$$

We now describe the general case. The construction is inductive. Assuming that relations for X_i with $r_k + 1 \leq i \leq m - 1$ have been determined, we proceed to derive relations for X_{r_k}.

First, let us assume for definiteness that $\alpha b_m - b_{m-1} - c > 0$, so that $X_{m-1} > 0$. The reader can readily furnish the modifications in the contrary case.

Case 1. Let

$$
\begin{aligned}
X_i &= 0 \qquad \text{for} \quad r_2 < i \leq r_1, \quad r_4 < i \leq r_3, \quad \ldots, \quad r_k < i \leq r_{k-1}, \\
X_i &= \frac{b_{r_2}}{b_{r_1}} \cdot \frac{b_{r_4}}{b_{r_3}} \cdots \frac{b_{r_k}}{b_{r_{k-1}}} \left(\frac{M_0 + \sum_{j=0}^{r_k} (b_{j+1} - b_j)X_j}{b_{r_k}} \right) \\
&\qquad\qquad\qquad\qquad \text{for} \quad r_1 < i \leq m - 1, \\
X_i &= \frac{b_{r_4}}{b_{r_3}} \cdots \frac{b_{r_k}}{b_{r_{k-1}}} \left(\frac{M_0 + \sum_{j=0}^{r_k} (b_{j+1} - b_j)X_j}{b_{r_k}} \right) \\
&\qquad\qquad\qquad\qquad \text{for} \quad r_3 < i \leq r_2, \\
&\vdots \\
X_i &= \frac{b_{r_k}}{b_{r_{k-1}}} \left(\frac{M_0 + \sum_{j=0}^{r_k} (b_{j+1} - b_j)X_j}{b_{r_k}} \right) \qquad \text{for} \quad r_{k-1} < i \leq r_{k-2},
\end{aligned}
\tag{5.12.8}
$$

with $m - 1 > r_1 > r_2 > \cdots > r_{k-1} > r_k$. Inserting these expressions in (5.12.4), we see that the coefficient of X_{r_k} is

$$
\begin{aligned}
\alpha^{r_k}(\alpha b_{r_{k+1}} - b_{r_k} - c) + \frac{b_{r_{k+1}} - b_{r_k}}{b_{r_k}} \Bigg\{ & \frac{b_{r_k}}{b_{r_{k-1}}} \sum_{r_{k-1} < i \leq r_{k-2}} \alpha^i(\alpha b_{i+1} - b_i - c) \\
& + \frac{b_{r_{k-2}}}{b_{r_{k-3}}} \frac{b_{r_k}}{b_{r_{k-1}}} \sum_{r_{k-4} < i \leq r_{k-3}} \alpha^i(\alpha b_{i+1} - b_i - c) + \cdots \\
& + \frac{b_{r_2}}{b_{r_1}} \frac{b_{r_4}}{b_{r_3}} \cdots \frac{b_{r_k}}{b_{r_{k-1}}} \sum_{r_1 < i \leq m-1} \alpha^i(\alpha b_{i+1} - b_i - c) \Bigg\}.
\end{aligned}
\tag{5.12.9}
$$

Therefore, if (5.12.9) is negative, the optimal value of X_{r_k} must be 0. In this circumstance, we examine the coefficient of $X_{r_{k-1}}$, after first reducing all the relations (5.12.8) and observing that they are again of the same form. If (5.12.9) is positive, then $b_{r_{k+1}} > b_{r_k}$ and X_{r_k} must equal its upper bound in (5.12.3). We are thus led to a second set of relations, which are valid when the last group of X's determined has satisfied the constraint relation (5.12.3) with equality in the upper bound. These are of the following form.

Case 2.

$$X_i = 0 \quad \text{for} \quad r_2 < i \leq r_1, \quad r_4 < i \leq r_3, \quad \ldots, \quad r_k < i \leq r_{k-1} \quad \text{and} \quad r_{k+1} < r_k,$$

$$X_i = \frac{b_{r_2}}{b_{r_1}} \cdot \frac{b_{r_4}}{b_{r_3}} \cdots \frac{b_{r_k}}{b_{r_{k-1}}} \frac{M_0 + \sum_{j=0}^{r_{k+1}} (b_{j+1} - b_j) X_j}{b_{r_{k+1}+1}} \quad \text{for} \quad r_1 < i \leq m - 1,$$

$$X_i = \frac{b_{r_4}}{b_{r_3}} \cdots \frac{b_{r_k}}{b_{r_{k-1}}} \frac{M_0 + \sum_{j=0}^{r_{k+1}} (b_{j+1} - b_j) X_j}{b_{r_{k+1}+1}} \quad \text{for} \quad r_3 < i \leq r_2, \tag{5.12.10}$$

$$\vdots$$

$$X_i = \frac{b_{r_k}}{b_{r_{k-1}}} \frac{M_0 + \sum_{j=0}^{r_{k+1}} (b_{j+1} - b_j) X_j}{b_{r_{k+1}+1}} \quad \text{for} \quad r_{k-1} < i \leq r_{k-2},$$

$$X_i = \frac{M_0 + \sum_{j=0}^{r_{k+1}} (b_{j+1} - b_j) X_j}{b_{r_{k+1}+1}} \quad \text{for} \quad r_{k+1} < i \leq r_k.$$

The decision about $X_{r_{k+1}}$ depends on the sign of its coefficient $L_{r_{k+1}}$, obtained by substituting the relations (5.12.10) into (5.12.4). The value of $L_{r_{k+1}}$ is

$$\begin{aligned} &\alpha^{r_{k+1}}(\alpha b_{r_{k+1}+1} - b_{r_{k+1}} - c) \\ &\quad + \frac{b_{r_{k+1}+1} - b_{r_{k+1}}}{b_{r_{k+1}+1}} \left\{ \sum_{r_{k+1} < i \leq r_k} \alpha^i (\alpha b_{i+1} - b_i - c) \right. \\ &\quad + \frac{b_{r_k}}{b_{r_{k-1}}} \sum_{r_{k-1} < i \leq r_{k-2}} \alpha^i (\alpha b_{i+1} - b_i - c) + \cdots \\ &\quad \left. + \frac{b_{r_2}}{b_{r_1}} \cdot \frac{b_{r_4}}{b_{r_3}} \cdots \frac{b_{r_k}}{b_{r_{k-1}}} \sum_{r_1 < i \leq m-1} \alpha^i (\alpha b_{i+1} - b_i - c) \right\}. \end{aligned} \tag{5.12.11}$$

To summarize: Relations for X_k are determined step by step, starting from X_{m-1}. For each X, we find that either (5.12.8) or (5.12.10) is satisfied. If (5.12.8) holds, the relation for the next X is determined after examining the sign of (5.12.9). If (5.12.10) holds, the relation for the next X is determined by examining the sign of (5.12.11).

The algorithm for the construction of the solution is now clear. By backward induction, a system of equations like (5.12.8) and (5.12.10) is satisfied by the X's. After each new X is related to the preceding X's, the previous equations are reduced, using the newly characterized X. The end result is a system of linear equations in the X's with a triangular coefficient matrix, so that the explicit values of the X's can be calculated recursively.

5.13 Problems.

1. (a) Two products A and B when manufactured must pass through four machine operations I, II, III, and IV. The machine time in hours per unit produced is as follows:

	I	II	III	IV
A	2	4	3	1
B	$\frac{1}{4}$	2	1	4

The total machine time available on machines I, II, III, and IV is 45, 100, 300, and 50 hours, respectively. Product A sells for \$6 per unit and product B for \$4 per unit. What combination of products A and B should be manufactured in order to maximize profit?

(b) Suppose market demands are such that at least 22 units of product A are required. Solve the preceding problem with this constraint.

2. A gasoline refinery has two types of crude oil, A and B, which may be used to produce fuel oil and gasoline in any one of the following three patterns:

(1) 1 unit of A + 2 units of $B \rightarrow$ 2 units of fuel oil + 3 units of gasoline.
(2) 2 units of A + 1 unit of $B \rightarrow$ 5 units of fuel oil + 1 unit of gasoline.
(3) 2 units of A + 2 units of $B \rightarrow$ 2 units of fuel oil + 1 unit of gasoline.

Assuming that fuel oil sells for \$1 a unit, and gasoline sells for \$10 a unit, what is the most profitable production scheme if we start out with 10 units of A and 15 units of B?

3. A man wishes to invest \$1000 in three common stocks, and no more than \$400 in any one of them. Stock A sells for \$50 a share and pays a dividend of \$2 a year; stock B sells for \$200 a share and pays a dividend of \$5 a year; stock C sells for \$20 a share and pays no dividend, but has an even chance (probability $\frac{1}{2}$) of appreciating to \$25 a share in one year. If it does not appreciate to \$25, it will stay unchanged. What amount of

each stock should be bought to maximize dividends plus expected appreciation over one year? (*Note:* Fractional shares may be purchased.)

4. Solve the linear programming problem

$$\min (x_1 + 2x_2 + 3x_3 + \cdots + nx_n)$$

subject to the constraints

$$x_i \geq 0, \qquad x_1 + x_2 + \cdots + x_i \geq i \qquad (i = 1, \ldots, n).$$

5. Construct an example in which there is no feasible vector satisfying (5.1.1).

6. Construct an example such that the constraint set (5.1.1) is nonvoid but such that Problem I has no solution.

7. There are n refineries with daily capacity D_i $(i = 1, 2, \ldots, n)$ and m oilfields with daily production rate Q_j $(j = 1, 2, \ldots, m)$. The value of sending one barrel of oil from the jth source to the ith refinery is p_{ij}. Assume that

$$\sum_{i=1}^{n} D_i \geq \sum_{j=1}^{m} Q_j.$$

Set up the linear programming problem determining the refining process on the daily oil yield which maximizes value subject to the natural constraints.

8. Prove that if $\mathbf{b}$ is a vector with all components nonnegative and there exists an admissible vector $\mathbf{y}$ for Problem II, then Problems I and II have solutions.

9. Prove that if $\mathbf{b}$ is a vector with all components positive and $\mathbf{A}$ is a matrix with all components positive, then Problems I and II have solutions.

10. (Generalized Weak Le Chatelier Principle.) Consider Problem I: $\max_x (\mathbf{c}, \mathbf{x})$ subject to the restrictions $\mathbf{Ax} \leq \mathbf{b}, \mathbf{x} \geq \mathbf{0}$, and let $\mathbf{x}_0$ denote a solution. Let $\boldsymbol{\delta}\mathbf{c}$ denote a variation of the vector $\mathbf{c}$ so that $(\mathbf{c} + \boldsymbol{\delta}\mathbf{c}, \mathbf{x})$ is maximized under the same restrictions as before by $\mathbf{x}_0 + \boldsymbol{\delta}\mathbf{x}_0$. Prove that $\mathbf{x}_0$ satisfies $(\boldsymbol{\delta}\mathbf{c}, \boldsymbol{\delta}\mathbf{x}_0) \geq 0$.

11. An airplane is able to carry M pounds. Its pilot is asked to load n objects; the ith object weighs α_i and its value is β_i. How should the plane be loaded so as to maximize the value of its cargo without exceeding the weight limit? Set up the problem mathematically and solve it.

12. Given two sets of numbers $a_1, a_2, \ldots, a_n$ and $b_1, b_2, \ldots, b_n$, find the arrangement of a_i which minimizes

$$\sum_{i=1}^{n} b_i a_{\nu_i},$$

where ν_i represents a permutation of the indices $1, 2, \ldots, n$.

13. A firm must schedule the production of n objects, and the ith object takes T_i time units to process. The total waiting time of a single object is the length of time until its processing is completed. Find the order of processing the n objects which minimizes the total waiting time of all objects.

14. There are n objects to be processed. Let T_i represent the processing time of the ith object, and let D_i be the due date of the ith object. If the objects are processed in numerical order, the tardiness of the ith object is $\delta_i = T_1 + T_2 + \cdots + T_i - D_i$. Let $\Delta = \max \delta_i$. Show that if $D_1 \leq D_2 \leq D_3 \leq \cdots \leq D_n$, the order $1, 2, 3, \ldots, n$ minimizes Δ.

15. Show that the system of linear equations $\mathbf{yA} = \mathbf{a}$ has no nonnegative solution if and only if the inequalities $\mathbf{Ax} \leq \mathbf{0}$ and $(\mathbf{a}, \mathbf{x}) > 0$ have a solution.

16. Let $\mathbf{A}$ and $\mathbf{B}$ be two $m \times n$ matrices, and $b_{ij} \geq 0$. We call a real number λ admissible if there exists $\mathbf{x} > \mathbf{0}$ such that $\mathbf{Ax} \geq \lambda\mathbf{Bx}$. Let Ω denote the set of all admissible λ, and assume that Ω is nonvoid and bounded above.

We call μ admissible if there exists $\mathbf{y} > \mathbf{0}$ such that $\mathbf{yA} \leq \mu\mathbf{yB}$. Let M denote all admissible μ, and assume that M is nonvoid and bounded below. Prove that

$$\lambda_0 = \sup_{\lambda \in \Omega} \lambda \geq \mu_0 = \inf_{\mu \in M} \mu.$$

17. Prove that if $((\mathbf{B} + \mathbf{A})\mathbf{x}, \mathbf{y}) > 0$ for every $\mathbf{x} > \mathbf{0}$ and $\mathbf{y} > \mathbf{0}$, then the assumptions of Problem 16 are satisfied and moreover $\lambda_0 = \mu_0$.

18. Let $v(\mathbf{b}) = \max_x (\mathbf{c}, \mathbf{x})$, where $\mathbf{x}$ satisfies $\mathbf{x} \geq \mathbf{0}, \mathbf{Ax} \leq \mathbf{b}$. Prove that $v(\mathbf{b})$ is a concave function of $\mathbf{b}$, and for the same function $v(\mathbf{b})$ show that

$$\left(\frac{\partial v}{\partial b_k}\right)_+ \leq y_k \leq \left(\frac{\partial v}{\partial b_k}\right)_-$$

for any set of shadow prices $y_1, y_2, \ldots, y_m$ where $+$ and $-$ denote right- and left-hand derivatives respectively.

19. Let $\mathbf{c}$ denote any vector of n components. Let I denote the indices i for which $c_i > 0$, J the indices j for which $c_j = 0$, and K the indices k for which $c_k < 0$.

For $i \in I$, define $\boldsymbol{\alpha}_i = \langle 0, 0, \ldots, 1, 0, \ldots, 0 \rangle$, with 1 only in the ith component; for $j \in J$, define $\boldsymbol{\beta}_j = \langle 0, 0, \ldots, 1, 0, \ldots, 0 \rangle$, with 1 only in the jth component; and for $i \in I$, $k \in K$, define

$$\boldsymbol{\nu}_{ik} = \left\langle 0, \ldots, \frac{1}{c_i}, 0, \ldots, -\frac{1}{c_k}, 0, \ldots, 0 \right\rangle,$$

with $1/c_i$ in the ith component, $-1/c_k$ in the kth component, and zero elsewhere.

Let $\mathbf{T}(\mathbf{c})$ denote the matrix whose column vectors are $\boldsymbol{\alpha}$'s, $\boldsymbol{\beta}$'s, and $\boldsymbol{\nu}$'s. Prove that $(\mathbf{c}, \mathbf{x}) \geq 0, \mathbf{x} \geq \mathbf{0}$ if and only if $\mathbf{x} = \mathbf{T}(\mathbf{c})\boldsymbol{\omega}$ for some $\boldsymbol{\omega} \geq \mathbf{0}$.

20. Let $\mathbf{x}_0$ and $\mathbf{y}_0$ represent solutions to Problems I and II, respectively, and let $\mathbf{x}_0 + \boldsymbol{\delta}\mathbf{x}_0$ and $\mathbf{y}_0 + \boldsymbol{\delta}\mathbf{y}_0$ represent solutions to Problems I and II where $\mathbf{c}$, $\mathbf{b}$, and $\mathbf{A}$ are replaced by $\mathbf{c} + \boldsymbol{\delta}\mathbf{c}$, $\mathbf{b} + \boldsymbol{\delta}\mathbf{b}$, and $\mathbf{A} + \boldsymbol{\delta}\mathbf{A}$, respectively. Prove that

$$-(\boldsymbol{\delta}\mathbf{c}, \boldsymbol{\delta}\mathbf{x}_0) + (\mathbf{y}_0 + \boldsymbol{\delta}\mathbf{y}_0, \boldsymbol{\delta}\mathbf{A}\, \boldsymbol{\delta}\mathbf{x}_0) + (\boldsymbol{\delta}\mathbf{y}_0, \mathbf{A}\, \boldsymbol{\delta}\mathbf{x}_0) \leq 0$$

and

$$-(\boldsymbol{\delta}\mathbf{y}_0, \boldsymbol{\delta}\mathbf{b}) + (\boldsymbol{\delta}\mathbf{y}_0, \boldsymbol{\delta}\mathbf{A}(\mathbf{x}_0 + \boldsymbol{\delta}\mathbf{x}_0)) + (\boldsymbol{\delta}\mathbf{y}_0, \mathbf{A}\, \boldsymbol{\delta}\mathbf{x}_0) \geq 0.$$

(*Hint:* Use the saddle-value theorem, Theorem 5.3.1.)

21. Let S be a bounded, closed, polyhedral, convex set in E^n such that $\mathbf{y} > 0$ for any $\mathbf{y}$ in S. Prove that there exists a vector $\mathbf{v}$ such that $\mathbf{v} \gg 0$ and $(\mathbf{v}, \mathbf{y}) \leq 0$ for all $\mathbf{y}$ in S.

Problems 22–26 are based on the following structure. Let $\mathbf{A}_1, \mathbf{A}_2, \ldots, \mathbf{A}_m$ be given vectors in n-space, with coordinates consisting of 1's and 0's such that (a) the 1's in each $\mathbf{A}_i$ are consecutive and each $\mathbf{A}_i$ contains at least one 1; (b) if the coordinates in $\mathbf{A}_i$ which are 1 are $a, a+1, \ldots, b$ and the coordinates in $\mathbf{A}_j$ which are 1 are $c, c+1, \ldots, d$, and if $a < c$, then $b \leq d$.

Let r_i $(i = 1, \ldots, n)$ denote positive constants and t_i $(i = 1, \ldots, n)$ given constants. Consider Problem A: Minimize

$$\lambda X + \sum_{i=1}^{n} z_i$$

with respect to X and z_i, subject to the constraints

$$\begin{aligned} &(\mathbf{A}_i, \mathbf{z}) \geq t_i - X && (i = 1, \ldots, n), \\ &0 \leq z_i \leq r_i && (i = 1, \ldots, n), \\ &\lambda \geq 0. \end{aligned}$$

22. Show that the caterer's problem of Section 5.11 is a special case of Problem A.

Assume in what follows that the constraint set is nonvoid.

23. For each fixed X describe a method for solving Problem A.

24. If $z(X) = z_1^0(X) + \cdots + z_n^0(X)$ is the solution of Problem A for fixed X, show that $z(X)$ is a convex, nonincreasing, polygonal function of X.

25. Solve Problem A completely if $\mathbf{A}_i = \langle 0, 0, \ldots, 1, 0, \ldots, 0\rangle$, with 1 occurring in the ith coordinate only.

26. On the basis of Problems 23 and 24, describe an algorithm for the solution of the general Problem A.

Notes and References to Chapter 5

§ 5.1. The formal theory of linear programming originated shortly after World War II, although special linear programming models have been considered over the past century. For example, results in the theory of linear inequalities can be traced back to the early works of Weyl [253], Farkas [79], and Motzkin [186]. See [163] for complete references.

Dantzig is commonly called the father of linear programming because it was he, in conjunction with von Neumann, who discovered and popularized the basic computational algorithm for solving linear programming problems. More recently, Charnes and Cooper have made a vast number of important contributions to the theory and its applications. An introduction to the theory and the simplex algorithm of linear programming is given in Charnes, Cooper, and Henderson [46].

An intermediate-level linear programming textbook by Gass [103] contains an account of many computing devices useful for different types of linear programming problems, and an extensive bibliography of linear programming applications.

Students with a moderate undergraduate mathematical training but oriented toward economic theory are referred to Dorfman, Samuelson, and Solow [65].

Among the numerous other practical linear programming texts are Ferguson and Sargent [82] and Reinfeld and Vogel [203].

The development of linear programming theory has also been substantially influenced by the study of activity analysis models, notably by Koopmans and the economists of the Cowles Commission. The reader is referred to two volumes on activity analysis edited by Koopmans [146], and also to Morgenstern [182] and Dorfman [64].

The first linear programming problem solved, although it was not so labeled, was a version of the transportation problem (Hitchcock [124]). The method used was a form of the simplex method. The earliest modern linear programming problem considered was the diet problem, which was first formulated and analyzed by Stigler [229]. Also of historical interest are the programming models of air force operations as set forth by Wood and Dantzig.

The most elaborate large-scale linear programming studies have been concerned with the workings of the oil industry. A review of the difficulties and accomplishments of linear programming in this industry is given in Manne [171].

The most common linear programming problem is the transportation problem, sometimes referred to as the Hitchcock distribution problem. There is an extensive literature devoted to the transportation problem and its variants (see, for example, Kantorovitch [131], Koopmans and Reiter [150], Flood [84], and Ford and Fulkerson [87]). The warehouse storage model cited in this chapter is due to Charnes and Cooper [45]. Various forms of activity analysis problems have been dealt with by Leontief [166], Samuelson [210], and Koopmans [146].

The methods and theory of linear programming have become a basic tool for operations researchers. Many operations research texts incorporate introductory summaries of the methodology of linear programming (see Chapters 11–13 of Churchman, Ackoff, and Arnoff [50] and Vajda [236]).

§ **5.2.** The interpretation of the dual linear programming problem is not too clearly understood. For variations of the meanings ascribed here to the dual solution, see Samuelson [212] and Uzawa [234] (see also Koopmans [147]).

§ **5.3.** The proof of Lemma 5.3.1 is new and represents a slight extension of the usual basis for deriving the duality and existence theorems of linear programming.

Theorem 5.3.1 is a special case of what is now known as the classical Kuhn-Tucker theorem of concave programming [164].

§ **5.4.** The duality and existence theorems are the principal theorems of linear programming; almost all other developments rest on these two propositions. Numerous slightly different versions of the two theorems have been published, notably by Dantzig [55] and Gale, Kuhn, and Tucker [99].

§ **5.5.** The equivalence of matrix-game problems and linear programming problems was recognized first by Dantzig [55] and von Neumann. The interaction between these two mathematical disciplines has accelerated the growth of each. What is more important is the common mode of thinking and methodology which underlies both subjects.

§ **5.6.** The matrix extensions of the duality and existence theorems of linear programming were first suggested by Gale, Kuhn, and Tucker [99]. They proved Theorem 5.6.2 as well as other generalizations. The proof of that theorem given here is new, and the proof of Theorem 5.6.1, as well as its formulation, appears also to be new.

§ **5.7.** The formulation and method of solution of the warehouse storage problem are due to Charnes and Cooper [45]. They did not give a rigorous proof of the optimality of the algorithm proposed for solving the problem. The formal arguments are indicated here. Charnes and Cooper have pursued many extensions of these models.

§ **5.8.** The optimal assignment problem with all its ramifications was investigated by von Neumann [240]. The more elegant treatment of these topics followed in this section was suggested by Dantzig.

§ **5.9.** More than half of the applications of linear programming described in the proceedings of a 1956 conference on the subject [167] were to transportation problems. Journals interested in programming have printed innumerable refinements and extensions of versions of the model described in this section. See Ford and Fulkerson [85, 86] and Kuhn [158] as typical.

§ **5.10.** The maximal-flow, minimal-cut theorem of network flow is a beautiful application of the duality theorem. This result was discovered and proved by Ford and Fulkerson [85]. At about the same time, other writers showed how to treat many set-theoretic combinatorial problems of mathematics as applications of the duality theorem. For example, let b be a finite partially ordered set consisting of elements $b_1, \ldots, b_m$. A chain of P is a set of one or more distinct $b_{i_1}, \ldots, b_{i_k}$ with $b_{i_1} \geq b_{i_2} \geq \cdots \geq b_{i_k}$. A set of disjunct chains covers P if every element of P belongs to some chain of the set. Find a minimal covering of P. (Dantzig and Hoffman [58].) For other applications of the duality theorem, see [90] and [87].

§ 5.11. The caterer's problem was proposed and partially solved by Jacobs [129]. A different attack on the general problem, yielding an algorithm for the solution, was undertaken by Gaddum, Hoffman, and Sokolowsky [92]. Their methods apply to a more general type of problem (see Problems 22–26).

Prager [199] recognized that the caterer's problem could be formulated as a transportation problem and was therefore susceptible to the numerous specialized procedures developed for the transportation problem.

§ 5.12. The price speculation model is due to Arrow and Karlin [15].

§ 5.13. The result of Problem 10 was first noted by Samuelson. An extension of this result, embodied in the statement of Problem 20, is due to Beckmann. The assertions of Problems 13 and 14 are taken from Smith [226]. Problem 18 is due to Samuelson [212]. Problem 19 was suggested by Uzawa (unpublished). Problems 22–26 are based on Gaddum, Hoffman, and Sokolowsky [92].

CHAPTER 6

COMPUTATIONAL METHODS FOR LINEAR PROGRAMMING AND GAME THEORY

We have seen that a wide variety of problems can be cast in game form if not in linear programming form. To one grappling with a specific problem, however, there is nothing so important as a numerical answer. In this chapter we discuss relatively simple, efficient procedures for obtaining numerical answers to linear programming and game problems. We concentrate on two such procedures: the Dantzig simplex method for solving linear programming problems and the Brown algorithm for determining the value of the game.

Both of these procedures have been programmed for machine computation, and are said to work far more efficiently than might be expected in the light of their known theoretical bounds. In many respects the practical importance of linear programming and game theory depends on these numerical procedures.

The first two sections of this chapter are devoted to a description and analysis of the simplex method; the third section illustrates the workings of this method in a product mix problem. As a sample of some other procedures used in connection with special linear programming problems, we present in Section 6.4 a technique for solving network flow problems. The final sections discuss the Brown algorithm and a differential-equations method of approximating the value of a game.

This chapter is intended only as an introduction to the computational methods used in solving decision problems. It should be remembered that often the special structure of a problem, in conjunction with the general philosophy of decision theory, leads more directly to its solution than the routine application of the standard computational procedure. This is necessarily true of nonlinear programming problems (see Chapter 7), for which no general numerical methods are available. Consequently, this chapter emphasizes the interaction between general theory and computational procedure.

6.1 The simplex method. Consider the linear programming problem I: Find $\mathbf{x}$ which maximizes $L(\mathbf{x}) = (\mathbf{c}, \mathbf{x})$, with $\mathbf{x}$ subject to the constraints

$$\sum_{j=1}^{n} a_{ij}x_j \leq b_i \qquad (i = 1, \ldots, m),\ x_j \geq 0. \tag{6.1.1}$$

The set of $\mathbf{x}$ satisfying the constraints will be denoted by S.

We now describe a computational procedure—the simplex method—for obtaining a solution $\mathbf{x}^0$ to Problem I. (The simplex algorithm for the dual problem is presented in Section 6.2.) Justifications of the various steps of the procedure are given in the course of the description.

The method consists, essentially, of a systematic way of determining the relevant extreme points of the constraint set S. Since the function being maximized is linear and the constraint set S is convex, it is clear that the maximum, if it exists, must be achieved for at least one extreme point of S. The simplex method begins by first determining any extreme point. In the case in which the constraints are described by linear inequalities, this is easy to do (see paragraph E of this section). Once an extreme point of S is known, a test can be applied to determine whether this is a solution. If it is not, we seek a new extreme point of S, one that is more likely to maximize the given linear form L. The direction in S in which we move to locate this new extreme point is in essence the direction of the gradient of L. After a finite number of steps either a solution is obtained, or we learn that the maximum is infinite and the problem has no solution.

The detailed elaboration of the procedure follows.

A. Introduction of "slack" variables. The first step of the simplex method is to convert the inequalities of (6.1.1) into equalities. This is done by introducing m new variables, $x_{n+1}, \ldots, x_{n+m}$, which are known as "slack" variables and are defined by

$$x_{n+k} = b_k - \sum_{j=1}^{n} a_{kj}x_j \qquad (k = 1, \ldots, m). \tag{6.1.2}$$

If we convert the matrix $\mathbf{A}$ into the matrix

$$\tilde{\mathbf{A}} = \left\| \begin{matrix} a_{11} & \ldots & a_{1n} & \overbrace{1 \ldots 0}^{m} \\ \cdot & & \cdot & \cdot\, 1\, \cdot \\ \cdot & & \cdot & \cdot\ \ \cdot\ \ \cdot \\ \cdot & & \cdot & \cdot\ \ \ \cdot\cdot \\ a_{m1} & \ldots & a_{mn} & 0 \ldots 1 \end{matrix} \right\|,$$

and $\mathbf{x}$ into $\tilde{\mathbf{x}} = \langle x_1, \ldots, x_n, x_{n+1}, \ldots, x_{n+m}\rangle$, the system of constraints $\mathbf{Ax} \leq \mathbf{b}, \mathbf{x} \geq \mathbf{0}$, in conjunction with (6.1.2), can be written as

$$\tilde{\mathbf{A}}\tilde{\mathbf{x}} = \mathbf{b}, \qquad \tilde{\mathbf{x}} \geq \mathbf{0}. \tag{6.1.3}$$

If we now replace $\mathbf{c}$ by $\tilde{\mathbf{c}} = \langle c_1, c_2, \ldots, c_n, c_{n+1}, \ldots, c_{n+m}\rangle$, where $c_{n+1} = \cdots = c_{n+m} = 0$, the linear programming problem becomes:

Find $\tilde{\mathbf{x}}_0$ such that

$$(\tilde{\mathbf{c}}, \tilde{\mathbf{x}}_0) = \max_{\tilde{x}} (\tilde{\mathbf{c}}, \tilde{\mathbf{x}})$$

among all $\tilde{\mathbf{x}}$ vectors satisfying the conditions $\tilde{\mathbf{A}}\tilde{\mathbf{x}} = \mathbf{b}$ and $\tilde{\mathbf{x}} \geq \mathbf{0}$.

The computational procedure deals only with this equivalent problem. Henceforth, we drop the tildes in (6.1.3) and assume that our original constraints for $\mathbf{A}$ were as described in (6.1.3). Restating the problem in new notation, we have: Find a solution to $\max_\lambda (\mathbf{c}, \boldsymbol{\lambda})$, subject to the conditions $\mathbf{A}\boldsymbol{\lambda} = \mathbf{b}$ and $\boldsymbol{\lambda} \geq \mathbf{0}$, where $\mathbf{A}$ is an $m \times n$ matrix.

Let $\mathbf{p}_j$ denote the jth column vector of $\mathbf{A}$, and let $\mathbf{p}_0$ denote the vector $\mathbf{b}$. Each $\mathbf{p}_j$ has m components.

In order to avoid certain inessential technical difficulties associated with the simplex method, we tentatively make the following assumption.

ASSUMPTION 6.1.1. The vector $\mathbf{p}_0$ cannot be expressed as a positive combination of fewer than r of the $\mathbf{p}_j$, where r is the rank of $\mathbf{A}$.

In most cases the rank of $\mathbf{A}$ is equal to its number of rows. This is always so when the original constraint set is composed entirely of inequalities and $\mathbf{A}$ is the result of enlarging the prescribed constraint matrix to include slack variables. In this circumstance the submatrix of $\mathbf{A}$ whose columns correspond to the slack variables is nonsingular. From here on, in order to simplify the exposition, we assume that $\mathbf{A}$ is of rank m, the number of its rows. We leave to the reader the task of modifying the procedures for the case in which $\mathbf{A}$ is of rank smaller than m.

Assumption 6.1.1 is known as the nondegeneracy assumption. After presenting a method for finding a solution under this assumption, we shall show how to circumvent it and still obtain a solution.

(When degeneracies are present, a lack of uniqueness appears in the process of obtaining new extreme points and a certain degree of arbitrariness is introduced. This is primarily a nuisance rather than a fundamental difficulty: it means that the procedure must be carried out with greater care and that some of its simplicity is sacrificed.)

B. Characterization of extreme points. Let Λ denote the set of admissible $\boldsymbol{\lambda}$'s which satisfy the constraints

$$\lambda_1 \mathbf{p}_1 + \lambda_2 \mathbf{p}_2 + \cdots + \lambda_n \mathbf{p}_n = \mathbf{p}_0, \qquad \lambda_i \geq 0. \tag{6.1.4}$$

▶ LEMMA 6.1.1. If $\boldsymbol{\lambda}$ is an extreme point of Λ, then $\boldsymbol{\lambda}$ has at most m nonzero components.

Proof. Suppose to the contrary that $\lambda_1, \lambda_2, \ldots, \lambda_{m+1} > 0$. Consider the matrix $\mathbf{A} = \|\mathbf{p}_1, \ldots, \mathbf{p}_{m+1}\|$ and the system of homogeneous linear equations $\mathbf{Az} = \mathbf{0}$. Since the number of unknowns clearly exceeds the rank of $\mathbf{A}$, there exists a nontrivial solution $\mathbf{z}^0 = \langle z_1^0, \ldots, z_{m+1}^0 \rangle$. Enlarge $\mathbf{z}^0$ to $\tilde{\mathbf{z}}^0 = \langle z_1^0, \ldots, z_{m+1}^0, 0, \ldots, 0 \rangle$ by adding $n - m - 1$ zeros. Select $\epsilon > 0$ sufficiently small so that the components of the vectors

$$\boldsymbol{\lambda}^1 = \boldsymbol{\lambda} + \epsilon \tilde{\mathbf{z}}^0 \quad \text{and} \quad \boldsymbol{\lambda}^2 = \boldsymbol{\lambda} - \epsilon \tilde{\mathbf{z}}^0$$

are all nonnegative. Since $\mathbf{A}\boldsymbol{\lambda}^1 = \mathbf{b}$ and $\mathbf{A}\boldsymbol{\lambda}^2 = \mathbf{b}$, it follows that $\boldsymbol{\lambda}^1, \boldsymbol{\lambda}^2 \in \Lambda$. But $\boldsymbol{\lambda}^1 \neq \boldsymbol{\lambda}^2$, and

$$\boldsymbol{\lambda} = \tfrac{1}{2}(\boldsymbol{\lambda}^1 + \boldsymbol{\lambda}^2).$$

This contradicts the assumption that $\boldsymbol{\lambda}$ is an extreme point, and the lemma follows.

Assumption 6.1.1 was not used in this lemma, but is used in the one that follows.

▶LEMMA 6.1.2. If $\boldsymbol{\lambda} \in \Lambda$ and $\lambda_1, \ldots, \lambda_m > 0, \lambda_{m+1} = \cdots = \lambda_n = 0$, then $\boldsymbol{\lambda}$ is an extreme point of Λ.

Proof. Suppose that there exist $\boldsymbol{\lambda}^1, \boldsymbol{\lambda}^2 \in \Lambda$ such that

$$\boldsymbol{\lambda} = \tfrac{1}{2}(\boldsymbol{\lambda}^1 + \boldsymbol{\lambda}^2).$$

Since the last $n - m$ components of $\boldsymbol{\lambda}$ are equal to zero, the last $n - m$ components of both $\boldsymbol{\lambda}^1$ and $\boldsymbol{\lambda}^2$ must equal zero. Then $\mathbf{A}\tilde{\boldsymbol{\lambda}}^1 = \mathbf{p}_0$ and $\mathbf{A}\tilde{\boldsymbol{\lambda}}^2 = \mathbf{p}_0$, where $\tilde{\boldsymbol{\lambda}}^i = \langle \lambda_1^i, \lambda_2^i, \ldots, \lambda_m^i \rangle$. Thus $\mathbf{A}(\tilde{\boldsymbol{\lambda}}^1 - \tilde{\boldsymbol{\lambda}}^2) = \mathbf{0}$. If $\tilde{\boldsymbol{\lambda}}^1 - \tilde{\boldsymbol{\lambda}}^2 \neq \mathbf{0}$, then ϵ can be chosen sufficiently small so that

$$\boldsymbol{\mu} = \tilde{\boldsymbol{\lambda}}^1 + \epsilon(\tilde{\boldsymbol{\lambda}}^1 - \tilde{\boldsymbol{\lambda}}^2) \geq \mathbf{0} \quad \text{and} \quad \mu_1\mathbf{p}_1 + \mu_2\mathbf{p}_2 + \cdots + \mu_n\mathbf{p}_n = \mathbf{p}_0.$$

If some component of $\tilde{\boldsymbol{\lambda}}^1 - \tilde{\boldsymbol{\lambda}}^2$ is negative, we can increase ϵ until at least one of the components of $\boldsymbol{\mu}$ vanishes while $\boldsymbol{\mu}$ persists as a nonnegative vector. We thus obtain a representation of $\mathbf{p}_0$ by fewer than m of the $\mathbf{p}_i$, contradicting Assumption 6.1.1. If some component of $\tilde{\boldsymbol{\lambda}}^1 - \tilde{\boldsymbol{\lambda}}^2$ is positive, we choose ϵ negative and a similar argument applies. Hence in all cases $\boldsymbol{\lambda}^1 = \boldsymbol{\lambda}^2$, and thus $\boldsymbol{\lambda}$ is an extreme point.

C. Getting a new extreme point from the old. Suppose that $\boldsymbol{\lambda}^0$ has $\lambda_1^0, \ldots, \lambda_m^0 > 0$ and $\lambda_{m+1}^0 = \cdots = \lambda_n^0 = 0$ and satisfies (6.1.4). (We shall discuss later how to obtain such a point.) Then $\boldsymbol{\lambda}^0$ is an extreme point of Λ (Lemma 6.1.2). It follows that the vectors $\mathbf{p}_1, \ldots, \mathbf{p}_m$ are linearly independent and hence constitute a basis of m-space. Otherwise, there exists a nonzero vector $\boldsymbol{\mu} = \langle \mu_1, \ldots, \mu_m \rangle$ such that

$$\sum_{i=1}^{m} \mu_i \mathbf{p}_i = \mathbf{0}.$$

Then

$$\boldsymbol{\lambda}^0 = \frac{(\boldsymbol{\lambda}^0 + \epsilon\boldsymbol{\mu}) + (\boldsymbol{\lambda}^0 - \epsilon\boldsymbol{\mu})}{2}$$

and $\boldsymbol{\lambda}^0 \pm \epsilon\boldsymbol{\mu}$ are nonzero distinct feasible vectors for ϵ sufficiently small, contradicting the extremal property of $\boldsymbol{\lambda}^0$. Consequently, for every vector $\mathbf{p}_i$ $(i = 1, 2, \ldots, n)$ there exist numbers x_{ji} $(j = 1, 2, \ldots, m)$ such that

$$\mathbf{p}_i = \sum_{j=1}^{m} \mathbf{p}_j x_{ji}.$$

We refer to $\mathbf{X} = \|x_{ji}\|$ $(j = 1, \ldots, m;\ i = 1, \ldots, n)$ as the "tableau" associated with the extreme-point vector $\boldsymbol{\lambda}^0$. The corresponding basis vectors $\mathbf{p}_1, \mathbf{p}_2, \ldots, \mathbf{p}_m$ used in the representation of $\mathbf{p}_0$ with weights λ_j^0 are called the basis vectors of the extreme point $\boldsymbol{\lambda}^0$.

We may obviously rewrite (6.1.4) as

$$\lambda_1^0\mathbf{p}_1 + \cdots + \lambda_m^0\mathbf{p}_m + \theta\left(\mathbf{p}_k - \sum_{j=1}^{m} \mathbf{p}_j x_{jk}\right) = \mathbf{p}_0,$$

where k denotes an arbitrary index between $m + 1$ and n. This equation may be rearranged to read

$$\sum_{j=1}^{m} (\lambda_j^0 - \theta x_{jk})\mathbf{p}_j + \theta\mathbf{p}_k = \mathbf{p}_0. \tag{6.1.5}$$

So long as the coefficients of $\mathbf{p}_k$ and $\mathbf{p}_j$ $(j = 1, \ldots, m)$ in (6.1.5) are nonnegative and θ is steadily increased from zero, we move in a direction of Λ where $\mathbf{p}_k$ is increasingly involved in the representation of $\mathbf{p}_0$. Let $\boldsymbol{\lambda}^1$ denote the point of Λ that yields the representation of $\mathbf{p}_0$ given in (6.1.5).

Consider for k fixed

$$\begin{aligned}
(\mathbf{c}, \boldsymbol{\lambda}^1) &= \sum_{j=1}^{m} (\lambda_j^0 - \theta x_{jk})c_j + \theta c_k \\
&= (\mathbf{c}, \boldsymbol{\lambda}^0) + \theta\left(c_k - \sum_{j=1}^{m} x_{jk}c_j\right) \\
&= (\mathbf{c}, \boldsymbol{\lambda}^0) - \theta(z_k - c_k),
\end{aligned} \tag{6.1.6}$$

where z_k is implicitly defined by the last equation. We now distinguish three cases.

Case 1. There exists a value k_0 $(m + 1 \leq k_0 \leq n)$ for which $z_{k_0} - c_{k_0} < 0$, where

$$z_k = \sum_{j=1}^{m} x_{jk}c_j$$

and $x_{jk_0} \leq 0$ for $j = 1, \ldots, m$. In this case $\lambda_j^0 - \theta x_{jk_0} \geq 0$, and hence $\boldsymbol{\lambda}^1$ qualifies as a member of Λ for every $\theta > 0$. Letting θ (the value of λ_{k_0}) tend to $+\infty$, we see that $(\mathbf{c}, \boldsymbol{\lambda}^1)$ increases beyond all limits and the problem has no solution; i.e., the maximum is infinite.

Case 2. There exists a k_0 $(m + 1 \leq k \leq n)$ for which $z_{k_0} - c_{k_0} < 0$ and $x_{jk_0} > 0$ for some j. Let

$$\theta = \min_j \frac{\lambda_j^0}{x_{jk_0}},$$

the minimum being taken over all j for which $x_{jk_0} > 0$. Clearly, $\theta > 0$ and $\lambda_j^0 - \theta x_{jk_0} \geq 0$ for $j = 1, \ldots, m$, with equality valid for at least one j. Two such terms cannot vanish together, however; if they do, $\mathbf{p}_0$ is expressed by a positive combination of fewer than m of the $\mathbf{p}_j$, in contradiction to Assumption 6.1.1. When two components simultaneously vanish, we say that a degeneracy occurs. With θ determined as indicated, the components of $\boldsymbol{\lambda}^1$ are $\lambda_j = \lambda_j^0 - \theta x_{jk_0}$ for $j = 1, \ldots, m$, $\lambda_{k_0} = \theta$, and zeros elsewhere. The vector $\boldsymbol{\lambda}^1$ possesses m nonzero components and therefore, by Lemma 6.1.2, is an extreme point of Λ. Moreover, $(\mathbf{c}, \boldsymbol{\lambda}^1) > (\mathbf{c}, \boldsymbol{\lambda}^0)$, so that the extreme point $\boldsymbol{\lambda}^1$ definitely yields a larger value than $\boldsymbol{\lambda}^0$.

Case 3. If $z_k - c_k \geq 0$ for all k, then $\boldsymbol{\lambda}^0$ is optimal. To prove this statement, we must show that

$$\sum_{i=1}^n \lambda_i \mathbf{p}_i = \mathbf{p}_0$$

implies $(\boldsymbol{\lambda}, \mathbf{c}) \leq (\boldsymbol{\lambda}^0, \mathbf{c})$. Note that $z_k = c_k$ for $1 \leq k \leq m$. From the condition of Case 3, it follows that

$$\sum_{i=1}^n \lambda_i c_i \leq \sum_{i=1}^n \lambda_i \sum_{j=1}^m x_{ji} c_j = \sum_{j=1}^m c_j \sum_{i=1}^n \lambda_i x_{ji}.$$

The proof will be finished if it is shown that

$$\lambda_j^0 = \sum_{i=1}^n \lambda_i x_{ji}.$$

Explicitly,

$$\mathbf{p}_0 = \sum_{i=1}^n \lambda_i \mathbf{p}_i = \sum_{i=1}^n \lambda_i \sum_{j=1}^m x_{ji} \mathbf{p}_j = \sum_{j=1}^m \mathbf{p}_j \sum_{i=1}^n \lambda_i x_{ji} = \sum_{j=1}^m \lambda_j^0 \mathbf{p}_j.$$

By virtue of Assumption 6.1.1 we deduce the equality of the coefficients, and the proof is complete.

D. Formalization. Suppose that we have an extreme point $\boldsymbol{\lambda}^0$ and a basis $\mathbf{p}_1, \ldots, \mathbf{p}_m$. Compute x_{ik} of the following matrix, where

$$\mathbf{p}_k = \sum_{j=1}^{m} \mathbf{p}_j x_{jk} \qquad (k = 0, 1, \ldots, n),$$

and the auxiliary quantities are as indicated:

	p_1	p_2		p_k		p_n	p_0	
p_1	x_{11}	x_{12}	$\cdots$	x_{1k}	$\cdots$	x_{1n}	λ_1^0	c_1
p_2	x_{21}	x_{22}	$\cdots$	x_{2k}	$\cdots$	x_{2n}	λ_2^0	c_2
$\vdots$								
p_m	x_{m1}	x_{m2}	$\cdots$	x_{mk}	$\cdots$	x_{mn}	λ_m^0	c_m
z:	z_1	z_2	$\cdots$	$z_k = \sum_{j=1}^{m} x_{jk} c_j$	$\cdots$	z_n		
$z - c$:	$z_1 - c_1$	$z_2 - c_2$	$\cdots$	$z_k - c_k$	$\cdots$	$z_n - c_n$		

Case 1. If $z_i - c_i \geq 0$ for all i, then $\boldsymbol{\lambda}^0 = \langle \lambda_1^0, \lambda_2^0, \ldots, \lambda_m^0 \rangle$ is the solution of the maximization problem.

Case 2. If $z_{k_0} - c_{k_0} < 0$ for some k_0, and $x_{ik_0} \leq 0$ for all i, then $\max\,(\mathbf{c}, \boldsymbol{\lambda}) = \infty$, the maximum being extended over all feasible λ.

Case 3. If $z_{k_0} - c_{k_0} < 0$ for some k_0, and $x_{ik_0} > 0$ for some i, then we can get a new extreme point $\boldsymbol{\lambda}^1$ for which $(\mathbf{c}, \boldsymbol{\lambda}^1) > (\mathbf{c}, \boldsymbol{\lambda}^0)$. In practice it is best to choose the index k_0 for which $z_{k_0} - c_{k_0}$ has the largest possible negative value; this tends to decrease the number of iterations needed for the simplex technique.

The components of $\boldsymbol{\lambda}^1$ are $\lambda_i^1 = \lambda_i^0 - \theta x_{ik_0}$ for $i = 1, \ldots, m$,

$$\lambda_{k_0} = \theta = \min_{\substack{i \\ x_{ik_0} > 0}} \frac{\lambda_i^0}{x_{ik_0}},$$

and zeros elsewhere. Corresponding to this extreme point are a new basis and a new tableau.

Define l_0 such that $\lambda_{l_0}^0 - \theta x_{l_0,k_0} = 0$. Then the basis vectors associated with $\boldsymbol{\lambda}^1$ are $\mathbf{p}_1, \mathbf{p}_2, \ldots, \mathbf{p}_{l_0-1}, \mathbf{p}_{l_0+1}, \ldots, \mathbf{p}_m, \mathbf{p}_{k_0}$. In other words, $\mathbf{p}_{k_0}$ has replaced $\mathbf{p}_{l_0}$.

The tableau $\mathbf{Y} = \|y_{ji}\|$ of $\boldsymbol{\lambda}^1$ may be readily computed. To this end, suppose for simplicity that $l_0 = 1$. Then the new basis consists of the vectors $\mathbf{p}_2, \ldots, \mathbf{p}_m, \mathbf{p}_{k_0}$.

Now

$$\mathbf{p}_i = \sum_{j=2}^{m} y_{ji}\mathbf{p}_j + y_{k_0i}\mathbf{p}_{k_0}. \tag{6.1.7}$$

These y_{ji} can be expressed in terms of old x_{ji}. Indeed, since x_{1k_0} cannot be zero by Assumption 6.1.1, we obtain

$$\mathbf{p}_1 = -\sum_{j=2}^{m} \frac{x_{jk_0}}{x_{1k_0}}\mathbf{p}_j + \frac{1}{x_{1k_0}}\mathbf{p}_{k_0},$$

$$\mathbf{p}_i = -x_{1i}\sum_{j=2}^{m} \frac{x_{jk_0}}{x_{1k_0}}\mathbf{p}_j + \frac{x_{1i}}{x_{1k_0}}\mathbf{p}_{k_0} + \sum_{j=2}^{m} x_{ji}\mathbf{p}_j \qquad (i \geq 2).$$

On comparison with (6.1.7), it follows that

$$y_{ji} = \begin{cases} x_{ji} - \dfrac{x_{1i}x_{jk_0}}{x_{1k_0}} & (j \neq k_0), \\ \dfrac{x_{1i}}{x_{1k_0}} & (j = k_0). \end{cases} \tag{6.1.8}$$

For the general case in which $\mathbf{p}_{k_0}$ replaces $\mathbf{p}_{r_0}$, the index 1 in (6.1.8) is to be replaced throughout by r_0. Once a tableau $\mathbf{Y}$ is explicitly known for the new extreme point $\boldsymbol{\lambda}^1$, we are ready to repeat the process and get a further extreme point, one which yields a still larger value of $(\mathbf{c}, \boldsymbol{\lambda})$, unless we find that the problem has no solution or that $\boldsymbol{\lambda}^1$ is the solution. After a finite number of steps, since there are only a finite number of distinct extreme points, we obtain an optimal extreme vector.

E. Obtaining an initial extreme point. For the linear programming problem I we have the restrictions $\mathbf{A}\boldsymbol{\lambda} \leq \mathbf{b}$, $\boldsymbol{\lambda} \geq \mathbf{0}$. The expanded matrix $\tilde{\mathbf{A}}$ including the slack variables is

$$\left\| \begin{matrix} a_{11} & \cdots & a_{1n} & 1 & \cdots & 0 \\ \cdot & & \cdot & & \cdot 1 & \cdot \\ \cdot & & \cdot & \cdot & \cdot & \cdot \\ \cdot & & \cdot & \cdot & & \cdot\cdot \\ a_{m1} & \cdots & a_{mn} & 0 & \cdots & 1 \end{matrix} \right\| = \|\mathbf{A}, \mathbf{I}\|.$$

(the last m columns, spanned by a brace labeled m)

When $\mathbf{b}$ is a nonnegative vector, the last m columns of $\tilde{\mathbf{A}}$ constitute a basis of the extreme point $\langle 0, \ldots, 0, b_1, \ldots, b_m \rangle$.

This is not the case if some of the components of $\mathbf{b}$ are negative, since then the linear combination of the vectors $\mathbf{p}_{n+1}, \mathbf{p}_{n+2}, \ldots, \mathbf{p}_{n+m}$ which represent $\mathbf{p}_0$ has coefficients $\langle b_1, b_2, \ldots, b_m \rangle$ not all nonnegative. In this case, to obtain an initial extreme point we introduce the concept of "artificial variables." Suppose for definiteness that the first r components $b_1, b_2, \ldots, b_r$ are negative and the remaining b's are nonnegative. Consider the augmented matrix $\mathbf{A}^* = \|\mathbf{A}, \mathbf{I}_m, \mathbf{P}_r\|$ with

$$\mathbf{P}_r = \begin{Vmatrix} -1 & \cdots & 0 \\ \vdots & \ddots & \vdots \\ 0 & & -1 \\ 0 & \cdots & 0 \\ \cdot & & \cdot \\ \cdot & & \cdot \\ \cdot & & \cdot \\ 0 & \cdots & 0 \end{Vmatrix}$$

that is, $\mathbf{P}_r$ is an $m \times r$ matrix such that $p_{ii} = -1$ $(1 \leq i \leq r)$ and $p_{ij} = 0$ otherwise. We choose $c_{n+m+1} = c_{n+m+2} = \cdots = c_{n+m+r} = -M$, with M positive and sufficiently large so that the last r components of $\boldsymbol{\lambda}$ in the maximization of $(\boldsymbol{\lambda}, \mathbf{c})$ are equal to zero, provided a solution exists.

For the enlarged matrix $\mathbf{A}^*$ the last m columns constitute a basis and thus we secure an initial extreme point. It is standard practice in implementing the computational algorithm to keep the vector $\mathbf{b}$ nonnegative. This means simply that suitable rows of the constraining equations are to be multiplied by -1, an operation that does not alter the choice of the starting basis.

This device of adding artificial variables is always applicable if no other initial extreme point is available. Its disadvantage is that it increases the number of iterations required by the simplex technique.

F. Circumventing the nondegeneracy assumption. It may be possible to represent $\mathbf{p}_0$ by fewer than m vectors $\mathbf{p}_i$. In this case

$$\theta = \min_{\substack{j \\ x_{jk_0} > 0}} \frac{\lambda_j^0}{x_{jk_0}}$$

may be achieved for two or more indices. It is conceivable at this point that a cycling among several extreme points occurs as the change of basis is effected. More specifically, if θ's of equal value are found, then on subsequent iterations one cannot be sure that the new extreme vector for a new basis furnishes a larger maximum.

This difficulty may be overcome by modifying the initial problem as follows: Keep the vectors $\mathbf{p}_i$ unchanged but replace $\mathbf{p}_0$ by

$$\mathbf{p}_0(\epsilon) = \mathbf{p}_0 + \sum_{j=1}^{n} \epsilon^j \mathbf{p}_j,$$

where ϵ is an unspecified positive parameter and sufficiently small. The method to be described is independent of the parameter ϵ and in essence involves only slight additional care in applying the basic simplex algorithm. Let us consider a basis with index set I and tableau matrix $\mathbf{X}$ (obviously all basis systems for the original and modified problems coincide). If

$$\sum_{i \in I} \lambda_i \mathbf{p}_i = \mathbf{p}_0,$$

then

$$\sum_{i \in I} \left(\lambda_i + \sum_{j=1}^{n} \epsilon^j x_{ij} \right) \mathbf{p}_i = \mathbf{p}_0(\epsilon).$$

Hence

$$\lambda'_r(\epsilon) = \lambda_r + \epsilon^r + \sum_{j \notin I} \epsilon^j x_{rj}.$$

Each $\lambda'_r(\epsilon)$ contains a power of ϵ not appearing in the representation of any other. The value of this feasible vector λ' is

$$(\mathbf{c}, \boldsymbol{\lambda}(\epsilon)) = f[\boldsymbol{\lambda}(\epsilon)] = \sum_{i \in I} c_i \lambda_i + \sum_{j=1}^{n} \epsilon^j z_j,$$

where the z_j are defined the same as before. Repeating the procedure of (6.1.5) for the modified problem, we have

$$\mathbf{p}_0(\epsilon) = \sum_{i \in I} \left(\lambda_i + \sum_{j=1}^{n} \epsilon^j x_{ij} - \theta x_{ik} \right) \mathbf{p}_i + \theta \mathbf{p}_k,$$

from which we get a new feasible vector $\lambda'(\epsilon)$ with value

$$f[\boldsymbol{\lambda}'(\epsilon)] = f[\boldsymbol{\lambda}(\epsilon)] + \theta(c_k - z_k).$$

There are three mutually exclusive possibilities to be considered, as in paragraph D:

(a) If $z_k - c_k \geq 0$ for all k, then $\boldsymbol{\lambda}$ is maximal.

(b) If $z_k - c_k < 0$ for some k, and $x_{ik} \leq 0$ for all i in I, then $f(\boldsymbol{\lambda}') \to \infty$ as $\theta \to \infty$ for every ϵ and the problem has no solution.

(c) If $z_{k_0} - c_{k_0} < 0$ for some k_0, and $x_{ik_0} > 0$ for some i in I, then let

$$\theta(\epsilon) = \min_i \frac{(\lambda_i + \sum_{j=1}^{n} \epsilon^j x_{ij})}{x_{ik_0}},$$

the minimum being taken over all i such that $x_{ik_0} > 0$. In practice, as mentioned earlier, it is customary to select the value of k_0 for which $z_{k_0} - c_{k_0}$ is most negative. Whenever

$$\min_{\substack{i \\ x_{ik_0}>0}} \frac{\lambda_i}{x_{ik_0}}$$

is achieved at a single index value i_0, then the same is true for $\theta(\epsilon)$ with ϵ sufficiently small. If $\min_i (\lambda_i/x_{ik_0})$ is achieved at two or more indices—call the set of such indices J—then we select the index value of J for which x_{i1}/x_{ik_0} is smallest (algebraically). If this minimum is attained for at least two indices of J, we proceed to examine the ratio $x_{i2}/x_{ik}, x_{i3}/x_{ik}, \ldots,$ determining their minimum over the successively smaller index sets at which the preceding minima were attained, and so on until we find a minimum that is achieved at a single index value i_0. The process must terminate, since the $\lambda_i'(\epsilon)$ are all distinct polynomials. The vector with index k_0 replaces the vector with index i_0, and the algorithm may be repeated.

It is never necessary to calculate $\boldsymbol{\lambda}^1(\epsilon)$, since all that is relevant for the next step of the algorithm is the components of λ_i for the new basis of the unmodified problem. The reader may observe that the nondegeneracy Assumption 6.1.1 applies to every modified problem with ϵ positive and sufficiently small, and hence that the modified procedure yields ultimately a single solution valid for every such problem. If we then put $\epsilon = 0$, this vector also provides a solution to the original problem.

***6.2 Auxiliary simplex methods.** In this section we discuss several consequences and modifications of the simplex method.

A. Solution of the dual problem. We show first how one can obtain a solution to Problem II, the dual problem, simultaneously with the solution to Problem I. This should help to clarify the dual simplex algorithm which is discussed in paragraph B below.

The simplex method solves the problem of finding a vector $\boldsymbol{\lambda}$ that maximizes $(\mathbf{c}, \boldsymbol{\lambda})$ subject to the constraints $\mathbf{A}\boldsymbol{\lambda} = \mathbf{b}$ and $\boldsymbol{\lambda} \geq \mathbf{0}$. The dual problem is to find a vector $\mathbf{y}$ that minimizes $(\mathbf{y}, \mathbf{b})$ subject to the constraints $\mathbf{yA} \geq \mathbf{c}$. This formulation is no less valid when $\mathbf{A}$ represents an augmented matrix obtained by adding suitable slack variables and defining the corresponding c's to be zero, since in that case the constraints $\mathbf{yA} \geq \mathbf{c}$ already express the necessary restrictions of the form $y_i \geq 0$.

At the end of the simplex algorithm, we arrive at a basis $\mathbf{B}$, consisting of a submatrix of $\mathbf{A}$ whose columns comprise the vectors of the basis, such that

$$\mathbf{B}\boldsymbol{\lambda}^0 = \mathbf{b}, \qquad \boldsymbol{\lambda}^0 = \mathbf{B}^{-1}\mathbf{b},$$

where $\boldsymbol{\lambda}^0$ is the solution of Problem I and the vector $\mathbf{b}$ is the same as $\mathbf{p}_0$. Let $\mathbf{c}_B$ denote a contraction of the vector $\mathbf{c}$ restricted to the components corresponding to the basis vectors of $\mathbf{B}$. Define $\mathbf{y}^0 = \mathbf{c}_B\mathbf{B}^{-1}$. Writing $\mathbf{A}$ in the form $\|\mathbf{B}, \mathbf{P}\|$, where $\mathbf{P}$ is the set of column vectors of $\mathbf{A}$ not included in $\mathbf{B}$, we have

$$\mathbf{y}^0\mathbf{A} = \mathbf{y}^0\|\mathbf{B}, \mathbf{P}\| = \langle \mathbf{c}_B, \mathbf{c}_B\mathbf{B}^{-1}\mathbf{P}\rangle. \tag{6.2.1}$$

Equation (6.2.1) states that $\mathbf{y}^0\mathbf{A}$ is the vector whose first m components are those of $\mathbf{c}_B$ and whose last $n - m$ components are those of $\mathbf{c}_B\mathbf{B}^{-1}\mathbf{P}$. We also note that the tableau associated with the basis $\mathbf{B}$, when the vectors of $\mathbf{B}$ are arranged first, is clearly $\|\mathbf{I}, \mathbf{B}^{-1}\mathbf{P}\|$. Hence the vector $\langle \mathbf{c}_B, \mathbf{c}_B\mathbf{B}^{-1}\mathbf{P}\rangle$ is identical to the vector $\mathbf{z}$ (see page 165) for the basis $\mathbf{B}$; and since $\mathbf{B}$ is optimal, $\mathbf{z} \geq \mathbf{c}$. Thus $\mathbf{y}^0$ is feasible for the dual problem. Moreover,

$$(\mathbf{y}^0, \mathbf{b}) = (\mathbf{c}_B\mathbf{B}^{-1}, \mathbf{b}) = (\mathbf{c}_B, \mathbf{B}^{-1}\mathbf{b}) = (\mathbf{c}, \boldsymbol{\lambda}^0). \tag{6.2.2}$$

It follows from Corollary 5.4.1 that $\mathbf{y}^0$ is optimal for the dual problem.

B. The dual simplex algorithm. We shall refer to the simplex method of Section 6.1 hereafter as the primal simplex algorithm, to distinguish it from the dual simplex algorithm described below. The dual simplex algorithm furnishes another algorithm for solving Problem I. It is referred to as the dual algorithm, since its underlying justification is bound intrinsically to the solution of the dual problem. We shall discuss briefly the merits and defects of each of these algorithms at the close of this section.

A solution $\boldsymbol{\lambda}^0$ to Problem I is determined whenever a matrix tableau $\mathbf{X}$ is achieved for a basis satisfying the property that $z_j - c_j \geq 0$ for all j, where

$$z_k = \sum_{j \in B} c_j x_{jk},$$

the summation extending over the indices corresponding to the basis vectors. The main idea of the transformations underlying a change of basis is to eliminate step by step the negative components of $z_j - c_j$ while maintaining $\boldsymbol{\lambda}$ as a feasible vector.

A dual method of calculation is to insist that the vector $\mathbf{z} - \mathbf{c}$ always be nonnegative but permit the vector $\boldsymbol{\lambda}$ used in the representation of $\mathbf{p}_0$ to possess components of arbitrary sign. This is equivalent to maintaining feasibility for the dual problem and proceeding until a feasible vector for the original problem is obtained. A new basis is formed by replacing a vector corresponding to a negative coordinate of $\boldsymbol{\lambda}$ by a new vector. The process is repeated and a solution is achieved after a finite number of steps.

Since we are concerned here only with the salient features of the dual algorithm, and not with its many special cases, we shall impose the following nondegeneracy assumption.

ASSUMPTION 6.2.1. For no feasible $\mathbf{y}$ of the dual problem can equality of $\mathbf{yA}$ and $\mathbf{c}$ be achieved at more than m coordinates.

The formal procedure is as follows. Suppose that we have a basis of column vectors $\mathbf{q}_1, \mathbf{q}_2, \ldots, \mathbf{q}_m$ with matrix $\mathbf{Q}$ such that $\mathbf{Qu} = \mathbf{b}$; the components of $\mathbf{u}$ are not necessarily all nonnegative. Let $\mathbf{X}$ represent the tableau associated with $\mathbf{Q}$, and let $\mathbf{c}_Q$ denote the vector $\mathbf{c}$ contracted to the components corresponding to the columns of $\mathbf{Q}$. We further assume that $z_j - c_j \geq 0$ for all j, with

$$z_j = \sum_{i \in Q} c_i x_{ij}.$$

As we see from paragraph A of this section, this is equivalent to saying that $\mathbf{y} = \mathbf{c}_Q \mathbf{Q}^{-1}$ is a feasible vector for the dual problem. Because of Assumption 6.2.1, equality holds for $\mathbf{yA} \geq \mathbf{c}$ at the components corresponding to the basis vectors, with strict inequality elsewhere.

If $u_i \geq 0$ for every component, then $\mathbf{u}$ is optimal by the primal simplex algorithm.

Let r_0 denote the index for which u_i is most negative. We now distinguish two cases.

Case 1. If $x_{r_0 j} \geq 0$ for all j, then there exists no feasible vector satisfying the constraints of Problem I.

Proof. Analogous to (6.1.5), we have

$$\mathbf{p}_0 = \sum_{l \in Q} (u_l - \theta x_{lk}) \mathbf{q}_l + \theta \mathbf{q}_k, \tag{6.2.3}$$

where k is arbitrary and θ is any real number. Suppose that there exists a representation

$$\mathbf{p}_0 = \sum_{j=1}^{m} \lambda_j \mathbf{q}_{k_j},$$

with $\lambda_j \geq 0$. Inserting $\theta = \lambda_j$ and $k = k_j$ successively into (6.2.3) and adding, we obtain

$$m\mathbf{p}_0 = \sum_{l \in Q} \left[m u_l - \sum_j \lambda_j x_{lk_j} \right] \mathbf{q}_l + \mathbf{p}_0$$

or

$$\mathbf{p}_0 = \sum_{l \in Q} \left[\frac{m}{m-1} u_l - \frac{\sum_j \lambda_j x_{lk_j}}{m-1} \right] \mathbf{q}_l. \tag{6.2.4}$$

Since the vectors $\mathbf{q}_l$ constitute a basis, the coefficient of $\mathbf{q}_l$ in (6.2.4) is equal to u_l, and in particular

$$u_{r_0} = \frac{m}{m-1} u_{r_0} - \frac{\sum_j \lambda_j x_{r_0 k_j}}{m-1}$$

or

$$0 = \frac{1}{m-1} u_{r_0} - \frac{\sum_j \lambda_j x_{r_0 k_j}}{m-1}.$$

But $u_{r_0} < 0$, $\lambda_j \geq 0$, and $x_{r_0 k} \geq 0$ (all k) contradict this relation. Therefore there exists no feasible primal vector. Since the dual problem has feasible vectors ($\mathbf{c}_Q \mathbf{Q}^{-1}$), this means that the dual form tends to $-\infty$.

Case 2. Let $u_{r_0} < 0$ and $x_{r_0 j} < 0$ for some j. (Necessarily, j is the index of a vector outside the basis.)

Define k_0 as the index value at which

$$\alpha = \min_{\substack{j \\ x_{r_0 j} < 0}} \left(\frac{z_j - c_j}{-x_{r_0 j}} \right)$$

is achieved. From the nondegeneracy assumption it readily follows that k_0 is unique. We now choose θ in (6.2.3) so that the coefficient of $\mathbf{q}_{r_0}$ vanishes, viz.,

$$\theta = \frac{u_{r_0}}{x_{r_0 k_0}}.$$

Since both u_{r_0} and $x_{r_0 k_0}$ are negative, we have θ positive. A new basis $\tilde{\mathbf{Q}}$ is obtained from the old by replacing $\mathbf{q}_{r_0}$ by $\mathbf{q}_{k_0}$. The coefficients in the representation of $\mathbf{p}_0$ are

$$\tilde{u}_l = u_l - \theta x_{l k_0} \qquad \text{and} \qquad \tilde{u}_{k_0} = \theta. \tag{6.2.5}$$

A straightforward computation, utilizing the transformation of basis (6.1.8) and the definition of α, shows that $\tilde{\mathbf{z}} - \tilde{\mathbf{c}}$ for the new basis $\tilde{\mathbf{Q}}$ is a nonnegative vector and hence that $\mathbf{c}_{\tilde{Q}} \tilde{\mathbf{Q}}^{-1}$ is a feasible vector for the dual. On comparison with (6.2.2), we get

$$(\mathbf{c}_{\tilde{Q}} \tilde{\mathbf{Q}}^{-1}, \mathbf{b}) = \sum_{l \in \tilde{Q}} \tilde{u}_l c_l = \sum_{l \in Q} u_l c_l - \theta z_{k_0} + \theta c_{k_0}$$

and

$$(\mathbf{c}_Q \mathbf{Q}^{-1}, \mathbf{b}) = \sum_l u_l c_l.$$

Therefore

$$(\mathbf{c}_Q \mathbf{Q}^{-1}, \mathbf{b}) - (\mathbf{c}_{\tilde{Q}} \tilde{\mathbf{Q}}^{-1}, \mathbf{b}) = \theta (z_{k_0} - c_{k_0}) > 0,$$

since $\theta > 0$ and $z_{k_0} - c_{k_0} > 0$; it follows from Assumption 6.2.1 that $\mathbf{c}_{\tilde{Q}} \tilde{\mathbf{Q}}^{-1}$ is an improvement over $\mathbf{c}_Q \mathbf{Q}^{-1}$ for the dual problem. Both vectors are extreme and feasible, and therefore the procedure terminates with u nonnegative in a finite number of steps.

The merit of the dual simplex method is that it involves far fewer iterations than the usual simplex method for certain kinds of problems. For example, consider the problem

$$\min\,(\mathbf{c}, \mathbf{y}) \tag{6.2.6}$$

subject to the constraints

$$\mathbf{y} \geq \mathbf{0}, \qquad \mathbf{yA} \geq \mathbf{b}. \tag{6.2.7}$$

This problem can be converted into a maximization problem by substituting $-\mathbf{A}$ for $\mathbf{A}$, $-\mathbf{b}$ for $\mathbf{b}$, and $-\mathbf{c}$ for $\mathbf{c}$. With this done, we can formulate the problem as

$$\max\,(\mathbf{c}, \mathbf{y})$$

subject to the constraints

$$\mathbf{y} \geq \mathbf{0}, \qquad \mathbf{yA} \leq \mathbf{b}.$$

Now if $\mathbf{c}$ and $\mathbf{b}$ of (6.2.6) and (6.2.7) are positive vectors, then in the changeover to a maximization problem they become negative vectors. In this case, in order to get started for the primal simplex algorithm, it seems necessary to introduce artificial variables on top of the slack variables, thus markedly increasing the number of iterations. In these circumstances, it appears that the dual method would yield a solution more quickly.

Indeed, suppose that the tableau $\mathbf{T}$ associated with the vectors of the slack variables is

$$\mathbf{T} = \|\mathbf{A}, \mathbf{I}\|.$$

Since the c_i attached to the slack variables are defined to be zero, we get for this basis

$$z_i - c_i = -c_i,$$

which is a positive vector if all $c_i \leq 0$. Hence, the dual method can be started immediately with an initial basis whose tableau is $\mathbf{T}$. The use of artificial variables in this circumstance would be entirely redundant.

The classical diet problem (see Section 5.1) is of the sort described in (6.2.6) and (6.2.7). It is more readily solved by the dual algorithm than by the primal.

Many variations of the simplex method have been introduced to take advantage of special forms of the constraint matrix. For example, the problem created by adding a single additional constraint to a problem already solved by the simplex method can be solved without having to repeat the full procedure. Another special case, the problem in which the variables are restricted by upper bounds in addition to the usual constraints, can be solved without doubling the number of variables. Descriptions of some of these special techniques are given in several elementary texts in linear programming (see notes to this chapter).

6.3 An illustration of the use of the simplex method. A manufacturer has the option of using one or more of four different production processes. The first and second processes yield items which we call A, and the third and fourth yield items B. The inputs for each of these processes are labor measured in man-weeks, pounds of raw material w, and boxes of raw material ω. Since each process has different input requirements, the profits yielded by the various processes differ, even for processes producing the same item. The manufacturer, in deciding on a week's production schedule, is limited in the range of possibilities by the available amounts of the two raw materials and of manpower. Table 6.1 describes the situation.

TABLE 6.1

	One item of A		One item of B		Availabilities
	Process 1	Process 2	Process 3	Process 4	
Man-weeks	1	1	1	1	15
Pounds of material w	7	5	3	2	120
Boxes of material ω	3	5	10	15	100
Unit profit \$	6	5.5	9	8	
Production levels	x_1	x_2	x_3	x_4	

Every production plan $\langle x_1, x_2, x_3, x_4 \rangle$ must obey the availability restrictions

$$\begin{aligned} x_1 + x_2 + x_3 + x_4 &\leq 15, \\ 7x_1 + 5x_2 + 3x_3 + 2x_4 &\leq 120, \\ 3x_1 + 5x_2 + 10x_3 + 15x_4 &\leq 100. \end{aligned} \tag{6.3.1}$$

Furthermore, we do not allow "negative production"; i.e., we require

$$x_1 \geq 0, \qquad x_2 \geq 0, \qquad x_3 \geq 0, \qquad x_4 \geq 0. \tag{6.3.2}$$

Our problem is to find a set of values for the unknown x's which satisfy the relations (6.3.1) and (6.3.2), and which maximize total profit r_0, where

$$r_0 = 6x_1 + 5.5x_2 + 9x_3 + 8x_4. \tag{6.3.3}$$

If we explicitly include unused materials or idle man-weeks as variables—

$$\begin{aligned} x_5 &= \text{slack (unused) man-weeks}, \\ x_6 &= \text{slack (unused) material } w, \\ x_7 &= \text{slack (unused) material } \omega \end{aligned} \tag{6.3.4}$$

—where again

$$x_5 \geq 0, \qquad x_6 \geq 0, \qquad x_7 \geq 0, \tag{6.3.5}$$

then we can convert the relations in (6.3.1) to equalities:

$$x_1 + x_2 + x_3 + x_4 + x_5 = 15,$$
$$7x_1 + 5x_2 + 3x_3 + 2x_4 + x_6 = 120,$$
$$3x_1 + 5x_2 + 10x_3 + 15x_4 + x_7 = 100,$$

without altering the profit function (6.3.3).

The various simplex iterations are described in Tables 6.2, 6.3, and 6.4, the starting basis (Table 6.2) consisting of the vectors corresponding to the slack variables. Note that the components of **z** are the dot products of the columns of the "tableau" with the column **c** on the left. The components of **λ** appear under $\mathbf{p}_0$.

TABLE 6.2

c			6	5.5	9	8	0	0	0
	Basis	$\mathbf{p}_0$	$\mathbf{p}_1$	$\mathbf{p}_2$	$\mathbf{p}_3$	$\mathbf{p}_4$	$\mathbf{p}_5$	$\mathbf{p}_6$	$\mathbf{p}_7$
0	$\mathbf{p}_5$	15	1	1	1	1	1	0	0
0	$\mathbf{p}_6$	120	7	5	3	2	0	1	0
0	$\mathbf{p}_7$	100	3	5	10	15	0	0	1
	z	0	0	0	0	0	0	0	0
	z − **c**		−6	−5.5	−9	−8	0	0	0

The largest negative component of **z** − **c** is −9 and the smallest ratio of λ_i^0/x_{ik_0} is 100/10. Hence vector $\mathbf{p}_3$ replaces $\mathbf{p}_7$ for the next basis (Table 6.3).

TABLE 6.3

c	Basis	$\mathbf{p}_0$	$\mathbf{p}_1$	$\mathbf{p}_2$	$\mathbf{p}_3$	$\mathbf{p}_4$	$\mathbf{p}_5$	$\mathbf{p}_6$	$\mathbf{p}_7$
0	$\mathbf{p}_5$	5	0.7	0.5	0	−0.5	1	0	−0.1
0	$\mathbf{p}_6$	90	6.1	3.5	0	−2.5	0	1	−0.3
9	$\mathbf{p}_3$	10	0.3	0.5	1	1.5	0	0	0.1
	z		2.7	4.5	9	13.5	0	0	0.9
	z − **c**		−3.3	−1	0	5.5	0	0	0.9

Here the largest negative component of **z** − **c** is −3.3 and the smallest ratio is 5/0.7. Hence vector $\mathbf{p}_1$ replaces $\mathbf{p}_5$.

TABLE 6.4

c	Basis	$\mathbf{p}_0$	$\mathbf{p}_1$	$\mathbf{p}_2$	$\mathbf{p}_3$	$\mathbf{p}_4$	$\mathbf{p}_5$	$\mathbf{p}_6$	$\mathbf{p}_7$
6	$\mathbf{p}_1$	7.143	1	0.714	0	−0.714	1.429	0	−0.143
0	$\mathbf{p}_6$	46.868	0	−0.855	0	1.855	8.717	1	0.572
9	$\mathbf{p}_3$	7.857	0	0.288	1	1.714	−0.429	0	0.143
$\mathbf{z}$			6	6.886	9	11.142	4.713	0	0.429
$\mathbf{z} - \mathbf{c}$			0	1.386	0	3.142	4.713	0	0.429

Since $z_i - c_i$ is nonnegative for all i (Table 6.4), we have secured a solution:

$$x_1^0 = 7.143, \qquad x_3^0 = 7.857, \qquad x_6^0 = 46.868.$$

The optimal production plan consistent with the constraints calls for 7.143 units of item A using process 1, and 7.857 units of item B using process 3. The total profit earned under this plan is

$$(7.143)6 + (7.857)9 = 113.571.$$

All available man-weeks and all available material ω are used in this plan, but since $x_6^0 > 0$, some pounds of material w are not used. In this example it can be shown that the solution is unique.

***6.4 Computation of network flow.** In this section we describe a special algorithm for the solution of maximal network flow problems (see Sections 5.9 and 5.10) in which there are capacity constraints on arcs only.

Consider a network consisting of nodes p_i $(i = 1, \ldots, n)$ and arcs p_ip_j (not necessarily directed) joining certain pairs of nodes. Let p_1 be the source, p_n the sink, and p_i $(i = 2, \ldots, n - 1)$ intermediate nodes. Associated with each arc p_ip_j is a nonnegative capacity c_{ij} representing an upper bound on flow from p_i to p_j. (Throughout what follows we have assumed $c_{ii} = 0$.) If x_{ij} is the flow from p_i to p_j, and v denotes the net flow leaving p_1 (the flow value), the problem may be stated as follows: Maximize v subject to

$$\sum_j (x_{1j} - x_{j1}) = v,$$

$$\sum_j (x_{ij} - x_{ji}) = 0 \qquad (i = 2, \ldots, n - 1), \tag{6.4.1}$$

$$\sum_j (x_{nj} - x_{jn}) = -v,$$

and

$$0 \leq x_{ij} \leq c_{ij}. \tag{6.4.2}$$

The constraints (6.4.1) state that the net flow leaving the source (or entering the sink) is v, and that the flow into any intermediate node is equal to the flow out; (6.4.2) states that the flow in any arc does not exceed its capacity. Arcs not present in the network may be assumed to have zero capacity, or alternatively we may think of the corresponding variables x_{ij} as suppressed.

This is, of course, a linear programming problem, which means that it may be solved by the simplex method. In fact, it can be shown that the simplex computation for this problem always involves solving triangular systems of equations, and hence is very efficient. In this section, however, we present an even simpler and faster algorithm for the solution of maximal flow problems. Like the simplex method, it yields solutions to both the primal and dual problems.

Before describing the computation in general terms, let us consider a simple example. Figure 6.1 is a network with source p_1, sink p_4, and arc capacities as indicated. We ask whether the flow shown in Fig. 6.2, namely, $x_{13} = x_{32} = x_{24} = 1$, $x_{12} = x_{23} = x_{34} = 0$ (with value $v =$ 1), is maximal. (In each arc the first number is the capacity and the second is the actual flow.) The answer is no, since we may, for example, increase x_{12}, x_{23}, and x_{34} each by one unit, yielding the flow shown in Fig. 6.3, which has value $v = 2$.

Is this new flow maximal? Again the answer is no, since we may increase x_{12}, *decrease* x_{32}, and increase x_{34}, each by one unit, obtaining the flow shown in Fig. 6.4, which has value $v = 3$.

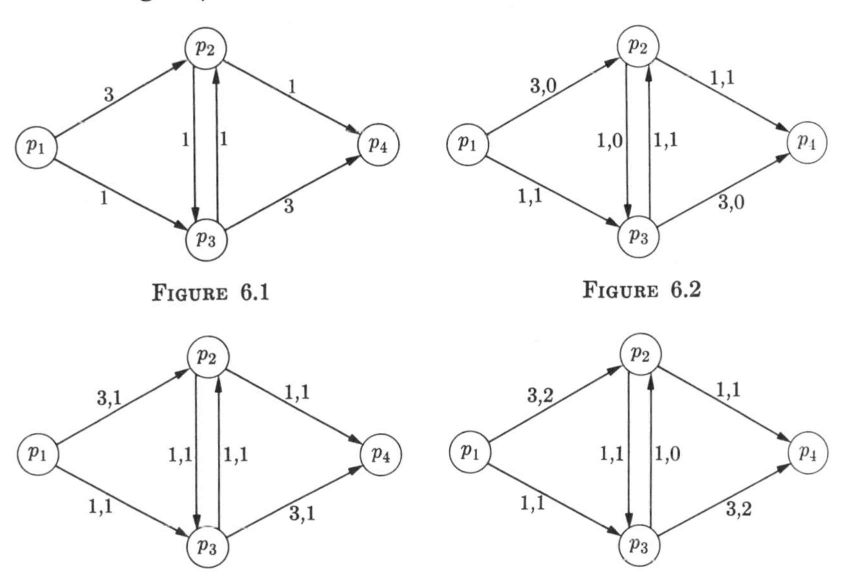

FIGURE 6.1

FIGURE 6.2

FIGURE 6.3

FIGURE 6.4

The flow is now maximal, since the cut consisting of arcs p_1p_3, p_2p_3, and p_2p_4 also has value $v = 3$.

Notice that each of the flow improvements in the example was accomplished by altering a sequence of variables x_{ij} along a "path" from source to sink, by either adding or subtracting a positive constant from each arc flow along the path. The algorithm which we now describe presents a systematic and efficient method of searching for flow improvements of this kind. If no such improvement exists, it will be shown that a maximal flow has been attained.

To ensure termination of the computation, we make the assumption that all the arc capacities are nonnegative integers. Since a problem with rational c_{ij} can be converted to one with integral c_{ij} simply by clearing fractions, this is not, computationally speaking, a drastic restriction.

The computation may start with any integral flow $\mathbf{x}$ (for example, the zero flow). For some of the nodes p_i, we define labels recursively as follows:

Step (a): Start by considering p_1 as labeled and unscanned.

Step (b): Take any labeled, unscanned node p_i, and scan p_i for all unlabeled nodes p_j such that $x_{ij} < c_{ij}$. Label such nodes p_i^+. Next scan p_i for all unlabeled nodes p_j such that $x_{ji} > 0$. Label these nodes p_i^-. Consider p_i as scanned.

Step (b) is repeated either until the sink p_n is labeled, or until no further labeling is possible and p_n is unlabeled. In the latter case, the computation terminates; in the former case, the flow $\mathbf{x}$ may be improved along a path from p_1 to p_n as follows. If p_n has been labeled p_j^+, tentatively add $\epsilon > 0$ to x_{jn}; if p_n has been labeled p_j^-, subtract ϵ from x_{nj}. Next look at the label on p_j. If this is p_i^+, add ϵ to x_{ij} (if p_i^-, subtract ϵ from x_{ji}) and proceed to p_i and its label. Continue in this fashion until the source has been reached. Since for any ϵ, the constraints (6.4.1) remain satisfied with a new flow value $v + \epsilon$, we take ϵ as large as possible consistent with (6.4.2). Clearly ϵ, being the minimum of certain positive integers, is itself a positive integer. Thus we arrive at a new feasible flow vector with a flow value in excess of the preceding by at least one unit.

The algorithm is then repeated with this new flow as a starting point and we obtain a new labeled process. If p_n is labeled, we may again construct a feasible flow vector with total flow value at least one unit more. Obviously, after a finite number of iterations the procedure must terminate with a flow whose corresponding labeled process does not include p_n as a labeled node.

It remains to show that if the labeling process ceases with p_n unlabeled, the flow $\mathbf{x}$ is maximal. Let S be the set of indices i for which p_i is labeled. Thus $1 \in S$, $n \notin S$. Now sum equations (6.4.1) over $i \in S$, obtaining

$$v = \sum_{\substack{i \in S \\ j \notin S}} (x_{ij} - x_{ji}),$$

the terms $j \in S$ having canceled out in summation. But $i \in S$, $j \notin S$ imply $x_{ij} = c_{ij}$ and $x_{ji} = 0$, for otherwise p_j would have received a label. Hence

$$v = \sum_{\substack{i \in S \\ j \notin S}} c_{ij}.$$

The set of arcs leading from S to its complement is a cut; otherwise there must exist a flow path from p_1 to p_n all of whose nodes belong to S, from which it follows that p_n is labeled, contrary to hypothesis. By Theorem 5.10.1 the flow $\mathbf{x}$ is maximal and the cut is minimal.

We conclude this section by indicating a type of problem which has constraining equations very similar to those of the transportation problem, but which, in fact, can also be identified as a network flow problem: viz., find $\mathbf{X} = \|x_{ij}\|$ which maximizes $\sum_{i,j} x_{ij}$ where

$$\sum_{j=1}^{m} x_{ij} \leq a_i, \qquad \sum_{i=1}^{m} x_{ij} \leq b_j, \tag{6.4.3}$$

and $x_{ij} = 0$ for a prescribed set Ω of pairs (i, j). To see this as a flow problem, set up the directed network of $m + n + 2$ nodes as shown in Fig. 6.5.

The capacity on the directed arc P_iQ_j is zero if $i, j \in \Omega$, and large otherwise; the capacities of the source and sink arcs are the a_i and b_j as shown, and x_{ij} is interpreted as the flow from P_i to Q_j.

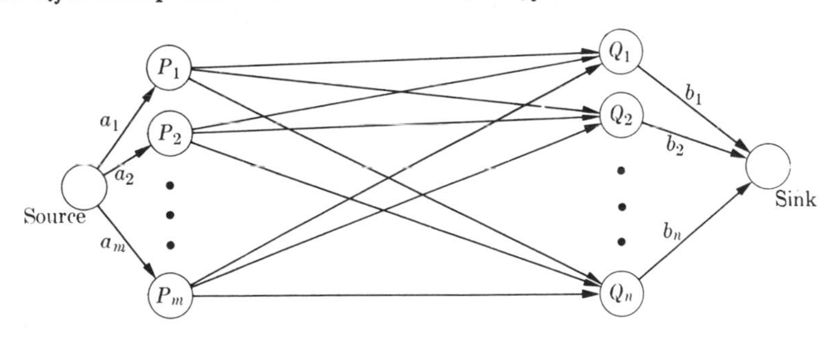

FIGURE 6.5

6.5 A method of approximating the value of a game. The computational methods discussed in the previous sections dealt directly with linear programming problems. The method described in this section provides a means of approximating the value of a game. This method enables us to construct a sequence of strategies for each player such that all limit strategies secured from this sequence are optimal strategies for the game. In particular, if both players have unique optimal strategies, these optimal strategies may be approximated with any prescribed degree of accuracy. If a linear programming problem and its dual problem have unique solutions, then by considering the equivalent game problem we can obtain another iterative scheme for solving the problem, and vice versa.

The method of this section is founded on the following intuitive considerations. Suppose that the game is played repeatedly, and that on each successive playing each player chooses a strategy on the assumption that his opponent's future actions will resemble the past. If what each player did on the first playing is known, this principle for selecting a strategy leads to a determinate sequence of plays, except for a slight arbitrariness which arises if several alternatives happen at an intermediate play (in this case any choice among the alternatives can be made without invalidating the convergence properties established below). From this sequence one can calculate upper and lower bounds for the value of the game and approximate the optimal strategy for each player.

The formal description of the iterations follows.

Let $\mathbf{A} = \|a_{ij}\|$ be the pay-off matrix. Let $\mathbf{a}_i$ denote the ith row and $\mathbf{b}_j$ the jth column of $\mathbf{A}$.

Assume that for the first play of the game Player II chooses a pure strategy corresponding to some column of $\mathbf{A}$ which we shall denote by $\mathbf{c}(1)$. In order to maximize his gain, Player I must then choose the pure strategy corresponding to the row index for which the maximum of $\mathbf{c}(1)$ is achieved. We denote this row of $\mathbf{A}$ by $\mathbf{r}(1)$. Thus, if $\mathbf{c}(1)$ and $\mathbf{r}(1)$ are the j_0th column and the i_0th row, respectively, we have

$$a_{i_0j_0} = \max_k a_{kj_0}, \qquad \text{or} \qquad c_{i_0}(1) = \max_k c_k(1).$$

Turning to the second play of the game, if Player I chooses the strategy whose yields are the components of the row $\mathbf{r}(1)$, Player II might counter with the pure strategy j_1, where

$$a_{i_0j_1} = \min_j a_{i_0j} = \min_j r_j(1).$$

However, for continuity we suppose that both players modify their past choice of a strategy only in the light of their experience. Actually, each plays the pure strategy that is optimal in terms of his total experience with his opponent, assigning equal weights to his opponent's previous actions.

Specifically, suppose that Player II chooses the mixed strategy $\langle \frac{1}{2}, \frac{1}{2} \rangle$ on j_0 and j_1.

Let $\mathbf{c}(2) = \mathbf{c}(1) + \mathbf{b}_{j_1}$.

Now Player I tries to maximize his gain again. He proceeds to determine $\max_i \frac{1}{2}[c_i(1) + b_{ij_1}]$, or equivalently $\max_i c_i(2)$. If this maximum occurs at i_1, he plays the average of $\mathbf{r}(1)$ and $\mathbf{a}_{i_1}$.

Let $\mathbf{r}(2) = \mathbf{r}(1) + \mathbf{a}_{i_1}$.

We continue this fictitious style of play in which each player in turn responds optimally to his opponent's previous plays. Next Player II finds a column index j_2 at which $\min_j r_j(2)$ is achieved and plays the mixed strategy $\langle \frac{1}{3}, \frac{1}{3}, \frac{1}{3} \rangle$ on j_0, j_1, j_2.

Let $\mathbf{c}(3) = \mathbf{c}(2) + \mathbf{b}_{j_2}$.

In answer to this strategy, Player I selects the largest component of $\frac{1}{3}\mathbf{c}(3)$. If the maximum occurs at i_2, he plays $\langle \frac{1}{3}, \frac{1}{3}, \frac{1}{3} \rangle$ on i_0, i_1, i_2, and so on. We thus obtain two sequences of vectors:

$$\mathbf{c}(k) = \mathbf{c}(k-1) + \mathbf{b}_{j_{k-1}}, \quad \text{where} \quad r_{j_{k-1}}(k-1) = \min_j r_j(k-1),$$

and

$$\mathbf{r}(k) = \mathbf{r}(k-1) + \mathbf{a}_{i_{k-1}}, \quad \text{where} \quad c_{i_k}(k) = \max_i c_i(k).$$

We shall demonstrate in the next section that the maximum and minimum pay-offs corresponding to the strategies $\mathbf{c}(n)/n$, $\mathbf{r}(n)/n$ converge to v, i.e., that

$$\lim_n \left[\max_i \frac{c_i(n)}{n} \right] = \lim_n \left[\min_j \frac{r_j(n)}{n} \right] = v.$$

First, however, we shall consider briefly the analogous computing scheme formulated directly in terms of mixed strategies.

We assume that Player II starts with some mixed strategy $\mathbf{y}_1$. Player I then chooses $\mathbf{x}_1$, where

$$\max_x (\mathbf{x}, \mathbf{A}\mathbf{y}_1) = (\mathbf{x}_1, \mathbf{A}\mathbf{y}_1).$$

Player II next chooses $\mathbf{y}_2$, where

$$\min_y (\mathbf{x}_1, \mathbf{A}\mathbf{y}) = (\mathbf{x}_1, \mathbf{A}\mathbf{y}_2).$$

Assume that Player II plays the average $\frac{1}{2}(\mathbf{y}_1 + \mathbf{y}_2)$ on this second round; Player I then selects $\mathbf{x}_2$, where

$$\max_x \left(\mathbf{x}, \mathbf{A}\left(\frac{\mathbf{y}_1 + \mathbf{y}_2}{2} \right) \right) = \left(\mathbf{x}_2, \mathbf{A}\left(\frac{\mathbf{y}_1 + \mathbf{y}_2}{2} \right) \right).$$

Each player continues to play the average of all his choices, and the opposing player tries to improve his chances against this average. For instance, if $\mathbf{x}_n$ is Player I's best choice against

$$\frac{1}{n} [\mathbf{y}_1 + \cdots + \mathbf{y}_n],$$

he will choose the strategy

$$\frac{1}{n}[\mathbf{x}_1 + \cdots + \mathbf{x}_n];$$

then Player II will play

$$\frac{1}{n+1}[\mathbf{y}_1 + \cdots + \mathbf{y}_{n+1}],$$

etc. It can be established that

$$\lim_n \left[\max_x \left(\mathbf{x}, \frac{1}{n}\mathbf{A}(\mathbf{y}_1 + \cdots + \mathbf{y}_n)\right)\right]$$
$$= \lim_n \left[\min_y \left(\frac{1}{n}(\mathbf{x}_1 + \cdots + \mathbf{x}_n), \mathbf{A}\mathbf{y}\right)\right] = v, \quad (6.5.1)$$

and that any limit point of

$$\frac{1}{n}\sum_{i=1}^{n} \mathbf{x}_i \quad \text{or} \quad \frac{1}{n}\sum_{i=1}^{n} \mathbf{y}_i$$

is an optimal strategy for the game. We shall prove the convergence of the scheme dealing with columns and rows, but a modification of our proof yields (6.5.1).

The following example illustrates the method. Consider the game defined by the pay-off matrix

$$\mathbf{A} = \begin{Vmatrix} 2 & 1 & 0 \\ 2 & 0 & 3 \\ -1 & 3 & -3 \end{Vmatrix}.$$

Suppose that Player I begins with i_0, Player II selects j_0, Player I comes back with i_1, etc., each player making alternate choices according to the scheme described above. Let

$$\underline{v}(N) = \frac{1}{N}\min_j r_j(N), \qquad \bar{v}(N) = \frac{1}{N}\max_i c_i(N).$$

The figures are tabulated in Table 6.5. The value of the game is 1.

***6.6 Proof of the convergence.** A sequence of vectors $\mathbf{c}(k)$ and $\mathbf{r}(k)$, of dimension n and m, respectively, is said to form a "vector system" with respect to an $n \times m$ matrix $\mathbf{A}$ if

(1) $\max_i c_i(0) = \min_j r_j(0)$;
(2) where $\max_i c_i(k)$ is achieved for the νth component,

$$\mathbf{r}(k+1) = \mathbf{r}(k) + \mathbf{a}_\nu;$$

TABLE 6.5

N	i_N	$r_1(N)$	$r_2(N)$	$r_3(N)$	$\underline{v}(N)$	j_N	$c_1(N)$	$c_2(N)$	$c_3(N)$	$\bar{v}(N)$
1	1	2	1	$\underline{0}$	0.000	3	0	3	−3	3.000
2	2	4	$\underline{1}$	3	0.500	2	1	3	0	1.500
3	2	6	$\underline{1}$	6	0.333	2	2	3	6	1.000
4	2	8	1	9	0.250	2	3	3	6	1.500
5	3	7	4	6	0.800	2	4	3	9	1.800
6	3	8	7	3	0.500	3	4	6	6	1.000
7	2	8	7	6	0.857	3	4	9	3	1.286
8	2	10	7	9	0.875	2	5	9	6	1.125
9	2	12	7	12	0.778	2	6	9	9	1.000
10	2	14	7	15	0.700	2	7	9	12	1.200
11	3	13	10	12	0.909	2	8	9	15	1.364
12	3	12	13	9	0.750	3	8	12	12	1.000
13	2	14	13	12	0.923	3	8	15	9	1.154
14	2	16	13	15	0.929	2	9	15	12	1.071
15	2	18	13	18	0.867	2	10	15	15	1.000
16	2	20	13	21	0.812	2	11	15	18	1.125
17	3	19	16	18	0.941	2	12	15	21	1.235
18	3	18	19	15	0.833	3	12	18	18	1.000

(3) where $\min_j r_j(k)$ is achieved for the μth component,

$$\mathbf{c}(k) = \mathbf{c}(k-1) + \mathbf{b}_\mu$$

($\mathbf{a}_\nu$ denotes the νth row and $\mathbf{b}_\mu$ the μth column of $\mathbf{A}$).

We can readily see that the vectors $\mathbf{c}(k)$ and $\mathbf{r}(k)$ of the scheme introduced in Section 6.5 form a vector system if $\mathbf{c}(0) \equiv \mathbf{r}(0) \equiv \mathbf{0}$.

▶THEOREM 6.6.1. For any vector system,

$$\lim_n \left[\max_i \frac{c_i(n)}{n}\right] = \lim_n \left[\min_j \frac{r_j(n)}{n}\right] = v.$$

It is conjectured that the process converges at a rate $1/\sqrt{k}$, where k is the number of iterations. In actual cases, it is found that the process is far more efficient than is expected theoretically.

▶LEMMA 6.6.1. For any vector system,

$$\underline{\lim}_{T\to\infty} \left[\frac{\max \mathbf{c}(T) - \min \mathbf{r}(T)}{T}\right] \geq 0.$$

By $\max \mathbf{c}(T)$ we mean the maximum component of the vector $\mathbf{c}(T)$.

Proof. Special case. $\mathbf{c}(0) = \mathbf{r}(0) = \mathbf{0}$. Then

$$\max \frac{\mathbf{c}(T)}{T} \geq v \geq \min \frac{\mathbf{r}(T)}{T},$$

and this implies the lemma.

General case. Let n_i, m_j denote the number of times that $\mathbf{a}_i$, $\mathbf{b}_j$ are added to $\mathbf{r}(0)$, $\mathbf{c}(0)$ in forming $\mathbf{r}(T)$, $\mathbf{c}(T)$, respectively. Then

$$\sum n_i = \sum m_j = T,$$

and

$$x_i = \frac{n_i}{T}, \qquad y_j = \frac{m_j}{T}$$

are strategies for the game. We observe that

$$\frac{c_i(T)}{T} = \frac{c_i(0)}{T} + \sum_j a_{ij} \frac{m_j}{T} = \frac{c_i(0)}{T} + (\mathbf{Ay})_i,$$

$$\frac{r_j(T)}{T} = \frac{r_j(0)}{T} + \sum_i a_{ij} \frac{n_i}{T} = \frac{r_j(0)}{T} + (\mathbf{xA})_j,$$

and

$$\frac{\max \mathbf{c}(T) - \min \mathbf{r}(T)}{T} = \max_i \frac{\mathbf{c}(0)}{T} - \min_j \frac{\mathbf{r}(0)}{T} + \max_i (\mathbf{Ay})_i - \min_j (\mathbf{xA})_j. \tag{6.6.1}$$

But

$$\lim_T \max_i \frac{c_i(0)}{T} = \lim_T \min_j \frac{r_j(0)}{T} = 0. \tag{6.6.2}$$

Also, by the min-max theorem and Lemma 1.4.1,

$$v = \min_y \max_x (\mathbf{x}, \mathbf{Ay}) = \min_y \max_i (\mathbf{Ay})_i \leq \max \mathbf{Ay}.$$

Similarly,

$$v \geq \min_j (\mathbf{xA})_j \rightarrow \max \mathbf{Ay} - \min \mathbf{xA} \geq 0. \tag{6.6.3}$$

Relations (6.6.1), (6.6.2), and (6.6.3) prove the lemma.

DEFINITION. A row $\mathbf{a}_\nu$ of $\mathbf{A}$ is said to be eligible in the interval $(M, M + T)$ if there exists R $(M \leq R \leq M + T)$ such that

$$\max \mathbf{c}(R) = c_\nu(R).$$

A column $\mathbf{b}_\mu$ is eligible if

$$\min \mathbf{r}(R) = r_\mu(R).$$

This definition does not imply that $\mathbf{a}_\nu$, $\mathbf{b}_\mu$ are actually used in $\mathbf{r}(R)$, $\mathbf{c}(R)$.

▶Lemma 6.6.2. If all rows and columns of $\mathbf{A}$ are eligible in $(M, M + T)$, then

$$\max \mathbf{c}(M + T) - \min \mathbf{c}(M + T) \leq 2aT \tag{6.6.4a}$$

and

$$\max \mathbf{r}(M + T) - \min \mathbf{r}(M + T) \leq 2aT, \tag{6.6.4b}$$

where

$$a = \max_{i,j} |a_{ij}|.$$

Proof. We shall prove only the first inequality. Let ν denote any row for which $\min \mathbf{c}(M + T)$ is achieved. Since $\mathbf{a}_\nu$ is eligible, there exists R such that

$$\max \mathbf{c}(R) = c_\nu(R). \tag{6.6.5}$$

Also, since $M \leq R \leq M + T$, we may obtain $\mathbf{c}(M + T)$ by adding at most T columns to $\mathbf{c}(R)$, recognizing that every time the components of the added vectors are bounded by a. Hence

$$c_j(M + T) \geq c_j(R) - aT, \qquad c_j(M + T) \leq c_j(R) + aT.$$

In particular,

$$\min \mathbf{c}(M + T) = c_\nu(M + T) \geq c_\nu(R) - aT. \tag{6.6.6}$$

Suppose ν_0 is such that $\max \mathbf{c}(M + T) = c_{\nu_0}(M + T)$. Then by equation (6.6.5)

$$c_{\nu_0}(M + T) \leq c_{\nu_0}(R) + aT \leq c_\nu(R) + aT. \tag{6.6.7}$$

Equations (6.6.6) and (6.6.7) imply

$$\max \mathbf{c}(M + T) - \min \mathbf{c}(M + T) \leq 2aT.$$

▶Lemma 6.6.3. If all rows and columns of $\mathbf{A}$ are eligible in $(M, M + T)$, then

$$\max \mathbf{c}(M + T) - \min \mathbf{r}(M + T) \leq 4aT. \tag{6.6.8}$$

Proof.

$$\begin{aligned}\max \mathbf{c}(M + T) - \min \mathbf{r}(M + T) &= [\max \mathbf{c}(M + T) - \min \mathbf{c}(M + T)] \\ &\quad + [\min \mathbf{c}(M + T) - \max \mathbf{r}(M + T)] \\ &\quad + [\max \mathbf{r}(M + T) - \min \mathbf{r}(M + T)] \\ &\leq 4aT + [\min \mathbf{c}(M + T) - \max \mathbf{r}(M + T)].\end{aligned}$$

It remains to show that the last term is negative.

Let k_i, l_j denote the number of times that $\mathbf{b}_i$, $\mathbf{a}_j$ are added to $\mathbf{c}(0)$, $\mathbf{r}(0)$ in order to form $\mathbf{c}(M + T)$ and $\mathbf{r}(M + T)$, respectively.

As in Lemma 6.6.1, we see that

$$\frac{1}{M + T} \Sigma k_i = \frac{1}{M + T} \Sigma l_j = 1.$$

Let $\mathbf{B} = \mathbf{A}'$ and thus $b_{ij} = a_{ji}$; let v^* denote the value of the game with matrix $\mathbf{B}$. Then

$$(M + T)\mathbf{x} = \langle k_1, \ldots, k_m \rangle, \qquad (M + T)\mathbf{y} = \langle l_1, \ldots, l_n \rangle$$

are strategies for Players I and II, respectively, of the game with matrix $\mathbf{B}$. Furthermore,

$$\frac{c_j(M + T)}{M + T} = \frac{c_j(0)}{M + T} + \sum_i b_{ij} \frac{k_i}{M + T} = \frac{c_j(0)}{M + T} + (\mathbf{xB})_j$$

and

$$\frac{r_i(M + T)}{M + T} = \frac{r_i(0)}{M + T} + \sum_j b_{ij} \frac{l_j}{M + T} = \frac{r_i(0)}{M + T} + (\mathbf{By})_i.$$

Then

$$\frac{\min \mathbf{c}(M + T) - \max \mathbf{r}(M + T)}{M + T} = \frac{\min \mathbf{c}(0) - \max \mathbf{r}(0)}{M + T} + \min \mathbf{xB} - \max \mathbf{By}.$$

Also, $\min \mathbf{c}(0) = \max \mathbf{r}(0)$ by definition of a vector system. Finally, arguing as in Lemma 6.6.1, we find that

$$\min \mathbf{xB} \leq v^* \leq \max \mathbf{By},$$

or

$$\min \mathbf{xB} - \max \mathbf{By} \leq 0.$$

Hence

$$\frac{\min \mathbf{c}(M + T) - \max \mathbf{r}(M + T)}{M + T} \leq 0.$$

This completes the proof of the lemma.

▶ LEMMA 6.6.4. For every $\epsilon > 0$, there exists $T^*(\epsilon)$ such that if $T > T^*$, then

$$\frac{\max \mathbf{c}(T) - \min \mathbf{r}(T)}{T} < \epsilon$$

for every vector system associated with a fixed matrix $\mathbf{A}$. The value of $T^*(\epsilon)$ is independent of the vector system.

Proof. We shall proceed by induction. The lemma is clearly valid for 1×1 matrices, since $r(0) = c(0)$ implies $r(n) \equiv c(n)$ for all n. We shall assume that the lemma is valid for all submatrices of $\mathbf{A}$ and then show that it is true for $\mathbf{A}$. Since the number of submatrices is finite, we may assume that $T^*(\epsilon/2)$ is the same for all proper submatrices of $\mathbf{A}$ for all vector systems. In the remainder of the proof T^* will denote this number.

SUB-LEMMA. If in the interval $(M, M + T^*)$ the νth row is ineligible, then

$$\max \mathbf{c}(M + T^*) - \min \mathbf{r}(M + T^*) \leq \max \mathbf{c}(M) - \min \mathbf{r}(M) + \frac{\epsilon}{2} T^*, \tag{6.6.9}$$

where $T^* = T^*(\epsilon/2)$. The same result holds if the μth column is ineligible.

Proof of the Sub-lemma. Define the new vector system:

$$\left.\begin{aligned} \bar{\mathbf{r}}(k) &= \mathbf{r}(M + k) + [\max \mathbf{c}(M) - \min \mathbf{r}(M)]\mathbf{u}, \\ &\qquad \text{where } \mathbf{u} = \langle 1, 1, \ldots, 1 \rangle \\ \bar{\mathbf{c}}(k) &= \text{projection of } \mathbf{c}(M + k) \text{ obtained by} \\ &\qquad \text{deleting the } \nu\text{th row} \end{aligned}\right\} \quad (k \leq T^*).$$

For $k > T^*$, use rules (2) and (3) of the definition of a vector system (page 182) to continue the system with respect to the submatrix obtained from $\mathbf{A}$ by deleting the νth row.

We can verify that $\bar{\mathbf{r}}(k)$ and $\bar{\mathbf{c}}(k)$ form a vector system. Rule (1) is satisfied as follows:

$$\begin{aligned} \min \bar{\mathbf{r}}(0) &= \min \mathbf{r}(M) + \max \mathbf{c}(M) - \min \mathbf{r}(M) \\ &= \max \mathbf{c}(M) = \max \bar{\mathbf{c}}(0), \end{aligned}$$

since the νth row is ineligible. Rules (2) and (3) are satisfied for $k \leq T^*$ because the νth row is ineligible, and for $k > T^*$ by construction.

Now we can prove (6.6.9). Since $\bar{\mathbf{r}}(k)$ and $\bar{\mathbf{c}}(k)$ are defined over a submatrix of $\mathbf{A}$, namely the submatrix obtained by deleting the νth row, we have by the induction hypothesis

$$\frac{\max \bar{\mathbf{c}}(T^*) - \min \bar{\mathbf{r}}(T^*)}{T^*} < \frac{\epsilon}{2}, \tag{6.6.10}$$

which is equivalent to (6.6.9).

Proof of the Lemma. Let $T = (\theta + q)T^*$ ($0 \leq \theta < 1$; q and T integers, with the value of q to be specified later). Let

$$I_s = [(\theta + s - 1)T^*,\ (\theta + s)T^*] \qquad (1 \leq s \leq q).$$

Case 1. There exists an s such that all rows and columns of $\mathbf{A}$ are eligible in I_s.

Let s' be the largest such s. If $s' \leq q - 1$, we can apply the sub-lemma to the interval $(M, M + T^*)$ with $M = (\theta + q - 1)T^*$ and $M + T^* = T$; then

$$\begin{aligned} \max \mathbf{c}(T) - \min \mathbf{r}(T) \leq{} & \max \mathbf{c}[(\theta + q - 1)T^*] \\ & - \min \mathbf{r}[(\theta + q - 1)T^*] + \frac{\epsilon}{2} T^*. \end{aligned} \tag{6.6.11}$$

If $s' \leq q - 2$, we can repeat the inequality by applying the sub-lemma to the right side of (6.6.11). Then

$$\begin{aligned} \max \mathbf{c}(T) - \min \mathbf{r}(T) \leq{} & \max \mathbf{c}[(\theta + q - 2)T^*] \\ & - \min \mathbf{r}[(\theta + q - 2)T^*] + 2\frac{\epsilon}{2} T^*. \end{aligned}$$

We continue iterating the inequalities until we reach $I_{s'}$. There the process must stop, since all rows and columns are eligible. Hence

$$\begin{aligned} \max \mathbf{c}(T) - \min \mathbf{r}(T) \leq{} & \max \mathbf{c}[(\theta + s')T^*] \\ & - \min \mathbf{r}[(\theta + s')T^*] + (q - s')\frac{\epsilon}{2} T^* \\ \leq{} & \max \mathbf{c}[(\theta + s')T^*] \\ & - \min \mathbf{r}[(\theta + s')T^*] + q\frac{\epsilon}{2} T^*. \end{aligned} \tag{6.6.12}$$

But in the interval $I_s = [(\theta + s' - 1)T^*,\ (\theta + s')T^*]$ of length T^* all rows and columns are eligible, so that Lemma 6.6.3 may be applied to yield

$$\max \mathbf{c}[(\theta + s')T^*] - \min \mathbf{r}[(\theta + s')T^*] \leq 4aT^*. \tag{6.6.13}$$

On substituting this result in (6.6.12), we obtain

$$\max \mathbf{c}(T) - \min \mathbf{r}(T) \leq 4aT^* + q\frac{\epsilon}{2} T^*.$$

Case 2. There exists no s such that all rows and columns of $\mathbf{A}$ are eligible in I_s. Then the sub-lemma may be applied as in (6.6.11) and the result iterated until $s = 1$. We obtain

$$\begin{aligned}\max \mathbf{c}(T) - \min \mathbf{r}(T) &\leq \max \mathbf{c}(\theta T^*) - \min \mathbf{r}(\theta T^*) + q\frac{\epsilon}{2}T^* \\ &\leq 2\theta a T^* + q\frac{\epsilon}{2}T^* + M,\end{aligned}$$

where $M = \max|\mathbf{c}(0)| + \max|\mathbf{r}(0)|$.

The last inequality holds because $\mathbf{c}(\theta T^*)$ is obtained by adding θT^* columns to $\mathbf{c}(0)$, and we know that $|a_{ij}| \leq a$. Hence

$$|\max \mathbf{c}(\theta T^*)| \leq a\theta T^* + \max|\mathbf{c}(0)|.$$

Similarly,

$$|\min \mathbf{r}(\theta T^*)| \leq a\theta T^* + \max|\mathbf{r}(0)|.$$

Finally, since $\theta < 1$, if we choose $T^* > M/2a$, we obtain

$$\max \mathbf{c}(T) - \min \mathbf{r}(T) \leq 4aT^* + q\frac{\epsilon}{2}T^*. \tag{6.6.14}$$

Clearly, equation (6.6.14) is valid for both cases. Then

$$\frac{\max \mathbf{c}(T) - \min \mathbf{r}(T)}{T} \leq \frac{4a + q(\epsilon/2)}{q} \leq \frac{4a}{q} + \frac{\epsilon}{2}.$$

If we choose q so large that $4a/q < \epsilon/2$, then for all $T > qT^*$

$$\frac{\max \mathbf{c}(T) - \min \mathbf{r}(T)}{T} < \epsilon.$$

Proof of Theorem 6.6.1. By Lemma 6.6.4

$$\overline{\lim}\left[\frac{\max \mathbf{c}(T) - \min \mathbf{r}(T)}{T}\right] \leq 0,$$

and by Lemma 6.6.1

$$\underline{\lim}\left[\frac{\max \mathbf{c}(T) - \min \mathbf{r}(T)}{T}\right] \geq 0.$$

Therefore

$$\lim\left[\frac{\max \mathbf{c}(T) - \min \mathbf{r}(T)}{T}\right] = 0. \tag{6.6.15}$$

But

$$\overline{\lim}\left[\min\frac{\mathbf{r}(T)}{T}\right] \leq v \leq \underline{\lim}\left[\max\frac{\mathbf{c}(T)}{T}\right],$$

showing that each limit in (6.6.15) exists separately; hence

$$\lim_T\left[\min\frac{\mathbf{r}(T)}{T}\right] = \lim_T\left[\max\frac{\mathbf{c}(T)}{T}\right] = v.$$

***6.7 A differential-equations method for determining the value of a game.** The method discussed in Sections 6.5 and 6.6 was discrete in concept and in operation. In contrast, our next approximation method involves continuous alteration of the strategies to take advantage of the opponent's errors. The strategies vary continuously in time and satisfy certain differential equations which express the way each player adjusts his play in reaction to his opponent's play. When solved, these differential equations furnish a one-parameter family $\mathbf{y}(t)$ of strategies for each player, with the property that as time goes to infinity all limit points of $\mathbf{y}(t)$ become optimal.

Unlike the discrete method, this continuous method enables us to establish precise bounds on the rates of convergence. This advantage, however, and the method itself are more of theoretical interest than of practical value, since in most cases the differential equations are difficult to solve.

Since any matrix game may be imbedded in an equivalent symmetric game (see Theorem 2.6.1), we shall consider here the symmetric game only.

Let Γ be a symmetric game with skew-symmetric $n \times n$ matrix $\mathbf{A}$. We consider a one-parameter family of strategies $\mathbf{y}(t)$ as functions of the real variable t. Let us form the following functions of $\mathbf{y}(t)$:

$$u_i[\mathbf{y}(t)] = \sum_{j=1}^{n} a_{ij}y_j(t), \qquad \phi(u_i) = \max(0, u_i),$$
$$\phi(\mathbf{y}) = \sum_{i=1}^{n} \phi(u_i), \qquad \psi(\mathbf{y}) = \sum_{i=1}^{n} \phi^2(u_i). \tag{6.7.1}$$

Let $\mathbf{y}^0$ be any fixed strategy for Γ. We now consider the following system of differential equations.

$$\left.\begin{aligned} \frac{dy_j(t)}{dt} &= \phi[u_j(t)] - \phi[\mathbf{y}(t)]y_j(t), \\ y_j(0) &= y_j^0, \end{aligned}\right\} \qquad (j = 1, \ldots, n). \tag{6.7.2}$$

The reason we choose these particular differential equations is as follows. Suppose that $\phi(u_j) > 0$; then by playing the jth row Player I obtains more than the value of the game, provided Player II plays $\{y_j\}$ (remember $v = 0$). If Player II now changes his strategy by increasing y_j to 1, the return to Player I becomes a_{jj}, which is zero. Hence it is to Player II's advantage to increase y_j whenever $\phi(u_j) > 0$. This situation is accounted for by the first term of the differential equation, which compels y_j to increase whenever $\phi(u_j) > 0$. The second term serves the role of a normalizing factor; it maintains the relations $\sum y_j(t) \equiv 1$, provided $\mathbf{y}^0$ was a strategy to start with.

Since the right side is continuous, it is known from the theory of differential equations that such a system has at least one solution. We shall prove that if $y_j(t)$ is any solution, and $\{t_i\}$ is a sequence converging to infinity, then any limit point of $\mathbf{y}(t_i)$ is an optimal strategy for Player II.

▶ LEMMA 6.7.1. Let $y_j(t)$ $(j = 1, \ldots, n)$ be a solution to the system of equations (6.7.2). Then $\langle y_j(t) \rangle$ is a strategy for any fixed t.

Proof. We prove first that $y_j(t) \geq 0$. Assume that this is false. Then there exists t_1 such that $y_j(t_1) < 0$. Let t_0 be the largest value of t less than t_1 for which $y_j(t_0) \geq 0$. Actually, $y_j(t_0) = 0$. Then if $t_0 < t \leq t_1$, it follows that $y_j(t) < 0$; but $\phi(u_j) \geq 0$ and $\phi(\mathbf{y}) \geq 0$ always. Hence $\phi(u_j) - \phi(\mathbf{y})y_j \geq 0$, and thus $dy_j/dt \geq 0$ by (6.7.2). Applying the mean value theorem, we obtain

$$y_j(t_1) = y_j(t_0) + y_j'(\tau)(t_1 - t_0) \geq y_j(t_0) = 0,$$

which contradicts the assumption $y_j(t_1) < 0$.

We prove next that

$$\sum_{j=1}^{n} y_j(t) = 1.$$

Observe that

$$\frac{d}{dt}\left(1 - \sum_{j=1}^{n} y_j(t)\right) = -\sum_{j=1}^{n} \frac{dy_j}{dt} = -\sum_{j=1}^{n} \phi(u_j) + \phi(\mathbf{y}) \sum_{j=1}^{n} y_j$$

$$= -\phi(\mathbf{y})\left(1 - \sum_{j=1}^{n} y_j(t)\right). \tag{6.7.3}$$

Integrating equation (6.7.3) from 0 to t, taking account of the initial condition, we obtain

$$1 - \sum_{j=1}^{n} y_j(t) = -\int_0^t \phi(\mathbf{y})\left(1 - \sum_{j=1}^{n} y_j(\tau)\right) d\tau.$$

The equation has an explicit integral

$$1 - \sum_{j=1}^{n} y_j(t) = c \exp\left[-\int_0^t \phi(y) dy\right],$$

and the initial conditions imply $c = 0$.

▶ LEMMA 6.7.2.

$$[\psi(\mathbf{y})]^{1/2} \leq \phi(\mathbf{y}) \leq n^{1/2}[\psi(\mathbf{y})]^{1/2}.$$

Proof. The first inequality is immediate, since

$$[\phi(\mathbf{y})]^2 = [\Sigma\phi(u_j)]^2 \geq \Sigma[\phi(u_j)]^2 = \psi(\mathbf{y}).$$

By Schwarz's inequality,

$$\phi(\mathbf{y}) = \sum_{j=1}^{n} 1 \cdot \phi(u_j) \leq \left[\sum_{j=1}^{n} 1\right]^{1/2} \left[\sum_{j=1}^{n} [\phi(u_j)]^2\right]^{1/2} = n^{1/2}[\psi(\mathbf{y})]^{1/2},$$

as desired.

▶THEOREM 6.7.1.

$$\psi[\mathbf{y}(t)] \leq \frac{C_0}{(C_2 + t)^2}; \qquad \phi[\mathbf{y}(t)] \leq \frac{C_1}{C_2 + t}.$$

Proof. If $\phi(u_i) > 0$ for a given t, then

$$\frac{d\phi(u_i)}{dt} = \sum_{j=1}^{n} a_{ij} \frac{dy_j}{dt} = \sum_{j=1}^{n} a_{ij}[\phi(u_j) - \phi(\mathbf{y})y_j]$$

$$= \sum_{j=1}^{n} a_{ij}\phi(u_j) - \phi(\mathbf{y})\phi(u_i).$$

If we multiply both sides by $\phi(u_i)$ and sum, we obtain the following equation, valid even if $\phi(u_i) = 0$:

$$\sum_{i=1}^{n} \phi(u_i) \frac{d\phi(u_i)}{dt} = \sum_{i,j} a_{ij}\phi(u_i)\phi(u_j) - \phi(\mathbf{y}) \sum_{i=1}^{n} \phi^2(u_i)$$

$$= -\phi(\mathbf{y})\psi(\mathbf{y}). \tag{6.7.4}$$

The sum over i, j vanishes because $\mathbf{A}$ is skew-symmetric; that is, $a_{ij} = -a_{ji}$. Equation (6.7.4) is equivalent to

$$\frac{1}{2} \frac{d\psi[\mathbf{y}(t)]}{dt} = -\phi[\mathbf{y}(t)]\psi[\mathbf{y}(t)]. \tag{6.7.5}$$

Therefore $\psi[\mathbf{y}(t)]$ is a decreasing function of t. By Lemma 6.7.2, we know also that it vanishes only when $\phi[\mathbf{y}(t)]$ vanishes, and further that

$$\frac{1}{2} \frac{d\psi}{dt} \leq -\psi^{1/2}\psi.$$

Then

$$\frac{1}{2} \frac{d\psi}{\psi^{3/2}} \leq -dt, \qquad -\psi^{-1/2}[\mathbf{y}(t)] + \psi^{-1/2}[\mathbf{y}(0)] \leq -t.$$

If $C_2 = \psi^{-1/2}[\mathbf{y}(0)]$, we have $\psi^{-1/2}[\mathbf{y}(t)] \geq C_2 + t$ or

$$\psi^{1/2}[\mathbf{y}(t)] \leq \frac{1}{C_2 + t}. \tag{6.7.6}$$

The last equation implies that for every i,

$$\phi(u_i) \leq \frac{n^{1/2}}{C_2 + t}.$$

Since $\phi(u_i) = \max(0, u_i)$, we see that

$$u_i = \sum a_{ij} y_j(t) \leq \frac{n^{1/2}}{C_2 + t}.$$

Hence, if a sequence $\{t_m\} \to \infty$, then $\sum a_{ij} y_j(t_m)$ has a limit superior bounded above by zero, which is the value of the game. Thus any limit point of the sequence $\mathbf{y}(t_m)$ is an optimal strategy for Player II, and also for Player I since the game is symmetric. We note that the yields

$$\sum a_{ij} y_j(t_m)$$

reduce to less than or equal to $v = 0$ at least as fast as $n^{1/2}/(C_2 + t)$.

6.8 Problems.

1. Use the simplex method to maximize

$$2\xi_1 + 4\xi_2 + \xi_3 + \xi_4$$

subject to the conditions

$$\begin{aligned} \xi_1 + 3\xi_2 \qquad\quad + \xi_4 &\leq 4, \\ 2\xi_1 + \xi_2 \qquad\qquad\qquad &\leq 3, \\ \xi_2 + 4\xi_3 + \xi_4 &\leq 3. \end{aligned}$$

2. Three processes are available for producing two final products, corn and hogs. Process A has hogs as a final product but uses corn as an intermediate product (for feeding the hogs); Process B is concerned only with growing corn; and Process C has corn and hogs as joint products. Corn sells for \$20 per unit (ton), and hogs also sell for \$20 per unit (head). Process A uses 100 unit-months of labor and 10 acres of land per unit output; Process B uses 25 unit-months of labor and 50 acres of land per unit output; and Process C uses 50 unit-months of labor and 40 acres of land

per unit output. There are 50 unit-months of labor available and 52.5 acres of land. Assuming constant returns to scale, find the levels at which Processes A, B, and C must be employed in order to achieve maximum profits.

3. Use the simplex method to solve the transportation problem for the following array:

Sources	*Destinations*
$a_1 = 3$	$b_1 = 3$
$a_2 = 4$	$b_2 = 3$
$a_3 = 2$	$b_3 = 6$
$a_4 = 8$	$b_4 = 2$
	$b_5 = 1$
	$b_6 = 2$

Unit shipping costs

	b_1	b_2	b_3	b_4	b_5	b_6
a_1	5	3	7	3	8	5
a_2	5	6	12	5	7	11
a_3	2	8	3	4	8	2
a_4	9	6	10	5	10	9

4. If a system of m linear equations in n nonnegative variables has a solution, then a solution exists in which k variables are positive and $n - k$ are zero, with $k \leq \min(m, n)$.

5. In the warehouse model of Section 5.7, solve explicitly the special case in which $b_i > c_i$, $b_{i+1} > c_i$, and $c_i \geq 0$ for all i.

6. Consider a transportation problem with availabilities at the sources given by $c_i > 0$ $(i = 1, \ldots, n)$ and demands at the destinations given by $b_j > 0$ $(j = 1, \ldots, m)$, such that

$$\sum_{i=1}^{n} c_i = \sum_{j=1}^{m} b_j.$$

Show that the rank of the constraint matrix $\mathbf{A}$ of page 137 is exactly $m + n - 1$.

7. Solve the diet problem

$$\min (\mathbf{c}, \mathbf{x}) \text{ subject to } \mathbf{A}\mathbf{x} \geq \mathbf{b}, \mathbf{x} \geq \mathbf{0},$$

where

$$\mathbf{A} = \begin{Vmatrix} 2 & 1 & 7 & 1 & 0 & 4 \\ 3 & 0 & 2 & 8 & 6 & 4 \\ 1 & 4 & 7 & 3 & 5 & 0 \end{Vmatrix}, \qquad \mathbf{b} = \begin{Vmatrix} 1 \\ 1 \\ 1 \end{Vmatrix},$$

and

$$\mathbf{c} = \langle 2 \quad 3 \quad 4 \quad 1 \quad 2 \quad 3 \rangle.$$

Use the method of artificial variables and then the dual simplex method.

8. Minimize

$$z_1 + z_2 + z_3 + \lambda x,$$

where

$$\begin{aligned} z_1 + z_2 + z_3 + x &\geq 10 \\ z_1 + z_2 \qquad + x &\geq 7 \qquad 0 \leq z_i \leq 5 \\ z_2 + z_3 + x &\geq 9 \qquad\quad x \geq 0, \end{aligned}$$

and λ is a positive parameter.

9. Prove that degeneracy in the simplex computation for the transportation problem can occur only if a partial sum of the c_i equals a partial sum of the b_j.

10. Set up the caterer's problem as a transportation problem.

11. A factory has 100 items on hand which may be shipped to an outlet at the cost of $1 apiece to meet an uncertain demand d which is uniformly distributed between 70 and 80. In the event that the demand exceeds the supply, it is necessary to meet the unsatisfied demand by purchases at the local market at $2 apiece. Set up the equations and find the shipping policy that minimizes expected costs.

12. Use the labeling algorithm to find a maximal flow and a minimal cut for the network diagrammed below. Reverse the arc orientations, interchange source and sink, and locate another minimal cut by labeling.

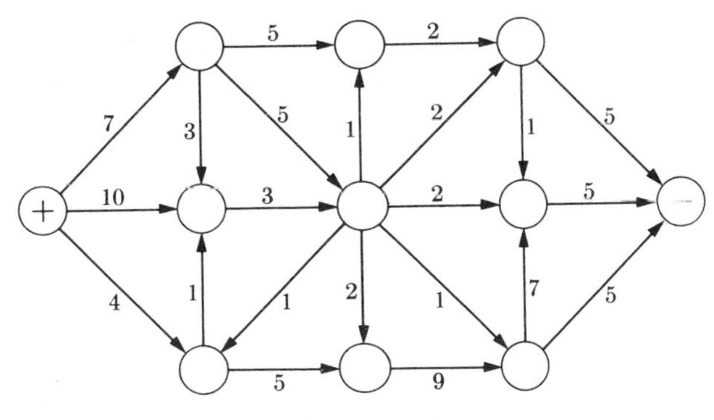

FIGURE 6.6

13. Consider an optimal assignment problem in which there are only two ratings: i.e., man i in job j is either good or bad. Show that such a problem is a maximal network flow problem.

14. Show how a flow problem with capacity constraints on nodes can be converted to one with capacity constraints on arcs only.

15. For a given unoriented network, the capacity constraint on an arc p_ip_j is of the form $x_{ij} + x_{ji} \leq c_{ij}$. Reduce this to the case of an oriented network with constraints $x_{ij} \leq c_{ij}$. How is the statement of the max-flow, min-cut theorem for the original network affected?

16. The matrix $\mathbf{A}$ of the constraints (6.4.1) is the incidence matrix of nodes *vs.* arcs of the network. Assuming that all arcs are present, prove that the rank of $\mathbf{A}$ is $n - 1$.

17. Write out the dual of the maximal flow problem as described in (6.4.1) and (6.4.2), and show that the labeling process yields an integral solution to the dual problem.

18. Suppose that $\{S_i, \overline{S}_i\}$, where $i = 1, 2$, represent two minimal cuts in partition form; i.e., the set of arcs leading from S_i to its complement $\overline{S}_i$ is a minimal cut. Show that $\{S_1 \cup S_2, \overline{S}_1 \cap \overline{S}_2\}$ and $\{S_1 \cap S_2, \overline{S}_1 \cup \overline{S}_2\}$ also represent minimal cuts.

19. Let $\mathbf{P}_j$ and $\mathbf{Q}$ denote $n + 1$ vectors in E^m $(m < n)$ satisfying the nondegeneracy Assumption 6.1.1. Let X describe the set of all $\mathbf{x} = \langle x_1, \ldots, x_n \rangle$ satisfying

$$\sum_{i=1}^{n} x_i\mathbf{P}_i = \mathbf{Q}$$

and $\mathbf{x} \geq \mathbf{0}$. Define

$$\phi_1 = \sum_{i=1}^{n} \alpha_i x_i, \qquad \phi(\mathbf{x}) = \phi_1(\mathbf{x}) + \beta \sum_{i=1}^{n} \delta(x_i)$$

for $\mathbf{x}$ in X, where $\beta \geq 0$ and $\delta(x_i)$ equals 1 or 0 according as $x_i > 0$ or $x_i = 0$. Prove that any extreme point of X which minimizes $\phi_1(\mathbf{x})$ also minimizes $\phi(\mathbf{x})$, and conversely that any minimum point of $\phi(\mathbf{x})$ is also a minimum point of $\phi_1(\mathbf{x})$.

20. Suppose that the nondegeneracy assumption of Problem 19 is removed, while the other hypotheses remain in effect. Show that if

$$\inf_{\mathbf{x} \in X} \sum_{j=1}^{n} [\alpha_j x_j + \beta_j \delta(x_j)] = \mu > -\infty \qquad (\beta_j \geq 0),$$

then there exists an extreme point $\mathbf{x}^*$ of X such that

$$\sum_{j=1}^{n} \alpha_j x_j^* + \beta_j \delta(x_j^*) = \mu.$$

21. Minimize

$$\sum_{i=1}^{N} u_i$$

subject to restrictions

$$\begin{aligned} u_i + u_{i+1} &\geq \alpha_i \qquad (i = 1, \ldots, N - 1), \\ u_N + u_1 &\geq \alpha_N. \end{aligned}$$

22. Use the labeling algorithm and Problem 16 to prove the max-flow, min-cut theorem (Theorem 5.10.1).

23. Let a_1 and a_2 be two arcs of a network flow with capacities c_1 and c_2, respectively, and let Δ_1 be the change in maximal flow value effected by increasing c_1 one unit while holding other capacities fixed. Similarly define Δ_2. Let Δ_{12} be the change in flow value effected by increasing c_1 and c_2 one unit each. Prove that

(a) if a_1 and a_2 both lead from the source, then $\Delta_1 + \Delta_2 \geq \Delta_{12}$;

(b) if a_1 leads from the source and a_2 leads into the sink, then $\Delta_1 + \Delta_2 \leq \Delta_{12}$.

Notes and References to Chapter 6

§ 6.1. The practical importance of linear programming theory depends to a large extent on the simplex algorithm described in this section. Dantzig attributes the method to the famous fertile imagination of von Neumann, who reportedly proposed it in a casual conversation between them. It was Dantzig, however, who exploited the suggestion and developed the method, which is now recognized as the forte of linear programming methodology.

A thorough elementary summary and illustration of the simplex algorithm and some of its modifications is contained in Charnes, Cooper, and Henderson [46]. A discussion of the method and further illustrations are included in Churchman, Ackoff, and Arnoff [50]. Some other expository articles on the algorithm are given by Eisemann [74] and Dantzig [56]. Comprehensive books on the subject are being written by Dantzig and by Charnes and Cooper; see also Gass [103]. Dorfman, Samuelson, and Solow [65] offers an exhaustive introductory discussion of the concepts and methodology of linear programming and related economic analysis.

The use of slack variables was initially proposed by Dantzig, and the device of artificial variables was also introduced by Dantzig. The problem of handling degeneracies in the simplex algorithm was solved by Charnes [44] and independently by Dantzig, Orden, and Wolfe [59]; our discussion follows Charnes.

§ 6.2. The dual simplex algorithm presented here is that of Lemke [165]. Wagner [243] gives an illuminating exposition of the algorithm and discusses the general usefulness of the dual theorem as a computational aid. There exist techniques which combine the primal and dual simplex algorithms. For a discussion of these techniques, which are useful for some problems, we direct the reader to Lemke [165] and Dantzig [57]. Special adjustments of the primal and dual algorithm to deal with secondary constraints and parametric programming problems are discussed by Gass [103], Wagner [243], and Dantzig [57].

§ 6.3. The illustration of this section was offered by Wagner.

§ 6.4. The elegant algorithm of this section, valid for network flow problems and numerous varieties of the transportation problem, is due to Ford and Fulkerson [85]. We have closely followed their presentation. For other transportation algorithms, the reader is referred to [90], [91], and [160].

§ **6.5.** The iterative convergence procedure for determining the value of a game was proposed by Brown [41]. It was Robinson [208] who rigorously established the convergence. Variations of the convergence procedure have been given by von Neumann [241] and Brown [40]. Continuous analogs of the scheme applying to infinite games are discussed by Danskin [51].

§ **6.6.** The proof of the convergence of the Brown method appears to be very delicate. The rate of convergence needs further study, along with such questions as whether the convergence is affected if the past strategies are assigned unequal weightings.

§ **6.7.** The differential-equations method, which is due to Brown and von Neumann [42], is introduced primarily to shed light on the discrete procedure of § 6.5. Modifications of this method have been suggested by Bellman (unpublished).

§ **6.8.** Problem 2 is due to Allen [2]. Problems 14–18 were suggested by Ford and Fulkerson [85]. Problems 19 and 20 are due to Hirsch and Dantzig [123]. Problem 21 is due to Gross [115]. The idea of Problem 23 was noted by Shapley.

CHAPTER 7

NONLINEAR PROGRAMMING

Nonlinear programming is an extension of the theory of linear programming to deal with problems of maximizing nonlinear functions subject to general constraints. Unfortunately, it seems difficult to obtain general qualitative characterizations of the solutions to nonlinear programming problems. Some success, however, has been achieved where the function to be maximized is concave (or the function to be minimized is convex) and the constraints limit the free variables to ranging over a convex set. Fortunately, many economic and management problems arising in practice involve functions which are naturally convex or concave. Heuristically, the convexity assumption refers to a situation of increasing returns per unit increase of input, and the concavity assumption to a situation of decreasing returns per unit increase of input. Furthermore, it is common in theoretical economic analysis to suppose that the utility functions representing individual desire are concave. (See Section 8.5.)

From a mathematical point of view, the principal reason that convex or concave functions yield to analysis hinges on the fact that such functions can be approached by linear functions with certain inequalities preserved; in geometric terms, there are supporting planes to the graph of the function which keep the whole graph of the function on one side of the hyperplane.

The primary object of this chapter is to demonstrate the analogs of the duality theorem and the equivalence of game problems to linear programming problems for the case of nonlinear programming. This is accomplished in Sections 7.1 and 7.8.

An elaborate interindustry nonlinear activity analysis model is described in Section 7.2. The model is of interest because of its interpretations and because of the special mathematical methods used in its analysis.

An approximate solution to the concave programming problem can be obtained by solving an associated system of gradient differential equations. The solutions to these equations may be regarded as expressing the perturbation mechanism of changing feasible vectors into feasible vectors with larger return. These ideas are discussed in Section 7.3.

The important theory of conjugate convex functions initiated by Fenchel is developed in Sections 7.5 through 7.7. This elegant and interesting mathematical theory is applied in the concluding sections to the duality theory of nonlinear programming and to the structure of convex sets.

7.1 Concave programming. Nonlinear programming deals with problems such as: Maximize a function $g(\mathbf{x})$ of n real variables constrained by m inequalities $f_j(\mathbf{x}) \geq 0$ $(j = 1, \ldots, m)$ and $\mathbf{x} \in X$, where X is a convex subset of E^n. In most applications, X is identified with the nonnegative orthant of E^n. The problem is called concave programming if $g(\mathbf{x})$ and $f_j(\mathbf{x})$ are all concave functions of $\mathbf{x}$. Unless there is a statement to the contrary, we assume from here on that the functions $g(\mathbf{x})$ and $f_j(\mathbf{x})$ are concave in the region $\mathbf{x} \in X$.

Kuhn and Tucker were the first to give a formal discussion of various mathematical aspects of concave programming problems. They distinguish a class of "proper" solutions, which are essentially solutions that can be recognized unambiguously by differential conditions; and their proofs, depending crucially on differential methods, apply only to this class of proper solutions. The methods and proofs given below do not involve such restrictions. Our results, however, are complementary to theirs rather than an extension of theirs; hence we shall also discuss the differential approach later in this section.

Under suitable assumptions the problem of characterizing the maximum points $\mathbf{x}^0$ of the concave programming problem described above can be transformed into an equivalent game or saddle value problem.

We shall prove first (Theorem 7.1.1 below) that a particular vector $\mathbf{x}^0$ maximizes $g(\mathbf{x})$ subject to the constraints $\{f_j(\mathbf{x})\} = \mathbf{F}(\mathbf{x}) \geq 0$ (the vector function as indicated) and $\mathbf{x} \in X$, if and only if there exists some $\mathbf{u}^0$ with nonnegative components which, for the function

$$\phi(\mathbf{x}, \mathbf{u}) = g(\mathbf{x}) + (\mathbf{u}, \mathbf{F}(\mathbf{x})) \tag{7.1.1}$$

satisfies

$$\phi(\mathbf{x}, \mathbf{u}^0) \leq \phi(\mathbf{x}^0, \mathbf{u}^0) \leq \phi(\mathbf{x}^0, \mathbf{u}) \tag{7.1.2}$$

for all $\mathbf{u} \geq \mathbf{0}$ and $\mathbf{x} \in X$. The pair $\{\mathbf{x}^0, \mathbf{u}^0\}$ will be called a saddle point of the kernel $\phi(\mathbf{x}, \mathbf{u})$. A feasible vector $\mathbf{x}^0$, i.e., a point at which $\mathbf{x}$ satisfies the constraints $\mathbf{x} \in X$ and $\mathbf{F}(\mathbf{x}) \geq \mathbf{0}$, is called optimal if $\mathbf{x}^0$ maximizes $g(\mathbf{x})$ among all feasible vectors.

The economic interpretation of the maximization as suggested by Kuhn and Tucker can best be stated in relation to the game with pay-off kernel $\phi(\mathbf{x}, \mathbf{u})$. The constraints are usually of the form $f_j(\mathbf{x}) = \xi_j - k_j(\mathbf{x})$, with $\boldsymbol{\xi}$ a constant vector and k_j a convex function of its argument in the region $\mathbf{x} \geq \mathbf{0}$. The vector $\mathbf{x}$ represents the activity level of the various operations. Interpret $\boldsymbol{\xi}$ as the available (vector) amount of primary resources and $\mathbf{k}(\mathbf{x})$ as the (vector) amount actually used, so that $\mathbf{F}(\mathbf{x})$ is the unused balance of primary resources. The function $g(\mathbf{x})$ is a measure of the resulting output of the desired commodity. The Lagrange multiplier $\mathbf{u}$ appearing in the definition of $\phi(\mathbf{x}, \mathbf{u})$ may be thought of as a price vector of the primary resources relative to a unit price for the desired commodity. The kernel $\phi(\mathbf{x}, \mathbf{u})$ can be interpreted as the combined output of the desired commodity and the unused balance of the primary resources. The existence of a saddle point expresses in a sense an equilibrium between prices of the available resources and the value of the output such that the use of the primary resources is controlled by the manager, while prices are generated or promulgated by some independent agency (see Chapter 8).

The first fundamental theorem may be stated as follows.

▶THEOREM 7.1.1. Let $g(\mathbf{x})$ and $\{f_j(\mathbf{x})\} = \mathbf{F}(\mathbf{x})$ all be concave functions defined on the convex set X of E^n, with the property that if $\mathbf{u}$ is a nonnegative vector not identically zero, then there exists some $\mathbf{x} \in X$ such that $(\mathbf{u}, \mathbf{F}(\mathbf{x})) > 0$. If $\mathbf{x}^0$ is a point at which $g(\mathbf{x})$ achieves its maximum subject to the constraints $\mathbf{x} \in X$ and $\mathbf{F}(\mathbf{x}) \geq \mathbf{0}$, that is, $\mathbf{x}^0$ is optimal, then there exists a choice of $\mathbf{u}^0 \geq \mathbf{0}$ such that for $\phi(\mathbf{x}, \mathbf{u}) = g(\mathbf{x}) + (\mathbf{u}, \mathbf{F}(\mathbf{x}))$,

$$\phi(\mathbf{x}, \mathbf{u}^0) \leq \phi(\mathbf{x}^0, \mathbf{u}^0) \leq \phi(\mathbf{x}^0, \mathbf{u}) \tag{7.1.2}$$

for all $\mathbf{x} \in X$, $\mathbf{u} \geq \mathbf{0}$, and $(\mathbf{u}^0, \mathbf{F}(\mathbf{x}^0)) = 0$. Conversely, if $\{\mathbf{x}^0, \mathbf{u}^0\}$ satisfies (7.1.2), then $\mathbf{x}^0$ maximizes $g(\mathbf{x})$ among all $\mathbf{x}$ satisfying $\mathbf{F}(\mathbf{x}) \geq \mathbf{0}$ and $\mathbf{x} \in X$.

Remark 7.1.1. The hypothesis of the theorem is satisfied if there exists a vector $\mathbf{x} \in X$ such that $\mathbf{F}(\mathbf{x})$ has all strictly positive components. (In this connection, see Problem 12.) In general, however, any $\mathbf{x} \in X$ satisfying the hypothesis for a given $\mathbf{u}$ need not satisfy the constraints $\mathbf{F}(\mathbf{x}) \geq \mathbf{0}$.

Proof. First we shall show that if $\{\mathbf{x}^0, \mathbf{u}^0\}$ is a saddle point of $\phi(\mathbf{x}, \mathbf{u})$, then $\mathbf{x}^0$ maximizes $g(\mathbf{x})$ subject to the constraints $\mathbf{x} \in X$, $\mathbf{F}(\mathbf{x}) \geq \mathbf{0}$.

Substituting (7.1.1) in (7.1.2), we have

$$g(\mathbf{x}) + (\mathbf{u}^0, \mathbf{F}(\mathbf{x})) \leq g(\mathbf{x}^0) + (\mathbf{u}^0, \mathbf{F}(\mathbf{x}^0)) \leq g(\mathbf{x}^0) + (\mathbf{u}, \mathbf{F}(\mathbf{x}^0)) \tag{7.1.3}$$

for all $\mathbf{x} \in X$, $\mathbf{u} \geq \mathbf{0}$.

Since the right-hand inequality holds for any $\mathbf{u} \geq \mathbf{0}$, it follows that

$$\mathbf{F}(\mathbf{x}^0) \geq \mathbf{0}, \qquad (\mathbf{u}^0, \mathbf{F}(\mathbf{x}^0)) = 0. \tag{7.1.4}$$

Then the left-hand inequality of (7.1.3) becomes

$$g(\mathbf{x}) + (\mathbf{u}^0, \mathbf{F}(\mathbf{x})) \leq g(\mathbf{x}^0) \qquad (\mathbf{x} \in X). \tag{7.1.5}$$

Therefore, for any $\mathbf{x} \in X$ such that $\mathbf{F}(\mathbf{x}) \geq \mathbf{0}$, we have

$$g(\mathbf{x}) \leq g(\mathbf{x}) + (\mathbf{u}^0, \mathbf{F}(\mathbf{x})) \leq g(\mathbf{x}^0),$$

which shows $\mathbf{x}^0$ to be optimal.

Next we shall prove that, subject to the hypothesis of the theorem, if $\mathbf{x}^0$ is optimal, then there exists a value $\mathbf{u}^0$ such that $\{\mathbf{x}^0, \mathbf{u}^0\}$ is a saddle point of $\phi(\mathbf{x}, \mathbf{u})$.

Consider two sets A and B, in $(m + 1)$-dimensional space, defined by

$$A = \left\{\begin{pmatrix} y_0 \\ \mathbf{y} \end{pmatrix} \middle| \begin{pmatrix} y_0 \\ \mathbf{y} \end{pmatrix} \leq \begin{pmatrix} g(\mathbf{x}) \\ \mathbf{F}(\mathbf{x}) \end{pmatrix} \quad \text{for some} \quad \mathbf{x} \in X \right\},$$

$$B = \left\{\begin{pmatrix} y_0 \\ \mathbf{y} \end{pmatrix} \middle| \begin{pmatrix} y_0 \\ \mathbf{y} \end{pmatrix} \gg \begin{pmatrix} g(\mathbf{x}^0) \\ \mathbf{0} \end{pmatrix} \right\},$$

where y^0 is obviously a scalar and $\mathbf{y}$ is a vector of m components.

Since $g(\mathbf{x})$ and $\mathbf{F}(\mathbf{x})$ are concave, the set A is convex. Also B is clearly convex, consisting of the interior of an orthant with vertex at

$$\begin{pmatrix} g(\mathbf{x}_0) \\ \mathbf{0} \end{pmatrix}$$

By the optimality of $\mathbf{x}^0$, sets A and B have no vector in common. Hence, by the separation theorem of convex sets (Theorem B.1.2), there is a vector

$$\begin{pmatrix} v_0 \\ \mathbf{v} \end{pmatrix} \neq 0$$

such that

$$v_0 y_0 + (\mathbf{v}, \mathbf{y}) \leq v_0 z_0 + (\mathbf{v}, \mathbf{z}) \quad \text{for all} \begin{pmatrix} y_0 \\ \mathbf{y} \end{pmatrix} \in A, \quad \begin{pmatrix} z_0 \\ \mathbf{z} \end{pmatrix} \in B. \tag{7.1.6}$$

By the definition of B, (7.1.6) implies

$$\begin{pmatrix} v_0 \\ \mathbf{v} \end{pmatrix} > \mathbf{0}. \tag{7.1.7}$$

Since

$$\begin{pmatrix} g(\mathbf{x}^0) \\ \mathbf{0} \end{pmatrix}$$

is on the boundary of B, we have, by the definition of A,

$$v_0 g(\mathbf{x}) + (\mathbf{v}, \mathbf{F}(\mathbf{x})) \leq v_0 g(\mathbf{x}^0) \qquad (\mathbf{x} \in X). \tag{7.1.8}$$

It follows that $v_0 > 0$. Otherwise, by (7.1.7) and (7.1.8)

$$\mathbf{v} > \mathbf{0} \quad \text{and} \quad (\mathbf{v}, \mathbf{F}(\mathbf{x})) \leq 0 \qquad (\mathbf{x} \in X),$$

which contradicts the hypothesis of the theorem.

Let $\mathbf{u}^0 = \mathbf{v}/v_0$. Then

$$\mathbf{u}^0 \geq \mathbf{0} \tag{7.1.9}$$

and

$$g(\mathbf{x}) + (\mathbf{u}^0, \mathbf{F}(\mathbf{x})) \leq g(\mathbf{x}^0) \qquad (\mathbf{x} \in X). \tag{7.1.10}$$

Putting $\mathbf{x} = \mathbf{x}^0$ in (7.1.10), we get

$$(\mathbf{u}^0, \mathbf{F}(\mathbf{x}^0)) \leq 0.$$

On the other hand, we have

$$\mathbf{F}(\mathbf{x}^0) \geq \mathbf{0}. \tag{7.1.11}$$

Hence

$$(\mathbf{u}^0, \mathbf{F}(\mathbf{x}^0)) = 0. \tag{7.1.12}$$

Equations (7.1.9–12) in conjunction show that $\{\mathbf{x}^0, \mathbf{u}^0\}$ is a saddle point of $\phi(\mathbf{x}, \mathbf{u})$.

The hypothesis of Theorem 7.1.1, weak as it is, may be further slightly weakened by introducing unnatural conditions. However, some sort of nondegeneracy condition must be expressed in the hypothesis or else the theorem is false. Consider the example

$$g(\mathbf{x}) = x,$$

$$\mathbf{F}(\mathbf{x}) = -x^2.$$

Clearly, $x = 0$ will be the optimal solution to the problem. But $\phi(\mathbf{x}, \mathbf{u}) = x - ux^2$ has no saddle point.

In the case in which the constraints are defined by linear inequalities, that is, $\mathbf{F}(\mathbf{x}) \geq \mathbf{0}$ is identical with $\mathbf{Ax} \leq \mathbf{b}$ for some matrix $\mathbf{A}$ and vector $\mathbf{b}$, it is possible to establish the conclusion of Theorem 7.1.1 without any extra hypothesis regarding the constraint set.

▶ THEOREM 7.1.2. Let $g(\mathbf{x})$ represent a concave function defined for all $\mathbf{x}$. If $\mathbf{x}^0$ is a point at which $g(\mathbf{x})$ achieves its maximum with respect to all $\mathbf{x}$ satisfying the restriction

$$\mathbf{x} \geq \mathbf{0}, \ \mathbf{Ax} \leq \mathbf{b},$$

then there exists a value $\mathbf{u}^0$ ($\mathbf{u}^0 \geq \mathbf{0}$) such that $\{\mathbf{x}^0, \mathbf{u}^0\}$ is a saddle point of the function

$$\phi(\mathbf{x}, \mathbf{u}) = g(\mathbf{x}) + (\mathbf{u}^0, \mathbf{b} - \mathbf{Ax}),$$

and conversely.

The proof is an adaptation of the arguments of Theorem 5.3.1, utilizing the natural extension of Lemma 5.3.1 to this case. The details are omitted.

If sufficient differentiability is assumed for the functions g and f_j so that $\partial\phi/\partial x$ exists for every nonnegative $\mathbf{x}$, then the conditions for the existence of a saddle point may be expressed in the form of differential inequalities.

▶THEOREM 7.1.3. Given that X is the positive orthant, a necessary condition in order that $\{\mathbf{x}^0, \mathbf{u}^0\}$ satisfy (7.1.2) for any continuously differentiable function ϕ is that

$$\boldsymbol{\phi}_x^0 = \left[\frac{\partial \phi}{\partial x_i}\right]_{\{x^0, u^0\}} \leq \mathbf{0}, \qquad (\boldsymbol{\phi}_x^0, \mathbf{x}^0) = 0 \qquad (\mathbf{x}^0 \geq \mathbf{0}) \quad (7.1.13)$$

and

$$\boldsymbol{\phi}_u^0 = \left[\frac{\partial \phi}{\partial u_j}\right]_{\{x^0, u^0\}} \geq \mathbf{0}, \qquad (\boldsymbol{\phi}_u^0, \mathbf{u}^0) = 0 \qquad (\mathbf{u}^0 \geq \mathbf{0}). \quad (7.1.14)$$

If $\phi(\mathbf{x}, \mathbf{u})$ is concave in $\mathbf{x}$ and convex in $\mathbf{u}$, then (7.1.13) and (7.1.14) are both necessary and sufficient to establish (7.1.2).

Proof. Necessity. The relations (7.1.13) and (7.1.14) assert that $\phi(\mathbf{x}, \mathbf{u}^0)$ has a local maximum at $\mathbf{x} = \mathbf{x}^0$ and that $\phi(\mathbf{x}^0, \mathbf{u})$ has a local minimum at $\mathbf{u} = \mathbf{u}^0$, respectively. This indeed must be the case if (7.1.2) holds.

Sufficiency. Since $\phi(\mathbf{x}, \mathbf{u})$ is concave in $\mathbf{x}$, we have

$$\phi(\mathbf{x}, \mathbf{u}^0) - \phi(\mathbf{x}^0, \mathbf{u}^0) \leq (\boldsymbol{\phi}_x^0, (\mathbf{x} - \mathbf{x}^0))$$

(property vii of Section B.4).

$$(\boldsymbol{\phi}_x^0, (\mathbf{x} - \mathbf{x}^0)) = (\boldsymbol{\phi}_x^0, \mathbf{x}) - (\boldsymbol{\phi}_x^0, \mathbf{x}^0) \leq 0.$$

Hence

$$\phi(\mathbf{x}, \mathbf{u}^0) \leq \phi(\mathbf{x}^0, \mathbf{u}^0) \qquad (\mathbf{x} \geq \mathbf{0}).$$

On the other hand, by the convexity of $\phi(\mathbf{x}, \mathbf{u})$ in $\mathbf{u}$ and by (7.1.14), we have

$$\phi(\mathbf{x}^0, \mathbf{u}) - \phi(\mathbf{x}^0, \mathbf{u}^0) \geq (\boldsymbol{\phi}_u^0, \mathbf{u} - \mathbf{u}^0) = (\boldsymbol{\phi}_u^0, \mathbf{u}) \geq 0 \qquad (\mathbf{u} \geq \mathbf{0}).$$

7.2 Examples of concave programming. In this section we describe two models (portfolio selection and interindustry activity analysis) that lead naturally to concave programming problems. More extensive forms of nonlinear programming problems arise in connection with production and inventory studies.

In general there is no standard method available for solving nonlinear programming problems—i.e., nothing corresponding to the routine simplex algorithm of linear programming. Nevertheless a modest classification of types of nonlinear programming problems susceptible to a common approach can be accomplished. For example, as we have seen in Section 7.1, concave programming problems can be reduced to an equivalent Lagrangian saddle value form. Moreover, it is known and evident that the solution of a convex programming problem (maximizing a convex function where the variables range over a convex set X) occurs at an extreme point of X.

Further particular qualitative results for special classes of problems may be obtained (see Chapter 8). It should be noted that since there is no single procedure for solving nonlinear programming problems—and this lack seems clearly inherent in the subject itself—the analysis of special examples is doubly important.

A. *Portfolio selection.* The process of selecting a portfolio of securities may be divided into two stages. The first stage is an appraisal of the future performances of available securities. The second stage starts with the relevant beliefs about future performances and ends with a choice of portfolio. We are concerned here with the second stage. Specifically, we seek to determine what rule the investor should follow in order to maximize discounted expected future returns.

Let R_i represent the discounted return of one unit of the ith security. The value R_i can be further decomposed as an infinite sum of the form

$$R_i = \sum_{t=1}^{\infty} d_{it} r_{it},$$

where r_{it} is the anticipated return at time t per dollar invested in security i, and d_{it} is the rate of return on the ith security at time t discounted back to the present.

Suppose that only securities $i = 1, 2, \ldots, N$ are under consideration. Let x_i describe a given investment policy which denotes the proportions of security i bought; i.e.,

$$\sum_{i=1}^{N} x_i = 1.$$

We further assume $x_i \geq 0$, which implies that short sales are excluded as a possibility. The relative cumulative return for such a policy is

$$\sum_{i=1}^{N} R_i x_i = P(\mathbf{x}). \tag{7.2.1}$$

If we suppose further that R_i consists of normal random variables with known mean values μ_i and covariance matrix $\|\sigma_{ij}\|$, then the mean value of $P(\mathbf{x})$ is

$$\sum_{i=1}^{N} \mu_i x_i$$

and its variance is $\sum_{i,j} x_i x_j \sigma_{ij}$. Consider the objective function

$$g(\mathbf{x}) = a \sum_{i=1}^{N} \mu_i x_i - b \sum_{i,j} x_i x_j \sigma_{ij}, \tag{7.2.2}$$

where a and b are positive parameters whose relative magnitudes describe the importance of each of the terms of $g(x)$. The first is the expected relative yield according to the policy, which obviously is to be made large. The second is the measure of variability of the return, which can be regarded as the risk attending a given portfolio selection.

The problem, then, is to maximize $g(x)$ subject to the constraints

$$x_i \geq 0, \qquad \sum_{i=1}^{N} x_i = 1.$$

This is clearly a concave programming problem to which the results of Section 7.1 are applicable. The quadratic programming algorithm is also relevant (see Problem 13 and the notes to this chapter).

B. *An interindustry nonlinear programming model.* Suppose that for each of n desired commodities there is a production activity, an import activity, and an export activity. Let the variables of production, export, and import be

$$\begin{aligned} X_j &= \text{amount of commodity } j \text{ produced},\\ M_j &= \text{amount of } j \text{ imported},\\ E_j &= \text{amount of } j \text{ exported}. \end{aligned}$$

The parameters of the model are as follows:

a_{ij} = input coefficient for commodity i used in production of j (i.e., the amount of i used in producing one unit of j),

w_j = input coefficient for labor used in production of j,

c_j = input coefficient for capital used in production of j,

g_j = import unit price of j,

h_j = export unit price of j,

Y_j = final demand for j (which we shall assume to be strictly positive),

L = available supply of labor,

D = maximum allowable trade deficit.

(All the parameters of the model are naturally assumed to be nonnegative.)

The programming problem may be stated as follows: Minimize the total capital required,

$$c = \sum_j c_j X_j, \tag{7.2.3}$$

subject to the following restrictions:

Nonnegativity:

$$X_j \geq 0, \qquad M_j \geq 0, \qquad E_j \geq 0 \qquad (j = 1, \ldots, n). \tag{7.2.4}$$

Production:

$$X_j + M_j - E_j - \sum_{r=1}^{n} a_{jr}X_r \geq Y_j \qquad (j = 1, \ldots, n); \tag{7.2.5}$$

i.e., if E_j units are exported and

$$\sum_{r=1}^{n} a_{jr}X_r$$

units are used up by domestic production out of a total of $X_j + M_j$ units produced and imported, enough should remain to satisfy the final demand Y_j.

Foreign exchange balance:

$$D + \sum_{j=1}^{n} h_jE_j - \sum_{j=1}^{n} g_jM_j \geq 0; \tag{7.2.6}$$

since the export revenue is

$$\sum_{j=1}^{n} h_jE_j$$

and the import cost is

$$\sum_{j=1}^{n} g_jM_j,$$

it follows that (7.2.6) expresses the allowable trade deficit.

Labor supply:

$$L - \sum_{j} w_jX_j \geq 0. \tag{7.2.7}$$

Two further natural assumptions are

$$a_{ij} \geq 0 \quad (i, j = 1, \ldots, n), \qquad \sum_{i} a_{ij} < 1 \quad (j = 1, \ldots, n) \tag{7.2.8}$$

and

$$h_j = \gamma_j + \rho_jE_j, \qquad \gamma_j > 0, \qquad \rho_j < 0 \qquad (j = 1, \ldots, n). \tag{7.2.9}$$

The first assumption is simply that the output exceeds the input for every commodity, a necessary condition for efficient production plans. The second assumption is one commonly made in economics: to wit, the larger the amount of export goods made available, the smaller the return per unit. The linear functional form of (7.2.9) expresses a special case of this kind of dependence.

The reader will observe that in view of (7.2.9), the only nonlinearity in this model is contained in (7.2.6).

**Formal statement of the method.* The function to be minimized, (7.2.3), is a convex function of X_j $(j = 1, \ldots, n)$, and the set $\{X, M, E\}$ satisfying restrictions (7.2.4–7) is a convex set. We shall now assume that there is a feasible $\{X^0, M^0, E^0\}$ with strict inequality for (7.2.5), (7.2.6), and (7.2.7). Then, according to Theorem 7.1.1,

$$\bar{\mathbf{x}} = \begin{pmatrix} \bar{X} \\ \bar{M} \\ \bar{E} \end{pmatrix}$$

is optimal for the problem if and only if there exists a value

$$\bar{\mathbf{p}} = \begin{pmatrix} \bar{P} \\ \bar{P}_w \\ \bar{P}_f \end{pmatrix}$$

such that $\{\bar{\mathbf{x}}, \bar{\mathbf{p}}\}$ is a saddle point of the Lagrangian form $\phi(\mathbf{x}, \mathbf{p})$ defined by

$$\begin{aligned}\phi(\mathbf{x}, \mathbf{p}) = &-\textstyle\sum_j c_j X_j + \sum_j P_j\,(X_j - \sum_r a_{jr} X_r + M_j - E_j - Y_j) \\ &+ P_w\,(L - \textstyle\sum_j w_j X_j) + P_f\,(D - \sum_j g_j M_j + \sum_j h_j E_j);\end{aligned} \tag{7.2.10}$$

that is,

$$\phi(\mathbf{x}, \bar{\mathbf{p}}) \leq \phi(\bar{\mathbf{x}}, \bar{\mathbf{p}}) \leq \phi(\bar{\mathbf{x}}, \mathbf{p}) \tag{7.2.11}$$

$$\text{for all} \quad \mathbf{x} = \begin{pmatrix} X \\ M \\ E \end{pmatrix} \geq \mathbf{0}, \qquad \mathbf{p} = \begin{pmatrix} P \\ P_w \\ P_f \end{pmatrix} \geq \mathbf{0}.$$

It will be noted that by virtue of (7.2.5) and $Y_j > 0$, for any feasible

$$\begin{pmatrix} X \\ M \\ E \end{pmatrix}, \text{ we have}$$

$$X_j + M_j > 0 \qquad (j = 1, \ldots, n). \tag{7.2.12}$$

Substituting (7.2.9) into (7.2.10) and rearranging, we get

$$\begin{aligned}\phi(\mathbf{x}, \mathbf{p}) = P_w L + P_f D - \textstyle\sum_j P_j Y_j + \sum_j(-c_j + P_j - \sum_i P_i a_{ij} - P_w w_j) X_j \\ + \textstyle\sum_j (P_j - P_f g_j) M_j + \sum_j [(-P_j + P_f \gamma_j) E_j + P_f \rho_j E_j^2].\end{aligned} \tag{7.2.13}$$

---

* The remainder of this section is of advanced nature and may be omitted at first reading without loss of continuity.

The saddle value problem will be solved in the following manner. We shall show that its $\bar{\mathbf{p}}$ component is uniquely determined and may be explicitly written out. With $\bar{\mathbf{p}}$ known, it is easy to secure the maximum $\bar{\mathbf{x}}$ of $\phi(\mathbf{x}, \bar{\mathbf{p}})$, which completes the solution of the problem.

In order for

$$\mathbf{p} = \begin{pmatrix} P \\ P_w \\ P_f \end{pmatrix}$$

to be the **p**-component of a saddle point, it is necessary (i) that

$$\max_{x \geq 0} \phi(\mathbf{x}, \mathbf{p})$$

be finite, and (ii) that there exist a value

$$\mathbf{x} = \begin{pmatrix} X \\ M \\ E \end{pmatrix}$$

which maximizes $\phi(\mathbf{x}, \mathbf{p})$ and for which (7.2.12) is satisfied.

For the finiteness of (7.2.13) the coefficients of X_j and M_j must be nonpositive. Thus (i) and (ii) are satisfied if and only if

$$P_j - P_f g_j \leq 0 \tag{7.2.14}$$

and

$$P_j - \sum_i P_i a_{ij} - P_w w_j - c_j \leq 0, \tag{7.2.15}$$

and for each $j = 1, \ldots, n$ equality holds for either (7.2.14) or (7.2.15) or both (since $X_j + M_j > 0$).

We can replace conditions (7.2.14) and (7.2.15) by

$$P_j = \min \left\{ P_f g_j, \sum_i P_i a_{ij} + P_w w_j + c_j \right\} \qquad (j = 1, \ldots, n). \tag{7.2.16}$$

The relations (7.2.16) possess at most one solution P_j for prescribed values of P_f and P_w. To demonstrate this, we suppose that P_j and $\tilde{P}_j$ $(j = 1, \ldots, n)$ both satisfy (7.2.16) for given P_f and P_w. We consider the case

$$P_j = P_f g_j \leq \sum_i P_i a_{ij} + P_w w_j + c_j,$$

$$\tilde{P}_j = \sum_i \tilde{P}_i a_{ij} + P_w w_j + c_j \leq P_f g_j.$$

Then

$$|P_j - \tilde{P}_j| \leq \left| \sum_i (P_i - \tilde{P}_i) a_{ij} \right| \leq \left(\max_i |P_i - \tilde{P}_i| \right) c \qquad (j = 1, \ldots, n),$$

where $c = \max_j \sum_i a_{ij} < 1$ by assumption (7.2.8). The same inequality ensues in all other circumstances, and we deduce that $P_j = \widetilde{P}_j$ $(j = 1, \ldots, n)$.

We now describe an iterative procedure for securing a solution of (7.2.16) with P_f and P_w specified.

The components $(P_1, \ldots, P_n)$ satisfying (7.2.16) will be determined in the order $\ldots, P_n^{(\nu)}, \ldots, P_1^{(\nu)}, P_n^{(\nu+1)}, \ldots, P_1^{(\nu+1)}, \ldots$ from the recursive formulas

$$P_j^{(\nu+1)} = \min \left\{P_f g_j, \sum_{i>j} P_i^{(\nu+1)} a_{ij} + \sum_{i \leq j} P_i^{(\nu)} a_{ij} + P_w w_j + c_j\right\}$$
$$(j = 1, \ldots, n; \nu = 0, 1, 2, \ldots), \quad (7.2.17)$$

with any nonnegative initial values $P_1^{(0)}, \ldots, P_n^{(0)}$.

▶ LEMMA 7.2.1. For any nonnegative (P_w, P_f) and any nonnegative initial values $(P_1^{(0)}, \ldots, P_n^{(0)})$, the values $(P_1^{(\nu)}, \ldots, P_n^{(\nu)})$ determined by (7.2.17) converge to a set of values $(P_1, \ldots, P_n)$ satisfying (7.2.16).

Proof. It will first be noted that

$$|P_j^{(\nu+1)} - P_j^{(\nu)}| \leq Kc^\nu, \quad (7.2.18)$$

where K is a positive constant and $c = \max_j \sum_i a_{ij}$. Relation (7.2.18) is established by induction. Let us assume that

$$|P_j^{(k+1)} - P_j^{(k)}| \leq Kc^k \qquad (k = 0, 1, \ldots, \nu - 1; j = 1, \ldots, n;$$
$$k = \nu; j > i).$$

Then by (7.2.17) we have

$$\begin{aligned}
|P_j^{(\nu+1)} - P_j^{(\nu)}| &\leq \left|\left(\sum_{i>j} P_i^{(\nu+1)} a_{ij} + \sum_{i \leq j} P_i^{(\nu)} a_{ij} + P_w w_j + c_j\right)\right. \\
&\quad \left. - \left(\sum_{i>j} P_i^{(\nu)} a_{ij} + \sum_{i \leq j} P_i^{(\nu-1)} a_{ij} + P_w w_j + c_j\right)\right| \\
&= \left|\sum_{i>j} (P_i^{(\nu+1)} - P_i^{(\nu)}) a_{ij} + \sum_{i \leq j} (P_i^{(\nu)} - P_i^{(\nu-1)}) a_{ij}\right| \\
&\leq \sum_{i>j} |P_i^{(\nu+1)} - P_i^{(\nu)}| a_{ij} + \sum_{i \leq j} |P_i^{(\nu)} - P_i^{(\nu-1)}| a_{ij} \\
&\leq Kc^\nu \sum_{i>j} a_{ij} + Kc^{\nu-1} \sum_{i \leq j} a_{ij} \\
&\leq Kc^{\nu-1} \sum_i a_{ij} \leq Kc^\nu,
\end{aligned}$$

since $c < 1$.

The convergence of $P_j^{(\nu)}$ to a P_j which obviously must satisfy (7.2.16) is now clear, and the proof of the lemma is complete.

For the triangular case, where

$$a_{ij} = 0, \qquad i \leq j,$$

the recursive formulas (7.2.17) are replaced by the exact formula

$$P_j = \min \left\{ P_f g_j, \sum_{i>j} P_i a_{ij} + P_w w_j + c_j \right\}, \tag{7.2.19}$$

determined in the order $j = n, \ldots, 1$.

A value

$$\hat{\mathbf{x}} = \begin{pmatrix} \hat{X} \\ \hat{M} \\ \hat{E} \end{pmatrix}$$

that maximizes $\phi(\mathbf{x}, \mathbf{p})$ subject to $\mathbf{x} \geq \mathbf{0}$ with $\mathbf{p}$ fixed will be now determined. Let

$$S = \left\{ j; P_j < \sum_{i>j} P_i a_{ij} + P_w w_j + c_j \right\},$$

$$T = \{ j; P_j < P_f g_j \}.$$

It follows from (7.2.19) that S and T are disjoint sets. Then

$$\hat{E}_j = \max \left\{ 0, \frac{P_f \gamma_j - P_j}{-2P_f \rho_j} \right\}, \qquad (j = 1, \ldots, n), \tag{7.2.20}$$

since the term involving E_j in (7.2.13) is a quadratic minimized by (7.2.20). Also, since the terms of X_j and M_j in (7.2.13) are linear it follows that $\hat{X}_j = 0$ for $j \in S$, and $\hat{M}_j = 0$ for $j \in T$. We may determine $\hat{X}_j$ and $\hat{M}_j$ as solutions to the following relations which maximize $\phi(\mathbf{x}, \mathbf{p})$. Explicitly,

$$\hat{X}_j - \sum_{k \notin S} a_{jk} \hat{X}_k = Y_j + \hat{E}_j \qquad \text{for} \quad j \notin S, \tag{7.2.21}$$

and

$$\hat{M}_j = Y_j + \hat{E}_j + \sum_{k \notin S} a_{jk} \hat{X}_k \qquad \text{for} \quad j \notin T. \tag{7.2.22}$$

With the aid of condition (7.2.8) it is easily proved that equations (7.2.21) possess a unique strictly positive solution. The values of $\hat{X}_j$ and $\hat{M}_j$, both function of P_f and P_w, calculated by means of (7.2.21) and (7.2.22) satisfy the constraints (7.2.5).

Define

$$L(P_w, P_f) = \sum_j w_j \hat{X}_j$$

and

$$D(P_w, P_f) = \sum_j g_j\hat{M}_j - \sum_j \hat{h}_j\hat{E}_j,$$

where

$$\hat{h}_j = \gamma_j + \rho_j\hat{E}_j \qquad (j = 1, \ldots, n).$$

An application of Theorem 7.1.1 shows that

▶ LEMMA 7.2.2. (P_w, P_f) determines optimal $(P_1, \ldots, P_n, P_w, P_f)$ if and only if

$$L(P_w, P_f) \leq L \quad \text{and} \quad D(P_w, P_f) \leq D, \tag{7.2.23}$$

with equality when $P_w > 0$ and $P_f > 0$, respectively.

Since $L(P_w, P_f)$ and $D(P_w, P_f)$ are approximately linear functions of P_w and P_f, optimal (P_w, P_f) can be approximated by linear interpolations.

If no choices of P_w and P_f suffice to satisfy (7.2.23), then (7.2.21) and (7.2.22) must be altered as follows:

$$\hat{X}_j + \hat{M}_j - \sum_R a_{jk}\hat{X}_k = Y_j + \hat{E}_j \quad \text{for } j \in R = \langle 1, 2, \ldots, n\rangle - S - T,$$

$$\hat{X}_j - \sum_{k \notin S} a_{jk}\hat{X}_k = Y_j + \hat{E}_j \quad \text{for } j \in T,$$

and

$$\hat{M}_j = Y_j + \hat{E}_j + \sum_{k \notin S} a_{jk}\hat{X}_k \quad \text{for } j \in S - T.$$

The problem is then reduced to a standard problem of linear programming. In most cases, however, the solutions of (7.2.21) and (7.2.22) are valid.

We are interested in this model for two reasons. One is its usefulness in determining optimal plans for the production, export, and import of commodities. The second and more important reason is its usefulness as a tool for the investigation of similar nonlinear programming problems.

***7.3 The Arrow-Hurwicz gradient method.** We have seen by Theorems 7.1.1 and 7.1.2 that concave programming problems can be solved by finding a saddle point of the Lagrangian form $\phi(\mathbf{x}, \mathbf{u})$, where $\phi(\mathbf{x}, \mathbf{u})$ is concave in $\mathbf{x}$ and linear in $\mathbf{u}$. In this section we shall describe a method of approaching a saddle point of $\phi(\mathbf{x}, \mathbf{u})$ by solving a system of classical gradient equa-

---

* Starred sections are advanced discussions that may be omitted at first reading without loss of continuity.

tions modified on the boundary to keep the variables in the positive orthant.

Let $\phi(\mathbf{x}, \mathbf{u})$ be strictly concave in $\mathbf{x} \geq \mathbf{0}$ and linear in $\mathbf{u} \geq \mathbf{0}$. For the first part of the theory it is enough to assume that $\phi(\mathbf{x}, \mathbf{u})$ is convex in $\mathbf{u}$.

Consider the following system of differential equations:

$$\dot{x}_i = \begin{cases} 0 & \text{if} \quad x_i = 0 \text{ and } \phi_{x_i} < 0, \\ \phi_{x_i}[\mathbf{x}(t), \mathbf{u}(t)] & \text{otherwise}, \end{cases} \qquad (i = 1, \ldots, n),$$

$$\dot{u}_k = \begin{cases} 0 & \text{if} \quad u_k = 0 \text{ and } \phi_{u_k} > 0, \\ -\phi_{u_k}[\mathbf{x}(t), \mathbf{u}(t)] & \text{otherwise}, \end{cases} \qquad (k = 1, \ldots, m), \tag{7.3.1}$$

where ϕ_{x_i} represents partial differentiation with respect to the x_i variable and ϕ_{u_k} has an analogous meaning. (Whenever we write ϕ_{x_i} or ϕ_{u_k} without indicating the points at which these derivatives are evaluated, the reader should interpret these derivatives in the obvious way.) In what follows we shall assume that the system (7.3.1) has a solution $\{\mathbf{x}(t), \mathbf{u}(t)\}$ which is uniquely determined by any initial position $\{\mathbf{x}^0, \mathbf{u}^0\} \geq 0$ and continuous with respect to $\{\mathbf{x}^0, \mathbf{u}^0\}$.

▶ THEOREM 7.3.1. Let $\phi(\mathbf{x}, \mathbf{u})$ be strictly concave in $\mathbf{x}$ and convex in $\mathbf{u}$ and possess a saddle point. Then the solution $\{\mathbf{x}(t), \mathbf{u}(t)\}$ of (7.3.1) with an arbitrary nonnegative initial position $\{\mathbf{x}^0, \mathbf{u}^0\}$ converges to a saddle point of $\phi(\mathbf{x}, \mathbf{u})$.

Proof. We observe first that the $\mathbf{x}$ component $\bar{\mathbf{x}}$ of a saddle point of $\phi(\mathbf{x}, \mathbf{u})$ is uniquely determined. Assume to the contrary that $\{\bar{\mathbf{x}}, \bar{\mathbf{u}}\}$ and $\{\bar{\bar{\mathbf{x}}}, \bar{\bar{\mathbf{u}}}\}$ are two saddle points of $\phi(\mathbf{x}, \mathbf{u})$ such that $\bar{\mathbf{x}} \neq \bar{\bar{\mathbf{x}}}$; then by the strict concavity of ϕ in $\mathbf{x}$, and by the definition of a saddle point, we have

$$\phi(\bar{\bar{\mathbf{x}}}, \bar{\mathbf{u}}) < \phi(\bar{\mathbf{x}}, \bar{\mathbf{u}}) \leq \phi(\bar{\mathbf{x}}, \bar{\bar{\mathbf{u}}})$$

and

$$\phi(\bar{\mathbf{x}}, \bar{\bar{\mathbf{u}}}) < \phi(\bar{\bar{\mathbf{x}}}, \bar{\bar{\mathbf{u}}}) \leq \phi(\bar{\bar{\mathbf{x}}}, \bar{\mathbf{u}}).$$

Hence

$$\phi(\bar{\bar{\mathbf{x}}}, \bar{\mathbf{u}}) < \phi(\bar{\bar{\mathbf{x}}}, \bar{\mathbf{u}}),$$

which is a contradiction.

Now consider the solution $\{\mathbf{x}(t), \mathbf{u}(t)\}$ with an initial position $\{\mathbf{x}^0, \mathbf{u}^0\}$, and put

$$V(t) = \tfrac{1}{2}\{|\mathbf{x}(t) - \bar{\mathbf{x}}|^2 + |\mathbf{u}(t) - \bar{\mathbf{u}}|^2\},$$

where $\{\bar{\mathbf{x}}, \bar{\mathbf{u}}\}$ is a saddle point of $\phi(\mathbf{x}, \mathbf{u})$. (The term $|\bar{\mathbf{x}}(t) - \bar{\mathbf{x}}|$ represents the Euclidean length of the vector $\bar{\mathbf{x}}(t) - \bar{\mathbf{x}}$, and $|\mathbf{u}(t) - \bar{\mathbf{u}}|$ is similarly defined.)

We shall show that

$$\dot{V}(t) \leq 0 \qquad \text{for any} \quad t,$$

and

$$\dot{V}(t) < 0 \qquad \text{for } t \text{ such that} \quad \mathbf{x}(t) \neq \bar{\mathbf{x}}.$$

We observe that

$$\begin{aligned} \dot{V}(t) &= (\dot{\mathbf{x}}(t), \mathbf{x}(t) - \bar{\mathbf{x}}) + (\dot{\mathbf{u}}(t), \mathbf{u}(t) - \bar{\mathbf{u}}) \\ &= (\boldsymbol{\phi}_{x(t)}, \mathbf{x}_{I(t)} - \bar{\mathbf{x}}_{I(t)}) - (\boldsymbol{\phi}_{u(t)}, \mathbf{u}_{I(t)} - \bar{\mathbf{u}}_{I(t)}), \end{aligned} \tag{7.3.2}$$

where

$$\mathbf{x}_{I(t)} = \boldsymbol{\delta}_{x(t)} \cdot \mathbf{x}(t), \qquad \mathbf{u}_{I(t)} = \boldsymbol{\delta}_{u(t)} \cdot \mathbf{u}(t),$$

$$\bar{\mathbf{x}}_{I(t)} = \boldsymbol{\delta}_{x(t)} \cdot \bar{\mathbf{x}}, \qquad \bar{\mathbf{u}}_{I(t)} = \boldsymbol{\delta}_{u(t)} \cdot \bar{\mathbf{u}},$$

$$\boldsymbol{\delta}_{x(t)} = \begin{Vmatrix} \delta_{x_1(t)} & & 0 \\ & \ddots & \\ 0 & & \delta_{x_n(t)} \end{Vmatrix}, \qquad \boldsymbol{\delta}_{u(t)} = \begin{Vmatrix} \delta_{u_1(t)} & & 0 \\ & \ddots & \\ 0 & & \delta_{u_m(t)} \end{Vmatrix},$$

$$\boldsymbol{\delta}_{x_i(t)} = \begin{cases} 0 & \text{if} \quad x_i(t) = 0 \text{ and } \phi_{x_i(t)} < 0, \\ 1 & \text{otherwise,} \end{cases}$$

$$\boldsymbol{\delta}_{u_k(t)} = \begin{cases} 0 & \text{if} \quad u_k(t) = 0 \text{ and } \phi_{u_k(t)} > 0, \\ 1 & \text{otherwise.} \end{cases}$$

Since $\phi(\mathbf{x}, \mathbf{u})$ is strictly concave in $\mathbf{x}$ and convex in $\mathbf{u}$, we have

$$\phi[\bar{\mathbf{x}}, \mathbf{u}(t)] - \phi[\mathbf{x}(t), \mathbf{u}(t)] < (\boldsymbol{\phi}_{x(t)}, \bar{\mathbf{x}} - \mathbf{x}(t)) \qquad \text{for} \quad \bar{\mathbf{x}} \neq \mathbf{x}(t), \tag{7.3.3}$$

and

$$\phi[\mathbf{x}(t), \bar{\mathbf{u}}] - \phi[\mathbf{x}(t), \mathbf{u}(t)] \geq (\boldsymbol{\phi}_{u(t)}, \bar{\mathbf{u}} - \mathbf{u}(t)) \tag{7.3.4}$$

(see Section B.4). Moreover, since $\{\bar{\mathbf{x}}, \bar{\mathbf{u}}\}$ is a saddle point of $\phi(\mathbf{x}, \mathbf{u})$, we have

$$\phi[\mathbf{x}(t), \bar{\mathbf{u}}] \leq \phi[\bar{\mathbf{x}}, \bar{\mathbf{u}}] \leq \phi[\bar{\mathbf{x}}, \mathbf{u}(t)].$$

Hence

$$0 \leq \phi[\bar{\mathbf{x}}, \mathbf{u}(t)] - \phi[\mathbf{x}(t), \bar{\mathbf{u}}] < (\boldsymbol{\phi}_{x(t)}, \bar{\mathbf{x}} - \mathbf{x}(t)) - (\boldsymbol{\phi}_{u(t)}, \bar{\mathbf{u}} - \mathbf{u}(t)). \tag{7.3.5}$$

Set

$$\mathbf{x}(t) = \mathbf{x}_{I(t)} + \mathbf{x}_{II(t)}, \qquad \mathbf{u}(t) = \mathbf{u}_{I(t)} + \mathbf{u}_{II(t)}.$$

From (7.3.2) and (7.3.5), after noticing that

$$(\boldsymbol{\phi}_{x(t)}, \mathbf{x}_{II(t)}(t)) = 0, \qquad (\boldsymbol{\phi}_{u(t)}, \mathbf{u}_{II(t)}(t)) = 0,$$

we obtain

$$\begin{aligned}\dot{V}(t) = {} & (\boldsymbol{\phi}_{x(t)}, \mathbf{x}(t) - \bar{\mathbf{x}}) - (\boldsymbol{\phi}_{x(t)}, \mathbf{x}_{II(t)} - \bar{\mathbf{x}}_{II(t)}) \\ & - (\boldsymbol{\phi}_{u(t)}, \mathbf{u}(t) - \bar{\mathbf{u}}) + (\boldsymbol{\phi}_{u(t)}, \mathbf{u}_{II(t)} - \bar{\mathbf{u}}_{II(t)}) \\ & < (\boldsymbol{\phi}_{x(t)}, \bar{\mathbf{x}}_{II(t)}) - (\boldsymbol{\phi}_{u(t)}, \bar{\mathbf{u}}_{II(t)}) \leq 0, \qquad \text{if} \quad \mathbf{x}(t) \neq \bar{\mathbf{x}}. \qquad (7.3.6)\end{aligned}$$

The last inequality is valid, since by definition $\bar{\mathbf{x}}_{II(t)}$ has its only nonzero components where $\boldsymbol{\phi}_{x(t)} < 0$, and $\bar{\mathbf{u}}_{II(t)}$ has its only nonzero components where $\boldsymbol{\phi}_{u(t)} > 0$. Similarly, we have $\dot{V}(t) \leq 0$ for any t.

Now we shall prove that the $\mathbf{x}$ component $\mathbf{x}(t)$ converges to the uniquely determined $\mathbf{x}$ component $\bar{\mathbf{x}}$ of saddle points.

Since $\dot{V}(t) \leq 0$ for any t and $V(t) \geq 0$, there exists a value

$$V^* = \lim_{t \to \infty} V(t).$$

Hence the set $\{\mathbf{x}(t) | t \geq 0\}$ is bounded. Let $\mathbf{x}^*$ be any accumulation point of $\{\mathbf{x}(t) | t \geq 0\}$. Then there is a sequence $\{t_\nu\}$ such that

$$\lim_{\nu \to \infty} \mathbf{x}(t_\nu) = \mathbf{x}^*$$

and

$$\lim_{\nu \to \infty} \mathbf{u}(t_\nu) = \mathbf{u}^*,$$

for a suitable $\mathbf{u}^*$. By the assumption of uniqueness the solution with initial position $\{\mathbf{x}(t_\nu), \mathbf{u}(t_\nu)\}$ is $\{\mathbf{x}(t + t_\nu), \mathbf{u}(t + t_\nu)\}$. Since the solution depends continuously on its initial position, we obtain that

$$\{\mathbf{x}^*(t), \mathbf{u}^*(t)\} = \lim_{\nu \to \infty} \{\mathbf{x}(t + t_\nu), \mathbf{u}(t + t_\nu)\},$$

where $\{\mathbf{x}^*(t), \mathbf{u}^*(t)\}$ represents the solution with starting position $\{\mathbf{x}^*, \mathbf{u}^*\}$. But

$$\begin{aligned}V^* &= \lim_{\nu \to \infty} \tfrac{1}{2}\{|\mathbf{x}(t + t_\nu) - \bar{\mathbf{x}}|^2 + |\mathbf{u}(t + t_\nu) - \bar{\mathbf{u}}|^2\} \\ &= \tfrac{1}{2}\{|\mathbf{x}^*(t) - \bar{\mathbf{x}}|^2 + |\mathbf{u}^*(t) - \bar{\mathbf{u}}|^2\}. \qquad (7.3.7)\end{aligned}$$

This is incompatible with (7.3.6) unless $\mathbf{x}^* = \bar{\mathbf{x}}$. Consequently, every accumulation point of $\{\mathbf{x}(t)\}$ is $\bar{\mathbf{x}}$ and we conclude that

$$\lim_{t \to \infty} \mathbf{x}(t) = \bar{\mathbf{x}}.$$

Lastly, we shall prove that the $\mathbf{u}$ component $\mathbf{u}(t)$ of the solution converges to the $\mathbf{u}$ component $\tilde{\mathbf{u}}$ of a saddle point $\{\bar{\mathbf{x}}, \tilde{\mathbf{u}}\}$, although $\tilde{\mathbf{u}}$ need not coincide with $\bar{\mathbf{u}}$. Since $V(t) \to V^*$ and $\mathbf{x}(t) \to \bar{\mathbf{x}}$, we find that $\mathbf{u}(t)$ converges and denote its limit by $\tilde{\mathbf{u}}$. Since $\boldsymbol{\phi}_u(\bar{\mathbf{x}}, \mathbf{u})$ is independent of $\mathbf{u}$ (ϕ is linear in $\mathbf{u}$), we infer that

$$\boldsymbol{\phi}_u(\bar{\mathbf{x}}, \tilde{\mathbf{u}}) = \boldsymbol{\phi}_u(\bar{\mathbf{x}}, \bar{\mathbf{u}}) \geq 0, \tag{7.3.8}$$

the second relation being valid because $\{\bar{\mathbf{x}}, \bar{\mathbf{u}}\}$ is a saddle point (Theorem 7.1.3). But $\dot{V}(t)$ tends to zero, and hence by (7.3.2)

$$(\boldsymbol{\phi}_{x(t)}, \mathbf{x}_{I(t)} - \bar{\mathbf{x}}_{I(t)}) - (\boldsymbol{\phi}_{u(t)}, \mathbf{u}_{I(t)} - \bar{\mathbf{u}}_{I(t)})$$

converges to zero. However, since $\mathbf{x}_{I(t)} \to \bar{\mathbf{x}}_{I(t)}$, it follows that

$$(\boldsymbol{\phi}_u(\bar{\mathbf{x}}, \tilde{\mathbf{u}}), \tilde{\mathbf{u}}_I - \bar{\mathbf{u}}_I) = 0$$

or

$$(\boldsymbol{\phi}_u(\bar{\mathbf{x}}, \tilde{\mathbf{u}}), \tilde{\mathbf{u}}_I) = (\boldsymbol{\phi}_u(\bar{\mathbf{x}}, \bar{\mathbf{u}}), \bar{\mathbf{u}}_I),$$

again exploiting the fact that $\boldsymbol{\phi}_u(\bar{\mathbf{x}}, \mathbf{u})$ is independent of $\mathbf{u}$. The indices of $\tilde{\mathbf{u}}_I$ include the terms in which $\tilde{u}_j > 0$ by the definition of I. However, $\boldsymbol{\phi}_u(\bar{\mathbf{x}}, \bar{\mathbf{u}}) \geq 0, \bar{\mathbf{u}} \geq \mathbf{0}$, and $(\boldsymbol{\phi}_u(\bar{\mathbf{x}}, \bar{\mathbf{u}}), \bar{\mathbf{u}}) = 0$ by Theorem 7.1.3. Therefore, by (7.3.8) and the definition of $\tilde{\mathbf{u}}_I$, we have

$$0 \leq (\boldsymbol{\phi}_u(\bar{\mathbf{x}}, \tilde{\mathbf{u}}), \tilde{\mathbf{u}}) \leq (\boldsymbol{\phi}_u(\bar{\mathbf{x}}, \bar{\mathbf{u}}), \bar{\mathbf{u}}) = 0,$$

which gives us

$$(\boldsymbol{\phi}_u(\bar{\mathbf{x}}, \tilde{\mathbf{u}}), \tilde{\mathbf{u}}) = 0. \tag{7.3.9}$$

Comparing (7.3.8) and (7.3.9) with the formulas of Theorem 7.1.3, we conclude that $\{\bar{\mathbf{x}}, \tilde{\mathbf{u}}\}$ is a saddle point as asserted.

7.4 The vector maximum problem. The nonlinear programming problem of Section 7.1 may be extended to deal with the problem of maximizing a vector function $\mathbf{G}(\mathbf{x})$ in E^r subject to the constraints $\mathbf{F}(\mathbf{x}) \geq \mathbf{0}, \mathbf{x} \in X$, where X is closed and convex. The components of $\mathbf{G}(\mathbf{x})$ and $\mathbf{F}(\mathbf{x})$ are assumed to be concave functions of $\mathbf{x}$. An $\mathbf{x}^0$ satisfying the constraints (i.e., a feasible vector) is said to be an "efficient point" if there exists no other vector $\mathbf{x}$ satisfying the constraints such that $\mathbf{G}(\mathbf{x}) > \mathbf{G}(\mathbf{x}^0)$. The vector maximum problem is solved when all efficient points have been determined.

The constraints $\mathbf{F}(\mathbf{x}) \geq \mathbf{0}$ and $\mathbf{x} \in X$ as before are assumed to satisfy the conditions of Theorem 7.1.1. An efficient point $\mathbf{x}^0$ satisfying the constraints is characterized by constructing an ordinary maximization problem with the same constraints such that $\mathbf{x}^0$ is a maximum point. For this purpose, the following lemma is fundamental.

▶ LEMMA 7.4.1. If $\mathbf{x}^0$ is an efficient point, then there exists a vector $\mathbf{v} = \langle v_1, \ldots, v_r \rangle$ with $v_i \geq 0$ and

$$\sum_{i=1}^{r} v_i = 1$$

such that the maximum of the function $g(\mathbf{x}) = (\mathbf{v}, \mathbf{G}(\mathbf{x}))$ in the set of all $\mathbf{x}$ satisfying $\mathbf{F}(\mathbf{x}) \geq \mathbf{0}$, $\mathbf{x} \in X$ is achieved for $\mathbf{x} = \mathbf{x}^0$.

Proof. The constraint set, which is clearly convex, will be denoted by S. Let

$$H = \{\mathbf{y} = \mathbf{G}(\mathbf{x}) - \mathbf{G}(\mathbf{x}^0) \in E^r | \mathbf{x} \in S\},$$

and let K be the convex set spanned by H. Since $\mathbf{x}^0$ is efficient, the set H cannot overlap with the interior of the positive orthant in E^r. It follows immediately from the concavity of G and the fact that $\mathbf{x}^0$ is efficient that the set K does not contain an interior point of the positive orthant. Let

$$\boldsymbol{\xi} = \langle \xi_1^0, \xi_2^0, \ldots, \xi_r^0, c \rangle \qquad (\boldsymbol{\xi} \neq 0)$$

represent a hyperplane which separates the positive orthant from K. The arguments of Lemma B.1.3 apply here, and we see that $\xi_i^0 \geq 0$ and $c = 0$. Set

$$v_i^0 = \frac{\xi_j^0}{\sum_j \xi_j^0} \cdot$$

Then it is readily verified that the vector $\mathbf{v}^0$ satisfies the assertion of the lemma.

Remark 7.4.1. Let us assume momentarily that G has continuous partial derivatives. If the set H is not tangent to any coordinate direction at zero, then the separating hyperplane $\boldsymbol{\xi}$ may be chosen so that it touches the first orthant only at zero. In this case, we see that $v_i^0 > 0$ for every i. Some conditions are now cited which guarantee the above property for H.

▶ COROLLARY 7.4.1. If $\mathbf{x}^0$ is an efficient interior point of the constraint set S with the property that $\mathbf{G}_{x^0} \cdot \mathbf{dx} \geq \mathbf{0}$ for no vector differential $\mathbf{dx}$, then the corresponding vector $\mathbf{v}$ of Lemma 7.4.1 may be chosen possessing all positive components.

The same conclusion holds if $\mathbf{x}^0$ is efficient and belongs to the boundary of the constraint set, provided $\mathbf{G}_{x^0} \cdot \mathbf{dx} \geq \mathbf{0}$ for no vector differential satisfying $\mathbf{F}_{x^0}\, \mathbf{dx} \geq \mathbf{0}$, $\mathbf{I}_1\, \mathbf{dx} \geq \mathbf{0}$, where the inequalities $\mathbf{F}(\mathbf{x}^0) \geq 0$ and $\mathbf{I}\mathbf{x}^0 \geq 0$ (I being the identity matrix) have been separated into the two sets of conditions $\mathbf{F}_1(\mathbf{x}^0) = \mathbf{0}$, $\mathbf{I}_1\mathbf{x}^0 = \mathbf{0}$ and $\mathbf{F}_2(\mathbf{x}^0) > \mathbf{0}$, $\mathbf{I}_2\mathbf{x}^0 > \mathbf{0}$.

The saddle point problem that is equivalent to the efficient point problem can now be given.

▶THEOREM 7.4.1. Let $\mathbf{G}$ be a concave vector function defined for $\mathbf{x} \in X$, and let $\mathbf{F}$ be the same as in Theorem 7.1.1. In order for $\mathbf{x}^0$ to be an efficient point, it is necessary that there exist two nonnegative vectors $\mathbf{v}^0$ ($\mathbf{v}^0 \neq \mathbf{0}$) and $\mathbf{u}^0$ with the same number of components as $\mathbf{G}$ and $\mathbf{F}$, respectively, such that $\{\mathbf{x}^0, \mathbf{u}^0\}$ is a saddle point of

$$\chi(\mathbf{x}, \mathbf{u}, \mathbf{v}^0) = (\mathbf{v}^0, \mathbf{G}(\mathbf{x})) + (\mathbf{u}, \mathbf{F}(\mathbf{x}));$$

that is,

$$\chi(\mathbf{x}, \mathbf{u}^0, \mathbf{v}^0) \leq \chi(\mathbf{x}^0, \mathbf{u}^0, \mathbf{v}^0) \leq \chi(\mathbf{x}^0, \mathbf{u}, \mathbf{v}^0) \tag{7.4.1}$$

for all nonnegative $\mathbf{u}$ and $\mathbf{x} \in X$. If $\mathbf{v}^0$ has all components strictly positive, then (7.4.1) is also sufficient.

Proof. Necessity. By Lemma 7.4.1 there exists a nonnegative vector $\mathbf{v}^0$ such that the maximum of $g(\mathbf{x}) = (\mathbf{v}^0, \mathbf{G}(\mathbf{x}))$ in the set S is achieved for $\mathbf{x}^0$. The conclusion of (7.4.1) is obtained by applying Theorem 7.1.1.

Sufficiency. Reversing the reasoning, it follows again by Theorem 7.1.1 that the maximum of $(\mathbf{v}^0, \mathbf{G}(\mathbf{x}))$ in S is achieved for $\mathbf{x}^0$. Since $\mathbf{v}^0$ is a strictly positive vector, it also follows that $\mathbf{x}^0$ is efficient.

The economic interpretation of the saddle-point problem (7.4.1) is analogous to that of (7.1.2). Since the desired output is now represented not by a real number but by a vector, the concept of maximum is replaced by that of efficient point.

***7.5 Conjugate functions.** A further extension of the results of linear programming to nonlinear maximization and minimization problems involving concave and convex functions can be effected by means of the theory of conjugate functions as set forth by Fenchel. Aside from its value in nonlinear programming, the theory is elegant and interesting in itself and has additional applications to the theory of inequalities and to the theory of the structure of convex sets. Both these topics embrace many of the fundamental working tools of finite game theory and programming problems. In the following sections we develop the basic properties of conjugate functions. Their relevance to nonlinear programming and convex set theory is described in Sections 7.8 and 7.9, respectively.

Let f be a convex function defined on a convex set C in E^n. Recall that f is continuous in the relative interior of C, and that if $\mathbf{x}^*$ is on the boundary of C and $f(\mathbf{x}^*)$ is defined, then

$$\underline{\lim}_{x \to x^*} f(\mathbf{x}) \leq f(\mathbf{x}^*)$$

(see Section B.4).

For the discussion that follows, whenever

$$\underline{\lim}_{x \to x^*} f(\mathbf{x})$$

is finite, the set C is redefined to include all points $\mathbf{x}^*$. Likewise, the definition of f is altered at such points so that

$$f(\mathbf{x}^*) = \underline{\lim}_{x \to x^*} f(\mathbf{x}).$$

This modified function and the enlarged set will be denoted by f^0 and C^0, respectively.

▶ DEFINITION 7.5.1. A convex function defined on a convex set S will be called closed if for any $\mathbf{x}^0$ where

$$\underline{\lim}_{x \to x^0} f(\mathbf{x})$$

is finite, $f(\mathbf{x}^0)$ is defined and is equal to

$$\underline{\lim}_{x \to x^0} f(\mathbf{x}).$$

The process of obtaining f^0 and C^0 from f and C will be referred to as "closing the function," and $[f^0, C^0]$ will be called the "closure" of $[f, C]$. The closure of $[f^0, C^0]$ is again $[f^0, C^0]$.

▶ LEMMA 7.5.1. If f is a convex function defined on a convex set C in E^n, then its closure f^0 is convex with domain of definition C^0.

Proof. Let $\mathbf{x}^* = t\mathbf{x}_1 + (1 - t)\mathbf{x}_2$, with $0 \leq t \leq 1$ and $\mathbf{x}_1, \mathbf{x}_2 \in C^0$. We can find sequences of interior points $\mathbf{x}_1^n \to \mathbf{x}_1$ and $\mathbf{x}_2^n \to \mathbf{x}_2$ such that

$$\lim_{n \to \infty} f^0(\mathbf{x}_1^n) = \lim_{n \to \infty} f(\mathbf{x}_1^n) = f^0(\mathbf{x}_1)$$

and

$$\lim_{n \to \infty} f(\mathbf{x}_2^n) = f^0(\mathbf{x}_2).$$

Then

$$f^0(\mathbf{x}^*) = \underline{\lim}_{x \to x^*} f(\mathbf{x}) \leq \underline{\lim}_{n \to \infty} f[t\mathbf{x}_1^n + (1 - t)\mathbf{x}_2^n] \leq tf^0(\mathbf{x}_1) + (1 - t)f^0(\mathbf{x}_2).$$

Henceforth we shall assume that the original convex function f defined on C is already closed.

We now construct the set

$$[f, C] = \{\{z, \mathbf{x}\} \in E^{n+1} | \mathbf{x} \in C \quad \text{and} \quad z \geq f(\mathbf{x})\}. \tag{7.5.1}$$

The lower boundary of this set is the graph of the function f. Since f is assumed to be closed and convex, it follows readily that $[f, C]$ is a closed convex set in E^{n+1}.

A convex set S of points in E^n can also be characterized by specifying the set of all its supporting hyperplanes (see Section B.1). It is therefore natural in the study of S to consider the set of all its supporting planes (determined by a normal unit vector and a real number) as a dual set S^* in E^n space. Starting with S^* and constructing its dual set leads essentially back to S. We now explore the consequences of this duality principle as applied to $[f, C]$. The *conjugate set* $[f, C]^*$ is defined as the set of all nonvertical hyperplanes in E^{n+1} with the property that the set $[f, C]$ lies above the plane. More precisely, a plane in E^{n+1} is determined by a vector $\boldsymbol{\xi}$ in E^n and a real number c such that $-c$ is the intercept of the plane and the vertical axis of E^{n+1}. The point $\{c, \boldsymbol{\xi}\}$ in E^{n+1} belongs to $[f, C]^*$ if and only if

$$z \geq (\mathbf{x}, \boldsymbol{\xi}) - c \qquad \text{for every} \quad \{z, \mathbf{x}\} \text{ in } [f, C]. \tag{7.5.2}$$

In order to verify that a point $\{c, \boldsymbol{\xi}\}$ is a member of $[f, C]^*$, it is enough to satisfy the inequalities

$$c \geq (\mathbf{x}, \boldsymbol{\xi}) - f(\mathbf{x}) \qquad \text{for every} \quad \mathbf{x} \text{ in } C. \tag{7.5.3}$$

Clearly, if $\{c_1, \boldsymbol{\xi}\}$ is in $[f, C]^*$ and $c_2 > c_1$, then $\{c_2, \boldsymbol{\xi}\}$ is in $[f, C]^*$. Consequently, for a given vector $\boldsymbol{\xi}$, the smallest admissible c is

$$\sup_{x \in C} [(\mathbf{x}, \boldsymbol{\xi}) - f(\mathbf{x})]. \tag{7.5.4}$$

Let

$$\Gamma = \{\boldsymbol{\xi} \text{ in } E^n | \sup_{x \in C} [(\mathbf{x}, \boldsymbol{\xi}) - f(\mathbf{x})] < \infty\}$$

and set

$$\phi(\boldsymbol{\xi}) = \sup_{x \in C} [(\mathbf{x}, \boldsymbol{\xi}) - f(\mathbf{x})]$$

for $\boldsymbol{\xi}$ in Γ. The function ϕ is referred to as the conjugate function of f. It is the largest vertical intercept of planes with normal $\{-1, \boldsymbol{\xi}\}$ lying below the graph of f.

▶ LEMMA 7.5.2. The set Γ is convex and nonvoid, and ϕ is a convex closed function with domain Γ.

Proof. If C has an interior in E^n, we construct a supporting hyperplane to $[f, C]$ at $\mathbf{x}^0$, an interior point of C. Such a plane is clearly nonvertical. If C has no interior in E^n, let $\mathbf{x}^0$ denote a relative interior point of C and $\widetilde{C}$ the linear extension of C in E^n. We may consider f defined in the affine†

† An affine space is a translation of a linear space.

space $\tilde{C} \supset C$. A supporting plane to $[f, C]$ at $\mathbf{x}^0$, defined only in the smaller space of the values of f and the set $\tilde{C}$, is a nonvertical hyperplane which may be extended in infinitely many ways to E^n while remaining a nonvertical hyperplane. Thus in any case $[\phi, \Gamma]$ is nonvoid. The proof that ϕ is convex is straightforward.

We indicate the proof that ϕ is also closed. Let $\xi^n \to \xi^*$; then $\phi(\xi^n) \geq (\mathbf{x}, \xi^n) - f(\mathbf{x})$, and hence

$$\varliminf_{n} \phi(\xi^n) \geq \lim_{n} (\mathbf{x}, \xi^n) - f(\mathbf{x}) = (\mathbf{x}, \xi^*) - f(\mathbf{x})$$

for all $\mathbf{x} \in C$. Therefore

$$\varliminf_{n} \phi(\xi^n) \geq \sup_{x \in C} [(\mathbf{x}, \xi^*) - f(\mathbf{x})] = \phi(\xi^*).$$

Since the opposite inequality always holds for convex functions, we must have equality.

Remark 7.5.1. The discussion preceding this last lemma implies that

$$[f, C]^* = [\phi, \Gamma]. \tag{7.5.5}$$

The striking conclusion of Lemma 7.5.2 and (7.5.5) is that the dual set to $[f, C]$ is again of the same type. Boundary points of $[\phi, \Gamma]$ correspond to points on the graph of $\phi(\xi)$ and may be viewed as planes of support to the graph of $f(\mathbf{x})$. If $\{f(\mathbf{x}), \mathbf{x}\}$ is a point of contact between the set $[f, C]$ and the plane $\{\phi(\xi), \xi\}$ with normal vector of components $\{-1, \xi\}$ and intercept $\phi(\xi)$ in the vertical direction, then $f(\mathbf{x}) + \phi(\xi) = (\mathbf{x}, \xi)$. The next lemma expresses the duality principle common to the study of convex sets.

▶ LEMMA 7.5.3. $[[f, C]^*]^* = [f, C]$.

Proof. We first show that $[f, C] \subset [\phi, \Gamma]^* = [[f, C]^*]^*$. Let $\mathbf{x}$ be in C; then $\phi(\xi) \geq (\mathbf{x}, \xi) - f(\mathbf{x})$ for any ξ in Γ. But $z \geq (\mathbf{x}, \xi) - \phi(\xi)$ whenever $z \geq f(\mathbf{x})$ for all ξ in Γ. Therefore $\{z, \mathbf{x}\}$ is in $[\phi, \Gamma]^*$.

In order to establish the converse, let $\{z^0, \mathbf{x}^0\}$ be a point in E^{n+1} but not in $[f, C]$. Since $[f, C]$ is closed and convex, there exists a nonvertical hyperplane that separates the point $\{z^0, \mathbf{x}^0\}$ from the set $[f, C]$. Let the plane be described by $\{c, \xi\}$. Then $f(\mathbf{x}) \geq (\mathbf{x}, \xi) - c$ for all $\mathbf{x}$ in C, and $z^0 < (\mathbf{x}^0, \xi) - c$. The first inequality implies that $\{c, \xi\}$ is in $[\phi, \Gamma]$, and the second that $\{z^0, \mathbf{x}^0\}$ is not in $[\phi, \Gamma]^*$. Hence $[f, C] \supset [\phi, \Gamma]^*$, which together with the first result proves the lemma.

Since $f(\mathbf{x})$ is conjugate to $\phi(\xi)$,

$$f(\mathbf{x}) = \sup_{\xi \in \Gamma} [(\mathbf{x}, \xi) - \phi(\xi)]. \tag{7.5.6}$$

Boundary points. If $\boldsymbol{\xi}^0$ is in Γ, then $\{\phi(\boldsymbol{\xi}^0), \boldsymbol{\xi}^0\}$ is a point on the boundary of $[\phi, \Gamma]$ and represents a supporting plane to $[f, C]$. Let the contact set be $C(\boldsymbol{\xi}^0)$. Any point $\{f(\mathbf{x}^0), \mathbf{x}^0\}$ in the contact set is a supporting hyperplane to $[\phi, \Gamma]$ at the point $\{\phi(\boldsymbol{\xi}^0), \boldsymbol{\xi}^0\}$. Thus if $\{f(\mathbf{x}^0), \mathbf{x}^0\}$ belongs to $C(\boldsymbol{\xi}^0)$, then $\{\phi(\boldsymbol{\xi}^0), \boldsymbol{\xi}^0\}$ is in the intersection of $[\phi, \Gamma]$ and the plane $\{f(\mathbf{x}^0), \mathbf{x}^0\}$. If the contact sets corresponding to $\mathbf{x}^0$ and $\boldsymbol{\xi}^0$ consist of unique interior points, and if partial derivatives exist for both $f(\mathbf{x})$ and $\phi(\boldsymbol{\xi})$ at $\mathbf{x}^0$ and $\boldsymbol{\xi}^0$, respectively, then, since $\phi(\boldsymbol{\xi}^0) + f(\mathbf{x}^0) = (\mathbf{x}^0, \boldsymbol{\xi}^0)$, we have

$$\left(\frac{\partial f}{\partial x_i}\right)_{x^0} = \xi_i^0, \qquad \left(\frac{\partial \phi}{\partial \xi_i}\right)_{\xi^0} = x_i^0. \tag{7.5.7}$$

***7.6 Composition of conjugate functions.** Three operations are commonly applied to convex sets in building new convex sets. They are: (1) forming direct sums of convex sets, (2) constructing the convex hulls of sets, and (3) taking the intersections of convex sets (see Appendix B). In this section, we investigate the application of these operations to the conjugate sets introduced above.

Let $[f_1, C_1]$ and $[f_2, C_2]$ be two closed convex sets of the type described in (7.5.1). Let their conjugate sets be $[\phi_1, \Gamma_1]$ and $[\phi_2, \Gamma_2]$, respectively. The vector sum $S = [\phi_1, \Gamma_1] + [\phi_2, \Gamma_2]$ is the set of all points of the form $\{a_1 + a_2, \boldsymbol{\xi}^1 + \boldsymbol{\xi}^2\}$, where $\{a_1, \boldsymbol{\xi}^1\}$ traverses $[\phi_1, \Gamma_1]$ and $\{a_2, \boldsymbol{\xi}^2\}$ traverses $[\phi_2, \Gamma_2]$. The closure of S will be denoted by $\overline{S}$.

▶THEOREM 7.6.1. If $C = C_1 \cap C_2$ is nonvoid and $f(\mathbf{x}) = f_1(\mathbf{x}) + f_2(\mathbf{x})$ for $\mathbf{x}$ in C, then $[f, C]^* = \overline{S} = [\phi, \Gamma]$, where $\Gamma_1 + \Gamma_2 \subset \Gamma \subset \overline{\Gamma_1 + \Gamma_2}$ and

$$\phi(\boldsymbol{\xi}) = \inf_{\substack{\boldsymbol{\xi}^1 \in \Gamma_1,\ \boldsymbol{\xi}^2 \in \Gamma_2 \\ \boldsymbol{\xi}^1 + \boldsymbol{\xi}^2 = \boldsymbol{\xi}}} \{\phi_1(\boldsymbol{\xi}^1) + \phi_2(\boldsymbol{\xi}^2)\} \tag{7.6.1}$$

for $\boldsymbol{\xi}$ in $\Gamma_1 + \Gamma_2$.

Proof. For clarity, the proof is divided into three separate parts.

(i) $$[f, C]^* \supset \overline{S}.$$

The conjugate of $[f, C]$ is the set of all nonvertical planes $\{a, \boldsymbol{\xi}\}$ for which $a \geq (\mathbf{x}, \boldsymbol{\xi}) - f(\mathbf{x})$ for all $\mathbf{x}$ in C. If $\{a_i, \boldsymbol{\xi}^i\}$ belongs to $[\phi_i, \Gamma_i]$, then $a_i \geq (\boldsymbol{\xi}^i, \mathbf{x}) - f_i(\mathbf{x})$ for $i = 1, 2$. By adding, we obtain that $a_1 + a_2 \geq (\boldsymbol{\xi}^1 + \boldsymbol{\xi}^2, \mathbf{x}) - f(\mathbf{x})$ for $\mathbf{x}$ in C, so that $\{a_1 + a_2, \boldsymbol{\xi}^1 + \boldsymbol{\xi}^2\}$ is in $[f, C]^*$. Hence $[f, C]^* \supset S$, and since $[f, C]^*$ is closed, it follows that $[f, C]^* \supset \overline{S}$.

(ii) $$\phi(\boldsymbol{\xi}) \quad \text{is convex and} \quad \overline{S} = [\phi, \Gamma].$$

Next we define $\phi(\boldsymbol{\xi})$ for $\boldsymbol{\xi}$ in $\Gamma_1 + \Gamma_2$ according to (7.6.1). In view of (i), $\phi(\boldsymbol{\xi})$ is finite for each $\boldsymbol{\xi}$ in $\Gamma_1 + \Gamma_2$. From its definition, we may verify that

$\phi(\xi)$ is convex. Moreover, by examining the construction of $\bar{S}$, we see that $\bar{S} = [\phi, \Gamma]$ as asserted.

(iii) $$\bar{S} \supset [f, C]^*.$$

It suffices to verify that $\bar{S}^* = [\phi, \Gamma]^* \subset [f, C]$. By definition the set $\bar{S}^*$ is composed of all nonvertical barriers to S; hence its elements can be represented in the form $\{z, \mathbf{x}\}$. A point $\{z, \mathbf{x}\}$ is in $\bar{S}^*$ if and only if for every $\{a_1, \xi^1\}$ and $\{a_2, \xi^2\}$ in $[\phi_1, \Gamma_1]$ and $[\phi_2, \Gamma_2]$, respectively,

$$z \geq (\mathbf{x}, \xi^1 + \xi^2) - a_1 - a_2.$$

Since $a_1 \geq \phi_1(\xi^1)$ and $a_2 \geq \phi_2(\xi^2)$, this last condition is equivalent to $z \geq (\mathbf{x}, \xi^1 + \xi^2) - \phi_1(\xi^1) - \phi_2(\xi^2)$. Hence

$$z \geq \sup_{\xi^1 \in \Gamma^1} [(\mathbf{x}, \xi^1) - \phi_1(\xi^1)] + \sup_{\xi^2 \in \Gamma^2} [(\mathbf{x}, \xi^2) - \phi_2(\xi^2)] = f_1(\mathbf{x}) + f_2(\mathbf{x}).$$

Consequently, $\{z, x\}$ is in $[f, C]$. This applies to every point in $\bar{S}^*$, and the proof of the last statement is therefore complete.

Let $[f_\alpha, C_\alpha]$ be a family of closed convex sets in E^{n+1} of the form given in (7.5.1). Let $C \subset \cap_\alpha C_\alpha$ be the set of all x for which $\sup_\alpha f_\alpha(\mathbf{x})$ is finite. On C, define $f(\mathbf{x}) = \sup_\alpha f_\alpha(\mathbf{x})$; then C is convex and $f(\mathbf{x})$ is a convex closed function defined over C. In terms of the point set $[f, C] = \cap_\alpha [f_\alpha, C_\alpha]$, we wish to characterize the set $[\phi, \Gamma] = [f, C]^*$. Let $[\phi_\alpha, \Gamma_\alpha] = [f_\alpha, C_\alpha]^*$ and let H be the convex hull of $\{\Gamma_\alpha\}$.

▶ **Theorem 7.6.2.** If C is nonvoid and $f(\mathbf{x}) = \sup_\alpha f_\alpha(\mathbf{x})$ for $\mathbf{x}$ in C, then $[f, C]^*$ is the closure of the convex hull G of $[\phi_\alpha, \Gamma_\alpha]$. Furthermore, $\bar{G} = [\phi, \Gamma]$, $H \subset \Gamma \subset \bar{H}$; and

$$\phi(\xi) = \inf_{\Sigma \lambda_i \xi^{\alpha_i} = \xi} \sum_{i=1}^{n+1} \lambda_i \phi_{\alpha_i}(\xi^{\alpha_i}), \tag{7.6.2}$$

the infimum operation being extended over all representations of ξ as a centroid of $n + 1$ points taken from any $n + 1$ of the sets Γ_α.

Proof. Since $[f, C] \subset [f_\alpha, C_\alpha]$, it follows that $[f, C]^* \supset [f_\alpha, C_\alpha]^* = [\phi_\alpha, \Gamma_\alpha]$. The fact that $[f, C]^*$ is closed and convex implies that it must contain all finite convex combinations of $[\phi_\alpha, \Gamma_\alpha]$ and their limits. Hence $[f, C]^* \supset \bar{G}$.

Since G in E^{n+1} is the convex hull of $[\phi_\alpha, \Gamma_\alpha]$, we may define $\phi(\xi)$ as the smallest ordinate (i.e., first component) with $\{\phi(\xi), \xi\}$ in $\bar{G}$. Since $[\phi_\alpha, \Gamma_\alpha]$ spans G convexly in E^{n+1}, every point of $\bar{G}$ can be approached by points of the form

$$\xi = \sum_{i=1}^{n+2} \lambda_i \xi^{\alpha_i} \qquad \text{and} \qquad c = \sum_{i=1}^{n+2} \lambda_i c_{\alpha_i}$$

with $\{c_{\alpha_i}, \xi^{\alpha_i}\}$ in $[\phi_{\alpha_i}, \Gamma_{\alpha_i}]$. (See Lemma B.2.1.) But $\phi(\xi)$ is the smallest admissible c for the vector ξ. Therefore

$$\phi(\xi) = \inf_{\Sigma \lambda_i \xi^{\alpha_i} = \xi} \sum_{i=1}^{n+2} \lambda_i \phi_{\alpha_i}(\xi^{\alpha_i}).$$

The point $\{\phi(\xi), \xi\}$ is on the boundary of $\overline{G}$; it is therefore located in a simplex of dimension at most n. Since extreme points of $\overline{G}$ must arise from extreme points of $[\phi_{\alpha_i}, \Gamma_{\alpha_i}]$, the corners of this simplex can be made to correspond to points $\{\phi_{\alpha_i}(\xi^{\alpha_i}), \xi^{\alpha_i}\}$. Hence, in evaluating $\phi(\xi)$ it suffices to consider convex combinations of at most $n + 1$ points. With the definition of $\phi(\xi)$ as proposed, we may easily verify that $\phi(\xi)$ is a closed function defined on Γ as indicated and that $\overline{G} = [\phi, \Gamma]$.

It remains to prove that $[f, C]^* = \overline{G}$. In view of the preceding, we need only establish that $[f, C]^* \subset \overline{G} = [\phi, \Gamma]$, or equivalently $[f, C] \supset [\phi, \Gamma]^*$. Let $\{z^0, \mathbf{x}^0\}$ be a point of $[\phi, \Gamma]^*$. Then for every α and ξ^α in Γ_α, we have $z^0 \geq (\xi^\alpha, \mathbf{x}^0) - \phi_\alpha(\xi^\alpha)$. Hence $\{z^0, \mathbf{x}^0\}$ is in $[\phi_\alpha, \Gamma_\alpha]^*$ for every α, and therefore $\{z^0, \mathbf{x}^0\} \in \cap_\alpha [f_\alpha, C_\alpha] = [f, C]$.

Another useful consequence of the definition of conjugate functions is the following theorem.

▶THEOREM 7.6.3. Let $f(\mathbf{x})$ in C be a closed convex function and $\phi(\xi)$ its conjugate function. Then the conjugate function of $f(\mathbf{x}) + k$ in C, where k is a constant, is $\phi(\xi) - k$ in Γ. The conjugate of $f(\mathbf{x} - \mathbf{v})$ in $C + \mathbf{v}$, where $\mathbf{v}$ is a constant vector, is $\phi(\xi) + (\mathbf{v}, \xi)$ in Γ.

Our next theorem will be of value in the theory of convex games as well as in the study of convex sets (see Volume II, Section 4.3).

▶THEOREM 7.6.4. Let $f_\alpha(\mathbf{x})$ be a family of convex functions defined over a bounded convex set C in E^n, such that $\sup_\alpha f_\alpha(\mathbf{x})$ is finite for each $\mathbf{x}$. Then for any $a \leq \inf_x \sup_\alpha f_\alpha(\mathbf{x})$ and $\epsilon > 0$, there exist functions f_{α_i} and constants $\lambda_i \geq 0$ $(i = 1, \ldots, n + 1)$ such that

$$\sum_{i=1}^{n+1} \lambda_i f_{\alpha_i}(\mathbf{x}) \geq a - \epsilon \qquad \text{for all} \quad \mathbf{x} \in C$$

and $\sum \lambda_i = 1$.

Proof. Let $f(\mathbf{x}) = \sup_\alpha f_\alpha(\mathbf{x})$, and $z_0 = \inf_x f(\mathbf{x})$. The horizontal plane of height z_0 is a supporting plane to the convex set $[f, C]$, and it is also a boundary point of $[f, C]^*$. For this plane, $\boldsymbol{\xi} = \mathbf{0}$ (the zero vector) and $\phi(\mathbf{0}) = -z_0$. It now follows from Theorem 7.6.2 that

$$-z_0 = \phi(\mathbf{0}) = \inf_{\Sigma\lambda_i\boldsymbol{\xi}^{\alpha_i}=\mathbf{0}} \sum_{i=1}^{n+1} \lambda_i\phi_{\alpha_i}(\boldsymbol{\xi}^{\alpha_i}),$$

where $\lambda_i \geq 0$ and $\sum\lambda_i = 1$. Therefore, for the given $\epsilon > 0$, we can find λ_i and $\boldsymbol{\xi}^{\alpha_i}$ such that

$$-z_0 \geq \sum_{i=1}^{n+1} \lambda_i\phi_{\alpha_i}(\boldsymbol{\xi}^{\alpha_i}) - \epsilon \qquad \text{and} \qquad \sum_{i=1}^{n+1} \lambda_i\boldsymbol{\xi}^{\alpha_i} = \mathbf{0}.$$

But for any $\mathbf{x} \in C$, we have $\phi_{\alpha_i}(\boldsymbol{\xi}^{\alpha_i}) \geq (\mathbf{x}, \boldsymbol{\xi}^{\alpha_i}) - f_{\alpha_i}(\mathbf{x})$. Then

$$z_0 \leq -\sum_{i=1}^{n+1} \lambda_i(\mathbf{x}, \boldsymbol{\xi}^{\alpha_i}) + \sum_{i=1}^{n+1} \lambda_i f_{\alpha_i}(\mathbf{x}) + \epsilon = \sum_{i=1}^{n+1} \lambda_i f_{\alpha_i}(\mathbf{x}) + \epsilon.$$

Since $a - \epsilon \leq z_0 - \epsilon$, the theorem is valid.

Support functions. Given any convex set $C \subset E^n$, we may define $f(\mathbf{x}) \equiv 0$ on C. The dual set to $[f, C]$ is of the form $[\phi, \Gamma]$, where

$$\phi(\boldsymbol{\xi}) = \sup_{x\in C} (\mathbf{x}, \boldsymbol{\xi})$$

and the domain Γ is the set of vectors for which

$$\sup_{x\in C} (\mathbf{x}, \boldsymbol{\xi})$$

is finite. We call $\phi(\boldsymbol{\xi})$ the support function of the set C.

▶ LEMMA 7.6.1. The support function $\phi(\boldsymbol{\xi})$ of a convex set C is a positively homogeneous function, and $[\phi, \Gamma]$ is a cone with vertex at the origin. Conversely, if a function ϕ defined over a closed convex cone of E^n with vertex at the origin is positively homogeneous, then $\phi(\boldsymbol{\xi})$ is the support function of some convex set C in E^n.

Proof. If $k > 0$, then

$$\phi(k\boldsymbol{\xi}) = \sup_{x\in C} (\mathbf{x}, k\boldsymbol{\xi}) = k \sup_{x\in C} (\mathbf{x}, \boldsymbol{\xi}) = k\phi(\boldsymbol{\xi}).$$

Hence Γ and $[\phi, \Gamma]$ are cones with vertex at the origin.

Conversely, let $\phi(\boldsymbol{\xi})$ be a positively homogeneous function defined over a closed convex cone Γ with vertex at the origin. Then $[\phi, \Gamma]$ is a cone of the same type, and the extreme points of $[\phi, \Gamma]^*$ must be planes through the origin. Hence $[\phi, \Gamma]^* = [f, C]$ where $f \equiv 0$ on C, and C is convex. Finally,

$$[\phi, \Gamma] = [\phi, \Gamma]^{**} = [f, C]^*,$$

so that ϕ is the support function of the set C.

The representation of the support function of an intersection of convex sets in terms of the support functions of the individual sets can be deduced from (7.6.2). Let C be a family of convex sets in E^n, and let ϕ be their support functions. If $C = \cap_\alpha C_\alpha$ and ϕ is the support function of C, then by Theorem 7.6.2

$$\phi(\xi) = \inf_{\Sigma\lambda_i\xi^{\alpha_i}=\xi} \sum_{i=1}^{n+1} \lambda_i\phi_{\alpha_i}(\xi_{\alpha_i}) = \inf_{\Sigma\lambda_i\xi^{\alpha_i}=\xi} \sum_{i=1}^{n+1} \phi_{\alpha_i}(\lambda_i\xi^{\alpha_i})$$

$$= \inf_{\Sigma\eta^{\alpha_i}=\xi} \sum_{i=1}^{n+1} \phi_{\alpha_i}(\eta^{\alpha_i}),$$

by the positive homogeneity of $\phi_{\alpha_i}(\xi)$.

***7.7 Conjugate concave functions.** A conjugate function theory for concave functions $g(\mathbf{x})$ defined over a convex set D can be developed analogously to the theory for conjugate convex functions. We define the convex set

$$[g, D] = (\{z, \mathbf{x}\} | \mathbf{x} \in D \quad \text{and} \quad z \leq g(\mathbf{x})).$$

By an appropriate redefinition of $g(\mathbf{x})$ and D, we can make the set $[g, D]$ closed (see Section 7.5). We may then define the conjugate set of $[g, D]$ in terms of nonvertical planes which keep $[g, D]$ on one side. A nonvertical plane is determined by a vector ξ and a number b, where $-b$ is the z intercept of the plane. The barrier plane is in $[g, D]^*$ if and only if $(\xi, \mathbf{x}) - b \geq g(\mathbf{x})$ or

$$b \leq (\xi, \mathbf{x}) - g(\mathbf{x}). \tag{7.7.1}$$

If b_1 is admissible for a vector ξ and if $b_2 < b_1$, then b_2 is also admissible by (7.7.1). If we denote by $\psi(\xi)$ the largest admissible b, then

$$\psi(\xi) = \inf_{x \in C} [(\xi, \mathbf{x}) - g(\mathbf{x})]. \tag{7.7.2}$$

The domain Δ of ψ is the set of vectors for which (7.7.2) is finite. It is easy to verify that Δ is convex, ψ is concave, and $[\psi, \Delta]$ is the conjugate of $[g, D]$. Furthermore, $[\psi, \Delta]^* = [g, D]$ and

$$g(\mathbf{x}) = \inf_{\xi \in \Delta} [(\mathbf{x}, \xi) - \psi(\xi)]. \tag{7.7.3}$$

All of the algebraic and geometric characteristics of the composition of conjugate concave functions carry over. In particular, the analog of Theorem 7.6.4 becomes:

▶THEOREM 7.7.1. Let $g_\alpha(\mathbf{x})$ be a family of concave functions defined over a bounded convex set C in E^n, such that $\inf_\alpha g_\alpha(\mathbf{x})$ is finite for each $\mathbf{x}$. Then for any $a \geq \sup_x \inf_\alpha g_\alpha(\mathbf{x})$ and $\epsilon > 0$, there exist functions g_{α_i} and constants $\lambda_i \geq 0$ $(i = 1, \ldots, n+1)$ such that

$$\sum_{i=1}^{n+1} \lambda_i g_{\alpha_i}(\mathbf{x}) \leq a + \epsilon \qquad \text{for all} \quad x \in C$$

and $\sum \lambda_i = 1$.

***7.8 A duality theorem of nonlinear programming.** Let $f(\mathbf{x})$ be convex over a convex set C and $g(\mathbf{x})$ be concave over a convex set D. Let $[f, C]^* = [\phi, \Gamma]$ and $[g, D]^* = [\psi, \Delta]$.

Hypothesis. We assume that $C \cap D$ and $\Gamma \cap \Delta$ have points which are relatively interior to C, D and Γ, Δ, respectively.

Consider the following extremum problems.

Problem I. Find a point $\mathbf{x}^0$ in $C \cap D$ that maximizes $g(\mathbf{x}) - f(\mathbf{x})$ in $C \cap D$.

If $g(\mathbf{x}) - f(\mathbf{x}) \geq 0$ in $C \cap D$, this problem, stated geometrically, is to find the maximal vertical chord of the convex set $[f, C] \cap [g, D]$ in E^{n+1}. If $f(\mathbf{x}) \equiv 0$ in C, this problem reduces to the programming problem, i.e., the problem of maximizing $g(\mathbf{x})$ for $\mathbf{x}$ in C.

Problem II. Find a point ξ^0 in $\Gamma \cap \Delta$ that minimizes $\phi(\xi) - \psi(\xi)$ in $\Gamma \cap D$.

If $\phi(\xi) - \psi(\xi) \geq 0$ in $\Gamma \cap \Delta$, this problem, stated geometrically, is to find the minimum vertical segment joining the sets $[\phi, \Gamma]$ and $[\psi, \Delta]$ in E^{n+1}.

These two problems may be considered as extensions of Problems I and II of linear programming to the case of nonlinear programming. However, the problems here are not equivalent to the saddle-value problem discussed in Section 7.1. In some situations the dual problem can be given an economic or industrial interpretation.

▶THEOREM 7.8.1. $\displaystyle\sup_{x \in C \cap D} [g(\mathbf{x}) - f(\mathbf{x})] = \inf_{\xi \in \Gamma \cap D} [\phi(\xi) - \psi(\xi)].$

Proof. Let $\xi \in \Gamma \cap \Delta$. Then

$$\begin{aligned} \phi(\xi) &\geq (\mathbf{x}, \xi) - f(\mathbf{x}) \qquad (\mathbf{x} \in C), \\ \psi(\xi) &\leq (\mathbf{x}, \xi) - g(\mathbf{x}) \qquad (\mathbf{x} \in D), \\ \phi(\xi) - \psi(\xi) &\geq g(\mathbf{x}) - f(\mathbf{x}) \qquad (\mathbf{x} \in C \cap D). \end{aligned}$$

Hence

$$\inf_{\xi \in \Gamma \cap D} [\phi(\xi) - \psi(\xi)] \geq \sup_{x \in C \cap D} [g(\mathbf{x}) - f(\mathbf{x})]. \tag{7.8.1}$$

Relation (7.8.1) shows that both quantities are finite.

In order to prove the opposite inequality, we set

$$\mu = \sup_{x \in C \cap D} [g(\mathbf{x}) - f(\mathbf{x})]$$

and $f_1(\mathbf{x}) = f(\mathbf{x}) + \mu$. The conjugate of $f_1(\mathbf{x})$ will be denoted by $\phi_1(\xi)$. The choice of μ allows us to deduce that the sets $[f_1, C]$ and $[g, D]$ come arbitrarily close to each other but do not overlap at any interior points. Therefore there exists a separating plane h which is a supporting plane to both sets. We distinguish two cases.

Case 1. If h is not a vertical plane, then h is determined by a vector ξ^0 and its z intercept $-\eta$. Consequently ξ^0 belongs to $\Gamma \cap \Delta$, $\eta = \psi(\xi^0)$, and

$$\eta = \phi_1(\xi^0) = \sup_{x \in C} [(\mathbf{x}, \xi^0) - f(\mathbf{x}) - \mu] = \phi(\xi^0) - \mu.$$

Hence

$$\sup_{x \in C \cap D} [g(\mathbf{x}) - f(\mathbf{x})] = \mu = \phi(\xi^0) - \psi(\xi^0) \geq \inf_{\xi \in \Gamma \cap \Delta} [\phi(\xi) - \psi(\xi)],$$

and equality holds in (7.8.1).

Case 2. If h is a vertical plane, we project $[f, C]$, $[g, D]$, and h vertically onto E^n. The two convex sets project onto C and D, and the projection of h is a hyperplane h_1 of E^n which must separate C and D. The last assertion contradicts the hypothesis that C and D have relative interior points in common. Thus Case 2 is impossible.

▶COROLLARY 7.8.1. The proof of the last theorem actually shows that

$$\inf_{\xi \in \Gamma \cap \Delta} [\phi(\xi) - \psi(\xi)] = \phi(\xi^0) - \psi(\xi^0),$$

with ξ^0 in $\Gamma \cap \Delta$. Therefore we may replace "infimum" by "minimum." Since the hypothesis of the theorem is symmetric in f, g and ϕ, ψ, we can also replace "supremum" by "maximum." Thus

$$\min_{\xi \in \Gamma \cap \Delta} [\phi(\xi) - \psi(\xi)] = \max_{x \in C \cap D} [g(\mathbf{x}) - f(\mathbf{x})]. \tag{7.8.2}$$

▶COROLLARY 7.8.2. If we replace $g(\mathbf{x})$ and $\phi(\xi)$ by their equivalents in terms of ψ and f, respectively,

$$g(\mathbf{x}) = \inf_{\xi \in \Delta} [(\mathbf{x}, \xi) - \psi(\xi)], \qquad \phi(\xi) = \sup_{x \in C} [(\mathbf{x}, \xi) - f(\mathbf{x})], \tag{7.8.3}$$

the theorem asserts that

$$\max_{x \in C \cap D} \inf_{\xi \in \Delta} [(\mathbf{x}, \boldsymbol{\xi}) - f(\mathbf{x}) - \psi(\boldsymbol{\xi})] = \min_{\xi \in \Gamma \cap \Delta} \sup_{x \in C} [(\mathbf{x}, \boldsymbol{\xi}) - f(\mathbf{x}) - \psi(\boldsymbol{\xi})].$$

If the given sets and functions have the additional properties that $\Gamma = \Delta$, $C = D$, and inf, sup may be replaced by min, max in (7.8.3), then

$$\max_{x \in C} \min_{\xi \in \Delta} [(\mathbf{x}, \boldsymbol{\xi}) - f(\mathbf{x}) - \psi(\boldsymbol{\xi})] = \min_{\xi \in \Delta} \max_{x \in C} [(\mathbf{x}, \boldsymbol{\xi}) - f(\mathbf{x}) - \psi(\boldsymbol{\xi})]. \tag{7.8.4}$$

This equation amounts to demonstrating the existence of a saddle point for the function $F(\mathbf{x}, \boldsymbol{\xi}) = (\mathbf{x}, \boldsymbol{\xi}) - f(\mathbf{x}) - \psi(\boldsymbol{\xi})$. Therefore, there exist $\boldsymbol{\xi}^0$ and $\mathbf{x}^0$ such that

$$F(\mathbf{x}^0, \boldsymbol{\xi}) \geq F(\mathbf{x}^0, \boldsymbol{\xi}^0) \geq F(\mathbf{x}, \boldsymbol{\xi}^0), \qquad \text{for all } \mathbf{x} \in C \text{ and } \boldsymbol{\xi} \in \Delta. \tag{7.8.5}$$

Conversely, whenever (7.8.5) is satisfied, Theorem 7.8.1 is valid.

If the expressions for ψ and g are substituted into (7.8.2), then, analogously to (7.8.4), we obtain

$$\max_{x \in C} \min_{\xi \in \Delta} [\phi(\boldsymbol{\xi}) + g(\mathbf{x}) - (\mathbf{x}, \boldsymbol{\xi})] = \min_{\xi \in \Delta} \sup_{x \in C} [\phi(\boldsymbol{\xi}) + g(\mathbf{x}) - (\mathbf{x}, \boldsymbol{\xi})]. \tag{7.8.6}$$

This amounts to demonstrating the existence of a saddle point for the function $\phi(\boldsymbol{\xi}, \mathbf{x}) = \phi(\boldsymbol{\xi}) + g(\mathbf{x}) - (\mathbf{x}, \boldsymbol{\xi})$.

A stronger form of Theorem 7.8.1 can be obtained which involves no special assumptions about the existence of relative interior points to $C \cap D$ and $\Gamma \cap \Delta$. All that is required is that both these last sets be nonvoid. The proof is based on the nature of the composition of conjugate functions. Let $\psi(\boldsymbol{\xi})$ be the conjugate of $f(\mathbf{x}) + [-g(\mathbf{x})]$ in $C \cap D$. It follows from Theorem 7.6.1 that $\psi(\boldsymbol{\xi})$ is defined on a set containing $\Gamma + (-\Delta)$. Since Γ and Δ have common points, $\Gamma - \Delta$ contains the origin. Hence $\psi(\mathbf{0})$ is defined and by Theorem 7.6.1

$$\begin{aligned} \psi(\mathbf{0}) &= \inf_{\substack{\xi^1 \in \Gamma,\, \xi^2 \in -\Delta \\ \xi^1 + \xi^2 = 0}} [\phi(\boldsymbol{\xi}^1) - \psi(-\boldsymbol{\xi}^2)] \\ &= \inf_{\xi \in \Gamma \cap \Delta} [\phi(\boldsymbol{\xi}) - \psi(\boldsymbol{\xi})]. \end{aligned}$$

On the other hand, the very definition of the conjugate of $f(\mathbf{x}) - g(\mathbf{x})$, when taken for $\boldsymbol{\xi} = \mathbf{0}$, yields

$$\psi(\mathbf{0}) = \sup_{x \in C \cap D} [g(\mathbf{x}) - f(\mathbf{x})].$$

This last argument suffers from the defect that it does not indicate when inf, sup can be replaced by min, max (see Corollary 7.8.1).

***7.9 Applications of the theory of conjugate functions to convex sets.** Theorem 7.6.4 points the way to a series of results about convex sets, approximation to function, fitting polynomials to prescribed loci, etc. These structural properties of convex sets are of constant use in the study of inequalities occurring in various aspects of programming problems. The following classical theorem, which expresses a strengthened form of the finite intersection property characterizing compactness, is fundamental.

▶THEOREM 7.9.1. (*Helly's Theorem*). If C_α is a collection of bounded, closed, convex sets in E^n such that every $n + 1$ sets intersect, then $\bigcap_\alpha C_\alpha$ is nonvoid.

Proof. Let $\phi_\alpha(\mathbf{y})$ be the function measuring the distance between $\mathbf{y}$ and C_α in E^n. It is trivial to verify that each ϕ_α is continuous and convex. Let S be a closed bounded sphere containing one of the C_α, say C_{α_0}. If $\bigcap_\alpha C_\alpha$ is void, then

$$\inf_{y \in S} \sup_\alpha \phi_\alpha(\mathbf{y}) > 0.$$

Since ϕ_α are all continuous functions, we can find a finite subfamily $\{C_{\alpha_i}\}$ such that

$$\inf_{y \in S} \sup_i \phi_{\alpha_i}(\mathbf{y}) > 0.$$

Since S is compact, we may write

$$\inf_{y \in S} \sup_i \phi_{\alpha_i}(\mathbf{y}) \geq \delta > 0.$$

It follows from Theorem 7.6.4 that there exists a convex combination (which for convenience of notation we label as the first $n + 1$) satisfying

$$\sum_{i=1}^{n+1} \lambda_i \phi_{\alpha_i}(\mathbf{y}) \geq \delta > 0 \qquad (\mathbf{y} \in S). \tag{*}$$

Therefore the $n + 1$ sets C_{α_i} must intersect outside S.

Let S' be a closed bounded sphere containing S, C_{α_0}, and the C_{α_i} $(i = 1, \ldots, n + 1)$ of equation (*). Then

$$\inf_{y \in S'} \sup_i \phi_{\alpha_i}(\mathbf{y}) > 0.$$

Repeating the previous argument for the family consisting of the $n + 2$ elements C_{α_i} $(i = 0, \ldots, n + 1)$, we conclude that some $n + 1$ of them do not intersect over S'. Since they are all contained in S', they do not intersect at all, contradicting the hypothesis.

▶ COROLLARY 7.9.1. If every $n + 1$ sets of the family C_α possesses a nonvoid common intersection, and if either

(a) C_α is a finite family of open convex sets (not necessarily bounded), or
(b) C_α is a finite family of closed convex sets (not necessarily bounded), or
(c) C_α is an infinite family of closed convex sets (not necessarily bounded, but possessing a finite subfamily with a bounded intersection),

then $\cap_\alpha C_\alpha$ is nonvoid.

In each case modifying the sets appropriately yields an equivalent problem to which Theorem 7.9.1 is applicable.

The theorem is not true for an arbitrary family of closed convex sets. In E^1, let $C_k = [k, \infty]$ $(k = 1, 2, \ldots)$. Any finite number of sets intersect, but there is no common point for the whole family.

An important application of Helly's Theorem is to the Polynomial Approximation Problem. Given $n + 1$ pairs of real numbers

$$\langle x_i, y_i \rangle \qquad (i = 0, \ldots, n; x_i < x_{i+1}),$$

we seek to find the polynomial P of degree at most $n - 1$ that minimizes the quantity $\max_i |y_i - P(x_i)|$. Let

$$\rho = \min_P \max_i |y_i - P(x_i)|.$$

That ρ is actually achieved by a polynomial P^* is easily proved by a routine compactness argument. We leave the details for the reader. Later, an explicit expression for ρ and P^* is noted.

▶ LEMMA 7.9.1. If P^* is a solution, then $|y_i - P^*(x_i)| = \rho$ for all i.

Proof. (By contradiction.) Assume that there exists some i, say $i = 0$, such that $|y_0 - P^*(x_0)| < \rho$. We shall show that this assumption would allow us to construct a better polynomial approximation, contradicting the choice of ρ.

For each $k \neq 0$ we may construct a polynomial $Q_k(x)$ of degree $n - 1$ such that $Q_k(x_i) = 0$ except for $i = 0$ and $i = k$. Furthermore, we can always choose $Q_k(x)$ such that $Q_k(x_k) > 0$.

Let $\epsilon_k = \text{sign}\,[y_k - P^*(x_k)]$. Select $a_i > 0$ $(i = 1, 2, \ldots, n)$ so small that

$$|y_0 - P^*(x_0) - a_i\epsilon_i Q_i(x_0)| < \rho.$$

This is possible in view of the supposition that $|y_0 - P^*(x_0)| < \rho$. But $a_i > 0$ and $|y_i - P^*(x_i)| \leq \rho$ imply that

$$|y_i - P^*(x_i) - a_i\epsilon_i Q_i(x_i)| < \rho, \tag{7.9.1}$$

while the values at the remaining x_j are not disturbed because Q_i vanishes there. Hence

$$\left| y_i - P^*(x_i) - \sum_{k=1}^{n} a_k \epsilon_k Q_k(x_i) \right| < \rho \qquad \text{for all} \quad i. \qquad (7.9.2)$$

But (7.9.2) is in contradiction to the definition of ρ.

▶LEMMA 7.9.2. If $P^*(x)$ is a solution, then the $y_i - P^*(x_i)$ alternate in sign.

Proof. If for some x_i and x_{i+1}, the values $y_i - P^*(x_i)$ are of the same sign, say positive, we can construct a polynomial of degree $n - 1$ that vanishes at the remaining x_i and is positive at x_i and at x_{i+1}. By subtracting an appropriate multiple of this polynomial we can reduce the error at x_i and x_{i+1} without disturbing the errors at the other points. This contradicts the assertion of Lemma 7.9.1.

It is clear from Lemmas 7.9.1 and 7.9.2 that if

$$P^*(x) = \sum_{i=0}^{n-1} a_i^* x^i$$

is a solution, it must satisfy the system of equations

$$(-1)^k \eta + \sum_{i=0}^{n-1} a_i^* x_k^i = y_k \qquad (k = 0, \ldots, n), \qquad (7.9.3)$$

where η is either $+\rho$ or $-\rho$.

The solution is actually unique. If two solutions exist with the same value of η, the difference $P_1^* - P_2^*$ must be a polynomial of degree at most $n - 1$ vanishing at $n + 1$ points $x_0, \ldots, x_n$. Hence $P_1^* = P_2^*$. On the other hand, if two solutions exist such that $\eta_1 = +\rho$ and $\eta_2 = -\rho$, their difference must vanish at some point between x_i and x_{i+1} for each i. Hence it vanishes for at least n points, and this means it is zero.

We may regard (7.9.3) as a system of $n + 1$ equations in the $n + 1$ unknowns $a_0^*, \ldots, a_n^*$. The value of η is determined uniquely by

$$\eta = \frac{\begin{vmatrix} y_0 & 1 & x_0 & \cdots & x_0^{n-1} \\ y_1 & 1 & x_1 & \cdots & x_1^{n-1} \\ \vdots & & & & \\ y_n & 1 & x_n & \cdots & x_n^{n-1} \end{vmatrix}}{\begin{vmatrix} 1 & 1 & x_0 & \cdots & x_0^{n-1} \\ -1 & 1 & x_1 & \cdots & x_1^{n-1} \\ \vdots & & & & \\ (-1)^n & 1 & x_n & \cdots & x_n^{n-1} \end{vmatrix}}.$$

The coefficients of the unique polynomial solution $P^*(x)$ can be determined by solving the system (7.9.3).

We now direct our attention to the problem of fitting a polynomial of degree $n - 1$ to a prescribed function in some best sense. Explicitly, let $f(x)$ be bounded over the interval $\alpha \leq x \leq \beta$ and let P denote a generic polynomial of degree at most $n - 1$. We seek to find the polynomial $P^*(x)$ of degree $n - 1$ for which

$$\min_P \max_{\alpha \leq x \leq \beta} |f(x) - P(x)|$$

is achieved.

Let $X = (x_1, \ldots, x_{n+1})$, $\alpha \leq x_i \leq \beta$, and $y_i = f(x_i)$. Then $\rho(X)$ will denote the solution to the approximation problem for the points x_i, y_i. If at least two of the x_i are equal, the corresponding y_i are equal and we can find a polynomial passing through all the points; hence in this case $\rho = 0$.

▶ **Theorem** 7.9.2. Let $\rho_0 = \sup_X \rho(X)$. Then

$$\min_P \max_{\alpha \leq x \leq \beta} |f(x) - P(x)| = \rho_0.$$

Proof. For any fixed P and X,

$$\min_P \max_{\alpha \leq x \leq \beta} |f(x) - P(x)| \geq \min_P \max_i |f(x_i) - P(x_i)| = \rho(X).$$

Therefore

$$\min_P \max_{\alpha \leq x \leq \beta} |f(x) - P(x)| \geq \sup_X \rho(X) = \rho_0.$$

In order to prove the opposite inequality, for each x in $[\alpha, \beta]$ we define the set

$$A_x = \left\{ \mathbf{a} = \langle a_0, \ldots, a_{n-1} \rangle \,\middle\|\, \left| f(x) - \sum_{i=0}^{n-1} a_i x^i \right| \leq \rho_0 \right\}.$$

Each A_x is a closed convex set, not necessarily bounded. By the definition of ρ_0 we know that any $n + 1$ of these sets intersect. If we take $n + 1$ distinct x_i's, the intersection of the corresponding sets is a parallelepiped, and therefore is bounded. It follows from Corollary 7.9.1 that they all intersect, and the theorem is established.

Suppose that in Theorem 7.9.2 we add the assumption that $f(x)$ is continuous on $\alpha \leq x \leq \beta$. It is not hard to show that $\rho(x_0, x_1, \ldots, x_n)$ is continuous on $\alpha \leq x_0 \leq x_1 \leq \cdots \leq x_n \leq \beta$. Thus we have the following corollary.

▶ **Corollary** 7.9.2. If $f(x)$ is continuous on $\alpha \leq x \leq \beta$, there exist $n + 1$ points $(x_0^*, x_1^*, \ldots, x_n^*)$ as described above such that

$$\min_{P_{n-1}} \max_{\alpha \leq x \leq \beta} |f(x) - P_{n-1}(x)| = \min_{P_{n-1}} \max_{x_\nu^*} |f(x_\nu^*) - P_{n-1}(x_\nu^*)|.$$

If we choose $f(x) = x^n$ ($\alpha = 0$, $\beta = 1$), then the solution of the problem is well known to be $-x^n + T_{n-1}(x)$, where $T_{n-1}(x) = \cos (n - 1)\theta$, with $\cos \theta = x$. In this case $x_{\nu-1}^*$ turns out to be those x for which $\theta = \pi/\nu$ ($\nu = 1, \ldots, n$) and $x_n = 1$.

Another application of Helly's Theorem is concerned with the covering of the unit sphere in E^n by closed hemispheres. A family of such hemispheres is called compact if the unit normals from the origin to the hyperplanes bounding the hemispheres, directed into the hemisphere, constitute a compact set.

▶THEOREM 7.9.3 (*Covering Theorem*). Let the surface of a sphere in E^n be covered by a compact family of closed hemispheres; then we can find m ($m \leq n + 1$) members of the family which cover the surface.

Proof. Let $\mathbf{l}_\alpha$ be the unit normal to the hemisphere H_α with the orientation as specified. A point $\mathbf{x}$ on the surface of the sphere is covered by H_α if and only if $(\mathbf{x}, \mathbf{l}_\alpha) \geq 0$.

Let $\{\mathbf{l}_i\}$ be a countable set dense in $\{\mathbf{l}_\alpha\}$, and let Γ_m be the compact set spanned within the unit sphere by $\mathbf{l}_1, \ldots, \mathbf{l}_m$. We wish to show that for m sufficiently large, Γ_m is arbitrarily close to the origin $\boldsymbol{\theta}$. Otherwise for some $\epsilon > 0$ the distance $d(\boldsymbol{\theta}, \Gamma_i)$ exceeds ϵ for all i. By choice of $\{\mathbf{l}_i\}$, the last statement implies that $d(\boldsymbol{\theta}, \Gamma) \geq \epsilon$, where Γ is the convex set spanned by $\{\mathbf{l}_\alpha\}$. Since Γ is convex, we can find a plane through the origin that does not pass within ϵ of Γ; if $\mathbf{x}_0$ is the unit normal to this plane directed away from Γ, then $(\mathbf{l}_\alpha, \mathbf{x}_0) \leq -\epsilon$ for all α. It follows that $\mathbf{x}_0$ is not covered by H_α for any α, which contradicts our hypothesis.

This contradiction implies that for each k there exists $m(k)$ such that $d(\boldsymbol{\theta}, \Gamma_{m(k)}) < 1/k$. Let $\mathbf{x}^k$ be a point of $\Gamma_{m(k)}$ of distance less than $1/k$ from the origin. By Lemma B.2.1, we have a convex representation

$$\mathbf{x}^k = \sum_{i=1}^{n+1} \xi_i^{(k)} \mathbf{l}_i^{(k)}.$$

Since $n + 1$ is fixed and $\xi_i^{(k)}$, $\mathbf{l}_i^{(k)}$ are drawn from compact sets, we may pass to the limit and obtain a representation

$$\boldsymbol{\theta} = \sum_{i=1}^{n+1} \xi_i \mathbf{l}_i,$$

with $\xi_i \geq 0$, $\sum \xi_i = 1$. The hemispheres H_i corresponding to the l_i of the representation must cover the full sphere.

The theorem is not true if the compactness requirement is removed; for example, the family of hemispheres in E^2 described by the angles π, $1, \frac{1}{2}, \ldots, 1/m, \ldots$ covers the full circle but no finite subfamily will cover it.

The following corollary shows that one of the hemispheres may be chosen at will.

▶ COROLLARY 7.9.3. Let a given hemisphere H on the surface of a sphere in E^n be covered by a compact family of closed hemispheres. Then we can find m ($m \leq n$) members of the family that cover H.

Proof. The given family, together with the closed complement H_0 of H, covers the sphere. The last theorem guarantees that we can find a subfamily of the augmented family consisting of at most $n + 1$ members which also covers the sphere. If this subfamily does not contain H_0, we consider the convex set C spanned within the unit sphere by the unit normals $\mathbf{l}_i$ of the subfamily. C must contain the origin $\boldsymbol{\theta}$, since the subfamily covers the sphere. Let $\mathbf{l}_0$ denote the unit normal to H_0, and $\mathbf{y}_0$ the intersection of the radius $[\boldsymbol{\theta}, -\mathbf{l}_0]$ with the boundary of C. Then $\mathbf{y}_0$ is a convex combination of n (or fewer) of the $\mathbf{l}_i$, and $\boldsymbol{\theta}$ is a convex combination of $\mathbf{y}_0$ and $\mathbf{l}_0$. (If $\mathbf{y}_0$ and $\boldsymbol{\theta}$ happen to coincide, then $\mathbf{l}_0$ will appear vacuously.) It follows that an $(n + 1)$-member subfamily containing $\mathbf{l}_0$ and covering the sphere can always be found. The closed complement of H_0 must be covered by the other n members of the family.

Most of these statements about convex sets are essentially based on the proposition that any point belonging to the convex hull of a set H in E^n can be represented as a convex combination of at most $n + 1$ points of H.

Another fundamental representation theorem which applies to interior points of convex sets is the following.

▶ THEOREM 7.9.4. Let C denote the convex hull of a finite set of points H in E^n. If $\mathbf{z}$ is a point of C such that there exists an open sphere about $\mathbf{z}$ in C, then there exists a subset H' of H consisting of at most $2n$ points and such that its convex hull has $\mathbf{z}$ as an interior point.

Proof. The proof is by induction on the dimension of E^n. Suppose that the result is true for convex sets in Euclidean spaces of dimension less than n. The case of E^1 is trivial. Choose m points $\mathbf{y}_i$ in H such that

$$\mathbf{z}_0 = \sum_{i=1}^{m} \lambda_i \mathbf{y}_i \qquad \left(\lambda_i > 0; \sum \lambda_i = 1\right)$$

and $2 \leq m \leq n + 1$. Since $\mathbf{z}_0$ is interior, it is certain that $m \geq 2$; the upper bound comes from the classical representation (Lemma B.2.1). The $\mathbf{y}_i$ may be selected so that the linear variety L determined by these points is $(m - 1)$-dimensional. Through $\mathbf{z}_0$ select any $(n - m + 1)$-dimensional linear manifold M independent of L. Project C perpendicularly in M. The image Proj (H) of H in M spans convexly the projection of C. Within the space M, the point $\mathbf{z}_0$ is also interior to Proj (C). By the induction hypothesis there exist at most $2(n - m + 1)$ points of Proj (H) whose convex span contains $\mathbf{z}_0$ on its interior (in the sense of the space M). To each such manifold M are associated at most $2(n - m + 1)$ points of Proj (H) with the properties specified above. The points may be chosen as continuous functions of M; hence, by a connectedness argument, we can obtain a single set of no more than $2(n - m + 1)$ points $\mathbf{z}_i$ which furnish the desired representation of $\mathbf{z}_0$ for any manifold M independent of $\mathbf{z}_0$. These $\mathbf{z}_i$, which together with $\mathbf{y}_i$ total at most $2(n - m + 1) + m = 2n - m + 2 \leq 2n$ (since $m \geq 2$), span a convex set C' with $\mathbf{z}_0$ on its interior. Otherwise, there must be a hyperplane Y through $\mathbf{z}_0$ that keeps C' on one side. Let $\mathbf{l}$ be the perpendicular direction away from Y not including C', and let M_0 be any linear manifold of dimension $2(n - m + 1)$ with $M_0 \cap L = \mathbf{z}_0$ containing the direction $\mathbf{l}$. Then the projection of $\mathbf{z}_i$ in M_0 spans a convex set with $\mathbf{z}_0$ noninterior, which contradicts the construction of $\mathbf{z}_i$.

A typical application of Theorem 7.9.4 is given in the following corollary.

▶Corollary 7.9.4. Let a unit sphere S in E^n be covered by a finite family of open hemispheres. There exists a subfamily consisting of at most $2n$ open hemispheres which also covers S.

The reader is referred to the problems at the close of this chapter for further applications.

We close this section with a constructive method of obtaining a finite representation. Let S_m denote the unit sphere of m-dimensional Euclidean space. We assume (1) that a closed set $X = \{\mathbf{x}_\alpha\}$ of points on S_m is given, and (2) that the origin is interior to the convex set spanned by the points $\mathbf{x}_\alpha$. Assumption (2) is equivalent to the statement (2′) that there exists a number $d > 0$ such that, for every point $\mathbf{y} \in S_m$, $(\mathbf{x}_\alpha, \mathbf{y}) \geq d$ for some α. What is desired is a way of choosing in a finite number of steps a finite subset $\{\alpha_i\}$ satisfying (2) or (2′).

The method. Assume (1) and (2′) as stated above, and proceed according to the following recursive instructions:

(0) Start with any point $\mathbf{x}_1 \in X$.

(1_k) Having chosen $\mathbf{x}_1, \mathbf{x}_2, \ldots, \mathbf{x}_k$, choose the point $\mathbf{y}_k \in S_m$ which minimizes $\max [(\mathbf{y}, \mathbf{x}_1), (\mathbf{y}, \mathbf{x}_2), \ldots, (\mathbf{y}, \mathbf{x}_k)]$.

(2_k) Choose $\mathbf{x}_{k+1}$ as a point in X that maximizes $(\mathbf{y}_k, \mathbf{x})$.

The purpose of (1_k) is to determine the point on the sphere farthest from those already chosen. So long as the origin is not in the interior of the convex set determined by $\mathbf{x}_1, \mathbf{x}_2, \ldots, \mathbf{x}_k$, this point will be uniquely defined, and will in fact be proportional to the (unique) nearest point of that set to the origin.

Step (2_k) selects a point of X nearest to $\mathbf{y}_k$. The way in which this is accomplished must depend on the way X is described. Since X is closed, the selection is always possible.

The process is stopped as soon as the minimum of (1_k) is positive. This will occur at some finite value of k.

Proof that the method terminates. Suppose that the method does not terminate, and define

$$m_k = \min_y \max\,[(\mathbf{y}, \mathbf{x}_1), (\mathbf{y}, \mathbf{x}_2), \ldots, (\mathbf{y}, \mathbf{x}_k)],$$

$$n_k = (\mathbf{y}_k, \mathbf{x}_{k+1}).$$

Then, clearly,

$$-1 = m_1 \leq m_2 \leq \cdots \leq m_k \leq \cdots \leq 0$$

and $n_k \geq d$ for all k. Making use of the compactness of S_m and X, select a subsequence $k_1, k_2, \ldots, k_l, \ldots$ for which y_{k_l} and x_{k_l+1} both converge. Then

$$\lim_{l\to\infty} m_{k_l} \geq \lim_{l\to\infty} (\mathbf{y}_{k_l}, \mathbf{x}_{k_{l-1}+1}) = \lim_{l\to\infty} n_{k_l} \geq d > 0.$$

Hence m_{k_0} is positive for some finite k_0, contrary to the hypothesis that the method did not terminate.

7.10. Problems.

1. Consider the following nonlinear programming problem: Maximize $g(\mathbf{x})$ with $\mathbf{x} \in E^n$ satisfying the vector constraints $\mathbf{F}(\mathbf{x}) \geq 0$ but otherwise unrestricted, $g(\mathbf{x})$ concave, and $\mathbf{F}(\mathbf{x})$ composed of concave functions. Assuming the condition of Theorem 7.1.1, show that the equivalent saddle-value problem has a kernel

$$\phi(\mathbf{x}, \mathbf{u}) = g(\mathbf{x}) + (\mathbf{u}, \mathbf{F}(\mathbf{x})),$$

where $\mathbf{u} \geq \mathbf{0}$ and $\mathbf{x}$ varies unrestrictedly.

2. Consider the problem

$$\max g(\mathbf{x}) \qquad \text{subject to} \qquad f(\mathbf{x}) \geq 0, \qquad \mathbf{x} \geq \mathbf{0},$$

where f and g are concave functions. Prove that $\mathbf{x}^0$ is optimal if and only if $\mathbf{x}^0$ is a solution to

$$\max_x\,(\mathbf{g}_{x^0}, \mathbf{x}) \qquad \text{subject to} \qquad f(\mathbf{x}) \geq 0, \qquad \mathbf{x} \geq \mathbf{0}$$

($\mathbf{g}_{x^0}$ represents the vector of partial derivatives evaluated at $\mathbf{x}^0$).

3. Consider the following problem: Maximize $(\mathbf{p}, \mathbf{x})$ subject to $\mathbf{x} \geq \mathbf{0}$, $\mathbf{F}(\mathbf{x}) \leq \mathbf{v}$, where F is a convex differentiable vector function and homogeneous of order one and $\mathbf{p}$ is a fixed constant vector. Prove that if Theorem 7.1.1 is applicable, then $(\mathbf{p}, \mathbf{x}^0) = (\mathbf{u}^0, \mathbf{v})$ for any saddle point $\{\mathbf{x}^0, \mathbf{u}^0\}$.

4. Prove that the linear programming problem

$$\max\,(\mathbf{c}, \mathbf{x}) \qquad \text{subject to} \qquad \mathbf{A}\mathbf{x} \leq \mathbf{b}, \mathbf{x} \geq \mathbf{0},$$

where $\mathbf{A}$ is a matrix, reduces to the quadratic programming problem

$$\max\,[(\mathbf{c}, \mathbf{x}) - \tfrac{1}{2}\epsilon(\mathbf{x}, \mathbf{x})] \qquad \text{subject to} \qquad \mathbf{A}\mathbf{x} \leq \mathbf{b}, \mathbf{x} \geq \mathbf{0}$$

when ϵ is a sufficiently small positive number.

5. Let $f(t)$ be a monotone-increasing convex function with $f(0) > 0$. Find the set of x's which minimize

$$\sum_{i=1}^{n} f(x_i)$$

subject to the restrictions

$$\sum_{i=1}^{n} x_i = T, \qquad x_i \geq 0, \qquad n \geq 1.$$

(Use Lagrange multipliers for each n.)

6. Solve the preceding problem with "convex" replaced by "concave."

7. Let $f(x)$ be a continuous, strictly increasing convex function for nonnegative real x. Let $0 \leq a_1 \leq a_2 \leq \cdots \leq a_n$, and define m as the largest index such that

$$\frac{a_m}{m} = \max_{1 \leq j \leq m} \frac{a_j}{j}.$$

Show that the solution of

$$\min_{x} \sum_{k=1}^{n} f(x_k), \qquad \text{where} \qquad \sum_{j=1}^{i} x_j \geq a_i \qquad \text{and} \qquad x_i \geq 0,$$

has the property that $x_1 = x_2 = \cdots = x_m = a_m/m$.

8. Determine inductively the complete solution of Problem 7.

9. Consider the following problem: Maximize $(\mathbf{a}, \mathbf{x}) - \frac{1}{2}(\mathbf{x}, \mathbf{A}\mathbf{x})$ subject to $\mathbf{B}\mathbf{x} \leq \mathbf{b}$, where $\mathbf{A}$ is a positive-definite matrix and $\mathbf{B}$ is an arbitrary matrix. Prove that a solution to this problem, if one exists, can be obtained by solving the problem

$$\max\,[(\mathbf{c}, \mathbf{y}) - \tfrac{1}{2}(\mathbf{y}, \mathbf{D}\mathbf{y})] \qquad \text{subject to} \qquad \mathbf{y} \geq \mathbf{0},$$

where $\mathbf{c} = \mathbf{D}\mathbf{A}^{-1}\mathbf{a} - \mathbf{b}$ and $\mathbf{D} = \mathbf{B}\mathbf{A}^{-1}\mathbf{B}'$.

10. Let $h(v) = \max_x \{g(\mathbf{x}) | f(\mathbf{x}) \leq v, \mathbf{x} \geq \mathbf{0}\}$, where f is convex and g is concave. Prove that $h(v)$ is concave and that for any shadow price vector $\langle u_1^0, \ldots, u_m^0 \rangle$

$$\left(\frac{\partial h}{\partial v_k}\right)_+ \leq u_k^0 \leq \left(\frac{\partial h}{\partial v_k}\right)_- \qquad (k = 1, \ldots, m).$$

(Assume that Theorem 7.1.1 can be applied.)

11. Let $\mathbf{G}(\mathbf{x})$ and $\mathbf{F}(\mathbf{x})$ denote two families of vector functions all of whose components consist of concave functions defined for all $\mathbf{x}$ in E^n. Consider the problem: Find $\mathbf{x}^0$ satisfying

$$\max_x \min_i \, (\mathbf{G}(\mathbf{x}))_i,$$

where $\mathbf{F}(\mathbf{x}) \geq \mathbf{0}, \mathbf{x} \geq \mathbf{0}$. We call this problem the minimum-component maximum problem. Prove that $\mathbf{x}^0$ cannot be a solution of the minimum-component maximum problem unless there exists a nonnegative vector $\mathbf{v}^0$ ($\sum v_i^0 = 1$) satisfying $(\mathbf{v}^0, \mathbf{G}(\mathbf{x}^0)) = \min_i [\mathbf{G}(\mathbf{x}^0)]_i$ and such that $\{\mathbf{x}^0, \mathbf{u}^0\}$, where $\mathbf{u}^0 > 0$, is a saddle point of

$$\phi(\mathbf{x}, \mathbf{u}) = (\mathbf{v}^0, \mathbf{G}(\mathbf{x})) + (\mathbf{u}, \mathbf{F}(\mathbf{x})).$$

12. In the statement of Theorem 7.1.1 the following condition was imposed: For each $\mathbf{u} > \mathbf{0}$ there exists $\mathbf{x}$ in X satisfying $(\mathbf{u}, \mathbf{F}(\mathbf{x})) > 0$. Prove that this is equivalent to the existence of an element $\mathbf{x}^0$ in X for which $\mathbf{F}(\mathbf{x}^0) \gg \mathbf{0}$.

13. Consider the problem

$$\max \, [(\mathbf{c}, \mathbf{x}) - \tfrac{1}{2}(\mathbf{x}, A\mathbf{x})] \qquad \text{subject to} \qquad \mathbf{x} \geq \mathbf{0},$$

where A is positive-definite. Prove that the iterative relation

$$x_i^{(u+1)} = \max \left\{0, \frac{1}{a_{ii}}\left(c_i - \sum_{j<i} a_{ij}x_j^{(u+1)} - \sum_{j \geq i} a_{ij}x_j^{(u)}\right)\right\}$$
$$(i = 1, \ldots, n; u = 0, 1, 2, \ldots),$$

where $\langle x_1^0, x_2^0, \ldots, x_n^0 \rangle \geq 0$, converges to a solution.

14. Consider the problem

$$\max g(\mathbf{x}) \qquad \text{subject to} \qquad \mathbf{F}(\mathbf{x}) \geq \mathbf{0}, \qquad \mathbf{x} \geq \mathbf{0},$$

where $\mathbf{F}$ is a vector function. Let $K = \{k \mid f_k(\mathbf{x}) = 0 \text{ for all feasible } x\}$. Assume (1) that $f_k(\mathbf{x})$ is linear in $\mathbf{x}$ for $k \in K$; (2) that df_k/dx is linearly independent for $k \in K$; (3) that for any i, there is a feasible vector $\mathbf{x}^i$ such that $x_i^i > 0$. Prove that the original problem reduces to the problem of finding a saddle point for $g(\mathbf{x}) + (\mathbf{u}, \mathbf{F}(\mathbf{x}))$ in $\mathbf{x} \geq \mathbf{0}, \mathbf{u} \geq \mathbf{0}$.

The next three problems employ the notation of Sections 7.5 and 7.6.

15. In Theorem 7.6.1, when is $V = [\phi_1, \Gamma_1] \cup [\phi_2, \Gamma_2]$ closed? Show that the set V will be closed whenever C_1 and C_2 have common relative interior points.

16. Show that $\sup f_\alpha(\mathbf{x})$ exists for some $\mathbf{x}$ if and only if $[f_\alpha, C_\alpha]$ have a common point or $[\phi_\alpha, \Gamma_\alpha]$ have a common nonvertical barrier.

17. Prove that whenever the directional derivatives of $\partial f/\partial x$ and $\partial g/\partial x$ are uniformly bounded, there exists a nonvertical hyperplane separating $[g, D]$ and $[f + \mu, C]$ of Theorem 7.8.1, even if C and D have no common relative interior points.

18. Deduce the min-max theorem of matrices (Theorem 1.4.1) from Theorem 7.8.1.

19. Using the theory of conjugate functions, prove that

$$\xi x \leq \frac{|x|^p}{p} + \frac{|\xi|^q}{q},$$

where ξ and x are real numbers and

$$1/p + 1/q = 1 \; (p > 1).$$

20. Let a unit sphere S in E^n be covered by a finite family of open hemispheres. Show that at most $2n$ open hemispheres of this family cover S (this is Corollary 7.9.4).

21. Let H_α denote a finite family of open half-spaces with boundary through the origin which covers all of E^n. Prove that at most $2n$ members of H_α cover all of E^n.

22. Let M_α denote a finite family of closed half-spaces in E^n with the origin on the boundary of each M_α such that every $2n$ members of M_α have a common ray. Prove that all members of M_α possess a common ray.

23. A finite family R_α of rays in E^n is said to be irreducible if for any hyperplane through the origin there is at least one ray properly on each side of the hyperplane. Prove that if R_α is a finite irreducible family of rays in E^n, then there exists a subfamily of R_α consisting of at most $2n$ rays which is also irreducible.

24. Let S and S' be two finite collections of points in E^n such that for every $n + 2$ points R of $S \cup S'$ there exists a hyperplane strictly separating $S' \cap R$ from $S \cap R$. Prove that there exists a hyperplane which strictly separates S from S'.

25. Let S and S' be two finite collections of points in E^n such that for any $2n + 2$ points T of $S \cup S'$, there exists a hyperplane which separates $S \cap T$ from $S' \cap T$. Show that there exists a hyperplane separating S from S'.

Notes and References to Chapter 7

§ 7.1. The first major development in the theory of nonlinear programming (N.L.P.) was the fundamental paper by Kuhn and Tucker [164]. Their work may more appropriately be called concave programming, which is the terminology accepted in this book.

The Kuhn-Tucker analogs of Theorems 7.1.1 and 7.1.2 depend on the differentiability properties of the functions, whereas our methods depend only on the separation properties of disjoint convex sets. Consequently, the hypotheses of Theorem 7.1.1 are not equivalent or comparable to those of Kuhn and Tucker. Theorem 7.1.3 is taken directly from their paper.

Other modifications of the Kuhn-Tucker theorems have been presented by Arrow and Hurwicz [9].

Two collections of contributions to the theory of N.L.P. and linear inequalities are [163] and [12], the first edited by Kuhn and Tucker and the second edited by Arrow, Hurwicz, and Uzawa.

§ 7.2. The mathematical problem of portfolio selection was discussed by Markowitz [172]. We partly follow his treatment. Markowitz is interested in the more general problem of determining efficient vector portfolio investment schemes in which the first component is the expected yield and the second component is the variance of the portfolio selection.

The blending of octane numbers of gasoline mixtures is a natural example of concave programming (see Manne [171]).

The theory of consumer buying abounds with examples of N.L.P. (see Samuelson [211]).

A N.L.P. version of foreign trade and exchange has been studied by Reiter [204].

The interindustry activity analysis model discussed in this section was formulated by Chenery [47] (see also Chenery and Kretchmer [48]). The iterative solution was proposed by Chenery and Uzawa [49].

Many important nonlinear programming models come from production and inventory studies in which the cost of production is a convex function of the amount produced. The original investigation of this type of problem was made by Hohn and Modigliani [125]. An extensive study of these models is carried out in Arrow, Karlin, and Scarf [16, Chapters 4–7].

§ 7.3. The gradient method of equations (7.3.1) was proposed by Arrow and Hurwicz [10], who proved the convergence to a pair of saddle points in the small. This means that if the starting pair of feasible vectors is sufficiently close to a saddle point, the solution of the gradient equations converges. Uzawa [235] established the convergence for any initial position. We have followed his proof. The gradient method of this section may be construed as a natural extension of the Brown-Von Neumann differential-equations scheme of Section 6.7.

With regard to computing procedures, we call attention to the simplex method for quadratic programming developed by Wolfe [256]. Alternative computational methods that are sometimes valid for quadratic programming problems have been proposed by Markowitz [173] and Houthakker [127].

§ 7.4. The vector maximum problem and the corresponding theorems are due originally to Kuhn and Tucker [164], who rely on differentiability arguments in contradistinction to the methods used here.

In connection with the vector maximum problem, the reader is referred to the work of Hurwicz [128] on programming in abstract spaces (see also Kretchmer [154]).

§ 7.5. The interesting theory of conjugate functions was discovered by Fenchel. An exposition of this theory is contained in his book *Convex Cones, Sets, and Functions* [81]. Fenchel develops the duality of conjugate functions from the algebraic point of view, using the notion of polar reciprocity. Our presentation of this subject is strictly geometric.

§ 7.6. Theorem 7.6.4 is due to Bohnenblust, Karlin, and Shapley [37]. The proof here is new and uses the ideas of conjugate functions.

§ 7.8. The results of this section are all due to Fenchel [81].

§ 7.9. Theorems on fitting points by polynomials of fixed degree have a long history. Theorem 7.9.2 is due to Rademacher and Schoenberg [202]. Most of the other results of this section can be found in Karlin and Shapley [140].

§ 7.10. In connection with Problems 5–8, see Charnes and Cooper [45] and Savage [214]. Problem 9 is due to Hildreth [122]. Problem 11 is based on results first obtained by Kuhn and Tucker [164]. The result of Problem 12 was noted by Uzawa. Problem 13 is due to Hildreth [122]. Problem 14 was suggested by Uzawa [12].

CHAPTER 8

MATHEMATICAL METHODS IN THE STUDY OF ECONOMIC MODELS

In the past two decades, impressive progress has been made in the mathematical analysis of economic models. It has been necessary to develop new methods to deal with such models, since they typically involve nondifferentiable functions and variables subject to inequality constraints, which cannot be handled by the classical calculus. Broadly speaking, the methods needed are those used in game theory—i.e., the theory of convex sets, topological fixed-point theorems, the theory of positive matrices, etc. In this chapter and the next, we explore the use of these techniques in the context of mathematical economics.

In this chapter we shall examine elements of production theory, consumption theory, and equilibrium theory. In Chapter 9 we shall consider welfare economics, models of the dynamic theory of balanced growth, and certain problems of stability theory associated with equilibrium prices. We do not formally consider macro-economic models, aggregation theory, and monopoly-oligopoly studies, since the mathematical aspects of these subjects are not too well developed (see Problems 1–6).

In short, we have not attempted to present an exhaustive study of mathematical economics. Rather, we have striven to give as complete a description as possible of the intrinsic mathematical methodology of economic models, and to point out the relevance of these techniques to the broader concept of decision processes.

In Section 8.1 we formulate various simple linear models of equilibrium and exchange. We arrive at solutions of these linear models by appealing to the theory of positive matrices, which is discussed at length in Section 8.2. Positive matrices, aside from occurring naturally in economic studies, appear also as part of the theory of finite stochastic processes in the investigation of some mechanical systems; hence the results of Section 8.2 are of fundamental importance. Also in this section, which may be read independently of the rest of the chapter, we describe some properties of matrices whose eigenvalues have negative real parts. These properties (see pp. 247–56) are relevant to our discussion of stability theory in Chapter 9.

Prior to a general investigation of the competitive equilibrium, it seems necessary to elaborate separately on the two main components of equilibrium theory, namely production theory and consumption theory. In Section 8.4 we describe the nature of the mathematical problems connected with production theory, and in Section 8.5 we examine the problem of characterizing efficient points for activity analysis models, i.e., the problem of selecting an optimal production process from among several alternative processes.

In Section 8.6 we discuss the theory of consumption. Our approach is to postulate a preference relation, from which we derive a demand function. Since this approach is independent of the specific form of the utility index function of the consumer, we can formulate Slutsky relations which describe the variations of the demand function solely in terms of price changes and changes in the level of the budget constraint for a given preference relation. The concept of revealed preference is also discussed in this section.

In the final sections of this chapter, production and consumption theory are combined in the analysis of general static equilibrium models. We treat several different versions of equilibrium, culminating in the Arrow-Debreu model. The mathematically oriented reader will be impressed by the depth of the topological techniques that are used in analyzing the equations of supply and demand.

8.1 Open and closed linear Leontief models. One of the simplest models of exchange for an economy can be described as follows:

There are n activities (or industries), each producing a product. Let X_i represent the total output of activity i, and let x_{ij} represent the total amount of the product of activity i used by the jth activity. If

$$Y_i = X_i - \sum_{j=1}^{n} x_{ij},$$

then Y_i expresses the difference between the total output of activity i and the amount of its product consumed by the n activities. Y_i is commonly referred to as the "final demand" of commodity i.

In a *closed Leontief model* $Y_i = 0$ for all i. There are no surpluses. If labor is a type of activity, the total labor force is employed; if steel is another type of activity, all of the steel produced is consumed by the system. A closed model typically includes such activities as services, management, and the production of raw materials.

In an *open Leontief model* $Y_i > 0$ for at least one i. From an economic viewpoint an open model implies the presence of exogenous material—some quantity of capital, labor, raw material, etc., that is supplied to the system from the outside. The essential characteristic of the open model is the existence of outside demand or supply or both.

We define

$$a_{ij} = \frac{x_{ij}}{X_j},$$

which may be interpreted as the amount of activity i's commodity needed to produce one unit of commodity j. The quantities a_{ij} are referred to as the "production coefficients," and are assumed in this section to be constants independent of the actual values of X_i. (This assumption is of doubtful validity and should be thought of only as a first approximation.) It follows that

$$(\mathbf{I} - \mathbf{A})\mathbf{X} = \mathbf{Y}, \tag{8.1.1}$$

where $\mathbf{A} = \|a_{ij}\|$, $\mathbf{X} = \langle X_1, X_2, \ldots, X_n \rangle$, and $\mathbf{Y} = \langle Y_1, Y_2, \ldots, Y_n \rangle$.

In practice the matrix $\mathbf{A}$ is assumed to be known or estimable. This is a severe assumption over which economists have fired off many a polemic.

The most frequent problem concerning the open Leontief model is as follows: Given that the Y_i are known, find X_i satisfying (8.1.1). Of course, the answer is meaningful only if the resulting solution $\mathbf{X} = (\mathbf{I} - \mathbf{A})^{-1}\mathbf{Y}$ exists and is a nonnegative vector.

Let $(\mathbf{I} - \mathbf{A})^{-1} = \|A_{ij}\|$. Then $X_i = \sum_j A_{ij} Y_j$ $(i = 1, 2, \ldots, n)$, so that A_{ij} can be interpreted as the amount by which the output of activity i must be increased to produce one additional unit of commodity j.

Another problem involving the open model is the following: Consider n countries $1, 2, \ldots, n$. Let a_{ij} denote country j's marginal propensity to import from country i (that is, the increase in imports from country i to country j per unit increase in income of country j), and let a_{ii} denote country i's marginal propensity to consume its own goods. Let X_i represent country i's national income, and let c_i represent country i's national expenditure, which is independent of income. Then

$$X_i = \sum_{j=1}^{n} a_{ij} X_j + c_i,$$

and hence $(\mathbf{I} - \mathbf{A})\mathbf{X} = \mathbf{c}$. To be economically meaningful, $\mathbf{X}$ and $\mathbf{c}$ must be positive vectors.

This example may also be interpreted as a closed model of trade between countries. Suppose that all income to country j comes from the sale of goods to other countries or to itself. Let α_{ij} represent the fraction of country j's income that is spent on goods from country i. Let X_j be the annual income of country j. The total value of exports from country i is clearly

$$\sum_{j=1}^{n} \alpha_{ij} X_j,$$

and this, by the definition of a closed system, is X_i. We thus have

$$\sum_{j=1}^{n} \alpha_{ij} X_j = X_i \qquad (i = 1, \ldots, n),$$

where

$$\sum_{i=1}^{n} \alpha_{ij} = 1, \qquad \alpha_{ij} \geq 0.$$

One problem to be solved in dealing with equations of this sort is to determine conditions which assure a positive solution.

In the case of a closed Leontief model, we seek a positive solution $\mathbf{X}$ of the system of equations $\mathbf{AX} = \mathbf{X}$. Nontrivial solutions exist if and only if the value 1 is an eigenvalue of the matrix $\mathbf{A}$. But even when 1 is an eigenvalue, to ensure the existence of a positive eigenvector requires something more. The theory of positive matrices serves to resolve such problems. Several conditions for the solvability of these problems are developed in the next section and applied in Section 8.3.

The usual dual problems in an economic model involving labor and prices make their appearance in the system of equations generated by the transpose matrix of $\mathbf{A}$. Suppose that the $(n + 1)$th activity represents labor. If there is no unemployment, then

$$X_{n+1} - \sum_{j=1}^{n} X_{n+1,j} = 0.$$

As before, we define

$$a_{n+1,j} = \frac{X_{n+1,j}}{X_j},$$

so that $a_{n+1,j}$ can be interpreted as the amount of labor used to produce a single unit of commodity j. It follows that $\sum_i a_{n+1,i} A_{ij}$ is the additional amount of labor needed to produce one more unit of commodity j.

Let p_i be the price of a unit of commodity i. What is usually assumed is that the price of a commodity is equal to the cost of producing it, i.e., that

$$p_j = \sum_{i=1}^{N} p_i a_{ij} + p_{n+1} a_{n+1,j} \qquad \text{or} \qquad \mathbf{p}(\mathbf{I} - \mathbf{A}) = p_{n+1}(\mathbf{a}_{n+1}).$$

Here $\mathbf{a}_{n+1}$ represents the vector $\langle a_{n+1,1}, a_{n+1,2}, \ldots, a_{n+1,n}\rangle$. This is a sort of equilibrium condition; it may be thought of as resulting from competition so complete as to make profit impossible. Ordinarily p_{n+1} is taken to be 1, so that the p_i represent relative prices.

†8.2 The theory of positive matrices. The study of matrices with positive elements is very important for a variety of applications. Such matrices often arise in a natural way—i.e., the nature of the model implies the posi-

† The results of this section will only be used in Sections 8.3 of this chapter and in Sections 9.3, 9.7, and 9.8 of Chapter 9.

tivity of the coefficients—in probability models (particularly the study of Markov chains), economic analyses, the study of vibrations of mechanical systems, etc. We accordingly digress in this section to display in separate form several propositions concerning eigenvalues of positive matrices that will be pertinent to our analysis of economic models.

Suppose that $\mathbf{A} = \|a_{ij}\|$ is a square matrix. If all a_{ij} are nonnegative, we write $\mathbf{A} \geq \mathbf{0}$; if $\mathbf{A} \geq \mathbf{0}$ and some $a_{ij} > 0$, we write $\mathbf{A} > \mathbf{0}$; and if all $a_{ij} > 0$, we write $\mathbf{A} \gg \mathbf{0}$. This is consistent with the notation and terminology adopted earlier for vectors.

We write $\mathbf{A} \geq \mathbf{B}$ if $\mathbf{A} - \mathbf{B} \geq \mathbf{0}$, etc. A positive matrix as a mapping clearly preserves the partial ordering defined by $>$. In particular, if $\mathbf{x}' \geq \mathbf{x}$, then $\mathbf{x}'\mathbf{A} \geq \mathbf{x}\mathbf{A}$ for $\mathbf{A} \geq \mathbf{0}$. Let

$$\Lambda = \{\lambda | \mathbf{x}\mathbf{A} \geq \lambda\mathbf{x} \quad \text{for some} \quad \mathbf{x} > \mathbf{0}\}. \tag{8.2.1}$$

The set always contains the value $\lambda = 0$ whenever $\mathbf{A} \geq \mathbf{0}$. The relation $\mathbf{x}\mathbf{A} \geq \lambda\mathbf{x}$ being homogeneous, we may without loss of generality assume that the nonnegative vector $\mathbf{x}$ corresponding to a value λ is normalized so that $\sum x_i = 1$. Such vectors $\mathbf{x}$ conveniently traverse a compact set. Let $\lambda_0 = \sup_\Lambda \lambda$. We obviously have $\lambda_0 < \infty$. We begin by an analysis of matrices with strictly positive elements and then pass to the modifications that are necessary in dealing with general positive matrices.

▶ THEOREM 8.2.1. If $\mathbf{A} \gg \mathbf{0}$, then (a) there exists $\mathbf{x}^0 \gg \mathbf{0}$ such that $\mathbf{x}^0\mathbf{A} = \lambda_0\mathbf{x}^0$; (b) if $\lambda \neq \lambda_0$ is any other eigenvalue of $\mathbf{A}$, it follows that $|\lambda| < \lambda_0$; (c) λ_0 is an eigenvalue of geometrical and algebraic multiplicity 1.

Remark 8.2.1. To clarify the meaning of the theorem, we make the following assertions, which are proved in Section A.4. If

$$N_\lambda^r = [\mathbf{x} | \mathbf{x}(\mathbf{A} - \lambda\mathbf{I})^r = \mathbf{0}] \quad \text{and} \quad M_\lambda^r = [\mathbf{x} = \mathbf{y}(\mathbf{A} - \lambda\mathbf{I})^r | \mathbf{y} \quad \text{arbitrary}],$$

then $N_\lambda^0 \subset N_\lambda^1 \subset N_\lambda^2 \subset \cdots \subset N_\lambda^k = N_\lambda^{k+1} = N_\lambda^{k+2} \cdots$ for any $\mathbf{A}$ and λ, where $\subset$ means "is contained in but not equal to." The integer k thus determined is called the index of λ. The dimension of N_λ^1 is the geometric multiplicity of λ, and the dimension of N_λ^k is the algebraic multiplicity of λ. Also $M_\lambda^0 \supset M_\lambda^1 \supset M_\lambda^2 \supset \cdots \supset M_\lambda^k = M_\lambda^{k+1} = M_\lambda^{k+2} = \cdots$ for the same k. For $\mathbf{A}$ self-adjoint, $k = 1$ for any eigenvalue λ, and hence its geometric and algebraic multiplicities are the same. In any case the geometric multiplicity is no larger than the algebraic multiplicity. When the algebraic and geometric multiplicities agree, the elementary divisors are simple.

Proof of the Theorem.

(a) A simple compactness argument implies the existence of a vector $\mathbf{x}^0 > \mathbf{0}$ such that $\mathbf{x}^0\mathbf{A} \geq \lambda_0\mathbf{x}^0$. Since $\mathbf{A} \gg \mathbf{0}$ implies that $\mathbf{x}\mathbf{A} \gg \mathbf{0}$ for

$\mathbf{x} > \mathbf{0}$, it follows that $\mathbf{x}^0\mathbf{A} = \mathbf{y}^0 \gg \mathbf{0}$ satisfies $\mathbf{y}^0\mathbf{A} \gg \lambda_0\mathbf{y}^0$ unless $\mathbf{x}^0\mathbf{A} = \lambda_0\mathbf{x}^0$. But if $\mathbf{y}^0\mathbf{A} \gg \lambda_0\mathbf{y}^0$, a positive ϵ can be found such that $\mathbf{y}^0\mathbf{A} \geq (\lambda_0 + \epsilon)\mathbf{y}^0$, thus contradicting the meaning of λ_0. Hence $\mathbf{x}^0\mathbf{A} = \lambda_0\mathbf{x}^0$ and $\mathbf{x}^0 \gg \mathbf{0}$.

(b) Let $\lambda \neq \lambda_0$ be an eigenvalue of $\mathbf{A}$ and let the corresponding nontrivial eigenvector $\mathbf{z}$ be such that $\mathbf{zA} = \lambda\mathbf{z}$. Let $|\mathbf{z}|$ denote the vector whose components are $\langle|z_i|\rangle$. Then the positivity of $\mathbf{A}$ implies that $|\mathbf{z}|\mathbf{A} \geq |\lambda||\mathbf{z}|$. From the construction of λ_0, it follows that $|\lambda| \leq \lambda_0$. In order to see that strict inequality persists, we consider the matrix $\mathbf{A}_\delta = \mathbf{A} - \delta\mathbf{I}$, where δ is chosen so small and positive that $\mathbf{A}_\delta$ is a strictly positive matrix. The largest real positive eigenvalue of $\mathbf{A}_\delta$ is $\lambda_0 - \delta$, which as we have seen bounds (in magnitude) all other eigenvalues of $\mathbf{A}_\delta$. If the eigenvalue satisfies $|\lambda| = |\lambda_0|$, with λ distinct from λ_0, then $|\lambda - \delta| > |\lambda_0 - \delta|$, contradicting the previous conclusion. The proof of the theorem is complete.

(c) Suppose that the set of all eigenvectors with eigenvalue λ_0 is at least two-dimensional. Since $\mathbf{A}$ is a real matrix, there exists a real eigenvector $\mathbf{z}$ that is linearly independent of $\mathbf{x}^0$. Since $\mathbf{x}^0 \gg \mathbf{0}$, a linear combination $\mathbf{w} = t\mathbf{x}^0 + \mathbf{z}^0$ exists ($t = \min_i - z_i^0/x_i^0$), with $\mathbf{w} > \mathbf{0}$ but not $\mathbf{w} \gg \mathbf{0}$. Since $\mathbf{wA} = \lambda_0\mathbf{w}$ is strictly positive and cannot be identically zero, we have an absurdity. Hence we conclude that the geometrical multiplicity of λ_0 is 1.

Suppose that $\mathbf{y}(\mathbf{A} - \lambda_0\mathbf{I}) = \mathbf{z}$ is such that $\mathbf{z}(\mathbf{A} - \lambda_0\mathbf{I}) = \mathbf{0}$. By the first part of this argument $\mathbf{z}$ is a multiple of $\mathbf{x}^0$. On multiplying $\mathbf{y}$ by a nonzero constant, we may assume that $\mathbf{z} = a\mathbf{x}^0$, with $a \geq 0$. Let $\mathbf{f}^0$ denote a positive eigenvector for the matrix $\mathbf{A}'$ corresponding to the eigenvalue λ_0; we may verify the existence of $\mathbf{f}^0$ by applying assertion (a) of the theorem to the strictly positive matrix $\mathbf{A}'$. Observe that

$$(a\mathbf{x}^0, \mathbf{f}^0) = (\mathbf{y}(\mathbf{A} - \lambda_0\mathbf{I}), \mathbf{f}^0) = (\mathbf{y}, (\mathbf{A}' - \lambda_0\mathbf{I})\mathbf{f}^0) = 0,$$

so that since $(\mathbf{x}^0, \mathbf{f}^0) > 0$ we must have $a = 0$. Hence if $\mathbf{y}(\mathbf{A} - \lambda_0\mathbf{I})^2 = \mathbf{0}$, then $\mathbf{y}(\mathbf{A} - \lambda_0\mathbf{I}) = \mathbf{0}$, and thus $N_{\lambda_0}^2 = N_{\lambda_0}^1$. This means that the algebraic multiplicity of λ_0 is equal to the geometric multiplicity of λ_0.

In order to explore in greater detail the structure of the eigenmanifold attached to the eigenvalue λ_0, we introduce a projection operation which projects vectors into the linear space of eigenvectors associated with λ_0. For this purpose, let the eigenvector $\mathbf{x}^0$, where $\mathbf{x}^0\mathbf{A} = \lambda_0\mathbf{x}^0$, and the eigenvector $\mathbf{f}^0 \gg \mathbf{0}$, where $\mathbf{Af}^0 = \lambda_0\mathbf{f}^0$, be normalized so that $\sum x_i^0 f_i^0 = 1$ Define $\mathbf{P}$ as the matrix $\|x_j^0 f_i^0\|$. Then $\mathbf{P}$ has the following properties:

(i) For any vectors $\mathbf{x}$ and $\mathbf{f}$, $\mathbf{xP} = (\mathbf{x}, \mathbf{f}^0)\mathbf{x}^0$ and $\mathbf{Pf} = (\mathbf{f}, \mathbf{x}^0)\mathbf{f}^0$.
(ii) $\mathbf{P}^2 = \mathbf{P}$.
(iii) $\mathbf{AP} = \mathbf{PA} = \lambda_0\mathbf{P}$.

The verification of these assertions is immediate.

For any matrix $\mathbf{B}$, it is known that $(\mathbf{B} - \lambda\mathbf{I})^{-1}$ represents an analytic function of λ (i.e., each component is an analytic function of λ) wherever the inverse makes sense. Each eigenvalue of $\mathbf{B}$ gives rise to a pole of $(\mathbf{B} - \lambda\mathbf{I})^{-1}$ (i.e., a pole for some component) whose order is related to the algebraic multiplicity of the eigenvalue. Specifically, the order of the pole at an eigenvalue $\tilde{\lambda}$ is equal to the index of $\tilde{\lambda}$ (see Remark 8.2.1 above and pp. 383–85 of Appendix A). Moreover,

$$(\mathbf{B} - \lambda\mathbf{I})^{-1} = -\sum_{n=0}^{\infty} \frac{1}{\lambda^{n+1}} \mathbf{B}^n$$

converges for $|\lambda| > r$, where r denotes the radius of the smallest circle containing the eigenvalues of $\mathbf{B}$, also called the spectral radius. (We frequently write $\lambda_0(\mathbf{A}) = r(\mathbf{A})$ to denote the spectral radius of $\mathbf{A}$.) But $(\mathbf{B} - \lambda\mathbf{I})^{-1}$ has a pole at some value λ^*, where $|\lambda^*| = \lambda_0$; hence

$$r = \overline{\lim_{n\to\infty}} \sqrt[n]{\max_{ij} |b_{ij}^{(n)}|},$$

where $\mathbf{B}^n = \|b_{ij}^{(n)}\|$. We are now prepared to prove the following theorem.

▶ THEOREM 8.2.2. For $\mathbf{A} \gg \mathbf{0}$, λ_0, and $\mathbf{P}$ as defined above,

$$\frac{1}{\lambda_0^n} \mathbf{A}^n \to \mathbf{P}.$$

Proof. The eigenvalues of the matrix $\mathbf{B} = \mathbf{A} - \lambda_0\mathbf{P}$ must all lie strictly within the circle of radius λ_0, since (i) any nonzero eigenvalue of $\mathbf{B}$ is an eigenvalue of $\mathbf{A}$, and (ii) λ_0 is not an eigenvalue of $\mathbf{B}$.

We now prove these statements:

Let us assume that these two assertions are false. Then $\mathbf{zB} = \lambda\mathbf{z}$ for some nonzero z. It follows that $\mathbf{zBP} = \lambda\mathbf{zP}$ and

$$\mathbf{zBP} = \mathbf{z}(\mathbf{AP} - \lambda_0\mathbf{P}^2) = \lambda_0\mathbf{z}(\mathbf{P} - \mathbf{P}) = \mathbf{0}.$$

Therefore, if $\lambda \neq 0$, it follows that $\mathbf{zP} = \mathbf{0}$ and $\mathbf{zB} = \lambda\mathbf{z}$ reduces to $\mathbf{zA} = \lambda\mathbf{z}$. In particular, if $\mathbf{zB} = \lambda_0\mathbf{z}$, then $\mathbf{zP} = \mathbf{0}$ and $\mathbf{zA} = \lambda_0\mathbf{z}$. But then $\mathbf{z}$ is a nonzero multiple of $\mathbf{x}^0$ and $\mathbf{zP}$ cannot be zero. This contradiction shows that the assertions are true. The spectral radius of $\mathbf{B}$ must be smaller than λ_0, and it follows that the maximum component of $\mathbf{B}^m$ satisfies the bound $|b_{ij}^{(m)}| < K\rho^m$ for all m with a suitable $\rho < \lambda_0$. (See the remarks preceding this theorem.) But $\mathbf{B}^m = \mathbf{A}^m - \lambda_0^m\mathbf{P}$; hence we infer that $\mathbf{A}^m/\lambda_0^m$ converges to $\mathbf{P}$ at the geometric convergence rate ρ/λ_0.

All of the foregoing theory holds if $\mathbf{A} \geq \mathbf{0}$ and $\mathbf{A}^m \gg \mathbf{0}$ for some m. It also holds if $\mathbf{A}$ maps some closed proper convex cone (i.e., contained properly in a half space) in E^n into its interior. The cone for positive matrices is the first quadrant.

Several of the above results hold whenever $\mathbf{A} \geq \mathbf{0}$, i.e., even when we do not insist that some iterate of $\mathbf{A}$ be strictly positive. Let $\mathbf{U}$ be the square matrix with every component 1; then $\mathbf{A} + \delta\mathbf{U}$ ($\delta > 0$) is strictly positive and the assertions of Theorem 8.2.1 apply. A standard limiting argument on δ will establish the validity of parts (a) and (b) of the following theorem.

▶THEOREM 8.2.3. Let $\mathbf{A} > \mathbf{0}$ and let λ_0 be defined as in (8.2.1). Then (a) λ_0 is an eigenvalue and there exists $\mathbf{x}^0 > \mathbf{0}$ such that $\mathbf{x}^0\mathbf{A} = \lambda_0\mathbf{x}^0$; (b) if λ is any other eigenvalue, $|\lambda| \leq \lambda_0$; (c)

$$\frac{1}{m}\sum_{i=1}^{m}\frac{\mathbf{A}^i}{\lambda_0^i}$$

converges whenever $\mathbf{x}^0 \gg \mathbf{0}$; and (d) if λ is an eigenvalue of $\mathbf{A}$ with $|\lambda| = \lambda_0$, it follows that $\eta = \lambda/\lambda_0$ is a root of unity and $\eta^m\lambda_0$ is an eigenvalue of $\mathbf{A}$ for $m = 0, 1, 2, \ldots$

Proof of (*c*). For convenience of exposition we assume that $\lambda_0 = 1$. We first note that the elements of $\mathbf{A}^m$ are uniformly bounded. In fact, $\mathbf{x}^0\mathbf{A}^m = \mathbf{x}^0$ implies

$$0 \leq a_{ij}^{(m)} \leq \frac{\max_i x_i^0}{\min_i x_i^0},$$

as desired.

Let $\mathcal{L}$ denote the space of elements for which $\mathbf{z}\mathbf{A} = \mathbf{z}$. Clearly, $\mathcal{L}$ is a closed linear space such that for $\mathbf{z}$ in $\mathcal{L}$

$$\mathbf{z}\,\frac{\mathbf{A} + \mathbf{A}^2 + \cdots + \mathbf{A}^m}{m} = \mathbf{z}$$

converges as m increases to infinity. Next define $\mathcal{K} = \{\mathbf{y}|\mathbf{y} = \mathbf{z}(\mathbf{I} - \mathbf{A})\}$; that is, $\mathcal{K}$ is the range of $\mathbf{I} - \mathbf{A}$. We assert that for $\mathbf{y}$ in $\mathcal{K}$, $\mathbf{y}\mathbf{S}_m$ tends to zero, where

$$\mathbf{S}_m = \frac{\mathbf{A} + \mathbf{A}^2 + \cdots + \mathbf{A}^m}{m}.$$

Explicitly, since $\mathbf{y} = \mathbf{z}(\mathbf{I} - \mathbf{A})$, it follows that $\mathbf{y}\mathbf{S}_m = (\mathbf{z}\mathbf{A} - \mathbf{z}\mathbf{A}^{m+1})/m$ tends to zero, by virtue of the fact that the coefficients of $\mathbf{A}^{m+1}$ are uniformly bounded.

We have thus established that $\mathbf{S}_m$ converges for all vectors contained in the sum $\mathcal{L} + \mathcal{K}$. It remains only to show that $\mathcal{L} + \mathcal{K}$ is all of the Euclidean n-space. To this end, we observe that every vector $\mathbf{x}$ can be written as

$$\mathbf{x} = (\mathbf{x} - \mathbf{x}\mathbf{S}_m) + \mathbf{x}\mathbf{S}_m = \mathbf{y}_m + \mathbf{z}_m.$$

Moreover, as m tends to infinity we may select a limit vector $\mathbf{z}_0$ of $\mathbf{z}_m$.

This is possible since $\mathbf{S}_m$ consists of a sequence of uniformly bounded matrices. It is immediately verified that $\mathbf{z}_0$ is in $\mathcal{L}$. Furthermore, it is readily verified that each $\mathbf{y}_m$ belongs to $\mathcal{K}$ and hence its limit belongs to $\mathcal{K}$. This finishes the proof of (c).

Proof of (d). We may assume without loss of generality that there exists some $\mathbf{f}^0 \gg \mathbf{0}$ for which $\mathbf{A}\mathbf{f}^0 = \mathbf{f}^0$. The reason for this is as follows: Suppose for definiteness that $\mathbf{f}^0$ has the form $\langle f_1^0, f_2^0, \ldots, f_r^0, 0, \ldots, 0\rangle$, where $f_1^0 > 0, \ldots, f_r^0 > 0$[$\mathbf{f}^0$ exists by (a) above applied to the matrix A']. Since $\mathbf{A}\mathbf{f}^0 = \mathbf{f}^0$, it follows that

$$\mathbf{A} = \begin{pmatrix} \mathbf{A}_1 & \mathbf{B} \\ 0 & \mathbf{A}_2 \end{pmatrix},$$

where $\mathbf{A}_1$ is a positive $r \times r$ matrix and $\mathbf{A}_2$ is a positive $(n-r) \times (n-r)$ matrix. To investigate the nature of the eigenvalues of $\mathbf{A}$ it is clearly sufficient to study the eigenvalues of $\mathbf{A}_1$ and $\mathbf{A}_2$ separately. By continuing in this way, we ultimately arrive at the problem for which the assumption made concerning $\mathbf{f}^0$ is valid.

We now proceed to the proof of (d). Suppose that λ is an eigenvalue of $\mathbf{A}$ of modulus 1 and that $\mathbf{x}\mathbf{A} = \lambda\mathbf{x}$ for some nonzero vector $\mathbf{x}$, that is,

$$\sum_{i=1}^{n} x_i a_{ij} = \lambda x_j.$$

Then $\sum_i |x_i| a_{ij} \geq |x_j|$ or $|\mathbf{x}|\mathbf{A} \geq |\mathbf{x}|$. But

$$(\mathbf{f}^0, |\mathbf{x}|) \leq (\mathbf{f}^0, |\mathbf{x}|\mathbf{A}) = (\mathbf{A}\mathbf{f}^0, |\mathbf{x}|) = (\mathbf{f}^0, |\mathbf{x}|),$$

with strict inequality yielding a contradiction unless $|\mathbf{x}|\mathbf{A} = |\mathbf{x}|$. Consequently,

$$\sum_{i=1}^{n} |x_i| a_{ij} = \left| \sum_{i=1}^{n} x_i a_{ij} \right|.$$

This implies the existence of numbers $u_j (|u_j| = 1)$ such that

$$x_i a_{ij} = |x_i| a_{ij} u_j \tag{*}$$

for each i and j. Let $\mathbf{x} \cdot \mathbf{y}$ denote the vector whose components are $\{x_i y_i\}$. Multiplying the preceding relation by u_i and summing shows that

$$(\mathbf{x} \cdot \mathbf{u})\mathbf{A} = (\mathbf{u} \cdot |\mathbf{x}|)\mathbf{A} \cdot \mathbf{u}$$

and from (*) by summation on i

$$|\mathbf{x}| \cdot \mathbf{u} = (|\mathbf{x}|\mathbf{A}) \cdot \mathbf{u} = \mathbf{x}\mathbf{A} = \lambda\mathbf{x}.$$

Consider the vector $\mathbf{x} \cdot \mathbf{u}$. Using the last two relations proves that

$$(\mathbf{x} \cdot \mathbf{u})\mathbf{A} = (\mathbf{u} \cdot |\mathbf{x}|)\mathbf{A} \cdot \mathbf{u} = \lambda \mathbf{x}\mathbf{A} \cdot \mathbf{u} = \lambda^2 \mathbf{x} \cdot \mathbf{u},$$

and hence $\mathbf{u} \cdot \mathbf{x}$ is an eigenvector corresponding to λ^2. Furthermore,

$$u_i^2 |x_i| a_{ij} u_j = u_i^2 x_i a_{ij},$$

and hence $(|\mathbf{x}| \cdot \mathbf{u}^2)\mathbf{A} \cdot \mathbf{u} = (\mathbf{x} \cdot \mathbf{u}^2)\mathbf{A}$. But

$$(|\mathbf{x}| \cdot \mathbf{u}^2)\mathbf{A} = \lambda(\mathbf{x} \cdot \mathbf{u})\mathbf{A} = \lambda^3 \mathbf{u} \cdot \mathbf{x},$$

and thus $(\mathbf{x} \cdot \mathbf{u}^2)\mathbf{A} = \lambda^3(\mathbf{x} \cdot \mathbf{u}^2)$. This shows that $\mathbf{x} \cdot \mathbf{u}^2$ is an eigenvector for λ^3. Proceeding in this way, we find that if λ is an eigenvalue of modulus 1, then λ is a root of unity and $\lambda, \lambda^2, \lambda^3, \ldots, \lambda^n, \ldots$ are all eigenvalues as asserted. An eigenvector of λ^{r+1} is $\mathbf{x} \cdot \mathbf{u}^r$, where $\mathbf{x}$ and $\mathbf{u}$ have the meaning indicated above. This completes the proof of Theorem 8.2.3.

The following characterization of the spectral radius $\lambda_0(\mathbf{A})$, which is implicit in the construction of Λ in (8.2.1), was used in the proofs of Theorems 8.2.1 and 8.2.3. It is useful enough in general to be worth highlighting here.

▶Corollary 8.2.1. The eigenvalue of $\mathbf{A}$ of largest magnitude $\lambda_0 = \lambda_0(\mathbf{A})$ is real and nonnegative and is characterized as $\lambda_0 = \max_\Lambda \lambda$, where

$$\Lambda = \{\lambda | \mathbf{x}\mathbf{A} \geq \lambda \mathbf{x} \quad \text{for some} \quad \mathbf{x} > \mathbf{0}\}. \tag{8.2.1}$$

Another useful consequence is as follows:

▶Corollary 8.2.2. If there exists $\mathbf{x}^0 \gg \mathbf{0}$ such that $\mathbf{x}^0\mathbf{A} \leq \mu \mathbf{x}^0$, then μ is an upper bound to the spectral radius of $\mathbf{A}$.

Proof. It quickly follows that

$$|a_{ij}^{(n)}| \leq \mu^n \frac{\max_i x_i^0}{\min_i x_i^0},$$

where $\mathbf{A}^n = \|a_{ij}^n\|$; hence

$$\overline{\lim} \sqrt[n]{|a_{ij}^{(n)}|} \leq \mu,$$

which leads easily to the desired result (see the discussion preceding Theorem 8.2.2).

The next corollary follows from Corollary 8.2.1.

▶Corollary 8.2.3. If $\mathbf{A} \geq \mathbf{B} \geq \mathbf{0}$ and $\lambda_0(\mathbf{A}), \lambda_0(\mathbf{B})$ are defined as in (8.2.1), corresponding to $\mathbf{A}$ and $\mathbf{B}$, respectively, then $\lambda_0(\mathbf{A}) \geq \lambda_0(\mathbf{B})$; i.e., the spectral radius of a positive matrix is a monotone-increasing function of its components.

The following corollary may be compared with the fundamental min-max theorem of game theory.

▶ COROLLARY 8.2.4. If $\mathbf{A} \geq 0$ and $\mu_0 = \inf \{\mu | \mathbf{Af} \leq \mu\mathbf{f}, \mathbf{f} > 0\}$, and if the eigenvector $\mathbf{x}^0$ is strictly positive, then

$$\mu_0 = \lambda_0 = \sup_{\substack{x>0 \\ \Sigma x_i=1}} \inf_{\substack{f>0 \\ \Sigma f_i=1}} \frac{(\mathbf{f}, \mathbf{xA})}{(\mathbf{f}, \mathbf{x})} = \inf_{\substack{f>0 \\ \Sigma f_i=1}} \sup_{\substack{x>0 \\ \Sigma x_i=1}} \frac{(\mathbf{f}, \mathbf{xA})}{(\mathbf{f}, \mathbf{x})},$$

where it is understood that if both $(\mathbf{f}, \mathbf{x}) = 0$ and $(\mathbf{f}, \mathbf{xA}) = 0$ hold, the meaning of $(\mathbf{f}, \mathbf{xA})/(\mathbf{f}, \mathbf{x})$ is taken to be λ_0.

Proof. Since there exists a positive eigenvector $\mathbf{f}^0$ (Theorem 8.2.3) corresponding to $\mathbf{A}'$ for the eigenvalue λ_0, it follows that $\mu_0 \leq \lambda_0$.

But for $\mathbf{f}$ corresponding to μ_0, which certainly exists, we obtain

$$\mu_0(\mathbf{f}, \mathbf{x}^0) \geq (\mathbf{Af}, \mathbf{x}^0) = (\mathbf{f}, \mathbf{x}^0\mathbf{A}) = \lambda_0(\mathbf{f}, \mathbf{x}^0).$$

The hypothesis on $\mathbf{x}^0$ compels $\mu_0 \geq \lambda_0$. Hence $\mu_0 = \lambda_0$. The remaining relations follow in the usual way (see Theorem 1.3.1).

The following theorem is useful for its applications to economics.

▶ THEOREM 8.2.4. If $\rho > \lambda_0(\mathbf{A})$ and $\mathbf{A} \geq \mathbf{0}$, then $(\rho\mathbf{I} - \mathbf{A})^{-1}$ transforms nonnegative vectors into nonnegative vectors.

Proof. For $\lambda > \lambda_0(\mathbf{A})$, where $\lambda_0(\mathbf{A})$ is the spectral radius, an explicit calculation yields the representation

$$(\lambda\mathbf{I} - \mathbf{A})^{-1} = \sum_{k=0}^{\infty} \frac{\mathbf{A}^k}{\lambda^{k+1}}.$$

Since each $\mathbf{A}^k$ maps the nonnegative orthant into itself, so does the infinite sum.

The converse of Theorem 8.2.4 is also valid: i.e., if $(\rho\mathbf{I} - \mathbf{A})^{-1}$ maps the nonnegative orthant into itself, then ρ is an upper bound to the spectral radius of $\mathbf{A}$. Explicitly, if $\mathbf{xA} \geq \lambda\mathbf{x}$ for $\mathbf{x} > \mathbf{0}$, then $\mathbf{z} = \mathbf{x}(\mathbf{A} - \rho\mathbf{I}) + (\rho - \lambda)\mathbf{x}$ is a nonnegative vector. Hence $(\rho\mathbf{I} - \mathbf{A})^{-1}$ applied to $\mathbf{z}$ yields a nonnegative vector, or

$$-\mathbf{x} + (\rho - \lambda)\mathbf{x}(\mathbf{A} - \rho\mathbf{I})^{-1} \geq \mathbf{0}.$$

But $\mathbf{x} > \mathbf{0}$ and $\mathbf{x}(\mathbf{A} - \rho\mathbf{I})^{-1} \geq \mathbf{0}$, which implies that $\rho > \lambda$. On the other hand, λ_0 is such an admissible λ value (Corollary 8.2.1), and hence $\rho > \lambda_0(\mathbf{A})$ as asserted.

We may express the conditions under which $(\rho\mathbf{I} - \mathbf{A})^{-1}$ transforms positive vectors into positive vectors in terms of restrictions on the principal minors of $(\rho\mathbf{I} - \mathbf{A})$ as follows: If every principal minor of $\rho\mathbf{I} - \mathbf{A}$ is positive, then $(\rho\mathbf{I} - \mathbf{A})^{-1}$ is positivity-preserving.

To prove this statement it is enough by virtue of Theorem 8.2.4 to show that $\rho > \lambda_0(\mathbf{A})$. This is done as follows: The relation $\rho = \lambda_0(\mathbf{A})$ is impossible, since by hypothesis $\det(\rho\mathbf{I} - \mathbf{A}) > 0$. In view of Theorem 8.2.4, it is enough to show that $\det(t\mathbf{I} - \mathbf{A}) \neq 0$ for $t > \rho$. Clearly,

$$\det(t\mathbf{I} - \mathbf{A}) = \det[(t - \rho)\mathbf{I} + \mathbf{B}(\rho)], \tag{8.2.2}$$

where $\mathbf{B}(\rho) = \rho\mathbf{I} - \mathbf{A}$. Expanding the second terms of (8.2.2), we see because of the hypothesis that $\det(t\mathbf{I} - \mathbf{A}) > 0$ for $t > \rho$ as desired.

The hypothesis can be weakened so that the assertion reads as follows: If a single sequence of n "bordered" principal minors $|\rho\mathbf{I} - \mathbf{A}|_1^p$ $(p = 1, \ldots, n)$ is positive, then every principal minor of $\rho\mathbf{I} - \mathbf{A}$ is positive, and hence, by what has been established above, ρ exceeds the greatest characteristic number of the matrix $\mathbf{A}$. The proof of this assertion is omitted (see Problem 11).

Matrix decomposability. Another set of ideas connected with positive matrices involves the notion of matrix indecomposability. We consider $\mathbf{A} = \|a_{ij}\|$, a nonnegative $n \times n$ matrix. A set S of indices will be said to be closed if $a_{pq} = 0$ for $q \in S$, $p \notin S$. If no proper closed set exists, the matrix $\mathbf{A}$ is said to be indecomposable. This will happen if and only if there exists no permutation matrix $\mathbf{\Pi}$† such that

$$\mathbf{\Pi}\,\mathbf{A}\,\mathbf{\Pi}^{-1} = \begin{bmatrix} \mathbf{A}_1 & \mathbf{A}_3 \\ 0 & \mathbf{A}_2 \end{bmatrix},$$

where $\mathbf{A}_1$ and $\mathbf{A}_2$ are square submatrices.

A matrix is decomposable, as indicated in the proof of Theorem 8.2.3, part (d), whenever it has a nonnegative eigenvector, not strictly positive, of a positive eigenvalue. For instance, if $\mathbf{A}\mathbf{y}^0 = \lambda_0\mathbf{y}^0$, where $\mathbf{y}^0 = \langle y_1^0, \ldots, y_n^0\rangle$, is arranged so that $y_1^0 > 0, \ldots, y_r^0 > 0$ while $y_{r+1}^0 = y_{r+2}^0 = \cdots = y_n^0 = 0$, then necessarily

$$\mathbf{A} = \begin{matrix} r\{ \\ n-r\{ \end{matrix} \begin{Vmatrix} \overbrace{\mathbf{A}_1}^{r} & \overbrace{\mathbf{A}_2}^{n-r} \\ \mathbf{A}_3 & \mathbf{A}_4 \end{Vmatrix},$$

with $\mathbf{A}_3 = \mathbf{0}$.

† A permutation matrix is a matrix such that each row and each column has a single nonzero entry of value 1.

The class of positive indecomposable matrices may further be divided into an aperiodic subclass and a periodic subclass. It is elementary to show that an indecomposable matrix $\mathbf{A}$ by permutation of rows and columns can be put in the form

$$\mathbf{\Pi A \Pi}^{-1} = \begin{bmatrix} \mathbf{0} & \mathbf{0} & \cdots & \mathbf{0} & \mathbf{G}_\omega \\ \mathbf{G}_1 & \mathbf{0} & \cdots & \mathbf{0} & \mathbf{0} \\ \mathbf{0} & \mathbf{G}_2 & \cdots & \mathbf{0} & \mathbf{0} \\ \vdots & \vdots & & \vdots & \vdots \\ \mathbf{0} & \mathbf{0} & \cdots & \mathbf{G}_{\omega-1} & \mathbf{0} \end{bmatrix}, \tag{8.2.3}$$

where the $\mathbf{0}$'s on the diagonal are square matrices. The number ω is related to the number of eigenvalues on the circle of radius λ_0 [see Theorem 8.2.3, part (d)]. If $\omega > 1$, the indecomposable matrix $\mathbf{A}$ is said to be periodic; if $\omega = 1$, $\mathbf{A}$ is said to be aperiodic.

Applications of some of these ideas to economic models will be found in the next section. Indecomposability is also an important concept related to the principle of communication between states of a Markov chain.

Metzler matrices. We close this section with a discussion of the eigenvalue structure of matrices $\mathbf{C} = \|c_{ij}\|$, where $c_{ij} \geq 0$ $(i \neq j)$. Such matrices are referred to by economists as Metzler matrices; we shall call them M-matrices. M-matrices may be treated as a special class of positive matrices, since by adding a sufficiently large positive multiple of the identity matrix we can achieve

$$\mathbf{A} = \mathbf{C} + \mu \mathbf{I}$$

such that $\mathbf{A} \geq \mathbf{0}$.

We now direct our analysis to the problem of determining criteria for M-matrices which ensure that all their eigenvalues have negative real parts. M-matrices of this form are of great importance in economic studies, notably in the study of stability characteristics associated with price systems converging to equilibrium, since local stability is guaranteed for a matrix system of differential equations whenever all eigenvalues have negative real parts (see Section 9.3).

Since we only translate $\mathbf{C}$, all the eigenvalues of $\mathbf{C}$ have negative real parts if and only if $\mathcal{R}(\lambda_i) < \mu$, where λ_i are the eigenvalues of $\mathbf{A}$. Because of Theorem 8.2.3, since $\mathbf{A} \geq \mathbf{0}$, this will be so if and only if $\lambda_0(\mathbf{A}) < \mu$. Analogous arguments relying heavily on the theory of positive matrices allow us to deduce more, namely:

(a) The eigenvalue of an M-matrix with largest real part is real and has an associated nonnegative eigenvector. We define $\sigma(\mathbf{C}) = \max_i \mathcal{R}(\alpha_i)$, where α_i are the eigenvalues of $\mathbf{C}$. (This use of σ will be maintained hereafter.)

(b) $\sigma(\mathbf{C}) < 0$ if and only if there exists $\mathbf{x} > \mathbf{0}$ such that $\mathbf{Cx} \ll \mathbf{0}$.
(c) $\sigma(\mathbf{C}) < 0$ if and only if $-\mathbf{C}^{-1}$ is positivity-preserving.

Assertions (a) and (b) follow immediately from Theorem 8.2.3. Assertion (c) follows from Theorem 8.2.4.

Assertion (c) can be expressed equivalently in terms of the minors of $\mathbf{C}$ (see the parallel argument on p. 254). Explicitly, $\sigma(\mathbf{C}) < 0$ if and only if

$$c_{11} < 0, \quad \begin{vmatrix} c_{11} & c_{12} \\ c_{21} & c_{22} \end{vmatrix} > 0, \quad \begin{vmatrix} c_{11} & c_{12} & c_{13} \\ c_{21} & c_{22} & c_{23} \\ c_{31} & c_{32} & c_{33} \end{vmatrix} < 0, \ldots, (-1)^n \begin{vmatrix} c_{11} & \cdots & c_{1n} \\ \vdots & & \vdots \\ c_{n1} & \cdots & c_{nn} \end{vmatrix} > 0. \tag{8.2.4}$$

The reader is referred to Problems 10–15 for further propositions concerning positive matrices and M-matrices and their applications.

8.3 Applications of the theory of positive matrices to the study of linear models of equilibrium and exchange. (1) In our discussion of closed Leontief models in Section 8.1, we did not prove the existence of positive solutions to the system of equations $\mathbf{AX} = \mathbf{X}$, where $\mathbf{A}$ represents the matrix of production coefficients. (Recall that $a_{ij} = x_{ij}/X_j$, where x_{ij} is the total amount of the ith product used by the jth activity.) To ensure the existence of positive solutions naturally requires some additional conditions. One reasonable condition suggested by economic considerations is that $\sum_i x_{ij} = X_j$, i.e., that the total value output of the jth activity is fully used in acquiring the various inputs used. This may be thought of as being part of the definition of a closed economic system: for example, in an economy consisting solely of farmers, weavers, and carpenters, our assumption says that the farmer spends all his income on clothing, housing, and food.

The problem is to determine the amounts to be produced consistent with the production coefficients. In terms of the coefficients, our assumption leads to the relation

$$\sum_{i=1}^{n} a_{ij} = 1$$

for every j.

As a consequence, the matrix $\mathbf{A}$ must have spectral radius 1. For if $\mathbf{u} = \langle 1, 1, 1, \ldots, 1 \rangle$, with n components, then $\mathbf{uA} = \mathbf{u}$; and if $\mathbf{Az} = \lambda\mathbf{z}$, with $\sum|z_i| = 1$, then $(\mathbf{u}, \mathbf{A}|\mathbf{z}|) \geq |\lambda|\sum|z_i|$. (Again $|\mathbf{z}|$ represents the vector with components $|z_i|$.) But $(\mathbf{uA}, |\mathbf{z}|) = \sum|z_i|$, from which it follows that $|\lambda| \leq 1$, and since the eigenvalues of $\mathbf{A}$ and $\mathbf{A}'$ coincide, $\lambda_0(\mathbf{A}) = 1$. It follows by Theorem 8.2.3 that there exists a nonzero, nonnegative vector $\mathbf{X}$ which solves $\mathbf{AX} = \mathbf{X}$.

We have proved the following theorem.

▶THEOREM 8.3.1. In a closed Leontief system with

$$\sum_{i=1}^{n} x_{ij} = X_j,$$

there exists a nontrivial positive solution to the equations $\mathbf{AX} = \mathbf{X}$. If $\mathbf{A}$ has a strictly positive iterate, then the solution is unique except for a multiplying factor and possesses only strictly positive components.

The last part of the theorem is inferred from the statement of Theorem 8.2.1.

In the case of an open system, our problem is to determine conditions such that $(\mathbf{I} - \mathbf{A})\mathbf{X} = \mathbf{Y}$ can be solved for any prescribed final demand vector $\mathbf{Y}$ such that the solution $\mathbf{X}$ is a nonnegative vector.

▶THEOREM 8.3.2. If there exists some output vector $\mathbf{x}^0$ such that $(\mathbf{I} - \mathbf{A})\mathbf{x}^0$ is a strictly positive vector, then $(\mathbf{I} - \mathbf{A})^{-1}$ exists and transforms nonnegative vectors into nonnegative vectors. (The expression $(\mathbf{I} - \mathbf{A})^{-1}$ may be viewed as a linear transformation acting on either column or row vectors, whichever is appropriate. The positivity-preserving character of a matrix and that of its transpose are obviously equivalent.)

Remark 8.3.1. The hypothesis of this theorem means that there is some combination of outputs which guarantees a surplus for every commodity.

Proof. In view of Theorem 8.2.4 it is sufficient to show that all eigenvalues of $\mathbf{A}$ are of magnitude smaller than 1. Let λ be such that $\mathbf{zA} = \lambda\mathbf{z}$, with $\mathbf{z}$ a nonzero vector. The positivity of $\mathbf{A}$ implies that $|\mathbf{z}|\mathbf{A} \geq |\lambda||\mathbf{z}|$. The hypothesis requires that

$$0 < ((\mathbf{I} - \mathbf{A})\mathbf{x}_0, |\mathbf{z}|) = (\mathbf{x}_0, |\mathbf{z}|) - (\mathbf{x}_0, |\mathbf{z}|\mathbf{A}),$$

and since $\mathbf{x}_0$ is a nonnegative vector, it follows that $0 < (\mathbf{x}_0, |\mathbf{z}|)(1 - |\lambda|)$. Hence $|\lambda| < 1$ as asserted.

The conclusion of Theorem 8.3.2 holds for any hypothesis strong enough to drive all the eigenvalues of $\mathbf{A}$ into the interior of the unit circle of the complex plane. The hypothesis of Theorem 8.3.2 accomplishes this.

Another theorem of this kind, whose proof is a paraphrase of the previous argument, is the following.

▶THEOREM 8.3.3. If there exists some price vector $\mathbf{p}^0$ such that $\mathbf{p}^0(\mathbf{I} - \mathbf{A})$ is strictly positive, then $(\mathbf{I} - \mathbf{A})^{-1}$ transforms nonnegative vectors into nonnegative vectors.

(2) As an example of a situation in economics in which the concepts of indecomposability are meaningful, consider the model introduced on p. 245

involving export-import relations between countries. If the import matrix $\mathbf{A}$ is indecomposable, then a dollar spent in any one country will eventually induce spending in every other country. In particular, if $\mathbf{A}$ has the form of (8.2.3), then countries in G_i will spend only in those in G_{i+1}; and after ω instances of spending, the original expenditure by G_i in G_{i+1} will have an influence in G_i, giving rise to a sort of spending cycle.

(3) We close this section with a short description of a dynamic situation whose analysis depends on Theorem 8.2.2.

We may think of the model of exchange introduced on p. 245 as happening over time. In the original analysis the equations $\mathbf{Ax} = \mathbf{x}$ describe a closed model in which the components x_i of $\mathbf{x}$ represent the total output (or activity level) of the ith activity with no reference to a time period over which this equilibrium is maintained. These equations may be expanded and given a dynamic interpretation as follows. Let $\mathbf{x}^1$ represent the initial vector of activity levels. Then $\mathbf{Ax}^1 = \mathbf{x}^2$ can be regarded as the outputs attributable to the activities of one time period. Similarly, $\mathbf{Ax}^2 = \mathbf{x}^3$ expresses the vector of the activity level required in the third time period, and $\mathbf{Ax}^n = \mathbf{x}^{n+1}$ has the analogous interpretation. Since $\mathbf{x}^{n+1} = \mathbf{A}^n\mathbf{x}^1$, we find that under suitable assumptions, namely where Theorem 8.2.2 is applicable, $\mathbf{x}^{n+1}$ converges to a limiting output vector $\mathbf{x}^0$ such that $\mathbf{x}^0 = \mathbf{Ax}^0$. It is easy to impose criteria which guarantee that $\mathbf{x}^0$ is a nonzero positive eigenvector of $\mathbf{A}$ and possesses suitable meaning in the context of the closed model. Other convergence and stability results of this sort will be developed in Sections 9.2 through 9.6.

8.4 The theory of production. In this section we generalize the Leontief assumptions to describe a nonlinear model involving production.

The theory of production is concerned first with the allocation of productive factors among various technological activities to produce goods for consumption, and then with the distribution of the value of the total product among the productive factors.

Consider a situation in which n finished products are produced by using a given amount of each of r factors of production, where labor, land, raw material, etc. are considered factors of production. (We shall later discuss a production problem in which we make no differentiation between finished products and factors of production.) The products will be denoted by $i = 1, 2, \ldots, n$, and the factors of production by $k = 1, 2, \ldots, r$.

The technological possibility of production will be explained in terms of *activity analysis*. The economy has a finite set of basic activities $j = 1, \ldots, m$, each of which has an "activity level" associated with it in any given set of circumstances. The activity level for the economy as a whole may be denoted by an m-vector $\mathbf{x} = \langle x_1, x_2, \ldots, x_m \rangle$, where each component x_j is nonnegative and stands for the level at which the basic activity j is operated. The technology of the system will now be characterized by specifying a pair of vector functions $\mathbf{f}(\mathbf{x})$ and $\mathbf{g}(\mathbf{x})$: $\mathbf{f}(\mathbf{x}) = \langle f_1(\mathbf{x}), \ldots, f_n(\mathbf{x})\rangle$ is an n-vector, the ith component of which stands for the amount of product i produced by operating the system at activity level $\mathbf{x}$, while $\mathbf{g}(\mathbf{x}) = \langle g_1(\mathbf{x}), \ldots, g_r(\mathbf{x})\rangle$ is an r-vector, the kth component of which stands for the amount of productive factor k required in order to operate at activity level $\mathbf{x}$.

Let us now assume that certain amounts of the productive factors, which are represented by an r-vector $\mathbf{v} = \langle v_1, v_2, \ldots, v_r\rangle$, are available to the system, and that the system is concerned only with getting the maximum value of the output as evaluated at given market prices

$$(\mathbf{p} = \langle p_1, p_2, \ldots, p_n\rangle,\ p_i \geq 0).$$

The problem of production can now be stated as follows: Find an activity level $\mathbf{x} = \langle x_1, x_2, \ldots, x_n\rangle$ that maximizes the value of output

$$(\mathbf{p}, \mathbf{f}(\mathbf{x})) = \sum_i p_i f_i(\mathbf{x}) \tag{8.4.1}$$

subject to the restrictions

$$\mathbf{x} \geq \mathbf{0}, \tag{8.4.2}$$

$$\mathbf{g}(\mathbf{x}) \leq \mathbf{v}. \tag{8.4.3}$$

The allocation of productive factors among various activities thus formulated is a special version of one of the linear (or nonlinear) programming problems (see Section 7.1).

The basic assumptions for the technology, which play an important role in the economic analysis of production and will be assumed in what follows, are the laws of constant returns to scale and of the diminishing marginal rates of transformation. The former is expressed by the positive homogeneity of $\mathbf{f}(\mathbf{x})$ and $\mathbf{g}(\mathbf{x})$, the latter by the concavity of $\mathbf{f}(\mathbf{x})$ and the convexity of $\mathbf{g}(\mathbf{x})$.

Invoking the theory of nonlinear programming (Section 7.1), we may reduce the problem of production to the problem of finding a saddle point $\{\bar{\mathbf{x}}, \bar{\boldsymbol{\omega}}\}$ of the Lagrangian form

$$\phi(\mathbf{x}, \boldsymbol{\omega}) = (\mathbf{p}, \mathbf{f}(\mathbf{x})) + (\boldsymbol{\omega}, \mathbf{v} - \mathbf{g}(\mathbf{x})) \qquad (\mathbf{x} \geq \mathbf{0},\ \boldsymbol{\omega} \geq \mathbf{0}).$$

The vector $\bar{\boldsymbol{\omega}}$ of the solution of the saddle-value problem may be interpreted as an imputed price vector; it can be construed as a relative measure of the distribution of the value of the total product among the productive factors (see p. 117).

A special case of the production problem occurs when there is no joint product, i.e., when the amount of each product produced is uniquely determined by the amounts of the productive factors used in the activity that produces it. By taking an appropriate scaling of the basic activities, we may state this special problem as follows: Find the distribution of the productive factors $\langle v_{ik} \rangle$ that maximizes

$$\sum_{i=1}^{n} p_i f_i(v_{i1}, \ldots, v_{ir})$$

subject to the restraints

$$v_{ik} \geq 0 \qquad (i = 1, \ldots, n; k = 1, \ldots, r),$$

$$\sum_{i=1}^{n} v_{ik} \leq v_k \qquad (k = 1, \ldots, r),$$

where v_{ik} denotes the amount of the productive factor k used in activity i.

A typical example is the Cobb-Douglas model:

$$f_i(v_{i1}, \ldots, v_{ir}) = \alpha_i v_{i1}^{\beta_{i1}}, \ldots, v_{ir}^{\beta_{ir}},$$

where α_i, β_{ik} are constants and $\beta_{ik} \geq 0$, $\sum_k \beta_{ik} = 1$.

Production possibility sets. The distinction between finished products (consumption goods) and factors of production is rather arbitrary. A finished product may be used in the production of other finished products, as in the classical Leontief model, and also a productive factor may be produced.

It is therefore of some purpose to formulate the model in such a way as to treat all goods (including productive factors as well as consumption goods) as symmetrical. Let us represent the goods by $i = 1, \ldots, n + r$, and denote by an $(n + r)$-vector the outputs of all goods (a negative amount here represents an input).

The technology of the system is described by specifying the production possibility set Z as a subset of the $(n + r)$-vector space such that an output vector $\mathbf{z}$ is technologically possible if and only if $\mathbf{z} \in Z$. The assumption corresponding to the laws of constant returns to scale and of the diminishing marginal rates of transformation will be that Z is a convex cone in the $(n + r)$-space.

The original setup of the production model can be represented by a production possibility set. Explicitly, if the technology is formulated in terms of $\mathbf{f}(\mathbf{x})$ and $\mathbf{g}(\mathbf{x})$, the corresponding Z is defined by

$$Z = \left\{ \begin{pmatrix} \mathbf{f}(\mathbf{x}) \\ -\mathbf{g}(\mathbf{x}) \end{pmatrix} \middle| \mathbf{x} \geq \mathbf{0} \right\}.$$

On the other hand, if the technology is given in terms of Z, the corresponding $\mathbf{f}(\mathbf{x})$ and $\mathbf{g}(\mathbf{x})$ are defined by taking a parametric representation of Z, if this is possible, namely $\mathbf{z} = \mathbf{z}(\mathbf{x})(\mathbf{x} \geq \mathbf{0})$, and defining

$$\mathbf{f}(\mathbf{x}) = \langle z_1(\mathbf{x}), \ldots, z_n(\mathbf{x})\rangle, \qquad \mathbf{g}(\mathbf{x}) = \langle -z_{n+1}(\mathbf{x}), \ldots, -z_{n+r}(\mathbf{x})\rangle.$$

Considered in terms of the production possibility sets, the production problem is a matter of determining and characterizing all efficient technologically possible production vectors. An instance of this will be examined in the following section.

A specific efficient production vector may be arrived at by taking account of market prices, as is done in (8.4.1).

The classical Leontief production model. The open Leontief production model corresponds in the above formulation to the case in which $r = 1$ (labor) and

$$Z = \left\{ \begin{pmatrix} (\mathbf{I} - \mathbf{A})\mathbf{x} \\ (-\mathbf{l}, \mathbf{x}) \end{pmatrix} \middle| \begin{array}{l} \mathbf{x} = \langle x_1, x_2, \ldots, x_n\rangle \\ \mathbf{x} \geq \mathbf{0} \end{array} \right\},$$

where $\mathbf{A} = \|a_{ik}\|$ stands for the input-output coefficient matrix and $\mathbf{l} = \langle l_1, l_2, \ldots, l_n\rangle$ represents the labor coefficient vector. Here x_i may be interpreted directly as the amount of the ith product produced, and a_{ik} measures the amount of the ith product used in producing one unit of the kth product. The negative of the last component of Z evaluates the quantity of labor consumed for a given output vector.

8.5 Efficient points of a Leontief-type model. One of the main limitations of the Leontief model as described in the preceding section is that it associates only one activity with the production of each product. We shall see in this section that even if many processes are available for each commodity, nevertheless all "efficient" modes of production involve a single production process for each activity, provided that no joint production is allowed and only one primary factor is available. This is the famous Samuelson substitution theorem. We illustrate this concept in the following model.

Suppose that we have n industries engaged in the production of n commodities. Let a_{ij} be the amount of the jth product produced (or consumed) by the ith industry when it is operating at the unit level of activity.

We assume that $a_{ij} \leq 0$ if $i \neq j$, but that $a_{ii} > 0$. A negative a_{ij} may be interpreted as material used up in the production process. We see that whereas each industry may use all commodities, it produces only one (this is what is meant by no joint production).

If we represent the total amount of available labor by unity, then the level of activity of the ith industry may be represented by $x_i \geq 0$, where

x_i is the proportion of the total available labor employed in this industry. The net product of commodity j is $\sum_i x_i a_{ij}$. Suppose that the ith industry has available alternative processes $a_{ij}^{\lambda_i}$ (Λ_i will index the set of processes available in the ith industry; $\lambda_i \in \Lambda_i$); then the net product of commodity j, corresponding to a choice of production processes $\lambda_1, \ldots, \lambda_n$ for industries $1, \ldots, n$, respectively, is $\sum_i x_i a_{ij}^{\lambda_i}$.

An alternative interpretation of this model is that the manufacturers vary the number of factories engaged in the production of a given commodity.

Let S_k $(k = 1, \ldots, n)$ denote the set of process vectors available to the kth industry. For fixed k, the set S_k consists of all possible process vectors $\mathbf{a}_k^{\lambda_k} = \langle a_{k1}^{\lambda_k}, \ldots, a_{kn}^{\lambda_k} \rangle$. We impose the requirement that S_k is convex and compact for each k; this is a natural assumption in terms of the alternative interpretation and can be justified as reasonable for many situations. By the set S we mean the subset of E^n which is the convex hull of the sets $S_1, \ldots, S_n$ (see Section B.2).

The set $\{\mathbf{A}^\lambda\} = \mathcal{A}$ of matrices representing possible processes consists of the matrices $\|a_{ij}^{\lambda_i}\|$ of order n; the ith row is an element of the set S_i.

A vector $\mathbf{c}$ in E^n is called feasible if $\mathbf{c} = \mathbf{x}\mathbf{A}$ for some $\mathbf{A}$ in $\mathcal{A}$ and $\mathbf{x}$ such that $x_i \geq 0$, $\sum x_i = 1$. (Vectors $\mathbf{x}$ of this form will be called admissible.)

▶LEMMA 8.5.1. The set of all possible feasible vectors coincides with the set S.

The verification of this assertion is direct and will be omitted.

A feasible point $\mathbf{c} = \mathbf{x}^0\mathbf{A}^{\lambda^0}$ is efficient if there exists no admissible $\mathbf{x}$ and $\mathbf{A}^\lambda$ in $\mathcal{A}$ such that $\mathbf{x}^0\mathbf{A}^{\lambda^0} \leq \mathbf{x}\mathbf{A}^\lambda$ with strict inequality in at least one component.

The efficient point concept introduced here is identical to the concept of an efficient point in nonlinear programming theory (see Section 7.4).

Our present objective is to characterize all efficient points for this Leontief model. For this purpose the following lemma will be needed.

▶LEMMA 8.5.2. Let $\mathbf{p}_i$ denote vectors belonging to S_i $(i = 1, \ldots, n)$. Form the plane generated by the n points $\mathbf{p}_i$, that is, the sets of all points $\sum_i y_i \mathbf{p}_i$, $\sum_i y_i = 1$. If this plane intersects the interior of the first quadrant for $\mathbf{y} \geq \mathbf{0}$, then the intersection of the plane and the first quadrant is contained in the set $\Gamma = \{\bar{\mathbf{p}} = \sum x_i \mathbf{p}_i | x_i \geq 0, \sum x_i = 1\}$.

Proof. (1) Let $\mathbf{A}^\lambda$ denote the process matrix constructed from the vectors $\mathbf{p}_i$. Suppose that $x_i \geq 0$, $\sum x_i = 1$, and $x_{i_0} = 0$ for some i_0. Then $\mathbf{x}\mathbf{A}^\lambda$ will have at least one nonpositive component, namely i_0, because $a_{ii_0}^{\lambda_i} \leq 0$ for $i \neq i_0$. Thus if $x_i \geq 0$, $\sum x_i = 1$, and $x_{i_0} = 0$, it follows that $\mathbf{x}\mathbf{A}^\lambda$ does not belong to the interior of the first quadrant.

(2) The set of all $\mathbf{y}$ such that $\sum y_i = 1$ and $\sum y_i \mathbf{p}_i \gg 0$ is clearly convex. By hypothesis and by (1) there exists an $\mathbf{x}^0$ such that $x_i^0 > 0$, $\sum x_i^0 = 1$, and $\sum x_i^0 \mathbf{p}_i \gg 0$. Suppose that there exists a $\mathbf{y}$ such that $\sum y_i = 1$ and $\sum y_i \mathbf{p}_i \gg 0$ but $y_i < 0$ for some i. Then we can find a value $\lambda (0 < \lambda < 1)$ such that the nonnegative vector $\lambda \mathbf{x}^0 + (1 - \lambda)\mathbf{y}$ has at least one component equal to zero. But it follows from (1) that $\lambda \mathbf{x}^0 + (1 - \lambda)\mathbf{y} \gg \mathbf{0}$, which is impossible. Hence $y_i \geq 0$ for all i.

The validity of Lemma 8.5.2 depends in a vital way on the hypothesis. The plane must intersect the interior of the positive orthant; otherwise the conclusion of the lemma is false. The reader can readily supply counterexamples in E^3; e.g., take

$$\mathbf{p}_1 = \langle \tfrac{1}{2}, -\tfrac{1}{2}, 0 \rangle, \qquad \mathbf{p}_2 = \langle -\tfrac{1}{2}, \tfrac{1}{2}, 0 \rangle, \qquad \mathbf{p}_3 = \langle 0, 0, 1 \rangle.$$

With the aid of Lemma 8.5.2 we are now able to prove the principal theorem of this section.

▶ THEOREM 8.5.1. If there exists a feasible point in the interior of the positive orthant, then the efficient points in the positive orthant belong to a hyperplane and every point of the intersection of the hyperplane with the positive orthant is efficient.

Proof. Let $\mathbf{p}_0$ be the feasible point, with strictly positive coordinates, and let l be the ray from the origin through the point $\mathbf{p}_0$. For every set of n points $\mathbf{p}_1, \ldots, \mathbf{p}_n$ ($\mathbf{p}_i \in S_i$), which determines a feasible point, construct the plane generated by these points. Consider the plane that intersects the ray l at the point of the positive orthant that is farthest from the origin. Let this plane be denoted by P. Then P is the desired hyperplane; its intersection P' with the positive orthant gives the set of efficient points in the positive orthant.

Suppose that this is not so; i.e., suppose that there exists a feasible point $\mathbf{g}$ in the positive orthant which dominates ($\geq$, with $>$ in at least one component) some point $\mathbf{p}$ of P'. But, P intersects the coordinate axes in the points $\langle a_1, 0, \ldots, 0 \rangle$, $\langle 0, a_2, 0, \ldots, 0 \rangle$, $\ldots$, $\langle 0, \ldots, 0, a_n \rangle$; and by Lemma 8.5.2 all these points are feasible points. The point $\mathbf{g}$ also belongs to a plane G generated by a set of points $\vec{\mathbf{g}}_1, \ldots, \vec{\mathbf{g}}_n$ ($\vec{\mathbf{g}}_i \in S_i$), and G intersects the coordinate axes in the points $\langle b_1, 0, \ldots, 0 \rangle$, $\langle 0, b_2, 0, \ldots, 0 \rangle$, $\ldots$, $\langle 0, \ldots, 0, b_n \rangle$, which are feasible points by the lemma. Since $\mathbf{g}$ dominates $\mathbf{p}$, at least one of the b_i's must be greater than one of the a_i's (say b_1). The points $\langle b_1, 0, \ldots, 0 \rangle$, $\langle 0, a_2, 0, \ldots, 0 \rangle$, $\ldots$, $\langle 0, \ldots, 0, a_n \rangle$ are all feasible; hence their convex hull is a subset of S in the positive orthant. The points of this subset dominate the points of P', and in particular this subset intersects the ray at a point farther from the origin than P'. But this contradicts our construction of P'. Therefore P' is the set of efficient points in the positive orthant.

There are three interesting consequences of this theorem. If there is a combination of processes and labor which yields a strictly positive surplus for each substance, then:

(1) There is one process for each industry which gives the nonnegative efficient points. Each industry need not substitute different processes to secure nonnegative efficient points. Each can always use the same process.

(2) The set of nonnegative efficient points is partially independent of labor, in that there are many distributions of labor which yield nonnegative efficient points.

(3) Prices are independent of demands, since the normal vector to the unique efficient hyperplane determines the same relative price system for all efficient modes of production.

†*Computational procedure for the Leontief model, and its connection with completely mixed games.* Theorem 8.5.1 shows that all efficient points of the positive orthant lie in a hyperplane, and the proof of the theorem also establishes that this hyperplane is constructed from a process matrix $\mathbf{A} \in \mathfrak{A}$. Since this plane is assumed to intersect the interior of the positive orthant, for this A there exists a vector $\mathbf{x}^0$ such that $\mathbf{x}^0\mathbf{A} \gg \mathbf{0}$.

Furthermore, the arguments of the theorem indicate that it is sufficient to find the matrix A for which xA is a maximum along any particular ray entering the first quadrant in E^n. We choose to maximize along the 45-degree line. Obviously, the problem is equivalent to determining the matrix A for which

$$\max_{A \in \mathfrak{A}} v(\mathbf{A})$$

is achieved, where $v(\mathbf{A}) = \max_x \min_j \sum_i x_i a_{ij}$.

The quantity $v(\mathbf{A})$ is recognized as the value of the game defined by the matrix $\mathbf{A}$. These matrix games are of the form referred to earlier as Minkowski-Leontief games (see Section 2.5). An important property they possess is stated in the following lemma.

▶ LEMMA 8.5.3. If $\mathbf{A} \in \mathfrak{A}$ and there exists a vector $\mathbf{x}^0$ such that $\mathbf{x}^0\mathbf{A} \gg \mathbf{0}$, then $\mathbf{A}$ is completely mixed.

Proof. The relation $\mathbf{x}^0\mathbf{A} \gg \mathbf{0}$ implies that $v(\mathbf{A}) > 0$. If any optimal x strategy had a zero component, the yield for the corresponding column would be negative, hence less than v. Thus every optimal strategy for Player I has all components strictly positive.

It follows from Theorem 2.4.1 that the kernel of any pair of optimal strategies must be the matrix $\mathbf{A}$, and the optimal $\mathbf{y}$ strategy must be unique.

---

† The remainder of this section may be passed over on first reading without loss of continuity.

The dimension relations also show that $\mathbf{y}$ must have all components positive.

Theorem 2.4.1 shows also that

$$v(\mathbf{A}) = \frac{1}{(\mathbf{u}, \mathbf{A}^{-1}\mathbf{u})} = \frac{1}{\sum_{i,j} A_{ij}},$$

where $\|A_{ij}\| = \mathbf{A}^{-1}$.

Let $\tilde{\mathcal{A}} = \{\mathbf{A} \in \mathcal{A} | v(\mathbf{A}) > 0\}$. To characterize the efficient points it is sufficient to find the matrix $\mathbf{A} \in \tilde{\mathcal{A}}$ for which $v(\mathbf{A})$ is a maximum, or equivalently to find $\mathbf{A}$ such that

$$\max_{A \in \mathcal{A}} \frac{1}{\sum_{i,j} A_{ij}}.$$

***8.6 The theory of consumer choice.** The general theory of equilibrium that describes the relationship of supply and demand to prices and other factors of the economy can be divided into two main parts: (1) production theory, which also embraces the theory of distribution of value (as embodied, for example, in the shadow price concept); and (2) consumption theory, which is concerned with individual buying preferences. In the previous sections we have reviewed the mathematical theory of production. In this section we concentrate on describing some of the problems arising in the mathematical theory of consumer choice. In Sections 8.7 and 8.8 we examine a global equilibrium model which recognizes simultaneously the production and consumption factors.

The theory of consumer choice is concerned essentially with the following question: Given a consumer with a limited budget and a definite set of preferences with respect to different commodity bundles, what quantities does he consume when confronted with given market prices for the various commodities? Another important question is this: By what rules does such a consumer's preference pattern change in response to variations in market prices or in his income?

Suppose that there are n commodities consumed by the consumer, and let the quantities of these commodities be represented by an n-vector $\mathbf{x} = \langle x_1, x_2, \ldots, x_n \rangle$, where x_i stands for the amount of the ith commodity consumed. We may assume, by taking an appropriate scale, that $x_i \geq 0$ $(i = 1, \ldots, n)$.

The consumer is assumed to have a preference relation P between vectors. That is to say, given any two vectors $\mathbf{x} = \langle x_1, x_2, \ldots, x_n \rangle$ and $\mathbf{y} = \langle y_1, y_2, \ldots, y_n \rangle$, either $\mathbf{x}$ is preferred to $\mathbf{y}$ ($\mathbf{x}\,\mathrm{P}\,\mathbf{y}$), or $\mathbf{y}$ is preferred to $\mathbf{x}$ ($\mathbf{y}\,\mathrm{P}\,\mathbf{x}$), or neither is preferred to the other ($\overline{\mathbf{x}\,\mathrm{P}\,\mathbf{y}}$ and $\overline{\mathbf{y}\,\mathrm{P}\,\mathbf{x}}$, where

---

*Starred sections are advanced discussions that may be omitted at first reading without loss of continuity.

$\overline{\mathbf{x}\,\mathrm{P}\,\mathbf{y}}$ stands for the negation of $\mathbf{x}\,\mathrm{P}\,\mathbf{y}$). In this last case we shall also say that $\mathbf{x}$ is indifferent to $\mathbf{y}$ and write $\mathbf{x}\,\mathrm{I}\,\mathbf{y}$. We do not claim transitivity for the indifference relation at present, although we shall see later that once postulates P–1 to P–5 below are accepted, the I relations indeed become transitive and reflexive. We postulate that P has the following properties:

P–1. Irreflexivity: $\overline{\mathbf{x}\,\mathrm{P}\,\mathbf{x}}$ for any $\mathbf{x} \geq \mathbf{0}$.

P–2. Transitivity: $\mathbf{x}\,\mathrm{P}\,\mathbf{y}$ and $\mathbf{y}\,\mathrm{P}\,\mathbf{z}$ imply $\mathbf{x}\,\mathrm{P}\,\mathbf{z}$.

P–3. Monotonicity; $\mathbf{x} > \mathbf{y}$ implies $\mathbf{x}\,\mathrm{P}\,\mathbf{y}$.

P–4. Convexity: $\mathbf{x}\,\mathrm{P}\,\mathbf{z}$ or $\mathbf{x}\,\mathrm{I}\,\mathbf{z}$ and $\mathbf{y}\,\mathrm{P}\,\mathbf{z}$ or $\mathbf{y}\,\mathrm{I}\,\mathbf{z}$ with $\mathbf{y} \neq \mathbf{x}$ imply $[t\mathbf{x} + (1 - t)\mathbf{y}]\mathrm{P}\,\mathbf{z}$ for any $0 < t < 1$.

P–5. Continuity: $P_x = \{\mathbf{y}|\mathbf{y}\,\mathrm{P}\,\mathbf{x}\}$ and $Q_x = \{\mathbf{y}|\mathbf{x}\,\mathrm{P}\,\mathbf{y}\}$ are open sets for any $\mathbf{x}$.

If there is a numerical function $U(\mathbf{x})$ ($x \geq 0$) such that

$$U(\mathbf{x}) > U(\mathbf{y}) \qquad \text{if and only if } \mathbf{x}\,\mathrm{P}\,\mathbf{y}, \tag{8.6.1}$$

(i.e., $U(\mathbf{x}) > U(\mathbf{y})$ and $\mathbf{x}\,\mathrm{P}\,\mathbf{y}$ are equivalent statements), then U is called a utility index of P. If U is an index and $F(U)$ is a monotone-increasing function of U, then $V = F(U)$ is also an index.

In order for the monotonicity and convexity assumptions to be satisfied by a utility index function U, it is necessary and sufficient that the set $\{\mathbf{x}|U(\mathbf{x}) \geq c\}$ defined for each constant c be a strictly convex set [see Section B.4, property (i)] which includes with each $\mathbf{x}$ all vectors $\mathbf{y} > \mathbf{x}$. The postulate of continuity is implied whenever U is continuous. Conversely,

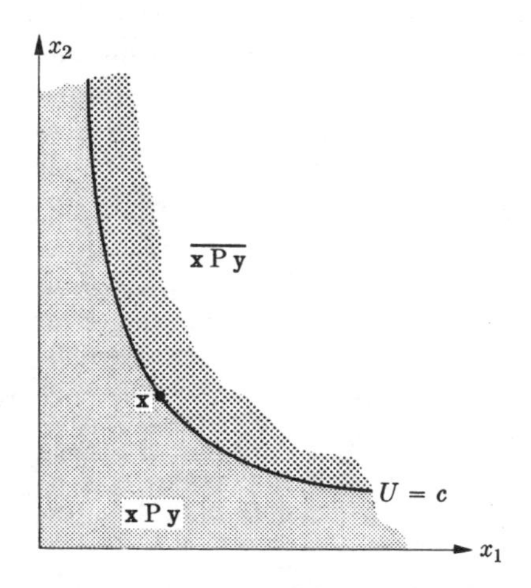

FIGURE 8–1

any real-valued function U with these properties defines a preference ordering by (8.6.1). Clearly $\mathbf{x} \text{ I } \mathbf{y}$ is synonymous with $U(\mathbf{x}) = U(\mathbf{y})$.

The preference relation induced by an index function is pictured in Fig. 8.1. The dotted region consists of all $\mathbf{y}$ to which $\mathbf{x}$ is preferred. This is an open set. The crosshatched region, which includes the boundary surface $\{\mathbf{y}|U(\mathbf{y}) \geq c\}$, represents all $\mathbf{y}$ to which $\mathbf{x}$ is not preferred. The same geometrical picture applies to the general preference relation P that satisfies postulates P–1 to P–5.

Demand function for the preference relation P. We shall assume that the consumer behaves in such a manner as to maximize his preference among all commodity vectors $\mathbf{x}$ satisfying his budget constraint. More precisely, given market prices $\mathbf{p} = \langle p_1, p_2, \ldots, p_n \rangle$ and his income M, he chooses a vector $\mathbf{x}$ for which

$$\mathbf{x} \geq \mathbf{0}, \qquad (\mathbf{p}, \mathbf{x}) \leq M, \tag{8.6.2}$$

where $(\mathbf{p}, \mathbf{x})$ evaluates the total cost to the consumer of the consumption vector when the market prices are given by $\mathbf{p}$, and

$$\mathbf{x} \text{ P } \mathbf{y} \qquad \text{for any} \quad \mathbf{y} \geq \mathbf{0}, \quad (\mathbf{p}, \mathbf{y}) \leq M, \tag{8.6.3}$$

where $\mathbf{p} \gg \mathbf{0}$ and $M \geq 0$.

By postulates P–1 through P–5, it is easily seen that the vector $\mathbf{x}$ satisfying (8.6.2) and (8.6.3) always exists and is uniquely determined by $\mathbf{p}$ and M.

The proof of uniqueness is an immediate consequence of transitivity and irreflexibility. Since the proof of existence is slightly more complicated, we present a formal proof.

For each $\mathbf{x}$ satisfying $(\mathbf{p}, \mathbf{x}) \leq M$, we consider the set

$$B_x = \{\mathbf{y}|(\mathbf{p}, \mathbf{y}) \leq M, \overline{\mathbf{x} \text{ P } \mathbf{y}}\}. \tag{8.6.4}$$

We prove first that $\Gamma = B_{x_1} \cap B_{x_2} \cap \cdots \cap B_{x_n}$ is nonvoid for any finite $\mathbf{x}_i$. To this end, let the set $\mathbf{x}_1, \mathbf{x}_2, \ldots, \mathbf{x}_n$ be separated into two groups X^1 and X^2, where some member of X^1 is preferred to each member of X^2, and where $\overline{\mathbf{x}_i \text{ P } \mathbf{x}_j}$ for each $\mathbf{x}_i$ and $\mathbf{x}_j$ in X^1. Such a decomposition is possible with X^1 nonvoid by virtue of the irreflexibility and transitivity postulated. All one has to do is to start systematically with $\mathbf{x}_1$ and $\mathbf{x}_2$ and compare them. If one of them is preferred to the other, the dominated member is assigned to X^2. We then compare the remaining member with $\mathbf{x}_3$, etc.

Now suppose that the $\mathbf{x}_i$ are labeled such that $\mathbf{x}_1, \ldots, \mathbf{x}_k$ belong to X^1 and $\mathbf{x}_{k+1}, \ldots, \mathbf{x}_n$ belong to X^2. We prove that $(\mathbf{x}_1 + \cdots + \mathbf{x}_k)/k$ is in Γ. By P–4,

$$\frac{\mathbf{x}_1 + \cdots + \mathbf{x}_k}{k} \text{ P } \mathbf{x}_i \qquad (i - 1, \ldots, k),$$

and by transitivity and the definition of X^2,

$$\frac{\mathbf{x}_1 + \cdots + \mathbf{x}_k}{k} \mathrm{P} \mathbf{x}_j \qquad (j = 1, \ldots, n).$$

Furthermore,

$$\overline{\mathbf{x}_j \mathrm{P} \frac{\mathbf{x}_1 + \cdots + \mathbf{x}_k}{k}},$$

for otherwise P–1 would be contradicted for the vector $(\mathbf{x}_1 + \cdots + \mathbf{x}_k)/k$. Since each of the sets B_x is a closed subset (P–5) of a compact set $T = \{\mathbf{y}|(\mathbf{p}, \mathbf{y}) \leq M, \mathbf{p} \gg \mathbf{0}\}$ and possesses the finite intersection property, it follows that

$$\Lambda = \bigcap_{x \in T} B_x$$

is nonvoid. Let $\mathbf{x}^0$ be in Λ; then $\mathbf{x}^0$ satisfies the budget constraint. Moreover, $\mathbf{x}^0 \mathrm{P} \mathbf{y}$ for all $\mathbf{y} \neq \mathbf{x}^0$. If we suppose the contrary, then $\overline{\mathbf{x}^0 \mathrm{P} \mathbf{y}^0}$ for some $\mathbf{y}^0$ in T, and $\overline{\mathbf{y}^0 \mathrm{P} \mathbf{x}^0}$ by the construction of Λ. Hence $\mathbf{y}^0 \mathrm{I} \mathbf{x}^0$. Also $\mathbf{x}^0 \mathrm{I} \mathbf{x}^0$, and hence $[t\mathbf{x}^0 + (1 - t)\mathbf{y}^0] \mathrm{P} \mathbf{x}^0$ for $0 < t < 1$ by P–4, contradicting the choice of $\mathbf{x}^0$.

We may write

$$\mathbf{x}^0 = \mathbf{f}(\mathbf{p}, M),$$

and f will be called the demand function derived from the preference relation P.

The demand function may also be characterized as the vector $\mathbf{x}^0$ satisfying

$$\mathbf{x}^0 \geq \mathbf{0}, \qquad (\mathbf{p}, \mathbf{x}^0) = M, \qquad \mathbf{x}^0 \mathrm{P} \mathbf{x} \tag{8.6.5}$$

for all $\mathbf{x}$ such that $(\mathbf{p}, \mathbf{x}) < M$ when $\mathbf{p} \gg \mathbf{0}$ and $M > 0$.

Proof. If $\mathbf{x}^0$ is the demand function for the parameters $\mathbf{p}$ and M and $(\mathbf{p}, \mathbf{x}^0) < M$, then we may increase all the components of $\mathbf{x}^0$ while maintaining $(\mathbf{p}, \mathbf{x}) \leq M$. But by P–3 the property (8.6.3) is contradicted.

Conversely, suppose that $\mathbf{x}^0$ satisfies (8.6.5). In this case we show that $\mathbf{x}^0$ is the demand function of the preference relation for the parameters $\mathbf{p}$ and M. Suppose that the demand for $\mathbf{p}$ and M is $\mathbf{y}^0 \neq \mathbf{x}^0$. Let

$$Q_{x^0} = \{\mathbf{y}|\mathbf{y} \mathrm{P} \mathbf{x}^0\}.$$

This set is open (by P–5) and includes $\mathbf{y}^0$. Hence there exists a $\mathbf{y}$ with the property $\mathbf{y} \mathrm{P} \mathbf{x}^0$ and $(\mathbf{p}, \mathbf{y}) < M$, which is impossible by (8.6.5). Hence $\mathbf{y}^0$ agrees with $\mathbf{x}^0$, as was to be proved.

It will be helpful for future discussions to diagram the position of $\mathbf{x} = \mathbf{f}(\mathbf{p}, M)$; this is done in Fig. 8.2.

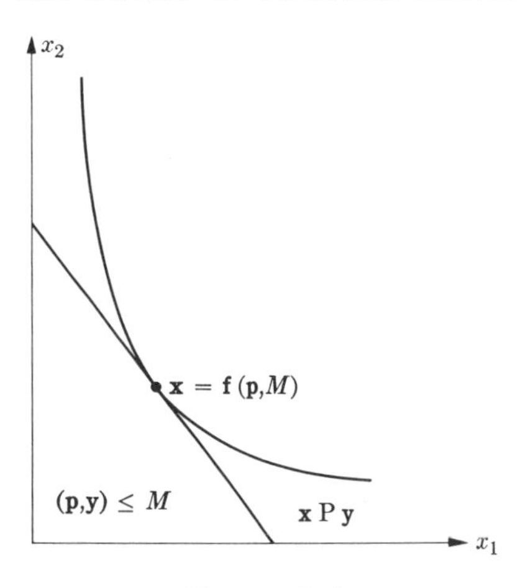

FIGURE 8–2

If the preference relation is described by a utility index U, then $\mathbf{x}^0$ is the vector for which $U(\mathbf{x})$ is maximized with respect to all $\mathbf{x}$ satisfying the restrictions $(\mathbf{p}, \mathbf{x}) \leq M$.

In view of (8.6.5), it is easily shown that $\mathbf{f}$ is continuous at $\mathbf{p} \gg \mathbf{0}$ and $M > 0$. In general, by a demand function $\mathbf{x} = \mathbf{f}(\mathbf{p}, M)$ will be meant a function $\mathbf{f}$, defined on $\mathbf{p} \gg \mathbf{0}$ and $M \geq 0$, such that

D–1. $\mathbf{f}(\mathbf{p}, M) \geq 0, (\mathbf{p}, \mathbf{f}(\mathbf{p}, M)) = M$.

D–2. $\mathbf{f}$ is continuous at $\mathbf{p} \gg \mathbf{0}$, $M > 0$.

In what follows we shall make two further assumptions:

D–3. For any $\mathbf{x} \geq \mathbf{0}$, there exist $\mathbf{p} \gg 0$ and $M \geq 0$ such that $\mathbf{x} = \mathbf{f}(\mathbf{p}, M)$.

D–4. $\mathbf{f}(\mathbf{p}, M)$ satisfies a Lipschitz condition with respect to $M > 0$; i.e., for any $M > 0$, there exist $K > 0$ and $\delta > 0$ such that

$$|\mathbf{f}(\mathbf{p}, M_1) - \mathbf{f}(\mathbf{p}, M)| \leq K|M_1 - M|$$

whenever $|M_1 - M| < \delta$.

If P is derived from a utility index U, where $\{\mathbf{x}|U(\mathbf{x}) \geq c\}$ is a strictly convex set with property P–3 and U is sufficiently differentiable, then it is easily verified that D–3 and D–4 are satisfied.

The axiom of revealed preference. With the aid of a given demand function $\mathbf{x} = \mathbf{f}(\mathbf{p}, M)$ fulfilling conditions D–1 through D–4, we now define the concept of "revealed preference." It is said that vector $\mathbf{x}^0 = \mathbf{f}(\mathbf{p}^0, M^0)$ is revealed-preferred to vector $\mathbf{x}^1 = \mathbf{f}(\mathbf{p}^1, M^1)$ if

$$(\mathbf{p}^0, \mathbf{x}^0) \geq (\mathbf{p}^0, \mathbf{x}^1) \qquad (\mathbf{x}^0 \neq \mathbf{x}^1).$$

This relationship will be designated by

$$\mathbf{x}^0 \mathrm{R}\, \mathbf{x}^1;$$

that is, $\mathbf{x}^0 \mathrm{R}\, \mathbf{x}^1$ if $\mathbf{x}^1$ belongs to the budget constraint set over which $\mathbf{x}^0$ is demanded. Naturally, to be consistent with the meaning of demand functions, it is impossible for $\mathbf{x}^0$ to be an admissible consumption vector for the budget constraint satisfied by $\mathbf{x}^1$.

We also define relations R^s and R^* as follows: $\mathbf{x}\, R^s\, \mathbf{y}$ if there exist vectors $\mathbf{x}^1, \mathbf{x}^2, \ldots, \mathbf{x}^{s-1}$ such that $\mathbf{x}\, \mathrm{R}\, \mathbf{x}^1, \mathbf{x}^1\, \mathrm{R}\, \mathbf{x}^2, \ldots, \mathbf{x}^{s-1}\, \mathrm{R}\, \mathbf{y}$; and $\mathbf{x}\, \mathrm{R}^*\, \mathbf{y}$ if $\mathbf{x}\, \mathrm{R}^s\, \mathbf{y}$ for some integer s. It is readily shown that the relation R^* is transitive, monotonic, and continuous. Moreover, whenever $\mathbf{x} = \mathbf{f}(\mathbf{p}, M)$ is derived from a preference relation P, the relation R^* is irreflexive: for since $\mathbf{x}\, \mathrm{R}\, \mathbf{y}$ implies $\mathbf{x}\, \mathrm{P}\, \mathbf{y}$, it follows that $\mathbf{x}\, \mathrm{R}^*\, \mathbf{y}$ implies $\mathbf{x}\, \mathrm{P}\, \mathbf{y}$ by the transitivity of P; hence the irreflexivity of R^* is a consequence of that of P.

Furthermore, if the induced demand function satisfies D–3 and D–4, then relation R^* coincides with P:

$$R^* = P.$$

The proof of this is involved and will be omitted (see the notes at the close of this chapter). However, for the case of $n = 2$ the orientation of the unit circle can be exploited and the proof is not difficult (see Problem 19).

It can also be shown that if the demand function $\mathbf{f}(\mathbf{p}, M)$ satisfies conditions D–1 through D–4 and R^* is irreflexive, then R^* is a preference relation satisfying postulates P–1 through P–5, and $\mathbf{f}(\mathbf{p}, M)$ is the demand function derived from R^*.

The irreflexivity of R^* may be characterized as follows: For any integer s, the relations

$$(\mathbf{p}^0, \mathbf{x}^0) \geq (\mathbf{p}^0, \mathbf{x}^1), (\mathbf{p}^1, \mathbf{x}^1) \geq (\mathbf{p}^1, \mathbf{x}^2), \ldots, (\mathbf{p}^{s-1}, \mathbf{x}^{s-1}) \geq (\mathbf{p}^{s-1}, \mathbf{x}^s)$$
$$(\mathbf{x}^i \neq \mathbf{x}^j \quad \text{for some} \quad i, j)$$

imply

$$(\mathbf{p}^s, \mathbf{x}^0) > (\mathbf{p}^s, \mathbf{x}^s), \tag{8.6.6}$$

where

$$\mathbf{x}^t = \mathbf{f}(\mathbf{p}^t, M^t) \qquad (t = 0, 1, \ldots, s).$$

This characterization is usually called the *strong axiom of revealed preference*. A related concept, the *weak axiom of revealed preference*, is formulated as follows:

$$(\mathbf{p}^0, \mathbf{x}^0) \geq (\mathbf{p}^0, \mathbf{x}^1) \quad (\mathbf{x}^0 \neq \mathbf{x}^1) \qquad \text{implies} \qquad (\mathbf{p}^1, \mathbf{x}^0) > (\mathbf{p}^1, \mathbf{x}^1). \tag{8.6.7}$$

The strong axiom obviously implies the weak axiom. We shall have occasion to apply the concept of revealed preference in our later discussion of stability (see Section 9.4).

Compensated income function for price $\mathbf{p}$. As an aid to our investigation of the properties of the demand function $\mathbf{f}(\mathbf{p}, M)$ derived from the preference relation P, we introduce the notion of "compensated income change."

Let $\mathbf{p}^0 \gg \mathbf{0}$, $M^0 > 0$, and $\mathbf{x}^0 = \mathbf{f}(\mathbf{p}^0, M^0)$ be given. We define the compensated income $M(\mathbf{p})$ for price $\mathbf{p}$ by

$$M(\mathbf{p}) = \inf \{M | \mathbf{f}(\mathbf{p}, M) \mathrm{P} \mathbf{x}^0\},$$

or, equivalently,

$$M(\mathbf{p}) = \sup \{M | \mathbf{x}^0 \mathrm{P} \mathbf{f}(\mathbf{p}, M)\}.$$

Figure 8.3 helps to indicate the meaning of $M(\mathbf{p})$. For a given utility value that corresponds to the utility of $\mathbf{x}^0$, the compensated income $M(\mathbf{p})$ can be interpreted as the minimum budget for a given price vector which will purchase the same utility as $\mathbf{x}^0$.

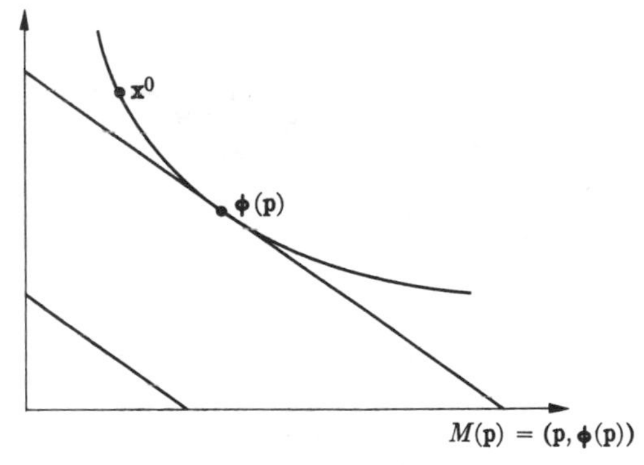

FIGURE 8-3

It is easy to show (by appealing to the geometrical description of $M(\mathbf{p})$ as indicated in Fig. 8.3) that $M(\mathbf{p})$ is a concave function of $\mathbf{p}$, and therefore that it is continuous at $\mathbf{p} \gg \mathbf{0}$ and differentiable almost everywhere.

We assume for simplicity of exposition that $M(\mathbf{p})$ is differentiable everywhere. We may also regard $M(\mathbf{p})$ as a support function of a convex set; explicitly,

$$M(\mathbf{p}) = \min (\mathbf{p}, \mathbf{x}),$$

where the minimum is extended over all $\mathbf{x}$ for which $\overline{\mathbf{x}_0 \mathrm{P} \mathbf{x}}$. $M(\mathbf{p})$ is a concave function, being a minimum of linear functions. Let

$$\phi(\mathbf{p}) = \mathbf{f}[\mathbf{p}, M(\mathbf{p})]; \tag{8.6.8}$$

then

$$(\mathbf{p}, \boldsymbol{\phi}(\mathbf{p})) = M(\mathbf{p}) \tag{8.6.9}$$

and

$$M(\mathbf{p}^0) = M^0, \qquad \mathbf{x}^0 = \boldsymbol{\phi}(\mathbf{p}^0).$$

According to the budget constraint $M(\mathbf{p})$, the vector $\boldsymbol{\phi}(\mathbf{p})$ is clearly the preferred consumption vector with respect to all $\mathbf{x}$ meeting the condition $(\mathbf{p}, \mathbf{x}) \leq M(\mathbf{p})$.

It is clear from the geometry of Fig. 8.3 and from the definition of $\phi(\mathbf{p})$ and $M(\mathbf{p})$ (a rigorous proof is easy) that the function $\sum \tilde{p}_i\phi_i(\mathbf{p}) = g(\mathbf{p})$, with $\tilde{\mathbf{p}} \gg \mathbf{0}$ as a parameter, is minimized as a function of $\mathbf{p}$ uniquely at $\mathbf{p} = \tilde{\mathbf{p}}$. Hence the usual necessary condition required of a minimum value of a function yields the relation

$$\sum_{i=1}^{n} p_i \frac{\partial \phi_i}{\partial p_j} = 0 \qquad (j = 1, 2, \ldots, n). \tag{8.6.10}$$

From (8.6.10), we have

$$\left(\frac{\partial M(\mathbf{p})}{\partial p}\right)_j = \phi_j(\mathbf{p}) + \sum_{i=1}^{n} p_i \frac{\partial \phi_i}{\partial p_j} = \phi_j(\mathbf{p}). \tag{8.6.11}$$

All these manipulations are performed while keeping the utility level constant and equal to that of $\mathbf{x}^0$.

We shall define

$$K_{ij}(\mathbf{p}^0) = \left(\frac{\partial \phi_i}{\partial p_j}\right)_{p^0} = \left(\frac{\partial f_i}{\partial p_j}\right)_{U(x)=U(x^0)} \tag{8.6.12}$$

as the residual variability of the ith commodity with respect to the compensated price change of the jth commodity. In other words, K_{ij} is a measure of the change of desire for consumption of the ith commodity as the price of the jth commodity is varied, always keeping the same utility level. The idea behind the study of the function $K_{ij}(\mathbf{p}^0)$ is as follows. Suppose that we attempt to assign meaning to the discrepancy of two distinct demand vectors, $\mathbf{x}^0 = \mathbf{f}(\mathbf{p}^0, M^0)$ and $\mathbf{x}' = \mathbf{f}(\mathbf{p}^1, M^1)$. The change of $\mathbf{x}^0$ to $\mathbf{x}^1$ can be decomposed into two components: (1) the change due to the change of prices in $\mathbf{x}^0$, keeping the same utility level; (2) the change for a fixed price resulting from a new budget size. $K_{ij}(\mathbf{p}^0)$, often referred to as the Slutsky term, can be regarded as a measure of the first change.

By (8.6.8) and (8.6.11), we have

$$\frac{\partial \phi_j}{\partial p_k} = \frac{\partial f_j}{\partial p_k} + \frac{\partial f_j}{\partial M}\frac{\partial M}{\partial p_k} = \frac{\partial f_j}{\partial p_k} + \frac{\partial f_j}{\partial M}\,\phi_k.$$

Evaluating (8.6.12) at $\mathbf{p} = \mathbf{p}^0$, we obtain

$$K_{jk}(\mathbf{p}^0) = \frac{\partial x_j^0}{\partial p_k} + \frac{\partial x_j^0}{\partial M} x_k^0,$$

where $\mathbf{x}^0 = \mathbf{f}(\mathbf{p}^0, M^0)$. Thus, if p_0 is the only price that changes, the change in demand can be written as the difference between the Slutsky term and the term measuring the effective loss in income due to the change in price.

THEOREM 8.6.1. K_{jk} satisfies

$$K_{jk} = K_{kj} \qquad (j, k = 1, \ldots, n), \tag{8.6.13}$$

$$\sum_{i,j=1}^{n} K_{ij}h_ih_j \leq 0 \qquad \text{for any real vector } \langle h_1, \ldots, h_n \rangle, \tag{8.6.14}$$

$$\sum_j p_j^0 K_{jk}(\mathbf{p}^0) = 0 \qquad (k = 1, \ldots, n). \tag{8.6.15}$$

Proof. Equation (8.6.13) is easily implied by (8.6.11). Explicitly,

$$K_{jk}(\mathbf{p}^0) = \left(\frac{\partial \phi_j}{\partial p_k}\right)_0 = \left(\frac{\partial^2 M(p)}{\partial p_j \partial p_k}\right)_0 = \left(\frac{\partial^2 M(p)}{\partial p_k \partial p_j}\right)_0 = \left(\frac{\partial \phi_k}{\partial p_j}\right)_0 = K_{kj}(\mathbf{p}^0). \tag{8.6.16}$$

Equation (8.6.15) has already been proved in (8.6.10). Relation (8.6.14) is implied by the fact that $M(\mathbf{p})$ is concave and by (8.6.11) and (8.6.12) (see Section B.4).

8.7 Nonlinear models of equilibrium. The state of an economic system under conditions of competition at any point in time can be formulated as the solution of a system of inequality relations expressing the demand for goods by consumers, the supply of goods by producers, and the equilibrium condition that supply exceeds demand in every market, it being assumed that if the supply of any commodity is overabundant, the price of that commodity is zero. From the study of the existence of equilibrium solutions, we acquire insight into descriptive and normative economic theory.

If competitive economic models are a reasonably accurate description of reality, then the fact that these models have solutions is one verification of their possible usefulness. Of equal importance is the relation of the existence of competitive-equilibrium solutions to the problem of welfare economics (see Section 9.1).

In this and the following section we discuss a series of similar nonoverlapping models describing phenomena of equilibrium and exchange. The key mathematical tool used in establishing the existence of solutions for each of these models is some form of the Kakutani fixed-point theorem, with additional slight trimmings.

The linear equations of the Leontief models of Sections 8.1 and 8.3 can be interpreted as expressing an equilibrium of exchange among various activities. In order to bring the model into closer conformity with reality, we must abandon the assumption of "linearity," i.e., the condition that the numbers a_{ij} are constants independent of X_i. The study of certain basic nonlinear models will be the subject of the remainder of this section.

By way of introduction, we first consider the following model. Suppose that we have n producers, and that each producer P_i produces some commodity G_i. Let $f_{ij}(x)$ $(i \neq j)$ represent the amount of money that P_i spends on G_j when his income is x. The functions $f_{ij}(x)$ may be regarded as demand functions in the sense of Section 8.6. If each producer spends his entire income in buying goods from the other producers, then

$$x = \sum_{j=1}^{n} f_{ij}(x) \tag{8.7.1}$$

for each choice of i and positive x value. The function $f_{ii}(x)$ is defined for convenience to be identically zero. The economic law states that the income to the ith producer x_i is determined such that the total amount of each commodity sold by a producer must equal the total amount of that commodity bought by the other producers. Mathematically, we seek to find values x_i which satisfy the system of equations

$$x_j = \sum_{i=1}^{n} f_{ij}(x_i). \tag{8.7.2}$$

This problem is a nonlinear analog of the problems of the closed Leontief model (see Theorem 8.3.1).

▶THEOREM 8.7.1. If the nonnegative functions f_{ij} are continuous, and if (8.7.1) is satisfied for all nonnegative numbers, then there exist nonnegative numbers $\bar{x}_1, \bar{x}_2, \ldots, \bar{x}_n$ which satisfy (8.7.2).

Proof. Let S be the set of all nonnegative vectors $\mathbf{y}$, with

$$\sum_{i=1}^{n} y_i = 1.$$

We construct a continuous mapping T of S into itself. Define T such that the jth component of Ty is

$$(Ty)_j = \sum_{i=1}^{n} f_{ij}(\mathbf{y}).$$

It follows from (8.7.1) that $\mathbf{Ty}$ is in S whenever $\mathbf{y}$ is chosen from S. The mapping T is continuous, and hence by the Brouwer fixed-point theorem (see Section C.2 of the Appendix) there exists a vector $\bar{\mathbf{x}}$ in S such that $\mathbf{T}\bar{\mathbf{x}} = \bar{\mathbf{x}}$. The vector $\bar{\mathbf{x}}$ obviously is a solution to the problem.

A model with variable prices. In the models considered up to now all prices were taken as fixed, and in order that supply should equal demand each producer was forced to control the volume of his output. We shall now take the opposite point of view and attempt to balance supply and demand by shifting prices rather than volume. To this end, we ask the question: Does there exist a set of prices such that the value of the goods sold by each producer is equal to the value of the goods he buys?

According to the classical law of supply and demand, the price of a commodity rises when the demand for that commodity exceeds the available supply and drops when the supply exceeds the demand. Economic equilibrium can therefore be achieved only when prices are such that supply and demand for each commodity are equal, taking into account the variability of prices. Let a_i be the amount of G_i produced by P_i in some fixed period of time, and let p_i be the price of one unit of G_i. If each producer sells his entire output, then his income will be $p_i a_i$. Let $D_{ij}(p_1, p_2, \ldots, p_n) = x_{ij}$ represent the amount of G_j demanded by P_i. The functions D_{ij} logically depend on a_i in the sense that the p_i are dependent on a_i. The assumption that each producer spends his entire income means that

$$p_i a_i = \sum_j p_j D_{ij}(\mathbf{p}) \qquad (i = 1, 2, \ldots, n). \tag{8.7.3}$$

(Set $D_{ii} = 0$ by definition.)

We assume that (8.7.3) holds for any price vector $\mathbf{p}$. The total demand of goods G_j is $\sum_i D_{ij}(\mathbf{p})$, and the total supply is a_j. The condition for equilibrium is

$$a_j - \sum_j D_{ij}(\mathbf{p}) = 0. \tag{8.7.4}$$

The problem is: Does there exist for prescribed a_i and demand functions D which satisfy (8.7.3) a nonnegative solution $\mathbf{p}$ of (8.7.4)? Unfortunately such a solution does not always exist. It is conceivable, for instance, that one of the producers may be supplying a commodity which none of the others want, in which case the left-hand side of (8.7.4) becomes positive. This consideration leads us to the following theorem.

▶THEOREM 8.7.2. If the nonnegative function $D_{ij}(\mathbf{p})$ satisfies (8.7.3) for every positive vector $\mathbf{p}$ and $a_i > 0$ for all i, then there exists $\langle p_1^*, p_2^*, \ldots, p_n^* \rangle$ such that

$$a_j - \sum_i D_{ij}(\mathbf{p}^*) \geq 0; \tag{8.7.5}$$

and if equality does not hold in the j_0th equation, then $p_{j_0}^* = 0$.

(The theorem states that there always exist prices for which supply is at least equal to demand, and that only goods which are oversupplied have price zero. This modification of the law of supply and demand is well known to economists. In this connection, see also Theorem 5.4.3.)

Proof. A real value μ is called admissible if there exists a positive vector $\mathbf{p} = \langle p_1, \ldots, p_n \rangle$ such that $\sum p_i = 1$ and

$$\mu a_j \geq F_j(\mathbf{p}), \tag{8.7.6}$$

where

$$F_j(\mathbf{p}) = \sum_i D_{ij}(\mathbf{p}) \qquad (j = 1, \ldots, n).$$

The set of all such normalized vectors $\mathbf{p}$ will be denoted by S. The set V of all admissible μ is clearly nonvoid and bounded below. Let $\mu_0 = \inf_V \mu$; then from the compactness and continuity of D_{ij} it follows that μ_0 is also admissible. Since $\mu_0 - \epsilon$ is not admissible, for every $\mathbf{p}$ in S ($S =$ the set of all nonnegative vectors whose components sum to one) there exists at least one vector $\mathbf{y}$ in S such that

$$((\mu_0 - \epsilon)\mathbf{a} - \mathbf{F}(\mathbf{p}), \mathbf{y}) \leq 0. \tag{8.7.7}$$

The set $S(\mathbf{p})$ of all such $\mathbf{y}$ in S satisfying (8.7.7) is convex, closed, and nonvoid. As a function of $\mathbf{p}$, the sets $S(\mathbf{p})$ are continuous in the usual sense of point-to-set mapping (see Section C.2).

Since the requirements of the Kakutani fixed-point theorem are fulfilled, it follows that there exists an element $\mathbf{p}^0$ in S such that

$$((\mu_0 - \epsilon)\mathbf{a} - \mathbf{F}(\mathbf{p}^0), \mathbf{p}^0) \leq 0. \tag{8.7.8}$$

This last equation, in conjunction with (8.7.3), implies that $\mu_0 - \epsilon \leq 1$, and hence $\mu_0 \leq 1$. If we multiply the jth equation of (8.7.6) by p_j and sum, it follows from (8.7.3) that $\mu \geq 1$ for any admissible μ. Therefore $\mu_0 \geq 1$, and thus $\mu_0 = 1$. Let $\mathbf{p}^*$ be the vector corresponding to $\mu_0 = 1$. Then

$$a_j \geq \sum_{i=1}^{n} D_{ij}(\mathbf{p}^*) \qquad (j = 1, \ldots, n).$$

Since $(\mathbf{a}_j, \mathbf{p}^*) - (\mathbf{F}(\mathbf{p}^*), \mathbf{p}^*) = 0$ by (8.7.3), it follows that $p_j^* = 0$ whenever

$$a_j > \sum_{i=1}^{n} D_{ij}(\mathbf{p}^*).$$

A model with variable prices and volumes. In our next model both the amount and the type of goods supplied vary with prices. For instance, the quality of services supplied will often depend on the remuneration received by the supplier. Consequently, we shall assume the existence of supply functions $S_{ij}(\mathbf{p})$ $(i \neq j)$ which represent the amounts of G_j which P_i will supply at prices $p_1, \ldots, p_n$. The demand functions $D_{ij}(\mathbf{p})$ are specified as previously. The equation

$$\sum_{j=1}^{n} p_j S_{ij}(\mathbf{p}) = \sum_{j=1}^{n} p_j D_{ij}(\mathbf{p}) \tag{8.7.9}$$

(again $S_{ii}(\mathbf{p})$ is defined to be identically zero) is the analog of (8.7.3). The law of supply and demand leads to the relation

$$\sum_{i} S_{ij}(\mathbf{p}) \geq \sum_{i} D_{ij}(\mathbf{p}). \tag{8.7.10}$$

We seek to find a price vector $\mathbf{p}$ which satisfies (8.7.10), and such that the price of any commodity for which the supply exceeds the demand is zero.

▶THEOREM 8.7.3. Let $S_{ij}(\mathbf{p})$ and $D_{ij}(\mathbf{p})$ be continuous nonnegative functions such that (8.7.9) is satisfied for each $\mathbf{p}$ and $\sum_i S_{ij}(\mathbf{p})$ is positive for each j. Then there exists a nonnegative vector

$$\mathbf{p}^* = \langle p_i^*, \ldots, p_n^* \rangle \qquad (\Sigma p_i^* = 1)$$

which satisfies (8.7.10), and in addition

$$\sum_j p_j^* \sum_i [S_{ij}(\mathbf{p}^*) - D_{ij}(\mathbf{p}^*)] = 0.$$

The proof parallels that of Theorem 8.7.2 and is therefore omitted.

† *A general theorem of Wald on the equilibrium equations.* Let

$$R_1, R_2, \ldots, R_m$$

be the factors of production which in different combinations can be used to produce n different products $S_1, S_2, \ldots, S_n$. Specifically, a_{ij} units of R_i are needed to produce one unit of S_j. Suppose that the producer has available r_1 units of R_1, r_2 units of $R_2, \ldots, r_m$ units of R_m. If the price of a unit of S_j is $f_j(s_1, \ldots, s_n)$ when s_i units of S_i $(i = 1, \ldots, n)$ are produced, then the equations for the unknown prices p_i for a unit of R_i and the unknown quantities s_j are

$$r_i = \sum_{j=1}^{n} a_{ij} s_j \qquad (i = 1, \ldots, m) \tag{8.7.11}$$

† The remainder of this section is of more special nature and may be omitted on first reading without loss of continuity.

and

$$\sigma_j = \sum_{i=1}^{m} a_{ij}p_i \qquad (j = 1, \ldots, n), \tag{8.7.12}$$

where $\sigma_j = f_j(s_1, s_2, \ldots, s_n)$.

The equations as they stand need not have economically meaningful solutions. It has accordingly been suggested that (8.7.11) be replaced by

$$r_i \geq \sum_{j=1}^{n} a_{ij}s_j, \tag{8.7.13}$$

with the convention that wherever strict inequality holds, the corresponding price p_i is zero. This conforms with the general economic dictum that factors of production in excess of need should be priced zero (see Theorems 8.7.2 and 8.7.3).

Equation (8.7.12), which states that the production cost of a commodity equals its price, is retained without change. The general cost functions f_j are assumed to be continuous and nonnegative, and to satisfy the following condition:

(A) If a sequence of nonnegative $s_1^k, \ldots, s_n^k$ tends to $s_1, \ldots, s_n$, respectively with $s_j^k \neq 0$ for every k and $s_j = 0$, then

$$\lim_{k\to\infty} f_j(s_1^k, \ldots, s_n^k) = \infty \qquad (j = 1, \ldots, n;\ k = 1, 2, \ldots),$$

and otherwise f_j remains bounded.

This assumption is hard to justify on economic grounds. It is necessary, however, for the mathematical analysis which follows. One possible argument in its behalf is in terms of the existence of a fixed setup cost associated with producing.

(B) $r_i > 0$ $(i = 1, \ldots, m)$.

(C) $a_{ij} \geq 0$ for all i, j, and for each j there exists at least one i such that $a_{ij} > 0$.

Conditions (B) and (C) guarantee that the set of nonnegative vectors $\mathbf{s} = \langle s_1, \ldots, s_n \rangle$ satisfying (8.7.13) is a compact, convex, nonvoid, n-dimensional bounded set. Let us denote this set by S. Let $\mathbf{c}$ be an interior point of S, and let S_ϵ be the set of all $\mathbf{s}$ in S which satisfy the property that $\mathbf{s} = \epsilon\mathbf{c} + (1 - \epsilon)\mathbf{s}'$ for $\mathbf{s}'$ in S and $0 < \epsilon < 1$. The set S_ϵ is an interior subset such that the boundary of S_ϵ is traversed as $\mathbf{s}'$ traverses the boundary of S. By the projection $\mathbf{P}_\epsilon$ of S into S_ϵ we shall mean

$$\mathbf{P}_\epsilon\mathbf{s} = \epsilon\mathbf{c} + (1 - \epsilon)\mathbf{s}.$$

Our objective is to establish the existence of nonnegative solutions $\langle p_i \rangle$

and $\langle s_j \rangle$ to the system of equations (8.7.12) and (8.7.13), with the added feature that

$$\sum_{i=1}^{m} p_i \left(r_i - \sum_{j=1}^{n} a_{ij} s_j \right) = 0.$$

The analysis depends upon the use of the Kakutani fixed-point theorem and the duality theorem of linear programming. For every $\mathbf{s}_0$ in S_ϵ, we construct the vector $\boldsymbol{\sigma} = \langle \sigma_j \rangle = \langle f_j(\mathbf{s}_0) \rangle$. Consider the following two problems:

I. For $\mathbf{Au} \leq \mathbf{r}$, where $\mathbf{u} \geq \mathbf{0}$, maximize $(\boldsymbol{\sigma}, \mathbf{u})$.

II. For $\mathbf{pA} \geq \boldsymbol{\sigma}$, where $\mathbf{p} \geq \mathbf{0}$, minimize $(\mathbf{r}, \mathbf{p})$.

Since the set of allowable vectors $\mathbf{u}$ traverses S, a bounded, closed, convex subset of the first orthant, a solution to Problem I exists. Let the set of all solutions be denoted by $\overline{T}(\mathbf{s}_0)$. The projection $\mathbf{P}_\epsilon[\overline{T}(\mathbf{s}_0)] = T(\mathbf{s}_0)$ is closed, nonvoid, and convex. The fundamental duality theorem (Theorem 5.4.1) tells us that there exists a solution $\mathbf{p}^0$ to Problem II which has the property that $(\mathbf{p}^0, \mathbf{r} - \mathbf{Au}^0) = 0$ for any $\mathbf{u}^0$ in $\overline{T}(\mathbf{s}_0)$. The point-to-set mapping $\mathbf{s}_0 \to T(\mathbf{s}_0)$ fulfills all the conditions needed to apply the Kakutani fixed-point theorem. Consequently, there exists an $\mathbf{s}_\epsilon^*$ in S_ϵ such that $\mathbf{s}_\epsilon^*$ is also in $T(\mathbf{s}_\epsilon^*)$.

Let $\mathbf{s}^*$ denote a limit point of $\mathbf{s}_\epsilon^*$ with the maximal number of zero coordinates. We first prove that no coordinate of $\mathbf{s}^* = \langle s_1^*, s_2^*, \ldots s_n^* \rangle$ can be zero. Suppose that the indices I_0 have $s_{i_0}^* = 0$ for i_0 in I_0; then the functions $f_j(\mathbf{s}_\epsilon^*)$ tend to ∞ as $\epsilon \to 0$ for every j in I_0, but remain bounded in the other coordinates. For ϵ sufficiently small the vector $\mathbf{s}_\epsilon^*$ is close to the set of u which maximizes $(\mathbf{f}, \mathbf{u})$, where $\mathbf{u}$ is subject to $\mathbf{Au} \leq \mathbf{r}$ and $\mathbf{r}$ is strictly positive. It follows, however, that some of the coordinates of $\mathbf{s}_\epsilon^*$ with indices in I_0 must be chosen bounded away from zero; for if the contrary were true and all the coordinates of $\mathbf{s}_\epsilon^*$ with i in I_0 converged to zero, then a larger value of $(\mathbf{f}(\mathbf{s}_\epsilon^*), \mathbf{u})$ could be obtained by choosing a $\mathbf{u}$ in S with the i_0 coordinate strictly greater than a fixed positive ϵ_0 for every i_0 in I_0, and the set $T(\mathbf{s}_\epsilon^*)$ could not contain $\mathbf{s}_\epsilon^*$ as constructed. Consequently, in order to avoid these absurdities, we must have every coordinate of $\mathbf{s}^*$ strictly bounded away from zero, and the corresponding $f_j(\mathbf{s}^*)$ are therefore uniformly bounded.

Since the associated solutions $\mathbf{p}_\epsilon^0$ to Problem II for the particular choice of $\boldsymbol{\sigma}$ corresponding to $\mathbf{s}_0 = \mathbf{s}_\epsilon^*$ minimize $(\mathbf{r}, \mathbf{p})$ and each r_j is strictly positive, it follows that the $\mathbf{p}_\epsilon^0$ lie in a bounded portion of the first quadrant, and hence a limit point $\mathbf{p}^0$ exists as ϵ tends to zero. The vectors $\mathbf{s}^*$ and $\mathbf{p}^0$ satisfy the following relations:

$$r_i \geq \sum_{j=1}^{n} a_{ij}s_j^* \qquad (i = 1, \ldots, m). \tag{8.7.14}$$

$$\sigma_j \leq \sum_{i=1}^{m} a_{ij}p_i^0 \qquad (j = 1, \ldots, n), \tag{8.7.15}$$

$$\sigma_j = f_j(s_1^*, \ldots s_n^*) \qquad (j = 1, \ldots, n). \tag{8.7.16}$$

The linear programming analysis shows that $\mathbf{s}^*$ solves Problem I, with $\sigma_j = f_j(\mathbf{s}^*)$, and $\mathbf{p}^0$ solves Problem II. Since each coordinate of $\mathbf{s}^*$ is positive, equality must hold in (8.7.15) (by Theorem 5.4.3). Again, by virtue of the same theorem, we obtain

$$\sum_{i=1}^{m} p_i^0 \left(r_i - \sum_{j=1}^{n} a_{ij}s_j^*\right) = 0. \tag{8.7.17}$$

The following theorem summarizes the preceding discussion.

▶THEOREM 8.7.4. If r_i, a_{ij}, and f_j satisfy conditions (A), (B), and (C), then there exist two positive vectors $\mathbf{s}^*$ and $\mathbf{p}^0$, the first strictly positive, satisfying (8.7.14), (8.7.16), and (8.7.17), with the equality sign for every equation of the set (8.7.15).

***8.8 The Arrow-Debreu equilibrium model of a competitive economy.** We suppose that there are a finite number r of distinct commodities (including services of all kinds) available. If the same commodity is bought and sold at m different locations or at m different times, it may be regarded as m distinct commodities.

The commodities are produced in production units (e.g., firms). We assume that for each production unit j ($j = 1, 2, \ldots, n$) there is a set Y_j of possible production plans. An element $\mathbf{y}_j$ of Y_j is a vector in E^r whose hth component y_{hj} designates the output of commodity h according to that plan. As usual, inputs are treated as negative components. $Y = \sum_j Y_j$ represents the set of all possible input-output schedules of the whole production sector. We make the following assumptions:

I(a). Y_j is a compact subset of E^r containing 0.

I(b). Y is convex.

The convexity assumption reflects the principle of nonincreasing marginal rates of transformation. Moreover, we assume the existence of a finite number of consumption units (individuals or institutions).

Let X_i ($i = 1, \ldots, m$) represent the set of consumption vectors in E^r available to consumer i. We now make another assumption:

II. X_i is a closed convex subset of E^r, with the additional property that whenever $\mathbf{x}_i^{(n)} \in X_i$ and some coordinate x_{hoi}^n of $\mathbf{x}_i^n$ tends to ∞, then $x_{hi}^n \to \infty$ for all coordinates $h = 1, 2, \ldots, r$.

We shall suppose that there is defined a utility index function $U_i(\mathbf{x}_i)$ for the ith consumer such that:

III(a). $U_i(\mathbf{x}_i)$ is a continuous differentiable function on X_i.

III(b). If $U_i(\mathbf{x}_i) \geq U_i(\mathbf{x}'_i)$ and $\mathbf{x}_i, \mathbf{x}'_i \in X_i$ ($\mathbf{x}_i \neq \mathbf{x}'_i; 0 < t < 1$), then $U_i[t\mathbf{x}_i + (1 - t)\mathbf{x}'_i] > U_i(\mathbf{x}'_i)$.

III(c). For any $\mathbf{x}_i$ in X_i, there is an $\mathbf{x}'_i$ in X_i such that $U_i(\mathbf{x}'_i) > U_i(\mathbf{x}_i)$.

Assumption III(b) guarantees that the set $\{\mathbf{x}|U_i(\mathbf{x}) \geq c\}$ is a strictly convex set [see Section B.4, property (ii)] for each constant c. The function U clearly induces a preference relation as defined in Section 8.6. Without any budgetary restrictions the ith consumer would make his selection from X_i strictly on the basis of his preference scale.

Assumption III(c) states that if we disregard budget considerations the desires of consumers are never completely satisfied.

Let $\boldsymbol{\xi}_i$ represent the commodities which the ith consumer holds for initial trading. We assume that:

IV. There exists an $\mathbf{x}_i$ in X_i such that $\boldsymbol{\xi}_i \gg \mathbf{x}_i$.

Although assumption IV is too strong and not altogether realistic, unfortunately it cannot be completely eliminated. It can be weakened slightly, but only at the expense of making the arguments more tedious and no more rewarding. (For extensions see the notes at the close of the chapter.)

Finally, we assume that the ith consumer has a contractual claim to the share α_{ij} of the profit of the jth production of each commodity such that

V. $\alpha_{ij} \geq 0$ for all i, j, and

$$\sum_{i=1}^{m} \alpha_{ij} = 1$$

for all j.

Let $\mathbf{p} = \langle p_1, \ldots, p_r \rangle$ represent a vector of prices of the commodities. Since only relative prices are important, we may normalize $\mathbf{p}$ as follows without restricting generality:

$$p_v \geq 0, \quad \sum_{v=1}^{r} p_v = 1.$$

Definition of a competitive equilibrium. The existence of a competitive equilibrium is equivalent to the existence of a set of vectors

$$\{\mathbf{p}^*; \mathbf{x}_1^*, \ldots, \mathbf{x}_m^*; \mathbf{y}_1^*, \ldots, \mathbf{y}_n^*\}$$

such that

(A) $$\mathbf{p}^* \in P = \left\{\mathbf{p}|\mathbf{p} \in E^r, \quad \mathbf{p} \geq \mathbf{0}, \quad \sum_{\nu=1}^{r} p_\nu = 1\right\};$$

(B) for each j

$$(\mathbf{p}^*, \mathbf{y}_j^*) = \max_{y_j \in Y_j} (\mathbf{p}^*, \mathbf{y}_j);$$

(C) for each i, $\mathbf{x}_i^*$ maximizes $U_i(\mathbf{x}_i)$ over the set

$$\left\{\mathbf{x}_i | \mathbf{x}_i \in X_i, \quad (\mathbf{p}^*, \mathbf{x}_i) \leq (\mathbf{p}^*, \boldsymbol{\xi}_i) + \sum_{j=1}^{n} \alpha_{ij}(\mathbf{p}^*, \mathbf{y}_j^*)\right\} ;$$

(D) $$\mathbf{x}^* \leq \mathbf{y}^* + \boldsymbol{\xi}^* \quad \text{and} \quad (\mathbf{p}^*, \mathbf{x}^* - \mathbf{y}^* - \boldsymbol{\xi}^*) = 0,$$

where

$$\mathbf{x}^* = \sum_{i=1}^{m} \mathbf{x}_i^*, \quad \mathbf{y}^* = \sum_{j=1}^{n} \mathbf{y}_j^*.$$

Condition (B) asserts that production is such as to maximize profits when the prices are the equilibrium market prices. Condition (C) states that the ith consumer seeks to maximize his utility with respect to all available consumption vectors, where a consumption vector is possible if it belongs to his availability set X_i and if its cost does not exceed his income. His income derives from two sources: first, the income associated with his initial holdings; and second, the income derived from his share of the profits of all production. In other words, condition (C) expresses the budget restraint under conditions of equilibrium. Condition (D) states that all demand is fulfilled, and further that any commodity supplied in excess of demand is priced zero.

The basic theorem is as follows:

▶THEOREM 8.8.1. For any system satisfying assumptions I–V, a competitive equilibrium exists.

The proof of this theorem is a simple consequence of the following topological result concerning point-to-set mapping, which also is of some independent interest.

▶LEMMA 8.8.1. Let S be a bounded, upper semicontinuous, set-valued function mapping the set P into E^r such that (a) $S(\mathbf{p})$ is a nonvoid convex set for all $\mathbf{p} \in P$, and (b) if $\mathbf{z} \in S(\mathbf{p})$, then $(\mathbf{p}, \mathbf{z}) \geq 0$. Then there exist values $\mathbf{p}$ in P and $\mathbf{z}$ in $S(\mathbf{p})$ such that $\mathbf{z} \geq \mathbf{0}$.

(The mapping $S(\mathbf{p})$ as upper semicontinuous is to be understood in the usual sense, i.e., if $\mathbf{z}_n \in S(\mathbf{p}_n)$ and $\mathbf{z}_n \to \mathbf{z}_0$, $\mathbf{p}_n \to \mathbf{p}_0$, then $\mathbf{z}_0 \in S(\mathbf{p}_0)$; see also Section C.2.)

Proof. Suppose that the lemma is false. This means that each $S(\mathbf{p})$ does not touch the nonnegative orthant. In that case we claim that there exists a positive ϵ_0 such that the family of convex sets

$$T(\mathbf{p}) = \{\mathbf{w} | \mathbf{w} = \mathbf{z} + \epsilon_0 \mathbf{p}, \qquad \mathbf{z} \in S(\mathbf{p})\}$$

for each $\mathbf{p}$ in P also does not touch the nonnegative orthant. Otherwise, there must exist a sequence $\mathbf{p}_n \to \tilde{\mathbf{p}}$ and $\mathbf{z}_n \in S(\mathbf{p}_n) \to \mathbf{z}$ such that $\mathbf{z}_n + \epsilon_n \mathbf{p}_n \geq 0$, with $\epsilon_n \to 0$ ($\mathbf{z}_n$ may be chosen to converge, since all $S(\mathbf{p}_n)$ are compact and lie in a bounded region); and it follows from the continuity of S that $\mathbf{z}$ is in $S(\tilde{\mathbf{p}})$ and $\mathbf{z} \geq \mathbf{0}$, in contradiction to our assumption. Hence $T(\mathbf{p})$ is disjoint from the nonnegative orthant as stated.

We notice next that a plane separating the positive orthant and one of the sets $T(\mathbf{p})$ may be identified with a vector $\mathbf{q} = \langle q_1, \ldots, q_r \rangle$ such that

$$q_i \geq 0, \quad \sum_{i=1}^{r} q_i = 1, \quad (\mathbf{q}, \mathbf{w}) \leq 0$$

for $\mathbf{w}$ in $T(\mathbf{p})$, and conversely (see Lemma B.1.3).

We now construct a point-to-set mapping as follows: For each $\mathbf{p}$ in P let $Q(\mathbf{p}) \subset P$ consist of all normalized vectors in P representing all separating planes of the positive orthant and $T(\mathbf{p})$. Since $T(\mathbf{p})$ does not touch the positive orthant, it follows that $Q(\mathbf{p})$ is nonvoid (Lemma B.1.3). Clearly, $Q(\mathbf{p})$ is convex and closed and describes an upper semicontinuous point-to-set mapping; the nature of the continuity of $Q(\mathbf{p})$ follows easily from the corresponding continuity of $T(\mathbf{p})$. By the Kakutani theorem (see Section C.2) there exists $\mathbf{p}_0 \in Q(\mathbf{p}_0)$. But for this $\mathbf{p}_0$, it follows from hypothesis (b) that $(\mathbf{p}_0, \mathbf{z}) \geq 0$ when $\mathbf{z}$ is in $S(\mathbf{p}_0)$, and hence $(\mathbf{p}_0, \mathbf{w}) > 0$ for $\mathbf{w}$ in $T(\mathbf{p}_0)$. This inference is incompatible with the meaning of $\mathbf{p}_0$ in $Q(\mathbf{p}_0)$. Hence our original supposition was false and the lemma is proved.

Proof of Theorem 8.8.1. For each $\mathbf{p}$ in P, we construct $S(\mathbf{p})$ with the properties of Lemma 8.8.1 as follows. First define $Y_j(\mathbf{p})$ as the set of all vectors in Y_j which maximize $(\mathbf{p}, \mathbf{y}_j)$. Let $Y(\mathbf{p}) = \sum Y_j(\mathbf{p})$, and let

$$\beta_j(\mathbf{p}) = \max_{y_j \in Y_j} (\mathbf{p}, \mathbf{y}_j).$$

It is easily established that $Y(\mathbf{p})$ is a closed, nonvoid, convex set and defines a bounded upper semicontinuous point-to-set mapping. Next let $X_i(\mathbf{p})$ designate the set of all $\mathbf{x}_i$ in X_i which maximize $U_i(\mathbf{x}_i)$ subject to the condition

$$(\mathbf{p}, \mathbf{x}_i) \leq (\mathbf{p}, \boldsymbol{\xi}_i) + \sum_{j=1}^{n} \alpha_{ij} \beta_j(\mathbf{p}). \tag{8.8.1}$$

An important observation, which we shall call on later, is that for $\mathbf{x}_i$ in $X_i(\mathbf{p})$ equality must prevail in (8.8.1). Let us suppose the contrary. Then by assumption III(c) there exists an $\mathbf{x}_i' \in X_i$ such that $U_i(\mathbf{x}_i') > U_i(\mathbf{x}_i)$, and by assumptions III(a) and III(b) we can find $\tilde{\mathbf{x}}_i = t\mathbf{x}_i + (1 - t)\mathbf{x}_i'$ for appropriate t ($0 < t < 1$) such that $U_i(\tilde{\mathbf{x}}_i) > U_i(\mathbf{x}_i)$ and $\tilde{\mathbf{x}}_i$ still obeys the constraints of (8.8.1), thus contradicting the meaning of $\mathbf{x}_i$ in $X_i(\mathbf{p})$. Hence, equality in (8.8.1) holds for such $\mathbf{x}_i$.

From the fact that $\beta_j(\mathbf{p}) \geq 0$ [recall that $\mathbf{0} \in Y_i$ by assumption I(a)] and from assumptions I, II, III, and IV, we see immediately that $X_i(\mathbf{p})$ is a nonvoid, convex, closed, bounded set. Define

$$X(\mathbf{p}) = \sum_{i=1}^{n} X_i(\mathbf{p}),$$

which is also nonvoid, convex, closed, and bounded. Again, in view of assumptions III and IV, it is readily verified that $X(\mathbf{p})$ is an upper semicontinuous point-to-set mapping. Finally, let us form

$$S(\mathbf{p}) = \{\mathbf{z}|\mathbf{z} = \mathbf{y} - \mathbf{x} + \boldsymbol{\xi}, \qquad \mathbf{y} \text{ in } Y(\mathbf{p}), \qquad \mathbf{x} \text{ in } X(\mathbf{p})\},$$

where

$$\boldsymbol{\xi} = \sum_{i=1}^{m} \boldsymbol{\xi}_i.$$

We now prove that $S(\mathbf{p})$ is a bounded upper semicontinuous point-to-set mapping from P to E^r such that (a) $S(\mathbf{p})$ is convex, nonvoid, and closed, and (b) $(\mathbf{p}, \mathbf{z}) = 0$ for each $\mathbf{z}$ in $S(\mathbf{p})$.

We have already noted the truth of (a). To check (b), we see that the terms of the sum of $\mathbf{y}$ in $Y(\mathbf{p})$ and $\mathbf{x}$ in $X(\mathbf{p})$ obey the relation (8.8.1) with equality, i.e.,

$$(\mathbf{p}, \mathbf{x}_i) = (\mathbf{p}, \boldsymbol{\xi}_i) + \sum_{j=1}^{n} \alpha_{ij}\beta_j(\mathbf{p}) = (\mathbf{p}, \boldsymbol{\xi}_i) + \sum_{j=1}^{n} \alpha_{ij}(\mathbf{p}, \mathbf{y}_j),$$

where

$$\mathbf{x} = \sum_{i=1}^{m} \mathbf{x}_i, \quad \mathbf{y} = \sum_{j=1}^{n} \mathbf{y}_j.$$

If we sum on i, it follows from assumption V that

$$(\mathbf{p}, \mathbf{x}) = (\mathbf{p}, \boldsymbol{\xi}) + (\mathbf{p}, \mathbf{y}),$$

which agrees with the statement of (b).

The preparations for invoking the result of Lemma 8.8.1 are here, and we deduce the existence of $\mathbf{p}^*$ and $\mathbf{z}^* = \mathbf{y}^* - \mathbf{x}^* + \boldsymbol{\xi}$ such that $\mathbf{z}^*$ belongs to $S(\mathbf{p}^*)$ and $\mathbf{z}^* \geq \mathbf{0}$. The vector system

$$\{\mathbf{p}^*; \mathbf{x}_1^*, \ldots, \mathbf{x}_m^*; \mathbf{y}_1^*, \ldots, \mathbf{y}_n^*\}$$

constitutes a competitive equilibrium where

$$\mathbf{x}^* = \sum_{i=1}^{m} \mathbf{x}_i^*, \quad \mathbf{y}^* = \sum_{j=1}^{n} \mathbf{y}_j^*.$$

Conditions (A), (B), and (C) (see p. 281) are satisfied by the definition of $\mathbf{x}_i^*$, $\mathbf{y}_j^*$ and the fact that $\mathbf{z}^*$ is in $S(\mathbf{p}^*)$. Condition (D) follows from the result $\mathbf{z}^* \geq \mathbf{0}$ and relation (b) above. This completes the proof of the theorem.

8.9 Problems. Problems 1–6 embody some simple formulations of macro-economic relations.

1. Suppose that the supply $S(t)$ and the demand $D(t)$ for a single commodity are related to its price $p(t)$ by the expressions

$$S(t) = \alpha + a\left(p(t) + \frac{dp}{dt}\right), \qquad D(t) = \beta + bp(t),$$

where α, a, β, and b are constants. Assume that in each instance the price is set so that $S(t) = D(t)$. Discuss the limiting properties of $p(t)$.

2. Let Q_t denote the stocks on hand at the end of period t. Suppose that supply and demand are known functions of price at time t of the form

$$D(t) = \beta + bP(t), \qquad S(t) = \alpha + aP(t).$$

Consider a model in which the price varies with the change in stocks according to the law

$$P(t) = P(t-1) - \lambda(Q_{t-1} - Q_{t-2}) \qquad (t \geq 2). \tag{*}$$

Show that $P_t = \overline{P} + [P(0) - \overline{P}]C^t$, where

$$\overline{P} = \frac{\alpha - \beta}{b - a}, \qquad C = 1 - \lambda(b - a).$$

3. Consider the model of Problem 2, taking time to be continuous. The law (*) becomes

$$\frac{dP}{dt} = -\lambda \frac{dQ}{dt},$$

where

$$Q(t) = Q(0) + \int_0^t [S(\tau) - D(\tau)]\, d\tau,$$

$D(t) = \beta + bP(t)$, and $S(t) = \alpha + aP(t)$. Determine $P(t)$ explicitly.

4. A well-known macro-economic relation is formulated as follows: national income = consumption + investment (savings) + autonomous investment, or

$$Y(t) = C(t) + I(t) + A(t).$$

Consider this relation at time t if $C(t) = cY(t)$ [$0 < c < 1$; here $1/(1 - c)$

is called the multiplier],

$$I(t) = v\,\frac{dY}{dt}$$

[$v > 0$ is called the accelerator], and $A = A_0 e^{rt}$. Determine $Y(t)$ for all t and interpret the solution.

5. In the macro-economic model of Problem 4, let

$$C(t) = \gamma Y + c_1\,\frac{dY}{dt} + c_1\,\frac{d^2Y}{dt^2}, \qquad I_t = v\,\frac{dY}{dt}, \qquad A = 0.$$

Determine the form of $Y(t)$ for all t and interpret the solution.

6. Consider the model of Problem 4 if time is taken to be discrete. Let $C(t) = c_1 Y_{t-1} + c_2 Y_{t-2}$, $I(t) = v(Y_{t-1} - Y_{t-2})$; and $A(t) = A =$ constant for all t. Determine $Y(t)$ for all t.

7. The problem of simple monopoly can be stated as follows: Suppose that it costs $\phi(\mathbf{x})$ to produce a commodity bundle $\mathbf{x}$, and that when $\mathbf{x}$ is offered on the market the price vector obtained per unit is $\mathbf{p} = h(\mathbf{x})$. The net profit on $\mathbf{x}$ is therefore

$$(\mathbf{p}, \mathbf{x}) - \phi(\mathbf{x}). \tag{**}$$

Determine a relationship satisfied by $\mathbf{x}$ which maximizes (**).

8. If $\mathbf{C}$ is a real or complex $n \times n$ matrix such that

$$S_j = \sum_{i=1}^{n} |C_{ij}| \leq 1 \qquad (j = 1, \ldots, n),$$

prove that

$$\prod_{j=1}^{n} \left(1 - \sum_{i=1}^{n} |C_{ij}|\right) \leq \det(\mathbf{I} - \mathbf{C}).$$

9. Let $\mathbf{A}$ be a matrix $\|a_{ij}\|$ $(i, j = 1, \ldots, n)$ such that

$$2 \det \left| \begin{pmatrix} a_{11}, a_{12} \\ a_{21}, a_{22} \end{pmatrix} \right| > \sum_{j<k} \left| \det \begin{pmatrix} a_{1j}, a_{1k} \\ a_{2j}, a_{2k} \end{pmatrix} \right|$$

and

$$2|a_{ii}| > \sum_{j} |a_{ij}| \qquad (i = 3, 4, \ldots, n).$$

Prove that $\det \mathbf{A} \neq 0$.

10. Use Corollary 8.2.1 to show that if $\mathbf{A}$ is a strictly positive matrix of order n and has more than one real characteristic root, then all the maximal characteristic numbers of the principal minors of $\mathbf{A}$ of order $n - 1$ lie between the two greatest real characteristic numbers of the matrix $\mathbf{A}$.

(The maximal characteristic number of a positive matrix is equal to $\lambda_0(\mathbf{A})$; see p. 252.)

11. Show that for the positive number λ to be an upper limit to the moduli of the characteristic numbers of the matrix $\mathbf{A}$ ($\mathbf{A} \geq \mathbf{0}$), it is necessary and sufficient that the first set of principal minors of the determinant $\mathbf{B}(\lambda) = \det(\lambda\mathbf{I} - \mathbf{A})|_1^n$, that is,

$$b_{11}, \quad \begin{vmatrix} b_{11}, & b_{12} \\ b_{21}, & b_{22} \end{vmatrix}, \quad \begin{vmatrix} b_{11}, & b_{12}, & b_{13} \\ b_{21}, & b_{22}, & b_{23} \\ b_{31}, & b_{32}, & b_{33} \end{vmatrix} \cdots \begin{vmatrix} b_{11}, \ldots, b_{1k} \\ b_{21}, \ldots, b_{2k} \\ \vdots \qquad \vdots \\ b_{k1}, \ldots, b_{kk} \end{vmatrix} \cdots,$$

be positive.

12. Consider a matrix $\mathbf{A}$ ($\mathbf{A} \gg \mathbf{0}$) having more than one real characteristic root. Show that between the two greatest real characteristic numbers of $\mathbf{A}$ there exists at least one principal minor of $\mathbf{A}$ of order $n - 2$ which has its maximal characteristic root lying in this interval. (*Hint:* Use Corollary 8.2.1.)

13. Suppose that the cofactors of a certain row or column in the determinant of $\mathbf{B}(\lambda) = (\lambda\mathbf{E} - \mathbf{A})$ ($\mathbf{A} \gg \mathbf{0}, \lambda > 0$) are all nonnegative but not all zero. Prove that $\lambda > \lambda_0(\mathbf{A})$ whenever $\det \mathbf{B}(\lambda) > 0$.

14. Let $\mathbf{A}$ be a positive matrix, and for fixed λ let all algebraic cofactors B_{ik} of the determinant $B(\lambda) = \det(\lambda\mathbf{I} - \mathbf{A})|_1^n$ be positive. Show that the inequalities $0 < \tilde{a}_{ij} \leq a_{ij}$ ($i, j = 1, \ldots, n$) imply the inequalities

$$\det(\lambda\mathbf{I} - \tilde{\mathbf{A}})|_1^k \geq \det(\lambda\mathbf{I} - \mathbf{A})|_1^k \qquad (k = 1, \ldots, n).$$

When does equality hold?

15. Let $\mathbf{B} = \|b_{ij}\|$ be an $r \times r$ matrix with $b_{ij} \geq 0$ for $i \neq j$. Suppose that there exists $\mathbf{z}$ such that $\mathbf{z} < \mathbf{0}$, $\mathbf{Bz} \gg \mathbf{0}$. Show that $\mathbf{B}$ is stable. (A matrix is called stable if all its eigenvalues have negative real parts.)

16. A product y is produced by the use of two factors x_1 and x_2 according to the linear homogeneous production function $y = kx_1x_2$, where k is a constant. Given the factor prices p_1 and p_2, find the optimum usage of the factors yielding a minimum cost for the product y. Show that at this optimum $x_1 = c_1 y$ and $x_2 = c_2 y$.

17. Prove that for any matrix $\mathbf{C}$ there exist nonnegative vectors $\mathbf{x}$ and $\mathbf{y}$ such that

$$\mathbf{xC} \geq \mathbf{0}, \qquad \mathbf{Cy} \leq \mathbf{0};$$

$$\mathbf{x} - \mathbf{Cy} \gg \mathbf{0}, \qquad \mathbf{y} + \mathbf{xC} \gg \mathbf{0}.$$

18. Using the Arrow-Debreu model and assuming that the technology is subject to constant returns to scale (i.e., each Y_j is a closed convex cone),

show that in an equilibrium the profit of each firm is zero—and hence that an equilibrium is determined without regard to α_{ij} (see p. 281).

19. (a) Prove that for $n = 2$ the strong axiom of revealed preference is implied by the weak axiom.

(b) Prove that if the demand vector function $\mathbf{f}(\mathbf{p}, M)$ satisfies the weak axiom of revealed preference, then $\mathbf{f}(\mathbf{p}, M)$ is positively homogeneous of order 0, that is, that $\mathbf{f}(\lambda\mathbf{p}, \lambda M) = \mathbf{f}(\mathbf{p}, M)$ for any $\lambda > 0$.

Notes and References to Chapter 8

It is generally agreed that the mathematical method began to play a significant role in economics with A. A. Cournot's *Recherches sur les principes mathématiques de la théorie des richesses*, which was published in 1838, although a few economists like Ceva and von Thunen contributed to the subject before Cournot. In the 1870's W. Stanley Jevons, Léon Walras, and Karl Menger created the subjective theory of value, which distinguishes the modern economics from the classical. Of the three, Walras, who was greatly influenced by Cournot, was most prominent; his theory of general equilibrium did for economics what Newton did for physics. Among those who may be classified as belonging to the Lausanne school founded by Walras are Pareto, Barone, Pantaleoni, Schumpeter, Leontief, and Hicks. Menger himself lacked the appropriate mathematical tools, but his marginal theory of value and much of the other work of the Austrian school of economics, which he helped to found, had a decisive influence on modern mathematical economics.

In England F. Y. Edgeworth's *Mathematical Psychics* (1881) proposed a detailed treatment of exchange. Some years later Alfred Marshall created the Cambridge school of economic analysis, based on a proper balance of intuitive knowledge and scientific precision; prominent later members of this school were Alphonse Pigou, John Maynard Keynes, and D. H. Robertson. It is accurate to say that the majority of contemporary research in economics derives ultimately from the massive contributions of the Lausanne, Austrian, and Cambridge schools.

The two main sources of current developments in mathematical economic theory are J. R. Hicks, *Value and Capital* [121], and P. A. Samuelson, *Foundations of Economic Analysis* [210], which was written at Harvard as a dissertation in 1937 and published in book form later.

Two very useful textbooks, both covering a multitude of topics of mathematical economics, are Allen [2] and Dorfman, Samuelson, and Solow [65].

The subject of mathematical economics divides into two main parts, macro-economic theory and micro-economic theory. The first is concerned chiefly with relations of economic variables aggregated over the system, the second with relations between individual economic variables, notably consumption units, production units, and market prices of commodity bundles.

Mathematical economics may also be divided into static theory and dynamic theory. A static problem is a problem in which time does not explicitly appear;

a dynamic problem is one in which the time variable is directly involved. Of course, even when we are confronted with a dynamic problem, it may happen that the solution is independent of time. We have then a stationary phenomenon. Frequently, but not always, a static situation may be interpreted as a stationary state of a dynamic system.

Research work in micro-economic theory has been concerned mainly with statics, while the majority of research work in macro-economic theory has been in dynamics. A typical procedure in macro theory is to propose models of aggregated consumption, investment, and national income as related functions of time, and then analyze the convergence or oscillatory character of, say, the income function over time.

We have not mentioned the well-developed statistical point of view of econometrics. From this point of view the statistical problem is to estimate the parameters underlying the dynamic system by methods embracing modifications of regression analysis. Econometric problems have been studied by J. Koavelma, T. Koopmans, A. Wald, T. W. Anderson, H. Rubin, L. Hurwicz, J. Marschak, N. Schultz, I. Timberger, and H. Wold. The leading textbook is L. R. Klein, *A Textbook of Econometrics.*

§ 8.1. The first linear model of equilibrium and exchange was proposed by Walras, who developed a system of equations to describe the simple phenomenon of equilibrium. However, he did not pursue the mathematical questions of the existence and uniqueness of solutions to these equations. With the aid of the theory of positive matrices this gap has been filled adequately by many workers in the field.

Until about 1948, the study of equilibrium models was largely restricted to the linear case.

The models of this section follow the formulation of open and closed systems due to Leontief (see [166] and Morgenstern [182]).

§ 8.2. The theory of matrices with nonnegative entries begins with Frobenius [89]. Since 1908 there have been innumerable extensions and applications of the Frobenius results. The generalizations have followed two main directions: (1) replacing the property of positivity by the requirement that the matrix A keep a cone invariant, and (2) pursuing the infinite-dimensional analogs of the Frobenius theorems involving positive kernels. An impressive collection of these extensions is found in Gantmacher and Krein [102], which in addition develops much of the finer structure of the theory of positive matrices. For abstract treatments of this subject see Krein and Rutman [153] and Karlin [139]. Unfortunately, economists continue to rediscover many of these theorems and to assign them thoroughly inaccurate priorities.

Direct interpretations of the theory of positive matrices with reference to economic theory, and various expositions of the properties of positive matrices and M-matrices, have been presented by Metzler [180], Mosak [185], Debreu and Herstein [63], Solow [227], Hawkins and Simon [119], and many others.

The simple proofs of this section represent a direct paraphrase of the abstract arguments of Krein and Rutman [153] and Karlin [139]. The characterization of the largest eigenvalue according to Corollary 8.2.1 seems to be new and appears very useful, as many of the later arguments indicate. The elegant argument proving part (d) of Theorem 8.2.3 is new and due to Bohnenblust (private communication).

Results related to Theorem 8.2.4 can be found in Kotelyonskii [151].

For a discussion of the classification of a matrix into its periodic and non-periodic components, see Feller [80, Chaps. 13 and 15].

The fact that results valid for M-matrices can be translated to positive matrices and vice versa is well known. The statement embodying (8.3.4) is known to economists as the Hawkins-Simon criterion [119] (see also Kotelyonskii [151]).

§ 8.3. For other results of the kind developed in this section, see Morgenstern [182].

§ 8.4. The modern development of the theory of production is attributed to Walras and Marshall. Samuelson devotes Chap. iv of his *Foundations* to production theory, see also Hicks [121] and Dorfman [64].

The current theory of production is formulated in terms of activity analysis models, utilizing the theorems of linear and nonlinear programming. As general references on activity analysis, we cite the compilations edited by Koopmans [146] and Morgenstern [182]. The nonlinear programming paper of Kuhn and Tucker [164] is basic to the finer mathematical analysis of production theory.

Inseparable from the theory of production is the theory of the distribution of wealth (imputed prices). Information on the distribution of wealth is obtained by solving the dual problem of the associated programming problem and suitably interpreting the concept of shadow price (see Uzawa [234]).

§ 8.5. Theorem 8.5.1 was suggested by Samuelson. Koopmans [148] proved the theorem for the case of three commodities, and Arrow [3] established the result in general. An alternative proof was advanced by Gale [94]. The proof given here is new.

§ 8.6. The modern theory of consumption originated with Walras and Jevons. Slutsky [225] and later Hicks [121] greatly elaborated this theory. Slutsky's analysis relies on the existence of a utility indicator function, which is used in deducing the Slutsky relations satisfied by the demand functions (see Theorem 8.6.1). McKenzie [177] derived the same conclusions without the use of a utility indicator function. Our proof is a modification of McKenzie's. The concept of the weak principle of revealed preference was introduced by Samuelson [211], who also showed that a demand function induced by an indicator function implies the validity of this principle.

Houthakker [126] proved that the strong axiom of revealed preference is equivalent to the existence of the utility function.

An axiomatic approach to the analysis of preference relations was carried out by Uzawa [232]. He established a one-to-one correspondence between a preference relation induced by the strong axiom of revealed preference and a general preference relation satisfying postulates P–1 through P–5. The discussion of this section is based on Uzawa's paper.

§ 8.7. The main advances to date in the study of equilibrium models are associated with the names of Walras, Wald, McKenzie, and Arrow and Debreu. To Walras [252] we owe the first sophisticated formulation of the equilibrium phenomenon (see notes to § 8.1). Wald [251] later carried out a comprehensive mathematical analysis of Walras's equations and established the existence and uniqueness of solutions to these equations under a variety of conditions. Finally, McKenzie, and Arrow and Debreu [7] independently formulated general models of equilibrium and proved the existence of a competitive equilibrium. None of the three models (Walras-Wald, McKenzie, Arrow-Debreu) subsumes any of the others as a special case; rather, all three should be viewed as variant descriptions of equilibrium in a competitive economy. Alternative analyses of the same models or other versions of equilibrium have been proposed independently by Gale, McKenzie, and Uzawa.

The statements of Theorems 8.7.1–3 were discovered by Gale [93]. The proofs evolved here appear to be more direct.

Our analysis of the Wald equations, which combines the Kakutani theorem and the duality theorem of linear programming, follows the analysis of Kuhn [159].

§ 8.8. The setup of the model here is that of Arrow and Debreu, with some small changes. The arguments of this section appear to be new and simpler than the original arguments, although it should be pointed out that the Arrow-Debreu formulation is slightly more general and undoubtedly more realistic.

Lemma 8.8.1 was stated by Gale [93], who appeals to the Knaster-Marcenkiewicz fixed-point theorem for its proof. Direct application of the Kakutani fixed-point theorem gives a much simpler proof.

CHAPTER 9

MATHEMATICAL METHODS IN THE STUDY OF ECONOMIC MODELS (Continued)

In this chapter we examine aspects of welfare economics, the stability of equilibrium price vector systems, and the theory of balanced growth.

Equilibrium models are essentially descriptive rather than normative; their function is simply to describe as accurately as possible the interactions of supply, demand, and prices in a competitive economy. By contrast, the theory of welfare economics is essentially normative; it seeks to determine production and consumption vectors which will yield optimal results from a definite point of view, that of the consumer. Equilibrium theory is concerned with things as they are, welfare theory with things as they might be.

The goal of welfare economics is a system in which all consumers uniformly achieve their maximal utility. When more than one consumer is involved, this rarely happens. In this case we characterize the form of all efficient consumption vectors which are so constituted that the utility of any one consumer cannot be increased without making another consumer worse off. This concept is known as the "Pareto optimum." Certainly, any consumption program should be formulated in terms of a Pareto optimum. However, since there exist a multitude of Pareto optimum consumption vectors, a further criterion must be imposed to single out a specific optimum consumption vector system. One such criterion is noted in Section 9.1.

The mathematics of the theory of welfare economics is in the main a nontrivial application of the vector maximum nonlinear programming problem of Section 7.4. The connections of welfare economics to equilibrium theory are also brought out in Section 9.1.

In the remainder of this chapter we consider various dynamic problems of mathematical economics. In Section 9.2 we formulate several versions of the process by which prices respond to changes from period to period in the excess demand vector functions (i.e., changes in the amount by which total demand exceeds total supply). The excess demand functions are themselves functions of the prices of the previous period. For the equilibrium price vector, the excess demand functions are essentially zero.

When time is considered continuous, the price adjustment process is described by a system of differential equations. The problem of determining whether this process is stable is essentially a matter of determining conditions which cause the price vector as a function of time to converge with increasing time to an equilibrium price vector. Mathematically speaking, the problem is one of determining criteria which ensure the stability of solutions of a special system of differential equations (or difference equations, depending on the model).

The nature of these criteria depends on economic considerations. For example, the concepts of "gross substitutes" and "gross complements" (see pp. 305 and 307) lead to natural conditions satisfied by the excess demand functions; among other things, they permit us to pair off commodities according to the manner in which a rise in the price of one commodity affects the demand for other commodities. The mathematics of the study of the gross substitutes concept involves some intriguing relations for special vector fields of functions (see in particular Section 9.4). Another meaningful economic concept which implies stability of the price adjustment process is the principle of revealed preference.

In Sections 9.2 through 9.5 we investigate the stability structure of the differential-equations formulation of the price adjustment process for a variety of conditions. In Section 9.6 we examine a difference-equation version of the same process. In Sections 9.7 and 9.8 we explore aspects of the theory of local stability of the adjustment process when the excess demand functions depend not only on current prices but also on the so-called "anticipated future prices" of the process. The two versions of stability (with and without the factor of anticipated future prices) are compared.

Sections 9.9 and 9.10 are concerned with a different kind of dynamic process: the growth of production and consumption in successive periods. One fundamental corollary shows that for a suitable hypothesis the rate of expansion is equal to the rate of interest. In Section 9.9 we describe the classical model of an expanding economy formulated by von Neumann. The final section presents a generalization of this model and an analysis of one of its special cases, the Leontief dynamic model.

***9.1 Welfare economics.** A descriptive study of the economy can be characterized mainly by the phenomenon of equilibrium. An equilibrium state is a situation such that with respect to a set of prices no consumer can make himself better off without spending more, and no producer can make a larger profit. A study of welfare economics is mainly concerned with the problem of describing the states of the economy where no consumer can be made better off without making another consumer worse off.

We turn now to a detailed analysis of welfare theory. Commodity bundles are represented by vectors of size r. To the ith consumer ($i = 1, 2, \ldots, m$) corresponds a set X_i of consumption vectors in E^r, and a utility indicator function $U_i(\mathbf{x}_i)$ defined for $\mathbf{x}_i$ in X_i which characterizes

---

* Starred sections are advanced discussions that may be omitted at first reading without loss of continuity.

the ith consumer's preference scale. (The preference relation induced by $U_i(\mathbf{x}_i)$ is assumed to satisfy all the usual properties enumerated in Section 8.6. For convenience of exposition we shall assume further that $U_i(\mathbf{x}_i)$ defines a strictly concave function for $\mathbf{x}_i$ in X_i.) The economy attains perfection when all consumers can simultaneously achieve their maximal utility. We shall see that the equations of a competitive economy generally possess no solution consistent with the ability of all consumers to achieve maximal utility.

A vector system $\{\mathbf{x}_i\}$ ($\mathbf{x}_i \in X_i$) is said to be a Pareto optimum if (a) $\{\mathbf{x}_i\}$ is possible (see below) and (b) there exists no other vector system $\{\mathbf{x}_i'\}$ ($\mathbf{x}_i' \in X_i$) such that

$$U_i\{\mathbf{x}_i'\} \geq U_i(\mathbf{x}_i) \qquad (i = 1, \ldots, m),$$

with strict inequality for at least one consumer. (This concept is in essence the same as the notion of "admissibility" in statistics.)

Our objective in this section is twofold: first, we seek to characterize all Pareto optimum consumption vector systems; second, we examine the relationship between a given Pareto optimum and the existence of a competitive equilibrium (see Section 8.8).

At the close of the section we discuss briefly the problem of how to select a Pareto optimum vector system from the set of all such efficient vector systems.

We first consider the question of when a Pareto optimum is a competitive equilibrium and conversely.

The case of a single consumer. Let Y_j ($j = 1, \ldots, n$) in E^r designate the set of production possibility vectors of firm j. Clearly,

$$Y = \sum_{j=1}^{n} Y_j$$

constitutes the vector of the total production of the economy.

Let X contained in the positive orthant of E^r represent the consumption set available to the consumer, and let $U(\mathbf{x})$ denote his utility index function, which we assume to be a strictly concave function. The total production set Y is assumed to be convex and compact. The consumer desires to select a vector $\mathbf{x}^*$ from X such that

$$U(\mathbf{x}^*) = \max U(\mathbf{x}),$$

where the maximum is extended over all $\mathbf{x}$ in X for which there exist vectors $\mathbf{y}$ in Y such that $\mathbf{x} \leq \mathbf{y}$.

Any vector $\mathbf{x}$ in X for which there exists $\mathbf{y}$ in Y such that $\mathbf{y} \geq \mathbf{x}$ is called feasible or possible. In other words, the consumer acts in a manner to maximize his utility, taking account of the technological possibilities.

Since Y is compact, it is clear that the range of possible $\mathbf{x}$ vectors ($\mathbf{x} \leq \mathbf{y}$) is also compact, and hence the operation max $U(\mathbf{x})$, extended only over possible consumption vectors, is well defined.

We shall also impose the following requirements:

Assumption I. There exist vectors $\mathbf{x}^0$ in X and $\mathbf{y}^0$ in Y such that $\mathbf{y}^0 - \mathbf{x}^0 \gg \mathbf{0}$.

Assumption II. For each $\mathbf{x}$ in X there exists a vector $\mathbf{x}'$ in X such that $U(\mathbf{x}') > U(\mathbf{x})$.

Assumption II states that if we disregard budget restrictions and feasibility, there is no consumption vector which is a point of saturation (see p. 281).

The problem of determining the optimum consumption vector can be put into the form of a concave programming problem as follows.

For $\{\mathbf{x}, \mathbf{y}\}$ consisting of any pair in $X \otimes Y$, we define

$$g(\mathbf{x}, \mathbf{y}) = U(\mathbf{x}).$$

Let $\mathbf{F}(\mathbf{x}, \mathbf{y}) = \langle \mathbf{y}_1 - \mathbf{x}_1, \mathbf{y}_2 - \mathbf{x}_2, \ldots, \mathbf{y}_r - \mathbf{x}_r \rangle$ denote a vector system of functions defined everywhere in $E^r \otimes E^r$. In this notation the problem of selecting a consumption vector of maximal utility becomes

$$\max g(\mathbf{x}, \mathbf{y}) = \max U(\mathbf{x}) \tag{9.1.1}$$

subject to the constraints

$$\mathbf{F}(\mathbf{x}, \mathbf{y}) \geq \mathbf{0} \qquad (\mathbf{x} \in X; \mathbf{y} \in Y). \tag{9.1.2}$$

The definition of a competitive equilibrium in the case of one consumer is as follows: A vector system $\{\mathbf{p}^*, \mathbf{x}^*, \mathbf{y}^*\}$, where $\mathbf{y}^*$ is in Y, $\mathbf{x}^*$ is in X, and $\mathbf{p}^*$ is a relative price vector

$$\left(p_i^* \geq 0, \qquad \sum_{i=1}^{r} p_i^* = 1\right),$$

is a competitive equilibrium if:

$$U(\mathbf{x}^*) = \max_{x \in \hat{X}} U(\mathbf{x}), \tag{9.1.3}$$

where $\hat{X} = \{\mathbf{x} | \mathbf{x} \in X, (\mathbf{p}^*, \mathbf{x}) \leq (\mathbf{p}^*, \mathbf{y}^*)\}$;

$$(\mathbf{p}^*, \mathbf{y}^*) = \max_{y \in Y} (\mathbf{p}^*, \mathbf{y}); \tag{9.1.4}$$

and

$$\mathbf{x}^* \leq \mathbf{y}^*, \qquad (\mathbf{p}^*, \mathbf{x}^* - \mathbf{y}^*) = 0. \tag{9.1.5}$$

The key theorem relating competitive equilibrium and consumption of maximal utility is as follows.

▶THEOREM 9.1.1. Let assumptions I and II be satisfied. A vector $\mathbf{x}^*$ maximizes $U(\mathbf{x})$ among all feasible vectors if and only if there exist a vector $\mathbf{y}^*$ in Y and a price vector $\mathbf{p}^*$ such that $\{\mathbf{p}^*, \mathbf{x}^*, \mathbf{y}^*\}$ is a competitive equilibrium.

Proof. (The proof of this theorem is an application of Theorem 7.1.1.) Suppose that $\mathbf{x}^*$ maximizes $U(\mathbf{x})$ among all feasible vectors. Choose any $\mathbf{y}^* \in Y$ such that $\mathbf{y}^* \geq \mathbf{x}^*$. Then $\{\mathbf{x}^*, \mathbf{y}^*\}$ solves the problem of (9.1.1) and (9.1.2). Moreover, assumption I ensures that the extra hypothesis of Theorem 7.1.1 is satisfied. Consequently, by virtue of Theorem 7.1.1, the Lagrangian form

$$\phi(\mathbf{x}, \mathbf{y}; \mathbf{u}) = g(\mathbf{x}, \mathbf{y}) + (\mathbf{u}, \mathbf{F}(\mathbf{x}, \mathbf{y})) = U(\mathbf{x}) + (\mathbf{u}, \mathbf{F}(\mathbf{x}, \mathbf{y})),$$

where $\{\mathbf{x}, \mathbf{y}\}$ traverses $X \otimes Y$ and $\mathbf{u} \geq \mathbf{0}$ has a saddle point $\{\mathbf{x}^*, \mathbf{y}^*; \mathbf{u}^0\}$ such that

$$U(\mathbf{x}) + (\mathbf{u}^0, \mathbf{F}(\mathbf{x}, \mathbf{y})) \leq U(\mathbf{x}^*) + (\mathbf{u}^0, \mathbf{F}(\mathbf{x}^*, \mathbf{y}^*)) \leq U(\mathbf{x}^*) + (\mathbf{u}, \mathbf{F}(\mathbf{x}^*, \mathbf{y}^*)) \tag{9.1.6}$$

for all $\mathbf{x}, \mathbf{y}$ in $X \otimes Y$ and $\mathbf{u} \geq \mathbf{0}$. Furthermore, the same theorem indicates that

$$\mathbf{F}(\mathbf{x}^*, \mathbf{y}^*) \geq \mathbf{0}, \qquad (\mathbf{u}^0, \mathbf{F}(\mathbf{x}^*, \mathbf{y}^*)) = 0. \tag{9.1.7}$$

We now deduce that $\mathbf{u}^0 > \mathbf{0}$, since otherwise the left-hand inequality of (9.1.6) contradicts assumption II. Let $\mathbf{p}^* = \lambda \mathbf{u}^0$, where

$$\frac{1}{\lambda} = \sum_{i=1}^{n} u_i^0.$$

We now establish that $\{\mathbf{x}^*, \mathbf{y}^*, \mathbf{p}^*\}$ constitutes a competitive equilibrium. First, note that (9.1.7) implies (9.1.5). Since

$$(\mathbf{u}^0, \mathbf{F}(\mathbf{x}, \mathbf{y})) = \frac{1}{\lambda} [(\mathbf{p}^*, \mathbf{y}) - (\mathbf{p}^*, \mathbf{x})],$$

we find with reference to (9.1.7) and the left-hand inequality of (9.1.6) that $U(\mathbf{x}^*) \geq U(\mathbf{x})$ whenever $\mathbf{x} \in \hat{X}$ in accordance with (9.1.3). Again, using the left-hand inequality of (9.1.6) for the special choice of $\{\mathbf{x}, \mathbf{y}\} = \{\mathbf{x}^*, \mathbf{y}\}$, with $\mathbf{y}$ arbitrary in Y, we arrive at

$$(\mathbf{u}^0, \mathbf{y}) \leq (\mathbf{u}^0, \mathbf{y}^*),$$

which is equivalent to (9.1.4). The first half of the theorem has now been proved.

Suppose, conversely, that $\{\mathbf{p}^*, \mathbf{x}^*, \mathbf{y}^*\}$ defines a competitive equilibrium in the sense of (9.1.3–5). We must prove that $U(\mathbf{x}^*) = \max U(\mathbf{x})$ for all feasible $\mathbf{x}$. It is clear that for every feasible $\mathbf{x}$ there exists some $\mathbf{y}$ in Y such that $\mathbf{y} \geq \mathbf{x}$. Consequently, $(\mathbf{p}^*, \mathbf{y}) \geq (\mathbf{p}^*, \mathbf{x})$. Also $(\mathbf{p}^*, \mathbf{y}^*) \geq (\mathbf{p}^*, \mathbf{x})$ by (9.1.4); and hence, by (9.1.3), $U(\mathbf{x}^*) \geq U(\mathbf{x})$ as desired. This completes the proof.

The case of many consumers. Let X_i represent the consumption set in E^r available to consumer i ($i = 1, 2, \ldots, m$), and let

$$X = \sum_{i=1}^{m} X_i$$

designate the cumulative consumption set of all consumers. The preference ordering of consumer i is described by a strictly concave utility function $U_i(\mathbf{x}_i)$. Each set X_i is assumed to be convex and closed, and lies in the nonnegative orthant of E^r. The situation of the production possibility vectors is the same as in the case of a single consumer.

Again we seek to determine conditions which imply that the Pareto optimum is an element of a competitive equilibrium. A system of vectors $\{\mathbf{x}_1^*, \mathbf{x}_2^*, \ldots, \mathbf{x}_m^*; \mathbf{y}_1^*, \mathbf{y}_2^*, \ldots, \mathbf{y}_n^*; \mathbf{p}^*\}$ is said to be a competitive equilibrium if $\mathbf{p}^*$ is a relative price vector

$$\left(p_i^* \geq 0, \quad \sum_{i=1}^{r} p_i^* = 1\right)$$

such that

$$(\mathbf{p}^*, \mathbf{y}_j^*) = \max_{y_j \in Y_j} (\mathbf{p}^*, \mathbf{y}_j); \tag{9.1.8}$$

$$U_i(\mathbf{x}_i^*) = \max_{x_i \in \hat{X}_i} U_i(\mathbf{x}_i), \tag{9.1.9}$$

where

$$\hat{X}_i = \left\{\mathbf{x}_i | \mathbf{x}_i \in X_i, (\mathbf{p}^*, \mathbf{x}_i) \leq \sum_{j=1}^{n} \alpha_{ij}(\mathbf{p}^*, \mathbf{y}_j^*)\right\};^*$$

and

$$\mathbf{x}^* \leq \mathbf{y}^*, \qquad (\mathbf{p}^*, \mathbf{x}^* - \mathbf{y}^*) = 0, \tag{9.1.10}$$

---

* The term α_{ij} is the share of the profit that is allotted to the ith consumer; it is assumed to satisfy the conditions

$$\sum_{i=1}^{m} \alpha_{ij} = 1 \qquad (j = 1, \ldots, m), \qquad \alpha_{ij} \geq 0.$$

For the full economic interpretation of α_{ij}, see Section 8.8.

where

$$\mathbf{x}^* = \sum_{i=1}^{m} \mathbf{x}_i^*, \qquad \mathbf{y}^* = \sum_{j=1}^{n} \mathbf{y}_j^*.$$

We may formulate the problem of characterizing Pareto optima as a vector maximum problem in the sense of Theorem 7.4.1. Explicitly, we define for each $\{\mathbf{x}_1, \ldots, \mathbf{x}_m; \mathbf{y}_1, \ldots, \mathbf{y}_n\}$ $(\mathbf{x}_i \in X_i;\ \mathbf{y}_j \in Y_j)$ the vector function

$$\mathbf{G}(\mathbf{x}_1, \ldots, \mathbf{x}_m; \mathbf{y}_1, \ldots, \mathbf{y}_n) = \{U_i(\mathbf{x}_i)\} = \mathbf{U}(\mathbf{x}). \tag{9.1.11}$$

Let

$$\mathbf{F}(\mathbf{x}_1, \ldots, \mathbf{x}_m; \mathbf{y}_1, \ldots, \mathbf{y}_n) = \mathbf{y} - \mathbf{x}, \tag{9.1.12}$$

where

$$\mathbf{y} = \sum_{j=1}^{n} \mathbf{y}_j, \qquad \mathbf{x} = \sum_{i=1}^{m} \mathbf{x}_i.$$

In accordance with the terminology of Section 7.4 the characterization of Pareto optima is equivalent to characterizing all efficient points of $\{U_i(\mathbf{x}_i)\}$ subject to the constraints $\mathbf{F} \geq \mathbf{0}$, $\mathbf{x}_i \in X_i$, and $\mathbf{y}_j \in Y_j$.

Again assumptions I and II hold, with the second extended in the natural manner to hold for each component U_i.

The relations between Pareto optimum vector systems and competitive equilibrium are derived from the statements of Theorem 7.4.1. The following theorem states these relations more precisely.

▶THEOREM 9.1.2. Let assumptions I and II be fulfilled. Let

$$\{\mathbf{x}_1^*, \mathbf{x}_2^*, \ldots, \mathbf{x}_m^*\}$$

be a Pareto optimum such that if $\mathbf{y}^* \geq \mathbf{x}^*$ there exists a consumption vector system $\{\tilde{\mathbf{x}}_i\}$ for each i_0, where $\tilde{\mathbf{x}}_i = \mathbf{x}_i^*$, $i \neq i_0$, and there is some $\tilde{\mathbf{y}}$ in Y such that $\tilde{\mathbf{y}} - \tilde{\mathbf{x}} \gg \mathbf{0}$. Then there exist vectors $\mathbf{y}_1^*, \mathbf{y}_2^*, \ldots, \mathbf{y}_n^*$ $(\mathbf{y}_i^* \in Y_i)$, a price vector $\mathbf{p}^*$, and a matrix $\|\alpha_{ij}^*\|$

$$\left(\alpha_{ij}^* \geq 0, \qquad \sum_{i=1}^{m} \alpha_{ij}^* = 1 \quad \text{for all } j\right)$$

such that

$$\left\{\mathbf{x}_1^*, \mathbf{x}_2^*, \ldots, \mathbf{x}_m^*;\ \ \mathbf{y}_1^*, \mathbf{y}_2^*, \ldots, \mathbf{y}_n^*;\ \ \mathbf{p}^*, \alpha_{ij}^*\right\}$$

is a competitive equilibrium.

Proof. Suppose that $\{\mathbf{x}_1^*, \ldots, \mathbf{x}_m^*\}$ is a Pareto optimum. Let $\mathbf{y}_1^*, \ldots, \mathbf{y}_n^*$ designate the respective terms of Y_i, giving

$$\mathbf{y}^* = \sum_{i=1}^{n} \mathbf{y}_i^*,$$

which is selected such that

$$\mathbf{y}^* \geq \mathbf{x}^* = \sum_{i=1}^{m} \mathbf{x}_i^*.$$

Clearly $\{\mathbf{x}_1^*, \ldots, \mathbf{x}_m^*\}$ is an efficient point of $\mathbf{G}$ [equation (9.1.11)] among all vectors $\mathbf{x}_i$, $\mathbf{y}_j$ satisfying the constraints $\mathbf{F} = \mathbf{y} - \mathbf{x} \geq \mathbf{0}$, $\mathbf{x}_i \in X_i$, and $\mathbf{y}_j \in Y_j$. Because of assumption I the hypothesis of Theorem 7.4.1 is satisfied. By Theorem 7.4.1 there exist vectors $\mathbf{u}^0$ and $\mathbf{v}^0$ ($\mathbf{u}^0 \geq \mathbf{0}$; $\mathbf{v}^0 > 0$) such that

$$(\mathbf{v}^0, \mathbf{U}(\mathbf{x})) + (\mathbf{u}^0, \mathbf{y} - \mathbf{x}) \leq (\mathbf{v}^0, \mathbf{U}(\mathbf{x}^*)) + (\mathbf{u}^0, \mathbf{y}^* - \mathbf{x}^*)$$
$$\leq (\mathbf{v}^0, \mathbf{U}(\mathbf{x}^*)) + (\mathbf{u}, \mathbf{y}^* - \mathbf{x}^*) \qquad (9.1.13)$$

for all $\mathbf{u} \geq \mathbf{0}$ and $\mathbf{x}_i \in X_i$, $\mathbf{y}_j \in Y_j$. Since $\mathbf{y}^* \geq \mathbf{x}^*$, we conclude from the right-hand inequality that $(\mathbf{u}^0, \mathbf{y}^* - \mathbf{x}^*) = 0$ (see the proof of Theorem 7.1.1). Next, observe that $\mathbf{u}^0 > \mathbf{0}$, since if $\mathbf{u}^0 = \mathbf{0}$ and $\mathbf{v}^0 > \mathbf{0}$ the left-hand inequality would be incompatible with assumption II. Finally, we observe that $\mathbf{v}^0 \gg \mathbf{0}$. For suppose to the contrary that $v_{i_0}^0 = 0$; then in the left-hand inequality we may assign $\tilde{\mathbf{x}}_i = \mathbf{x}_i^*$ for $i \neq i_0$ (as guaranteed by the hypothesis) such that $\tilde{\mathbf{x}} \ll \tilde{\mathbf{y}}$. Inserting these choices of $\mathbf{x}$ and $\mathbf{y}$ into (9.1.13), we see plainly that the left-hand inequality is contradicted.

Define

$$\mathbf{p}^* = \frac{1}{\sum_{i=1}^{r} u_i^0}\, \mathbf{u}^0$$

and $\alpha_{ij}^* = a_i$, where

$$a_i = \frac{(\mathbf{p}^*, \mathbf{x}_i^*)}{(\mathbf{p}^*, \mathbf{y}^*)}.$$

Note that

$$a_i \geq 0, \qquad \sum_{i=1}^{m} a_i = 1,$$

since

$$\sum_{i=1}^{m} (\mathbf{p}^*, \mathbf{x}_i^*) = (\mathbf{p}^*, \mathbf{y}^*), \qquad (\mathbf{p}^*, \mathbf{x}_i) \geq 0.$$

We are now prepared to establish the claim that

$$\{\mathbf{x}_1^*, \mathbf{x}_2^*, \ldots, \mathbf{x}_m^*; \quad \mathbf{y}_1^*, \mathbf{y}_2^*, \ldots, \mathbf{y}_n^*; \quad \mathbf{p}^*, \alpha_{ij}^*\}$$

is a competitive equilibrium.

Equation (9.1.10) follows immediately from

$$\mathbf{y}^* \geq \mathbf{x}^* \qquad \text{and} \qquad (\mathbf{u}^0, \mathbf{y}^* - \mathbf{x}^*) = 0.$$

Next, set $\mathbf{x}_i = \mathbf{x}_i^*$ for all i, and $\mathbf{y}_j = \mathbf{y}_j^*$ for $j \neq j_0$. Then the left-hand inequality of (9.1.13) yields

$$(\mathbf{u}^0, \mathbf{y}_{j_0}) \leq (\mathbf{u}^0, \mathbf{y}_{j_0}^*),$$

which gives (9.1.8) since j_0 is arbitrary. Finally, set $\mathbf{y}_j = \mathbf{y}_j^*$ for all j and $\mathbf{x}_i = \mathbf{x}_i^*$ for $i \neq i_0$, and choose $\mathbf{x}_{i_0}$ in $\hat{X}_{i_0}$ [equation (9.1.9)]. The left-hand inequality of (9.1.13) then reduces to

$$v_{i0}^0 U_{i0}(\mathbf{x}_{i0}) + \left(\sum_{i=1}^{r} u_i^0\right)(\mathbf{p}^*, \mathbf{y}^* - \mathbf{x}) \leq v_{i0}^0 U_{i0}(\mathbf{x}_{i0}^*). \qquad (9.1.14)$$

Now $\mathbf{x}_i^*$ belongs to $\hat{X}_i$, by the construction of α_{ij}^*. Then,

$$(\mathbf{p}^*, \mathbf{x}_i^*) = a_i(\mathbf{p}^*, \mathbf{y}^*) \qquad (i \neq i_0),$$
$$(\mathbf{p}^*, \mathbf{x}_{i_0}) \leq a_{i_0}(\mathbf{p}^*, \mathbf{y}^*),$$

and by summation

$$(\mathbf{p}^*, \mathbf{x}) \leq (\mathbf{p}^*, \mathbf{y}^*).$$

From this observation, combined with $v_{i_0}^0 > 0$, we deduce that $U_{i_0}(\mathbf{x}_{i_0}) \leq U_{i_0}(\mathbf{x}_{i_0}^*)$ for $\mathbf{x}_{i_0}$ in $\hat{X}_{i_0}$, the statement of (9.1.9). This completes the proof of the theorem.

The restriction stated in the theorem was used in the preceding proof only in showing that $v_i^0 > 0$ for all i. The reader will note that all conclusions of Theorem 9.1.2 follow whenever we can guarantee that $\mathbf{v}^0 \gg \mathbf{0}$. In this connection we refer to the discussion of Section 7.4, where alternative conditions are indicated which yield $\mathbf{v}^0 \gg \mathbf{0}$.

We turn to the converse of Theorem 9.1.2.

▶ THEOREM 9.1.3. If $\{\mathbf{x}_1^*, \ldots, \mathbf{x}_m^*; \mathbf{y}_1^*, \ldots, \mathbf{y}_n^*; \mathbf{p}^*, \alpha_{ij}^*\}$ is a competitive equilibrium, then $\{\mathbf{x}_1^*, \ldots, \mathbf{x}_m^*\}$ is a Pareto optimum.

Proof. Suppose to the contrary that there exists a feasible vector system $\{\mathbf{x}_i'\}$ such that $U_i(\mathbf{x}_i') \geq U_i(\mathbf{x}_i^*)$ $(i = 1, \ldots, m)$, with inequality for at least one i. Let us arrange the indices such that

$$\mathbf{x}_i' = \mathbf{x}_i^* \quad (i = 1, \ldots, s), \qquad \mathbf{x}_i' \neq \mathbf{x}_i^* \quad (i = s+1, \ldots, m).$$

Clearly $s < m$, since $\mathbf{x}_i'$ dominates $\mathbf{x}_i^*$ in the sense of utility by assumption. Since U_i is strictly concave, it follows that

$$\tilde{\mathbf{x}}_i = t\mathbf{x}_i' + (1-t)\mathbf{x}_i^* \qquad (i = 1, \ldots, m; 0 < t < 1)$$

has

$$U_i(\tilde{\mathbf{x}}_i) > U_i(\mathbf{x}_i^*) \qquad (i = s+1, \ldots, m).$$

Moreover, $\tilde{\mathbf{x}}_i$ is feasible, and hence there exists $\tilde{\mathbf{y}}_j$ such that

$$\tilde{\mathbf{y}} = \sum_{j=1}^{n} \tilde{\mathbf{y}}_j \geq \tilde{\mathbf{x}} = \sum_{i=1}^{m} \tilde{\mathbf{x}}_i.$$

Therefore $(\mathbf{p}^*, \tilde{\mathbf{y}}) \geq (\mathbf{p}^*, \tilde{\mathbf{x}})$, and by (9.1.8)

$$(\mathbf{p}^*, \mathbf{y}^*) \geq (\mathbf{p}^*, \tilde{\mathbf{x}}). \tag{9.1.15}$$

Taking into account $\tilde{\mathbf{x}}_i = \mathbf{x}_i^*$ for $i = 1, \ldots, s$ and (9.1.15), we infer the existence of i_0 $(m \geq i_0 \geq s + 1)$ such that

$$(\mathbf{p}^*, \tilde{\mathbf{x}}_{i_0}) \leq \sum_{j=1}^{n} \alpha_{i_0 j}(\mathbf{p}^*, \mathbf{y}_j^*).$$

However, $U_{i_0}(\tilde{\mathbf{x}}_{i_0}) > U_{i_0}(\mathbf{x}_{i_0}^*)$, in contradiction to (9.1.9). This completes the proof.

Now that we have essentially characterized the Pareto optimum consumption vector systems, the problem remains of how to single out a specific Pareto optimum. To do this, it is necessary to impose some further criterion. Many such criteria have been proposed. One of them, known as the principle of equity, amounts to assigning equal weights $\lambda_i = 1/m$ and determining $\mathbf{x}_i \in X_i$ such that

$$\sum_{i=1}^{m} \lambda_i U_i(\mathbf{x}_i)$$

is maximized, where the $\mathbf{x}_i$ are subject to the constraints $\mathbf{x}_i \in X_i$ and there exists some $\mathbf{y}$ in Y for which

$$\mathbf{x} = \sum_{i=1}^{m} \mathbf{x}_i \leq \mathbf{y}.$$

The $\mathbf{x}_i^*$ so obtained obviously constitutes a Pareto optimum. The economic merits of such a selection process have been argued cogently by Edgeworth and Marshall, among others.

In general, selecting a Pareto optimum represents a formidable philosophical problem very much akin to the task of justifiably specifying an admissible statistical procedure from a complete class of such procedures.

9.2 The stability of a competitive equilibrium. In the previous chapter and the preceding section, we were concerned primarily with investigating the existence, uniqueness, and optimality of a competitive equilibrium. These concepts are static. Of equal importance are the dynamics of equilibrium, and particularly the concept of stability. In this section we shall describe a formal dynamic model whose characteristics reflect the nature of the competitive process, and examine its stability properties in the light of certain assumptions as to the properties of the individual units or of the aggregate excess demand functions.

To begin with, let us assume as before that there exist m consumers with available consumption sets X_i in E^{r+1} $(i = 1, \ldots, m)$.† Consumer i seeks to choose a possible consumption vector $\mathbf{x}_i$ from X_i which maximizes his utility (see p. 267). Let $\mathbf{x} = \sum \mathbf{x}_i$ denote an aggregate demand consumption vector.

Let Y represent the aggregate production sector of commodity bundles in E^{r+1}. Each of the production units (firms) seeks to produce in a manner which maximizes profits. Consequently, with any price vector $\mathbf{p} = \langle p_0, p_1, \ldots, p_r \rangle$ of the commodity bundle there are associated an aggregate demand vector $\mathbf{x}(\mathbf{p})$ and an aggregate production vector $\mathbf{y}(\mathbf{p})$, determined according to conditions (B) and (C) of p. 281, respectively, which possess the desired maximality characteristics. Subject to the usual conditions underlying equilibrium (primarily the dictum of unsaturated demand),

$$(\mathbf{p}, \mathbf{x}(\mathbf{p}) - \mathbf{y}(\mathbf{p})) = (\mathbf{p}, \mathbf{F}(\mathbf{p})) = 0 \tag{9.2.1}$$

for each price vector $\mathbf{p}$. For the derivation of (9.2.1) the reader should consult p. 284. We shall refer to $\mathbf{F}(\mathbf{p})$ as the (aggregate) excess demand vector function for the price vector $\mathbf{p}$. The excess demand vector for the ith consumer is denoted by $\mathbf{F}^i(\mathbf{p})$;

$$\mathbf{F}(\mathbf{p}) = \sum_{i=1}^{m} \mathbf{F}^i(\mathbf{p}).$$

It follows from the same considerations that $(\mathbf{p}, \mathbf{F}^i(\mathbf{p})) = 0$.

Throughout what follows we assume that $\mathbf{F}(\mathbf{p})$ is continuous-differentiable and single-valued for $\mathbf{p} > \mathbf{0}$. Relation (9.2.1) is classically known as the Walras law. We emphasize that it operates universally, no matter what prices prevail in the market.

An equilibrium price vector $\mathbf{p}^*$ is distinguished further by the property that

$$\mathbf{F}(\mathbf{p}^*) \leq \mathbf{0}, \tag{9.2.2}$$

which ensures that the consumption vector $\mathbf{x}(\mathbf{p}^*)$ can be achieved by the production vector $\mathbf{y}(\mathbf{p}^*)$.

For mathematical convenience and to help clarify the dynamic price adjustment process, we shall assume in what follows that a strictly positive

---

† We are considering a commodity vector with $r + 1$ components instead of r in order to be able to identify the zeroth component as the numéraire whenever it is important to distinguish it from the others.

equilibrium price vector $\mathbf{p}^*$ exists, i.e., that $\mathbf{p}^* \gg \mathbf{0}$. This implies, in view of (9.2.1) and (9.2.2), that $\mathbf{F}(\mathbf{p}^*) = \mathbf{0}$.

We shall also assume that the excess demand function $\mathbf{F}(\mathbf{p})$ is a positively homogeneous function of degree zero, i.e.,

$$\mathbf{F}(\lambda p_0, \lambda p_1, \ldots, \lambda p_r) = \mathbf{F}(p_0, p_1, \ldots, p_r) \qquad (\lambda > 0). \qquad (9.2.3)$$

This is the natural assumption that only the relative prices of the various commodities are relevant in influencing fluctuations of the excess demand functions.

We are now prepared to propose a dynamic model which describes how prices adjust over time to variations of supply and demand in a competitive economy.

One such dynamic model is based on the system of differential equations

$$\frac{dp_j}{dt} = F_j(p_0, p_1, \ldots, p_r) \qquad (j = 0, 1, \ldots, r), \qquad (9.2.4)$$

or in vector form

$$\frac{d\mathbf{p}}{dt} = \mathbf{F}(\mathbf{p}). \qquad (9.2.5)$$

For such a system the price of a commodity increases (decreases) when the excess demand for that commodity is positive (negative). The rate of increase or decrease is proportional to the amount of the excess demand or supply of each commodity. This appealing functional relationship between supply, demand, and prices is often postulated and represents the simplest and most direct characterization of the price adjustment process.

By a suitable choice of units of measurement for each of the commodities we can convert (9.2.4) into

$$\frac{dp_j}{dt} = k_j F_j \qquad (j = 0, 1, \ldots, r),$$

with the proportionality constants k_j positive and arbitrary. Thus there is no loss of generality in supposing $k_j = 1$.

A discrete-time analog of the dynamic system (9.2.4) is

$$p_j(t + 1) = p_j(t) + \rho F_j[p_0(t), \ldots, p_r(t)] \qquad (j = 0, 1, \ldots, r). \qquad (9.2.6)$$

In view of the physical interpretation of the model, it is necessary for accuracy to indicate some restrictions in (9.2.5) and (9.2.6) which ensure that $\mathbf{p}(t)$ is a nonnegative vector for all t.

In the case of discrete time we modify (9.2.6) to the system of equations

$$\mathbf{p}(t + 1) = \max \{0, \mathbf{p}(t) + \rho\mathbf{F}[\mathbf{p}(t)]\}, \qquad (9.2.7)$$

thus preserving the positivity of $\mathbf{p}(t)$ for all t while still reflecting the dynamics of the price adjustment process. We postpone the analysis of (9.2.7) until Section 9.6.

For the case of continuous time we shall not bother to check whether $\mathbf{p}(t)$ remains a positive vector at all times. The independent reader should be able to check this condition without too much difficulty.

We shall have occasion to study stability for two systems emerging from (9.2.5): (1) the normalized price adjustment process, (2) the nonnormalized price adjustment process. In the normalized process a single commodity (say the zeroth) is distinguished, and the price vector is normalized so that $\mathbf{q} = \langle 1, q_1, \ldots, q_r \rangle$. The commodity singled out is frequently referred to as the numéraire. For the normalized process we consider the dynamics in terms of the relative prices $q_j = p_j/p_0$ $(j = 1, \ldots, r)$. Here we reformulate the price adjustment process and postulate that again it is described by a system of differential equations

$$\frac{dq_j}{dt} = G_j(\mathbf{q}) \qquad (j = 1, \ldots, r), \tag{9.2.8}$$

where $G_j(\mathbf{q}) = F_j(1, q_1, \ldots, q_r)$ and F_j has the same meaning as previously.

The nonnormalized adjustment process treats all commodities symmetrically, and the dynamics are expressed by means of (9.2.5).

Stability criteria will be established simultaneously for the normalized and nonnormalized processes. From a mathematical point of view the nonnormalized process is complicated by the multitude of equilibrium price vectors, since if $\mathbf{p}^*$ is an equilibrium point, all points on the ray $\lambda\mathbf{p}^*$ $(\lambda > 0)$ are equilibrium points. However, this model is the more appealing from an economic point of view because of its symmetry. The normalized process is more meaningful in many special situations, notably those in which the commodity "money" is not a free commodity and may therefore be distinguished as the numéraire.

A solution of (9.2.5) is denoted by $\mathbf{p}(t) = \psi(t, \mathbf{p}^0)$, where $\mathbf{p}^0$ designates the initial price vector. The excess demand functions $F_j(\mathbf{p})$ are henceforth assumed to possess enough smoothness properties to guarantee that the solution $\psi(t, \mathbf{p}^0)$ is uniquely determined by, and changes continuously with, the initial price vector $\mathbf{p}^0$. A solution of (9.2.8) is denoted by $\mathbf{q}(t) = \phi(t, \mathbf{q}^0)$. Here we make similar assumptions regarding the uniqueness and continuity of solutions. Clearly, any strictly positive equilibrium vector

$$\mathbf{p}^*(t) = \psi(t, \mathbf{p}^*) = \mathbf{p}^*$$

is a fixed-point solution. With this background, the main problem of Sections 9.2 through 9.5 can be stated as follows: We seek to determine criteria which imply that any solution $\psi(t, \mathbf{p}^0)$ of (9.2.5) or solution $\phi(t, \mathbf{p}^0)$ of (9.2.8) converges to an equilibrium vector.

Stability phenomena can be investigated on two levels. First, we can determine conditions for local stability, i.e., conditions implying convergence to the equilibrium vector $\mathbf{p}^*$ when the initial price vector $\mathbf{p}^0$ is sufficiently close to $\mathbf{p}^*$. Second, and more significantly, we can seek to determine conditions for global stability, i.e., conditions implying convergence to an equilibrium price vector regardless of the nature of the initial price vector $\mathbf{p}^0$.

At this point it seems worth while to summarize and discuss briefly the interpretation of various conditions which yield stability, reserving the detailed analysis for later sections.

Let

$$\mathbf{A}(\mathbf{p}) = \left\| \frac{\partial F_i}{\partial p_j} \right\|_{i,j=0,\ldots,r} \quad \text{and} \quad \mathbf{B}(\mathbf{q}) = \left\| \frac{\partial G_i}{\partial q_j} \right\|_{i,j=1,\ldots,r}. \tag{9.2.9}$$

We say that gross substitutability (strictly) prevails for the nonnormalized process if

$$\frac{\partial F_i}{\partial p_j} \geq 0 \quad (> 0) \qquad \text{for every} \quad i \neq j. \tag{9.2.10}$$

An analogous definition applies to the normalized process. Relation (9.2.10) has the property of gross substitutability for each $\mathbf{p}$ if and only if $\mathbf{A}(\mathbf{p})$ is an M-matrix (see p. 255). The economic justification underlying this property is as follows. If the price of the jth commodity rises while the prices of the other commodities remain unchanged, then an increase in the demand for every other commodity may be expected and hence (9.2.10) holds. Commodities related in this way are called "substitutes." This relationship is not always valid: if the price of butter goes up, for example, it may well happen that the demand for bread decreases, since many people cannot enjoy bread without butter.

Nevertheless, it is of some interest to study the price adjustment process in the situation of gross substitutability. Of particular interest is the principal theorem of Section 9.4, which states that if strict gross substitutability prevails for either the normalized or the nonnormalized process, and $\mathbf{p}^*$ is a strictly positive equilibrium vector, then the process is globally stable.

Another proposition of some economic interest is that the normalized process (9.2.8) is globally stable whenever $\mathbf{B}(\mathbf{q})$ defines a real negative quasi-definite matrix. (A real matrix $\mathbf{C}$ is said to be negative quasi-definite if $\mathbf{C} + \mathbf{C}'$ is negative definite.) The fact that $\mathbf{B}(\mathbf{q})$ is a negative quasi-definite matrix is closely related to the concept of revealed preference [see Section 8.6 and equation (8.6.4)].

Other detailed results concerning stability are presented in the following sections.

9.3 Local stability. *Normalized process.* In studying local stability, we may replace the system (9.2.8) by a system of linear differential equations. Explicitly, by expanding the right-hand side of (9.2.8) about the equilibrium point $\mathbf{q} = \mathbf{q}^*$ (throughout this section, unless there is an explicit statement to the contrary, $\mathbf{q}^*$ is assumed to be $\gg \mathbf{0}$) we obtain

$$\frac{d(\mathbf{q} - \mathbf{q}^*)}{dt} = \frac{d\mathbf{q}}{dt} \sim \mathbf{G}(\mathbf{q}^*) + \mathbf{B}(\mathbf{q}^*)(\mathbf{q} - \mathbf{q}^*) = \mathbf{B}(\mathbf{q}^*)(\mathbf{q} - \mathbf{q}^*), \tag{9.3.1}$$

where

$$\mathbf{B}(\mathbf{q}) = \left\| \frac{\partial G_i(\mathbf{q})}{\partial q_j} \right\|_{i,j=1,\ldots,r.}$$

Local stability for the system (9.2.8) is obviously equivalent to the stability of the linear differential system

$$\frac{d\mathbf{z}}{dt} = \mathbf{C}\mathbf{z}, \tag{9.3.2}$$

where $\mathbf{C} = \mathbf{B}(\mathbf{q}^*)$, which in this circumstance reduces to the proposition that all solutions converge to zero. Clearly, the system (9.3.2) will be stable if and only if all characteristic roots of $\mathbf{C}$ have negative real parts, since every solution of (9.3.2) is a linear combination of exponential functions $e^{\lambda_i t}$, where the λ_i represent eigenvalues of $\mathbf{C}$. A matrix $\mathbf{C}$ whose eigenvalues all have negative real parts is called a stable matrix.

We now indicate a series of conditions which imply local stability. The hypotheses all possess some economic relevance.

▶THEOREM 9.3.1. If $\mathbf{B}(\mathbf{q}^*)$ is an M-matrix and either (a) $\det \mathbf{B}(\mathbf{q}^*) \neq 0$ and $\partial G_0/\partial q_i \geq 0$ or (b) $\partial G_0/\partial q_i > 0$ $(i = 1, 2, \ldots, r)$, then the normalized adjustment process (9.2.8) is locally stable.

(As already noted, the first assumption is synonymous with the statement that the normalized process locally (at $\mathbf{q}^*$) has the property of gross substitutability.)

Proof. Differentiation of the Walras law [equation (9.2.1)] and evaluation at $\mathbf{q}^*$ gives

$$\frac{\partial G_0(\mathbf{q}^*)}{\partial q_i} + \sum_{j=1}^{r} q_j^* \frac{\partial G_j(\mathbf{q}^*)}{\partial q_i} = 0. \tag{9.3.3}$$

Hence, by the hypothesis of our theorem, $\mathbf{q}^*\mathbf{C} \leq \mathbf{0}$ and by assumption $\mathbf{q}^* \gg \mathbf{0}$. Since $\mathbf{C}$ is an M-matrix, it is enough to show that $\sigma(\mathbf{C}) < 0$, where $\sigma(\mathbf{C}) = \max \mathcal{R}(\lambda_i)$ (λ_i are the eigenvalues of $\mathbf{C}$). By adding a positive multiple of the identity we may suppose that $\mathbf{D} = \mathbf{C} + \mu\mathbf{I}$ is a nonnegative matrix. Then

$$\mathbf{q}^*\mathbf{D} \leq \mu\mathbf{q}^*.$$

To complete the proof of the theorem it is necessary to verify that $\lambda_0(\mathbf{D}) < \mu$, where $\lambda_0(\mathbf{D})$ is the spectral radius of $\mathbf{D}$.

Let λ_0 denote the largest positive eigenvalue of $\mathbf{D}$, and suppose that $\lambda_0 \geq \mu$. In case (a), $\lambda_0 > \mu$. From the characterization of λ_0 as given in Corollary 8.2.1, there exists a vector $\mathbf{x} > \mathbf{0}$ such that $\mathbf{Dx} > (\lambda_0 - \epsilon)\mathbf{x}$ for $\lambda_0 - \epsilon > \mu$ and ϵ small enough. But since $\mathbf{q}^* \gg \mathbf{0}$, it follows that

$$\mu(\mathbf{q}^*, \mathbf{x}) \geq (\mathbf{q}^*\mathbf{D}, \mathbf{x}) = (\mathbf{q}^*, \mathbf{Dx}) > (\lambda_0 - \epsilon)\,(\mathbf{q}^*, \mathbf{x}),$$

which is impossible.

In case (b), by (9.3.3)

$$\mathbf{q}^*\mathbf{D} \ll \mu\mathbf{q}^*.$$

Invoking the result of Corollary 8.2.2, we conclude that $\lambda_0(\mathbf{D}) < \mu$ as desired. The proof of Theorem 9.3.1 is complete.

A further condition of the same kind which guarantees local stability is as follows.

▶ THEOREM 9.3.2. Let $\mathbf{T} = \mathbf{B}(\mathbf{q}^*) + \mu\mathbf{I}$ be a positive matrix such that some power of $\mathbf{T}$ is strictly positive. If $\partial G_0/\partial q_i \geq 0$ $(i = 1, 2, \ldots, r)$, with strict inequality for some i, then $\mathbf{B}(\mathbf{q}^*)$ is stable.

(Note that the hypothesis is satisfied if $\mathbf{B}(\mathbf{q}^*)$ has the property of strict gross substitutability.)

Proof. Relation (9.3.3) and the hypothesis together yield $\mathbf{q}^*\mathbf{B}(\mathbf{q}^*) < \mathbf{0}$, and hence $\mathbf{q}^*\mathbf{T} < \mu\mathbf{q}^*$. Some iterate of $\mathbf{T}$ has $\mathbf{q}^*\mathbf{T}^k \ll \mu^k\mathbf{q}^*$. Consequently, by Corollary 8.2.2, the spectral radius of $\mathbf{T}^k$ is smaller than μ^k. A familiar result (see p. 383) tells us that the spectral radius of $\mathbf{T}$ lies within the circle of radius μ, and the theorem follows.

Other analytical criteria implying stability for matrices abound in the literature (see Problem 3 as typical).

We have concentrated here on developing theorems which indicate when an M-matrix is stable because of the natural appeal of the concept of "gross substitutes" and its transparent economic interpretations.

On the other end, two commodities labeled i and j are called complementary if $\partial F_i/\partial p_j < 0$ and $\partial F_j/\partial p_i < 0$, e.g., pen and ink. If the price of ink rises, the demand for pens drops, and conversely.

Let us suppose that the indices of the matrix $\mathbf{B}(\mathbf{q}^*)$ can be divided into two groups J and J' such that for two distinct indices k and k' in the same group (either J or J')

$$b_{k,k'} \geq 0 \qquad \text{and} \qquad b_{k',k} \geq 0,$$

whereas for k in one group and k' in the other,

$$b_{k,k'} \leq 0 \qquad \text{and} \qquad b_{k',k} \leq 0.$$

We call such a matrix an $\widetilde{M}$-matrix. For definiteness, let us arrange the rows and columns so that the indices of J precede the indices of J'. Clearly, if $\mathbf{C}$ is an $\widetilde{M}$-matrix, then $\mathbf{SCS}$ is an M-matrix, where

$$\mathbf{S} = \begin{pmatrix} \mathbf{I} & 0 \\ 0 & -\mathbf{I} \end{pmatrix},$$

with the identity matrix $\mathbf{I}$ of the same order as the number of indices in J and $-\mathbf{I}$ of the same order as the number of indices in J'.

▶ THEOREM 9.3.3. If $\mathbf{B}(\mathbf{q}^*)$ is an $\widetilde{M}$-matrix and there exists a vector $\mathbf{u}$ such that $\mathbf{v} = \mathbf{Su} > \mathbf{0}$ and $\mathbf{SB}(\mathbf{q}^*)\mathbf{u} \ll \mathbf{0}$, then $\mathbf{B}(\mathbf{q}^*)$ is stable.

Proof. Since $\mathbf{SS} = \mathbf{I}$, we have

$$\mathbf{0} \gg \mathbf{SB}(\mathbf{q}^*)\mathbf{u} = \mathbf{SB}(\mathbf{q}^*)\mathbf{SSu} = \mathbf{Dv},$$

where $\mathbf{D}$ is an M-matrix and $\mathbf{v} > \mathbf{0}$. It follows from Problem 15 of Chapter 8 that $\mathbf{D}$ is stable; and therefore $\mathbf{B}(\mathbf{q}^*)$, being obtained from $\mathbf{D}$ by a similarity transformation, is also stable. (See also Problem 4.)

Nonnormalized process. Stability for the nonnormalized process must be interpreted, on account of the homogeneity of the excess demand functions $F_i(\mathbf{p})$; as noted earlier, the whole ray $\Gamma = \{\lambda \mathbf{p}^*\}$ $(\lambda > 0)$ consists of equilibrium vectors. In the following discussion, local stability shall mean that if $\mathbf{p}$ is a price vector near the ray Γ, then $\mathbf{p}(t)$ tends to some equilibrium vector situated on the ray Γ.

▶ THEOREM 9.3.4. Let $\mathbf{A}(\mathbf{p}^*)$ be a strict M-matrix (that is, $a_{ij} > 0; i \neq j$), with $i, j = 0, 1, \ldots, r$. If $\mathbf{p}^* \gg \mathbf{0}$ is an equilibrium vector and $\mathbf{p}$ is sufficiently close to $\Gamma = \{\lambda \mathbf{p}^* | \lambda > 0\}$, then $\mathbf{p}(t)$ converges to an equilibrium point of Γ.

The proof depends on the following lemma.

LEMMA 9.3.1. Given the hypothesis of Theorem 9.3.4, if $\mathbf{p}$ is sufficiently close to Γ, then

$$(\mathbf{p}^*, \mathbf{F}(\mathbf{p})) > 0 \qquad \text{for} \qquad \mathbf{p} \neq \lambda \mathbf{p}^*.$$

Proof. Let $h(\theta) = (\mathbf{p}^*, \mathbf{F}[\mathbf{p}(\theta)])$, where $\mathbf{p}(\theta) = \mathbf{p}^* + \theta(\mathbf{p} - \mathbf{p}^*)$; then we shall prove that $h(0) = h'(0) = 0$ and $h''(0) > 0$ unless $\mathbf{p} = \lambda \mathbf{p}^*$ for some positive λ.

The conclusion $h(0) = 0$ is immediate from the Walras law (9.2.1). By differentiation of (9.2.1), since $\mathbf{p}^*$ is an equilibrium vector $[\mathbf{F}(\mathbf{p}^*) = \mathbf{0}]$, we deduce that $h'(0) = 0$. Explicitly,

$$h'(\theta) = \sum_{i=0}^{r} p_i^* \sum_{j=0}^{r} \frac{\partial F_i[\mathbf{p}(\theta)]}{\partial p_j} (p_j - p_j^*). \tag{9.3.4}$$

By differentiation of (9.2.1) with respect to p_j we obtain

$$\sum_{i=0}^{r} p_i \frac{\partial F_i(\mathbf{p})}{\partial p_j} + F_j(\mathbf{p}) = 0. \tag{9.3.5}$$

At $\mathbf{p} = \mathbf{p}^*$ this reduces to $\mathbf{p}^*\mathbf{A}(\mathbf{p}^*) = \mathbf{0}$. Using this in (9.3.4), after changing the order of summation, shows that $h'(0) = 0$.

By differentiation of (9.3.4) and the same manipulations with (9.3.5), we obtain

$$h''(0) = -\sum_{j,k=0}^{r} (p_j - p_j^*)\left[\frac{\partial F_j(\mathbf{p}^*)}{\partial p_k} + \frac{\partial F_k(\mathbf{p}^*)}{\partial p_j}\right](p_k - p_k^*), \tag{9.3.6}$$

or in vector inner product notation

$$h''(0) = -(\mathbf{p} - \mathbf{p}^*, [\mathbf{A}(\mathbf{p}^*) + \mathbf{A}'(\mathbf{p}^*)](\mathbf{p} - \mathbf{p}^*)) = -(\mathbf{p} - \mathbf{p}^*, \mathbf{T}(\mathbf{p} - \mathbf{p}^*)),$$

where $\mathbf{A}'$ as usual denotes the transpose of $\mathbf{A}$. We have already noted that $\mathbf{p}^*\mathbf{A}(\mathbf{p}^*) = \mathbf{0}$. The homogeneity of $F_i(\mathbf{p})$, which is equivalent to the Euler relation

$$\sum_{j=0}^{r} \frac{\partial F_i(\mathbf{p})}{\partial p_j} p_j = 0,$$

gives in particular $\mathbf{A}(\mathbf{p}^*)\mathbf{p}^* = \mathbf{0}$. It follows that $\mathbf{p}^*[\mathbf{A}(\mathbf{p}^*) + \mathbf{A}'(\mathbf{p}^*)] = \mathbf{p}^*\mathbf{T} = \mathbf{0}$. But $\mathbf{T}$ is a real symmetric M-matrix, and since $\mathbf{p}^* \gg \mathbf{0}$ we may invoke property (b) of M-matrices to conclude that any other solution $\mathbf{x}$ of $\mathbf{xT} = \mathbf{0}$ is a scalar multiple of $\mathbf{p}^*$, and that all nonzero eigenvalues of $\mathbf{T}$ are negative. Thus, T is negative semidefinite. Hence $h''(0) > 0$ unless $\mathbf{p} = \lambda\mathbf{p}^*$, again by virtue of the fact that F_i is homogeneous. The result of Lemma 9.3.1 is now clear, provided that we alter $\mathbf{p}^*$ to a suitable $\lambda\mathbf{p}^*$ ($\lambda > 0$) in $(\mathbf{p}^*, (\mathbf{F}(\mathbf{p}))$ and form a Taylor expansion about $\lambda\mathbf{p}^*$.

Proof of Theorem 9.3.4. Define

$$V(t) = \tfrac{1}{2}|\mathbf{p}(t) - \mathbf{p}^*|^2 = \tfrac{1}{2}\sum_{i=0}^{r} [p_i(t) - p_i^*]^2,$$

where $\mathbf{p}(t) = \psi(t, \mathbf{p}^0)$ and $\mathbf{p}^0$ is sufficiently close to the ray $\lambda\mathbf{p}^*$. We calculate

$$\frac{dV(t)}{dt} = (\mathbf{p}(t) - \mathbf{p}^*, \mathbf{F}[\mathbf{p}(t)]) = -(\mathbf{p}^*, \mathbf{F}[\mathbf{p}(t)]),$$

the second equality resulting from the Walras law. By Lemma 9.3.1, $dV(t)/dt < 0$ unless $\mathbf{p}(t) = \lambda\mathbf{p}^*$. Since $V(t)$ decreases, we may select a limit point of $\mathbf{p}(t) = \tilde{\mathbf{p}}$. If $\tilde{\mathbf{p}}$ is not on Γ, an easy contradiction ensures, as follows. There exists a sequence $t^\nu \to \infty$ such that

$$\lim_{\nu\to\infty} \psi(t^\nu, \mathbf{p}^0) = \tilde{\mathbf{p}}.$$

The continuity of solutions with respect to the initial position vector clearly implies that

$$\lim_{\nu\to\infty} \psi(t^\nu + t, \mathbf{p}^0) = \psi(t, \tilde{\mathbf{p}}).$$

Now $\dfrac{dV(t^\nu + t)}{dt}$ tends to zero, but if $\tilde{\mathbf{p}}$ is not on Γ,

$$\lim_{\nu\to\infty} \frac{dV(t^\nu + t)}{dt} = \frac{d}{dt}\,(\tfrac{1}{2}|\psi(t, \tilde{\mathbf{p}}) - \mathbf{p}^*|^2) < 0.$$

Consequently, if $\lim V(t) = a$, then all limit points of $\mathbf{p}(t)$ lie on the ray Γ at distance $2a$ from $\mathbf{p}^*$. Any $\tilde{\mathbf{p}}$ consists of at most one of two points. However, the differential equation (9.2.5) shows, with the aid of (9.2.1) that

$$\sum_{i=0}^{r} p_i(t)\,\frac{dp_i}{dt} = \sum_{i=0}^{r} p_i(t)F_i[\mathbf{p}(t)] = 0,$$

and hence

$$\sum_{i=0}^{r} p_i^2(t)$$

is a constant b independent of t. Clearly, $\tilde{\mathbf{p}} = \lambda\mathbf{p}^*$, where

$$\lambda = \sqrt{\frac{b}{\sum_{i=0}^{r} p_i^{*2}}}.$$

***9.4 Global stability of price adjustment processes.** In this section we investigate various conditions which imply stability in the large. The criteria are mostly based on economic considerations. We concentrate on the nonnormalized process whose dynamics are described by the differential-equations system (9.2.5).

In this section again (unless there is an explicit statement to the contrary) we assume the validity of the Walras law, the homogeneity of the excess demand functions F_i, and the strict positivity of an equilibrium price vector $\mathbf{p}^*$, which is postulated to exist (see also Problem 13). The following lemma is crucial to most of our discussion of global stability.

▶ LEMMA 9.4.1. If $(\mathbf{p}^*, \mathbf{F}(\mathbf{p})) > 0$ for all vectors $\mathbf{p}$ for which $\mathbf{F}(\mathbf{p}) \neq \mathbf{0}$, then the nonnormalized price adjustment process is globally stable; i.e., as t tends to infinity the solution $\psi(t, \mathbf{p}^0)$ converges to a fixed-point solution.

After proving Lemma 9.4.1, we shall devote the remainder of this section to establishing natural conditions which imply the hypothesis of this lemma. Theorems 9.4.1 and 9.4.2 state two such conditions.

The literature on the global stability of differential-equations systems is based largely on one fundamental method of analysis. This method consists in introducing some appropriate norm of the distance between the solution and the fixed point. If it can be shown that the values of this norm strictly decrease with time whenever the distance is not zero, then stability in the large follows. The norm we use in this section is the ordinary Euclidean distance. In the following section we employ two other norms that are suitable for the hypothesis proposed there (see also the notes at the close of the chapter). A more general technique for establishing stability is to consider functions $V[\mathbf{p}(t)]$ which are not necessarily norms but possess appropriate contraction properties relative to equilibrium points. We exploit this concept on pp. 316–325.

Proof of Lemma 9.4.1. Let $\mathbf{p}^0$ be any initial vector and let $\mathbf{p}(t)$ designate the solution of (9.2.5) with initial position $\mathbf{p}^0$; that is, $\mathbf{p}(t) = \psi(t, \mathbf{p}^0)$. Consider the norm deviation measure

$$V(t) = \tfrac{1}{2}|\mathbf{p}(t) - \mathbf{p}^*|^2 = \tfrac{1}{2}(\mathbf{p}(t) - \mathbf{p}^*, \mathbf{p}(t) - \mathbf{p}^*).$$

It follows from (9.2.5) that

$$V'(t) = (\mathbf{p}(t) - \mathbf{p}^*, \mathbf{F}[\mathbf{p}(t)]) = -(\mathbf{p}^*, \mathbf{F}[\mathbf{p}(t)]), \tag{9.4.1}$$

the last simplification resulting from an application of the Walras law. By the hypothesis, $V'(t) < 0$ unless $\mathbf{p}(t)$ satisfies $\mathbf{F}(\mathbf{p}(t)) = \mathbf{0}$. The proof from here proceeds as in the argument of Theorem 9.3.4.

We now discuss a series of applications of Lemma 9.4.1.

The aggregate excess demand functions are said to satisfy the weak axiom of revealed preference if

$$(\mathbf{p}', \mathbf{F}(\mathbf{p}'') - \mathbf{F}(\mathbf{p}')) \leq 0 \qquad \text{and} \qquad \mathbf{F}(\mathbf{p}'') \neq \mathbf{F}(\mathbf{p}') \tag{9.4.2}$$

imply

$$(\mathbf{p}'', \mathbf{F}(\mathbf{p}'') - \mathbf{F}(\mathbf{p}')) < 0 \qquad \text{(see also p. 269).}$$

▶ THEOREM 9.4.1. If the excess demand functions satisfy the weak axiom of revealed preference, then the process described by (9.2.5) is globally stable.

Proof. Let $\mathbf{p}'' = \mathbf{p}^*$ and $\mathbf{p}' = \mathbf{p}$. Then the first inequality of (9.4.2) is fulfilled because of the Walras law and by the definition of an equilibrium point $[\mathbf{F}(\mathbf{p}^*) = \mathbf{0}]$. Also, if $\mathbf{p}$ is not an equilibrium price vector, we have

$\mathbf{F}(\mathbf{p}) \neq \mathbf{0}$ and hence the second inequality of (9.4.2) is also satisfied. It follows that $-(\mathbf{p}^*, \mathbf{F}(\mathbf{p})) < 0$, which means that the hypothesis of Lemma 9.4.1 holds. Consequently, the process is globally stable and the theorem is established.

The weakness of Theorem 9.4.1 lies in the fact that the axiom of revealed preference is usually verified only in the neighborhood of an equilibrium point and need not hold in the large. The following result is deeper and also has considerable economic appeal.

▶THEOREM 9.4.2. If the excess demand functions satisfy the property of strict gross substitutability in the interior of the positive orthant, then the process described by (9.2.5) is globally stable. Moreover, $(\mathbf{p}^*, \mathbf{F}(\mathbf{p})) > 0$ for all $\mathbf{p} \neq \lambda\mathbf{p}^*$ (λ real and positive), where $\mathbf{p}^* \gg \mathbf{0}$ is an equilibrium vector.

Remarks on the hypothesis of Theorem 9.4.2.† It must be emphasized that the restriction of strict gross substitutability to the interior of the positive orthant, as opposed to the full closed positive orthant, is necessary. Indeed, the property of strict gross substitutability and the requirement of homogeneity are contradictory. For let us assume to the contrary that the excess demand function $\mathbf{F}(\mathbf{p})$ is defined and homogeneous and satisfies the property of strict gross substitutes for all

$$\mathbf{p} = \langle p_0, p_1, \ldots, p_r \rangle \qquad (p_i \geq 0),$$

including the boundary. Now, let $\mathbf{p}^0 = \langle 0, p_1^0, \ldots, p_r^0 \rangle$, with $\mathbf{p}^0 > \mathbf{0}$. Then $F_0(\lambda\mathbf{p}^0) = F_0(\mathbf{p}^0)$ is a constant in λ. But at least one of the components $\lambda p_1^0, \ldots, \lambda p_r^0$ strictly increases, and hence F_0 should strictly increase, a contradiction.

If strict gross substitutability is assumed merely for the interior of the positive orthant, then the assumptions of homogeneity, the validity of the Walras law, and the property of strict gross substitutability are fully consistent.

Examples of vector functions $\mathbf{F}$ obeying the stated restrictions are constructed as follows. For $\mathbf{p} \gg \mathbf{0}$, we define

$$F_k(\mathbf{p}) = \frac{\sum_{i=0}^{r} a_{ik} p_i}{p_k} \qquad (k = 0, \ldots, r). \tag{9.4.3}$$

The further conditions $a_{ik} > 0$ for $i \neq k$ and

$$\sum_{k=0}^{r} a_{ik} = 0$$

† It is indicated in Problem 7 that subject to the conditions of the theorem, any solution of (9.2.5) with $\mathbf{p}(0) \gg \mathbf{0}$ necessarily has $\mathbf{p}(t) \gg \mathbf{0}$ for all $t > 0$, and thus $\mathbf{F}[\mathbf{p}(t)]$ is well defined.

imply trivially that the property of strict gross substitutability prevails and that the Walras law holds. That $F_k(\mathbf{p})$ are homogeneous functions is also immediately clear. A simple special instance of (9.4.3) is obtained by setting $a_{ik} = 1$ for $i \neq k$ and $a_{kk} = -r$, $(i, k = 0, 1, \ldots, r)$.

It is interesting to note that an excess demand function of the structure (9.4.3) can be derived by performing an appropriate consumer utility maximization. Explicitly, the problem

$$U(\mathbf{x}^*) = \max_{x} U(\mathbf{x}) = \max_{x_i} \sum_{i=0}^{r} \alpha_i \log (x_i + x_i^0 + c_i)$$

subject to the constraint

$$\sum_{i=0}^{r} p_i x_i = 0,$$

which is easily resolved by the standard Lagrange multiplier methods, leads to $x_i^*(\mathbf{p}) = F_i(\mathbf{p})$ of the form (9.4.3). (The α_i are fixed positive constants and $\mathbf{x}^0 \gg \mathbf{0}$ is interpreted as a fixed vector of initial holdings for an individual consumer.)

If to the hypotheses of the theorem we add the condition that $F_j(\mathbf{p})$ tends to infinity whenever $\mathbf{p}$ approaches a vector $\mathbf{p}^0 > \mathbf{0}$ for which $p_j^0 = 0$ and otherwise approaches a finite limit, we may secure the existence of an equilibrium price vector $\mathbf{p}^* \gg \mathbf{0}$ (see Problems 12 and 13 and the discussion of solutions to the problems on p. 357).

Proof of Theorem 9.4.2. We shall prove that if $\mathbf{p} \neq \lambda \mathbf{p}^*$, then

$$(\mathbf{p}^*, \mathbf{F}(\mathbf{p})) > 0, \tag{9.4.4}$$

from which, in view of Lemma 9.4.1, the theorem follows.

We may without loss of generality assume that

$$\mathbf{p}^* = \mathbf{u}^* = \langle 1, 1, \ldots, 1 \rangle.$$

To demonstrate this, we let

$$H_i(\mathbf{p}) = H_i(p_0, p_1, \ldots, p_r) = p_i^* F_i(p_0^* p_0, p_1^* p_1, p_2^* p_2, \ldots, p_r^* p_r).$$

It is readily verified that the vector function $\mathbf{H}$ satisfies all the properties that $\mathbf{F}$ satisfies, including gross substitutability; but now $\mathbf{H}(\mathbf{u}^*) = \mathbf{0}$ by the Walras law, and thus $\mathbf{u}^*$ is an equilibrium vector for $\mathbf{H}$. Finally, we observe that $(\mathbf{u}^*, \mathbf{H}(\mathbf{p})) > 0$ for $\mathbf{p} \neq \lambda \mathbf{u}^*$ if and only if $(\mathbf{p}^*, \mathbf{F}(\mathbf{p})) > 0$ for $\mathbf{p} \neq \lambda \mathbf{p}^*$.

At this point it is convenient to revert to our old notation involving the vector function $\mathbf{F}$ and the equilibrium vector $\mathbf{p}^* = \langle 1, 1, \ldots, 1 \rangle$.

Let $\mathbf{p} = \langle p_0, p_1, \ldots, p_r \rangle$ denote any vector not a multiple of $\mathbf{p}^*$. Without loss of generality we may so number the commodities that

$$p_0 \leq p_1 \leq \cdots \leq p_r. \tag{9.4.5}$$

Since $\mathbf{p}$ is not a multiple of $\mathbf{p}^*$, there must be at least one case of strict inequality in (9.4.5), say $p_\nu < p_{\nu+1}$ for some $0 \leq \nu < r$. We now define a sequence of vectors $\mathbf{p}^0, \mathbf{p}^1, \ldots, \mathbf{p}^r$ by the relations

$$\begin{aligned}
\mathbf{p}^0 &= \langle p_0, p_0, p_0, \ldots, p_0 \rangle, \\
\mathbf{p}^1 &= \langle p_0, p_1, p_1, \ldots, p_1 \rangle, \\
&\cdots \\
\mathbf{p}^s &= \langle p_0, p_1, \ldots, p_{s-1}, p_s, p_s, \ldots, p_s \rangle, \\
&\cdots \\
\mathbf{p}^r &= \langle p_0, p_1, \ldots, p_r \rangle,
\end{aligned} \tag{9.4.6}$$

where the p_j are the components of $\mathbf{p}$. Obviously $\mathbf{p}^r = \mathbf{p}$ and $\mathbf{p}^0$ is a multiple of $\mathbf{p}^*$. We shall prove that

$$(\mathbf{p}^*, \mathbf{F}(\mathbf{p}^{s+1}) - \mathbf{F}(\mathbf{p}^s)) \geq 0 \qquad (s = 0, 1, \ldots, r-1), \tag{9.4.7}$$

with strict inequality for $s = \nu$. Adding the equations (9.4.7) and recognizing that $\mathbf{F}(\mathbf{p}^0) = \mathbf{0}$, we deduce the desired inequality (9.4.4).

It remains to prove the correctness of (9.4.7). Because of the assumption of strict gross substitutability, since only the components greater than s increase, it follows that

$$F_i(\mathbf{p}^{s+1}) - F_i(\mathbf{p}^s) \geq 0 \qquad (i = 0, 1, \ldots, s), \tag{9.4.8}$$

where strict equality holds when $s = \nu$.

Moreover, the homogeneity of $\mathbf{F}$, in conjunction with the fact that $p_s/p_{s+1} \leq 1$, shows that

$$\begin{aligned}
F_j(\mathbf{p}^{s+1}) - F_j(\mathbf{p}^s) &= F_j\left(\frac{p_s}{p_{s+1}}\, p_0, \frac{p_s}{p_{s+1}}\, p_1, \ldots, \frac{p_s}{p_{s+1}}\, p_s, p_s, p_s, \ldots, p_s\right) \\
&\quad - F_j(p_0, p_1, \ldots, p_{s-1}, p_s, p_s, \ldots, p_s) \leq 0
\end{aligned} \tag{9.4.9}$$

for $j = s+1, \ldots, r$. This follows from the hypothesis of strict gross substitutability, on the assumption that F_j is first evaluated at a price vector whose initial $s+1$ components do not exceed in magnitude the respective values of the first $s+1$ components of the second vector at which F_j is evaluated, while the last $r-(s+1)$ coordinates are identical. Again strict inequality prevails in (9.4.9) if $s = \nu$.

For fixed j, iterating (9.4.9) to the zero index shows that

$$F_j(\mathbf{p}^{s+1}) \leq F_j(\mathbf{p}^s) \leq \cdots \leq F_j(\mathbf{p}^0) = 0 \qquad (j = s+1, \ldots, r), \tag{9.4.10}$$

with strict inequality at the point $\mathbf{p}^\nu$ in the sequence (9.4.10), provided that $\mathbf{p}^\nu$ appears.

Consider any fixed index s and partition all vectors of $r + 1$ components into two smaller vectors, the first consisting of the first $s + 1$ ordered components and the second consisting of the remaining components. If $\mathbf{x}$ is a general vector of $r + 1$ components, we write $\mathbf{x}_*$ and $\mathbf{x}_{**}$ for the two associated vectors constructed by the rules just set forth. Note that

$$\mathbf{p}_*^{s+1} = \mathbf{p}_*^s \tag{9.4.11}$$

and

$$\mathbf{p}_{**}^{s+1} = (p_r^{s+1})\mathbf{p}_{**}^* = \mathbf{p}_{**}^s + (p_{s+1} - p_s)\mathbf{p}_{**}^*, \tag{9.4.12}$$

where $\mathbf{p}_{**}^*$ is a vector of appropriate size with all components 1. Decomposing the expression $(\mathbf{p}^{s+1}, \mathbf{F}(\mathbf{p}^{s+1})) - (\mathbf{p}^s, \mathbf{F}(\mathbf{p}^s))$ (which is zero by the Walras law) in accordance with the partitioning scheme suggested above, we obtain

$$\begin{aligned}
0 &= (\mathbf{p}_*^{s+1}, \mathbf{F}(\mathbf{p}^{s+1})_*) + (\mathbf{p}_{**}^{s+1}, \mathbf{F}(\mathbf{p}^{s+1})_{**}) \\
&\qquad\qquad - (\mathbf{p}_*^s, \mathbf{F}(\mathbf{p}^s)_*) - (\mathbf{p}_{**}^s, \mathbf{F}(\mathbf{p}^s)_{**}) \\
&= (\mathbf{p}_*^s, \mathbf{F}(\mathbf{p}^{s+1})_* - \mathbf{F}(\mathbf{p}^s)_*) + (\mathbf{p}_{**}^{s+1}, \mathbf{F}(\mathbf{p}^{s+1})_{**}) - (\mathbf{p}_{**}^s, \mathbf{F}(\mathbf{p}^s)_{**}) \\
&\qquad\qquad [\text{by } (9.4.11)] \\
&= (\mathbf{p}_*^s, \mathbf{F}(\mathbf{p}^{s+1})_* - \mathbf{F}(\mathbf{p}^s)_*) + (\mathbf{p}_{**}^{s+1}, \mathbf{F}(\mathbf{p}^{s+1})_{**} - \mathbf{F}(\mathbf{p}^s)_{**}) \\
&\qquad\qquad + (p_{s+1} - p_s)(\mathbf{p}_{**}^*, \mathbf{F}(\mathbf{p}^s)_{**}) \\
&\qquad\qquad [\text{by } (9.4.12) \text{ and rearrangement}] \\
&= (\mathbf{p}_*^s, \mathbf{F}(\mathbf{p}^{s+1})_* - \mathbf{F}(\mathbf{p}^s)_*) + p_r^{s+1}(\mathbf{p}_{**}^*, \mathbf{F}(\mathbf{p}^{s+1})_{**} - \mathbf{F}(\mathbf{p}^s)_{**}) \\
&\qquad\qquad + (p_{s+1} - p_s)(\mathbf{p}_{**}^*, \mathbf{F}(\mathbf{p}^s)_{**}) \qquad [\text{by } (9.4.12)] \\
&\leq p_r^{s+1}(\mathbf{p}^*, \mathbf{F}(\mathbf{p}^{s+1}) - \mathbf{F}(\mathbf{p}^s)) + (p_{s+1} - p_s)(\mathbf{p}_{**}^*, \mathbf{F}(\mathbf{p}^s)_{**}) \\
&\qquad\qquad [\text{by } (9.4.5) \text{ and } (9.4.11)].
\end{aligned}$$

This may be written as

$$(\mathbf{p}^*, \mathbf{F}(\mathbf{p}^{s+1}) - \mathbf{F}(\mathbf{p}^s)) \geq -\frac{p_{s+1} - p_s}{p_r^{s+1}}(\mathbf{p}_{**}^*, \mathbf{F}(\mathbf{p}^s)_{**}). \tag{9.4.13}$$

The right-hand side of (9.4.13) is nonnegative because

$$\frac{p_{s+1} - p_s}{p_r^{s+1}} \geq 0$$

and $(\mathbf{p}^*_{**}, \mathbf{F}(\mathbf{p}^s)_{**}) \leq 0$. Furthermore, strict inequality holds for $s = \nu$, in view of (9.4.10) and the fact that $p_{\nu+1} > p_\nu$. Hence (9.4.7) follows and the proof is complete.

The global stability of the system (9.2.5) under the conditions of Theorem 9.4.2 can be established more simply as indicated in Theorem 9.4.3 below. However, the interesting supplementary inequality (9.4.4) does not appear to be susceptible to easy verification except by the reasoning of the above proof.

We next consider a price adjustment process defined by the system of differential equations

$$\frac{dp_i}{dt} = H_i[F_i(\mathbf{p})], \qquad (i = 0, 1, \ldots, r), \tag{9.4.14}$$

where $H_i(x)$ is continuous and sign-preserving, i.e., sign $H_i(x) =$ sign x for all real x. The usual assumptions regarding uniqueness and continuity of the solution of (9.4.14) as a function of the initial position are made.

▶THEOREM 9.4.3. Under the conditions of Theorem 9.4.2, with the single change that the Walras law is not required to hold, any solution $\mathbf{p}(t)$ satisfying (9.4.14) with $\mathbf{p}(0) \gg \mathbf{0}$ is stable.

Proof. As demonstrated in the proof of Theorem 9.4.2, we may without loss of generality normalize any strict positive equilibrium vector to the form $\mathbf{p}^* = \langle 1, \ldots, 1\rangle$. We suppose henceforth that this is done.

The following auxiliary result is needed.

(I) If $\mathbf{p} = \langle p_0, \ldots, p_r\rangle$ is strictly positive and satisfies

$$p_{i_0} = \max_{0\leq j\leq r} p_j > \min_{0\leq j\leq r} p_j = p_{k_0},$$

then $F_{i_0}(\mathbf{p}) < 0$ and $F_{k_0}(\mathbf{p}) > 0$.

We prove only the first inequality, since the second is obtained by analogous reasoning. To this end, we increase all components of p_j $(j \neq i_0)$ if necessary, and thus achieve the price vector $\mathbf{p}' = \langle p_{i_0}, p_{i_0}, \ldots, p_{i_0}\rangle$. Because of the property of strict gross substitutability, $0 = F_{i_0}(\mathbf{p}') > F_{i_0}(\mathbf{p})$, as was to be proved. The same argument also implies that the set $\{\lambda\mathbf{p}^*\}$ $(\lambda > 0)$ spans the totality of equilibrium vectors.

We now continue with the proof of the theorem. Let $\mathbf{p}(t)$ be the solution of (9.4.14) for an initial vector $\mathbf{p}(0) \gg \mathbf{0}$. We shall establish in (9.4.18) that $\mathbf{p}(t) \gg \mathbf{0}$, thus guaranteeing that $\mathbf{F}[\mathbf{p}(t)]$ is meaningful for all t. For the present we proceed as if this has been done. (The reader may check afterwards that the arguments are all consistent.) We now introduce the functions

$$\Lambda(t) = \max_{0 \leq j \leq r} p_j(t), \qquad \lambda(t) = \min_{0 \leq j \leq r} p_j(t). \tag{9.4.15}$$

The definition implies the existence of an index i such that

$$\overline{\lim_{h \to 0+}} \frac{\Lambda(t+h) - \Lambda(t)}{h} \leq \overline{\lim_{h \to 0+}} \frac{p_i(t+h) - p_i(t)}{h} = \frac{dp_i(t)}{dt}$$

($h \to 0+$ means that h approaches zero through positive values only) and

$$p_i(t) = \max_{0 \leq j \leq r} p_j(t).$$

Hence by (I),

$$\overline{\lim_{h \to 0+}} \frac{\Lambda(t+h) - \Lambda(t)}{h} \leq H_i(F_i[\mathbf{p}(t)]) \leq 0, \tag{9.4.16}$$

with strict inequality if $\mathbf{p}(t)$ is not of the form $\lambda\mathbf{p}^*$ for some positive λ. In a similar manner, we obtain

$$\underline{\lim_{h \to 0+}} \frac{\lambda(t+h) - \lambda(t)}{h} \geq 0, \tag{9.4.17}$$

with strict inequality except when $\mathbf{p}(t)$ is of the form $\lambda\mathbf{p}^*$ ($\lambda > 0$). The inequalities (9.4.16) and (9.4.17) show that $\Lambda(t)$ is nonincreasing and $\lambda(t)$ is nondecreasing since each is continuous. In particular,

$$0 < \lambda(0) \leq \lambda(t) \leq \Lambda(t) \leq \Lambda(0); \tag{9.4.18}$$

hence the solution $\mathbf{p}(t)$ is always bounded above and below, and each component of $p_j(t)$ strictly avoids zero. Let

$$\tilde{\Lambda} = \lim_{t \to \infty} \Lambda(t),$$

and let $\tilde{\mathbf{p}}$ be a limit point of the sequence $\mathbf{p}(t)$; i.e.,

$$\tilde{\mathbf{p}} = \lim_{\mu \to \infty} \mathbf{p}(t_\mu)$$

for some sequence $t_\mu \to \infty$. Let $\tilde{\mathbf{p}}(t)$ be the solution of (9.4.14) with initial vector $\tilde{\mathbf{p}}$, and define

$$\tilde{\Lambda}(t) = \max_{0 \leq j \leq r} \tilde{p}_j(t).$$

Then

$$\tilde{\Lambda}(t) = \lim_{\mu \to \infty} \Lambda(t + t_\mu) = \tilde{\Lambda},$$

which implies

$$\overline{\lim_{h\to 0}} \frac{\tilde{\Lambda}(h) - \tilde{\Lambda}(0)}{h} = 0.$$

By (9.4.16), $\tilde{\mathbf{p}}$ is a multiple of $\mathbf{p}^*$ and hence an equilibrium vector. Since $\tilde{\Lambda}$ is uniquely determined, we infer that $\tilde{\mathbf{p}} = \tilde{\Lambda} \cdot \mathbf{p}^*$ is also uniquely determined. Therefore $\mathbf{p}(t)$ itself converges to $\tilde{\mathbf{p}}$, and the proof of the theorem is complete.

Note that the proof of this theorem does not refer to the Walras law. By contrast, if weak rather than strict gross substitutability is assumed, i.e., if we require only that $\partial F_i/\partial p_i \geq 0$ for $i \neq j$ and $\mathbf{p} \gg \mathbf{0}$, the validity of the Walras law must be postulated.

▶THEOREM 9.4.4. If the excess demand functions $\mathbf{F}(\mathbf{p})$ are defined, differentiable for $\mathbf{p} \gg \mathbf{0}$, and homogeneous, and if they satisfy the Walras law and possess the property of weak gross substitutability, then the price adjustment process characterized by (9.4.14) is stable.

(As usual, we assume the existence of a strict positive equilibrium vector.)

Proof. Similarly to the proof of Theorem 9.4.3, the relation (9.4.18) holds for the present case; hence the solution $\mathbf{p}(t)$ of the system (9.4.14) with any initial vector $\mathbf{p}(0) \gg \mathbf{0}$ is bounded and positive.

Now consider the function

$$R(t) = \max_{i \in S(t)} \left\{ \frac{F_i(t)}{p_i(t)} \right\} \tag{9.4.19}$$

($F_i(t)$ is an abbreviation for $F_i[\mathbf{p}(t)]$), where

$$S(t) = \left\{ i \,\middle|\, \frac{H_i[F_i(t)]}{p_i(t)} = \max_{0 \leq j \leq r} \frac{H_j[F_j(t)]}{p_j(t)} \right\}.$$

It is noted that $R(t) = 0$ or $R(t) > 0$ according as $\mathbf{p}(t)$ is or is not an equilibrium vector. For if $R(t) \leq 0$, it follows from the definition of $R(t)$ that $H_i(F_i[\mathbf{p}(t)]) \leq 0$, where $i = 0, 1, \ldots, r$; and since H_i is sign-preserving, $F_i[\mathbf{p}(t)] \leq 0$. From these inequalities and the Walras law, it follows that $F_i[\mathbf{p}(t)] = 0$ and $\mathbf{p}(t)$ is an equilibrium vector.

Consider

$$W(t) = \overline{\lim_{h\to 0}} \frac{R(t+h) - R(t)}{h} \leq \frac{d}{dt}\left(\frac{F_i[\mathbf{p}(t)]}{p_i(t)}\right), \tag{9.4.20}$$

where the last inequality holds for an appropriate index i in $S(t)$. We obtain

$$\frac{d}{dt}\left(\frac{F_i(t)}{p_i(t)}\right) = -\frac{F_i(t)}{p_i^2(t)}\frac{dp_i(t)}{dt} + \frac{1}{p_i(t)}\sum_{j=0}^{r}\frac{\partial F_i}{\partial p_j}\frac{dp_j}{dt}$$

$$= -\frac{F_i(t)H_i[F_i(t)]}{p_i^2(t)} + \frac{1}{p_i(t)}\sum_{j=0}^{r}\frac{\partial F_i}{\partial p_j}H_j[F_j(t)].$$

Applying the property of weak gross substitutability, and taking account of the meaning of the index i and the homogeneity of $\mathbf{F}$, we obtain

$$\frac{1}{p_i(t)}\sum_{j=0}^{r}\frac{\partial F_i}{\partial p_j}H_j[F_j(t)] = \frac{H_i[F_i(t)]}{p_i(t)}\frac{\partial F_i}{\partial p_i} + \frac{1}{p_i}\sum_{j\neq i}\frac{\partial F_i}{\partial p_j}H_j[F_j(t)]$$

$$\leq \frac{H_i[F_i(t)]}{p_i(t)}\frac{\partial F_i}{\partial p_i} + \frac{H_i[F_i(t)]}{p_i^2(t)}\sum_{j\neq i}\frac{\partial F_i}{\partial p_j}p_j(t)$$

$$= \frac{H_i[F_i(t)]}{p_i^2(l)}\sum_{j=0}^{r}\frac{\partial F_i}{\partial p_j}p_j(t) = 0.$$

Hence the inequality (9.4.20) may be written as

$$W(t) \leq -\frac{F_i(t)H_i[F_i(t)]}{p_i^2(t)}. \tag{9.4.21}$$

Since H_i is sign-preserving, we infer from (9.4.21), together with the observation following (9.4.19), that

$$W(t) \leq 0, \tag{9.4.22}$$

with equality if and only if $F[\mathbf{p}(t)] = 0$.

Thus the function $R(t)$ is nonincreasing and nonnegative (the latter by virtue of the Walras law); and hence $R(t)$ converges to $\tilde{R}$ as t tends to infinity. Now let $\tilde{\mathbf{p}}$ be a limit point of $\mathbf{p}(t_v)$ for some sequence t_v approaching infinity. Define $\tilde{R}(t)$ and $\tilde{W}(t)$ for the solution $\tilde{\mathbf{p}}(t)$ with initial vector $\tilde{\mathbf{p}}$ as $R(t)$ and $W(t)$ are defined for $\mathbf{p}(t)$. Then

$$\tilde{R}(t) = \lim_{v\to\infty} R(t+t_v) = \tilde{R},$$

and hence $\tilde{W}(t) = 0$ at $t = 0$.

By (9.4.22), $\tilde{\mathbf{p}}$ is an equilibrium price vector. We may now define for the solution $\mathbf{p}(t)$,

$$\Lambda(t) = \max_j \frac{p_j(t)}{\tilde{p}_j} \quad \text{and} \quad \lambda(t) = \min_j \frac{p_j(t)}{\tilde{p}_j}. \tag{9.4.23}$$

Although the reasoning in Theorem 9.4.3 was carried out with respect to the equilibrium vector $\mathbf{p}^*$, it is readily shown that it remains correct for $\Lambda(t)$ and $\lambda(t)$ as defined in (9.4.23). Hence

$$\lim_{t\to\infty} \Lambda(t) \qquad \text{and} \qquad \lim_{t\to\infty} \lambda(t)$$

both exist. Moreover, since $\lim \mathbf{p}(t_v) = \tilde{\mathbf{p}}$, we have

$$\lim_{t\to\infty} \Lambda(t) = \lim_{t\to\infty} \lambda(t) = 1.$$

From this, in conjunction with (9.4.23), it follows that

$$\lim_{t\to\infty} \mathbf{p}(t) = \tilde{\mathbf{p}},$$

and the proof of the theorem is complete.

Valid adaptations of Theorems 9.4.1–4 can be formulated for the normalized price adjustment process.

***9.5. Global stability (continued).** In the last section we established global stability in one case by showing that the norm measured by the Euclidean distance between the solution and the equilibrium price vector decreased with time. We subsequently achieved sharper results by using other norms. In this section we again employ norms other than the Euclidean distance for the same purpose under different types of hypothesis. The three principal theorems of this section are formulated in terms of the normalized process. By suitable modifications, these theorems may be translated into propositions that are relevant for the nonnormalized process.

▶ THEOREM 9.5.1. The normalized process (9.2.8) is globally stable if there exists a set of positive constants $\langle c_1, c_2, \ldots, c_r\rangle$ such that for each $j = 1, 2, \ldots, r$

$$G_{jj} < 0, \qquad c_j|G_{jj}| > \sum_{k\neq j} c_k|G_{kj}|, \tag{9.5.1}$$

where $G_{kj} = \partial G_k/\partial q_j$.

Remark 9.5.1. The condition of a dominant diagonal says that the price of the kth commodity influences the demand for that commodity more than it influences the demand for other commodities.

Proof. Consider the function

$$V(t) = \sum_{j=1}^{r} c_j|G_j|. \tag{9.5.2}$$

It is easily shown that

$$\lim_{h \to 0+} \frac{V(t+h) - V(t)}{h} = W(t)$$

exists. This value of $W(t)$ may be calculated as

$$W(t) = \sum_{J} c_j \operatorname{sign} G_j(t) \sum_{k=1}^{r} \frac{\partial G_j}{\partial q_k} G_k + \sum_{J'} c_j \left| \sum_{k=1}^{r} \frac{\partial G_j}{\partial q_k} G_k \right|, \qquad (9.5.3)$$

the first sum being extended over the indices j for which $G_j(t) \neq 0$ and the second sum over the indices j for which $G_j(t) = 0$. The usual meaning is attached to sign, namely

$$\operatorname{sign} a = \begin{cases} +1 & a > 0, \\ 0 & a = 0, \\ -1 & a < 0. \end{cases}$$

We shall prove that $-W(t)$ is positive unless $G_j = 0$ for all j (i.e., the solution is at an equilibrium price vector). We consider the coefficient $L_{k_0}(t)$ in $-W(t)$ of $G_{k_0}(t)$, which we suppose is not zero: Specifically,

$$L_{k_0}(t) = -c_{k_0} \frac{\partial G_{k_0}}{\partial q_{k_0}} \operatorname{sign} G_{k_0}(t) - \sum_{\substack{j \neq k_0 \\ J}} c_j \operatorname{sign} G_j(t) \frac{\partial G_j}{\partial q_{k_0}}$$
$$- \sum_{\substack{j \neq k_0 \\ J'}} c_j \operatorname{sign} \left(\sum_{k=1}^{r} \frac{\partial G_j}{\partial q_k} G_k \right) \cdot \frac{\partial G_j}{\partial q_{k_0}}.$$

Hence

$$G_{k_0}(t) L_{k_0}(t) \geq c_{k_0} \left| \frac{\partial G_{k_0}}{\partial q_{k_0}} \right| \left| G_{k_0}(t) \right| - \left(\sum_{j \neq k_0} c_j \left| \frac{\partial G_j}{\partial q_{k_0}} \right| \right) |G_{k_0}(t)|, \qquad (9.5.4)$$

and by the hypothesis the right-hand side is strictly positive. Adding the inequalities (9.5.4) for the different k_0 (note that the terms in which $G_{k_0}(t) = 0$ make no contribution to the value of $W(t)$ and are thus of no relevance), we obtain $W(t) < 0$ except when $G_k(t) = 0$ for all k. The norm $V(t)$, as defined by (9.5.2), strictly decreases, and global stability results as in Theorem 9.3.4.

A dual result to Theorem 9.5.1 is as follows.

▶Theorem 9.5.2. The normalized process (9.2.8) is globally stable if there exists a set of positive constants $\langle c_1, c_2, \ldots, c_r \rangle$ such that for each $j = 1, 2, \ldots, r$

$$G_{jj} < 0, \qquad c_j |G_{jj}| > \sum_{k \neq j} c_k |G_{jk}|. \qquad (9.5.5)$$

Remark 9.5.2. The hypothesis may be loosely interpreted to mean that the price of the kth commodity is the major influence on the excess demand function of that commodity.

The difference between Theorem 9.5.1 and Theorem 9.5.2 is that (9.5.1) pertains to the column sums and (9.5.5) to the row sums of the matrix $\|G_{jk}\|$. These sets of conditions may be construed as dual criteria, each yielding global stability. The proofs likewise proceed in a dual fashion.

Proof of Theorem 9.5.2. Consider the norm function

$$V(t) = \max_j \frac{|G_j|}{c_j}. \tag{9.5.6}$$

From the point of view of linear space theory, (9.5.6) may be regarded as conjugate (dual) to (9.5.2). Here also we shall prove that $dV/dt < 0$ except at equilibrium, wherever the derivative exists. Suppose that

$$V(t) = \frac{G_{j_0}[\mathbf{q}(t)]}{c_{j_0}} \operatorname{sign} G_{j_0}[\mathbf{q}(t)]$$

and $G_{j_0}[\mathbf{q}(t)] \neq 0$. Then

$$\frac{dV}{dt} = \frac{\operatorname{sign} G_{j_0}[\mathbf{q}(t)]}{c_{j_0}} \sum_k \frac{\partial G_{j_0}}{\partial p_k} G_k \leq -\frac{|G_{j_0}|}{c_{j_0}} \left(-\frac{\partial G_{j_0}}{\partial p_{j_0}}\right) + \frac{1}{c_{j_0}} \sum_{k \neq j_0} c_k \left|\frac{\partial G_{j_0}}{\partial p_k}\right| \frac{|G_k|}{c_k}.$$

But by the definition of j_0

$$\frac{|G_j|}{c_j} \leq \frac{|G_{j_0}|}{c_{j_0}},$$

and hence we have

$$\frac{dV}{dt} \leq -\frac{|G_{j_0}|}{c_{j_0}} \left|\frac{\partial G_{j_0}}{\partial p_{j_0}}\right| + \frac{|G_{j_0}|}{c_{j_0}^2} \sum_{k \neq j_0} c_k \left|\frac{\partial G_{j_0}}{\partial p_k}\right| < 0, \tag{9.5.7}$$

the last inequality following from the hypothesis of the theorem and the fact that $|G_{j_0}(t)| \neq 0$. Of course, when $\mathbf{G} = \mathbf{0}$ we are at an equilibrium price vector and we stay there for all t. When dV/dt does not exist, we estimate

$$\lim_{h \to 0+} \frac{V(t+h) - V(t)}{h},$$

which always exists, and modify the arguments accordingly (see Theorem 9.5.1). Global stability follows from (9.5.7) in the usual way.

Another criterion of a different sort which also yields global stability is the requirement that $\mathbf{B}(\mathbf{q}) = \|\partial G_i(\mathbf{q})/\partial q_j\|$ be negative quasi-definite.

(See p. 305 for the definition of a negative quasi-definite matrix.) A slight generalization of this criterion is embodied in the following theorem.

▶ THEOREM 9.5.3. Let $\mathbf{S}$ denote a positive definite matrix of order r. If $\mathbf{B}(\mathbf{q})'\mathbf{S} + \mathbf{S}\mathbf{B}(\mathbf{q})$ is negative definite for every price vector $\mathbf{q}$, then the normalized process is globally stable.

(Note that when $\mathbf{S}$ is the identity matrix, the hypothesis reduces to the statement that $\mathbf{B}(\mathbf{q})$ is negative quasi-definite.)

Proof. Here we introduce the metric

$$V(t) = (\mathbf{G}, \mathbf{S}\mathbf{G}) \tag{9.5.8}$$

for $\mathbf{G}[\mathbf{q}(t)]$. Differentiation of $V(t)$ produces

$$V'(t) = (\mathbf{G}, [\mathbf{B}(\mathbf{q})'\mathbf{S} + \mathbf{S}\mathbf{B}(\mathbf{q})]\mathbf{G}),$$

which is strictly negative unless $\mathbf{G} = \mathbf{0}$. Global stability now follows in a routine fashion.

We close this section by outlining a general approach to the problem of ascertaining the stability of dynamic processes.

Consider the system of differential equations

$$\frac{dx_i}{dt} = f_i(\mathbf{x}) \qquad (i = 1, \ldots, r), \tag{9.5.9}$$

where $\langle f_1(\mathbf{x}), \ldots, f_r(\mathbf{x})\rangle$ is a continuous vector function defined on a set $X \subset E^r$.

It is assumed that the set X is closed and has a nonvoid interior X^0; and that the solution $\mathbf{x}(t; \mathbf{x}^0)$ to the system with any initial position $\mathbf{x}^0$ in X^0 is uniquely determined, is continuous with respect to $\mathbf{x}^0$, and remains in X.

Let E be the set of all equilibrium vectors in X, i.e.,

$$E = \{\bar{\mathbf{x}} | \bar{\mathbf{x}} \text{ in } X, \mathbf{f}(\bar{\mathbf{x}}) = \mathbf{0}\},$$

and define the distance from a point $\mathbf{x}$ in X to the set E by

$$V(\mathbf{x}) = \inf_{\bar{x} \text{ in } E} |\mathbf{x} - \bar{\mathbf{x}}|^2.$$

The system (9.5.9) is called *quasi-stable* if, for any initial position $\mathbf{x}^0$ in X^0,

$$\lim_{t\to\infty} V[\mathbf{x}(t; \mathbf{x}^0)] = 0.$$

We state without proof the following general theorem.

▶ THEOREM 9.5.4. If for any initial position $\mathbf{x}^0$ in X^0, the solution to the system (9.5.9) is bounded, and if there exists a function $\phi(\mathbf{x})$ defined on X

which satisfies a Lipschitz condition in X such that $D_{\phi,f}(\mathbf{x}) < 0$ for any $\mathbf{x}$ in $X - E$, then the system (9.5.9) is quasi-stable.

The function $D_{\phi,f}$ is defined as follows: In general,

$$D_{\phi,f}(\mathbf{x}) = \overline{\lim_{\Delta x \to 0}} \sum \frac{\Delta_i \phi(\mathbf{x})}{\Delta x_i} f_i(\mathbf{x}),$$

where

$$\Delta_i \phi(\mathbf{x}) = \phi(x_1 + \Delta x_1, \ldots, x_i + \Delta x_i, x_{i+1}, \ldots, x_r)$$
$$- \phi(x_1 + \Delta x_1, \ldots, x_{i-1} + \Delta x_{i-1}, x_i, x_{i+1}, \ldots, x_r).$$

If ϕ is differentiable, then

$$D_{\phi,f} = \sum_i \frac{\partial \phi}{\partial x_i} f_i.$$

The theorem can be proved by a suitable adaptation of the reasoning used in proving Theorems 9.4.3 and 9.4.4 of the previous section.

As a special application of this theorem, consider the situation in which $\mathbf{f}(\mathbf{x})$ has a symmetrical matrix of the partial derivatives, i.e., $\partial f_i/\partial x_j = \partial f_j/\partial x_i$ for all i and j. From the theory of exact differentials, we infer the existence of a (potential) function ϕ such that

$$-\frac{\partial \phi}{\partial x_i} = f_i(\mathbf{x}).$$

Therefore

$$\sum_{i=1}^{r} \frac{\partial \phi}{\partial x_i} f_i(\mathbf{x}) = -\sum_{i=1}^{r} f_i(\mathbf{x})^2 < 0, \qquad \text{if} \quad \mathbf{f}(\mathbf{x}) \neq \mathbf{0},$$

and the theorem may be applied. In this connection the reader may consult the notes at the close of the chapter (see also Problem 14).

In line with these concepts we give a general result on quasi-stability which applies in the case of an excess demand vector function induced by a utility maximization for a single consumer.

Let $U(\mathbf{x})$ be a utility function defined for all $\mathbf{x} > \mathbf{0}$ in E^{r+1} and obeying the inequality

$$\frac{\partial U}{\partial x_i} > 0 \tag{9.5.9}$$

for at least one index i. (This essentially means that if budget considerations are disregarded, then no consumption vector exists which will saturate the individual's desires.)

The individual consumer selects the consumption vector which maximizes his utility function subject to his budget constraint. Formally, we postulate the existence of a unique $\mathbf{x}^*(\mathbf{p})$ for which

$$U(\mathbf{x}^*) = \max_x U(\mathbf{x})$$

under the condition

$$\sum_{i=0}^{r} p_i x_i = \sum_{i=0}^{r} p_i x_i^0$$

and $\mathbf{x} \geq \mathbf{0}$. Here, $\mathbf{p} > \mathbf{0}$ represents a fixed price vector and $\mathbf{x}^0 \gg \mathbf{0}$ is interpreted as the fixed vector of initial holdings. The maximization can be carried out by the techniques of Lagrange multipliers. There result the necessary conditions

$$\frac{\partial U}{\partial x_i}(\mathbf{x}^*) \leq \lambda p_i, \qquad (i = 0, 1, \ldots, r), \tag{9.5.10}$$

with equality in the ith relation if $x_i^* > 0$. Assumption (9.5.9) implies $\lambda > 0$.

For this setup we define the price adjustment process by (9.4.14), where $\mathbf{F}(\mathbf{p}) = \mathbf{x}^0 - \mathbf{x}^*(\mathbf{p})$. We can now prove that this process is quasi-stable. To this end, consider the function $\phi[\mathbf{p}(t)] = U(\mathbf{x}^*(\mathbf{p}))$. We have

$$\dot{\phi}[\mathbf{p}(t)] = \sum_{i=0}^{r} \frac{\partial U}{\partial x_i}(x^*)\frac{dx_i^*}{dt} \leq \lambda \sum_{i=0}^{r} p_i \frac{dx_i^*}{dt}, \tag{9.5.11}$$

since either $x_i^* > 0$ and equality holds in (9.5.10) or $x_i^* = 0$ and then clearly $\dot{x}_i^* \geq 0$.

With the aid of (9.4.14) and the identity

$$\sum_{i=0}^{r} p_i \frac{\partial x_i^*}{\partial p_j} = -x_j^* + x_j^0 \qquad (j = 0, 1, \ldots, r)$$

resulting from differentiation of the budget constraint, we can reduce (9.5.11) to

$$\dot{\phi}(t) \leq -\lambda \sum_{j=0}^{r} (x_j^0 - x_j^*) H_j(F_j(t)) \leq 0, \tag{9.5.12}$$

with strict inequality unless $F_j(\mathbf{p}(t)) = 0$. Quasi-stability is a consequence of the inequality (9.5.12) in the usual way.

For further stability results we refer the reader to Problems 5 through 15 and to a discussion of their solutions on page 355.

***9.6 A difference-equations formulation of global stability.** In this section we describe the price adjustment mechanism over time by means of a system of difference equations. Explicitly,

$$\mathbf{p}(t+1) = \max\{\mathbf{0}, \mathbf{p}(t) + \rho\mathbf{F}[\mathbf{p}(t)]\} \qquad (t = 0, 1, 2, \ldots), \qquad (9.6.1)$$

where $\mathbf{p}(t)$ is the price vector at time t and $\mathbf{F}$ represents the excess demand function and ρ is a fixed positive constant.

This formulation is harder to analyze than the differential-equations formulation. Nevertheless, it will permit us to prove the existence of an equilibrium price vector in a simpler manner. Furthermore, in the circumstance of strict gross substitutability, where the equilibrium vector is unique, we shall establish stability in the large.

We assume as always that the Walras law,

$$\sum_{i=0}^{r} p_i F_i(p_0, p_1, \ldots, p_r) = 0, \qquad (9.6.2)$$

is valid for any nonnegative price vector.

The system investigated here corresponds to the nonnormalized process, and in this case it is customary to suppose that the excess demand functions are homogeneous functions of degree zero.

First we shall simply prove the existence of an equilibrium price vector. Recall that an equilibrium price is a price vector $\mathbf{p}^* = \langle p_0^*, p_1^*, \ldots, p_r^* \rangle$ for which

$$F_i(\mathbf{p}^*) \leq 0. \qquad (9.6.3)$$

Let every nonzero price vector be normalized as

$$\sum_{i=0}^{r} p_i = 1.$$

We denote the set of all normalized price vectors by $\mathcal{Y}$. We observe that for $\mathbf{p} \in \mathcal{Y}$

$$\sum_{i=0}^{r} \max[0, p_i + \rho F_i(\mathbf{p})] = \lambda(\mathbf{p}) > 0.$$

Otherwise $p_i + \rho F_i(\mathbf{p}) \leq 0$ for all i, and thus

$$\sum_{i=0}^{r} p_i^2 = \sum_{i=0}^{r} p_i^2 + \rho \sum_{i=0}^{r} p_i F_i(\mathbf{p}) \leq 0,$$

the first equality being true by virtue of the Walras law. But then $p_i = 0$ for all i, which is contradictory to the supposition that $\mathbf{p} \in \mathcal{Y}$.

We now define a mapping $\mathbf{T}$ of $\mathcal{Y}$ into $\mathcal{Y}$ by

$$\mathbf{T}(\mathbf{p}) = \frac{1}{\lambda(\mathbf{p})} \max [\mathbf{0}, \mathbf{p} + \rho \mathbf{F}(\mathbf{p})]$$

(the maximum operation is taken componentwise). It is obvious that $\mathbf{T}(\mathbf{p})$ is a continuous mapping of $\mathcal{Y}$ into $\mathcal{Y}$. We may therefore apply the Brouwer fixed-point theorem and infer the existence of $\bar{\mathbf{p}} \in \mathcal{Y}$ such that $\bar{\mathbf{p}} = \mathbf{T}(\bar{\mathbf{p}})$, that is,

$$\lambda(\bar{\mathbf{p}})\bar{\mathbf{p}} = \max [\mathbf{0}, \bar{\mathbf{p}} + \rho F(\bar{\mathbf{p}})].$$

Then

$$\lambda(\bar{\mathbf{p}})\bar{p}_i \geq \bar{p}_i + \rho F_i(\bar{\mathbf{p}}) \qquad (i = 0, 1, \ldots, r), \tag{9.6.4}$$

with equality for all i such that $\bar{p}_i > 0$. Hence, multiplying by $\bar{p}_i$ and adding the resulting equations,

$$\lambda(\bar{\mathbf{p}}) \sum_{i=0}^{r} \bar{p}_i^2 = \sum_{i=0}^{r} \bar{p}_i^2 + \rho \sum_{i=0}^{r} \bar{p}_i F_i(\bar{\mathbf{p}}) = \sum_{i=0}^{r} \bar{p}_i^2.$$

Therefore $\lambda(\bar{\mathbf{p}}) = 1$, and (9.6.4) shows that $\bar{\mathbf{p}}$ is an equilibrium vector.

The existence of equilibrium having been established, we turn to a discussion of the dynamics of stability. The following theorem gives one criterion for stability of the discrete time system (9.6.1).

▶THEOREM 9.6.1. Let $\mathbf{p}^*$ denote a strictly positive equilibrium vector, and suppose that the excess demand vector function $\mathbf{F}$ satisfies the property of strict gross substitutability in the large. Then there exists a positive ρ_0 such that for $\rho < \rho_0$ the price system $\mathbf{p}(t)$ satisfying (9.6.1) converges to an equilibrium price vector.

Remark 9.6.1. The conclusion of this theorem can be interpreted as follows: If only relative prices are relevant, then the discrete-time price adjustment process, as defined by the system (9.6.1) is, globally stable.

Proof. Consider the norm function

$$V(t) = \sum_{i=0}^{r} [p_i(t) - p_i^*]^2.$$

A direct calculation, using (9.6.1), shows that

$$V(t) - V(t+1) = \sum_{i=0}^{r} p_i^2(t) - p_i^2(t+1) - 2[p_i(t) - p_i(t+1)]p_i^*$$

$$\geq \sum_{i=0}^{r} p_i^2(t) - p_i^2(t+1) + 2\rho \sum_{i=0}^{r} p_i^* F_i[\mathbf{p}(t)]. \tag{9.6.5}$$

Let R denote these indices of the set $\{1, 2, \ldots, r\}$ for which $p_i(t+1) = p_i(t) + \rho F_i[\mathbf{p}(t)]$, and R' the complementary set of indices, those for which $p_i(t+1) = 0$. We note that

$$\sum_{i \in R} [p_i^2(t) - p_i^2(t+1)] = -\rho^2 \sum_{i \in R} F_i^2[\mathbf{p}(t)] - 2\rho \sum_{i \in R} p_i(t) F_i[\mathbf{p}(t)]$$

$$= -\rho^2 \sum_{i \in R} F_i^2[\mathbf{p}(t)] + 2\rho \sum_{i \in R'} p_i(t) F_i[\mathbf{p}(t)],$$

the last equality resulting by virtue of the Walras law (9.6.2). Thus

$$\sum_{i=1}^{r} [p_i^2(t) - p_i^2(t+1)] = -\rho^2 \sum_{i=1}^{r} F_i^2[\mathbf{p}(t)] + \sum_{i \in R'} \{[p_i(t)] + \rho F_i[\mathbf{p}(t)]\}^2$$

$$\geq -\rho^2 \sum_{i=1}^{r} F_i^2[\mathbf{p}(t)].$$

We choose ρ satisfying

$$0 < \rho < \min_{p \neq \lambda p^*} \frac{2\sum_{i=0}^{r} p_i^* F_i(\mathbf{p})}{\sum_{i=0}^{r} F_i^2(\mathbf{p})}. \tag{9.6.6}$$

That this is feasible is proved as follows. We expand the numerators and denominators of the right-hand side of (9.6.6) in the neighborhood of $\mathbf{p}^*$, and simplify to

$$S(\mathbf{p}) = \frac{\sum_{j,k=0}^{r} p_j \left(\frac{\partial F_j(\mathbf{p}^*)}{\partial p_k} + \frac{\partial F_k(\mathbf{p}^*)}{\partial p_j} \right) p_k}{\sum_{j,k=0}^{r} p_k p_j \sum_{i=0}^{r} \frac{\partial F_i(\mathbf{p}^*)}{\partial p_j} \frac{\partial F_i(\mathbf{p}^*)}{\partial p_k}}$$

[see (9.3.6)]. A further expansion of the two quadratic forms occurring in the numerator and denominator of $S(\mathbf{p})$ will show that since strict gross substitutability prevails,

$$\inf_{p \neq \lambda p^*} S(\mathbf{p}) > 0.$$

We next observe by virtue of Theorem 9.4.2 that

$$\sum_{i=0}^{r} p_i^* F_i(\mathbf{p}) > 0, \qquad \sum_{i=0}^{r} F_i^2(\mathbf{p}) > 0$$

for all $\mathbf{p} \neq \lambda \mathbf{p}^*$.

The continuity and homogeneity of $F_i(\mathbf{p})$, plus the fact that $S(\mathbf{p})$ is bounded away from zero, enable us to deduce the validity of (9.6.6). For any selection of ρ satisfying (9.6.6), it follows from (9.6.5) that

$$V(t) > V(t+1) \tag{9.6.7}$$

unless $\mathbf{p}(t) = \lambda\mathbf{p}^*$. Consequently,

$$\lim_{t\to\infty} V(t)$$

exists and any limit point of $\mathbf{p}(t)$ is a multiple of $\mathbf{p}^*$, as was shown in the proof of Theorem 9.3.4. Since it is now easy to establish the convergence of $\mathbf{p}(t)$, we omit the details; the proof of the theorem is hereby complete.

9.7 Stability and expectations (Model I). The previous models involved the adjustment of prices over time in response to changes in the excess demand functions, which are functions of the current market prices only. In this section and the next we examine the influence of anticipated future prices in addition to actual prices on the dynamic stability of the market system.

With this end in view, we propose to write the general dynamic equilibrium system as

$$\frac{d\mathbf{p}}{dt} = \mathbf{KF}, \tag{9.7.1}$$

where $\mathbf{p}$ is the price vector of the commodity bundle of r components, $\mathbf{K}$ is an $r \times r$ matrix of constants which determines how rapidly prices adjust to excess demands, and $\mathbf{F}$ designates, as usual, the excess demand vector. Here, in general, each component of $\mathbf{F}$ is a function of the current price vector $\mathbf{p}$ and the expected future price vector $\mathbf{p}^f$. In discrete time the price vector $\mathbf{p}^f$ refers to the anticipated prices for the next time period. In continuous time $\mathbf{p}^f$ corresponds to anticipated prices in the next infinitesimal time period.

In our previous discussions we implicitly assumed that $\mathbf{p}^f = \mathbf{p}$, i.e., that expectations are static, and we accordingly took the matrix $\mathbf{K}$ as in essence a diagonal matrix with positive components. Here, by contrast, we permit the expected future prices to deviate from current prices according to a determinative law and in this way influence the fluctuation of current prices. We shall show later that a stable dynamic system can absorb the effects of some extrapolation of price movements, together with the corresponding expected future price changes, and remain stable. The effects of expected future prices will be considered only with regard to the structure of local stability.

At equilibrium $\mathbf{p} = \mathbf{p}^*$, it is customary to suppose that $\mathbf{p}^f = \mathbf{p}^*$, i.e., that the expectations and the actual prices agree. Hence, by a Taylor expansion we may approximate $F_i(\mathbf{p}, \mathbf{p}^f) = F_i(p_1, \ldots, p_r; p_1^f, \ldots, p_r^f)$ about an equilibrium price vector $\mathbf{p}^*$ by the linear expression

$$F_i = \sum_{j=1}^{r} a_{ij}(p_j - p_j^*) + \sum_{j=1}^{r} b_{ij}(p_j^f - p_j^*), \tag{9.7.2}$$

which is valid in the neighborhood of $\mathbf{p}^*$.

The problem of ensuring the local stability of the system (9.7.1) can thus be reduced to the problem of stability of the linear system of differential equations

$$\frac{d\mathbf{z}}{dt} = \mathbf{K}(\mathbf{A}\mathbf{z} + \mathbf{B}\mathbf{w}), \tag{9.7.3}$$

where $\mathbf{z}(t) = \mathbf{p}(t) - \mathbf{p}^*$ and $\mathbf{w}(t) = \mathbf{p}^f(t) - \mathbf{p}^*$. (The proof of this assertion is essentially the same as the argument of p. 306.)

The linear system (9.7.3) involves the two functions $\mathbf{z}(t)$ and $\mathbf{w}(t)$. Therefore, to completely specify the dynamics of the price system it is necessary to indicate how $\mathbf{w}(t)$ is related to current prices. The simplest assumption is that changes in the expected future price of the ith commodity are induced by changes in its actual price according to the relationship

$$p_i^f = p_i + \eta_i \frac{dp_i}{dt}, \tag{9.7.4}$$

where η_i is a constant. When $\eta_i = 0$, current prices are expected to persist. When $\eta_i > 0$, some multiple of the change in prices is added to the current price in arriving at the expected price; hence the expected price may be described as extrapolative. When $\eta_i < 0$, expected prices do not change as much as actual prices.

In vector notation, (9.7.4) may be written as

$$\mathbf{p}^f = \mathbf{p} + \boldsymbol{\eta} \frac{d\mathbf{p}}{dt}, \tag{9.7.5}$$

where $\boldsymbol{\eta}$ is a diagonal matrix. We generalize immediately and allow $\boldsymbol{\eta}$ to be an arbitrary constant matrix. In the notation of (9.7.3),

$$\mathbf{w}(t) = \mathbf{z}(t) + \boldsymbol{\eta} \frac{d\mathbf{z}(t)}{dt}. \tag{9.7.6}$$

Inserting (9.7.6) into (9.7.3), we obtain

$$\frac{d\mathbf{z}}{dt} = [\mathbf{I} - \mathbf{K}\mathbf{B}\boldsymbol{\eta}]^{-1}\mathbf{K}(\mathbf{A} + \mathbf{B})\mathbf{z}, \tag{9.7.7}$$

provided that the matrix $(\mathbf{I} - \mathbf{K}\mathbf{B}\boldsymbol{\eta})^{-1}$ exists. We shall assume for convenience in what follows that the inverse matrix is well defined. In the case of static expectations ($\mathbf{p}^f = \mathbf{p}$), the corresponding dynamic system is

$$\frac{d\mathbf{z}}{dt} = \mathbf{K}(\mathbf{A} + \mathbf{B})\mathbf{z}. \tag{9.7.8}$$

We shall now develop conditions on the matrices appearing in the systems (9.7.7) and (9.7.8) that imply the stability or instability of the two systems together.

The problem discussed above may be stated formally as follows:

Let $\mathbf{T}$ denote a stable matrix (all eigenvalues of $\mathbf{T}$ have negative real parts). Two questions are of interest. First, under what conditions on the matrix $\mathbf{Q}$ is $\mathbf{QT}$ again stable? Second, what properties must $\mathbf{T}$ and $\mathbf{Q}$ have for $\mathbf{T}$ and $\mathbf{QT}$ to be simultaneously stable? Theorems 9.7.1 and 9.7.2 provide partial answers to these questions (see also Problems 1–4).

▶THEOREM 9.7.1. Let $\mathbf{T}$ be a stable M-matrix (see p. 255), and let $\mathbf{Q}$ be a real diagonal matrix. Then $\mathbf{QT}$ is stable if and only if the diagonal elements of $\mathbf{Q}$ are strictly positive.

Proof. (Sufficiency.) The hypothesis asserts that the diagonal elements of $\mathbf{Q}$ are strictly positive. Clearly $\mathbf{QT}$ is an M-matrix. Moreover, $-(\mathbf{QT})^{-1} = -\mathbf{T}^{-1}\mathbf{Q}^{-1}$ is positivity-preserving (by property (c) of p. 256), and hence $\mathbf{QT}$ is stable.

The proof of necessity turns upon the following lemma.

▶LEMMA 9.7.1. Let $\mathbf{T}$ be a stable M-matrix of order r. If $\mathbf{D}$ is a nonsingular diagonal matrix, then $\mathbf{DT}$ has no eigenvalues on the imaginary axis.

Proof. Suppose to the contrary that $\omega\lambda$ is an eigenvalue of $\mathbf{Q} = \mathbf{DT}$. (Here $\omega^2 = -1$, and obviously $\lambda \neq 0$ since $\mathbf{T}$ is nonsingular.) Consequently, there exists a nonzero vector $\mathbf{x}$ such that

$$d_{ii}\sum_{j=1}^{r} t_{ij}x_j = \omega\lambda x_i \qquad (i = 1, 2, \ldots, r). \tag{9.7.9}$$

Adding a sufficiently large positive multiple μ of the identity makes it certain that $\mathbf{T} + \mu\mathbf{I} = \mathbf{S}$ is a nonnegative matrix. Equation (9.7.9) becomes

$$\sum_{j=1}^{r} S_{ij}x_j = \left(\mu + \frac{\omega\lambda}{d_{ii}}\right)x_i. \tag{9.7.10}$$

Taking absolute values and identifying $y_i = |x_i|$, we obtain

$$\mathbf{Sy} \geq \left(\min_i \left|\mu + \frac{\omega\lambda}{d_{ii}}\right|\right)\mathbf{y} \geq \mu\mathbf{y}$$

and $\mathbf{y} > \mathbf{0}$. From the characterization of the spectral radius of nonnegative matrices as given in Corollary 8.2.1 it follows that there exists an eigenvalue $\lambda_0(\mathbf{S}) \geq \mu$. Consequently $\mathbf{T}$ has a nonnegative eigenvalue, contrary to the hypothesis.

Proof of Theorem 9.7.1. (Necessity.) We first observe that since $\mathbf{QT}$ is stable, $\mathbf{Q}$ is nonsingular.

Let θ represent a real parameter varying between 0 and 1, and consider the family of matrices

$$\mathbf{U}(\theta) = \mathbf{Q}[\theta\mathbf{T} - (1 - \theta)\mathbf{I}].$$

For each θ $(0 \leq \theta \leq 1)$, the bracketed term is clearly a stable M-matrix; therefore, by the lemma, $\mathbf{U}(\theta)$ never has an imaginary eigenvalue. But the eigenvalues vary continuously, and hence for each θ $(0 \leq \theta \leq 1)$ they always lie interior to the left half complex plane, since that is the situation for $\theta = 1$. Using this fact for $\theta = 0$, we infer that the diagonal elements of $\mathbf{Q}$ are all positive, as was to be proved.

▶ THEOREM 9.7.2. If $\mathbf{S}$ is symmetric and $\mathbf{T}$ is a real negative quasi-definite matrix, $\mathbf{ST}$ is stable if and only if $\mathbf{S}$ is positive definite.

Remark 9.7.1. We remind the reader that a matrix $\mathbf{T}$ is negative quasi-definite if and only if $(\mathbf{x}, \mathbf{Tx})$ is negative for every nonzero real vector $\mathbf{x}$.

A real negative quasi-definite matrix is stable. We prove this assertion as follows. Suppose that

$$\mathbf{Tx} = \lambda \mathbf{x}, \tag{9.7.11}$$

where $\mathbf{x} = \mathbf{x}_1 + i\mathbf{x}_2$ and $\lambda = \lambda_1 + i\lambda_2$, with $\mathbf{x}_1$ and $\mathbf{x}_2$ real vectors not both zero. Rewriting the real and imaginary parts of (9.7.11), we obtain

$$\mathbf{Tx}_1 = \lambda_1\mathbf{x}_1 - \lambda_2\mathbf{x}_2, \qquad \mathbf{Tx}_2 = \lambda_1\mathbf{x}_2 + \lambda_2\mathbf{x}_1.$$

Hence

$$(\mathbf{x}_1, \mathbf{Tx}_1) + (\mathbf{x}_2, \mathbf{Tx}_2) = \lambda_1[(\mathbf{x}_1, \mathbf{x}_1) + (\mathbf{x}_2, \mathbf{x}_2)],$$

and we conclude that $\lambda_1 < 0$.

Proof of Theorem 9.7.2. (Sufficiency.) If $\mathbf{S}$ is positive definite, $\mathbf{S}^{1/2}\mathbf{T}\mathbf{S}^{1/2}$ is real negative quasi-definite and hence stable, where $\mathbf{S}^{1/2}$ is real and positive definite. But $\mathbf{R}\mathbf{S}^{1/2}\mathbf{T}\mathbf{S}^{1/2}\mathbf{R}^{-1} = \mathbf{ST}$ is stable for $\mathbf{R} = \mathbf{S}^{1/2}$, since it arises from a stable matrix by a similarity transformation.

(Necessity.) We show that if $\mathbf{S}$ is nonsingular symmetric and $\mathbf{T}$ is real negative quasi-definite, then $\mathbf{ST}$ has no purely imaginary eigenvalue. To this end, suppose that there exists a nonnull vector $\mathbf{x}$ such that $\mathbf{STx} = \lambda\mathbf{x}$. Premultiply both sides by $\bar{\mathbf{x}}\mathbf{S}^{-1}$, where $\bar{\mathbf{x}}$ denotes the conjugate of $\mathbf{x}$:

$$(\bar{\mathbf{x}}\mathbf{S}^{-1}, \mathbf{STx}) = (\bar{\mathbf{x}}, \mathbf{Tx}) = \lambda(\bar{\mathbf{x}}, \mathbf{S}^{-1}\mathbf{x}). \tag{9.7.12}$$

Write $\mathbf{x} = \mathbf{y} + i\mathbf{z}$ and $\lambda = \alpha + i\beta$, where $\mathbf{y}$, $\mathbf{z}$, α, and β are all real. Equating real parts of (9.7.12) gives

$$(\mathbf{y}, \mathbf{Ty}) + (\mathbf{z}, \mathbf{Tz}) = \alpha[(\mathbf{y}, \mathbf{S}^{-1}\mathbf{y}) + (\mathbf{z}, \mathbf{S}^{-1}\mathbf{z})]. \tag{9.7.13}$$

The left-hand side is strictly negative, and hence $\alpha \neq 0$. The remainder of the proof follows the proof of Theorem 9.7.1. We omit the details.

The economic significance of Theorem 9.7.1 is plain. The relation of M-matrices and gross substitutes has been duly discussed. The fact that the matrix $\mathbf{T} = \mathbf{A} + \mathbf{B}$ has nonnegative off-diagonal elements means that the system contains no complementary commodities. If in addition the matrices $\mathbf{K}$, $\mathbf{B}$, and $\boldsymbol{\eta}$ are all diagonal, then Theorem 9.7.1 asserts that the expectationless system is stable if and only if all the k_{ii} are positive. Moreover, for the same $\mathbf{T} = \mathbf{A} + \mathbf{B}$, stability remains unaffected by the introduction of the expectational coefficients if and only if $1 > k_{ii}b_{ii}\eta_{ii}$ for all i. Theorem 9.7.2 deals with another case, in which the income effects are essentially symmetric. Here we no longer require that $\mathbf{K}$, $\mathbf{B}$, and $\boldsymbol{\eta}$ be diagonal.

Further consequences along the lines of these theorems can be obtained by applying similarity transformations to the matrices $\mathbf{A}$ and $\mathbf{T}$ (see Problem 1).

9.8 Stability and expectations (Model II). A relationship between expected future prices $\mathbf{p}^f$ and actual prices was expressed by (9.7.4) and more generally by (9.7.5).

Another such relationship of some economic merit is

$$\frac{dp_i^f}{dt} = \beta_i[p_i(t) - p_i^f(t)] \qquad (i = 1, 2, \ldots, r). \tag{9.8.1}$$

When $\beta_i = 0$, changes in current prices have no effect on expected prices. The condition for static expectations is now $\beta_i = +\infty$.

For a given time path of prices $p_i(t)$, equation (9.8.1) has the solution

$$p_i^f(t) = p_i^f(0)e^{-\beta_i t} + e^{-\beta_i t}\int_0^t \beta_i p_i(u)e^{\beta_i u}\,du, \tag{9.8.2}$$

where $p_i^f(0)$ is the initial value of the expected price of the ith commodity. If the origin is in the sufficiently distant past, the term in (9.8.2) involving $p_i^f(0)$ is negligible; hence $p_i^f(t)$ may be taken as an exponentially weighted average of past prices. Consequently, (9.8.1) indicates that the expected price as formed is a properly weighted average of actual past prices [exponential in the case of (9.8.2)].

Still another rationale, attributable to Hicks, is as follows. If we think of time as a discontinuous variable, then the notion of "elasticity of expectations" may be formulated as

$$\frac{p_i^f(t) - p_i^f(t-1)}{p_i(t-1) - p_i^f(t-1)} = \beta_i, \tag{9.8.3}$$

where the prices are expressed in logarithms and β_i is a constant. When $\beta_i = 0$, changes in actual prices have no effect on expected price; when $\beta_i = 1$, current price is projected forward as people's expectations of the level of future prices. The value of β_i can be considered as a measure of the elasticity of people's expectations between the extreme values of $\beta = 0$ and $\beta = 1$.

If we drop the assumption that prices are expressed in logarithms, the differential-equations analog of (9.8.3) is (9.8.1).

We turn now to a discussion of the local stability of (9.7.1) in terms of the relationship (9.8.1). We seek to compare the conditions of stability for the expectationless system

$$\frac{d\mathbf{z}}{dt} = \mathbf{K}(\mathbf{A} + \mathbf{B})\mathbf{z}, \tag{9.8.4}$$

where $\mathbf{K}$ is a diagonal matrix and $\mathbf{A}$ and $\mathbf{B}$ are as in (9.7.2), with the conditions of stability for the dynamic system with expected future prices accounted for. This system is represented by

$$\frac{d\mathbf{z}}{dt} = \mathbf{K}(\mathbf{A}\mathbf{z} + \mathbf{B}\mathbf{w}), \qquad \frac{d\mathbf{w}}{dt} = \boldsymbol{\beta}\mathbf{z} - \boldsymbol{\beta}\mathbf{w}, \tag{9.8.5}$$

where $\mathbf{z} = \mathbf{p}(t) - \mathbf{p}^*$, $\mathbf{w} = \mathbf{p}^f(t) - \mathbf{p}^*$, and $\boldsymbol{\beta}$ is a diagonal matrix with the elements β_i along the diagonal and 0 elsewhere. The coefficient matrix of the system (9.8.5) is of the form

$$\mathbf{C} = \begin{pmatrix} \mathbf{KA} & \mathbf{KB} \\ \boldsymbol{\beta} & -\boldsymbol{\beta} \end{pmatrix},$$

where $\tilde{\mathbf{A}} = \mathbf{KA}$ is the matrix $\|k_{ii}a_{ij}\|$ and $\tilde{\mathbf{B}} = \mathbf{KB}$ is the matrix $\|k_{ii}b_{ij}\|$. The coefficient matrix for the system (9.8.4) is of the form

$$\mathbf{D} = \mathbf{K}(\mathbf{A} + \mathbf{B}) = \tilde{\mathbf{A}} + \tilde{\mathbf{B}}.$$

In order to simplify matters, we assume that all present commodities are gross substitutes for one another and also that each future commodity is a gross substitute for each present commodity. That is, we assume that $a_{ij} \geq 0$ $(i \neq j)$ and $b_{ij} \geq 0$ $(i, j = 1, \ldots, r)$. Since $\mathbf{K}$ and $\boldsymbol{\beta}$ are diagonal nonnegative matrices, we see that $\mathbf{C}$ and $\mathbf{D}$ are both M-matrices.

▶ THEOREM 9.8.1. The system (9.8.5) with matrix $\mathbf{C}$ is stable if and only if the system (9.8.4) with matrix $\mathbf{D}$ is stable.

Remark 9.8.1. The principal argument employed in the proof of the theorem hinges on two observations indicated earlier: (a) an M-matrix $\mathbf{T}$ is stable if and only if the spectral radius of the positive matrix $\mathbf{S} = (\mathbf{T} + \mu\mathbf{I})$ is smaller than μ, where μ is positive and sufficiently large to make $\mathbf{T} + \mu\mathbf{I}$ a nonnegative matrix (see p. 255); (b) the spectral radius $\lambda_0(\mathbf{S})$ of a nonnegative matrix $\mathbf{S}$ is characterized as the supremum of all λ for which there exists a positive vector $\mathbf{x}$ satisfying $\mathbf{S}\mathbf{x} \geq \lambda\mathbf{x}$.

Proof of Theorem 9.8.1. Let $\mathbf{U} = \mathbf{C} + \mu\mathbf{I}_{2r}$ and $\mathbf{V} = \mathbf{D} + \mu\mathbf{I}_r$, where μ is large enough so that $\mathbf{U}$ and $\mathbf{V}$ are both nonnegative matrices. (Of course, $\mathbf{U}$ is a matrix of order $2r$ and $\mathbf{V}$ is a matrix of order r.) Let $\lambda_0(\mathbf{U})$ and $\lambda_0(\mathbf{V})$ denote the spectral radii of $\mathbf{U}$ and $\mathbf{V}$, respectively. Suppose that μ is such that $\mathbf{Vx} \geq \mu\mathbf{x}$, where $\mathbf{x} = \langle x_1, \ldots, x_r \rangle$ and $\mathbf{x} > \mathbf{0}$, which is certainly the case if $\mathbf{D}$ is not zero. Define the vector $\mathbf{y}$ of $2r$ coordinates by $\mathbf{y} = \langle x_1, \ldots, x_r; x_1, \ldots, x_r \rangle$. Then, by immediate check, we have $\mathbf{Uy} \geq \mu\mathbf{y}$, and hence

$$\lambda_0(\mathbf{U}) \geq \lambda_0(\mathbf{V}) \geq \mu. \tag{9.8.6}$$

Thus, if $\mathbf{C}$ is stable we must have $\mathbf{D}$ stable, since in the contrary event we obtain $\lambda_0(\mathbf{U}) \geq \mu$, in contradiction to the assumption that $\mathbf{C}$ is stable. Next, consider μ such that

$$\mathbf{Uy}' \geq \mu\mathbf{y}' \tag{9.8.7}$$

for some $\mathbf{y}' > \mathbf{0}$. The last r inequalities of (9.8.7) reduce to

$$\beta_i(y'_i - y'_{i+r}) \geq 0,$$

and hence $y'_i \geq y'_{i+r} \geq 0$. Define $\mathbf{x}' = \langle y'_1, \ldots, y'_r \rangle$. Since $\mathbf{U}$ is nonnegative, we may increase the values of the last r components to the values y'_i, and the first r inequalities of (9.8.7) become

$$\mathbf{Vx}' = \tilde{\mathbf{A}}\mathbf{x}' + \tilde{\mathbf{B}}\mathbf{x}' + \mu\mathbf{Ix}' \geq \mathbf{Uy}' \geq \mu\mathbf{x}'. \tag{9.8.8}$$

Clearly, $\mathbf{x}' > \mathbf{0}$, and therefore

$$\lambda_0(\mathbf{V}) \geq \lambda_0(\mathbf{U}) \geq \mu. \tag{9.8.9}$$

Using the same reasoning as in the analysis of (9.8.6) we may infer from (9.8.9) the assertion that if $\mathbf{D}$ is stable, then $\mathbf{C}$ is stable. The proof is now complete.

It can be argued that the stability of the dynamic system under static expectations is plausible; hence, by the theorem proved, stability in general is plausible no matter what the inertia of the system or the elasticities of expectations.

9.9 The von Neumann model of an expanding economy. As we have seen, the equilibrium relations for the laws of supply and demand reduce mathematically to certain systems of equations. The problem is then to establish the existence of economically meaningful solutions to these equations, and it is solved by imposing various kinds of assumptions on the nature of the supply and demand functions. All theory of this sort—i.e., the sort discussed in Chapter 8—is concerned with the statics of the economic process; it yields no insight into the dynamic workings of the economy.

In this chapter (Sections 9.2–9.7) we have formulated the dynamic price adjustment process and investigated the stability of the system with respect to convergence of the price vector system to equilibrium. The price adjustment process can be regarded as an aspect of the short-run dynamics of the economy, in the sense that price changes upsetting the equilibrium situation are followed by readjustments toward equilibrium in response to the natural fluctuation of supply and demand.

We can also view the dynamics of the economy from a long-run perspective, in the sense of keeping account of actual amounts produced and consumed in each period, where the economy is assumed to operate in an efficient manner for an infinite time. A natural growth of production and consumption capacity appears to be associated with the economic process; we seek to determine the rate of this growth. This expansion process is known as the phenomenon of balanced growth.

We begin with a description of the classical model of an expanding economy proposed by von Neumann. Consider an economic system engaged in the production of n commodities ($i = 1, \ldots, n$). Let $j = 1, \ldots, m$ denote the activities that are carried out in the system. If the jth activity is carried out at the unit level of intensity, then a_{ij} and b_{ij} denote respectively the input and the output of the ith commodity in the jth activity. We make the natural assumption that $a_{ij} \geq 0$, $b_{ij} \geq 0$. Let $\mathbf{A} = \|a_{ij}\|$, $\mathbf{B} = \|b_{ij}\|$.

The level of intensity of the jth activity at time t will be denoted by $y_j(t)$. The price of the ith commodity will be denoted by $p_i(t)$, and the total amount of the ith commodity that is available by $x_i(t)$.

The basic assumptions imposed on the economic system are as follows:

(a) $x_i(t) \geq \sum_j a_{ij} y_j(t)$ for all i and t.

This is the obvious restriction that the amount of any commodity used at a particular time must not exceed the amount of that commodity available at that time.

(b) $x_i(t + 1) = x_i(t) + \sum_j (b_{ij} - a_{ij}) y_j(t)$ for all i and t.

This difference equation is the dynamic equation of the system. A continuous-time formulation of the dynamics of the system would involve replacing (b) by an analogous differential equation, but the essential features of the analysis would remain unchanged. There are, of course, prescribed initial conditions.

(c) $y_j(t) = y(t) y_j$, where $y(t) = e^{\lambda t}$.

The proportions between the intensities of the activities remain the same; only the total output increases with time. This assumption can be justified as an equilibrium condition for the economy.

(d) $\sum_{i,j} p_i(t+1)b_{ij}y_j(t) - \sum_{i,j} p_i(t)a_{ij}y_j(t) \leq 0$ for all $y_j(t)$.

There is no positive return from the system; it is a closed system.

(e) $p_i(t) = p(t)p_i$, where $p(t) = e^{-\mu t}$.

The λ in assumption (c) is the expansion rate, and the μ here is the discount rate of the system. Alternatively, $e^{\mu t}$ can be regarded as the rate at which the interest of a unit of money accumulates with time.

By some manipulations of the assumptions we now develop various relations satisfied by the expansion rate and interest rate. If we substitute the values of $p_i(t)$ and $y_j(t)$ in (d), we obtain

$$\sum_{i,j} p_i e^{-\mu(t+1)} b_{ij} y_j e^{\lambda t} - \sum_{i,j} p_i e^{-\mu t} a_{ij} y_j e^{\lambda t} \leq 0,$$

or

$$e^{-\mu} \sum_{i,j} p_i b_{ij} y_j \leq \sum_{i,j} p_i a_{ij} y_j$$

for all $\mathbf{y}$. Hence

$$e^{-\mu}\mathbf{pB} \leq \mathbf{pA}. \tag{9.9.1}$$

If we substitute the value of $y(t)$ in (b), we obtain

$$x_i(t+1) = x_i(t) + \sum_{j=1}^{m} (b_{ij} - a_{ij}) y_j e^{\lambda t}.$$

The solution of this first-order difference equation is given by

$$x_i(t) = x_i(0) + \frac{\sum_{j=1}^{m} (b_{ij} - a_{ij}) y_j}{1 - e^{\lambda}} + \frac{e^{\lambda t}}{e^{\lambda} - 1} \sum_j (b_{ij} - a_{ij}) y_j$$

$$= c_0 + \frac{1}{e^{\lambda} - 1} e^{\lambda t} \sum_{j=1}^{m} (b_{ij} - a_{ij}) y_j. \tag{9.9.2}$$

Using this result in (a), we obtain

$$c_0 + \frac{1}{e^{\lambda} - 1} e^{\lambda t} \sum_j (b_{ij} - a_{ij}) y_j \geq \sum_j a_{ij} y_j e^{\lambda t},$$

$$e^{-\lambda t}(e^{\lambda} - 1)c_0 + \sum_j b_{ij} y_j \geq e^{\lambda} \sum_j a_{ij} y_j.$$

Since this inequality holds for all values of t, we permit t to become infinite and the inequality reduces to

$$\sum b_{ij} y_j \geq e^{\lambda} \sum a_{ij} y_j, \qquad \text{or} \qquad \mathbf{By} \geq e^{\lambda}\mathbf{Ay} \tag{9.9.3}$$

In the following section, by imposing some additional mild positivity restrictions on $\mathbf{A}$ and $\mathbf{B}$, we shall establish the existence of nonnegative vectors $\mathbf{p}^0$, $\mathbf{y}^0$ fulfilling (9.9.3) and (9.9.1) when $\lambda = \mu > 0$. In this case we say that the economy undergoes a balanced growth such that the expansion rate and interest rate are equal. For other applications of this model the reader is referred to Problems 22 and 23.

9.10 A general model of balanced growth. In this section we formulate a general model of production which includes the von Neumann model as a particular case, and prove the existence of a unique rate of uniform expansion and its coincidence with the rate of interest. The dynamic Leontief model (see Section 8.6) may also be considered as a special case of the present model, and these consequences will also be indicated.

Consider an economy in which n distinct commodities are produced, and in which the technology of production obeys the laws of nonincreasing marginal rates of transformation (i.e., the production possibility set is convex) and of constant returns to scale (i.e., the production possibility set is a cone).

The technological possibility of production is described by the transformation set T of the economy, where T is a set of pairs of n vectors $\{\mathbf{x}, \mathbf{y}\}$ such that the production of output $\mathbf{y} = \langle y_1, \ldots, y_n \rangle$ is technologically possible from input $\mathbf{x} = \langle x_1, \ldots, x_n \rangle$ if and only if $\{\mathbf{x}, \mathbf{y}\} \in T$.

The transformation set T is assumed to possess the following properties:

(T_1) T is a closed convex cone in the nonnegative orthant of the $2n$-vector space.

(T_2) $\{\mathbf{x}, \mathbf{y}\} \in T$ and $\mathbf{x}' \geq \mathbf{x}$, $\mathbf{0} \leq \mathbf{y}' \leq \mathbf{y}$ imply that $\{\mathbf{x}', \mathbf{y}'\} \in T$. This assumption states that the disposal activity is costless. A smaller output is certainly possible with a larger input.

(T_3) $\{\mathbf{0}, \mathbf{y}\} \in T$ implies that $\mathbf{y} = \mathbf{0}$. That is, it is impossible to produce something from nothing.

(T_4) For any $i = 1, \ldots, n$ there exists $\{\mathbf{x}^i, \mathbf{y}^i\} \in T$ such that $y_i^i > 0$. That is, every commodity can be produced. (T_4), in view of (T_1), is equivalent to the following assertion:

(T_4') There exists $\{\mathbf{x}^0, \mathbf{y}^0\} \in T$ such that $\mathbf{y}^0 \gg \mathbf{0}$.

For a possible nonzero input-output relation $\{\mathbf{x}, \mathbf{y}\} \in T$, the rate of expansion $\lambda(\mathbf{x}, \mathbf{y})$ will be defined by

$$\lambda(\mathbf{x}, \mathbf{y}) = \max \{\lambda | \mathbf{y} \geq \lambda \mathbf{x}\}. \tag{9.10.1}$$

By (T_1) and (T_3)

$$0 \leq \lambda(\mathbf{x}, \mathbf{y}) < \infty. \tag{9.10.2}$$

▶THEOREM 9.10.1. There exists $\{\mathbf{x}^*, \mathbf{y}^*\} \in T$ such that

$$\mathbf{y}^* = \lambda^* \mathbf{x}^*, \qquad \lambda^* = \lambda(\mathbf{x}^*, \mathbf{y}^*) \qquad (\mathbf{x}^* > \mathbf{0}), \tag{9.10.3}$$

and

$$\lambda^* \geq \lambda(\mathbf{x}, \mathbf{y}) \tag{9.10.4}$$

for any $\{\mathbf{x}, \mathbf{y}\} \in T$ with $\mathbf{x} > \mathbf{0}$.

Proof. Let

$$\lambda^* = \sup_{\substack{\{x,y\} \in T \\ x > 0}} \lambda(\mathbf{x}, \mathbf{y}). \tag{9.10.5}$$

Then $0 < \lambda^* < \infty$. The left-hand inequality is easily demonstrated: by taking $\{\mathbf{x}^0, \mathbf{y}^0\} \in T$ with $\mathbf{y}^0 \gg \mathbf{0}$, we obtain

$$\lambda^* \geq \lambda(\mathbf{x}^0, \mathbf{y}^0) > 0.$$

As for the right-hand inequality, if λ^* is not finite, then there exists a sequence $\{\mathbf{x}^\nu, \mathbf{y}^\nu\} \in T$ such that $\mathbf{y}^\nu > 0$, $\mathbf{y}^\nu \geq \lambda^\nu \mathbf{x}^\nu$, and $\lim \lambda^\nu = \infty$. Normalizing $\mathbf{y}^\nu$ such that

$$\sum_{i=1}^n y_i^\nu = 1$$

and taking a limit point $\bar{\mathbf{y}}$ of $(\mathbf{y}^\nu)$, we are led to the existence of $\{\mathbf{0}, \bar{\mathbf{y}}\} \in T$ such that $\bar{\mathbf{y}} > \mathbf{0}$. This contradicts ($T_3$).

Since T is a closed cone, there exists a vector pair $\{\bar{\mathbf{x}}, \bar{\mathbf{y}}\} \in T$ with $\lambda^* = \lambda(\bar{\mathbf{x}}, \bar{\mathbf{y}})$, where λ^* is defined by (9.10.5). By (T_2) we can find $\{\mathbf{x}^*, \mathbf{y}^*\}$ which satisfies (9.10.3) and (9.10.4). This completes the proof of the theorem.

The vector pair $\{\mathbf{x}^*, \mathbf{y}^*\}$ is referred to as a *balanced growth* and λ^* as the *rate* of *balanced growth.*

Associated with the balanced growth we have also an equilibrium price system, as indicated in the following theorem.

▶THEOREM 9.10.2. There exists a vector $\mathbf{p}^*$ such that $\mathbf{p}^* > \mathbf{0}$ and

$$(\mathbf{p}^*, \mathbf{y}) \leq \lambda^*(\mathbf{p}^*, \mathbf{x}) \tag{9.10.6}$$

for all $\{\mathbf{x}, \mathbf{y}\} \in T$.

Proof. Consider a set V in E^n defined as follows:

$$V = \left\{\mathbf{y} - \lambda^*\mathbf{x} | \{\mathbf{x}, \mathbf{y}\} \in T, \quad \sum_{i=1}^n (x_i + y_i) \leq 1\right\}.$$

Since T is a closed convex set, V is also a closed convex set. Moreover, from the very meaning of λ^* we note that V does not overlap the interior of the positive orthant O_n of E^n. Hence there exists a hyperplane through the origin which separates O_n and V (Theorem B.1.3). That is, there exists a nonzero vector $\mathbf{p}^* = \langle p_1^*, \ldots, p_n^* \rangle$ such that

$$(\mathbf{p}^*, \mathbf{y} - \lambda^*\mathbf{x}) \leq 0 \tag{9.10.7}$$

for all $\{\mathbf{x}, \mathbf{y}\} \in T$ (the normalization being irrelevant) and

$$(\mathbf{p}^*, \mathbf{z}) \geq 0 \tag{9.10.8}$$

for all $\mathbf{z}$ in O_n. It follows readily from (9.10.8) that $\mathbf{p}^* > \mathbf{0}$. Relation (9.10.7) is exactly the statement of the theorem.

The von Neumann model. The von Neumann model of an expanding economy may be considered as a special case of the present model. We conform to the notation of the preceding section. Let the number of activities be labeled by j ($j = 1, \ldots, m$). Suppose that the operation of the jth activity at unit level requires an input $\langle a_{1j}, \ldots, a_{nj} \rangle$ of the n commodities and produces $\langle b_{1j}, \ldots, b_{nj} \rangle$. It will be assumed that

(i) $a_{ij} \geq 0$, $b_{ij} \geq 0$ ($i = 1, \ldots, n; j = 1, \ldots, m$).

The transformation set T is now defined by

$$T = [\{\mathbf{x}, \mathbf{y}\} \mid \mathbf{x} \geq \mathbf{Az}, \mathbf{y} \leq \mathbf{Bz} \quad \text{for some} \quad \mathbf{z} \geq \mathbf{0}, \mathbf{z} = \langle z_1, \ldots, z_m \rangle],$$

where

$$\mathbf{A} = \|a_{ij}\|, \quad \mathbf{B} = \|b_{ij}\| \quad (i = 1, \ldots, n; j = 1, \ldots, m).$$

Assumptions (T_1) and (T_2) are always satisfied. (T_3) is equivalent to:

(ii) For any j there is at least one i such that $a_{ij} > 0$. (Every activity uses some commodity as input.)

(T_4) is equivalent to:

(iii) For any i, there is at least one j such that $b_{ij} > 0$. (Every commodity can be produced by some activity.)

Theorem 9.10.1 combined with Theorem 9.10.2 implies the following theorem.

►THEOREM 9.10.3. If $\mathbf{A}$ and $\mathbf{B}$ satisfy conditions (i)–(iii), then there exist an activity level and a price system $\mathbf{p}^*$ such that

$$\mathbf{z}^* > \mathbf{0}, \mathbf{p}^* > \mathbf{0}; \qquad \mathbf{Bz}^* \geq \lambda^* \mathbf{Az}^*; \qquad \mathbf{p}^*\mathbf{B} \leq \lambda^* \mathbf{p}^* \mathbf{A}.$$

The Leontief dynamic model. Suppose that there are n commodities as before. Let $\mathbf{A}$ be the input-output coefficient matrix; an element a_{kj} of $\mathbf{A}$ represents the amount of commodity k required for one unit of output of commodity j. Let $\mathbf{B}$ denote the capital coefficient matrix. Here b_{kj} can be interpreted as the amount of commodity k the producer must have on hand in order to produce one unit of commodity j.

A commodity vector $\mathbf{x} = \langle x_1, \ldots, x_n \rangle$ at the beginning of the time period will produce a commodity vector $\mathbf{y} = \langle y_1, \ldots, y_n \rangle$ at the end of the period if and only if for some activity level $\mathbf{z} = \langle z_1, \ldots, z_n \rangle$,

$$\mathbf{y} \leq (\mathbf{I} - \mathbf{A})\mathbf{z} + \mathbf{x}, \qquad \mathbf{x} \geq \mathbf{Bz}, \qquad \mathbf{z} > \mathbf{0}. \tag{9.10.9}$$

We have here identified activity level with output. (This merely amounts to an appropriate choice of units of measurement.)

There is an exogenous source of input—for example, raw material or land—such that the stock of commodities grows with each time period. Relations (9.10.9) express in a sense this growth process within the framework of the natural economic limitations of availability and producibility.

Relations (9.10.9) can be elucidated as follows. Stock on hand of the kth commodity plus output of the kth commodity in a given time period is given by the formula $x_k + z_k$. During the same period an amount $\sum_j a_{kj}z_j$ of the kth commodity is used up; hence $x_k + z_k - \sum a_{kj}z_j$ remains available for final consumption. The capital constraints $\mathbf{x} \geq \mathbf{Bz}$ do not affect this amount, since the investment in equipment remains a capital asset.

In order to deal with the mathematical aspects of this model, it seems convenient to assume that there exists a vector $\mathbf{z}^0$ ($\mathbf{z}^0 > \mathbf{0}$) such that

$$(\mathbf{I} - \mathbf{A})\mathbf{z}^0 \gg \mathbf{0}. \tag{9.10.10}$$

This assumption asserts the existence of an activity vector which produces a surplus of each commodity; it is vital to the correctness of the relations (9.10.12a,b) below, which in fact need not hold if we do not require (9.10.10). (See p. 257 for a further discussion of the relevance of this kind of assumption.)

We shall further assume that for each j there exists an i such that

$$b_{ij} > 0. \tag{9.10.11}$$

The transformation set T in this case is defined by

$$T = [\{\mathbf{x}, \mathbf{y}\} | \mathbf{0} \leq \mathbf{y} \leq (\mathbf{I} - \mathbf{A})\mathbf{z} + \mathbf{x}, \quad \mathbf{x} \geq \mathbf{Bz} \quad \text{for some} \quad \mathbf{z} \geq \mathbf{0}].$$

With (9.10.10) and (9.10.11) assumed to hold, it becomes evident that assumptions (T_1)–(T_4) are satisfied. It is also easily seen that Theorems 9.10.1 and 9.10.2 imply the existence of $\lambda^* > 1$, $\mathbf{p}^* > \mathbf{0}$, $\mathbf{z}^* > \mathbf{0}$ such that

$$(\mathbf{I} - \mathbf{A})\mathbf{z}^* \geq (\lambda^* - 1)\mathbf{Bz}^*, \tag{9.10.12a}$$

$$\mathbf{p}^*(\mathbf{I} - \mathbf{A}) \leq (\lambda^* - 1)\mathbf{p}^*\mathbf{B}. \tag{9.10.12b}$$

These last two inequalities can be converted to equalities by imposing further assumptions on the matrix $\mathbf{A}$. We indicate one such extension. Two indices i and j are said to communicate (in symbols, $i \sim j$; the relation may be nonreflexive) if there exists a chain of indices $i_1, i_2, \ldots, i_r$, such that $a_{i_\nu i_{\nu+1}} > 0$ for $\nu = 1, \ldots, r - 1$, where $i_1 = i$ and $i_r = j$. If all indices communicate, then the relations (9.10.12) may be written with equality, and $\mathbf{z}^*$ and $\mathbf{p}^*$ are strictly positive vectors. To begin with, we prove that $\mathbf{z}^* \gg \mathbf{0}$. Let us suppose to the contrary that $z^*_{i_0} = 0$ for some i_0. Examining the i_0th equation of (9.10.12a), we conclude that $z^*_i = 0$ for all i which communicate with i_0 for a chain of length 1. Iteration shows that $z^*_i = 0$ for all i which communicate with i_0. This gives $\mathbf{z}^* \equiv \mathbf{0}$, contrary to the construction of (9.10.12a); hence $\mathbf{z}^* \gg \mathbf{0}$.

It follows that equality holds in (9.10.12b). For otherwise

$$(\mathbf{p}^*(\mathbf{I} - \mathbf{A}), \mathbf{z}^*) < (\lambda^* - 1)(\mathbf{p}^*\mathbf{B}, \mathbf{z}^*). \tag{9.10.13}$$

But premultiplication of (9.10.12a) by $\mathbf{p}^*$ gives

$$(\mathbf{p}^*(\mathbf{I} - \mathbf{A}), \mathbf{z}^*) \geq (\lambda^* - 1)(\mathbf{p}^*\mathbf{B}, \mathbf{z}^*), \tag{9.10.14}$$

which is obviously incompatible with (9.10.13).

Since equality holds in (9.10.12b) we may argue as above and show that $\mathbf{p}^* \gg \mathbf{0}$. By a symmetrical argument this yields

$$(\mathbf{I} - \mathbf{A})\mathbf{z}^* = (\lambda^* - 1)\mathbf{B}\mathbf{z}^*,$$

and the proof is complete. (For further results of this kind, see Problems 20 and 21.)

Efficient production. In any model in which the transformation set T satisfies assumptions (T_1)–(T_4) the efficiency of a production plan is easily characterized by the following definition:

A possible production plan $\{\bar{\mathbf{x}}, \bar{\mathbf{y}}\}$ is said to be efficient if $\{\bar{\mathbf{x}}, \bar{\mathbf{y}}\} \in T$, $\bar{\mathbf{x}} > \mathbf{0}$ and there is no $\{\mathbf{x}, \mathbf{y}\} \in T$ distinct from $\{\bar{\mathbf{x}}, \bar{\mathbf{y}}\}$ such that $\mathbf{x} \leq \bar{\mathbf{x}}$, $\bar{\mathbf{y}} \leq \mathbf{y}$.

▶ THEOREM 9.10.4. A production plan $\{\bar{\mathbf{x}}, \bar{\mathbf{y}}\}$ is efficient if and only if there exist price systems $\bar{\mathbf{p}}$ and $\bar{\mathbf{q}}$ with positive components ($\bar{\mathbf{p}}, \bar{\mathbf{q}} \gg \mathbf{0}$) such that

$$(\bar{\mathbf{q}}, \mathbf{y}) - (\bar{\mathbf{p}}, \mathbf{x}) \leq (\bar{\mathbf{q}}, \bar{\mathbf{y}}) - (\bar{\mathbf{p}}, \bar{\mathbf{x}}) \tag{9.10.15}$$

for all $\{\mathbf{x}, \mathbf{y}\} \in T$.

Remark 9.10.1. The left-hand expression evaluates the net profit—i.e., the value of the output $(\bar{\mathbf{q}}, \mathbf{y})$ minus the cost of the input $(\bar{\mathbf{p}}, \mathbf{x})$—according to the price system $\{\bar{\mathbf{q}}, \bar{\mathbf{p}}\}$. The theorem asserts that the principle of profit maximization applies with some appropriate price vector for each efficient production plan.

Proof of Theorem 9.10.4. We consider the set Γ in E^{2n} defined as follows:

$$\Gamma = \langle \lambda\bar{\mathbf{x}} - \mathbf{x}, \quad \mathbf{y} - \lambda\bar{\mathbf{y}} \rangle$$

for $\{\mathbf{x}, \mathbf{y}\} \in T$ and $\lambda \geq 0$. The set Γ is a closed convex cone which touches the nonnegative orthant O_{2n} only at the origin. Moreover, the set Γ cannot be tangent at the origin to any direction in the positive orthant. Otherwise, since Γ is a closed cone, the ray in that direction of the first orthant would belong to Γ, which is impossible by the definition of Γ. Consequently, we can find a hyperplane through the origin, containing no direction of O_{2n}, which separates Γ and O_{2n}. We denote the components of the normal to this hyperplane by $\langle \bar{p}_1, \bar{p}_2, \ldots, \bar{p}_n; \bar{q}_1, \bar{q}_2, \ldots, \bar{q}_n \rangle$. Since the hyperplane supports O_{2n}, it follows easily that $\bar{p}_i \geq 0$ and $\bar{q}_i \geq 0$. Actually, $\bar{p}_i > 0$ and $\bar{q}_i > 0$, since the hyperplane contains no segment of O_{2n}. Finally,

$$(\bar{\mathbf{p}}, \bar{\mathbf{x}} - \mathbf{x}) + (\bar{\mathbf{q}}, \mathbf{y} - \bar{\mathbf{y}}) \leq 0$$

for $\{\mathbf{x}, \mathbf{y}\} \in T$, since the hyperplane separates Γ from O_{2n}. This establishes (9.10.15).

On the other hand, if there exist systems $\bar{\mathbf{p}} \gg \mathbf{0}$ and $\bar{\mathbf{q}} \gg \mathbf{0}$ satisfying (9.10.15), the efficiency of $\{\bar{\mathbf{x}}, \bar{\mathbf{y}}\}$ is evident.

9.11 Problems.

1. Let $\mathbf{D}$ be a diagonal matrix, and let $\mathbf{A}$ be a stable matrix (i.e., all eigenvalues of $\mathbf{A}$ have negative real parts) such that $\mathbf{DA}$ is stable if and only if the diagonal elements of $\mathbf{D}$ are all positive ($d_{ii} > 0$). Let $\mathbf{C} = \mathbf{EAE}^{-1}$, where $\mathbf{E}$ is any nonsingular diagonal matrix. Prove that $\mathbf{C}$ is stable, and that $\mathbf{DC}$ is stable if and only if $d_{ii} > 0$ for all i.
2. Let $\mathbf{A}$ be any stable matrix. If $\mathbf{D}$ is diagonal and $\mathbf{DA}$ is stable, prove that at least one characteristic root of $\mathbf{D}$ is positive.
3. Show that if $\mathbf{A} = \|a_{ij}\|$ is a matrix such that $a_{ii} < 0$ and

$$|a_{ii}| > \sum_{j \neq i} |a_{ij}| \qquad \text{for all} \quad i,$$

then $\mathbf{A}$ is stable.
4. Consider a matrix $\mathbf{D} = \|d_{ij}\|$ such that sign $d_{ij} =$ sign d_{ji} $(i \neq j)$ and sign $(d_{ij}, d_{jk}) =$ sign d_{ik} $(i \neq j \neq k)$ provided $d_{ij}d_{jk} \geq 0$. Show that $\mathbf{D}$ is an $\widetilde{M}$-matrix (see p. 308).

The next two problems apply to the normalized price adjustment process, and the following five problems refer to the nonnormalized price adjustment process. We keep the notation of Section 9.2.

5. Consider the case of $n = 2$ commodities. At any point $\mathbf{p}^*$ of stable equilibrium show that

$$\frac{\partial F_0}{\partial p_1}(\mathbf{p}^*) \geq 0, \qquad \frac{\partial F_1}{\partial p_0}(\mathbf{p}^*) \geq 0.$$

6. Show that for any number of commodities it is impossible to divide all the commodities (including the numéraire) into two nonvoid sets R and S such that $\partial F_r/\partial p_s < 0$ for all $r \in R$, $s \in S$ and all price vectors $\mathbf{p}$.

In Problems 7–11 we assume that the excess demand functions $\mathbf{F}(\mathbf{p})$ are continuously differentiable for $\mathbf{p} \gg \mathbf{0}$ and homogeneous of degree zero.

7. Suppose that $\mathbf{F}$ has the property of strict gross substitutability. Prove that if $\mathbf{F}(\mathbf{p}^0) \gg \mathbf{0}$, then $\mathbf{F}[\mathbf{p}(t)] \gg \mathbf{0}$, where $\mathbf{p}(t) = \psi(t, \mathbf{p}^0)$.

8. Let strict gross substitutability prevail for the nonnormalized price adjustment process of (9.2.4). Prove that

$$V(t) = \max_{0 \le i \le n} |p_i(t)/p_i^* - 1|$$

is strictly decreasing unless $p_i(t) = \lambda p_i^*$ for all i and some positive λ.

9. For the case of two commodities, prove that

$$V(t) = [p_0(t) - p_0^*]^{2k} + [p_1(t) - p_1^*]^{2k}$$

is strictly decreasing in t for any positive integer k, provided that $p_i(t) \neq \lambda p_i^*$ for $i = 0, 1$ and some positive λ. (Assume strict gross substitutability.)

10. If strict gross substitutability prevails for two commodities, show that

$$\sum_{i,j=0}^{1} p_i^* \frac{\partial F_i}{\partial p_j} (\mathbf{p}) F_j(\mathbf{p}) < 0$$

unless $\mathbf{p} = \lambda \mathbf{p}^*$ $(\lambda > 0)$.

11. Under the same conditions as in Problem 10, prove that

$$V(t) = \sum_{i=0}^{1} [p_i(t) - p_i^*]^2$$

is a strictly decreasing convex function of t, where $\mathbf{p}(t) = \psi(t, \mathbf{p}^0)$ and $\mathbf{p}^0 \neq \lambda \mathbf{p}^*$.

12. Let $\mathbf{F}(\mathbf{p})$ be homogeneous, finite in the interior of the positive orthant I, but continuous in the extended sense (i.e., infinite values allowed) in the closure of I with the origin removed. Moreover, assume that $\partial F_i/\partial p_j$ $(i \neq j)$ is strictly positive and continuous over the same domain. Prove that if $\mathbf{p} > \mathbf{0}$ and $p_k = 0$, then $F_k(\mathbf{p}) = \infty$.

13. Under the same conditions as in Problem 12, with the added restrictions that $\mathbf{F}(\mathbf{p})$ is uniformly bounded from below and satisfies the Walras law, prove the existence of a strictly positive equilibrium vector.

14. Let strict gross substitutability prevail for the nonnormalized process (9.2.4), and assume the existence of an equilibrium vector $\mathbf{p}^* \gg \mathbf{0}$. Show that Theorem 9.5.4 applies if we take for $\phi(\mathbf{p})$ any one of the following:

(a) $\max_{1 \leq i \leq r} \left\{ \frac{p_i(t)}{p_i^*} \right\},$

(b) $\sum_{i=1}^{r} \max\,(0, p_i(t) F_i[\mathbf{p}(t)]),$

(c) $\max_{1 \leq i \leq r} \left\{ \frac{F_i[\mathbf{p}(t)]}{p_i(t)} \right\}.$

15. Consider the case of the nonnormalized process of two commodities such that the Walras law is satisfied and the excess demand functions are homogeneous of degree zero. Assume that $F_1(p_1, 1) < 0$ if p_1 is sufficiently large. Define

$$\phi(p_1, p_2) = -\int_0^{p_1/p_2} F_1(v, 1)\, dv \qquad \text{for} \quad p_1, p_2 > 0.$$

Show that Theorem 9.5.4 applies and hence that the process (9.2.5) is *quasi-stable.*

16. Let F_i denote the excess demand functions and suppose that $\partial F_i / \partial p_j$ is symmetric at an equilibrium price vector $\mathbf{p}^* \gg \mathbf{0}$. Show that local stability of the system is invariant with respect to the choice of numéraire $i = 0, 1, \ldots, r$.

17. Prove that if the excess demand function $\mathbf{F}(\mathbf{p})$ of Section 9.6 is continuous, satisfies the weak axiom of revealed preference, and is homogeneous of order zero, the system

$$\mathbf{p}(t + 1) = \max\,\{0, \mathbf{p}(t) + \rho \mathbf{F}[\mathbf{p}(t)]\} \qquad (t = 0, 1, 2, \ldots),$$

with nonnegative initial price $\mathbf{p}(0)$, is stable provided ρ is a sufficiently small positive number.

18. Let $\mathbf{F}(\mathbf{p})$ be an excess demand function. Assume that $\mathbf{F}(\mathbf{p})$ is differentiable, is homogeneous of order zero, satisfies the gross substitutability condition

$$\frac{\partial F_i}{\partial p_j} > 0 \qquad (i \neq j, \mathbf{p} \gg \mathbf{0}),$$

and has an equilibrium vector $\mathbf{p}^* \gg \mathbf{0}$ such that $\mathbf{F}(\mathbf{p}^*) = \mathbf{0}$. Consider $\mathbf{p}(t)$ $(t = 1, 2, \ldots)$ defined recursively by

$$F_i[p_1(t + 1), \ldots, p_i(t + 1), p_i(t), \ldots, p_r(t)] = 0$$
$$(t = 0, 1, 2, \ldots; i = 1, 2, \ldots, r),$$

where $\mathbf{p}(0) \gg \mathbf{0}$. Show first that $\mathbf{p}(t)$ is well defined, and then that $\mathbf{p}(t)$ converges to an equilibrium vector.

19. Consider a single commodity whose output in each period (time t) is split into two parts: consumption for the next period, $c(t + 1)$; and input for the next period's output, $x(t + 1)$. Suppose that one unit of input gives rise in the next period to a units of output and that the law of constant returns to scale prevails. Suppose that at time $t = 0$ we start with K units of the commodity. Prove the formula

$$c(n) + x(n) \leq a^n K - \sum_{r=0}^{n-1} a^{n-r} c(r),$$

and set up the problem of maximizing $c(n)$ as a linear programming problem.

20. Show that (9.10.12a,b) become equalities if there exists a power n_0 such that $\mathbf{A}^{n_0} \gg \mathbf{0}$.

21. Show that (9.10.12a,b) become equalities if there exists a power n_0 such that $\mathbf{B}^{n_0} \gg \mathbf{0}$.

22. Consider the von Neumann model of an expanding economy as described in Section 9.9. Prove that if $a_{ij} + b_{ij} > 0$ for all $i, j = 1, \ldots, n$, then

$$\sup_{\mu} e^{-\mu} = \inf_{\lambda} e^{-\lambda},$$

where $e^{-\mu}$ satisfies (9.9.1) for some price vector $\mathbf{p}$ and $e^{-\lambda}$ satisfies (9.9.3) for some commodity vector $\mathbf{y}$.

23. Consider the model of technological production described by a set of pairs of vectors $\{\mathbf{x}, \mathbf{y}\} \in T$, with $\mathbf{x}, \mathbf{y} \in E^n$ satisfying postulates (T_1)–(T_4) of Section 9.10. Suppose in addition that $x_i + y_i > 0$ $(i = 1, \ldots, n)$ for each $\{\mathbf{x}, \mathbf{y}\} \in T$. Let $\mathcal{P}$ denote the set of all normalized price vectors $(\mathbf{p} = \langle p_1, \ldots, p_n \rangle, p_i \geq 0, \sum p_i = 1)$, and define

$$\phi(\mathbf{p}; \mathbf{x}, \mathbf{y}) = \begin{cases} \dfrac{\sum_{i=1}^{n} x_i p_i}{\sum_{i=1}^{n} y_i p_i} & \sum y_i p_i > 0, \\ 0 & \sum y_i p_i = 0 \end{cases}$$

for $\mathbf{p} \in \mathcal{P}$ and $\{\mathbf{x}, \mathbf{y}\} \in T$. Prove that there exist a vector $\mathbf{p}_0 \in \mathcal{P}$ and a pair $\{\mathbf{x}^0, \mathbf{y}^0\} \in T$ such that

$$\phi(\mathbf{p}; \mathbf{x}^0, \mathbf{y}^0) \geq \phi(\mathbf{p}^0; \mathbf{x}^0, \mathbf{y}^0) \geq \phi(\mathbf{p}^0; \mathbf{x}, \mathbf{y})$$

for every $\{\mathbf{x}, \mathbf{y}\} \in T$ and $\mathbf{p}^0 \in \mathcal{P}$.

NOTES AND REFERENCES TO CHAPTER 9

§ 9.1. A good review of the nature of the problem of welfare economics can be found in Samuelson [210]. A critical essay discussing the relations of welfare economics and equilibrium theory is included in Koopmans [147]. Arrow [4] is another important contribution to the area of welfare economics.

The fundamental articles of Arrow [4] and Debreu [62] both offer detailed mathematical analyses of the connections between Pareto optima and equilibrium theory. Debreu's presentation is in very abstract form and applies to infinite-dimensional analogs. Arrow's approach is geometric, appealing frequently to the structure theorems of convex sets. We take still another line in this section: that of reducing the relationship between a given Pareto optimum and the existence of an equilibrium to the vector maximum problem of nonlinear programming discussed in Section 7.4.

Both Arrow and Debreu develop the characterization of Pareto optima for the more general case in which the utility indicator functions U_i are assumed to be quasi-concave. The proofs in this section are valid when the U_i are quasi-like concave (see [4]).

§ 9.2. Stability of the price adjustment process was studied in the nineteenth century by Walras [252]. In modern form, stability of the dynamic process has been discussed by Hicks [121], Samuelson [210], Metzler [180], Allais [1], and others.

In 1958 Arrow and Hurwicz [11] made a comprehensive survey of stability theory and considerably extended the scope of the then existing results. This paper was in the main superseded by Arrow, Block, and Hurwicz [6], in which the results obtained are considerably sharper. Much of our formulation of the process is based on their presentation, and on several elegant refinements and extensions obtained by Uzawa.

The setup of a discrete-time price adjustment process referred to on p. 326 was suggested by Uzawa.

§ 9.3. Part (a) of Theorem 9.3.1 is due originally to Arrow and Hurwicz [11], who gave a rather elaborate proof of its validity. Hahn [117] offered a simpler proof. Our proof is still simpler. Part (b) of this same theorem may be new.

Theorem 9.3.2 is new.

The concepts of "gross substitutes" and "complementary commodities" were introduced by Mosak [185] and Metzler [180].

The relations between M-matrices and $\widetilde{M}$-matrices were worked out by Morishima [183].

Theorem 9.3.4 is an application to a local situation of a more general theorem formulated by Arrow, Block, and Hurwicz [6].

§ 9.4. Theorems 9.4.1 and 9.4.2 are due to Arrow, Block, and Hurwicz [6]. The elegant analysis of Theorems 9.4.3 and 9.4.4 was devised by Uzawa. A weaker version of Theorem 9.4.3 is contained in the first reference.

§ 9.5. Theorem 9.5.1 is new; it is the dual of Theorem 9.5.2, which is due to Arrow, Block, and Hurwicz [6]. Theorem 9.5.3 represents a slight extension of a result of Arrow and Hurwicz [11, Theorem 4]. The concept of quasi-stability originated with Uzawa. This concept makes possible a unified approach to the problem of deriving stability under a variety of assumptions on the structure of the price adjustment process.

§ 9.6. The model of this section was proposed by Uzawa [233]. He established the existence of an equilibrium vector, and we have followed his proof. The normalized discrete-time process was shown by Uzawa to be stable whenever the weak axiom of revealed preference is satisfied everywhere. Theorem 9.6.1 establishes the same conclusion for the nonnormalized discrete-time process when strict gross substitutability prevails.

§ 9.7. The discussion of this section follows Enthoven and Arrow [75] and Arrow and McManus [13].

§ 9.8. The material of this section stems from Arrow and Nerlove [14]. However, the proof of Theorem 9.8.1 is new and more direct.

§ 9.9. The model of an expanding economy was constructed by von Neumann [239] as a sweeping formulation of the dynamics of production and consumption. Its solution was closely related to the development of the theory of games. Unfortunately, in von Neumann's original analyses he was obliged to assume that $a_{ij} + b_{ij} > 0$ for all i and j—i.e., that every commodity is either consumed or produced, which on economic grounds is not easily justified. Kemeny, Thompson, and Morgenstern [142] and Gale [94] later substantially weakened the hypothesis and were still able to obtain the quantitative link between the rate of expansion and the rate of interest.

Other generalizations have been offered by Georgescu-Roegen [104] and Nikaido [192]. The last model is nonlinear in form.

Another version of balanced growth was presented by Solow and Samuelson [228]. Neisser gave an interpretation to this model and Suits [230] provided a simple mathematical proof.

Morishima [184] discusses the theory of balanced growth in terms of dynamic Leontief models.

§ 9.10. The general balanced-growth model examined here was proposed by Gale [94]. Our discussion is a simplification of his treatment.

§ 9.11. Problems 5, 6, and 8 are based on Arrow and Hurwicz [11]. Problems 12 and 13 are due to Arrow, Block, and Hurwicz [6]. Problems 14–18 were suggested by Uzawa.

SOLUTIONS TO PROBLEMS OF CHAPTERS 5–9

CHAPTER 5

Problem 2. Let x, y, z denote the quantities of total crude oil used in the three production patterns. The problem reduces to that of maximizing $32x + 15y + 12z$ subject to $x, y, z \geq 0$, and the availability constraints become

$$x + 2y + 2z \leq 10,$$
$$2x + y + 2z \leq 15.$$

The solution is $x = 15/2$, $y = 0$, $z = 0$ and the maximum profit is 240.

Problem 4. $x_1 = n$, $x_2 = x_3 = \cdots = x_n = 0$.

Problem 8. Since $\mathbf{b} \geq \mathbf{0}$, it follows that $\mathbf{x} = \mathbf{0}$ is feasible for Problem I. Now use the existence theorem (Theorem 5.4.2).

Problem 11. The problem can be stated mathematically as follows: Maximize $\sum_i \beta_i x_i$ subject to $x_i \geq 0$ and $\sum \alpha_i x_i \leq M$. Let

$$\frac{\beta_{i_0}}{\alpha_{i_0}} = \max \frac{\beta_i}{\alpha_i}\cdot$$

The solution has the form $x_{i_0} = M/\alpha_{i_0}$ and $x_j = 0$ for $j \neq i_0$.

Problem 12. Arrange the a's and b's in order as follows: $a_{\mu_1} < \cdots < a_{\mu_n}$ and $b_{\nu_1} > \cdots > b_{\nu_n}$. Then the minimum is

$$\sum_{i=1}^{n} a_{\mu_i} b_{\nu_i}.$$

(This is known as Tchebycheff's inequality.)

Problem 16. Consider the set $\Gamma = \{(\mathbf{A} - \lambda_0 \mathbf{B})\mathbf{x} | \mathbf{x} \geq \mathbf{0}\}$. The definition of λ_0 implies that Γ does not intersect the interior of the positive orthant. By Lemma B.1.3 there exists $\mathbf{y}^0 > \mathbf{0}$ such that $(\mathbf{y}^0, (\mathbf{A} - \lambda_0\mathbf{B})\mathbf{x}) \leq 0$ for all $\mathbf{x} \geq \mathbf{0}$ or $\lambda_0 \mathbf{y}^0 \mathbf{B} \geq \mathbf{y}^0 \mathbf{A}$. Hence $\mu_0 \leq \lambda_0$, as claimed.

Problem 18. That $v(\mathbf{b})$ is a concave function of $\mathbf{b}$ is easily shown. Next suppose that $\{\bar{\mathbf{x}}, \bar{\mathbf{y}}\}$ is a saddle point of $\phi(\mathbf{x}, \mathbf{y}) = (\mathbf{c}, \mathbf{x}) + (\mathbf{y}, \mathbf{b} - \mathbf{A}\mathbf{x})$, and that $\{\bar{\mathbf{x}} + \delta\mathbf{x}, \bar{\mathbf{y}} + \delta\mathbf{y}\}$ is a saddle point of $\tilde{\phi}(\mathbf{x}, \mathbf{y}) = (\mathbf{c}, \mathbf{x}) + (\mathbf{y}, \mathbf{b} + \Delta\mathbf{b} - \mathbf{A}\mathbf{x})$. Then

$$(\mathbf{c}, \bar{\mathbf{x}} + \delta\mathbf{x}) + (\bar{\mathbf{y}}, \mathbf{b} - \mathbf{A}(\bar{\mathbf{x}} + \delta\mathbf{x})) \leq (\mathbf{c}, \bar{\mathbf{x}}) + (\bar{\mathbf{y}}, \mathbf{b} - \mathbf{A}\bar{\mathbf{x}}),$$

and hence

$$\Delta v = (\mathbf{c}, \delta\mathbf{x}) \leq (\bar{\mathbf{y}}, \mathbf{A}\,\delta\mathbf{x}).$$

But $\mathbf{A}\bar{\mathbf{x}} + \mathbf{A}\,\delta\mathbf{x} \leq \mathbf{b} + \Delta\mathbf{b}$ and $(\bar{\mathbf{y}}, \mathbf{A}\bar{\mathbf{x}}) = (\bar{\mathbf{y}}, \mathbf{b})$ [equation (5.3.10)] implies $(\bar{\mathbf{y}}, \mathbf{A}\,\delta\mathbf{x}) \leq (\bar{\mathbf{y}}, \Delta\mathbf{b})$. Collecting, we have $\Delta v \leq (\bar{\mathbf{y}}, \Delta\mathbf{b})$, from which the desired conclusion follows by an appropriate choice of $\Delta\mathbf{b}$.

Problem 19. The following sequence of easily proved equivalent statements establishes the result:

(i) $\mathbf{x} \geq 0, (\mathbf{c}, \mathbf{x}) \geq 0$;

(ii) $x_i \geq 0 \ (i = 1, \ldots, n), \sum_{i \in I} c_i x_i \geq \sum_{k \in K} (-c_k) x_k$;

(iii) there exists $\alpha_{ik} \geq 0$ such that

$$c_i x_i \geq \sum_{k \in K} \alpha_{ik} \quad \text{for} \quad i \in I, \qquad -c_k x_k = \sum_{i \in I} \alpha_{ik} \quad \text{for} \quad k \in K$$

(this is proved by induction on the number of elements in I and K);

(iv) there exists γ_i such that

$$x_i = \gamma_i + \sum_{k \in K} \frac{\alpha_{ik}}{c_i} \quad \text{for} \quad i \in I, \qquad x_k = \sum_{i \in I} \frac{\alpha_{ik}}{-c_k} \quad \text{for} \quad k \in K;$$

(v) $\mathbf{x} = \sum_{i \in I} \gamma_i \boldsymbol{\alpha}_i + \sum_{\substack{i \in I \\ k \in K}} \alpha_{ik} \boldsymbol{\nu}_{ik} + \sum_{j \in J} x_j \boldsymbol{\beta}_j,$

with γ_i, α_{ik}, and x_j nonnegative.

Problem 20. By virtue of Theorem 5.3.1 and Equation (5.3.10) we have the following relations:

$$(\mathbf{c}, \mathbf{x}_0 + \boldsymbol{\delta}\mathbf{x}_0) + (\mathbf{y}_0, \mathbf{b} - \mathbf{A}(\mathbf{x}_0 + \boldsymbol{\delta}\mathbf{x}_0)) \leq (\mathbf{c}, \mathbf{x}_0) \quad \text{and} \quad (\mathbf{y}_0, \mathbf{b}) = (\mathbf{y}_0, \mathbf{A}\mathbf{x}_0),$$

$$(\mathbf{c} + \boldsymbol{\delta}\mathbf{c}, \mathbf{x}_0) + (\mathbf{y}_0 + \boldsymbol{\delta}\mathbf{y}_0, \mathbf{b} + \boldsymbol{\delta}\mathbf{b} - (\mathbf{A} + \boldsymbol{\delta}\mathbf{A})\mathbf{x}_0) \leq (\mathbf{c} + \boldsymbol{\delta}\mathbf{c}, \mathbf{x}_0 + \boldsymbol{\delta}\mathbf{x}_0),$$

$$(\mathbf{y}_0 + \boldsymbol{\delta}\mathbf{y}_0, \mathbf{b} + \boldsymbol{\delta}\mathbf{b}) = (\mathbf{y}_0 + \boldsymbol{\delta}\mathbf{y}_0, (\mathbf{A} + \boldsymbol{\delta}\mathbf{A})(\mathbf{x}_0 + \boldsymbol{\delta}\mathbf{x}_0)).$$

By combining these relations we obtain the first inequality of the problem. The second is established by using the other half of the saddle-value inequalities.

Problem 21. The solution follows an argument parallel to that of Lemma 5.3.1.

Problem 22. Compare with Problem C″ of Section 5.11.

Problem 25. The problem is to minimize

$$c = \sum_{i=1}^{n} z_i + \lambda X$$

subject to the constraints $z_i \geq t_i - X$ and $0 \leq z_i \leq r_i$. A solution exists only if $r_i \geq t_i - X$, a condition that we assume is satisfied. For fixed X, c is minimized when $z_i = \max \{0, t_i - X\}$ $(i = 1, \ldots, n)$, and in this case

$$c(X) = \sum_{k=1}^{n} \max \{0, t_k - X\} + \lambda X.$$

Suppose for definiteness that $t_1 \geq t_2 \geq \cdots \geq t_n$. If $t_m > X \geq t_{m+1}$, then

$$c(X) = \sum_{k=1}^{m} t_k + (\lambda - m)X,$$

which is minimized when

$$X = \begin{cases} t_m, & \lambda - m < 0, \\ t_{m+1}, & \lambda - m \geq 0. \end{cases}$$

Hence

$$c(X) = \begin{cases} \sum_1^m t_k + (\lambda - m)t_{m+1}, & \lambda - m \geq 0, \\ \sum_1^m t_k + (\lambda - m)t_m, & \lambda - m < 0. \end{cases}$$

Let $[\lambda] = m_0$, where $m_0 \leq \lambda < m_0 + 1$; then the optimum $X = t_{m_0+1}$ and

$$z_i = \begin{cases} t_i - t_{m_0+1}, & i \leq m_0, \\ 0, & i > m_0. \end{cases}$$

CHAPTER 6

Problem 2. Operate Process A at level 0.25 and Process B at level 1.0; do not operate Process C.

Problem 4. Use Lemma 6.1.1.

Problem 6. Consider $\mathbf{zA} = \mathbf{0}$, with the vector $\mathbf{z}$ represented as

$$\mathbf{z} = \langle x_1, x_2, \ldots, x_n, y_1, \ldots, y_m \rangle.$$

It follows that $x_i = -y_j$ for all $i = 1, \ldots, n$ and $j = 1, \ldots, m$. Hence the null space of $\mathbf{A}$ is one-dimensional and the rank of $\mathbf{A}$ is $n + m - 1$.

Problem 8.

	$0 \leq \lambda < 1$	$1 \leq \lambda < 2$	$2 \leq \lambda$
x	10	1	0
z_1	0	1	2
z_2	0	5	5
z_3	0	3	4

Problem 10. See Prager [199].

Problem 11. Ship 75.

Problem 13. Compare with the model of Fig. 6.5 and set c_{ij} equal to 1 or 0 according to whether (i, j) is good or bad. If sufficiently large a's and b's are chosen, their capacity constraints are not relevant.

Problem 14. Let p be a node with arcs $A_1, \ldots, A_r$ directed into the node and arcs $B_1, \ldots, B_s$ directed outwards and limited by a capacity restraint c. Replace the node by two nodes labeled p and p' such that $A_1, \ldots, A_r$ is directed into p and $B_1, \ldots, B_s$ is directed out of p', and add an arc directed from p to p' with capacity constraint c.

Problem 16. The relation $\mathbf{xA} = \mathbf{0}$ holds if and only if $x_i = x_j$ for all $i, j = 1, \ldots, n$. It follows that the null space of $\mathbf{A}$ is one-dimensional, and hence the rank of $\mathbf{A}$ is $n - 1$.

Problem 19. By Lemma 6.1.1 every extreme point $\mathbf{x}$ in X has exactly m positive components, and conversely. The result now follows from the further observation that $\sum_i \delta(x_i)$ is a constant over the set of extreme points of X.

Problem 20. Since $\varphi(\mathbf{x})$ is concave.

Problem 21. See Gross [115].

CHAPTER 7

Problem 1. Theorem 7.1.1 with $X = E^m$.

Problem 2. By virtue of the concavity of g, we have

$$g(\mathbf{x}) \leq g(\mathbf{x}^0) + (\mathbf{g}_{x^0}, \mathbf{x} - \mathbf{x}^0) \qquad \text{(property vii of Section B.4).}$$

Now if $\mathbf{x}^0$ maximizes $(\mathbf{g}_{x^0}, \mathbf{x})$ subject to $f(\mathbf{x}) \geq \mathbf{0}, \mathbf{x} \geq \mathbf{0}$, it follows that $(\mathbf{g}_{x^0}, \mathbf{x}) \leq (\mathbf{g}_{x^0}, \mathbf{x}^0)$ for $\mathbf{x}$ satisfying the constraints, and consequently $g(\mathbf{x}) \leq g(\mathbf{x}^0)$. On the other hand, if $g(\mathbf{x}^0) \geq g(\mathbf{x})$ for all $\mathbf{x}$ satisfying $\mathbf{x} \geq \mathbf{0}$ and $f(\mathbf{x}) \geq \mathbf{0}$, then $\mathbf{x}^0 + t(\mathbf{x} - \mathbf{x}^0)$ $(0 \leq t \leq 1)$ also satisfies the constraints. The maximum of $g[\mathbf{x}^0 + t(\mathbf{x} - \mathbf{x}^0)] = I(t)$ where $0 \leq t \leq 1$ is achieved at $t = 0$. Since $I(t)$ is concave, this can occur if and only if $I'(0) \leq 0$ or $(\mathbf{g}_{x^0}, \mathbf{x} - \mathbf{x}^0) \leq 0$, as was to be shown.

Problem 3. By Theorem 7.1.1,

$$(\mathbf{p}, \mathbf{x}) + (\mathbf{u}^0, \mathbf{v} - \mathbf{F}(\mathbf{x})) \leq (\mathbf{p}, \mathbf{x}^0) + (\mathbf{u}^0, \mathbf{v} - \mathbf{F}(\mathbf{x}^0)) \leq (\mathbf{p}, \mathbf{x}^0) + (\mathbf{u}, \mathbf{v} - \mathbf{F}(\mathbf{x}^0))$$

for any $\mathbf{x}, \mathbf{u} \geq \mathbf{0}$. Hence $(\mathbf{u}^0, \mathbf{v} - \mathbf{F}(\mathbf{x}^0)) = 0$, and thus $(\mathbf{p}, \mathbf{x}) - (\mathbf{u}^0, \mathbf{F}(\mathbf{x})) \leq (\mathbf{p}, \mathbf{x}^0) - (\mathbf{u}^0, \mathbf{F}(\mathbf{x}^0))$. Note that $\mathbf{F}(\mathbf{x})$ is homogeneous; hence for $\mathbf{x} = \lambda\mathbf{x}^0$, with λ an arbitrary positive scalar, we have $(\mathbf{p}, \mathbf{x}^0) = (\mathbf{u}^0, \mathbf{F}(\mathbf{x}^0)) = (\mathbf{u}^0, \mathbf{v})$.

Problem 4. Theorem 7.1.1 shows that $\mathbf{x}^*$ is optimum for the quadratic problem if and only if for some $\mathbf{u}^* \geq \mathbf{0}$

$$\begin{aligned} \mathbf{c} - \epsilon\mathbf{x}^* - \mathbf{u}^*\mathbf{A} &\leq \mathbf{0}, \\ (\mathbf{c} - \epsilon\mathbf{x}^* - \mathbf{u}^*\mathbf{A}, \mathbf{x}^*) &= 0, \\ \mathbf{A}\mathbf{x}^* \leq \mathbf{b}, \qquad \mathbf{x}^* &\geq 0, \\ (\mathbf{u}^*, \mathbf{b} - \mathbf{A}\mathbf{x}^*) &= 0. \end{aligned}$$

It follows from these relations that $\mathbf{x}^*$ is an optimum for the problem: Maximize $(\mathbf{c} - \epsilon\mathbf{x}^*, \mathbf{x})$ subject to $\mathbf{A}\mathbf{x} \leq \mathbf{b}, \mathbf{x} \geq \mathbf{0}$. If ϵ is sufficiently small, $\mathbf{x}^*$ is an optimum for the original linear programming problem (by Theorem 5.3.1).

Problem 5. For fixed n the optimum $\mathbf{x}^n$ is given by $x_i^{(n)} = T/n$ and

$$u(n) = \sum_{i=1}^{n} f(x_i^{(n)}) = nf\left(\frac{T}{n}\right).$$

The maximum of $u(n)$ is achieved where n is determined by the condition $u'(n) = 0$, or equivalently by

$$\frac{T}{n} f'\left(\frac{T}{n}\right) = f\left(\frac{T}{n}\right).$$

Problem 6. Since f is concave, the optimum is located at an extreme point. Hence $\min_i \sum f(x_i) = f(T)$.

Problem 7. Use the solution of Problem 5, making sure that the constraints are satisfied.

Problem 9. By Theorem 7.1.1, $\mathbf{x}^*$ is optimum for the first problem if and only if for some $\mathbf{u}^* \geq \mathbf{0}$

$$\begin{aligned} \mathbf{a} - \mathbf{A}\mathbf{x}^* &= \mathbf{u}^*\mathbf{B}, \\ \mathbf{B}\mathbf{x}^* &\leq \mathbf{b}, \\ (\mathbf{u}^*, \mathbf{b} - \mathbf{B}\mathbf{x}^*) &= 0. \end{aligned}$$

These conditions are equivalent to

$$\begin{aligned} &\mathbf{x}^* = \mathbf{A}^{-1}\mathbf{a} - \mathbf{A}^{-1}\mathbf{B}'\mathbf{u}^* \qquad (\mathbf{B}' \text{ is the transpose of } \mathbf{B}), \\ &\mathbf{B}\mathbf{A}^{-1}\mathbf{a} - \mathbf{b} - \mathbf{B}\mathbf{A}^{-1}\mathbf{B}'\mathbf{u}^* \leq 0, \\ &(\mathbf{B}\mathbf{A}^{-1}\mathbf{a} - \mathbf{b} - \mathbf{B}\mathbf{A}^{-1}\mathbf{B}'\mathbf{u}^*, \mathbf{u}^*) = 0. \end{aligned}$$

These last conditions are equivalent to saying that $\mathbf{u}^*$ maximizes

$$(\mathbf{B}\mathbf{A}^{-1}\mathbf{a} - \mathbf{b}, \mathbf{u}) - \tfrac{1}{2}(\mathbf{u}, \mathbf{B}\mathbf{A}^{-1}\mathbf{B}'\mathbf{u})$$

subject to $\mathbf{u} \geq \mathbf{0}$, again by Theorem 7.1.1.

Problem 10. See Problem 17 of Chapter 5.

Problem 11. Use Theorem 7.4.1.

Problem 12. Suppose the contrary; then $\mathbf{F}(\mathbf{x})$ has some component nonpositive for each $\mathbf{x}$ in X. Let H be the image in E^n of the vector function $\mathbf{F}(\mathbf{x})$ as $\mathbf{x}$ ranges over X. Since $\mathbf{F}$ is concave, the convex hull of H does not intersect the interior of the first orthant. Hence a separating plane $\mathbf{u}$ exists and $(\mathbf{u}, \mathbf{F}(\mathbf{x})) \leq 0$ for all $\mathbf{x}$ in X, contradicting the hypothesis.

Problem 13. Let $g(\mathbf{x}) = (\mathbf{c}, \mathbf{x}) - \frac{1}{2}(\mathbf{x}, \mathbf{A}\mathbf{x})$. Then x_i^{v+1} uniquely maximizes $g(x_1^{(v+1)}, \ldots, x_{i-1}^{v+1}, x_i, x_{i+1}^{(v)}, \ldots, x_n^{(v)})$ subject to $x_i \geq 0$. Therefore

$$\begin{aligned} g(x_1^{(v+1)}, \ldots, x_{i-1}^{(v+1)}, x_i^{(v+1)}, x_{i+1}^{(v)}, \ldots, x_n^{(v)}) \\ \geq g(x_1^{(v+1)}, \ldots, x_{i-1}^{(v+1)}, x_i^{(v)}, x_{i+1}^{(v)}, \ldots, x_n^{(v)}), \end{aligned}$$

with equality only if $x_i^{v+1} = x_i^v$. Thus $g(x_1^{(v+1)}, \ldots, x_n^{(v+1)}) \geq g(x_1^{(v)}, \ldots, x_n^{(v)})$, with strict inequality unless $x_1^{(v)}, \ldots, x_n^{(v)}$ is the maximum of $g(\mathbf{x})$. But the set $\{\mathbf{x} | g(\mathbf{x}) \geq g(\mathbf{x}^{(1)})\}$ is bounded, and the sequence $g(\mathbf{x}^{(1)}), g(\mathbf{x}^{(2)}), \ldots$ is also bounded. Hence $\mathbf{x}^{(1)}, \mathbf{x}^{(2)}, \ldots$ converges to the unique optimum.

Problem 16. There exists a vector $\mathbf{x}^0$ such that $c = \sup_\alpha f_\alpha(\mathbf{x}^0) < \infty$ if and only if $\{c, \mathbf{x}^0\} \in [f_\alpha, C_\alpha]$ for all α, which in turn is equivalent to

$$\sup_{\xi \in \Gamma_\alpha} [(\mathbf{x}^0, \xi) - \varphi_\alpha(\xi)] = f_\alpha(\mathbf{x}^0) \leq c.$$

Problem 18. Set $f(\mathbf{x}) = \min_y (\mathbf{x}, \mathbf{Ay})$ and $g(\mathbf{y}) = \max_x (\mathbf{x}, \mathbf{Ay})$, where $\mathbf{x}$ and $\mathbf{y}$ traverse the usual sets of simplicities. Apply Theorem 7.8.1.

Problem 20. Let $\mathbf{t}_\alpha$ denote the set of unit vectors through the origin orthogonal to the bounding hyperplanes of O_α and directed into O_α. Let $\mathbf{p}_\alpha$ denote the terminal points of $\mathbf{t}_\alpha$ on the unit sphere. The hypothesis implies that the convex span of $\mathbf{p}_\alpha$ has the origin on its interior. Apply Theorem 7.9.4.

Problem 22. Consider the set of open half spaces complementary to the specified closed half-spaces M_α, and apply Problem 21.

Problem 23. Identify the rays $\mathbf{r}_\alpha$ with the points $\mathbf{p}_\alpha$ at which they pierce the unit sphere centered on the origin. The hypothesis implies that $\mathbf{p}_\alpha$ spans a convex set with the origin as an interior point. Apply Theorem 7.9.4.

Problem 24. A vector $\mathbf{p} \in S$ will be denoted by $\langle x_1, \ldots, x_n \rangle$, a vector $\mathbf{p}' \in S'$ by $\langle y_1, \ldots, y_n \rangle$. Vectors in E^{n+1} will be denoted by $\mathbf{a} = \langle a_0, a_1, \ldots, a_n \rangle$. Let

$$H_p = \{\mathbf{a} | a_0 + a_1 x_1 + \cdots + a_n x_n < 0\},$$

$$H_{p'} = \{\mathbf{a} | a_0 + a_1 y_1 + \cdots + a_n y_n > 0\}.$$

We have thus generated a finite subfamily of half-spaces in E^{n+1}; any $n + 2$ of them correspond to $n + 2$ points from $S \cup S'$. By assumption they intersect. Apply Corollary 7.9.1.

Problem 25. Use the method of Problem 24 and Theorem 7.9.4.

CHAPTER 8

Problem 1. $p(t) = \left(p(0) + \dfrac{\alpha - \beta}{a - b}\right) \exp[-(a - b)t/a] - \dfrac{\alpha - \beta}{a - b}.$

If $a > b$, then $p(t)$ is convergent; if $a = b$, then $p(t)$ is a constant; and if $a < b$, then $p(t)$ is divergent.

Problem 4. $Y(t) = cY(t) + v\dot{Y}(t) + A_0 e^{rt}$. Then

$$Y(t) = \frac{A_0/v}{\dfrac{1 - c}{v} - r} e^{rt} + ke^{(1-c)t/v}.$$

Problem 6. $Y(t) = c_1Y(t-1) + c_2Y(t-2) + v(Y_{t-1} - Y_{t-2}) + A$. Let $\overline{Y} = A/(1 - c_1 - c_2)$. Then $Y(t) = \overline{Y} + k_1\lambda_1^t + k_2\lambda_2^t$, where λ_1 and λ_2 are roots of $x^2 - (c_1 + v)x + (v - c_2) = 0$. If multiple roots or complex roots occur, the form of the solution is modified accordingly.

Problem 8. (Proof by induction.) Consider the equation $\mathbf{D}(\mathbf{I} - \mathbf{C}) = \mathbf{E}$ for $c_{11} \neq 1$, where $d_{11} = 1$, $d_{j1} = c_{j1}/(1 - c_{11})$ for $j = 2, \ldots, n$, and $d_{ij} = \delta_{ij}$ otherwise. Then

$$\mathbf{E} = \begin{Vmatrix} 1 - c_{11} & -c_{12} & \cdots & -c_{1n} \\ 0 & 1 - \tilde{c}_{22} & \cdots & -\tilde{c}_{2n} \\ \vdots & \vdots & & \vdots \\ 0 & -\tilde{c}_{n2} & \cdots & 1 - \tilde{c}_{nn} \end{Vmatrix}$$

where

$$\tilde{c}_{ij} = c_{ij} + \frac{c_{i1}c_{1j}}{1 - c_{11}} \qquad \text{for} \quad i, j = 2, \ldots, n.$$

Observe that

$$\sum_{i=2}^{n} |\tilde{c}_{ij}| \leq 1.$$

Hence it follows from the induction hypothesis that

$$\det(\mathbf{I} - \mathbf{C}) = (1 - c_{11}) \det(\mathbf{I} - \tilde{\mathbf{C}}) \geq (1 - |c_{11}|) \prod_{j=2}^{n} \left(1 - \sum_{i=2}^{n} |\tilde{c}_{ij}|\right)$$

$$\geq \left(1 - \sum_{i=1}^{n} |c_{i1}|\right) \prod_{j=2}^{n} \left(1 - \sum_{i=1}^{n} |c_{ij}|\right).$$

The modification of this argument where $c_{11} = 1$ is trivial.

Problem 10. Let $\mathbf{x}^0 \gg \mathbf{0}$ be defined by $\mathbf{A}\mathbf{x}^0 = \lambda_0(\mathbf{A})\mathbf{x}^0$ and $\mathbf{A}\mathbf{y} = \mu\mathbf{y}$, with $\mathbf{y}$ a real vector and μ real. If the kth coordinate of $\mathbf{y}$ is zero, then $\mathbf{A}_{kk}|\mathbf{y}| \geq |\mu|\,|\mathbf{y}|$, where $\mathbf{A}_{kk}$ is the matrix obtained by deleting the kth row and the kth column. By Corollary 8.2.1, we have $\lambda_0(\mathbf{A}) \geq \lambda_0(\mathbf{A}_{kk}) \geq |\mu| \geq \mu$. If the kth coordinate of $\mathbf{y}$ is not zero, we determine a such that $ay_k + x_k^0 = 0$. Since $\mathbf{y}$ and $\mathbf{x}$ are linearly independent, $a\mathbf{y} + \mathbf{x}^0 \neq 0$. Now $\mathbf{A}_{kk}(a\mathbf{y} + \mathbf{x}^0) = \mathbf{A}(a\mathbf{y} + \mathbf{x}^0) \geq \mu(a\mathbf{y} + \mathbf{x}^0)$, and the conclusion follows as before, since k is arbitrary.

Problem 11. See Gantmacher [101, chap. 13].

Problem 13. Suppose for definiteness that the first column vector $\mathbf{b}$ of $\mathbf{B}(\lambda)$ is nonnegative and not all zeros. Since $(\lambda\mathbf{I} - \mathbf{A})(\lambda\mathbf{I} - \mathbf{A})^{-1} = \mathbf{I}$, we have $(\lambda\mathbf{I} - \mathbf{A})\mathbf{b} > \mathbf{0}$. By Theorem 8.2.1 there exists a vector $\mathbf{y}^0 \gg \mathbf{0}$ such that $\mathbf{y}^0\mathbf{A} = \lambda_0(\mathbf{A})\mathbf{y}^0$; hence $\lambda(\mathbf{y}^0, \mathbf{b}) > \lambda_0(\mathbf{A})(\mathbf{y}^0, \mathbf{b})$ and $\lambda > \lambda_0(\mathbf{A})$, as was to be shown.

Problem 15. Let λ be the eigenvalue with largest real part. Then $\mathbf{x}\mathbf{B} = \lambda\mathbf{x}$ for some $\mathbf{x} > \mathbf{0}$. Hence $0 < (\mathbf{x}, \mathbf{B}\mathbf{z}) = \lambda(\mathbf{x}, \mathbf{z})$. Since $(\mathbf{x}, \mathbf{z}) \leq 0$, we have $\lambda < 0$.

Problem 17. Assume the contrary. Then the set Γ in E^{2n} where

$$\Gamma = \left\{ \begin{pmatrix} \mathbf{I}, & -\mathbf{C} \\ \mathbf{C}', & \mathbf{I} \end{pmatrix} \begin{pmatrix} \mathbf{x} \\ \mathbf{y} \end{pmatrix} \middle\| \begin{matrix} \mathbf{C}' & 0 \\ 0 & -\mathbf{C} \\ \mathbf{I} & 0 \\ 0 & \mathbf{I} \end{matrix} \middle\| \begin{pmatrix} \mathbf{x} \\ \mathbf{y} \end{pmatrix} \geq 0 \right\},$$

has no point in common with the interior of the positive cone. The remainder of the argument is standard.

Problem 18. Let $\{\mathbf{p}^*, \mathbf{x}_i^*, \mathbf{y}_j^*\}$ be an equilibrium. Then

$$(\mathbf{p}^*, \mathbf{y}_j^*) = \max_{y_j \in Y_j} (\mathbf{p}^*, \mathbf{y}_j) \qquad (j = 1, \ldots, n).$$

Since Y_j is a cone, we deduce that $(\mathbf{p}^*, \mathbf{y}_j^*) = 0$ for $j = 1, \ldots, n$, as claimed.

Problem 19. (a) This is easily proved by a geometric argument.

(b) Let $\mathbf{p}^1 = \lambda \mathbf{p}^0$, $M^1 = \lambda M^0$, $\lambda > 0$, and $\mathbf{x}^0 = \mathbf{f}(\mathbf{p}^0, M^0)$, $\mathbf{x}^1 = \mathbf{f}(\mathbf{p}^1, M^1)$. Then $(\mathbf{p}^1, \mathbf{x}^0) = \lambda(\mathbf{p}^0, \mathbf{x}^0) = \lambda M^0 = M^1 = (\mathbf{p}^1, \mathbf{x}^1)$. If $\mathbf{x}^0 \neq \mathbf{x}^1$, then by the weak axiom of revealed preference $(\mathbf{p}^0, \mathbf{x}^0) < (\mathbf{p}^0, \mathbf{x}^1)$. But $(\mathbf{p}^0, \mathbf{x}^0) = M^0$ and

$$(\mathbf{p}^0, \mathbf{x}^1) = \frac{1}{\lambda}(\mathbf{p}^1, \mathbf{x}^1) = \frac{1}{\lambda} M^1 = M^0,$$

which is a contradiction.

CHAPTER 9

Problem 1. **DC** stable $\rightleftarrows$ $\mathbf{E}^{-1}\mathbf{DCE} = \mathbf{E}^{-1}\mathbf{DEA}$ stable (**D** and **E** commute since both are diagonal matrices) $\rightleftarrows$ $\mathbf{E}^{-1}\mathbf{DE}$ has all positive diagonal elements $\rightleftarrows d_{ii} > 0$.

Problem 4. Classify $\{1, \ldots, n\}$ according to the equivalence relation

$$i \sim j \rightleftarrows \operatorname{sign} d_{ij} \geq 0.$$

Then $i \sim j$ implies $j \sim i$ and $i \sim j$, $j \sim k$ imply $i \sim k$.

Problem 7. Let

$$\lambda(t) = \min_{1 \leq i \leq r} \frac{p_i(t)}{p_i^*},$$

where $\mathbf{p}^*$ is a strict positive equilibrium vector. Then

$$\underline{\lim}_{h \to 0+} \frac{\lambda(t+h) - \lambda(t)}{h} \geq 0,$$

and hence $\lambda(t) \geq \lambda(0)$. See also the proof of Theorem 9.4.3.

Problem 8.

$$\overline{\lim_{h\to 0+}} \frac{V(t+h)-V(t)}{h} \le \begin{cases} \dfrac{1}{p_i^*}\dfrac{dp_i^*}{dt} \text{ for } V(t) = -1 + p_i(t)/p_i^* \\ -\dfrac{1}{p_i^*}\dfrac{dp_i^*}{dt} \text{ for } V(t) = 1 - (p_i(t)/p_i^*) \end{cases} = \begin{matrix} F_i(t)/p_i^*, \\ \\ -F_i(t)/p_i^*. \end{matrix}$$

for an appropriate index i.

If $p_i(t) > p_i^*$, it follows from assertion (I) of the proof of Theorem 9.4.3 that $F_i(t) \le 0$, with equality only at an equilibrium point since the sequence $p_j(t)/p_j^*$ $(j = 0, 1, \ldots, n)$ is maximized for the index $j = i$. Similarly, if i is such that $V(t) = -(p_i(t) - p_i^*)$, then i is the index for which $p_j(t)/p_j^*$ $(0 \le j \le n)$ is minimized, and therefore $F_i(t) \ge 0$, with equality only if $\mathbf{p}(t)$ is an equilibrium vector. In either case, we obtain

$$\overline{\lim_{h\to 0+}} \frac{V(t+h)-V(t)}{h} \le 0,$$

and the remainder of the proof proceeds as in Theorem 9.4.3.

Problem 12. Consider $\mathbf{p}^0 > 0$ of the form $\mathbf{p}^0 = \langle 0, p_2^0, p_3^0, \ldots, p_r^0\rangle$. We show that $F_1(\mathbf{p}^0) = \infty$. Let $\mathbf{p}_\lambda = \langle \lambda, p_2^0, p_3^0, \ldots, p_r^0\rangle$, with $\lambda > 0$. From Euler's identity (homogeneity) we obtain

$$\lambda \frac{\partial F_1}{\partial p_1}(\mathbf{p}_\lambda) = -\sum_{k=2}^{r} p_k^0 \frac{\partial F_1}{\partial p_k}(\mathbf{p}_\lambda) \le -c < 0,$$

or

$$\lambda \frac{dF_1}{d\lambda} \le -c < 0 \qquad \text{for all} \quad 0 \le \lambda \le \lambda_0.$$

Integrating this relation from ϵ to λ_0 gives

$$F_1(\mathbf{p}_\epsilon) > F_1(\mathbf{p}_{\lambda_0}) - c \log \epsilon + c \log \lambda_0,$$

and as $\epsilon \to 0$ the conclusion follows.

Problem 13. We restrict attention to the set Δ of all vectors $\mathbf{p}$ satisfying $p_i \ge 0$ and

$$\sum_{i=1}^{r} p_i = 1.$$

Consider the functions $g_k(\mathbf{p}) = e^{-F_k(p)}$ for $\mathbf{p}$ in Δ $(1 \le k \le r)$. By Problem 12, $g_k(\mathbf{p})$ are continuous on Δ and $g_k(\mathbf{p}) = 0$ whenever $p_k = 0$ (and conversely). Let

$$h_k(\mathbf{p}) = \max_{1\le j\le r} \{g_j(\mathbf{p})\} - g_k(\mathbf{p}).$$

Suppose there exists no $\mathbf{p}$ in Δ for which $h_k(\mathbf{p}) = 0$ $(k = 1, \ldots, r)$. In that case, consider the mapping of Δ into the boundary of Δ, defined as follows:

$$\langle p_k \rangle \to \left\langle \frac{h_k(\mathbf{p})}{\sum_{i=1}^{r} h_i(\mathbf{p})} \right\rangle.$$

By virtue of the Brouwer fixed-point theorem (Section C.2) we secure a fixed-point vector $\mathbf{p}^*$ having at least one zero component. If $h_{k_0}(\mathbf{p}^*) = \mathbf{p}^*_{k_0} = 0$, then $g_{k_0}(\mathbf{p}^*) = 0$ and

$$\max_{1 \le j \le r} g_j(\mathbf{p}^*) = 0,$$

which means that $\mathbf{p}^* = 0$, an absurdity. Consequently, there exists a choice $\mathbf{p}^*$ with the property $g_1(\mathbf{p}^*) = g_2(\mathbf{p}^*) = \cdots = g_r(\mathbf{p}^*) \neq 0$. Hence $F_1(\mathbf{p}^*) = F_2(\mathbf{p}^*) = \cdots = F_r(\mathbf{p}^*) \neq \infty$, which by virtue of the Walras law reduces to $F(\mathbf{p}^*) = 0$ and $\mathbf{p}^* \gg \mathbf{0}$.

Problem 16. We may assume without loss of generality that $\mathbf{p}^* = \langle 1, \ldots, 1 \rangle$. Let

$$a_{ij} = \left(\frac{\partial F_i}{\partial p_j} \right)_{p^*} \qquad (i, j = 0, 1, \ldots, r).$$

Assume that the system

$$\frac{dp_i}{dt} = \sum_{j=1}^{r} a_{ij} p_j \qquad (i = 1, \ldots, r)$$

is stable (the zeroth component is the numéraire). Since the matrix $\|a_{ij}\|_1^r$ is symmetric it is negative definite. Consider another commodity, say r, as the numéraire. By homogeneity,

$$a_{i0} = -\sum_{j=1}^{r} a_{ij} \qquad (i = 1, \ldots, r)$$

and since the numéraire is constant

$$a_{00} = -\sum_{i=1}^{r} a_{i0} = \sum_{i,j=1}^{r} a_{ij}.$$

Now, for any $\mathbf{x} = \langle x_0, \ldots, x_{r-1} \rangle \neq 0$ we obtain by some simple calculations that

$$\sum_{i,j=0}^{r-1} a_{ij} x_i x_j = \sum_{i,j=1}^{r} a_{ij} x'_i x'_j,$$

where $x'_i = x_0 - x_i$ for $i = 1, \ldots, r-1$ and $x'_r = x_0$. Hence the matrix $\|a_{ij}\|_0^{r-1}$ is stable.

Problem 18. Let

$$\lambda(t) = \min_{1 \le i \le r} \left\{ \frac{p_i(t)}{p_i^*} \right\}, \qquad \Lambda(t) = \max_{1 \le i \le r} \left\{ \frac{p_i(t)}{p_i^*} \right\}.$$

Then $-\lambda(t)$ and $\Lambda(t)$ are nonincreasing functions (strictly if $\mathbf{p}(t)$ is not an equilibrium vector). From here on, an argument analogous to the proof of Theorem 9.4.3 is applicable.

Problem 19. The relation of output and consumption over two successive periods obeys the law $x(t) + c(t) \leq ax(t - 1)$. By iteration,

$$\begin{aligned} &a^{n-t}x(t) + a^{n-t}c(t) \leq a^{n-(t-1)}x(t-1) \\ &\vdots \\ &a^{n}x(0) + a^{n}c(0) \leq a^{n}K. \end{aligned}$$

Hence

$$c(n) + x(n) \leq a^{n}K - \sum_{t=0}^{n-1} a^{n-t}c(t).$$

Stated as a linear programming problem: Maximize $c(n)$ subject to the constraints

$$a^{t}K - \sum_{v=0}^{t} a^{n-v}c(v) \geq \upsilon \qquad (t = 0, 1, 2, \ldots, n)$$

and $c(v) \geq 0$ for all v.

Problem 20. In this case $\mathbf{z}^* \gg \mathbf{0}$ and $\mathbf{p}^* \gg \mathbf{0}$.

APPENDIXES

In order to make this volume as complete in itself as possible and accessible to the mature student who has only a modicum of mathematical training beyond the calculus, we devote these appendixes to a brief review of those topics of upper-class undergraduate mathematics which play a fundamental role in the studies of this book. Where an analysis requires more advanced mathematics, the essentials are usually directly incorporated in the text.

In Appendix A we summarize the salient features of the theory of vector spaces and matrices. Our approach to this theory, unlike the usual approach, is primarily geometric rather than algebraic. We do not develop matrix theory abstractly in terms of general rings and fields, but restrict the scalar field of our vector spaces to the real or complex numbers. Proofs that are not easily accessible in the literature are indicated in detail, and in all circumstances we have at least sketched the arguments verifying the main statements of the theory.

Appendix B treats some aspects of the theory of convex sets and convex functions. Much of the theory of finite games, linear programming, and mathematical economics is based on operations with sets and functions derived from linear and convex inequalities. We summarize here the appropriate versions of the separation theorems for convex sets and discuss simple forms of the duality theory of convex cones. Some advanced topics in the theory of convex sets—for example, the development of the theory of conjugate convex functions and certain convexity results connected with moment spaces—are presented in the text.

Appendix C reviews several miscellaneous facts regarding functions, integrals, and fixed-point theorems which are explicitly called upon in making some of the analyses in this volume. The concepts of semi-continuity and equi-continuity are defined precisely, and some of the properties of these functions are listed. The important Kakutani fixed-point theorem is noted in Section C.2. The relevant properties of Lebesgue-Stieltjes integrals are recorded for reference in Section C.3.

The following books and articles may profitably be consulted in connection with the topics covered in these appendixes:

Appendix A: Perlis [195], Birkoff and McLane [26], Gantmacher [101].

Appendix B: Fenchel [81], Hardy, Littlewood, and Pólya [118, Chap. 2].

Appendix C: Graves [112], Rudin [209].

APPENDIX A

VECTOR SPACES AND MATRICES

A.1 Euclidean and unitary spaces. A real vector space may be defined as follows: We have a set V; an associative and commutative addition, $+$, of pairs of its elements; a unique zero element $\boldsymbol{\theta}$; and a multiplication of elements of V by real numbers, for which, with α, β real and $\mathbf{x}$, $\mathbf{y}$ in V

$$\mathbf{x} + \boldsymbol{\theta} = \mathbf{x} \qquad 1 \cdot \mathbf{x} = \mathbf{x}, \qquad 0 \cdot \mathbf{x} = \boldsymbol{\theta},$$

$$\alpha(\mathbf{x} + \mathbf{y}) = \alpha\mathbf{x} + \alpha\mathbf{y}, \qquad (\alpha + \beta)\mathbf{x} = \alpha\mathbf{x} + \beta\mathbf{x}, \qquad \alpha(\beta\mathbf{x}) = (\alpha\beta)\mathbf{x}.$$

As a prime example we have the Euclidean n-dimensional space E^n, which consists of the set of all ordered n-tuples $\mathbf{x} = \langle x_1, x_2, \ldots, x_n \rangle$ of real numbers $x_1, x_2, \ldots, x_n$, with addition and multiplication by a real number γ defined by

$$\langle x_1, x_2, \ldots, x_n \rangle + \langle y_1, y_2, \ldots, y_n \rangle = \langle x_1 + y_1, x_2 + y_2, \ldots, x_n + y_n \rangle,$$

$$\gamma\langle x_1, x_2, \ldots, x_n \rangle = \langle \gamma x_1, \gamma x_2, \ldots, \gamma x_n \rangle.$$

Here $\boldsymbol{\theta}$ (which we shall henceforth call $\mathbf{0}$) is $\langle 0, 0, \ldots, 0 \rangle$; each $\mathbf{x} = \langle x_1, x_2, \ldots, x_n \rangle$ has a negative $-\mathbf{x} = \langle -x_1, -x_2, \ldots, -x_n \rangle$ for which $\mathbf{x} + (-\mathbf{x}) = \mathbf{0}$ (the same is true abstractly: $\mathbf{x} + (-1) \cdot \mathbf{x} = 1 \cdot \mathbf{x} + (-1) \cdot \mathbf{x} = [1 + (-1)] \cdot \mathbf{x} = 0 \cdot \mathbf{x} = \boldsymbol{\theta}$); and we may define subtraction by $\mathbf{y} - \mathbf{x} = \mathbf{y} + (-\mathbf{x})$. It should be noted that in this example the real numbers may be replaced by the complex numbers; in that case the corresponding vector space U^n is referred to as the unitary n-dimensional space.

Actually E^n (and U^n) are not merely vector spaces; we can perform the additional operation of forming the inner product $(\mathbf{x}, \mathbf{y})$ of two elements $\mathbf{x}$, $\mathbf{y}$, which has a value

$$(\mathbf{x}, \mathbf{y}) = \sum_{i=1}^{n} x_i \bar{y}_i,$$

where $\bar{y}_i$ is the complex conjugate of y_i. (Two elements $\mathbf{x}$, $\mathbf{y}$ are said to be *orthogonal* if $(\mathbf{x}, \mathbf{y}) = 0$.) By means of this operation we define the usual distance function

$$d(\mathbf{x}, \mathbf{y}) = (\mathbf{x} - \mathbf{y}, \mathbf{x} - \mathbf{y})^{\frac{1}{2}} = \left(\sum_{i=1}^{n} |x_i - y_i|^2\right)^{\frac{1}{2}},$$

where $d(\mathbf{x}, \mathbf{y}) = d(\mathbf{y}, \mathbf{x}) \geq 0$ (with equality holding if and only if $\mathbf{x} = \mathbf{y}$), and the triangle inequality

$$d(\mathbf{x}, \mathbf{y}) \leq d(\mathbf{x}, \mathbf{z}) + d(\mathbf{z}, \mathbf{y}).$$

Conventionally, x_i is called the ith *component* of $\mathbf{x} = \langle x_1, x_2, \ldots, x_n \rangle$, and $d(\mathbf{0}, \mathbf{x})$ is called the *length* of $\mathbf{x}$. In all that follows we shall be considering E^n and U^n as the vector spaces involved, and any statements made concerning E^n may be applied with equal force to U^n (but not conversely). To eliminate reference to real and complex numbers we shall refer to the numbers involved as *scalars* (α, β, γ, etc.), and to the elements of a vector space as *vectors* ($\mathbf{x}$, $\mathbf{y}$, $\mathbf{z}$, etc.).

Exercises

1. Let $\mathcal{P}^n$ be the set of all polynomials $p(t)$ defined on $-\infty < a \leq t \leq b < \infty$, with real (complex) coefficients such that the degree of $p(t)$ is at most $n - 1$. Verify that under the ordinary notions of addition of polynomials and multiplication by a real (complex) scalar, $\mathcal{P}^n$ is a real (complex) vector space. Show that an inner product for p_1 and p_2 in $\mathcal{P}^n$ may be defined by

$$(p_1, p_2) = \int_a^b p_1(t)\bar{p}_2(t)\, dt.$$

2. Let C be the set of complex numbers, addition and multiplication by *real* scalars being defined as usual. Show that C is a real vector space, which possesses an inner product for $x = a + bi$ and $y = c + di$ given by $(x, y) = ac + bd$.

3. In E^3, for instance, we can associate each vector $\langle x, y, z \rangle$ with the point in three-dimensional space having Cartesian coordinates $\{x, y, z\}$. Thus in referring to sets of vectors we may speak of points, lines, and planes. Interpret this statement.

4. For E^n let

$$d(\mathbf{x}, \mathbf{y}) = \max_{1 \leq i \leq n} |x_i - y_i|.$$

Show that $d(\mathbf{x}, \mathbf{y})$ satisfies the usual requirements for a distance function; namely,

$$d(\mathbf{x}, \mathbf{y}) = d(\mathbf{y}, \mathbf{x}) \geq 0 \qquad \text{(with equality if and only if } \mathbf{x} = \mathbf{y}\text{)},$$
$$d(\mathbf{x}, \mathbf{y}) \leq d(\mathbf{x}, \mathbf{z}) + d(\mathbf{z}, \mathbf{y}).$$

5. Using the usual distance function for E^n and noticing the relationships between "orthogonal" and the usual definition of "perpendicular," state and prove a general Pythagorean theorem for E^n.

A.2 Subspaces, linear independence, basis, direct sums, orthogonal complements. A subspace V_1 of a vector space V is a subset which is closed under addition and scalar multiplication: if $\mathbf{x}$ and $\mathbf{y}$ are in V_1, and α is a scalar, then $\mathbf{x} + \mathbf{y}$ and $\alpha\mathbf{y}$ are in V_1. Under these operations V_1 is a vector space in its own right. For example, the set of all linear combinations

$$\sum_{i=1}^{k} \alpha_i \mathbf{x}^i = \alpha_i \mathbf{x}^1 + \cdots + \alpha_k \mathbf{x}^k$$

of a fixed set $\{\mathbf{x}^1, \mathbf{x}^2, \ldots, \mathbf{x}^k\}$ of *vectors* forms a subspace called the span of $\{\mathbf{x}^1, \mathbf{x}^2, \ldots, \mathbf{x}^k\}$. Alternatively, if V_1 is the span of $\{\mathbf{x}^1, \ldots, \mathbf{x}^k\}$, then $\{\mathbf{x}^1, \ldots, \mathbf{x}^k\}$ *spans* V_1.

A subset $\{\mathbf{x}^1, \ldots, \mathbf{x}^k\}$ is *linearly independent* if

$$\sum_{i=1}^{k} \alpha_i \mathbf{x}^i = \mathbf{0}$$

implies $\alpha_i = 0$ $(i = 1, \ldots, k)$. A *linearly independent* subset $\{\mathbf{x}^1, \ldots, \mathbf{x}^k\}$ which *spans* V is a *basis* for V. In E^n,

$$\mathbf{e}^1 = \langle 1, 0, 0, \ldots, 0\rangle, \qquad \mathbf{e}^2 = \langle 0, 1, 0, \ldots, 0\rangle, \qquad \ldots,$$
$$\mathbf{e}^n = \langle 0, 0, \ldots, 0, 1\rangle$$

clearly forms a basis. (Not all vector spaces have finite bases; a trivial example is the case in which V has exactly one element, $\mathbf{0}$, where no subset is independent. In this case we shall not usually regard V as a vector space.)

If $\{\mathbf{x}^1, \ldots, \mathbf{x}^k\}$ spans V, then a subset forms a basis for V. This is clear if the set is linearly independent; otherwise we have

$$\sum_{i=1}^{k} \alpha_i \mathbf{x}^i = \mathbf{0},$$

with, say, $\alpha_1 \neq 0$. It follows that

$$\mathbf{x}^1 = \sum_{i=2}^{k} (-\alpha_i \cdot \alpha_1^{-1})\mathbf{x}^i,$$

and clearly $\{\mathbf{x}^2, \ldots, \mathbf{x}^k\}$ spans V. If this subset is not independent, we repeat the argument until we arrive at one that is. The advantage of recognizing a basis $\{\mathbf{y}^1, \ldots, \mathbf{y}^l\}$ for V is that each $\mathbf{x}$ in V is *uniquely* expressible in the form

$$\sum_{i=1}^{l} \alpha_i \mathbf{y}^i,$$

where we refer to the unique scalars α_i as the coordinates of $\mathbf{x}$ relative to the basis $\{\mathbf{y}_1, \ldots, \mathbf{y}_l\}$.

Some important facts about bases are the following, which we state without proof. Each nontrivial subspace V_1 of E^n has a basis, and each basis for V_1 has the same number of elements, the *dimension* (abbreviated dim) of V_1. If $V_1 = \{\mathbf{0}\}$, then dim $V_1 = 0$. If $V_1 \neq E^n$ (i.e., V_1 is a *proper* subspace), then each basis $\{\mathbf{x}^1, \mathbf{x}^2, \ldots, \mathbf{x}^k\}$ of V_1 can be enlarged to a basis $\{\mathbf{x}^1, \mathbf{x}^2, \ldots, \mathbf{x}^k, \mathbf{y}^1, \ldots, \mathbf{y}^l\}$ of E^n (thus V_1 has dimension $k < n =$ dimension E^n), and indeed any linearly independent subset can be so enlarged, being a basis of its span. Consequently, any linearly independent subset of E^n with n elements is already a basis for E^n. It should be noted that V_1 as a vector space is essentially E^k imbedded in E^n, since the one-to-one (1-1) mapping

$$\langle \alpha_1, \alpha_2, \ldots, \alpha_k \rangle \to \sum_{i=1}^{k} \alpha_i \mathbf{x}^i$$

(α_i scalars) of E^k onto V_1 preserves all sums and scalar multiples, as does its inverse.

In E^n or U^n a basis is *orthonormal* if

$$(\mathbf{y}^i, \mathbf{y}^j) = \begin{cases} 0 & i \neq j, \\ 1 & i = j, \end{cases}$$

i.e., if the distinct elements of the basis are orthogonal and each is of length 1. The components of a vector relative to an orthonormal basis are computed from the inner product:

$$\mathbf{x} = \sum_{i=1}^{k} (\mathbf{x}, \mathbf{y}^i)\mathbf{y}^i$$

for $\mathbf{x}$ in the span of $\{\mathbf{y}^1, \ldots, \mathbf{y}^k\}$, since the difference of the two sides is orthogonal to each $\mathbf{y}^i$, and hence to itself. *Every subspace of E^n (or U^n) has an orthonormal basis.*

If V_1 and V_2 are subspaces of V and

$$V = V_1 + V_2 = \{\mathbf{x} + \mathbf{y} | \mathbf{x} \in V_1, \mathbf{y} \in V_2\}$$

while $V_1 \cap V_2 = \{\mathbf{0}\}$, then V is the *direct sum* of V_1 and V_2; symbolically $V = V_1 \oplus V_2$. Each $\mathbf{z}$ in V is then uniquely expressible as $\mathbf{x} + \mathbf{y}$, where $\mathbf{x} \in V_1, \mathbf{y} \in V_2$; for $\mathbf{x} + \mathbf{y} = \mathbf{x}' + \mathbf{y}'$ implies that $\mathbf{x} - \mathbf{x}' = \mathbf{y}' - \mathbf{y}$ is in both V_1 and V_2, and hence equals zero. Consequently, if $\{\mathbf{x}^1, \ldots, \mathbf{x}^k\}$ is a basis for V_1 and $\{\mathbf{y}^1, \ldots, \mathbf{y}^m\}$ is a basis for V_2, then

$$\{\mathbf{x}^1, \ldots, \mathbf{x}^k, \mathbf{y}^1, \ldots, \mathbf{y}^m\}$$

is a basis for V; for it clearly spans V, while $\sum \gamma_i \mathbf{x}^i + \sum \delta_1 \mathbf{y}^j = \mathbf{0}$ or $\sum \gamma_i \mathbf{x}^i = -\sum \delta_1 \mathbf{y}^j$ implies that each is zero, and hence that all γ_i and δ_j are 0.

For any subspace V of E^n (or U^n) the set

$$V^{\perp} = \{\mathbf{x} | (\mathbf{x}, \mathbf{y}) = 0 \quad \text{for} \quad \mathbf{y} \text{ in } V\}$$

is called the *orthogonal complement* of V. It is clearly a subspace, and $E^n = V \oplus V^{\perp}$; also $V^{\perp\perp} = (V^{\perp})^{\perp} = V$. Since clearly $\dim (V_1 \oplus V_2) = \dim V_1 + \dim V_2$, it follows that $\dim V^{\perp} = n - \dim V$.

Exercises

1. In $\mathcal{P}^n$ (see Problem 1 of Section A.1) show that the polynomials $1, t, \ldots, t^{n-1}$ comprise a basis, and that $\mathcal{P}^n$ has dimension n.
2. With C as specified in Problem 2 of Section A.1, prove that an orthogonal basis is formed from the elements 1 and i, and thus that C is two-dimensional.
3. Prove that any nonvoid proper subspace of E^3 corresponds to a line or plane passing through the origin, and that its orthogonal complement corresponds to the perpendicular plane or line also passing through the origin.
4. Let $\mathbf{a}$ be a nonzero vector in E^n. Let $H = \{\mathbf{x} | \mathbf{x} \in E^n, (\mathbf{x}, \mathbf{a}) = 0\}$. Show that H, called a hyperplane, is a subspace of E^n and find its dimension.
5. Find an orthogonal basis for $\mathcal{P}^2$ and express the polynomials 1 and t in terms of this basis.

A.3 Linear transformations, matrices, and linear equations. A linear transformation $\mathbf{T}$ mapping E^n into E^m is a function with domain E^n and values in E^m which preserves sums and scalar multiples: $\mathbf{T}(\mathbf{x} + \mathbf{y}) = \mathbf{T}\mathbf{x} + \mathbf{T}\mathbf{y}$, and $\mathbf{T}(\alpha\mathbf{x}) = \alpha\mathbf{T}\mathbf{x}$. Clearly, $\mathbf{T0} = \mathbf{0}$, since $\mathbf{T0} = \mathbf{T}(\mathbf{0} + \mathbf{0}) = \mathbf{T0} + \mathbf{T0}$; similarly, $\mathbf{T}(\mathbf{x} - \mathbf{y}) = \mathbf{T}\mathbf{x} - \mathbf{T}\mathbf{y}$, since

$$\mathbf{T}(\mathbf{x} - \mathbf{y}) + \mathbf{T}\mathbf{y} = \mathbf{T}(\mathbf{x} - \mathbf{y} + \mathbf{y}) = \mathbf{T}\mathbf{x}.$$

The null space of $\mathbf{T}$, $\mathfrak{N}(\mathbf{T}) = \{\mathbf{x} | \mathbf{T}\mathbf{x} = \mathbf{0}\}$, is a subspace of the domain E^n (that is, $\mathbf{x}$, $\mathbf{y}$ in $\mathfrak{N}(\mathbf{T})$ yields $\alpha\mathbf{x} + \beta\mathbf{y}$ in $\mathfrak{N}(\mathbf{T})$ for any scalars α and β); however, $\mathfrak{N}(\mathbf{T})$ can reduce to the trivial subspace $\{\mathbf{0}\}$. This occurs exactly when $\mathbf{T}$ is one-to-one, and $\mathbf{T}$ being one-to-one is equivalent to $\mathfrak{N}(\mathbf{T}) = \{\mathbf{0}\}$, or $\dim \mathfrak{N}(\mathbf{T}) = 0$.

Similarly, the *range* of $\mathbf{T}$, $\mathcal{R}(\mathbf{T}) = \{\mathbf{T}\mathbf{x} | \mathbf{x} \in E^n\}$, is a subspace of E^m, and the following dimensional relationship holds:

$$\dim \mathfrak{N}(\mathbf{T}) + \dim \mathcal{R}(\mathbf{T}) = n. \tag{A.3.1}$$

We demonstrate this relationship as follows. If we select a basis

$$\{\mathbf{x}^1, \ldots, \mathbf{x}^k\}$$

of $\mathfrak{N}(\mathbf{T})$ and enlarge it to a basis $\{\mathbf{x}^1, \ldots, \mathbf{x}^k, \mathbf{y}^{k+1}, \ldots, \mathbf{y}^n\}$ of E^n, then, with $V = \text{span}\,\{\mathbf{y}^{k+1}, \ldots, \mathbf{y}^n\}$, we have $E^n = \mathfrak{N}(\mathbf{T}) \oplus V$. We have only to check $\mathfrak{N}(\mathbf{T}) \cap V = \{\mathbf{0}\}$. But $\mathbf{z}$ in $\mathfrak{N}(\mathbf{T}) \cap V$ implies

$$\mathbf{z} = \sum_{i=1}^{k} \alpha_i \mathbf{x}^i = \sum_{j=k+1}^{n} \beta_j \mathbf{y}^j;$$

hence all $\alpha_i, \beta_j = 0$, since $\{\mathbf{x}^1, \ldots, \mathbf{x}^k, \mathbf{y}^{k+1}, \ldots, \mathbf{y}^n\}$ is linearly independent, and thus $\mathbf{z} = \mathbf{0}$. But the image of the general element $\sum \alpha_i \mathbf{x}^i + \sum \beta_j \mathbf{y}^j$ of E^n is $\mathbf{T}(\sum \beta_j \mathbf{y}^j) = \sum \beta_j \mathbf{T}\mathbf{y}^j$, and hence $\{\mathbf{T}\mathbf{y}^{k+1}, \ldots, \mathbf{T}\mathbf{y}^n\}$ spans

$\mathcal{R}(\mathbf{T})$. On the other hand, the set is linearly independent, since $\sum\beta_j\mathbf{T}\mathbf{y}^j = \mathbf{0}$ implies that $\mathbf{T}(\sum\beta_j\mathbf{y}^j) = \mathbf{0}$, which means that $\sum\beta_j\mathbf{y}^j \in \mathfrak{N}(\mathbf{T}) \cap V = \{\mathbf{0}\}$, and hence all $\beta_j = 0$. Thus dim $\mathcal{R}(\mathbf{T}) = n - k$, and our relation holds. It should be noted that V is in no way unique.

As consequences, note that dim $\mathcal{R}(\mathbf{T}) \leq n$, with equality holding if and only if $\mathbf{T}$ is 1-1. Thus when $m = n$, the transformation $\mathbf{T}$ maps E^n onto itself if and only if it is 1-1.

Matrices. A rectangular array of numbers

$$\mathbf{A} = \|a_{ij}\| = \begin{Vmatrix} a_{11} & a_{12} & \cdots & a_{1n} \\ a_{21} & a_{22} & \cdots & a_{2n} \\ \vdots & \vdots & & \vdots \\ a_{m1} & a_{m2} & \cdots & a_{mn} \end{Vmatrix}$$

is called a *matrix*—in this case an $m \times n$ (m-row by n-column) matrix—with *entries* a_{ij}. If $m = n$, we call the matrix *square*.

$\mathbf{A}$ determines a linear transformation $\mathbf{T}$ of E^n into E^m defined by setting

$$\mathbf{T}\langle x_1, x_2, \ldots, x_n\rangle = \left\langle \sum_{j=1}^{n} a_{1j}x_j, \quad \sum_{j=1}^{n} a_{2j}x_j, \ldots, \quad \sum_{j=1}^{n} a_{mj}x_j \right\rangle.$$

Conversely, if we denote the usual orthonormal basis of E^n by $\{\mathbf{e}^1, \ldots, \mathbf{e}^n\}$ and that of E^m by $\{\mathbf{f}^1, \ldots, \mathbf{f}^m\}$, each linear transformation $\mathbf{T}$ is so determined. For $\mathbf{T}$ is determined by its values at the $\mathbf{e}^j$, since

$$\mathbf{T}\left(\sum_{j=1}^{n} x_j\mathbf{e}^j\right) = \sum_{j=1}^{n} x_j\mathbf{T}\mathbf{e}^j;$$

and since each $\mathbf{T}\mathbf{e}^j$ is expressible as

$$\sum_{i=1}^{m} a_{ij}\mathbf{f}^i$$

for some unique $a_{1j}, a_{2j}, \ldots, a_{mj}$, we obtain

$$\mathbf{T}\mathbf{x} = \sum_{j=1}^{n} x_j\left(\sum_{i=1}^{m} a_{ij}\mathbf{f}^i\right) = \sum_{j=1}^{n}\sum_{i=1}^{m} x_ja_{ij}\mathbf{f}^i = \sum_{i=1}^{m}\left(\sum_{j=1}^{n} a_{ij}x_j\right)\mathbf{f}^i,$$

and the components of $\mathbf{T}\mathbf{x}$ are obtained as before from $\mathbf{A}$. Moreover, it is clear that we may replace $\{\mathbf{e}^1, \ldots, \mathbf{e}^n\}$ and $\{\mathbf{f}^1, \ldots, \mathbf{f}^m\}$ by any other fixed ordered bases in E^n and E^m and set up a similar one-to-one correspondence between linear transformations mapping E^n into E^m and $m \times n$ matrices. The exact correspondence will of course depend greatly on the bases chosen (and on their indexing). However, when we consider a matrix as a transformation we shall have in mind our original correspondence.

Corresponding to the linear transformations **T** and **S** mapping E^n into E^m there is a natural sum $\mathbf{T} + \mathbf{S}$, again a linear transformation, such that $(\mathbf{T} + \mathbf{S})(\mathbf{x}) = \mathbf{Tx} + \mathbf{Sx}$. Similarly, $c\mathbf{T}$ at $\mathbf{x}$ is defined to be $c \cdot \mathbf{Tx}$ (c a scalar). The corresponding operations on matrices are clearly the addition of corresponding entries:

$$\mathbf{A} + \mathbf{B} = \left\| \begin{matrix} a_{11} + b_{11} & a_{12} + b_{12} & \cdots \\ a_{21} + b_{21} & \cdots\cdots\cdots & \cdots \\ \vdots & & \\ a_{m1} + b_{m1} & \cdots\cdots\cdots & \cdots \end{matrix} \right\|$$

and

$$c\mathbf{A} = \left\| \begin{matrix} ca_{11} & ca_{12} & \cdots \\ \vdots & \vdots & \\ ca_{m1} & ca_{m2} & \cdots \end{matrix} \right\|$$

(each entry multiplied by c). Clearly the $m \times n$ matrices thus form a vector space with an obvious basis $\mathbf{e}_{ij}$ ($i = 1, \ldots, m; j = 1, \ldots, n$), where $\mathbf{e}_{ij}$ has all entries 0 except for a single 1 in the ijth place. Thus the space is mn-dimensional.

If **T** maps E^n into E^m and **S** maps E^m into E^l, we can compose the mappings in such a way that **ST** maps E^n into E^l. Explicitly, we set $\mathbf{ST}(\mathbf{x}) = \mathbf{S}(\mathbf{Tx})$. The linear transformation **ST** is called the product of **S** and **T**; the corresponding matrix product is easily computed: with **S** corresponding to $\mathbf{A}(l \times m)$ and **T** corresponding to $\mathbf{B}(m \times n)$, **AB** is an $l \times n$ matrix with ijth entry

$$\sum_{k=1}^{m} a_{ik}b_{kj} \qquad (i = 1, \ldots, l; \quad j = 1, \ldots, n).$$

If we view the elements of E^n as $n \times 1$ matrices

$$\left\| \begin{matrix} x_1 \\ x_2 \\ \vdots \\ x_n \end{matrix} \right\|$$

(a *column* vector), our computation of the components of **Tx** (where **T** corresponds to $\|b_{ij}\|$) is simply the formation of the matrix product

$$\left\| \begin{matrix} b_{11} & b_{12} & \cdots & b_{1n} \\ b_{21} & b_{22} & \cdots & b_{2n} \\ \vdots & \vdots & & \vdots \\ b_{m1} & b_{m2} & \cdots & b_{mn} \end{matrix} \right\| \left\| \begin{matrix} x_1 \\ x_2 \\ \vdots \\ x_n \end{matrix} \right\| = \left\| \begin{matrix} \sum_{j=1}^{k} b_{1j}x_j \\ \vdots \\ \sum_{j=1}^{n} b_{mj}x_j \end{matrix} \right\|$$

giving the image elements of E^m as column vectors $(m \times 1)$. The resulting column vector can also be expressed as

$$x_1 \begin{Vmatrix} b_{11} \\ b_{21} \\ \vdots \\ b_{m1} \end{Vmatrix} + x_2 \begin{Vmatrix} b_{12} \\ b_{22} \\ \vdots \\ b_{m2} \end{Vmatrix} + \cdots + x_n \begin{Vmatrix} b_{1n} \\ b_{2n} \\ \vdots \\ b_{mn} \end{Vmatrix}$$

so that the *dimension of* $\mathcal{R}(\mathbf{T})$, *called the rank of* $\mathbf{T}$ *or* $\mathbf{B}$, *is the number of linearly independent columns of* $\mathbf{B}$ (viewed as vector elements of E^m). One can obtain the rank by eliminating dependent columns, or by determinants; the rank is k if after deleting rows and columns of $\mathbf{B}$, the largest square submatrices with nonzero determinant are $k \times k$.

Finally, a linear transformation from E^n into E^1, which we may regard as the real line, is called a *linear functional.* As we have seen, it is given (relative to the bases $\{\mathbf{e}^1, \ldots, \mathbf{e}^n\}$ and $\{\mathbf{f}^1\}$, where $\mathbf{f}^1 = \langle 1 \rangle$) by a $1 \times n$ matrix, and its value at $\langle x_1, x_2, \ldots, x_n \rangle$ is of the form $c_1x_1 + c_2x_2 + \cdots + c_nx_n$, with c_i fixed. Thus each linear functional is obtained from the inner product and maps $\mathbf{x}$ into $(\mathbf{x}, \mathbf{y})$ for some fixed $\mathbf{y}$ in E^n.

Linear equations. The correspondence between matrices and linear transformations can be applied to systems of linear equations. Consider the following system of m equations in n unknown (real or complex) numbers $x_1, x_2, \ldots, x_n$:

$$\begin{aligned} a_{11}x_1 + a_{12}x_2 + \cdots + a_{1n}x_n &= c_1 \\ a_{21}x_1 + a_{22}x_2 + \cdots + a_{2n}x_n &= c_2 \\ &\vdots \\ a_{m1}x_1 + a_{m2}x_2 + \cdots + a_{mn}x_n &= c_m. \end{aligned} \tag{A.3.2}$$

Solving the system is equivalent to finding an $\mathbf{x} = \langle x_1, x_2, \ldots, x_n \rangle$ in E^n mapping onto $\mathbf{c} = \langle c_1, c_2, \ldots, c_m \rangle$. A solution can be obtained for every c if and only if the rank of $\mathbf{A} = \|a_{ij}\|$ is m.

If each $c_i = 0$, the system is called *homogeneous* (otherwise inhomogeneous) and has the trivial solution $\langle 0, 0, \ldots, 0 \rangle$; nontrivial solutions exist whenever $\mathfrak{N}(\mathbf{A}) \neq \{\mathbf{0}\}$. If $n > m$ (more unknowns than equations), this must be the case, for otherwise $\dim \mathcal{R}(\mathbf{A}) = n$, which is greater than the dimension of E^m, while $\mathcal{R}(\mathbf{A}) \subset E^m$. If $n = m$, (A.3.2) is solvable for each $\mathbf{c}$ if and only if each solution is unique, or indeed if the solution for $\mathbf{c} = \mathbf{0}$ is unique; this amounts to saying that $\mathbf{A}$ maps E^n onto itself if and only if $\mathfrak{N}(\mathbf{A}) = \{\mathbf{0}\}$.

Transformations from E^n *into* E^n. Let $L(E^n)$ be the set of all linear transformations $\mathbf{T}$ mapping E^n into E^n. If $\mathbf{S}$ and $\mathbf{T}$ are in $L(E^n)$, then $\mathbf{S} + \mathbf{T}$, $\mathbf{ST}$, and $c\mathbf{T}$ are all elements of $L(E^n)$, and $L(E^n)$ forms a vector space with a (noncommutative) associative and distributive multiplication. The zero element maps each $\mathbf{x}$ into $\mathbf{0}$, while the identity transformation $\mathbf{I}$ mapping $\mathbf{x}$ into $\mathbf{x}$ yields $\mathbf{TI} = \mathbf{IT} = \mathbf{T}$. In terms of matrices, the zero transformation corresponds to the matrix with all entries 0, and

$$\mathbf{I} = \begin{Vmatrix} 1 & 0 & \cdots & 0 \\ 0 & 1 & \cdots & 0 \\ \vdots & \vdots & \ddots & \vdots \\ 0 & 0 & \cdots & 1 \end{Vmatrix}$$

(with 1's only on the main diagonal). We know that $\mathbf{T}$ is 1-1 if and only if $\mathbf{T}$ is onto; in this case the inverse transformation $\mathbf{T}^{-1}$ (which has as its domain all of E^n) is also linear and thus in $L(E^n)$; $\mathbf{T}^{-1}\mathbf{y} = \mathbf{x}$ is equivalent by definition to $\mathbf{y} = \mathbf{Tx}$. The 1-1 transformations in $L(E^n)$ are just those with a multiplicative inverse $\mathbf{T}^{-1}$: $\mathbf{TT}^{-1} = \mathbf{T}^{-1}\mathbf{T} = \mathbf{I}$. They are called *nonsingular* or invertible; all others are *singular*. The same terminology is applied to matrices. In particular, $\mathbf{A}$ is nonsingular if $\mathbf{AB} = \mathbf{I}$ (or $\mathbf{BA} = \mathbf{I}$) for some matrix $\mathbf{B}$ (for $\mathbf{A}$, considered as a linear transformation, cannot lower dimension, and therefore is onto and 1-1). $\mathbf{B}$ of course must be the (unique) inverse matrix, denoted by $\mathbf{A}^{-1}$, and

$$\sum_{j=1}^{n} a_{ij}b_{jk} = 0$$

if $i \neq k$. If $\mathbf{A}$ is singular, we have numbers $x_1, x_2, \ldots, x_n$, not all zero, with $\mathbf{A}\langle x_1, \ldots, x_n\rangle = \langle 0, \ldots, 0\rangle$.

Finally note that a product of nonsingular matrices or transformation is nonsingular: $(\mathbf{ST}) \cdot (\mathbf{T}^{-1}\mathbf{S}^{-1}) = \mathbf{T}^{-1}\mathbf{S}^{-1} \cdot \mathbf{ST} = \mathbf{I}$.

Exercises

1. In E^2 with basis $\{\mathbf{e}^1, \mathbf{e}^2\}$ let the transformation $\mathbf{A}_\theta$ describe the rotation of all points θ degrees counterclockwise about the origin. Show that the matrix of $\mathbf{A}_\theta$ is given by

$$\mathbf{A}_\theta = \begin{pmatrix} \cos\theta & \sin\theta \\ -\sin\theta & \cos\theta \end{pmatrix}.$$

Show further that $\mathbf{A}_{\theta_1+\theta_2} = \mathbf{A}_{\theta_1}\mathbf{A}_{\theta_2}$ and $\mathbf{A}_\theta^{-1} = \mathbf{A}_{-\theta}$.

2. In the space $\mathcal{P}^n$ show that

$$\int_a^b p(t)\, dt$$

is a linear functional and express it in the form $(\mathbf{q}, \mathbf{p})$, where $\mathbf{q}$ is some element in $\mathcal{P}^n$ and (,) is the inner product specified in Problem 1 of Section A.1.

3. On the space $\mathcal{P}^n$ with the basis $\{1, t, t^2, \ldots, t^{n-1}\}$, define the linear transformation $\mathbf{A}_k$ ($k \geq 1$) as follows:

$$\mathbf{A}_k\mathbf{p} = \frac{d^k}{dt^k}\, p(t).$$

Verify that $\mathbf{A}_k$ is a singular transformation whose matrix representation in terms of this basis is

$$\left\| \begin{matrix} 0 & \cdots & 0 & \frac{k!}{0!} & 0 & \cdots & \cdots & 0 \\ 0 & \cdots & \cdots & 0 & \frac{(k+1)!}{1!} & 0 & \cdots & 0 \\ \vdots & & & & & & & \\ 0 & \cdots & \cdots & \cdots & \cdots & \cdots & \cdots & \frac{(n-1)!}{(n-k-1)!} \\ \vdots & & & & & & & \vdots \\ 0 & \cdots & \cdots & \cdots & \cdots & \cdots & \cdots & 0 \end{matrix} \right\|$$

Show also that $\mathcal{R}(\mathbf{A}_k)$ is the span of $\{t^{n-k}, t^{n-k+1}, \ldots, t^{n-1}\}$, and that $\mathfrak{N}(\mathbf{A}_k)$ is the span of $\{1, t, \ldots, t^{n-k-1}\}$.

A.4 Eigenvalues, eigenvectors, and the Jordan canonical form. If $\mathbf{T} - \lambda\mathbf{I}$ is singular for some scalar λ, then λ is called an *eigenvalue* of $\mathbf{T}$. Associated with λ we have some $\mathbf{x} \neq \mathbf{0}$ such that $(\mathbf{T} - \lambda\mathbf{I})\mathbf{x} = \mathbf{0}$ or $\mathbf{Tx} = \lambda\mathbf{x}$; this vector $\mathbf{x}$ is called an *eigenvector*, corresponding to the eigenvalue λ.

Each $\mathbf{T}$ in $L(U^n)$ has eigenvalues, and hence eigenvectors. The corresponding statement for $L(E^n)$, where λ is taken to be real, is false. However, any real matrix corresponds to an element of $L(U^n)$ and thus has complex eigenvalues and eigenvectors.

The existence of eigenvalues is established as follows. Note that since $L(U^n)$ is n^2-dimensional, the $n^2 + 1$ elements $\mathbf{I}, \mathbf{T}, \mathbf{T}^2, \ldots, \mathbf{T}^{n^2}$ are linearly dependent; we have $a_0, a_1, \ldots, a_{n^2}$ not all zero, with $a_0\mathbf{I} + a_1\mathbf{T} + \cdots + a_{n^2}\mathbf{T}^{n^2} = \mathbf{0}$. The corresponding polynomial $p(z) = a_0 + a_1 z + \cdots + a_{n^2}z^{n^2}$ can of course be factored by the fundamental theorem of algebra:

$$p(z) = \prod_{i=1}^{k} (z - \lambda_i).$$

Since the identity of the polynomials says exactly that we obtain the coefficients of p by multiplying out the right side, and the same coefficients are obtained in multiplying out

$$\prod_{i=1}^{k} (\mathbf{T} - \lambda_i\mathbf{I}),$$

we have

$$\prod_{i=1}^{k} (\mathbf{T} - \lambda_i \mathbf{I}) = a_0\mathbf{I} + a_1\mathbf{T} + \cdots + a_{n^2}\mathbf{T}^{n^2} = p(\mathbf{T}) = 0;$$

consequently one of the $\mathbf{T} - \lambda_i\mathbf{I}$ is singular.

For any basis $\{\mathbf{y}^1, \ldots, \mathbf{y}^n\}$ of U^n, by writing

$$\mathbf{x} = \sum_{i=1}^{n} x_i \mathbf{y}^i$$

and similarly expressing $\mathbf{Tx}$ in terms of its components relative to $\{\mathbf{y}^1, \ldots, \mathbf{y}^n\}$, we can represent $\mathbf{T}$ by a matrix $\mathbf{A}$ mapping $\langle x_1, \ldots, x_n\rangle$ into the vector of components of $\mathbf{Tx}$ relative to the basis. For an appropriate choice of basis, $\mathbf{A}$ has the Jordan *canonical form*

$$\mathbf{A} = \begin{Vmatrix} \mathbf{A}_1 & \mathbf{0} & \cdots & \mathbf{0} \\ \mathbf{0} & \mathbf{A}_2 & \cdots & \mathbf{0} \\ \vdots & \vdots & \ddots & \vdots \\ \mathbf{0} & \mathbf{0} & \cdots & \mathbf{A}_k \end{Vmatrix}$$

where $\mathbf{0}$ represents appropriate rectangular matrices with all entries zero, and each $\mathbf{A}_i$ is a square matrix which can similarly be written (in so-called block-diagonal form) with diagonal blocks of the form

$$\mathbf{B} = \begin{Vmatrix} \lambda_i & 0 & 0 & \cdots & 0 \\ 1 & \lambda_i & 0 & \cdots & 0 \\ 0 & 1 & \lambda_i & \cdots & 0 \\ \vdots & \vdots & \vdots & & \vdots \\ 0 & 0 & 0 & \cdots & \lambda_i \end{Vmatrix}$$

(λ_i on the main diagonal, 1's below, 0's elsewhere). The same λ_i appears in all sub-blocks of $\mathbf{A}_i$, but $\lambda_i \neq \lambda_j$, $i \neq j$, and $\lambda_1, \ldots, \lambda_k$ comprise the set of all eigenvalues of $\mathbf{T}$. The effect of this decomposition of $\mathbf{A}$ into blocks is the following: Any vector in the span of the basis elements

$$\{\mathbf{y}^j, \mathbf{y}^{j+1}, \ldots, \mathbf{y}^k\}$$

corresponding to the block $\mathbf{B}$ maps into a vector in this same subspace; moreover, on this subspace $\mathbf{T}$ acts as the sum of two simple transformations: the transformation that maps $\mathbf{x}$ into $\lambda_i\mathbf{x}$ and the transformation that maps $\mathbf{y}^j$ into $\mathbf{y}^{j+1}$, $\mathbf{y}^{j+1}$ into $\mathbf{y}^{j+2}, \ldots, \mathbf{y}^{k-1}$ into $\mathbf{y}^k$, $\mathbf{y}^k$ into $\mathbf{0}$ (thus $\mathbf{T} - \lambda_i\mathbf{I}$ reduces to the latter transformation on the subspace). In effect, we have a decomposition of E^n into (a direct sum of) pieces, and $\mathbf{T}$ has a simple form on each piece. In many important cases each block $\mathbf{B}$ is 1×1; the matrix $\mathbf{A}$ assumes the pure diagonal form

$$\begin{Vmatrix} \lambda_1 & 0 & \cdots & 0 \\ 0 & \lambda_2 & \cdots & 0 \\ \vdots & \vdots & \ddots & \vdots \\ 0 & 0 & \cdots & \lambda_n \end{Vmatrix},$$

where now the λ_i need not all be distinct; and the eigenvectors of $\mathbf{T}$ span U^n. Moreover, if $\mathbf{T}$ *has n distinct eigenvalues*, it is clear that each block $\mathbf{B}$ is 1×1 and $\mathbf{A}$ has the pure diagonal form.

We now indicate how the basis $\{\mathbf{y}^1, \ldots, \mathbf{y}^n\}$ is constructed. For a singular $\mathbf{S}$ in $L(U^n)$, or indeed $L(E^n)$, consider $\mathfrak{N}(\mathbf{S}), \mathfrak{N}(\mathbf{S}^2), \ldots$. Clearly, $\mathfrak{N}(\mathbf{S}) \subset \mathfrak{N}(\mathbf{S}^2) \subset \cdots$, and since proper subspaces have smaller dimensions, there is some integer k ($\leq n$) for which

$$\mathfrak{N}(\mathbf{S}) \subsetneqq \mathfrak{N}(\mathbf{S}^2) \subsetneqq \cdots \subsetneqq \mathfrak{N}(\mathbf{S}^k) = \mathfrak{N}(\mathbf{S}^{k+1}).$$

Then $\mathfrak{N}(\mathbf{S}^k) = \mathfrak{N}(\mathbf{S}^{k+j})$ for all $j \geq 0$. For since $\mathfrak{N}(\mathbf{S}^k) = \mathfrak{N}(\mathbf{S}^{k+1})$, if $\mathbf{S}^{k+1}\mathbf{x} = \mathbf{0}$ it follows that $\mathbf{S}^k\mathbf{x} = \mathbf{0}$; and thus $\mathbf{S}^{k+j}\mathbf{x} = \mathbf{0} = \mathbf{S}^{k+1}(\mathbf{S}^{j-1}\mathbf{x})$ implies $\mathbf{S}^k(\mathbf{S}^{j-1}\mathbf{x}) = \mathbf{S}^{k+j-1}\mathbf{x} = \mathbf{0}$, from which it follows that

$$\mathbf{S}^{k+j-1}\mathbf{x} = \mathbf{S}^{k+j-2}\mathbf{x} = \cdots = \mathbf{S}^k\mathbf{x} = 0.$$

By the dimension relation (A.3.1),

$$\mathfrak{R}(\mathbf{S}) \supsetneqq \mathfrak{R}(\mathbf{S}^2) \supsetneqq \cdots \supsetneqq \mathfrak{R}(\mathbf{S}^k) = \mathfrak{R}(\mathbf{S}^{k+1}) = \cdots$$

Thus $\mathbf{S}$ restricted to $\mathfrak{R}(\mathbf{S}^k)$ is 1-1 because it is onto. But, since $\mathbf{S}^k$ is 1-1, we also have $\mathfrak{N}(\mathbf{S}^k) \cap \mathfrak{R}(\mathbf{S}^k) = \{\mathbf{0}\}$. If $\{\mathbf{x}^1, \ldots, \mathbf{x}^l\}$ is a basis for $\mathfrak{N}(\mathbf{S}^k)$ and $\{\mathbf{y}^1, \ldots, \mathbf{y}^m\}$ is a basis for $\mathfrak{R}(\mathbf{S}^k)$, it follows by (A.3.1) that $l + m = n$. The conjunction $\{\mathbf{x}^1, \ldots \mathbf{x}^l; \mathbf{y}^1, \ldots, \mathbf{y}^m\}$ forms a basis for U^n, and $U^n = \mathfrak{N}(\mathbf{S}^k) \oplus \mathfrak{R}(\mathbf{S}^k)$. Corresponding to our basis, $\mathbf{S}$ would have the matricial structure

$$\begin{pmatrix} \mathbf{A} & \mathbf{0} \\ \mathbf{0} & \mathbf{B} \end{pmatrix},$$

where $\mathbf{A}$ is $l \times l$ and $\mathbf{B}$ is $(n - l) \times (n - l)$.

Now suppose that $\mathbf{T}$ is any element of $L(U^n)$. We know that there is an eigenvalue λ_1; hence we may let $\mathbf{S} = \mathbf{T} - \lambda_1\mathbf{I}$ and let

$$\mathfrak{N}_1 = \mathfrak{N}[(\mathbf{T} - \lambda_1\mathbf{I})^{k_1}],$$

with k_1 the k obtained for $\mathbf{S}$ above. Accordingly, $\mathfrak{R}_1 = \mathfrak{R}[(\mathbf{T} - \lambda_1\mathbf{I})^{k_1}]$. We have $U^n = \mathfrak{N}_1 \oplus \mathfrak{R}_1$, with $(\mathbf{T} - \lambda_1\mathbf{I})\mathbf{x}$ in $\mathfrak{N}_1$ or $\mathfrak{R}_1$ according as $\mathbf{x}$ is in $\mathfrak{N}_1$ or $\mathfrak{R}_1$. Since $\lambda_1\mathbf{x}$ is in a given subspace whenever $\mathbf{x}$ is, $\mathbf{Tx}$ is in $\mathfrak{N}_1$ if $\mathbf{x}$ is in $\mathfrak{N}_1$, and $\mathbf{Tx}$ is in $\mathfrak{R}_1$ if $\mathbf{x}$ is in $\mathfrak{R}_1$. Let $\mathbf{T}_1$ be the restriction of $\mathbf{T}$ to

$\mathcal{R}_1$, mapping $\mathcal{R}_1$ into itself. Since we can identify $\mathcal{R}_1$ with some U^m, it follows that $\mathbf{T}_1$ has an eigenvalue λ_2 such that $\mathbf{Tx} = \mathbf{T}_1\mathbf{x} = \lambda_2\mathbf{x}$ for $\mathbf{x} \neq \mathbf{0}$ in $\mathcal{R}_1$. Since $(\mathbf{T} - \lambda_1\mathbf{I})\mathbf{x} \neq \mathbf{0}$ ($\mathbf{T} - \lambda_1\mathbf{I}$ being 1-1 on $\mathcal{R}_1$), we deduce that $\lambda_2 \neq \lambda_1$. Defining $\mathfrak{N}_2$ and $\mathcal{R}_2$ in the obvious fashion, we obtain $\mathcal{R}_1 = \mathfrak{N}_2 \oplus \mathcal{R}_2$ (hence $U^n = \mathfrak{N}_1 \oplus \mathfrak{N}_2 \oplus \mathcal{R}_2$), with

$$\mathbf{T}\mathfrak{N}_2 = \mathbf{T}_1\mathfrak{N}_2 = \{\mathbf{T}_1\mathbf{x} | \mathbf{x} \in \mathfrak{N}_2\} \subset \mathfrak{N}_2, \qquad \mathbf{T}\mathcal{R}_2 \subset \mathcal{R}_2.$$

We continue and obtain $U^n = \mathfrak{N}_1 \oplus \cdots \oplus \mathfrak{N}_r$, with associated distinct eigenvalues $\lambda_1, \ldots, \lambda_r$ such that $(\mathbf{T} - \lambda_i\mathbf{I})^{k_i}\mathbf{x} = 0$ for all $\mathbf{x}$ in $\mathfrak{N}_i$ and some appropriate $k_i \geq 1$. Choosing any basis for U^n which is formed by combining bases of the $\mathfrak{N}_i$ now yields a matrix $\mathbf{A}$ in block-diagonal form, since this amounts to saying $\mathbf{T}\mathfrak{N}_i \subset \mathfrak{N}_i$.

The further decomposition into sub-blocks is delicate: we shall only note that if we choose an $\mathbf{x}_1$ in $\mathfrak{N}_1$ for which $(\mathbf{T} - \lambda_1\mathbf{I})^{k_1-1}\mathbf{x}_1 \neq \mathbf{0}$ (i.e., an element of $\mathfrak{N}[(\mathbf{T} - \lambda_1\mathbf{I})^{k_1}]$ not in $\mathfrak{N}[(\mathbf{T} - \lambda_1\mathbf{I})^{k_1-1}]$, a choice that is made possible by the meaning of k_1), then for $\mathbf{S} = \mathbf{T} - \lambda_1\mathbf{I}$ the elements $\mathbf{x}_1$, $\mathbf{Sx}_1, \ldots, \mathbf{S}^{k-1}\mathbf{x}_1$ are independent: for if

$$\gamma_1\mathbf{x}_1 + \gamma_2\mathbf{Sx}_1 + \cdots + \gamma_{k_1-1}\mathbf{S}^{k_1-1}\mathbf{x}_1 = \mathbf{0},$$

then by applying $\mathbf{S}^{k_1-1}$ we obtain $\gamma_1\mathbf{S}^{k_1-1}\mathbf{x}_1 = \mathbf{0}$ and hence $\gamma_1 = 0$; we then apply $\mathbf{S}^{k_1-2}, \mathbf{S}^{k_1-3}, \ldots$ successively to deduce $\gamma_2 = \gamma_3 = \cdots = 0$. These vectors need not constitute a basis for $\mathfrak{N}_1$, but it can be shown that there exists a finite set $\mathbf{x}_1, \mathbf{x}_2, \ldots$ such that $\mathbf{x}_1, \mathbf{Sx}_1, \ldots, \mathbf{S}^{k_1-1}\mathbf{x}_1$; $\mathbf{x}_2, \mathbf{Sx}_2, \ldots, \mathbf{S}^{j_2-1}\mathbf{x}_2$; $\mathbf{x}_3, \mathbf{Sx}_3, \ldots, \mathbf{S}^{j_3-1}\mathbf{x}_3$; ... form a basis for $\mathfrak{N}_1$ where $\mathbf{S}^{j_2}\mathbf{x}_2 = \mathbf{0}, \mathbf{S}^{j_3}\mathbf{x}_3 = \mathbf{0}, \ldots j_2 \leq k_1, j_3 \leq j_2$, etc.* It is easily seen that on the span V_0 of $\mathbf{x}_1, \mathbf{Sx}_1, \ldots, \mathbf{S}^{k_1-1}\mathbf{x}_1$ we can write $\mathbf{S}$ as a matrix of the form

$$\begin{Vmatrix} 0 & 0 & \ldots & 0 & 0 \\ 1 & 0 & \ldots & 0 & 0 \\ 0 & 1 & \ldots & 0 & 0 \\ \vdots & \vdots & \ddots & \vdots & \vdots \\ 0 & 0 & \ldots & 1 & 0 \end{Vmatrix};$$

hence on V_0

$$\mathbf{T} = \lambda_1\mathbf{I} + \mathbf{S} = \begin{Vmatrix} \lambda_1 & 0 & \ldots & 0 \\ 1 & \lambda_1 & \ldots & 0 \\ 0 & 1 & \ldots & 0 \\ \vdots & \vdots & & \vdots \\ 0 & 0 & \ldots & \lambda_1 \end{Vmatrix}.$$

* For the sake of preserving the continuity here, the proof of this assertion has been placed at the end of the argument (see p. 376).

Clearly, each such sub-block of our final matrix $\mathbf{A}$ yields exactly one eigenvector (here $\mathbf{S}^{k_1-1}\mathbf{x}_1$). Similar statements apply to the span of

$$\mathbf{x}_2, \mathbf{S}\mathbf{x}_2, \ldots, \mathbf{S}^{j_2-1}\mathbf{x}_2, \text{ etc.}$$

The number of eigenvectors we thus obtain corresponding to λ_1 is called the *geometric multiplicity* of λ_1; it is precisely the dimension of the subspace spanned by the eigenvectors corresponding to λ_1. Since $\mathbf{T}$ maps

$$\begin{aligned}
&\mathbf{x}_1 \to \lambda_1\mathbf{x}_1 + \mathbf{S}\mathbf{x}_1,\\
&\mathbf{S}\mathbf{x}_1 \to \lambda_1\mathbf{S}\mathbf{x}_1 + \mathbf{S}^2\mathbf{x}_1,\\
&\vdots\\
&\mathbf{S}^{k_1-2}\mathbf{x}_1 \to \lambda_1\mathbf{S}^{k_1-2}\mathbf{x}_1 + \mathbf{S}^{k_1-1}\mathbf{x}_1,\\
&\mathbf{S}^{k_1-1}\mathbf{x}_1 \to \lambda_1\mathbf{S}^{k_1-1}\mathbf{x}_1,
\end{aligned}$$

$\mathbf{T} - \lambda\mathbf{I}$ maps

$$\begin{aligned}
&\mathbf{x}_1 \to (\lambda_1 - \lambda)\mathbf{x}_1 + \mathbf{S}\mathbf{x}_1,\\
&\mathbf{S}\mathbf{x}_1 \to (\lambda_1 - \lambda)\mathbf{S}\mathbf{x}_1 + \mathbf{S}^2\mathbf{x}_1,\\
&\vdots\\
&\mathbf{S}^{k_1-2}\mathbf{x}_1 \to (\lambda_1 - \lambda)\mathbf{S}^{k_1-2}\mathbf{x}_1 + \mathbf{S}^{k_1-1}\mathbf{x}_1,\\
&\mathbf{S}^{k_1-1}\mathbf{x}_1 \to (\lambda_1 - \lambda)\mathbf{S}^{k_1-1}\mathbf{x}_1.
\end{aligned}$$

Consequently, if $\lambda \neq \lambda_1$, it follows that $\mathcal{R}(\mathbf{T} - \lambda\mathbf{I})$ contains $\mathbf{S}^{k_1-1}\mathbf{x}_1$, hence $\mathbf{S}^{k_1-2}\mathbf{x}_1$, and thus all the elements spanning V_0. Therefore $\mathbf{T} - \lambda\mathbf{I}$ maps V_0 onto itself for any $\lambda \neq \lambda_i$ and all i, yielding $\mathfrak{N}_i \subset \mathcal{R}(\mathbf{T} - \lambda\mathbf{I})$. Hence $\mathbf{T} - \lambda\mathbf{I}$ is onto and $(\mathbf{T} - \lambda\mathbf{I})^{-1}$ exists, and thus $\lambda_1, \ldots, \lambda_k$ is the full set of eigenvalues.

Finally, note that if $\mathbf{S}_0$ is the unique linear transformation mapping $\mathbf{e}^i$ onto $\mathbf{y}^i$—that is, if

$$\mathbf{S}_0 \sum_{i=1}^{n} \gamma_i\mathbf{e}^i = \sum_{i=1}^{n} \gamma_i\mathbf{y}^i$$

—then for $\mathbf{z} = \langle\gamma_1, \gamma_2, \ldots, \gamma_n\rangle$ we have

$$\mathbf{T}\mathbf{S}_0\mathbf{z} = \sum_{i=1}^{n}\left(\sum_{j=1}^{n} a_{ij}\gamma_j\right)\mathbf{y}^i;$$

and therefore (since $\mathbf{S}_0$, being onto, is nonsingular)

$$\mathbf{S}_0^{-1}\mathbf{T}\mathbf{S}_0\mathbf{z} = \sum_{i=1}^{n}\left(\sum_{j=1}^{n} a_{ij}\gamma_j\right)\mathbf{e}^i \qquad (\mathbf{S}_0^{-1}\mathbf{y}^i = \mathbf{e}^i).$$

Thus if $\mathbf{T}$ corresponds to the matrix $\mathbf{B}$ relative to the basis $\{\mathbf{e}^1, \ldots, \mathbf{e}^n\}$, and $\mathbf{S}_0$, $\mathbf{S}_0^{-1}$ to $\mathbf{C}$, $\mathbf{C}^{-1}$, respectively, we have $\mathbf{C}^{-1}\mathbf{BC} = \mathbf{A}$, or equivalently $\mathbf{B} = \mathbf{C}\,(\mathbf{C}^{-1}\mathbf{BC})\mathbf{C}^{-1} = \mathbf{CAC}^{-1}$. Two matrices $\mathbf{A}$ and $\mathbf{B}$ for which $\mathbf{B} = \mathbf{CAC}^{-1}$ for some $\mathbf{C}$ are called *similar*, and any matrix is therefore similar to one of the (block-diagonal) Jordan canonical form. Any two matrices $\mathbf{A}$ and $\mathbf{B}$ are similar if and only if their Jordan canonical forms differ only in the order in which the blocks and sub-blocks appear.

We now present the formal proof referred to in the footnote on p. 374.

(i) Suppose that we have a linearly independent set of vectors of $\mathfrak{N}_1$ of the form

$$\begin{array}{llll} \mathbf{x}_1, & \mathbf{Sx}_1, & \ldots, & \mathbf{S}^{j_1-1}\mathbf{x}_1; \\ \mathbf{x}_2, & \mathbf{Sx}_2, & \ldots, & \mathbf{S}^{j_2-1}\mathbf{x}_2; \\ \vdots & & & \\ \mathbf{x}_r, & \mathbf{Sx}_r, & \ldots, & \mathbf{S}^{j_r-1}\mathbf{x}_r, \end{array}$$

where $k_1 = j_1 \geq j_2 \geq \cdots \geq j_r$ and $\mathbf{S}^{j_i}\mathbf{x}_i = 0$ $(i = 1, 2, \ldots, r)$.

(ii) Suppose also that for each $l = 1, 2, \ldots, r$ there exists no $\mathbf{x}$ satisfying $\mathbf{S}^{k_1}\mathbf{x} = \mathbf{0}$ and such that $\mathbf{x}, \mathbf{Sx}, \ldots, \mathbf{S}^{j_l}\mathbf{x}$ together with the first $l - 1$ rows of vectors in (i) are linearly independent.

Then either the vectors in (i) form a basis for $\mathfrak{N}_1$ or they can be extended so as to satisfy both (i) and (ii).

Proof. Let j_{r+1} be the largest integer satisfying the properties that for some $\mathbf{x}_{r+1}$, $\mathbf{S}^{k_1}\mathbf{x}_{r+1} = \mathbf{0}$ and $\mathbf{x}_{r+1}, \mathbf{Sx}_{r+1}, \ldots, \mathbf{S}^{j_{r+1}-1}\mathbf{x}_{r+1}$ together with the vectors in (i) are linearly independent. Then, since $j_{r+1} \leq j_r$ by definition, if $\mathbf{S}^{j_{r+1}}\mathbf{x}_{r+1} = \mathbf{0}$ we are through.

Suppose, then, that $\mathbf{0} \neq \mathbf{S}^{j_{r+1}}\mathbf{x}_{r+1} = \sum \alpha_i \mathbf{S}^{v_i}\mathbf{x}_{\mu_i}$, with all $\alpha_i \neq 0$ and $v_1 \leq v_2 \leq \ldots$ Clearly, we must have all $\mu_i \leq r$. For otherwise we could repeatedly multiply both sides by $\mathbf{S}$, substitute the above equality for $\mathbf{S}^{j_{r+1}}\mathbf{x}_{r+1}$ when it appears on the right, and obtain all $\mathbf{S}^{j_{k+1}+r}\mathbf{x}_{r+1} \neq \mathbf{0}$ for $r = 0, 1, 2, \ldots$ (since $\alpha_i \neq 0$), contradicting $\mathbf{S}^{k_1}\mathbf{x}_{r+1} = \mathbf{0}$. If $v_1 \leq j_{r+1} - 1$, then $j_{r+1} + \mu_1 - v_1 - 1 \geq \mu_1$. But

$$\mathbf{S}^{j_{r+1}+\mu_1-v_1-1}\mathbf{x}_{r+1} = \alpha_1\mathbf{S}^{\mu_1-1}\mathbf{x}_{\mu_1} + \cdots$$

together with $\mathbf{x}_{r+1}, \mathbf{Sx}_{r+1}, \ldots, \mathbf{S}^{j_{r+1}+\mu_1-v_1-2}\mathbf{x}_{r+1}$, and the first $\mu_1 - 1$ rows of vectors in (i) are independent, contradicting (ii). Hence $\mathbf{S}^{j_{r+1}}\mathbf{x}_{r+1}$ has the representation $\mathbf{S}^{j_{r+1}}\mathbf{x}_{r+1} = \mathbf{S}^{j_{r+1}}\,(\sum \alpha_i \mathbf{S}^{v'_i}\mathbf{x}_{\mu_i})$, with all $\mu_i \leq r$. Hence the vector $\mathbf{x}_{r+1} - \sum \alpha_i \mathbf{S}^{v'_i}\mathbf{x}_{\mu_i}$ satisfies all the necessary requirements.

The case of distinct eigenvalues. If a matrix $\mathbf{B}$ has n distinct eigenvalues, it is similar to a diagonal matrix $\mathbf{A}$ given by its canonical form $\mathbf{C}^{-1}\mathbf{BC} = \mathbf{A}$. In this case $\mathbf{C}$ can be constructed directly from the eigenvectors of $\mathbf{B}$ (which form the basis $\{\mathbf{y}^1, \ldots, \mathbf{y}^n\}$ of the preceding section). For $\mathbf{B}\mathbf{y}^i = \lambda_i \mathbf{y}^i$, and if $\mathbf{y}^i = \langle y_1^i, y_2^i, \ldots, y_n^i \rangle$, then, writing $\mathbf{y}^i$ as a column vector, we have

$$\begin{Vmatrix} b_{11} & b_{12} & \cdots & b_{1n} \\ b_{21} & b_{22} & \cdots & b_{2n} \\ \vdots & \vdots & & \vdots \\ b_{n1} & b_{n2} & \cdots & b_{nn} \end{Vmatrix} \begin{Vmatrix} y_1^i \\ y_2^i \\ \vdots \\ y_n^i \end{Vmatrix} = \lambda_i \begin{Vmatrix} y_1^i \\ y_2^i \\ \vdots \\ y_n^i \end{Vmatrix},$$

whence, by quick computation

$$\begin{Vmatrix} b_{11} & b_{12} & \cdots & b_{1n} \\ b_{21} & b_{22} & \cdots & b_{2n} \\ \vdots & \vdots & & \vdots \\ b_{n1} & b_{n2} & \cdots & b_{nn} \end{Vmatrix} \begin{Vmatrix} y_1^1 & y_1^2 & \cdots & y_1^n \\ y_2^1 & y_2^2 & \cdots & y_2^n \\ \vdots & \vdots & & \vdots \\ y_n^1 & y_n^2 & \cdots & y_n^n \end{Vmatrix}$$

$$= \begin{Vmatrix} y_1^1 & y_1^2 & \cdots & y_1^n \\ y_2^1 & y_2^2 & \cdots & y_2^n \\ \vdots & \vdots & & \vdots \\ y_n^1 & y_n^2 & \cdots & y_n^n \end{Vmatrix} \begin{Vmatrix} \lambda_1 & 0 & \cdots & 0 \\ 0 & \lambda_2 & \cdots & 0 \\ \vdots & \vdots & & \vdots \\ 0 & 0 & \cdots & \lambda_n \end{Vmatrix}.$$

Since the matrix $\mathbf{Y} = \|y_j^i\|$ maps the basic column vectors $\langle 1, 0, \ldots, 0 \rangle$, $\langle 0, 1, 0, \ldots, 0 \rangle, \ldots$ onto $\mathbf{y}^1, \mathbf{y}^2, \ldots$, respectively, it maps U^n onto itself; hence $\mathbf{Y}^{-1}$ exists and $\mathbf{Y}^{-1}\mathbf{B}\mathbf{Y} = \mathbf{\Lambda}$, where $\mathbf{\Lambda}$ denotes the above diagonal matrix of the eigenvalues.

Exercises

1. Find the Jordan forms for the matrices

$$\begin{Vmatrix} 1 & 0 & 2 \\ 0 & 3 & 1 \\ 2 & -1 & 0 \end{Vmatrix} \quad \text{and} \quad \begin{Vmatrix} 1 & -1 \\ 1 & 1 \end{Vmatrix}.$$

2. A common type of square matrix is one in which all elements are nonnegative and the sum of the elements in each row is unity. For such a matrix $\mathbf{A}$ prove that 1 is an eigenvalue, that all eigenvalues are less than or equal to 1 in absolute value, and that the index of the eigenvalue 1 is 1; that is, the null space of $\mathbf{A} - \mathbf{I}$ and $(\mathbf{A} - \mathbf{I})^2$ coincide.

3. Construct an $n \times n$ matrix $\mathbf{A}$ with distinct eigenvalues such that $\mathbf{A}^n = \mathbf{I}$.

4. An equivalence relation on transformations is one which satisfies the following three postulates:

(1) Any transformation is equivalent to itself.
(2) If **A** is equivalent to **B**, then **B** is equivalent to **A**.
(3) If **A** is equivalent to **B** and **B** is equivalent to **C**, then **A** is equivalent to **C**.

Prove that the notion of "similarity" is an equivalence relation.

5. In E^3 with basis $\mathbf{e}^1, \mathbf{e}^2, \mathbf{e}^3$ let $\mathbf{a}$ be a fixed vector with $(\mathbf{a}, \mathbf{a}) = 1$. Consider the hyperplane $H = \{\mathbf{x} | (\mathbf{a}, \mathbf{x}) = 0\}$. By the orthogonal projection $\mathbf{y}'$ of a vector onto H we mean the unique vector $\mathbf{y}'$ in H such that $\mathbf{y}' = \mathbf{y} - \alpha\mathbf{a}$ for some scalar α. Define $\mathbf{Py} = \mathbf{y}'$. Then $\mathbf{Py} = \mathbf{y} - (\mathbf{y}, \mathbf{a})\mathbf{a} \cdot \mathcal{R}(P) = H$. The null space of P is the space spanned by the vector $\mathbf{a}$. The eigenvalues of $\mathbf{P}$ are 0 and 1, and the Jordan form of $\mathbf{P}$ is given by

$$\begin{Vmatrix} 0 & 0 & 0 \\ 0 & 1 & 0 \\ 0 & 0 & 1 \end{Vmatrix}$$

Verify all these statements.

A.5 Transposed, normal, and hermitian matrices; orthogonal complement. If $\mathbf{A}$ is any complex $m \times n$ matrix, the *conjugate transpose* $\mathbf{A}^*$ of $\mathbf{A}$ is the $n \times m$ matrix with ijth entry

$$a_{ij}^* = \bar{a}_{ji} \qquad (i = 1, \ldots, n; j = 1, \ldots, m).$$

Clearly $(\mathbf{A} + \gamma\mathbf{B})^* = \mathbf{A}^* + \bar{\gamma}\mathbf{B}^*$, and $(\mathbf{AB}^*) = \mathbf{B}^*\mathbf{A}^*$ is easily verified when $\mathbf{AB}$ makes sense. For $\mathbf{x}$ in U^n and $\mathbf{y}$ in U^m

$$(\mathbf{Ax}, \mathbf{y}) = \sum_{i=1}^{m} \sum_{j=1}^{n} a_{ij} x_j \bar{y}_i = \sum_{j=1}^{n} x_j \sum_{i=1}^{m} \bar{a}_{ij} y_i = (\mathbf{x}, \mathbf{A}^*\mathbf{y}).$$

As a consequence, $\mathcal{R}(\mathbf{A})$ is orthogonal to $\mathfrak{N}(\mathbf{A}^*)$; that is, $(\mathbf{Ax}, \mathbf{y}) = 0$ if $\mathbf{A}^*\mathbf{y} = \mathbf{0}$. Indeed, $(\mathbf{Ax}, \mathbf{y}) = 0$ for all $\mathbf{x}$ if and only if $\mathbf{A}^*\mathbf{y} = \mathbf{0}$. Hence, if we denote by $\mathcal{R}(\mathbf{A})^\perp$ the set of all elements of U^m orthogonal to the subspace $\mathcal{R}(\mathbf{A})$, we may write $\mathcal{R}(\mathbf{A})^\perp = \mathfrak{N}(\mathbf{A}^*)$; similarly $\mathcal{R}(\mathbf{A}^*)^\perp = \mathfrak{N}(\mathbf{A})$. $\mathcal{R}(\mathbf{A})^\perp$ is known as the *orthogonal complement of* $\mathcal{R}(\mathbf{A})$. (The orthogonal complement of any subset of U^n is always a subspace; if V is a subspace, it is easily shown that $V^{\perp\perp} = (V^\perp)^\perp = V$.) When we are dealing with the real vector space E^n, the conjugate transpose matrix $\mathbf{A}^*$ reduces to the ordinary transpose matrix $\mathbf{A}'$(obtained from $\mathbf{A}$ by interchanging corresponding rows and columns), and all the corresponding operations and relations of $\mathbf{A}'$, $\mathcal{R}(\mathbf{A}')^\perp$, and $\mathfrak{N}(\mathbf{A}')$ remain valid.

When $\mathbf{A}$ is $n \times n$, $\mathbf{A}$ is called *normal* if $\mathbf{A}^*\mathbf{A} = \mathbf{AA}^*$ (in general, $\mathbf{A}^*\mathbf{A}$ is $n \times n$ and $\mathbf{AA}^*$ is $m \times m$). One form of normal matrix is the hermitian matrix: $\mathbf{A}$ is *hermitian* if $\mathbf{A} = \mathbf{A}^*$. (A real hermitian matrix is called symmetric; here $\mathbf{A} = \mathbf{A}'$.) Since $(\mathbf{AB})^* = \mathbf{B}^*\mathbf{A}^*$, and clearly $\mathbf{A}^{**} = \mathbf{A}$, each $n \times n$ matrix $\mathbf{A}$ has $\mathbf{A}^*\mathbf{A}$ (or $\mathbf{AA}^*$) hermitian.

Normal matrices have elegant properties compared with general matrices. As we shall see, *the eigenvectors of a normal matrix* $\mathbf{A}$ *span* U^n *and those corresponding to distinct eigenvalues are orthogonal, so that we may obtain an orthonormal basis of eigenvectors for* U^n; moreover, $\mathbf{Ax} = \lambda\mathbf{x}$ $(\mathbf{x} \neq \mathbf{0})$ *is equivalent to* $\mathbf{A}^*\mathbf{x} = \bar{\lambda}\mathbf{x}$, that is, $\mathbf{A}$ *and* $\mathbf{A}^*$ *share the same eigenvectors with conjugate eigenvalues.*

To obtain these facts, we begin by noting that when $\mathbf{A}$ is normal, $\mathbf{B} = \mathbf{A} - \lambda\mathbf{I}$ is also normal. Thus if $\mathbf{B}^k\mathbf{x} = \mathbf{0}$, we have $\mathbf{B}^{*k}\mathbf{B}^k = \mathbf{0}$; and since $\mathbf{B}^*$ and $\mathbf{B}$ commute, $(\mathbf{B}^*\mathbf{B})^k\mathbf{x} = \mathbf{0}$, and surely $(\mathbf{B}^*\mathbf{B})^{2^k}\mathbf{x} = \mathbf{0}$. Let $\mathbf{C} = \mathbf{B}^*\mathbf{B} = \mathbf{C}^*$. Then

$$(\mathbf{C}^{2^k}\mathbf{x}, \mathbf{x}) = 0 = (\mathbf{C}^{2^{k-1}}\mathbf{x}, \mathbf{C}^{2^{k-1}}\mathbf{x}),$$

and thus $\mathbf{C}^{2^{k-1}}\mathbf{x} = \mathbf{0}$; similarly $\mathbf{C}^{2^{k-2}}\mathbf{x} = \mathbf{0}$, and continuing, we find ultimately that $\mathbf{Cx} = \mathbf{0}$, or $\mathbf{B}^*\mathbf{Bx} = \mathbf{0}$. But again $(\mathbf{B}^*\mathbf{Bx}, \mathbf{x}) = 0 = (\mathbf{Bx}, \mathbf{Bx})$, and thus $\mathbf{Bx} = \mathbf{0}$. Consequently each element of $\mathbf{x}_i$ obtained in our Jordan decomposition is an eigenvector of $\mathbf{A}$, and U^n is spanned by eigenvectors. Further, if $\mathbf{Bx} = \mathbf{0}$, then

$$(\mathbf{Bx}, \mathbf{Bx}) = 0 = (\mathbf{B}^*\mathbf{Bx}, \mathbf{x}) = (\mathbf{BB}^*\mathbf{x}, \mathbf{x}) = (\mathbf{B}^*\mathbf{x}, \mathbf{B}^*\mathbf{x}),$$

and thus $\mathbf{B}^*\mathbf{x} = \mathbf{0}$. In terms of $\mathbf{A} - \lambda\mathbf{I}$, if $\mathbf{Ax} = \lambda\mathbf{x}$, then $\mathbf{A}^*\mathbf{x} = \bar{\lambda}\mathbf{x}$; applying this to $\{\mathbf{A}^*, \bar{\lambda}\}$ in place of $\{\mathbf{A}, \lambda\}$ yields the converse.

Consequently, if $\mathbf{Ax} = \lambda\mathbf{x}$ and $\mathbf{Ay} = \mu\mathbf{y}$, with $\lambda \neq \mu$ and $\mathbf{x}, \mathbf{y} \neq \mathbf{0}$, we have

$$(\mathbf{Ax}, \mathbf{y}) = (\lambda\mathbf{x}, \mathbf{y}) = \lambda(\mathbf{x}, \mathbf{y}) = (\mathbf{x}, \mathbf{A}^*\mathbf{y}) = (\mathbf{x}, \bar{\mu}\mathbf{y}) = \mu(\mathbf{x}, \mathbf{y}).$$

It follows that $(\lambda - \mu)(\mathbf{x}, \mathbf{y}) = 0$; thus $(\mathbf{x}, \mathbf{y}) = 0$, and distinct eigenvalues have orthogonal eigenvectors. Since the set of eigenvectors corresponding to λ_i forms the subspace Y_i when $\mathbf{0}$ is adjoined, we can find an orthonormal basis for each Y_i; obviously the union of these bases is the required orthonormal basis for U^n.

Applying the construction of the last section, we find that $\mathbf{Y}^{-1}\mathbf{AY} = \mathbf{\Lambda}$, a diagonal matrix, where the column vectors of $\mathbf{Y}$ are orthonormal (being our basis of orthonormal eigenvectors). Since this amounts to saying

$$\sum_{j=1}^{n} y_j^i \bar{y}_j^k = \delta_{ik},$$

we have $\mathbf{YY}^* = \mathbf{I}$, and thus $\mathbf{Y}^{-1} = \mathbf{Y}^*$. Such a matrix is called *unitary*, and $\mathbf{A}$ is said to be unitarily equivalent (rather than unitarily similar) to $\mathbf{\Lambda}$. A unitary matrix preserves inner products (hence length and orthogonality), for $(\mathbf{Yx}, \mathbf{Yy}) = (\mathbf{Y^*Yx}, \mathbf{y}) = (\mathbf{x}, \mathbf{y})$.

Hermitian matrices have the further property that each λ_i is real, for the equal matrices $\mathbf{A}$ and $\mathbf{A}^*$ have conjugate eigenvalues; moreover, any normal matrix with this property is hermitian, for then

$$\mathbf{A} = \mathbf{Y\Lambda Y^*} = (\mathbf{Y\Lambda^*Y^*})^* = (\mathbf{Y\Lambda Y^*})^* = \mathbf{A}^*.$$

A unitary matrix is of course normal, and each eigenvalue is of unit modulus ($|\lambda| = 1$; for $\mathbf{Ux} = \lambda\mathbf{x}$ has the same length as $\mathbf{x}$). Conversely, any normal matrix with unimodular eigenvalues is unitary. A real unitary matrix is called *orthogonal*; any real eigenvalue is ± 1 and has corresponding real eigenvectors, which either are fixed points of the corresponding transformation or are reflected into their negatives. (No real eigenvalues need exist if n is even.) An orthogonal matrix also preserves angles.

Projections. If $U^n = V \oplus W$, then the unique expression of $\mathbf{x}$ as $\mathbf{v} + \mathbf{w}$ ($\mathbf{v} \in V$; $\mathbf{w} \in W$) leads to a linear transformation $\mathbf{P}$ mapping $\mathbf{x}$ into $\mathbf{v}$, and $\mathbf{P}^2\mathbf{x} = \mathbf{Pv} = \mathbf{v} = \mathbf{Px}$, that is, $\mathbf{P}^2 = \mathbf{P}$. If V and W are orthogonal subspaces ($W = V^\perp$), then $\mathbf{P} = \mathbf{P}^2 = \mathbf{P}^*$. Conversely, any transformation $\mathbf{P}$ such that $\mathbf{P} = \mathbf{P}^2 = \mathbf{P}^*$ (called an *orthogonal projection*, or *projection*) yields a direct-sum decomposition $U^n = \mathcal{R}(\mathbf{P}) + \mathcal{N}(\mathbf{P})$. As a normal matrix, $\mathbf{P}$ can be represented (by an appropriate choice of orthonormal basis) as a diagonal matrix with 0's and 1's on the main diagonal and 0's elsewhere.

Exercises

1. Interpret and prove the statement that any orthogonal transformation may be regarded as a rotation of axes about a fixed origin.

2. We have defined $\mathbf{A}^*$ with respect to a particular basis of U^n, but it may also be defined directly from the inner product. Prove that for each $\mathbf{y} \in U^n$, $\mathbf{A}^*\mathbf{y}$ is the unique vector $\mathbf{y}'$ such that $(\mathbf{Ax}, \mathbf{y}) = (\mathbf{x}, \mathbf{y}')$ for all $\mathbf{x} \in U^n$.

3. On the space $\mathcal{P}^n$ of polynomials with real coefficients over the interval $[a, b]$, let $f(t)$ be a specific even polynomial. Define the linear transformation $\mathbf{A}$ on $\mathcal{P}^n$ as follows:

$$(\mathbf{A}p)(t) = \int_a^b f(t - y)p(y)\, dy.$$

Show that $\mathbf{A}$ is a symmetric transformation.

4. Prove that a matrix is unitary if and only if it transforms orthonormal vectors into orthonormal vectors.

5. Let $\mathbf{A}$ be a normal transformation with eigenvalues $\lambda_1, \lambda_2, \ldots, \lambda_r$ (eigenvalues of multiplicity higher than one appearing only once). Let $\mathbf{E}_k$ be the per-

pendicular projection on the subspace of eigenvectors of λ_k. Show that if $i \neq j$, then $\mathbf{E}_i\mathbf{E}_j = \mathbf{0}$. Show also that $\mathbf{E}_1 + \mathbf{E}_2 + \cdots + \mathbf{E}_k$ is a perpendicular projection for all $k \leq r$, and that $\mathbf{E}_1 + \mathbf{E}_2 + \cdots + \mathbf{E}_r = \mathbf{I}$.

6. For the same setup as in Problem 5, prove the matrix relation

$$\mathbf{A} = \lambda_1\mathbf{E}_1 + \lambda_2\mathbf{E}_2 + \cdots + \lambda_r\mathbf{E}_r.$$

A.6 Quadratic form. Any expression

$$\sum_{i,j=1}^{n} a_{ij}x_ix_j$$

in the n real variables $x_1, \ldots, x_n$ (where a_{ij} is a real constant) is called a quadratic form. Since

$$Q(\mathbf{x}) = \sum_{i,j=1}^{n} a_{ij}x_ix_j = \sum \frac{1}{2}(a_{ij} + a_{ji})x_ix_j,$$

one can replace the original a_{ij} by a symmetric array with $a_{ij} = a_{ji}$. The corresponding matrix $\mathbf{A}$ is then a hermitian transformation of U^n into itself and has only real eigenvalues. Let λ denote one of these eigenvalues. Then since $\mathbf{A}(\mathbf{x} + i\mathbf{y}) = \lambda(\mathbf{x} + i\mathbf{y})$ implies $\mathbf{A}\mathbf{x} = \lambda\mathbf{x}$ and $\mathbf{A}\mathbf{y} = \lambda\mathbf{y}$, and one of these real vectors is nonzero, λ has a corresponding eigenvector in E^n. Repeating our former arguments (applying them now to E^n instead of U^n), we obtain a real orthonormal basis $\{\mathbf{y}^1, \ldots, \mathbf{y}^n\}$ of eigenvectors corresponding to the eigenvalues $\lambda_1, \ldots, \lambda_n$.

Further, the matrix $\mathbf{Y}$ we obtain from $\mathbf{y}^1, \ldots, \mathbf{y}^n$ is now a real unitary (i.e., orthogonal) matrix, which yields $\mathbf{Y}^*\mathbf{A}\mathbf{Y} = \mathbf{\Lambda}$ (diagonal), or $\mathbf{A} = \mathbf{Y}\mathbf{\Lambda}\mathbf{Y}^*$. Consequently, if we write

$$\mathbf{x} = \sum_{i=1}^{n} x'_i\mathbf{y}^i \qquad (\mathbf{x} \in E^n),$$

from which it follows that

$$\mathbf{Y}\mathbf{x}' = \sum_{i=1}^{n} x'_i\mathbf{y}^i = \mathbf{x},$$

then $\mathbf{Y}\overset{*}{\mathbf{x}} = \mathbf{Y}^{-1}\mathbf{x} = \mathbf{x}'$ and we have

$$Q(\mathbf{x}) = (\mathbf{A}\mathbf{x}, \mathbf{x}) = (\mathbf{Y}\mathbf{\Lambda}\mathbf{Y}^*\mathbf{x}, \mathbf{x}) = (\mathbf{\Lambda}\mathbf{Y}^*\mathbf{x}, \mathbf{Y}^*\mathbf{x}) = \sum_{i=1}^{n} \lambda_i(x'_i)^2.$$

Thus, by selecting an appropriate orthonormal basis we may express Q in diagonal form, with no mixed products of components appearing. Geometrically, then, the locus of all $\mathbf{x}$ satisfying $Q(\mathbf{x}) = 1$ is an ellipsoid if all

$\lambda_i > 0$. This occurs if and only if $Q(\mathbf{x})$ is nonnegative and $Q(\mathbf{x}) = 0$ implies $\mathbf{x} = \mathbf{0}$, i.e., when Q is *positive definite.*

Two quadratic forms Q and Q_1 are said to be unitarily equivalent if one arises from the other via a unitary transformation of E^n into itself, i.e., if $Q(\mathbf{x}) = Q_1(\mathbf{Ux})$. (In our discussion $\mathbf{U}$ is of course an orthogonal matrix, but similar results are available for complex quadratic forms.) This occurs if and only if $\mathbf{A}$ and $\mathbf{A}_1$ are unitarily equivalent, for then $(\mathbf{Ax}, \mathbf{x}) = (\mathbf{A}_1\mathbf{Ux}, \mathbf{Ux}) = (\mathbf{U}^*\mathbf{A}_1\mathbf{Ux}, \mathbf{x})$ and it follows that $\mathbf{A} = \mathbf{U}^*\mathbf{A}_1\mathbf{U}$. Thus Q and Q_1 are unitarily equivalent if and only if their diagonal forms feature the same eigenvalues (of identical multiplicities, but possibly in a different arrangement).

If Q is a positive definite quadratic form, then one can replace the usual inner product by

$$[\mathbf{x}, \mathbf{y}] = (\mathbf{Ax}, \mathbf{y}),$$

where $\mathbf{A}$ is the corresponding hermitian matrix; for $[\mathbf{x}, \mathbf{y}]$ will have all the required properties. Geometrically, this amounts to replacing the unit sphere $(\mathbf{x}, \mathbf{x}) = 1$ by an ellipsoid. In terms of the components $x_1', x_2', \ldots, x_n'$ of $\mathbf{x}$ relative to this basis, the new length of $\mathbf{x}$ is computed as the ordinary length of $(\sqrt{\lambda_1}\, x_1', \ldots, \sqrt{\lambda_n}\, x_n')$.

The new inner product will of course give rise to a new version of the conjugate transpose of a matrix $\mathbf{A}$ if one insists that $[\mathbf{Ax}, \mathbf{y}] = [\mathbf{x}, \mathbf{A}^*\mathbf{y}]$. Thus one obtains an analogous concept of normal and hermitian linear transformations relative to the new inner product; these are of course related to the old by virtue of the obvious change of scale (in each x_i') which transforms our new inner product into the old.

Exercises

1. Show that for any real matrix transformation $\mathbf{A}$ on E^n,

$$2(\mathbf{Ax}, \mathbf{y}) = (\mathbf{A}(\mathbf{x} + \mathbf{y}), \mathbf{x} + \mathbf{y}) - (\mathbf{Ax}, \mathbf{x}) - (\mathbf{Ay}, \mathbf{y}).$$

With the aid of this identity prove that if $(\mathbf{Ax}, \mathbf{x}) = (\mathbf{Bx}, \mathbf{x})$ for all $\mathbf{x}$ in E^n, then $\mathbf{A} = \mathbf{B}$.

2. Prove that $[\mathbf{x}, \mathbf{y}]$ as defined in this section satisfies the requirements for an inner product. Is it necessary that $\mathbf{A}$ be hermitian?

3. Prove that if

$$\sum_{i=1}^{n} x_i^2 = Q_1 + Q_2,$$

where Q_1 and Q_2 are quadratic forms of $\mathbf{x} = \langle x_1, x_2, \ldots, x_n \rangle$ of ranks r and $n - r$, respectively, then there is an orthogonal transformation $\mathbf{C}$ such that

$$\mathbf{x} = \mathbf{Cy}, \qquad Q_1 = \sum_{i=1}^{r} y_i^2, \qquad Q_2 = \sum_{i=r+1}^{n} y_i^2.$$

4. Prove that if Q is positive definite, then the locus of all $\mathbf{x}$ such that $Q(\mathbf{x}) =$ constant is an ellipsoid with semi-axes proportional to $1/\sqrt{\lambda_r}$, where the λ_r's are the eigenvalues of A. Any direction corresponding to an eigenvector of λ_r is a semi-axis of the ellipsoid.

A.7 Matrix-valued functions. For any $n \times n$ matrix $\mathbf{A}$ and polynomial $p(t) = a_0 + a_1 t + \cdots + a_k t^k$ one may clearly form $a_0\mathbf{I} + a_1\mathbf{A} + \cdots + a_k\mathbf{A}^k = p(\mathbf{A})$; thus we obtain a matrix-valued function of $\mathbf{A}$; that is, $\mathbf{A}$ maps into $p(\mathbf{A})$. Similarly, for appropriate power series

$$f(t) = \sum_{i=0}^{\infty} a_i t^i$$

the corresponding matrix series

$$f(\mathbf{A}) = \sum_{i=0}^{\infty} a_i \mathbf{A}^i$$

(where $\mathbf{A}^0 = \mathbf{I}$) will be convergent [in the sense that each entry of the partial sums converges, with its limit the value of the corresponding entry in $f(\mathbf{A})$]. For example, if f is an entire function, $f(\mathbf{A})$ will be defined for every square matrix; in particular,

$$e^{\mathbf{A}} = \mathbf{I} + \mathbf{A} + \frac{1}{2!}\mathbf{A}^2 + \cdots$$

is well defined. *If $\lambda_1, \ldots, \lambda_k$ are the eigenvalues of $\mathbf{A}$, then $f(\lambda_1), \ldots, f(\lambda_k)$ are the eigenvalues of $f(\mathbf{A})$.* For $\mathbf{A}\mathbf{x} = \lambda\mathbf{x}$ implies $\mathbf{A}^2\mathbf{x} = \lambda\mathbf{A}\mathbf{x} = \lambda^2\mathbf{x}$, $\mathbf{A}^3\mathbf{x} = \lambda^3\mathbf{x}, \ldots$; hence $f(\mathbf{A})\mathbf{x} = f(\lambda)\mathbf{x}$, so that each $f(\lambda_i)$ is an eigenvalue of $f(\mathbf{A})$ whose associated eigenvectors are those belonging to the eigenvalue λ_i of $\mathbf{A}$. To see that all eigenvalues of $f(\mathbf{A})$ are so obtained, consider the Jordan canonical form $\mathbf{B}^{-1}\mathbf{A}\mathbf{B}$ of $\mathbf{A}$, noting that $\mathbf{B}^{-1}\mathbf{A}^k\mathbf{B} = (\mathbf{B}^{-1}\mathbf{A}\mathbf{B})^k$ and thus $\mathbf{B}^{-1}f(\mathbf{A})\mathbf{B} = f(\mathbf{B}^{-1}\mathbf{A}\mathbf{B})$; the computation of $f(\mathbf{B}^{-1}\mathbf{A}\mathbf{B})$ can be made sub-block by sub-block, and its eigenvalues become apparent.

A particularly important example of a matrix-valued function is furnished by $(\lambda\mathbf{I} - \mathbf{A})^{-1}$. When $|\lambda|$ exceeds the modulus of each eigenvalue, the formal power series expansion

$$\begin{aligned}(\lambda\mathbf{I} - \mathbf{A})^{-1} &= \lambda^{-1}\left(\mathbf{I} - \frac{1}{\lambda}\mathbf{A}\right)^{-1} \\ &= \lambda^{-1}\left(\mathbf{I} + \frac{1}{\lambda}\mathbf{A} + \frac{1}{\lambda^2}\mathbf{A}^2 + \cdots\right) = \sum_{i=0}^{\infty} \frac{\mathbf{A}^i}{\lambda^{i+1}}\end{aligned}$$

is valid. Moreover, by fixing $\mathbf{A}$ we obtain an analytic function of the complex variable λ; that is, λ maps into $(\lambda\mathbf{I} - \mathbf{A})^{-1}$, defined for all $\lambda \neq \lambda_1, \ldots, \lambda_k$. The function is analytic in the sense that

$$\frac{d}{d\lambda}(\lambda\mathbf{I} - \mathbf{A})^{-1} = \lim_{\lambda' \to \lambda} \frac{1}{\lambda' - \lambda}[(\lambda'\mathbf{I} - \mathbf{A})^{-1} - (\lambda\mathbf{I} - \mathbf{A})^{-1}]$$

$$= -(\lambda\mathbf{I} - \mathbf{A})^{-2}$$

exists where $\lambda \neq \lambda_1, \ldots, \lambda_k$. One may also apply the Cauchy integral theorem, its validity being obtained in the usual fashion; here, however, we can interpret the function's analyticity as meaning that it has a power series expansion about each $\lambda \neq \lambda_1, \ldots, \lambda_k$, with matrix coefficients. Each λ_i is a pole of the function; for if $\mathbf{C} = \mathbf{B}^{-1}\mathbf{AB}$ is again the Jordan canonical form, we have $\mathbf{A} = \mathbf{BCB}^{-1}$ and

$$\lambda\mathbf{I} - \mathbf{A} = \lambda\mathbf{I} - \mathbf{BCB}^{-1} = \mathbf{B}(\lambda\mathbf{I} - \mathbf{C})\mathbf{B}^{-1};$$

hence $(\lambda\mathbf{I} - \mathbf{A})^{-1} = \mathbf{B}(\lambda\mathbf{I} - \mathbf{C})^{-1}\mathbf{B}^{-1}$, and it suffices to consider $(\lambda\mathbf{I} - \mathbf{C})^{-1}$, or indeed $(\mathbf{C} - \lambda\mathbf{I})^{-1}$. Since $\mathbf{C} - \lambda\mathbf{I}$ has the same block-diagonal form as $\mathbf{C}$, and since block-diagonal matrices (with blocks of similar size) multiply block by block—i.e., since

$$\begin{pmatrix} \mathbf{A} & \mathbf{0} \\ \mathbf{0} & \mathbf{B} \end{pmatrix}\begin{pmatrix} \mathbf{C} & \mathbf{0} \\ \mathbf{0} & \mathbf{D} \end{pmatrix} = \begin{pmatrix} \mathbf{AC} & \mathbf{0} \\ \mathbf{0} & \mathbf{BD} \end{pmatrix}$$

if $\mathbf{A}$ and $\mathbf{C}$ are $k \times k$, and $\mathbf{B}$ and $\mathbf{D}$ are $l \times l$—it is sufficient to compute the inverse of each sub-block

$$\begin{Vmatrix} \lambda_i - \lambda & 0 & \cdots & \cdots & 0 \\ 1 & \lambda_i - \lambda & \cdots & \cdots & 0 \\ 0 & 1 & \cdots & \cdots & 0 \\ \vdots & \vdots & & & \vdots \\ 0 & 0 & \cdots & \cdots & \lambda_i - \lambda \end{Vmatrix}$$

of $\mathbf{C} - \lambda\mathbf{I}$. Writing

$$\mathbf{D} = (\lambda_i - \lambda)\mathbf{I} + \mathbf{E} = (\lambda_i - \lambda)\left(\mathbf{I} + \frac{1}{\lambda_i - \lambda}\mathbf{E}\right),$$

where $\mathbf{E}$ is the $k \times k$ matrix with 1's below the main diagonal and 0's elsewhere, and formally expanding in analogy with $(1 + x)^{-1} = 1 - x + x^2 - x^3 + \cdots$, we obtain

$$\mathbf{D}^{-1} = (\lambda_i - \lambda)^{-1}\left(\mathbf{I} + \frac{1}{\lambda_i - \lambda}\mathbf{E}\right)^{-1}$$

$$= (\lambda_i - \lambda)^{-1}\left(\mathbf{I} - \frac{1}{\lambda_i - \lambda}\mathbf{E} + \frac{1}{(\lambda_i - \lambda)^2}\mathbf{E}^2 - \cdots\right),$$

which converges, since $\mathbf{E}^k = \mathbf{0}$. Moreover, for the same reason we may check that the value is the appropriate inverse by multiplying by

$$\mathbf{I} + \frac{1}{\lambda_i - \lambda}\mathbf{E}.$$

Thus the highest power of $(\lambda_i - \lambda)^{-1}$ which appears is the kth, and $\mathbf{D}^{-1}$ appears as a Laurent expansion (with constant matrix coefficients) corresponding to a pole of order k at λ_i. Consequently the order of the pole we obtain at λ_i for $(\lambda\mathbf{I} - \mathbf{C})^{-1}$ [or $(\lambda\mathbf{I} - \mathbf{A})^{-1}$] is the size of the largest sub-block featuring λ_i; recall that this is the least k for which

$$\mathfrak{N}[(\lambda\mathbf{I} - \mathbf{A})^k] = \mathfrak{N}[(\lambda\mathbf{I} - \mathbf{A})^{k+1}].$$

Exercises

1. Show that if $\mathbf{AB} = \mathbf{BA}$, then $e^{A+B} = e^A e^B$.
2. Find a Laurent series expansion for $(\lambda\mathbf{I} - \mathbf{A})^{-n}$ and show that it converges when λ exceeds the modulus of each eigenvalue. Under these conditions, show that

$$\frac{d}{d\lambda}(\lambda\mathbf{I} - \mathbf{A})^{-n} = -n(\lambda\mathbf{I} - \mathbf{A})^{-n-1}.$$

3. Let $\lambda_1, \lambda_2, \ldots, \lambda_r$ and $\mathbf{E}_1, \mathbf{E}_2, \ldots, \mathbf{E}_r$ (as defined in Problem 5 of Section A.5) be the eigenvalues and perpendicular projections corresponding to a normal matrix $\mathbf{A}$. Let $f(\lambda)$ be a convergent power series. Show that

$$f(\mathbf{A}) = \sum_{j=1}^{r} f(\lambda_j)\mathbf{E}_j.$$

4. Show that if $f(d) = \mathbf{A}_0 + \mathbf{A}_1 d + \mathbf{A}_2 d^2 + \cdots$ converges absolutely in some interval about d_0, then $f'(d_0) = \mathbf{A}_1 + 2\mathbf{A}_2 d_0 + 3\mathbf{A}_3 d_0^2 + \cdots$.
5. If $\mathbf{A}$ is a normal matrix with nonzero eigenvalues, how can you define $\log \mathbf{A}$?

A.8 Determinants; minors, cofactors. The determinant $|\mathbf{A}|$ (sometimes written $\det \mathbf{A}$) of an $n \times n$ matrix $\mathbf{A}$ is given by the sum

$$\sum_{\pi} \delta(\pi) a_{1,\pi(1)} a_{2,\pi(2)} \cdots a_{n,\pi(n)} = \sum_{\pi} \delta(\pi) a_{\pi(1),1} a_{\pi(2),2} \cdots a_{\pi(n),n}, \tag{A.8.1}$$

where π runs over all permutations of $\{1, 2, \ldots, n\}$, and $\delta(\pi)$ is 1 if π is an even permutation and -1 if π is an odd permutation. Interchange of any two rows (or columns) of $\mathbf{A}$ produces a matrix $\mathbf{B}$ such that $|\mathbf{B}| = -|\mathbf{A}|$, since the corresponding terms in (A.8.1) occur with permutations of opposite parity; thus if $\mathbf{A}$ has two identical rows (or columns), $|\mathbf{A}| = 0$. The determinant of the matrix obtained by deleting the ith row and jth column of $\mathbf{A}$ is the *minor* m_{ij} of a_{ij}; the *cofactor* c_{ij} of a_{ij} is then defined by $c_{ij} = (-1)^{i+j} m_{ij}$.

It follows directly from (A.8.1) that

$$|\mathbf{A}| = \sum_{k=1}^{n} a_{ik} c_{ik} = \sum_{k=1}^{n} a_{kj} c_{kj} \tag{A.8.2}$$

for all i, j; these are the so-called row and column expansions of $|\mathbf{A}|$. Further, if $i \neq j$,

$$\sum_{k=1}^{n} a_{ik} c_{jk} = 0 = \sum_{k=1}^{n} a_{ki} c_{kj},$$

since these expressions are the row and column expansions of $|\mathbf{B}|$ where $\mathbf{B}$ is a matrix with two identical rows or columns. Consequently

$$|\mathbf{A}| = \sum_{k=1}^{n} a_{ik} c_{ik} - \lambda \sum_{k=1}^{n} a_{jk} c_{ik} = \sum_{k=1}^{n} (a_{ik} - \lambda a_{jk}) c_{ik}$$

if $i \neq j$, so that $|\mathbf{A}|$ coincides with the value of the determinant of the matrix obtained from $\mathbf{A}$ by subtracting λ times the jth row from the ith row. Writing $|\mathbf{A}|$ in the descriptive form

$$\begin{vmatrix} a_{11} & a_{12} & \cdots & a_{1n} \\ a_{21} & a_{22} & \cdots & a_{2n} \\ \vdots & \vdots & & \vdots \\ a_{n1} & a_{n2} & \cdots & a_{nn} \end{vmatrix},$$

we may thus subtract any multiple of a row (or column) of the determinant from any other without changing its value.

Let

$$\mathbf{A}\begin{pmatrix} i_1, i_2, \ldots, i_r \\ j_1, j_2, \ldots, j_r \end{pmatrix}$$

denote the determinant of the matrix obtained from $\mathbf{A}$ by deleting all rows but $i_1, i_2, \ldots, i_r$ and all columns but $j_1, \ldots, j_r$, where $i_1 < i_2 < \cdots < i_r, j_1 < j_2 < \cdots < j_r$; that is,

$$\mathbf{A}\begin{pmatrix} i_1, i_2, \ldots, i_r \\ j_1, j_2, \ldots, j_r \end{pmatrix} = \begin{vmatrix} a_{i_1 j_1} & a_{i_1 j_2} & \cdots & a_{i_1 j_r} \\ a_{i_2 j_1} & \cdots & \cdots & \cdots \\ \vdots & & & \\ a_{i_r j_1} & \cdots & \cdots & a_{i_r j_r} \end{vmatrix}.$$

Such subdeterminants of a product **AB** are related to those of **A** and **B** by the expansion

$$\mathbf{AB}\begin{pmatrix} i_1, i_2, \ldots, i_r \\ j_1, j_2, \ldots, j_r \end{pmatrix} = \sum_{k_1<k_2<\cdots<k_r} \mathbf{A}\begin{pmatrix} i_1, i_2, \ldots, i_r \\ k_1, k_2, \ldots, k_r \end{pmatrix} \mathbf{B}\begin{pmatrix} k_1, k_2, \ldots, k_r \\ j_1, j_2, \ldots, j_r \end{pmatrix}, \quad \text{(A.8.3)}$$

where the sum extends over all $\{k_1, k_2, \ldots, k_r\}$ satisfying the indicated relation.

From (A.8.2) and our succeeding remarks, if adj (**A**) denotes the matrix with ijth entry

$$b_{ij} = c_{ji},$$

the *adjoint matrix* of **A**, then $\mathbf{A} \cdot \operatorname{adj}(\mathbf{A}) = |\mathbf{A}|\mathbf{I} = \operatorname{adj}(\mathbf{A}) \cdot \mathbf{A}$. Thus if $|\mathbf{A}| \neq 0$,

$$\frac{1}{|\mathbf{A}|} \operatorname{adj}(\mathbf{A})$$

is the inverse of **A**. But $|\mathbf{AB}| = |\mathbf{A}||\mathbf{B}|$ by (A.8.3); thus if $\mathbf{A}^{-1}$ exists, $|\mathbf{A}||\mathbf{A}^{-1}| = |\mathbf{I}| = 1$ and $|\mathbf{A}| \neq 0$. *Hence the nonsingular matrices are those with nonzero determinants.*

If $\mathbf{B}^{-1}\mathbf{AB}$ is the Jordan canonical form of **A**, then, since $|\mathbf{B}^{-1}\mathbf{AB}| = |\mathbf{B}^{-1}||\mathbf{A}||\mathbf{B}| = |\mathbf{A}|$, it follows that $|\mathbf{A}|$ is the product of the powers of its eigenvalues. Consequently, if 0 appears among the eigenvalues (say $\lambda_i = 0$), we may slightly change λ_i so as to have nonzero eigenvalues and a nonsingular matrix; the corresponding changes in the entries of **A** are small, and thus *each singular* **A** *is a limit of nonsingular matrices.*

Further, since

$$|\mathbf{B}^{-1}\mathbf{AB} - \lambda\mathbf{I}| = |\mathbf{B}^{-1}(\mathbf{A} - \lambda\mathbf{I})\mathbf{B}| = |\mathbf{A} - \lambda\mathbf{I}|,$$

with **B** as above, it follows that

$$|\mathbf{B}^{-1}\mathbf{AB} - \lambda\mathbf{I}| = \prod_{i=1}^{k} (\lambda_i - \lambda)^{n_i}.$$

Thus the λ_i are exactly the roots of the polynomial equation

$$|\mathbf{A} - \lambda\mathbf{I}| = 0$$

in λ, the *characteristic equation* of **A**. The multiplicity of λ_i as a root is called the *algebraic multiplicity* of λ_i (the algebraic multiplicity $\geq$ the geometric multiplicity = number of linearly independent eigenvectors corresponding to λ_i); it is of course the number of times λ_i appears on the main diagonal of the Jordan form.

Exercises

1. Show that if $\langle x_1, y_1\rangle$ and $\langle x_2, y_2\rangle$ are distinct points on the real plane, the equation

$$\begin{vmatrix} x & y & 1 \\ x_1 & y_1 & 1 \\ x_2 & y_2 & 1 \end{vmatrix} = 0$$

represents a straight line through the points.

2. Prove the following statement:

Given a nonsingular $n \times n$ matrix $\mathbf{A}$ and a vector $\mathbf{y}$ in U^n, let $\mathbf{A}_{iy}$ be the matrix in which column i has been replaced by y and all other columns remain unchanged. Then the solution of the system of equations $\mathbf{Ax} = \mathbf{y}$, where $\mathbf{x} = \langle x_1, x_2, \ldots, x_n\rangle$, is given by

$$x_i = \frac{|\mathbf{A}_{iy}|}{|\mathbf{A}|} \qquad (i = 1, \ldots, n).$$

3. Show that if all the elements in a matrix above the main diagonal are zero, the diagonal elements are the eigenvalues.

4. Let $\mathbf{A}$ have eigenvalues $\lambda_1, \lambda_2, \ldots, \lambda_n$, with repetitions if necessary. Find $|(I - \lambda\mathbf{A})^{-1}|$.

5. By $|\mathbf{x}|$ we mean $(\mathbf{x}, \mathbf{x})^{\frac{1}{2}}$. We say that a matrix or transformation $\mathbf{A}$ is bounded on U if there is a constant K such that $\|\mathbf{Ax}\| \leq K\|\mathbf{x}\|$ for all $\mathbf{x}$ in U. The greatest lower bound, K_0, of all such K's is the bound of $\mathbf{A}$. Prove that all transformations on U^n are bounded. Show that if $\mathbf{A}$ is the limit of nonsingular matrices, then $\mathbf{A}$ is singular if and only if the bounds of the inverses of these matrices approach infinity.

A.9 Some identities. Let $\mathbf{U}$ be the $n \times n$ matrix with all entries 1, and let $\mathbf{u} = \langle 1, 1, \ldots, 1\rangle$ be an n-component vector of $\mathbf{U}$ with all entries 1. The following identities relating the determinants and adjoints of $\mathbf{A}$ and $\mathbf{A} + a\mathbf{U}$ are used in Section 2.4:

$$\begin{aligned} |\mathbf{A} + a\mathbf{U}| &= |\mathbf{A}| + a\,(\mathbf{u}, \text{adj}\,(\mathbf{A})\,\mathbf{u}), \\ \mathbf{u}\,\text{adj}\,(\mathbf{A} + a\mathbf{U}) &= \mathbf{u}\,\text{adj}\,(\mathbf{A}), \\ \text{adj}\,(\mathbf{A} + a\mathbf{U})\mathbf{u} &= \text{adj}\,(\mathbf{A})\mathbf{u}. \end{aligned}$$

Here $(\mathbf{u}, \text{adj}\,(\mathbf{A})\mathbf{u})$ when calculated yields

$$\sum_{i,j=1}^{n} A_{ij},$$

where A_{ij} is the cofactor of a_{ij}. If we regard $|\mathbf{A}|$ as a function of the n^2 variables a_{ij}, the row expansion

$$|\mathbf{A}| = \sum_{k=1}^{n} a_{ik}A_{ik}$$

implies $\partial|\mathbf{A}|/\partial a_{ij} = A_{ij}$. Thus, regarding $|\mathbf{A} + a\mathbf{U}|$ as a function of a, setting $a'_{ij} = a_{ij} + a$, and denoting by A'_{ij} the cofactor of a'_{ij} in $\mathbf{A} + a\mathbf{U}$, we have

$$\frac{d}{da}|\mathbf{A} + a\mathbf{U}| = \sum_{i,j=1}^{n} \frac{\partial}{\partial a'_{ij}}|\mathbf{A} + a\mathbf{U}| \frac{d}{da}(a'_{ij}) = \sum_{i,j=1}^{n} A'_{ij}.$$

But since $\sum_{j=1}^{n} A'_{ij} \cdot 1$ is evidently the expansion about the ith row of a matrix obtained from $\mathbf{A} + a\mathbf{U}$ by replacing the ith row by $1, 1, \ldots, 1$, that is, the matrix

$$\begin{vmatrix} a_{11} + a & a_{12} + a & \cdots & a_{1n} + a \\ \vdots & & & \\ a_{i-1,1} + a & a_{i-1,2} + a & \cdots & a_{i-1,n} + a \\ 1 & 1 & \cdots & 1 \\ a_{i+1,1} + a & a_{i+1,2} + a & \cdots & a_{i+1,n} + a \\ \vdots & & & \\ a_{n1} + a & a_{n2} + a & \cdots & a_{nn} + a \end{vmatrix},$$

we can subtract a times the ith row from all others to obtain

$$\sum_{j=1}^{n} A'_{ij} = \sum_{j=1}^{n} A_{ij}. \tag{A.9.1}$$

Thus

$$\frac{d}{da}|\mathbf{A} + a\mathbf{U}| = \sum_{i,j=1}^{n} A_{ij},$$

which is independent of a. It follows that $|\mathbf{A} + a\mathbf{U}|$ is the linear function

$$|\mathbf{A}| + a \cdot \sum_{i,j=1}^{n} A_{ij},$$

which is our first identity. Further, (A.9.1) states precisely that

$$\begin{Vmatrix} A'_{11} & A'_{12} & \cdots & A'_{1n} \\ A'_{21} & A'_{22} & \cdots & A'_{2n} \\ \vdots & \vdots & & \vdots \\ A'_{n1} & A'_{n2} & \cdots & A'_{nn} \end{Vmatrix} \begin{Vmatrix} 1 \\ 1 \\ \vdots \\ 1 \end{Vmatrix} = \begin{Vmatrix} \sum A'_{1j} \\ \sum A'_{2j} \\ \vdots \\ \sum A'_{nj} \end{Vmatrix} = \begin{Vmatrix} A_{11} & A_{12} & \cdots & A_{1n} \\ A_{21} & A_{22} & \cdots & A_{2n} \\ \vdots & \vdots & & \vdots \\ A_{n1} & A_{n2} & \cdots & A_{nn} \end{Vmatrix} \begin{Vmatrix} 1 \\ 1 \\ \vdots \\ 1 \end{Vmatrix},$$

that is, our third identity. The second follows by the original argument concerning

$$\sum_{j=1}^{n} A'_{ij} \qquad \text{applied to} \qquad \sum_{i=1}^{n} A'_{ij}$$

as a column expansion.

It follows directly from the definition that $|\lambda\mathbf{A}| = \lambda^n|\mathbf{A}|$ for any $n \times n$ matrix $\mathbf{A}$. Consequently, since

$$\mathbf{A}^{-1} = \frac{1}{|\mathbf{A}|} \operatorname{adj}(\mathbf{A})$$

when $\mathbf{A}$ is nonsingular, we have

$$|\mathbf{A}^{-1}| = \frac{1}{|\mathbf{A}|} = \frac{1}{|\mathbf{A}|^n} |\operatorname{adj}(\mathbf{A})|,$$

and thus $|\operatorname{adj}(\mathbf{A})| = |\mathbf{A}|^{n-1}$. Since this last relation holds for any limit of nonsingular matrices, it holds for all $\mathbf{A}$.

For any nonsingular matrix $\mathbf{A}$,

$$\mathbf{A}^{-1}\begin{pmatrix} i_1, i_2, \ldots, i_p \\ j_1, j_2, \ldots, j_p \end{pmatrix} = \frac{\mathbf{A}\begin{pmatrix} j'_1, j'_2, \ldots, j'_{n-p} \\ i'_1, i'_2, \ldots, i'_{n-p} \end{pmatrix}}{|\mathbf{A}|} (-1)^{\Sigma_{\nu=1}^{p}(i_\nu + j_\nu)} \tag{A.9.2}$$

where $\{j'_1, j'_2, \ldots, j'_{n-p}\}$ forms the complementary set of column indices to $\{j_1, \ldots, j_p\}$, and similarly for $\{i'_1, \ldots, i'_{n-p}\}$. Equation (A.9.2) is known as Sylvester's identity.

Consider the case in which $i_1 = j_1 = 1$, $i_2 = j_2 = 2$, $\ldots$, $i_p = j_p = p$, so that the identity becomes

$$\mathbf{A}^{-1}\begin{pmatrix} 1, 2, \ldots, p \\ 1, 2, \ldots, p \end{pmatrix} = \frac{1}{|\mathbf{A}|}\mathbf{A}\begin{pmatrix} p+1, p+2, \ldots, n \\ p+1, p+2, \ldots, n \end{pmatrix}.$$

Noting that the transpose of a matrix has the same determinant as the matrix and that $|\mathbf{A}||\mathbf{B}| = |\mathbf{AB}|$, and denoting by A_{ij} the cofactor of a_{ij}, we have the product

$$\begin{vmatrix} A_{11} & A_{12} & \cdots & A_{1p} & A_{1,p+1} & \cdots & \cdots & A_{1n} \\ A_{21} & A_{22} & \cdots & A_{2p} & A_{2,p+1} & \cdots & \cdots & A_{2n} \\ \vdots \\ A_{p1} & A_{p2} & \cdots & A_{pp} & A_{p,p+1} & \cdots & \cdots & A_{pn} \\ 0 & 0 & \cdots & 0 & 1 & 0 & \cdots & 0 \\ 0 & 0 & \cdots & 0 & 0 & 1 & \cdots & 0 \\ \vdots & \vdots & & \vdots & \vdots & & & \vdots \\ 0 & 0 & \cdots & 0 & 0 & \cdots & \cdots & 1 \end{vmatrix}$$

$$\times \begin{vmatrix} a_{11} & a_{21} & \cdots & a_{n1} \\ a_{12} & a_{22} & \cdots & a_{n2} \\ \vdots \\ a_{1p} & a_{2p} & \cdots & a_{n,p} \\ a_{1,p+1} & a_{2,p+1} & \cdots & a_{n,p+1} \\ \vdots & \vdots & & \vdots \\ a_{1n} & a_{2n} & \cdots & a_{nn} \end{vmatrix}$$

(where the first p rows of the first determinant coincide with those of adj ($\mathbf{A}$), while the last $n - p$ feature only 1's on the main diagonal as non-zero entries), equal to

$$\begin{vmatrix} |\mathbf{A}| & 0 & \cdots & \cdots & \cdots & 0 & 0 & \cdots & 0 \\ 0 & |\mathbf{A}| & 0 & \cdots & \cdots & \cdots & \cdots & \cdots & \cdots \\ \vdots \\ 0 & \cdots & \cdots & \cdots & 0 & |\mathbf{A}| & 0 & \cdots & 0 \\ a_{1,p+1} & \cdots & \cdots & \cdots & \cdots & a_{p,p+1} & a_{p+1,p+1} & \cdots & a_{n,p+1} \\ a_{1,p+2} & \cdots & \cdots & \cdots & \cdots & a_{p,p+2} & a_{p+1,p+2} & \cdots & a_{n,p+2} \\ \vdots \\ a_{1,n} & \cdots & \cdots & \cdots & \cdots & a_{p,n} & a_{p+1,n} & \cdots & a_{n,n} \end{vmatrix}$$

$$= |\mathbf{A}|^p \mathbf{A}\begin{pmatrix} p+1, \ldots, n \\ p+1, \ldots, n \end{pmatrix}$$

(expanding by rows, using the first p rows). On the other hand, if we expand the first factor of our product by rows (using the last n $-$ p rows), it becomes, with $\mathbf{C} = \text{adj}\ (\mathbf{A})$,

$$\mathbf{C}\begin{pmatrix} 1, 2, \ldots, p \\ 1, 2, \ldots, p \end{pmatrix} |\mathbf{A}|;$$

hence

$$\mathbf{C}\begin{pmatrix} 1, 2, \ldots, p \\ 1, 2, \ldots, p \end{pmatrix} = |\mathbf{A}|^{p-1} \mathbf{A}\begin{pmatrix} p+1, \ldots, n \\ p+1, \ldots, n \end{pmatrix}.$$

But

$$\mathbf{A}^{-1} = \frac{1}{|\mathbf{A}|}\mathbf{C},$$

and thus

$$\mathbf{A}^{-1}\begin{pmatrix} 1, \ldots, p \\ 1, \ldots, p \end{pmatrix} = \frac{1}{|\mathbf{A}|^p}\mathbf{C}\begin{pmatrix} 1, \ldots, p \\ 1, \ldots, p \end{pmatrix} = \frac{1}{|\mathbf{A}|}\mathbf{A}\begin{pmatrix} p+1, \ldots, n \\ p+1, \ldots, n \end{pmatrix},$$

which is exactly the identity in our special case.

By further interchanging appropriate rows and columns we may achieve (A.9.2). In terms of $\mathbf{C} = \operatorname{adj}(\mathbf{A})$ the identity (A.9.2) becomes

$$\frac{1}{|\mathbf{A}|^p}\mathbf{C}\begin{pmatrix} i_1, \ldots, i_p \\ j_1, \ldots, j_p \end{pmatrix} = (-1)^{\Sigma_{\nu=1}^{p}(i_\nu + j_\nu)}\frac{1}{|\mathbf{A}|}\mathbf{A}\begin{pmatrix} j_1', \ldots, j_{n-p}' \\ i_1', \ldots, i_{n-p}' \end{pmatrix},$$

or

$$\mathbf{C}\begin{pmatrix} i_1, \ldots, i_p \\ j_1, \ldots, j_p \end{pmatrix} = (-1)^{\Sigma_{\nu=1}^{p}(i_\nu + j_\nu)}|\mathbf{A}|^{p-1}\mathbf{A}\begin{pmatrix} j_1', \ldots, j_{n-p}' \\ i_1', \ldots, i_{n-p}' \end{pmatrix},$$

since

$$\frac{1}{|\mathbf{A}|}\mathbf{C} = \mathbf{A}^{-1}.$$

The determinants

$$\mathbf{A}\begin{pmatrix} i_1, \ldots, i_p \\ j_1, \ldots, j_p \end{pmatrix}$$

are clearly generalized minors of $\mathbf{A}$; but in particular the ijth minor of $\mathbf{A}$ is

$$\mathbf{A}\begin{pmatrix} 1, 2, \ldots, i-1, i+1, \ldots, n \\ 1, 2, \ldots, j-1, j+1, \ldots, n \end{pmatrix},$$

while of course

$$\mathbf{A}\begin{pmatrix} 1, \ldots, n \\ 1, \ldots, n \end{pmatrix} = |\mathbf{A}|.$$

Let $p < n$ be fixed, and set

$$d_{ij} = \mathbf{A}\begin{pmatrix} 1, \ldots, p, i \\ 1, \ldots, p, j \end{pmatrix} \qquad (i, j \geq p+1),$$

so that the $(n-p) \times (n-p)$ array $\mathbf{D} = \|d_{ij}\|$ forms a matrix. Then (*Sylvester's theorem*)

$$\mathbf{D}\begin{pmatrix} i_1, \ldots, i_q \\ j_1, \ldots, j_q \end{pmatrix} = \mathbf{A}\begin{pmatrix} 1, \ldots, p \\ 1, \ldots, p \end{pmatrix}^{q-1} \mathbf{A}\begin{pmatrix} 1, \ldots, p, i_1, \ldots, i_q \\ 1, \ldots, p, j_1, \ldots, j_q \end{pmatrix}. \quad \text{(A.9.3)}$$

Since we shall be concerned only with the matrix obtained from $\mathbf{A}$ by deleting all rows but $1, 2, \ldots, p, i_1, \ldots, i_q$ and all columns but $1, 2, \ldots, p, j_1, \ldots, j_q$, we can assume without loss of generality that this matrix forms our original $\mathbf{A}$ and thus prove the identity in the one case in which $q = n - p$:

$$\mathbf{D}\begin{pmatrix} p+1, \ldots, n \\ p+1, \ldots, n \end{pmatrix} = \mathbf{A}\begin{pmatrix} 1, \ldots, p \\ 1, \ldots, p \end{pmatrix}^{n-p-1} \mathbf{A}\begin{pmatrix} 1, \ldots, n \\ 1, \ldots, n \end{pmatrix}$$
$$= \mathbf{A}\begin{pmatrix} 1, \ldots, p \\ 1, \ldots, p \end{pmatrix}^{n-p-1} \cdot |\mathbf{A}|.$$

Let $\mathbf{C} = \text{adj}\,(\mathbf{A})$. Then

$$c_{ij} = (-1)^{i+j}\mathbf{A}\begin{pmatrix} 1, \ldots, j-1, j+1, \ldots, n \\ 1, \ldots, i-1, i+1, \ldots, n \end{pmatrix}.$$

In view of our first identity (written in terms of the adjoint), we have

$$\mathbf{C}\begin{pmatrix} p+1, \ldots, r-1, r+1, \ldots, n \\ p+1, \ldots, s-1, s+1, \ldots, n \end{pmatrix}$$
$$= (-1)^{r+s}|\mathbf{A}|^{n-p-2}\mathbf{A}\begin{pmatrix} 1, \ldots, p, s \\ 1, \ldots, p, r \end{pmatrix} \quad \text{(A.9.4)}$$
$$= (-1)^{r+s}|\mathbf{A}|^{n-p-2}d_{sr},$$

and

$$\mathbf{C}\begin{pmatrix} p+1, \ldots, n \\ p+1, \ldots, n \end{pmatrix} = |\mathbf{A}|^{n-p-1}\mathbf{A}\begin{pmatrix} 1, \ldots, p \\ 1, \ldots, p \end{pmatrix}. \quad \text{(A.9.5)}$$

But

$$b_{sr} = (-1)^{s+r}\mathbf{C}\begin{pmatrix} p+1, \ldots, r-1, r+1, \ldots, n \\ p+1, \ldots, s-1, s+1, \ldots, n \end{pmatrix} \quad (r, s \geq p+1)$$

defines the elements of the adjoint $\mathbf{B}$ of the $(n-p) \times (n-p)$ matrix

$$\left\| \begin{matrix} c_{p+1,p+1} & c_{p+1,p+2} & \cdots & c_{p+1,n} \\ c_{p+2,p+2} & \cdots & \cdots & c_{p+2,n} \\ \vdots & & & \vdots \\ c_{n,p+1} & \cdots & \cdots & c_{n,n} \end{matrix} \right\|,$$

and thus

$$\mathbf{B}\begin{pmatrix} p+1, \ldots, n \\ p+1, \ldots, n \end{pmatrix} = |\mathbf{B}| = \mathbf{C}\begin{pmatrix} p+1, \ldots, n \\ p+1, \ldots, n \end{pmatrix}^{n-p-1}$$

since

$$\mathbf{C}\begin{pmatrix} p+1, \ldots, n \\ p+1, \ldots, n \end{pmatrix}$$

is just the determinant of this matrix and $|\text{adj } (\mathbf{A}_0)| = |\mathbf{A}_0|^{k-1}$ for any $k \times k$ matrix $\mathbf{A}_0$. But clearly $b_{sr} = |\mathbf{A}|^{n-p-2} d_{sr}$; hence

$$|\mathbf{B}| = |\mathbf{A}|^{(n-p-2)(n-p)} \mathbf{D}\begin{pmatrix} p+1, \ldots, n \\ p+1, \ldots, n \end{pmatrix}$$

and

$$\mathbf{C}\begin{pmatrix} p+1, \ldots, n \\ p+1, \ldots, n \end{pmatrix}^{n-p-1} = |\mathbf{A}|^{(n-p)(n-p-2)} \mathbf{D}\begin{pmatrix} p+1, \ldots, n \\ p+1, \ldots, n \end{pmatrix}.$$

Thus by (A.9.5)

$$|\mathbf{A}|^{(n-p-1)^2} \mathbf{A}\begin{pmatrix} 1, \ldots, p \\ 1, \ldots, p \end{pmatrix}^{n-p-1} = |\mathbf{A}|^{(n-p)(n-p-2)} \mathbf{D}\begin{pmatrix} p+1, \ldots, n \\ p+1, \ldots, n \end{pmatrix},$$

or

$$\mathbf{D}\begin{pmatrix} p+1, \ldots, n \\ p+1, \ldots, n \end{pmatrix} = |\mathbf{A}| \mathbf{A}\begin{pmatrix} 1, \ldots, p \\ 1, \ldots, p \end{pmatrix}^{n-p-1},$$

completing the proof.

A.10 Compound matrices. From an $n \times n$ matrix $\mathbf{A}$ we form the pth associated (compound) matrix $\mathfrak{A}_p$, whose elements comprise all $p \times p$ subdeterminants of $\mathbf{A}$, arranged in alphabetical order (viz., $\{1, 3\}$ precedes $\{1, 8\}$; $\{2, 6\}$ precedes $\{3, 1\}$). For example, the following matrix $\mathfrak{A}_2$ is a square matrix of size $\binom{n}{2} \times \binom{n}{2}$:

$$\left\| \begin{matrix} A\begin{pmatrix}1,2\\1,2\end{pmatrix} & A\begin{pmatrix}1,2\\1,3\end{pmatrix} & \cdots & A\begin{pmatrix}1,2\\1,n\end{pmatrix} & A\begin{pmatrix}1,2\\2,3\end{pmatrix} & \cdots & A\begin{pmatrix}1, & 2\\n-1, & n\end{pmatrix} \\ A\begin{pmatrix}1,3\\1,2\end{pmatrix} & A\begin{pmatrix}1,3\\1,3\end{pmatrix} & \cdots & A\begin{pmatrix}1,3\\1,n\end{pmatrix} & A\begin{pmatrix}1,3\\2,3\end{pmatrix} & \cdots & A\begin{pmatrix}1, & 3\\n-1, & n\end{pmatrix} \\ \vdots & & & & & & \vdots \\ A\begin{pmatrix}n-1, & n\\1, & 2\end{pmatrix} & \cdots & \ddots & \cdots & \cdots & \cdots & A\begin{pmatrix}n-1, n\\n-1, n\end{pmatrix} \end{matrix} \right\|$$

Similarly, $\mathfrak{A}_p$ is a matrix of size $\binom{n}{p} \times \binom{n}{p}$.

It follows immediately from formula (A.8.3) that if $\mathbf{A} \cdot \mathbf{B} = \mathbf{C}$, then $\mathfrak{A}_p \cdot \mathfrak{B}_p = \mathfrak{C}_p$, where $\mathfrak{A}_p$, $\mathfrak{B}_p$, and $\mathfrak{C}_p$ are the pth compound matrices of $\mathbf{A}$, $\mathbf{B}$, and $\mathbf{C}$, respectively. In particular, if $\mathbf{B} = \mathbf{A}^{-1}$, then $\mathfrak{B}_p$ is the inverse matrix of $\mathfrak{A}_p$.

Let $\lambda_1, \lambda_2, \ldots, \lambda_n$ denote a complete system of eigenvalues (allowing repetitions when required) of the matrix $\mathbf{A}$, and suppose that $\mathbf{A} = \mathbf{PTP}^{-1}$, where

$$\mathbf{T} = \left\| \begin{matrix} \lambda_1 & & & & \bigcirc \\ \cdot\cdot & \lambda_2 & & & \\ & & \cdot & & \\ & & & \cdot & \\ & & & & \cdot \\ \cdot & \cdot & & & \lambda_n \end{matrix} \right\|$$

represents the Jordan canonical form of $\mathbf{A}$. Then manifestly

$$\mathfrak{A}_p = \mathfrak{P}_p \mathfrak{J}_p \mathfrak{P}_p^{-1}, \tag{A.10.1}$$

where $\mathfrak{J}_p$ is a triangular matrix whose diagonal elements are

$$\mathbf{T}\begin{pmatrix} i_1, \ldots, i_p \\ i_1, \ldots, i_p \end{pmatrix} = \lambda_{i_1}\lambda_{i_2} \cdots \lambda_{ip}.$$

Thus the eigenvectors of $\mathfrak{A}_p$ are apparent, being the diagonal entries of $\mathfrak{J}_p$.

In the important special case in which the eigenvalues are distinct, we have verified (p. 376) that the similarity transformation $\mathbf{P}$ can be formed as a matrix whose column vectors are the eigenvectors of $\lambda_1, \lambda_2, \ldots, \lambda_n$, respectively. But we see by inspection of (A.10.1) that since $\mathfrak{J}_p$ is diagonal, the eigenvector corresponding to the eigenvalue $\lambda_{k_1}\lambda_{k_2} \ldots \lambda_{k_p}$ is the vector

$$\mathbf{P}\begin{pmatrix} \alpha_1, \alpha_2, \ldots, \alpha_p \\ k_1, k_r, \ldots, k_p \end{pmatrix},$$

where $(\alpha_1, \ldots, \alpha_p)$ designates the variable component index.

A matrix **A** is said to be totally positive (T.P.) of order p if

$$\mathbf{A}\begin{pmatrix} i_1, i_2, \ldots, i_p \\ j_1, j_2, \ldots, j_p \end{pmatrix} \geq 0$$

for every pair of index sets $1 \leq i_1 < i_2 < \cdots < i_p \leq n$ and $1 \leq j_1 < j_2 < \cdots < j_p \leq n$. Because of (A.8.3) the product of two totally positive matrices of order p is again totally positive of order p. A matrix is said to be weakly totally positive if

$$(-1)^{\Sigma_{\nu=1}^{p}(i_\nu + j_\nu)} \mathbf{A}\begin{pmatrix} i_1, \ldots, i_p \\ j_1, \ldots, j_p \end{pmatrix} \geq 0 \qquad \text{for}$$

$$1 \leq \begin{matrix} i_1 < i_2 < \cdots < i_p \\ j_1 < j_2 < \cdots < j_p \end{matrix} \leq n.$$

We may immediately deduce from Sylvester's identity (A.9.2) that if **A** is T.P. of order p, then $\mathbf{B} = \mathbf{A}^{-1}$ (provided the inverse exists) is weakly T.P. of order p.

By way of illustration, we suppose that $\mathbf{A} = \|a_{ij}\|$, where

$$a_{ij} = \begin{cases} b_i c_j & i \leq j, \\ b_j c_i & i \geq j, \end{cases}$$

and all b_i and c_i have the same sign and are nonzero. A calculation will show that A is T.P. of all orders if and only if

$$\frac{b_1}{c_1} \leq \frac{b_2}{c_2} \leq \cdots \leq \frac{b_n}{c_n}.$$

Exercises

1. If **A** is normal or unitary, show that the same is true for $\mathcal{A}_p$.
2. Prove that if **A** is symmetric, $\mathcal{A}_p$ is symmetric; and that if the quadratic form for **A** is positive definite, the same holds for $\mathcal{A}_p$.
3. Prove that if $|\mathbf{A}| > 0$ and **A** is weakly T.P. of order p, then $\mathbf{A}^{-1}$ is T.P. of order p.
4. Find $|\mathcal{A}_p|$ in terms of $|\mathbf{A}|$.

APPENDIX B

CONVEX SETS AND CONVEX FUNCTIONS

B.1 Convex sets in $\mathbf{E}^n$. A set X of vectors in E^n is called a convex set if, for any two vectors $\mathbf{x}$ and $\mathbf{y}$ in X, the following relations hold: $\alpha\mathbf{x}+\beta\mathbf{y} \in X$; $\alpha, \beta \geq 0$; $\alpha + \beta = 1$. Among the simpler convex sets are linear subspaces, spheres, and triangles.

For a finite number of vectors $\mathbf{a}^1, \ldots, \mathbf{a}^r$ and real numbers $\alpha_1, \ldots, \alpha_r$, the set H defined by

$$H = \{\mathbf{x} | (\mathbf{a}^i, \mathbf{x}) = \alpha_i, \qquad i = 1, \ldots, r\}$$

is a closed convex set. In particular, for a nonzero vector $\mathbf{a}$ and a real number α, the set $H = \{\mathbf{x} | (\mathbf{a}, \mathbf{x}) = \alpha\}$ is called a hyperplane [or an $(n - 1)$-dimensional linear variety].

A hyperplane H determines two *closed half-spaces*

$$H_1 = \{\mathbf{x} | (\mathbf{a}, \mathbf{x}) \geq \alpha\}, \qquad H_2 = \{\mathbf{x} | (\mathbf{a}, \mathbf{x}) \leq \alpha\},$$

both of which are closed, convex sets.

A hyperplane H is said to be a *supporting hyperplane* to a convex set X if X is contained in one of the half-spaces of H and the boundary of X has a point in common with H. More precisely, a hyperplane $\{\mathbf{a}, \alpha\}$ is a supporting hyperplane to X if

$$\inf_{x \in X} (\mathbf{a}, \mathbf{x}) = \alpha.$$

(Here the appropriate infinite point must be admitted as a possible point.)

▶ LEMMA B.1.1. Let X be a convex set and $\mathbf{y}$ be a point exterior to the closure of X. Then there exists a vector $\mathbf{a}$ such that

$$\inf_{x \in X} (\mathbf{a}, \mathbf{x}) > (\mathbf{a}, \mathbf{y}).$$

Proof. Let $\mathbf{x}^0$ be a boundary point of X such that

$$\sqrt{(\mathbf{y} - \mathbf{x}^0, \mathbf{y} - \mathbf{x}^0)} = |\mathbf{y} - \mathbf{x}^0| = \inf_{x \in X} |\mathbf{y} - \mathbf{x}|. \qquad \text{(B.1.1)}$$

If $\mathbf{x} \in X$, then $\mathbf{x}^0 + t(\mathbf{x} - \mathbf{x}^0) \in \overline{X}$ $(0 \leq t \leq 1)$, where $\overline{X}$ denotes the closure of X.

By (B.1.1)

$$|\mathbf{x}^0 + t(\mathbf{x} - \mathbf{x}^0) - \mathbf{y}|^2 \geq |\mathbf{x}^0 - \mathbf{y}|^2. \qquad \text{(B.1.2)}$$

Expanding (B.1.2), we obtain

$$2t(\mathbf{x} - \mathbf{x}^0, \mathbf{x}^0 - \mathbf{y}) + t^2(\mathbf{x} - \mathbf{x}^0, \mathbf{x} - \mathbf{x}^0) \geq 0 \qquad (0 \leq t \leq 1).$$

Hence $(\mathbf{x} - \mathbf{x}^0, \mathbf{x}^0 - \mathbf{y}) \geq 0$, or

$$(\mathbf{x}, \mathbf{x}^0 - \mathbf{y}) \geq (\mathbf{x}^0, \mathbf{x}^0 - \mathbf{y}) > (\mathbf{y}, \mathbf{x}^0 - \mathbf{y}).$$

Let $\mathbf{a} = \mathbf{x}^0 - \mathbf{y}$. Then $\mathbf{a} \neq \mathbf{0}$ and

$$\inf_{x \in X} (\mathbf{x}, \mathbf{a}) \geq (\mathbf{x}^0, \mathbf{a}) > (\mathbf{y}, \mathbf{a}),$$

as asserted.

▶ LEMMA B.1.2. Let X be a convex set and $\mathbf{y}$ be on the boundary of X. Then there exists a supporting plane through $\mathbf{y}$; i.e., there exists a nonzero vector $\mathbf{a}$ such that

$$\inf_{x \in X} (\mathbf{a}, \mathbf{x}) = (\mathbf{a}, \mathbf{y}).$$

Proof. Consider a sequence $\{\mathbf{y}^\nu\}$, with $\mathbf{y}^\nu$ exterior to the closure of X and such that $\lim_\nu \mathbf{y}^\nu = \mathbf{y}$.

By Lemma B.1.1, there is a sequence $\{\mathbf{a}^\nu\}$ which may be normalized such that

$$(\mathbf{a}^\nu, \mathbf{a}^\nu) = 1, \qquad \inf_{x \in X} (\mathbf{x}, \mathbf{a}^\nu) > (\mathbf{y}^\nu, \mathbf{a}^\nu) \qquad (\nu = 1, 2, \ldots).$$

Taking a limiting point $\mathbf{a}$ of $\{\mathbf{a}^\nu\}$, we have, for any $\mathbf{x} \in X$,

$$(\mathbf{x}, \mathbf{a}) = \lim_\nu (\mathbf{x}, \mathbf{a}^\nu) \geq \lim_\nu (\mathbf{y}^\nu, \mathbf{a}^\nu) = (\mathbf{y}, \mathbf{a}),$$

as asserted.

The preceding two lemmas in conjunction yield the following theorem.

▶ THEOREM B.1.1. A closed convex set is the intersection of all its supporting half-spaces, and every boundary point of the set lies on a supporting hyperplane.

▶ THEOREM B.1.2. If X possesses interior points and X and Y are two convex sets with no interior point in common, then there is a hyperplane H that separates X and Y; that is, there exist a nonzero vector $\mathbf{a}$ and a scalar α such that $(\mathbf{a}, \mathbf{x}) \geq \alpha$ for all $\mathbf{x} \in X$ and $(\mathbf{a}, \mathbf{y}) \leq \alpha$ for all $\mathbf{y} \in Y$.

Proof. Consider the set

$$X - Y = \{\mathbf{x} - \mathbf{y} | \mathbf{x} \in X, \mathbf{y} \in Y\},$$

which is convex and does not contain **0** in the interior. By Lemmas B.1.1 and B.1.2 such a set has a vector **a** (**a** $\neq$ **0**) such that

$$(\mathbf{a}, \mathbf{x} - \mathbf{y}) \geq (\mathbf{a}, \mathbf{0}) = 0 \qquad (\mathbf{x} \in X, \mathbf{y} \in Y).$$

The assertion of the theorem follows immediately.

▶THEOREM B.1.3. If X and Y are closed convex sets having no point in common and at least one of them is bounded, then there exists a hyperplane H determined by the parameters $\{\mathbf{a}, \alpha\}$ that strictly separates X and Y: that is, $(\mathbf{a}, \mathbf{x}) > \alpha$ for all $\mathbf{x} \in X$ and $(\mathbf{a}, \mathbf{y}) < \alpha$ for all $\mathbf{y} \in Y$.

Proof. In this case $X - Y$ is closed and does not contain **0**; hence the theorem follows immediately from Lemma B.1.1.

The following result is frequently used in the text.

▶LEMMA B.1.3. Let X be a convex set which has no point in common with the nonnegative orthant. Then there exists a vector $\mathbf{a} > \mathbf{0}$ (every component of **a** is nonnegative and at least one component is positive) such that $(\mathbf{a}, \mathbf{x}) \leq 0$ for all $\mathbf{x} \in X$.

Proof. Apply Theorem B.1.2 to X and Y, where Y denotes the nonnegative orthant. Then there exists a vector $\mathbf{a} \neq \mathbf{0}$ such that $(\mathbf{a}, \mathbf{x}) \leq 0$ for all $\mathbf{x} \in X$ and $(\mathbf{a}, \mathbf{y}) \geq 0$ for all $\mathbf{y} \in Y$. It is easy to show that $\mathbf{a} > \mathbf{0}$.

EXERCISES

1. Theorem B.1.3 is not necessarily true if both X and Y are unbounded. Show that a counterexample can be constructed for $E^2 = \{\langle x_1, x_2\rangle\}$ by taking $X = \{\langle x_1, x_2\rangle | x_1 \leq 0\}$ $Y = \{\langle x_1, x_2\rangle | x_2 \geq e^{-x_1}\}$.
2. Prove that every set and its closure have the same exterior points. Show also that for convex sets it is also true that every convex set and its closure have the same inner points, hence also the same boundary points.
3. Prove that in E^2 every polygon is convex if and only if its interior angles are less than 180 degrees.
4. Prove that the intersection of any number of convex sets is convex. Prove that the closure of a convex set is convex.
5. Prove that every set in E^n can be embedded uniquely in a subspace of lowest possible dimension.
6. To obtain the convex hull of a closed bounded set in E^2 we can "wrap a string around the set and pull it tight." The area determined by the string and its interior is the convex hull. Show that this method works for finding the convex hull of two triangles, or two circles. What does Lemma B.2.2 say about the convex hull of two triangles (or circles)?

B.2 Convex hulls of sets and extreme points of convex sets. The smallest convex set containing a set X is called the *convex hull* of X. It is usually denoted by $[X]$, sometimes by Co (X). The set $[X]$ may be constructively formed as follows:

$$[X] = \left\{\sum_{k=1}^{r} \alpha_k \mathbf{x}^k | \mathbf{x}^k \in X, \sum_{k=1}^{r} \alpha_k = 1, \alpha_k \geq 0, k = 1, \ldots, r\right\},$$

where r is an arbitrary positive integer. If in particular $X = \{\mathbf{a}^1, \ldots, \mathbf{a}^r\}$, then $[\mathbf{a}^1, \ldots, \mathbf{a}^r]$ is referred to as the polyhedral convex set spanned by $\mathbf{a}^i$. It is easily shown that if X is bounded, $[X]$ is bounded, and that if X is compact, $[X]$ is compact. The set $[\mathbf{a}^1, \ldots, \mathbf{a}^r]$ is clearly closed and bounded.

The following well-known representation is of fundamental value.

▶ LEMMA B.2.1. Let X be a subset of E^n, and let $[X]$ be the convex hull of X. Then every point of $[X]$ can be represented as a convex combination of at most $n + 1$ points of X.

Proof. It suffices to show that if

$$\mathbf{x} = \sum_{k=1}^{r} \alpha_k \mathbf{x}^k \;(\mathbf{x}^k \in X, \alpha_k > 0), \qquad \text{and} \qquad \sum_{k=1}^{r} \alpha_k = 1 \qquad (r > n + 1),$$

then the number of vectors $\mathbf{x}^1, \ldots, \mathbf{x}^r$ of X used in the representation of $\mathbf{x}$ can be reduced. Since $\mathbf{x}^1, \ldots, \mathbf{x}^r$ $(r > n + 1)$ are linearly dependent, there exist real numbers $\beta_1, \ldots, \beta_r$ not all zero such that

$$\sum_{k=1}^{r} \beta_k \mathbf{x}^k = \mathbf{0}, \sum_{k=1}^{r} \beta_k = 0.$$

Let ϵ be a real number such that $\alpha_k + \epsilon\beta_k \geq 0$ $(k = 1, \ldots, r)$ and $\alpha_k + \epsilon\beta_k = 0$ for some k, say k_0. Then

$$\mathbf{x} = \mathbf{x} + \sum_{k=1}^{r} \epsilon\beta_k \mathbf{x}^k = \sum_{k=1}^{r} (\alpha_k + \epsilon\beta_k)\mathbf{x}^k,$$

where $\alpha_k + \epsilon\beta_k \geq 0$ $(k = 1, \ldots, r)$, $\alpha_{k_0} + \epsilon\beta_{k_0} = 0$; and

$$\sum_{k=1}^{r} (\alpha_k + \epsilon\beta_k) = 1,$$

as was to be proved.

By imposing further restrictions on X we may sharpen the preceding result as follows.

▶ LEMMA B.2.2. If X has at most n connected components, then the number $n + 1$ in Lemma B.2.1 may be decreased to n.

Proof. Consider the set S spanned by a fixed set of $n + 1$ points of X, and assume that S has an interior point $\mathbf{x}^0$ that cannot be represented by a combination of n or fewer points of X. Corresponding to each $(n - 2)$-dimensional face L of S, let

$$T_L = \{\mathbf{x} = (1 - \lambda)\mathbf{x}^0 + \lambda\mathbf{y} | \lambda \leq 0, \mathbf{y} \in L\}.$$

The set T_L cannot intersect X, because if $\mathbf{c} = (1 - \lambda)\mathbf{x}^0 + \lambda\mathbf{y}$ were in X, then $\lambda \leq 0$ and

$$\mathbf{x}^0 = \frac{1}{1 - \lambda}\mathbf{c} + \frac{-\lambda}{1 - \lambda}\mathbf{y}$$

would represent $\mathbf{x}^0$ as a combination of n or fewer points of X. The sets T_L obtained by traversing the $(n - 2)$-dimensional faces of S form the boundaries of $n + 1$ nonoverlapping cones whose vertices are at $\mathbf{x}^0$ and whose union is all of E^n. Furthermore, each vertex $\mathbf{x}^i$ of S belongs to a different cone. Since X cannot intersect the boundaries of these cones, X must have at least $n + 1$ connected components, thus contradicting the hypothesis.

▶ LEMMA B.2.3. If X is a bounded set in E^n and

$$\mathbf{x} = \sum_{i=1}^{\infty} \alpha_i \mathbf{x}^i,$$

where

$$\mathbf{x}^i \in X, \alpha_i > 0, \sum_{i=1}^{\infty} \alpha_i = 1,$$

then $\mathbf{x}$ belongs to the convex hull of X.

This is easily proved by employing Theorem B.1.1 plus an induction argument based on the dimension of the smallest linear space containing the point $\mathbf{x}$. Suppose that the result has been proved for sets in E^n. Then let X be a bounded set in E^{n+1}. If $\mathbf{x}$ is a boundary point of the closure of the convex span Γ of $\{\mathbf{x}^i\}$, then any supporting plane p to Γ at $\mathbf{x}$ necessarily contains all $\mathbf{x}^i$. But the part common to p and Γ lies in a Euclidian space of dimension at most n, and therefore the induction hypothesis applies. Finally, if $\mathbf{x}$ is interior to Γ, the result is immediate.

A point $\mathbf{x}$ in X is called an *extreme point* of X if there are no points $\mathbf{x}^1$ and $\mathbf{x}^2$ in X such that $\mathbf{x} = \lambda\mathbf{x}^1 + (1 - \lambda)\mathbf{x}^2$ for some λ $(0 < \lambda < 1)$, where $\mathbf{x}^1 \neq \mathbf{x}^2$.

▶ LEMMA B.2.4. A closed, bounded, convex set X is spanned by its extreme points. That is, every $\mathbf{x}$ in X can be represented in the form

$$\mathbf{x} = \sum_{k=1}^{r} \lambda_k \mathbf{x}^k \ (\lambda_k \geq 0; k = 1, \ldots, r; \sum_k \lambda_k = 1),$$

where $\mathbf{x}^1, \ldots, \mathbf{x}^r$ *are extreme points* of X.

The proof again follows by an induction argument on the dimension of the smallest linear space containing X. We omit the formal details.

Exercises

1. Let S be a convex subset of E^n containing the origin and n linearly independent vectors. Prove that S has inner points relative to E^n.
2. Apply Lemma B.2.1 to E^2 to show that the convex hull of a set S is obtained by taking the union of all possible triangles in E^2 with vertices in S.
3. Let $\mathbf{e}$ be the point in the closed bounded convex set S which is farthest from the origin. Prove that $\mathbf{e}$ is an extreme point of S.
4. Show that any closed convex set in E^n is the intersection of a denumerable number of closed half-spaces.
5. Prove that a closed convex set in E^n has a finite number of extreme points if and only if it is the intersection of a finite number of closed half-spaces.
6. Establish that a closed, bounded, convex set has extreme points in every supporting hyperplane.

B.3 Convex cones. A set X in E^n is called a *convex cone* if, for any $\mathbf{x}$ and $\mathbf{y}$ in X,

$$\alpha\mathbf{x} + \beta\mathbf{y} \in X \qquad (\alpha, \beta \geq 0).$$

A convex cone, of course, is a *convex set.* Linear subspaces and half-spaces are examples of convex cones.

The smallest convex cone that contains a set X is called the *convex cone spanned by* X and is denoted by $\mathcal{P}(X)$. Equivalently,

$$\mathcal{P}(X) = \left\{\sum_{k=1}^{r} \alpha_k \mathbf{x}^k \middle| \alpha_k \geq 0, \mathbf{x}^k \in X, k = 1, \ldots, r; r \text{ arbitrary}\right\}.$$

The polar cone X^+ is defined as the set of all directed separating hyperplanes to a set X which contain the origin. In symbols,

$$X^+ = \{\mathbf{a} | (\mathbf{a}, \mathbf{x}) \geq 0 \quad \text{for all} \quad \mathbf{x} \in X\}. \tag{B.3.1}$$

Clearly X^+ is a closed convex cone.

For any sets X and Y, the following four properties are simple consequences of the definition (B.3.1) of polar cones.

(i) $X \subset Y$ implies $X^+ \supset Y^+$;

(ii) $X \subset X^{++}$;

(iii) $X^+ = X^{+++}$;

(iv) $(X + Y)^+ = X^+ \cap Y^+ \quad$ if $\mathbf{0} \in X, Y$.

▶ THEOREM B.3.1. For any closed convex cones X and Y

$$\text{(I)}\quad X = X^{++} \qquad \text{(duality)},$$

$$\left.\begin{aligned}\text{(II)}\quad (X + Y)^+ &= X^+ \cap Y^+,\\ \text{(III)}\quad (X \cap Y)^+ &= \overline{X^+ + Y^+},\end{aligned}\right\}\quad \text{(modularity)}.$$

(Recall that a bar symbol over a set denotes closure.)

Proof. (I) Let $\mathbf{y} \notin X$. Then by Theorem B.1.3 there exists a vector $\mathbf{a}$ such that $(\mathbf{a}, \mathbf{x}) > (\mathbf{a}, \mathbf{y})$ for all $\mathbf{x} \in X$. Since X is a cone, $(\mathbf{a}, \mathbf{x}) \geq 0 > (\mathbf{x}, \mathbf{y})$ for all $\mathbf{x} \in X$. Hence $\mathbf{a} \in X^+$, $\mathbf{y} \notin X^{++}$, as was to be shown.

(II) This assertion was noted in property (iv) above.

(III) $(X \cap Y)^+ = (X^{++} \cap Y^{++})^+ = (X^+ + Y^+)^{++} = \overline{X^+ + Y^+}$.

The following theorem can be proved in a similar manner.

▶ THEOREM B.3.2. For any set X,

$$\overline{p(X)} = X^{++}.$$

Convex polyhedral cones. The convex cone spanned by a finite set of vectors $\mathbf{a}^1, \ldots, \mathbf{a}^r$ is called a *convex polyhedral cone* and denoted by

$$\mathcal{P}(\mathbf{a}^1, \ldots, \mathbf{a}^r) = \left\{\sum_{k=1}^{r} \lambda_k \mathbf{a}^k \middle| \lambda_k \geq 0,\, k = 1, \ldots, r\right\}.$$

Using matrix notation, we may write

$$\mathcal{P}(\mathbf{a}^1, \ldots, \mathbf{a}^r) = \mathcal{P}(\mathbf{A}),$$

where $\mathbf{A}$ is the $n \times r$ matrix whose column vectors are $\mathbf{a}^1, \ldots, \mathbf{a}^r$. In particular, for a nonzero vector $\mathbf{a}$, we have $\mathcal{P}(\mathbf{a}) = \{\lambda \mathbf{a} | \lambda \geq 0\}$. The set $\mathcal{P}(\mathbf{a})$ is referred to as the *ray* spanned by $\mathbf{a}$.

A convex polyhedral cone is closed. This is a consequence of the following more general theorem.

▶ THEOREM B.3.3. A convex polyhedral cone is represented as the intersection of a finite set of supporting half-spaces. Equivalently, for any finite set of vectors $\mathbf{a}^1, \ldots, \mathbf{a}^r$, there exist vectors $\mathbf{b}^1, \ldots, \mathbf{b}^s$ such that

$$\mathcal{P}(\mathbf{a}^1, \ldots, \mathbf{a}^r) = (\mathbf{b}^1)^+ \cap \ldots \cap (\mathbf{b}^s)^+,$$

and conversely.

The vectors $\mathbf{b}^j$ may be found by taking all supporting planes passing through the origin to the convex space spanned by the vectors $\mathbf{a}^i$.

The duality theorem for convex polyhedral cones may be expressed in matrix notation as follows.

▶THEOREM B.3.4. The inner product $(\mathbf{x}, \mathbf{y})$ is nonnegative for all $\mathbf{yA} \geq \mathbf{0}$ if and only if $\mathbf{x} = \mathbf{Au}$ for some $\mathbf{u} \geq \mathbf{0}$.

The following theorem is useful in our study of linear programming.

▶THEOREM B.3.5. Let X be a bounded convex polyhedral set or a convex closed cone having no point but $\mathbf{0}$ in common with the nonnegative orthant of E^n. Then there exists a vector $\mathbf{a}$ with strictly positive components ($\mathbf{a} \gg \mathbf{0}$) such that $(\mathbf{a}, \mathbf{x}) \leq 0$ for all $\mathbf{x} \in X$.

Proof. It is enough to prove the case when X is a cone. Let us suppose that $\mathbf{a} \geq \mathbf{0}$ and $(\mathbf{a}, \mathbf{x}) \leq 0$ for all $\mathbf{x} \in X$ imply that $a_i = 0$ for some component index i, say $i = 1$. Then

$$(-X)^+ \cap (I)^+ \subset H = \{\langle u_1, \ldots, u_n \rangle | u_1 = 0\},$$

where I represents the nonnegative orthant. By the duality theorem (Theorem B.3.1),

$$\overline{-X + I} \supset H^+,$$

and it is readily shown that $-X + I$ is already closed.

Since $\langle \alpha, 0, 0, \ldots, 0 \rangle \in H^+$ for all α, there exist vectors $\mathbf{x} \in X$ and $\mathbf{b} \geq \mathbf{0}$ such that

$$-\mathbf{x} + \mathbf{b} = \langle \alpha, 0, \ldots, 0 \rangle = \mathbf{a}.$$

If we take $\alpha < 0$, it is clear that $\mathbf{x} = \mathbf{b} - \mathbf{a}$ is a nonnegative vector belonging to X and with a positive first component, in contradiction to the hypothesis of the theorem.

EXERCISES

1. In E^3 let $X = \{(x, y, 1) | x^2 + y^2 = 1\}$. Find $\mathcal{P}(x)$.

2. In E^2 let $X = \{(x_1, x_2) | x_1 = \sqrt{x_2^2 + 1}\}$. Then $\mathcal{P}(X) = \{(0, 0)\} + \{(x_1, x_2) | 0 < |x_2| < x_1\}$. Thus X and its convex hull may be closed without the same being true of $\mathcal{P}(X)$. Verify this statement.

3. Prove that $\overline{\mathcal{P}(X)} \supset \mathcal{P}(\overline{X})$.

4. Let C be a convex set. Show that $\mathcal{P}(C)$ consists of the union of all rays spanned by points of C.

5. Prove that if C is a closed bounded convex set, $\mathcal{P}(C)$ is the convex hull of the union of all rays spanned by extreme points of C.

6. Let $\mathbf{x}_0$ be a vector in X, a bounded convex polyhedral set. Prove that there exists a vector $\mathbf{a} \gg \mathbf{0}$ such that $(\mathbf{a}, \mathbf{x}) \leq (\mathbf{a}, \mathbf{x}_0)$ for all $\mathbf{x}$ in X, if and only if there is no vector $\mathbf{x}_1$ in X such that $\mathbf{x}_1 - \mathbf{x}_0 > \mathbf{0}$.

B.4 Convex and concave functions. A function $f(\mathbf{x})$ defined on a convex set X of E^n is called a *convex* function if

$$f[t\mathbf{x} + (1 - t)\mathbf{y}] \leq tf(\mathbf{x}) + (1 - t)f(\mathbf{y}) \qquad (\mathbf{x}, \mathbf{y} \in X; 0 \leq t \leq 1). \tag{B.4.1}$$

If (B.4.1) holds with strict inequality whenever $\mathbf{x} \neq \mathbf{y}$ and $0 < t < 1$, then $f(\mathbf{x})$ is called *strictly convex*. A function $f(\mathbf{x})$ is called *concave* (*strictly concave*) if $-f(\mathbf{x})$ is convex (strictly convex).

Examples of convex functions are such functions as e^x and $x_1^2 + x_2^2$. The condition (B.4.1) implies that

$$f\left(\sum_{k=1}^{r} t_k \mathbf{x}^k\right) \leq \sum_{k=1}^{r} t_k f(\mathbf{x}^k), \tag{B.4.2}$$

where $\mathbf{x}^k \in X$, $\sum t_k = 1$, $t_k \geq 0$, $k = 1, \ldots, r$.

We list the following important facts concerning convex functions, each of which is proved in a straightforward fashion.

(i) Let

$$[X, f] = \left\{\begin{pmatrix}\mathbf{x}\\ \xi\end{pmatrix} \middle| \mathbf{x} \in X, \xi \geq f(\mathbf{x})\right\}.$$

Then $f(\mathbf{x})$ is a convex function defined on X if and only if $[X, f]$ is a convex set in E^{n+1}.

If f is strictly convex, then $[X^0, f]$ is a strictly convex set S; that is, the midpoint of any two points of the set S lies in the relative interior of S. Here X^0 denotes the interior of X.

(ii) If $f_k(\mathbf{x})$ $(k = 1, \ldots, r)$ are convex functions defined on X, and $\lambda_k \geq 0$ $(k = 1, \ldots, r)$, then $\sum_k \lambda_k f_k(\mathbf{x})$ is also a convex function on X.

(iii) Let $f_v(\mathbf{x})$ be a set of convex functions on X; then

$$X_0 = \{\mathbf{x} | \mathbf{x} \in X, \sup_v f_v(\mathbf{x}) < \infty\}$$

is a convex set, and $\sup_v f_v(\mathbf{x})$ is convex on X_0.

(iv) If $f(\mathbf{x})$ is strictly convex on X, then $f(\mathbf{x})$ has at most one local minimum. Any local minimum is an absolute minimum.

(v) If $f(\mathbf{x})$ is convex on X, it is continuous in the relative interior of X.

(vi) For a boundary point $\mathbf{y}$ of X,

$$\varliminf_{x \to y} f(\mathbf{x}) \leq f(\mathbf{y}).$$

(vii) If X is an open convex set and $f(\mathbf{x})$ is differentiable in X, we have another characterization of the convexity of $f(\mathbf{x})$. Specifically, $f(\mathbf{x})$ is convex if and only if, for $\mathbf{x}^0$ and $\mathbf{x}$ in X,

$$f(\mathbf{x}) - f(\mathbf{x}^0) \geq \sum_i \left(\frac{\partial f}{\partial x_i}\right)^0 (x_i - x_i^0).$$

(viii) If $f(\mathbf{x})$ is convex on an open convex set X, then $f(\mathbf{x})$ has second-order partial derivatives existing *almost everywhere* (in the sense of the Lebesgue measure).

(ix) If X is one-dimensional and f is convex on X, then f has a derivative except for at most a countable set of points. Moreover, the right-hand and left-hand derivatives exist everywhere and are right and left continuous, respectively.

(x) If $f(\mathbf{x})$ is a twice-differentiable function, and X is a general convex set in E^n, then $f(\mathbf{x})$ is convex if and only if the matrix of second derivatives

$$\left(\frac{\partial^2 f}{\partial x_i \partial x_j}\right)_{i,j}$$

is positive semidefinite for every x in X. If

$$\left(\frac{\partial f}{\partial x_i \partial x_j}\right)_{i,j}$$

is positive definite, then $f(\mathbf{x})$ is strictly convex.

Exercises

1. Consider the following examples: Prove that $(\mathbf{a}, \mathbf{x})$ is a convex function on E^n, but never strictly convex. For $\mathbf{x} = \langle x_1, x_2, \ldots, x_n \rangle$ in E^n, $f(\mathbf{x}) = \sum_1^r \lambda_i |x_i|^{\alpha_i}$ $(\alpha_i \geq 1)$ is convex on E^n if all $\lambda_i \geq 0$.

2. Let $f(\mathbf{x})$ be a convex function on E^n. Show by applying the theorem on supporting hyperplanes that we obtain the result that for any point $\mathbf{x}_0$ in E^n there is a linear function $l(\mathbf{x})$ (that is, $(\mathbf{l}, \mathbf{x})$) such that $l(\mathbf{x}) \leq f(\mathbf{x})$ and $l(\mathbf{x}_0) = f(\mathbf{x}_0)$. Compare this with (vii).

3. If $f_k(x)$, $k = 1, 2, \ldots$ are nonnegative monotonically increasing convex functions on E^1, so is $\Pi_{k=1}^n f_k(x)$.

4. If in E^n, $f(\mathbf{x})$ is convex (strictly convex) and $\mathbf{A} \not\equiv \mathbf{0}$, then so is $f(\mathbf{Ax} + \mathbf{b})$, where $\mathbf{A}$ and $\mathbf{b}$ are a matrix and a vector of appropriate dimensions.

5. Let $f(\mathbf{x})$ be a continuous function on a convex set X of E^n. Then f is convex if and only if

$$f\left(\frac{\mathbf{x} + \mathbf{y}}{2}\right) \leq \frac{f(\mathbf{x}) + f(\mathbf{y})}{2} \qquad \mathbf{x}, \mathbf{y} \in X.$$

6. Construct a convex function on [0, 1] whose derivative fails to exist for an infinity of points.

APPENDIX C

MISCELLANEOUS TOPICS

C.1 Semicontinuous and equicontinuous functions. *Semicontinuous functions.* A real-valued function $f(\mathbf{x})$ defined on a set X in E^n is said to be continuous at $\mathbf{x}_0$ in X if whenever $\mathbf{x}_n \to \mathbf{x}_0$ (i.e., whenever the Euclidian distance $d(\mathbf{x}_n, \mathbf{x}_0)$ tends to zero), $f(\mathbf{x}_n) \to f(\mathbf{x}_0)$. An equivalent definition expressed in terms of neighborhoods runs as follows: For any $\epsilon > 0$ there exists some $\delta(\epsilon, \mathbf{x}_0) > 0$ such that if $d(\mathbf{x}, \mathbf{x}_0) < \delta$, then

$$-\epsilon \leq f(\mathbf{x}) - f(\mathbf{x}_0) \leq \epsilon. \tag{C.1.1}$$

A function is continuous in X if it is continuous at every point of X.

If δ may be chosen independent of $\mathbf{x}_0$ in X, then f is said to be uniformly continuous in X.

A function f is said to be upper semicontinuous at $\mathbf{x}_0$ if the right-hand inequality of (C.1.1) holds for any arbitrary prescribed ϵ ($\epsilon > 0$) and all values $\mathbf{x}$ satisfying $d(\mathbf{x}, \mathbf{x}_0) \leq \delta$ for a suitable positive δ. Similarly, f is lower semicontinuous at $\mathbf{x}_0$ if the left-hand inequality of (C.1.1) holds for any prescribed $\epsilon > 0$ and the corresponding δ. Equivalently, f is upper (lower) semicontinuous at $\mathbf{x}_0$ if and only if

$$\overline{\lim_{x \to x_0}}\, f(\mathbf{x}) \leq f(\mathbf{x}_0)\Big(\underline{\lim_{x \to x_0}}\, f(\mathbf{x}) \geq f(\mathbf{x}_0)\Big).$$

If f is upper semicontinuous on X (i.e., at every point of X), then the set $L = \{\mathbf{x} | f(\mathbf{x}) < \alpha\}$ for each real α is relatively open in X, and conversely. For if $\mathbf{x}_0$ belongs to L, then there exists an ϵ such that $f(\mathbf{x}_0) + \epsilon < \alpha$. By the definition of an upper semicontinuous function, there exists about $\mathbf{x}_0$ an open sphere of radius δ all of whose members, also in X, belong to L; hence L is relatively open. The converse is proved similarly. Analogously, a function f is lower semicontinuous on X if $\{\mathbf{x} | f(\mathbf{x}) > \alpha\}$ is relatively open for each real α.

With the aid of these characterizations we may readily establish the following properties:

(i) The set of all upper (lower) semicontinuous functions is closed under addition. The negative of an upper semicontinuous function is lower semicontinuous, and conversely.

(ii) If f_α are upper (lower) semicontinuous on X and bounded below (bounded above), then $\inf_\alpha f_\alpha$ ($\sup_\alpha f_\alpha$) is upper (lower) semicontinuous. (Here the last equivalent definition of upper semicontinuity applies most easily.)

(iii) An upper (lower) semicontinuous function defined on a compact set achieves its maximum (minimum).

(iv) An upper (lower) semicontinuous function defined on a compact set may be approached by a decreasing (increasing) sequence of continuous functions, and conversely.

Equicontinuous functions. A family of functions f_α defined on X is said to be equicontinuous at $\mathbf{x}_0$ if for any $\epsilon > 0$ there exists some $\delta(\epsilon, \mathbf{x}_0) > 0$ (independent of α) such that $|f_\alpha(\mathbf{x}) - f_\alpha(\mathbf{x}_0)| < \epsilon$ for $d(\mathbf{x}, \mathbf{x}_0) \leq \delta$ and all α. The functions f_α are said to be equicontinuous on X if they are equicontinuous at each point of X. When δ may be chosen independent of $\mathbf{x}_0$ in X, we speak of uniform equicontinuity for f_α. To illustrate this concept, we exhibit two families of equicontinuous functions.

(1) Let $K(s, t)$ be a uniformly continuous function defined on the unit square, and let $f_\alpha(t)$ constitute a family of uniformly bounded functions $(0 \leq t \leq 1)$. Then

$$g_\alpha(s) = \int_0^1 K(s, t) f_\alpha(t)\, dt$$

represents a uniform equicontinuous family of functions. The estimate

$$|g_\alpha(s_1) - g_\alpha(s_2)| \leq M \max_{0 \leq t \leq 1} |K(s_1, t) - K(s_2, t)|,$$

where M is the bound of f_α, clearly implies the assertion.

(2) Let $f_\alpha(z)$ constitute a family of uniformly bounded functions that are analytic in the unit circle C of the complex plane. Then on X (any interior circle of C) the functions f_α are uniformly equicontinuous. We prove this statement by the same reasoning as in Example (1), using the Cauchy integral formula to represent f_α for points of X.

For our purposes the most important property of equicontinuous functions is given by the classical Ascoli theorem, which may be stated as follows: *Let f_α be a uniformly bounded equicontinuous family of real functions defined on a compact separable set X; then we can select a subsequence f_{α_i} which converges uniformly.*

We sketch the idea of the proof. We specify a countable dense subset $\{\mathbf{x}_i\}$ of X. By the diagonal procedure we may select a subsequence f_{α_i} which converges for all $\mathbf{x}_i$. Owing to the equicontinuity, it follows that f_{α_i} converges uniformly as desired.

C.2 Fixed-point theorems. The following theorems are reviewed with no proofs included. For detailed discussion of these results, see [130].

Brouwer's fixed-point theorem. Let $\phi(\mathbf{x})$ be a continuous point-to-point mapping of a closed simplex X into itself. Then there exists a point $\mathbf{x}_0$ in

X such that $\mathbf{x}_0 = \phi(\mathbf{x}_0)$. (For our purposes, we define a simplex in E^n as a convex set spanned by the origin and n linearly independent points.)

The theorem is actually correct for any set which is homeomorphic to an n-dimensional simplex.

The fixed-point theorem was generalized by Kakutani to cover point-to-set mappings. Prior to stating this result, we clarify a few terms. A mapping ϕ which transforms every point $\mathbf{x}$ in X into a subset of X is called a *point-to-set* mapping. A point-to-set mapping ϕ is called *upper semicontinuous* if $\mathbf{x}_n \to \mathbf{x}_0$, $\mathbf{y}_n \to \mathbf{y}_0$, $\mathbf{x}_n \in X$, $\mathbf{y}_n \in \phi(\mathbf{x}_n)$ imply $\mathbf{y}_0 \in \phi(\mathbf{x}_0)$.

It is easily verified that a mapping ϕ is upper semicontinuous if and only if the graph $\{\langle \mathbf{x}, \mathbf{y}\rangle | \mathbf{x} \in X, \mathbf{y} \in \phi(\mathbf{x})\}$ is closed in E^{2n}.

A point-to-set mapping $\phi(\mathbf{x})$ is said to be lower semicontinuous if for every $\mathbf{y} \in \phi(\mathbf{x}_0)$ whenever $\mathbf{x}_n \to \mathbf{x}_0$ there exists $\mathbf{y}_n \in \phi(\mathbf{x}_n)$ such that $\mathbf{y}_n \to \mathbf{y}$.

A point-to-set mapping is continuous if it is both lower and upper semicontinuous. The definition of upper semicontinuous for a point-to-set mapping ϕ reduces to that of Section C.1 when ϕ is a point function, provided we define the containing relation $\mathbf{y} \in \phi(\mathbf{x})$ as $\mathbf{y} \leq \phi(\mathbf{x})$. It is also possible to express the properties of point-to-set mappings in terms of neighborhood concepts. For example, $\phi(\mathbf{x})$ is upper semicontinuous at $\mathbf{x}_0$ if corresponding to any open set U containing $\phi(\mathbf{x}_0)$ there exists some $\delta(U) > 0$ such that $d(\mathbf{x}, \mathbf{x}_0) < \delta$ implies $\phi(\mathbf{x}) \subset U$.

Kakutani's fixed-point theorem. Let X be a closed simplex, and let ϕ represent an upper semicontinuous mapping which maps each point of X into a closed convex subset of X. Then there exists a point $\mathbf{x}_0 \in X$ such that $\mathbf{x}_0 \in \phi(\mathbf{x}_0)$.

This fixed-point theorem has been further generalized by Eilenberg and Montgomery [73] and Begle [18].

C.3 Set functions and probability distributions. We review very briefly, without proofs, some of the theory behind the Lebesgue-Stieltjes integral. For full discussions, see Graves [112].

The class of Borel sets. The set of all points that belong to some member of a sequence $\{S_n\}$ is the union of the sequence, and the set of all points that belong to every member of the sequence is called the intersection. These two sets are denoted by

$$\bigcup_{n=1}^{\infty} S_n \quad \text{and} \quad \bigcap_{n=1}^{\infty} S_n,$$

respectively. The set of points that belong to S_1 but do not belong to S_2 is called the difference, and it is denoted by $S_1 - S_2$.

A class $\mathcal{S}$ of subsets of E^k is said to be an additive class of sets if it satisfies the following conditions:

(1) E^k belongs to $\mathcal{S}$.

(2) If $S_1, \ldots, S_n, \ldots$ all belong to $\mathcal{S}$, then

$$\bigcup_{n=1}^{\infty} S_n$$

belongs to $\mathcal{S}$.

(3) If S_1 and S_2 belong to $\mathcal{S}$, then $S_1 - S_2$ belongs to $\mathcal{S}$.

The smallest additive class of sets that contains all the "rectangles" in E^k (or all the intervals if $k = 1$) is called the class of Borel sets of E^k.

Nonnegative set functions. A nonnegative set function P is a real-valued function that is defined on a class of Borel sets $\mathcal{S}$ and satisfies the following conditions:

(1) $P(S) \geq 0 \qquad \text{for all} \quad S \in \mathcal{S}.$

(2) $P\left(\bigcup_{n=1}^{\infty} S_n\right) = \sum_{n=1}^{\infty} P(S_n) \;\; \text{if} \;\; S_n \cap S_m = \phi \;\; \text{for} \;\; n \neq m.$

(3) $P(S) < \infty \qquad \text{if } S \text{ is bounded.}$

The class of all nonnegative set functions defined on the class of Borel sets of real numbers is equivalent to the class of all monotone-increasing functions (defined on the real line) that are continuous from the right; this equivalence is unique if we can identify two increasing functions that differ everywhere by a fixed constant. The equivalence is constructed as follows: Given P and any real number α, define

$$F(x, \alpha) = \begin{cases} P\{\xi | \alpha < \xi \leq x\} & x > \alpha, \\ 0 & x = \alpha, \\ -P\{\xi | x < \xi \leq \alpha\} & x < \alpha. \end{cases}$$

$F(x, \alpha)$ is an increasing function for each α, and it is continuous from the right; furthermore, if $\alpha_1 < \alpha_2$, then

$$F(x, \alpha_1) - F(x, \alpha_2) = P\{\xi | \alpha_1 < \xi \leq \alpha_2\}.$$

That is, $F(x, \alpha_1)$ and $F(x, \alpha_2)$ differ by a constant, independent of x. On the other hand, if $F(x)$ is any monotone-increasing function that is also continuous from the right, one may define

$$P\{\xi | a < \xi \leq b\} = F(b) - F(a).$$

The resulting set function defined on intervals can be extended to a nonnegative set function defined on all Borel sets.

If $P\{\xi|-\infty < \xi < \infty\} = 1$, the set function is called a probability measure, and the corresponding point function F is commonly normalized so that $F(-\infty) = 0$ and $F(\infty) = 1$. The function F, when normalized, is referred to as the (cumulative) distribution function. If $F'(x) = f(x)$ exists, then $f(x)$ is called the probability density function. Clearly,

$$f(x) \geq 0, \qquad \int_{-\infty}^{\infty} f(x)\, dx = 1.$$

Any function f that satisfies the last two conditions is the density function of some probability distribution.

These concepts can be extended to E^n. The probability distributions are the nonnegative set functions P with the property that $P(E^n) = 1$. To each P there corresponds a cumulative distribution function F defined by

$$F(x_1, \ldots, x_n) = P\{\xi \in E^n | \xi_1 \leq x_1, \ldots, \xi_n \leq x_n\},$$

etc. This function F is monotone-increasing in each variable;

$$F(-\infty, x_2, \ldots, x_n) = \cdots = F(x_1, x_2, \ldots, x_{n-1}, -\infty) = 0$$

and

$$F(\infty, \infty, \ldots, \infty) = 1.$$

Furthermore, F is continuous from the right in each variable; and the difference $\Delta_1 \Delta_2 \cdots \Delta_n F$, where by definition

$$\begin{aligned} \Delta_i F = &\ F(a_1, \ldots, a_{i-1}, a_i + \delta_i, a_{i+1}, \ldots, a_n) \\ &- F(a_1, \ldots, a_{i-1}, a_i, a_{i+1}, \ldots, a_n), \end{aligned}$$

is nonnegative since this difference is

$$P\{\xi | a_1 < \xi_1 \leq a_1 + \delta_1, \ldots, a_n < \xi_n \leq a_n + \delta_n\}.$$

Conversely, any function with the last four properties is the cumulative distribution function of some probability distribution.

Marginal distributions. If $F(x_1, \ldots, x_n)$ is a cumulative distribution function defined on E^n, then $F(x_1, \ldots, x_{n-1}, \infty)$ is also a distribution function defined on E^{n-1}. In general, if k of the variables are assigned the value $+\infty$, then F induces a distribution function of the remaining $n - k$ variables, which is called the marginal distribution.

Lebesgue-Stieltjes integral. Let g be a bounded function defined on E^n, and let P be a probability distribution defined on the class of Borel sets of

E^n. In order to define the upper and lower integrals of g with respect to P, we consider first any subdivision of E^n into disjoint Borel sets $E_1, \ldots, E_k$ and define

$$m_i = \inf_{x \in E_i} g(x), \qquad M_i = \sup_{x \in E_i} g(x),$$

$$s = \sum_{i=1}^{k} m_i P(E_i), \qquad S = \sum_{i=1}^{k} M_i P(E_i).$$

The numbers s and S are called the lower and upper Darboux sums of g for the given subdivision. By varying the subdivision one obtains different values for the Darboux sums; if a subdivision is a refinement of another, the lower Darboux sum increases and the upper Darboux sum decreases. The supremum of the lower Darboux sums is called the lower integral of g with respect to p; the infimum of the upper Darboux sums is called the upper integral. If both integrals are equal, then the function g is said to be integrable with respect to P, and its integral is the common value of the upper and lower integrals; this value is usually denoted by

$$\int g(x)\, dP(x).$$

If F is the point function associated with the set function P, the previous notation is commonly replaced by

$$\int g(x)\, dF(x).$$

Besides the obvious additive properties of integrals, we record for reference the Lebesgue criterion of dominated convergence. Let $|f_n| \leq g$ and let g possess an integral with respect to P. If f_n converges everywhere to f (actually a.e. with respect to P would suffice), then

$$\lim_{n \to \infty} \int f_n\, dP = \int f\, dP. \tag{C.3.1}$$

If f_n forms a monotone-increasing sequence of integrable functions with no other restrictions and P is a positive measure, then the same conclusion holds, provided that the value infinity is allowed.

Sequences of distributions. Weak convergence.* A sequence $\{F_n\}$ of distribution functions converges weak* to a distribution function F if $F_n(x) \to F_0(x)$ at every point of continuity of F_0. Every sequence of distribution functions contains a subsequence F_{n_k} that converges weak* to a function F, but F need not be a distribution function. However, if

g is continuous $g(\pm\infty) = 0$, and $F_{n_k} \to F$ (weak*), then

$$\lim_{k\to\infty} \int g(x)\, dF_{n_k}(x) = \int g(x)\, dF(x). \tag{C.3.2}$$

If, in addition, F is a distribution function, the convergence of (C.3.2) is valid for any g which is bounded and continuous.

These results are known as *Helly's convergence theorem.*

BIBLIOGRAPHY

Each reference work is listed once under the name of the first author as the article or book appears in the literature. At the end of the bibliography, each co-author is cited for the articles or books not previously listed directly under his name. In this case, the relevant reference is indicated by the appropriate numbered article in the bibliography.

[1] ALLAIS, M., *Traité d'economie pure,* Imprimerie Nationale, Paris, (1952), 852 pp.

[2] ALLEN, R. G. D., *Mathematical Economics,* London: Macmillan, 1956.

[3] ARROW, K. J., "Alternative proof of the substitution theory for Leontief models in the general case," Chap. 9 of [146].

[4] ARROW, K. J., "An extension of the basic theorems of classical welfare economics," in [200], pp. 507–32.

[5] ARROW, K. J., E. W. BARANKIN, and D. BLACKWELL, "Admissible points of convex sets," in [162], pp. 87–91.

[6] ARROW, K. J., D. BLOCK, and L. HURWICZ, "On the stability of the competitive equilibrium, II," *Econometrica,* **27** (1959), 82–109.

[7] ARROW, K. J., and G. DEBREU, "Existence of an equilibrium for a competitive economy," *Econometrica,* **22** (1954), 265–90.

[8] ARROW, K. J., T. E. HARRIS, and J. MARSCHAK, "Optimal inventory policy," *Econometrica,* **19** (1951), 250–72.

[9] ARROW, K. J., and L. HURWICZ, "Reduction of constrained maxima to saddle-point problems," in [201], V, 1–20.

[10] ARROW, K. J., and L. HURWICZ, "The gradient method for concave programming," Chap. 6 of [12].

[11] ARROW, K. J., and L. HURWICZ, "On the stability of the competitive equilibrium, I," *Econometrica,* **26** (1958), 522–552.

[12] ARROW, K. J., L. HURWICZ, and H. UZAWA (eds.), *Studies in Linear and Non-Linear Programming,* Stanford, Calif.: Stanford University Press, 1958.

[13] ARROW, K. J., and M. MCMANUS, "A note on dynamic stability," *Econometrica,* **26** (1958), 448–54.

[14] ARROW, K. J., and M. NERLOVE, "A note on expectation and stability," *Econometrica,* **26** (1958), 297–305.

[15] ARROW, K. J., and S. KARLIN, "Price speculation under certainty," Chap. 13 of [12].

[16] ARROW, K. J., S. KARLIN, and H. SCARF, *Studies in the Mathematical Theory of Inventory and Production,* Stanford, Calif.: Stanford University Press, 1958.

[17] BALDWIN, R. R., W. E. CANTEY, H. MAISEL, and J. P. MCDERMOTT, "The Optimum Strategy in Blackjack," *J. Amer. Stat. Assoc.,* **51** (1956), 429–39.

[18] BEGLE, E. G., "A fixed point theorem," *Ann. Math.,* **51** (1950), 544–50.

[19] BELLMAN, R., *Dynamic Programming,* Princeton, N. J.: Princeton University Press, 1957.

[20] BELLMAN, R., "On games involving bluffing," *Rendiconti del Circolo Mathematico di Palermo*, Series 2, Vol. 1 (1952), 139–56.

[21] BELLMAN, R., and M. A. GIRSHICK, "An extension of results on duels with two opponents, one bullet each, silent guns, equal accuracy," The RAND Corporation, D-403, 1949.

[22] BELLMAN, R., I. GLICKSBERG, and O. GROSS, "On some variational problems occurring in the theory of dynamic programming," *Rendiconti del Circolo Mathematico di Palermo*, Series 2, Vol. 3 (1954), 1–35.

[23] BELLMAN, R., and M. SHIFFMAN, "On the Min-Max of

$$\int_0^1 f(x)a(x)\, d(x)\, dt(x),\text{"}$$

The RAND Corporation, RM-308-1, 1949.

[24] BELZER, R. L., "Silent duels, specified accuracies, one bullet each," The RAND Corporation, RAD(L)-301, 1948.

[25] BERGE, C., "Sur une théorie ensembliste des jeux alternatifs," *Journal de mathématiques pures et appliquées*, **32** (1953), 129–84.

[26] BIRKOFF, G., and S. MACLANE, *A Survey of Modern Algebra*, New York: Macmillan, 1941.

[27] BLACKETT, D. W., "Some blotto games," *Naval Research Logistics Quarterly*, **1** (1954), 55–60.

[28] BLACKWELL, D., "The silent duel, one bullet each, arbitrary accuracy," The RAND Corporation, RM-302, 1948.

[29] BLACKWELL, D., "The noisy duel, one bullet each, arbitrary, non-monotone accuracy," The RAND Corporation, D-442, 1949.

[30] BLACKWELL, D., and M. A. GIRSHICK, *Theory of Games and Statistical Decisions*, New York: Wiley, 1954.

[31] BLACKWELL, D., and M. A. GIRSHICK, "A loud duel with equal accuracy where each duelist has only a probability of possessing a bullet," The RAND Corporation, RM-219, 1949.

[32] BLACKWELL, D., and M. SHIFFMAN, "A bomber-fighter duel," The RAND Corporation, D-509, 1949.

[33] BLACKWELL, D., and M. SHIFFMAN, "A bomber-fighter duel," The RAND Corporation, RM-193, 1949.

[34] BOHNENBLUST, H. F., "The Theory of Games," E. F. BECKENBACH (ed.), *Modern Mathematics for the Engineer*, New York: McGraw-Hill, 1956.

[35] BOHNENBLUST, H. F., and S. KARLIN, "On a theorem of Ville," in [161], pp. 155–61.

[36] BOHNENBLUST, H. F., S. KARLIN, and L. S. SHAPLEY, "Solutions of discrete two-person games," in [161], pp. 51–73.

[37] BOHNENBLUST, H. F., S. KARLIN, and L. S. SHAPLEY, "Games with continuous, convex pay-off," in [161], pp. 181–92.

[38] BOREL, E., "Applications aux jeux de hasard," *Traité du calcul des probabilités et de ses applications*, Paris: Gauthier-Villars, 1938.

[39] BOREL, E., "The theory of play and integral equations with skew symmetrical kernels," "On games that involve chance and the skill of the players,"

and "On systems of linear forms of skew symmetric determinants and the general theory of play," trans. L. J. Savage, *Econometrica*, **21** (1953), 97–117.

[40] Brown, G. W., "Some notes on computation of games solutions," The RAND Corporation, D-436, 1949.

[41] Brown, G. W., "Iterative solution of games by fictitious play," Chap. 24 of [146].

[42] Brown, G. W., and J. von Neumann, "Solutions of games by differential equations," in [161], pp. 73–79.

[43] Brown, R. H., "The solution of a certain two-person zero-sum game," *Operations Research*, **5** (1957), 63–67.

[44] Charnes, A., "Optimality and degeneracy in linear programming," *Econometrica*, **20** (1952), 160–70.

[45] Charnes, A., and W. W. Cooper, "Generalizations of the warehousing model," *Operations Research Quarterly*, **6** (1955), 131–72.

[46] Charnes, A., W. W. Cooper, and A. Henderson, *Introduction to Linear Programming*, New York: Wiley, 1953.

[47] Chenery, H. B., "The role of industrialization in development programs," *Amer. Econ. Rev.*, **45** (1955), 40–56.

[48] Chenery, H. B., and K. S. Kretchmer, "Resource allocation for economic development," *Econometrica*, **24** (1956), 365–99.

[49] Chenery, H. B., and H. Uzawa, "Non-linear programming in economic development," Chap. 15 of [12].

[50] Churchman, C. W., R. L. Ackoff, and E. L. Arnoff, *Introduction to Operations Research*, New York: Wiley, 1957.

[51] Danskin, J. M., "Fictitious play for continuous games," *Naval Research Logistics Quarterly*, **1** (1954), 313–20.

[52] Danskin, J. M., "Mathematical treatment of a stockpiling problem," *Naval Research Logistics Quarterly*, **2** (1955), 99–109.

[53] Danskin, J. M., and L. Gillman, "Explicit solution of a game over function space," The RAND Corporation, P-235, 1951.

[54] Danskin, J. M., and L. Gillman, "A game over function space," *Rivista Mat. Univ. Parma*, **4** (1953), 83–94.

[55] Dantzig, G. B., "A proof of equivalence of the programming problem and the game problem," Chap. 20 of [146].

[56] Dantzig, G. B., "Maximization of a linear function of variables subject to linear inequalities," Chap. 21 of [146].

[57] Dantzig, G. B., *The Theory of Mathematical Programming*, notes, The RAND Corporation.

[58] Dantzig, G. B., and A. J. Hoffman, "Dilworth's theorem on partially ordered sets," in [163], pp. 207–14.

[59] Dantzig, G. B., A. Orden, and P. Wolfe, "The generalized simplex method for minimizing a linear form under linear inequality restraints," *Pac. J. Math.*, **5** (1955), 183–95.

[60] Dantzig, G. B., and A. Wald, "On the fundamental lemma of Neyman and Pearson," *Ann. Math. Stat.*, **22** (1951), 87–93.

[61] Debreu, G., "A social equilibrium existence theorem," *Proc. Nat. Acad. Sci.*, **38** (1952), 886–93.

[62] DEBREU, G., "Valuation equilibrium and Pareto optimum," *Proc. Nat. Acad. Sci.*, **40** (1954), 588–92.

[63] DEBREU, G., and I. N. HERSTEIN, "Non-negative square matrices," *Econometrica*, **21** (1953), 597–607.

[64] DORFMAN, R., *Application of Linear Programming to the Theory of the Firm*, Berkeley and Los Angeles: University of California Press, 1951.

[65] DORFMAN, R., P. A. SAMUELSON, and R. M. SOLOW, *Linear Programming and Economic Analysis*, New York: McGraw-Hill, 1958.

[66] DRESHER, M., *Theory and Applications of Games of Strategy*, The RAND Corporation, R-216 (1951).

[67] DRESHER, M., "Methods of solution in game theory," *Econometrica*, **18** (1950), 179–81.

[68] DRESHER, M., "Games of strategy," *Mathematics Magazine*, **25** (1951), 93–99.

[69] DRESHER, M., "Le Her," RAND report (unpublished).

[70] DRESHER, M., and S. KARLIN, "Solutions of convex games as fixed points," in [162], pp. 75–86.

[71] DRESHER, M., S. KARLIN, and L. S. SHAPLEY, "Polynomial games," in [161], pp. 161–180.

[72] DRESHER, M., A. W. TUCKER, and P. WOLFE (eds.), *Contributions to the theory of Games, III*, (*Ann. Math. Studies*, Vol. 39), Princeton, N. J.: Princeton University Press, 1957.

[73] EILENBERG, S., and D. MONTGOMERY, "Fixed point theorems for multi-valued transformations," *Amer. J. Math.*, **68** (1946), 214–22.

[74] EISEMANN, K., "Linear programming," *Quart. Appl. Math.*, **13** (1955), 209–32.

[75] ENTHOVEN, A. C., and K. J. ARROW, "A theorem on expectations and the stability of equilibrium," *Econometrica*, **24** (1956), 288–93.

[76] FAN, K., "Fixed point and minimax theorems in locally convex topological linear spaces," *Proc. Nat. Acad. Sci. U. S. A.*, **38** (1952), 121–26.

[77] FAN, K., "Minimax theorems," *Proc. Nat. Acad. Sci. U. S. A.*, **39** (1953), 42–47.

[78] FAN, K., I. GLICKSBERG, and A. J. HOFFMAN, "Systems of inequalities involving convex functions," *Proc. Amer. Math. Soc.*, **8** (1957), 617–22.

[79] FARKAS, J., "Theorie der einfachen Ungleichungen," *Journal für reine und angewandte Mathematik*, **124** (1901), 1–27.

[80] FELLER, W., *An Introduction to Probability Theory and its Applications, I*, New York: Wiley, 1950.

[81] FENCHEL, W., "Convex Cones, Sets, and Functions," Lecture notes, Department of Mathematics, Princeton University, 1953.

[82] FERGUSON, R. O., and L. F. SARGENT, *Linear Programming*, New York: McGraw-Hill, 1958.

[83] FLEMING, W. H., "On a class of games over function space and related variation problems," *Ann. Math.*, **60** (1954), 578–94.

[84] FLOOD, M. M., "On the Hitchcock distribution problem," *Pac. J. Math.*, **3** (1953), 369–86.

[85] FORD, L. R., JR., and D. R. FULKERSON, "Maximal flow through a network," *Canadian J. Math.*, **8** (1956), 399–404.

[86] FORD, L. R., JR., and D. R. FULKERSON, "A simple algorithm for finding maximal network flows and an application to the Hitchcock Problem," *Canadian J. Math.*, **9** (1957), 210–18.

[87] FORD, L. R., JR., and D. R. FULKERSON, "A primal dual algorithm for the capacitated Hitchcock problem," *Naval Research Logistics Quarterly*, **4** (1957), 47–54.

[88] FRANK, M., and P. WOLFE, "An algorithm for quadratic programming," *Naval Research Logistics Quarterly*, **3** (1956), 95–110.

(89) FROBENIUS, G., "Über Matrizen aus positiven Elementen," *Sitzungsberichte*, **8** (1908), 471–76.

[90] FULKERSON, D. R., "Hitchcock transportation problem," The RAND Corporation, P-890, 1956.

[91] FULKERSON, D. R., "Notes on linear programming. Part XLV. A network-flow feasibility theorem and combinatorial applications," The RAND Corporation, RM-2159, 1958.

[92] GADDUM, J. W., A. J. HOFFMAN, and D. SOKOLOWSKY, "On the solution of the caterer problem," *Naval Research Logistics Quarterly*, **1** (1954), 223–29.

[93] GALE, D., "The law of supply and demand," *Mathematica Scandinavica*, **3** (1955), 155–69.

[94] GALE, D., "The closed linear model of production," in [163], pp. 285–303.

[95] GALE, D., "Information in games with finite resources," in [72], pp. 141–45.

[96] GALE, D., Lecture notes, Brown University (1957).

[97] GALE, D., and O. GROSS, "A note on polynomial and separable games," The RAND Corporation, P-1216, 1957.

[98] GALE, D., H. W. KUHN, and A. W. TUCKER, "On symmetric games," in [161], pp. 81–87.

[99] GALE, D., H. W. KUHN, and A. W. TUCKER, "Linear programming and the theory of games," Chap. 19 of [146].

[100] GALE, D., and S. SHERMAN, "Solutions of finite two-person games," in [161], pp. 37–49.

[101] GANTMACHER, F., *Theory of Matrices* (in Russian), Moscow, 1954.

[102] GANTMACHER, F., and M. KREIN, *Oscillatory Matrices and Kernels and Small Vibrations of Mechanical Systems*, 2d ed. (in Russian), Moscow, 1950.

[103] GASS, S. I., *Linear Programming, Methods and Applications*, New York: McGraw-Hill, 1958.

[104] GEORGESCU-ROEGEN, N., "The aggregate linear production function and its application to von Neumann's economic model," Chap. 4 of [146].

[105] GILLIES, D. B., J. P. MAYBERRY, and J. VON NEUMANN, "Two variants of poker," in [162] pp. 13–50.

[106] GILLMAN, L., "Operations analysis and the theory of games: an advertising example," *J. Amer. Stat. Assoc.*, **45** (1950), 541–46.

[107] GLICKSBERG, I., "Noisy duel, one bullet each, with simultaneous fire and unequal worths," The RAND Corporation, RM-474, 1950.

[108] GLICKSBERG, I., "A further generalization of the Kakutani fixed point theorem, with application to Nash equilibrium points," *Proc. Amer. Math. Soc.*, **3** (1952), 170–74.

[109] GLICKSBERG, I., and O. GROSS, "Butterfly games," The RAND Corporation, RM-655, 1951.

[110] GLICKSBERG, I., and O. GROSS, "Notes on games over the square," in [162], pp. 173–84.

[111] GOLDMAN, A. J., and A. W. TUCKER, "Theory of linear programming," in [163], pp. 53–97.

[112] GRAVES, L. M., *The Theory of Functions of Real Variables*, New York: McGraw-Hill, 1946.

[113] GROSS, O., "A rational payoff characterization of the Cantor distribution," The RAND Corporation, D-1349, 1952.

[114] GROSS, O., "The derivatives of the value of a game," The RAND Corporation, RM-1286, 1954.

[115] GROSS, O., "A simple linear programming problem explicitly solvable in integers," The RAND Corporation, RM-1560, 1955.

[116] GROSS, O., "A rational game on the square," in [72], pp. 307–11.

[117] HAHN, F. H., "Gross substitutes and the dynamic stability of general equilibrium," *Econometrica*, **26** (1958), 169–70.

[118] HARDY, G. H., J. E. LITTLEWOOD, and G. PÓLYA, *Inequalities*, Cambridge, England: Cambridge University Press, 1952.

[119] HAWKINS, D., and H. A. SIMON, "Note: Some conditions of macroeconomic stability," *Econometrica*, **17** (1949), 245–48.

[120] HELMER, O., "Open problems in game theory," *Econometrica*, **20** (1952), 90 abstract.

[121] HICKS, J. R., *Value and Capital*, 2d ed., Oxford: Oxford University Press, 1953.

[122] HILDRETH, C., "A quadratic programming procedure," *Naval Research Logistics Quarterly*, **4** (1957), 79–85.

[123] HIRSCH, W., and G. B. DANTZIG, "The fixed charge problem," The RAND Corporation, RM-1383, 1954.

[124] HITCHCOCK, F. L., "Distribution of a product from several sources to numerous localities," *J. Math. Phys.*, **20** (1941), 224–30.

[125] HOHN, F. E., and F. MODIGLIANI, "Production planning over time and the nature of the expectation and planning horizon," *Econometrica*, **23** (1955), 46–66.

[126] HOUTHAKKER, H. S., "Revealed preference and the utility function," *Economica*, **17** (1950), 159–74.

[127] HOUTHAKKER, H. S., "La forme des courbes d'Engel," *Cahiers du Séminaire d'Économétrie*, (1953), 59–66.

[128] HURWICZ, L., "Programming in linear spaces," Chap. 4 of [12].

[129] JACOBS, W. W., "The caterer problem," *Naval Research Logistics Quarterly*, **1** (1954), 154–65.

[130] KAKUTANI, S., "A generalization of Brouwer's fixed point theorem," *Duke Math. Journal*, **8** (1941), 457–58.

[131] KANTOROVITCH, L., "On the transformation of masses," *Dokl. Akad. Nauk USSR*, **37** (1942), 199–201.

[132] KAPLANSKY, I., "A contribution to von Neumann's theory of games," *Ann. Math.*, **46** (1945), 474–79.

[133] KARLIN, S., "Operator treatment of minmax principle," in [161], pp. 133–54.

[134] KARLIN, S., "Reduction of certain classes of games to integral equations," in [162], pp. 125–58.

[135] KARLIN, S., "The theory of infinite games," *Ann. Math.*, **58** (1953), 371–401.

[136] KARLIN, S., "Polya type distributions, II," *Ann. Math. Stat.*, **28** (1957), 281–308.

[137] KARLIN, S., "On games described by bell-shaped kernels," in [72], pp. 365–91.

[138] KARLIN, S., and R. RESTREPO, "Multistage poker models," in [72], pp. 337–63.

[139] KARLIN, S., "Positive Operators," *Journal of Math. and Mech.*, (1960).

[140] KARLIN, S., and L. S. SHAPLEY, "Some applications of a theorem on convex functions," *Ann. Math.*, **52** (1950), 148–53.

[141] KARLIN, S., and L. S. SHAPLEY, "Geometry of Moment Spaces," *Memoirs of the American Mathematics Society*, **12** (1953), 1–93.

[142] KEMENY, J. G., O. MORGENSTERN, and G. L. THOMPSON, "A generalization of the von Neumann model of an expanding economy," *Econometrica*, **24** (1956), 115–35.

[143] KIEFER, J., "Invariance, minimax sequential estimation and continuous time processes," *Ann. Math. Stat.*, **28** (1957), 573–601.

[144] KNIGHT, F. H., "A note on Professor Clark's illustration of marginal productivity," *Journal of Political Economics*, **33** (1925), 550–53.

[145] KOOPMAN, B. O., "The theory of search, I: kinematic bases," *Operations Research*, **4** (1956), 324–46.

[146] KOOPMANS, T. C. (ed.), *Activity Analysis of Production and Allocation*, Cowles Commission Monograph No. 13, New York: Wiley, 1951.

[147] KOOPMANS, T. C., *Three Essays on the State of Economic Analysis*, New York: McGraw-Hill, 1957.

[148] KOOPMANS, T. C., "Alternative proof of the substitution theorem for Leontief models in the case of three industries," Chap. 8 of [146].

[149] KOOPMANS, T. C., "Water storage policy in a simplified hydroelectric system," in *Proc. International Conference on Operations Research*, Bristol, England: John Wright and Sons, 1958.

[150] KOOPMANS, T. C., and S. REITER, "A model of transportation," Chap 14 of [146].

[151] KOTELYANSKII, D. M., "On some properties of matrices with positive elements," (in Russian), *Mat. Sbornik* (*N. S.*), **31** (1952), 497–506.

[152] KOTELYANSKII, D. M., "On a property of sign-symmetric matrices," *Uspehi Matem. Nauk* (*N. S.*), **8** (1953), 163–67.

[153] KREIN, M. G., and M. A. RUTMAN, "Linear operators leaving invariant a

cone in a Banach space," *Uspehi Matem. Nauk* (*N. S.*), **3** (1948), 3–95 (Amer. Math. Soc. Translation No. 26, New York, 1950).

[154] KRETCHMER, K. S., "Linear programming in locally convex spaces and its use in analysis," Ph.D. thesis, Carnegie Inst. of Tech.

[155] KUDO, H., "On minimax invariant estimates of the transformation parameter," *National Science Report of the Ochanomizu University*, **6** (1955), 31–73.

[156] KUHN, H. W., "A simplified two-person poker," in [161], pp. 97–103.

[157] KUHN, H. W., "A combinatorial algorithm for the assignment problem," Issue 11, Logistics Papers, George Washington University Logistics Research Project, 1954.

[158] KUHN, H. W., "The Hungarian method for the assignment problem," *Naval Research Logistics Quarterly*, **2** (1955), 83–97.

[159] KUHN, H. W., "On a theorem of Wald," in [163], pp. 265–73.

[160] KUHN, H. W., "Variants of the Hungarian method for assignment problems," *Naval Research Logistics Quarterly*, **3** (1956), 253–58.

[161] KUHN, H. W., and A. W. TUCKER (eds.), *Contributions to the Theory of Games, I*, (*Ann. Math. Studies*, Vol. 24), Princeton, N. J.: Princeton University Press, 1950.

[162] KUHN, H. W., and A. W. TUCKER (eds.), *Contributions to the Theory of Games, II*, (*Ann. Math. Studies*, Vol. 28), Princeton, N. J.: Princeton University Press, 1953.

[163] KUHN, H. W., and A. W. TUCKER (eds.), *Linear Inequalities and Related Systems*, (*Ann. Math. Studies*, Vol. 38), Princeton, N. J.: Princeton University Press, 1956.

[164] KUHN, H. W., and A. W. TUCKER, "Non-linear programming," in [200], pp. 481–92.

[165] LEMKE, C. E., "The dual method of solving the linear programming problem," *Naval Research Logistics Quarterly*, **1** (1954), 36–47.

[166] LEONTIEF, W. W., *The Structure of the American Economy*, Cambridge, Mass: Harvard University Press, 1941.

[167] *Linear Programming*, Vols. I and II, Second Symposium, National Bureau of Standards, U. S. Department of Commerce, January 27–29, 1955.

[168] LOOMIS, L. H., "On a theorem of von Neumann," *Proc. Nat. Acad. Sci. U. S. A.*, **32** (1946), 213–15.

[169] LUCE, R. D., and H. Raiffa, *Games and Decisions*, New York: Wiley, 1957.

[170] LUCE, R. D., and A. W. TUCKER (eds.), *Contributions to the Theory of Games, IV* (*Ann. Math. Studies*, Vol. 40), Princeton, N. J.: Princeton University Press, 1959.

[171] MANNE, A. S., *Scheduling of Petroleum Refinery Operations*, Cambridge, Mass.: Harvard University Press, 1956.

[172] MARKOWITZ, H. M., "Portfolio selection," *Journal of Finance*, **7** (1952), 77–91.

[173] MARKOWITZ, H. M., "The optimization of a quadratic function subject to linear constraints," *Naval Research Logistics Quarterly*, **3** (1956), 111–13.

[174] MARKOWITZ, H. M., and A. S. MANNE, "On the solution of discrete programming problems," *Econometrica*, **25** (1957), 84–110.

[175] McCloskey, J. F., and F. N. Trefethen, *Operations Research for Management*, Baltimore: Johns Hopkins University Press, 1954.

[176] McDonald, J., *Strategy in Poker, Business and War*, New York: Norton, 1950.

[177] McKenzie, L. W., "Competitive equilibrium with dependent consumer preferences," in [167], 277–94.

[178] McKenzie, L. W., "Demand theory without a utility index," *Review of Economic Studies*, **24** (1956–57), 185–89.

[179] McKinsey, J. C. C., *Introduction to the Theory of Games*, New York: McGraw-Hill, 1952.

[180] Metzler, L. A., "Stability of multiple markets: the Hicks conditions," *Econometrica*, **13** (1945), 277–92.

[181] Mills, H. D., "Marginal values of matrix games and linear programs," in [163], pp. 183–93.

[182] Morgenstern, O. (ed.), *Economic Activity Analysis*, New York: Wiley, 1954.

[183] Morishima, M., "On the laws of change of the price system in an economy which contains complementary commodities," *Osaka Economics Papers*, **1** (1952), 101–13.

[184] Morishima, M., "Prices, interest and profits in a dynamic Leontief system," *Econometrica*, **26** (1958), 358–80.

[185] Mosak, J. L., *General Equilibrium Theory in International Trade*, Bloomington, Indiana: The Principia Press, 1944.

[186] Motzkin, T. S., *Beiträge zur Theorie der linearen Ungleichungen*, [Inaugural Dissertation] Jerusalem, 1936.

[187] Motzkin, T. S., H. Raiffa, G. L. Thompson, and R. M. Thrall, "The double description method," in [162], pp. 51–73.

[188] Nash, J. F., "Equilibrium points in N-person games," *Proc. Nat. Acad. Sci., U. S. A.*, **36** (1950), 48–49.

[189] Nash, J. F., and L. S. Shapley, "A simple three-person poker game," in [161], pp. 105–16.

[190] Neyman, J., and E. Pearson, "On the problem of the most efficient tests of statistical hypotheses," *Philosophical Transactions of the Royal Society*, **A-231** (1933), 289–337.

[191] Nikaido, H., "On a minimax theorem and its applications to functional analysis," *Journal of the Mathematical Society of Japan*, **5** (1953), 86–94.

[192] Nikaido, H., "Note on the general economic equilibrium for non-linear production functions," *Econometrica*, **22** (1954), 49–53.

[193] Nikaido, H., "On von Neumann's minimax theorem," *Pac. J. Math.*, **4** (1954), 65–72.

[194] Peisakoff, M. P., "Transformation parameters," Ph.D. thesis, Princeton University, 1950.

[195] Perlis, S., *Theory of Matrices*, Cambridge, Mass.: Addison-Wesley, 1952.

[196] Pitman, E. J. G., "The estimation of location and scale parameters of a continuous population of any given form," *Biometrika*, **30** (1939), 391–421.

[197] Pitman, E. J. G., "Tests of hypotheses concerning location and scale parameters," *Biometrika*, **31** (1939), 200–215.

[198] Prager, W., "On the role of congestion in transportation problems," *Zeit. für angewandte Math. Mech.*, **35** (1955), 381–96.

[199] Prager, W., "On the caterer problem," *Management Science*, **3** (1956), 15–23.

[200] *Proceedings of the Second Berkeley Symposium on Mathematical Statistics and Probability*, Berkeley and Los Angeles: University of California Press, 1951.

[201] *Proceedings of the Third Berkeley Symposium on Mathematical Statistics and Probability*, Berkeley and Los Angeles: University of California Press, 1955. 5 vols.

[202] Rademacher, H., and I. J. Schoenberg, "Helly's theorems on convex domains and Tchebycheff's approximation problem," *Canadian Math.*, **2** (1950), 245–56.

[203] Reinfeld, N. V., and W. R. Vogel, *Mathematical Programming*, Englewood Cliffs, N. J.: Prentice-Hall, 1958.

[204] Reiter, S., "Efficiency and prices in the theory of an international economy," Technical Report No. 13, Stanford University, 1954.

[205] Restrepo, R., "Tactical problems involving several actions," in [72], pp. 313–35.

[206] Restrepo, R., "Games with a random move," unpublished.

[207] Riley, V., and S. I. Gass, *A Bibliography of Linear Programming and Associated Techniques*, Baltimore: Johns Hopkins University Press, 1958.

[208] Robinson, J., "An iterative method of solving a game," *Ann. Math.*, **54** (1951), 296–301.

[209] Rudin, W., *Principles of Mathematical Analysis*, New York: McGraw-Hill, 1953.

[210] Samuelson, P. A., *Foundations of Economic Analysis*, Cambridge, Mass.: Harvard University Press, 1955.

[211] Samuelson, P. A., "A note on the pure theory of consumers' behavior," *Economica*, **18** (1938), 61–71, 353–54.

[212] Samuelson, P. A., "Frank Knight's theorem in linear programming," The RAND Corporation, D-782, 1950 (hectographed).

[213] Samuelson, P. A., "Abstract of a theorem concerning substitutability in open Leontief models," Chap. 7 of [146].

[214] Savage, I. R., "Cycling," *Naval Research Logistics Quarterly*, **3** (1956), 163–75.

[215] Schoenberg, I. J., "On smoothing operations and their generating functions," *Bull. Amer. Math. Soc.*, **59** (1953), 199–230.

[216] Shapley, L. S., *The Theory of n-Person Games*, unpublished notes.

[217] Shapley, L. S., "The silent duel, one bullet *vs.* two, equal accuracy," The RAND Corporation, RM-445, 1950.

[218] Shapley, L. S., "An example of an infinite non-constant-sum game," The RAND Corporation, RM-898, 1952.

[219] Shapley, L. S., and R. N. Snow, "Basic solutions of discrete games," in [161], pp. 27–37.

[220] Sherman, S., "Games and subgames," *Proc. Amer. Math. Soc.*, **2** (1951), 186–87.

[221] Shiffman, M., "On the equality of min max = max min, and the theory of games," The RAND Corporation, RM-243, 1941.

[222] Shiffman, M., "Games of timing," in [162], pp. 97–123.

[223] Sion, M., and P. Wolfe, "On a game without a value," in [72], pp. 299–305.

[224] Sion, M., "On general minimax theorems," *Pac. J. Math.*, **8** (1958), 171–76.

[225] Slutsky, E., "Sulla teoria del bilancio del consumatore," *Giornale degli Economisti*, **51** (1915), 19–23.

[226] Smith, W. E., "Various optimizers for single-stage production," *Naval Research Logistics Quarterly*, **3** (1956), 59–66.

[227] Solow, R. M., "On the structure of linear models," *Econometrica*, **20** (1952), 29–46.

[228] Solow, R. M., and P. A. Samuelson, "Balanced growth under constant returns to scales," *Econometrica*, **21** (1953), 412–24.

[229] Stigler, G. J., "The cost of substance," *Journal of Farm Economics*, **27** (1945), 303–14.

[230] Suits, D. B., "Dynamic growth under diminishing returns to scales," *Econometrica*, **22** (1954), 496–501.

[231] Thompson, G. L., "On the solution of a game-theoretic problem," in [163], pp. 275–84.

[232] Uzawa, H., "On the logical relation between preference and revealed preference," Technical Report No. 38, Department of Economics, Stanford University, 1956.

[233] Uzawa, H., "A note on the stability of equilibrium," Technical Report No. 44, Department of Economics, Stanford University, 1957.

[234] Uzawa, H., "A note on the Menger-Wieser theory of imputation," Technical Report No. 45, Department of Economics, Stanford University, 1957.

[235] Uzawa, H., "The gradient method for concave programming, II. Global results," Chap. 7 of [12].

[236] Vajda, S., *Theory of Games and Linear Programming*, New York: Wiley, 1956.

[237] Ville, J., "Note sur la théorie générale des jeux où intervient l'habilité des jouers," in "Applications aux jeux de hasard," by E. Borel and J. Ville, Tome IV, Fascicule II, in Borel [1938].

[238] Von Neumann, J., "Zur Theorie der Gesellschaftsspiele," *Mathematische Annalen*, **100** (1928), 295–320.

[239] Von Neumann, J., "Über ein ökonomisches Gleichungssystem und eine Verallgemeinerung des Brouwerschen Fixpunktsatzes," *Ergebnisse eines Mathematischen Kolloquiums*, **8** (1937), 73–83.

[240] Von Neumann, J., "A certain zero-sum two-person game equivalent to the optimum assignment problem," in [162], pp. 5–12.

[241] Von Neumann, J., "A numerical method to determine optimum strategy," *Naval Research Logistics Quarterly*, **1** (1954), 109–15.

[242] Von Neumann, J., and O. Morgenstern, *Theory of Games and Economic Behavior*, 2d ed., Princeton, N. J.: Princeton University Press, 1947 (1st ed., 1944).

[243] Wagner, H., "A practical guide to the dual theorem," *Operations Research*, **6** (1958), 364–84.

[244] Wald, A., *Sequential Analysis*, New York: Wiley, 1947.

[245] Wald, A., *Statistical Decision Functions*, New York: Wiley, 1950.

[246] Wald, A., "Über die eindeutige positive Lösbarkeit der neuen Produktionsgleichungen," *Ergebnisse eines mathematischen Kolloquiums*, **6** (1935), 12–20.

[247] Wald, A., "Über die Produktionsgleichungen der ökonomischen Wertlehre, II," *Ergebnisse eines mathematischen Kolloquiums*, **7** (1936), 1–6.

[248] Wald, A., "Statistical decision functions which minimize the maximum risk," *Ann. Math.*, **46** (1945), 265–80.

[249] Wald, A., "Statistical decision theory," *Ann. Math. Stat.*, **20** (1949), 165–205.

[250] Wald, A., "Note on zero-sum two-person games," *Ann. Math.*, **52** (1950), 739–42.

[251] Wald, A., "On some systems of equations of mathematical economics," *Econometrica*, **19** (1951), 368–403.

[252] Walras, L., *Elements of Pure Economics, or the Theory of Social Wealth*, trans. W. Jaffe, Homewood, Ill.: Irwin, 1954.

[253] Weyl, H., "The elementary theory of convex polyhedra," in [161], pp. 3–18.

[254] Williams, J. D., *The Compleat Strategyst*, New York: McGraw-Hill, 1954.

[255] Wolfe, P., "Determinateness of polyhedral games," in [163], pp. 195–98.

[256] Wolfe, P., "The simplex method for quadratic programming," The RAND Corporation, P-1295, 1957.

[257] Wood, M. K., and G. B. Dantzig, "Programming of interdependent activities, I. General discussion," *Econometrica*, **17** (1949), 193–99. (Reprinted in revised form as Chap. 1 of [146].)

Ackoff, R. L., see [50]
Arnoff, E. L., see [50]
Arrow, K. J., see [75]
Barankin, E. W., see [5]
Blackwell, D., see [5]
Block, D., see [5]
Cantey, W. E., see [17]
Cooper, W. W., see [45], [46]
Dantzig, G. B., see [123], [257]
Debreu, G., see [7]
Fulkerson, D. R., see [85], [86], [87]
Gass, S. I., see [207]
Gillman, L., see [53], [54]
Girschick, M. A., see [21], [30], [31], [78]
Glicksberg, I., see [22], [78]
Gross, O., see [22], [97], [109], [110]
Harris, T. E., see [8]
Henderson, A., see [46]
Hoffman, A. J., see [58], [78], [92]
Hurwicz, L., see [6], [9], [10], [11], [12]
Karlin, S., see [15], [16], [35], [36], [37], [70], [71]
Krein, M., see [102]
Kretchmer, K. S., see [48]
Kuhn, H. W., see [98], [99]
Littlewood, J. E., see [118]
MacLane, S., see [26]
Maisel, H., see [17]
Manne, A. S., see [174]
Marschak, J., see [8]
Mayberry, J. P., see [105]
McDermott, J. P., see [17]
McManus, M., see [13]
Modigliani, F., see [125]
Morgenstern, O., see [142], [242]

INDEX

INDEX

VOLUME II:
THE THEORY OF INFINITE GAMES

NOTE TO THE READER

This volume differs from Volume I in only one essential respect: the dimensionality of the strategy spaces available to the players of the game. In Volume I our strategy spaces were finite-dimensional; in this volume they are infinite-dimensional.

In order that this volume may be studied independently of Volume I, the essential background material from Volume I is reproduced here in its entirety: i.e., Chapter 1, which presents the underlying concepts of game theory in their simplest form and introduces the basic notation; and the appendixes, which review matrix theory, the properties of convex sets, and miscellaneous topics of function theory.

The organization of this volume is similar in all respects to that of Volume I.

SAMUEL KARLIN

Stanford, California
August 1959

LOGICAL INTERDEPENDENCE OF THE VARIOUS CHAPTERS

(*applies only to unstarred sections of each chapter*)

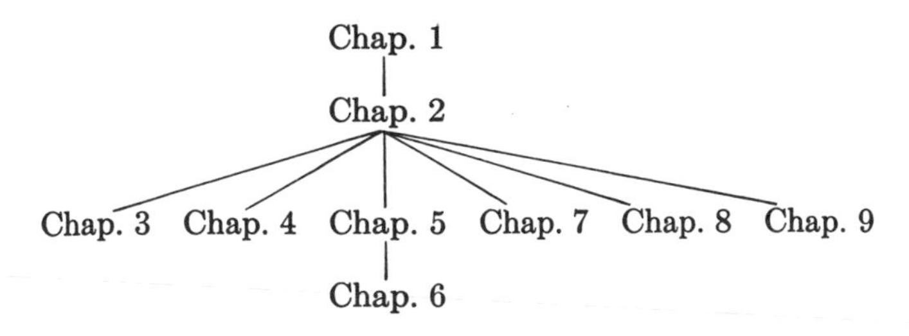

CONTENTS

NOTATION

Vectors

A vector $\mathbf{x}$ with n components is denoted by

$$\mathbf{x} = \langle x_1, x_2, \ldots, x_n \rangle.$$

$\mathbf{x} \geq \mathbf{0}$ shall mean $x_i \geq 0$ $(i = 1, 2, \ldots, n)$.

$\mathbf{x} > \mathbf{0}$ shall signify $x_i \geq 0$ and at least one component of $\mathbf{x}$ strictly positive.

$\mathbf{x} \gg \mathbf{0}$ denotes that $x_i > 0$ $(i = 1, 2, \ldots, n)$ (all components are positive).

The inner product of two real vectors $\mathbf{x}$ and $\mathbf{y}$ in E^n is denoted by

$$(\mathbf{x}, \mathbf{y}) = \sum_{i=1}^{n} x_i y_i.$$

In the complex case

$$(\mathbf{x}, \mathbf{y}) = \sum_{i=1}^{n} x_i \bar{y}_i,$$

where $\bar{y}_i$ denotes conjugate complex.

The distance between two vectors $\mathbf{x}$ and $\mathbf{y}$ is denoted by $|x - y|$.

Matrices

A matrix $\mathbf{A}$ in terms of its components is denoted by $\|a_{ij}\|$.

A matrix $\mathbf{A}$ applied to vector $\mathbf{x}$ in the manner $\mathbf{xA}$ gives the vector

$$\langle \sum x_i a_{i1}, \sum x_i a_{i2}, \ldots, \sum x_i a_{in} \rangle.$$

Similarly,

$$\mathbf{Ay} = \langle \sum a_{1j} y_j, \sum a_{2j} y_j, \ldots, \sum a_{nj} y_j \rangle \cdot$$

$(\mathbf{Ax})_j$ denotes the jth component of the vector $\mathbf{Ax}$.

The transpose of $\mathbf{A}$ is denoted by $\mathbf{A}'$.

The determinant of $\mathbf{A}$ is usually designated by $|\mathbf{A}|$ and alternatively by $\det \mathbf{A}$.

Convex Sets

The convex set spanned by a set S is denoted alternatively by Co $(S) = [S]$.

The convex cone, spanned by S, is denoted by P_S or $\mathscr{P}_S$.

Distributions

The symbols

$$\mathbf{x}_{t_0}, \mathbf{y}_{t_0} \quad \text{and} \quad \mathbf{I}_{t_0} \quad (0 \leq t_0 \leq 1)$$

are used interchangeably to represent a distribution defined on the unit interval which concentrates its full mass at the point t_0, that is,

$$\mathbf{x}_{t_0}(\xi) = \begin{cases} 0 & \xi < t_0, \\ 1 & \xi \geq t_0. \end{cases}$$

The symbol

$$\sum_{i=1}^{k} \lambda_i I_{\xi_i}$$

represents the probability distribution function with jumps λ_i located at ξ_i.

CHAPTER 1

THE DEFINITION OF A GAME AND THE MIN-MAX THEOREM

1.1 Introduction. Games in normal form. The mathematical theory of games of strategy deals with situations involving two or more participants with conflicting interests. The outcome of such games is usually controlled partly by one side and partly by the opposing side or sides; it depends to some extent on chance, but primarily on the intelligence and skill employed by the participants. Aside from games proper, such as poker and chess, there are many conflicting situations to which the theory of games can be applied, notably in certain areas of operations research, economics, politics, and military science.

We shall first consider only two-person zero-sum games, i.e., games with only two participants (competing persons, teams, firms, nations) in which one participant wins what the other loses. It should be noted that the terms of this situation exclude the possibility of any bargaining between the participants. We shall also be concerned with two-person constant-sum games, in which the two players compete irreconcilably for the greatest possible share of the kitty. By suitable renormalization such a game may be converted into a zero-sum game.

A fundamental concept in game theory is that of *strategy*. A strategy for Player I is a complete enumeration of all actions Player I will take for every contingency that might arise, whether the contingency be one of chance or one created by a move of the opposing player. What a strategy is should not be interpreted too naively. It might seem that once Player I has chosen a strategy, every move he makes at any stage is determined, in the sense that he has in mind at the beginning of the game a sequence of moves which he will carry out no matter what his opponent does. However, what we mean by "strategy" is a rule which in determining Player I's ith moves takes into account everything that has happened before his ith turn. The reason a player does not change a strategy during a game is not that the strategy has committed him to a sequence of moves he must make no matter what his opponent does, but that it gives him a move to make in any circumstances that may arise.

It is usual, in describing a game, to regard all possible procedures, good or bad, as possible strategies. Even in simple games the number of possible strategies is often forbidding. Consider the game of ticktacktoe.* Suppose Player I makes the first move. There are nine possible positions for his first cross. Player II then has eight possible moves he can make, and what Player I does at his second opportunity to move will depend on the preceding move of Player II. In this situation there are seven possible moves he may make. Player I's third move will, of course, depend on all the preceding moves of both players; and so on.

A strategy might start off as follows: Player I's first cross is to be made in the upper right-hand square. If Player II marks either the square below or the square to the left, Player I makes his second cross in the center square; if Player II marks the center square, Player I makes his second cross in the lower left-hand square; if Player II marks any of the other five squares, Player I makes his second cross in the square immediately below the first; and thus the description goes on. Player I might even embody in his strategy the possibility of randomizing, according to a fixed probability distribution, among alternatives at a given move.

Clearly, an enormous number of possible strategies present themselves even for such a simple game as ticktacktoe. Although many of them intuitively seem to be poor strategies, we are obliged to include all the possibilities in order to give a complete description of the game. In the course of this book mathematical tools for manipulating and analyzing these large sets of strategies will be developed.

A second fundamental concept in game theory is that of the *pay-off*. The pay-off is the connecting link between the set of strategies open to Player I and the set open to Player II. Specifically, it is a rule that tells how much Player I may be expected to win from Player II if Player I chooses any particular strategy from his set of strategies and Player II chooses any particular strategy from his set. The pay-off function is always evaluated in terms of the appropriate utility units (see the notes to this section on p. 20).

We are now ready for a formal definition of a game.

A game is defined to be a *triplet* $\{X, Y, K\}$, where X denotes the space of strategies for Player I, Y signifies the space of strategies of Player II, and K is a real-valued function of X and Y. Player I chooses a strategy $\mathbf{x}$ from X and Player II chooses a strategy $\mathbf{y}$ from Y. For the pair $\{\mathbf{x}, \mathbf{y}\}$ the pay-off to Player I is $K(\mathbf{x}, \mathbf{y})$ and the pay-off to Player II is $-K(\mathbf{x}, \mathbf{y})$. We shall call K the pay-off kernel.

In the absence of a statement to the contrary, the following conditions are assumed to be satisfied throughout this chapter:

---

* Ticktacktoe is played on a 3×3 matrix grid. Players move alternately and on each turn are allowed to capture one of the remaining free squares. The first player who takes possession of three squares which are on a single horizontal, vertical, or diagonal line wins.

(a) X is a convex, closed, bounded set in Euclidean n-space E^n.

(b) Y is a convex, closed, bounded set in Euclidean m-space E^m.

(c) The pay-off kernel K is a convex linear function of each variable separately. Explicitly,

$$K[\lambda \mathbf{x}_1 + (1 - \lambda)\mathbf{x}_2, \mathbf{y}] = \lambda K(\mathbf{x}_1, \mathbf{y}) + (1 - \lambda)K(\mathbf{x}_2, \mathbf{y})$$

and

$$K[\mathbf{x}, \lambda \mathbf{y}_1 + (1 - \lambda)\mathbf{y}_2] = \lambda K(\mathbf{x}, \mathbf{y}_1) + (1 - \lambda)K(\mathbf{x}, \mathbf{y}_2),$$

where λ is a real number satisfying $0 \leq \lambda \leq 1$.

Several of these limitations will be relaxed in later chapters.

The property that is essential in Chapters 1–4 is that X and Y are convex sets and have the character of finite dimensionality. (A representation of a game as a triplet involving strategy spaces which are finite-dimensional is necessarily a restriction; its justification rests on the fact that numerous actual games are of this kind.) The identification of strategies with points in Euclidean n-space is a convenience that simplifies the mathematical analysis.

An important special class of games is obtained where X is taken as the simplex S^n in E^n, defined as the set of all $\mathbf{x} = \langle x_1, x_2, \ldots, x_n \rangle$ where $x_i \geq 0$ and

$$\sum_{i=1}^{n} x_i = 1,$$

and the space Y is the corresponding simplex T^m in E^m. The pay-off kernel then takes the form

$$K(\mathbf{x}, \mathbf{y}) = \sum_{j=1}^{m} \sum_{i=1}^{n} x_i a_{ij} y_j = (\mathbf{x}, \mathbf{A}\mathbf{y}),$$

where $\mathbf{A}$ is the matrix $\|a_{ij}\|$. In the case of such matrix games we shall often denote the pay-off corresponding to strategies $\mathbf{x}$ and $\mathbf{y}$ as $A(\mathbf{x}, \mathbf{y})$ in place of $K(\mathbf{x}, \mathbf{y})$ to suggest that these games are matrix games, i.e., games in which $K(\mathbf{x}, \mathbf{y}) = (\mathbf{x}, \mathbf{A}\mathbf{y})$.

Certain special strategies consisting of vertex points of X are denoted by $\boldsymbol{\alpha}_i = \langle 0, \ldots, 0, 1, 0, \ldots, 0 \rangle$ $(i = 1, \ldots, n)$, where the 1 occurs in the ith component. These are Player I's *pure strategies*. Similarly, the strategies $\boldsymbol{\beta}_j = \langle 0, \ldots, 0, 1, 0, \ldots, 0 \rangle$ of Y $(j = 1, \ldots, m)$ are referred to as Player II's pure strategies. Since $A(\boldsymbol{\alpha}_i, \boldsymbol{\beta}_j) = K(\boldsymbol{\alpha}_i, \boldsymbol{\beta}_j) = a_{ij}$, we see that the i, j element of the matrix array A expresses the yield to Player I when Player I uses the pure strategy $\boldsymbol{\alpha}_i$ and Player II employs the pure strategy $\boldsymbol{\beta}_j$.

A strategy $\mathbf{x} = \langle x_1, x_2, \ldots, x_n \rangle$ with no component equal to 1 is called a *mixed strategy*. In view of the relationship

$$\sum_{i=1}^{n} x_i K(\boldsymbol{\alpha}_i, \mathbf{y}) = K(\mathbf{x}, \mathbf{y}), \tag{1.1.1}^*$$

the mixed strategy $\mathbf{x}$ can be effected as follows. An experiment is conducted with n possible outcomes such that the probability of the ith outcome is x_i. The ith pure strategy is used by Player I if and only if the ith outcome has resulted, and the yield to Player I becomes the expected yield from this experiment, namely

$$\sum_{i=1}^{n} x_i K(\boldsymbol{\alpha}_i, \mathbf{y}).$$

This amounts to playing each pure strategy with a specified probability. We can therefore interpret a mixed strategy as a probability distribution defined on the space of pure strategies, and conversely. Similar interpretations apply to the strategy space T^m of Player II.

The usual introductory approach to game theory is to enumerate the spaces of pure strategies for both players and to specify the pay-off matrix $\mathbf{A}$ corresponding to these pure strategies. The concept of mixed strategy is introduced subsequently, and the pay-off function is then replaced in a natural way by the expected pay-off function. In this book we have started with the more general formulation of a game, in terms of a triplet specifying the complete strategy spaces for both players and the pay-off kernel. In this formulation, the distinction between pure and mixed strategies does not exist. Nevertheless, in many special classes of games it is natural to single out the pure strategies, which span convexly the space of all strategies. This will be done wherever it is advantageous, e.g., whenever pure strategies occur in a natural manner and possess special significance. Broadly speaking, however, we lose no generality by starting directly with two-person zero-sum games in normal form, where the strategy spaces are finite-dimensional.

Again, it has been common in the literature of game theory to make a distinction between games in extensive form and games in normal form and to take the first as a point of departure. A game formulated in extensive form is developed in terms of more primitive concepts such as "play," "chance move," "personal move," and "information structure." A strategy

---

* Although we have postulated this formula, it is possible to derive it by appealing to a suitable axiom system satisfied by a preference pattern which selects among alternative probability distributions over the space of outcomes resulting from the choice of a pure strategy (see the notes to this section).

is then defined within the framework of these notions, and the analysis of optimal strategies proceeds from this point. Finally, a theorem is proved to the effect that any game in extensive form may be in fact reduced to an equivalent game in normal form.

In contrast, our definition of a game as a triplet begins immediately with the concepts of strategy and pay-off, and is flexible and general enough to encompass all forms of finite game theory, including in particular the structure of games in extensive form. By carefully defining strategies and specifying completely X, Y and $K(\mathbf{x}, \mathbf{y})$ we are able to handle all forms of information patterns that arise. This will become clear as we study specific games.

It may happen that in special instances one can construct two apparent strategy spaces which are in fact equivalent in terms of pay-off. Whenever necessary we shall demonstrate this equivalence. For purposes of mathematical consistency, however, whenever any two given games differ in a specific component, the two games are taken to be distinct. For example, if we enlarge the X strategy space while the components Y and K remain unchanged, we create a new game. This is so even if the added strategies are obviously inferior and cannot affect either player's ultimate choice of an optimal strategy. In practice, as will be seen, the strategy spaces exhibited constitute an exhaustive class of procedures, i.e., a class of procedures that take into account all the fine structure of the model.

1.2 Examples. In dealing with finite matrix games it is sufficient to specify the pure strategies for both players and the corresponding pay-off matrix $\|a_{ij}\|$. The pay-off kernel for arbitrary mixed strategies $\mathbf{x}$ and $\mathbf{y}$ is given by the expression

$$K(\mathbf{x}, \mathbf{y}) = \sum_{j=1}^{m} \sum_{i=1}^{n} x_i a_{ij} y_j. \tag{1.2.1}$$

Example 1. Matching pennies. Players I and II each display simultaneously a single penny. If I matches II, i.e., if both are heads or both are tails, I takes II's penny. Otherwise, II takes I's penny. The pay-off kernel is represented in matrix form as follows:

		Player II	
		H	T
Player I	H	1	−1
	T	−1	1

The first pure strategy for I would be to display heads and the second tails. One possible mixed strategy would be to randomize equally between showing heads and tails (i.e., $\mathbf{x} = \langle \frac{1}{2}, \frac{1}{2} \rangle$).

Example 2. Two-finger Morra. Each player displays either one or two fingers and simultaneously guesses how many the opposing player will show. If both players guess correctly or both guess incorrectly, the game is a draw. If only one guesses correctly, he wins an amount equal to the total number of fingers shown by both players.

In this case each pure strategy will have two components: (a) the number of fingers to show, and (b) the number of fingers to guess. Thus, each strategy can be represented by a pair $\langle a, b \rangle$, where a denotes the first component and b the second. For example, the strategy $\langle 2, 1 \rangle$ for Player I is to show two fingers and guess one. There will be four such pure strategies for each player: $\langle 1, 1 \rangle$, $\langle 1, 2 \rangle$, $\langle 2, 1 \rangle$, and $\langle 2, 2 \rangle$. The pay-off matrix is shown below.

		Player II			
		$\langle 1, 1 \rangle$	$\langle 1, 2 \rangle$	$\langle 2, 1 \rangle$	$\langle 2, 2 \rangle$
Player I	$\langle 1, 1 \rangle$	0	2	−3	0
	$\langle 1, 2 \rangle$	−2	0	0	3
	$\langle 2, 1 \rangle$	3	0	0	−4
	$\langle 2, 2 \rangle$	0	−3	4	0

Example 3. Poker model. In this poker model there are only three possible hands, as compared with $\binom{52}{5}$ in regular poker, and all three are considered equally likely to occur. One hand is dealt to each player. There is a preference ordering among the hands: Hand 1 wins over hands 2 and 3, and hand 2 wins over hand 3. The ante is a units. Player I can either pass or bet b units. If Player I bets then Player II can either fold or call. If Player I passes and Player II bets, then Player I again has the option of folding or calling. The three possible courses of play can be diagrammed as follows:

Player I		Player II		Player I
pass	—	pass		
pass	—	bet	—	{fold / call}
bet	—	{fold / call}		

If a pass follows a bet, the bet wins. If both contestants pass or if one contestant calls, the hands are compared and the player with the better hand wins the pot.

A strategy for Player I must provide a set of actions to be taken for each of the three possible hands that he might draw and for all possible contingencies of the play of a hand. We can denote such a strategy by a triplet of pairs $[\{\alpha, \beta\}_1, \{\alpha, \beta\}_2, \{\alpha, \beta\}_3]$. The component α of $\{\alpha, \beta\}_k$ represents for Player I the probability of betting on the first round if he draws hand k. The component β represents the probability of his betting on the second round if he holds hand k. The second component is relevant only if there is a second round, but since a strategy must take all contingencies into account, a strategy in this game must include the values of β.

Similarly, Player II's strategies can be represented by a triplet of pairs $[\{\lambda, \mu\}_1, \{\lambda, \mu\}_2, \{\lambda, \mu\}_3]$. The component λ of $\{\lambda, \mu\}_k$ is the probability of Player II's contesting a bet by Player I if Player II has drawn hand k. The component μ represents the probability of Player II's betting if Player I has previously passed.

In this example the mixed strategies have been introduced directly. In view of the fact that the components of such a strategy are formulated in terms of the behavior at each move of both players, strategies of this sort are often called "behavior strategies." This specific example is fully discussed in Section 4.3. For other examples see Problems 1 through 4 at the end of this chapter.

1.3 Choice of strategies. Suppose that Player II is compelled to announce to Player I what strategy he is going to use, and that he announces $\mathbf{y}_0$. Then Player I, seeking to maximize his pay-off or yield, naturally chooses his strategy $\mathbf{x}_0$ so that $K(\mathbf{x}_0, \mathbf{y}_0) = \max_x K(\mathbf{x}, \mathbf{y}_0)$. The best thing for Player II to do under these circumstances would be to announce $\mathbf{y}_0$ such that

$$\max_x K(\mathbf{x}, \mathbf{y}_0) = \min_y \max_x K(\mathbf{x}, \mathbf{y}) = \bar{v},$$

where $\bar{v}$ (the upper value) can be interpreted as the most that Player I can achieve if Player II employs the strategy $\mathbf{y}_0$.

Suppose the tables are turned and Player I has to announce his strategy $\mathbf{x}_0$. Since Player II is sure to choose $\mathbf{y}_0$ such that

$$K(\mathbf{x}_0, \mathbf{y}_0) = \min_y K(\mathbf{x}_0, \mathbf{y}),$$

Player I can best protect himself by choosing $\mathbf{x}_0$ such that

$$\min_y K(\mathbf{x}_0, \mathbf{y}) = \max_x \min_y K(\mathbf{x}, \mathbf{y}) = \underline{v},$$

where $\underline{v}$ (the lower value) can be interpreted as the most that Player I can guarantee himself independent of Player II's choice of strategy. In order to continue this line of reasoning, we first establish the following simple result.

▶LEMMA 1.3.1. (a) Let $f(\mathbf{x}, \mathbf{y})$ denote a real-valued function defined on $X \times Y$; then

$$\inf_y \sup_x f(\mathbf{x}, \mathbf{y}) \geq \sup_x \inf_y f(\mathbf{x}, \mathbf{y}) \qquad (\pm\infty \text{ are possible values}). \quad (1.3.1)$$

(b) If X and Y are compact and f is continuous, then sup and inf in (1.3.1) may be replaced by max and min, respectively.

Proof. By definition, for any $\mathbf{y}$, $\sup_x f(\mathbf{x}, \mathbf{y}) \geq f(\mathbf{x}, \mathbf{y})$. Hence $\inf_y \sup_x f(\mathbf{x}, \mathbf{y}) \geq \inf_y f(\mathbf{x}, \mathbf{y})$, and (1.3.1) follows immediately. The simple proof of (b) is omitted.

Lemma 1.3.1 for $f = K$ gives

$$\min_{y \in Y} \max_{x \in X} K(\mathbf{x}, \mathbf{y}) \geq \max_{x \in X} \min_{y \in Y} K(\mathbf{x}, \mathbf{y}).$$

Expressed in terms of the upper and lower values of a game, we obtain

$$\bar{v} \geq \underline{v}.$$

Our principal aim in this chapter is to establish that for a matrix game $\bar{v} = \underline{v} = v$. By an appropriate choice of strategy Player I can guarantee himself the value $\underline{v} = v$, and by judicious play Player II can prevent Player I from achieving more than $\bar{v} = \underline{v}$. Thus, unless Player I has additional information about Player II's mode of behavior—for example, expects him to choose a strategy in a wild manner—Player I should play so as to achieve v. If he departs from the course of action that assures him the value v, his ultimate yield might be less than v. Thus, it is reasonable to call this common value v the value of the game to Player I. Of course, $-v$ is the value to Player II.

All strategies $\mathbf{x}_0$ and $\mathbf{y}_0$ such that $K(\mathbf{x}_0, \mathbf{y}) \geq v$ for all $\mathbf{y}$ and $K(\mathbf{x}, \mathbf{y}_0) \leq v$ for all $\mathbf{x}$ will be referred to as optimal strategies for Players I and II, respectively.

The following simple criterion is often very useful in determining when $\bar{v} = \underline{v}$; moreover, it points up the connection between the existence of optimal strategies and the equality of the lower and upper values.

▶THEOREM 1.3.1. If there exist $\mathbf{x}_0 \in X$, $\mathbf{y}_0 \in Y$ and a real number v such that

$$K(\mathbf{x}_0, \mathbf{y}) \geq v \qquad \text{for all} \quad \mathbf{y} \in Y$$

and

$$K(\mathbf{x}, \mathbf{y}_0) \leq v \qquad \text{for all} \quad \mathbf{x} \in X,$$

then

$$\bar{v} = \min_y \max_x K(\mathbf{x}, \mathbf{y}) = v = \max_x \min_y K(\mathbf{x}, \mathbf{y}) = \underline{v},$$

and conversely.

Proof. (a) Since

$$K(\mathbf{x}_0, \mathbf{y}) \geq v \qquad \text{for all} \quad \mathbf{y},$$

it follows that

$$\min_y K(\mathbf{x}_0, \mathbf{y}) \geq v \qquad \text{and} \qquad \max_x \min_y K(\mathbf{x}, \mathbf{y}) \geq v.$$

Similarly,

$$\max_x K(\mathbf{x}, \mathbf{y}_0) \leq v;$$

hence

$$\min_y \max_x K(\mathbf{x}, \mathbf{y}) \leq v.$$

This gives

$$\max_x \min_y K(\mathbf{x}, \mathbf{y}) \geq \min_y \max_x K(\mathbf{x}, \mathbf{y}).$$

But

$$\min_y \max_x K(\mathbf{x}, \mathbf{y}) \geq \max_x \min_y K(\mathbf{x}, \mathbf{y})$$

by Lemma 1.3.1. Therefore

$$\min_y \max_x K(\mathbf{x}, \mathbf{y}) = \max_x \min_y K(\mathbf{x}, \mathbf{y}).$$

(b) Conversely, choose $\mathbf{y}_0$ such that

$$\max_x K(\mathbf{x}, \mathbf{y}_0) = \min_y \max_x K(\mathbf{x}, \mathbf{y}) = v,$$

and $\mathbf{x}_0$ such that

$$\min_y K(\mathbf{x}_0, \mathbf{y}) = \max_x \min_y K(\mathbf{x}, \mathbf{y}) = v.$$

Then $K(\mathbf{x}, \mathbf{y}_0) \leq v$ and $K(\mathbf{x}_0, \mathbf{y}) \geq v$. This completes the proof.

▶Corollary 1.3.1. A necessary and sufficient condition that

$$\min_y \max_x K(\mathbf{x}, \mathbf{y}) = \max_x \min_y K(\mathbf{x}, \mathbf{y})$$

is that there exist strategies $\{\mathbf{x}_0, \mathbf{y}_0\}$ such that for all $\mathbf{y}$ in Y and all $\mathbf{x}$ in X

$$K(\mathbf{x}_0, \mathbf{y}) \geq K(\mathbf{x}_0, \mathbf{y}_0) \geq K(\mathbf{x}, \mathbf{y}_0).$$

Obviously, $v = K(\mathbf{x}_0, \mathbf{y}_0)$.

1.4 The min-max theorem for finite matrix games. Throughout this section we shall be concerned exclusively with matrix games. For this special case the strategy spaces X and Y of Players I and II are identifiable with the simplexes S^n and T^m, respectively, and thus each set of all strategies is spanned convexly by the pure strategies. Our objective is to establish the fundamental min-max theorem for these matrix games.

The number of different proofs of this theorem is very large. Von Neumann constructed two proofs: one based on the Brouwer fixed-point theorem (see Section C.2 of the Appendix), the other on the proposition that nonoverlapping convex sets lying in a Euclidean space can be separated by hyperplanes. Other proofs have been given which involve algebraic inequalities and the properties of convex functions. Three of these proofs are reviewed in this section and the next. We have singled out those proofs that are of use in obtaining further information concerning optimal strategies (see Remark 1.4 below).

The first proof presented is based on the idea of separating convex sets by supporting hyperplanes. Specific geometric representations of optimal strategies are developed in the course of the argument. Similar representations will be used later in analyzing the nature of the sets of optimal strategies.

▶ Lemma 1.4.1. If S^n denotes the collection of all vectors in E^n satisfying $x_i \geq 0$ $(i = 1, \ldots, n)$, $\sum_{i=1}^n x_i = 1$, and $\mathbf{b} = \langle b_1, \ldots, b_n \rangle$ is a fixed vector, then

$$\max_{x \in S^n} \sum_{i=1}^{n} x_i b_i = \max_i b_i. \tag{1.4.1}$$

Proof. Clearly, $\max_i b_i = \sum_{j=1}^n (x_j \max_i b_i) \geq \sum_{j=1}^n x_j b_j$. On the other hand, $\max_i b_i = b_{i_0} \leq \max_{x \in S^n} \sum_{i=1}^n x_i b_i$.

We next develop a geometrical interpretation for strategies and a schematic representation for the value of a game. To this end, we consider the pay-off matrix

$$\mathbf{A} = \begin{Vmatrix} a_{11} & \cdots & a_{1m} \\ \vdots & & \vdots \\ a_{n1} & \cdots & a_{nm} \end{Vmatrix} = \|\mathbf{a}_1, \ldots, \mathbf{a}_m\|,$$

where $\mathbf{a}_i$ stands for the column vector $\langle a_{1i}, \ldots, a_{ni} \rangle$. The $\mathbf{a}_i$'s can be viewed as m points in n-dimensional space.

Let Γ denote the convex set in E^n-space spanned by the $\mathbf{a}_i$. Equivalently, Γ is the intersection of all convex sets containing all $\mathbf{a}_i$. Formally,

$$\Gamma = \left\{ \sum_{j=1}^{m} y_j \mathbf{a}_j \,\middle|\, \sum_{j=1}^{m} y_j = 1, y_j \geq 0 \right\}.$$

Admissible values. We call a real number λ admissible if there exists a vector $\mathbf{p} \in \Gamma$ such that $\mathbf{p} \leq \boldsymbol{\lambda}$. (Recall that $\mathbf{p} \leq \boldsymbol{\lambda}$ means that every component p_i is smaller than or equal to λ; see p. x in the front of the book.)

If Ω denotes the set of admissible numbers, then (i) Ω is nonvoid since it contains N, provided $N > a_{ij}$ for all i, j, and (ii) Ω is bounded from below by $-N$, for N sufficiently large.

Let

$$\lambda_0 = \inf_{\lambda \in \Omega} \lambda.$$

In order to show that λ_0 is admissible—i.e., that $\lambda_0 \in \Omega$—we select a sequence $\{\lambda_n\}$ in Ω such that $\lambda_n \to \lambda_0$.

To every λ_n there corresponds at least one vector $\mathbf{q}_n$ in Γ such that $\mathbf{q}_n \leq \boldsymbol{\lambda}_n$. Since Γ is compact and the sequence $\{\mathbf{q}_n\}$ must have at least one limit point $\mathbf{q}_0$, it follows readily that $\mathbf{q}_0 \leq \boldsymbol{\lambda}_0$, and thus λ_0 is admissible.

Consider the translation of the negative orthant O_{λ_0} consisting of all vectors having no component greater than λ_0. The set O_{λ_0} touches Γ with no overlap. In fact, $O_{\lambda_0} \cap \Gamma \supset \mathbf{q}_0$, but if the interior of O_{λ_0} were to contain points of Γ, the meaning of λ_0 would be contradicted.

Supporting planes to O_{λ_0}. We shall now characterize the form of all supporting planes to O_{λ_0} that are oriented so that their normals are directed into the half-space not containing O_{λ_0} (see also Lemma B.1.3 in Appendix B).

Let us consider a hyperplane defined by

$$\sum_{i=1}^{n} \mu_i \xi_i + \mu_0 = 0, \qquad \sum_{i=1}^{n} |\mu_i| \neq 0 \tag{1.4.2}$$

whose normal is directed away from O_{λ_0}. This provides a supporting plane to O_{λ_0} if and only if for every $\boldsymbol{\xi}$ in O_{λ_0}

$$\sum_{i=1}^{n} \mu_i \xi_i + \mu_0 \leq 0, \tag{1.4.3}$$

and there exists a $\boldsymbol{\xi}^0$ in O_{λ_0} for which equality holds. We now show that μ_i can be characterized as follows:

$$\mu_i \geq 0 \quad (i = 1, \ldots, n), \qquad \sum_{i=1}^{n} \mu_i > 0, \qquad \mu_0 = -\lambda_0 \sum_{i=1}^{n} \mu_i. \tag{1.4.4}$$

That μ_i $(1 \leq i \leq n)$ is nonnegative is seen by applying (1.4.3) to the vector $\boldsymbol{\xi} = \langle \xi_1, \ldots, \xi_n \rangle$ with components $\xi_j = \lambda_0$ for $j \neq i$, $\xi_i = -N$, and N sufficiently large. Since (1.4.2) defines a hyperplane, it follows that

$$\sum_{i=1}^{n} \mu_i > 0.$$

Finally, any supporting plane to O_{λ_0} clearly must contain the vertex point $\boldsymbol{\xi} = \langle \lambda_0, \lambda_0, \ldots, \lambda_0 \rangle$. Therefore,

$$O = \left(\sum_{i=1}^{n} \mu_i \right) \lambda_0 + \mu_0$$

as asserted. Conversely, any set of μ_i satisfying the conditions of (1.4.4) defines a supporting hyperplane to O_{λ_0} whose normal is directed into the half-space not containing O_{λ_0}.

If we change notation from μ_i to x_i, supporting planes to O_{λ_0} whose normal is directed away from O_{λ_0} can be characterized by the relation

$$\sum_{i=1}^{n} x_i \xi_i - \lambda_0 = 0,$$

where $\sum_{i=1}^{n} x_i = 1$ and $x_i \geq 0$.

Geometrical representation of strategies. Strategies $\mathbf{x} = \langle x_1, \ldots, x_n \rangle$ in S^n, i.e.,

$$\sum_{i=1}^{n} x_i = 1, \qquad x_i \geq 0,$$

generate supporting planes to O_{λ_0} that are defined by the equation

$$\sum_{i=1}^{n} x_i \xi_i - \lambda_0 = 0,$$

and conversely.

Strategies $\mathbf{y} = \langle y_1, \ldots, y_m \rangle$ in T^m for Player II correspond to points in Γ. To $\mathbf{y}$ we associate the point $\mathbf{p}$ with coordinates

$$p_i = \sum_{j=1}^{m} a_{ij} y_j.$$

Unfortunately, the correspondence of Y to Γ is not one to one, since subsets of Y map into points of Γ if and only if $\mathbf{A}$ has rank less than m.

Finally, the expected pay-off $A(\mathbf{x}, \mathbf{y})$ can be interpreted as the inner product of the normal to the plane determined by $\mathbf{x}$, and the point of Γ determined by $\mathbf{y}$. Explicitly,

$$(\mathbf{x}, \mathbf{A}\mathbf{y}) = \sum_{i=1}^{n} \sum_{j=1}^{m} x_i a_{ij} y_j.$$

▶THEOREM 1.4.1 (*Min-Max Theorem*). If $\mathbf{x}$ and $\mathbf{y}$ range over S^n and T^m, respectively, then

$$\min_y \max_x A(\mathbf{x}, \mathbf{y}) = \max_x \min_y A(\mathbf{x}, \mathbf{y}) = v.$$

Proof. We construct O_{λ_0} and Γ as described above.

Since O_{λ_0} and Γ are nonoverlapping convex sets by Lemma B.1.2 in Appendix B, there exists a separating plane. Since O_{λ_0} and Γ touch, the separating plane must also be a supporting plane, which may be oriented so that its normal is directed away from O_{λ_0}.

Let the direction numbers of its normal be x_i^0. Then, since

$$\sum_{i=1}^{n} x_i^0 = 1$$

and $x_i^0 \geq 0$, as required by (1.4.4),

$$\sum_{i=1}^{n} x_i^0 \xi_i - \lambda_0 \leq 0 \qquad \text{for all} \quad \boldsymbol{\xi} \in O_{\lambda_0},$$

and

$$\sum_{i=1}^{n} x_i^0 \xi_i - \lambda_0 \geq 0 \qquad \text{for all} \quad \boldsymbol{\xi} \in \Gamma.$$

If $\mathbf{y}$ is any given strategy, then the vector $\boldsymbol{\xi}$, where

$$\xi_i = \sum_{j=1}^{m} a_{ij} y_j,$$

is in Γ, and thus

$$A(\mathbf{x}^0, \mathbf{y}) = \sum_{i=1}^{n} x_i^0 \sum_{j=1}^{m} a_{ij} y_j \geq \lambda_0 \tag{1.4.5}$$

for any strategy $\mathbf{y}$.

If $\mathbf{y}^0$ denotes a strategy corresponding to a point Γ which also belongs to O_{λ_0}, then

$$\sum_{j=1}^{m} a_{ij} y_j^0 \leq \lambda_0 \qquad \text{for all} \quad i, \tag{1.4.6}$$

and hence $A(\mathbf{x}, \mathbf{y}^0) \leq \lambda_0$ for all $\mathbf{x}$ strategies (by Lemma 1.4.1).

The proof of the theorem is complete by virtue of Theorem 1.3.1.

Remark 1.4. We have observed that strategies for I can be viewed as supporting planes, and strategies for II as points in Γ. We further see that optimal strategies for I correspond to separating planes between O_{λ_0} and Γ, and that optimal strategies for II correspond to those points of Γ which have contact with O_{λ_0}. This geometric characterization of optimal strategies will be exploited in Chapter 3.

The interpretations used in the arguments of Theorem 1.4.1 are unsymmetrical, in that the $\mathbf{x}$ strategies correspond to hyperplanes and the $\mathbf{y}$

strategies to points of Euclidean n-space. However, a completely parallel analysis can be developed which associates $\mathbf{y}$ strategies with hyperplanes and $\mathbf{x}$ strategies with points. In this analysis the analog of Γ is the convex set spanned by the row vectors of $\mathbf{A}$, and so on.

It is often of interest to know when there exist optimal pure strategies. This will be the case when there exist pure strategies $\{\boldsymbol{\alpha}_{i_0}, \boldsymbol{\beta}_{j_0}\}$ such that

$$\begin{aligned} a_{i_0,j} &\geq a_{i_0,j_0} \qquad \text{for all} \quad j, \\ a_{i_0,j_0} &\geq a_{i,j_0} \qquad \text{for all} \quad i. \end{aligned} \tag{1.4.7}$$

The first inequality implies that for any mixed strategy $\mathbf{y}$,

$$\sum_{j=1}^{m} a_{i_0,j} y_j \geq a_{i_0,j_0} = v,$$

or $A(\boldsymbol{\alpha}_{i_0}, \mathbf{y}) \geq v$. Similarly, we obtain $v \geq A(\mathbf{x}, \boldsymbol{\beta}_{j_0})$, and thus $\boldsymbol{\alpha}_{i_0}$ and $\boldsymbol{\beta}_{j_0}$ are optimal strategies for Players I and II, respectively.

Where pure optimal strategies exist, the matrix $\mathbf{A}$ is said to have a *saddle point* at (i_0, j_0). A saddle point is an element of a matrix which is both a minimum of its row and a maximum of its column.

***1.5 General min-max theorem.** In this section two alternative proofs of the min-max theorem for general finite games are given. The first proof is based on the Kakutani fixed-point theorem. The second proof relies only on the properties of convex functions.

▶THEOREM 1.5.1. Let $f(\mathbf{x}, \mathbf{y})$ be a real-valued function of two variables $\mathbf{x}$ and $\mathbf{y}$ which traverse C and D, respectively, where both C and D are closed, bounded, convex sets. If f is continuous, convex in $\mathbf{y}$ for each $\mathbf{x}$, and concave in $\mathbf{x}$ for each $\mathbf{y}$, then $\min_y \max_x f(\mathbf{x}, \mathbf{y}) = \max_x \min_y f(\mathbf{x}, \mathbf{y})$.

Proof. For each $\mathbf{x}$, define $\phi_x(\mathbf{y}) = f(\mathbf{x}, \mathbf{y})$. Let

$$B_x = \{\mathbf{y} | \phi_x(\mathbf{y}) = \min_{y \in D} \phi_x(\mathbf{y})\}.$$

It is easily verified that B_x is nonvoid, closed, and convex. The fact that B_x is convex is a consequence of the convexity of the function f in the variable $\mathbf{y}$. For each $\mathbf{y}$, define $\psi_y(\mathbf{x}) = f(\mathbf{x}, \mathbf{y})$. Let

$$A_y = \{\mathbf{x} | \psi_y(\mathbf{x}) = \max_{x \in C} \psi_y(\mathbf{x})\}.$$

Again, A_y is nonvoid, closed, and convex.

---

* Starred sections are advanced discussions that may be omitted at first reading without loss of continuity.

Let the set E be the direct product, $E = C \times D$, of the convex sets C and D, and recall that the direct product of two convex sets is itself convex.

Let g be the mapping which maps the point $\{\mathbf{x}, \mathbf{y}\} \in E$ into the set $(A_y \times B_x)$. Since A_y and B_x are nonvoid, closed, convex subsets of C and D, respectively, $(A_y \times B_x)$ is a nonvoid, closed, convex subset of E. Hence, condition (i) of the Kakutani theorem is fulfilled (see Section C.2 in the Appendix). It is also easy to verify that the continuity requirement of the Kakutani theorem is satisfied. Therefore, invoking the Kakutani theorem, there exists an $\{\mathbf{x}^0, \mathbf{y}^0\} \in (A_{y^0} \times B_{x^0})$, that is,

$$f(\mathbf{x}^0, \mathbf{y}^0) \geq f(\mathbf{x}, \mathbf{y}^0) \qquad \text{for all} \quad \mathbf{x} \in C,$$

$$f(\mathbf{x}^0, \mathbf{y}^0) \leq f(\mathbf{x}^0, \mathbf{y}) \qquad \text{for all} \quad \mathbf{y} \in D.$$

It then follows from Corollary 1.3.1 that

$$\min_{y \in D} \max_{x \in C} f(\mathbf{x}, \mathbf{y}) = \max_{x \in C} \min_{y \in D} f(\mathbf{x}, \mathbf{y}). \tag{1.5.1}$$

Whenever (1.5.1) is satisfied, we say that the game defined by (f, C, D) possesses a value or, alternatively, that the game is determined.

The other proof proceeds as follows. Assume first that $f(\mathbf{x}, \mathbf{y})$ is a strictly convex function of $\mathbf{y}$, for fixed $\mathbf{x}$, and $f(\mathbf{x}, \mathbf{y})$ is a strictly concave function of $\mathbf{x}$, for fixed $\mathbf{y}$. By virtue of the strict convexity, there exists a unique $\mathbf{y}(\mathbf{x})$ for which

$$f[\mathbf{x}, \mathbf{y}(\mathbf{x})] = \min_y f(\mathbf{x}, \mathbf{y}) = m(\mathbf{x}).$$

From the uniform continuity of f and the uniqueness of $\mathbf{y}(\mathbf{x})$ it follows readily that $m(\mathbf{x})$ and $\mathbf{y}(\mathbf{x})$ are continuous. Also, since the minimum of a family of concave functions remains concave, $m(\mathbf{x})$ is concave. Let $\mathbf{x}^*$ be a point where

$$m(\mathbf{x}^*) = \max_x m(\mathbf{x}) = \max_x \min_y f(\mathbf{x}, \mathbf{y}).$$

For any $\mathbf{x}$ in C and any t in $0 < t < 1$, we have by the hypothesis that

$$\begin{aligned} f[(1 - t)\mathbf{x}^* + t\mathbf{x}, \mathbf{y}] &> (1 - t)f(\mathbf{x}^*, \mathbf{y}) + tf(\mathbf{x}, \mathbf{y}) \\ &\geq (1 - t)m(\mathbf{x}^*) + tf(\mathbf{x}, \mathbf{y}). \end{aligned}$$

Put $\mathbf{y} = \mathbf{y}[(1 - t)\mathbf{x}^* + t\mathbf{x}] = \bar{\mathbf{y}}$, so that

$$m[(1 - t)\mathbf{x}^* + t\mathbf{x}] \geq (1 - t)m(\mathbf{x}^*) + tf(\mathbf{x}, \bar{\mathbf{y}}).$$

But since $m(\mathbf{x}^*) \geq m(\mathbf{x})$ for all $\mathbf{x}$ in C, it follows that

$$m(\mathbf{x}^*) \geq m[(1 - t)\mathbf{x}^* + t\mathbf{x}],$$

and hence

$$f(\mathbf{x}, \bar{\mathbf{y}}) \leq m(\mathbf{x}^*) = f[\mathbf{x}^*, \mathbf{y}(\mathbf{x}^*)].$$

Now let $t \to 0$, so that $(1 - t)\mathbf{x}^* + t\mathbf{x} \to \mathbf{x}^*$ and $\bar{\mathbf{y}} \to \mathbf{y}(\mathbf{x}^*)$; from this we deduce that

$$f[\mathbf{x}, \mathbf{y}(\mathbf{x}^*)] \leq f[\mathbf{x}^*, \mathbf{y}(\mathbf{x}^*)] \tag{1.5.2}$$

for any $\mathbf{x}$. If we let $\mathbf{y}(\mathbf{x}^*) = \mathbf{y}^*$ and $v = f(\mathbf{x}^*, \mathbf{y}^*)$, then it is clear from (1.5.2) and the construction of $\mathbf{y}(\mathbf{x})$ that

$$f(\mathbf{x}, \mathbf{y}^*) \leq v \leq f(\mathbf{x}^*, \mathbf{y}).$$

This is equivalent to the min-max theorem. It remains only to relax the conditions of strictness imposed above. To this end, define

$$f_\epsilon(\mathbf{x}, \mathbf{y}) = f(\mathbf{x}, \mathbf{y}) - \epsilon f(\mathbf{x}) + \epsilon g(\mathbf{y}),$$

where

$$f(\mathbf{x}) = \sum_{i=1}^{n} x_i^2, \qquad g(\mathbf{y}) = \sum_{j=1}^{m} y_j^2.$$

Each f_ϵ satisfies the strict convexity-concavity requirements used in the previous argument, and consequently

$$f(\mathbf{x}, \mathbf{y}_\epsilon) - \epsilon f(\mathbf{x}) \leq f_\epsilon(\mathbf{x}, \mathbf{y}_\epsilon) \leq v_\epsilon \leq f_\epsilon(\mathbf{x}_\epsilon, \mathbf{y}) \leq f(\mathbf{x}_\epsilon, \mathbf{y}) + \epsilon g(\mathbf{y})$$

for all $\{\mathbf{x}, \mathbf{y}\}$ in $C \times D$. Take a sequence of ϵ's approaching zero and consider a subsequence $\mathbf{x}_\epsilon \to \mathbf{x}^*$, $\mathbf{y}_\epsilon \to \mathbf{y}^*$, $v_\epsilon \to v^*$. We obtain in the limit

$$f(\mathbf{x}, \mathbf{y}^*) \leq v^* \leq f(\mathbf{x}^*, \mathbf{y}),$$

and the proof is complete.

1.6 Problems.

1. "Stone, Paper, and Scissors." Both players must simultaneously name one of these objects; "paper" defeats "stone," "stone" defeats "scissors," and "scissors" defeats "paper." The player who chooses the winning object wins one unit; if both players choose the same object, the game is a draw. Set up the matrix for this game and describe the optimal strategies.

2. Two cards marked Hi and Lo are placed in a hat. Player I draws a card at random and inspects it. He may pass immediately, in which case he pays Player II the amount $a > 0$; or he may bet, in which case Player II may either pass, paying amount a to Player I, or call. If he calls, he

receives from or pays to Player I the amount $b > a$, according as Player I holds Lo or Hi. Set up the game and solve by the methods of Theorem 1.4.1.

3. Player I draws a card Hi or Lo from a hat. He then has the option of either betting or passing. If he bets, Player II may either fold, in which case he pays Player I an amount a, or call, in which case he wins or loses $b > a$ according as Player I holds Hi or Lo. If Player I passes initially, Player II draws Hi or Lo from a new hat. Player II then has the symmetrical choice of actions, to which Player I must respond with the corresponding pay-off. If Player II passes on the second round, the pay-off is zero. Describe the game and the strategy spaces.

4. Each player must choose a number between 1 and 9. If a player's choice is one unit higher than his opponent's, he loses two dollars; if it is at least two units higher, he wins one dollar; if both players choose the same number, the game is a draw. Set up the matrix for this game.

5. Show that the $n \times n$ matrix $\|a_{ij}\|$ with $a_{ij} = i - j$ has a saddle point, and describe the corresponding optimal strategies.

6. Use Theorem 1.4.1 to solve the game with matrix

$$\begin{Vmatrix} 3 & -1 & 3 & 7 \\ -1 & 9 & 3 & 0 \end{Vmatrix}.$$

7. Two games have matrices $\mathbf{A} = \|a_{ij}\|$ and $\mathbf{B} = \|b_{ij}\|$, respectively; the value of the first game is v, and $b_{ij} = a_{ij} + a$. Show that the value of $\mathbf{B}$ is $v + a$, and that the optimal strategies are the same for both games.

8. Show that each of the two matrices

$$\mathbf{A}_1 = \begin{Vmatrix} 0 & x \\ 1 & 2 \end{Vmatrix}, \qquad \mathbf{A}_2 = \begin{Vmatrix} 2 & 1 \\ x & 0 \end{Vmatrix}$$

has a saddle point, and find values x_1 and x_2 such that

$$v(\mathbf{A}_1 + \mathbf{A}_2) < v(\mathbf{A}_1) + v(\mathbf{A}_2)$$

and

$$v(\mathbf{A}_1 + \mathbf{A}_2) > v(\mathbf{A}_1) + v(\mathbf{A}_2),$$

respectively.

9. Give an example to show that

$$(\mathbf{x}^*, \mathbf{A}\mathbf{y}^*) = \min_{y \in Y} \max_{x \in X} (\mathbf{x}, \mathbf{A}\mathbf{y}) = \max_{x \in X} \min_{y \in Y} (\mathbf{x}, \mathbf{A}\mathbf{y}),$$

is not sufficient proof that $\mathbf{x}^*$ and $\mathbf{y}^*$ are optimal.

10. Let $\mathbf{A} = \|a_{ij}\|$ be an $n \times m$ matrix. Show that either there exists

$$\mathbf{u} = \langle u_1, \ldots, u_n \rangle \qquad \left(\text{with } u_i \geq 0, \sum_{i=1}^{n} u_i = 1\right)$$

such that

$$\sum_{i=1}^{n} u_i a_{ij} \geq 0 \qquad \text{for all} \quad j,$$

or there exists

$$\mathbf{w} = \langle w_1, \ldots, w_m \rangle \qquad \left(\text{with } w_j \geq 0, \sum_{j=1}^{m} w_j = 1\right)$$

such that

$$\sum_{j=1}^{m} a_{ij} w_j < 0 \qquad \text{for all} \quad i.$$

11. Let $\mathbf{B} = \|b_{ij}\|$ be a fixed $n \times m$ game, and let $f(t)$ be a convex increasing function of t. Let Γ be the game with strategies

$$\mathbf{x} = \langle x_1, \ldots, x_n \rangle, \qquad \mathbf{y} = \langle y_1, \ldots, y_m \rangle$$

and pay-off

$$\sum_{i=1}^{n} x_i f\left(\sum_{j=1}^{m} b_{ij} y_j\right).$$

Show that the optimal strategies for $\mathbf{B}$ are also optimal for the new game.

12. Show that a game $\sum x_i a_{ij} y_j$ played over closed sets of strategies S and T has a value if and only if the game played over the convex hulls of S and T has two optimal strategies $\mathbf{x}^* \in S$ and $\mathbf{y}^* \in T$.

13. Let $\mathbf{A}$, $\mathbf{B}$, and $\mathbf{D}$ be matrix games of orders $n_1 \times m_1$, $n_1 \times m_2$, $n_2 \times m_2$ and values a, b, d, respectively. Let $\{\mathbf{x}^0, \mathbf{y}^0\}$ and $\{\mathbf{x}^*, \mathbf{y}^*\}$ be optimal for $\mathbf{A}$ and $\mathbf{D}$, respectively, and let $\langle \xi, 1 - \xi \rangle$ and $\langle \eta, 1 - \eta \rangle$ be optimal for the game

$$\begin{Vmatrix} a & b \\ 0 & d \end{Vmatrix}.$$

Show that if $a < 0 < d$, the strategies

$$\mathbf{x} = \langle \xi x_1^0, \ldots, \xi x_{n_1}^0, (1 - \xi) x_{n_1+1}^*, \ldots, (1 - \xi) x_{n_1+n_2}^* \rangle,$$

$$\mathbf{y} = \langle \eta y_1^0, \ldots, \eta y_{m_1}^0, (1 - \eta) y_{m_1+1}^*, \ldots, (1 - \eta) y_{m_1+m_2}^* \rangle$$

are optimal for the game

$$\begin{Vmatrix} \mathbf{A} & \mathbf{B} \\ 0 & \mathbf{D} \end{Vmatrix}.$$

14. Show that the value of a matrix game is a nondecreasing continuous function of the components of the matrix.

15. Let X and Y be convex compact sets, and let $f(\mathbf{x}, \mathbf{y})$ be concave in $\mathbf{x}$ and convex in $\mathbf{y}$. Show that for any finite set $\mathbf{x}_1, \ldots, \mathbf{x}_n, \mathbf{y}_1, \ldots, \mathbf{y}_m$ there exists a pair $\{\mathbf{x}_0, \mathbf{y}_0\}$ such that

$$\max_{1 \leq i \leq n} f(\mathbf{x}_i, \mathbf{y}_0) \leq \min_{1 \leq j \leq m} f(\mathbf{x}_0, \mathbf{y}_j).$$

16. Let X and Y be convex sets, with Y also compact. Let $f(\mathbf{x}, \mathbf{y})$ be a function defined on $X \times Y$ with the property that f is lower-semicontinuous in $\mathbf{y}$, concave in $\mathbf{x}$, and convex in $\mathbf{y}$. Show that for each real α, either there exists $\mathbf{y}_0$ in Y such that $f(\mathbf{x}, \mathbf{y}_0) \leq \alpha$ for all $\mathbf{x}$ in X, or there exists $\mathbf{x}_0$ in X such that $f(\mathbf{x}_0, \mathbf{y}) > \alpha$ for all $\mathbf{y}$ in Y.

17. Show that if every 2×2 submatrix of a matrix $\mathbf{A}$ has a saddle point, then $\mathbf{A}$ has a saddle point.

18. Let $\mathbf{A}$ be a nonsingular $n \times m$ matrix with $m \geq 3$. Show that if every $n \times m - 1$ submatrix has a strict saddle point, then $\mathbf{A}$ has a saddle point.

19. Let X and Y denote polyhedral convex sets in E^n and E^m space, respectively. Let $\mathbf{A}$ be an $n \times m$ matrix. Define

$$v_2 = \inf_{y \in Y} \sup_{x \in X} (\mathbf{x}, \mathbf{Ay}), \qquad v_1 = \sup_{x \in X} \inf_{y \in Y} (\mathbf{x}, \mathbf{Ay}).$$

Prove that if $v_1 \neq v_2$, then $v_1 = -\infty$ and $v_2 = +\infty$. Construct examples satisfying $v_1 = -\infty$; $v_2 = +\infty$; $v_1 = v_2 = +\infty$; $v_1 = v_2 = -\infty$. (A polyhedral convex set is defined as the intersection of a finite number of closed half spaces.)

Notes and References to Chapter 1

§ **1.1.** Several popular and entertaining discussions of discrete game theory and its applications are available for the nonmathematically-oriented reader, notably Williams [254] and McDonald [176]. A comprehensive critical survey of the concepts of game theory, utility theory, and related subjects is contained in Luce and Raiffa [169].

Historically, the modern mathematical approach to game theory is attributed to von Neumann in his papers of 1928 [238] and 1937 [239], although some game problems were dealt with earlier by Borel [39]. The primary stimulation for the present vigorous research in game theory was the publication in 1944 of the fundamental book on games and economic behavior by von Neumann and Morgenstern [242]. For introductory mathematical expositions in game theory, the reader is directed to McKinsey [179], Gale [96], and Vajda [236].

To the more advanced student we recommend the volumes edited by Kuhn and Tucker [161, 162]. Another volume of studies in game theory has been compiled by Dresher, Tucker, and Wolfe [72]. To the student interested primarily in n-person game theory we call attention to Shapley [216] and a volume edited by Luce and Tucker [170].

Several texts on related subjects incorporate some aspects of game theory. These include Blackwell and Girshick [30], Churchman, Ackoff, and Arnoff [50], McCloskey and Trefethen [175], and Allen [2].

Luce and Raiffa, in addition to their discussions of the underlying concepts of games and decisions, provide in eight appendixes a moderate survey of several of the techniques of the subject.

A handbook of important results in game theory has been compiled by Dresher [66].

Although a knowledge of utility theory is not essential to the analytical study of the structure of games and their solutions, a few words on the subject may be helpful.

Von Neumann and Morgenstern in their classic book introduced an axiom system satisfied by a preference pattern $\succ$ defined on a finite set of outcomes and the space $\mathcal{P}$ of probability distributions of outcomes. That is to say, of any two probability distributions P_1 and P_2 on the space of outcomes, the preference relation selects one as preferable to the other. (In symbols, either $P_1 \succ P_2$ or $P_2 \succ P_1$.)

The axioms, stated loosely, are as follows:

I. If $P_1 \succ P_2$ and $Q_1 \succ Q_2$, then any probability mixture of P_1 and Q_1 is preferred to the same probability mixture of P_2 and Q_2.

II. (Continuity Axiom.) There is no probability distribution on the space of outcomes that is infinitely desirable or infinitely undesirable. Formally, we may express this axiom as follows: If $P_1 \prec P_2 \prec P_3$, there are numbers $\lambda < 1$ and $\mu > 0$ such that

$$\lambda P_1 + (1 - \lambda)P_3 \prec P_2, \qquad \mu P_1 + (1 - \mu)P_3 \succ P_2.$$

When the preference relation obeys these axioms, it can be expressed as a numerical utility function U defined for probability distributions on the set of outcomes. Furthermore, U is linearly convex with respect to the operation of forming convex combinations of elements of $\mathcal{P}$. It is also shown by these authors that U is uniquely determined up to a positive linear affine transformation. That is, if U is a utility function, then $aU + b$ (with $a > 0$ and b real) is the only other possible utility function. The proof is given in Chapter i of [242].

Blackwell and Girshick [30] have extended this result to the case in which the set of outcomes may contain a countable number of elements. Luce and Raiffa [169] subject the von Neumann and Morgenstern axioms to a penetrating review. We recommend their critical discussion to the reader interested in the foundations of game theory.

In the classical approach to the theory of games, it is necessary when defining a strategy to specify decisions for both players which appropriately take account of the complete present state of information. Because of chance moves which may occur in the course of the play the outcomes of a pair of strategies may take the form of a random variable with a known probability distribution. Generally, in choosing among strategies, it is necessary to specify a preference relation which chooses among probability distributions over the set of outcomes. If the von Neumann and Morgenstern axioms are valid for this preference relation, then the preference associated with any pair of pure strategies can be expressed by a number obtained by applying utility notions to distributions over outcomes. In this way the pay-off function can be so constructed as to be consistent with the preference pattern. Moreover, the pay-off operates additively (i.e., is linearly convex) when computing expected values with respect to probability distributions on the set of pure strategies. Hence the formula (1.1.1) obtains.

§ 1.2. The poker example of this section was first introduced and studied by Kuhn [156]. A general technique for the analysis of parlor games is developed in Volume II Chapter 9 of the present work.

§ 1.3. The concept of the upper and lower values was proposed by von Neumann, who refers to corresponding situations as the majorant game and minorant game, respectively.

§ 1.4. Proofs of the fundamental min-max theorems have been given by von Neumann [239], Ville [237], Loomis [168], Karlin [133], Fan [77], Glicksberg [108], and others. All proofs except Loomis' utilize in essence some form of convexity and compactness of the strategy spaces.

§ 1.5. The first proof in this section is due to Kakutani [130], who obtained, in the course of establishing the min-max theorem, a generalization of the classical Brouwer fixed-point theorem. Kakutani's method depends on certain convexity properties of the mapping function considered. Eilenberg and Montgomery [73] and Begle [18] have shown that this generalized fixed-point theorem is valid when the requirement of simple connectedness for certain sets is substituted for convexity. Further extensions along these lines have been given by Debreu [61]. The second proof in this section was proposed by Shiffman [221].

§ 1.6. Problems 2 and 3 are based on Borel [39]. The statement of Problem 12 was first observed by I. Glicksberg. The conclusions of Problems 15 and 16 were used by Fan [77]. Problems 17 and 18 were suggested by L. Shapley. Problem 19 is due to Wolfe [255].

CHAPTER 2

THE NATURE AND STRUCTURE OF INFINITE GAMES

This chapter is descriptive in nature; its purpose is to introduce the reader, in broad terms, to the concepts and techniques of infinite games. The chapter is essential for what follows, but it need not be studied in detail, since formal proofs and particulars of the results it discusses are presented in Chapters 3–9.

2.1. Introduction. In Chapter 1 we defined a game as a triplet $\{X, Y, K\}$, where X is the strategy space for Player I, Y is the strategy space for Player II, and K is a function defined for $X \otimes Y$ whose values are interpreted as the pay-off to Player I. This same definition applies to the theory of infinite games, the only difference being that each of the components of an infinite game has a more complicated structure. In all circumstances, we shall require that X and Y be convex sets and that $K(x, y)$ define a real-valued function that is convex in y and concave in x. In most examples, K is in fact bilinear.

In Chapter 1 we specified that the sets X and Y must be identifiable with subsets of finite-dimensional Euclidean spaces. This is always the situation in finite matrix games, where X is specifically composed of all vectors $\mathbf{x} = \langle \xi_1, \xi_2, \ldots, \xi_n \rangle$ which lie in the intersection of the positive orthant of E^n and the hyperplane $\xi_1 + \xi_2 + \cdots + \xi_n = 1$, and Y is analogously described.

A natural extension of the theory of games is to the case in which the spaces X and Y are infinite-dimensional. In this event we must distinguish two categories: either (1) X and Y are small enough to be identified with subsets of a classical function space, or (2) X and Y cannot be represented in such terms. Throughout Chapters 2–9 we concentrate exclusively on games of the first category. Games of the second category will be discussed in a future volume.

Games of the first category are chiefly of two types: (a) X consists of a set of probability measures defined over a region lying in Euclidean n-space; (b) X consists of a set of n-tuples, each component of which is a function defined on a prescribed set of E^n with values restricted between 0 and 1. Specific cases of each kind will be described in Sections 2.3 and 2.4.

In representing a game we assume the strategy spaces to be exhaustive; i.e., we take into account all possible modes of procedure, good and bad. It will be clear to the reader that in each specific application of game theory in these pages, the formal game as constructed completely describes the given conflict of interest.

As with the matrix games, in order to ascribe a proper interpretation to a game and to delimit a rational mode of behavior for each of the antagonists, it is necessary that the relationship

$$\min_{y \in Y} \max_{x \in X} K(x, y) = \max_{x \in X} \min_{y \in Y} K(x, y) \tag{2.1.1}$$

be valid. We call this common number the "value" of the game. It follows from (2.1.1) that there exist strategies x_0 in X and y_0 in Y satisfying

$$K(x, y_0) \leq K(x_0, y_0) \leq K(x_0, y) \tag{2.1.2}$$

for all y in Y and x in X. The proof of this statement is very simple and resembles that of Theorem 1.3.1. We omit the formal details.

Any pair of strategies x_0 and y_0 satisfying (2.1.2) are said to be optimal strategies (solutions) for Players I and II, respectively. Relation (2.1.2) states that if Player I uses x_0, he can ensure himself a yield of at least $v = K(x_0, y_0)$ regardless of the strategy used by Player II, and that Player II can prevent Player I from attaining more than v by using the strategy y_0. If Player II does not employ an optimal strategy, Player I by a judicious choice of strategy can secure more than v, so that unless Player II is unwise or desires to be excessively charitable (contrary to the conflict-of-interest motivation) he will in general play optimally. Thus, as in the finite case, v ultimately measures the average pay-off accruing to Player I when both players act rationally.

In the theory of infinite games, the truth of relation (2.1.1) is a deep question, requiring some kind of assumption of continuity for $K(x, y)$ and the restriction that at least one of the spaces X and Y is a compact space in some suitable sense. It suffices, for example, to have $K(x, y)$ continuous in both variables and X compact. See the notes at the close of the chapter for a discussion of an abstract version of the problem (see also problems 1–3). In the remainder of this volume we shall assume that (2.1.1) holds for the class of games with which we are concerned, unless there is an explicit statement to the contrary. Actually, we shall on most occasions verify that the inequalities of (2.1.2) hold, and hence that (2.1.1) holds for the games considered.

If the conditions on X, Y, and K are relaxed slightly, it is sometimes possible to prove a weaker form of (2.1.1), namely

$$\sup_{x \in X} \inf_{y \in Y} K(x, y) = \inf_{y \in Y} \sup_{x \in X} K(x, y) = v. \tag{2.1.3}$$

This is essentially equivalent to guaranteeing the existence of ϵ-optimal strategies for an arbitrary positive value ϵ. A pair of strategies x_0 and y_0

are said to be ϵ-optimal if

$$K(x_0, y) \geq v - \epsilon \qquad \text{for all} \quad y \text{ in } Y$$

and

$$K(x, y_0) \leq v + \epsilon \qquad \text{for all} \quad x \text{ in } X.$$

In most cases of interest, if (2.1.3) holds, then (2.1.2) also holds.

Our definition of a game here, as in Chapter 1, is given directly in terms of the full strategy spaces, rather than initially in terms of pure strategies alone. Nevertheless, for many classes of games it is advantageous and natural to single out the set of pure strategies, which span convexly the set of all strategies. In particular, this distinction is relevant to games played over the unit square (see Section 2.2).

Within the framework of a game as a triplet $\{X, Y, K\}$, following the extension from pure to mixed strategies, we may enlarge X to a set $\overline{X}$ comprising all regular probability measures defined over the Borel field generated by the open subsets of X. We form $\overline{Y}$ from Y similarly, and put

$$K(\mu, \nu) = \int_Y \int_X K(x, y)\, d\mu(x)\, d\nu(y)$$

for μ in $\overline{X}$ and ν in $\overline{Y}$. The function K must be measurable with respect to the Borel field of subsets as specified.

The points of X may be identified as the degenerate probability measures, labeled μ_x, which concentrate their full mass at the point x. That is,

$$\mu_x(B) = \begin{cases} 0 & B \not\supset \{x\}, \\ 1 & B \supset \{x\}, \end{cases}$$

where B is a measurable set in X. A similar meaning is attached to the symbol ν_y. This extension process is analogous to the procedure of embedding a space of pure strategies in a space of mixed strategies. If, however, the min-max property (2.1.1) already holds for the game $\{X, Y, K\}$, the above extension is of no significance. In this event, any pair $\{x_0, y_0\}$ of optimal strategies for the original game is also optimal for the extended game with the same value. Suppose

$$K(x_0, y) \geq v \qquad \text{and} \qquad K(x, y_0) \leq v \tag{2.1.4}$$

for all y and x, respectively. Then, for ν in Y and μ in X, by integrating the inequalities of (2.1.4) we obtain

$$K(\mu_{x_0}, \nu) \geq v \qquad \text{and} \qquad K(\mu, \nu_{y_0}) \leq v$$

for every ν in $\overline{Y}$ and μ in $\overline{X}$, respectively.

In relation to playing the game, the interpretation of the preceding remarks is as follows: To carry out a μ in $\bar{X}$ we select an x in X by randomization on the basis of the probability measure μ, and use x in playing the game. The above arguments show that such randomization devices cannot guarantee on the average a larger yield than the optimal strategy x_0 in X.

To sum up: In setting up a game the strategy spaces X and Y are usually taken to describe all possible courses of action. If (2.1.1) holds, then any further extension of the concept of strategy to that of mixed strategy as indicated above does not alter the value.

We close this section by noting the following useful property of optimal strategies.

▶ LEMMA 2.1.1. If X^0 (Y^0) denotes the set of optimal strategies for Player I (II), then X^0 (Y^0) is convex.

Proof. The proof follows by virtue of the hypothesis that X is convex and K is concave in the variable x. Explicitly, if x_1 and x_2 belong to X^0, then

$$K(\lambda x_1 + (1 - \lambda)x_2, y) \geq \lambda K(x_1, y) + (1 - \lambda)K(x_2, y) \geq v$$

for all y, $0 \leq \lambda \leq 1$, and hence $\lambda x_1 + (1 - \lambda)x_2$ is in X^0.

2.2 Games on the unit square. One of the simplest versions of the continuous analog of finite matrix games is the class of games played over the unit square. In such games, the triplet $\{X, Y, K\}$ consists of a kernel function $K(\xi, \eta)$ of the two variables ξ and η, each of which ranges over the closed unit interval [0, 1], and strategy spaces X and Y, consisting of all cumulative distribution functions $x(\xi)$ and $y(\eta)$, respectively. (The reader is referred to Section C.3 in the Appendix for a discussion of distributions on the real line.) The pay-off corresponding to a choice of x by Player I and y by Player II is

$$K(x, y) = \int_0^1 \int_0^1 K(\xi, \eta)\, dx(\xi)\, dy(\eta). \tag{2.2.1}$$

The special strategies

$$x_{\xi_0}(\xi) = \begin{cases} 0 & \xi < \xi_0, \\ 1 & \xi \geq \xi_0 \end{cases} \quad \text{and} \quad y_{\eta_0}(\eta) = \begin{cases} 0 & \eta < \eta_0, \\ 1 & \eta \geq \eta_0 \end{cases}$$

are called "pure strategies"; i.e., in this example the pure strategies for each player are identified with points in the unit interval of the real line. (We sometimes use the notation I_γ for the degenerate distribution which represents the pure strategy where only the value γ can be observed.) A general strategy $x(\xi)$ in X is a probability mixture of pure strategies. $K(x, y)$ represents the expected yield to Player I for the pure strategies ξ and η, respectively, where ξ is randomized according to x and η is independently randomized according to y.

We shall frequently use, for simplicity, the following notation: $K(x, \eta)$ for $K(x, y_\eta)$; $K(\xi, y)$ for $K(x_\xi, y)$, and $K(\xi, \eta)$ for $K(x_\xi, y_\eta)$. In words, $K(x, \eta)$ is the expected yield to Player I if he adopts the strategy x while Player II uses the pure strategy y_η.

There are certain points of correspondence between games on the unit square and matrix games. In matrix games there is a finite number of pure strategies, whereas in the present case there is a one-dimensional continuum of pure strategies identified with the unit interval. In both cases, the complete strategy spaces are generated from the pure strategies; i.e., they consist of all possible probability mixtures of the pure strategies. Clearly, games of increasing degrees of complexity can be obtained by enlarging the dimensionality and cardinality of the space of pure strategies. Applications of generalized games in which the pure strategy spaces coincide with n-dimensional unit cubes are discussed in Sections 4.4 and 5.4.

As remarked in Section 2.1, the existence of optimal strategies for infinite games and the rational interpretation of these games hinge on the validity of (2.1.1). If $K(\xi, \eta)$ is a continuous function of both variables, and if $0 \leq \xi, \eta \leq 1$, then it can be shown that (2.1.1) holds. As we have noted, this fact will be assumed hereafter.

One objective in this volume is to determine ways of calculating explicit optimal strategies for both players for any game on the unit square. Of almost equal importance and relevance is the goal of characterizing properties of solutions as a function of the form of the game. The following discussion indicates the inherent obstacles confronting the analysis of games on the unit square.

Specify an arbitrary pair of strategies x_0 and y_0 for Players I and II, respectively. Then there exists a continuous kernel $K(\xi, \eta)$ on the unit square such that x_0 and y_0 are the unique optimal strategies for this game. The kernel $K(\xi, \eta)$ can also be constructed as an analytic function in each of its variables. By way of illustration, take x_0 and y_0 both equal to the infamous Cantor distribution (see Section 7.1); then there exists a rational function $K(\xi, \eta)$ that is continuous for $0 \leq \xi, \eta \leq 1$, with x_0 and y_0 the unique optimal strategies.

In view of the weird nature of optimal strategies for seemingly reasonable kernels, it would appear to be intrinsically difficult to characterize simply the solutions associated with a kernel. It seems proper, therefore, to investigate special classes of kernels which arise naturally in the study of examples, in an effort to find properties of the solutions that are functionally dependent on the nature of the kernel. From a mathematical point of view, the problem is one of classification; in effect, we seek to relate assumptions about the special characteristics of the kernel to the form of the solution. In the following section we shall describe a series of special classes of continuous kernels on the unit square, together with results that are valid for them.

We first introduce some appropriate terminology and establish some simple facts for general infinite games on the unit square. The definitions are given in terms of x strategies; the corresponding notation and terminology apply for y strategies.

For any distribution f on the real line, and in particular for members of X and Y, the support or spectrum of the measure f is defined as the complement of the largest open set in which f vanishes (see Section C.3). Clearly, if ξ is in the support of $f(\xi)$, then for every open interval $I_{a,b}$ about ξ we have $f(b) - f(a) > 0$.

A pure strategy ξ is said to be "essential" if there is an optimal strategy x in which ξ is used, i.e., if ξ belongs to the spectrum of x.

A strategy x is said to be of finite type if x is a finite convex combination of pure strategies; that is to say, the support of x consists of a finite number of points and x has the representation

$$x = \sum_{i=1}^{n} \lambda_i x_{\xi_i}, \qquad \text{where} \quad \lambda_i > 0, \qquad \sum_{i=1}^{n} \lambda_i = 1.$$

The pure strategies $\xi_1, \xi_2, \ldots, \xi_n$ involved in x are said to be "relevant" or "essential" to x with weights $\lambda_1, \lambda_2, \ldots, \lambda_n$, respectively.

A finite strategy is effected by choosing a single number j from 1 to n according to the probabilities λ_j $(j = 1, \ldots, n)$, and then following the dictates of the pure strategy ξ_j. Games which possess optimal strategies of finite type are significant because of their simple form.

Another class of strategies that we encounter frequently is the class of completely mixed (c.m.) strategies. These are the strategies whose spectrum is the full unit interval. For example, if x is a distribution with a density that is positive on the unit interval [0, 1], then x is c.m.

The following lemma will prove useful.

▶ **Lemma 2.2.1.** Let x_0 and y_0 be optimal. If η_0 is in the support of y_0, then $K(x_0, \eta_0) = v$, where v is the value of the game. (This is the direct analog of Lemma 2.1.2 of Volume I.)

Proof. Since x_0 is optimal, $K(x_0, \eta) \geq v$ for $0 \leq \eta \leq 1$. If $K(x_0, \eta_0)$ were strictly greater than v, inequality would hold for an interval about η_0 (recall that $K(\xi, \eta)$ is assumed to be continuous with respect to both variables). But η_0 is in the support of y_0, which means that this interval has positive y_0 measure. Hence if we integrate $K(x_0, \eta)$ with respect to

dy_0, we obtain $K(x_0, y_0) > v$, which contradicts the assumption that y_0 is optimal. The same result is valid for the other player.

A strategy $\tilde{x}$ is called an "equalizer" if $K(\tilde{x}, \eta) = c$ for $0 \leq \eta \leq 1$ and some constant c. From Lemma 2.2.1 we obtain the following corollaries.

▶ COROLLARY 2.2.1. If x^0 is a completely mixed optimal strategy, then every optimal strategy for Player II is an equalizer strategy.

▶ COROLLARY 2.2.2. If every ξ in the unit interval is essential for Player I, then every optimal strategy for Player II is an equalizer.

A game with kernel $K(\xi, \eta)$ is said to be symmetric if $K(\eta, \xi) = -K(\xi, \eta)$. This concept is completely analogous to the concept of symmetric matrix games. Moreover, the results are fully parallel: the value of a symmetric game is zero and the sets of optimal strategies for the players coincide. Since these statements can be proved routinely, we omit the formal arguments.

One final definition: A pair of pure strategies $\{\xi^0, \eta^0\}$ is said to constitute a saddle point for the game if

$$K(\xi^0, \eta) \geq K(\xi^0, \eta^0) \geq K(\xi, \eta^0) \qquad \text{for} \quad 0 \leq \xi, \eta \leq 1.$$

2.3 Classes of games on the unit square. We describe in this section the structure of certain important classes of games defined on the unit square, together with certain characteristics of each class which are useful in determining solutions to games in that class. By way of illustrating the general scope of these games, we relate our examples to situations in politics, military tactics, and business.

Proofs of all the assertions relating to these classes of games will be found in Chapters 2–7.

Games with separable kernels. It is possible to reduce certain infinite games defined on the unit square to finite games. This can be done if the kernel $K(\xi, \eta)$ has the form

$$K(\xi, \eta) = \sum_{i=1}^{n} \sum_{j=1}^{m} a_{ij} r_i(\xi) s_j(\eta),$$

where $r_i(\xi)$ and $s_j(\eta)$ are continuous functions. Such a game is called a separable game. A direct calculation shows that

$$\begin{aligned} K(x, y) &= \sum_{i,j} a_{ij} \int_0^1 r_i(\xi)\, dx(\xi) \int_0^1 s_j(\eta)\, dy(\eta) \\ &= \sum_{i,j} a_{ij} r_i s_j, \end{aligned} \tag{2.3.1}$$

with $\mathbf{r} = \langle r_1, r_2, \ldots, r_n \rangle$ and $\mathbf{s} = \langle s_1, s_2, \ldots, s_m \rangle$ defined in the obvious way. Consequently, if the pay-off values accruing from the use of strategies are the guide to the actions undertaken, then the separable game with kernel (2.3.1) may be redefined as the triplet $\{R, S, A\}$. Here R consists of all possible vectors $\mathbf{r} = \langle r_1, r_2, \ldots, r_n \rangle$ such that

$$r_i = \int_0^1 r_i(\xi)\, dx(\xi)$$

for x in X; S is similarly formed, consisting of all vectors

$$\mathbf{s} = \langle s_1, s_2, \ldots, s_m \rangle$$

such that

$$s_j = \int_0^1 s_j(\eta)\, dy(\eta)$$

for y in Y; and $\mathbf{A}$ is the matrix $\|a_{ij}\|$. The pay-off for a given $\mathbf{r}$ and $\mathbf{s}$ is $(\mathbf{r}, \mathbf{As}) = A(\mathbf{r}, \mathbf{s})$. Although several distinct elements of X may give rise to the same point of R, because of (2.3.1) they can influence the pay-off only through the component values of the vector $\mathbf{r}$, and thus all $\mathbf{x}$ with the same $\mathbf{r}$ produce identical contributions to the pay-off. Finally we note that the sets R and S are convex, bounded, and closed, since $r_i(\xi)$ and $s_j(\eta)$ are continuous functions.

By applying the general min-max theorem of Section 1.5 to the game $\{R, S, A\}$ we deduce the existence of $\mathbf{r}^0$ and $\mathbf{s}^0$ (not necessarily unique) such that

$$A(\mathbf{r}^0, \mathbf{s}) \geq A(\mathbf{r}^0, \mathbf{s}^0) \geq A(\mathbf{r}, \mathbf{s}^0)$$

for all $\mathbf{r}$ in R and $\mathbf{s}$ in S. Any x^0 in X whose image of X in E^n by the mapping $Ux = \{\int r_i(\xi)\, dx(\xi)\}$ is $\mathbf{r}^0$, is a solution for the game $\{X, Y, K\}$. Similarly, optimal strategies for Player II are those y^0 in Y which have the image $\mathbf{s}^0$ by the linear transformation $Vy = \{\int s_j(\eta)\, dy(\eta)\}$. The converse is also true: any optimal strategies $\{x^0, y^0\}$ for the separable game $\{X, Y, K\}$ give rise to optimal strategies $\{\mathbf{r}^0, \mathbf{s}^0\}$ for the reduced game $\{R, S, A\}$.

Therefore, solving a separable game is equivalent to determining all solutions of an associated finite game, provided we can recognize the inverse image of the mappings U and V for any given $\mathbf{r}$ and $\mathbf{s}$, respectively. This problem is examined in the following chapter.

An important separable game is the game with kernel

$$K(\xi, \eta) = \sum_{i=0}^{n} \sum_{j=0}^{m} a_{ij}\xi^i\eta^j \qquad (0 \leq \xi \leq 1, 0 \leq \eta \leq 1).$$

Since any game with a continuous kernel defined on the unit square can be uniformly approximated by polynomial kernels, the corresponding optimal strategies possess limit strategies which are optimal for the original game (see Theorem 3.3.1). However, the limiting procedure in general entails great complexities, and consequently polynomial games are not very useful in deducing the nature of solutions for general continuous games on the unit square. Nevertheless, polynomial games are of interest to us here for two reasons. First, a fairly complete analysis of the properties of the solutions to such games can be given, and this analysis helps indicate how to approach general separable games. Second, explicit computational methods exist for determining solutions to polynomial games (see Section 3.2 and Problems 1–3, of Chapter 3).

We shall prove in Chapter 3 that both players in a separable game possess optimal strategies of finite type. Furthermore, it is possible to establish the existence of solutions having bounds on the size of their spectrums determined by the number of terms in the representation of $K(\xi, \eta)$. More precise statements relating to the spectrums of optimal strategies for both players can be given in the case of polynomial games.

Games with convex kernels. Convex kernels are kernels such that

$$\frac{\partial^2 K(\xi, \eta)}{\partial \eta^2} \geq 0$$

for each ξ and η in the unit interval. In games defined by convex kernels, Player II possesses a pure optimal strategy while Player I has an optimal strategy involving at most two points in its spectrum. Since we know this much about the form of the optimal strategies, games of this sort are readily solved (see Section 4.2).

The analysis of these games can be generalized to yield results for kernels which satisfy

$$\frac{\partial^r K(\xi, \eta)}{\partial \eta^r} \geq 0$$

for some r, and also to the case in which ξ ranges over a compact connected set R in E^n, η ranges over a convex compact set S in E^m, and $K(\xi, \eta)$, continuous in both variables, is a convex function of η for each ξ. The reader should note that since S is m-dimensional, we are speaking here of a convex function of several variables (see Section B.4).

We now describe a problem which gives rise to a game of this type. The model is a military situation of area defense. Let Player II be the defender of n targets $T_1, \ldots, T_n$, whose relative values are $k_1, \ldots, k_n$. We order the targets in such a way that $k_1 \geq k_2 \geq \cdots \geq k_n$; i.e., target 1 is the most important, target 2 the next most important, etc. The total defending force is D. A pure strategy for Player II is an n-tuple of nonnegative numbers $\langle \eta_1, \eta_2, \ldots, \eta_n \rangle$ satisfying $\sum \eta_i = D$, where η_i represents the part of force assigned to defend T_i. Let Player I be the attacker, and denote his total force by A. A pure strategy for Player I is a vector $\langle \xi_1, \ldots, \xi_n \rangle$ with $\xi_i \geq 0$ and $\sum \xi_i = A$, such that the portion of his force attacking T_i is ξ_i.

We assume that at T_i the value to the attacker is proportional to $\xi_i - \eta_i$, provided the attacking force is greater than the defending force; otherwise the value is zero. Hence

$$K(\xi, \eta) = \sum_{i=1}^{n} k_i \max(0, \xi_i - \eta_i). \tag{2.3.2}$$

The function is convex in each variable ξ and η and jointly continuous; ξ and η vary over simplexes with vertices $\mathbf{d}_1, \ldots, \mathbf{d}_n$ and $\mathbf{a}_1, \ldots, \mathbf{a}_n$, respectively, where

$$\mathbf{d}_1 = \langle D, 0, \ldots, 0 \rangle, \mathbf{d}_2 = \langle 0, D, 0, \ldots, 0 \rangle, \ldots, \mathbf{d}_n = \langle 0, \ldots, 0, D \rangle,$$

and

$$\mathbf{a}_1 = \langle A, 0, \ldots, 0 \rangle, \ldots, \mathbf{a}_n = \langle 0, \ldots, 0, A \rangle.$$

This example may also be interpreted in terms of a political campaign, with two opposing political parties each trying to decide in which of several localities to invest campaign money. The details of this example are discussed in Chapter 4, and a complete solution is presented.

The key theorem underlying the analysis of the example is to the effect that for a generalized convex game—a game for which $K(\xi, \eta)$ is convex in η, ξ traverses R, η traverses S, and S is n-dimensional—Player II has an optimal pure strategy and Player I has an optimal strategy of finite type with a spectrum consisting of at most $n + 1$ points.

Games of timing. Games of timing are games in which the choice of a pure strategy represents the choice of a time to perform a specific action.

Suppose, for example, that two large mail-order companies are intending to issue their yearly catalogs. When to issue the catalog is a big question. The company that comes out first with its catalog has the advantage of getting a jump on its competitor. On the other hand, the company that waits until later can capitalize on weaknesses in its competitor's catalog.

Or suppose that two opposing political parties are trying to decide when to send out their campaign pamphlets. Some people are influenced early and never change their minds. Others are influenced most by what has reached them most recently.

The following time problem is common in the book publishing business. Two competing outfits are trying to sign up a prospective author. The timing—i.e., the time for the publisher's representative to appear ready with the contract and various inducements—represents the set of pure strategies. Too early a contact may meet with failure, since the author is not far enough along in his work to be interested in a publishing decision. On the other hand, a late bid may be too late; the competitor may have gotten there first. The right psychological moment is a well-known and much debated concept in this game.

The reader can readily think of many other examples of games of timing. Such games need not involve only one possible action (as in the above examples), but may involve several repeated actions.

From the mathematical viewpoint the game of timing is described as a game defined over the unit square, whose pay-off kernel has the following properties:

(i)
$$K(\xi, \eta) = \begin{cases} L(\xi, \eta) & \text{if} \quad \xi < \eta, \\ \phi(\xi) & \text{if} \quad \xi = \eta, \\ M(\xi, \eta) & \text{if} \quad \xi > \eta. \end{cases} \tag{2.3.3}$$

(ii) Each of the functions $K(\xi, \eta)$ and $M(\xi, \eta)$ is jointly continuous in ξ and η.

(iii) $L(\xi, \eta)$ is monotone-increasing in ξ for each η,
$M(\xi, \eta)$ is monotone-increasing in ξ for each η,
$L(\xi, \eta)$ is monotone-decreasing in η for each ξ,
$M(\xi, \eta)$ is monotone-decreasing in η for each ξ.

The variables ξ and η are to be viewed as the times when the two opponents act. The fact that they range over the unit interval is a normalization condition; it involves no real restriction of the model. The monotonicity of the functions $L(\xi, \eta)$ and $M(\xi, \eta)$ and the discontinuity of the kernel at $\xi = \eta$ have the following interpretation: If Player II is going to act at a fixed time η, Player I improves his chances of success by waiting as long as possible, provided he acts before Player II does. If Player I waits until after Player II acts, however, he may lose if Player II is successful; hence the discontinuity at $\xi = \eta$. Once Player II has acted, Player I's chances of success increase with time. This is expressed by the monotonicity of $M(\xi, \eta)$ as a function of ξ. The analogous statements apply for Player II, since he benefits whenever the pay-off to Player I decreases.

Here we encounter for the first time a pay-off kernel $K(\xi, \eta)$ which may be discontinuous on its diagonal. This is permissible, since the abstract theory of games can be invoked to demonstrate that kernels possessing discontinuities only on the diagonal satisfy

$$\sup_x \inf_y \iint K(\xi, \eta)\, dx(\xi)\, dy(\eta) = \inf_y \sup_x \iint K(\xi, \eta)\, dx(\xi)\, dy(\eta). \quad (2.3.4)$$

If (2.3.4) is valid, there exist ϵ-optimal strategies, and conversely. In all specific cases under consideration ϵ-optimal strategies will be explicitly exhibited, thus establishing the truth of (2.3.4) in a direct way. Even more, in most instances we may replace sup by max and inf by min in (2.3.4).

Such games have not only practical significance but theoretical interest, since they can be thoroughly analyzed and their optimal strategies can be fully characterized, both qualitatively and quantitatively. It turns out that the optimal strategies are composed of discrete jumps plus a density part. The density part either is obtained by solving a standard eigenvalue problem of a linear integral equation, or is expressible as a Neumann series (see Chapter 5, where these notions are elaborated). Frequently the payoff kernel has a separable form, which allows us to replace the integral equation by a more readily resolved system of differential equations. In almost all cases the optimal strategies are unique, which simplifies their determination and adds to their attractiveness.

There are two classes of games of timing. In class I, the simpler class, $L(\xi, \eta)$ is a function of ξ alone and $M(\xi, \eta)$ is a function of η alone. We shall refer to such a game as a game of timing of "complete information." The idea of complete information is to be understood here in the sense that when either player acts, his action and its consequences become known to his opponent. The mail-order catalog example may be considered as a game of complete information for obvious reasons.

Patterns of complete information as they occur in matrix games usually imply that optimal pure strategies may be found. Randomization of pure strategies arises most often because of the uncertainties surrounding the problem. Optimal strategies are almost necessarily mixed strategies whenever some form of uncertainty is present in the model. In view of these heuristic remarks, a further justification for the appelation "games of timing of complete information" rests in the observation that games of this class usually possess optimal pure strategies.

Class II consists of all games of timing not in class I, i.e., games in which L or M or both explicitly depend on both ξ and η. In such games a player's acts are unknown to his opponent. So-called silent duels are of this type (see Chapter 5); so is the competition between two publishing houses described above. Optimal strategies for games in this class generally involve a genuine randomization of pure strategies.

Chapter 5 contains a complete analysis of games of timing of class II and a discussion of several interesting examples which illustrate methods of solving such games. Chapter 6 is devoted primarily to a study of games of timing of class I; an introduction to the theory of games of timing involving several actions is also presented in this chapter. The integral-equation methods that are used to solve games of timing are applied to kernels of the form given in (2.3.3), where the kernels are assumed to be continuous throughout the unit square except for a discontinuity in the first derivative along the principal diagonal (Section 6.5).

Bell-shaped games. The following game is a bell-shaped game. A submarine S seeks to escape a depth bomb to be dropped by an airplane A. During the time before the bomb is discharged the submarine (Player II) can locate its position anywhere in the interval $-a \leq \eta \leq a$, where a is a fixed positive constant. The plane is aware of the possible positions of the submarine. If S locates itself at η and A aims the bomb at ξ, we assume the damage done is proportional to the error function $e^{-b(\xi-\eta)^2}$ (b is a positive constant). With appropriate renormalization the game may be formulated as a continuous game played on the unit square such that

$$K(\xi, \eta) = e^{-\lambda(\xi-\eta)^2},$$

where λ is a parameter depending on the constants b and a. The submarine chooses η, the airplane selects ξ, and the pay-off is K as given above.

A kernel K is said to be bell-shaped if $K(\xi, \eta) = \phi(\xi - \eta)$, where ϕ has the following properties: (i) $\phi(u)$ is an analytic function defined for all real u; (ii) for every n and every set of values ξ_i and η_j such that $\xi_1 < \xi_2 < \cdots < \xi_n$ and $\eta_1 < \eta_2 < \cdots < \eta_n$, the determinant of the matrix $\|\phi(\xi_i - \eta_j)\|$ is positive; and (iii)

$$\int_{-\infty}^{\infty} \phi(u)\, du < \infty.$$

The function $\phi(u) = e^{-\lambda u^2}$ satisfies these conditions (see Section 7.2).

We can prove that bell-shaped games possess unique optimal strategies of finite type (Chapter 7). In the case of $\phi(u) = e^{-\lambda u^2}$ the explicit optimal strategies may be computed by a recursive scheme with λ serving as a varying parameter.

Miscellaneous games. Limited results can be obtained for other games as a function of assumptions about the kernel of the game. Two such games that deserve mention here are analytic games and games that are invariant with respect to groups of transformations.

If $K(\xi, \eta)$ is an analytic function of each of the variables and obeys suitable growth properties, then optimal strategies of finite type exist. In general, however, it is very difficult to discern properties for solutions of analytic kernels. To indicate the difficulty, we need only recall the proposition cited previously (see also Section 7.1) to the effect that any two strategies x and y may qualify as the unique solutions for an analytic kernel. It may be argued that since games characterized purely by the property of analyticity do not arise naturally, it is not surprising that no elegant distinctive theory exists for such games. This argument cannot be lightly dismissed. There is indeed some reason to believe that in order for a pattern to exist between solution and kernel, there must be a physical model which reflects the given kernel in a natural way. This is true in any event for convex games, games of timing, and bell-shaped games.

Unlike analytic games, games that are invariant with respect to groups of transformations are often readily solved. Examples of these are games with kernels of the form $K(\xi, \eta) = \phi(\xi - \eta)$, where ϕ is a periodic function of period 1. Here it is easy to see that the uniform distribution $d\xi$ is an optimal strategy for each player (Section 7.5). If a game is so constructed that a group of transformations operating on the kernel induces a group of transformations on the strategy spaces such that the pay-offs are unchanged, we might reasonably expect to find optimal strategies that are also invariant with respect to the elements of the transformation groups. Heuristically speaking, this states that operations which change the labels of the strategies should not affect the solution of the game. In the example above, the relevant group of transformations is the translation group on the real line, and the only invariant distribution with respect to all translations modulo 1 is the uniform distribution.

The theory of invariance groups and decision problems is basic to the solution of problems involving statistical analysis; such matters, however, are beyond the scope of this book. In Section 7.5 we note several simple applications of invariance concepts to the theory of games.

For other instructive examples of miscellaneous games the reader should consult the problems of this chapter and those of Chapter 7.

2.4 Infinite games whose strategy spaces are known function spaces. In the preceding two sections we proposed for analysis various classes of games defined on the unit square whose strategy spaces X and Y are identified with the set of all probability distributions on the unit interval. In this section we describe some typical examples of games whose strategy spaces are subsets of recognizable function spaces, but which cannot be regarded as games played over the unit square.

Fighter-bomber duel. A fighter F provided with a single round of ammunition tries to destroy a bomber B. Since F can fire only one blast, his strategy is a matter of choosing when to discharge his blast. Let $a(t)$ be the accuracy of the fighter, i.e., the probability that B will be destroyed if F fires at time (distance) t; the variable t is normalized to vary between 0 and 1. A mixed strategy for F (Player I) is a probability distribution over [0, 1]. We assume in this model that once F has fired his salvo, successfully or otherwise, he flies off (if he can) and the duel is ended.

The bomber is equipped with a machine gun and in defending itself seeks to destroy the fighter. A strategy for B is specified by a function $p(t)$, which measures the intensity with which the machine gun is being fired at time t. We assume that B's total amount of ammunition, if fired at maximum intensity, will last for a period $\delta < 1$. Therefore nonrandomized strategies $p(t)$ for B consist of all functions satisfying the restrictions

$$0 \leq p(t) \leq 1 \quad \text{and} \quad \int_0^1 p(t)\,dt = \delta. \tag{2.4.1}$$

The accuracy $r(t)$ of the bomber is given by a continuous monotone function with $r(0) = 0$, $r(1) = 1$. This function measures the accuracy in the sense that if B fires with intensity $p(t)$ over a small interval of time $[t, t + h]$ then the probability that F will be destroyed is $p(t)r(t)h + o(h)$.

Suppose $\phi(t, p)$ is the probability that F will survive up to time t when B has chosen to fire with intensity $p(t)$. Clearly,

$$\phi(t + h, p) = \phi(t, p)\,[1 - p(t)r(t)h] + o(h)$$

for h small and positive, and hence

$$\frac{\phi(t + h, p) - \phi(t, p)}{h} = -\phi(t, p)p(t)r(t) + o(1).$$

In the limit we find that $\phi(t, p)$ satisfies the differential equation

$$\phi'(t, p) = -\phi(t, p)r(t)p(t),$$

with the initial condition $\phi(0, p) = 1$, whose solution is

$$\phi(t, p) = \exp\left[-\int_0^t p(u)r(u)\,du\right]\cdot \tag{2.4.2}$$

The pay-off function for the game $K(\xi, p)$ when F uses a pure strategy ξ (time of firing the single burst of ammunition) and B uses the strategy p is evaluated as follows:

Denote the pay-off to F by α if F destroys B, by $-\beta$ if B destroys F, and by zero if neither is destroyed. Then the kernel given as the expected pay-off accruing to F (Player I) is

$$K(\xi, p) = \alpha a(\xi)\phi(\xi, p) - \beta[1 - \phi(\xi, p)]. \tag{2.4.3}$$

The pay-off for a mixed strategy x for Player I and the choice of p for Player II is obtained by averaging (2.4.3) with respect to $dx(\xi)$:

$$K(x, p) = \int_0^1 K(\xi, p)\,dx(\xi). \tag{2.4.4}$$

We may summarize this game as the triplet $\{X, Y, K\}$, where X is the space of probability distributions on the unit interval, Y consists of all p satisfying (2.4.1), and K is computed according to the formula (2.4.4).

The differences in form between this game and games defined on the unit square are apparent. The strategy space Y in this case is of a different character and requires new methods of analysis. In deriving the form of the optimal strategy for F, we make use for the first time of the classical Neyman-Pearson lemma (see Section 8.4), a fundamental proposition of statistics. The use of the Neyman-Pearson lemma is natural in view of the nature of the constraints (2.4.1).

In Chapter 8 we characterize the solution to this problem and several of its variants, including the two-machine-gun duel.

Poker game. One rich source of examples of nontrivial games is the class of parlor games. Variants of poker games, in particular, provide an inexhaustible flow of interesting and challenging game models. Over and above their place in game theory, poker problems are interesting because of their marked similarity to the statistical decision problem of testing hypotheses.

Poker games may be described generally as follows. Strategies for Players I and II are identified with vectors of functions

$$\boldsymbol{\phi}(\xi) = \langle \phi_1(\xi), \phi_2(\xi), \ldots, \phi_n(\xi) \rangle \qquad (0 \leq \phi_i(\xi) \leq 1)$$

and

$$\boldsymbol{\psi}(\eta) = \langle \psi_1(\eta), \psi_2(\eta), \ldots, \psi_m(\eta) \rangle \qquad (0 \leq \psi_j(\xi) \leq 1).$$

These vectors may be subject to additional linear constraints. We assume that ξ and η are random variables whose distribution functions are F and G, respectively, and which are observed privately by Players I and II, respectively. The strategy $\boldsymbol{\phi}$ for Player I is a function of the observation ξ and Player II's possible actions; the strategy $\boldsymbol{\psi}$ for Player II is a function of the observation η and Player I's possible actions. The pay-off function is of the form

$$K(\boldsymbol{\phi}, \boldsymbol{\psi}) = \iint P(\xi, \eta; \boldsymbol{\phi}(\xi), \boldsymbol{\psi}(\eta))\, dF(\xi)\, dG(\eta),$$

where P is a function specified by the model.

In a typical model the unit interval is taken as the representation of all possible hands that can be dealt to a player. Each hand is considered equally likely, and therefore the operation of dealing a hand to a player may be considered as equivalent to selecting a random number from the unit interval according to the uniform distribution. Of course, a hand ξ_1 is inferior to a hand ξ_2 if and only if $\xi_1 < \xi_2$. The game proceeds as follows: Players I and II select points (draw hands) ξ and η, respectively, from the unit interval (the full pack) according to the uniform distribution (the equal likelihood of all possible hands). Both players ante one unit. Player I, knowing his value ξ, acts first and has the option of either folding immediately, thus forfeiting the ante to Player II, or betting any one of the amounts $a_1, a_2, \ldots, a_n$, where $1 < a_1 < a_2 < \cdots < a_n$. Player II then must respond by either passing immediately or seeing the bet. In the first circumstance Player I wins B's ante. If B sees the bet, the hands ξ and η are compared and the player with the better hand wins the kitty. If $\xi = \eta$, no payment is made.

A strategy for Player I can be described as an n-tuple of functions

$$\phi(\xi) = \langle \phi_1(\xi), \phi_2(\xi), \ldots, \phi_n(\xi) \rangle,$$

where $\phi_i(\xi)$ expresses the probability of his betting the amount a_i when his hand is ξ. The $\phi_i(\xi)$ must satisfy

$$\sum_{i=1}^{n} \phi_i(\xi) \leq 1.$$

The probability that Player I will fold immediately is

$$1 - \sum_{i=1}^{n} \phi_i(\xi).$$

A strategy for Player II can be represented by the n-tuple

$$\psi(\eta) = \langle \psi_1(\eta), \psi_2(\eta), \ldots, \psi_n(\eta) \rangle,$$

where $\psi_i(\eta)$ expresses the probability of his seeing a bet of a_i units when he holds the hand η. The probability that Player II will fold after Player I has bet a_i is $1 - \psi_i(\eta)$. Each $\psi_i(\eta)$ is subject only to the condition that $0 \leq \psi_i(\eta) \leq 1$.

If Player I uses the strategy ϕ and Player II uses the strategy ψ, the expected gain to Player I is denoted by $K(\phi, \psi)$. Enumerating all the possibilities, we find that

$$K(\phi, \psi) = (-1)\int_0^1 \left[1 - \sum_{i=1}^{n} \phi_i(\xi)\right] d\xi$$

$$+ \int_0^1 \int_0^1 \sum_{i=1}^{n} \phi_i(\xi)(1 - \psi_i(\eta))\, d\xi\, d\eta$$

$$+ \sum_{i=1}^{n} (a_i + 1) \int_0^1 \int_0^1 \phi_i(\xi)\psi_i(\eta) L(\xi, \eta)\, d\xi\, d\eta,$$

where

$$L(\xi, \eta) = \begin{cases} 1 & \text{if} \quad \xi > \eta, \\ 0 & \text{if} \quad \xi = \eta, \\ -1 & \text{if} \quad \xi < \eta. \end{cases}$$

The validity of relation (2.1.1) for such games is easily demonstrated by appealing to the appropriate abstract min-max theorems. What is more important is that in spite of the complicated form of the model, a constructive methodology exists which leads to explicit optimal strategies for each player. Chapter 9 is devoted to an elaboration of this technique and to the solution of a number of interesting and significant examples.

2.5 How to solve infinite games. Solving games is like solving differential equations, a combination of good guesswork (intuition), adroit use of perturbation arguments, and an ability to exploit the special features of a problem.

Whenever possible, one must make use of the unique features of the kernel. In polynomial games, for example, the detailed structure of the theory of moments is very useful for supplying bounds on the size of the spectrum of solutions. Consider as an illustration the kernel

$$K(\xi, \eta) = \xi\eta^2 - \xi^2 + 2\xi\eta - \eta^2.$$

The theory tells us that there exist solutions for each player with spectrums either confined to the points $\{0, 1\}$ or consisting of a single point on the interior of the unit interval (see Section 3.8). With these facts in mind, some easy juggling produces the optimal strategies

$$x^0(\xi) = x_{\frac{1}{3}}(\xi), \qquad y^0(\eta) = \tfrac{7}{9}y_0(\eta) + \tfrac{2}{9}y_1(\eta),$$

and the value is $-1/9$. The same answer could also have been obtained by appealing to the theory of convex games (more precisely, in this case, to the corresponding concave analog).

In games of timing, convex games, and bell-shaped games, the theory itself provides a routine for determining the optimal strategies. This routine is not as formal as the routines for solving finite games or linear programming problems. Often the computation is reduced to an integral or differential equation, which can be handled, if necessary, by the techniques of numerical analysis. This is to be the interpretation of "routine" in the present context. We add once again that the calculating procedures in all these cases can often be immeasurably speeded up by exploiting the characteristics of the specific kernel.

A problem frequently encountered is what to do about kernels for which the form of the solution is not even partially known. One possibility is the use of analogy to suggest an initial guess for the solution. By appropriate perturbations and modifications of this guess it may be possible to approach the solution. We illustrate this method in the following paragraph.

For bell-shaped games, a substantial theory is available to us. Since finite-type solutions exist for such games, we can set up a system of equations from which the solutions may be determined by recursive means (see Section 7.2). Suppose we digress for a moment and seek to solve the game with kernel

$$K(\xi, \eta) = \frac{1}{1 + \lambda(\xi - \eta)^2}.$$

This kernel does not quite fit the hypothesis of bell-shaped games, but the graph of $1/(1 + \lambda u^2)$ is closely akin to that of $e^{-\lambda u^2}$. Hence a good guess would be to apply the theory of bell-shaped games to the game defined above. This gives us the optimal strategies

$$x^0(\xi) = x_{\frac{1}{2}}(\xi) \qquad \text{and} \qquad y^0(\eta) = \tfrac{1}{2}y_0(\eta) + \tfrac{1}{2}y_1(\eta)$$

for $0 \leq \lambda \leq 4/3$. Similarly, for $\lambda = 2$ it can be shown that the optimal strategy for each player involves a support of two points. The analysis for different values of λ continues in this same vein, fully parallel to the case of bell-shaped games.

If no clear analogy is available, we may try to find an optimal strategy x^0 for which $K(x^0, y) = c$ identically for all y in Y. An x^0 with this property is called an equalizer. (This is consistent with the notion of an equalizer introduced in connection with optimal strategies for games on the unit square.) If $K(x^0, y) = c$ has a solution x^0 in X and $K(x, y^0) \equiv b$ has a solution y^0 in Y, then x^0, y^0 trivially constitute a pair of optimal strategies and necessarily $b = c$. This trick of seeking out a strategy which yields identically v applies more often than might be expected. We cite one example where it works.

Suppose

$$K(\xi, \eta) = \frac{(1 + \xi)^k(1 + \eta)^k(1 - \xi\eta)}{(1 + \xi\eta)^{k+1}} \qquad (0 \leq \xi, \eta \leq 1, k > 0). \qquad (2.5.1)$$

We seek a strategy $x(\xi)$ satisfying

$$\int_0^1 K(\xi, \eta)\, dx(\xi) = c \qquad (0 \leq \eta \leq 1) \qquad (2.5.2)$$

for some constant c.

By expanding in suitable power series, we may write (2.5.2) as

$$\int_0^1 (1+\xi)^k \left[\sum_{r=0}^{\infty} (-1)^r \binom{r+k}{r} \xi^r \eta^r\right] (1-\xi\eta)\, dx(\xi) = c \sum_{n=0}^{\infty} (-1)^n \binom{n+k-1}{n} \eta^n. \quad (2.5.3)$$

Set

$$a_r = \int_0^1 (1+\xi)^k \xi^r \, dx(\xi) \qquad (r = 0, 1, 2, \ldots).$$

Equating coefficients of η^n on both sides of (2.5.3) leads to the relation

$$a_n\left[\binom{n+k}{n} + \binom{n+k-1}{n-1}\right] = c\binom{n+k-1}{n},$$

or

$$a_n = c\frac{k}{2n+k}.$$

But

$$\int_0^1 \xi^n \xi^{(k/2)-1}\, d\xi = \frac{2}{2n+k}.$$

Hence, by setting

$$dx_0(\xi) = M\frac{\xi^{(k/2)-1}}{(1+\xi)^k}\, d\xi,$$

where M is a normalization constant, we secure (2.5.2) for an appropriate constant c. By symmetry, with $dy_0(\eta)$ chosen the same as $dx_0(\xi)$, we also get

$$\int_0^1 K(\xi, \eta)\, dy_0(\eta) = c.$$

It follows that $dx_0(\xi)$ and $dy_0(\eta)$ as specified above are both equalizers and hence optimal (see also Problem 10).

The device of determining equalizer strategies plays a large part in the theory of games of timing. For such games full equalizers are not required; instead we seek part equalizers—i.e., strategies that are equalizers for a subinterval of ξ and η values. Because of the inherent monotonicity characteristics of kernels of games of timing, this requirement is sufficient to ensure the calculation of optimal strategies. The same idea of seeking part equalizers works for many other examples (see the problems of Chapter 5).

A final method of getting at a solution embodies a fixed-point idea. For any pair x^*, y^* consider the problems

$$\max_{x \in X} K(x, y^*), \qquad \min_{y \in Y} K(x^*, y), \tag{2.5.4}$$

and let X_{y^*} (Y_{x^*}) denote the set of all strategies x (y) in which the maximum (minimum) of (2.5.4) is attained. If $x^* \in X_{y^*}$ and $y^* \in Y_{x^*}$, then according to (2.1.2) x^* and y^* comprise a pair of optimal strategies.

Surprisingly enough, this simple fixed-point notion can often be used effectively in solving certain classes of games. The operations of maximum and minimum in (2.5.4) lead to consistency relations which may suggest the nature of the solutions. This approach works particularly well in solving parlor games for which good intuitive strategies based on centuries of experience are available. With such time-tested or intuitively sound strategies used initially in (2.5.4) for x^* and y^*, it is possible with suitable slight alterations of x^* and y^* to arrive at a pair x_0 and y_0 which are indeed fixed points and hence optimal (see Chapter 9).

In closing this chapter we urge the reader to attempt to solve the problems at the ends of Chapters 2–9. Facility in analyzing games cannot be acquired except by matching one's wits against explicit game problems, the more the better.

2.6 Problems.

1. Let $K(\xi, \eta)$ be a continuous function defined on the unit square, and let K_n be the $(n + 1) \times (n + 1)$ matrix whose (i, j) component is $K(i/n, j/n)$ for $i, j = 0, \ldots, n$. Let $\boldsymbol{\lambda}^n = \langle \lambda_0^n, \ldots, \lambda_n^n \rangle$ and $\boldsymbol{\mu}^n = \langle \mu_0^n, \ldots, \mu_n^n \rangle$ be optimal strategies for the game with matrix $\mathbf{K}_n$, and let v_n be the value of this game. Show that there exists a sequence ϵ_n that tends to zero, such that

$$\int K(\xi, \eta) d\left(\sum_{i=0}^{n} \lambda_i^n x_{i/n}(\xi)\right) \geq v_n - \epsilon_n \qquad \text{for} \quad 0 \leq \eta \leq 1$$

and

$$\int K(\xi, \eta) d\left(\sum_{j=0}^{n} \mu_j^n y_{j/n}(\eta)\right) \leq v_n + \epsilon_n \qquad \text{for} \quad 0 \leq \xi \leq 1.$$

2. Let $\mathfrak{F}$ and $\mathfrak{G}$ denote the sets of probability distributions on the intervals $0 \leq \xi \leq 1$ and $0 \leq \eta \leq 1$. Use the results of Problem 1 to show that if K is continuous, then

$$\sup_{x \in \mathfrak{F}} \inf_{y \in \mathfrak{G}} \iint K(\xi, \eta)\, dx(\xi)\, dy(\eta) = \inf_{y \in \mathfrak{G}} \sup_{x \in \mathfrak{F}} \iint K(\xi, \eta)\, dx(\xi)\, dy(\eta).$$

3. Using Helly's selection theorem (Section C.3), show that sup and inf in Problem 2 can be replaced by max and min, so that a continuous game with continuous kernel $K(\xi, \eta)$ has a value and optimal strategies.

4. Let

$$K(\xi, \eta) = \begin{cases} -\dfrac{1}{\xi^2} & \xi > \eta, \\ 0 & \xi = \eta, \\ \dfrac{1}{\eta^2} & \xi < \eta. \end{cases}$$

Show that if the game with kernel K is played on the closed unit square, the point $\{0, 0\}$ is a saddle point. Prove also that if the game is played on the open unit square, then the game has no value, by verifying that

$$\sup_x \inf_\eta \int K(\xi, \eta)\, dx(\xi) \geq +1$$

and

$$\inf_y \sup_\xi \int K(\xi, \eta)\, dy(\eta) \leq -1.$$

5. Let

$$K(\xi, \eta) = \begin{cases} -1 & \xi = 1 \quad \text{and} \quad 0 \leq \eta < 1, \\ 1 & 0 \leq \eta < \xi < 1, \\ 0 & \xi = \eta, \\ -1 & 0 \leq \xi < \eta < 1, \\ 1 & \eta = 1 \quad \text{and} \quad 0 \leq \xi < 1. \end{cases}$$

Show that the game with kernel $K(\xi, \eta)$ has no value.

6. Show that the game

$$K(\xi, \eta) = \begin{cases} -1 & \xi < \eta < \xi + \frac{1}{2}, \\ 0 & \xi = \eta \quad \text{or} \quad \eta = \xi + \frac{1}{2}, \\ 1 & \text{otherwise} \end{cases}$$

has no value. Prove that

$$\sup_x \inf_y K(\xi, \eta) = \tfrac{1}{3} \qquad \text{and} \qquad \inf_y \sup_x K(\xi, \eta) = \tfrac{3}{7}.$$

7. Let $\mathbf{K} = \|K_{ij}\|$ be any $(n + 1) \times (n + 1)$ matrix with unique optimal strategies $\langle \lambda_0^0, \ldots, \lambda_n^0 \rangle$ and $\langle \mu_0^0, \ldots, \mu_n^0 \rangle$. Let $K(\xi, \eta)$ be a function that is linear in ξ and η separately in each square

$$\frac{i}{n} \leq \xi \leq \frac{i+1}{n}, \qquad \frac{j}{n} \leq \eta \leq \frac{j+1}{n},$$

and such that $K(i/n, j/n) = K_{ij}$ for each i and j. Show that the strategies

$$x^0 = \sum_{i=0}^{n} \lambda_i^0 x_{i/n}(\xi) \qquad \text{and} \qquad y^0 = \sum_{j=0}^{n} \mu_j^0 y_{j/n}(\eta)$$

are optimal for the continuous game with kernel $K(\xi, \eta)$ defined on the unit square.

8. Let

$$L_{ij}(\xi, \eta) = \epsilon\left[\left(\xi - \frac{i}{n} - \frac{1}{2n}\right)^2 - \left(\eta - \frac{j}{n} - \frac{1}{2n}\right)^2\right],$$

with $\epsilon > 0$. Let $K^*(\xi, \eta) = K(\xi, \eta) + L_{ij}(\xi, \eta)$ for

$$\frac{i}{n} \leq \xi \leq \frac{i+1}{n}, \qquad \frac{j}{n} \leq \eta \leq \frac{j+1}{n},$$

where K is defined as in Problem 7. Show that the *only* optimal strategies for the game with kernel $K^*(\xi, \eta)$ defined on the unit square are the strategies x^0 and y^0 of Problem 7.

9. Let $K(\xi, \eta)$ be any continuous function defined on the unit square. Show that for each $\epsilon > 0$ there exists a continuous function $K_\epsilon(\xi, \eta)$ with $|K(\xi, \eta) - K_\epsilon(\xi, \eta)| < \epsilon$ and such that the optimal strategies for $K_\epsilon(\xi, \eta)$ are unique. (*Hint*: Use the result of Problem 8 above.)

10. The game with kernel

$$\frac{(1 + \xi)^a(1 + \eta)^b}{(1 + \xi\eta)^c}$$

defined on the unit square has an optimal equalizing strategy with density

$$K\,\frac{\xi^m(1 - \xi)^n}{(1 + \xi)^p}$$

for some constants K, m, n, p. Find the value of these constants in terms of a, b, c. Assume that $c > a, b$.

11. Solve the game

$$K(\xi, \eta) = \begin{cases} 1 & \xi + \eta < 1, \\ \lambda & 1 < \xi + \eta < \frac{3}{2}, \\ 1 & \frac{3}{2} < \xi + \eta \leq 1 \end{cases}$$

in the following two cases:

(a) $K(\xi, \eta) = \lambda$ if $\xi + \eta = 1$ and
$K(\xi, \eta) = 1$ if $\xi + \eta = \frac{3}{2}$,

(b) $K(\xi, \eta) = \lambda$ if $\xi + \eta = 1$ and $\xi + \eta = \frac{3}{2}$.

12. Let B denote any positive number and let the intervals $[0, B + 1]$ and $[0, B]$ be the spaces of pure strategies for Players I and II, respectively. Let $0 < \theta < 1$ and

$$K(\xi, \eta) = \begin{cases} 1 - \theta & \xi \leq \eta, \\ 1 & \eta < \xi \leq \eta + 1, \\ 0 & \eta + 1 < \xi. \end{cases}$$

Find a pair of optimal strategies, and show that the value of the game is $(1 - \theta)/(1 - \theta^{n+1})$, where n is the greatest integer less than or equal to B. (*Hint*: There is an equalizing optimal strategy for Player I.)

Notes and References to Chapter 2

§ 2.1. The theory of infinite games is at most twenty years old, whereas the theory of discrete finite games can be traced back to the beginning of the twentieth century in the writings of Borel. However, it was the appearance of the classic book of von Neumann and Morgenstern [242] in 1944 which first stimulated vigorous activity among mathematicians in the study of all kinds of games, finite and infinite. Almost simultaneously with the publication of this book Wald advanced his first major writings [245, 249] on statistical decision theory. The significance of game theory for statistical theory and practice is incontestable. Furthermore, the thinking processes of game theory and statistical decision theory are so akin that statisticians or mathematicians interested in either discipline are naturally attracted to both. Fundamental to both is the general min-max theorem.

The earliest proof of an infinite version of the min-max theorem is due to Ville [237], who concentrated on games arising from continuous kernels defined on the unit square. He was also the first to exhibit a counterexample eradicating the possibility of general validity for the min-max theorem of bounded pay-off kernels.

The min-max theorem as an abstract result for functions and linear operators defined on subsets of an abstract linear topological space was analyzed by Karlin in 1948 at the RAND Corporation. This analysis appeared in 1950 [133]. The principal relevant notions are appropriate convexity assumptions and the compactness of the strategy spaces with respect to the appropriate topology imposed on them. Wald independently and simultaneously [248] presented other sufficient conditions which ensure the correctness of the min-max theorem. Wald's hypotheses derive from a statistical background. Many abstract extensions of these results have appeared since. We mention in this connection Fan [77, 76], Glicksberg [108], Berge [25], Sion [224], and others (see also [133, 135], and references on p. 19ff.). All these studies are concerned with what we have labeled games of the first category. Fleming [83] and Gross (unpublished) have been concerned to some extent with the min-max theorem for games of the second category.

For general reference books covering some aspects of infinite games, see McKinsey [179], Blackwell and Girshick [30, Chaps. 1 and 2], and Luce and Raiffa [169]. The main source of research articles on infinite games is Volumes 1–3 of *Contributions to the Theory of Games* [161, 162, 72].

§ 2.2. Scattered examples of infinite games can be found in von Neumann and Morgenstern [242], including a simplified poker example (see also Ville [237]). However, the real pioneering work on games played over the unit square was done by RAND mathematicians and consultants.

The intensive work in game theory done at the RAND Corporation in the late 1940's and early 1950's had many repercussions in the fields of statistics, mathematical economics, linear programming, multistage decision problems, etc., and greatly accelerated their development. Most of the contributions to infinite game theory since about 1952 have been in the form of solutions of special models.

General results of the kind described in this section cannot be ascribed to any single person but may be regarded as a product of the RAND Corporation program.

§ 2.3. The min-max theorem is now reasonably well understood. Additional classes of games possessing the min-max property are discovered from time to time, but only as a result of refinements of the concepts of convexity and compactness which constitute the salient elements of the theorem. What is far less understood is the relationship of the form of the pay-off kernel to the form of the solutions; this would seem to be one of the most important general problems now confronting game theorists. Some beginnings in the classification of games and solutions are presented in Chapters 3–9.

For a complete historical perspective on the various types of games discussed in these last seven chapters, we direct the reader to the remarks at the close of each chapter. Here, we list briefly some of the principal contributions to the understanding of the types of games covered in this section.

The theory of games with separable kernels (or more specifically, polynomial kernels) was proposed in [71]. Finite convex games are discussed extensively in [70].

The properties of games with convex kernels are explored in [37].

Examples of games of timing were first conceived and dealt with by Blackwell and Girschick [31]. Further particular examples were investigated by Shapley [217], Belzer [24], Glicksberg [107], and others. Shiffman [222] later showed how to solve the general symmetric game of timing, and Karlin [134] solved the one-action game of timing with no restrictions.

The theory of bell-shaped games is due to Karlin [137]. Analytic games have been treated by Glicksberg and Gross [110]. The concept of games that are invariant with respect to groups of transformations is an outgrowth of developments in statistical decision theory.

§ 2.4. The fighter-bomber duel was partly treated by Blackwell and Shiffman [32]. In Chapter 8 a complete solution of this model is indicated.

Very possibly because of poker's popular appeal to the natural gambling instincts, variants of poker models have received attention from numerous authors, notably Gillies, Mayberry, and von Neumann [105] and Bellman [20] (see also [138]).

§ 2.5. The use of equalizing strategies to determine solutions originated with Girschick, who exploited this technique primarily in solving games of timing. The methods of moments utilized in discussing the rational game example of this section may be ascribed to Gross (in conversation).

The fixed-point trick is inherent in the Kakutani theorem as applied to game theory.

§ 2.6. Problem 4 is due to Shapley. Problem 6 was suggested by Sion and Wolfe [223]. Problem 12 is discussed by Brown [43].

CHAPTER 3*

SEPARABLE AND POLYNOMIAL GAMES

The class of games whose kernels consist of polynomials

$$K(\xi, \eta) = \sum_{i=0}^{n} \sum_{j=0}^{m} a_{ij}\xi^i\eta^j$$

can be regarded, by reason of their familiarity, as the simplest class of infinite games. Since any continuous game on the unit square can be approximated by polynomial games, it seems reasonable to expect that insight into the structure of solutions to polynomial games will shed light on the structure of solutions to general continuous games played over the unit square. Unfortunately, the special properties of polynomial games do not translate into results concerning general continuous games. However, the theory of polynomial games does extend to the class of all separable games.

There are very few actual situations that can be represented in terms of separable games. Such games are interesting primarily because we know so much about the properties of their solutions.

The separable game is in reality a finite game (Chapter 1) which is played as follows: Player I selects a point $\mathbf{r}$ from a set R in E^n, and Player II chooses a point $\mathbf{s}$ from a set S in E^m. The pay-off to Player I is a linear function of both variables, $A(\mathbf{r}, \mathbf{s})$. We shall show in Section 3.1 that every separable game is a finite game of this type. More precise knowledge of the structure of convex sets in finite Euclidean spaces is essential to the investigation of separable games. Accordingly, in Sections 3.6 and 3.7 the duality theory of convex cones and the facial dimensions of convex sets are examined, and the results obtained in these sections are then applied in deducing the form of solutions for separable games. Some computational methods are also indicated.

3.1 General finite convex games. This and the following section are devoted to a discussion of the theory of finite convex games. This theory is later applied to the analysis of separable games (see also Section 2.3). We begin with a description of what is meant by a finite convex game.

An $n \times m$ matrix $\|a_{ij}\|$ is prescribed. Player I chooses as a strategy a point $\mathbf{r} = \langle r_1, \ldots, r_n \rangle$ from a set R in E^n, and Player II chooses a point

* This entire chapter is an advanced discussion that may be omitted at first reading without loss of continuity.

$\mathbf{s} = \langle s_1, \ldots, s_m \rangle$ from a set S in E^m. The pay-off to Player I is given by

$$A(\mathbf{r}, \mathbf{s}) = \sum_{i=1}^{n} \sum_{j=1}^{m} a_{ij} r_i s_j.$$

The sets R and S are assumed to be closed, bounded, and convex. Such a game hereafter will be referred to as a finite convex game defined by the triplet $\{R, S, A\}$. In particular, a matrix game (Chapter 1) is a finite convex game.

Another important class of finite games is defined in the following manner. Consider a matrix game in which the strategies satisfy certain further convex restrictions in addition to the usual restrictions. Then the choice of a strategy no longer corresponds to selecting a point out of the simplex Δ (the set of $\mathbf{x}$ vectors with $x_i \geq 0$, $\sum x_i = 1$); the selection must now be made from a convex closed subset R of Δ. If the constraints are defined by linear inequalities, then R describes a convex polyhedral subset of the simplex Δ. It is possible to relabel the vertices of the new polyhedron in such a manner that the new convex game is reconverted into an ordinary matrix game. Usually, however, this considerably enlarges the size of the matrix and makes it inaccessible for computation.

Still another class of finite convex games consists of games on the unit square with separable pay-off kernels, i.e., kernels of the form

$$K(\xi, \eta) = \sum_{i=1}^{n} \sum_{j=1}^{m} a_{ij} r_i(\xi) s_j(\eta), \tag{3.1.1}$$

where $r_i(\xi)$ and $s_j(\eta)$ are continuous functions. If we set

$$r_i = \int_0^1 r_i(\xi)\, dx(\xi), \qquad s_j = \int_0^1 s_j(\eta)\, dy(\eta), \tag{3.1.2}$$

where x and y are strategies, the pay-off becomes

$$K(x, y) = \int_0^1 \int_0^1 \sum_{i,j} a_{ij} r_i(\xi) s_j(\eta)\, dx(\xi)\, dy(\eta) = \sum_{i,j} a_{ij} r_i s_j. \tag{3.1.3}$$

This is clearly a game of the form $\{R, S, A\}$. Here the sets R and S can be identified as consisting of all points obtained by the relations (3.1.2) as x and y vary over all possible distributions (strategies). We shall refer in this case to the sets R and S as the finite moment spaces of the functions $\{r_i(\xi)\}$ and $\{s_j(\eta)\}$, respectively, since if $r_i(\xi)$ are powers ξ^i, the values r_i are the classical moments of x. In general, the r_i may be considered as generalized moments with respect to the functions $r_i(\xi)$. To sum up, a separable game may be regarded as a finite convex game $\{R, S, A\}$, where R and S are the reduced moment spaces and the pay-off is calculated in accordance with (3.1.3). The sets R and S may be characterized more simply, as indicated in Theorem 3.1.1.

▶THEOREM 3.1.1. The set R of (3.1.2) is the convex set spanned by a curve C whose parametric representation is $r_i = \{r_i(\xi)\}$ for $0 \leq \xi \leq 1$ and $i = 1, 2, \ldots, n$.

Proof. Let D be the convex set spanned by C (i.e., the convex hull of C).

(i) Suppose that $\mathbf{r}^0$ is in R but not in D. Since $r_i(\xi)$ are continuous functions, D is closed and convex, so that there exists a hyperplane h strictly separating $\mathbf{r}^0$ from D. That is, for some fixed $\delta > 0$,

$$\sum_{i=1}^{n} h_i r_i^0 - \sum_{i=1}^{n} h_i r_i(\xi) \geq \delta \tag{3.1.4}$$

for $0 \leq \xi \leq 1$. Since $\mathbf{r}^0$ is in R, there exists a distribution function $x(\xi)$ such that

$$\int_0^1 r_i(\xi)\, dx(\xi) = r_i^0.$$

By integrating (3.1.4) with respect to $dx(\xi)$, we obtain

$$\sum h_i r_i^0 \int_0^1 dx(\xi) - \sum h_i \int_0^1 r_i(\xi)\, dx(\xi) \geq \delta \int_0^1 dx(\xi),$$

or

$$\sum h_i r_i^0 - \sum h_i r_i^0 \geq \delta > 0,$$

an absurdity. Hence $\mathbf{r}^0$ must be in D.

(ii) Let $\mathbf{r}^0$ be in D. Then

$$r_i^0 = \sum_{k=1}^{n+1} \alpha_k r_i(\xi_k), \qquad \alpha_k \geq 0, \qquad \sum \alpha_k = 1$$

(Lemma B.2.1). If x_{ξ_k} denotes the degenerate distribution that has full weight at the point ξ_k, then

$$x(\xi) = \sum_{k=1}^{n+1} \alpha_k x_{\xi_k}$$

is a distribution with a discrete spectrum located at the set of points ξ_k. Furthermore,

$$\int_0^1 r_i(\xi)\, dx(\xi) = \sum_{k=1}^{n+1} r_i(\xi_k)\alpha_k = r_i^0 \qquad (i = 1, \ldots, n).$$

Hence $D \subset R$ and $R = D$.

For polynomial games,

$$r_i(\xi) = \xi^i \quad (i = 0, \ldots, n), \qquad s_j(\eta) = \eta^j \quad (j = 0, \ldots, m),$$

and C denotes the curve described parametrically by the equations $x_i = \xi^i$.

In the remaining sections of this chapter we develop the properties of solutions to finite convex games.

3.2 The fixed-point method for finite convex games. A pair of strategies $\mathbf{r}^0$ and $\mathbf{s}^0$ are optimal for Players I and II, respectively, if

$$\max_{r \in R} A(\mathbf{r}, \mathbf{s}^0) = \min_{s \in S} A(\mathbf{r}^0, \mathbf{s}) = v = A(\mathbf{r}^0, \mathbf{s}^0). \tag{3.2.1}$$

The validity of (3.2.1) for finite convex games can be established by two methods. One method appeals to the properties of convex sets, specifically to the fact that two nonoverlapping convex sets can be separated by a hyperplane. This idea will be basic in the development of the theory of dual cones discussed in Section 3.4. The second method depends on the Kakutani fixed-point theorem (see Section C.2). It is along the lines of the fixed-point theorem that a method of obtaining the optimal strategies of finite convex games will now be suggested.

The optimal strategies can be regarded as generalized fixed points of the mapping described as follows: Define for $\mathbf{r}^0$ in R the set $S(\mathbf{r}^0) \subset S$, where

$$\min_{s \in S} A(\mathbf{r}^0, \mathbf{s})$$

is assumed. $S(\mathbf{r}^0)$, which either is the intersection of a hyperplane with the boundary of S or agrees fully with S (since $A(\mathbf{r}^0, \mathbf{s})$ is linear in S), is obviously a convex set. Let $\mathbf{s}^0$ be in S and let the image of $\mathbf{s}^0$ in R be the set of points $R(\mathbf{s}^0) \subset R$, where

$$\max_{r \in R} A(\mathbf{r}, \mathbf{s}^0)$$

is assumed. If $\mathbf{r}^0$ is in $R(\mathbf{s}^0)$ and $\mathbf{s}^0$ is in $S(\mathbf{r}^0)$, then $\mathbf{r}^0$ and $\mathbf{s}^0$ satisfy (3.2.1), and they are therefore optimal strategies. It is clear that if R^0 and S^0 are the sets of optimal strategies of the two players, then every point of R^0 is an image of every point of S^0 and every point of S^0 is an image of every point of R^0.

We can also express the solutions as fixed points of the point-to-set mapping F which takes a point $\{\mathbf{r}, \mathbf{s}\}$ in the product space $R \otimes S$ into the nonvoid set $\{R(\mathbf{s}), S(\mathbf{r})\} = F(\mathbf{r}, \mathbf{s})$ in $R \otimes S$, where $R(\mathbf{s})$ and $S(\mathbf{r})$ are as defined above. A point $\{\mathbf{r}, \mathbf{s}\}$ is said to be a fixed point if $\{\mathbf{r}, \mathbf{s}\}$ belongs to its image set $F(\mathbf{r}, \mathbf{s})$.

The assertion that generalized fixed points for the mapping F exist is precisely a restatement of the fundamental min-max theorem. Here the special properties of the mapping $F(\mathbf{r}, \mathbf{s})$ will be exploited to obtain a method of characterizing the solution sets as well as a computational procedure for securing solutions.

In analyzing the finite convex game it is convenient to express the payoff in two equivalent forms

$$A(\mathbf{r}, \mathbf{s}) = \sum_{i,j=1}^{n,m} a_{ij} r_i s_j = \sum_{j=1}^{m} f_j(\mathbf{r}) s_j + f_0(\mathbf{r}) = \sum_{i=1}^{n} g_i(\mathbf{s}) r_i + g_0(\mathbf{s}), \tag{3.2.2}$$

where $f_j(\mathbf{r})$ and $g_i(\mathbf{s})$ are linear functions of $\mathbf{r}$ and $\mathbf{s}$, respectively, and $\mathbf{r}$ and $\mathbf{s}$ range over R and S, respectively. When the variables $\mathbf{r}$ and $\mathbf{s}$ are separated, the mapping properties of $A(\mathbf{r}, \mathbf{s})$ are more easily discernible.

Before proceeding with a description of the mapping system, we interpret geometrically two general types of optimal strategies which are of practical importance: (1) optimal strategies interior to R and S; (2) optimal strategies which yield identically v to a player independent of the strategies used by the other player.

These two categories of strategies are interrelated. If Player I has an interior optimal strategy $\mathbf{r}^0$, consider some optimal strategy $\mathbf{s}^0$ of Player II. The point $\mathbf{s}^0$ maps onto a set $R(\mathbf{s}^0)$ containing no interior point of R, since $A(\mathbf{r}, \mathbf{s}^0)$ is linear in $\mathbf{r}$, unless $g_i(\mathbf{s}^0) = 0$ for all i. Therefore every optimal strategy of Player II is on the intersection of the planes $g_i(\mathbf{s}) = 0$ and yields identically v. The interior solution $\mathbf{r}^0$ of Player I need not have any special position relative to the planes $f_j(\mathbf{r}) = 0$. However, an optimal strategy $\mathbf{r}^0$ yields identically v if and only if it lies on the intersection of all the planes $f_j(\mathbf{r}) = 0$.

The computation of fixed points. The mapping procedure suggested in proving the min-max theorem for finite convex games can also be developed into an algorithm for obtaining the solutions to games in this class. The method consists primarily of systematically keeping track of the image sets $S(\mathbf{r})$ and $R(\mathbf{s})$ until the fixed points are found. It is particularly effective in dealing with finite convex games in which R and S are both two-dimensional. The procedure consists of dividing the space R into a finite number of convex subsets $R_1, R_2, \ldots, R_\alpha$ by means of the hyperplanes $f_j(\mathbf{r}) = 0$ $(j = 1, 2, \ldots, m)$ and the boundaries of S, and similarly dividing the space S into a finite number of convex subsets $S_1, S_2, \ldots, S_\beta$ by means of the hyperplanes $g_i(\mathbf{s}) = 0$ $(i = 1, 2, \ldots, n)$ and the boundaries of R. The division of R is such that each of its subsets R_i maps onto some boundary set of S and each of the subsets S_j maps onto some boundary set of R. The subsets R_i may overlap, but their union is the full strategy space R. From the previous discussion, it follows that R^0 and S^0 are the fixed points in the mappings of $R_1, R_2, \ldots, R_\alpha$ onto S and $S_1, S_2, \ldots, S_\beta$ onto R, respectively. Often $R^0 = R_i$ for some i, and $S^0 = S_j$ for some j. Sometimes R^0 coincides with the set of $\mathbf{r}$ values in R defined by the relations $f_{i_1}(\mathbf{r}) = f_{i_2}(\mathbf{r}) = \cdots = f_{i_k}(\mathbf{r})$ and $f_i(\mathbf{r}) = 0$ $(i \neq i_v)$ for a suitable set of indices i_v. Similar considerations apply to S^0.

We illustrate the method by two examples: first, a game with a finite number of pure strategies, and second, a game with a continuum of pure strategies and a separable pay-off kernel.

Example 1. Consider the game in which Player I picks a strategy from the polyhedral convex set R defined in two-dimensional space by

$$\tfrac{1}{10} \leq r_1 \leq \tfrac{4}{5}, \quad \tfrac{1}{20} \leq r_2 \leq \tfrac{1}{2}, \quad r_3 \geq \tfrac{6}{5}(1 - \tfrac{20}{9}r_1), \quad r_1 + r_2 + r_3 = 1,$$

and Player II picks a strategy from S defined by

$$\tfrac{1}{10} \leq s_2 \leq 2s_1, \qquad s_3 \geq \tfrac{1}{6}, \qquad s_1 + s_2 + s_3 = 1.$$

Let the pay-off matrix be

$$||a_{ij}|| = \frac{1}{3}\begin{Vmatrix} 3 & 39 & 30 \\ 33 & 9 & 0 \\ 28 & 4 & 25 \end{Vmatrix}, \qquad \text{with } A(\mathbf{r}, \mathbf{s}) = \sum_{j=1}^{3} \sum_{i=1}^{3} r_i s_j a_{ij}.$$

Using the relations $r_1 + r_2 + r_3 = 1$ and $s_1 + s_2 + s_3 = 1$, we may eliminate the variables r_3 and s_3. The analysis is then carried out in terms of the remaining variables r_1, r_2 and s_1, s_2, subject to the appropriate constraints. R is the pentagon $ABCDE$ and S is the triangle XYZ in Fig. 3.1.

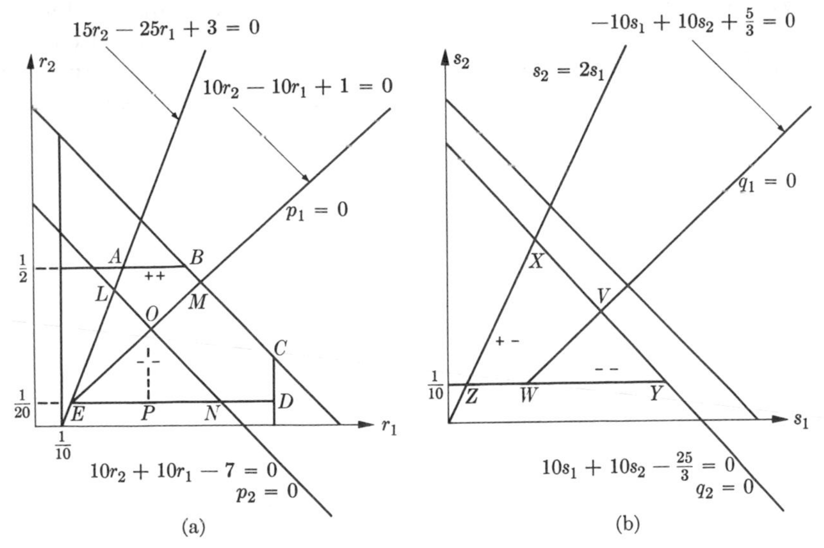

FIGURE 3.1

A direct calculation with r_3 replaced by $1 - r_1 - r_2$ gives

$$\begin{aligned} A(\mathbf{r}, \mathbf{s}) &= (10r_2 - 10r_1 + 1)s_1 + (10r_2 + 10r_1 - 7)s_2 + \tfrac{1}{3}(5r_1 - 25r_2 + 25) \\ &= (-10s_1 + 10s_2 + \tfrac{5}{3})r_1 + (10s_1 + 10s_2 - \tfrac{25}{3})r_2 + (s_1 - 7s_2 + \tfrac{25}{3}) \\ &= p_1s_1 + p_2s_2 + p_3 = q_1r_1 + q_2r_2 + q_3. \end{aligned}$$

The strategy space R is divided into four regions ($OMCDN$, etc.), and the strategy space S into two regions, by means of the pay-off function. In region VWY, $q_1 \leq 0$ and $q_2 \leq 0$.

We map each of the four regions of R into S by means of $\min_S A$ and each of the two regions of S into R by means of $\max_R A$. For example, consider a point $\mathbf{r}^*$ in R belonging to the interior of $OLABM$ (++ region) and examine $A(\mathbf{r}^*, \mathbf{s})$ as a function of $\mathbf{s}$ in S. Since $A(\mathbf{r}^*, \mathbf{s})$ is a linear expression in $\mathbf{s}$ with positive coefficients, it achieves its minimum at Z of S. Let $\mathbf{s}^*$ represent the point Z of S and consider $A(\mathbf{r}, \mathbf{s}^*)$, which is a linear function of the form $q_1r_1 + q_2r_2 + q_3$ with $q_1 > 0$ and $q_2 < 0$. The maximum of this for $\{r_1, r_2\}$ in R is attained by making r_1 as large as possible and r_2 as small as possible consistent with belonging to R. Both these criteria are satisfied by the single point D of R. This implies that points of the ++ region R cannot satisfy the fixed-point property. In a similar manner it is easy to see that points of the interior of ELO (+− of R) map into a subset of ZX which in turn has the image D. Hence, for the same reasons as before, no interior point of ELO is optimal.

Proceeding in this way, inspecting the separate regions of R, we arrive ultimately at the fixed points of R and S. It is readily verified that the line OP and the point V are the only fixed points. For instance, O maps into the full set S and V maps into the full set R. The points of OP are precisely those for which $A(\mathbf{r}, \mathbf{s})$ is a linear function in s_1 and s_2 with equal negative coefficients whose image is the line XY. The complete solution is given by

$$r_1 = \tfrac{2}{5}, \qquad \tfrac{1}{20} \leq r_2 \leq \tfrac{3}{10}$$

for Player I and

$$s_1 = \tfrac{1}{2}, \qquad s_2 = \tfrac{1}{3}, \qquad s_3 = \tfrac{1}{6}$$

for Player II; and the value of the game is 6.5.

Example 2. As an example of an infinite game—a type of game for which this method is very suitable—consider the separable game whose pay-off is given by

$$\begin{aligned} K(\xi, \eta) = {} & \eta^2 \left(\cos \frac{\pi}{2}\, \xi + \sin \frac{\pi}{2}\, \xi - 1\right) \\ & + \frac{4}{3}\, \eta \left(\cos \frac{\pi}{2}\, \xi - 3 \sin \frac{\pi}{2}\, \xi\right) \\ & + \frac{1}{3} \left(5 \sin \frac{\pi}{2}\, \xi - 3 \cos \frac{\pi}{2}\, \xi\right) \cdot \end{aligned}$$

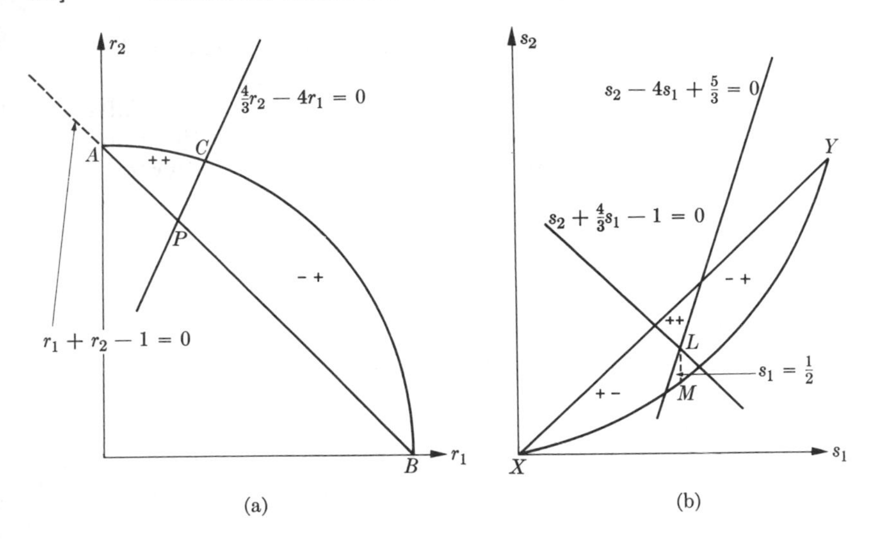

FIGURE 3.2

This is equivalent, according to Theorem 3.1.1, to a finite convex game in which R is the convex hull of the curve

$$r_1(\xi) = \sin\frac{\pi}{2}\xi \quad (0 \le \xi \le 1), \qquad r_2(\xi) = \cos\frac{\pi}{2}\xi \quad (0 \le \xi \le 1),$$

and S is the convex hull of the curve

$$s_1(\eta) = \eta \qquad (0 \le \eta \le 1), \qquad s_2(\eta) = \eta^2 \qquad (0 \le \eta \le 1).$$

The pay-off becomes

$$K(\mathbf{r}, \mathbf{s}) = A(\mathbf{r}, \mathbf{s}) = s_2(r_1 + r_2 - 1) + \tfrac{4}{3}s_1(r_2 - 3r_1) + \tfrac{1}{3}(5r_1 - 3r_2).$$

The convex set R for Player I is $ACBA$ in Fig. 3.2, and the convex set S for Player II is $XYMX$. Writing the pay-off function in two different ways, we obtain a division of the strategy spaces into two regions for Player I and four regions for Player II. In mapping the regions, we note that P maps into every point of S, and in particular into LM. Similarly, LM maps into the line AB, and in particular into P. Therefore P and LM are the fixed points and the solutions of the game. The value of the game is $-1/3$.

3.3 Dimension relations for solutions of finite convex games. The following theorem establishes the continuity properties of the solution sets for finite convex games. This result will be needed in the further study of such sets.

▶ THEOREM 3.3.1. The set of optimal strategies or solution set of a finite convex game is an upper semicontinuous function of the pay-off matrix **A**. (See Section C.2 for the definition of upper semicontinuous point-to set mapping.) If the strategy spaces R_n converge to R, then the set of solutions is upper semicontinuous. (A set of spaces R_n is said to converge to R if every point of R is a limit point of points of R_n and there exist no other limit points of points of R_n.)

Proof. Let G be an open set containing the optimal strategies $\widetilde{R}$ of Player I in the game with pay-off $A = A(\mathbf{r}, \mathbf{s})$. Suppose that each element of $\mathbf{A}$ is perturbed by at most ϵ. Let $\mathbf{A}_\epsilon$ denote the resulting pay-off and R_ϵ denote the set of optimal strategies of Player I. We assert that for all sufficiently small ϵ, the set R_ϵ is in G. To prove this assertion, we assume the contrary: that there exists a sequence $\{\epsilon_n\}$ tending to zero with a corresponding sequence of pay-offs $\mathbf{A}_{\epsilon_n}$ such that each R_{ϵ_n} is not contained in G. Let $\mathbf{r}_n$ be in R_{ϵ_n} but not in G. Since R is compact, we may suppose that $\mathbf{r}_n \to \mathbf{r}$ as $n \to \infty$, where $\mathbf{r}$ is not in G but in R. Further, it is readily verified that a limit point of optimal strategies for the pay-offs $\mathbf{A}_{\epsilon_n}$ is an optimal strategy for the pay-off A. But $\mathbf{r}$ is not in G and hence not in $\widetilde{R}$, and so we arrive at a contradiction.

The second half of the theorem is proved by an analogous argument.

Finite convex games with unique solutions. We derive here dimension relations for finite convex games with unique solutions. These are extensions of the structure theorems for solution sets of matrix games (Chapter 3 of Vol. I) to finite convex games, in which the strategy spaces of both players are polyhedral. It is shown in this section that the relevant property of the polyhedral face containing the optimal strategy is the dimension of the face rather than the number of vertices.

The method of analysis utilizes the mapping procedure involving the image set $S(\mathbf{r})$ for any $\mathbf{r}$ in R and the image set $R(\mathbf{s})$ for any $\mathbf{s}$ in S.

Let $\mathbf{r}^0$, $\mathbf{s}^0$ be the unique solution of the game with pay-off $A(\mathbf{r}, \mathbf{s})$ over the strategy space $\{R, S\}$. Assume that R and S are polyhedral convex sets. Let $\mathbf{r}^0$ be interior to a k-dimensional face R^0 of R, and $\mathbf{s}^0$ be interior to an l-dimensional face S^0 of S. R^0 and S^0 are polyhedral faces. Since $\mathbf{s}^0$ is optimal, it maps onto some face $R' \supseteq R^0$. Similarly, $\mathbf{r}^0$ maps onto some face $S' \supseteq S^0$.

▶ LEMMA 3.3.1. The game with pay-off $A(\mathbf{r}, \mathbf{s})$ over the reduced strategy spaces $\{R', S'\}$ has the unique solution $\mathbf{r}^0$, $\mathbf{s}^0$.

(The polyhedral nature of R and S is essential to the validity of this lemma.)

Proof. Since $\mathbf{r}^0$ maps onto $S' \supseteq S^0 \supseteq \{\mathbf{s}^0\}$ and $\mathbf{s}^0$ maps onto $R' \supseteq R^0 \supseteq \{\mathbf{r}^0\}$, it follows that $\mathbf{r}^0$, $\mathbf{s}^0$ is a solution of the game over the strategy space $\{R', S'\}$. To show that it is unique, let us assume that $\mathbf{s}' \in S'$ is another optimal strategy for Player II. Then $\mathbf{s}'$ maps onto some face containing R^0. Let $\bar{\mathbf{s}} = \epsilon\mathbf{s}' + (1 - \epsilon)\mathbf{s}^0$ be a solution close to $\mathbf{s}^0$, so that $\bar{\mathbf{s}}$ maps onto a face containing R^0. Now for the original game over $\{R, S\}$ we defined R' as the maximal face onto which $\mathbf{s}^0$ maps. Hence a point $\bar{\mathbf{s}}$ sufficiently close to $\mathbf{s}^0$ will map *into* some part of R' containing $\mathbf{r}^0$. Therefore $\bar{\mathbf{s}}$ is another solution of the full game taken over $\{R, S\}$, which contradicts the uniqueness assumption.

We may now confine ourselves to the reduced polyhedral game over $\{R', S'\}$. In this game $\mathbf{r}^0$ is interior to R^0 and maps onto S'. Similarly, $\mathbf{s}^0$ is interior to S^0 and maps onto R'.

The reduced polyhedral game is analogous to the "essential part" of a discrete game.

▶LEMMA 3.3.2. If $\mathbf{r}^0$, $\mathbf{s}^0$ is the unique solution of the polyhedral game over $\{R', S'\}$, then $R^0 = R'$ and $S^0 = S'$.

Proof. Suppose that $\mathbf{s}^0$ is on the boundary of S', or $S^0 \subset S'$. Consider a sequence S_n of polyhedrons interior to S' excluding S^0 which expand to fill all of S' as n increases. We can construct S_n by using any inner point $\mathbf{c}$ of S' and taking the set of all points on the segment

$$\lambda\mathbf{c} + (1 - \lambda)\mathbf{s} \qquad (\epsilon_n \leq \lambda \leq 1),$$

where $\mathbf{s}$ is any point of S' and $\epsilon_n \to 0$.

Consider the game over the spaces $\{R', S_n\}$. Let $\{X_n, Y_n\}$ denote the solutions to this game. Then from the upper semicontinuity of the solutions (Theorem 3.3.1) it follows that every sequence of solutions $\mathbf{s}_n$ tends to $\mathbf{s}^0$. Now since $\mathbf{r}^0$ is interior to R^0, it follows that for n sufficiently large the solutions $\mathbf{r}_n$ lie interior to a polyhedron having R^0 as a face, and hence $\mathbf{s}_n$ maps into a polyhedron containing R^0. But since $\mathbf{r}^0$ maps into all of S', it follows that $\{\mathbf{r}^0, \mathbf{s}_n\}$ is a solution of the game over $\{R', S_n\}$, and of the game over $\{R', S'\}$ as well. This contradicts the hypothesis, since $\mathbf{s}_n$ is interior to S'. Therefore $S' = S^0$, and similarly $R' = R^0$.

Remark 3.3.1. Using a similar argument, we can extend Lemma 3.3.2 to include games with nonunique solutions. Let R^0 be the smallest polyhedral face containing Player I's set of optimal strategies. Now all of Player II's optimal strategies will map onto some polyhedral face of R, and some of them will map onto R^0. Let R' be the maximal intersection of these polyhedral faces. Then, by an argument identical to the proof of Lemma 3.3.2, it can readily be shown that $R' = R^0$. Similarly, if S^0 is the smallest polyhedral face containing Player II's optimal strategies and S' is the intersection of all polyhedral faces into which Player I's optimal strategies are mapped, we can show that $S' = S^0$. Again we may confine ourselves to the reduced game over the space $\{R^0, S^0\}$.

▶THEOREM 3.3.2. If a polyhedral game has a unique solution, then the two optimal strategies lie in polyhedrons of the same dimension.

Proof. Since $\mathbf{r}^0$ maps into an l-dimensional polyhedron in S, we must have l linear independent relations satisfied by

$$f_j(\mathbf{r}^0) \qquad (j = 1, 2, \ldots, m).$$

These relations determine a manifold of points in R mapping onto S^0 and having dimension at least $m - l$. Now the manifold and the k-dimensional polyhedron have only $\mathbf{r}^0$ in common, since otherwise the uniqueness of the solution would be contradicted. Therefore $m - l + k \leq m$, or $k \leq l$. Similarly, we can show that $l \leq k$, and therefore $k = l$.

The dimension theorem for polyhedral games with arbitrary solution sets can be obtained by a suitable reduction to the case of a unique solution. The result is stated and the proof left as an exercise for the reader (see Problems 15 and 16 of this chapter).

For the next theorem, we suppose that X, the set of optimal strategies for Player I, is k-dimensional and interior to a polyhedron R^0 which is μ-dimensional; and that Y, the set of optimal strategies of Player II, is l-dimensional and interior to a polyhedron S^0 which is ν-dimensional.

▶THEOREM 3.3.3. For any polyhedral game, the set of solutions and their containing polyhedrons satisfy the dimension relation

$$\mu - k = \nu - l.$$

3.4 The method of dual cones. So far our analysis of finite convex games has been based on the workings of the Kakutani fixed-point theorem. Another approach is by way of the separation theorems of convex sets (which we also used in our analysis of matrix games), notably the method of dual convex cones. This is the familiar interplay between points of a convex cone and its planes of support. (We urge the reader not acquainted with this theory to review Section B.3.) The duality principle of convex cones was applied in connection with the theory of conjugate functions (Section 7.5 of Vol. I) and was also seen to be fundamental in the development of the duality results of linear programming. In the present context the duality relations between convex cones will be useful in furnishing a geometrical representation of strategies for finite convex games from which we can deduce bounds on the spectrums of the optimal strategies.

Let the finite convex game be defined as before by the triplet $\{R, S, A\}$, where

$$A(\mathbf{r}, \mathbf{s}) = \sum_{i=1}^{n} \sum_{j=1}^{m} a_{ij} r_i s_j \qquad (\mathbf{r} \in R, \mathbf{s} \in S).$$

Denote the value of the game by v. We enlarge the matrix $\|a_{ij}\|$ by defining $a_{00} = -v$, $a_{i0} = 0$ $(i = 1, \ldots, n)$, and $a_{0j} = 0$ $(j = 1, \ldots, m)$. The new matrix will again be denoted by $\mathbf{A}$. The set of strategies R in E^n is replaced by the set $\overline{R}$ in E^{n+1} consisting of all vectors of the form $\langle 1, r_1, r_2, \ldots, r_n \rangle$, where $\langle r_1, r_2, \ldots, r_n \rangle$ is in R. In a similar manner we replace S by $\overline{S}$. We are clearly justified in again allowing a value $\mathbf{r}$ to represent a point of $\overline{R}$, and a value $\mathbf{s}$ to represent a point of $\overline{S}$. The pay-off for the augmented game is

$$\sum_{i=0}^{n} \sum_{j=0}^{m} a_{ij} r_i s_j = -v + \sum_{i=1}^{n} \sum_{j=1}^{m} a_{ij} r_i s_j,$$

and the value of the new game is obviously 0. Of course, the sets of optimal strategies for the two games coincide. For our discussion of finite convex games we adopt the standard notation.

$$(\mathbf{rA})_j = \sum_i r_i a_{ij} \qquad \text{and} \qquad (\mathbf{As})_i = \sum_j a_{ij} s_j.$$

Consequently, the pay-off function can be written in inner product language as $A(\mathbf{r}, \mathbf{s}) = (\mathbf{rA}, \mathbf{s}) = (\mathbf{r}, \mathbf{As})$.

The cones P_R, P_S and the dual cones P_R^*, P_S^*. We embed the convex sets $\overline{R}$ and $\overline{S}$ in the convex cones P_R and P_S, where

$$P_R = \{\lambda \mathbf{r} | \mathbf{r} \in \overline{R}, \lambda \geq 0\}, \qquad P_S = \{\lambda \mathbf{s} | \mathbf{s} \in \overline{S}, \lambda \geq 0\}.$$

The dual cones P_R^*, P_S^* are defined as follows:

$$P_R^* = \{\mathbf{h} | \mathbf{h} \text{ in } E^{n+1}, (\mathbf{r}, \mathbf{h}) \geq 0 \quad \text{for all} \quad \mathbf{r} \text{ in } P_R\},$$

$$P_S^* = \{\mathbf{h} | \mathbf{h} \text{ in } E^{m+1}, (\mathbf{h}, \mathbf{s}) \geq 0 \quad \text{for all} \quad \mathbf{s} \text{ in } P_S\}.$$

(See Section B.3.)

We now summarize for immediate use some of the properties of dual cones whose proofs can be found in Section B.3.

(1) The elements of the convex cone P_R^* in E^{n+1} may be identified as the set of oriented hyperplanes passing through the origin of E^{n+1}, where by an orientation we mean to distinguish the half-space associated with each element $\mathbf{h}$ of P_R^* in which

$$\sum_{i=0}^{n} h_i r_i \geq 0$$

for all $\mathbf{r}$ in P_R. In other words, P_R^* may be regarded as the set of all closed half-spaces containing the origin on their boundary and enclosing P_R.

These half-spaces are represented analytically as points in E^{n+1} in terms of the coordinates of the normal vectors to the boundary hyperplane directed into the distinguished half-space. In particular, the boundary points of P_R^* are associated with supporting hyperplanes to P_R which touch P_R along a full ray. Hence $\mathbf{h}$ is a boundary point of P_R^* if and only if there exists a point $\mathbf{r}^0 \neq \mathbf{0}$ in P_R such that $(\mathbf{r}^0, \mathbf{h}) = 0$.

(2) If $(P_R^*)^*$ represents the dual cone to P_R^*, then $(P_R^*)^* = P_R$ (see Theorem B.3.1).

In terms of the geometry of the convex cones P_R^* and P_S^* we can now study the structure of the solutions to the game determined by $A(\mathbf{r}, \mathbf{s})$, $\overline{R}$, and $\overline{S}$. The two following lemmas, like the corresponding remarks of Chapter 1 (see especially Remark 1.1.4), embrace a geometrical representation of optimal strategies for Players I and II by points and hyperplanes, respectively.

▶ Lemma 3.4.1. Let $\overline{R}\mathbf{A}$ denote the image of $\overline{R}$ under $\mathbf{A}$. Then

$$\overline{R}\mathbf{A} \cap P_S^* = R^0\mathbf{A},$$

where R^0 denotes the set of optimal strategies for Player I. Furthermore, $\overline{R}\mathbf{A}$ does not overlap P_S^* on its interior. (Interior is defined relative to E^{m+1}.)

Proof. Assume to the contrary that the two sets overlap. Then there exists a point $\mathbf{r}_0$ in $\overline{R}$ such that $\mathbf{r}_0\mathbf{A}$ is an interior point of P_S^*. It follows that $(\mathbf{r}_0\mathbf{A}, \mathbf{s}) > 0$ for all $\mathbf{s}$ in $\overline{S}$. Since $\overline{S}$ is compact, we have $(\mathbf{r}_0\mathbf{A}, \mathbf{s}) \geq \delta > 0$ uniformly in $\mathbf{s}$, which implies that the value of the game is at least δ, contradicting the fact that the value is zero. The last statement of the lemma has thus been established.

Furthermore, since optimal strategies exist, $R^0\mathbf{A}$ is not empty. Let $\mathbf{r}^0 \in R^0$; then clearly $\mathbf{r}^0\mathbf{A} \in \overline{R}\mathbf{A}$, and also, by optimality, $(\mathbf{r}^0\mathbf{A}, \mathbf{s}) \geq 0$ for all $\mathbf{s}$ in $\overline{S}$. Consequently $\mathbf{r}^0\mathbf{A} \in P_S^*$, and so $R^0\mathbf{A} \subset \overline{R}\mathbf{A} \cap P_S^*$.

Conversely, if $\mathbf{r}\mathbf{A} \in P_S^*$ for any $\mathbf{r}$ in $\overline{R}$, then $(\mathbf{r}\mathbf{A}, \mathbf{s}) \geq 0$ for all $\mathbf{s}$ in $\overline{S}$, and therefore $\mathbf{r}$ must be optimal. Hence $R^0\mathbf{A} \supset \overline{R}\mathbf{A} \cap P_S^*$. This completes the proof of the lemma.

In view of Lemma 3.4.1 the convex sets $\overline{R}\mathbf{A}$ and P_S^* may be separated by at least one hyperplane $\mathbf{p}$. A separating plane of $\overline{R}\mathbf{A}$ and P_S^* is in particular a supporting plane to P_S^*, and any plane of support to a cone must necessarily contain its vertex 0. Thus $\mathbf{p}$ can be represented as a linear form defined in E^{m+1} determined by the $(m + 1)$-tuple $\mathbf{p} = \mathbf{s} = \langle s_0, s_1, s_2, \ldots, s_m \rangle$, with some $s_j \neq 0$. Since $\mathbf{p}$ supports the cone P_S^*, on multiplying all the components s_j by the same factor ± 1, we may assume without loss of generality that $\mathbf{s}$ is in $(P_S^*)^* = P_S$. Identical reasoning shows that any hyperplane through the origin which keeps P_S^* entirely on one side can be associated with points of P_S. Moreover, all points of

P_S corresponding to the same plane bounding P_S^* must lie on a ray through the origin. Therefore it is convenient to take a single point of that ray to represent the hyperplane. This is accomplished by multiplying $\mathbf{p}$ by a suitable positive constant. Since $\overline{S}$ was a natural cross section used to span P_S, all hyperplanes containing 0 which keep P_S^* on one side are in one-to-one correspondence with the section $\overline{S}$ of P_S.

The particular supporting planes to P_S^* which separate $\overline{R}\mathbf{A}$ and P_S^* play a distinguished role.

▶LEMMA 3.4.2. The separating planes of $\overline{R}\mathbf{A}$ and P_S^* are in one-to-one correspondence with the optimal strategies for Player II.

Proof. Let S^0 denote the set of all optimal strategies for Player II. Every $\mathbf{s}^0$ in S^0 satisfies

$$\begin{aligned} (\mathbf{rA}, \mathbf{s}^0) &\leq 0 \qquad (\text{all} \quad \mathbf{r} \in \overline{R}), \\ (\mathbf{h}, \mathbf{s}^0) &\geq 0 \qquad (\text{all} \quad \mathbf{h} \in P_S^*). \end{aligned} \tag{3.4.1}$$

Hence $\mathbf{s}^0$ represents a separating hyperplane.

Conversely, since $\overline{R}\mathbf{A}$ and P_S^* are in contact, any separating plane $\mathbf{s}^*$ must be a supporting plane to both. In view of the discussion preceding this lemma, we may multiply $\mathbf{s}^*$ by an appropriate constant and write

$$(\mathbf{rA}, \mathbf{s}^*) \leq 0 \qquad (\text{all} \quad \mathbf{rA} \in \overline{R}\mathbf{A}) \tag{3.4.2}$$

and

$$(\mathbf{h}, \mathbf{s}^*) \geq 0 \qquad (\text{all} \quad \mathbf{h} \in P_S^*), \tag{3.4.3}$$

where $\mathbf{s}^*$ is in $\overline{S}$. It follows from (3.4.2) that $\mathbf{s}^*$ is an optimal strategy for Player II.

3.5 Structure of solution sets of separable games. To sum up the geometry developed in the previous section: (a) Player I's strategies can be regarded as the image points of $\overline{R}$ under the mapping $\mathbf{A}$, and his optimal strategies correspond to the contact points of $\overline{R}\mathbf{A}$ and P_S^*; (b) Player II's strategies can be represented as supporting hyperplanes to P_S^*, and his optimal strategies correspond to the supporting planes to P_S^* which also separate P_S^* and $\overline{R}\mathbf{A}$.

With the aid of these geometrical characterizations of strategies, we are now prepared to derive dimension relations and properties of optimal strategies for finite convex games, and in particular for separable games.

The next few theorems imply that for separable games, there exist optimal strategies having finite spectrums with known upper bounds. The surprising feature of these results stems from the fact that separable games are after all infinite games allowing for arbitrary mixed strategies. Nevertheless, the finite character of the kernel implies the existence of optimal strategies of finite type.

Results which establish the existence of optimal strategies of finite type are of value for two reasons. First, such strategies can be simulated easily for computation. Second, these are the simplest types of mixed strategies that occur in infinite games, and in practice the most easily effected. Consider the separable game

$$K(\xi, \eta) = \sum_{i=1}^{n} \sum_{j=1}^{m} a_{ij} r_i(\xi) s_j(\eta), \tag{3.5.1}$$

where $r_i(\xi)$ $(i = 1, \ldots, n)$ and $s_j(\eta)$ $(j = 1, \ldots, m)$ are continuous functions. Let C be a curve lying in E^{n+1} and defined parametrically as $r_i = r_i(\xi)$ $(0 \leq \xi \leq 1; i = 0, 1, \ldots, n)$, where $r_0(\xi)$ is defined as identically 1. As we have seen in Section 3.1, the game with kernel K can be reduced to a finite convex game $\{R, S, \mathbf{A}\}$, where $\mathbf{A}$ is the matrix $\|a_{ij}\|$ and R and S are characterized as in Theorem 3.1.1.

▶THEOREM 3.5.1. In a separable game, if the dimensions (in R and S) of the sets of optimal strategies are μ and ν, respectively, and if ρ is the rank of the original matrix $\|a_{ij}\|$ $(i = 1, \ldots, n; j = 1, \ldots, m)$, then Player I has an optimal strategy with at most $\min(\rho, m - \nu + 1)$ points in its spectrum and Player II has an optimal strategy with at most $\min(\rho, n - \mu + 1)$ points in its spectrum.

Proof. We recall that an optimal strategy $\mathbf{r}^0$ has its image $\mathbf{r}^0\mathbf{A}$ in $\overline{R}\mathbf{A} \cap P_S^*$.

The convex set $\overline{R}\mathbf{A}$ is at most ρ-dimensional. Hence by Lemma B.2.2 every point is spanned by at most ρ points of $(C)\mathbf{A}$. (The term $(C)\mathbf{A}$ designates the set in E^{m+1} that is obtained by transforming the points of the curve C by the matrix $\mathbf{A}$.)

Furthermore, any point of $\overline{R}\mathbf{A} \cap P_S^*$ must lie in the intersection L of all the hyperplanes separating the two sets, as Lemma 3.4.1 asserts. By Lemma 3.4.2, the dimension of L is $m - \nu$. Thus $\mathbf{r}^0\mathbf{A}$ must be contained in the set spanned by $(C)\mathbf{A} \cap L$. Since this set need not be connected, by Lemma B.2.1 we may conclude that $\mathbf{r}^0\mathbf{A}$ can be expressed as a convex combination involving at most $\min(\rho, m - \nu + 1)$ points of $(C)\mathbf{A}$. Hence $\mathbf{r}^0$ involves at most $\min(\rho, m - \nu + 1)$ points of C.

The second assertion of the theorem is similarly proved.

▶COROLLARY 3.5.1. In a separable game with kernel $K(\xi, \eta)$ as in (3.5.1) and $r_i(\xi)$, $s_j(\eta)$ continuous, both players have optimal strategies with spectrums of size at most $\min(m, n)$.

This follows from the well-known inequality $\rho \leq \min(m, n)$.

The bounds of Theorem 3.5.2 may be improved upon for special finite games. For games defined by a polynomial kernel very precise bounds for the existence of optimal strategies of given size can be established (see Section 3.8). In that case we must invoke knowledge of the precise nature of the boundaries of the classical moment space.

The next theorem is not concerned with the number of spectral points of optimal strategies; rather, it parallels the statement of dimension relations between sets of optimal strategies in matrix games.

▶THEOREM 3.5.2. If μ, ν, and ρ are defined as in Theorem 3.5.1, then

$$\mu + \nu \leq m + n - \rho.$$

Proof. The dimension of the null space of $\mathbf{A}'$ is $n - \rho$ (recall that $\mathbf{A}'$ designates the transpose matrix of $\mathbf{A}$). If we view $\mathbf{A}'$ as a linear mapping on subspaces of E^n, it follows that the image space cannot diminish in dimension by more than $n - \rho$. In particular, the dimension of $R^0\mathbf{A}$ is at least $\mu - (n - \rho)$, since the dimension of R^0 is μ.

On the other hand, $R^0\mathbf{A}$ is the contact set of $\overline{R}\mathbf{A}$ and P_S^*. Consequently, it lies in $\nu + 1$ linearly independent hyperplanes in E^{m+1} whose intersection has dimension $m - \nu$. Then

$$\mu - (n - \rho) \leq \dim R^0\mathbf{A} \leq m - \nu$$

and

$$\mu + \nu \leq m + n - \rho.$$

▶COROLLARY 3.5.2. If the matrix $\|a_{ij}\|$ $(i = 1, \ldots, n; j = 1, \ldots, m)$ is of full rank, then

$$\mu + \nu \leq \max(m, n).$$

3.6 General remarks on convex sets in E^n.

In the following sections we shall make a more thorough investigation of the nature of optimal strategies for separable games whose kernels are polynomials. In games of this class the strategy spaces R and S are recognized as the classical finite moment spaces. Our further observations about the form of solutions to polynomial games require an exact description of the boundaries of the moment spaces. We therefore digress to consider the classical finite moment spaces and the boundary structure of convex sets in Euclidean spaces.

The idea is to associate with any convex set several numerical or integer-valued functions which measure the facial points on the boundary of the set, and thus fully determine and describe it. By means of these index functions we can distinguish between various kinds of convex sets. Those associated with polyhedral convex sets, for example, often satisfy very simple relations. The reduced classical moment spaces also possess interesting dimension relations in terms of these indices.

Given any closed, bounded, convex set Γ, we define the dimension of Γ to be the dimension of the smallest linear manifold containing Γ. If $\mathbf{r}$ is a point on the boundary of Γ, $L(\mathbf{r})$ will denote the intersection of all the supporting hyperplanes to Γ at $\mathbf{r}$. The contact set $C(\mathbf{r})$ of a point $\mathbf{r}$ is defined to be the intersection of $L(\mathbf{r})$ and Γ. We define two indices of dimension:

$$a(\mathbf{r}) = \dim L(\mathbf{r}) \qquad \textit{facial dimension of } \mathbf{r},$$

$$c(\mathbf{r}) = \dim C(\mathbf{r}) \qquad \textit{contact dimension of } \mathbf{r}.$$

For an interior point $\mathbf{u}$ of Γ, for formal completeness, we set

$$a(\mathbf{u}) = c(\mathbf{u}) = n,$$

where n is the dimension of Γ.

It can be seen easily that in a bounded polyhedral set (a convex set spanned by a finite number of points) $a(\mathbf{r}) = c(\mathbf{r})$ always. In general, the difference $a(\mathbf{r}) - c(\mathbf{r})$ describes quantitatively the curvature of the boundary of Γ at $\mathbf{r}$. In contrast, intuitively speaking, at a round point on the boundary of a convex set like the point on the boundary of the unit sphere, $c(\mathbf{r}) = 0$.

If we embed Γ in E^{n+1} in the manner in which we embedded R in E^{n+1} in Section 3.4, we may then construct the cones P_Γ and P_Γ^*.

If P_Γ possesses no interior, then P_Γ^* is identical to all of E^{n+1} by definition and this case is of no interest. Consequently, throughout our discussion we shall assume, unless we state otherwise, that Γ is such that P_Γ possesses an interior. Since P_Γ is a closed convex cone, $(P_\Gamma^*)^* = P_\Gamma$ (see Theorem B.3.1). Any bounded cross section Γ^* of P_Γ^* is called a dual of Γ. Analytically, a cross section of P_Γ^* can be represented as the set of all $\mathbf{s}$ in P_Γ^* satisfying $(\mathbf{u}, \mathbf{s}) = 1$, where $\mathbf{u}$ is an interior point of P_Γ. Points in the boundary of Γ^* may be regarded as the supporting hyperplanes to Γ, and vice versa.

We say that a point $\mathbf{s}$ on the boundary of Γ^* is conjugate to a point $\mathbf{r}$ on the boundary of Γ if $\mathbf{s}$ is an interior point to the set of supporting planes to Γ at $\mathbf{r}$. In the same manner we define the points $\mathbf{r}$ in Γ that are conjugate to a point $\mathbf{s}$ in Γ^*.

The relationship is not always symmetric. In Fig. 3.3, $\mathbf{s}$ is conjugate to $\mathbf{r}$, $\mathbf{r}$ is a supporting plane to Γ^* at $\mathbf{s}$, but $\mathbf{r}$ is not conjugate to anything.

Of course, $\mathbf{s}$ is the only supporting plane at $\mathbf{r}$, and hence it is interior to the set of planes of support to Γ at $\mathbf{r}$.

▶ Lemma 3.6.1. Let $\mathbf{r}$ be in the boundary of Γ and $\mathbf{s}$ in the boundary of Γ^*. If $\mathbf{s}$ is conjugate to $\mathbf{r}$ or $\mathbf{r}$ is conjugate to $\mathbf{s}$, then

$$a(\mathbf{r}) + c(\mathbf{s}) = c(\mathbf{r}) + a(\mathbf{s}) = n - 1. \tag{3.6.1}$$

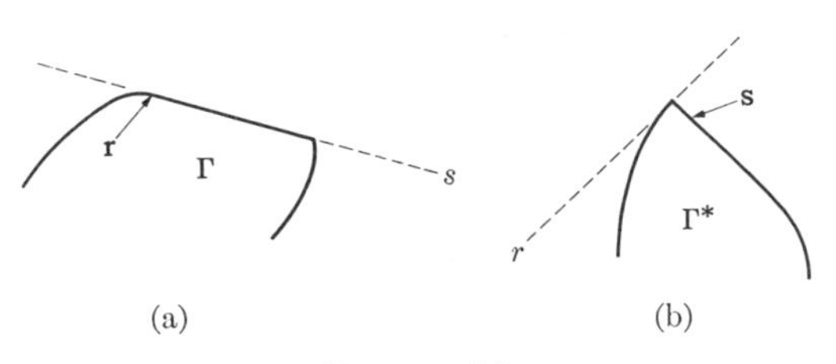

FIGURE 3.3

Remark 3.6.1. The lemma expresses in a sense a duality between the dimension of the "face" of the convex set containing a given point and the dimension of the family of all possible supporting planes at the given point. Formula (3.6.1) enables us to translate dimension relations for the set Γ into corresponding dimension relations for Γ^*, and conversely.

Proof. We take any $\mathbf{r}$ in the boundary of Γ, and let $\mathbf{s}$ in Γ^* be a supporting plane to Γ at $\mathbf{r}$. Then $(\mathbf{r}, \mathbf{s}) = 0$.

Let A be the set of supporting planes to Γ^* at $\mathbf{s}$:

$$A = \{\mathbf{r}' \in \Gamma | (\mathbf{r}', \mathbf{s}) = 0\}.$$

Since $a(\mathbf{s})$ is the dimension of the intersection of these planes, there must be $n - a(\mathbf{s})$ linearly independent planes in A. The dimension of A, considered as a convex subset of the boundary of Γ, is therefore $n - a(\mathbf{s}) - 1$.

Let $B = L(\mathbf{r}) \cap \Gamma$. By definition, the dimension of B is $c(\mathbf{r})$. We shall prove that $A = B$.

If $\mathbf{r}' \in B$, then $\mathbf{r}'$ is in all the supporting planes to Γ at $\mathbf{r}$. In particular, $\mathbf{r}'$ is in the plane that corresponds to $\mathbf{s}$; hence $(\mathbf{r}', \mathbf{s}) = 0$. This implies that $\mathbf{r}' \in A$ and $B \subset A$.

Conversely, to prove $A \subset B$ we consider two cases.

Case 1. If $\mathbf{s}$ is conjugate to $\mathbf{r}$, then $\mathbf{s}$ is an inner supporting plane to Γ at $\mathbf{r}$† Hence $\mathbf{s}$ supports Γ in precisely the set B. Thus every $\mathbf{r}'$ in Γ satisfying $(\mathbf{r}', \mathbf{s}) = 0$ is in B, and $A \subset B$.

Case 2. If $\mathbf{r}$ is conjugate to $\mathbf{s}$, then $\mathbf{r}$ is an inner point of A. Hence every supporting plane through $\mathbf{r}$ must contain all of A. Otherwise A would have points on both sides of the plane, and the plane would not be a supporting plane. Hence the intersection $L(\mathbf{r})$ of all these planes must contain all of A. Since Γ also contains A by definition, we have that $L(\mathbf{r}) \cap \Gamma \supset A$, or $B \supset A$.

† An "inner" point is interior relative to the defining set; an "interior" point is interior relative to the full space E^n.

In either case $B \supset A$, and therefore $B = A$. Then $\dim B = \dim A$; that is,

$$c(\mathbf{r}) = n - a(\mathbf{s}) - 1$$

or

$$c(\mathbf{r}) + a(\mathbf{s}) = n - 1.$$

Since the hypothesis of the lemma is symmetric in $\mathbf{r}$ and $\mathbf{s}$,

$$c(\mathbf{s}) + a(\mathbf{r}) = n - 1.$$

A final important dimension associated with points of a convex set is related to the fundamental fact that convex sets are spanned by their extreme points (see Lemma B.2.4). Since every member of Γ can be expressed as a finite convex combination of extreme points, we may denote by $b(\mathbf{r})$ the least number of extreme points of Γ required to span $\mathbf{r}$.

In view of Lemma B.2.1, for points in a closed, bounded, m-dimensional convex set

$$b(\mathbf{r}) \leq m + 1,$$

and also $b(\mathbf{r}) \leq c(\mathbf{r}) + 1$.

3.7 The reduced moment spaces. The preceding relations for general finite-dimensional convex sets possess interesting interpretations and refinements when applied to the classical moment spaces. The classical n-dimensional moment space D^n consists of all vectors $\langle r_1, r_2, \ldots, r_n \rangle$ such that there exists a distribution $x(\xi)$ on the interval $[0, 1]$ and

$$r_i = \int_0^1 \xi^i \, dx(\xi) \qquad (i = 1, 2, \ldots, n).$$

For $n \geq 2$ the curve C described parametrically as $r_i = \xi^i$ $(0 \leq \xi \leq 1)$ comprises the extreme points of D^n (see Lemma 3.7.1 below).

Figures 3.4 and 3.5 show the two- and three-dimensional moment spaces. In Fig. 3.5 the solid lines connecting the curve C^3 to the point $\langle 1, 1, 1 \rangle$ span the upper boundary of D^3. The dotted lines which connect $\langle 0, 0, 0 \rangle$ to the curve span the lower boundary of D^3.

Our object now is to analyze D^n in terms of the concepts of the facial dimensions introduced for general convex sets. Like polyhedral convex sets, which are easily distinguished, the classical moment spaces D^n are rich in structure. We begin by studying the spanning properties of C^n with respect to D^n. The screw nature of the curve C^n $(n \geq 2)$ readily implies that no point of C^n can be represented as a convex combination of other points of D^n and hence that every point of C^n constitutes an extreme point of D^n. By Theorem 3.1.1, all other points can be expressed in terms of

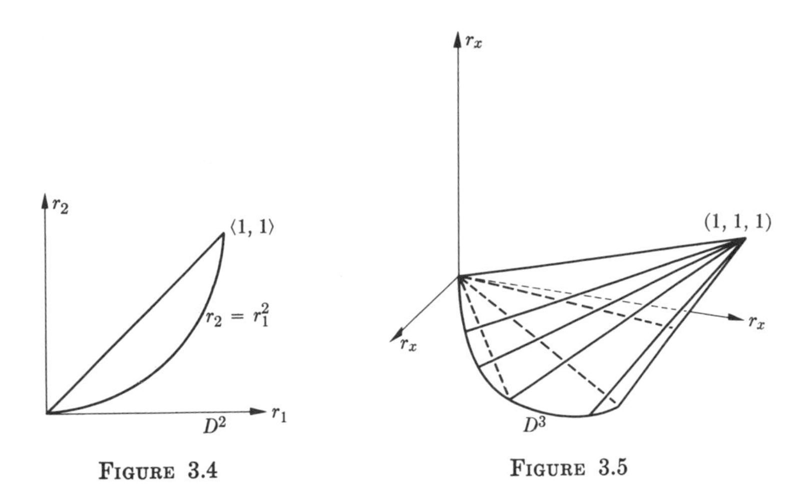

FIGURE 3.4

FIGURE 3.5

finite convex combinations of points of C^n. A more precise statement of this representation is given in the following lemma.

▶LEMMA 3.7.1. Points on the boundary of D^n have unique representations as convex combinations of extreme points. For each interior point of D^n there exist infinitely many representations in terms of points of C^n.

Proof. We present the proof of the first assertion only, leaving the second as an exercise for the reader.

We observe first that if H is any supporting plane to D^n, it has the characteristic property

$$(\mathbf{h}, \mathbf{r}) = \sum_{i=1}^{n} h_i r_i + h_0 \geq 0 \tag{3.7.1}$$

for all $\mathbf{r}$ in D^n. In particular,

$$h_0 + \sum_{i=1}^{n} h_i \xi^i \geq 0 \qquad (0 \leq \xi \leq 1), \tag{3.7.2}$$

and equality must hold for at least one value of ξ, since the plane is a supporting plane. But not all h_i are zero; hence equality cannot hold for more than n values of ξ, since the polynomial is of degree at most n. Let the number of zeros (contact points of C^n) be $k \leq n$. These points are linearly independent, since their matrix is the Vandermonde matrix whose rank is k. (A collection of n points is linearly independent if the $n - 1$ vectors from any one of the points to each of the others are linearly independent.)

Therefore every supporting hyperplane H meets D^n on a simplex whose vertices are linearly independent. Any point in the simplex may be represented in a unique way as a convex combination of its vertices. Since every point on the boundary of D^n is touched by some supporting plane, the first statement of the lemma is true.

In pursuing further the investigation of the explicit manner in which C^n spans D^n, it is convenient to assign different weightings to points of C^n. A point $\mathbf{r}$ in C^n is given weight 1 unless it corresponds to the parameter values 0 or 1, in which case the weight is taken to be 1/2. The justification of this will appear later.

For any point $\mathbf{r}$ in D^n, the index $b(\mathbf{r})$ denotes the minimum number of points of C^n that span $\mathbf{r}$ (see Section 3.6). We now let $b'(\mathbf{r})$ denote the same number of points, but with the endpoints $\langle 0, 0, \ldots, 0\rangle$, $\langle 1, 1, \ldots, 1\rangle$ counted as half-points; in other words, $b'(\mathbf{r})$ measures the number of extreme points used in the representation of $\mathbf{r}$ in terms of the new weightings introduced above.

We may say that $b'(\mathbf{r})$ is the weight of the point $\mathbf{r}$, and it can clearly take only half-integral or integral values. Obviously

$$b(\mathbf{r}) - 1 \leq b'(\mathbf{r}) \leq b(\mathbf{r}). \tag{3.7.3}$$

The next two lemmas exhibit some of the various relations that exist between the facial dimension indices of moment spaces. In addition, Lemma 3.7.3 shows, in effect, that the boundary of D^n has sufficient curvature to cut the dimension of $L(\mathbf{r})$ in half. By means of values of the index $b'(\mathbf{r})$ it is possible to decompose the boundary into various components. Within any component the dimension of the set of supporting planes is the same.

Many other interesting features of the nature of the classical moment spaces can be obtained. Some of these features will be mentioned in our interpretation of the cone generated by D^n and its dual cone. First, we establish the following two important results.

▶ LEMMA 3.7.2. If $\mathbf{r}$ is a point on the boundary of D^n,

$$b(\mathbf{r}) = c(\mathbf{r}) + 1, \tag{3.7.4}$$

and

$$2b'(\mathbf{r}) = a(\mathbf{r}) + 1. \tag{3.7.5}$$

If $\mathbf{r}$ is in the interior of D^n,

$$2b'(\mathbf{r}) = n + 1 = a(\mathbf{r}) + 1. \tag{3.7.6}$$

Proof. In the proof of the previous lemma we saw that the number of linearly independent elements in $L(\mathbf{r}) \cap D^n$ is $b(\mathbf{r})$. But the number of linearly independent elements is one more than the dimension; hence

$$b(\mathbf{r}) = \dim (L(\mathbf{r}) \cap D^n) + 1 = c(\mathbf{r}) + 1.$$

To establish (3.7.5) for the boundary point $\mathbf{r}$ of D^n with index number $b'(\mathbf{r})$ we construct a polynomial P_0 of degree $2b'(\mathbf{r})$ as follows. Each interior point of C in the representation of $\mathbf{r}$ is taken to be a root of multiplicity 2, and each boundary point of C (i.e., 0 or 1) involved in representing $\mathbf{r}$ is taken to be a simple root. The sign of the polynomial P_0 is chosen so that P_0 is nonnegative throughout the unit interval. Next consider the family of all polynomials Q of degree $n - 2b'(\mathbf{r})$, which are also so restricted as to be nonnegative on the unit interval. The implied fact that $n \geq 2b'(\mathbf{r})$ is easily proved. The coefficients $h = (h_i)$ $(i = 0, 1, \ldots, n)$ of any polynomial in the set of polynomials QP_0 generate a supporting plane to D^n at $\mathbf{r}$. The fact that QP_0 is nonnegative for all $0 \leq \xi \leq 1$ means that the curve C^n lies on one side of the plane, and hence $D^n = \operatorname{Co} C^n$ (Co = convex span of) lies on the same side. Moreover, QP_0 vanishes for all ξ involved in the representation of $\mathbf{r}$, and hence $(\mathbf{h}, \mathbf{r}) = 0$, that is, $\mathbf{r}$ belongs to the hyperplane $\mathbf{h}$ whose components are the coefficients of QP_0.

Conversely, any supporting plane to D^n at $\mathbf{r}$ produces a polynomial obeying the inequalities (3.7.2) and necessarily having P_0 as a factor (see the proof of Lemma 3.7.1).

The dimension of all supporting planes (all nonnegative polynomials Q) is precisely $n - 2b'(\mathbf{r}) + 1$, and hence the dimension of their common intersection $a(\mathbf{r})$ is $a(\mathbf{r}) = 2b'(\mathbf{r}) - 1$. This completes the proof of formula (3.7.5). Since (3.7.6) is demonstrated along similar lines, we omit the details.

▶ **Lemma** 3.7.3. On the boundary of D^n

$$a(\mathbf{r}) - 1 \leq 2c(\mathbf{r}) \leq a(\mathbf{r}) + 1. \tag{3.7.7}$$

Proof. From (3.7.3) we obtain

$$2b'(\mathbf{r}) \leq 2b(\mathbf{r}) \leq 2b'(\mathbf{r}) + 2,$$

or

$$2b'(\mathbf{r}) \leq 2[c(\mathbf{r}) + 1] \leq 2b'(\mathbf{r}) + 2;$$

then by (3.7.5) in Lemma 3.7.2

$$a(\mathbf{r}) + 1 \leq 2c(\mathbf{r}) + 2 \leq a(\mathbf{r}) + 3.$$

The cone P_D. If the notion of representation by convex combinations is extended to permit infinite sets of points, then the convex representation of any $\mathbf{r}^0$ in D^n corresponds one-to-many to the distribution functions $x^0(\xi)$ whose moments r_i are the coordinates of $\mathbf{r}^0$.

The moment

$$r_0 = \int_0^1 1\, dx(\xi)$$

corresponds to the coordinate r_0. The cone P_D is obtained by considering distributions with arbitrary nonnegative total weight r_0. Equivalently, a point of P_D is an $(n + 1)$-tuple $\langle r_0, r_1, \ldots, r_n\rangle$, where

$$r_i = \int_0^1 \xi^i \, dx(\xi) \qquad (i = 0, 1, \ldots, m)$$

and where the normalization of $x(\xi)$ is not restricted to be 1.

The dual cone P_D^*. The correspondence between the linear form in (3.7.1) and the polynomial in (3.7.2) indicates that P_D^* may be viewed as the set of polynomials of degree less than or equal to n which are nonnegative in the interval $0 \leq \xi \leq 1$. From the proof of Lemma 3.7.1, we see that the boundary of P_D^* comprises just those polynomials of P_D^* with at least one root in the unit interval. It can be shown that the extreme points (of a bounded cross section of the cone) are those with all n roots in the interval.

The cross section of P_D^* is defined as the set of all possible

$$\left\{h_i \middle| i = 0, 1, \ldots, n; (\mathbf{h}, \mathbf{r}) \geq 0 \quad \text{for} \quad r \text{ in } P_D, \text{ with } \sum_{i=0}^{n} \frac{h_i}{i+1} = 1\right\}.$$

This is illustrated graphically for $n = 2$ in Fig. 3.6.

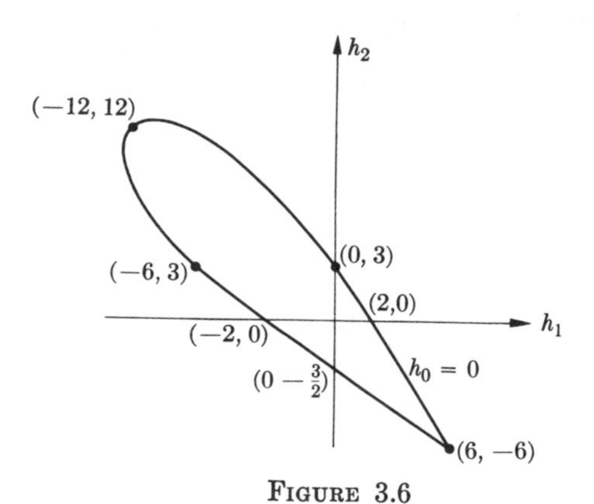

FIGURE 3.6

3.8 Polynomial games. A polynomial game is defined by a separable kernel of the form

$$K(\xi, \eta) = \sum_{i=0}^{n} \sum_{j=0}^{m} a_{ij} \xi^i \eta^j.$$

The geometric representation of strategies as points of convex sets and as supporting planes to suitable convex cones was of help in deducing dimension relations for the sets of optimal strategies of separable games. In particular, the use of the theory of polar cones applied to separable games enabled us to deduce the existence of optimal strategies of finite type. Similarly, our more complete knowledge of the geometry of the classical moment spaces should enable us to discover more about the form of optimal strategies for polynomial games. We proceed now to a detailed investigation of polynomial games.

Let μ and ν denote as before the dimensions of the sets R^0 and S^0 of optimal strategies for Players I and II, respectively. Let ρ denote the rank of the $n \times (m + 1)$ matrix obtained from $\mathbf{A}$ by omitting the row $i = 0$; let σ denote the rank of the $(n + 1) \times m$ matrix obtained by omitting the column $j = 0$. (Observe that a change in a_{00} to make $v = 0$ does not change these ranks.)

Assumption. We consider only the case in which $\rho = \min(n, m + 1)$ and $\sigma = \min(n + 1, m)$, so that both matrices have full rank. For definiteness, we shall study only the case $n \leq m + 1$, so that $\rho = n$.

The assumption $\rho = n$ implies that the set $(R)\mathbf{A}$ has dimension n. Under these circumstances, the dimension indices are preserved:

$$a(\mathbf{rA}) = a(\mathbf{r}), \qquad c(\mathbf{rA}) = c(\mathbf{r}).$$

(We emphasize that $a(\mathbf{r})$ and $c(\mathbf{s})$ are defined with respect to the sets R and S, respectively.)

To help clarify the arguments of the following theorems, we recall the definition of essential strategy and remark on its relationship to mixed strategies in polynomial games.

We define as essential a pure strategy corresponding to a parameter value of ξ if there is an optimal mixed strategy $x(\xi)$ in which ξ is used. If $\mathbf{r}^0$ is an interior point of R, then there exists a representation of $\mathbf{r}^0$ in which ξ may be used, so that in this case every pure strategy is essential (Lemma 3.7.1). On the other hand, if $\mathbf{r}^0$ is on the boundary of R, then there are exactly $b(\mathbf{r}^0)$ essential pure strategies in the unique representation of $\mathbf{r}^0$. In other words, any mixed strategy $x(\xi)$ which gives rise to the moment point $\mathbf{r}^0$ on the boundary of R is made up of mass concentrating at precisely $b(\mathbf{r}^0)$ points of the unit interval. In fact, we can say more. By virtue of Lemma 3.7.1, there is a unique distribution which corresponds to $\mathbf{r}^0$, and it is of finite type as described. Further, if $\mathbf{r}^0$ is on the boundary of R but $\mathbf{r}^0\mathbf{A}$ is an inner point of $R^0\mathbf{A}$ (see footnote, p. 65), all of R^0 must be on the boundary of R. Then all of R^0 must be spanned by the same points that span $\mathbf{r}^0$; otherwise $\mathbf{r}^0\mathbf{A}$ would not be an inner point of $R^0\mathbf{A}$. Hence the number of essential pure strategies is $b(\mathbf{r}^0)$.

The next series of theorems provides a detailed characterization of optimal strategies.

▶THEOREM 3.8.1. In the polynomial game with kernel

$$K(\xi, \eta) = \sum_{i=0}^{n} \sum_{j=0}^{m} a_{ij}\xi^i\eta^j \qquad (n \le m + 1),$$

in which the $n \times (m + 1)$ matrix $\|a_{ij}\|$ $(i \neq 0)$ has rank $\rho = n$, let $\mathbf{r}^0$ be an inner point of the μ-dimensional convex set of optimal strategies for Player I and let $\mathbf{s}^0$ be an inner point of the ν-dimensional set of optimal strategies for Player II. Then

$$\mu \le m + 1 - 2b'(\mathbf{s}^0). \tag{3.8.1}$$

Proof.

Case 1. Suppose that $\mathbf{s}^0$ is on the boundary of S. Let $\mathbf{s}^*$ be any point of P_S^* that is conjugate to $\mathbf{s}^0$. We shall prove first that every supporting plane to P_S^* at $\mathbf{s}^*$ contains the contact set $P_S^* \cap R\mathbf{A}$. Let h denote one such supporting plane, and assume that there exists a point $\mathbf{r}'\mathbf{A}$ that is in $R^0\mathbf{A}$ but not in h. Let T denote the set of supporting planes to S at $\mathbf{s}^0$. Since $\mathbf{r}'\mathbf{A}$ is in P_S^* (Lemma 3.4.1) but not in h, we must have $(\mathbf{r}'\mathbf{A}, \mathbf{h}) > 0$. However, $\mathbf{s}^0$ and $\mathbf{r}'$ are optimal, so that $(\mathbf{r}'\mathbf{A}, \mathbf{s}^0) = 0$. Consequently, $\mathbf{r}'\mathbf{A}$ may be regarded as an element of the set T. But $\mathbf{s}^*$ is an inner point of the set T by the definition of a conjugate point. Hence we can find $\mathbf{t}$ in T such that $\mathbf{s}^* = \alpha\mathbf{r}'\mathbf{A} + \beta\mathbf{t}$, where $0 < \alpha \le 1$, $\alpha + \beta = 1$. Forming the inner product with $\mathbf{h}$,

$$0 = (\mathbf{s}^*, \mathbf{h}) = \alpha(\mathbf{r}'\mathbf{A}, \mathbf{h}) + \beta(\mathbf{t}, \mathbf{h}).$$

Since $\mathbf{t} \in T \subset P_S^*$ implies $(\mathbf{t}, \mathbf{h}) \ge 0$, the last equation contradicts the assumption $\mathbf{r}'\mathbf{A} \notin h$.

Hence the intersection of all supporting planes to P_S^* at the point $\mathbf{s}^*$ contains the contact set $R^0\mathbf{A}$, which has dimension μ. Allowing for contact along one of the rays of P_S^*, we have

$$c(\mathbf{s}^*) + 1 \ge \mu.$$

By Lemma 3.6.1, $c(\mathbf{s}^*) + a(\mathbf{s}^0) = m - 1$. Hence $m - a(\mathbf{s}^0) \ge \mu$, and it follows from Lemma 3.7.2 that $m + 1 - 2b'(\mathbf{s}^0) \ge \mu$ as desired.

Case 2. Let $\mathbf{s}^0$ be an interior point of S. Then $a(\mathbf{s}^0) = m$, and we must show that $\mu = 0$.

Let $\mathbf{r}'$ be in R^0. Then $\mathbf{r}'\mathbf{A}$ is in $R\mathbf{A} \cap P_S^*$. Moreover, since $\mathbf{r}'\mathbf{A}$ is in P_S^* and $\mathbf{s}^0$ is interior to S, we have $(\mathbf{r}'\mathbf{A}, \mathbf{s}^0) > 0$, unless $\mathbf{r}'\mathbf{A} = \mathbf{0}$. But the fact that $\mathbf{r}'$ is optimal implies that $(\mathbf{r}'\mathbf{A}, \mathbf{s}^0) = 0$. Hence $\mathbf{r}'\mathbf{A} = \mathbf{0}$, and $R^0\mathbf{A}$ contains only one element, the origin. Since $\mathbf{A}'$ is a one-to-one transformation when applied to R, we have $\mu = \dim R^0 = \dim R^0\mathbf{A} = 0$. This concludes the proof of the theorem.

Relation (3.8.1) implies that the larger the dimension of Player I's set of optimal strategies, the more likely it is that Player II will have optimal strategies of finite type involving few pure strategies. More precisely, if Player I's optimal strategies span a μ-dimensional set, then Player II has an optimal strategy $\mathbf{s}_0$ with weight

$$b'(\mathbf{s}_0) \leq \frac{m + 1 - \mu}{2}.$$

The next theorem formulates the corresponding relationship between the dimension of Player II's set of optimal strategies and the weight of an optimal strategy of Player I.

▶ THEOREM 3.8.2. Under the conditions of Theorem 3.8.1,

$$\nu \leq m + 1 - 2b'(\mathbf{r}^0). \tag{3.8.2}$$

Proof. Let $L(\mathbf{r}^0\mathbf{A})$ denote the intersection of the supporting planes to $R\mathbf{A}$ at $\mathbf{r}^0\mathbf{A}$. Then $a(\mathbf{r}^0) = a(\mathbf{r}^0\mathbf{A}) = \dim L(\mathbf{r}^0\mathbf{A})$. Hence the number of linearly independent planes must be $m + 1 - a(\mathbf{r}^0)$. Since the dimension of a set is one less than the number of linearly independent elements in the set (taking linear independence in the sense of p. 67), the dimension of the set of supporting planes is $m - a(\mathbf{r}^0)$. This set contains the set of separating planes S^0. Hence

$$\nu = \dim S^0 \leq m - a(\mathbf{r}^0),$$

which by Lemma 3.7.2 gives (3.8.2).

It is interesting to note the difference between the arguments used in proving Theorems 3.8.1 and 3.8.2. The proof of Theorem 3.8.1 amounts to applying a counting procedure to the contact set of $R\mathbf{A}$ and P_S^*, whereas Theorem 3.8.2 is proved by enumerating the number of independent separating planes between $R\mathbf{A}$ and P_S^*. The reader may wonder why a parallel symmetrical argument cannot be applied to prove both theorems simultaneously, rather than in one case analyzing the contact set and in the other investigating the set of separating hyperplanes. The two separate arguments are necessitated by the unsymmetrical assumption that $\rho = n$. In this connection, we should remark that neither (3.8.1) nor (3.8.2) is symmetrical, since both involve the dimension number m.

▶ THEOREM 3.8.3. Under the conditions of Theorem 3.8.1, if $\mathbf{r}^0$ is on the boundary of R, then

$$\mu \leq b(\mathbf{r}^0) - 1 \tag{3.8.3}$$

and the number of essential pure strategies for Player I is exactly $b(\mathbf{r}^0)$.

Similarly, if $\mathbf{s}^0$ is on the boundary of S, then

$$\nu \leq b(\mathbf{s}^0) - 1 \tag{3.8.4}$$

and the number of essential pure strategies for Player II is exactly $b(\mathbf{s}^0)$.

Proof. It is clear from examining (3.8.3) and (3.8.4) with reference to Lemma 3.7.2 that we need only prove $\mu \leq c(\mathbf{r}^0)$ and $\nu \leq c(\mathbf{s}^0)$. If $\mathbf{r}^0$ is on the boundary of R but $\mathbf{r}^0\mathbf{A}$ is an inner point of the set $R^0\mathbf{A}$, and h is any supporting plane to R at $\mathbf{r}^0$, then h contains all of R^0. Otherwise, we could find a point $\mathbf{r}^1$ in R^0 which is not in h. Then, since $\mathbf{r}^0\mathbf{A}$ is inner to $R^0\mathbf{A}$, there exists a point $\mathbf{r}^2\mathbf{A}$ in $R^0\mathbf{A}$ such that $\mathbf{r}^0\mathbf{A} = \alpha_1\mathbf{r}^1\mathbf{A} + \alpha_2\mathbf{r}^2\mathbf{A}$, where $\alpha_1 + \alpha_2 = 1, 0 < \alpha_1, \alpha_2 \leq 1$. Since $\mathbf{A}$ has rank n, it follows that $\mathbf{r}^0 = \alpha_1\mathbf{r}^1 + \alpha_2\mathbf{r}^2$. Hence

$$0 = (\mathbf{h}, \mathbf{r}^0) = \alpha_1(\mathbf{h}, \mathbf{r}^1) + \alpha_2(\mathbf{h}, \mathbf{r}^2). \tag{3.8.5}$$

Since h is a supporting plane, it follows that $(\mathbf{h}, \mathbf{r}^1) \geq 0, (\mathbf{h}, \mathbf{r}^2) \geq 0$. Hence, by (3.8.5), $(\mathbf{h}, \mathbf{r}^1) = 0$ and $\mathbf{r}^1$ is in h.

Therefore the intersection $L(\mathbf{r}^0)$ of all supporting planes to R at $\mathbf{r}^0$ contains R^0, and we have

$$c(\mathbf{r}^0) = \dim (L(\mathbf{r}^0) \cap R) \geq \dim R^0 = \mu.$$

In a similar manner we see that if $\mathbf{s}^0$ is on the boundary of S, the intersection $L(\mathbf{s}^0)$ contains all S^0. The rest of the proof for ν is the same as for μ.

If more precise information is known about the form of the optimal strategies for one of the players (say Player II), then some of the inequalities of Theorems 3.8.1–3 can be converted into equalities. Such a situation occurs if, for example, $\mathbf{s}^0$ is interior to S.

▶THEOREM 3.8.4. In the polynomial game of Theorem 3.8.1, if $\mathbf{s}^0$ is interior to S, then Player I has a unique optimal strategy $\mathbf{r}^0$, and

$$\nu = m + 1 - 2b'(\mathbf{r}^0).$$

Proof. If $\mathbf{s}^0$ is interior, then $\mathbf{s}^0$ is a separating plane of P_S^* and $R\mathbf{A}$ which touches P_S^* only at the vertex. Hence $R\mathbf{A}$ has only one point in common with P_S^*; and since $\mathbf{A}$ is of rank n, $\mathbf{r}^0$ is unique.

Since $R\mathbf{A}$ touches P_S^* only at the origin and any sufficiently small tilt of the hyperplane $\mathbf{s}^0$ about the vertex of P_S^* remains a supporting plane to P_S^*, we see that the dimension of the set of separating planes of $R\mathbf{A}$ and P_S^* is precisely the dimension of the set of all supporting planes to $R\mathbf{A}$ at $\mathbf{r}^0\mathbf{A}$.

Recall that the dimension of the set of all supporting planes to $R\mathbf{A}$ at $\mathbf{r}^0\mathbf{A}$ is $m - a(\mathbf{r}^0)$. Thus

$$\nu = m - a(\mathbf{r}^0) = m + 1 - 2b'(\mathbf{r}^0),$$

the last identity being valid by Lemma 3.7.2.

Retaining the one-sided hypothesis $\rho = n$, we now suppose that Player I has an interior optimal strategy. In this case n must be strictly less than $m + 1$, so that $\mathbf{r}^0\mathbf{A}$ may be on the boundary of $R\mathbf{A}$. We now have $a(\mathbf{r}^0), c(\mathbf{r}^0) = n$, but the substitution of these values in the lemmas does not lead beyond the results of Theorems 3.8.1–2. Thus, whereas the assumption of an interior optimal strategy for the player with the larger choice of strategies is of interest and leads to Theorem 3.8.4, a similar assumption for the other player is hardly restrictive at all.

Another theorem along the same lines is the following.

▶THEOREM 3.8.5. In the square case $\rho = n = m$, if $\mathbf{s}^0$ is interior and unique, then $\mathbf{r}^0$ is also interior and unique, and

$$b'(\mathbf{r}^0) = b'(\mathbf{s}^0) = \frac{n+1}{2}.$$

Proof. Since $\mathbf{s}^0$ is interior, it follows from Theorem 3.8.4 that $\mathbf{r}^0$ is unique. Since $\mathbf{s}^0$ is unique, it also follows from Theorem 3.8.4 that

$$0 = \nu = m + 1 - 2b'(\mathbf{r}^0).$$

Hence

$$b'(\mathbf{r}^0) = \frac{m+1}{2} = \frac{n+1}{2}.$$

It follows from the theory of the geometry of moment spaces that any point having a representation with $b' = (n + 1)/2$ is interior (see Lemma 3.7.2); hence $\mathbf{r}^0$ is interior. Also since $\mathbf{s}^0$ is interior,

$$b'(\mathbf{s}^0) = \frac{m+1}{2} = \frac{n+1}{2}$$

by Lemma 3.7.2. Hence

$$b'(\mathbf{s}^0) = \frac{n+1}{2} = b'(\mathbf{r}^0).$$

Equalizer optimal strategies for polynomial games. A strategy for one player that makes the outcome of the game independent of his opponent's action is called an equalizing strategy or an equalizer. Equalizing strategies are not necessarily optimal, but when they are, their computation is greatly simplified. For example, the strategies ξ^0 and η^0 are equalizing optimal strategies if the pay-off may be written in the form

$$K(\xi, \eta) = P(\xi)Q(\eta)R(\xi, \eta) + c,$$

where ξ^0 is a root of $P(\xi)$ and η^0 is a root of $Q(\eta)$. If either strategy is used, the outcome is c, and clearly c is the value of the game.

For the polynomial game with value 0, the following geometrical interpretation may be given. If Player I has an optimal equalizer $\mathbf{r}^0$, then $(\mathbf{r}^0\mathbf{A}, \mathbf{s}) = 0$ for all $\mathbf{s}$ in S; hence $\mathbf{r}^0\mathbf{A} = \langle 0, \ldots 0 \rangle$ and the origin lies in $R\mathbf{A}$. If Player II has an optimal equalizer $\mathbf{s}^0$, then $(\mathbf{rA}, \mathbf{s}^0) = 0$ for all $\mathbf{r}$ in R. Hence the supporting plane to P_S^* corresponding to $\mathbf{s}^0$ contains all of $R\mathbf{A}$.

If one player has an interior solution, all his opponent's optimal strategies must be equalizers. For if $\mathbf{r}^0$ is an interior optimal strategy, any pure strategy ξ may be used with positive weight in a mixed strategy $x(\xi)$ representing $\mathbf{r}^0$ (Lemma 3.7.1). Consequently, if $\mathbf{s}^0$ is optimal for the other player, $(\mathbf{r}^0\mathbf{A}, \mathbf{s}^0) = 0$ and any element in the convex representation must yield a nonpositive result. Since their convex sum must be zero, every yield must be zero—in particular, $(\mathbf{rA}, \mathbf{s}^0) = 0$. The converse is not true: it does not follow from the fact that if all of one player's optimal strategies are equalizers, then the other player has an interior optimal strategy. (Consider as an example $K(\xi, \eta) = \xi\eta - \xi^2$.) In this respect polynomial games differ from finite matrix games. In finite games only the essential pure strategies yield the value of the game against every opposing optimal strategy (Theorem 3.1.1 of Vol. I).

3.9 Problems.

1. Use the mapping technique to solve the game with matrix

$$\begin{Vmatrix} -2 & 2 & -4 \\ 1 & -1 & 2 \\ 1 & 0 & 1 \end{Vmatrix}$$

subject to the added constraints $\eta_3 \geq 4/15$ and $\eta_1 \leq 1/5$, apart from the usual constraints of matrix games.

2. Use the mapping technique to solve the game with separable kernel

$$K(\xi, \eta) = 3 \cos 7\xi \cos 8\eta + 5 \cos 7\xi \sin 8\eta + 2 \sin 7\xi \cos 8\eta + \sin 7\xi \sin 8\eta.$$

3. Solve the separable game with kernel $K(\xi, \eta) = \xi(1 - \xi^2) \sin \pi\eta$, where $0 \leq \xi, \eta \leq 1$.

4. If $K(\xi, \eta) = \xi^2\eta - \xi\eta^2$, find the value of the game and two optimal strategies for each player.

5. Find the value and a pair of optimal strategies for the following kernels:

$$\text{(a)}\ \xi\eta^2 - 3\xi\eta - \xi + 2\eta,$$
$$\text{(b)}\ \xi^3\eta - \eta^2.$$

6. Find all solutions for the game with kernel $K(\xi, \eta) = \xi^3\eta - \eta^2$.

7. Find at least two optimal strategies for Player I in the second example of Section 3.2.

8. For the game with kernel

$$K(\xi, \eta) = 2\xi + 3\xi^2 + 3\eta - 3\xi\eta - \tfrac{9}{2}\xi^2\eta + 4\eta^2 - 4\xi\eta^2 - 6\xi^2\eta^2$$

find at least one equalizing strategy for each player.

9. Using the mapping technique of Section 3.2, find the first three moments of the optimal strategies for the game with kernel

$$K(\xi, \eta) = 21\xi + 18\xi^2 - 24\xi^3 - 16\eta - 36\xi\eta - 9\xi^2\eta + 18\xi^3\eta + 60\eta^2 - 36\eta^3.$$

10. Find a pair of optimal distributions $x(\xi)$ and $y(\eta)$ for Problem 9.

11. Consider the polynomial game with kernel $K(\xi, \eta) = \xi\eta - \xi^2$. Show that both players have unique pure optimal strategies located at $\xi = 0$ and $\eta = 0$, respectively. Show also that for the game with perturbed kernel $K_\epsilon(\xi, \eta) = \xi\eta - \xi^2 - \epsilon\xi$ $(\epsilon > 0)$, the optimal strategies are no longer unique.

12. Prove that for any finite convex game (not necessarily polyhedral), if both players have unique interior optimal strategies, all games resulting from small perturbations of the pay-off kernel have unique optimal strategies.

13. Show that if C is any closed bounded set in E^n, then the convex hull of C is also closed and bounded.

14. Find the dual cone P^* where P is the convex cone consisting of all vectors $\langle x_1, \ldots, x_n\rangle$ with $x_1 \geq 0$ and

$$x_1^2 \geq \sum_{k=2}^{n} x_k^2$$

(see Section 3.6).

15. Let Γ be a finite convex game with polyhedral convex strategy spaces R and S. Let μ and ν be the dimensions of the sets R^0 and S^0 of optimal strategies for Players I and II, respectively, and let k and l be the dimensions of the smallest polyhedral faces of R and S that contain R^0 and S^0. Show that

$$k - \mu = l - \nu.$$

16. Let Γ be a finite convex game with polyhedral convex strategy spaces. Show that if both players possess unique solutions, this uniqueness is not disturbed by sufficiently small changes in the pay-off kernel.

17. Let $\mathbf{r}^0$ and $\mathbf{s}^0$ be interior points of the n-dimensional moment space D^n, and let v be any number. Construct a matrix $\mathbf{A}$ such that $\mathbf{As}^0 = \langle v, 0, \ldots, 0\rangle$, $\mathbf{r}^0 A = \langle v, 0, \ldots, 0\rangle$. Show that $\mathbf{r}^0$ and $\mathbf{s}^0$ are the unique optimal strategies for $\mathbf{A}$. (This example illustrates Theorem 3.8.5.)

18. Consider a set of continuous functions $h_0(\xi)$, $h_1(\xi)$, $h_2(\xi)$, $\ldots$, $h_n(\xi)$, all defined for $0 \leq \xi \leq 1$, which satisfy

$$\begin{vmatrix} h_0(\xi_1) & h_1(\xi_1) & \cdots & h_n(\xi_1) \\ h_0(\xi_2) & h_1(\xi_2) & \cdots & h_n(\xi_2) \\ \vdots & & & \\ h_0(\xi_{n+1}) & \cdots & \cdots & h_n(\xi_{n+1}) \end{vmatrix} \neq 0$$

whenever $0 \leq \xi_1 < \xi_2 < \cdots < \xi_{n+1} \leq 1$. Show that there exists a distribution x which maximizes

$$\int_0^1 h_0(\xi)\, dx(\xi)$$

subject to the constraints

$$\int_0^1 h_i(\xi)\, dx(\xi) = a_i$$

and possesses a spectrum of at most $(n + 1)/2$ points, where 0 and 1 are each assigned the weight 1/2. (The constraint set is assumed to be nonempty.)

Notes and References to Chapter 3

§ 3.1. It was thought in the early stages of the development of infinite game theory that the solution of polynomial games would go far toward clearing the way for the ultimate analysis of all infinite games. This hope was rapidly shattered. It was soon realized that polynomial games of large degree possess solutions of great complexity which are impossible to so much as describe in qualitative terms, let alone calculate. Moreover, the slow convergence of polynomials to continuous kernels means that polynomial games are impractical as approximations to continuous games. Nevertheless, the desire to solve polynomial games led to the partially successful investigation of the more important general separable games. Among the earliest to examine particular instances of separable and polynomial games were Helmer [120] and Dresher [67].

§ 3.2. The fixed-point numerical method of handling finite convex games was first proposed by Dresher.

§ 3.3. Elaborations on the structure theorems of solution sets for finite polyhedral games based on the fixed-point technique are presented in Dresher and Karlin [70]. Other uses of the fixed-point method have been advanced elsewhere. For a discussion of these, the reader may consult McKinsey [179] and Dresher [67].

§ 3.4. The theory of convex cones is classical and possesses a vast literature (see Fenchel [81]). The relation of the theory of dual convex cones to separable games was developed by Dresher, Karlin, and Shapley [71]. Gale and Sherman [100] independently pointed out the relevance of polyhedral convex cones to the analysis of finite discrete games.

The duality theory of cones is most extensively used in connection with the theory of linear and nonlinear programming. (See Chapters 5–7 of Vol. I.) It must be emphasized again that in the case of linear programming the cones considered are polyhedral, whereas in the separable finite games of this chapter we deal with general closed convex cones, which are more complex and harder to characterize.

One special and important class of cones is the class formed by the classical reduced moment spaces. Most of the boundary structure of the moment cone can be discerned in simple elegant analytical forms. The geometry of moment spaces has been developed in Karlin and Shapley [141]. Here is an instance of how game theory impinged upon the development of an area of pure mathematics. While attempting to solve polynomial games, the authors found it necessary to study in minute detail the structure of the boundaries of moment spaces. This study led to an extensive study of moment spaces, [141], in which the subject was discussed without reference to game theory.

§ 3.5. The account of polynomial and separable games given here is taken from [71]. The analysis of such games depends upon a precise knowledge of the properties of the boundary components of the moment spaces. A knowledge of these properties enables us to obtain accurate descriptions of the dimensions of solution sets for polynomial games as well as the minimal sizes of the solution spectrum. Even more precise results have been obtained for discrete games, but this is because polyhedral cones are the simplest cones to characterize.

Gale and Gross [97] have shown how to construct a polynomial game whose unique solutions are any two prescribed distributions of finite type.

§ 3.6 and **§ 3.7.** The results of these sections are based on [141].

§ 3.8. This section is an elaboration of [71].

CHAPTER 4

GAMES WITH CONVEX KERNELS AND GENERALIZED CONVEX KERNELS

4.1 Introduction. In general, the analysis of continuous games defined by a kernel $K(\xi, \eta)$ on the unit square is very complicated. Often a further restriction of monotonicity is dictated by the nature of the model describing the game. Such a condition is the following: We say that a game defined by a kernel K is a generalized convex game if there exists some integer $n > 0$ such that

$$\frac{\partial^n K(\xi, \eta)}{\partial \eta^n} \geq 0 \qquad (0 \leq \xi, \eta \leq 1) \tag{4.1.1}$$

or some integer $m > 0$ such that

$$\frac{\partial^m K(\xi, \eta)}{\partial \xi^m} \leq 0 \qquad (0 \leq \xi, \eta \leq 1). \tag{4.1.2}$$

The tacit assumption that K possesses a sufficient number of continuous derivatives is made only for convenience of exposition. We have chosen to impose differentiability conditions in order to avoid a more detailed and tedious consideration of cases. The principal ingredients of the arguments and results are the same whether we use the assumption (4.1.1) or the corresponding difference inequality. Throughout what follows we restrict ourselves to games satisfying (4.1.1). The parallel proofs needed to prove the corresponding results for kernels obeying (4.1.2) are left as exercises for the reader.

Any kernel which is a polynomial of degree at most $n - 1$ in η obviously satisfies (4.1.1). However, one cannot expect results achieved for generalized convex games to be as sharp as those obtained by direct investigation of polynomial games. Yet, the assumption (4.1.1) alone yields substantial information concerning the form of optimal strategies of generalized convex games.

The cases $n = 1$ and $n = 2$ play an important role from both the theoretical and the practical point of view; moreover, of course, they yield to simpler analysis. We proceed to a discussion of the case $n = 1$, and in the next section we study the case $n = 2$.

Relation (4.1.1) for $n = 1$ states that K is monotone-increasing in η for each ξ. Since Player II wishes to minimize the return to Player I, he tries to minimize $K(\xi, \eta)$ for every ξ. By choosing the pure strategy y_0 that concentrates at the point $\eta = 0$, Player II simultaneously makes $K(\xi, \eta)$ as small as possible regardless of the choice of ξ. In other words, there is obviously no reason for Player II to consider any other choice of η, since with any fixed ξ the pure strategy $\eta = 0$ gives the best yield. Player I, realizing that Player II will choose $\eta = 0$, will do best to choose the pure strategy x_{ξ_0} satisfying $K(\xi_0, 0) = \max_\xi K(\xi, 0)$. It is immediately clear that the pure strategies ξ_0 and 0 for Players I and II, respectively, provide a saddle point for the kernel $K(\xi, \eta)$. The value of the game is $v = \max_\xi K(\xi, 0)$.

Note that only the requirement of monotonicity for $K(\xi, \eta)$ in η was used; smoothness conditions, e.g., the existence of derivatives, are irrelevant. However, the continuity of the kernel—or more generally, the fact that $\sup_\xi K(\xi, 0)$ is attained—is necessary to guarantee the existence of optimal strategies.

4.2 Convex continuous games. Relation (4.1.1) for $n = 2$ implies that $K(\xi, \eta)$ is a convex function of η for each ξ. Again, for the analysis of such kernels we can dispense with the differentiability assumption, since the concept of convexity may be defined independent of such assumptions. We need only require that $K(\xi, \eta)$ be convex in η for each ξ, and that K be continuous jointly in both ξ and η (the usual requirement). Throughout the remainder of this section we assume that this is the case. We can further generalize our results by allowing ξ to vary over an arbitrary compact set.

Our approach to such games resembles the analysis of $m \times 2$ games in Section 2.3 of Vol. I. In fact, the pay-off kernel of an $m \times 2$ matrix game can be regarded as a bilinear function $K(\xi_1, \xi_2, \ldots, \xi_{m-1}, \eta) = K(\boldsymbol{\xi}, \eta)$, where η traverses the unit interval, $\boldsymbol{\xi} = \langle \xi_1, \ldots, \xi_{m-1} \rangle$ with $\xi_i \geq 0$ and

$$\sum_{i=1}^{m-1} \xi_i \leq 1,$$

and K is convex (indeed linear) in η for each $\boldsymbol{\xi}$ belonging to Δ. It is therefore plausible to suppose that a generalization of the analysis of $m \times 2$ matrix games might yield results for convex continuous games. Proceeding by analogy, we plot the family of convex functions $\phi_\xi(\eta) = K(\xi, \eta)$, where ξ is a parameter, and form the envelope from above

$$\phi(\eta) = \max_{0 \leq \xi \leq 1} K(\xi, \eta).$$

In view of property (iv) of Section B.4, $\phi(\eta)$ also defines a continuous convex function of η as η traverses the unit interval. Hence $\phi(\eta)$ achieves its minimum. Let η_0 be any point at which this minimum is attained. We shall prove that y_{η_0} is optimal for Player II, and that $\phi(\eta_0)$ is the value of the game.

This is completely parallel to the construction of Section 2.3 of Vol. I. Once the optimal η strategy has been discovered, we determine two values of ξ which by appropriate combination yield an optimal strategy for Player I. The following lemma sets forth this procedure.

▶LEMMA 4.2.1. If the function $K(\xi, \eta)$ defined over the unit square is jointly continuous in ξ and η, and if $K(\xi, \eta)$ is convex in η for each ξ, then $\phi(\eta_0) = \inf_\eta \sup_\xi K(\xi, \eta)$ is achieved. Moreover, for every point η_0 at which the infimum is achieved, we can find values ξ_1 and ξ_2 such that

$$\begin{aligned} K(\xi_1, \eta) &\geq K(\xi_1, \eta_0) = \phi(\eta_0) \qquad (\eta \geq \eta_0), \\ K(\xi_2, \eta) &\geq K(\xi_2, \eta_0) = \phi(\eta_0) \qquad (\eta \leq \eta_0). \end{aligned} \tag{4.2.1}$$

If $\eta_0 = 0$ or $\eta_0 = 1$, then only one inequality is well defined and valid.

Proof. We need only establish the second half of the lemma. We present the arguments of the theorem only for the case in which η_0 is an interior point. Let $T \equiv \{\xi_\alpha | K(\xi_\alpha, \eta_0) = \phi(\eta_0)\}$. Let $t_\alpha(\eta)$ denote the right tangent line to $K(\xi_\alpha, \eta)$ at the point $\eta = \eta_0$, and let r_α denote the slope of this tangent. It is readily seen that the set T is compact and that the corresponding set of real numbers $\{r_\alpha\}$ achieves its maximum. Moreover, $\max_\alpha r_\alpha \geq 0$, for otherwise $r_\alpha \leq \delta < 0$ uniformly in α, which violates the hypothesis that $\phi(\eta_0) = \min_\eta \phi(\eta)$. Select any $\xi_1 = \xi_\alpha$ for which $r_\alpha \geq 0$. Since a convex function always lies above its tangent, we obtain for every $\eta > \eta_0$

$$K(\xi_1, \eta) \geq t_\alpha(\eta) \geq \phi(\eta_0).$$

The point ξ_2, satisfying the other half of (4.2.1), is similarly constructed, using left tangents.

Construction of optimal strategies for convex continuous games. To apply the results of the preceding lemma to the convex game under consideration, we distinguish three cases.

(a) If η_0 is an interior point, then there exist both ξ_1 and ξ_2 (not necessarily distinct) satisfying (4.2.1). If $t_1(\eta)$ denotes a right tangent line to $K(\xi_1, \eta)$ at $\eta = \eta_0$ and $t_2(\eta)$ a left tangent line to $K(\xi_2, \eta)$ at $\eta = \eta_0$, and if r_1 and r_2 are the slopes of these right and left tangents, respectively, then $r_1 \geq 0$ and $r_2 \leq 0$. We can consequently choose α satisfying $0 \leq \alpha \leq 1$ such that

$$\alpha t_1(\eta) + (1 - \alpha) t_2(\eta) = \phi(\eta_0).$$

Set $x_0(\xi) = \alpha x_{\xi_1} + (1 - \alpha)x_{\xi_2}$. Then

$$\int K(\xi, \eta)\, dx_0(\xi) = \alpha K(\xi_1, \eta) + (1 - \alpha)K(\xi_2, \eta)$$
$$\geq \alpha t_1(\eta) + (1 - \alpha)t_2(\eta) = \phi(\eta_0),$$

since any convex function always lies above its tangent. The last equation shows that $v \geq \phi(\eta_0)$.

On the other hand,

$$\int K(\xi, \eta)\, dy_{\eta_0}(\eta) = K(\xi, \eta_0) \leq \sup_\xi K(\xi, \eta_0) = \phi(\eta_0). \tag{4.2.2}$$

Hence $v \leq \phi(\eta_0)$. Therefore $v = \phi(\eta_0)$, and $\{x_0(\xi), y_{\eta_0}(\eta)\}$ are optimal strategies for the game.

(b) If $\eta_0 = 0$, we select ξ_1 such that

$$K(\xi_1, \eta) \geq K(\xi_1, \eta_0) = \phi(\eta_0).$$

Set $x(\xi) = x_{\xi_1}$; then

$$\int K(\xi, \eta)\, dx_{\xi_1}(\xi) = K(\xi_1, \eta) \geq \phi(\eta_0).$$

Equation (4.2.2) is still valid; $v = \phi(\eta_0)$, and $\{x(\xi), y_{\eta_0}(\eta)\}$ are again optimal.

(c) If $\eta_0 = 1$, we select ξ_2 and proceed as in (b). In any case Player I has an optimal strategy involving at most two pure strategies, and Player II has an optimal pure strategy.

It may happen that for each fixed η, the function $K(\xi, \eta)$ is also convex in ξ. Then

$$K(\xi, \eta) \leq \xi K(1, \eta) + (1 - \xi)K(0, \eta) \leq \max\{K(1, \eta), K(0, \eta)\}.$$

It follows that $\sup_\xi K(\xi, \eta) = \max\{K(1, \eta), K(0, \eta)\}$. It is then sufficient to plot these two curves and find the minimum point of their envelope. If the minimum occurs at one of the endpoints, one of the pure strategies will be optimal. If the minimum occurs at an interior point, some convex combination of x_ξ with $\xi = 0$ and $\xi = 1$ will be optimal for Player I (see the example below).

If $K(\xi, \eta)$, in addition to being convex in η, is also concave in ξ for each η, then Player I will have a pure optimal strategy x_{ξ_0} corresponding to a point ξ_0 at which $\max_\xi \min_\eta K(\xi, \eta)$ is achieved.

Example. The following example is a special case of the model described in Section 2.3.

We assume that Player II has the task of defending two cities, A and B, against Player I. We assume that in each city, if the attacking force is greater than the defending force, the defender's losses are proportional to the difference between the two forces; but that if the attacking force is smaller, the defender suffers no loss. Furthermore, losses in B are valued at λ times losses in A. We further assume that the total force available to each player is unity. If pure strategies ξ and η are identified with the proportion of the total forces of Players I and II, respectively, that are assigned to A, the yield to Player I becomes

$$K(\xi, \eta) = \max \{ \xi - \eta, \lambda(1 - \xi - 1 + \eta)\} = \max \{ \xi - \eta, \lambda(\eta - \xi)\},$$

and $K(\xi, \eta)$ is convex in each variable. Furthermore, $K(0, \eta) = \lambda\eta$ and $K(1, \eta) = 1 - \eta$. These two lines intersect for $\eta_0 = 1/(1 + \lambda)$; hence, if we choose according to the preceding construction, a pair of optimal strategies is

$$x^*(\xi) = \frac{1}{1 + \lambda} x_0(\xi) + \frac{\lambda}{1 + \lambda} x_1(\xi) \qquad \text{and} \qquad y^*(\eta) = y_{1/(1+\lambda)}(\eta).$$

(x_1 corresponds to attacking A.)

The defender keeps his force divided in the ratio $1:\lambda$ between A and B. The attacker should always attack a city in full force, randomizing his attacks in the ratio λ:1. It is interesting that

$$\frac{\lambda}{1 + \lambda} y_0(\eta) + \frac{1}{1 + \lambda} y_1(\eta)$$

is also optimal for Player II, although Player I has no optimal pure strategy.

***4.3 Generalized convex games.** An extension of the theory of convex continuous games to generalized convex games leads to the characterization that such games possess optimal strategies of finite type. Unfortunately, in the general case this characterization is not constructive: i.e., it does not show us how to determine the optimal strategies.

We assume in what follows that for generalized convex games the variable ξ ranges over a connected compact set C. The principal theorem for generalized convex games follows.

▶ THEOREM 4.3.1. If for some $n > 0$

$$\frac{\partial^n K(\xi, \eta)}{\partial \eta^n} \geq 0,$$

---

* Starred sections are advanced discussions that may be omitted at first reading without loss of continuity.

then Player I has an optimal strategy involving at most n pure strategies and Player II has an optimal strategy involving at most $n/2$ pure strategies, where each pure strategy η interior to the unit interval has a unit count and the endpoints $\eta = 0$ and $\eta = 1$ are weighted as half-counts.

Proof. For simplicity we assume $v = 0$. To wit, change K by subtracting a suitable constant.

(a) It suffices to prove the theorem in the case

$$\frac{\partial^n K(\xi, \eta)}{\partial \eta^n} > 0;$$

for if the theorem is true for these special games, and if

$$\frac{\partial^n K(\xi, \eta)}{\partial \eta^n} \geq 0,$$

we can clearly find a sequence of games with kernels $K^{(m)}(\xi, \eta)$ uniformly convergent to $K(\xi, \eta)$ with

$$\frac{\partial^n K^{(m)}(\xi, \eta)}{\partial \eta^n} > 0.$$

By assumption, these games have optimal strategies for Player I of the form

$$\sum_{k=1}^{n} \alpha_k^{(m)} x_{\xi_k}^{(m)}.$$

We can find a subsequence of indices such that the corresponding subsequences for $\alpha_k^{(m)}$, $\xi_k^{(m)}$ converge for every k and their limit points α_k, ξ_k still satisfy the conditions $\xi_k \in C$, $0 \leq \alpha_k \leq 1$, $\sum \alpha_k = 1$. It is immediately verified that

$$\sum_{k=1}^{n} \alpha_k x_{\xi_k}$$

is an optimal strategy for Player I for the game with kernel $K(\xi, \eta)$. The same argument applies for Player II.

In view of the preceding argument we suppose throughout the rest of this section that

$$\frac{\partial^n K(\xi, \eta)}{\partial \eta^n} > 0$$

for $\xi \in C, 0 \leq \eta \leq 1$.

(b) We now prove that Player II has an optimal strategy using at most $n/2$ points. Let $x^*(\xi)$ be optimal for Player I. Set

$$h(\eta) = \int K(\xi, \eta)\, dx^*(\xi);$$

then $d^n h(\eta)/d\eta^n > 0$.

The function $h(\eta)$ can vanish at most n times in the unit interval (a root with multiplicity r is counted as r roots). In fact, if m denotes the number of roots of $h(\eta)$ counting multiplicities, then $h'(\eta)$ vanishes at least $m - 1$ times, $h''(\eta)$ vanishes at least $m - 2$ times, etc., and finally $h^{(n)}(\eta)$ vanishes at least $m - n$ times, with proper multiplicities assigned to the roots. Since $h^{(n)} > 0$, it follows that $m \leq n$. Furthermore, since $x^*(\xi)$ is optimal and $v = 0$, we see that $h(\eta) \geq 0$ in the unit interval. Therefore any interior root must have even multiplicity. It follows that the maximum possible number of distinct roots is $n/2$, provided that the endpoints $\eta = 0$ and $\eta = 1$ are counted as half-points.

Now let y^* be optimal for Player II. Then

$$\int h(\eta)\, dy^*(\eta) = \iint K(\xi, \eta)\, dx^*(\xi)\, dy^*(\eta) = v = 0.$$

Since $h(\eta)$ is nonnegative, y^* may have mass only at the points where $h(\eta)$ vanishes, of which there are at most $n/2$.

(c) Player I has an optimal strategy with spectrum consisting of at most n points. As a step toward proving this statement we establish the following lemma.

▶LEMMA 4.3.1. If in the unit interval the function h satisfies $h(\eta) \geq 0$, $h^{(n)}(\eta) > 0$, then there exists a polynomial $P \geq 0$ of degree at most $n - 1$ such that $h(y) - P(y)$ is nonnegative on the unit interval and has exactly n roots counting multiplicities.

Proof. We know that $h(\eta)$ may have at most n roots. If it has exactly n roots, we choose $P(\eta) \equiv 0$; if it has fewer than n roots, the polynomial $P(\eta)$ is constructed as follows. Let η_i be roots of $h(\eta)$, and let $Q_0(\eta)$ be a polynomial with the same roots as $h(\eta)$ and the same multiplicities. Then $Q_0(\eta)$ is of degree at most $n - 1$, $Q_0 \geq 0$, and $Q_0{}^{(n)}(\eta) = 0$. Furthermore, if r_i is the multiplicity of the root η_i,

$$\lim_{\eta \to \eta_i} \frac{h(\eta)}{Q_0(\eta)} = \lim_{\eta \to \eta_i} \frac{h^{(r_i)}(\eta)}{Q_0^{(r_i)}(\eta)} = \delta_i > 0.$$

Hence we can find open sets E_i about the roots η_i such that

$$\frac{h(\eta)}{Q_0(\eta)} \geq \frac{\delta_i}{2}$$

in E_i. The rest of the unit interval is a compact set; hence h/Q_0 is the ratio of two continuous, nonvanishing functions, and consequently it achieves its minimum δ_0. Hence for $\epsilon \leq \min(\delta_0, \delta_i/2)$,

$$h(\eta) - \epsilon Q_0(\eta) \geq 0. \tag{4.3.1}$$

Let ϵ_0 be the supremum of all ϵ for which (4.3.1) is satisfied. Then $h(\eta) - \epsilon_0 Q_0(\eta)$ must have at least one more root; otherwise $h(\eta) - \epsilon_0 Q_0(\eta)$ would satisfy all the conditions that $h(\eta)$ satisfies, and we could find $\epsilon' > 0$ such that

$$[h(\eta) - \epsilon_0 Q_0(\eta)] - \epsilon' Q_0(\eta) \geq 0,$$

which contradicts the assumption that ϵ_0 is the supremum of the admissible ϵ's. Therefore, either we have at least one new root or the multiplicity of a former root has increased.

The function $h - \epsilon_0 Q_0 \geq 0$ satisfies the condition

$$\frac{\partial^n}{\partial \eta^n} (h - \epsilon Q_0) > 0.$$

We may continue the process until we arrive at

$$P(\eta) = \sum_k \epsilon_k Q_k(\eta),$$

satisfying the conditions of the lemma.

▶COROLLARY 4.3.1. If $\eta_1, \ldots, \eta_n$ are the roots of the function $h(\eta) - P(\eta)$, then $P(\eta)$ is the unique polynomial of degree $n - 1$ that satisfies the conditions of Lemma 4.3.1 with the fixed set of roots.

For if there exists another polynomial $R(\eta)$ such that $h(\eta) - R(\eta)$ has roots $\eta_1, \ldots, \eta_n$, then $P(\eta) - R(\eta)$ is a polynomial of degree at most $n - 1$ with n roots $\eta_1, \ldots, \eta_n$, and consequently $P(\eta) - R(\eta) \equiv 0$.

Continuing with the proof of Theorem 4.3.1, we define

$$h(\eta) = \int K(\xi, \eta)\, dx^*(\xi)$$

and construct the polynomial P of the previous lemma. We again denote the n roots of $h(\eta) - P(\eta)$ by $\eta_1, \ldots, \eta_n$. We define

$$\phi_\xi(\eta) = K(\xi, \eta) - \sum_{i=0}^{n-1} a_i(\xi)\eta^i, \tag{4.3.2}$$

where the coefficients $a_i(\xi)$ are determined in such a manner that the roots of $\phi_\xi(\eta)$ are exactly the numbers $\eta_1, \ldots, \eta_n$ with the appropriate multiplicities. Since every interior root of $h(\eta) - P(\eta)$ has even multiplicity, $\phi_\xi(\eta)$ is either nonpositive or nonnegative for each ξ. Moreover, $\phi_\xi(\eta)$ is either nonpositive or nonnegative for each η. To prove this assertion, we assume the contrary. Then $\eta \neq \eta_i$ exists and ξ_1, ξ_2 can be found such that $\phi_{\xi_1}(\eta) > 0$ and $\phi_{\xi_2}(\eta) < 0$. But since the $a_i(\xi)$ are continuous, there must exist a value ξ_0 such that $\phi_{\xi_0}(\eta) = 0$, and in this case $\phi_{\xi_0}(\eta)$ would have more than the prescribed number of roots. Hence, either

$$K(\xi, \eta) \geq \sum_{i=0}^{n-1} a_i(\xi)\eta^i \qquad (\text{all} \quad \xi, \eta \text{ in } [0, 1])$$

or

$$K(\xi, \eta) \leq \sum_{i=0}^{n-1} a_i(\xi)\eta^i \qquad (\text{all} \quad \xi, \eta \text{ in } [0, 1]).$$

If we integrate both sides of (4.3.2) with respect to $x^*(\xi)$, the function

$$h(\eta) - \sum_{i=0}^{n-1} \eta^i \int a_i(\xi)\, dx^*(\xi) \tag{4.3.3}$$

is always of one sign, and its roots are exactly $\eta_1, \ldots, \eta_n$ with the appropriate multiplicities. Since the polynomial

$$R(\eta) = \sum_{i=0}^{n-1} \eta^i \int a_i(\xi)\, dx^*(\xi)$$

is of degree $n - 1$, it follows from Corollary 4.3.1 that $R(\eta)$ must be identical to $P(\eta)$. Hence (4.3.3) is nonnegative, and therefore

$$K(\xi, \eta) \geq \sum_i a_i(\xi)\eta^i.$$

If we set $L(\xi, \eta) = \sum a_i(\xi)\eta^i$, then $L(\xi, \eta)$ is the kernel of a separable game. We shall show that if x^* and y^* are optimal for the game with kernel $K(\xi, \eta)$, they are also optimal for L, and that the value $v(L) = 0$ [$v(L)$ denotes the value of the game with kernel L].

Since $v(K) = 0$ and y^* is optimal,

$$0 \geq \int K(\xi, \eta)\, dy^*(\eta) \geq \int L(\xi, \eta)\, dy^*(\eta).$$

Thus Player II in the game L can prevent Player I from getting more than zero, and therefore $v(L) \leq 0$. On the other hand,

$$\int L(\xi, \eta)\, dx^*(x) = \sum \int a_i(\xi)\eta^i\, dx^*(\xi) = R(\eta) \geq 0.$$

Hence $v(L) \geq 0$, from which it follows that $v(L) = 0$.

By the results of Theorem 3.5.1, we know that the separable game L has some optimal strategy $x_0(\xi)$ which uses at most n points $\xi_1, \ldots, \xi_n$. Then

$$\int K(\xi, \eta)\, dx_0(\xi) \geq \int L(\xi, \eta)\, dx_0(\xi) \geq 0.$$

Hence $x_0(\xi)$ is also optimal for the game with kernel $K(\xi, \eta)$, and the theorem is established.

4.4 Games with convex pay-off in E^n. Let $K(\xi, \eta)$ be a function defined for all η in some convex compact set S in E^n, and for all ξ in some compact set R. We assume that $K(\xi, \eta)$ is jointly continuous in ξ and η, and that $K(\xi, \eta)$ is convex in η for each ξ (see Section B.4). We consider a game with kernel $K(\xi, \eta)$ and mixed strategy spaces X and Y consisting of all distributions over R and S, respectively. We seek to characterize and determine the optimal strategies of this game.

Games of this sort are of practical as well as theoretical interest; i.e., many actual problems give rise to kernels $K(\xi, \eta)$ which are convex in one or both variables. We illustrate this with an example which was previously described in Section 2.3. Our interpretation there was in terms of attack and defense; here, by way of emphasizing the flexibility and the range of possible interpretations of game problems, we describe the same example in terms of a political campaign (see also Section 4.4 of Vol. I).

Party II desires to match its previous vote-getting record in n precincts which normally return a majority vote for this party. The n precincts $T_1, T_2, \ldots, T_n$ have varying values $k_1, k_2, \ldots, k_n$, respectively, the labeling being done in such a way that $k_1 \geq k_2 \geq \cdots \geq k_n > 0$. That is, T_1 (precinct 1) is the most valuable (e.g., has the largest number of voters), T_2 is next in value, etc. Party II has available a total of D units of money to use in campaigning for its nominees in the various precincts. The opposing party, Party I, has a total of A units of money for the same purpose. The model is such that in any given precinct the yield to Party I is proportional to the difference between the amount of money it invests and the amount Party II invests, assuming Party I invests more than Party II. If Party I invests less than Party II, the yield to Party I is zero.

A pure strategy for Party II is a vector of nonnegative numbers $\boldsymbol{\eta} = \langle \eta_1, \eta_2, \ldots, \eta_n \rangle$ satisfying

$$\sum_{i=1}^{n} \eta_i = D.$$

A pure strategy for Party I is a vector $\boldsymbol{\xi} = \langle \xi_1, \xi_2, \ldots, \xi_n \rangle$, with $\xi_i \geq 0$ and

$$\sum_{i=1}^{n} \xi_i = A.$$

We impose the condition $A \geq D$; i.e., we assume that Party I is as rich as Party II or richer. According to the model, the total value accruing to Party I for each choice of pure strategies is

$$K(\boldsymbol{\xi}, \boldsymbol{\eta}) = \sum_{i=1}^{n} k_i \max\{0, \xi_i - \eta_i\}, \tag{4.4.1}$$

where η and ξ range over the simplexes with vertices $\mathbf{d}_1, \mathbf{d}_2, \ldots, \mathbf{d}_n$ and

$\mathbf{a}_1, \mathbf{a}_2, \ldots, \mathbf{a}_n$, respectively, with $\mathbf{d}_i = \langle 0, 0, \ldots, D, 0, \ldots, 0\rangle$, the component D occurring in the ith place, and $\mathbf{a}_i = \langle 0, 0, \ldots, A, 0, \ldots, 0\rangle$, the component A occurring in the ith place ($i = 1, 2, \ldots, n$). The pay-off kernel $K(\boldsymbol{\xi}, \boldsymbol{\eta})$ is easily seen to be convex in each variable separately and jointly continuous.

At the close of this section we obtain the explicit solution to this game, a solution that possesses several striking aspects. First, however, since the process of arriving at a solution typifies a general technique in game theory, and since in this case the general theory is especially useful as a starting point, we turn to a description of the optimal strategies for general games with convex pay-off in E^n. The analysis of such games depends on the following basic theorem.

▶THEOREM 4.4.1. Let $\{\phi_\alpha\}$ be a family of continuous convex functions defined over a compact, convex, n-dimensional set S. Then $\sup_\alpha \phi_\alpha(\boldsymbol{\eta})$ attains its minimum value c at some point $\boldsymbol{\eta}_0$ of S. Moreover, given any $\delta > 0$, there exist indices α_i and real numbers λ_i such that

$$\sum_{i=1}^{n+1} \lambda_i \phi_{\alpha_i}(\boldsymbol{\eta}) \geq c - \delta$$

for all $\boldsymbol{\eta}$ in S, where $\lambda_i \geq 0$ and $\sum \lambda_i = 1$.

This theorem has already appeared in our discussion of the theory of conjugate functions (Theorem 7.6.4 of Vol. I). An alternative proof is presented in Section 4.5, following.

In the present context we observe that the family of functions $\phi_{\boldsymbol{\xi}}(\boldsymbol{\eta}) = K(\boldsymbol{\xi}, \boldsymbol{\eta})$ satisfies the conditions of Theorem 4.4.1. Consequently, the function $\phi(\boldsymbol{\eta}) = \sup_{\boldsymbol{\xi}} K(\boldsymbol{\xi}, \boldsymbol{\eta})$ achieves its minimum at some point $\boldsymbol{\eta}_0$ in S, and for every $\epsilon > 0$ we can choose $\{\lambda_i^\epsilon\}$, $\{\boldsymbol{\xi}_i^\epsilon\}$ such that

$$\sum_{i=1}^{n+1} \lambda_i^\epsilon K(\boldsymbol{\xi}_i^\epsilon, \boldsymbol{\eta}) \geq \phi(\boldsymbol{\eta}_0) - \epsilon$$

for all $\boldsymbol{\eta}$ in S, where λ_i^ϵ, $\boldsymbol{\xi}_i^\epsilon$ vary over compact sets; more precisely, where $\lambda_i^\epsilon \geq 0$ and

$$\sum_{i=1}^{n+1} \lambda_i^\epsilon = 1.$$

Therefore as ϵ tends to zero, we may select for each i limit points λ_i^0 and $\boldsymbol{\xi}_i^0$ ($i = 1, \ldots, n + 1$) satisfying the properties $\lambda_i^0 \geq 0$,

$$\sum_{i=1}^{n+1} \lambda_i^0 = 1,$$

$\xi_i^0 \in R$, and

$$\sum_{i=1}^{n+1} \lambda_i^0 K(\xi_i^0, \eta) \geq \phi(\eta_0) \tag{4.4.2}$$

for all η in S. Relation (4.4.2) shows that if Player I employs the finite mixed strategy

$$x^*(\xi) = \sum_{i=1}^{n+1} \lambda_i^0 x_{\xi_i^0}(\xi),$$

his return is at least $\phi(\eta_0)$. On the other hand, if Player II uses the pure strategy η_0, then for any ξ the yield to Player I is

$$K(\xi, \eta_0) \leq \sup_{\xi} K(\xi, \eta_0) = \phi(\eta_0).$$

Thus $\phi(\eta_0)$ is the value of the game and $\{x^*, y_{\eta_0}\}$ are optimal strategies.

Remark 4.4.1. If R is a simplex with vertices $\mathbf{a}_i$ $(i = 1, \ldots, k)$, then every ξ in R has a representation of the form

$$\sum_{i=1}^{k} \mu_i \mathbf{a}_i, \qquad \mu_i \geq 0, \qquad \sum_{i=1}^{k} \mu_i = 1.$$

If $K(\xi, \eta)$ is also convex in ξ for each η, then

$$K(\xi, \eta) = K\left(\sum_{i=1}^{k} \mu_i \mathbf{a}_i, \eta\right) \leq \sum_{i=1}^{k} \mu_i K(\mathbf{a}_i, \eta),$$

so that

$$\sup_{\xi} K(\xi, \eta) = \sup_{1 \leq i \leq k} K(\mathbf{a}_i, \eta).$$

Hence there exist optimal strategies for Player I which are composed of the pure strategies corresponding to the vertices of R.

Solution of the campaign example. The solution to the convex pay-off kernel defined by (4.4.1) will now be given. In view of Remark 4.4.1, Party I possesses an optimal strategy of the form

$$x^* = \sum_{i=1}^{n} \lambda_i x_{a_i}, \tag{4.4.3}$$

where $\lambda_i \geq 0$, $\sum \lambda_i = 1$, and $\mathbf{a}_i$ are the vertices of R.

Let $\lambda_{i_1}, \ldots, \lambda_{i_r}$ be the strictly positive λ's in (4.4.3). If Party II selects the pure strategies associated with the vertices $\mathbf{d}_{i_1}, \ldots, \mathbf{d}_{i_r}$, the respective expected yields to Party I are

$$\begin{aligned}
&k_{i_1}\lambda_{i_1}(A - D) + k_{i_2}\lambda_{i_2}A + \cdots + k_{i_r}\lambda_{i_r}A \geq v, \\
&k_{i_1}\lambda_{i_1}A + k_{i_2}\lambda_{i_2}(A - D) + \cdots + k_{i_r}\lambda_{i_r}A \geq v, \\
&\vdots \\
&k_{i_1}\lambda_{i_1}A + k_{i_2}\lambda_{i_2}A + \cdots + k_{i_r}\lambda_{i_r}(A - D) \geq v.
\end{aligned} \tag{4.4.4}$$

Each one of these yields is equal to or exceeds the value v because Party I is presumably employing the optimal strategy x^*.

On the other hand, if Party II selects a pure strategy corresponding to any of the remaining $n - r$ vertices, Party I's expected return is

$$\sum_{j=1}^{r} \lambda_{i_j} k_{i_j} A,$$

which is in excess of v, as can be seen by comparing this quantity with that of (4.4.4). Hence any optimal strategy for Party II must be composed of pure strategies contained in the $(r - 1)$-dimensional face of S with vertices at $\mathbf{d}_{i_1}, \ldots, \mathbf{d}_{i_r}$. The general theory informs us that a pure optimal strategy exists; it is necessarily located in the indicated face of S.

Assume first that equality prevails in each of the equations (4.4.4). This assumption will be justified later. Subtracting the jth equation from the first, we obtain

$$\lambda_{i_j} = \frac{k_{i_1}}{k_{i_j}} \lambda_{i_1}.$$

Using the normalization $\sum \lambda_{i_j} = 1$, we obtain

$$\lambda_{i_j} = \frac{1}{k_{i_j} L}, \qquad \text{where} \qquad L = \sum_{j=1}^{r} \frac{1}{k_{i_j}}, \tag{4.4.5}$$

and

$$v = \sum_{j=1}^{r} k_{i_j}\lambda_{i_j}A - k_{i_1}\lambda_{i_1}D = \frac{rA - D}{\sum_{j=1}^{r} (1/k_{i_j})} = \frac{rA - D}{L}. \tag{4.4.6}$$

We notice that in (4.4.6) v increases as k_{i_j} increases. Since Party I desires the value to be as large as possible, the k_{i_j} should be chosen to correspond to the r most valuable precincts, $T_1, \ldots, T_r$. The precise number r of precincts to be campaigned in by Party I is to be determined afterwards so that (4.4.6) is a maximum. Since the values $k_1, \ldots, k_n$ were arranged in decreasing scale of magnitude, we can write

$$v = \frac{rA - D}{\sum_{j=1}^{r} (1/k_j)} = \max_{1 \leq s \leq n} \frac{sA - D}{\sum_{j=1}^{s} (1/k_j)},$$

moreover, $v \geq 0$ as a consequence of the hypothesis $A \geq D$.

We must now verify that the number v so obtained is indeed the value of the game, and that the equality assumption in (4.4.4) used to evaluate v is correct.

In order to show that the value is at least v, it is sufficient to show that Party I can obtain at least v, regardless of what pure strategy Party II chooses. In fact, if Party II chooses $\boldsymbol{\eta} = \langle \eta_1, \ldots, \eta_n \rangle$ ($\sum \eta_j = D$) and Party I chooses

$$x^* = \sum_{i=1}^{r} \lambda_i x_{a_i},$$

with λ_i as in (4.4.5), the pay-off becomes

$$\int K(\boldsymbol{\xi}, \boldsymbol{\eta})\, dx^*(\boldsymbol{\xi}) = \sum_{i=1}^{r} \lambda_i K(\mathbf{a}_i, \boldsymbol{\eta}) = \sum_{i=1}^{r} \lambda_i k_i (A - \eta_i)$$

$$= \frac{rA - \sum_{i=1}^{r} \eta_i}{\sum_{i=1}^{r} (1/k_i)} \geq \frac{rA - D}{\sum_{i=1}^{r} (1/k_i)} = v.$$

To establish that the value is at most v, we first show that if $m > r$, then $k_m A \leq v$. Since $m \geq r + 1$ implies $k_m \leq k_{r+1}$, it is enough to verify that $k_{r+1} A \leq v$. According to the definition of v,

$$\frac{rA - D}{\sum_{i=1}^{r} (1/k_i)} \geq \frac{(r + 1)A - D}{\sum_{i=1}^{r+1} (1/k_i)}.$$

Hence

$$\frac{1}{k_{r+1}} (rA - D) + \left(\sum_{i=1}^{r} \frac{1}{k_i}\right) (rA - D) \geq \left(\sum_{i=1}^{r} \frac{1}{k_i}\right) (rA - D + A),$$

$$v = \frac{rA - D}{\sum_{i=1}^{r} (1/k_i)} \geq k_{r+1} A. \tag{4.4.7}$$

A similar argument based on

$$\frac{rA - D}{\sum_{i=1}^{r} (1/k_i)} \geq \frac{(r - 1)A - D}{\sum_{i=1}^{r-1} (1/k_i)}$$

shows that $k_r A \geq v$, and hence

$$k_j A \geq v \qquad (j = 1, 2, \ldots, r). \tag{4.4.8}$$

Relation (4.4.7) indicates that it is not profitable for Party I to campaign in the least important $n - r$ precincts even if there is no campaigning there by Party II.

We now construct a vector $\boldsymbol{\eta}^0 = \langle \eta_1^0, \ldots, \eta_r^0, 0, \ldots, 0 \rangle$ that prevents Party I from obtaining a yield higher than v in the first r precincts. This requires (Lemma 2.2.1)

$$k_j(A - \eta_j^0) = v \quad \text{or} \quad \eta_j^0 = A - \frac{v}{k_j} \qquad (j = 1, \ldots, r). \tag{4.4.9}$$

Furthermore,

$$\sum_{j=1}^{r} \eta_j^0 = rA - v \sum_{j=1}^{r} \frac{1}{k_j} = D,$$

and it follows from (4.4.8) that $\eta_j^0 \geq 0$, so that $\boldsymbol{\eta}^0$ is indeed a strategy. Equation (4.4.9) states that if Party II chooses $\boldsymbol{\eta}^0$ and Party I responds with the pure strategy identified by the vertex $\mathbf{a}_i$ $(i = 1, \ldots, r)$, the return to Party I is v. For any $\boldsymbol{\xi}$, since $K(\boldsymbol{\xi}, \boldsymbol{\eta})$ is convex and

$$\boldsymbol{\xi} = \sum_{1}^{n} \alpha_i \mathbf{a}_i,$$

we obtain

$$K(\boldsymbol{\xi}, \boldsymbol{\eta}^0) \leq \sum_{i=1}^{n} \alpha_i K(\mathbf{a}_i, \boldsymbol{\eta}^0) = \sum_{i=1}^{n} \alpha_i k_i (A - \eta_i^0) \leq \sum_{i=1}^{n} \alpha_i v = v.$$

Thus by using $\boldsymbol{\eta}^0$ Party II can prevent Party I from obtaining a return higher than v. It follows that v is the value of the game and that the strategies $\{x^*, y_{\eta^0}\}$ are optimal.

Uniqueness. The number r of vertices out of which the optimal strategy of Party I is built need not necessarily be unique; in the game with $A = 2$, $D = 1$, $k_1 = 2$, $k_2 = 1$, for example, r can be either 1 or 2. However, the optimal strategy $\boldsymbol{\eta}^0$ for Party II is unique among the class of pure strategies, for if $\bar{\boldsymbol{\eta}} = \langle \bar{\eta}_1, \ldots, \bar{\eta}_n \rangle$ is optimal, it must prevent Party I from obtaining a return greater than v when Party I uses a pure strategy corresponding to any of the first r vertices. Hence

$$\bar{\eta}_j \geq A - \frac{v}{k_j} \qquad (j = 1, \ldots, r). \tag{4.4.10}$$

But

$$\sum_{j=1}^{r} \left(A - \frac{v}{k_j} \right) = D,$$

and therefore equality must obtain in (4.4.10). The last statement does not contradict the lack of uniqueness for r; for if Party I may use either r or $r + 1$ vertices, then (4.4.7) and (4.4.8) require $k_{r+1}A = v$, and hence

$$\eta_{r+1}^0 = A - \frac{v}{k_{r+1}} = 0.$$

The explicit form for the optimal strategy η^0 is

$$\begin{aligned} \eta_j^0 &= A - \frac{rA - D}{k_j \sum_{j=1}^r (1/k_j)} \qquad (j = 1, \ldots, r), \\ \eta_j^0 &= 0 \qquad\qquad\qquad\qquad\quad (j = r + 1, \ldots, n). \end{aligned} \tag{4.4.11}$$

In general, Party II possesses additional optimal mixed strategies: e.g., use $\mathbf{d}_i$ with probability η_i^0/D $(i = 1, \ldots, n)$ and probability 0 for $i > r$.

We summarize the results:

(1) Choose r such that

$$\frac{rA - D}{\sum_{j=1}^r (1/k_j)}$$

is maximized.

(2) The defender (Party II) has a unique pure optimal strategy, described in equation (4.4.11), which shows that he campaigns only in the r most valuable precincts.

(3) The attacker (Party I) employs his entire capacities in a single precinct, which he chooses from the r most valuable. Which precinct to concentrate on is determined by randomization, with the probability λ_i of pouring campaign effort into area i $(i = 1, \ldots, r)$ being $(1/k_i)\,(1/L)$, where

$$L = \sum_{j=1}^r \frac{1}{k_j}.$$

The problem in which

$$n = 3, \qquad A = 16, \qquad D = 12, \qquad k_1 = 6, \qquad k_2 = 2, \qquad k_3 = 1$$

has the following solution. Using all resources, Player I attacks precinct 1 with probability 1/4 and precinct 2 with probability 3/4. Player II defends precinct 1 with 11 units and precinct 2 with 1 unit. The value of the game is 30.

***4.5 A theorem on convex functions.** In this section we present a proof of Theorem 4.4.1 (see also Theorem 7.6.4 of Vol. I).

Let S be a compact convex region in an n-dimensional space whose elements will be denoted by $\boldsymbol{\eta}$. A function $f(\boldsymbol{\eta})$ is linear (not necessarily homogeneous) if

$$f(\sum \lambda_i \boldsymbol{\eta}_i) = \sum \lambda_i f(\boldsymbol{\eta}_i)$$

whenever $\sum \lambda_i = 1$; the function $f(\boldsymbol{\eta}) \equiv 1$, denoted by u, is linear. The set of linear functions forms an $(n + 1)$-dimensional space E. Elements of E will be denoted by $\mathbf{f}$; elements of its conjugate space E^* (i.e., all possible linear forms defined for elements in E) will be denoted by F. Clearly every η in S gives rise to an element of E^*. Lemma 4.5.1 states that subject to proper normalization E^* is identical with all of $E^n \supset S$.

▶LEMMA 4.5.1. If $F(u) = 1$, then there exists η in E^n such that $F(f) = f(\eta)$ for all f in E.

Proof. Let $f_1, \ldots, f_{n+1}$ be linearly independent. Since these functions span the space E, it is sufficient to show that the system of equations

$$F(f_i) = f_i(\eta) \qquad (i = 1, \ldots, n + 1)$$

has a solution. If we take $f_1 = u$, the first equation becomes an identity in η. The remaining n independent equations have a solution, since η is an n-dimensional vector.

Our next lemma again expresses a fundamental duality.

▶LEMMA 4.5.2. The set of all f in E which are nonnegative over S forms a closed convex cone $P \subset E$ having its vertex at the origin and containing the function u as an interior point. Moreover, the region in which $P(\eta) \geq 0$ (see the following note) is precisely S.

Note. The notation $P(\eta)$ means $f(\eta)$ for all f in P; $f(S)$ means $f(\eta)$ for all η in S.

Proof. The first part of the lemma is obvious. In order to prove the last assertion we remark that if η is not in S, then η and S can be separated by a hyperplane; thus there exists a linear functional f such that $f(\eta) < c \leq f(S)$; then $f - cu$ is in P, and $f - cu$ is negative for η. Hence η is not in the region in which $P(\eta) \geq 0$.

▶LEMMA 4.5.3. Let Q be a convex compact set of E which does not intersect the set P. Then we can find η in S and $\delta > 0$ such that $Q(\eta) \leq -\delta$.

Proof. Since Q and P are closed convex sets and do not intersect (and Q is bounded), they are separated by some F in E^*; that is, for some $\delta > 0$,

$$F(Q) + \delta \leq F(P).$$

This relation shows that $F(P)$ is bounded from below. Thus, since P is a cone with vertex at the origin and F is linear, $F(P)$ must be nonnegative. Since u is an interior point, $F(u) > 0$; by an appropriate normalization we can make $F(u) = 1$. It now follows from Lemma 4.5.1 that there exists a vector η in E^n such that $F(Q) = Q(\eta)$, $F(P) = P(\eta)$. Then, since $F(P) \geq 0$,

$$Q(\eta) + \delta \leq 0 \leq P(\eta).$$

The last equation, together with Lemma 4.5.2, shows that η is in S.

▶ LEMMA 4.5.4. If for a family $\{f_\alpha\}$ and for each $\boldsymbol{\eta}$ in S, $\sup_\alpha f_\alpha(\boldsymbol{\eta}) > 0$, then for a suitable choice of $\lambda_i \geq 0$, $\sum \lambda_i = 1$, and α_i $(i = 1, \ldots, n+1)$ the function

$$f = \sum_{i=1}^{n+1} \lambda_i f_{\alpha_i}$$

is in P; that is, $f(S) \geq 0$.

Proof. If $f_\alpha(\boldsymbol{\eta}) > 0$ for fixed α and $\boldsymbol{\eta}$, then we can find an open set containing $\boldsymbol{\eta}$ in which strict inequality will hold. Hence, by the Heine-Borel theorem, we can find a finite subfamily $\{f_{\alpha_j}\}$ such that $\max_j f_{\alpha_j}(\boldsymbol{\eta}) > 0$ for each $\boldsymbol{\eta}$ in S.

Let Q be the convex set spanned by this family. By Lemma 4.5.3, Q intersects P. Since Q is bounded while P is not, some boundary point f of Q must lie in P. This boundary point must lie on a polyhedral face of Q whose dimension is at most n. Hence, by Lemma B.2.1,

$$f = \sum_{i=1}^{n+1} \lambda_i f_{\alpha_i},$$

with $\sum \lambda_i = 1$, $\lambda_i \geq 0$; and $f \in P$ implies

$$\textstyle\sum_i \lambda_i f_{\alpha_i}(S) = f(S) \geq 0.$$

Proof of Theorem 4.4.1. For all $\boldsymbol{\eta}$ in S of Theorem 4.4.1, let

$$b > \inf_{\eta \in S} \sup_\alpha \phi_\alpha(\boldsymbol{\eta}) = c.$$

The set of $\boldsymbol{\eta} \in S$ for which $\phi_\alpha(\boldsymbol{\eta}) \leq b$ for all α is a nonvoid closed set. Furthermore, this set decreases as b decreases. Hence if we choose $b_k \to c$, the intersection of the corresponding sets must be nonvoid, by the principle of nested intervals. Any point $\boldsymbol{\eta}_0$ in this intersection satisfies the first part of the theorem.

For the second part of the theorem we let f_β be the family of all linear functions satisfying

$$[\phi_\alpha - f_\beta](S) \geq 0 \tag{4.5.1}$$

for some α. Any tangent plane to ϕ_α satisfies this condition, so that the family f_β contains all tangent planes to ϕ_α. Therefore $\sup_\beta f_\beta(\boldsymbol{\eta}) = \sup_\alpha \phi_\alpha(\boldsymbol{\eta}) \geq c$ for all $\boldsymbol{\eta}$.

For a fixed $\delta > 0$, we consider the family

$$\mathfrak{f}_\delta = \{f_\beta - (c - \delta)u\}.$$

Since $\sup f_\beta(S) \geq c$, we have $\sup_\beta [f_\beta - (c - \delta)u]\,(S) \geq \delta > 0$. We apply Lemma 4.5.4 to the family f_δ, and select $\beta_1, \ldots, \beta_{n+1}, \lambda_1, \ldots, \lambda_{n+1}$. Let ϕ_{α_i} be a function corresponding to f_{β_i} in (4.5.1). Then

$$\sum_{i=1}^{n+1} \lambda_i \phi_{\alpha_i}(\boldsymbol{\eta}) \geq \sum_{i=1}^{n+1} \lambda_i f_{\beta_i}(\boldsymbol{\eta}) \geq c - \delta$$

for all $\boldsymbol{\eta}$ in S, as asserted by the theorem.

4.6 Problems. In Problems 1–11, $K(\xi, \eta)$ denotes a continuous function defined on the square $0 \leq \xi, \eta \leq 1$.

1. If $K_\xi > 0$ and $K_\eta < 0$, indicate how to solve the game with kernel $K(\xi, \eta)$.

2. Show that the condition $K_\xi(\xi, \eta) \leq 0$ is sufficient to guarantee the existence of a saddle point of the kernel $K(\xi, \eta)$, and indicate how to find it.

3. Find the optimal strategies for the game with kernel

$$K(\xi, \eta) = (\xi - \eta)^2.$$

4. Solve the game on the unit square whose kernel is

$$K(\xi, \eta) = \begin{cases} \xi(1 - \eta) & \xi \leq \eta, \\ \eta(1 - \xi) & \xi \geq \eta. \end{cases}$$

5. Let $K(\xi, \eta)$ be strictly convex in each variable separately, and let η_0 and η_1 denote the points at which $K(0, \eta)$ and $K(1, \eta)$, respectively, achieve their minimum values. Show that:

(a) If $K(0, \eta_0) \geq K(1, \eta_0)$, then the optimal strategies for Players I and II are $x_{\xi_0}(\xi)$ (where $\xi_0 = 0$) and y_{η_0}, respectively.

(b) If $K(1, \eta_1) \geq K(0, \eta_1)$, then the optimal strategies are x_{ξ_1} (where $\xi_1 = 1$) and y_{η_1}, respectively.

(c) Otherwise, the optimal minimizing strategy is y_{η^*}, where η^* is the unique solution between η_0 and η_1 of the equation $K(0, \eta) = K(1, \eta)$.

6. Using the results of Problem 5, solve the game with kernel

$$\xi^2\eta^2 - \xi - \eta.$$

7. Solve the generalized convex game with kernel

$$K(\xi, \eta) = \xi^3\eta - \eta^2 - \xi\eta + \eta.$$

8. Find optimal strategies for the game with kernel $K(\xi, \eta) = \sin[\pi(\xi + \eta)/2]$. (*Hint:* K is concave with respect to ξ.)

9. Solve the game with kernel

$$K(\xi, \eta) = \sqrt{1 - \left(\frac{\xi}{3} - \eta\right)^2}.$$

10. Show that if $K_{\eta\eta\eta} > 0$ and $K_{\xi\xi\xi} < 0$, then both players have optimal strategies of weight at most 3/2.

11. Show that if $K_{\eta\eta\eta} > 0$, then any optimal minimizing strategy of weight 3/2 (where end points are weighted 1/2 and interior points 1) cannot be composed of some interior η_0 and 1.

12. Consider the campaign example of Section 4.4. Suppose that there are 5 precincts with values

$$k_1 = \tfrac{1}{4}, \qquad k_2 = \tfrac{1}{5}, \qquad k_3 = \tfrac{1}{7}, \qquad k_4 = \tfrac{1}{9}, \qquad k_5 = \tfrac{1}{12}$$

and $A = D = 10$. Solve the game.

13. Player I picks a point $\boldsymbol{\xi}$ in the unit circle S, and Player II simultaneously picks a point in the same unit circle. The pay-off to Player I is $|\boldsymbol{\xi} - \boldsymbol{\eta}|^2$ (the square of the distance from $\boldsymbol{\xi}$ to $\boldsymbol{\eta}$). Determine the form of the solutions to this game.

14. Determine the form of the solutions to the game of Problem 13 if S designates any compact convex set in E^n.

15. Consider the game in which the pure strategies Y of Player II range over the unit sphere in E^n and X consists of all $\boldsymbol{\lambda} = \langle \lambda_1, \lambda_2, \ldots, \lambda_m \rangle$ subject to

$$\lambda_i \geq 0, \qquad \sum_{i=1}^{m} \lambda_i = 1.$$

Let $\mathbf{a}^i$ denote m points on the surface of the unit sphere (that is, $|\mathbf{a}^i| = 1$) such that there exists a value $\boldsymbol{\lambda}^0$ satisfying

$$\sum_{i=1}^{m} \lambda_i^0 \mathbf{a}^i = \mathbf{0}.$$

Define the pay-off kernel

$$K(\lambda, y) = \int_{|\eta| \leq 1} \sum_{i=1}^{m} \lambda_i |\mathbf{a}^i - \boldsymbol{\eta}| \, dy(\eta).$$

Prove that $\boldsymbol{\lambda}^0$ is an optimal strategy for Player I, and that the pure strategy in which y concentrates at $\boldsymbol{\eta} = \mathbf{0}$ is optimal for Player II. Show also that in all optimal strategies $\boldsymbol{\lambda}$ for Player I

$$\textstyle\sum_i \lambda_i \mathbf{a}^i = \mathbf{0}.$$

16. Let the kernel $K(\boldsymbol{\xi}, \boldsymbol{\eta})$ of a continuous game be convex in $\boldsymbol{\eta}$ for each $\boldsymbol{\xi}$, where $\boldsymbol{\eta}$ ranges over a convex compact set B in E^n, and $\boldsymbol{\xi}$ ranges over a compact set A. Assume further that the set of supporting planes at every boundary point of B has at most n extreme points. Show that if Player II has a pure optimal strategy located at $\boldsymbol{\eta}_0$ on the boundary of B, then

Player I has an optimal strategy with spectrum of size not exceeding n points of increase.

17. Prove that the hypothesis concerning the number of supporting planes at boundary points of B in Problem 16 is always satisfied if $n = 2$.

18. Show that if $\partial^n K/\partial \eta^n > 0$, and if the minimizing player has a k-dimensional set of optimal strategies, then the maximizing player has an optimal strategy with spectrum of size at most $n - k$ points. (K is a kernel defined on the unit square.) The concept of dimension is well defined in view of Theorem 4.3.1.

Notes and References to Chapter 4

§ **4.1.** Bohnenblust was the first to solve a convex game. He considered kernels on the unit square of the form $K(\xi, \eta) = f(\xi - \eta)$, where f is a convex function, and arrived at the solution given in Section 4.2.

§ **4.2.** The theory of this section is due to Karlin and Martin (unpublished). The example discussed here was formulated by Paxson, who appealed to the results of Bohnenblust cited above in obtaining its solution.

§ **4.3.** The class of games known as generalized convex games was initially treated by Karlin [162, Chap 8], the proofs of this section follow a more elegant analysis due to Glicksberg.

§ **4.4.** The theory of this section is based on Bohnenblust, Karlin, and Shapley [37]. The elaborate convex game here interpreted in terms of a political campaign was proposed originally as a military area defense game. This game was resolved first by Gross (unpublished).

§ **4.5.** The theorem on families of convex functions in this section is from [37]. Another proof of this result, based on the duality theory of conjugate convex functions, was given by Fenchel [81]. For applications of this theorem, see Fan, Glicksberg, and Hoffman, [78]; see also [140] and Section 7.9 of Vol. I.

CHAPTER 5

GAMES OF TIMING OF ONE ACTION FOR EACH PLAYER

An important class of games defined on the unit square consists of games in which the pure strategy spaces $0 \leq \xi \leq 1$ for Player I and $0 \leq \eta \leq 1$ for Player II represent the possible times during which a certain action can be taken. The normalization $0 \leq \xi, \eta \leq 1$ is for convenience only and imposes no real restrictions on the models. In the present chapter we limit ourselves to the case in which only a single action is possible for each player; in Chapter 6 we shall consider games in which an action may be performed repeatedly a fixed number of times.

Consider the following idealized model: Each of two antagonists possesses one unit of firepower. That is, each player has one opportunity to strike at his opponent, with the knowledge that his accuracy—that is to say, his chance of firing successfully—increases with time. In a game of this class the pay-off depends fundamentally on the order in which the players act, and on their respective degrees of success. Each player obviously wishes to delay action as long as possible to increase his opportunity for success, but at the same time does not wish to delay so long that his opponent can precede him with effectiveness. The optimal strategy expresses the proper balance between the desire for delay and the danger of delaying.

Numerous versions of military games of timing can be formulated in terms of what information a player has about his opponent's action. A series of such games is described in Section 5.1. Although we use the language of tactical duels, the reader can interpret most of our models in terms of business duels, advertising campaigns, etc.

At this point the reader is urged to reread Section 2.3, which presents the necessary background for the analysis of games of timing. We shall make free use of the terminology introduced there. In particular, as noted in Section 2.3, a mathematical game of timing will be associated with a pay-off kernel $K(\xi, \eta)$ defined on the unit square and satisfying the following properties:

(i)

$$K(\xi, \eta) = \begin{cases} L(\xi, \eta) & \xi < \eta, \\ \Phi(\xi) & \xi = \eta, \\ M(\xi, \eta) & \xi > \eta. \end{cases} \tag{5.0.1}$$

(ii) Each of the functions $L(\xi, \eta)$ and $M(\xi, \eta)$ is jointly continuous in ξ and η.

(iii) $L(\xi, \eta)$ is monotone-increasing in ξ for each η,

$M(\xi, \eta)$ is monotone-increasing in ξ for each η,

$L(\xi, \eta)$ is monotone-decreasing in η for each ξ,

$M(\xi, \eta)$ is monotone-decreasing in η for each ξ.

Finally, we recall that a game of timing is said to be of class I if $L(\xi, \eta)$ is a function of ξ alone and $M(\xi, \eta)$ is a function of η alone, and of class II in all other circumstances.

5.1 Examples of games of timing. Although the general theory of Section 5.2 includes these examples as special cases, a discussion of examples is useful on two counts: (1) it exhibits a general method of determining optimal strategies in simple, lucid terms without having to take into account the morass of cases occurring in the general situation; (2) it helps to illustrate the ideas underlying the general theory of games of timing.

Example 1. Noisy duel. Two duelists are each allowed to fire only once. Assume that both have "noisy" guns, so that each knows when his opponent has fired. This is clearly a game of timing of class I (see Section 2.3). The term "noisy" generally will be used to signify a state of complete information—i.e., a situation such that each player knows when the other acts. The accuracy function $P_1(\xi)$ (the probability of success) for Player I is assumed to be continuous and monotone-increasing in ξ, with $P_1(0) = 0$, $P_1(1) = 1$; the accuracy of Player II's fire is similarly described by a continuous increasing function $P_2(\eta)$, with $P_2(0) = 0$, $P_2(1) = 1$. If I hits II, the value to I is assumed to be $+1$; if II hits I, the value to I is taken to be -1; the pay-off kernel $K(\xi, \eta)$ will be the expected value to I when the two players use pure strategies ξ and η.

If $\xi < \eta$, the probability of I's hitting II is $P_1(\xi)$ and the value to I is $1 \cdot P_1(\xi)$; the probability of I's failing to hit II is $1 - P_1(\xi)$. The structure of information in this game (the fact that the guns are noisy) is now taken into account in the formulation of the pay-off kernel. Clearly, when II has not yet fired but knows that I cannot fire again, II will increase his chances of success by waiting until $\eta = 1$ before he fires. Thus if I misses at ξ, he is certain to be hit by II if $\xi < \eta$; hence

$$L(\xi, \eta) = P_1(\xi) + (-1)[1 - P_1(\xi)] \qquad (\xi < \eta).$$

Similarly,

$$M(\xi, \eta) = P_2(\eta)\,(-1) + [1 - P_2(\eta)](1) \qquad (\xi > \eta)$$

and

$$\Phi(\xi) = P_1(\xi)[1 - P_2(\xi)](1) + P_2(\xi)[1 - P_1(\xi)](-1) \qquad (\xi = \eta).$$

In the last formula it is assumed that if both opponents fire simultaneously with success or failure, the value is zero. Simplifying, we obtain

$$K(\xi, \eta) = \begin{cases} 2P_1(\xi) - 1 & \xi < \eta, \\ P_1(\xi) - P_2(\xi) & \xi = \eta, \\ 1 - 2P_2(\eta) & \xi > \eta, \end{cases} \tag{5.1.1}$$

so that K defines a game of timing of class I. Note that the value at $\eta = \xi$ is the average of the values approached by the two functions L and M. Figure 5.1 pictures various cases of the pay-off kernel for $P_1(\xi) = P_2(\xi) = \xi$.

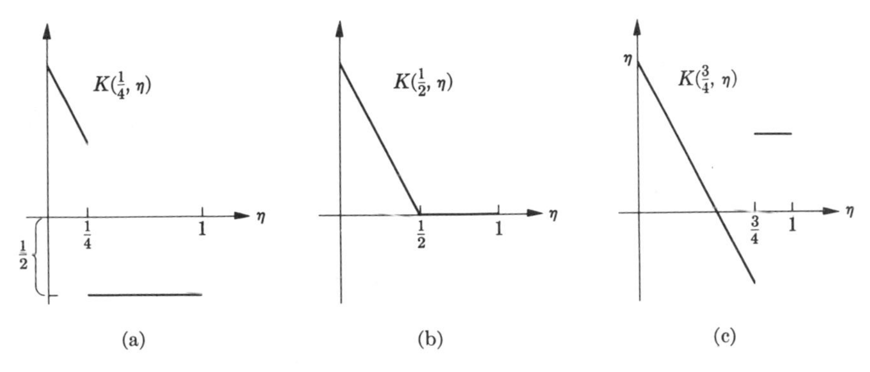

FIGURE 5.1

An examination of the yield $K(\xi, \eta)$ as a function of η for special choices of ξ (see Fig. 5.1) suggests that Player I will have an optimal pure strategy when the constant portion of the yield [i.e., $2P_1(\xi) - 1$] equals the smallest value of $1 - 2P_2(\eta)$ for $\eta < \xi$. This requires

$$2P_1(\xi) - 1 = 1 - 2P_2(\xi). \tag{5.1.2}$$

Since the right-hand side of (5.1.2) decreases continuously from $+1$ to -1 as ξ traverses $[0, 1]$, while the left-hand side increases continuously from -1 to $+1$, at least one solution z_0 to (5.1.2) can be found. Furthermore, if the two functions are strictly monotonic, then a single solution z_0 exists. Let the common value of (5.1.2) for z_0 be denoted by v. The monotonicity properties of $K(\xi, \eta)$ yield at once that where $\xi_0 = \eta_0 = z_0$,

$$K(\xi_0, \eta) \geq v \geq K(\xi, \eta_0) \tag{5.1.3}$$

for all η and ξ, and hence the pair $\{\xi_0, \eta_0\}$ are optimal strategies.

Example 2. Silent duel. Again two duelists are each allowed to fire only once, but this time both guns are fitted with silencers so that neither duelist can determine whether or not his opponent has fired. "Silent," in contrast to "noisy," will be used to designate an action game in which some knowledge of the play of the game is denied to at least one of the adversaries.

Assume for simplicity that the accuracy functions are given by $P_1(\xi) = P_2(\xi) = \xi$. The pay-off kernel describing this game is given by

$$K(\xi, \eta) = \begin{cases} \xi - (1 - \xi)\eta & \xi < \eta, \\ 0 & \xi = \eta, \\ -\eta + (1 - \eta)\xi & \xi > \eta. \end{cases} \tag{5.1.4}$$

The derivation of (5.1.4) is similar to that of (5.1.1), except that in the present case neither player can determine the time of his opponent's action, unless, of course, the other player has succeeded.

Observe that $K(\xi, \eta) = -K(\eta, \xi)$; hence this silent duel is a symmetric game in the sense of Section 2.2. The value of the game is therefore zero, and any strategy that is optimal for one player is also optimal for the other.

The kernel $K(\xi, \eta)$ of (5.1.4) defines a game of timing of class II; it has the form of (5.0.1), where

$$L(\xi, \eta) = \xi - \eta + \xi\eta, \qquad M(\xi, \eta) = \xi - \eta - \xi\eta$$

and

$$\left.\begin{aligned} L_\xi(\xi, \eta) &= 1 + \eta > 0, & L_\eta(\xi, \eta) &= -1 + \xi < 0, \\ M_\xi(\xi, \eta) &= 1 - \eta > 0, & M_\eta(\xi, \eta) &= -1 - \xi < 0, \end{aligned}\right\} 0 \leq \xi, \eta < 1.$$

(L_ξ and M_ξ designate partial differentiation with respect to ξ; L_η and M_η have the analogous meaning.) Therefore, the functions $L(\xi, \eta)$, $M(\xi, \eta)$ are strictly increasing in ξ and strictly decreasing in η on the interior of the unit interval.

Intuitively, the fact that the players have no knowledge of each other's actions suggests randomization, and conceivably each player might wait the full time interval to assure certain success. Hence a reasonable first guess has the optimal strategy of each player composed of a density with support consisting of an interval of the form $[a, 1]$ and a possible discrete mass at 1. To begin with, we discard the possibility of a jump at 1 and seek optimal strategies of the form

$$x^0(\xi) = \int_a^\xi f(t)\, dt, \qquad y^0(\eta) = \int_a^\eta g(t)\, dt.$$

We invoke the principle of strategies identically equal to v (see Section 2.5). However, the requirement of a constant yield of v is only to be met on the interval $[a, 1]$. Lemma 2.2.1 insists on this property as a necessary condition for optimality.* Hence

$$\int K(\xi, \eta)\, dx^0(\xi) = \int_a^1 K(\xi, \eta) f(\xi)\, d\xi = v = 0 \tag{5.1.5}$$

for all η in $[a, 1]$. In terms of the explicit expression of $K(\xi, \eta)$, equation (5.1.5) becomes

$$\int_a^\eta (\xi - \eta + \xi\eta) f(\xi)\, d\xi + \int_\eta^1 (\xi - \eta - \xi\eta) f(\xi)\, d\xi \equiv 0. \tag{5.1.6}$$

Recalling that

$$\int_a^1 f(\xi)\, d\xi = 1,$$

we write (5.1.6) in the form

$$\int_a^1 \xi f(\xi)\, d\xi - \eta + \eta \int_a^\eta \xi f(\xi)\, d\xi - \eta \int_\eta^1 \xi f(\xi)\, d\xi \equiv 0, \tag{5.1.7}$$

which may be regarded as an integral equation for the function $r(\xi) = \xi f(\xi)$. By performing two differentiations, we convert this into the differential equation

$$2\eta r'(\eta) + 4r(\eta) = 0,$$

whose general solution is $r(\eta) = k\eta^{-2}$; and therefore $f(\xi) = k\xi^{-3}$. Inserting $f(\xi) = k\xi^{-3}$ into (5.1.7) leads to the identity

$$\eta\left(-1 + \frac{k}{a} + k\right) + k\left(-3 + \frac{1}{a}\right) \equiv 0 \qquad (a \leq \eta \leq 1), \tag{5.1.8}$$

which is possible only if $a = 1/3$ and $k = 1/4$. Fortunately, these same values also satisfy the normalization restraint

$$\int_a^1 f(\xi)\, d\xi = k\left(\frac{1}{2a^2} - \frac{1}{2}\right) = 1. \tag{5.1.9}$$

To sum up, the optimal strategy x^0 has the density

$$f(\xi) = \begin{cases} 0 & 0 \leq \xi < \frac{1}{3}, \\ \frac{1}{4}\xi^{-3} & \frac{1}{3} \leq \xi \leq 1, \end{cases}$$

* In the statement of Lemma 2.2.1 the continuity of $K(\xi, \eta)$ was assumed. However, the essential thing is that $K(x^0, \eta)$ should be continuous, as it is here.

provided it is shown that

$$\int_a^1 K(\xi, \eta)f(\xi)\, d\xi > 0$$

for $\eta < a$. This fact follows easily as a result of the monotonicity properties of the kernel. Thus for $\eta < a$,

$$\frac{\partial}{\partial \eta}\int_a^1 K(\xi, \eta)f(\xi)\, d\xi = \int_a^1 M_\eta(\xi, \eta)f(\xi)\, d\xi \leq 0.$$

But

$$\int_a^1 K(\xi, \eta)f(\xi)\, d\xi$$

is continuous for all η and vanishes for $\eta = a$; hence the desired result. By symmetry, the same strategy is optimal for Player II. It can also be shown in this example that each player's optimal strategy is unique (see Problems 16–19 of this chapter).

Example 3. Silent-noisy duel. We assume that Player I has a silent gun and Player II a noisy gun, so that Player I knows whether Player II has fired or not, but not conversely. Again the accuracy functions are taken to be $P_1(\xi) = P_2(\xi) = \xi$. Then

$$K(\xi, \eta) = \begin{cases} \xi - \eta + \xi\eta & \xi < \eta, \\ 0 & \xi = \eta, \\ 1 - 2\eta & \xi > \eta. \end{cases}$$

Guided by the heuristic principles set forth in the analysis of Example 2, we shall search for strategies $x^0(\xi)$ consisting of part density $f(\xi)$ on the interval $[a, 1]$ with a weight α at $\xi = 1$, and $y^0(\eta)$ consisting of part density $g(\eta)$ on the same interval with a weight β at $\eta = 1$. Analogously to (5.1.5), the appropriate integral equations are

$$\int_a^1 K(\xi, \eta)f(\xi)\, d\xi + \alpha K(1, \eta) \equiv v \qquad (a \leq \eta \leq 1),$$

$$\int_a^1 K(\xi, \eta)g(\eta)\, d\eta + \beta K(\xi, 1) \equiv v \qquad (a \leq \xi \leq 1),$$

or

$$\int_a^t (\xi - t + \xi t)f(\xi)\, d\xi + \int_t^1 (1 - 2t)f(\xi)\, d\xi + \alpha(1 - 2t) \equiv v \qquad (a \leq t < 1),$$

$$\int_a^t (1 - 2\eta)g(\eta)\, d\eta + \int_t^1 (t - \eta + t\eta)g(\eta)\, d\eta + \beta(2t - 1) \equiv v \qquad (a \leq t < 1). \tag{5.1.10}$$

Two differentiations produce

$$3(1+\xi)f(\xi)+(\xi^2+2\xi-1)f'(\xi)=0,$$

$$(1-2\eta-\eta^2)g'(\eta)-3(1+\eta)g(\eta)=0.$$

Therefore, the two density functions for the optimal strategies are

$$f(\xi)=k_1(\xi^2+2\xi-1)^{-3/2}, \qquad g(\eta)=k_2(\eta^2+2\eta-1)^{-3/2}.$$

In order to complete the determination of the solutions, we must evaluate the unknown constants a, v, α, β, k_1, k_2. Equations (5.1.10) represent identities in ξ and η; if we substitute the values of $f(\xi)$, $g(\eta)$ and integrate, using

$$\int \frac{\xi+1}{(\xi^2+2\xi-1)^{3/2}}\,d\xi=-\frac{1}{\sqrt{\xi^2+2\xi-1}},$$

$$\int \frac{d\xi}{(\xi^2+2\xi-1)^{3/2}}=-\frac{1}{2}\,\frac{\xi+1}{\sqrt{\xi^2+2\xi-1}},$$

we obtain

$$k_1\left\{-\eta\left(\frac{a}{P(a)}+\frac{2\alpha}{k_1}-\frac{2}{\sqrt{2}}\right)+\frac{1-a}{2P(a)}+\frac{\alpha}{k_1}-\frac{1}{\sqrt{2}}\right\}\equiv v,$$

$$k_2\left\{\frac{3a-1}{2P(a)}-\frac{\xi}{\sqrt{2}}+\frac{\beta}{k_2}\,(2\xi-1)\right\}\equiv v, \tag{5.1.11}$$

where $P(a)=(a^2+2a-1)^{1/2}$. The second equation requires that the coefficient of ξ be zero; hence

$$\beta=\frac{k_2}{2\sqrt{2}}.$$

The first equation requires that the coefficient of η be zero; this can be achieved with $\alpha=0$ by letting

$$\frac{a}{P(a)}=\frac{2}{\sqrt{2}}.$$

Next, solving for a, we have

$$a=\sqrt{6}-2,\quad P(a)=\sqrt{3}-\sqrt{2}.$$

The normalizing equations

$$\int_a^1 f(\xi)\,d\xi=1-\alpha \qquad \text{and} \qquad \int_a^1 g(\eta)\,d\eta=1-\beta$$

result in

$$\frac{1}{k_1} = \frac{(a+1)}{2P(a)} - \frac{1}{\sqrt{2}} = \frac{\sqrt{3}+\sqrt{2}}{2},$$

$$\frac{1}{k_2} = \frac{(a+1)}{2P(a)} - \frac{1}{2\sqrt{2}} = \frac{3\sqrt{2}+2\sqrt{3}}{4}.$$

Finally, in order to justify the choice $\alpha = 0$, we must verify that equations (5.1.11) are consistent with the indicated values of α, β, a, k_1, k_2, and that the specified strategies are indeed optimal. The consistency is easily checked and leads to the value $v = 5 - 2\sqrt{6}$. The monotonicity properties of $K(\xi, \eta)$ guarantee that $\int K(\xi, \eta)\, dx^0(\xi) > v$ for $0 \leq \eta < a$, so that x^0 is optimal as asserted. A similar argument applies to y^0.

To be complete, it is necessary to check the yield for the strategies x^0 and y^0 at $\eta = 1$ and $\xi = 1$, respectively. This is easily checked for η, since $\alpha = 0$, and also for ξ, since $K(1, 1) = 0 < (2\xi - 1)|_{\xi=1}$. Note that in this case the conclusion of Lemma 2.2.1 does not hold: that is, $\xi = 1$ is in the spectrum of x^0, and yet $K(1, y^0) < v$ (see also the footnote on p. 105).

Finally, we note without proof that the optimal strategies x^0 and y^0 exhibited above are unique.

We contrast the solutions of the silent-noisy duel and the full silent duel. First, we notice that the spectrum begins in the silent duel at $a = 1/3$, whereas in the silent-noisy duel the spectrum starts at $a = \sqrt{6} - 2 > 1/3$. In other words, when one of the players has the advantage of being able to hear his opponent's shot, his optimal strategy calls for firing at a later time than under the ordinary circumstances of the silent duel. The second and most interesting difference in the solution of the silent-noisy duel is to the effect that Player II (the player with the noisy gun) must maintain a threat of firing until the very end. This is reflected in the jump of $y^0(\eta)$ at $\eta = 1$. Actually, this behavior is not too surprising, in view of the unequal information patterns available to the two players.

Example 4. Consider a silent duel defined not over the unit square but over the square $b \leq \xi \leq 1$, $b \leq \eta \leq 1$, with $b > 0$. The accuracy functions will be $P_1(\xi) = P_2(\xi) = \xi$. This example may be viewed as a version of the silent duel in which each player has a positive initial accuracy.

The pay-off function is

$$K(\xi, \eta) = \begin{cases} \xi - \eta + \xi\eta & \xi < \eta, \\ 0 & \xi = \eta, \\ \xi - \eta - \xi\eta & \xi > \eta, \end{cases} \qquad (b \leq \xi, \eta \leq 1).$$

The value of the game is zero, and any optimal strategy for one player is also optimal for the other. If $b \leq 1/3$, the solution of Example 2 is valid for this example also. For an arbitrary b we shall attempt to discover an optimal strategy consisting of a density part f with support $[a, 1]$ and a discrete mass α at the point b.

We shall show later (Section 5.2), as part of our general theorem, that it is never possible in any type of silent duel to have a jump at the end point 1 common to the optimal strategies of both players. In this example, since the game is symmetric, we infer that 1 is excluded as a jump in any optimal strategy.

However, we introduce here a strategy which entails the possibility that both players might want to fire immediately. This possibility follows from our assumption of a positive initial accuracy for both players. For such a strategy, the yield to Player I when Player II uses η $(a \leq \eta \leq 1)$ is

$$\int_a^\eta (\xi - \eta + \xi\eta) f(\xi)\, d\xi + \int_\eta^1 (\xi - \eta - \xi\eta) f(\xi)\, d\xi + \alpha(b - \eta + b\eta) \equiv 0 \qquad (a \leq \eta \leq 1).$$

Since this equation differs from (5.1.6) by a linear term only, it leads to the same differential equation with solution $f(\xi) = k\xi^{-3}$. The equations corresponding to (5.1.8) and (5.1.9) are clearly

$$k\eta\left(-\frac{1}{2a^2} + \frac{1}{a} + \frac{3}{2}\right) + k\left(\frac{1}{a} - 3\right) + \alpha(b - \eta + b\eta) = 0$$

and

$$\frac{k}{2a^2} - \frac{k}{2} = 1 - \alpha.$$

The solutions of these equations, assuming $b > 1/3$, are

$$a = \frac{b}{2 - 3b}, \qquad \alpha = \frac{3b - 1}{2b^2}, \qquad k = \frac{1}{4}.$$

Since $0 \leq \alpha \leq 1$, we must have in the second equation $b \leq 1/2$; it follows that if $b > 1/2$, we cannot have an optimal strategy involving a density of the prescribed form. In that case, the pure strategy $x_b(\xi)$ is optimal.

The results may be summarized as follows:

Case 1: $b \leq 1/3$. The optimal strategies are the same as in Example 2.

Case 2: $1/3 \leq b \leq 1/2$. Optimal strategy:

$$x^*(\xi) = \alpha x_b(\xi) + \int_a^\xi f(t)\, dt,$$

where $\alpha = (3b - 1)/2b^2$, $a = b/(2 - 3b)$, and

$$f(\xi) = \begin{cases} 0 & 0 \leq \xi < a, \\ \frac{1}{4}\xi^{-3} & a \leq \xi \leq 1. \end{cases}$$

Case 3: $1/2 \leq b$. Optimal strategy: $x_b(\xi)$.

5.2 The integral equations of games of timing and their solutions. In this section we shall consider kernels of the form

$$K(\xi, \eta) = \begin{cases} L(\xi, \eta) & \xi < \eta, \\ \Phi(\xi) & \xi = \eta, \\ M(\xi, \eta) & \xi > \eta \end{cases}$$

which satisfy the following conditions:

(a) The functions $L(\xi, \eta)$ and $M(\xi, \eta)$ are defined over the closed triangles $0 \leq \xi \leq \eta \leq 1$ and $0 \leq \eta \leq \xi \leq 1$, respectively. Furthermore, they possess continuous second partial derivatives defined in their respective closed triangles.

(b) The value of $\Phi(1)$ lies between $L(1, 1)$ and $M(1, 1)$; the value of $\Phi(0)$ lies between $L(0, 0)$ and $M(0, 0)$.

(c) $L_\xi(\xi, \eta) > 0$ and $M_\xi(\xi, \eta) > 0$ over their respective closed triangles, with the possible exception that $M_\xi(1, 1) = 0$; and $L_\eta(\xi, \eta) < 0$ and $M_\eta(\xi, \eta) < 0$ over their respective domains of definition, with the possible exception that $L_\eta(1, 1) = 0$.

Assumption (c) implies that $L(\xi, \eta)$ and $M(\xi, \eta)$ are strictly increasing in ξ and strictly decreasing in η over their respective domains of definition. This restricts our analysis to a subclass of kernels corresponding to games of timing of class II. Nevertheless, many of the theorems developed below are valid for all games of timing of class II. Our proofs will include the salient features of the analysis of the general game of timing of class II, while remaining at a relatively simple level of exposition.

Strategies. If the strategy x consists of a discrete mass α at $\xi = 0$ together with a density part whose support is $[a, b]$ and a discrete mass β at $\xi = 1$, we shall use the notation $x = (\alpha I_0, f_{ab}, \beta I_1)$. If, in addition, $b = 1$, we simply write $x = (\alpha I_0, f_a, \beta I_1)$. For any strategy x of this form, the yield to Player I is

$$\int_0^1 K(\xi, \eta)\, dx(\xi) = \alpha L(0, \eta) + \int_a^\eta L(\xi, \eta) f_{ab}(\xi)\, d\xi + \int_\eta^b M(\xi, \eta) f_{ab}(\xi)\, d\xi + \beta M(1, \eta), \tag{5.2.1}$$

provided $0 < \eta < 1$. Of course, if $\eta < a$, the first integral does not appear, and if $\eta > b$, the second integral does not appear. If $\eta = 0$, we must replace $L(0, 0)$ by $\Phi(0)$; if $\eta = 1$, we replace $M(1, 1)$ by $\Phi(1)$.

▶LEMMA 5.2.1. If both players possess optimal strategies of the form $x = (\alpha I_0, f_{ab}, \beta I_1)$, $y = (\gamma I_0, g_{cd}, \delta I_1)$, then $a = c$ and $b = d = 1$. In other words, the absolutely continuous part of both distributions has the same support extending from a to 1.

Proof. For any distribution of the type indicated above, (5.2.1) shows that

$$\int_0^1 K(\xi, \eta)\, dx(\xi)$$

is a continuous function of η (each term is continuous) for $0 < \eta < 1$. If $f_{ab}(\xi)$ is identically zero over some interval, then

$$\int_0^1 K(\xi, \eta)\, dx(\xi)$$

is strictly decreasing as η varies over that same interval. This follows from the fact that the effective limits of integration remain constant in (5.2.1) while the integrand is strictly decreasing by condition (c). Since the point $\eta = c$ is in the support of strategy y,

$$\int_0^1 K(\xi, c)\, dx(\xi) = v$$

by Lemma 2.2.1. Also, since x is optimal,

$$\int_0^1 K(\xi, \eta)\, dx(\xi) \geq v.$$

The last two equations, together with the decreasing properties of the left-hand side, require that the support of f_{ab} must contain the interval $[c, d]$. A similar argument applies to y and establishes the fact that $a = c$ and $b = d$. Now if $b < 1$,

$$\int_0^1 K(\xi, \eta)\, dx(\xi) < v$$

for $\eta > d$ and we obtain a contradiction.

It is not accidental that the spectrum begins at the same value a for both players. For the moment, think of the game as a general tactical duel involving silent guns. If one player has not fired by the time a, the other player is certainly better off firing at a than before a. Consequently, the initial firing time must be the same in the optimal strategies of both players.

Lemma 5.2.1 is not used directly in developing the general theory, but rather helps explain the analysis that follows. We shall try to construct solutions in the manner suggested by the lemma, i.e., by choosing suitable constants α, β, γ, δ, a and by solving an associated pair of integral equations for the functions f_a and g_a.

The integral equations. It will be shown in Lemma 5.4.1 that if $L(1, 1) \leq M(1, 1)$, the point $\xi = 1$, $\eta = 1$ is a saddle point for $K(x, y)$. In order to consider the more difficult case in which $L(1, 1) > M(1, 1)$, we first observe that under this assumption the continuity of $L(\xi, \xi)$ and $M(\xi, \xi)$ furnishes a nonvoid interval $a \leq \xi \leq 1$ for which $L(\xi, \xi) > M(\xi, \xi)$. Over any such interval we consider the linear integral equations

$$f(t) - \int_a^1 T(\xi, t) f(\xi)\, d\xi = \alpha p_0(t) + \beta p_1(t) \tag{5.2.2}$$

and

$$g(u) - \int_a^1 U(u, \eta) g(\eta)\, d\eta = \gamma q_0(u) + \delta q_1(u), \tag{5.2.3}$$

where

$$T(\xi, t) = \begin{cases} \dfrac{-L_\eta(\xi, t)}{L(t, t) - M(t, t)} & a \leq \xi < t \leq 1, \\[2ex] \dfrac{-M_\eta(\xi, t)}{L(t, t) - M(t, t)} & a \leq t \leq \xi \leq 1, \end{cases}$$

$$p_0(t) = \frac{-L_\eta(0, t)}{L(t, t) - M(t, t)}, \qquad p_1(t) = \frac{-M_\eta(1, t)}{L(t, t) - M(t, t)},$$

$$U(u, \eta) = \begin{cases} \dfrac{M_\xi(u, \eta)}{L(u, u) - M(u, u)} & a \leq \eta < u \leq 1, \\[2ex] \dfrac{L_\xi(u, \eta)}{L(u, u) - M(u, u)} & a \leq u \leq \eta \leq 1, \end{cases}$$

$$q_0(u) = \frac{M_\xi(u, 0)}{L(u, u) - M(u, u)}, \qquad q_1(u) = \frac{L_\xi(u, 1)}{L(u, u) - M(u, u)}.$$

The integral equations follow from Lemma 5.2.1. If there exist optimal strategies x, y of the type described in the lemma, then x must satisfy (5.2.1), with the right-hand side identically v when η ranges over the interval $a \leq \eta < 1$. Differentiation with respect to η followed by division by the nonzero factor $L(\eta, \eta) - M(\eta, \eta)$ yields the first integral equation. The second equation comes from performing the corresponding operations on the optimal strategy y.

In order to simplify the notation, let T_a denote the integral operator with kernel $T(\xi, t)$ and lower limit a, and U_a denote the operator with kernel $U(u, \eta)$ and lower limit a. The identity operator will be signified by I. The integral equations may be written compactly as

$$(I - T_a)f = \alpha p_0 + \beta p_1, \tag{5.2.4}$$

$$(I - U_a)g = \gamma q_0 + \delta q_1. \tag{5.2.5}$$

Such equations, classically known as integral equations of the second kind, have given rise to a substantial literature. Our concern here, however, is with the special positivity properties of the kernel of the integral operators, rather than with the general theory as such.

In order to make the analysis of these equations sufficiently rigorous, we must specify the domain of functions to which the linear operators apply. Specifically, we shall seek for a solution within the set of all continuous functions defined on the interval $a \leq \xi \leq 1$. These functions will span the domain of definition D_a of T_a and U_a unless there is an explicit statement to the contrary. The parameter a emphasizes that we deal here with a one-parameter family of linear transformations, and furnishes sufficient flexibility for the subsequent analysis.

Whenever a function f in the domain of the transformation T_a is extended to a larger interval, it is understood that $f = 0$ outside of $a \leq \xi \leq 1$. We emphasize again that (5.2.2) and (5.2.3) are defined only for those intervals $[a, 1]$ on which $L(\xi, \xi) > M(\xi, \xi)$. Since the operators U_a and T_a are used fundamentally in the characterization of the optimal strategies, a review of their basic properties will now be given, the proofs of which can be found in the next section.

First, U_a and T_a are strictly positive, i.e., the functions T_af and U_af into which nonnegative, nontrivial piecewise-continuous functions f are transformed by T_a and U_a, respectively, are strictly positive continuous functions on $a \leq \xi \leq 1$. The strict positivity of T_a and U_a follows from property (c) of the kernel $K(\xi, \eta)$, together with the fact that $L(\xi, \xi) > M(\xi, \xi)$ on $[a, 1]$.

Second, U_a and T_a are completely continuous. In the present context this means that any family of functions f_α that is uniformly bounded on the interval $[a, 1]$ is transformed into a family of functions $h_\alpha = T_af_\alpha$ which is equicontinuous (see Section C.1). A similar statement applies to the operator U_a. Consequently, from the family of image functions h_α a uniformly convergent subsequence can be selected (see p. 366).

The strict positivity and complete continuity of T_a and U_a imply that these operators have the following additional properties:

(1) If $\lambda(a)$ denotes the spectral radius of T_a (that is, $\lambda(a)$ is the radius of the smallest circle in the complex plane centered at the origin that contains all the eigenvalues of T_a), then $\lambda(a)$ is an eigenvalue of T_a and has a positive eigenfunction $f^{(a)}$. Furthermore, the strict positivity of T_a guarantees that the eigenfunction of $\lambda(a)$ is unique, up to a multiplicative constant factor.

(2) The spectral radius $\lambda(a)$ is a continuous strictly monotone function of a. As $a \to 1$, $\lambda(a) \to 0$. Furthermore, as $a \to a_0$, for any preassigned positive δ the eigenfunction solution $f^{(a)}$ converges uniformly on $a_0 + \delta \leq \xi \leq 1$ to $f^{(a_0)}$.

(3) Finally, if $\lambda > \lambda(a)$, then

$$\left(I - \frac{T_a}{\lambda}\right)^{-1}$$

exists and can be evaluated by means of the series

$$\sum_{n=0}^{\infty} \left(\frac{T_a}{\lambda}\right)^n.$$

This expression for the inverse is known as the abstract Neumann series. When it is applied to any continuous function, the resulting series of functions converges uniformly in the interval $[a, 1]$; furthermore, the operator series defines a strictly positive linear transformation which carries continuous positive functions into strictly positive continuous functions. Moreover,

$$\left(I - \frac{T_a}{\lambda}\right)^{-1}$$

changes continuously with a for fixed λ in the sense described in (2), i.e.,

$$\left(I - \frac{T_a}{\lambda}\right)^{-1} p \to \left(I - \frac{T_{a_0}}{\lambda}\right)^{-1} p$$

as $a \to a_0$ uniformly on $a_0 + \delta \leq \xi \leq 1$.

We let $\mu(a)$ denote the spectral radius of U_a and $g^{(a)}$ the associated eigenfunction. Properties (1), (2), and (3) apply equally to $\mu(a)$ and $g^{(a)}$, *mutatis mutandis*.

The reader may recognize these results as an extension of the theory of positive matrices to linear mappings defined for continuous functions which are positivity-preserving (see Theorem 8.2.1 of Vol. I). The proof of these assertions can be obtained by simple modifications of the reasoning employed in the matrix case (see Section 5.3).

With the summary of the relevant theory and properties of positive transformations complete, we return to our discussion of the integral equations of games of timing.

Since the solution of the integral equations (5.2.4) and (5.2.5) is very closely tied to the existence of optimal strategies of the special form $(\alpha I_0, f_a, \beta I_1)$, it is important to know whether it is possible to solve (5.2.4) for f_a with f_a positive. In view of properties (1) and (2) of T_a and the fact that $L(a, a) > M(a, a)$, this is indeed possible provided $\lambda(a) < 1$.

Our next lemma provides a criterion for determining whether or not there exists a value $a > 0$ such that $\lambda(a) = 1$. Note that property (2) of T_a enables us to conclude that for a' larger than a, $\lambda(a') < 1$.

▶Lemma 5.2.2. If $L(1, 1) > M(1, 1)$ and there exists a value ξ_0 $(0 \leq \xi_0 < 1)$ such that $L(\xi_0, \xi_0) = M(\xi_0, \xi_0)$, then there exists a value $a > \xi_0$ such that $\lambda(a) = 1$; similarly there exists a value $a^* > \xi_0$ such that $\mu(a^*) = 1$.

Proof. Let ξ_0 denote the largest ξ for which $M(\xi_0, \xi_0) = L(\xi_0, \xi_0)$. Then since $L(\xi, \eta)$ and $M(\xi, \eta)$ are continuous and $L(1, 1) > M(1, 1)$, it follows that $L(\xi, \xi) > M(\xi, \xi)$ for $\xi_0 < \xi \leq 1$.

For each $a > \xi_0$ the operators T_a, U_a are well defined; we let $f^{(a)}$ be the eigenfunction associated with $\lambda(a)$. For a fixed $\epsilon > 0$ such that $\xi_0 < 1 - \epsilon$ and a such that $\xi_0 < a < 1 - \epsilon$, we normalize $f^{(a)}$ so that

$$\int_a^{1-\epsilon} f^{(a)}(\xi)\, d\xi = 1.$$

This is possible since $f^{(a)}(\xi)$ is strictly positive on the interval $[a, 1 - \epsilon]$.

By condition (c) there exists a constant $c > 0$ for which

$$-L_\eta(\xi, \eta) > c \qquad (\xi \leq \eta \leq 1 - \epsilon),$$
$$-M_\eta(\xi, \eta) > c \qquad (\eta \leq \xi \leq 1 - \epsilon).$$

Hence, if $\xi_0 < a \leq \eta \leq 1 - \epsilon$,

$$\begin{aligned} 0 < c = c\int_a^{1-\epsilon} f^{(a)}(\xi)\, d\xi &\leq \int_a^\eta - L_\eta(\xi, \eta) f^{(a)}(\xi)\, d\xi + \int_\eta^{1-\epsilon} - M_\eta(\xi, \eta) f^{(a)}(\xi)\, d\xi \\ &\leq \int_a^\eta - L_\eta(\xi, \eta) f^{(a)}(\xi)\, d\xi + \int_\eta^1 - M_\eta(\xi, \eta) f^{(a)}(\xi)\, d\xi. \end{aligned}$$

Since $T_a f^{(a)} = \lambda(a) f^{(a)}$, the last expression is $[L(\eta, \eta) - M(\eta, \eta)]\lambda(a) f^{(a)}(\eta)$. Hence

$$0 < \frac{c}{[L(\eta, \eta) - M(\eta, \eta)]} \leq \lambda(a) f^{(a)}(\eta). \tag{5.2.6}$$

Furthermore, since the partial derivatives of L and M are bounded, there exists a constant C such that

$$L(\eta, \eta) - M(\eta, \eta) = [L(\eta, \eta) - M(\eta, \eta)] - [L(\xi_0, \xi_0) - M(\xi_0, \xi_0)] \leq C(\eta - \xi_0).$$

Then it follows from (5.2.6) that

$$0 < \frac{c'}{\eta - \xi_0} \leq \lambda(a) f^{(a)}(\eta).$$

Integrating the last equation over $[a, 1 - \epsilon]$, we obtain

$$c' \int_a^{1-\epsilon} \frac{d\eta}{\eta - \xi_0} \leq \lambda(a) \int_a^{1-\epsilon} f^{(a)}(\eta)\, d\eta = \lambda(a).$$

Thus

$$\lambda(a) \geq c' \log \frac{1 - \epsilon - \xi_0}{a - \xi_0},$$

which becomes unbounded as $a \to \xi_0$. Since we know that $\lambda(a) \to 0$ as $a \to 1$ and that $\lambda(a)$ is a continuous monotonic function, there exists some a for which $\lambda(a) = 1$, $\xi_0 < a < 1$. A similar argument establishes the existence of a^* for which $\mu(a^*) = 1$, and completes the proof of the lemma.

Our aim, of course, is to give an explicit representation of the optimal strategies for the game with kernel $K(\xi, \eta)$. In the description of solutions the eigenfunctions of the operator T_a associated with the eigenvalue $\lambda(a) = 1$ play a very important role. However, it may be possible that there is no value of a $(0 \leq a \leq 1)$ for which either $\lambda(a) = 1$ or $\mu(a) = 1$. In view of Lemma 5.2.2 and the monotonicity of $\lambda(a)$, $\mu(a)$, this can happen only if $L(\xi, \xi) > M(\xi, \xi)$ for all ξ $(0 \leq \xi \leq 1)$ and $\lambda(0) < 1$, $\mu(0) < 1$. In this case a new class of functions serves our needs; it consists of functions of the form

$$g_a = (I - U_a)^{-1}(\gamma q_0 + \delta q_1) = \gamma(I - U_a)^{-1} q_0 + \delta(I - U_a)^{-1} q_1,$$

and

$$f_a = (I - T_a)^{-1}(\alpha p_0 + \beta p_1),$$

where p_0, q_0, p_1, q_1 are the functions defined in (5.2.2) and (5.2.3) and α, β, γ, δ are nonnegative parameters to be determined.

Description of the optimal strategies for $K(\xi, \eta)$. There are many types of solutions to games of timing, depending on the magnitudes of the values of $L(\xi, \xi)$ and $M(\xi, \xi)$ and on the values of $K(\xi, \eta)$ at all the corner points of

TABLE 5-1

Kernel	Optimal F	Optimal G
Group 1:		
A $L(1, 1) \leq M(1, 1)$	(I_1)	(I_1)
Group 2:		
B $L(\xi, \xi) > M(\xi, \xi)$ $a \leq \xi \leq 1$ $\lambda(a) = 1, \mu(a) < 1$	(f_a)	$(g_a, \delta I_1)$
C $L(\xi, \xi) > M(\xi, \xi)$ $a \leq \xi \leq 1$ $\lambda(a) = 1, \mu(a) = 1$	(f_a)	(g_a)
D $L(\xi, \xi) > M(\xi, \xi)$ $a \leq \xi \leq 1$ $\lambda(a) < 1, \mu(a) = 1$	$(f_a, \beta I_1)$	(g_a)
If there exists ξ_0 $(0 \leq \xi \leq 1)$ such that $L(\xi_0, \xi_0) = M(\xi_0, \xi_0)$, either B, C, or D occurs.		
Group 3:		
$L(\xi, \xi) > M(\xi, \xi)$ $0 \leq \xi \leq 1$ $\mu(0) < 1; \lambda(0) < 1$ $L(0, 1) < M(1, 0)$		
E $\Phi(0) = L(0, 0)$	$(\alpha I_0, f_0)$	$(g_0, \delta I_1)$
F $L(0, 0) > \Phi(0) > S_0$	$(\alpha I_0, f_a)$	$(\gamma I_0, g_a, \delta I_1)$
G $\Phi(0) = S_0$	$(\alpha I_0, f_a)$	$(\gamma I_0, g_a)$
H $S_0 > \Phi(0) > M(0, 0)$	$(\alpha I_0, f_a, \beta I_1)$	$(\gamma I_0, g_a)$
I $\Phi(0) = M(0, 0)$	$(f_0, \beta I_1)$	$(\gamma I_0, g_0)$
Group 4:		
$L(\xi, \xi) > M(\xi, \xi)$ $0 \leq \xi \leq 1$ $\mu(0) < 1; \lambda(0) < 1$ $L(0, 1) \geq M(1, 0)$		
J $L(0, 1) \geq \Phi(0) \geq M(1, 0)$	(I_0)	(I_0)
K $L(0, 0) > \Phi(0) > L(0, 1)$	$(\alpha I_0, f_a)$	$(\gamma I_0, g_a, \delta I_1)$
L $L(0, 0) = \Phi(0)$	$(\alpha I_0, f_0)$	$(g_0, \delta I_1)$
M $M(1, 0) > \Phi(0) > M(0, 0)$	$(\alpha I_0, f_a, \beta I_1)$	$(\gamma I_0, g_a)$
N $\Phi(0) = M(0, 0)$	$(f_0, \beta I_1)$	$(\gamma I_0, g_0)$

the unit square. It is necessary to give a mutually exclusive classification of all possibilities simultaneously with their enumeration; this has been done in Table 5–1. The first group of kernels $K(\xi, \eta)$ (Case A in Table 5–1) have $L(1, 1) \leq M(1, 1)$; in this case the solutions form a saddle point. For the remaining kernels we have shown (Lemma 5.2.2) that the operators T_a and U_a can be defined. From here on, our classification is most easily made in terms of special values attained by the spectral radii $\lambda(a)$ of T_a and $\mu(a)$ of U_a. Thus the second group of kernels (Cases B, C, and D) are those for which a value a exists such that $\lambda(a) = 1$ or $\mu(a) = 1$ or both. Lemma 5.2.2 has shown that any kernel $K(\xi, \eta)$ for which there exists ξ_0 $(0 \leq \xi_0 < 1)$ such that $L(\xi_0, \xi_0) = M(\xi_0, \xi_0)$ will belong to the second group. If such a ξ_0 does not exist, then $\lambda(a)$ and $\mu(a)$ are well defined for every $0 \leq a \leq 1$ and represent monotonic functions decreasing to zero as $a \to 1$. Even here, if the kernel is such that either $\lambda(0) \geq 1$ or $\mu(0) \geq 1$ is satisfied, then the game belongs to the second group.

The remaining kernels $K(\xi, \eta)$ must satisfy the conditions $L(\xi, \xi) > M(\xi, \xi)$ $(0 \leq \xi \leq 1)$ and $\lambda(0) < 1, \mu(0) < 1$. We subdivide these kernels into two classes: Group 3 consists of kernels for which $L(0,1) < M(1, 0)$, Group 4 of kernels for which $L(0, 1) \geq M(1, 0)$. The forms that the solutions take in the various groups are exhibited in Table 5–1.

As a guide to the understanding of Table 5–1, we remind the reader that

$$x = (\alpha I_0, f_a, \beta I_1)$$

means that x is a distribution made up of a density f_a spread on the interval $[a, 1]$ and two jumps of magnitude α and β located at 0 and 1, respectively; and that

$$y = (\gamma I_0, g_a, \delta I_1)$$

has a similar interpretation. If any of the factors α, β, f_a, γ, δ, and g_a are missing, the missing factor or factors are understood to be zero. For example, $x = (\alpha I_0, f_a)$ implies that $\beta = 0$, while the remaining components have the interpretation as before. A new quantity S_0 appears in the description of the optimal strategies of Group 3. The significance of this symbol is discussed in Section 5.4.

▶THEOREM 5.2.1. The optimal strategies for the pay-off kernel $K(\xi, \eta)$ in games of timing of incomplete information are unique and take the forms indicated in Table 5–1. Furthermore, the densities f_a are either solutions of the equation $T_a f_a = f_a$ or expressible as Neumann series of the form

$$\sum_{n=0}^{\infty} T_a^n(\alpha p_0 + \beta p_1);$$

similarly for g_a.

The existence and explicit construction of these solutions will be discussed in Section 5.4. The uniqueness property of these solutions is summarized by means of Problems 16–19.

***5.3 Integral equations with positive kernels.** We digress in this section in order to verify the properties of integral operators set forth in the previous section.

Consider the general integral operator

$$\int_a^b A(\xi, \eta)f(\eta)\, d\eta,$$

or, in operator form, Af. (Throughout this section, all integral signs will be understood to be between the fixed limits a, b unless there is an explicit statement to the contrary.) We shall suppose that the kernel satisfies

$$A(\xi, \eta) \geq 0, \tag{5.3.1a}$$

where the points at which $A = 0$ form on each line ξ (= constant) or η (= constant) a point set containing no set of positive Lebesgue measure. We also assume that

$$A(\xi, \eta) \leq C, \tag{5.3.1b}$$

and that

$$\begin{aligned} \int |A(\xi_2, \eta) - A(\xi_1, \eta)|\, d\eta \to 0, \\ \int |A(\xi, \eta_2) - A(\xi, \eta_1)|\, d\xi \to 0, \end{aligned} \tag{5.3.1c}$$

if $|\xi_2 - \xi_1| \to 0$ or $|\eta_2 - \eta_1| \to 0$, respectively.

Let $\mathcal{P}$ represent the set of all bounded functions f defined on the interval $[a, b]$. The domain of A in all that follows is restricted to $\mathcal{P}$.

If, in particular, f is bounded on $[a, b]$, it follows from (5.3.1a–c) that

$$\int A(\xi, \eta)f(\eta)\, d\eta \qquad \text{and} \qquad \int f(\xi)A(\xi, \eta)\, d\xi$$

are continuous functions of ξ and η, respectively; and if in addition f is nonnegative, continuous, and not identically zero, then these expressions have a positive minimum. The first claim is an easy consequence of the Lebesgue convergence criterion (see Section C.3); the second is true because of (5.3.1a). We formalize these statements as a theorem.

▶ **Theorem 5.3.1.** The mapping A satisfying (5.3.1a–c) transforms nonzero, nonnegative continuous functions into strictly positive continuous functions.

---

*Starred sections are advanced discussions that may be omitted at first reading without loss of continuity.

Another useful property of such integral transformations is the fact that they are completely continuous mappings when defined on $\mathcal{P}$. The precise statement of this property is embodied in the following theorem.

▶ THEOREM 5.3.2. The operator A maps a family of functions that is uniformly bounded into an equicontinuous family of functions (see Section C.1 for the definition of equicontinuity).

The proof is immediate from (5.3.1c).

▶ COROLLARY 5.3.1. If f_α is a family of uniformly bounded functions defined on $[a, b]$, then we may select a subsequence from $g_\alpha = Af_\alpha$ which converges uniformly to a continuous function.

This is a direct consequence of the classical Ascoli theorem pertaining to equicontinuous families of functions (see Section C.1).

One final observation, which we shall use later, is that if $f \geq 0$ and $\int f = 1$, then

$$\int g \geq \min_{\xi} \int A(\xi, \eta)\, d\eta > 0, \tag{5.3.2}$$

where $g = Af$.

The principal theorem on the existence of positive eigenvalues and strictly positive eigenfunctions follows.

▶ THEOREM 5.3.3. Let $\lambda_0(A)$ denote the spectral radius of A. Then (a) $\lambda_0(A)$ is a strictly positive eigenvalue with an associated strictly positive continuous eigenfunction f_0; (b) if $Af = \lambda_0(A)f$, then f is a multiple of f_0.

Compare this theorem with Theorem 8.2.1 of Vol. I. The proof is accomplished by adapting the finite-dimensional analysis of positive matrix mappings to the case of the integral operator A.

Proof. A real value λ is said to be admissible if there exists a continuous, nonzero, nonnegative function f defined on $[a, b]$ such that $Af \geq \lambda f$, where the inequality is to be interpreted as holding for every value of the argument ξ in the interval $[a, b]$. The function f of the inequality is referred to as the qualifying or test function of λ.

Since the relation $Af \geq \lambda f$ is homogeneous, we may without loss of generality normalize all f which qualify for admissible λ by the condition $\int f = 1$. Clearly,

$$\mu = \min_{a \leq \xi \leq b} \int A(\xi, \eta)\, d\eta$$

is admissible with the test function $f \equiv 1/(b - a)$. Let

$$M = \max_{a \leq \xi \leq b} \int A(\xi, \eta)\, d\eta;$$

then it is clear that every admissible λ is bounded by M.

Let $\lambda_0 = \sup \lambda$, where the supremum is taken over all possible admissible λ. The value λ_0 can also be shown to be admissible. The compactness needed to carry out this argument is available, since A is completely continuous. Let $\lambda_n \to \lambda_0$, and let f_n be a nonzero nonnegative function satisfying $\int f_n = 1$ and $Af_n \geq \lambda_n f_n$. Clearly, f_n are uniformly bounded, by (5.3.1b). Moreover, since A preserves positivity and linearity, we have $Ag_n \geq \lambda_n g_n$, where $g_n = Af_n$ is an equicontinuous family of functions (Theorem 5.3.2) whose integrals are uniformly bounded from below in accordance with (5.3.2). Any limit point of g_n will clearly qualify to show that λ_0 is admissible.

Suppose that f_0 is a nonzero, nonnegative, continuous function such that $Af_0 \geq \lambda_0 f_0$. We shall show that equality holds everywhere in this inequality. Suppose the contrary; then as A transforms nonzero, nonnegative continuous functions into strictly positive continuous functions, we obtain $Ag_0 \geq (\lambda_0 + \epsilon)g_0$ for $g_0 = Af_0$ and $\epsilon > 0$. This contradicts the definition of λ_0.

We next show that λ_0 is equal to the spectral radius. It suffices to show that any other eigenvalue λ^* is smaller in magnitude than λ_0. Now if $Af^* = \lambda^* f^*$, it follows that $A|f^*| \geq |\lambda^*||f^*|$, where $|f^*| = |f^*(\xi)|$, so that from the very definition of λ_0 we may conclude that $\lambda_0 \geq |\lambda^*|$. Finally, we show that any other eigenfunction $\bar{f}$ associated with the eigenvalue λ_0 is a multiple of f_0. Without loss of generality, we may assume that $\bar{f}$ is real. Since f_0 is strictly positive, we may find a constant γ such that $f_0 + \gamma\bar{f}$ is nonnegative for $a \leq \xi \leq b$ and 0 somewhere. Since

$$A(f_0 + \gamma\bar{f}) = \lambda_0(f_0 + \gamma\bar{f})$$

is strictly positive or identically zero, we deduce that $\bar{f}$ is a multiple of f_0 as asserted. This completes the proof.

We are now prepared to deduce the abstract Neumann series of the integral equation of the second kind formed from the kernel A.

▶ THEOREM 5.3.4. If $\lambda > \lambda_0(A)$, then

$$\left(I - \frac{A}{\lambda}\right)^{-1} = \sum_{n=0}^{\infty} \frac{A^n}{\lambda^n},$$

where $A^0 = I$.

Proof. It is enough to show that

$$\sum_{n=0}^{\infty} \frac{A^n}{\lambda^n} f = g$$

represents a uniformly convergent series for any bounded positive function f, and that

$$\left(I - \frac{A}{\lambda}\right) g = f.$$

Let f_0 be a strictly positive continuous eigenfunction associated with the eigenvalue $\lambda_0(A)$. Such an eigenfunction exists by virtue of Theorem 5.3.3. Since f_0 is a strictly positive continuous function, we can determine a constant c such that

$$f \leq cf_0.$$

But A preserves positivity and hence

$$A^n f \leq c\lambda_0^n f_0,$$

from which we see that the series is uniformly and absolutely bounded, apart from a constant factor, by a convergent geometric series. The direct verification of $(I - A/\lambda)g = f$ offers no difficulties, and the proof of Theorem 5.3.4 is complete.

Our next result pertains to the variation of $\lambda_0(A)$ as a function of A. Specifically, let $\widetilde{A}$ define another integral operator generated by a kernel function $\widetilde{A}(\xi, \eta)$ defined on the square $a' \leq \xi, \eta \leq b'$, $(a' \leq a, b' \geq b)$ and satisfying (5.3.1a–c) on the larger interval. Suppose, in addition, that

$$\widetilde{A}(\xi, \eta) \geq A(\xi, \eta) \qquad (a \leq \xi, \eta \leq b).$$

We show that

$$\lambda_0(\widetilde{A}) \geq \lambda_0(A). \tag{5.3.3}$$

It will be recalled that $\lambda_0(A)$ was characterized in the course of the proof of Theorem 5.3.3 as the largest admissible value λ for which $Af \geq \lambda f$, with $f \geq 0$, f continuous, and

$$\int_a^b f = 1.$$

A slight modification of our analysis permits the same characterization of $\lambda_0(A)$ for all functions f satisfying the same normalization condition but allowed to possess discontinuities of the first kind only.

The quantity $\lambda_0(\widetilde{A})$ can be similarly characterized. Let λ be admissible for A, and let f denote the corresponding qualifying function. Extend f to the interval $[a', b']$ by setting $f = 0$ outside $[a, b]$. Then

$$\int_{a'}^{b'} \widetilde{A}(\xi, \eta)f(\eta)\, d\eta \geq \int_a^b A(\xi, \eta)f(\eta)\, d\eta \geq \lambda f(\xi) \qquad (a \leq \xi \leq b),$$

and the same inequality trivially prevails for $[a', b']$. Hence any admissible λ for A is also admissible for $\widetilde{A}$, which implies (5.3.3).

By the same methods we can examine the continuity properties of $\lambda_0(A)$ and its eigenfunction f_0 as a function of A. The formal details of these studies are omitted.

We close this section by noting that T_a and U_a as defined in (5.2.2) and (5.2.3) satisfy (5.3.1a–c). The condition (5.3.1a) is met by virtue of condition (c) of Section 5.2; the condition (5.3.1b) is immediate; and the condition (5.3.1c) is shown to be met as follows: Suppose that $\xi_2 > \xi_1$ and $\xi_2 - \xi_1 < \epsilon$; then

$$\int_a^b |T(\xi_1, \eta) - T(\xi_2, \eta)|\, d\eta \leq \int_a^{\xi_1 - \epsilon} |T(\xi_1, \eta) - T(\xi_2, \eta)|\, d\eta + C\epsilon + \int_{\xi_2+\epsilon}^b |T(\xi_1, \eta) - T(\xi_2, \eta)|\, d\eta,$$

where C is a fixed constant. But T is uniformly continuous in the respective triangular regions of the right-hand integral, and hence

$$\int_a^b |T(\xi_1, \eta) - T(\xi_2, \eta)|\, d\eta \to 0 \qquad \text{as} \qquad |\xi_1 - \xi_2| \to 0.$$

By applying Theorem 5.3.3, we deduce the validity of $T_a f^{(a)} = \lambda(a) f^{(a)}$, where $\lambda(a)$ is the largest eigenvalue of T_a and $f^{(a)}$ is a positive continuous eigenfunction normalized so that $\int f^{(a)}(\xi)\, d\xi = 1$.

The other assertions—i.e., properties (2) and (3) of p. 114—follow from Theorems 5.3.1–4. For instance, we show now that $\lambda(a)$ is a continuous, monotone-decreasing function of a. The monotonicity of $\lambda(a)$ follows easily from the discussion of (5.3.3). It remains to show that $\lambda(a) \leq \lambda(a') + \epsilon$ for prescribed $\epsilon > 0$, when a' is sufficiently close to a and $a' > a$. Consider

$$\begin{aligned} \lambda(a) f^{(a)}(\eta) &= \int_a^1 T(\xi, \eta) f^{(a)}(\xi)\, d\xi \\ &\leq \int_{a'}^1 T(\xi, \eta) f^{(a)}(\xi)\, d\xi + \delta \end{aligned} \tag{5.3.4}$$

(since $f^{(a)}(\xi)$ is uniformly bounded for $a' < \xi < a$), where clearly $\delta > 0$ approaches zero as $a' - a \to 0$. Since $f^{(a)}(\eta)$ are uniformly bounded (in a and a') from below for all $a' \leq \eta \leq 1$ when $a' - a \leq \epsilon$ (otherwise the strict positivity of the kernel T can be contradicted), we may rewrite (5.3.4) as

$$[\lambda(a) - \epsilon] f^{(a)}(\eta) \leq \int_{a'}^1 T(\xi, \eta) f^{(a)}(\xi)\, d\xi \qquad (a' \leq \eta \leq 1)$$

for appropriate ϵ, where $\epsilon \to 0$ as $a' \to a$. Hence $\lambda(a) - \epsilon$ is admissible for $T_{a'}$ and therefore $\lambda(a') \geq \lambda(a) - \epsilon$. A more careful examination of this proof reveals that $\lambda(a)$ is strictly monotonic. The remaining parts of properties (2) and (3) are proved by similar methods. These proofs are left as exercises for the reader.

*5.4 Existence proofs.

Optimal strategies for Group 1 kernels.

▶LEMMA 5.4.1. If $L(1, 1) \leq M(1, 1)$, then the point $\{1, 1\}$ is a saddle point (a pair of pure optimal strategies) of $K(\xi, \eta)$ and both players have the optimal strategy (I_1).

Proof. Condition (b) (p. 110) implies that $M(1, 1) \geq \Phi(1) \geq L(1, 1)$, and condition (c) implies that $M(1, \eta) \geq M(1, 1)$ and $L(\xi, 1) \leq L(1, 1)$. Hence

$$K(1, \eta) \geq \Phi(1) \geq K(\xi, 1) \qquad (\text{all} \quad \xi, \eta)$$

and

$$K(1, \eta) \geq K(1, 1) \geq K(\xi, 1) \qquad (\text{all} \quad \xi, \eta).$$

The last equation shows that the point $\{1, 1\}$ is a saddle point. The value of the game is $\Phi(1)$.

Optimal strategies for Group 2 kernels.

▶LEMMA 5.4.2. If $L(1, 1) > M(1, 1)$ and there exists ξ_0 $(0 \leq \xi_0 < 1)$ such that $L(\xi_0, \xi_0) = M(\xi_0, \xi_0)$, then there exist optimal strategies of the following form: densities f_a, g_a over a common interval $[a, 1]$ and a possible jump at 1 for one of the two players.

Proof. Lemma 5.2.2 implies that there exist a and a^* in the interval $(\xi_0, 1)$ such that $\lambda(a) = 1$, $\mu(a^*) = 1$. Furthermore, since the spectral radius is a strictly decreasing function (property (2), p. 114), if $a > a^*$, then $\mu(a) < 1$; similarly, if $a = a^*$, then $\lambda(a) = \mu(a) = 1$; and if $a < a^*$, then $\mu(a^*) = 1$ and $\lambda(a^*) < 1$.

We begin with the case $\lambda(a) = 1$, $\mu(a) < 1$. Since $\lambda(a) = 1$, the equation $T_a f_a = f_a$ has a strictly positive solution f_a. We normalize this function so that

$$\int_a^1 f_a(\xi)\, d\xi = 1,$$

and define

$$u(a) = \int_a^\eta L(\xi, \eta) f_a(\xi)\, d\xi + \int_\eta^1 M(\xi, \eta) f_a(\xi)\, d\xi. \qquad (5.4.1)$$

Clearly $u(a)$ is independent of η $(a \leq \eta \leq 1)$, since differentiation yields

$$[L(\eta, \eta) - M(\eta, \eta)]\{f_a(\eta)\} + \int_a^\eta L_\eta(\xi, \eta) f_a(\xi)\, d\xi + \int_\eta^1 M_\eta(\xi, \eta) f_a(\xi)\, d\xi = 0,$$

which is the same as

$$f_a - T_a f_a = 0.$$

We extend $f_a(\xi)$ beyond the interval $[a, 1]$ by defining $f_a(\xi) = 0$ outside that interval. If Player I uses the strategy f_a and Player II uses the pure strategy that concentrates at η $(0 \leq \eta \leq a)$, the return to Player I is

$$\int_0^1 K(\xi, \eta) f_a(\xi)\, d\xi = \int_a^1 M(\xi, \eta) f_a(\xi)\, d\xi \geq \int_a^1 M(\xi, a) f_a(\xi)\, d\xi = u(a),$$

the inequality emerging as a consequence of condition (c) on the kernel $K(\xi, \eta)$. On the other hand, if $a \leq \eta \leq 1$, the yield is given by (5.4.1). Hence Player I can obtain a yield of at least $u(a)$.

To define a strategy for Player II, set

$$g = (I - U_a)^{-1} q_1,$$

where q_1 is the function in (5.2.3). Since $\mu(a) < 1$, it follows that g exists and may be calculated by the series

$$\sum_{n=0}^{\infty} (U_a)^n q_1.$$

Let

$$c = \int_a^1 g(\eta)\, d\eta$$

and $\delta = 1/(1 + c)$. Finally, set $g_a(\eta) = \delta g(\eta)$ $(a \leq \eta \leq 1)$; elsewhere $g_a(\eta) = 0$.

By this definition $y^0 = (g_a, \delta I_1)$ is a strategy for Player II, and we shall show that it is optimal. The definition of g_a implies

$$(I - U_a) g_a = \delta q_1 \qquad (a \leq \xi \leq 1).$$

Define

$$w(a, \delta) = \delta L(\xi, 1) + \int_a^\xi M(\xi, \eta) g_a(\eta)\, d\eta + \int_\xi^1 L(\xi, \eta) g_a(\eta)\, d\eta$$
$$(a \leq \xi \leq 1); \qquad (5.4.2)$$

then by differentiation $w(a, \delta)$ is seen to be a constant independent of ξ. An analysis parallel to that of (5.4.1) yields

$$\int_0^1 K(\xi, \eta)\, dy^0(\eta) < w(a, \delta) \qquad (0 \leq \xi < a),$$

since equation (5.4.2) represents the yield to Player I for $a \leq \xi \leq 1$. For $\xi = 1$ the yield is

$$w(a, \delta) + \delta[\Phi(1) - L(1, 1)] \leq w(a, \delta).$$

Player II by using y^0 can prevent Player I from securing a yield larger than $w(a, \delta)$. But

$$w(a, \delta) = w(a, \delta)\int_a^1 dx^0(\xi) = \int_a^1\int_a^1 K(\xi, \eta)\, dx^0(\xi)\, dy^0(\eta)$$
$$= u(a)\int_a^1 dy^0(\eta) = u(a),$$

where x^0 is the strategy with density $f_a(\xi)$. Hence the value of the game is $u(a)$, and the strategies x^0 and y^0 are optimal.

In order to complete the proof of the lemma we must discuss the cases $\lambda(a) = \mu(a) = 1$ and $\lambda(a^*) < 1, \mu(a^*) = 1$. Clearly, if $\lambda(a^*) < 1$, $\mu(a^*) = 1$, an argument identical to the previous one will show that the solution g_{a^*} of the eigenvalue problem $U_{a^*}g_{a^*} = g_{a^*}$ is optimal for Player II, while a strategy of the form $(f_{a^*}, \beta I_1)$ is optimal for Player I. If $\lambda(a) = \mu(a) = 1$, the argument is more direct and simpler: the optimal strategies consist exclusively of densities obtained as the solutions of the eigenvalue problems $T_a f_a = f_a$ and $U_a g_a = g_a$.

Note that Example 2 (pp. 103–06) is of Group 2, Case B, with $a = 1/3$.

▶LEMMA 5.4.3. If $L(\xi, \xi) > M(\xi, \xi)$ $(0 \leq \xi \leq 1)$ and there exists a value a $(0 \leq a < 1)$ for which either $\lambda(a) = 1$ or $\mu(a) = 1$, then Case B, C, or D of Table 5–1 obtains and there exist optimal strategies of the form indicated in the table.

Proof. Since $\lambda(a)$ and $\mu(a)$ decrease monotonically and continuously to zero as $a \to 1$, it is clear that there exists a value a' $(0 \leq a' \leq 1)$ for which either $\lambda(a') = 1$ and $\mu(a') < 1$, or $\lambda(a') = \mu(a') = 1$, or $\lambda(a') < 1$ and $\mu(a') = 1$. The rest of the proof is now the same as the proof of the previous lemma.

We have now proved the existence, and characterized the form, of optimal strategies for the kernels of Group 2 in Table 5–1.

We turn next to the games whose kernels belong to Group 3. Since we have exhausted all other possibilities, we may assume hereafter that $L(\xi, \xi) > M(\xi, \xi)$ $(0 \leq \xi \leq 1)$ and $\lambda(0) < 1, \mu(0) < 1$.

Subject to these assumptions, the operators T_a and U_a are well-defined strictly positive mappings for every a $(0 \leq a \leq 1)$. Furthermore $(I - T_a)^{-1}$ and $(I - U_a)^{-1}$ exist for every a and transform positive continuous functions into strictly positive functions. Since $\lambda(a) < 1$, it follows that $(I - T_a)^{-1}p = \sum T_a^n p$, and that $(I - T_a)^{-1}$ varies continuously with a (see property (3), p. 114); hence if $f_a = (I - T_a)^{-1}p$,†

† Note that $(I - T_a)^{-1}p$ is defined for all η in [0, 1] for the purpose of this argument.

then

$$\lim_{a\to 1} (I - T_a)^{-1}p = I^{-1}p = p$$

uniformly in ξ, and thus

$$\lim_{a\to 1} \int_a^1 f_a(\xi)\, d\xi = \lim_{a\to 1} \int_a^1 p(\xi)\, d\xi = 0 \tag{5.4.3}$$

for every continuous function p. Analogous remarks apply to $(I - U_a)^{-1}$.

We now construct a family of strategies for the game as follows: Set

$$\phi_0 = (I - T_a)^{-1}p_0, \qquad \phi_1 = (I - T_a)^{-1}p_1,$$

$$f_a = (I - T_a)^{-1}(\alpha p_0 + \beta p_1) = \alpha\phi_0 + \beta\phi_1,$$

where $\alpha, \beta \geq 0$ and p_0, p_1 are the functions defined in (5.2.2). The parameters α, β are assumed always to be connected by the relation

$$\alpha\int_a^1 \phi_0 + \beta\int_a^1 \phi_1 = 1 - \alpha - \beta. \tag{5.4.4}$$

It is clear that with this normalization $x = (\alpha I_0, f_a, \beta I_1)$ is a strategy for the game. For each η $(a \leq \eta \leq 1)$, we form

$$u(a, \alpha, \beta) = \alpha L(0, \eta) + \beta M(1, \eta) + \int_a^\eta L(\xi, \eta) f_a(\xi)\, d\xi + \int_\eta^1 M(\xi, \eta) f_a(\xi)\, d\xi, \tag{5.4.5}$$

where the function $u(a, \alpha, \beta)$ is independent of η, since differentiating this expression with respect to η yields $\alpha p_0 + \beta p_1 = (I - T_a)f_a$. Moreover, by the monotonicity and continuity of $L(\xi, \eta)$ and $M(\xi, \eta)$ (condition (c), p. 110),

$$u(a, \alpha, \beta) < \alpha L(0, \eta) + \beta M(1, \eta) + \int_a^1 M(\xi, \eta) f_a(\xi)\, d\xi \qquad (0 \leq \eta < a). \tag{5.4.6}$$

In a similar manner we define

$$g_a = (I - U_a)^{-1}(\gamma q_0 + \delta q_1) = \gamma\psi_0 + \delta\psi_1$$

and normalize in such a way that

$$\gamma\int_a^1 \psi_0 + \delta\int_a^1 \psi_1 = 1 - \gamma - \delta, \tag{5.4.7}$$

so that $y = (\gamma I_0, g_a, \delta I_1)$ is a strategy. The same kind of argument shows that

$$w(a, \gamma, \delta) = \gamma M(\xi, 0) + \delta L(\xi, 1) + \int_a^\xi M(\xi, \eta) g_a(\eta)\, d\eta$$
$$+ \int_\xi^1 L(\xi, \eta) g_a(\eta)\, d\eta \qquad (a \leq \xi \leq 1) \qquad (5.4.8)$$

is independent of ξ and that

$$w(a, \gamma, \delta) > \gamma M(\xi, 0) + \delta L(\xi, 1) + \int_a^1 L(\xi, \eta) g_a(\eta)\, d\eta \qquad (0 < \xi < a).$$

For $0 < \eta < 1$, equations (5.4.5) and (5.4.6) describe the yield to Player I when he uses the strategy x and Player II uses the pure strategy η; we have

$$\int_0^1 K(\xi, \eta)\, dx(\xi) \geq u(a, \alpha, \beta) \qquad (0 < \eta < 1).$$

In a similar manner,

$$\int_0^1 K(\xi, \eta)\, dy(\eta) \leq w(a, \gamma, \delta) \qquad (0 < \xi < 1).$$

We now have available the five parameters $\alpha, \beta, \gamma, \delta, a$ (subject only to the restrictions imposed by (5.4.4) and (5.4.7) and the conditions $\alpha, \beta, \gamma, \delta \geq 0$ and $0 \leq a \leq 1$) which we shall show may be chosen, in each of the cases E-I of Group 3, to satisfy the following conditions:

$$u(a; \alpha, \beta) = w(a; \gamma, \delta),$$

$$\int_0^1 K(0, \eta)\, dy(\eta) \leq w(a; \gamma, \delta), \qquad \int_0^1 K(1, \eta)\, dy(\eta) \leq w(a; \gamma, \delta),$$

and

$$\int_0^1 K(\xi, 0)\, dx(\xi) \geq u(a; \alpha, \beta), \qquad \int_0^1 K(\xi, 1)\, dx(\xi) \geq u(a; \alpha, \beta).$$

When these conditions are fulfilled, we have

$$\int_0^1 K(\xi, \eta)\, dx(\xi) \geq u(a, \alpha, \beta) = v = w(a, \gamma, \delta) \geq \int_0^1 K(\xi, \eta)\, dy(\eta)$$

for all ξ and η, so that v is the value of the game and x, y are optimal strategies.

In the remaining part of this section it is to be understood that $\alpha, \beta, \gamma, \delta \geq 0$ and $0 \leq a \leq 1$. We note specifically that we may take α or β to be zero; in this case

$$\beta \int_a^1 \phi_1 = 1 - \beta \qquad \text{or} \qquad \beta = \left(1 + \int_a^1 \phi_1\right)^{-1},$$

which is a continuous function of a. We may, more generally, require that α and β have a fixed ratio r $(0 \leq r \leq \infty)$; in this case α is a continuous function of a and hence $u(a, \alpha, \beta)$ is a continuous function of a. Similar remarks apply to γ and δ.

▶Lemma 5.4.4. If

$$L(\xi, \xi) > M(\xi, \xi) \qquad (0 \leq \xi \leq 1)$$

and

$$\lambda(0) < 1, \mu(0) < 1,$$

then

$$u(0, \alpha, \beta) - w(0, \gamma, \delta) = \alpha\gamma[L(0, 0) - M(0, 0)] + \beta\delta[M(1, 1) - L(1, 1)].$$

Proof. By substituting $a = 0$ in equations (5.4.5) and (5.4.8) and evaluating these equations for $\eta = 0$, $\xi = 0$, $\eta = 1$, $\xi = 1$, we obtain

$$u(0, \alpha, \beta) = \alpha L(0, 0) + \beta M(1, 0) + \int_0^1 M(\xi, 0) f_0(\xi)\, d\xi,$$

$$w(0, \gamma, \delta) = \gamma M(0, 0) + \delta L(0, 1) + \int_0^1 L(0, \eta) g_0(\eta)\, d\eta,$$

$$u(0, \alpha, \beta) = \alpha L(0, 1) + \beta M(1, 1) + \int_0^1 L(\xi, 1) f_0(\xi)\, d\xi,$$

$$w(0, \alpha, \beta) = \gamma M(1, 0) + \delta L(1, 1) + \int_0^1 M(1, \eta) g_0(\eta)\, d\eta.$$

If we multiply (5.4.5) by g_0 and (5.4.8) by f_0 and integrate, we obtain

$$(1 - \gamma - \delta) u(0, \alpha, \beta) = \alpha \int_0^1 L(0, \eta) g_0(\eta)\, d\eta + \beta \int_0^1 M(1, \eta) g_0(\eta)\, d\eta + \int_0^1 \int_0^1 K(\xi, \eta) f_0(\xi) g_0(\eta)\, d\eta\, d\xi,$$

$$(1 - \alpha - \beta) w(0, \gamma, \delta) = \gamma \int_0^1 M(\xi, 0) f_0(\xi)\, d\xi + \delta \int_0^1 L(\xi, 1) f_0(\xi)\, d\xi + \int_0^1 \int_0^1 K(\xi, \eta) f_0(\xi) g_0(\eta)\, d\eta\, d\xi.$$

Multiplying the first four equations by γ, α, δ, and β, respectively, we can eliminate the integrals in the last two expressions; subtracting these, we finally arrive at

$$u(0, \alpha, \beta) - w(0, \gamma, \delta) = \alpha\gamma[L(0, 0) - M(0, 0)] + \beta\delta[M(1, 1) - L(1, 1)]. \qquad (5.4.9)$$

Optimal strategies for Group 3 kernels.

We consider kernels which in addition to the properties listed above satisfy the condition $L(0, 1) < M(1, 0)$.

▶ LEMMA 5.4.5. For every ratio r there exists an a for which

$$u(a, \alpha, 0) = w(a, \gamma, r\gamma).$$

Similarly, there exists an a' for which $u(a', \alpha, r\alpha) = w(a', \gamma, 0)$.

Proof. By substituting $\eta = 1$ in (5.4.5) and $\xi = 1$ in (5.4.8), we obtain

$$u(a, \alpha, \beta) - \alpha L(0, 1) - \beta M(1, 1) - \int_a^1 L(\xi, 1) f_a(\xi)\, d\xi = 0,$$

$$w(a, \gamma, \delta) - \gamma M(1, 0) - \delta L(1, 1) - \int_a^1 M(1, \eta) g_a(\eta)\, d\eta = 0.$$

Then for any $\epsilon > 0$ we can choose a sufficiently close to 1 so that

$$\left| u(a, \alpha, \beta) - \alpha L(0, 1) - \beta M(1, 1) - L(1, 1) \int_a^1 f_a(\xi)\, d\xi \right| < \frac{\epsilon}{3},$$

$$\left| w(a, \gamma, \delta) - \gamma M(1, 0) - \delta L(1, 1) - M(1, 1) \int_a^1 g_a(\eta)\, d\eta \right| < \frac{\epsilon}{3}.$$

We have seen in (5.4.3) that as $a \to 1$,

$$\int_a^1 f_a \to 0 \qquad \text{and} \qquad \int_a^1 g_a \to 0;$$

it follows that for a sufficiently close to 1,

$$\begin{aligned} |u(a, \alpha, \beta) &- w(a, \gamma, \delta) \\ &- \{\alpha L(0, 1) + \beta M(1, 1) - \gamma M(1, 0) - \delta L(1, 1)\}| < \epsilon \end{aligned} \tag{5.4.10}$$

for any $\alpha, \beta, \gamma, \delta$ satisfying (5.4.4) and (5.4.7).

By assumption, $L(0, 1) < M(1, 0)$; and by monotonicity,

$$L(0, 1) < L(1, 1), \qquad M(1, 1) < M(1, 0).$$

Hence

$$\begin{aligned} \alpha L(0, 1) + \beta M(1, 1) &- \gamma M(1, 0) - \delta L(1, 1) \\ &\leq (\alpha - \delta) L(0, 1) - (\gamma - \beta) M(1, 0). \end{aligned}$$

Moreover,

$$1 - \alpha - \beta = \int_a^1 f_a \to 0 \qquad \text{as} \qquad a \to 1,$$

so that $\alpha + \beta \to 1$. Similarly, $\gamma + \delta \to 1$ as $a \to 1$; therefore $\alpha - \delta \to \gamma - \beta$. But we have assumed that $L(0, 1) < M(1, 0)$, which implies that

$$\alpha L(0, 1) + \beta M(1, 1) - \gamma M(1, 0) - \delta L(1, 1) < 0$$

for a sufficiently close to 1. If we insert this relation into (5.4.10), we obtain the inequality

$$u(a, \alpha, \beta) - w(a, \gamma, \delta) < 0$$

for a near 1.

On the other hand, Lemma 5.4.4 asserts that

$$u(0, \alpha, \beta) - w(0, \gamma, \delta) = \alpha\gamma[L(0, 0) - M(0, 0)] + \beta\delta[M(1, 1) - L(1, 1)].$$

If $\beta = 0$, since $L(0, 0) > M(0, 0)$, then

$$u(0, \alpha, 0) - w(0, \gamma, \delta) \geq 0.$$

Finally, since $u(a, \alpha, 0) - w(a, \gamma, r\gamma)$ is a continuous function of a, the last two inequalities guarantee the existence of a value a' for which

$$u(a', \alpha, 0) = w(a', \gamma, r\gamma).$$

The other assertion can be established by setting $\delta = 0$ above.

▶ LEMMA 5.4.6. Given the conditions of Lemma 5.4.4 and $\Phi(0) = L(0, 0)$, Player I has an optimal strategy $x = (\alpha I_0, f_0)$ and Player II has an optimal strategy $y = (g_0, \delta I_1)$.

Proof. If we set $\beta = \gamma = 0$, Lemma 5.4.4 shows that $u(0, \alpha, 0) = w(0, 0, \delta) = v$. It remains to verify

$$\int K(\xi, 0)\, dx(\xi) \geq v, \qquad \int K(\xi, 1)\, dx(\xi) \geq v,$$

$$\int K(0, \eta)\, dy(\eta) \leq v, \qquad \int K(1, \eta)\, dy(\eta) \leq v,$$

where $x = (\alpha I_0, f_0)$ and $y = (g_0, \delta I_1)$. By assumption, $\Phi(0) = L(0, 0)$. Then

$$\int K(\xi, 0)\, dx(\xi) = \alpha\Phi(0) + \int_0^1 M(\xi, 0) f_0(\xi)\, d\xi$$

$$= \alpha L(0, 0) + \int_0^1 M(\xi, 0) f_0(\xi)\, d\xi = u(0, \alpha, 0)$$

by (5.4.5); since $\beta = 0$, (5.4.5) yields

$$\int_0^1 K(\xi, 1)\, dx(\xi) = v.$$

Also $\gamma = 0$ in conjunction with (5.4.8) shows that

$$\int_0^1 K(0, \eta)\, dy(\eta) = v.$$

Finally, since $M(1, 1) \leq \Phi(1) \leq L(1, 1)$,

$$\int_0^1 K(1, \eta)\, dy(\eta) = \delta\Phi(1) + \int_0^1 M(1, \eta) g_0(\eta)\, d\eta$$

$$\leq \delta L(1, 1) + \int_0^1 M(1, \eta) g_0(\eta)\, d\eta = w(0, 0, \delta)$$

as desired.

The following lemma is proved in a similar manner.

▶ LEMMA 5.4.7. Given the conditions of Lemma 5.4.4 and $\Phi(0) = M(0, 0)$, Player I has an optimal strategy $x = (f_0, \beta I_1)$ and Player II has an optimal strategy $(\gamma I_0, g_0)$.

▶ LEMMA 5.4.8. There exists a value S_0 with $M(0, 0) < S_0 < L(0, 0)$ such that if $\Phi(0) = S_0$, then there exist optimal strategies of the form indicated in Case G of Table 5–1.

Proof. We take $\beta = \delta = 0$, and determine a value of a for which $w(a, \gamma, 0) = u(a, \alpha, 0)$. The strategies $x = (\alpha I_0, f_a)$ and $y = (\gamma I_0, g_a)$ are now completely specified. We now produce a value S_0 (depending on L and M) such that if $\Phi(0) = S_0$, then the strategies x and y are optimal, and $v = u(a, \alpha, 0) = w(a, \gamma, 0)$ is the value of the game. To this end, we observe first that for $\beta = \delta = 0$ equations (5.4.5–8) imply that

$$v = u(a, \alpha, 0) \leq \int_0^1 K(\xi, \eta)\, dx(\xi) \qquad (0 < \eta \leq 1),$$

$$v = w(a, \gamma, 0) \geq \int_0^1 K(\xi, \eta)\, dy(\eta) \qquad (0 < \xi \leq 1),$$

with strict inequality in the subintervals $0 < \eta < a$ and $0 < \xi < a$. We now consider the yield for $\eta = 0$ as a function of $\Phi(0)$. If $\Phi(0) = L(0, 0)$, then

$$\int_0^1 K(\xi, 0)\, dx(\xi) = \alpha\Phi(0) + \int_a^1 M(\xi, 0) f_a(\xi)\, d\xi$$

$$= \alpha L(0, 0) + \int_a^1 M(\xi, 0) f_a(\xi)\, d\xi > u(a, \alpha, 0) = v, \qquad (5.4.11)$$

the last inequality being a consequence of (5.4.6). On the other hand, if $\Phi(0) = M(0, 0)$, then

$$\int_0^1 K(0, \eta)\, dy(\eta) = \gamma\Phi(0) + \int_a^1 L(0, \eta) g_a(\eta)\, d\eta$$
$$= \gamma M(0, 0) + \int_a^1 L(0, \eta) g_a(\eta)\, d\eta < w(a, \gamma, 0) = v;$$

and since $\alpha > 0$, we infer that

$$\int_0^1 \int_0^1 K(\xi, \eta)\, dx(\xi)\, dy(\eta) = \alpha \int_0^1 K(0, \eta)\, dy(\eta)$$
$$+ \int_0^1 \int_a^1 K(\xi, \eta) f_a(\xi)\, d\xi\, dy(\eta) < \alpha v + (1 - \alpha)v = v.$$

Reversing the order of integration, we obtain

$$\gamma \int_0^1 K(\xi, 0)\, dx(\xi) + \int_0^1 \int_a^1 K(\xi, \eta) g_a(\eta)\, d\eta\, dx(\xi)$$
$$= \gamma \int_0^1 K(\xi, 0)\, dx(\xi) + (1 - \gamma)v < v.$$

Therefore, when $\Phi(0) = M(0, 0)$,

$$\int_0^1 K(\xi, 0)\, dx(\xi) < v. \tag{5.4.12}$$

Since

$$\int_0^1 K(\xi, 0)\, dx(\xi)$$

is clearly a continuous monotonic function of $\Phi(0)$, (5.4.11) and (5.4.12) combined show that there exists a value S_0 for $\Phi(0)$ for which

$$\int_0^1 K(\xi, 0)\, dx(\xi) = v.$$

It remains to verify that when $\Phi(0) = S_0$,

$$\int_0^1 K(0, \eta)\, dy(\eta) = v.$$

This follows readily from

$$v = \int_0^1 \int_0^1 K(\xi, \eta)\, dx(\xi)\, dy(\eta)$$
$$= \alpha \int_0^1 K(0, \eta)\, dy(\eta) + \int_a^1 \int_0^1 K(\xi, \eta) f_a(\xi)\, dy(\eta)\, d\xi$$
$$= \alpha \int_0^1 K(0, \eta)\, dy(\eta) + (1 - \alpha)v.$$

▶LEMMA 5.4.9. Given the conditions of Lemma 5.4.4 and $L(0, 0) > \Phi(0) > S_0$, Players I and II have optimal strategies $x = (\alpha I_0, f_a)$ and $y = (\gamma I_0, g_a, \delta I_1)$, respectively; given the conditions of Lemma 5.4.4 and $S_0 > \Phi(0) > M(0, 0)$, Players I and II have optimal strategies $(\alpha I_0, f_a, \beta I_1)$ and $(\gamma I_0, g_a)$, respectively.

Proof. We now assume that the value a_0 of a used in the previous lemma is the smallest a for which $u(a, \alpha, 0) = w(a, \gamma, 0)$. Hence if $0 < a < a_0$, it follows from Lemma 5.4.4 that $u(a, \alpha, 0) > w(a, \gamma, 0)$. In order to prove the first assertion (Case F in Table 5–1), we first notice that if β is taken to be zero,

$$\alpha = \left(1 + \int_a^1 \phi_0\right)^{-1}.$$

Set

$$h(a) = \frac{1}{\alpha}\left(u(a, \alpha, 0) - \int_a^1 M(\xi, 0)f_a(\xi)\, d\xi\right).$$

By (5.4.5), $h(0) = L(0, 0)$ and by Lemma 5.4.8, $h(a_0) = S_0$. Therefore, since $h(a)$ is continuous if $S_0 < \Phi(0) < L(0, 0)$, there exists an a $(0 < a < a_0)$ for which $h(a) = \Phi(0)$. For this value of a, take $x = (\alpha I_0, f_a)$; then

$$\int_0^1 K(\xi, 0)\, dx(\xi) = \alpha\Phi(0) + \int_a^1 M(\xi, 0)f_a(\xi)\, d\xi = u(a, \alpha, 0).$$

Since $\beta = 0$, we also have

$$\int_0^1 K(\xi, 1)\, dx(\xi) = u(a, \alpha, 0).$$

If we take $\gamma = 0$ and $y = (g_a, \delta I_1)$, we get

$$\begin{aligned} u(a, \alpha, 0) &= \int_0^1\int_0^1 K(\xi, \eta)\, dx(\xi)\, dy(\eta) \\ &= \alpha\int_0^1 K(0, \eta)\, dy(\eta) + \int_a^1\int_0^1 K(\xi, \eta)f_a(\xi)\, dy(\eta)\, d\xi \\ &= \alpha\int_0^1 K(0, \eta)\, dy(\eta) + (1 - \alpha)w(a, 0, \delta). \end{aligned}$$

But

$$\int_0^1 K(0, \eta)\, dy(\eta) = \delta L(0, 1) + \int_a^1 L(0, \eta)g_a(\eta)\, d\eta < w(a, 0, \delta).$$

Hence $u(a, \alpha, 0) < w(a, 0, \delta)$. On the other hand, it follows from the choice of a_0 in Lemma 5.4.8 and $a < a_0$ that $u(a, \alpha, 0) > w(a, \gamma, 0)$. It is then clear that we can find a strategy $(\bar{\gamma} I_0, g_a, \bar{\delta} I_1) = \bar{y}$ for which $w(a, \gamma, \delta) = u(a, \alpha, 0)$, where a is the same as before and $0 < \bar{\gamma} < \gamma$, $0 < \bar{\delta} < \delta$. For simplicity we denote the new strategy also by $y = (\gamma I_0, g_a, \delta I_1)$. We shall now verify that x, y are optimal and that $v = u(a, \alpha, 0)$ is the value of the game. The equation

$$\int_0^1 K(\xi, \eta)\, dx(\xi) \geq u(a, \alpha, 0) = v$$

has already been established, and (5.4.8) shows that

$$\int_0^1 K(\xi, \eta)\, dy(\eta) \leq w = v$$

for $0 < \xi < 1$. Finally,

$$\int K(1, \eta)\, dy(\eta) = \gamma M(1, 0) + \delta \Phi(1) + \int_a^1 M(1, \eta) g_a(\eta)\, d\eta$$

$$\leq \gamma M(1, 0) + \delta L(1, 1) + \int_a^1 M(1, \eta) g_a(\eta)\, d\eta = w = v,$$

and since

$$\int_0^1 \int_0^1 K(\xi, \eta)\, dx(\xi)\, dy(\eta) = v,$$

it follows that

$$v = \alpha \int_0^1 K(0, \eta)\, dy(\eta) + (1 - \alpha) v.$$

Hence

$$\int_0^1 K(0, \eta)\, dy(\eta) = v,$$

and the proof is complete.

A similar argument establishes the optimal strategies for Case H in Table 5–1, completing the description of the optimal strategies for the Group 3 kernels.

The only kernels that have not been discussed are those for which

$$L(\xi, \xi) > M(\xi, \xi) \qquad (0 \leq \xi \leq 1),$$

$$\lambda(0) < 1 \quad \text{and} \quad \mu(0) < 1,$$

and

$$L(0, 1) \geq M(1, 0),$$

that is, Group 4. The necessary arguments in this case are modifications of those employed in analyzing Group 3 kernels; the details are omitted. We remind the reader that the uniqueness of the optimal strategies displayed in Table 5–1 is the subject of Problems 16–19.

5.5 The silent duel with general accuracy functions. As an application of the general theory, we shall solve the silent duel with arbitrary accuracy functions. Consider two opponents, each allowed to fire once. Player I fires with accuracy $P_1(\xi)$ and Player II with accuracy $P_2(\eta)$. We assume that P_1 and P_2 are continuous; that $P_1'(\xi) > 0$, $P_2'(\eta) > 0$ in $[0, 1]$; and that $P_1(0) = P_2(0) = 0$, $P_1(1) = P_2(1) = 1$. The value to Player I is $+1$ if he hits his opponent, -1 if he himself is hit, and 0 in case both fail or both succeed. The expected return to Player I is

$$K(\xi, \eta) = \begin{cases} P_1(\xi) - P_2(\eta) + P_1(\xi)P_2(\eta) & \xi < \eta, \\ P_1(\xi) - P_2(\xi) & \xi = \eta, \\ P_1(\xi) - P_2(\eta) - P_1(\xi)P_2(\eta) & \xi > \eta. \end{cases}$$

This kernel clearly satisfies conditions (b) and (c), p. 110.

The conditions of Lemma 5.2.2 are satisfied by $\xi_0 = 0$. Hence there exist a and b such that $\lambda(a) = 1$ and $\mu(b) = 1$. We proceed to determine a and f satisfying $T_a f = f$. The equation becomes

$$\begin{aligned} f(t) &= \int_a^t \left\{\frac{-P_2'(t)[-1 + P_1(\xi)]}{2P_1(t)P_2(t)}\right\} f(\xi)\, d\xi \\ &\quad + \int_t^1 \left\{\frac{-P_2'(t)[-1 - P_1(\xi)]}{2P_1(t)P_2(t)}\right\} f(\xi)\, d\xi \\ &= \frac{P_2'(t)}{2P_1(t)P_2(t)} \left[\int_a^1 f(\xi)\, d\xi - \int_a^t P_1(\xi) f(\xi)\, d\xi + \int_t^1 P_1(\xi) f(\xi)\, d\xi\right]. \end{aligned} \tag{5.5.1}$$

If we normalize f so that

$$\int_a^1 f(\xi)\, d\xi = 1$$

and define $h(\xi) = P_1(\xi)f(\xi)$, the last equation may be written in the form

$$\frac{2h(\xi)}{1 - \int_a^\xi h(t)\, dt + \int_\xi^1 h(t)\, dt} = \frac{P_2'(\xi)}{P_2(\xi)}. \tag{5.5.2}$$

Integrating, we obtain

$$1 - \int_a^\xi h(t)\, dt + \int_\xi^1 h(t)\, dt = \frac{k}{P_2(\xi)}.$$

Differentiation with respect to ξ yields

$$2h(\xi) = 2f(\xi)P_1(\xi) = \frac{kP_2'(\xi)}{[P_2(\xi)]^2},$$

and hence

$$f(\xi) = \frac{k_1 P_2'(\xi)}{P_1(\xi)[P_2(\xi)]^2}, \tag{5.5.3}$$

where $2k_1 = k$.

The function f is a solution of the differential equation obtained by differentiating (5.5.2). We now determine the number a that will make f a solution of the original integral equation. Inserting (5.5.3) in (5.5.2), we obtain

$$\frac{k_1}{P_2(\xi)} = \frac{1}{2}\left[1 + k_1\left\{\frac{2}{P_2(\xi)} - \frac{1}{P_2(a)} - \frac{1}{P_2(1)}\right\}\right] \quad \text{or} \quad \frac{1}{k_1} = 1 + \frac{1}{P_2(a)}.$$

The last formula, in conjunction with the normalization requirement on f, leads to a relation fulfilled by a, viz.,

$$\frac{1}{k_1} = 1 + \frac{1}{P_2(a)} = \int_a^1 \frac{P_2'(\xi)\,d\xi}{P_1(\xi)[P_2(\xi)]^2}. \tag{5.5.4}$$

The value a lying in the unit interval satisfying (5.5.4) is clearly the same as the value a for which $\lambda(a) = 1$.

A similar argument shows that the value b for which $\mu(b) = 1$ and the associated normalized eigenfunction g are determined from the equations

$$g(\eta) = \frac{k_2 P_1'(\eta)}{P_2(\eta)[P_1(\eta)]^2} \tag{5.5.5}$$

and

$$\frac{1}{k_2} = 1 + \frac{1}{P_1(b)} = \int_b^1 \frac{P_1'(\eta)\,d\eta}{P_2(\eta)[P_1(\eta)]^2}. \tag{5.5.6}$$

It should be pointed out that the equations that determine a and b have unique solutions in the interval $0 \le \xi \le 1$; to demonstrate the validity of this claim, it is enough to show that

$$r(z) = \int_z^1 \frac{P_1'(\xi)}{P_2(\xi)P_1^2(\xi)}\,d\xi - \frac{1}{P_1(z)}$$

strictly decreases from ∞ to -1 as z traverses the unit interval. Clearly,

$$r'(z) = -\frac{1 - P_2(z)}{P_2(z)}\,\frac{P_1'(z)}{P_1^2(z)} < 0,$$

and on direct examination

$$r(0) = +\infty \qquad \text{and} \qquad r(1) = -1.$$

The procedure for obtaining the optimal strategies is outlined in the discussion of Lemma 5.4.2. We follow its prescriptions.

First evaluate a and b by means of (5.5.4) and (5.5.6) and single out the larger of the two numbers. (If a and b agree, then the optimal strategy is a density for each player. This case is dismissed as the simplest to discuss and left for the reader.) Suppose for definiteness in what follows that a is the larger of the two. Then $f(\xi)$ as described in (5.5.3) is indeed an optimal strategy for Player I. To obtain an optimal strategy for Player II, set $g_a = (I - U_a)^{-1}\delta q_1$; thus g_a is a solution of the nonhomogeneous equation

$$(I - U_a)g_a = \delta q_1.$$

Writing this out explicitly, we have

$$g_a(\eta) - \int_a^\eta \frac{P_1'(\eta)[1 - P_2(t)]}{2P_1(\eta)P_2(\eta)} g_a(t)\,dt - \int_\eta^1 \frac{P_1'(\eta)[1 + P_2(t)]}{2P_1(\eta)P_2(\eta)} g_a(t)\,dt = \delta\,\frac{2P_1'(\eta)}{2P_1(\eta)P_2(\eta)}.$$

A rearrangement of the terms of this expression, utilizing the normalization

$$\int_a^1 g_a(\eta)\,d\eta = 1 - \delta$$

as in Lemma 5.4.2, gives

$$g_a(\eta) = \frac{P_1'(\eta)}{2P_1(\eta)P_2(\eta)} \left(2\delta + 1 - \delta - \int_a^\eta P_2(t)g_a(t)\,dt + \int_\eta^1 P_2(t)g_a(t)\,dt\right).$$

Comparison of this equation with (5.5.1) reveals instantly that

$$g_a(\eta) = \frac{kP_1'(\eta)}{P_2(\eta)[P_1(\eta)]^2}.$$

It is a remarkable phenomenon that the density part of the optimal strategy for Player II differs from the solution of the equation $U_b g = g$ only in the normalizing constant k, and in the domain of definition which is now the interval $[a, 1]$. If we substitute the explicit expression of g_a in the integral equation and integrate, we find that k and δ are connected by the relation

$$k\left(\frac{1}{P_1(a)} + 1\right) = 1 + \delta.$$

The last equation, combined with the normalizing equation

$$\int_a^1 \frac{kP_1'(t)\,dt}{P_2(t)[P_1(t)]^2} = 1 - \delta,$$

completely determines the constants k and δ. An appeal to Lemma 5.4.2 or a direct verification convinces us that (f_a) and $(g_a, \delta I_1)$ are optimal strategies. The value of the game as computed from (5.4.1) is

$$v = u(a) = \frac{1 - 3P_2(a)}{P_2(a)\displaystyle\int_a^1 \frac{P_2'(t)\,dt}{P_1(t)P_2^2(t)}}.$$

As a special example we take $P_1(\xi) = \xi$ and $P_2(\xi) = \xi^n$ $(n \geq 1)$. If we substitute these functions into (5.5.4) and (5.5.6), integrate, and simplify, we find that a and b are the roots of the equations

$$(2n + 1)a^{n+1} + (n + 1)a - n = 0,$$

$$(n + 2)b^{n+1} + (n + 1)b^n - 1 = 0.$$

We have already shown that each of these equations must have exactly one root between 0 and 1. Furthermore, if $n > 1$ then in the interval $0 \leq \xi \leq 1$ the graph of the first equation always lies below that of the second equation, the difference between the two being

$$(n - 1)\xi^{n+1} - (n + 1)\xi^n + (n + 1)\xi - n + 1.$$

This quantity is negative at $\xi = 0$. Moreover, by the Descartes rule of signs we know that it has at most three positive roots, which are clearly located at $\xi = 1$ as a triple root. Hence it follows that the root of the first equation is greater than the root of the second equation. If $n = 1$, both roots are equal and $a = 1/3$ (Example 2, Section 5.1).

It might be of some interest to particularize the accuracy functions assigned to the two players. Take $P_1(\xi) = \xi$ and $P_2(\xi) = P(\xi)$. Let α be the positive probability of Player I's firing at time 1, and let β be the corresponding probability for Player II; v, as usual, denotes the value of the game. The interval containing the spectrum of the optimal strategies is given by $[a, 1]$.

In the following tabulation, the P's are arranged in descending order; that is to say, $P(\xi)$ in (b) is less than $P(\xi)$ in (a) for $0 \leq \xi \leq 1$, etc. There are two exceptions to this rule: $P(\xi)$ in (c) is less than $P(\xi)$ in (b) only for $3 - \sqrt{6} < \xi < 1$, and $P(\xi)$ in (f) intersects each of the other $P(\xi)$ once.

	$P(\xi)$	a	α	β	v
(a)	ξ	0.333	0	0	0
(b)	$\frac{2\xi^2}{1+\xi^2}$	0.409	0	0.0764	0.1021
(c)	$\frac{\xi(3-\xi)}{2(2-\xi)}$	0.372	0.0063	0	0.0838
(d)	$\frac{\xi}{2-\xi}$	0.414	0	0	0.172
(e)	ξ^2	0.481	0	0.0729	0.2481
(f)	$\frac{2\xi^3}{4\xi^2-3\xi+1}$	0.415	0	0.1741	0.028

5.6 Problems.

1. Solve the silent duel in which each player has the same continuous strictly monotone-increasing accuracy function $P(\xi)$, where $P(0) = 0$ and $P(1) = 1$. [*Hint:* Transform the game into Example 2 of Section 5.1 by introducing new variables $\tilde{\xi} = P(\xi)$, $\tilde{\eta} = P(\eta)$.]

2. Find the critical point a and the mass β for the silent duels with $P_1(t) = t$, $P_2(t) = t^3$ and $P_1(t) = t$, $P_2(t) = t^4$.

3. Verify the table of solutions given above.

4. A beautiful girl A is to arrive at a railroad station at a time t $(0 \leq t \leq 1)$ chosen at random from a uniform distribution. Each of two players, Player I and Player II, goes to the station once and must leave immediately. If A is at the station when either player arrives, then A leaves with this player. Whoever leaves with A wins one unit from his opponent. If neither leaves with A the pay-off is zero. When should the players arrive at the station? Find the optimal strategies. (One could provide an interpretation of this problem in terms of the book publishing example discussed on p. 31.)

5. Solve the silent duel with equal accuracy function

$$(P_1(t) = P_2(t) = t)$$

assuming that each player has a probability p_i $(i = 1, 2)$ of having a bullet.

6. Solve Problem 5 on the assumption that the value to Player I is α if he alone survives, β if Player II alone survives, $\alpha + \beta$ if no one survives, 0 if both survive. Assume $\alpha > 0$ and $\beta < 0$.

7. Consider the game

$$K(\xi, \eta) = \begin{cases} \xi + \eta & \xi \leq \eta, \\ \xi - \eta & \xi > \eta. \end{cases}$$

Solve this game by the methods of this chapter, even though it does not fully qualify as a game of timing.

8. A large industrial concern is competing with a small company for a large contract. At time t, the company that spends m dollars for advertising has a chance $mp(t)$ of getting the contract where $p'(t) > 0$ for all t and $p(0) = 0$. The companies have l and s dollars for advertising. We assume $lp(t) \leq 1$ and $sp(t) \leq 1$ for all t. Each must plan its campaign at time $t = 0$, and must spend its full advertising budget at one time. The time interval over which the duel occurs is normalized to be in [0, 1]. The value to the large company is +1 if he alone gets the contract. If neither or both get the contract, the pay-off is zero. If the small company gets the contract, the value to the large company is −1. Set up the game and solve it.

9. A number t is chosen at random from a uniform distribution on (0, 1). Simultaneously the players choose numbers ξ and η, $(0 \leq \xi, \eta \leq 1)$ respectively. The pay-off to Player I is $K(\xi, \eta)$, where

$$K(\xi, \eta) = \begin{cases} 1 & \xi + t + \eta \leq \frac{3}{2}, \\ -1 & \xi + t + \eta > \frac{3}{2}. \end{cases}$$

Find the optimal strategies.

10. Describe the solution for the silent duel with monotone-increasing accuracy functions $P_1(t)$ and $P_2(t)$, assuming that $P_1(0) > 0$, $P_2(0) > 0$, and $P_1(1) = P_2(1) = 1$.

11. Assuming that all the inequalities in condition (c) of p. 110 are reversed, describe how the form of the optimal strategies is altered. (*Hint:* Replace ξ by $1 - \xi$ and η by $1 - \eta$ as new variables.)

12. Show that in the general game of timing of class II the point 1 cannot be a jump for the optimal strategies of both players unless $\{1, 1\}$ is a saddle point.

13. Let $A(\xi, \eta)$ possess continuous second partials defined on the triangle $0 \leq \xi \leq \eta \leq 1$, and such that $A_\xi(\xi, \eta) > 0$ and $A_\eta(\xi, \eta) < 0$. Let

$$K(\xi, \eta) = \begin{cases} A(\xi, \eta) & 0 \leq \xi < \eta \leq 1, \\ 0 & \xi = \eta, \\ -A(\eta, \xi) & 0 \leq \eta < \xi \leq 1. \end{cases}$$

Show that if $A(1, 1) \leq 0$, then I_1 is optimal for both players. Show also that if $A(0, 1) \geq 0$, then I_0 is optimal for both players.

14. Show that if $A(1, 1) > 0$ and $A(0, 1) < 0$, the optimal strategies for the kernel of Problem 13 are of the form $(\alpha I_0, f_a)$, with $\alpha \geq 0$ and f_a a continuous density on some interval $a \leq t \leq 1$.

15. Find the integral equations satisfied by the density f_a of Problem 14, and show that if

$$A(\xi, \eta) = \sum_{i=1}^{n} p_i(\xi)q_i(\eta),$$

then the integral equations can be replaced by a system of differential equations.

The next four problems are concerned with uniqueness results for games of timing of class II.

16. Prove that if x represents any optimal strategy for Player I which is not a component of a saddle point, then x is continuous on the interior of the unit interval. (*Hint:* Make use of the discontinuity of $K(\xi, \eta)$ on the diagonal, and of the fact that Player II has an optimal strategy of the form $(\gamma I_0, g_a, \delta I_1)$ which is an equalizer on the interval $[a, 1]$.)

17. Using Problem 16, prove that if $x(\xi)$ is an optimal strategy for Player I, then $x(\xi)$ is absolutely continuous on $a < \xi < 1$ and $x'(\xi)$ is continuous on this interval.

18. Using the results of Problems 16 and 17, prove that the optimal strategy for Player I is unique.

19. For the game discussed in the previous three problems, prove that the optimal strategy for Player II is unique.

20. In a game of timing of class II, assume that $\Phi(0)$ and $\Phi(1)$ do not lie between the respective values of L and M but K satisfies the other conditions of Section 5.2. Prove that the game has a value, although attainable optimal strategies may not exist.

21. Show that if $K(\xi, \eta)$ is a game of timing of class II satisfying (5.0.1), with $L_{\xi\xi}(\xi, \eta) \leq 0$, $M_{\xi\xi}(\xi, \eta) \leq 0$, $L_{\xi}(t, t) > M_{\xi}(t, t)$, and $L'(t, t) > M'(t, t)$, then the density part of the solution for Player II is a continuous decreasing function.

NOTES AND REFERENCES TO CHAPTER 5

The study of games of tactical duels began in 1948. The RAND Corporation at that time collected a team of mathematicians, statisticians, economists, and social scientists to analyze the meaning and structure of "the uncertainties of war" and to construct a blueprint for the optimal future operation of the economies of attack and defense in war. One of the by-products of the study of tactical warfare, a component of the program, was the solution of numerous examples of games of timing. The name "games of timing" for this class of tactical duels was actually coined late in 1950 by Shiffman [222], who recognized the wide scope of their possible applications. Two other members of this group who were instrumental in formulating and solving several versions of tactical duels were Blackwell and Girschick. As a result of this conference, it was felt in some quarters that the prevalent view of military tactical maneuvers should be revised in the light of considerations suggested by the solutions of certain games of timing.

This was the origin of an interesting, rich class of games requiring advanced applications of the theory of positive linear transformations for their complete solution. The early studies of these games pointed the way in turn to a broader class of games on the unit square whose kernels (or their derivatives) have some sort of discontinuity along the diagonal, a characteristic that distinguishes games of timing.

§ 5.1. The individual examples of games of timing that have been solved during the brief history of this class of games are too numerous to recount. Some of the more prolific contributors to the theory of games of timing include Blackwell, Girschick, Shapley, Bellman, Glicksberg, Belzer, and others. The first three examples described in this section are due to Blackwell and Girschick [30, Chap. 2]. The fourth is from Bellman and Girschick [21].

§ 5.2. The general symmetric one-action game of timing was solved by Shiffman [222], who showed that the solution, which is the same for each player, can be obtained by solving a single integral equation of the second kind. Karlin [134] extended these methods in order to determine the solution for the general game of timing. The assumption of differences in the two players' resources and opportunities leads to new solutions which cannot be surmised from the study of symmetric kernels. Whereas the symmetric game of timing involves four categories of solutions, the unsymmetrical case entails fourteen. The method of solving games of timing in both cases appeals fundamentally to the theory of positive integral transformations.

§ 5.3. The relevant theory of linear integral transformations with positive kernel is summarized in this section. This theory has a long history. Frobenius [89] investigated the discrete case, which corresponds to a positive matrix transformation acting on a finite-dimensional space into itself. Some integral extensions of this were first given by Jentzsch. An abstract treatment of these ideas, dealing with linear operators in Banach spaces which leave a cone invariant, has been presented by Krein and Rutman [153], who also furnish a substantial bibliography on the subject.

The characterization of the largest eigenvalue of a positive linear operator, as described in the proof of Theorem 5.3.3, may be new. It is due jointly to Bohnenblust and Karlin. (See also Section 8.2 of Vol. I.)

§ 5.4. The proofs of this section are amplifications of those contained in Karlin [134].

§ 5.5. The solution of the general silent duel with arbitrary accuracy functions is due to Blackwell [28]. He solves this game by employing a suitable extension of the equalizer strategy technique (see Section 2.5). In our discussion, this example emerges as an application of the general theory.

The tabulation on p. 140 is due to Belzer [24].

§ 5.6. Problem 6 was suggested by Blackwell and Girschick [30]. Problems 13–15 are due to Shiffman [222].

CHAPTER 6

GAMES OF TIMING (Continued)

6.1 Games of timing of class I. In the previous chapter we considered games of timing in which only partial information is available to each player. For instance, in the silent duel neither player knows whether he has been fired at unless he is hit. Because of this lack of information, one can expect that the players will have to randomize to some extent within the set of pure strategies, and indeed we have seen this to be the case. On the other hand, in the noisy duel each player knows when his opponent fires. The optimal strategies are pure strategies for each player, and the kernel of the game is of the form

$$K(\xi, \eta) = \begin{cases} l(\xi) & \xi < \eta, \\ \phi(\xi) & \xi = \eta, \\ m(\eta) & \xi > \eta, \end{cases} \tag{6.1.1}$$

where $c \geq l'(\xi) \geq \delta > 0$ and $-c \leq m'(\eta) \leq -\delta < 0$, $l(1) > m(1)$ and $l(0) < m(0)$, $\phi(\xi)$ is bounded but otherwise arbitrary, and $l(\xi)$ and $m(\eta)$ are assumed to have continuous second derivatives. Henceforth, any game of the type (6.1.1) will be called a game of timing of complete information or, in accordance with our previous terminology, a game of timing of class I.

The monotonicity properties of $l(\xi)$ and $m(\eta)$ obviously imply that there exists a unique value a_0 with $0 < a_0 < 1$ and $m(a_0) = l(a_0)$. The principal theorem concerning games of timing of class I follows.

▶ THEOREM 6.1.1. *The game* (6.1.1) *has a value* $v = l(a_0) = m(a_0)$, *and the optimal strategies* x^* *and* y^* *are described as follows. If* $\phi(a_0) > v$, *then* $x^*(\xi) = x_{a_0}$, *and Player II has no attainable optimal strategy. An* ϵ*-effective optimal strategy for Player II has the form of a density whose spectrum is a sufficiently small interval just to the right of* a_0. *If* $\phi(a_0) = v$, *then* $x^* = y^* = I_{a_0}$ (*a degenerate distribution concentrating at* a_0). *If* $\phi(a_0) < v$, *then* $y^*(\eta) = y_{a_0}$, *and Player I has no attainable optimal strategy. An* ϵ*-effective optimal strategy for Player I has the form of a density spread over an arbitrarily small interval to the right of* a_0.

The proofs of these statements will be discussed in Section 6.3. The idea is to treat the kernel (6.1.1) as a limiting case of kernels having the structure of games of timing of class II. Since the limiting procedure is handicapped by the discontinuities of the kernel along the diagonal, moderate care must be exercised in the analysis.

The conditions $l'(\xi) > 0$ and $m'(\eta) < 0$ are natural for games of timing (see the following section). The conditions $l(1) > m(1)$ and $l(0) < m(0)$, however, may appear offhand to be rather arbitrary. Hence, in order to cover some of the other possibilities, we describe the solutions with no restrictions imposed on the values of the kernel at the end points.

Let $K(\xi, \eta)$ satisfy the conditions of Theorem 6.1.1 except that instead of requiring $l(1) > m(1)$ and $l(0) < m(0)$ we assume that $\phi(0)$ and $\phi(1)$ lie between the respective values of l and m; then the optimal strategies are given as follows:

(a) If $l(1) < m(1)$, the point $\{1, 1\}$ is clearly a saddle point and I_1 is optimal for both players.

(b) If $l(0) > m(0)$, the point $\{0, 0\}$ is a saddle point and $x = y = I_0$.

(c) If neither (a) nor (b) is satisfied, the kernel fulfills the conditions of Theorem 6.1.1.

In cases (a), (b), and (c), if no conditions are imposed on the relative size of $\phi(0)$ and $\phi(1)$, it can be shown that the game defined by $K(\xi, \eta)$ need not have a value.

6.2 Examples. *Example 1.* We consider a noisy duel between two opponents whose respective accuracies are given by $P_1(\xi)$ and $P_2(\eta)$, defined on the interval $[0, 1]$. Both functions are assumed to be strictly increasing with continuous positive derivatives, and $P_1(0) = P_2(0) = 0$, $P_1(1) = P_2(1) = 1$. At the end of the duel Player I will benefit by an amount w as follows: If only Player I survives, $w = \alpha$; if only Player II survives, $w = \beta$; if both survive, $w = 0$; if neither survives, $w = \gamma$. We assume $\alpha > \beta$. The kernel $K(\xi, \eta)$ is evaluated as the expected value of w. Hence

$$K(\xi, \eta) = \begin{cases} (\alpha - \beta)P_1(\xi) + \beta & \xi < \eta, \\ \alpha P_1(\xi) + \beta P_2(\xi) + (\gamma - \alpha - \beta)P_1(\xi)P_2(\xi) & \xi = \eta, \\ \alpha - (\alpha - \beta)P_2(\eta) & \xi > \eta. \end{cases}$$

It follows from the assumption $\alpha > \beta$ that $K(\xi, \eta)$ satisfies the conditions of Theorem 6.1.1. The number a_0 is the solution of the equation

$$(\alpha - \beta)P_1(\xi) + \beta = \alpha - (\alpha - \beta)P_2(\xi) = v.$$

Hence a_0 is the solution of the equation $P_1(\xi) + P_2(\xi) = 1$, and

$$\begin{aligned} v = (\alpha - \beta)P_1(a_0) + \beta &= (\alpha - \beta)P_1(a_0) + \beta[P_1(a_0) + P_2(a_0)] \\ &= \alpha P_1(a_0) + \beta P_2(a_0). \end{aligned}$$

Player I will have an optimal strategy if and only if $\phi(a_0) \geq v$ (by Theorem 6.1.1) or

$$\alpha P_1(a_0) + \beta P_2(a_0) + (\gamma - \alpha - \beta)P_1(a_0)P_2(a_0) \geq \alpha P_1(a_0) + \beta P_2(a_0).$$

Equivalently, x_{a_0} is optimal for Player I if and only if $\gamma - \alpha - \beta \geq 0$, and y_{a_0} is optimal for Player II if and only if $\gamma - \alpha - \beta \leq 0$.

Example 2. We consider a similar noisy duel with $\alpha = 1$, $\beta = -1$, and $\gamma = 0$. However, we no longer require that P_1 and P_2 be monotonic functions; we require only that they be continuous. In these circumstances, if Player I fires at a time ξ before Player II has fired, Player II will not necessarily wait until $\eta = 1$ to fire; he will obviously fire when his accuracy is a maximum as η ranges in the interval $[\xi, 1]$. The optimal moment for Player II to fire is any time θ in $[\xi, 1]$ at which

$$P_2(\theta) = M_2(\xi) = \max_{t \geq \xi} P_2(t).$$

The expected return to Player I is

$$K(\xi, \eta) = \begin{cases} P_1(\xi) - [1 - P_1(\xi)]M_2(\xi) & \xi < \eta, \\ P_1(\xi) - P_2(\xi) & \xi = \eta, \\ -P_2(\eta) + [1 - P_2(\eta)]M_1(\eta) & \xi > \eta, \end{cases}$$

where

$$M_1(\eta) = \max_{t \geq \eta} P_1(t).$$

Simplifying notation,

$$K(\xi, \eta) = \begin{cases} f(\xi) & \xi < \eta, \\ \phi(\xi) & \xi = \eta, \\ g(\eta) & \xi > \eta, \end{cases}$$

where $f(\xi) = P_1(\xi) - [1 - P_1(\xi)]M_2(\xi)$, etc.

To solve this game it is convenient to define two auxiliary functions

$$F(\xi) = \max_{t \leq \xi} f(t), \qquad G(\eta) = \min_{t \leq \eta} g(t).$$

Note that $F(\xi)$ is increasing and $G(\eta)$ is decreasing, but not necessarily strictly. We distinguish three possibilities.

Case 1. There exists a value a_0 $(0 \leq a_0 \leq 1)$ for which $F(a_0) = G(a_0)$. We claim that the number $v = F(a_0) = G(a_0)$ is the value of the game whose kernel is $K(\xi, \eta)$ in the sense that inf sup = sup inf. The proof of this assertion proceeds as follows.

Let b be the smallest ξ for which $F(\xi) = F(a_0)$, and let c be the smallest η for which $G(\eta) = G(a_0)$. Clearly, $f(b) = F(b) = F(a_0)$ and $g(c) = G(c) = G(a_0)$. We shall now evaluate the yield to Player I when he adopts the strategy x_b and Player II uses a pure strategy η:

(i) If $\eta < b$, then

$$K(b, \eta) = g(\eta) \geq \min_{t \leq \eta} g(t) = G(\eta) \geq G(b) \geq G(a_0) = v.$$

(ii) If $\eta = b$, then $K(b, b) = \phi(b)$.
(iii) If $\eta > b$, then $K(b, \eta) = f(b) = F(a_0) = v$.

Similarly, if Player II adopts the strategy y_c and Player I uses the pure strategy ξ, the yield to Player I is as follows:

(i) If $\xi < c$, then

$$K(\xi, c) = f(\xi) \leq \max_{t \leq \xi} f(t) = F(\xi) \leq F(c) \leq F(a_0) = v.$$

(ii) If $\xi = c$, then $K(c, c) = \phi(c)$.
(iii) If $\xi > c$, then $K(\xi, c) = g(c) = G(a_0) = v$.

It is immediately clear that if $\phi(b) \geq v$ and $\phi(c) \leq v$, the pure strategies x_b and y_c are optimal and v is the value. These solutions are, in general, not unique.

For other values of $\phi(b)$ and $\phi(c)$, attainable optimal strategies need not exist. However, as indicated in Theorem 6.1.1, ϵ-effective strategies can be constructed. In fact, it is seen that by using a uniform distribution over a small interval $[b - \delta, b + \delta]$ Player I can secure at least $v - \epsilon$ uniformly in η, for any $\epsilon > 0$. A similar analysis shows that Player II can prevent Player I from obtaining a yield larger than $v + \epsilon$. Thus v is the value.

Case 2. $G(t) < F(t)$ $(0 \leq t \leq 1)$. Then $g(0) < f(0)$, and the point $\{0, 0\}$ is a saddle point if and only if $g(0) \leq \phi(0) \leq f(0)$. Explicitly,

$$K(0, \eta) = \begin{cases} \phi(0) & \eta = 0, \\ f(0) \geq \phi(0) & \eta > 0, \end{cases}$$

and

$$K(\xi, 0) = \begin{cases} \phi(0) & \xi = 0, \\ g(0) \leq \phi(0) & \xi > 0. \end{cases}$$

If such a saddle point exists, it is clear that $v = \phi(0)$, and I_0 (the degenerate distribution concentrated at zero) is optimal for both players.

Case 3. $G(t) > F(t)$ $(0 \leq t \leq 1)$. Then

$$g(1) \geq \min_{t \leq 1} g(t) = G(1) > F(1) = \max_{t \leq 1} f(t) \geq f(1).$$

The same methods yield that $\{1, 1\}$ is a saddle point if and only if $\phi(1)$ lies between $G(1)$ and $F(1)$. Then $v = \phi(1)$, and I_1 is optimal for both players.

***6.3 Proof of Theorem 6.1.1.** For clarity of exposition, we consider two lemmas.

▶LEMMA 6.3.1. The game $K(\xi, \eta)$ of (6.1.1) has a value v, in the sense that

$$\inf_y \sup_x \int K(\xi, \eta)\, dx(\xi)\, dy(\eta) = v = \sup_x \inf_y \int K(\xi, \eta)\, dx(\xi)\, dy(\eta).$$

Proof. Consider the family of games with kernels defined by

$$K_n(\xi, \eta) = K(\xi, \eta) + \frac{1}{n} K^*(\xi, \eta),$$

where $K^*(\xi, \eta)$ satisfies the conditions of Theorem 5.2.1. Explicitly,

$$K_n(\xi, \eta) = \begin{cases} l(\xi) + \dfrac{1}{n} L^*(\xi, \eta) & \xi < \eta, \\ \phi(\xi) + \dfrac{1}{n} K^*(\xi, \xi) & \xi = \eta, \\ m(\eta) + \dfrac{1}{n} M^*(\xi, \eta) & \xi > \eta, \end{cases}$$

with $n = 1, 2, 3, \ldots$. Clearly, $K_n(\xi, \eta)$ possesses the same characteristics as $K^*(\xi, \eta)$, and the optimal strategies are as described in Table 5–1. The assumptions on $l(\xi)$ and $m(\eta)$ readily imply that for each n sufficiently large there exists a value ξ_n $(0 < \xi_n < 1)$ such that

$$l(\xi_n) + \frac{1}{n} L^*(\xi_n, \xi_n) = m(\xi_n) + \frac{1}{n} M^*(\xi_n, \xi_n)$$

---

* Starred sections are advanced discussions that may be omitted at first reading without loss of continuity.

and $\xi_n \to a_0$. Therefore, the only cases of Table 5–1 (p. 117) that may occur are Cases B, C, and D. In these cases condition (b) of p. 110 is not relevant. The optimal strategies for the game $K_n(\xi, \eta)$ are of the form $(f_{a_n}, \beta_n I_1)$, $(g_{a_n}, \delta_n I_1)$, where $\xi_n < a_n < 1$ and either β_n or δ_n is zero. Thus for an infinite number of indices n, either β_n or δ_n is zero; for definiteness let us suppose that $\beta_n = 0$ for such indices. Then, by restricting ourselves to an appropriate subsequence, we may assume that $\beta_n = 0$, $a_n \to a$, and $v_n \to v$, where v_n is the value of the game $K_n(\xi, \eta)$ and a and v are the limit points of the corresponding sequences.

We now verify that this number v satisfies the conditions of the lemma. Since $v_n \to v$, for any $\epsilon > 0$ there exists $n_1(\epsilon)$ such that for all $n > n_1(\epsilon)$

$$\int K_n(\xi, \eta)\, dx_n(\xi) \geq v_n \geq v - \frac{\epsilon}{2}\cdot$$

Also, since $K_n(\xi, \eta)$ converges uniformly to $K(\xi, \eta)$, for $n \geq n_2(\epsilon)$

$$\int [K(\xi, \eta) - K_n(\xi, \eta)]\, dx_n(\xi) \geq -\frac{\epsilon}{2}\cdot$$

Hence for n sufficiently large, $\int K(\xi, \eta)\, dx_n(\xi) \geq v - \epsilon$, and similarly $\int K(\xi, \eta)\, dy_n(\eta) \leq v + \epsilon$. But the last two equations yield

$$\sup_x \inf_y \int K(\xi, \eta)\, dx(\xi)\, dy(\eta) \geq v \geq \inf_y \sup_x \int K(\xi, \eta)\, dx(\xi)\, dy(\eta).$$

Since the opposite inequality (inf sup $\geq$ sup inf) always holds, the lemma is established.

▶ LEMMA 6.3.2. $v = m(a_0) = l(a_0)$.

Proof. The optimal strategies $x_n = (f_{a_n})$ of the game $K_n(\xi, \eta)$ satisfy the integral equation (5.2.2). This integral equation after some easy manipulations becomes

$$m'(\eta)\int_\eta^1 f_{a_n}(\xi)\, d\xi + [l(\eta) - m(\eta)]f_{a_n}(\eta) + \rho_n(\eta) + \lambda_n(\eta) f_{a_n}(\eta) = 0, \tag{6.3.1}$$

where $\rho_n(\eta)$ and $\lambda_n(\eta)$ tend uniformly to zero as $n \to \infty$. By appealing to Helly's selection theorem (Section C.3) we may assume (again replacing our sequence by a subsequence if necessary) that there exists a distribution x_0 for which $x_n(\xi) \to x_0(\xi)$ for every continuity point ξ of x_0. Consequently, by integrating (6.3.1) over the range $[\eta, 1]$, where

$$1 > \eta \geq a + \delta \quad (\delta > 0) \qquad \text{and} \qquad a \geq a_0,$$

we get for n sufficiently large

$$\int_\eta^1 \frac{-m'(t)}{l(t) - m(t)} [1 - x_n(t)]\, dt = 1 - x_n(\eta) + \epsilon_n(\eta),$$

where $\epsilon_n(\eta)$ tends uniformly to zero.

Since $\xi_n < a_n < 1$ and $\xi_n \to a_0$, it follows that $a_0 \leq a$; and therefore $m'(t)/[l(t) - m(t)]$ is bounded for $t \geq a + \delta$. Hence

$$1 - x_0(\eta) = \int_\eta^1 \frac{-m'(t)}{l(t) - m(t)} [1 - x_0(t)]\, dt$$

at every continuity point η of x_0. This last observation, combined with a standard argument, implies that $1 - x_0(\eta) = 0$ for $1 > \eta \geq a + \delta$. We observe first that

$$1 - x_0(\eta) \leq M \int_\eta^1 [1 - x_0(t)]\, dt \leq M(1 - \eta)$$

for all η which are points of continuity of x_0. Iteration of this inequality (which is permissible since the points of discontinuity of x_0 form a set of measure zero) results in

$$1 - x_0(\eta) \leq \frac{M^n(1 - \eta)^n}{n!} \to 0.$$

Thus $x_0(\eta) = 1$ at every point of continuity $1 > \eta \geq a + \delta$. Since $x_0(\eta)$ is monotone-increasing, $x_0(1) = 1$; and since the points of discontinuity are denumerable, we conclude that $x_0(\eta) = 1$ for all $\eta \geq a + \delta$. But δ is arbitrary, and hence $x_0(\eta) = 1$ if $\eta > a$. On the other hand, if $\eta < a$, it follows that $x_n(\eta) = 0$ for n large. Hence, in the range $\eta < a$, we have $x_0(\eta) = 0$ and therefore $x_0 = I_a$.

We now show that $a = a_0$. The alternative is $a > a_0$ and $l(\eta) > m(\eta)$ for $\eta \geq a > a_0$. But the last inequality would allow us to infer from equation (6.3.1) that the functions f_{a_n} are uniformly bounded for large n, so that x_n could not converge to $x_0 = I_a$.

Finally, for $\eta > a_0$, $\int K(\xi, \eta)\, dx_n(\xi)$ tends to v. Hence

$$v = \int K(\xi, \eta)\, dx_0(\xi) = K(a_0, \eta) = l(a_0) = m(a_0).$$

In the course of the proof of this lemma we have actually verified that I_{a_0} is an optimal strategy for Player I if $\phi(a_0) \geq m(a_0)$. An analogous statement applies to an optimal strategy for Player II.

To complete the proof of Theorem 6.1.1 it remains to show that the only possible optimal strategy for each player is a pure strategy located at a_0.

Now suppose that x is optimal for the game $K(\xi, \eta)$. Then we assert that $x(\xi) = 0$ for $\xi < a_0$. Let us assume to the contrary that $x(\xi_0) > 0$ for some $\xi_0 < a_0$. Then since $a_n \to a_0$, we have $\int K(\xi, \eta)\, dy_n(\eta) = l(\xi) \leq l(\xi_0)$ for $\xi \leq \xi_0$ and $n \geq n_0$. Moreover, since $K_n(\xi, \eta)$ converges uniformly to $K(\xi, \eta)$ and $v_n \to v$, for any $\epsilon > 0$ and $n \geq n(\epsilon)$ we have

$$\int K(\xi, \eta)\, dy_n(\eta) \leq v + \epsilon$$

for all ξ. Since $l'(\xi) \geq \delta$, it follows that $\delta \cdot (a_0 - \xi_0) \leq l(a_0) - l(\xi_0)$; and if we set

$$\epsilon = \frac{\delta}{2} x(\xi_0) \cdot (a_0 - \xi_0) > 0,$$

we have

$$\int K(\xi, \eta)\, dy_n(\eta) \leq l(\xi_0) \leq l(a_0) - \delta(a_0 - \xi_0) = v - \delta \cdot (a_0 - \xi_0)$$

for $\xi \leq \xi_0$ and $n \geq n_0$. Consequently, for $n \geq \max\{n(\epsilon), n_0\}$,

$$\begin{aligned}\iint K(\xi, \eta)\, dy_n(\eta)\, dx(\xi) &\leq x(\xi_0)[v - \delta \cdot (a_0 - \xi_0)] + [1 - x(\xi_0)](v + \epsilon)\\ &< x(\xi_0)[v + \epsilon - \delta \cdot (a_0 - \xi_0)]\\ &\quad + [1 - x(\xi_0)](v + \epsilon)\\ &< v + \epsilon - x(\xi_0) \cdot (a_0 - \xi_0)\delta\\ &< v - \frac{\delta}{2} x(\xi_0) \cdot (a_0 - \xi_0) < v,\end{aligned}$$

so that x could not be optimal.

Since $x(\xi) = 0$ for $\xi < a_0$, we may write $x = \alpha I_{a_0} + \tilde{x}$, where $\tilde{x}$ is continuous at a_0, $\tilde{x}(a_0) = 0$, and $\tilde{x}(1) = 1 - \alpha$. For $\eta > a_0$ and η in the continuity set of $\tilde{x}$,

$$\begin{aligned}\int K(\xi, \eta)\, dx(\xi) &= \alpha l(a_0)\\ &\quad + \int_{a_0}^{\eta} l(\xi)\, d\tilde{x}(\xi) + m(\eta)[1 - \alpha - \tilde{x}(\eta)] \geq v = l(a_0) = m(a_0),\end{aligned}$$

so that

$$\int_{a_0}^{\eta} [l(\xi) - l(a_0)]\, d\tilde{x}(\xi) + [m(\eta) - m(a_0)][1 - \alpha - \tilde{x}(\eta)] \geq 0.$$

Since $c \cdot (\eta - a_0) \geq l(\eta) - l(a_0) \geq \delta \cdot (\eta - a_0)$ and $\tilde{x}$ is continuous at a_0, we have

$$\int_{a_0}^{\eta} [l(\xi) - l(a_0)]\, d\tilde{x}(\xi) = o(\eta - a_0);$$

since $m(\eta) - m(a_0) \leq -\delta \cdot (\eta - a_0) < 0$, we have

$$0 \leq \delta \cdot (\eta - a_0)[1 - \alpha - \tilde{x}(\eta)] \leq -[m(\eta) - m(a_0)][1 - \alpha - \tilde{x}(\eta)]$$

$$\leq \int_{a_0}^{\eta} [l(\xi) - l(a_0)]\, d\tilde{x}(\xi) = o(\eta - a_0),$$

so that $1 - \alpha - \tilde{x}(\eta) \to 0$ as $\eta \to a_0$ through the continuity set of $\tilde{x}$. Thus $\tilde{x}(\eta) = 1 - \alpha$ for $\eta > a_0$, and $\tilde{x} = (1 - \alpha)I_{a_0}$. Since $\tilde{x}$ was assumed continuous at a_0, it follows that $\alpha = 1$.

A completely analogous argument applies to Player II and shows that only I_{a_0} can serve as a solution. The remaining statements of Theorem 6.1.1 now follow readily.

***6.4 Games of timing involving several actions.** The problems considered in this section belong to the general class of silent duels, with the new feature that each player may act more than once during the course of the game if the opportunity arises. Although these games, like other silent duels, are applicable to a large number of competitive problems, it will be convenient to describe them in military terms.

Consider a combat between two airplanes that come within range of each other at time $t = 0$ and continue to approach until $t = 1$. The first plane carries n rounds of ammunition, and if one of these rounds is fired at time t, the enemy will be destroyed with probability $P(t)$. The second plane carries m rounds of ammunition, and the corresponding probability is $Q(t)$. Both players know the numbers m and n and the functions P and Q, but during the course of the game neither knows how many times his opponent has fired and missed. The pay-off for each player is 1 if he alone survives, 0 if both or none survive, and -1 if the enemy alone survives. It is convenient, as previously, to make the intuitively meaningful assumptions that P and Q are strictly increasing and continuously differentiable, and that $P(0) = Q(0) = 0$ and $P(1) = Q(1) = 1$.

Strategies and pay-off. If Player I fires his ith round of ammunition at $t = \xi_i$, his strategy is described by the vector $\xi = \langle \xi_1, \ldots, \xi_n \rangle$; this vector satisfies the restriction $0 \leq \xi_1 \leq \cdots \leq \xi_n \leq 1$, and any vector that satisfies these inequalities represents a pure strategy. Similarly, the pure strategies for Player II correspond to vectors $\eta = \langle \eta_1, \ldots, \eta_m \rangle$, with $0 \leq \eta_1 \leq \cdots \leq \eta_m \leq 1$. The pay-off function $K(\xi, \eta)$ can be constructed by the methods used in Section 5.1. However, in the analysis of any particular game it is convenient to use not one such function but a family of functions $K(\xi, \eta)$ depending on any number of variables $\xi_1, \ldots, \xi_r$ and $\eta_1, \ldots, \eta_s$. In all that follows, the variables ξ and η are associated with Players I and II, respectively, and are treated as essentially different symbols. With this convention in mind, $K(\xi_1, \ldots, \xi_r; \eta_1, \ldots, \eta_s)$ will always denote the pay-off to Player I when he fires at times $\xi_1, \ldots, \xi_r$ while Player II fires at times $\eta_1, \ldots, \eta_s$. For example, $K(\xi_1) = P(\xi_1)$, $K(\eta_1) = -Q(\eta_1)$; and $K(\xi_1; \eta_1)$ is the pay-off of the silent duel of equation (5.5.1) with P_1 replaced by P and P_2 by Q. In terms of these functions the pay-off kernel can be defined recursively by

$$K(\xi_1, \ldots, \xi_n; \eta_1, \ldots, \eta_m) = \begin{cases} P(\xi_1) + [1 - P(\xi_1)]K(\xi_2, \ldots, \xi_n; \eta_1, \ldots, \eta_m) & \xi_1 < \eta_1, \\ -Q(\eta_1) + [1 - Q(\eta_1)]K(\xi_1, \ldots, \xi_n; \eta_2, \ldots, \eta_m) & \eta_1 < \xi_1. \end{cases}$$

When $\eta_1 = \xi_1$, one may take the average value of the two functions. The consideration of the family of functions K not only simplifies the definition of the pay-off, but also allows us to isolate the contribution made by any component to the total pay-off. This property is illustrated by the following equation, which isolates the contribution of η to $K(\xi_1, \xi_2; \eta)$; the validity of the formula can be verified by direct computation:

$$K(\xi_1, \xi_2; \eta) - K(\xi_1, \xi_2) = \begin{cases} \prod_{i=1}^{2} [1 - P(\xi_i)]Q(\eta) & \xi_1 < \xi_2 < \eta, \\ -[1 - P(\xi_1)]Q(\eta)[1 + K(\xi_2)] & \xi_1 < \eta < \xi_2, \\ -Q(\eta)[1 + K(\xi_1, \xi_2)] & \eta < \xi_1 < \xi_2. \end{cases} \tag{6.4.1}$$

Characterization of a solution. We consider now the yield obtained by the two players when they choose strategies x and y that satisfy the following conditions:

(i) $x(\xi) = \prod_{i=1}^{n} x_i(\xi_i)$ and $y(\eta) = \prod_{j=1}^{m} y_j(\eta_j)$.

This is equivalent to seeking an optimal strategy which has the property that the times of firing are independent. In other words, x_1 and x_2 represent the marginal distributions of the times of discharging round 1 and round 2, respectively.

It is plausible that such optimal strategies should exist, in view of each player's complete lack of knowledge of his opponent's firing times.

(ii) The support of each x_i is an interval $[a_i, a_{i+1}]$, and the support of each y_j is an interval $[b_j, b_{j+1}]$. Moreover, $a_1 > 0$, $b_1 > 0$ and $a_{n+1} = b_{m+1} = 1$. We must necessarily have $a_1 = b_1$, for if $a_1 > b_1$ Player II would only improve his chances by waiting until a_1 before firing, and similarly for Player I.

(iii) All these distributions are continuous except for x_n and y_m, which may have discrete masses α and β at $\xi_n = 1$ and $\eta_m = 1$, respectively.

With these strategies we associate the quantities D_i and E_j defined by

$$D_i = \int_{a_i}^{a_{i+1}-} P(t)\, dx_i(t) \qquad (i = 1, \ldots, n),$$

$$E_j = \int_{b_j}^{b_{j+1}-} Q(t)\, dy_j(t) \qquad (j = 1, \ldots, m).$$

The fundamental properties of these strategies are given by the next two lemmas. For the sake of simplicity, the statements and the proofs of these lemmas are restricted to the case where $n = 2$ and $m = 1$. This case already exhibits all the characteristic features of the general problem, but it avoids most of the intricacies of notation. The proofs have been chosen so that they can be immediately generalized by iteration or induction. The consideration of the identities $\int K(\xi, \eta)\, dx \equiv \underline{v}$ and $\int K(\xi, \eta)\, dy \equiv \bar{v}$ is suggested by Lemma 2.2.1.

▶ LEMMA 6.4.1. The distributions x and y satisfy the identities

$$\int K(\xi_1, \xi_2; \eta)\, dx_1(\xi_1)\, dx_2(\xi_2) \equiv \underline{v} \qquad (b \leq \eta < 1)$$

and

$$\int K(\xi_1, \xi_2; \eta)\, dy(\eta) \equiv \bar{v} \qquad (a_i \leq \xi_i < a_{i+1};\, i = 1, 2)$$

if and only if the following conditions are satisfied:

(1) The distributions x_1, x_2 and y are absolutely continuous except at $\xi_2 = 1$, $\eta = 1$, and have densities

$$x_i'(t) = h_i \frac{Q'(t)}{Q^2(t)P(t)} \qquad (a_i \leq t < a_{i+1};\, i = 1, 2),$$

$$y'(t) = k_i \frac{P'(t)}{P^2(t)Q(t)} \qquad (a_i \leq t < a_{i+1};\, i = 1, 2).$$

(2) The constants a_1, a_2, b_1, b_2 $(b_2 = 1)$, h_1, h_2, k_1, k_2, α, and β satisfy the following system of equations:

$$a_1 = b_1,$$

$$2h_2 = 2\alpha + 1 - D_2, \qquad h_1 = [1 - D_1]h_2,$$

$$2k_2 = 2\beta + 1 - E, \qquad k_1 = [1 - P(a_2)]k_2.$$

Proof. Integrating both sides of (6.4.1) with respect to x and simplifying, we obtain

$$\int K(\xi_1, \xi_2; \eta)\, dx_1(\xi_1)\, dx_2(\xi_2) - \int K(\xi_1, \xi_2)\, dx_1(\xi_1)\, dx_2(\xi_2)$$

$$= \begin{cases} -(1 - D_1)Q(\eta)\left[2\int_\eta^1 P(\xi_2)\, dx_2(\xi_2) + 1 - D_2\right] & a_2 \leq \eta < 1, \\ -Q(\eta)\left[2\int_\eta^{a_2} P(\xi_1)\, dx_1(\xi_1) + (1 - D_1)(1 + D_2)\right] & a_1 \leq \eta \leq a_2, \\ -Q(\eta)[1 + D_1 + (1 - D_1)D_2] & 0 \leq \eta \leq a_1, \\ (1 - D_1)(1 - D_2 + \alpha) & \eta = 1. \end{cases} \tag{6.4.2}$$

The right-hand side is continuous except at $\eta = 1$ (remember that x_2 may have mass at $\xi = 1$), and the dependence in η is explicit. Obviously, $\int K(\xi_1, \xi_2; \eta)\, dx_1(\xi_1)\, dx_2(\xi_2)$ cannot be constant for $0 < \eta < a_1$. We may therefore conclude that $a_1 \leq b$. An examination of the other two terms shows that $\int K\, dx$ is constant for $b \leq \eta < 1$ if and only if there are constants h_1 and h_2 such that

$$Q(\eta)\left[2\int_\eta^1 P(\xi_2)\, dx_2(\xi_2) + 1 - D_2\right] = 2h_2 \qquad a_2 \leq \eta < 1,$$

$$Q(\eta)\left[2\int_\eta^{a_2} P(\xi_1)\, dx_1(\xi_1) + (1 - D_1)(1 + D_2)\right] = 2h_1 \qquad a_1 \leq \eta \leq a_2, \tag{6.4.3}$$

and $h_2(1 - D_1) = h_1$.

Integrating by parts, we see immediately that x_1 and x_2 are absolutely continuous with a continuous density on $a_1 \leq \eta \leq a_2$ and $a_2 \leq \eta < 1$, respectively. Next, dividing by $Q(\eta)$ and then differentiating both sides of (6.4.3), we deduce that

$$x_i'(t) = h_i Q'(t)[Q^2(t)P(t)]^{-1} \qquad (a_i \leq t \leq a_{i+1}; i = 1, 2).$$

When the form of x_i' is inserted back into (6.4.3), the integrals can be evaluated explicitly, showing that (6.4.3) holds only if $2h_2 = 2\alpha + 1 - D_2$ and $h_1 = [1 - D_1]h_2$.

The characterization of y is quite similar, but instead of (6.4.1) one must use the equations that relate $K(\xi_1, \xi_2; \eta)$ and $K(\xi_1; \eta)$ to exhibit the dependence on ξ_2 and thus characterize $y(t)$ for $a_2 \leq t < 1$. A further application of the same technique displays the dependence on ξ_1 and characterizes $y(t)$ in the interval $a_1 \leq t \leq a_2$.

A direct enumeration of cases yields

$$K(\xi_1, \xi_2; \eta) = K(\xi_1; \eta) + \begin{cases} [1 - P(\xi_1)]P(\xi_2)[1 + Q(\eta)] & \xi_2 < \eta, \\ [1 - P(\xi_1)][1 - Q(\eta)]P(\xi_2) & \eta < \xi_2. \end{cases}$$

Fixing $a_1 \leq \xi_1 \leq a_2$ and $a_2 \leq \xi_2 < 1$, we obtain for the strategy y

$$\int K(\xi_1, \xi_2; \eta)\, dy(\eta) + E = P(\xi_2)[1 - P(\xi_1)]\left[1 - E + 2\int_{\xi_2}^{1} Q(\eta)\, dy(\eta)\right] + P(\xi_1)\left[1 - E + 2\int_{\xi_1}^{1} Q(\eta)\, dy(\eta)\right]. \tag{6.4.4}$$

If we require that (6.4.4) be identically a constant in the region indicated, we deduce easily with the aid of appropriate differentiations that

$$y_i'(t) = k_i \frac{P'(t)}{P^2(t)Q(t)} \qquad (a_i \leq t \leq a_{i+1}).$$

Substituting this back into (6.4.4) leads to the equivalent conditions $2k_2 = 2\beta + 1 - E$ and $k_1 = [1 - P(a_2)]k_2$ as desired. The proof of Lemma 6.4.1 is hereby complete.

▶ Lemma 6.4.2. If $\int K(\xi_1, \xi_2; \eta)\, dx_1(\xi_1)\, dx_2(\xi_2) \equiv \underline{v}$ for $b \leq \eta < 1$, then $\int K(\xi_1, \xi_2; \eta)\, dx_1(\xi_1)\, dx_2(\xi_2) \geq \underline{v}$ for all η. And if $\int K(\xi_1, \xi_2; \eta)\, dy(\eta) = \bar{v}$ for $a_1 \leq \xi_1 \leq a_2 \leq \xi_2 < 1$, then $\int K(\xi_1, \xi_2; \eta)\, dy(\eta) \leq \bar{v}$ for all ξ_1 and ξ_2.

Proof. In the proof of the preceding lemma it was shown that if

$$0 < \eta \leq a_1,$$

then

$$\int K\, dx = \int K(\xi_1, \xi_2)\, dx_1(\xi_1)\, dx_2(\xi_2) - Q(\eta)[1 + D_1 + (1 - D_1)D_2].$$

This is clearly a decreasing function of η, and by assumption it achieves the value v at $\eta = a_1$. Therefore $\int K(\xi_1, \xi_2; \eta)\, dx_1(\xi_1)\, dx_2(\xi_2)$ exceeds $\underline{v}$ for all $\eta < a_1$. Finally, the yield achieved by using the strategy x against the pure strategy $\eta = 1$ is clearly

$$\int K(\xi_1, \xi_2)\, dx_1(\xi_1)\, dx_2(\xi_2) - (1 - D_1)(1 - D_2 + \alpha).$$

But letting $\eta \to 1$ in (6.4.2), we obtain

$$\underline{v} = \int K(\xi_1, \xi_2)\, dx_1(\xi_1)\, dx_2(\xi_2) - (1 - D_1)(2\alpha + 1 - D_2),$$

and hence the first inequality holds everywhere. The proof of the other inequality is similar. It should be noticed that strict inequality holds everywhere except in the regions $b \leq \eta \leq 1$ and $a_1 \leq \xi_1 \leq a_2 \leq \xi_2 \leq 1$.

An immediate consequence of Lemma 6.4.2 is that the strategies x and y are optimal if and only if one of them is continuous at 1. For if x is continuous at 1, then the identity $\int K\,dx \equiv \underline{v}$ holds in the closed interval $b \leq \eta \leq 1$, and it follows that $\int\int K\,dx\,dy \equiv \underline{v}$. Furthermore, if x is continuous at 1, the identity $\int K\,dy \equiv \bar{v}$ holds a.e. with respect to x, so that $\int K\,dx\,dy \equiv \bar{v}$. Thus $\underline{v} = \bar{v}$, and the conclusion of Lemma 6.4.2 is the statement that x and y are optimal. The same argument applies when y is continuous.

Existence of a solution. The last two lemmas have reduced the problem of finding optimal strategies to the problem of finding a solution for a system of equations. In the analysis of this system, nothing is gained by the restrictions $n = 2$ and $m = 1$; on the contrary, this case makes the subsequent notation appear rather artificial, so that it is desirable now to go back to arbitrary values of n and m. In this general form, Lemma 6.4.1 asserts that the desired densities $x_i'(t)$ and $y_j'(t)$ are always multiples of $Q'(t)[Q^2(t)P(t)]^{-1}$ and $P'(t)[P^2(t)Q(t)]^{-1}$, respectively. Moreover, the constants h and k that multiply these functions change only at the points $a_1, \ldots, a_n$ and $b_1, \ldots, b_m$. At the point a_i the constant multiplier in x changes by the factor $1 - D_{i-1}$ and the multiplier in y changes by the factor $1 - P(a_i)$. Similarly, at any point b_j the constant multiplier in y changes by the factor $1 - E_{j-1}$ and the multiplier in x changes by the factor $[1 - Q(b_j)]$. Thus, if we consider $a_1, \ldots, a_n$ and $b_1, \ldots, b_m$ as arbitrary parameters and define

$$f^*(t) = \prod_{b_j > t} [1 - Q(b_j)] \frac{Q'(t)}{Q^2(t)P(t)}, \tag{6.4.5a}$$

$$g^*(t) = \prod_{a_i > t} [1 - P(a_i)] \frac{P'(t)}{P^2(t)Q(t)}, \tag{6.4.5b}$$

condition (1) of Lemma 6.4.1 asserts that the densities x_i' and y_j' have the form

$$x_i'(t) = h_i f^*(t) \qquad (a_i \leq t < a_{i+1};\, i = 1, \ldots, n), \tag{6.4.6a}$$

$$y_j'(t) = k_j g^*(t) \qquad (b_j \leq t < b_{j+1};\, j = 1, \ldots, m). \tag{6.4.6b}$$

The equations in condition (2) of Lemma 6.4.1 are replaced by the equations

$$2h_n = 2\alpha + 1 - D_n, \tag{6.4.7a}$$

$$2k_m = 2\beta + 1 - E_m, \tag{6.4.7b}$$

$$h_i = [1 - D_i]h_{i+1} \qquad (i = 1, \ldots, n-1), \tag{6.4.8a}$$

$$k_j = [1 - E_j]k_{j+1} \qquad (j = 1, \ldots, m-1), \tag{6.4.8b}$$

$$a_1 = b_1, \tag{6.4.9}$$

and the continuity condition at 1 is given by

$$\alpha\beta = 0. \tag{6.4.10}$$

The previous results can now be summarized as follows: Any pair of optimal strategies x and y that satisfies conditions (i), (ii), and (iii) on p. 154 must satisfy equations (6.4.5) through (6.4.10); and conversely, any solution of these equations that is a strategy is also an optimal strategy. Furthermore, since the densities given by (6.4.6) are always positive, any solution of this system of equations is a strategy if and only if α and β are nonnegative and

$$h_n \int_{a_n}^{1} f^*(t)\, dt + \alpha = 1, \tag{6.4.11a}$$

$$k_m \int_{b_m}^{1} g^*(t)\, dt + \beta = 1, \tag{6.4.11b}$$

$$h_i \int_{a_i}^{a_{i+1}} f^*(t)\, dt = 1 \qquad (i = 1, \ldots, n-1), \tag{6.4.12a}$$

$$k_j \int_{b_j}^{b_{j+1}} g^*(t)\, dt = 1 \qquad (j = 1, \ldots, m-1). \tag{6.4.12b}$$

In (6.4.7) and (6.4.8) it is convenient to write D_i and E_j explicitly in terms of f^*, g^*, α, and β, and to eliminate h_i and k_j by means of (6.4.11) and (6.4.12). The resulting equations are as follows:

$$2(1 - \alpha) = \int_{a_n}^{1} [1 + \alpha - (1 - \alpha)P(t)]f^*(t)\, dt, \tag{6.4.13a}$$

$$2(1 - \beta) = \int_{b_m}^{1} [1 + \beta - (1 - \beta)Q(t)]g^*(t)\, dt, \tag{6.4.13b}$$

$$\frac{1}{h_{i+1}} = \int_{a_i}^{a_{i+1}} [1 - P(t)]f^*(t)\,dt \qquad (i = 1, \ldots, n-1), \quad (6.4.14a)$$

$$\frac{1}{k_{j+1}} = \int_{b_j}^{b_{j+1}} [1 - Q(t)]g^*(t)\,dt \qquad (j = 1, \ldots, m-1). \quad (6.4.14b)$$

The existence of a solution for the complete system of equations (6.4.9) through (6.4.14) can be established easily. To this effect, it is important to notice that equations (6.4.5a, b) make sense regardless of the number of parameters $a_1, \ldots, a_n$ and $b_1, \ldots, b_m$—even in the case of no parameters. Moreover, so long as the parameters are strictly interior to the unit interval, all the integrals in (6.4.13a) through (6.4.14b) diverge as their lower limits approach zero. Hence, if α ($\alpha < 1$) and $b_1, \ldots, b_m$ ($b_m < 1$) are prescribed parameters, then a_n can be obtained from (6.4.13a), h_n from (6.4.11a), a_{n-1} from (6.4.14a), etc. Moreover, if α is fixed, then a_n is a monotone-decreasing function of b_m (the integrand decreases), and $a_n \to 0$ as $b_m \to 1$; also if $b_1, \ldots, b_m$ are fixed, then a_n is an increasing function of α, and $a_n \to 1$ as $\alpha \to 1$.

Now let α and β be any numbers that satisfy the conditions $0 \leq \alpha < 1$, $\alpha\beta = 0$, and $0 \leq \beta < 1$. Using no parameters in the definitions of f^* and g^*, compute a_n and b_m from (6.4.13a) and (6.4.13b). Denote these solutions by a_n^0 and b_m^0, and assume for definiteness that $a_n^0 > b_m^0$. Then define $a_n^* = a_n^0$, neglect b_m^0, and compute the solution h_n of (6.4.11a). Using a_n^* as the only parameter in the definition of f^* and g^*, compute next the solutions b_m and a_{n-1} of (6.4.13b) and (6.4.14a), and denote these solutions by a_{n-1}^1 and b_m^1; it is easy to verify that $b_m^1 < b_m^0 < a_n^*$. For definiteness assume that $b_m^1 < a_{n-1}^1$, and define $a_{n-1}^* = a_{n-1}^1$. The process can be continued, using the previously starred letters as parameters and computing at each stage a new a_i^k and a new b_j^k. If $a_i^k \leq b_j^k$, one defines $b_j^* = b_j$; and if $a_i^k > b_j^k$, one defines $a_i^* = a_i^k$. In this procedure each parameter is smaller than those computed previously, and it cannot influence their values because it does not affect the values of f^* or g^* that are relevant for their computation. Hence the resulting numbers are consistent, in the sense that if the a_i^*'s are used as parameters, one obtains the same b_j^*'s, and conversely. Thus these numbers produce a solution of the complete system of equations except for the equation $a_1^* = b_1^*$.

Finally, for α close to 1 ($\beta = 0$), $a_n^* \to 1$, $b_m^* \to 0$ and $b_1^* < b_m^* < a_1^*$; similarly, for β close to 1, $a_1^* < b_1^*$. Since b_1^* and a_1^* are continuous functions of α and β, there is a solution for the complete system. If $\alpha = \beta = 0$ leads to the inequality $a_1^* < b_1^*$, the solution has $\alpha > 0$, $\beta = 0$; and if $\alpha = \beta = 0$ leads to the inequality $b_1^* < a_1^*$, the solution has $\alpha = 0, \beta > 0$.

Example 1. If $P(t) = Q(t) = t$ and $n = 2$, $m = 1$, then

$$a_1 = b_1 = \frac{1}{1 + 2\sqrt{1 + \sqrt{1/3}}}, \qquad a_2 = \frac{1}{1 + 2\sqrt{1/3}},$$

$$\alpha = 2 - \sqrt{3}, \qquad \beta = 0.$$

Example 2. If $P(t) = Q(t) = t$ and $n = m$, then

$$a_{n-k} = \frac{1}{2k + 3} = b_{n-k}.$$

***6.5 Butterfly-shaped kernels.** The one-action game of timing was characterized by a kernel of the form

$$K(\xi, \eta) = \begin{cases} L(\xi, \eta) & \xi < \eta, \\ \Phi(\xi) & \xi = \eta, \\ M(\xi, \eta) & \xi > \eta, \end{cases} \tag{6.5.1}$$

where $L(\xi, \eta)$ and $M(\xi, \eta)$ are suitable monotonic functions in each of the variables and $K(\xi, \eta)$ may have a discontinuity along the diagonal. We determined the optimal strategies in general form and found that they consisted in each case of a density and some discrete masses at the ends of the interval. The method of analysis revolved about solving suitable integral equations of the second kind. In this section we show how to extend the integral-equation technique to solve games of the form (6.5.1) which are continuous throughout but possess a discontinuity in the first derivative along the diagonal, provided that the functions L and M are both suitably concave (but not necessarily monotonic). Precisely, we assume that $K(\xi, \eta)$ is as in (6.5.1) and continuous for all $0 \leq \xi, \eta \leq 1$; that is, that $L(\xi, \xi) = \Phi(\xi) = M(\xi, \xi)$. In addition, we make the following assumptions:

(a) In their respective domains both $L(\xi, \eta)$ and $M(\xi, \eta)$ have continuous third partial derivatives.

(b) $L_{\xi\xi}(\xi, \eta)$, $L_{\eta\eta}(\xi, \eta)$ are strictly negative for $\xi \leq \eta$, and $M_{\xi\xi}(\xi, \eta)$, $M_{\eta\eta}(\xi, \eta)$ are strictly negative for $\xi \geq \eta$. In other words, each of the functions $L(\xi, \eta)$ and $M(\xi, \eta)$ is strictly concave in each variable over its domain of definition.

We note specifically that since we may have a discontinuity in the first partial derivatives, the kernel $K(\xi, \eta)$ is not necessarily concave in each variable throughout the unit square. As an alternative to condition (b) we could require that $M_{\xi\xi}$, $M_{\eta\eta}$, $L_{\xi\xi}$, and $L_{\eta\eta}$ be positive in their respective triangles. This requirement gives rise to an analogous theory, the precise conclusions of which are left to the reader.

(c) $L_\eta(t, t) - M_\eta(t, t) > 0$ and $M_\xi(t, t) - L_\xi(t, t) > 0$.

For M and L convex, condition (c) is replaced by

$$M_\eta(t, t) - L_\eta(t, t) > 0 \quad \text{and} \quad L_\xi(t, t) - M_\xi(t, t) > 0.$$

Condition (c) may be visualized as giving the kernel a butterfly shape in the neighborhood of the diagonal.

Before stating the principal theorem of this section, we recall some of the notation used for the theory of games of timing of class II which will also be of convenience here. A strategy x consisting of a density f on the interval $[a, b]$ and discrete masses of magnitude α and β at two points $c \leq a$ and $d \geq b$, respectively, will be written $x = (\alpha I_c, f_{ab}, \beta I_d)$. In a similar way we interpret $y = (\gamma I_c, g_{ab}, \delta I_d)$. The symbol I_c will again be used to represent the degenerate distribution located at c.

▶THEOREM 6.5.1. Under the conditions listed above, all the optimal strategies of butterfly-shaped games have the following form: There exists a unique interval $[a, b]$ such that any optimal maximizing strategy $x = (\alpha I_a, f_{ab}, \beta I_b)$ and any optimal minimizing strategy $y = (\gamma I_0, g_{ab}, \delta I_1)$. Furthermore, if $0 < a < 1$, then $\alpha > 0$, and if $0 < b < 1$, then $\beta > 0$; in general, $\alpha, \beta, \gamma, \delta \geq 0$. The densities f_{ab} and g_{ab} are obtained as Neumann series or as eigenfunctions of integral operators.

The existence of optimal strategies for the game with kernel $K(\xi, \eta)$ follows from Problem 3 of Chapter 2, which asserts that a game with a continuous kernel defined over the unit square will always have a solution. We concentrate here on characterizing the form of the optimal strategies as described in the theorem. The value of the game will be denoted as usual by v. The proof of Theorem 6.5.1 is divided into a series of lemmas.

▶LEMMA 6.5.1. There exists a unique interval $[a, b]$ which is the support of any optimal strategy for the maximizing player. If this interval is nondegenerate, it is (with the possible addition of 0 and 1) the support of any optimal strategy for the minimizing player as well. If $a = b$, then any optimal strategy for the minimizing player concentrates on the points a, 0, and 1.

Proof. We shall let x and y represent optimal strategies for the maximizing and minimizing players, respectively. If (c, d) is an open interval of x-measure zero with $0 < c < d < 1$, it is easily verified that $\int K(\xi, \eta)\, dx(\xi)$ is strictly concave over this interval. Hence

$$\int K(\xi, \eta)\, dx(\xi) > v \qquad (c < \eta < d).$$

This relation, in conjunction with Lemma 2.2.1, shows that the support of any optimal y cannot meet the interval (c, d). Hence the support of y must be contained in the support of x, with the possible addition of 0 and 1.

We now suppose that c and d are in the support of x, that $c < d$, and that (c, d) is of x-measure zero. Then (c, d) is of y-measure zero for any optimal y, and it is easily verified that $\int K(\xi, \eta)\, dy(\eta)$ is strictly concave over this interval. Since this integral is a continuous function of ξ and must assume the value v at the end points c and d, it follows that

$$\int K(\xi, \eta)\, dy(\eta) > v \qquad (c < \xi < d),$$

which contradicts the assumption that y is optimal. This contradiction implies that if c and d are in the support of x, the open interval (c, d) cannot have x-measure zero. Therefore we conclude that the support of x must be a closed interval $[a, b]$.

Suppose that for some optimal x this interval is nondegenerate; i.e., suppose $a < b$. Then for any optimal y,

$$\int K(\xi, \eta)\, dy(\eta) \equiv v \qquad (a \leq \xi \leq b).$$

The argument of the previous paragraph shows that no subinterval of $[a, b]$ can be of y-measure zero (otherwise the pay-off would be strictly concave and greater than v). Hence $[a, b]$ is contained in the support of any optimal y. Combining this result with the fact that the support of any optimal y (disregarding 0 and 1) is contained in the support of x, we deduce that the support of any optimal y consists exactly of the interval $[a, b]$, with the possible addition of 0 and 1; and hence (by virtue of our conclusions two paragraphs back) that the support of any optimal x must contain the nondegenerate interval $[a, b]$. Since it contains a nondegenerate interval, it must be contained in the support of any optimal y, so that the support of every optimal x is precisely $[a, b]$.

If $[a, b]$ were degenerate for every optimal x, it would still be unique. For if two different pure distributions I_a and $I_{a'}$ were both optimal, then the support of y would be contained in $\{0, a, 1\}$ and also in $\{0, a', 1\}$. The support of y would accordingly be in the point set $\{0, 1\}$; hence $\int K(\xi, \eta)\, dy(\eta)$ would be strictly concave for $0 \leq \xi \leq 1$, equal to at most v everywhere, and exactly equal to v at $\xi = a$ and $\xi = a'$. These conditions are clearly absurd; hence $a = a'$.

Our next two lemmas establish that on the interior of $[a, b]$ any optimal strategy comes from a continuous density.

►LEMMA 6.5.2. If x is optimal for Player I, then x is continuous in the interior of $[a, b]$.

Proof. Let η_0 be any interior point of $[a, b]$. Since x is assumed to be optimal, and since we know from Lemma 6.5.1 that the support of any optimal y contains $[a, b]$, we infer that $\int K(\xi, \eta)\, dx(\xi) \equiv v$ for all η in $[a, b]$. Hence

$$\int \frac{K(\xi, \eta) - K(\xi, \eta_0)}{\eta - \eta_0}\, dx(\xi) = 0 \qquad (\eta \neq \eta_0 \text{ and each in } [a, b]).$$

By assumption, the right-hand derivative $K_{\eta+}$ exists everywhere and is bounded; then, with the aid of bounded convergence, we obtain

$$\int_0^1 K_{\eta+}(\xi, \eta_0)\, dx(\xi) = 0;$$

similarly, we obtain

$$\int_0^1 K_{\eta-}(\xi, \eta_0)\, dx(\xi) = 0.$$

Finally, since $K_{\eta+}(\xi, \eta_0) - K_{\eta-}(\xi, \eta_0) = 0$ except for $\xi = \eta_0$, we deduce that

$$0 = \int [K_{\eta+}(\xi, \eta_0) - K_{\eta-}(\xi, \eta_0)]\, dx(\xi) = [L_\eta(\eta_0, \eta_0) - M_\eta(\eta_0, \eta_0)]\sigma_{\eta_0},$$

where σ_{η_0} is the mass of $x(\xi)$ at the point η_0. By condition (c), $\sigma_{\eta_0} = 0$, and hence x is continuous at η_0 as claimed.

▶ LEMMA 6.5.3. If x is optimal for Player I, then x is absolutely continuous on the interior of $[a, b]$ and its density x' is continuous for $a < \xi < b$.

Proof. The last lemma allows us to write $x = (\alpha_0 I_a, x_0, \beta_0 I_b)$, where x_0 is the continuous part of x. Since x is optimal,

$$v = \int_0^1 K(\xi, \eta)\, dx(\xi) = \alpha_0 L(a, \eta) + \beta_0 M(b, \eta) + \int_a^\eta L(\xi, \eta)\, dx_0(\xi) + \int_\eta^b M(\xi, \eta)\, dx_0(\xi) \tag{6.5.2}$$

for $a \leq \eta \leq b$. The continuity of x_0 permits a simple integration by parts, which yields

$$\int_a^\eta L\, dx_0 + \int_\eta^b M\, dx_0 = M(b, \eta)x_0(b) - \int_a^\eta L_\xi(\xi, \eta)x_0(\xi)\, d\xi - \int_\eta^b M_\xi(\xi, \eta)x_0(\xi)\, d\xi,$$

since by continuity $M(\eta, \eta) = L(\eta, \eta)$ and also $x_0(a) = 0$. If we substitute the resulting expression in (6.5.2) and differentiate, we obtain

$$0 = \alpha_0 L_\eta(a, \eta) + \beta_0 M_\eta(b, \eta) + x_0(b) M_\eta(b, \eta) - \int_a^\eta L_{\xi\eta}(\xi, \eta) x_0(\xi)\, d\xi$$

$$- \int_\eta^b M_{\xi\eta}(\xi, \eta) x_0(\xi)\, d\xi - [L_\xi(\eta, \eta) - M_\xi(\eta, \eta)] x_0(\eta).$$

But $L_\xi - M_\xi \leq -\epsilon < 0$ implies from the last equation and assumption (a) that $x_0(\eta)$ is absolutely continuous. If we differentiate the last equation once more—this step is valid by assumption (a)—we deduce that x_0' is continuous. In a similar manner we prove the next lemma.

▶ Lemma 6.5.4. If y is optimal for Player II, then y is absolutely continuous with continuous density on the interior of $[a, b]$.

▶ Lemma 6.5.5. The maximizing optimal strategies have the form $x = (\alpha I_a, f_{ab}, \beta I_b)$, with $\alpha > 0$ when $0 < a < 1$, and $\beta > 0$ when $0 < b < 1$; otherwise $\alpha, \beta \geq 0$.

Proof. Lemmas 6.5.1–3 put together imply that any optimal x has the form indicated. It remains to show that if $0 < a < 1$, then x possesses a discrete mass at the point $\xi = a$. We assume the contrary; then the derivative of $h(\eta) = \int K(\xi, \eta)\, dx(\xi)$ exists at $\eta = a$. But if the derivative exists at $\eta = a$, it must be zero there, since $h(\eta) = v$ for $a \leq \eta \leq b$. But $h(\eta)$ is strictly concave for $0 < \eta < a$; hence the last assertion implies that $\int K(\xi, \eta)\, dx(\xi) < v$ for $0 < \eta < a$, which is contrary to the optimal character of x. A similar argument applies at the end point b if it is interior to the unit interval.

▶ Lemma 6.5.6. Any minimizing strategy y has the form

$$y = (\gamma I_0, g_{ab}, \delta I_1), \quad \text{with} \quad \gamma, \delta \geq 0.$$

Proof. Lemma 6.5.4 asserts that y is absolutely continuous over the open interval (a, b). It remains only to show that y is continuous at $\xi = a$ if $0 < a < 1$, and continuous at $\xi = b$ if $0 < b < 1$. Arguing as in Lemma 6.5.2, we deduce first that $\int K_{\xi+}(a, \eta)\, dy(\eta) = 0$. Also

$$\int K(a, \eta)\, dy(\eta) = v, \qquad \int K(\xi, \eta)\, dy(\eta) \leq v \qquad (\xi < a).$$

Hence

$$\int \frac{K(a, \eta) - K(\xi, \eta)}{a - \xi}\, dy(\eta) \geq 0 \qquad (\xi < a),$$

and proceeding to the limit, $\int K_{\xi-}(a, \eta)\, dy(\eta) \geq 0$. Hence

$$0 \geq \int [K_{\xi+}(a, \eta) - K_{\xi-}(a, \eta)]\, dy(\eta) = [M_\xi(a, a) - L_\xi(a, a)]\sigma_a,$$

where σ_a is the y-measure of the point a. Since $\sigma_a \geq 0$ and we have assumed $M_\xi(a, a) - L_\xi(a, a) > 0$, we must conclude that $\sigma_a = 0$. A similar argument applies to the point b if $0 < b < 1$.

The integral equations. We now wish to exhibit the densities f_{ab} and g_{ab} as solutions of integral equations. If we use the notation $f = f_{ab}$ and $g = g_{ab}$, the identities

$$v \equiv \int_0^1 K(\xi, \eta)\, dx(\xi) \qquad (a \leq \eta \leq b),$$

$$v \equiv \int_0^1 K(\xi, \eta)\, dy(\eta) \qquad (a \leq \xi \leq b)$$

become

$$v \equiv \alpha L(a, \eta) + \int_a^\eta L(\xi, \eta) f(\xi)\, d\xi + \int_\eta^b M(\xi, \eta) f(\xi)\, d\xi + \beta M(b, \eta),$$

$$v \equiv \gamma M(\xi, 0) + \int_a^\xi M(\xi, \eta) g(\eta)\, d\eta + \int_\xi^b L(\xi, \eta) g(\eta)\, d\eta + \delta L(\xi, 1).$$

Two successive differentiations of the first identity yield respectively

$$0 \equiv \alpha L_\eta(a, \eta) + \int_a^\eta L_\eta(\xi, \eta) f(\xi)\, d\xi + \int_\eta^b M_\eta(\xi, \eta) f(\xi)\, d\xi + \beta M_\eta(b, \eta)$$

and

$$0 \equiv \alpha L_{\eta\eta}(a, \eta) + \int_a^\eta L_{\eta\eta}(\xi, \eta) f(\xi)\, d\xi + \int_\eta^b M_{\eta\eta}(\xi, \eta) f(\xi)\, d\xi + \beta M_{\eta\eta}(b, \eta)$$
$$+ [L_\eta(\eta, \eta) - M_\eta(\eta, \eta)] f(\eta).$$

If we divide the last identity by the nonzero coefficient of $f(\eta)$ and rearrange the terms, we obtain

$$f - T_{ab} f = \alpha p_a + \beta p_b, \tag{6.5.3}$$

where

$$T_{ab} f = \int_a^\eta \frac{-L_{\eta\eta}(\xi, \eta) f(\xi)\, d\xi}{(L_\eta - M_\eta)(\eta, \eta)} + \int_\eta^b \frac{-M_{\eta\eta}(\xi, \eta) f(\xi)\, d\xi}{(L_\eta - M_\eta)(\eta, \eta)},$$

$$p_a = \frac{-L_{\eta\eta}(a, \eta)}{(L_\eta - M_\eta)(\eta, \eta)}, \qquad p_b = \frac{-M_{\eta\eta}(b, \eta)}{(L_\eta - M_\eta)(\eta, \eta)}.$$

A similar differentiation of the identity satisfied by g shows that g must satisfy the integral equation

$$g - U_{ab} g = \gamma q_0 + \delta q_1, \tag{6.5.4}$$

where

$$U_{ab}g = \int_a^\xi \frac{-M_{\xi\xi}(\xi, \eta)g(\eta)\,d\eta}{(M_\xi - L_\xi)(\xi, \xi)} + \int_\xi^b \frac{-L_{\xi\xi}(\xi, \eta)g(\eta)\,d\eta}{(M_\xi - L_\xi)(\xi, \xi)},$$

$$q_0 = \frac{-M_{\xi\xi}(\xi, 0)}{(M_\xi - L_\xi)(\xi, \xi)}, \qquad q_1 = \frac{-L_{\xi\xi}(\xi, 1)}{(M_\xi - L_\xi)(\xi, \xi)}.$$

The integral equations (6.5.3) and (6.5.4) are to be used to determine the functions f and g. It is easy to verify that the operators T_{ab} and U_{ab} are strictly positive and completely continuous linear transformations. (The general theory of positive operators as discussed in Section 5.3 provide various characterizations of the spectral radius $\lambda(T_{ab})$. Another such characterization, valid only for strictly positive linear transformations, is

$$\lambda(T_{ab}) = \inf \{\lambda | T_{ab}f \leq \lambda f \qquad \text{for some} \quad f > 0\}.)†$$

The existence of an optimal strategy satisfying (6.5.3), together with the positivity of p_a and p_b, implies that $\lambda(T_{ab}) \leq 1$. If $0 < a < 1$ or $0 < b < 1$, Lemma 6.5.5 implies that $\lambda(T_{ab}) < 1$. Similarly, $\mu(U_{ab}) \leq 1$, where $\mu(U_{ab})$ is the spectral radius of U_{ab}. In general, both $\mu(U_{ab}) = 1$ and $\mu(U_{ab}) < 1$ may occur.

The final proofs of the statements of Theorem 6.5.1 are mostly contained in Lemmas 6.5.1–5. The assertion that f_{ab} and g_{ab} are expressible as Neumann series or eigenfunctions is equivalent to the assertion that $\lambda(T_{ab}) \leq 1$ and $\mu(U_{ab}) \leq 1$ (see Theorem 5.3.4). The proof of Theorem 6.5.1 is hereby complete.

We now indicate how the various parameters of the optimal strategies may be computed. The constants α, β, a, b can be determined from the following necessary relations:

$$\alpha + \beta + \int_a^b f_{ab}(t)\,dt = 1,$$

$$\frac{\partial}{\partial\eta^+}\left\{\int K(\xi, \eta)\,dx(\xi)\right\}\Bigg|_{\eta=a} = 0 \qquad \left(\frac{\partial}{\partial\eta^+} \text{ refers to a right-hand derivative}\right),$$

$$\int K(\xi, 0)\,dx(\xi) = \int K(\xi, 1)\,dx(\xi) = \int K(\xi, a)\,dx(\xi).$$

Of course f_{ab} is calculated as a Neumann series based on the operator T_{ab} if $\lambda(T_{ab}) < 1$. Explicitly,

$$f_{ab} = \sum_{n=0}^{\infty} T_{ab}^n(\alpha p_a + \beta p_b).$$

---

† The proof is easily obtained by utilizing the eigenfunction of the linear transformation whose kernel is the transpose of the kernel that defines T_{ab}.

When $\lambda(T_{ab}) = 1$, f_{ab} is a positive eigenfunction corresponding to the eigenvalue 1.

The constants γ and δ are calculated by the relations

$$\gamma + \delta + \int_a^b g_{ab}(t)\,dt = 1,$$

$$\frac{\partial}{\partial \xi}\left\{\int K(\xi, \eta)\,dy(\eta)\right\}\bigg|_{\xi=a} = 0,$$

and g_{ab} is obtained by the infinite series

$$g_{ab} = \sum_{n=0}^{\infty} U_{ab}^n(\gamma q_0 + \delta q_1)$$

when $\mu(U_{ab}) < 1$. In the case of $\mu(U_{ab}) = 1$, g_{ab} is the corresponding positive eigenfunction. The validity of these expressions has already been established in the proof of Theorem 6.5.1.

We illustrate the theory with the following example.

Example. Consider the kernel

$$K(\xi, \eta) = \begin{cases} L(\xi, \eta) = \eta - \xi - \lambda(\xi - \eta)^2 & \xi \leq \eta, \\ M(\xi, \eta) = \xi - \eta - \lambda(\xi - \eta)^2 & \xi \geq \eta, \end{cases}$$

with $\lambda > 0$. It is easily verified that

$$L_{\xi\xi} = L_{\eta\eta} = M_{\xi\xi} = M_{\eta\eta} = -2\lambda$$

and

$$L_\eta = 1 + 2\lambda(\xi - \eta), \qquad M_\eta = -1 + 2\lambda(\xi - \eta),$$

$$\frac{-L_{\eta\eta}}{L_\eta - M_\eta} = \frac{-M_{\eta\eta}}{L_\eta - M_\eta} = \frac{2\lambda}{2} = \lambda.$$

Furthermore, it is easy to show that the kernel is invariant under the transformations $\xi \to 1 - \xi$ and $\eta \to 1 - \eta$. Consequently we shall search for an interval of length c symmetric about 1/2 to play the role of $[a, b]$, and by the same symmetry considerations we assume that $\alpha = \beta$; the integral equation (6.5.3) becomes

$$2\alpha\lambda = f(\xi) - \int_{1/2-c/2}^{1/2+c/2} \lambda f(t)\,dt,$$

and hence $f(\xi) \equiv k$ and $2\alpha\lambda = k - c\lambda k$. The normalization condition requires $ck = 1 - 2\alpha$. If we eliminate k from the last two equations, we obtain $\alpha = (1 - \lambda c)/2$. Substituting this result into the previous equation yields $k = \lambda$. The condition

$$\int K(\xi, 0)\, dx(\xi) = \int K(\xi, a)\, dx(\xi)$$

implies after a simple calculation that either $c = 1$ or $c = -1 + 2/\lambda$. Clearly, if $c = 1$, then $a = 0$ and $b = 1$. If $c = -1 + 2/\lambda$, then $a = 1 - 1/\lambda$ and $b = 1/\lambda$; it is clear that these values are contained in the unit interval only if $\lambda > 1$, and in order that $c \geq 0$ it is necessary that $\lambda \leq 2$. We distinguish three cases.

Case 1. $0 < \lambda \leq 1$. We have shown that for this case $c = 1$, $a = 0$, and $b = 1$. Since $\alpha = (1 - \lambda c)/2$, we have $\alpha = (1 - \lambda)/2$. An optimal strategy for Player I is

$$x = \left(\frac{1 - \lambda}{2} I_0, f, \frac{1 - \lambda}{2} I_1\right),$$

where $f = \lambda$ on $[0, 1]$. Since any optimal strategy y for the minimizing player has the same density λ extending over the same interval, we deduce that $\gamma + \delta = \alpha + \beta = 2\alpha$. But the kernel of the game is symmetric about the point $\xi = 1/2$ and hence $\gamma = \delta$, which means $y = x$. Finally, we assert that these strategies are unique: their uniqueness follows from the uniqueness of $f_{ab} = g_{ab} = \lambda$, and from the symmetry that entails $\alpha = \beta$ and $\gamma = \delta$.

Case 2. $1 < \lambda < 2$. We have already shown that in this case $c = -1 + 2/\lambda$, $a = 1 - 1/\lambda$, and $b = 1/\lambda$; since $\alpha = (1 - \lambda c)/2$, we obtain $\alpha = (\lambda - 1)/2$. The strategies

$$x = \left(\frac{\lambda - 1}{2} I_{1-1/\lambda}, f, \frac{\lambda - 1}{2} I_{1/\lambda}\right),$$

$$y = \left(\frac{\lambda - 1}{2} I_0, f, \frac{\lambda - 1}{2} I_1\right),$$

where $f \equiv \lambda$ on $[1 - 1/\lambda, 1/\lambda]$, are optimal, and it is easy to show that they are unique.

Case 3. $2 \leq \lambda$. It is easy to verify directly that the optimal strategies are

$$x = (I_{1/2}), \qquad y = (\tfrac{1}{2}I_0, \tfrac{1}{2}I_1).$$

The value is now $1/2 - \lambda(1/2)^2$, and these strategies are also unique.

Butterfly kernels. A kernel is called a butterfly kernel if conditions (a) and (b) are unchanged but condition (c) is strengthened so that $L(\xi, \eta)$ is strictly increasing in η and strictly decreasing in ξ, and $M(\xi, \eta)$ is strictly increasing in ξ and strictly decreasing in η. In other words, the kernel is monotonic away from the diagonal.

Butterfly kernels are an important special class of the general kernels of this section. Two notable examples are

$$K(\xi, \eta) = \begin{cases} \phi(\xi)\psi(\eta) & \xi \leq \eta, \\ \phi(\eta)\psi(\xi) & \xi \geq \eta, \end{cases}$$

where

$$\phi(\xi) \geq k, \quad \psi(\xi) \geq k, \quad \phi'(\xi) \leq -k, \quad \psi'(\xi) \geq k,$$
$$\phi''(\xi) \leq -k, \quad \psi''(\xi) \leq -k \quad (k > 0)$$

and ϕ and ψ have continuous third derivatives; and

$$K(\xi, \eta) = |\xi - \eta| - \lambda(\xi - \eta)^2 \qquad (0 < \lambda \leq \tfrac{1}{2}).$$

We state without proof the following theorem for butterfly kernels.

▶THEOREM 6.5.2. If the kernel $K(\xi, \eta)$ is a butterfly kernel, then Players I and II possess *unique* optimal strategies of the form $x = (\alpha I_0, f_0, \beta I_1)$ and $y = (\gamma I_0, g_0, \delta I_1)$ respectively, where f_0 and g_0 are absolutely continuous over the full interval and are obtained as the solutions of a pair of integral equations

$$f_0 - Tf_0 = \alpha p_0 + \beta p_1,$$
$$g_0 - Ug_0 = \gamma q_0 + \delta q_1,$$

where $T = T_{01}$ and $U = U_{01}$ of (6.5.3) and (6.5.4), respectively, and p_0, p_1, q_0, q_1 are defined in these same equations.

Furthermore, the solution of these integral equations may be calculated as the Neumann series

$$f_0 = \sum_{n=0}^{\infty} T^n(\alpha p_0 + \beta p_1),$$

$$g_0 = \sum_{n=0}^{\infty} U^n(\gamma q_0 + \delta q_1).$$

The constants α and β are determined from the normalization condition on x and the relation

$$\frac{\partial}{\partial \eta} \int K(\xi, \eta)\, dx(\xi) \bigg|_{\eta=0} = 0.$$

The constants γ and δ are determined from the normalization condition on y and the relation

$$\frac{\partial}{\partial \xi} \int K(\xi, \eta)\, dy(\eta) \bigg|_{\xi=0} = 0.$$

6.6 Problems.

1. Solve the noisy duel with one bullet for each player and accuracy functions $P_1(t) = |\frac{1}{3} - t|$ and $P_2(t) = \frac{1}{2} - |\frac{1}{2} - t|$, using the method of Example 2, Section 6.1.

2. Solve the noisy duel between two opponents with probabilities p_1 and p_2 of having one bullet each. Assume that the accuracy function for each player is t, and that the value to Player I is α if he alone survives, β if Player II alone survives, $\alpha + \beta$ if neither survive and 0 if both survive. Assume that $\alpha > 0$ and $\beta < 0$. (*Hint*: This problem must be dealt with by the methods of Chapter 5.)

3. Consider the game in which Player I picks a point from either A_1 or A_2 and Player II picks a point from B with pay-off kernel as described. Prove that this game has no value. Interpret the game as a noisy duel.

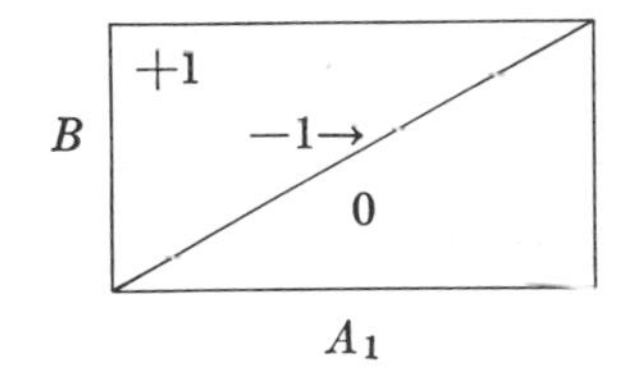

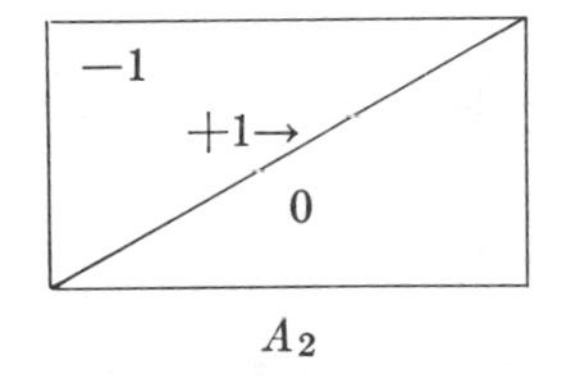

4. Verify that the noisy duel with equal accuracies $P(t) = t$ and m and n bullets for Players I and II, respectively, has the following solution:

(a) The value is $(m - n)/(m + n)$.

(b) As soon as a bullet is fired, the players should play optimally for the game with the remaining number of bullets. If no bullet is fired before $t = 1/(m + n)$, the player with the largest number of bullets should fire once at that time. If he fails to fire then, his opponent should fire immediately.

5. Solve Problem 4 with an arbitrary, monotone-increasing accuracy function $P(t)$, where $P(0) = 0$ and $P(1) = 1$.

6. Solve the game whose kernel, defined on the unit square, is

$$K(\xi, \eta) = \begin{cases} 1 & \xi > \eta \quad \text{or} \quad \eta > \xi + \lambda, \\ \lambda & \xi + \lambda \geq \eta \geq \xi, \end{cases}$$

where $\lambda < 1$.

[This game has the following interpretation. An obstacle is placed by Player I in the interval [0, 1] (which may represent time). Player II has discharged a missile which travels from 1 to 0 at a uniform rate 1 except for an interval of length λ, where it travels at an increased rate. The mechanism of the missile is such that the burst of speed on the interval of length λ can be arranged to start at any given instant. This is the element of pure strategy available to Player II. With probability 1 the obstacle pushes the missile off course when it is traveling at rate 1, and with probability λ the missile is diverted when traveling at the increased speed. The pay-off to Player I is the probability of diverting the missile.]

7. Give a proof of Theorem 6.5.2 by using the theory of positive transformations and alternatively by appealing to the methods of Theorem 6.5.1.

8. Players I and II submit sealed bids on a contract that will cost them respectively c_1 and c_2 units to fulfill. The contract is awarded to the lower bidder, or at random if the bids are equal. Set up the pay-offs to the two players, and show that if $c_1 = c_2$ there are pure equilibrium strategies.

[The game is not zero-sum. Let $K_1(\xi, \eta)$ and $K_2(\xi, \eta)$ denote the pay-off kernels for Players I and II, respectively. A pair of strategies $x(\xi)$ and $y(\eta)$ are said to be optimal in the sense of equilibrium points if

$$\int K_1(\xi, \eta)\, dx(\xi) \geq \int K_1(\xi, \eta)\, dx(\xi)\, dy(\eta) \qquad 0 < \eta,$$

and

$$\int K_2(\xi, \eta)\, dy(\eta) \geq \int K_2(\xi, \eta)\, dx(\xi)\, dy(\eta) \qquad 0 < \xi.$$

This is the equilibrium strategy concept. It reduces to the usual notion of optimal strategy when $K_2(\xi, \eta) = -K_1(\xi, \eta)$. Note that in this example we assume the region over which both ξ and η may vary to be the full positive real axis.]

9. In Problem 8, assume that $c_1 \neq c_2$ and that all bids are necessarily restricted to the interval $[a, \infty)$. Show that for any $a > \max(c_1, c_2)$ there exist optimal strategies (i.e., an equilibrium pair) with continuous positive densities on the interval $[a, \infty)$. (*Hint*: The methods of silent duels apply here. Try to determine a pair x and y which make the integrals of Problem 8 constant on an interval $[a, \infty)$.)

Notes and References to Chapter 6

§ **6.1.** The origins of noisy duel game problems or games of timing of class I were similar to those of silent duels (see Chapter 5). The general noisy duel, however, as formulated in this section, was first studied and solved by Glicksberg [107]. An independent and more complete determination of the solution was given by Karlin [134], using a different approach. The generalized noisy duel is solved by approximating a solution uniformly by a sequence of games of timing of class II and then invoking a standard limiting process. The uniqueness of the optimal strategies is established afterwards. The analysis is complicated by the fact that one of the players need not possess an attainable optimal strategy, although ϵ-effective optimal strategies do exist for every positive ϵ. Because of this the concept of uniqueness in this context must be defined with care.

In view of the fact that one player need not possess an attainable optimal strategy, it is not surprising to discover relatively simple noisy duels which are indeterminate in the sense that they have no value. Shapley (Problem 4 of Chapter 2) has constructed several such disconcerting examples (see also Sion and Wolfe [223]).

§ **6.2.** The first example of this section is a special case of the general game of timing of class I. The extension of the noisy duel which permits the possibility of nonmonotonic accuracy functions is due to Blackwell [29].

§ **6.3.** We follow here the proof of Karlin [134].

§ **6.4.** Duels involving several opportunities for action were originally introduced by Blackwell and Girschick (see notes to Chapter 5). They successfully treated the n-bullets-vs.-m-bullets noisy game for the case of equal accuracy functions [30] (see Problem 4). However, their methods do not extend to the asymmetrical situation, nor has anyone else devised methods covering this situation. The problem is complex, since attainable optimal strategies need not exist even though the game has a value.

More progress has been made with the multiple-action silent duel, namely the natural generalization of the model of Section 5.2. The solution of the equal-accuracy, equal-number, n-bullet silent duel was uncovered initially by good guesswork. Later the investigation of the equal-accuracy, 2-vs.-1 silent duel was attempted by Pinney while consulting for RAND. His results, although inaccurate in minor details, were close enough to enable Shapley to arrive at a valid solution [217], in which he ingeniously pointed out some of the underlying features of the solution to the general silent duel.

It was Restrepo [205], in his Ph.D. dissertation, who cracked the problem in general. The solution is vastly detailed—in particular, an enormous number of cumbersome transcendental equations are needed to be solved in order to determine the parameters of the solution. The complexity of the analysis may prevent some readers from sensing the importance of Restrepo's work; actually, however, an elaborate apparatus is required by the nature of the problem, and Restrepo's handling of it would be hard to improve on. In this section we discuss Restrepo's result, specialized in the proofs of Lemmas 6.4.1–2 to the 2-vs.-1 case and allowing arbitrary monotonic accuracy functions.

It should be mentioned at this point that the n-vs.-m silent duel does not represent a full extension of the general game of timing of class II. This class of one-action games includes many games that are not expressible as duels (kernels in separable form). The full generalization of games of timing to higher dimensions is still an open problem.

§ 6.5. Examples of butterfly games and several of their general properties were first uncovered by Glicksberg and Gross [109], who are responsible also for the apt and picturesque name. The general theory and its complete solution are due to Karlin [134]. The proof of Theorem 6.5.2 can be found in [134].

§ 6.6. Problem 3 is due to Shapley (unpublished). Problem 4 is due to Blackwell and Girschick [30]. The concept of an equilibrium point for nonzero-sum games originated with Nash [188]. Problems 8 and 9 were suggested by Shapley [218].

CHAPTER 7*

MISCELLANEOUS GAMES

We have seen that in order to secure information concerning the optimal strategies of a continuous game, it is necessary to impose special assumptions on the kernel of the game. Usually the necessary assumptions are suggested naturally by the model on which the game is founded. In such cases, the solutions can be guessed approximately from intuitive considerations. However, if we seek to develop a fuller understanding of infinite games in general, it is important to pursue the classification of games on the unit square without any reference to the physical meaning of the game. The reader is referred to Section 2.3 for a preliminary survey discussion of the miscellaneous games covered in this chapter.

In view of the relative accessibility of polynomial games, we shall examine first those games whose kernels are analytic or rational functions. These games are studied in Section 7.1. While it is possible to obtain some information about their optimal strategies, this information is necessarily limited. There are some rational games whose unique solutions are extremely complex distributions (for instance, the Cantor distribution).

In Section 7.2 we examine games whose kernels are bell-shaped. In this case, we can give a fairly complete description of the optimal strategies. The kernels are then generalized to so-called Pólya-type (P.T.) kernels, whose solutions are like those of butterfly-shaped games (see Section 6.5).

The final section of this chapter treats briefly the important concept of games that are invariant with respect to a group of transformations. Here we attempt to relate the symmetries of the kernel of the game to the form of the optimal strategy. The invariance principle is of some use in reducing complex game problems to simpler situations, but on the whole its applications appear to be more pertinent to the study of statistical decision problems. Our discussion of invariant games may be regarded as an introduction to the larger theory examined in the context of games on the unit square.

7.1 Games with analytic kernels. A pay-off kernel $K(\xi, \eta)$ on the unit square which is jointly continuous in (ξ, η) is said to be an analytic kernel if (i) for each fixed η it can be expanded about each ξ $(0 \leq \xi \leq 1)$ into

* This entire chapter is of more special nature and may be omitted at first reading without loss of continuity.

a convergent Taylor series and (ii) for each ξ it can be expanded about each η $(0 \leq \eta \leq 1)$ into a convergent Taylor series. Each power series possesses a positive radius of convergence.

Many of the theorems about analytic kernels describe the support of the optimal strategies. In this respect these theorems resemble those obtained for polynomial kernels, but the results are less precise. Typical of these theorems is the following one.

▶THEOREM 7.1.1. If the kernel is analytic and if one of the players has an optimal strategy whose support contains an infinite number of points, then every strategy that is optimal for the other player is an equalizing strategy.

Proof. Let y be any optimal minimizing strategy, and define $h(\xi) = \int K(\xi, \eta)\, dy(\eta)$. The function h is analytic and assumes the value v at every point ξ that is in the support of any optimal strategy (Lemma 2.2.1). If the number of these points is infinite, the analyticity of h implies that $h(\xi) \equiv v$. This is precisely the assertion that y is an equalizing strategy.

A similar argument can be used when the optimal minimizing strategies have a support that contains an infinite number of points.

Further explicit results can be obtained by imposing additional restrictions on the kernel. Theorems 7.1.2 and 7.1.3 describe two such results. Other theorems are suggested by Problems 3 and 4 at the end of the chapter.

▶THEOREM 7.1.2. Let $Q(\xi, \eta)$ be the reciprocal of $K(\xi, \eta)$. If $K(\xi, \eta)$ is analytic on the unit square, and if there exists some ξ_0 such that $Q_\xi(\xi_0, \eta) > 0$ for $0 \leq \eta \leq 1$, then the support of all the optimal maximizing strategies is a finite set.

Proof. If this theorem were false, the previous theorem would imply that for any optimal $y(\eta)$,

$$\int_0^1 \frac{1}{Q(\xi, \eta)}\, dy(\eta) = v.$$

Then, by differentiation,

$$\int_0^1 \frac{Q_\xi(\xi, \eta)}{[Q(\xi, \eta)]^2}\, dy(\eta) = 0.$$

But this identity is impossible for $\xi = \xi_0$, because the integrand is positive. Hence the theorem is true.

▶THEOREM 7.1.3. Let $K(\xi, \eta)$ be an analytic kernel that is strictly positive over the unit square. Let ξ be a complex variable, and assume that there exists in the complex plane a curve C that connects 0 to ∞ such that, when

ξ ranges over C and $0 \leq \eta \leq 1$, the kernel $K(\xi, \eta)$ remains analytic and uniformly bounded and there exists $\xi_n \in C$ such that

$$\lim_{\xi_n \to \infty} |K(\xi_n, \eta)| = 0.$$

Then the optimal strategies for Player I concentrate on a finite number of points.

Proof. We assume that the theorem is false. Then Theorem 7.1.1 asserts that for every optimal $y(\eta)$,

$$\int K(\xi, \eta)\, dy(\eta) \equiv v > 0 \qquad (0 \leq \xi \leq 1).$$

Since the left-hand side is an analytic function for $\xi \in C$, the last identity prevails for all $\xi \in C$. Then

$$\begin{aligned} v &= \lim_{\xi_n \to \infty} \int K(\xi_n, \eta)\, dy(\eta) \\ &= \int \lim_{\xi_n \to \infty} K(\xi_n, \eta)\, dy(\eta) = 0. \end{aligned}$$

But this is impossible, since

$$\min_{0 \leq \xi, \eta \leq 1} K(\xi, \eta) > 0$$

implies $v > 0$.

Example 1. Let

$$K(\xi, \eta) = \frac{1}{1 + \lambda(\xi - \eta)^2},$$

with $\lambda > 0$.

For the curve C we may take the real positive ξ-axis. Then Theorem 7.1.3 implies that the optimal strategies for Player I concentrate on a finite number of points.

In contrast to the results obtained above, it is possible to construct games with simple kernels and complex solutions. One of these games is described in Example 2. This example illustrates some of the difficulties that arise in any theory of continuous games. It also illustrates some of the techniques employed in the analysis of these games.

The kernel is a rational function, but the unique optimal maximizing strategy is the classical Cantor function $C(\xi)$. This function has some unusual properties: it is continuous everywhere, and differentiable except on a noncountable set of Lebesgue measure zero. Moreover, the derivative vanishes wherever it is defined, but $C(\xi)$ is a cumulative distribution function. The graph of $C(\xi)$ can be constructed as follows: Define $C(0) = 0$ and $C(1) = 1$. Then divide the interval $[0, 1]$ into three equal subintervals, and in the middle one $(\frac{1}{3} \leq \xi \leq \frac{2}{3})$ set $C(\xi) = [C(1) + C(0)]/2$. The process is iterated as follows: Divide each maximal interval $[a, b]$ in which $C(\xi)$ has not been defined into three equal subintervals, and in the middle subinterval set $C(\xi) = [C(a) + C(b)]/2$. For example, $C(\xi) = \frac{1}{4}$ for $\frac{1}{9} \leq \xi \leq \frac{2}{9}$, and $C(\xi) = \frac{3}{4}$ for $\frac{7}{9} \leq \xi \leq \frac{8}{9}$, etc. A formal definition of $C(\xi)$ is given below.

Example 2. In this example we consider a game with the Cantor distribution as its unique optimal strategy. The Cantor distribution as a distribution function is characterized by the functional equations $C(\xi/3) = \frac{1}{2}C(\xi)$, $C(1 - \xi) = 1 - C(\xi)$, together with the boundary conditions $C(0) = 0$, $C(1) = 1$. We shall show that $C(\xi)$ is the unique optimal maximizing strategy for the game whose kernel is the rational function

$$K(\xi, \eta) = (\eta - \tfrac{1}{2}) \left\{ \frac{1 + (\xi - \frac{1}{2})(\eta - \frac{1}{2})^2}{1 + (\xi - \frac{1}{2})^2(\eta - \frac{1}{2})^4} - \frac{1}{1 + (\xi/3 - \frac{1}{2})^2(\eta - \frac{1}{2})^4} \right\}.$$

In fact, $K(\xi, \eta)$ is an analytic function of ξ and η in a neighborhood of the origin.

This kernel is skew-symmetric about $\eta = \frac{1}{2}$. Hence, by symmetry, any y such that $y(1 - \eta) = 1 - y(\eta)$ is an equalizing strategy, with

$$\int_0^1 K(\xi, \eta)\, dy(\eta) \equiv 0. \tag{7.1.1}$$

To verify that $C(\xi)$ is optimal, it is sufficient to show that

$$\int_0^1 K(\xi, \eta)\, dC(\xi) = 0. \tag{7.1.2}$$

The term of $K(\xi, \eta)$ that involves

$$(\xi - \tfrac{1}{2})\,(\eta - \tfrac{1}{2})^2$$

is also skew-symmetric about $\xi = \frac{1}{2}$ and does not contribute to the integral of (7.1.2). Therefore, it is sufficient to show that

$$\int_0^1 \left\{ \frac{1}{1 + (\xi - \frac{1}{2})^2(\eta - \frac{1}{2})^4} - \frac{1}{1 + (\xi/3 - \frac{1}{2})^2(\eta - \frac{1}{2})^4} \right\} dC(\xi) \equiv 0.$$

The validity of this equation can be verified easily, using the substitution $\xi' = \xi/3$ and the properties of $C(\xi)$:

$$\int_0^1 \frac{dC(\xi)}{1+(\xi/3-\frac{1}{2})^2(\eta-\frac{1}{2})^4}$$

$$= \int_0^{1/3} \frac{dC(3\xi)}{1+(\xi-\frac{1}{2})^2(\eta-\frac{1}{2})^4} = 2\int_0^{1/3} \frac{dC(\xi)}{1+(\xi-\frac{1}{2})^2(\eta-\frac{1}{2})^4}$$

$$= \int_0^{1/3} \frac{dC(\xi)}{1+(\xi-\frac{1}{2})^2(\eta-\frac{1}{2})^4} + \int_{2/3}^{1} \frac{dC(\xi)}{1+(\xi-\frac{1}{2})^2(\eta-\frac{1}{2})^4}$$

$$= \int_0^{1} \frac{dC(\xi)}{1+(\xi-\frac{1}{2})^2(\eta-\frac{1}{2})^4}.$$

The fact that $C(\xi)$ is constant over the interval $\frac{1}{3} \leq \xi \leq \frac{2}{3}$ has been used in the last step.

Consider now the question of uniqueness. Equation (7.1.1) shows that y is optimal if $1 - y(\eta) = y(1-\eta)$ for $0 \leq \eta \leq 1$. Therefore, any optimal x must have the property that

$$\int_0^1 K(\xi, \eta)\, dx(\xi) \equiv 0 \qquad (0 \leq \eta \leq 1).$$

In the last equation $K(\xi, \eta)$ can be expressed as a uniformly convergent series of powers of $(\xi - \frac{1}{2})$, $(\xi/3 - \frac{1}{2})$ and $(\eta - \frac{1}{2})$. In this form, if x is optimal, then

$$\int_0^1 \sum_{n=0}^{\infty} (-1)^n \{(\xi-\tfrac{1}{2})^{2n}(\eta-\tfrac{1}{2})^{4n+1} + (\xi-\tfrac{1}{2})^{2n+1}(\eta-\tfrac{1}{2})^{4n+3}\}\, dx(\xi)$$
$$- \int_0^1 \sum_{n=0}^{\infty} (-1)^n (\xi/3-\tfrac{1}{2})^{2n}(\eta-\tfrac{1}{2})^{4n+1}\, dx(\xi) \equiv 0.$$

The left-hand side of this equation is a power series in $(\eta - \frac{1}{2})$, and this series vanishes identically in the interval $0 \leq \eta \leq 1$. Hence the coefficient of each power of $\eta - \frac{1}{2}$ must vanish; and since $4n + 1 \neq 4m + 3$ for any choice of integers n and m, it follows that

$$\int_0^1 \{(\xi-\tfrac{1}{2})^{2n} - (\xi/3-\tfrac{1}{2})^{2n}\}\, dx(\xi) = 0 \tag{7.1.3}$$

and

$$\int_0^1 (\xi-\tfrac{1}{2})^{2n+1}\, dx(\xi) = 0. \tag{7.1.4}$$

Denoting the moments of x by $\mu_0, \mu_1, \mu_2, \ldots$ and recalling that for every distribution $\mu_0 = 1$, we can successively determine $\mu_1, \mu_2, \ldots$ from (7.1.3) and (7.1.4) in terms of the moments computed previously. This procedure works in each equation, since the coefficient of the highest power of ξ is different from zero. Since all the moments are computed by solving algebraic equations that do not depend on x, it follows that all the optimal maximizing strategies have the same moments; and since the moments determine the strategy, there is only one optimal strategy. This strategy is, of course, $C(\xi)$.

Example 2 shows that there are games whose kernels are analytic functions in which one of the players possesses no simple solution. A rational game on the unit square can be constructed for which the Cantor distribution is the unique optimal strategy for both players (see Problem 11). More pathological examples are obtained by weakening the restrictions on the kernel. Glicksberg and Gross have constructed a game with a rational pay-off function and with unique optimal strategies that concentrate in countable dense sets (see the notes at the close of this chapter).

Example 3. In this example we consider a game whose unique optimal strategies are two given distributions, neither of which is a finite step function, but which are otherwise completely arbitrary.

Let x and y be the given distributions, and let μ_n and ν_n denote the nth moments of x and y, respectively. Define

$$K(\xi, \eta) = \sum_{n=0}^{\infty} \frac{1}{2^n} (\xi^n - \mu_n)(\eta^n - \nu_n) \qquad (0 \leq \xi, \eta \leq 1).$$

Each term of this series is smaller in absolute value than the corresponding term of the series $1/2^n$. Therefore the series is absolutely and uniformly convergent, and

$$\int_0^1 K(\xi, \eta)\, dx(\xi) = \sum_{n=1}^{\infty} \frac{1}{2^n} (\eta^n - \nu_n) \int_0^1 (\xi^n - \mu_n)\, dx(\xi) \equiv 0,$$

$$\int_0^1 K(\xi, \eta)\, dy(\eta) = \sum_{n=0}^{\infty} \frac{1}{2^n} (\xi^n - \mu_n) \int_0^1 (\eta^n - \nu_n)\, dy(\eta) \equiv 0.$$

In particular, x and y are optimal, and the value of the game is zero. Furthermore, since x and y are not finite-step functions and K is analytic, Theorem 7.1.1 can be applied to this game. Thus, if x^* is any optimal maximizing strategy, then

$$0 \equiv \int K(\xi, \eta)\, dx^*(\xi) = \sum_{n=0}^{\infty} \frac{1}{2^n} (\mu_n^* - \mu_n)(\eta^n - \nu_n),$$

where μ_n^* is the nth moment of x^*. Since the series vanishes identically, the coefficient of each power of η must vanish. Thus $\mu_n^* = \mu_n$ and $x^* = x$. Similarly, the optimal minimizing strategy is unique.

7.2 Bell-shaped kernels. Bell-shaped kernels include and generalize the properties of the well-known normal probability function $e^{-\xi^2}$, whose characteristic shape resembles the vertical cross section of a bell and whose common occurrence more than justifies the study of these kernels. This function is a suitable kernel for pursuit games in which the pursuer does not know the exact location of the quarry. For instance, a bomber that knows the course followed by an enemy submarine may be unaware of its exact location; if the bomber aims at a point ξ while the submarine is at another point η, the expected damage to the submarine may be proportional to $e^{-(\xi-\eta)^2}$ (see Section 2.4). Although this game can be solved explicitly, it is also possible to abstract from it the relevant properties of its kernel and thus construct a general theory that is applicable to a larger class of games. This abstraction leads to the concept of a bell-shaped kernel.

A kernel K is said to be bell-shaped if $K(\xi, \eta) = \varphi(\xi - \eta)$, where φ has the following properties:

(i) $\varphi(u)$ is defined and continuous for all u.

(ii) For every n and every set of values ξ_i and η_j such that

$$\xi_1 < \cdots < \xi_n \qquad \text{and} \qquad \eta_1 < \cdots < \eta_n,$$

the determinant of the matrix $\|\varphi(\xi_i - \eta_j)\|$ is nonnegative.

(iii) For each set $\{\xi_i\}$ with $\xi_1 < \cdots < \xi_n$ there exists a set $\{\eta_j\}$ with $\eta_1 < \cdots < \eta_n$ such that the determinant of $\|\varphi(\xi_i - \eta_j)\|$ is strictly positive; and the corresponding condition must hold if the η_j's are prescribed first.

(iv)
$$\int_{-\infty}^{\infty} \varphi(u)\, du < \infty.$$

Any function that satisfies properties (i), (ii), and (iv) is called a Pólya frequency function (P.F.F.), and any function that satisfies all four properties is called a regular P.F.F. When conditions (ii) and (iii) can be replaced by the stronger condition that det $\|\varphi(\xi_i - \eta_j)\| > 0$, the resulting function is called a *proper* Pólya frequency function (P.P.F.F.). Functions such as e^{-u^2} and e^{u-e^u} are P.P.F.F. (see Problem 5); $e^{-|u|}$ is a regular P.F.F. It can be shown (see references) that if φ_1 is a regular P.F.F. and φ_2 is a P.P.F.F., then the convolution

$$\varphi_1 * \varphi_2 = \int_{-\infty}^{\infty} \varphi_1(u)\varphi_2(x - u)\, du$$

is a P.P.F.F. In particular, if

$$\varphi_2 = \frac{1}{\sqrt{2\pi\sigma}} e^{-u^2/2\sigma},$$

we see, by letting $\sigma \to 0$, that any regular P.F.F. can be approximated by means of a P.P.F.F.

The sign variation of a function f, denoted by $V(f)$, is defined to be the supremum of the number of changes of sign of the sequence $f(\xi_1), f(\xi_2), \ldots, f(\xi_n)$, where the ξ_i's are chosen arbitrarily but arranged in increasing order, and where the number n is also allowed to vary arbitrarily. The fundamental proposition, established by Schoenberg, concerning Pólya frequency functions is that the transformation

$$g(u) = \int_{-\infty}^{\infty} \varphi(u - t) f(t)\, dt$$

is sign-variation-diminishing [that is, $V(g) \leq V(f)$] for all bounded f with a finite number of discontinuities of the first kind, if and only if φ is a P.F.F. The properties of the regular P.F.F. that are relevant to the theory of games are established in the following sequence of lemmas.

▶LEMMA 7.2.1. If φ is a regular P.F.F., then for any choice of real numbers a_j and η_j the function

$$f(\xi) = \sum_{j=1}^{n} a_j \varphi(\xi - \eta_j)$$

has at most $n - 1$ changes of sign.

Proof. Suppose to the contrary that the lemma is false, so that there exist numbers

$$\xi_1, \ldots, \xi_{n+1} \qquad \text{with} \qquad \xi_i < \xi_{i+1}$$

and

$$f(\xi_i) f(\xi_{i+1}) < 0 \qquad (i = 1, \ldots, n).$$

Then, by the definition of f,

$$\begin{vmatrix} f(\xi_1) & f(\xi_2) & \cdots & f(\xi_{n+1}) \\ \varphi(\xi_1 - \eta_1) & \varphi(\xi_2 - \eta_1) & \cdots & \varphi(\xi_{n+1} - \eta_1) \\ \vdots & \vdots & & \vdots \\ \varphi(\xi_1 - \eta_n) & \varphi(\xi_2 - \eta_n) & \cdots & \varphi(\xi_{n+1} - \eta_n) \end{vmatrix} = 0. \qquad (7.2.1)$$

Expanding this determinant by the elements of the first row, we obtain

$$\sum_{i=1}^{n+1} (-1)^{i+1} f(\xi_i) U \begin{pmatrix} \xi_1, \ldots, \xi_{i-1}, & \xi_{i+1}, \ldots, \xi_{n+1} \\ \eta_1, \ldots & \ldots \quad \ldots \ldots, \eta_n \end{pmatrix} = 0, \qquad (7.2.2)$$

where U denotes the appropriate minor in the determinant.† If φ is a P.P.F.F., equation (7.2.2) readily leads to a contradiction; indeed, since there is no loss of generality in assuming that $\eta_1 < \eta_2 < \cdots < \eta_n$, all the U's are strictly positive while all the products $(-1)^i f(\xi_i)$ are of the same sign. Thus, the left-hand side of (7.2.2) cannot vanish. If φ is not a P.P.F.F., it is sufficient to approximate φ by a P.P.F.F. φ^* whose corresponding function f^* would have as many changes of sign as f.

Remark 7.2.1. A more refined analysis shows that when φ is a P.P.F.F., the number of zeros of $f(\xi)$ (counting zeros of even multiplicities twice) is bounded by the number of changes of sign of the sequence $\{a_j\}$, provided that the corresponding η_j's are arranged in order of magnitude $\eta_1 < \eta_2 < \cdots < \eta_n$ and that the a_j's are not all zero. If, in addition, φ is infinitely continuously differentiable, then the number of zeros of $f(\xi)$ counting multiplicities in the usual sense is bounded by the number of changes of sign of the sequence $\{a_j\}$.

▶LEMMA 7.2.2. Any real, analytic, positive function which is a regular P.F.F. must be a P.P.F.F.

Proof. By assumption, $\varphi(\xi - \eta) > 0$, and one may proceed by induction on the size of the determinants. Assuming that all $n \times n$ determinants are strictly positive and that $\xi_1 < \xi_2 < \cdots < \xi_n$ and $\eta_1 < \eta_2 < \cdots < \eta_{n+1}$, let ξ be a variable, and let

$$f(\xi) = \begin{vmatrix} \varphi(\xi_1 - \eta_1) \cdots \varphi(\xi_1 - \eta_{n+1}) \\ \vdots \qquad\qquad \vdots \\ \varphi(\xi_n - \eta_1) \cdots \varphi(\xi_n - \eta_{n+1}) \\ \varphi(\xi - \eta_1) \cdots \varphi(\xi - \eta_{n+1}) \end{vmatrix}. \tag{7.2.3}$$

Expanding this determinant by the last row, we obtain

$$f(\xi) = \sum_{k=1}^{n+1} a_k \varphi(\xi - \eta_k),$$

where by virtue of the induction hypotheses the coefficients a_k are all nonzero and alternate in sign. If $\hat{f}$ and $\hat{\varphi}$ denote the Fourier transforms of f and φ, then

$$\hat{f}(s) = \hat{\varphi}(s) \sum_{k=1}^{n+1} a_k e^{is\eta_k}.$$

† See page 344 for further details concerning the notation customarily used in describing minors of determinants.

Clearly $\hat{f}(s)$ cannot vanish identically, since neither $\hat{\varphi}(s)$ nor the exponential polynomial vanishes except for isolated values. Therefore $f(\xi)$ does not vanish identically. Also, since $f(\xi)$ is analytic, it has only isolated zeros, and these zeros include the points $\xi_1, \xi_2, \ldots, \xi_n$.

We next show that $f(\xi)$ changes sign at the points $\xi_1, \ldots, \xi_n$. For $\xi > \xi_n$, property (ii) of P.F.F.'s shows that $f(\xi) \geq 0$. For $\xi_{n-1} < \xi < \xi_n$, interchanging the last two rows of (7.2.3), we see that $f(\xi) \leq 0$. The argument can be extended to the remaining intervals. Thus $f(\xi)$ alternates in sign in the intervals $(-\infty, \xi_1), (\xi_1, \xi_2), \ldots, (\xi_n, \infty)$, and hence it has the maximum number of changes of sign allowed by Lemma 7.2.1.

Finally, if $f(\xi)$ vanishes at some point ξ distinct from $\xi_1, \ldots, \xi_n$, then ξ must be a root of even multiplicity. We assume for definiteness that $\xi_k < \xi < \xi_{k+1}$ and that $f(\xi) \geq 0$ in this interval; then the function $g(\xi) = f(\xi) - \epsilon\varphi(\xi - \eta_i)$ (i is arbitrary) satisfies the conditions of Lemma 7.2.1. For ϵ sufficiently small, $g(\xi)$ has at least all the changes of sign of $f(\xi)$ and two more changes of sign near ξ. This contradicts Lemma 7.2.1 and shows that $f(\xi)$ can vanish only at $\xi_1, \ldots, \xi_n$.

▶ Lemma 7.2.3. If φ is a regular P.F.F. and φ' is continuous, then the function

$$g(\xi) = \sum_{i=1}^{n} a_i\varphi'(\xi - \eta_i)$$

has at most $2n - 1$ changes of sign.

Proof. Since φ' is continuous, the functions

$$g_h(\xi) = \frac{1}{h} \sum_{i=1}^{n} a_i\{\varphi(\xi + h - \eta_i) - \varphi(\xi - \eta_i)\}$$

converge uniformly to $g(\xi)$ over any finite interval. Suppose g changes sign at least $2n$ times. Thus, for h sufficiently small, g_h changes sign at least $2n$ times. But this is impossible, for since g_h is clearly of the form

$$\sum_{j=1}^{2n} b_j\varphi(\xi - z_j),$$

with $b_j = \pm a_i/h$ and z_j equal to either η_i or $\eta_i - h$, Lemma 7.2.1 shows that g_h can have at most $2n - 1$ changes of sign.

7.3 Bell-shaped games. A game with kernel K defined over the unit square is said to be bell-shaped if $K(\xi, \eta) = \varphi(\xi - \eta)$, where φ is a positive, analytic, regular P.F.F. Although the graph of $\varphi(t)$ is indeed bell-shaped, the above definition may eventually appear to be too restrictive: for example, it does not include functions like $\psi(t) = 1/(1 + t^2)$, whose graph is also bell-shaped and whose associated kernel $\psi(\xi - \eta)$ gives rise to a game whose solution has most of the properties described below. Nevertheless, some of the following proofs depend heavily on the sign-variation-diminishing properties of P.P.F.F.'s, and at the present time there are no proofs that are applicable to more general classes of bell-shaped functions. The results that have been obtained so far concern the nature of the spectrums of optimal strategies of bell-shaped games.

▶ LEMMA 7.3.1. The value v of a bell-shaped game is strictly positive, and the support of the optimal strategies of each player is a finite set of points.

Proof. The value of the game is bounded below by the minimum value of $K(\xi, \eta)$ in the unit square. Since φ is continuous, it actually achieves this minimum value δ, and since φ is strictly positive, $\delta > 0$. Thus $v \geq \delta > 0$.

If the support of the optimal strategies of one of the players (say Player I) has an infinite number of points, then for any optimal y,

$$\int_0^1 \varphi(\xi - \eta)\, dy(\eta) = v$$

for an infinite number of ξ's in the interval [0, 1]. Since the integral is an analytic function of ξ, it follows that

$$\int_0^1 \varphi(\xi - \eta)\, dy(\eta) \equiv v$$

for all ξ. But any P.F.F., by (iv), vanishes exponentially fast as $\xi \to \infty$, so that as $\xi \to \infty$,

$$\int_0^1 \varphi(\xi - \eta)\, dy(\eta) \to 0,$$

contradicting the fact that $v > 0$.

It should be noticed that the lemma is not valid if the assumption of analyticity is discarded; for instance, the kernel $e^{-|\xi-\eta|}$ satisfies all other assumptions but does not possess any optimal strategies with finite support. This game will be analyzed in Section 7.4.

▶ LEMMA 7.3.2. If there exists an optimal minimizing strategy whose support contains precisely n points, then the support of any optimal maximizing strategy consists of n or $n - 1$ fixed points.

Proof. Let y_{η_j} denote as usual the strategy that concentrates exclusively at the point η_j. Then any minimizing strategy with n points of support can be written in the form

$$\sum_{j=1}^{n} a_j y_{\eta_j},$$

and the corresponding pay-off to Player I is

$$h(\xi) = \sum_{j=1}^{n} a_j \varphi(\xi - \eta_j).$$

If the strategy is optimal, then $h(\xi) \leq v$ and $h(\xi) = v$ only at isolated points that include the points of support of the optimal maximizing strategies.

Obviously, if a point ξ_i belongs to the support of an optimal maximizing strategy and also lies interior to the unit interval, then ξ_i is a maximum of $h(\xi)$, and $h'(\xi)$ must change sign at ξ_i. Furthermore, between any two such points there must be a minimum of $h(\xi)$, and therefore $h'(\xi)$ will also change sign between each two successive relative maxima. Therefore, if the support of the optimal maximizing strategies contains $n + 1$ points interior to the unit interval, it follows that $h'(\xi)$ must change sign at least $2n + 1$ times, contradicting Lemma 7.2.3. The same contradiction results if the support contains $n + 1$ points including 0 or 1 or both. In this case the end point need not be a maximum, but since

$$\lim_{\xi \to \pm\infty} h(\xi) = 0,$$

there must be a maximum between $-\infty$ and 0 or between 1 and ∞.

A similar argument can be used to show that the support of any optimal maximizing strategy must contain at least $n - 1$ points. Indeed, if

$$\sum_{i=1}^{m} b_i x_{\xi_i}$$

is optimal, then the corresponding pay-off to Player I is

$$r(\eta) = \sum_{i=1}^{m} b_i \varphi(\xi_i - \eta).$$

As before, $r(\eta) \geq v$, $r(\eta) \not\equiv v$, and $r(\eta_j) = v$ $(j = 1, \ldots, n)$. Also, between any two successive η_j's there is at least one maximum of $r(\eta)$, and therefore $r'(\eta)$ changes sign in this interval; furthermore, each η_j interior to [0, 1] is also a minimum of $r(\eta)$, and therefore $r'(\eta)$ changes sign. Hence the total number of changes in sign is at least $(n - 2) + (n - 1) = 2n - 3$. Since the maximum number of changes of sign allowed by Lemma 7.2.3 is $2m - 1$, we may infer that $m \geq n - 1$.

The next lemma can be applied only to those bell-shaped games that are normalized so that $\varphi'(0) = 0$. At first glance, this normalization appears to be inconsequential. Since φ is positive and analytic and vanishes at

$\pm\infty$, it follows that φ' must have at least one zero. Further, suppose that φ' vanishes at ξ_1 and ξ_2; then

$$\varphi(\xi) - \frac{\varphi(\xi_1)}{\varphi(\xi_2)}\,\varphi(\xi - \xi_1 + \xi_2)$$

has a double zero at ξ_1. But this is impossible by Remark 7.2.1; hence φ' has only one zero. The normalization $\varphi'(0) = 0$ requires a translation of the axis, and such a translation alters the strategy spaces. Fortunately, many games naturally arise in normalized form.

▶ LEMMA 7.3.3. If $\varphi'(0) = 0$, then the points 0 and 1 belong to the support of every optimal minimizing strategy but do not belong to the support of any optimal maximizing strategy.

Proof. Let

$$x(\xi) = \sum_{i=1}^{m} b_i x_{\xi_i} \qquad \text{and} \qquad y(\eta) = \sum_{j=1}^{n} a_j y_{\eta_j}$$

be optimal strategies with a_i, $b_j > 0$ and ξ_i and η_j arranged in increasing order. To show that $\eta_1 = 0$, it will be shown that any other possibility leads to a contradiction.

Case 1. $0 \leq \xi_1 < \eta_1$. Let

$$h(\xi) = \sum_{j=1}^{n} a_j \varphi(\xi - \eta_j).$$

Since x and y are optimal, $h(\xi) \leq v$ for any ξ in the unit interval, and $h(\xi_1) = v$. These conditions and the fact that $h(-\infty) = 0$ imply that there exists some $\bar{\xi} \leq \xi_1$ for which $h'(\bar{\xi}) = 0$. However, such a $\bar{\xi}$ cannot exist because the normalization of φ would require that $\varphi'(\bar{\xi} - \eta_j) > 0$ for each j, contradicting the fact that $h'(\bar{\xi}) = 0$.

Case 2. $0 < \eta_1 \leq \xi_1$. Let

$$r(\eta) = \sum_{i=1}^{m} b_i \varphi(\xi_i - \eta).$$

Then $r(\eta) \geq v$ for all η in the unit interval, and $r(\eta_1) = v$. These conditions and the fact that $r(\eta) \to 0$ as $\eta \to -\infty$ imply that there exists some $\bar{\eta} < \eta_1$ for which $r'(\bar{\eta}) = 0$. However, it follows that $\xi_i - \bar{\eta} > 0$ and $\varphi'(\xi_i - \bar{\eta}) < 0$ for each i, in contradiction to $r'(\bar{\eta}) = 0$; hence such an $\bar{\eta}$ cannot exist.

Since all other possibilities have led to absurdities, we may conclude that $\eta_1 = 0$. A similar argument shows that $\eta_n = 1$.

The results obtained above can be improved slightly; indeed, we can show by analogous arguments that

$$0 = \eta_1 < \xi_1 < \eta_2 \qquad \text{and} \qquad \eta_{n-1} < \xi_m < \eta_n = 1.$$

▶ LEMMA 7.3.4. If $\varphi'(0) = 1$, the optimal strategies of a bell-shaped game are unique.

Proof. According to Lemma 7.3.2, two cases must be considered, depending upon whether the spectrums of all x and y optimal strategies total the same number or differ by one. Suppose first that the spectrum of some optimal x strategy is composed of $\xi_1, \ldots, \xi_n$ with all these ξ_i relevant, and that every optimal y strategy is composed of the points $\eta_1, \ldots, \eta_n$. Let

$$x = \sum_{i=1}^{n} a_i x_{\xi_i} \qquad \text{and} \qquad x^* = \sum_{i=1}^{n} a_i^* x_{\xi_i}$$

denote two optimal strategies for Player I. Then

$$\sum_{i=1}^{n} a_i \varphi(\xi_i - \eta_j) = v \quad \text{and} \quad \sum_{i=1}^{n} a_i^* \varphi(\xi_i - \eta_j) = v \quad (j = 1, \ldots, n).$$

Subtracting, we obtain

$$\sum_{i=1}^{n} (a_i - a_i^*) \varphi(\xi_i - \eta_j) = 0 \qquad (j = 1, \ldots, n).$$

This result and the fact that $\det \|\varphi(\xi_i - \eta_j)\| > 0$ immediately imply that $a_i^* - a_i = 0$; hence the maximizing strategy is unique. Under these same circumstances we can prove that the optimal y strategy is unique.

In order to complete the proof we must consider the remaining case, where the spectrum of all optimal x strategies is confined to $\xi_1, \xi_2, \ldots, \xi_{n-1}$ and the optimal y strategies are composed of the points $\eta_1, \ldots, \eta_n$. The uniqueness of the optimal x strategy follows as above. Suppose that

$$y = \sum_{j=1}^{n} b_j y_{\eta_j} \qquad \text{and} \qquad y^* = \sum_{j=1}^{n} b_j^* y_{\eta_j}$$

are both optimal. In this case, as in the previous one, the functions

$$h(\xi) = \sum_{j=1}^{n} b_j \varphi(\xi - \eta_j) \qquad \text{and} \qquad h^*(\xi) = \sum_{j=1}^{n} b_j^* \varphi(\xi - \eta_j)$$

assume the value v at the points $\xi_1, \ldots, \xi_{n-1}$, and their derivatives vanish at these points since none of the ξ_i can ever coincide with the end points 0 or 1 (see the remark following Lemma 7.3.3). We find that the function

$$\psi^*(\xi) = \sum_{j=1}^{n} (b_j - b_j^*)\varphi(\xi - \eta_j)$$

assumes the value 0 at $\xi_1, \ldots, \xi_{n-1}$, and its derivative vanishes at these points. Thus ψ^* has at least $2(n - 1)$ zeros, counting multiplicities. But if $\psi^*(\xi) \not\equiv 0$, then the number of zeros, counting multiplicities, that ψ^* can have is at most $n - 1$ (see Remark 7.2.1). Thus $\psi^*(\xi) \equiv 0$, and the optimal strategy is unique. (The reader should consult Problem 10 dealing with the question of uniqueness when the normalization condition is dropped.)

At this point it seems worth while to summarize our previous lemmas in the form of a theorem.

▶**Theorem 7.3.1.** If $K(\xi, \eta) = \varphi(\xi - \eta)$ is a bell-shaped game with $\varphi'(0) = 0$, then the optimal strategies for both players are unique finite-step distributions such that:

(1) If the minimizing optimal strategy consists of n steps, then the maximizing optimal strategy is composed of either n or $n - 1$ steps.

(2) The points 0 and 1 belong to the spectrum of the optimal y strategy.

(3) The first point ξ_1 in the optimal x strategy separates 0 and the next optimal η point. The last value of the optimal x strategy has a similar property relative to the point 1.

▶**Theorem 7.3.2.** If $\varphi(u) = \varphi(-u)$, then each optimal strategy is symmetric about the point $\frac{1}{2}$. That is, if the point $\xi(\eta)$ is in the support of an optimal strategy, then the point $1 - \xi\ (1 - \eta)$ is also in that support, and the corresponding weights associated with these points are equal.

Proof. Applied to Player I, the theorem asserts that if

$$\sum_{i=1}^{n} a_i x_{\xi_i}$$

is the optimal strategy, then

$$\sum_{i=1}^{n} a_i x_{1-\xi_i}$$

represents the same strategy. In view of the uniqueness of the solution, the present theorem can be proved by showing that

$$\sum_{i=1}^{n} a_i x_{1-\xi_i}$$

is optimal. The optimal property follows from the fact that the pay-off for this strategy is

$$\sum_{i=1}^{n} a_i\varphi(1 - \xi_i - \eta) = \sum_{i=1}^{n} a_i\varphi(-\xi_i + 1 - \eta)$$

$$= \sum_{i=1}^{n} a_i\varphi[\xi_i - (1 - \eta)] \geq v.$$

A similar argument establishes the symmetry of the minimizing strategy.

The two final lemmas of this section apply to families of bell-shaped games depending on a parameter. The kernels of these games will be of the form $K_\lambda(\xi, \eta) = \varphi[\lambda(\xi - \eta)]$ $(\lambda > 0)$, and the lemmas are concerned with the number $n(\lambda)$ of points contained in the support of the optimal minimizing strategy. The function φ will denote again a positive, analytic, regular P.F.F. The value of the game is denoted by $v(\lambda)$.

▶ LEMMA 7.3.5. In any bell-shaped game with pay-off $\varphi[\lambda(\xi - \eta)]$, the numbers $n(\lambda)$ and $v(\lambda)$ are related by the inequality $v(\lambda)n(\lambda) \geq \varphi(0)$.

Proof. In any strategy of the form

$$\sum_{j=1}^{n} b_j y_{\eta_j},$$

there exists some index, say j_0, such that $b_{j_0} > 1/n$. If the strategy is optimal for $\varphi[\lambda(\xi - \eta)]$, then

$$v(\lambda) \geq \sum_{j=1}^{n} b_j\varphi[\lambda(\xi - \eta_j)] \geq \frac{1}{n}\,\varphi[\lambda(\xi - \eta_{j_0})] \qquad (0 \leq \xi \leq 1).$$

Setting $\xi = \eta_{j_0}$, we obtain the desired result.

▶ LEMMA 7.3.6. As $\lambda \to \infty$, the value of the game approaches zero and the number of steps increases without bounds.

Proof. For each $\epsilon > 0$ there exists some λ_0 such that $\varphi[\lambda(\xi - \eta)] < \epsilon$ whenever

$$|\xi - \eta| \geq \epsilon \qquad \text{and} \qquad \lambda > \lambda_0.$$

Also, if

$$|\xi - \eta| < \epsilon, \qquad \text{then} \qquad \varphi[\lambda(\xi - \eta)] \leq \varphi_0,$$

where $\varphi_0 = \max_t \varphi(t)$. Thus we obtain that

$$\int_0^1 \varphi[\lambda(\xi - \eta)]\,d\eta < \epsilon(2\varphi_0 + 1) \qquad (\lambda > \lambda_0).$$

The last equation shows that for $\lambda > \lambda_0$ Player II can prevent Player I from obtaining a yield greater than $\epsilon(2\varphi_0 + 1)$. Since ϵ is arbitrary, $v(\lambda) \to 0$ and $n(\lambda) \to \infty$.

It is possible to obtain more precise results about the behavior of $n(\lambda)$ when $\varphi'(0) = 0$. For λ sufficiently small, $\varphi[\lambda(\xi - \eta)]$ is strictly concave in each variable for $0 \leq \xi, \eta \leq 1$, and these games are known to possess optimal strategies of the form $x = x_{\xi_1}$, $y = \alpha y_0 + (1 - \alpha)y_1$ (Section 4.2). For larger values of λ both players have optimal strategies that involve at least two points, etc. A precise evaluation of the critical points, where the size of the spectrum $n(\lambda)$ changes with increasing λ, is possible. (See Problems 7 and 8.)

7.4 Other types of continuous games. *Pólya-type kernels.* A direct generalization of bell-shaped kernels leads to the consideration of Pólya-type (P.T.) kernels. These are kernels $K(\xi, \eta)$ that satisfy the following conditions:

(i) $K(\xi, \eta)$ is defined and continuous in the square $0 \leq \xi, \eta \leq 1$.

(ii) For every n and every set of numbers ξ_i and η_i such that

$$0 \leq \xi_1 < \cdots < \xi_n \leq 1 \qquad \text{and} \qquad 0 \leq \eta_1 < \cdots < \eta_n \leq 1,$$

the determinant of the matrix $\|K(\xi_i, \eta_j)\|$ is nonnegative.

From the concept of a P.T. kernel one can proceed to the concepts of regular and proper P.T. kernels, whose definitions are analogous to the definitions of regular and proper Pólya frequency functions. Many of the results of Sections 7.2 and 7.3 can be extended to P.T. kernels, but it should be pointed out that Lemma 7.2.3 cannot be generalized. Thus there are games with P.T. kernels whose optimal strategies do not possess those properties of optimal strategies of bell-shaped games that depend on the validity of this lemma. Some properties of P.T. kernels are outlined in Problems 24 and 25.

Green-type kernels. A Green-type (G.T.) kernel is simply the negative of a butterfly kernel; that is, $K(\xi, \eta)$ is G.T. if and only if $-K(\xi, \eta)$ is a butterfly kernel (see Section 6.5, particularly Theorem 6.5.2, and Problem 24 of this chapter). Since the optimal strategies for butterfly kernels are always equalizing strategies, it is clear that strategies of the same type are also optimal for the G.T. kernel.

Green-type kernels are so named because they are actually the Green's functions of suitable differential equations. There is a large class of games with kernels of this type. Some of these games are described in the following example, which also illustrates a fundamental relationship between the optimal strategies for the game and the coefficients of the differential equations. In this example we shall consider a class of differential equations, and we shall show that the corresponding Green's functions give rise to kernels whose games possess easily discernible optimal strategies which can be expressed analytically in terms of the coefficients of the differential equations.

Example. Let $p(t)$ be positive and continuously differentiable, and let $q(t)$ be positive and continuous on some interval $a \leq t \leq b$ that contains the unit interval in its interior. Let φ and ψ be two linearly independent solutions of the differential equation

$$(pu')' - qu = 0 \qquad (a \leq t \leq b) \tag{7.4.1}$$

that satisfy homogeneous boundary conditions (for example, $\varphi(a) = \psi(b) = 0$). Then the Green's function

$$K(\xi, \eta) = \begin{cases} \varphi(\xi)\psi(\eta) & 0 \leq \xi \leq \eta \leq 1, \\ \psi(\xi)\varphi(\eta) & 0 \leq \eta \leq \xi < 1 \end{cases} \tag{7.4.2}$$

can be interpreted as the kernel of a game defined on the unit square. If $\varphi(t) > 0$, $\varphi'(t) > 0$, $\varphi''(t) > 0$, $\psi(t) > 0$, $\psi'(t) < 0$, and $\psi''(t) > 0$ for $0 \leq t \leq 1$, then it is easy to verify that $K(\xi, \eta)$ is indeed a G.T. kernel. Under these circumstances it follows from the general theory of G.T. kernels, completely analogous to the theory of butterfly kernels, that the support of the optimal strategies is the full unit interval; and hence, by Lemma 2.2.1, each optimal strategy must yield identically v. Therefore, since $K(\eta, \xi) = K(\xi, \eta)$, any strategy that is optimal for one player is also optimal for the other player, and whoever uses it always secures the value v. Finally, using our knowledge of butterfly games, since the game is G.T., we seek to determine optimal strategies composed of jumps α and β, at $\xi = 0$ and $\xi = 1$, respectively, and a positive density f spread on the unit interval. If such an optimal strategy exists, it must be an equalizer; hence

$$v \equiv \alpha\varphi(0)\psi(\eta) + \int_0^\eta \varphi(\xi)\psi(\eta)f(\xi)\,d\xi + \int_\eta^1 \psi(\xi)\varphi(\eta)f(\xi)\,d\xi + \beta\psi(1)\varphi(\eta).$$

(In this connection see also Theorem 6.5.2.) Differentiating this equation twice and eliminating the terms that involve integrals, we obtain

$$\alpha = v\,\frac{\varphi'(0)}{\varphi(0)} \cdot \frac{1}{\varphi'(0)\psi(0) - \varphi(0)\psi'(0)},$$

$$\beta = -v\,\frac{\psi'(1)}{\psi(1)} \cdot \frac{1}{\varphi'(1)\psi(1) - \varphi(1)\psi'(1)},$$

$$f(t) = v\,\frac{\varphi'(t)\psi''(t) - \psi'(t)\varphi''(t)}{[\varphi'(t)\psi(t) - \psi'(t)\varphi(t)]^2}.$$

By direct verification it can be shown that this specific strategy is in fact an equalizer and hence optimal for both players.

Although the general theory has successfully produced an explicit formula for the optimal strategy, it may be worth while to use the specific form of (7.4.1) to obtain simpler expressions for f, α, and β. To this end, we can verify directly that if φ and ψ are solutions of the differential equation, then the derivative of $p[\varphi'\psi - \psi'\varphi]$ is zero, so that $\varphi'\psi - \psi'\varphi = kp^{-1}$ for some constant k. Furthermore, multiplying the equations

$$p\varphi'' + p'\varphi' - q\varphi = 0 \qquad \text{and} \qquad p\psi'' + p'\psi' - q\psi = 0$$

by ψ' and φ', respectively, we can also verify that

$$\psi''\varphi' - \varphi''\psi' = p^{-1}[\varphi'\psi - \psi'\varphi]q = kp^{-2}q.$$

Thus, by substitution we obtain

$$\alpha = \frac{v}{k}\frac{\varphi'(0)}{\varphi(0)}\,p(0), \qquad f(t) = \frac{v}{k}\,q(t), \qquad \beta = \frac{-v}{k}\frac{\psi'(1)}{\psi(1)}\,p(1).$$

The number v/k can be computed directly by means of the normalizing condition

$$\int_0^1 dx(\xi) = 1.$$

Two specific examples of G.T. kernels may be obtained by considering the differential equations $u'' - u = 0$ and $u'' - (1 + t^2)u = 0$ over the full real line $-\infty < t < \infty$. The solutions of the first equations include the functions e^t and e^{-t}, which lead, in accordance with (7.4.2), to the kernel $K(\xi, \eta) = e^{-|\xi-\eta|}$. This kernel has appeared before as an example of a regular P.F.F. that is not analytic. In the present context it is evident that the spectrum of the optimal strategies is the full unit interval. This example demonstrates that the analyticity assumption of Lemma (7.3.1) cannot be discarded. The solutions of the second differential equation include the functions

$$\varphi(t) = e^{t^2/2}\int_{-\infty}^{t} e^{-u^2}\,du \qquad \text{and} \qquad \psi(t) = e^{t^2/2}\int_{t}^{\infty} e^{-u^2}\,du,$$

and these satisfy all the conditions required by the present theory. Therefore, if

$$K(\xi, \eta) = \begin{cases} \exp\left[\frac{1}{2}(\xi^2 + \eta^2)\right]\int_{-\infty}^{\xi} e^{-t^2}\,dt\int_{\eta}^{\infty} e^{-t^2}\,dt & \xi \leq \eta, \\ \exp\left[\frac{1}{2}(\xi^2 + \eta^2)\right]\int_{\xi}^{\infty} e^{-t^2}\,dt\int_{-\infty}^{\eta} e^{-t^2}\,dt & \xi \geq \eta, \end{cases}$$

the density part of the solution is $f(\xi) = k(1 + \xi^2)$.

It is interesting to note that the differential equation $(tu')' - n^2u/t = 0$ has fundamental solutions $\varphi(t) = t^n$ and $\psi(t) = t^{-n} - t^n$. These solutions violate the positivity requirements at 0 and 1, and the optimal strategies for the corresponding kernels cannot be of the requisite type because $q(t)$ is not integrable. However, if the game is restricted to the square with sides $[\delta, 1 - \delta]$, then all conditions are satisfied and the strategies have the desired form.

7.5 Invariant games. In this section we consider applications of the concept of invariance to the classification of games. We assume the existence of two groups of transformations which map the spaces of strategies onto themselves. These transformations, defined initially for the pure strategies ξ and η, affect also the mixed strategies $x(\xi)$ and $y(\eta)$. Indeed, any measurable transformation τ that maps ξ into $\tau\xi$ automatically induces a transformation $x \to \tau x$, where $\tau x(E) = x(\tau E)$ for any measurable set E (where $\tau E = \{\xi | \xi = \tau\xi', \xi' \in E\}$). In applications of invariance to the theory of games, it is necessary to assume that the induced transformations are such that they map probability measures into probability measures. It is easy to verify that the induced transformations having this property form a group of transformations of the space of mixed strategies into itself. A group of transformations mapping X onto itself will be denoted by T, and an associated group of transformations mapping Y onto itself will be denoted by T'. The following discussion is restricted to games defined on the unit square.

Definition 7.5.1. A game with kernel $K(\xi, \eta)$ is said to be invariant under T with respect to T' if for every $\tau \in T$ there exists some $\tau' \in T'$ such that $K(\tau\xi, \eta) = K(\xi, \tau'\eta)$. And the game is said to be invariant under T' with respect to T if for every $\tau' \in T'$ there exists some $\tau \in T$ such that $K(\xi, \tau'\eta) = K(\tau\xi, \eta)$.

This definition is illustrated by the following examples.

Example 1. Let T and T' be groups of transformations of the unit interval into itself, where each group contains precisely two elements: the identity transformation $\xi \to \xi$ (or $\eta \to \eta$) and the reflection $\xi \to 1 - \xi$ (or $\eta \to 1 - \eta$). Then any game with kernel $K(\xi, \eta) = \varphi(\xi - \eta)$, where φ is an even function, is invariant under T and T'. We have discussed this example in connection with bell-shaped games (p. 189).

Example 2. Any game with kernel $K(\xi, \eta) = \varphi(\xi - \eta)$, where φ is periodic with period 1, is called a convolution game. These games, played over the unit square, are invariant under the group of translations $\tau_a\xi = \xi + a$ (reduced modulo 1).

The next theorems describe some invariance properties of the solutions of invariant games. In this context, $T(E)$ will denote the set of all distributions of the form τx with $\tau \in T$ and $x \in E$, where E is a subset of X. The set E is said to be invariant under T if $T(E) = E$. Here the equality asserts that the sets $T(E)$ and E contain the same distributions. Similarly, a strategy x is said to be invariant under a transformation τ if $\tau x = x$. Here the equality means that τx and x define the same measure. Over the interval [0, 1] any strategy is completely determined by its moments, and thus in order to verify that x and τx define the same strategy, it is sufficient to verify that $\int \xi^n \, dx = \int \xi^n \, d\tau x$ for every n.

▶ LEMMA 7.5.1. If $K(\xi, \eta)$ is invariant under T (or under T'), then the set X_0 (or Y_0) of optimal strategies is also invariant under T (or under T').

Proof. Let x_0 be optimal and let K be invariant under T. Choosing the strategy τx_0, Player I obtains the yield

$$\int K(\xi, \eta) \, d\tau x_0(\xi) = \int K(\tau^{-1}\xi, \eta) \, dx_0(\xi) = \int K[\xi, (\tau^{-1})'\eta] \, dx_0(\xi) \geq v.$$

Thus τx_0 is optimal, and $T(X_0) \subset X_0$. Also, since T contains an identity element, it is clear that $T(X_0) \supset X_0$. Hence $T(X_0) = X_0$.

▶ THEOREM 7.5.1. If T is a compact topological group and K is continuous and invariant under T, then there exists an optimal strategy x_0 that is invariant under every transformation $\tau \in T$.

Proof. Let $\mu(\tau)$ denote the Haar measure of the group T, normalized so that $\mu(T) = 1$, and let x be any optimal strategy. For every measurable set E, define $x_0(E)$ by the equation

$$x_0(E) = \int_T x(\tau E) \, d\mu(\tau).$$

Since x and μ are positive, $x_0(E)$ is always positive. Since the x-measure of the unit interval is 1, it follows that the x_0-measure of the unit interval is also 1, and hence x_0 is a probability measure. The invariant character of μ implies that

$$x_0(\tau_0 E) = \int_T x(\tau\tau_0 E) \, d\mu(\tau) = \int_T x(\tau\tau_0 E) \, d\mu(\tau\tau_0) = x_0(E);$$

thus x_0 is invariant under T. Finally, in order to show that x_0 is also optimal, it is sufficient to observe that the change of variable $\zeta = \tau\xi$ implies that

$$\begin{aligned}\int_0^1 K(\xi, \eta)\, dx_0(\xi) &= \int_0^1 \int_T K(\xi, \eta)\, d\mu(\tau)\, dx(\tau\xi)\\ &= \int_0^1 \int_T K(\tau^{-1}\xi, \eta)\, d\mu(\tau)\, dx(\xi)\\ &= \int_T \int_0^1 K[\xi, (\tau^{-1})'\eta]\, dx(\xi)\, d\mu(\tau)\\ &\geq \int_T v\, d\mu(\tau) = v.\end{aligned}$$

An alternative approach to the study of invariant strategies can be based on a limiting process. This approach, described below, requires that the transformations τ be continuous. More precisely, given a sequence x_n such that for every continuous $h(\xi)$

$$\int h(\xi)\, dx_n(\xi) \to \int h(\xi)\, dx_0(\xi),$$

the transformations τ must have the property that

$$\int h(\xi)\, d\tau x_n(\xi) \to \int h(\xi)\, d\tau x_0(\xi).$$

This condition may be difficult to check directly. However, since

$$\int h(\xi)\, d\tau x(\xi) = \int h(\tau^{-1}\xi)\, dx(\xi),$$

it is sufficient to verify that the transformation $h(\xi) \to h(\tau^{-1}\xi)$ preserves the continuity of h. Hence it is sufficient that τ^{-1} be a continuous function on [0, 1].

▶ LEMMA 7.5.2. If τ_0 is continuous in the sense described above and K is continuous and invariant under T, then there exists an optimal strategy x_0 such that $\tau_0 x_0 = x_0$.

Proof. Let x be any optimal strategy and construct a sequence of strategies $\{x_n\}$ defined by

$$x_n = \frac{1}{n+1} \sum_{k=0}^{n} (\tau_0)^k x.$$

Lemma 7.5.1 implies that the strategies $(\tau_0)^k x$ are all optimal, and the convexity of the set of optimal strategies implies that each x_n is also optimal. Furthermore, by Helly's theorem (Section C.3), there exists a subsequence x_{n_i} that converges to a strategy x_0 in the sense that

$$\lim_{i\to\infty} \int h(\xi)[dx_{n_i}(\xi) - dx_0(\xi)] = 0$$

for every continuous function $h(\xi)$. Hence the strategy x_0 is optimal, since K is continuous and x_{n_i} is optimal. Also, to show that x_0 and $\tau_0 x_0$ define the same probability distribution, it is sufficient to show that for every continuous function h, the left-hand side of the following expression can be made arbitrarily small. In fact,

$$\left|\int h(\xi)\, d[x_0(\xi) - \tau_0 x_0(\xi)]\right| \le \left|\int h(\xi)\, d[x_0(\xi) - x_{n_i}(\xi)]\right|$$
$$+ \left|\int h(\xi)\, d[x_{n_i}(\xi) - \tau_0 x_{n_i}(\xi)]\right|$$
$$+ \left|\int h(\xi)\, d\tau_0[x_{n_i}(\xi) - x_0(\xi)]\right|.$$

The first and last terms in the right-hand side can be made arbitrarily small by choosing i sufficiently large. The other term also tends to zero because

$$x_{n_i}(\xi) - \tau_0 x_{n_i}(\xi) = \frac{1}{n_i + 1}\,[x(\xi) - \tau_0^{n_i+1} x(\xi)] \to 0$$

in variation. Thus $\int hd(x_0 - \tau_0 x_0) = 0$, as was to be proved.

▶**Theorem 7.5.2.** If T is Abelian and each τ is continuous, and if K is continuous and invariant under T, then there exists an optimal strategy x_0 that is invariant under every transformation $\tau \in T$.

Proof. For each τ in T, let $X(\tau)$ denote the set of all optimal strategies that are invariant under τ. The intersection of any finite number of these sets, say $X(\tau_1), \ldots, X(\tau_n)$, is the set of all optimal strategies that are invariant under $\tau_1, \ldots, \tau_n$. The existence of these strategies can be established by a finite iteration of the arguments used in Lemma 7.5.2 with the help of the Abelian character of T. We first produce an optimal strategy that is invariant with respect to τ_1. Then since $X(\tau_1)$ is nonvoid, convex, and closed in the sense of weak* convergence, we can repeat the argument of Lemma 7.5.2 with $X(\tau_1)$ replacing X_0 and thus obtain an optimal strategy that is invariant with respect to τ_1 and τ_2. The continuation of this procedure is clear.

Hence the sets $X(\tau)$ are weak* closed and nonvoid, and the intersection of any finite number of them is nonvoid. Since the space of all strategies is weak* compact, the intersection of all the sets $X(\tau)$ must contain at least one strategy x_0, and this strategy satisfies all the conditions of the theorem.

Theorem 7.5.2 may be extended to apply to the case in which T is a solvable group.

We close this chapter by describing the totality of solutions for games with convolution kernels which are also periodic on the unit interval.

Since the translation group modulo 1 is compact and leaves the game invariant, we find immediately that the Haar measure $d\xi$ (the uniform distribution) is an optimal equalizing strategy. The following theorem provides a necessary and sufficient criterion for determining when the Haar measure is the unique optimal strategy.

▶ THEOREM 7.5.3. Let $K(\xi, \eta) = \phi(\xi - \eta)$, where ϕ is continuous and periodic of period 1. Then $d\xi$ is the unique optimal strategy for both players of the game K if and only if for every $n \neq 0$,

$$\int_0^1 \phi(t)e^{2\pi int}\, dt \neq 0.$$

Proof. If for some $n \neq 0$,

$$\int_0^1 \phi(t)e^{2\pi int}\, dt = 0,$$

then, since ϕ is real-valued,

$$\int_0^1 \phi(t) \cos 2\pi nt\, dt = 0 \qquad \text{and} \qquad \int_0^1 \phi(t) \sin 2\pi nt\, dt = 0.$$

Furthermore, the function $x(\xi) = 1 + \cos 2\pi n\xi$ is nonnegative and has total integral 1, so that it is the density of a probability distribution. It is also optimal, since

$$\int_0^1 \phi(\xi - \eta)(1 + \cos 2\pi n\xi)\, d\xi = \int_0^1 \phi(\xi - \eta)\, d\xi$$
$$+ \int_0^1 \phi(\xi - \eta) \cos 2\pi n\xi\, d\xi = \int_0^1 \phi(t)\, dt = v.$$

Conversely, if we assume that $\int \phi(t)e^{2\pi int}dt \neq 0$ for $n \neq 0$, we now show that the optimal strategy is unique. Without loss of generality we may assume that the value $v = 0$; otherwise the game with kernel $\phi(\xi - \eta) - v$ has value zero and the same set of optimal strategies, and still satisfies the assumption $\int [\phi(t) - v]e^{2\pi int}dt \neq 0$ for $n \neq 0$. Since $d\xi$ has for its support the full unit interval, any optimal minimizing $y(\eta)$ strategy satisfies the equation

$$\int_0^1 \phi(\xi - \eta)\, dy(\eta) = 0.$$

Therefore, for every $n \neq 0$,

$$\int_0^1 e^{-2\pi in\xi} \left(\int_0^1 \phi(\xi - \eta)\, dy(\eta)\right) d\xi = 0$$

and

$$\int_0^1 \int_0^1 e^{-2\pi in(\xi-\eta)}\phi(\xi - \eta)e^{-2\pi in\eta}\, dy(\eta)\, d\xi = 0.$$

Therefore

$$\int_0^1 e^{-2\pi inu}\phi(u)\, du \int_0^1 e^{-2\pi in\eta}\, dy(\eta) = 0.$$

By assumption,

$$\int_0^1 e^{-2\pi inu}\phi(u)\, du \neq 0$$

if $n \neq 0$; hence

$$\int_0^1 e^{-2\pi in\eta}\, dy(\eta) = 0$$

for $n \neq 0$, and we conclude that $dy(\eta) = c\, d\eta$. Since $y(\eta)$ is a probability distribution, $c = 1$. A similar proof establishes the uniqueness of the optimal maximizing strategy.

7.6 Problems. In the following problems, the kernels $K(\xi, \eta)$ are always defined on the unit square $0 \leq \xi, \eta \leq 1$ unless there is an explicit statement to the contrary.

1. The game with kernel

$$K(\xi, \eta) = \frac{2(\xi + \eta)}{(2\xi + 1)(2\eta + 1)}$$

has an optimal equalizing strategy $x(t) = y(t)$ for each player. Show that $x(t)$ must satisfy the equation

$$\int_0^1 \frac{dx(\xi)}{2\xi + 1} = \frac{1}{2},$$

and find a strategy solution of this equation.

2. Show that the strategies

$$x(\xi) = \frac{\log (1 + \xi)}{\log 2} \qquad \text{and} \qquad y(\eta) = \frac{\log (1 + \eta)}{\log 2}$$

are optimal for the kernel

$$K(\xi, \eta) = \frac{(1 + \xi)(1 + \eta)}{(1 + \xi\eta)^2}.$$

Find the value of the game, and show that the game cannot have any optimal strategies with finite support.

3. Let $P(\xi, \eta)$ and $Q(\xi, \eta)$ be two polynomials with $Q(\xi, \eta) > 0$ for $0 \leq \xi, \eta \leq 1$. Let

$$K(\xi, \eta) = \frac{P(\xi, \eta)}{Q(\xi, \eta)}$$

and assume that the value of the game with this kernel is $v \neq 0$. Assume also that if ξ is considered as a complex variable, then there exists a curve C (in the complex plane) that connects to ∞ in such a way that $K(\xi, \eta)$ remains analytic when ξ ranges over C and η ranges over $[0, 1]$. Assume furthermore that $|K(\xi, \eta)| \leq M < \infty$ for $\xi \in C$, and finally suppose that for each η the degree in ξ of $Q(\xi, \eta)$ is greater than the degree in ξ of $P(\xi, \eta)$. Prove that Player I has an optimal strategy of finite type.

4. Let

$$K(\xi, \eta) = \frac{P(\xi, \eta)}{Q(\xi, \eta)}$$

be as in Problem 3 except that (i) v is arbitrary; and (ii) for each η the degree in ξ of $P(\xi, \eta)$ exceeds the degree in ξ of $Q(\xi, \eta)$. Prove that every optimal maximizing strategy is of finite type.

5. Show that

$$\frac{1}{\cosh u}, \quad e^{u-e^u}, \quad \begin{cases} u^n e^{-u} & u > 0, \\ 0 & u \leq 0, \end{cases} \quad (n \text{ is a positive integer})$$

are all Pólya frequency functions.

6. Solve explicitly the bomber-submarine problem (p. 181) where the pay-off is $e^{-3(\xi-\eta)^2}$. Here ξ is the point bombed, η is the actual position of the submarine, and $e^{-3(\xi-\eta)^2}$ is the expected damage to the submarine.

7. In the game

$$K(\xi, \eta) = \frac{1}{1 + \lambda(\xi - \eta)^2}$$

for λ sufficiently small Player II has an optimal strategy involving 0 and 1 and Player I has an optimal pure strategy confined to $\xi = \frac{1}{2}$. Determine the largest value of λ for which this is true.

8. Determine the largest value of λ for which the optimal strategy for Player I of the kernel $e^{-\lambda(\xi-\eta)^2}$ is pure. Find a relation satisfied by λ which implicitly determines the largest value of λ for which the spectrum of the optimal strategy of Player II consists of at most two points.

9. Show that if φ is a regular P.F.F. and φ'' is continuous, then the function

$$h(\xi) = \sum_{i=1}^{n} a_i \varphi''(\xi - \eta_i)$$

has at most $2n$ changes of sign, provided $a_i > 0$.

10. Show that the optimal strategy for Player I is always unique in a bell-shaped game (i.e., that it is unique even if $\varphi'(0) \neq 0$), and that the optimal strategy for Player II is unique if its support contains at least three points. Give an example of a bell-shaped game in which the optimal strategy of Player II is not unique.

11. (a) Prove that if h is a continuous function from the reals into the reals, then

$$\int_0^1 \left[2h(\xi) - h\left(1 - \frac{\xi}{3}\right) - h\left(\frac{\xi}{3}\right)\right] dC(\xi) = 0,$$

where C is the classical Cantor distribution.

(b) Use part (a) to show that if

$$K(\xi, \eta) = \sum_{n=0}^{\infty} \frac{1}{2^n}\left[2\xi^n - \left(1 - \frac{\xi}{3}\right)^n - \left(\frac{\xi}{3}\right)^n\right] \times \left[2\eta^n - \left(1 - \frac{\eta}{3}\right)^n - \left(\frac{\eta}{3}\right)^n\right],$$

then the Cantor distribution is the unique optimal strategy for each player. (*Hint:* Use the technique of moments of Example 2 in Section 7.1.)

12. Let $K(\xi, \eta)$ be a continuous kernel with unique solutions x and y and value 0, and let U and V be countable dense sets with $\int_U dx = \int_V dy = 0$. Put

$$K^*(\xi, \eta) = \begin{cases} K(\xi, \eta) & \xi \in U, \quad \eta \in V, \\ 0 & \text{otherwise.} \end{cases}$$

Show that $K^*(\xi, \eta) = 0$ a.e., and that x and y are the unique optimal strategies for $K^*(\xi, \eta)$.

13. Consider a game in which the pure strategy spaces for Players I and II are composed of the pairs $\langle \xi_1, \xi_2 \rangle$ and $\langle \eta_1, \eta_2 \rangle$, respectively, satisfying $\xi_1^2 + \xi_2^2 \leq 1$ and $\eta_1^2 + \eta_2^2 \leq 1$. Let the kernel $K(\xi_1, \xi_2; \eta_1, \eta_2)$ have the form $h[(\xi_1 - \eta_1)^2 + (\xi_2 - \eta_2)^2]$. Show that both players have optimal strategies that are invariant with respect to all rotations about the origin.

14. By considering the game

$$K(\xi, \eta) = \sum_{n=1}^{\infty} \frac{1}{n^2} \cos 2\pi n(\xi - \eta),$$

show that the set of all games with continuous kernels and unique optimal strategies for each player is not an open set. (*Hint:* Make use of Theorem 7.5.3.)

15. Show that the strategies

$$x = y = \sum_{n=1}^{\infty} \frac{1}{2^n} x_{2^{-n}}(\xi)$$

are optimal for the kernel

$$K(\xi, \eta) = \frac{2}{2 - \xi\eta} - \frac{2}{4 - \xi\eta} - \frac{1}{2 - \xi} - \frac{1}{2 - \eta}.$$

16. Let μ_n denote the moments of an arbitrary distribution x, and let $y = I_0$. Show that x and y are the unique optimal strategies for the kernel

$$K(\xi, \eta) = \sum_{n=1}^{\infty} 2^{-n}(\xi^n - \mu_n)\eta^n \sin\frac{1}{\eta} + e^{-\eta^{-2}}.$$

17. Let $x^0(\xi)$ and $y^0(\eta)$ denote arbitrary distributions on the unit interval. Verify that x^0 and y^0 are optimal for Players I and II, respectively, for the game whose kernel is

$$K(\xi, \eta) = M(\xi, \eta) - \int_0^1 M(\xi, \eta)\, dx^0(\xi) - \int_0^1 M(\xi, \eta)\, dy^0(\eta)$$

for any continuous $M(\xi, \eta)$.

18. Let $h(t) > 0$ be normalized so that

$$\int_0^1 \int_0^1 h(\xi\eta)\, d\xi\, d\eta = 1.$$

Consider the game

$$K(\xi, \eta) = \frac{h(\xi\eta)}{\int_0^1 h(\xi t)\, dt \int_0^1 h(\eta t)\, dt}.$$

Prove that

$$x'(\xi) = \int_0^1 h(\xi t)\, dt \qquad \text{and} \qquad y'(\eta) = \int_0^1 h(\eta t)\, dt$$

are densities of optimal strategies for Players I and II, respectively.

19. If in Problem 18 $h(t)$ is analytic in a neighborhood of the origin, i.e.,

$$h(t) = \sum_{n=0}^{\infty} a_n t^n$$

for $|t| \leq \delta$ and $a_{n_i} \neq 0$, where $\sum 1/n_i = \infty$, show that the solution described in Problem 18 is unique.

20. Let

$$\varphi(t) = \sum_{n=0}^{\infty} a_n t^n \qquad (a_n > 0, 0 \leq t \leq 1).$$

Let $y^0(\eta)$ be any distribution. Prove that $y^0(\eta)$ is the unique optimal strategy for Player II for the game with kernel

$$K(\xi, \eta) = \varphi(\xi\eta) - \int_0^1 \varphi(\xi\eta)\, dy^0(\xi) - \int_0^1 \varphi(\xi\eta)\, dy^0(\eta).$$

21. Show that the optimal strategies of Problem 2 are unique.

22. (Needed for Problem 23.) Let μ_n and μ_n^* denote the nth moments of two distributions x and x^* defined on the unit interval. Show that if $\mu_{2n} \geq \mu_{2n}^*$ and $\mu_{2n+1} \leq \mu_{2n+1}^*$ $(n = 0, 1, 2, \ldots)$, then $x = x^*$.

23. Consider the game with kernel

$$K(\xi, \eta) = \begin{cases} (-1)^n(\xi^n - \mu_n) & \eta = \dfrac{1}{n} \quad (n = 1, 2, \ldots), \\ 0 & \text{otherwise}, \end{cases}$$

which is 0 except for a countable number of verticals where

$$\mu_n = \int \xi^n dx^0(\xi).$$

Prove that x^0 is the unique optimal strategy for Player I.

24. Consider a P.T. kernel $K(\xi, \eta)$ whose values in a neighborhood of the diagonal decrease away from the diagonal. Suppose, moreover, that $K_{\xi\xi}$, $K_{\eta\eta}$ are strictly positive on each side of the diagonal $\xi \leq \eta$ and $\eta \geq \xi$, but not necessarily convex in each variable throughout the unit square. Prove that K is a G.T. kernel.

25. Suppose that $K(\xi, \eta) = \psi(|\xi - \eta|)$ (ψ is assumed three times continuously differentiable) is a P.T. kernel which is strictly convex on each side of the diagonal and whose values decrease away from the diagonal. Prove that K is a G.T. kernel.

26. Construct a kernel $K(\xi, \eta)$ whose unique optimal strategies are two-step functions x and y, each having at least one step interior to the unit interval.

27. Construct a kernel whose unique optimal strategies are

$$x = \alpha I_0 + (1 - \alpha)I_1 \qquad \text{and} \qquad y = \beta I_0 + (1 - \beta)I_1,$$

with

$$0 < \alpha, \beta < 1.$$

28. Show that for every nonvoid, convex, weak* closed set $\mathfrak{F}$ of distributions, there exist a sequence $\{\varphi_n\}$ of continuous functions and a sequence $\{a_n\}$ of real numbers such that

$$\mathfrak{F} = \bigcap_{n=1}^{\infty} \{x | \textstyle\int \varphi_n\, dx \geq a_n\} \cdot$$

29. Using the notation of the previous problem, let $K(\xi, 0) = 0$ and

$$K\left(\xi, \frac{1}{n}\right) = \frac{1}{n}\,[\varphi_n(\xi) - a_n],$$

and let $K(\xi, \eta)$ be defined by linear interpolation for $1/(n+1) \leq \eta \leq 1/n$. Show that $x(\xi)$ is optimal for $K(\xi, \eta)$ if and only if $x \in \mathfrak{F}$.

Notes and References to Chapter 7

The need to solve additional classes of games is compelling for two reasons. First, it furthers the task of classifying games in relating the properties of the kernel to the form of the solution. This is analogous to the objectives involved, for example, in classifying differential equations. Second, every specific advance makes easier and more manifest the ultimate practical application of the theory of infinite games.

§ 7.1. The beginning results in analytic games touched on here are due primarily to Gross and Glicksberg.

The remarkable example of the rational game whose unique solution is the Cantor distribution is due to Gross [113] (see also [116]). For other weird solutions to seemingly nice games, see Glicksberg and Gross [110]. These authors have also shown that for any specified pair of distributions, there exists a game with analytic kernel for which the prescribed distributions are the unique optimal strategies. Example 3 of this section is a special case of this construction.

§ 7.2. One day in 1951 a Lockheed employee consulted me on a version of the submarine game problem described in Section 2.4. In analyzing this example it became clear that this was part of a larger class of games—namely, bell-shaped games. The kernel $1/[1 + \lambda(\xi - \eta)^2]$ has a form analogous to that of the Gaussian kernel, and the nature of the solutions is likewise parallel. Some special cases of this kind can be found in McKinsey [179, p. 217].

The P.F.F.'s were first studied by Pólya in connection with the problem of characterizing functions that can be approximated in a given interval by polynomials having only real zeros. The sign-variation diminishing character of these kernels has been extensively investigated by Schoenberg [215] and Gantmacher and Krein [102]. The significance of Pólya-type kernels to statistical decision theory is developed in Karlin [136].

§ 7.3. All the results of this section are elaborations from Karlin [137].

§ 7.4. The theory of Green and Pólya kernels was introduced in [137].

§ 7.5. The importance of the invariance principle as related to statistical problems goes back to Pitman [196]. The most penetrating applications of these concepts were developed and announced jointly by Hunt and Stein while relaxing from their duties as meteorologists for the U. S. Army during World War II. These observations have not yet appeared in print. Nevertheless, they became well known and attracted numerous statistical theorists, who worked out many of their implications with regard to statistical procedures. (See Kiefer [143], Kudo [155], Blackwell and Girschick [30], and Peisakoff [194].)

Theorem 7.5.3 is due to L. J. Savage (unpublished). The application of invariance theory to games involving two rational adversaries is discussed in [135] in an abstract setting.

§ 7.6. Problems 3, 4, and 15–23 are due to Glicksberg and Gross (internal documents of the RAND Corporation). Problem 11 is due to Gross [113].

CHAPTER 8

INFINITE CLASSICAL GAMES NOT PLAYED OVER THE UNIT SQUARE

In most of the games examined in Chapters 3–7 a set of pure strategies could be singled out for each player and identified, usually with the unit interval of the real line. The set of all strategies in such games was recognized as the set of probability distributions defined on the space of pure strategies. The pure strategies usually possessed a special interpretation which was exploited in solving the game. For example, in games of timing the pure strategies represented the possible times that could be chosen for performing a certain act. In certain convex games a pure strategy denoted the proportion of resources assigned to each of two or more alternative tasks.

In this and the succeeding chapter we consider a class of games which cannot be described by a kernel function defined on subsets of Euclidean spaces, but for which the strategy space of each player can be embedded in a classical function space. Such games are still of category I according to our classification in Chapter 2, p. 22.

We defer the investigation of games of category II, whose strategy spaces are considerably larger than those studied here. The analysis of such games is complex: it obliges us, for example, to come to grips with the mathematically unresolved problem of clarifying the concept of measures over function spaces. Games of this class will be discussed in a projected volume.

The games studied in this chapter may be interpreted as tactical duels involving continuous firing. In the games of timing considered in earlier chapters, each player had one opportunity, or at most several opportunities, to act. In this chapter we allow at least one of the players to act in a continuous manner. The reader is referred to the example given in Section 2.4 as one instance of this class of games. In the present chapter we give the solution to this example and examine a generalization of this game, which may be interpreted as a two-machine-gun duel.

Such isolated examples do not comprise a general theory. Unfortunately, however, no general theory exists for handling these games—only a more or less useful technique that is often effective in solving specific continuous-fire duels. This technique will be further formalized in our discussions of poker-type models in the next chapter; we shall concentrate in this chapter on exhibiting its workings for the two examples outlined above. The challenge of developing a unifying theory, one that will include these examples as special cases, remains to be met.

Because of the nature of the constraints on firepower, etc., we shall find the well-known Neyman-Pearson lemma an indispensable tool in our analysis.

8.1 Preliminary results (the Neyman-Pearson lemma). In this chapter we prove the Neyman-Pearson lemma in its simplest form, and show how it may be used to solve two simple problems. Its application to a special nonlinear variational problem is discussed in the next section, as an illustration of the method to be used in solving the game problems of Sections 8.3 and 8.5.

Neyman-Pearson problem. Let $f(t)$ and $g(t)$ be two nonnegative functions that are integrable with respect to a sigma-finite measure μ defined on the interval $0 \leq t \leq T$, and let $g(t)$ be normalized so that

$$\int_0^T g(t)\, d\mu(t) = 1.$$

From the class $\mathcal{A}$ of measurable functions $\phi(t)$ subject to the restrictions $0 \leq \phi(t) \leq M$ almost everywhere and

$$\int \phi(t)g(t)\, d\mu(t) \leq \alpha \qquad (0 < \alpha < M), \tag{8.1.1}$$

we wish to determine the function $\phi_0(t)$ for which

$$\int_0^T \phi_0(t)f(t)\, d\mu(t) = \max_{\phi} \int_0^T \phi(t)f(t)\, d\mu(t).$$

Remark 8.1.1. The existence of a maximizing function follows from the fact that the sphere of radius M in the Banach space of bounded measurable functions is weak* compact. This remark is not essential to what follows, since we shall exhibit the solution explicitly and verify its maximal property directly.

We first note that if $\phi_0(t)$ is a solution, then any function which is equal to $\phi_0(t)$ almost everywhere, with respect to the μ-measure, is likewise a solution. To avoid ambiguities, throughout this chapter any two functions which differ only on a set of μ-measure zero are considered to represent the same function. Finally, for convenience we adopt the inner product notation

$$\int \phi(t)f(t)\, d\mu(t) = (f, \phi).$$

Solution. We map each measurable member function $\phi(t)$ satisfying $0 \leq \phi \leq M$ a.e. into a point $\langle \xi, \eta \rangle$ in E^2, where

$$\xi = (g, \phi), \qquad \eta = (f, \phi).$$

The image set in E^2 (which we shall denote by Γ) is clearly a convex set because the mapping is linear. The set Γ contains the points $\langle 0, 0\rangle$ and $\langle M, (f, M)\rangle$, where M stands for the function $\phi(t)$ identically equal to M; and furthermore, if $\langle \xi, \eta\rangle \in \Gamma$, then obviously

$$\eta = \int \phi(t) f(t)\, d\mu(t) \leq (f, M).$$

Therefore Γ is a convex set whose largest ordinate corresponds to its largest abscissa, so that in order to solve the problem it is sufficient to find the maximum of (f, ϕ) among those members of $\mathfrak{A}$ for which $(g, \phi) = \alpha$.

For convenience, we shall assume that $g(t)$ is positive wherever $f(t)$ is positive. In the more general case, where this assumption is dropped, the essential ideas of the analysis are the same but certain pathologies develop which must be treated separately.

Let $\eta = k\xi + b$ denote a supporting plane to Γ at the point whose abscissa is $(g, \phi) = \alpha$ and whose ordinate is the largest possible value for this abscissa. The previous remarks show that $k \geq 0$, and that $\eta \leq k\xi + b$ for all $(\xi, \eta) \in \Gamma$, with equality holding for the image of the solution ϕ_0 if one exists. We proceed as if a solution existed, and we seek to characterize its form. Once ϕ_0 is explicitly specified, its maximality property can be checked in a routine fashion (see below). Thus

$$(f, \phi_0) = k(g, \phi_0) + b,$$

$$(f, \phi) \leq k(g, \phi) + b;$$

and hence

$$(f - kg, \phi - \phi_0) \leq 0.$$

By taking first $\phi(t) \equiv M$ on the set where $f - kg > 0$ and $\phi(t) \equiv \phi_0$ otherwise, then $\phi(t) \equiv 0$ where $f - kg < 0$ and $\phi(t) \equiv \phi_0$ otherwise, we conclude that

$$\text{if} \quad \frac{f(t)}{g(t)} \begin{cases} > k, & \text{then } \phi_0(t) = M; \\ < k, & \text{then } \phi_0(t) = 0; \\ = k, & \text{then } \phi_0 \text{ is not determined.} \end{cases} \tag{8.1.2}$$

If g is zero, then by hypothesis $f = 0$, and f/g is to be interpreted as equal to k.

The last condition of (8.1.2) means that any determination of ϕ on the set where $f = kg$ does not alter the value of (f, ϕ).

Computation. In actual practice the method of computing the value of k and the function ϕ_0 is as follows. Let K be the set of real numbers k'

such that

$$v(k') = \int_{f/g>k'} Mg(t)\,d\mu(t) \geq \alpha,$$

where $v(k')$ is a nonincreasing right-continuous function of k', and let

$$k = \sup_{k' \in K} k' \geq 0.$$

If ϵ is defined by

$$\epsilon = \alpha - \int_{f/g>k} Mg(t)\,d\mu(t),$$

then $\epsilon \geq 0$ by the right continuity of v and $\epsilon \leq v(k-) - v(k)$.

We define ϕ_0 as in (8.1.2) and such that

$$\int_{f/g=k} \phi_0(t)g(t)\,d\mu(t) = \epsilon,$$

this equality clearly being possible because of the given bounds on ϵ.

In order to show that the function ϕ_0 defined in terms of this number k is the desired function, we may readily verify that for any $\phi(t)$ with $0 \leq \phi(t) \leq M$ a.e.,

$$(f - kg, \phi - \phi_0) \leq 0,$$

and therefore

$$(f, \phi) - (f, \phi_0) \leq k[(g, \phi) - (g, \phi_0)] = k[(g, \phi) - \alpha].$$

Hence $(f, \phi_0) \geq (f, \phi)$ whenever $(g, \phi) \leq \alpha$.

A similar problem. Let $g(t)$ be defined as before, and let $f(t)$ be an integrable function, not necessarily positive. The problem is to find a function $\phi(t)$ subject to the constraints

$$0 \leq \phi(t) \leq M, \qquad \int_0^T g(t)\phi(t)\,d\mu(t) = \alpha$$

for which (f, ϕ) is a maximum. We assume that $0 < \alpha < M$. The solution is trivial if $\alpha = 0$ or $\alpha = M$.

Note that the constraint (8.1.1) has been replaced by the stronger requirement that $(g, \phi) = \alpha$. This is necessitated by the fact that f is no longer assumed to be nonnegative; hence we cannot conclude, as previously, that the maximal ϕ has $(g, \phi) = \alpha$ unless we restrict ourselves by hypothesis to such ϕ.

The solution of this problem is similar to that of the previous one. The mapping $\phi \to [(g, \phi), (f, \phi)]$ maps the set of measurable functions ϕ $(0 \leq \phi \leq M)$ into a convex set Γ in E^2. This set has a supporting plane from above at the point with abscissa $\xi = \alpha$, and the equation of the plane is $\eta = k\xi + b$, but k is not necessarily positive. In terms of k the solution ϕ_0 is described as follows:

$$\text{if } \frac{f(t)}{g(t)} \begin{cases} > k, & \phi_0(t) = M; \\ < k, & \phi_0(t) = 0. \end{cases}$$

Where $f(t) = kg(t)$, the values of $\phi_0(t)$ are subject only to the restrictions

$$\int_0^T \phi_0(t)g(t)\,d\mu(t) = \alpha, \qquad 0 \leq \phi_0(t) \leq M.$$

Where $g(t) = 0$, we assume as before that $f(t) = 0$, so that such points are not relevant to the solution.

***8.2 Application of the Neyman-Pearson lemma to a variational problem.** The characterization of the solution to the Neyman-Pearson problem as given in (8.1.2) can be used to solve nonlinear variational problems of a special form. Since the technique involved will be needed for our analysis of machine-gun duels, we shall depart momentarily from our purely game-theoretic orientation and discuss the use of this technique on a nonlinear variational problem of some independent interest.

We summarize the idea of the method as follows. The problem is to determine a function $\phi(t)$ which maximizes (minimizes) an integral expression involving a concave (convex) function of ϕ, where ϕ is subject to linear inequality constraints. By manipulation of the integral functional of ϕ we reduce the problem to one to which the Neyman-Pearson solution applies, with one important reservation: in the notation of Section 8.1 the function f now involves $\phi_0(t)$, the solution. Thus the solution to the Neyman-Pearson problem is determined implicitly, with no assurance that it will be free of inconsistencies.

However, by careful analysis and ingenuity it is often possible to arrive at a consistent solution that satisfies the constraints. This is done by introducing suitable parameters derived from the natural relations of the problem. The consistency of the solution must afterwards be checked. We illustrate this process on the following variational problem.

Problem. Let $0 \leq F(x) \leq A$ be twice continuously differentiable, $F'(0) > 0$, and $F''(x) < 0$. Let $v(t)$ be positive, strictly decreasing, continuous, and integrable on $[0, \infty)$. Let $\phi(t)$ be integrable and subject to the condition

$$\int_0^\infty \phi(t)\,dt \leq \alpha \qquad (\alpha > 0,\ \phi(t) \geq 0). \tag{8.2.1}$$

* Starred sections are advanced discussions that may be omitted at first reading without loss of continuity.

We wish to characterize the function $\phi_0(t)$ in this class for which

$$\int_0^\infty F[\phi_0(t)]v(t)\,dt \qquad \text{is a maximum.} \tag{8.2.2}$$

The problem may be provided with the following physical setting. Let $\phi(t)$ be the rate of depletion at time t of an oilfield of known limited reserves α. $F[\phi(t)]$ is the profit rate associated with the extraction rate $\phi(t)$. F can be argued to be concave, since diseconomies of scale result in a convex cost function. Let $F[\phi(t)]v(t)$ denote the present value of future profits, after discounting by the function $v(t)$.

Our problem is to choose an optimal extraction function $\phi_0(t)$ satisfying (8.2.2) and such that

$$\int_0^\infty \phi_0(t)\,dt \le \alpha \qquad (\phi_0(t) \ge 0).$$

Solution. We first note that if a solution $\phi(t)$ exists, then this solution is unique. Let $\phi_1(t)$ be any other function that maximizes the integral, and let

$$B = \max_{\phi} \int_0^\infty F[\phi(t)]v(t)\,dt.$$

By strict concavity,

$$F[\lambda\phi_0(t) + (1-\lambda)\phi_1(t)] > \lambda F[\phi_0(t)] + (1-\lambda)F[\phi_1(t)]$$

for all t such that $\phi_0(t) \neq \phi(t)$, and hence

$$\int_0^\infty F[\lambda\phi_0(t) + (1-\lambda)\phi_1(t)]v(t)\,dt \ge \lambda\int_0^\infty F[\phi_0(t)]v(t)\,dt$$
$$+ (1-\lambda)\int_0^\infty F[\phi_1(t)]v(t)\,dt = B. \tag{8.2.3}$$

Since B is the maximum, we have equality in (8.2.3). By the previous inequality, we conclude that $\phi_0(t) = \phi_1(t)$ a.e.

We proceed by proving the following lemma, which we shall use later. (Note that since F' is strictly decreasing and continuous, its inverse function F'^{-1} exists and is also strictly decreasing and continuous.)

▶LEMMA 8.2.1. Let

$$J(t_0) = \int_0^{t_0} F'^{-1}\left(\frac{F'(0)v(t_0)}{v(t)}\right) dt,$$

where F, v satisfy the conditions of the problem. Then

$$\lim_{t_0\to 0} J(t_0) = 0, \qquad \lim_{t_0\to\infty} J(t_0) = \infty.$$

Proof. The fact that

$$\lim_{t_0 \to 0} J(t_0) = 0$$

is an immediate consequence of the fact that the integrand is a decreasing continuous function which itself approaches zero. To prove the other half of the lemma we first choose ϵ $(0 < \epsilon < 1)$ fixed so that

$$F'^{-1}[F'(0)\,(1 - \epsilon)] = \delta\ (\delta > 0).$$

Then, given n arbitrarily large, choose $\bar{t} = n/\delta$. Choose any $t_0 > \bar{t}$ such that

$$\frac{v(t_0)}{v(\bar{t})} \le 1 - \epsilon.$$

Then

$$F'^{-1}\left(\frac{F'(0)v(t_0)}{v(t)}\right) \ge \delta$$

for $t < \bar{t}$, and

$$\int_0^{t_0} F'^{-1}\left(\frac{F'(0)v(t_0)}{v(t)}\right) dt \ge \int_0^{\bar{t}} F'^{-1}\left(\frac{F'(0)v(t_0)}{v(t)}\right) dt \ge n.$$

We now consider an equivalent problem whose solution is obtained using the Neyman-Pearson lemma. Let

$$I(\lambda) = \int_0^\infty F[\lambda\phi_0(t) + (1 - \lambda)\phi(t)]v(t)\,dt,$$

$$I'(\lambda) = \int_0^\infty F'[\lambda\phi_0(t) + (1 - \lambda)\phi(t)]v(t)[\phi_0(t) - \phi(t)]\,dt,$$

$$I''(\lambda) = \int_0^\infty F''[\lambda\phi_0(t) + (1 - \lambda)\phi(t)]v(t)[\phi_0(t) - \phi(t)]^2\,dt < 0,$$

where we have assumed for ease of exposition that the second interchange of differentiation and integration is valid.

Thus $I(\lambda)$ is a strictly concave function of λ. If ϕ_0 is a solution of the original problem, then $I(\lambda)$ achieves its maximum at $\lambda = 1$. This is possible *if and only if* $I'(1) \ge 0$. Explicitly, $\phi_0(t)$ is a solution if and only if for every other $\phi(t)$ such that

$$\int_0^\infty \phi(t)\,dt \le \alpha \qquad (\alpha > 0,\ \phi(t) \ge 0), \tag{8.2.4}$$

we have

$$\int_0^\infty F'[\phi_0(t)]v(t)\phi_0(t)\,dt \ge \int_0^\infty F'[\phi_0(t)]v(t)\phi(t)\,dt. \tag{8.2.5}$$

Suppose that $\phi_0(t)$ satisfying (8.2.4) is such that for a fixed function $\psi(t)$

$$\int_0^\infty F'[\psi(t)]v(t)\phi_0(t)\,dt \geq \int_0^\infty F'[\psi(t)]v(t)\phi(t)\,dt \tag{8.2.6}$$

for all $\phi(t)$ satisfying (8.2.4). Then if in fact $\phi_0(t) \equiv \psi(t)$, it follows that $\phi_0(t)$ is a solution to the original problem, since (8.2.4) and (8.2.5) are satisfied.

Let us now restrict our attention to functions $\phi(t)$ such that $0 \leq \phi(t) \leq M$ for $0 \leq t \leq T$ and $\phi(t) = 0$ for $t > T$. We may solve this restricted problem by using the Neyman-Pearson result.

The solution $\phi_0(t)$ has the following characterization:

$$\text{if } F'[\psi(t)]v(t) \begin{cases} > k, & \text{then} \quad \phi_0(t) = M, \\ < k, & \text{then} \quad \phi_0(t) = 0, \\ = k, & \text{then} \quad \phi_0(t) \text{ is arbitrary but such} \\ & \qquad\quad \text{that } 0 \leq \phi_0(t) \leq M, \end{cases} \tag{8.2.7}$$

where k is determined so that

$$\int_0^T \phi_0(t)\,dt = \alpha.$$

We now proceed in a somewhat artistic manner to arrive at the function $\psi(t)$ which agrees with the Neyman-Pearson solution $\phi_0(t)$, keeping in mind the above characterization of ϕ_0.

Because of the restriction

$$\int_0^T \phi_0(t)\,dt \leq \alpha,$$

we suspect that $\phi_0(t)$ never attains the value M if M is chosen large enough to begin with. Furthermore, on any interval where $F'[\psi(t)]v(t) = k$ holds, we must have $\psi(t) = F'^{-1}[k/v(t)]$. Thus as a candidate for $\psi(t)$ we consider the function

$$\psi(t) = \begin{cases} F'^{-1}[k/v(t)] & 0 \leq t < t_0, \\ 0 & t_0 \leq t \leq T. \end{cases} \tag{8.2.8}$$

Then

$$F'[\psi(t)]v(t) = \begin{cases} k & 0 \leq t < t_0, \\ F'(0)v(t) & t_0 \leq t \leq T. \end{cases}$$

If we take $k = F'(0)v(t_0)$, then

$$F'[\psi(t)]v(t) = \begin{cases} F'(0)v(t_0) = k & 0 \leq t \leq t_0, \\ F'(0)v(t) < k & t_0 \leq t \leq T. \end{cases}$$

This choice of k is not surprising, since it amounts to making $\psi(t)$ continuous at t_0. It can be shown that any other choice of k leads to an inconsistency. To guarantee obvious consistency requirements we take M, as prescribed in (8.2.7), to be a constant greater than $\psi(t)$.

By virtue of Lemma 8.2.1 we choose t_0 such that

$$\int_0^{t_0} F'^{-1}\left(\frac{F'(0)v(t_0)}{v(t)}\right) dt = \alpha.$$

Then, taking $T \geq t_0$, we have

$$\phi_0(t) = \begin{cases} F'^{-1}\left(\dfrac{F'(0)v(t_0)}{v(t)}\right) & 0 \leq t \leq t_0, \\ 0 & t_0 \leq t \leq T. \end{cases} \tag{8.2.9}$$

We may verify that $\phi_0(t)$ as defined in (8.2.9) satisfies the conditions of (8.2.7) with $\psi(t) = \phi_0(t)$. Hence this ϕ_0 is a solution of the original problem if the admissible functions $\phi(t)$ are restricted further to be uniformly bounded by M and vanish outside a fixed interval $[0, T]$.

However, any function $\phi(t)$ for which

$$\int_0^\infty \phi(t)\, dt = \alpha < \infty$$

may be approximated by a sequence of bounded functions $\phi_n(t)$ that vanish outside compact sets, the approximation made so that

$$\int_0^\infty \{F[\phi(t)] - F[\phi_n(t)]\} v(t)\, dt \to 0.$$

It follows that $\phi(t)$, as specified in (8.2.9), is the solution of the variational problem with respect to all possible ϕ satisfying (8.2.1).

8.3 The fighter-bomber duel. In this and the following section we present the solution of the game introduced in Section 2.4, which describes a duel between a fighter capable of firing a single shot and a bomber capable of maintaining a continuous fire. Or rather we shall examine a slight variation of this game, which we shall interpret as an advertising battle. The salient features of the two solutions are identical. The model may not be perfectly true to life, but it does indicate what kind of intelligent recommendations for a course of action emerge from the analysis of game problems.

Consider a competition between two businessmen, Mr. Big (B) and Mr. Little (L). B is prosperous and has a stable supply of customers; L is on the verge of bankruptcy.

Now at time 0 a potential customer arrives on the scene. He may decide not to buy at all, but if he does buy he will place a large order with only one of the concerns. The order is sufficiently large to put L back on his feet, but to remain in business L must obtain the order by a certain time—call it time 1. B and L now engage in advertising competition. Since L will undoubtedly be forced out of business unless he secures this order, B would regard it as a victory even if the customer should decide not to buy at all. L wins if he secures the customer within the allotted time; B wins if the customer does not buy or, *a fortiori*, if B receives the order. By proper interpretation of the utility objectives of the participants the game is zero-sum.

We assume that only B or L can win the order, and that advertising will be the decisive factor in determining the customer's decision. The conditions are these: it is to B's advantage to spread his advertising over the full period, whereas L will do best to concentrate his resources in one elaborate appeal. Each in his own way must decide when to advertise.

We assume that the customer's resistance to buying decreases with time. Moreover, we suppose that the customer's psychology is such that all previous ads that have appeared exert no influence on him when he reads a given ad.

Finally, we formulate the campaign as a mathematically continuous affair. L selects a time t in $[0, 1]$ at which to advertise, while B selects a rate of spending $p(t)$ which is subject to the restrictions

$$0 \leq p(t) \leq 1, \qquad \int_0^1 p(t)\, dt = \delta < 1, \tag{8.3.1}$$

where δ is related to the money B has allotted to his campaign; i.e., if B is spending at maximum rate, then the money will last a duration δ of the time interval 1.

Let $a(t)$ be the probability of success for L when he advertises at time t, and let $r(t)$ represent the probability that a one-dollar investment at time t will secure the order for B if it has not been placed by that time. [More precisely, $r(t)p(t)$ is the probability density associated with the spending rate $p(t)$.] As stated before, we assume that $r(t)$ and $a(t)$ increase with time.

For fixed choices of these strategies, a time t for L and a spending rate $p(t)$ for B, the probability of L's winning is equal to the probability that the customer will resist B until time t and then succumb to L. Let $\phi(t, p)$ denote the probability of the customer's resisting B during the interval $[0, t]$; then from the independence assumptions we have

$$\phi(t, p) = \exp\left[-\int_0^t p(u)r(u)\, du\right]. \tag{8.3.2}$$

(For the proof of this formula the reader is referred to Section 2.4.)

The pay-off kernel $K[t, p(t)]$ when B (Player II) uses $p(t)$ and L (Player I) acts at time t is

$$\begin{aligned} K[t, p(t)] &= \alpha a(t)\phi[t, p(t)] - \beta\{1 - \phi[t, p(t)]\} \\ &\quad - \beta\phi[t, p(t)]\,[1 - a(t)] \\ &= [(\alpha + \beta)a(t)]\phi(t, p) - \beta, \end{aligned} \tag{8.3.3}$$

where α is the value of the pay-off to L if L wins the order and β is the value of the pay-off to B if L goes bankrupt. The coefficients of these factors are obviously the probabilities of the desired event according to the strategies t for L and $p(t)$ for B.

The pay-off for a mixed strategy x for L and the choice p for B is obtained by averaging over the space of pure strategies, i.e.,

$$K(x, p) = \int_0^1 K(t, p)\, dx(t). \tag{8.3.4}$$

We now display the solutions to this game of kernel $K(x, p)$, reserving the proofs for the next section. For this purpose we define

$$w(t) = \min\left(1, \frac{A'(t)}{A(t)r(t)}\right),$$

where $A(t) = (\alpha + \beta)a(t)$. (In the case of the fighter-bomber duel, $A(t) = \alpha a(t) + \beta$; see p. 36.)

The solution given below applies only when

$$\frac{a'(t)}{a(t)r(t)} = b(t)$$

is strictly decreasing in t. This will be true, for example, whenever $\log a(t)$ is concave. We shall see later that only the set of points for which $b(t) < 1$ may be relevant to an optimal strategy for Player II; our assumption guarantees that such a set is at worst a single interval. The extension to the general case can be made, and the analysis is the same except for tedious details, notably a more cumbersome type of strategy for Player II, consisting of a spending rate spread over several disjoint intervals of t values where $b(t) < 1$. We have made our assumption simply to avoid these irrelevant complexities; all the essential features of the general answer have been preserved, as well as the main elements of the arguments. It should be added that most of the common probability functions $a(t)$ and $r(t)$ which qualify are such that the assumption of $b(t)$ decreasing is indeed fulfilled.

We now introduce two further parameters. Let d be the unique value in $[0, 1]$ satisfying

$$\frac{A'(d)}{A(d)r(d)} = 1,$$

if such a value exists. If no solution of $A'(d)/A(d)r(d) = 1$ exists in the unit interval, we define $d = 1$ when

$$\frac{A'(t)}{A(t)r(t)} > 1$$

everywhere in $[0, 1]$ and $d = 0$ when the opposite inequality holds everywhere. Throughout what follows we shall consider only the case in which d exists. The other cases are left as exercises for the reader.

If

$$\int_0^1 w(t)\, dt = \delta_0 > \delta,$$

let c be chosen so that

$$\int_c^1 w(t)\, dt = \delta.$$

The solution of the game can now be described in terms of the quantities $w(t)$, c, and d. We consider three cases.

Case 1. $\delta_0 > \delta$, $c < d$. The optimal strategy for Player II is

$$p_0(t) = \begin{cases} 0 & 0 \le t < c, \\ w(t) & c \le t \le 1; \end{cases}$$

the optimal strategy for Player I is

$$x_0(t) = \begin{cases} 0 & 0 \le t < d, \\ 1 - \dfrac{r(c)}{r(t)} & d \le t < 1, \\ 1 & t = 1. \end{cases}$$

Case 2. $\delta_0 > \delta$, $d < c$. The optimal strategy for Player II is

$$p_0(t) = \begin{cases} 0 & 0 \le t < c, \\ w(t) & c \le t < 1; \end{cases}$$

the optimal strategy for Player I is

$$x_0(t) = \begin{cases} 0 & 0 \le t \le c, \\ 1 - \dfrac{r(c)}{r(t)} & c \le t < 1, \\ 1 & t = 1. \end{cases}$$

Case 3. $\delta_0 \leq \delta$. The optimal strategy for Player II is any $p_0(t)$ such that

$$\text{(a)} \qquad p_0(t) = 1 \qquad (0 \leq t \leq d),$$

$$\text{(b)} \qquad 1 \geq p_0(t) \geq \frac{A'(t)}{A(t)r(t)} \qquad (d \leq t \leq 1),$$

$$\text{(c)} \qquad \int p_0(t)\,dt = \delta;$$

the optimal strategy for Player I is a degenerate distribution concentrating at $t = d$.

In Cases 1 and 2 the optimal strategy for each player is unique, whereas in Case 3 Player II has a whole family of strategies as described above, and others as well.

If $b(t)$ is not monotonic, then the optimal strategy for Player II is composed of intervals in which $p_0(t) = w(t)$ alternated with intervals in which $p_0(t) = 0$.

As an example, consider the case $\beta = 0$, $\alpha = 1$, $r(t) = k^2t (k > 1)$, $a(t) = t$. Then

$$b(t) = (kt)^{-2}, \qquad d = \frac{1}{k}, \qquad \delta_0 = \frac{2k-1}{k^2};$$

if

$$\frac{k-1}{k^2}\begin{cases} \leq \delta \leq \delta_0, & \text{then} \quad c = \delta_0 - \delta \leq d, \\ > \delta, & \text{then} \quad c = \dfrac{1}{1+k^2\delta} > d. \end{cases}$$

The various solutions can now be recorded by appealing to the above tabulation.

***8.4 Solution of the fighter-bomber duel.** Since the constant β of (8.3.3) does not affect the set of optimal strategies, we study a general game with pay-off

$$V(t, p) = A(t)\exp\left[-\int_0^t p(u)r(u)\,du\right],$$

where $0 \leq t \leq 1$ and $p(u)$ is subject to the conditions expressed in (8.3.1). We assume that $A(t) > 0$, $A'(t) > 0$, and $r(u) > 0$, $r'(u) > 0$, and that

$$b(t) \equiv \frac{A'(t)}{A(t)r(t)}$$

is a strictly decreasing function of t. The first two sets of assumptions are satisfied in our model; the last is an extra assumption on the accuracy functions $r(t)$ and $a(t)$ whose meaning we have discussed in the last section.

Strategies. Strategies for Player II will be functions $p(t)$ satisfying (8.3.1), and mixed strategies for Player I will be probability distributions $x(t)$ on the unit interval. The corresponding expected return will be

$$\int V(t, p)\, dx(t) = \int A(t) \exp\left[-\int_0^t p(u)r(u)\, du\right] dx(t).$$

The fact that this game has a value and optimal strategies can be deduced from the abstract min-max theorem. The actual optimal strategies for this example are exhibited later in this section. The following lemma will be fundamental.

▶ LEMMA 8.4.1. If there exist two admissible strategies $p_1(t)$ and $p_2(t)$ such that $V(t, p_1) \geq V(t, p_2)$ for all t, then $p_1(t) = p_2(t)$ a.e.

Proof. By assumption,

$$\int_0^t r(u)p_1(u)\, du \leq \int_0^t r(u)p_2(u)\, du.$$

Therefore, if we define $f(u) = p_2(u) - p_1(u)$, it follows that

$$\psi(t) = \int_0^t r(u)f(u)\, du \geq 0;$$

also $\psi'(u) = r(u)f(u)$ a.e. But

$$\int_0^1 \frac{d\psi(u)}{r(u)} = \int_0^1 f(u)\, du = \int_0^1 [p_2(u) - p_1(u)]\, du = \delta - \delta = 0.$$

On the other hand, integrating by parts,

$$\lim_{\epsilon \to 0} \left[\frac{\psi(1)}{r(1)} - \frac{\psi(\epsilon)}{r(\epsilon)} + \int_\epsilon^1 \frac{\psi(u)}{[r(u)]^2} r'(u)\, du\right] = 0, \tag{8.4.1}$$

while

$$\left|\frac{\psi(\epsilon)}{r(\epsilon)}\right| = \frac{1}{r(\epsilon)}\left|\int_0^\epsilon r(u)f(u)\, du\right| \leq \int_0^\epsilon |f(u)|\, du \to 0.$$

Since the remaining terms in (8.4.1) are both nonnegative, we conclude that both must approach zero, or

$$\lim_{\epsilon \to 0} \int_\epsilon^1 \frac{\psi(t)}{[r(t)]^2} r'(t)\, dt = 0.$$

Since $\psi \geq 0$, it follows that $\psi(t) = 0$ a.e. Thus $f = 0$ a.e.

We define

$$w(t) = \min\left(1, \frac{A'(t)}{A(t)r(t)}\right),$$

and the constants c and d as previously.

Case 1. $\delta_0 > \delta$, $c < d$. Set

$$p_0(t) = \begin{cases} 0 & 0 \leq t < c, \\ w(t) & c \leq t \leq 1. \end{cases}$$

The expected return for this strategy is as follows:

(i) For $0 \leq t \leq c$,

$$V(t, p_0) = A(t).$$

(ii) For $c \leq t \leq d$,

$$V(t, p_0) = A(t) \exp\left[-\int_c^t r(u)\, du\right].$$

Also, in the domain of (ii)

$$V'(t, p_0) = \exp\left[-\int_c^t r(u)\, du\right][A'(t) - A(t)r(t)].$$

But

$$1 \leq \frac{A'(t)}{A(t)r(t)},$$

so that $V'(t, p_0)$ is strictly positive for $t < d$ and $V(t, p_0)$ is increasing. Its maximum value is achieved at d and will be denoted by M_1.

(iii) For $d \leq t \leq 1$,

$$\begin{aligned} V(t, p_0) &= A(t) \exp\left[-\int_c^d r(u)\, du\right] \exp\left[-\int_d^t \frac{A'(u)}{A(u)}\, du\right] \\ &= A(d) \exp\left[-\int_c^d r(u)\, du\right] = M_1. \end{aligned} \tag{8.4.2}$$

These results may be summarized as

$$\max_{0 \leq t \leq 1} V(t, p_0) = M_1,$$

and this value is attained only for $d \leq t \leq 1$.

▶ LEMMA 8.4.2. For any other strategy p,

$$\max_t V(t, p) > \max_t V(t, p_0),$$

and therefore p_0 is the optimal strategy for Player II (in Case 1).

Proof. By Lemma 8.4.1, if $p \neq p_0$ there exists a value t^* such that $V(t^*, p) > V(t^*, p_0)$. We first notice that t^* cannot fall between 0 and c, since in this range p_0 compels $\phi(t, p)$ to be as large as possible. But, if $c < t^* \leq d$, then

$$V(d, p) = A(d) \exp\left[-\int_0^{t^*} r(u)p(u)\,du\right] \exp\left[-\int_{t^*}^{d} r(u)p(u)\,du\right]$$

$$= \frac{A(d)}{A(t^*)} V(t^*, p) \exp\left[-\int_{t^*}^{d} r(u)p(u)\,du\right].$$

Similarly,

$$V(d, p_0) = \frac{A(d)}{A(t^*)} V(t^*, p_0) \exp\left[-\int_{t^*}^{d} r(u)\,du\right].$$

Clearly, since $p \leq 1$ and $V(t^*, p) > V(t^*, p_0)$, we must have $V(d, p) > V(d, p_0) = M_1$. If $d < t^* \leq 1$, then $V(t^*, p) > V(t^*, p_0)$ implies $V(t^*, p) > M_1$. Thus in any case $V(t, p) > M_1 = \max_t V(t, p_0)$ for some t.

Optimal strategy for Player I. If x_0 is optimal and v is the value, then for any p

$$\int_0^1 V(t, p)\,dx_0(t) \geq v,$$

while

$$\int_0^1 V(t, p_0)\,dx_0(t) = v.$$

We define

$$I(\lambda) = \int_0^1 V[t, \lambda p_0 + (1 - \lambda)p]\,dx_0(t) \qquad (0 \leq \lambda \leq 1).$$

Then

$$I'(\lambda) = \int_0^1 V[t, \lambda p_0 + (1 - \lambda)p] \times \left[-\int_0^t r(u)[p_0(u) - p(u)]\,du\right] dx_0(t),$$

$$I''(\lambda) = \int_0^1 V[t, \lambda p_0 + (1 - \lambda)p] \times \left[-\int_0^t r(u)[p_0(u) - p(u)]\,du\right]^2 dx_0(t) > 0.$$

Thus $I(\lambda)$ is a convex function of λ which achieves its minimum at $\lambda = 1$. This is possible if and only if

$$I'(\lambda)|_{\lambda=1} \leq 0,$$

i.e., if

$$\int_0^1 V(t, p_0) \left(\int_0^t r(u)[p(u) - p_0(u)]\,du\right) dx_0(t) \leq 0.$$

Changing the order of integration, we deduce that

$$\max_{p} \int_0^1 p(u)r(u) \left(\int_u^1 V(t, p_0)\, dx_0(t)\right) du \tag{8.4.3}$$

is achieved for $p = p_0$. Since the set of functions over which we maximize (8.4.3) is subject to the constraints (8.3.1), we may invoke the Neyman-Pearson lemma, which shows that $p_0(u)$ is characterized as follows: for some k,

$$\text{if} \quad r(u) \int_u^1 V(t, p_0)\, dx_0(t) \begin{cases} > k, & \text{then} \quad p_0(u) = 1, \\ = k, & \text{then} \quad 0 \leq p_0(u) \leq 1, \\ < k, & \text{then} \quad p_0(u) = 0. \end{cases}$$

Since $A'(t)/A(t)r(t)$ is strictly decreasing, we know that $0 < p_0(t) < 1$ if $d < t < 1$, so that in this range equality must hold above. Consequently, by (8.4.2) we have $r(u)M_1[1 - x_0(u)] = k$ for $d < u < 1$, so that

$$x_0(u) = 1 - \frac{k}{M_1} \frac{1}{r(u)} \qquad (d < u < 1).$$

Furthermore, $x_0(t)$ will concentrate in the interval $[d, 1]$ in which $V(t, p_0)$ is largest, so that $x_0(u) = 0 \quad (0 \leq u < d)$.

In order to evaluate the constant k, we notice that the definition of p_0 implies that

$$r(u) \int_u^1 V(t, p_0)\, dx_0(t) < k$$

for $u < c$, while for $u > c$ the opposite inequality holds. Since the above quantity is continuous at $u = c$, equality must be maintained at $u = c$. Then, since $V(t, p_0) = M_1$ in $[d, 1]$, where x_0 concentrates its full mass, it follows that $r(c)M_1 = k$, and hence

$$x_0(u) = 1 - \frac{r(c)}{r(u)} \qquad (d < u < 1).$$

In particular, $x_0(d^+) = 1 - r(c)/r(d)$, so that we have a mass $1 - r(c)/r(d)$ at the point $u = d$. The jump at $u = 1$ is of magnitude $r(c)/r(1)$.

Case 2. $\delta_0 > \delta$, $c > d$. Set

$$p_0(t) = \begin{cases} 0 & 0 \leq t < c, \\ w(t) & c \leq t \leq 1. \end{cases}$$

The expected return is as follows:

(i) For $0 \leq t < c$,

$$V(t, p_0) = A(t) < A(c).$$

(ii) For $c \leq t \leq 1$,

$$V(t, p_0) = A(c),$$

and in particular $\max_t V(t, p_0) = A(c) = M_2$.

▶ LEMMA 8.4.3. If $p(t) \neq p_0(t)$, then $\max_t V(t, p) > \max_t V(t, p_0)$, and hence $p_0(t)$ is the optimal strategy for Player II.

Proof. By Lemma 8.4.1, there exists a value t^* such that $V(t^*, p) > V(t^*, p_0)$. But t^* cannot be in the interval $[0, c]$, since in this interval p_0 makes the return as large as possible. Therefore t^* is in $[c, 1]$, so that $V(t^*, p) > M_2 = \max_t V(t, p_0)$.

Optimal strategy for Player I. The method used in Case 1 shows now that if x_0 is the optimal strategy for Player I, then p_0 has the following characterization:

$$\text{if} \quad r(u) \int_u^1 V(t, p_0)\, dx_0(t) \begin{cases} > k, & \text{then} \quad p_0(u) = 1, \\ = k, & \text{then} \quad 0 \leq p_0(u) \leq 1, \\ < k, & \text{then} \quad p_0(u) = 0. \end{cases}$$

But, from explicit knowledge of the form of p_0, we know that the first possibility never occurs (except possibly at c if $c = d$) while the second possibility is valid for $c < u < 1$, where $V(t, p_0) = M_2$. Then

$$r(u)M_2[1 - x_0(u)] = k \qquad (c < u),$$

or

$$x_0(u) = 1 - \frac{k}{M_2 r(u)} \qquad (c < u). \tag{8.4.4}$$

In order to evaluate k, we first notice that $p_0(t) = 0$ if $0 \leq t < c$, so that

$$r(u) \int_u^1 V(t, p_0)\, dx_0(t) < k \qquad (0 \leq u < c).$$

But $x_0(t)$ must concentrate in the interval $[c, 1]$, where the yield is largest for p_0. Since $V(t, p_0) = M_2$ for $t \geq c$, we obtain

$$r(u) \int_u^1 V(t, p_0)\, dx_0(t) = r(u)M_2 \leq k \qquad (0 \leq u < c).$$

Therefore $r(c)M_2 \leq k$. But if we suppose that strict inequality holds, then in view of (8.4.4) it follows that

$$x_0(c) < 1 - \frac{M_2 r(c)}{M_2 r(c)} = 0,$$

which is absurd. Therefore $k = M_2 r(c)$ and

$$x_0(u) = 1 - \frac{r(c)}{r(u)} \qquad (c \leq u < 1).$$

Finally, since $x_0(1) = 1$, it is clear that x_0 has a jump at $u = 1$ of magnitude $r(c)/r(1)$.

Case 3. $\delta_0 \leq \delta$. We shall show that the optimal strategy for Player I is I_d, a degenerate distribution located fully at $t = d$, and that any strategy $p_0(t) \geq w(t)$ is optimal for Player II. Such a strategy possesses the properties

(i) $$p_0(t) = 1 \qquad (0 \leq t \leq d),$$

(ii) $$1 \geq p_0(t) \geq \frac{A'(t)}{A(t)r(t)} \qquad (d \leq t \leq 1),$$

(iii) $$\int p_0(t)\, dt = \delta.$$

The expected return $V(t, p_0)$ is as follows:

(i) For $0 \leq t < d$,

$$V(t, p_0) = A(t) \exp\left[-\int_0^t r(u)\, du\right],$$

$$V'(t, p_0) = \exp\left[-\int_0^t r(u)\, du\right][A'(t) - A(t)r(t)] > 0.$$

Then

$$V(t, p_0) \leq V(d, p_0) = A(d) \exp\left[-\int_0^d r(u)\, du\right] = M_3.$$

(ii) For $d \leq t \leq 1$,

$$V(t, p_0) \leq A(t) \exp\left[-\int_0^d r(u)\, du\right] \exp\left[-\int_d^t \frac{A'(u)}{A(u)}\, du\right] = M_3.$$

In particular, $\max_t V(t, p_0) = M_3$.

▶LEMMA 8.4.4. If $p \neq p_0$, then $\max_t V(t, p) \geq \max_t V(t, p_0)$, so that p_0 is optimal for Player II. Also, I_d is optimal for Player I, and the value is M_3.

Proof. In the interval $[0, d]$, we have $p(t) \leq 1 = p_0(t)$. Hence

$$V(d, p) = A(d) \exp\left[-\int_0^d r(u)p(u)\,du\right]$$
$$\geq A(d) \exp\left[-\int_0^d r(u)\,du\right] = M_3.$$

Therefore, in order to prevent Player I from obtaining a yield greater than M_3, Player II must employ a strategy that is 1 a.e. in $[0, d]$. In order to show that any strategy of the same form as p_0 is optimal, we notice that if Player I uses d he can secure a yield at least equal to M_3. Thus M_3 is the value, and p_0 and I_d are optimal.

***8.5 The two-machine-gun duel.** Players I and II each possess a quantity of ammunition denoted by α and β, respectively. Their probability of scoring a hit when they fire at each other at a distance s is described by the functions $\xi(s)$ for Player I and $\eta(s)$ for Player II. We assume that these two functions are continuous and strictly decreasing, and moreover that $\xi(0) = \eta(0) = 1$, $\xi(\infty) = \eta(\infty) = 0$.

The strategies for the game are integrable functions $x(s)$ and $y(s)$. These functions, which describe the intensity of firing when the two opponents are a distance s apart, are subject to the obvious constraints

$$0 \leq x(s), \quad y(s) \leq 1$$

and

$$\int_0^\infty x(s)\,ds = \alpha, \qquad \int_0^\infty y(s)\,ds = \beta.$$

[The expression $x(s)\,\xi(s)\,ds$ may be construed as the probability of success for Player I when firing over the interval $(s, s + ds)$ if he adopts the firing policy $x(s)$.]

If the two opponents choose the strategies x and y and approach each other without retreat, then the probability that both are alive when they are a distance s apart will be denoted by $Q_{xy}(s)$. The probability that both will be alive at s equals the product of the probability that both were alive at $s + h$ and the conditional probability that they will survive from $s + h$ to s.

For small h, the probability that Player I will be destroyed in $[s + h, s]$ is approximately $y(s)\eta(s)h$, and the probability that Player I will survive is therefore $1 - y(s)\eta(s)h + 0(h)$. Since a similar argument is valid for Player II, we conclude that for h small

$$Q_{xy}(s) = Q_{xy}(s + h)[1 - x(s)\xi(s)h][1 - y(s)\eta(s)h] + o(h).$$

Then

$$\frac{Q'_{xy}(s)}{Q_{xy}(s)} = x(s)\xi(s) + y(s)\eta(s).$$

If we integrate this equation from ∞ to s and set $Q_{xy}(\infty) = 1$, we obtain

$$Q_{xy}(s) = \exp\left[-\int_s^\infty x(t)\xi(t)\,dt\right]\exp\left[-\int_s^\infty y(t)\eta(t)\,dt\right]. \qquad (8.5.1)$$

In order to simplify the notation, we shall write

$$Q_x(s) = \exp\left[-\int_s^\infty x(t)\xi(t)\,dt\right], \qquad Q_y(s) = \exp\left[-\int_s^\infty y(t)\eta(t)\,dt\right],$$

and hence $Q_{xy}(s) = Q_x(s)Q_y(s)$.

Pay-off. Player I will be the maximizing player; if Player II is destroyed, the value to Player I will be 1; otherwise the value will be zero. If the players choose the strategies $x(s)$ and $y(s)$, the pay-off reduces to the probability that Player II will be destroyed, which is clearly

$$\int_0^\infty Q_{xy}(s)\xi(s)x(s)\,ds = \int_0^\infty Q_y(s)\,dQ_x(s) = K(x, y). \qquad (8.5.2)$$

We shall solve this game for some important cases.

Other versions of duels may be constructed by assigning different pay-off values corresponding to various events. For instance, we can imagine a symmetric situation in which Player I wins 1 if his opponent is destroyed, loses 1 if he himself does not survive, and receives nothing otherwise; in this case the pay-off kernel is

$$L(x, y) = \int Q_x(s)Q_y(s)[\xi(s)x(s) - \eta(s)y(s)]\,ds$$

for strategies x and y.

Unfortunately, the methods which solve the game (8.5.2) are not applicable to the game $L(x, y)$, since L is not a concave function of x and a convex function of y. Hence the value of the game $L(x, y)$ in terms of the sets of strategies prescribed above is usually not determined. Actually, this game belongs properly to category II.

The game (8.5.2) may also be interpreted as an advertising conflict, along the lines of the advertising interpretation in Section 8.3, except that here both business concerns adopt a strategy of distributing their advertising expenditures. Player I has available resources totaling α dollars, Player II has β dollars, etc.; the rest of the interpretation should be clear. We now turn back to the analysis of the game (8.5.2).

Necessary conditions on optimal strategies. The existence of optimal strategies for the game (8.5.2) is a consequence of a general existence theorem. In this section we begin by assuming that optimal strategies x_0 and y_0 exist; we later exhibit them explicitly, relying on the Neyman-Pearson technique. To this end, we define

$$J(\lambda) = \int_0^\infty Q_{\lambda y_0 + (1-\lambda)y}(s)\, dQ_{x_0}(s) \qquad (0 \leq \lambda \leq 1).$$

Then

$$\begin{aligned} J'(\lambda) &= \int_0^\infty Q_{\lambda y_0 + (1-\lambda)y}(s) \left(- \int_s^\infty [y_0(t) - y(t)]\eta(t)\, dt \right) dQ_{x_0}(s), \\ J''(\lambda) &= \int_0^\infty Q_{\lambda y_0 + (1-\lambda)y}(s) \left(- \int_s^\infty [y_0(t) - y(t)]\eta(t)\, dt \right)^2 dQ_{x_0}(s). \end{aligned} \tag{8.5.3}$$

Clearly $J''(\lambda) \geq 0$; hence $J(\lambda)$ is a convex function of λ. Since y_0 is an optimal strategy for the minimizing player, $J(\lambda)$ achieves its minimum at $\lambda = 1$. For a convex function this is possible if and only if

$$J'(\lambda)|_{\lambda=1} \leq 0.$$

Hence, by (8.5.3),

$$\int_0^\infty Q_{y_0}(s) \left(\int_s^\infty y(t)\eta(t)\, dt \right) dQ_{x_0}(s)$$

attains its maximum for $y = y_0$. Interchanging the order of integration, we may write that

$$\int_0^\infty y(t)\eta(t) \left(\int_0^t Q_{y_0}(s)\, dQ_{x_0}(s) \right) dt \tag{8.5.4a}$$

achieves its maximum when $y(t) = y_0(t)$, where the maximum is obtained with respect to functions $y(t)$ which are subject to the constraints

$$0 \leq y(t) \leq 1, \qquad \int_0^\infty y(t)\, dt = \beta. \tag{8.5.4b}$$

In order to obtain a similar characterization for $x_0(t)$, we first note that integrating by parts gives

$$\int_0^\infty Q_y(s)\, dQ_x(s) = 1 - \left(Q_x(0)Q_y(0) + \int_0^\infty Q_x(s)\, dQ_y(s) \right).$$

We now define

$$I(\lambda) = Q_{\lambda x_0 + (1-\lambda)x}(0)Q_{y_0}(0) + \int_0^\infty Q_{\lambda x_0 + (1-\lambda)x}(s)\, dQ_{y_0}(s).$$

An analysis parallel to the previous one reveals that $I(\lambda)$ is a convex function of λ; and since x_0 is optimal, $I(\lambda)$ achieves its minimum at $\lambda = 1$. This assertion is equivalent to

$$I'(\lambda)|_{\lambda=1} \leq 0,$$

and this in turn implies that the maximum of

$$\int_0^\infty Q_{x_0}(0)Q_{y_0}(0)x(t)\xi(t)\,dt + \int_0^\infty x(t)\xi(t)\left(\int_0^t Q_{x_0}(s)\,dQ_{y_0}(s)\right)dt \quad (8.5.5a)$$

is achieved when $x(t) = x_0(t)$, where the maximum is taken over the set of functions $x(t)$ subject to the constraints

$$0 \leq x(t) \leq 1, \qquad \int_0^\infty x(t)\,dt = \alpha. \quad (8.5.5b)$$

Since y_0 and x_0 maximize (8.5.4a, 5a) subject to (8.5.4b, 5b), respectively, the Neyman-Pearson method gives the following characterization of these functions: Let

$$H(t) = \int_0^t Q_{y_0}(s)\,dQ_{x_0}(s) \quad (8.5.6)$$

and

$$K(t) = Q_{x_0}(0)Q_{y_0}(0) + \int_0^t Q_{x_0}(s)\,dQ_{y_0}(s); \quad (8.5.7)$$

then there exist positive constants h and k such that

$$\text{if}\quad H(t)\begin{cases} > \dfrac{h}{\eta(t)}, & \text{then} \quad y_0(t) = 1, & \text{(i)} \\ = \dfrac{h}{\eta(t)}, & \text{then} \quad 0 \leq y_0(t) \leq 1, & \text{(ii)} \\ < \dfrac{h}{\eta(t)}, & \text{then} \quad y_0(t) = 0; & \text{(iii)} \end{cases}$$

$$\text{if}\quad K(t)\begin{cases} > \dfrac{k}{\xi(t)}, & \text{then} \quad x_0(t) = 1, & \text{(iv)} \\ = \dfrac{k}{\xi(t)}, & \text{then} \quad 0 \leq x_0(t) \leq 1, & \text{(v)} \\ < \dfrac{k}{\xi(t)}, & \text{then} \quad x_0(t) = 0. & \text{(vi)} \end{cases}$$

We note specifically that since the functions $H(t)$ and $K(t)$ depend on the solutions $x_0(t)$ and $y_0(t)$ it is necessary to verify that the characterized functions ensuing from the Neyman-Pearson method are consistent with the x_0 and y_0 used in calculating $H(t)$ and $K(t)$.

More precisely, we must discover a pair of strategies x_0 and y_0 obeying (8.5.4b, 5b) such that when $H(t)$ and $K(t)$ are computed from x_0 and y_0, there exist constants h and k such that the x_0 and y_0 determined in accordance with relations (i-vi) agree with the original x_0 and y_0. Remarkably enough, judicious juggling of the Neyman-Pearson relations enable us to arrive at a solution. The technique involved has been demonstrated on a smaller scale in solving a variational problem (Section 8.2).

Some properties of $H(t)$ and $K(t)$.

(1) Since

$$H(t) = \int_0^t Q_{y_0}(s)\, dQ_{x_0}(s)$$
$$= Q_{y_0}(t)Q_{x_0}(t) - Q_{y_0}(0)Q_{x_0}(0) - \int_0^t Q_{x_0}(s)\, dQ_{y_0}(s),$$

it follows that

$$H(t) + K(t) = Q_{x_0y_0}(t).$$

(2) $H(t)$, $K(t)$, $Q_{x_0}(t)$, and $Q_{y_0}(t)$ are all differentiable at any point at which $x_0(t)$ and $y_0(t)$ are continuous. Furthermore, they are all absolutely continuous and nondecreasing in t. Finally, since both Q_{x_0} and Q_{y_0} are nonnegative and H and K are also nonnegative, it follows from (1) that $H(t), K(t) \leq 1$.

The next two lemmas and the remarks following them summarize several general properties of the optimal strategy which are deduced merely by manipulation of relations (i-vi). Such arguments are typical of the Neyman-Pearson technique when applied to nonlinear problems.

▶ LEMMA 8.5.1. There exists a nondegenerate maximal interval $[0, t_1]$ on which $y_0(t) = 0$.

Proof. It suffices to verify the existence of a nondegenerate interval with this property. Suppose that such an interval does not exist. Then for any $a > 0$ there exists a point in $[0, a]$ such that $H(t) \geq h/\eta(t)$. Since H and η are continuous, it now follows that $H(0) \geq h/\eta(0) > 0$, which contradicts the obvious fact that $H(0) = 0$.

▶ LEMMA 8.5.2. On $[0, t_1]$, $x_0(t) = 1$.

Proof. Lemma 8.5.1, in conjunction with (8.5.7), implies that $K(t) = Q_{x_0}(0)Q_{y_0}(0)$ on $[0, t_1]$.

If

$$Q_{x_0}(0)Q_{y_0}(0) \leq \frac{k}{\xi(0)},$$

then, since $\xi(t)$ is decreasing,

$$K(t_1) = Q_{x_0}(0)Q_{y_0}(0) < \frac{k}{\xi(t_1)}.$$

The continuity of the two functions now implies that inequality must hold also on an interval $[t_1, t_1 + \delta]$ with $\delta > 0$. Then on $[0, t_1 + \delta]$, $x_0(t) = 0$ by (vi), and therefore $H(t) = 0$ by (8.5.6). Since h must be greater than zero, $H(t) < h/\eta(t)$ and $y_0(t) = 0$ on $[0, t_1 + \delta]$, contradicting the choice of t_1. Thus $K(0) > k/\xi(0)$.

Since the functions are continuous, there exists a maximum interval $[0, t_2]$ on which $K(t) > k/\xi(t)$, and therefore $x(t) = 1$ on $[0, t_2]$ by (iv). The lemma will be established if we can show that $t_2 \geq t_1$. But if we assume $t_2 < t_1$, then $K(t) < k/\xi(t)$ on (t_2, t_1), since $K(t_2) = k/\xi(t_2)$ and $\xi(t)$ is decreasing. Hence $x_0(t) = 0$ on (t_2, t_1). Then

$$\frac{h}{\eta(t_1)} > \frac{h}{\eta(t_2)} \geq H(t_2) = H(t_1).$$

The last inequality requires by (iii) that $y_0(t) = 0$ in a neighborhood of t_1, contradicting the choice of t_1. This completes the proof of the lemma.

It is clear from the hypothesis that since $\eta(t)$ and $\xi(t)$ tend to zero and H and K are bounded, both $x_0(t)$ and $y_0(t)$ are identically zero for at least an interval of the form $[t_0, \infty]$.

Finally if $H(t) \equiv h/\eta(t)$ over an interval, then by differentiating we obtain

$$Q_{y_0}(t)Q_{x_0}(t)x_0(t)\xi(t) = -\frac{h\eta'(t)}{\eta^2(t)} \tag{8.5.8}$$

wherever η' exists, which is a.e. Thus it follows that $x_0(t) > 0$ in that interval. If

$$-\frac{\eta'(t)}{\eta(t)\xi(t)} < 1$$

in that interval, then $x_0(t)$ cannot be 1. For if $x_0(t) = 1$, then

$$K(t) \geq k/\xi(t),$$

and hence

$$K(t) + H(t) = Q_{x_0}(t)Q_{y_0}(t) \geq \frac{k}{\xi(t)} + \frac{h}{\eta(t)}.$$

But by (8.5.8)

$$Q_{y_0}Q_{x_0} = \frac{-\eta'(t)h}{\xi(t)\eta^2(t)} < \frac{h}{\eta(t)},$$

which is the desired contradiction. Thus when

$$-\frac{\eta'(t)}{\eta(t)\xi(t)} < 1,$$

if $H(t) \equiv h/\eta(t)$, then $K(t) \equiv k/\xi(t)$ and

$$Q_{x_0}Q_{y_0}y_0(t)\eta(t) = -\frac{k\xi'(t)}{\xi^2(t)}$$

where ξ' exists. This, together with (8.5.8) and the fact that $Q_{x_0}Q_{y_0} = h/\eta + k/\xi$, produces

$$x_0 = -\frac{\lambda\eta'(t)}{\eta(\lambda\xi + \eta)}, \qquad y_0 = -\frac{\xi'(t)}{(\lambda\xi + \eta)\xi},$$

where η' and ξ', respectively, exist and $\lambda = h/k$.

Solution. The analysis of the game (8.5.2) divides into numerous cases, each of which has a different type of solution. The form of the solution is heavily dependent on the nature of $\xi(s)$ and $\eta(s)$. We shall describe here one type of solution and determine sufficient conditions on ξ and η which assure its optimality.

We are interested in solutions of the following form:

$$\text{on } [0, t_1] \begin{cases} x_0(t) = 1, \\ y_0(t) = 0, \end{cases} \qquad \text{on } (t_1, t_2) \begin{cases} x_0 \neq 0, \\ y_0 \neq 0, \end{cases} \qquad \text{on } [t_2, \infty] \begin{cases} x_0 = 0, \\ y_0 = 0. \end{cases} \tag{8.5.9}$$

Such a proposed solution is entirely consistent with Lemmas 8.5.1 and 8.5.2. If a solution of the form of (8.5.9) exists, then necessarily

$$H(t_2) = \frac{h}{\eta(t_2)} \qquad \text{and} \qquad K(t_2) = \frac{k}{\xi(t_2)},$$

and hence, in view of property (1) and the nature of the strategies in $[t_2, \infty)$,

$$\frac{k}{\xi(t_2)} + \frac{h}{\eta(t_2)} = 1.$$

Another necessary condition is that $H(t_1) = h/\eta(t_1)$.

We shall need the formula

$$\begin{aligned} K(t_1) = Q_{x_0y_0}(0) &= Q_{x_0}(t_1)Q_{y_0}(t_1) \exp\left[-\int_0^{t_1} \xi(s)\, ds\right] \\ &= [K(t_1) + H(t_1)] \exp\left[-\int_0^{t_1} \xi(s)\, ds\right]. \end{aligned}$$

By requiring also that $x_0 < 1$ in (t_1, t_2), we obtain the additional necessary condition $K(t_1) = k/\xi(t_1)$, and the last equation may be written as

$$\frac{k}{\xi(t_1)} = \left(\frac{k}{\xi(t_1)} + \frac{h}{\eta(t_1)}\right) \exp\left[-\int_0^{t_1} \xi(s)\, ds\right].$$

Hence

$$\lambda = \frac{h}{k} = \frac{\eta(t_1)}{\xi(t_1)}\left\{\exp\left[\int_0^{t_1} \xi(s)\,ds\right] - 1\right\}. \tag{8.5.10}$$

We now impose one final restriction on $\xi(t)$ and $\eta(t)$. We postulate that

$$\int_0^\infty \frac{-\xi'(t)}{\xi(t)\eta(t)}\,dt > \beta, \tag{8.5.11}$$

and that

$$\frac{-\xi'(t)}{\eta(t)\xi(t)} < 1, \qquad \frac{-\eta'(t)}{\eta(t)\xi(t)} < 1 \tag{8.5.12}$$

on the infinite interval $[0, \infty]$; e.g.,

$$\eta(t) = \xi(t) = \frac{1}{(1+t)^\alpha} \qquad (0 < \alpha < 1).$$

[These assumptions are probably more restrictive than necessary in order to guarantee a solution of the form given in (8.5.9). Nevertheless, they furnish one explicit set of sufficient conditions which lead to a pair of optimal strategies; see also Problem 8.]

Subject to the hypothesis stated, we shall construct constants t_1 and t_2 so that the optimal strategies have the form

$$\begin{array}{lll} y_0 = 0, & x_0 = 1 & \text{for } t \text{ in } [0, t_1], \\ y_0 = \dfrac{-\xi'(t)}{\xi(t)(\lambda\xi + \eta)}, & x_0 = \dfrac{-\lambda\eta'(t)}{\eta(t)(\lambda\xi + \eta)} & \text{for } t \text{ in } (t_1, t_2), \\ y_0 = 0, & x_0 = 0 & \text{for } t \text{ in } [t_2, \infty], \end{array} \tag{A}$$

where λ is calculated by (8.5.10).

A solution of this form must satisfy the following constraints:

$$t_1 + \int_{t_1}^{t_2} \frac{-\eta'(t)\lambda}{\eta(\lambda\xi + \eta)}\,dt = \alpha, \tag{8.5.13}$$

$$\int_{t_1}^{t_2} \frac{-\xi'(t)}{\xi(\lambda\xi + \eta)}\,dt = \beta. \tag{8.5.14}$$

To verify that there are values t_1 and t_2 satisfying these equations, we proceed as follows: For each choice of $t_1 < \alpha$, we determine λ by (8.5.10); we then compute t_2 so that (8.5.13) is satisfied. In particular, if $t_1 = \alpha$, then $t_2 = \alpha$; and if $t_1 \to 0$, then $\lambda \to 0$ and $t_2 \to \infty$. Consequently, by continuity, for suitable t_1 in the interval $[0, \alpha]$ there exists a value of t_2

that will satisfy both (8.5.13) and (8.5.14). The quantities h and k are then determined by the conditions $H(t_1) = h/\eta(t_1)$ and $K(t_1) = k/\xi(t_1)$.

It remains to verify that x_0 and y_0 as prescribed above with these parameters lead back to the same x_0 and y_0 in relations (i-vi). Since $K(t) \equiv$ constant for $0 \leq t < t_1$, we find that $K(t) > k/\xi(t)$ over this interval, with equality at t_1.

It is an easy matter to show that $H(t) < h/\eta(t)$ on this same interval, with equality at t_1. The proof of this runs as follows. Certainly $H(0) = 0 < h/\eta(0)$. Suppose that there exists a $t^* < t_1$ such that $H(t^*) \geq h/\eta(t^*)$. Then

$$H(t^*) + K(t^*) \geq \frac{h}{\eta(t^*)} + \frac{k}{\xi(t_1)} \geq \frac{h}{\eta(t^*)} + \frac{k}{\xi(t^*)}. \tag{8.5.15}$$

But

$$\begin{aligned} H'(t^*) &= (H + K)'(t^*) \geq \left(\frac{h}{\eta(t^*)} + \frac{k}{\xi(t^*)}\right)\xi(t^*) > \frac{-h\eta'(t^*)}{\eta^2(t^*)} \\ &= \frac{d}{dt}\left(\frac{h}{\eta(t)}\right)\Big|_{t^*}, \end{aligned}$$

the sequence of inequalities resulting successively from the use of property (1), p. 229, (8.5.15), and (8.5.12). We infer as a consequence of the two outside inequalities that if $H(t^*) \geq h/\eta(t^*)$, strict inequality takes over for $t_1 \geq t > t^*$, in contradiction to $H(t_1) = h/\eta(t_1)$. Hence $H(t) < h/\eta(t)$ for t in $[0, t_1]$, as asserted.

On the interval $[t_1, t_2]$ we have identically $H(t) = h/\eta(t)$ and $K(t) = k/\xi(t)$. Finally, since both $H(t)$ and $K(t)$ are constant beyond t_2, we find that for $t > t_2$

$$H(t) < \frac{h}{\eta(t)}, \qquad K(t) < \frac{k}{\xi(t)}.$$

We have demonstrated that (A) is consistent with relations (i-vi) as regards x_0 and y_0; hence x_0 and y_0 are optimal. Moreover, it is not difficult to prove that these optimal strategies are unique.

The above analysis is always valid so long as x_0 and y_0 lie between 0 and 1 on the interval (t_1, t_2) and the constraints (8.5.13–14) are met. When these requirements are not fulfilled, then the same lines of reasoning can be applied with proper care (see Problem 7).

8.6 Problems.

1. Prove that

$$\mu = 2\int_0^1 (1 - \xi)\phi(\xi)\, d\xi - \left(\int \phi(\xi)\, d\xi\right)^2 \geq 0,$$

where ϕ is monotone-increasing, $\phi(0) = 0$, and $\phi(1) = 1$. (*Hint:* Use the Neyman-Pearson lemma to minimize

$$2\int_0^1 (1 - \xi)\phi(\xi)\, d\xi,$$

where ϕ is subject to the additional constraint that $\int\phi(\xi)\, d\xi = \alpha$ has a prescribed value, and then allow α to vary.)

2. Use the Neyman-Pearson method to solve the variational problem

$$\min_L \int_0^1 [L(t) - t]^2\, dt,$$

where $0 \leq L(t) \leq 1$ and

$$\int_0^1 L(t)\, dt = p \qquad (0 < p < 1).$$

3. Use the technique of Section 8.2 to solve

$$\max_\phi \int_{-\infty}^{\infty} p(t)[1 - e^{-\phi(t)}]\, dt \qquad \left(p(t) > 0, \quad \int p(t)\, dt < \infty\right),$$

where ϕ is subject to the conditions

$$\phi(t) \geq 0, \qquad \int_{-\infty}^{\infty} \phi(t)\, dt = \alpha \qquad (\alpha > 0).$$

4. Solve Problem 3 where the constraint set is

$$\phi(t) \geq 0, \qquad \int_{-\infty}^{\infty} h(t)\phi(t)\, dt = \alpha \qquad (\alpha > 0),$$

where $h(t) > 0$ and integrable on bounded sets and the functional to be maximized is

$$\int_{-\infty}^{\infty} p(t)[1 - e^{-g(t)\phi(t)}]\, dt,$$

with $g(t)$ bounded. Assume that the set $\{t|g(t) > 0\}$ has positive Lebesgue measure.

5. *Search problem.* An object is to arrive in box i ($i = 1, 2, \ldots, k$) with probability p_i at time t where t is uniformly distributed over $[0, N]$. We are allowed to look for the object N times, opening any box each time, i.e., a search plan consists of specifying N times and the box to be looked at each time. The gain is $g(t)$ if time t elapses from arrival to detection, $g'(t) < 0$.

(a) Show that in the optimum search plan (the plan that maximizes expected gain) the openings of a given box are equally spaced.

(b) Let m_i be the time interval between openings of the ith box and n_i be the number of times the ith box is opened; that is, $m_i = N/n_i$.

Prove that the expected gain from such a search policy is

$$F = \sum_{i=1}^{k} p_i \frac{G(m_i)}{m_i}, \tag{*}$$

where $G(t) = \int_0^t g(u)\,du$ and the m_i satisfy

$$\sum_{i=1}^{k} \frac{1}{m_i} = 1. \tag{**}$$

(This problem may be interpreted and related to the problem of determining an optimal sky scanning strategy which detects an oncoming missile in the earliest possible time.)

6. Use the method of Section 8.2 to determine the m_i which maximize (*) where m_i are subject to the inequality constraint $1/N \leq 1/m_i \leq 1$ and satisfy (**). This, of course, will admit solutions where the n_i are not integers, but it gives an upper bound for the gain possible in a "realizable" search plan.

7. Find conditions on α and β which guarantee a pair of optimal strategies of the two-machine-gun duel of the following form:

$$x_0(t) = \begin{cases} 1 & 0 < t < \alpha, \\ 0 & t > \alpha, \end{cases} \qquad y_0(t) = \begin{cases} 0 & 0 < t < t_0, \\ 1 & t_0 < t < t_0 + \beta, \\ 0 & t_0 + \beta < t, \end{cases}$$

where $t_0 < \alpha$, $t_0 + \beta > \alpha$ and $\xi(t) = \eta(t) = e^{-t}$.

8. Show that the solution of the two-machine-gun duel as described in (A) of Section 8.5 is always valid if there exists a positive value $\bar{t}$ satisfying the inequalities

$$\bar{t} > \alpha, \qquad \xi(\bar{t}) < \xi(0) \exp\left[-\alpha\xi(0) - \alpha\eta(0) - \beta\eta(0)\right]$$

and on $[0, \bar{t}]$

$$\frac{-\xi'(t)}{\xi(t)\eta(t)} < 1 \qquad \text{and} \qquad \frac{-\eta'(t)}{\xi(t)\eta(t)} < 1.$$

In Problems 9–11 we consider the game whose kernel is

$$K(\phi, \psi) = \int_0^1 f(t)\phi(t)\psi(t)\,dt$$

(assume $f(t) > 0$ continuous and strictly increasing), where the strategy spaces are defined by the inequalities

$$0 \leq \phi(t) \leq 1, \qquad 0 \leq \psi(t) \leq 1$$

and

$$\int_0^1 \phi(t)\,dt = B, \qquad \int_0^1 \psi(t)\,dt = A \quad (0 \le A, B \le 1).$$

Let Q and P $(Q \ge P)$ be real numbers defined by the relations

$$Q + f(P)\int_Q^1 \frac{dt}{f(t)} = A, \qquad Q - P + f(Q)\int_Q^1 \frac{dt}{f(t)} = B. \tag{+}$$

9. Show that if A and B are such that (+) has a solution $1 > Q > P > 0$, then the optimal strategies for the game $K(\phi, \psi)$ are

$$\psi^*(t) = \begin{cases} 1 & 0 < t < Q, \\ \dfrac{l}{f(t)} & Q < t < 1, \end{cases} \qquad \phi^*(t) = \begin{cases} 0 & 0 \le t < P, \\ 1 & P < t < Q, \\ \dfrac{k}{f(t)} & Q < t \le 1, \end{cases}$$

where $l = f(P)$ and $k = f(Q)$. (*Hint:* Use methods similar to those of Section 8.5; it is easier in this case because of the linearity of the kernel.)

10. If A and B are such that

$$P + f(P)\int_P^1 \frac{dt}{f(t)} = A, \qquad f(Q)\int_P^1 \frac{dt}{f(t)} = B \tag{++}$$

have a solution as in Problem 9 where $0 \le Q \le P \le 1$, show that $K(\phi, \psi)$ has a pair of optimal strategies of the form

$$\psi^*(t) = \begin{cases} 1 & 0 \le t \le P, \\ \dfrac{f(P)}{f(t)} & P \le t \le 1, \end{cases} \qquad \phi^*(t) = \begin{cases} 0 & 0 \le t \le P, \\ \dfrac{f(Q)}{f(t)} & P < t \le 1. \end{cases}$$

11. Show that the cases of equations (+) and (++) above are mutually exclusive.

12. Use Jensen's inequality on convex functions to verify directly that the optimal strategy proposed for Player I in the game of Section 8.4 is indeed optimal.

Notes and References to Chapter 8

Substantial and important classes of games on the unit square—notably games of timing and convex games—have been solved. Nevertheless, there are many lacunae in the general theory, and in the methods associated with solving arbitrary games on the unit square. Even less is known about classical games not played over the unit square.

Among such games are machine-gun duels (Section 8.5), reconnaissance games, and general games played over function spaces. At the present time, a coherent theory of these games does not exist. It seems proper, therefore, to try to gain insight into the nature of the solutions of these games by treating individual examples at some length. The content of this chapter may be regarded as a first step in this direction.

§ 8.1. There are some who claim that credit for the Neyman-Pearson lemma actually belongs to Gibbs, but by all accounts it was Neyman and Pearson [190] who truly accentuated the significance of this now classical result.

The geometric proof given in this section resembles an argument of Dantzig and Wald [60].

§ 8.2. Several authors have used the Neyman-Pearson lemma in solving nonlinear variational problems subject to linear constraints, notably Bellman, Glicksberg, and Gross [22], Danskin [52], Koopmans [149], F. Proschan (unpublished), and Arrow and Karlin [16, Chaps. 4–7].

§ 8.3. The first to examine the fighter-bomber duel as formulated here was Weiss, in the course of munition studies conducted at the Aberdeen Proving Ground. The interpretation of this problem as an advertising campaign was conceived by Gillman [106]. Weiss's results were corroborated independently by Bellman and Blackwell (unpublished), who further constructed an optimal strategy for the fighter (L in the present interpretation). Later, Blackwell and Shiffman [32, 33] produced an optimal strategy for B (the Bomber).

§ 8.4. Our way of arriving at an optimal strategy for L follows in the main that of Bellman and Blackwell. The methods for obtaining the optimal strategies of Player II seem to be new, relying heavily on the Neyman-Pearson characterization.

§ 8.5. The two-machine-gun duel was formulated and partially solved by Danskin and Gillman [54]. We treat here a case not studied by the original authors.

§ 8.6. Problems 3 and 4 are based on Koopman [145]. Problem 5 is due to Proschan. Problems 9 and 10 are based on Bellman and Shiffman [23].

CHAPTER 9

POKER AND GENERAL PARLOR GAMES

It is the considered opinion of many expert card players that poker requires the most skill and depth of all card games. Oswald Jacoby, a champion bridge player, makes a distinction between poker, which in his opinion concerns the management of money, and other card games, which emphasize the management of cards. In most card games, if the opponent holds a good hand, nothing can be done about it; but this is not necessarily the case in poker. The elements of bluff and threat are the very spirit of the game.

A knowledge of mathematical probability is necessary but not sufficient to make a good poker player. Correct poker is a proper combination of the tools of probability analysis, bluff, sandbagging, risk, psychology, etc. The timidity or daring of one's personality, the inconsistencies or deadly consistency of one's character, are manifest in the play of the game.

The many respects in which business, politics, and war resemble poker should be evident. Hence, any progress in our mathematical understanding of poker games has its counterpart interpretation in many relevant circumstances of life.

The impulse to gamble remains an eternal aspect of the apparent irrationality of man. Card games are one outlet for this impulse. When a game like poker is played over a long period of time, the experience acquired by the players tends to crystallize into a set of heuristic rules of play—a collection of practices and gambits that have been tested under fire and found effective. With the aid of these heuristic rules, with which all devotees of poker are more or less familiar, we shall be able to unravel in formal mathematical terms the nature of the optimal strategies in certain simple versions of poker games.

We are restricted in this chapter to two-person poker. This is less of a limitation than it may seem at first glance, since in any game of multi-person poker any player can be considered as concerned with one opponent, namely the totality of all the other players.

The components of a simple two-person poker game may be described as follows. We identify all possible hands which each player may receive with the unit interval. A hand ξ is superior to a hand η if and only if $\xi > \eta$. Each of the players is dealt a hand at random, in the following sense: Player I receives a random value ξ (his hand) according to the distribution $F(\xi)$, while Player II simultaneously receives an independent random value η (his hand) according to the distribution $G(\eta)$, which may or may not be the same as $F(\xi)$ $(0 \leq \xi, \eta \leq 1)$. At the beginning, each player knows only what is in his own hand. Following the rules, the players perform certain moves of the game (i.e., "folding," "seeing," "betting," "raising") in some order, and possibly several times if the bets are appropriately met. The hands are then compared, and the player with the superior hand wins the pot (i.e., the accumulated bets).

There are many variations of this game. Unfortunately, the form of the optimal strategies appears to be sensitive to the special features of each model. For example, we obtain different solutions depending on whether the moves in the game are simultaneous or alternate, on how many rounds of bets are allowed, on whether raises are permissible, etc. Each case necessitates a separate analysis. Nevertheless, a pronounced pattern shows up in the solutions of poker games. Moreover, implicit in the techniques used here to solve poker games is a general method which can be applied to a great variety of games.

The method is applicable if the pay-off is the expected value of a function P of two random variables ξ and η, provided that the strategies for Players I and II can be identified with the vectors

$$\boldsymbol{\phi}(\xi) = \langle \phi_1(\xi), \ldots, \phi_m(\xi) \rangle \qquad \text{and} \qquad \boldsymbol{\psi}(\eta) = \langle \psi_1(\eta), \ldots, \psi_n(\eta) \rangle,$$

respectively. (Of course two strategies which agree a.e. are hereafter considered identical.) These vectors may be subject to arbitrary constraints. The cumulative distributions F and G of ξ and η can be arbitrary, but implicit in the definition of the strategies is the assumption that Player I knows the value of ξ and Player II knows the value of η. In these games the pay-off will be of the form

$$K(\boldsymbol{\phi}, \boldsymbol{\psi}) = \iint P[\xi, \eta; \boldsymbol{\phi}(\xi), \boldsymbol{\psi}(\eta)]\, dF(\xi)\, dG(\eta).$$

In most of the subsequent examples it will be possible to write this pay-off in two different forms:

$$K(\boldsymbol{\phi}, \boldsymbol{\psi}) = M_1(\boldsymbol{\psi}) + \sum_{i=1}^{m} \int \phi_i(\xi) C_{\phi_i}(\boldsymbol{\psi}; \xi)\, dF(\xi)$$

and

$$K(\boldsymbol{\phi}, \boldsymbol{\psi}) = M_2(\boldsymbol{\phi}) + \sum_{j=1}^{n} \int \psi_j(\eta) C_{\psi_j}(\boldsymbol{\phi}; \eta)\, dG(\eta),$$

where M_1 and M_2 are independent of $\boldsymbol{\phi}$ and $\boldsymbol{\psi}$, respectively, and where C_{ϕ_i} and C_{ψ_j} denote the coefficients of ϕ_i and ψ_j in the expressions for $K(\boldsymbol{\phi}, \boldsymbol{\psi})$. The notation emphasizes the fact that C_{ϕ_i} and C_{ψ_j} are functions of $\boldsymbol{\psi}$ and $\boldsymbol{\phi}$, respectively.

In order to solve the game it is necessary and sufficient to find two strategies ϕ^* and ψ^* such that

$$K(\phi^*, \psi^*) = \max_{\phi} \left\{ M_1(\psi^*) + \sum_{i=1}^{m} \int \phi_i(\xi) C_{\phi_i}(\psi^*; \xi)\, dF(\xi) \right\}$$

and

$$K(\phi^*, \psi^*) = \min_{\psi} \left\{ M_2(\phi^*) + \sum_{j=1}^{n} \int \psi_j(\eta) C_{\psi_j}(\phi^*; \eta)\, dG(\eta) \right\}.$$

Each equation represents a simple maximum or minimum problem, subject to constraints. Any *a priori* information about some property of an optimal strategy, say ψ^*, can be used to obtain information about $C_{\phi_i}(\psi^*; \xi)$, and this in turn gives information about the optimal ϕ_i^*. The process can be iterated until the optimal strategies are obtained. This iteration process is the basic technique in our method. The details are illustrated in the course of this chapter by several examples of different degrees of complexity.

The principles of this method were discussed previously in Section 2.5, where it was called the fixed-point method. Its applicability is of substantial scope, as we shall see. This same kind of juggling was used in our analysis of the two-machine-gun duel in Chapter 8.

The games of kernel $K(\phi, \psi)$ given above belong to category I. Although they are not equivalent to standard games played over the unit square, the strategy spaces for each player, as indicated, are tractable and can be formulated in terms of recognized function spaces.

For all these games it is not difficult to prove the existence of optimal strategies and the validity of the min-max theorem. However, for all games considered in this chapter, we explicitly exhibit the optimal strategies.

In the following sections we discuss a series of poker-type examples which are amenable to analysis by suitable variations of the fixed-point method described earlier. Aside from the intrinsic interest of card games, the importance of this chapter lies in the exemplification of this method, and in pointing up its evident applicability to general game problems.

We begin in Section 9.1 with a simple card game which does not involve two opposing players in the usual sense. The game is an idealized version of blackjack in which the dealer's course of action is fixed by the probability of the card sequences, so that he cannot employ independent judgment or intelligence.

In Section 9.2 we tackle a simple poker model which entails only one round of betting and one size of bet. In the following section this one-round model is generalized to permit bets of n different amounts. These games do not embrace the possibility of raises.

In Section 9.4 we consider a poker model which involves two rounds of bets and raises, with only one size of bet permissible in each round. The solution of a poker model involving k rounds of betting which allows each player several opportunities to fold, see, or raise is presented in Section 9.5.

In all the preceding models the players move alternately. In Section 9.6 we treat a poker model in which moves are simultaneous.

To emphasize the applicability of these methods to many kinds of parlor games, we analyze the classical Le Her game in Section 9.7 and a game described as "High Hand Wins" in Section 9.8. (For other examples see Problems 11–12.)

9.1 A simplified blackjack game. The model we discuss in this section is not a game in the usual sense, since one of the players' strategies is fixed. Nevertheless, this example is a serviceable introduction to the fixed-point technique.

The model. Each of two players, a dealer (D) and a bettor (B), receives a card (a value from the unit interval), and before seeing it each must bet one unit of money. After seeing his card B has three choices: he may bet one more unit, he may bet M more units ($M > 1$ is fixed), or he may fold and forfeit his initial bet. D, however, has no choice: he must always cover B's bet. If B instead of folding bets either 1 or M additional units, then the two players compare hands and the one with the losing hand pays the other an amount equal to B's bet (which includes the initial bet of one unit).

The card that B draws (which we shall denote by ξ) is chosen at random according to a probability distribution $F(\xi)$. The card η that D draws is chosen according to a probability distribution $G(\eta)$. Both $F(\xi)$ and $G(\eta)$ are defined on the unit interval and are assumed, for convenience only, to be continuous and strictly increasing.

Strategies for B can be described in terms of the following functions: $\phi_1(\xi)$ is the probability that he will bet one additional unit; $\phi_2(\xi)$ is the probability that he will bet M additional units; and $1 - \phi_1(\xi) - \phi_2(\xi)$, is therefore the probability that he will fold where ξ represents the hand drawn by B. Denoting by $L(\xi, \eta)$ the sign of $\xi - \eta$, that is,

$$L(\xi, \eta) = \begin{cases} 1 & \xi > \eta, \\ 0 & \xi = \eta, \\ -1 & \xi < \eta, \end{cases}$$

we deduce that the expected return to B is

$$\begin{aligned} K(\phi_1, \phi_2) = (-1)\int [1 - \phi_1(\xi) - \phi_2(\xi)]\, dF(\xi) \\ + 2\iint \phi_1(\xi) L(\xi, \eta)\, dF(\xi)\, dG(\eta) \\ + (M + 1)\iint \phi_2(\xi) L(\xi, \eta)\, dF(\xi)\, dG(\eta). \end{aligned}$$

Since

$$\int L(\xi, \eta)\, dG(\eta) = 2G(\xi) - 1,$$

we may write

$$K(\phi_1, \phi_2) = -1 + \int \phi_1(\xi)[4G(\xi) - 1]\, dF(\xi)$$
$$+ \int \phi_2(\xi)\{1 + (M + 1)[2G(\xi) - 1]\}\, dF(\xi).$$

Since D has no choice of strategies, the optimal procedure for B is to select the functions $\phi_1(\xi)$ and $\phi_2(\xi)$ that maximize $K(\phi_1, \phi_2)$ subject to the restrictions $\phi_i(\xi) \geq 0$ and $0 \leq \phi_1(\xi) + \phi_2(\xi) \leq 1$. Clearly, we should set $\phi_i(\xi) = 1$ for those values of ξ for which the coefficient of ϕ_i is largest and positive. The two coefficients are equal when

$$4G(\xi) - 1 = 1 + (M + 1)[2G(\xi) - 1],$$

or

$$2[2G(\xi) - 1] = (M + 1)[2G(\xi) - 1].$$

Since $M + 1 > 2$, equality holds if and only if $2G(\xi) - 1 = 0$. Let ξ_0 be the root of this equation; if $\xi > \xi_0$, then

$$1 + (M + 1)[2G(\xi) - 1] \geq 4G(\xi) - 1 > 0.$$

Hence, if $\xi > \xi_0$, the optimal strategy requires $\phi_2(\xi) = 1$.

Next define a_0 as the unique solution of $4G(\xi) = 1$. Clearly, $a_0 < \xi_0$. Moreover, for $a_0 < \xi < \xi_0$ the coefficient of ϕ_1 is larger than that of ϕ_2 and is furthermore positive. Therefore the optimal strategy calls for $\phi_1(\xi) = 1$ when $a_0 < \xi < \xi_0$. Finally, for $\xi < a_0$ the bettor should fold.

For the particular case of $F(\xi) = G(\xi) = \xi$ (the uniform distribution) the optimal strategy for B reduces to the following: if $\xi < \frac{1}{4}$, he folds; if $\frac{1}{4} \leq \xi < \frac{1}{2}$, he bets 1 unit; if $\frac{1}{2} \leq \xi < 1$, he bets M units.

An interesting general characteristic of the optimal strategy is that it is independent of M so long as M is greater than 1.

A generalization. A more general version of the preceding model is as follows. The game begins when each player makes an initial bet of one unit; then they receive cards ξ_1 and η_1, respectively. B can either fold and forfeit his initial bet, or bet an amount $f(\xi_1)$, restricted by $1 \leq f(\xi_1) \leq M$; if B bets, D has no choice but to cover the bet. After this they receive new cards ξ_2 and η_2, respectively. B now has a choice of betting an amount $f(\xi_2)$ $[1 \leq f(\xi_2) \leq M]$ or folding and losing $1 + f(\xi_1)$. If B bets, D must again cover. The game continues in this way for N steps and then, if B has not yet folded, the hands are compared. B wins the total amount

$$1 + \sum_{i=1}^{N} f(\xi_i) \qquad \text{if} \quad \sum_{i=1}^{N} \xi_i > \sum_{i=1}^{N} \eta_i.$$

If the opposite inequality holds, B loses the same amount. The problem is to maximize the expected return to B.

The results may be summarized qualitatively as follows: B folds on hands which are too bad; B bets the minimum 1 on fair hands; B bets the maximum M on good hands; the classification of bad, fair, and good hands is independent of the upper limit M.

The proof and analysis of the conclusions of this model will not be included here; we shall, however, analyze a similar game resembling the classical game of Red Dog.

Red Dog is played as follows. There are N players ($N > 1$). The dealer deals himself and the other players four cards each. Before the play begins each player makes an initial bet of one unit so that there are N units in the pot. Upon looking at his cards each player in turn has the option of folding or betting an amount f ($1 \leq f \leq N$) that he can beat, in its suit, the next card that the dealer turns up from the deck. If he can, he wins an amount f from the pot; if not, he contributes f to the pot. The next player has a similar choice, with the difference that the upper limit is the size of the pot, so that if the previous player has bet and lost f, the new upper limit is $N + f$. Whenever the pot falls beneath N in size each player contributes one more unit; hence if the first player has bet and won f, the new upper limit is $2N - f$.

Each player sees at most one card from any other player's hand, for the rules are that a losing player throws his hand away face down, and a winning player need only show the majorizing card. We shall in the discussion below ignore the additional information which may be gained this way, and the fact that the upper limit of the wagers may become considerably higher than N.

We consider the following simple abstract version of the game. There are two players, B and D. B is given a "hand," i.e., a point $\mathbf{c} = \langle \xi_1, \xi_2, \ldots, \xi_n \rangle$ in the positive unit cube of E^n, where the distribution functions of $\xi_1, \ldots, \xi_n$ are known. D is given a "random card" z ($0 \leq z \leq 1$), whose distribution function is G, which has a probability p_i of being compared to ξ_i in order to determine the outcome of the game. B and D both make an initial bet of one unit, and then B is allowed to bet an amount $f(\mathbf{c})$ [$1 \leq f(\mathbf{c}) \leq M$], winning if $z \leq \xi_i$, losing if $z > \xi_i$. The problem is to determine the optimal procedure for B.

A strategy for B is to fold if $\mathbf{c}$ is in a certain region R and to bet if $\mathbf{c}$ is outside R. If the distribution of ξ_i is F_i, then the expected return is

$$E(B) = -\int_{c \in R} \prod_{i=1}^{n} dF_i + \sum_{i=1}^{n} p_i \int_{\substack{c \in I-R \\ \xi_i \geq z}} [1 + f(\mathbf{c})] \prod_{i=1}^{n} dF_i(\xi_i)\, dG(z)$$
$$- \sum_{i=1}^{n} p_i \int_{\substack{c \in I-R \\ \xi_i < z}} [1 + f(\mathbf{c})] \prod_{i=1}^{n} dF_i(\xi_i)\, dG(z) = -\int_{c \in R} \prod_{i=1}^{n} dF_i$$
$$+ \int_{c \in I-R} [1 + f(\mathbf{c})] \left\{ \sum_{i=1}^{n} p_i[2G(\xi_i) - 1] \right\} \prod_{i=1}^{n} dF_i, \tag{9.1.1}$$

where I denotes the unit cube and $\mathbf{c} \in R$ means that the point $\mathbf{c}$ is contained in R. But

$$-\int_{c \in R} \prod_{i=1}^{n} dF_i = -1 + \int_{c \in I-R} \prod_{i=1}^{n} dF_i.$$

Inserting this into (9.1.1), we obtain

$$E(B) = -1 + \int_{c \in I-R} \left[[1 + f(\mathbf{c})] \sum_{i=1}^{n} p_i[2G(\xi_i) - 1] + 1 \right] \prod_{i=1}^{n} dF_i.$$

Since the smallest bet that can be made is $f(\mathbf{c}) = 1$, we may conclude that the region of betting should consist of the set of points for which

$$2 \left\{ \sum_{i=1}^{n} p_i[2G(\xi_i) - 1] \right\} + 1 \geq 0. \tag{9.1.2}$$

Having determined the region of betting $I - R$, we must now determine the amount of betting over this region. With R fixed it is clear from (9.1.1) that in order to maximize the expected return, B should bet $f(\mathbf{c}) = M$ in the region in which the coefficient of $f(\mathbf{c})$ is positive. This is the subregion S of $I - R$ in which

$$\sum_{i=1}^{n} p_i[2G(\xi_i) - 1] > 0. \tag{9.1.3}$$

In the complementary portion of $I - R$ the bet should be $f(\mathbf{c}) = 1$.

Application to the game of Red Dog. We consider the special case in which the ξ_i's are uniformly distributed, z is uniformly distributed, and $p_i = \frac{1}{4}$ corresponding to the four suits. The folding region R by (9.1.2) is the region in which

$$0 \geq 1 + 2 \sum_{i=1}^{4} p_i[2G(\xi_i) - 1] = \sum_{i=1}^{4} \xi_i - 1.$$

The part of the complementary region in which the maximum should be bet is the subregion in which

$$0 \leq \sum_{i=1}^{4} p_i[2G(\xi_i) - 1] = \frac{1}{2}\sum_{i=1}^{4} \xi_i - 1.$$

In the actual game of Red Dog $dF_i = 0$ except at the points

$$k/13 \quad (k = 1, 2, \ldots, 13).$$

We have neglected in our analysis the possibility that a hand may be void of some suit.

9.2 A poker model with one round of betting and one size of bet. The model examined in this section is a special case of the model treated in Section 9.3, which permits n possible sizes of bet.

The model. Two players, A and B, ante one unit each at the beginning of the game. After each draws a card, A acts first: he may either bet a more units or fold and forfeit his initial bet. If A bets, B has two choices: he may either fold (losing his initial bet) or bet a units and "see" A's hand. If B sees, the two players compare hands, and the one with the better hand wins the pot.

We shall denote A's hand by ξ, whose distribution is assumed to be the uniform distribution on the unit interval, and B's hand by η, also distributed uniformly on the unit interval. We shall write $L(\xi, \eta) = \text{sign}(\xi - \eta)$ as before.

Strategies and pay-off. The strategies are composed as follows: Let

$$\begin{aligned}
\phi(\xi) &= \text{probability that if } A \text{ draws } \xi \text{ he will bet } a,\\
1 - \phi(\xi) &= \text{probability that if } A \text{ draws } \xi \text{ he will fold},\\
\psi(\eta) &= \text{probability that if } B \text{ draws } \eta \text{ he will see},\\
1 - \psi(\eta) &= \text{probability that if } B \text{ draws } \eta \text{ he will fold}.
\end{aligned}$$

If the two players follow these strategies, the expected net return $K(\phi, \psi)$ is the sum of the returns corresponding to three mutually exclusive possibilities: A folds; A bets a units and B sees; A bets and B folds. Thus

$$K(\phi, \psi) = (-1)\int [1 - \phi(\xi)]\, d\xi + (a + 1)\iint \phi(\xi)\psi(\eta)L(\xi, \eta)\, d\xi\, d\eta + \iint \phi(\xi)[1 - \psi(\eta)]\, d\xi\, d\eta.$$

The yield to A may also be more transparently written as

$$K(\phi, \psi) = -1 + \int_0^1 \phi(\xi)\left(2 + a\int_0^{\xi} \psi(\eta)\, d\eta - (a + 2)\int_{\xi}^1 \psi(\eta)\, d\eta\right) d\xi \tag{9.2.1}$$

or

$$K(\phi, \psi) = -1 + 2\int_0^1 \phi(\xi)\,d\xi + \int_0^1 \psi(\eta)\left(-(a+2)\int_0^\eta \phi(\xi)\,d\xi + a\int_\eta^1 \phi(\xi)\,d\xi\right)d\eta. \tag{9.2.2}$$

Method of analysis. We begin by observing that the existence of a pair of optimal strategies is equivalent to the existence of two functions ϕ^* and ψ^* satisfying the inequalities

$$K(\phi, \psi^*) \leq K(\phi^*, \psi^*) \leq K(\phi^*, \psi) \tag{9.2.3}$$

for all strategies ϕ and ψ, respectively. Thus ϕ^* maximizes $K(\phi, \psi^*)$ while ψ^* minimizes $K(\phi^*, \psi)$. We shall therefore search for the strategy ϕ^* that maximizes (9.2.1) with ψ replaced by ψ^*; and we shall also search for the strategy ψ^* that minimizes (9.2.2) with ϕ replaced by ϕ^*. Since the constant terms are not important, the problem is to find

$$\max_\phi \int_0^1 \phi(\xi)\left(2 + a\int_0^\xi \psi^*(\eta)\,d\eta - (a+2)\int_\xi^1 \psi^*(\eta)\,d\eta\right)d\xi \tag{9.2.4}$$

and

$$\min_\psi \int_0^1 \psi(\eta)\left(-(a+2)\int_0^\eta \phi^*(\xi)\,d\xi + a\int_\eta^1 \phi^*(\xi)\,d\xi\right)d\eta. \tag{9.2.5}$$

The crux of the argument is to verify that our results are consistent, i.e., that the function ϕ^* that maximizes (9.2.4) is the same function ϕ^* that appears in (9.2.5), and similarly for ψ^*; if these assertions are valid, then (9.2.3) is satisfied and we have found a solution.

At this point intuitive considerations suggest what type of solution we search for. Since B has no chance to bluff, $\psi^*(\eta) = 1$ for η greater than some critical number c, and $\psi^*(\eta) = 0$ otherwise; also, since B is minimizing, $\psi^*(\eta)$ should be equal to 1 when the coefficient of $\psi(\eta)$ in (9.2.5) is negative. But this coefficient expresses a decreasing function of η, and thus c is the value at which it first becomes zero. Hence

$$-(a+2)\int_0^c \phi^*(\xi)\,d\xi + a\int_c^1 \phi^*(\xi)\,d\xi = 0. \tag{9.2.6}$$

With this choice for $\psi^*(\eta)$, we find that the coefficient of $\phi(\xi)$ in (9.2.4) is constant for $\xi \leq c$. If we assume that this constant is 0, we obtain at $\xi = c$

$$2 + 0 - (a+2)(1-c) = 0,$$

or

$$c = \frac{a}{a+2}. \tag{9.2.7}$$

The reason we determine the constant c so as to make the coefficient zero in the interval $[0, c]$ is as follows. In maximizing (9.2.4) we are obviously compelled to make $\phi^*(\xi) = 1$ whenever its coefficient is positive, and $\phi^*(\xi) = 0$ whenever its coefficient is negative. The only arbitrariness allowed in the values of $\phi^*(\xi)$ occurs when its coefficient is zero. But we expect A to attempt some partial bluffing on low hands, which means that probably $0 < \phi^*(\xi) < 1$ for these hands. As pointed out, this is feasible if the coefficient of $\phi(\xi)$ is zero. With the determination of c according to (9.2.7), the coefficient of $\phi(\xi)$ in (9.2.4) is zero for $\xi \leq c$ and positive for $\xi > c$. Under these circumstances it follows from (9.2.4) that the maximizing player is obligated to have $\phi^*(\xi) = 1$ for $\xi > c$ while the values of $\phi^*(\xi)$ for $\xi \leq c$ are irrelevant, in the sense that they do not contribute to the pay-off. However, in order to satisfy (9.2.6) with this choice of ϕ^* we must have

$$-(a + 2)\int_0^c \phi^*(\xi)\, d\xi + a(1 - c) = 0,$$

or

$$\int_0^c \phi^*(\xi)\, d\xi = \frac{a(1 - c)}{a + 2} = \frac{2a}{(a + 2)^2},$$

and this can be accomplished with $\phi^*(\xi) < 1$. It is easy to verify now that if

$$\psi^*(\eta) = \begin{cases} 0 & 0 \leq \eta \leq \dfrac{a}{a + 2}, \\ 1 & \dfrac{a}{a + 2} < \eta \leq 1 \end{cases}$$

and

$$\phi^*(\xi) = \begin{cases} \text{arbitrary between 0 and 1} & \\ \text{but satisfying} & \\ \displaystyle\int_0^c \phi^*(\xi)\, d\xi = \frac{2a}{(a + 2)^2} & 0 \leq \xi \leq \dfrac{a}{a + 2}, \\ 1 & \dfrac{a}{a + 2} < \xi < 1, \end{cases}$$

then ϕ^* maximizes (9.2.4) for ψ^* and ψ^* minimizes (9.2.5) for ϕ^*.

The interpretation of the solution is of some interest:

(1) Both players bet or see on high hands. What is of special significance is that both players use the identical critical number $a/(a + 2)$ to distinguish high and low hands.

(2) The element of bluffing shows up for Player I only to the extent that the proportion of hands on which he should bluff is determined; he may

choose the actual hands in an arbitrary manner from $[0, a/(a+2)]$ subject only to the restriction

$$\int_0^{a/a+2} \phi^*(\xi)\, d\xi = \frac{2a}{(a+2)^2}.$$

9.3 A poker model with several sizes of bet. The model examined here is an extension of the one just analyzed. The same example was previously described in Section 2.4.

As before, the unit interval is taken as the representation of all possible hands that can be dealt to a player. Each hand is considered equally likely, and therefore the operation of dealing a hand to a player may be considered as equivalent to selecting a random number from the unit interval according to the uniform distribution. Of course, a hand ξ_1 is inferior to a hand ξ_2 if and only if $\xi_1 < \xi_2$. The game proceeds as follows. Two players A and B select points ξ and η, respectively, from the unit interval according to the uniform distribution. Both players ante one unit. A, knowing his value ξ, acts first and has the option of either folding immediately, thus forfeiting his ante to B, or betting any one of the amounts $a_1, a_2, \ldots, a_n$, where $1 < a_1 < a_2 < \cdots < a_n$. B must then respond by either passing immediately or seeing. In the first circumstance A wins B's ante. If B sees, the hands ξ and η are compared and the player with the better hand wins the pot. If $\xi = \eta$, no payment is made.

A strategy for A can be described as an n-tuple of functions

$$\boldsymbol{\phi}(\xi) = \langle \phi_1(\xi), \phi_2(\xi), \ldots, \phi_n(\xi) \rangle,$$

where $\phi_i(\xi)$ expresses the probability that A will bet the amount a_i when his hand is ξ. The $\phi_i(\xi)$ must satisfy $\phi_i(\xi) \geq 0$ and

$$\sum_{i=1}^{n} \phi_i(\xi) \leq 1.$$

The probability that A will fold immediately is

$$1 - \sum_{i=1}^{n} \phi_i(\xi).$$

A strategy for B can be represented by the n-tuple

$$\boldsymbol{\psi}(\eta) = \langle \psi_1(\eta), \psi_2(\eta), \ldots, \psi_n(\eta) \rangle,$$

where $\psi_i(\eta)$ expresses the probability that B will see a bet of a_i units when he holds the value η. The probability that B will pass after A has bet a_i is $1 - \psi_i(\eta)$. Each $\psi_i(\eta)$ is subject only to the condition that

$$0 \leq \psi_i(\eta) \leq 1.$$

If A uses the strategy ϕ and B uses the strategy ψ, the expected gain to A is denoted by $K(\phi, \psi)$. Enumerating all the possibilities, we find that

$$K(\phi, \psi) = (-1)\int_0^1 \left[1 - \sum_{i=1}^{n} \phi_i(\xi)\right] d\xi + \int_0^1\int_0^1 \sum_{i=1}^{n} \phi_i(\xi)[1 - \psi_i(\eta)]\, d\xi\, d\eta + \sum_{i=1}^{n} (a_i + 1)\int_0^1\int_0^1 \phi_i(\xi)\psi_i(\eta)L(\xi, \eta)\, d\xi\, d\eta,$$

where $L(\xi, \eta) = \text{sign } (\xi - \eta)$.

Any pair of optimal strategies ϕ^* and ψ^* satisfy the inequalities

$$K(\phi^*, \psi) \geq K(\phi^*, \psi^*) \qquad \text{for all} \quad \psi \tag{9.3.1}$$

and

$$K(\phi, \psi^*) \leq K(\phi^*, \psi^*) \qquad \text{for all} \quad \phi. \tag{9.3.2}$$

Conversely, if the inequalities are satisfied, the * strategies are optimal. Thus, ϕ^* maximizes $K(\phi, \psi^*)$ and ψ^* minimizes $K(\phi^*, \psi)$. In this case, rearranging the expression for $K(\phi, \psi)$, we can write

$$K(\phi, \psi^*) = -1 + \sum_{i=1}^{n} \int \phi_i(\xi)\left[2 + a_i\int_0^\xi \psi_i^*(\eta)\, d\eta - (a_i + 2)\int_\xi^1 \psi_i^*(\eta)\, d\eta\right] d\xi \tag{9.3.3}$$

and

$$K(\phi^*, \psi) = -1 + 2\sum_{i=1}^{n} \int \phi_i^*(\xi)\, d\xi + \sum_{i=1}^{n} \int_0^1 \psi_i(\eta)\left[-(a_i + 2)\int_0^\eta \phi_i^*(\xi)\, d\xi + a_i\int_\eta^1 \phi_i^*(\xi)\, d\xi\right] d\eta. \tag{9.3.4}$$

Thus (9.3.3) and (9.3.4) have the form

$$K(\phi, \psi^*) = C_1 + \sum_{i=1}^{n} \int_0^1 \phi_i(\xi)L_i(\xi)\, d\xi \tag{9.3.5}$$

and

$$K(\phi^*, \psi) = C_2 + \sum_{i=1}^{n} \int_0^1 \psi_i(\eta) K_i(\eta)\, d\eta, \tag{9.3.6}$$

where C_1 and C_2 are independent of ϕ and ψ, respectively, and L_i and K_i stand for the bracketed expressions in (9.3.3) and (9.3.4), respectively. In view of the constraints on $\phi_1, \ldots, \phi_n$, it is clear that in order to maximize (9.3.3) or (9.3.5), A must choose $\phi_i(\xi) = 1$ wherever $L_i(\xi)$ is positive and greater than all $L_j(\xi)$ with $j \neq i$; furthermore, he must choose $\phi_i(\xi) = 0$ wherever $L_i(\xi) < 0$; and finally, if $L_i(\xi) = 0$ and if the remaining coefficients $L_j(\xi)$ $(j \neq i)$ are nonpositive, he can maximize $K(\phi, \psi^*)$ by choosing $\phi_i(\xi)$ arbitrary consistent with $0 \leq \phi_i(\xi) \leq 1$ [or, if more than one $L_i(\xi)$ is zero, $\sum \phi_i(\xi) \leq 1$, where the sum is extended over those indices corresponding to $L_i(\xi) = 0$.] Similarly, in order to minimize (9.3.4) or (9.3.6), B must choose $\psi_i(\eta) = 1$ wherever $K_i(\eta) < 0$, and $\psi_i(\eta) = 0$ wherever $K_i(\eta) > 0$. Where $K_i(\eta) = 0$, the values of $\psi_i(\eta)$ will not affect the pay-off.

Guided to some extent by intuitive considerations, we shall construct two strategies ϕ^* and ψ^*. It will be easy to verify that these strategies satisfy (9.3.1) and (9.3.2) and are therefore optimal. The main problem in the construction is to make sure that the strategies are consistent—i.e., that the function ϕ^* that maximizes (9.3.3) is the same function that appears in (9.3.4), and similarly for ψ^*. This is another illustration of the fixed-point method. We now proceed with the details.

Since B has no opportunity to bluff, we may expect that

$$\psi_i^*(\eta) = \begin{cases} 0 & \eta < b_i, \\ 1 & \eta > b_i \end{cases} \tag{9.3.7}$$

for some b_i. This is in fact the case, since each $K_j(\eta)$ is nonincreasing.

On the other hand, we may expect that A will sometimes bluff when his hand is low. In order to allow for this possibility, we determine the critical numbers b_i which define $\psi_i^*(\eta)$ so that the coefficient $L_i(\xi)$ of ϕ_i is zero for $\xi < b_i$. This can be accomplished, since $L_i(\xi)$ is constant on this interval. Hence we choose

$$b_i = \frac{a_i}{2 + a_i}, \tag{9.3.8}$$

and thus $b_1 < b_2 < \cdots < b_n < 1$. The coefficient $L_i(\xi)$ of ϕ_i is zero in the interval $(0, b_i)$, and thereafter it is a linear function of ξ such that

$$L_i(1) = 4\left(1 - \frac{1}{a_i + 2}\right).$$

From this we deduce that the functions $L_i(\xi)$ and $L_j(\xi)$ intersect at the point

$$c_{ij} = 1 - \frac{2}{(2 + a_i)(2 + a_j)} \cdot \tag{9.3.9}$$

Clearly, c_{ij} is a strictly increasing function of i and j. Define $c_1 = b_1$ and $c_i = c_{i-1,i}$ for $i = 2, \ldots, n$ and $c_{n+1} = 1$. For ξ in the interval (c_i, c_{i+1}), it is clear that $L_i(\xi) > L_j(\xi) \geq 0$ for $j \neq i$. Consequently, according to our previous discussion, if ϕ^* maximizes $K(\phi, \psi^*)$, then $\phi_i^*(\xi) = 1$ for $c_i < \xi < c_{i+1}$. For definiteness we also set $\phi_i^*(c_i) = 1$; this is of no consequence, since if a strategy is altered only at a finite number of points (or on a set of Lebesgue measure zero), the yield $K(\phi, \psi)$ remains unchanged.

Summarizing, we have shown that if ψ^* is defined as in (9.3.7), with

$$b_i = \frac{a_i}{2 + a_i},$$

then $K(\phi, \psi^*)$ is maximized by any strategy ϕ^* of the form

$$\phi_i^*(\xi) = \begin{cases} \text{arbitrary} & \xi < c_1 = b_1, \\ 0 & c_1 \leq \xi < c_i, \\ 1 & c_i \leq \xi < c_{i+1}, \\ 0 & c_{i+1} \leq \xi \leq 1, \end{cases} \tag{9.3.10}$$

where

$$\sum_{i=1}^{n} \phi_i^*(\xi) \leq 1, \qquad \phi_i^*(\xi) \geq 0.$$

The values of $\phi_i^*(\xi)$ in the interval $0 \leq \xi < c_1$ are still undetermined because of the relations $L_i(\xi) \equiv 0$ which are valid for that same interval.

It remains to show that ψ^* as constructed actually minimizes $K(\phi^*, \psi)$. In order to guarantee this property for ψ^*, it might be necessary to impose some further conditions on the ϕ_i^*; for this purpose, we shall utilize the flexibility present in the definition of ϕ^* as ξ ranges over the interval $[0, c_1)$. In order to show that ψ^* minimizes $K(\phi^*, \psi)$, we must show that the coefficient $K_i(\eta)$ of $\psi_i(\eta)$ is nonnegative for $\eta < b_i$ and nonpositive for $\eta > b_i$. Since $K_i(\eta)$ is a continuous monotone-decreasing function, the last condition is equivalent to the relation

$$-(a_i + 2)\int_0^{b_i} \phi_i^*(\xi)\, d\xi + a_i\int_{b_i}^1 \phi_i^*(\xi)\, d\xi = 0. \tag{9.3.11}$$

Inserting the special form (9.3.10) of ϕ^* into (9.3.11) leads to the equations

$$2\int_0^{b_1} \phi_1^*(\xi)\, d\xi = b_1(1 - b_1)(b_2 + 1),$$

$$2\int_0^{b_1} \phi_n^*(\xi)\, d\xi = b_n(1 - b_n)(1 - b_{n-1}), \tag{9.3.12}$$

and

$$2\int_0^{b_1} \phi_i^*(\xi)\, d\xi = b_i(1 - b_i)(b_{i+1} - b_{i-1}) \qquad (i = 2, \ldots, n - 1). \tag{9.3.13}$$

Since

$$\sum_{i=1}^{n} \phi_i^*(\xi) \leq 1,$$

these equations can be satisfied if and only if

$$2\int_0^{b_1} \sum_{i=1}^{n} \phi_i^*(\xi)\, d\xi \leq 2b_1. \tag{9.3.14}$$

But the sum of the right-hand sides of (9.3.12) and (9.3.13) is at most $(2 + b_n - b_1)/4$, since always $b_i(1 - b_i) \leq \frac{1}{4}$. Since $b_1 \geq \frac{1}{3}$, we get

$$\tfrac{1}{4}(2 + b_n - b_1) \leq \tfrac{1}{4}(3 - b_1) \leq \tfrac{2}{3} \leq 2b_1.$$

Thus the requirements in (9.3.12–14) can be fulfilled. We have consequently established the inequalities (9.3.1) and (9.3.2) for the strategies ϕ^* and ψ^*. In summary, we display the optimal strategies as follows:

$$\psi_i^*(\eta) = \begin{cases} 0 & \eta < b_i, \\ 1 & \eta \geq b_i, \end{cases} \tag{9.3.15}$$

$$\phi_i^*(\xi) = \begin{cases} \text{arbitrary but satisfying} & \\ \int_0^{b_1} \phi_i^*(\xi)\, d\xi = b_i(c_{i+1} - c_i) & 0 \leq \xi < b_1, \\ 0 & \begin{cases} b_1 \leq \xi < c_i & \text{or} \\ c_{i+1} \leq \xi \leq 1, \end{cases} \\ 1 & c_i \leq \xi < c_{i+1}, \end{cases} \tag{9.3.16}$$

where

$$b_i = \frac{a_i}{2 + a_i}, \qquad c_1 = b_1,$$

$$c_i = 1 - \frac{2}{(2 + a_i)(2 + a_{i-1})} \quad (i = 2, \ldots, n), \qquad c_{n+1} = 1$$

and

$$\sum_{i=1}^{n} \phi_i^*(\xi) \leq 1.$$

Extensions. Instead of assuming that the hands correspond to the unit interval and are all equally likely (i.e., that a value ξ is selected according to the uniform distribution), we may make the broader assumption that A receives a random variable ξ from the unit interval according to the cumulative distribution function (c.d.f.) $F(\xi)$ and B receives a random variable η from the unit interval according to the c.d.f. $G(\eta)$. If both $F(\xi)$ and $G(\eta)$ are continuous throughout, the preceding analysis can be carried through and the form of the optimal strategies remains essentially unaltered. The critical numbers can be easily calculated in terms of the size of the bets and the distributions F and G. In the case in which the distributions are discrete, an additional feature enters. We illustrate this by describing the results for the case $F = G =$ the uniform distribution concentrating at N equally spaced points in the unit interval. Explicitly, the distribution F concentrates mass $1/N$ at each of the points $t_i = i/N$ $(i = 1, 2, \ldots, N)$. For convenience, we shall assume that the number of hands is sufficiently large so that all the numbers

$$\frac{2N}{a_1 + 2}, \ldots, \frac{2N}{a_n + 2}, \frac{2N}{(a_i + 2)(a_j + 2)} \qquad (i, j = 1, \ldots, n)$$

differ by at least one unit, and we shall also assume that $a_i > 2$. The analysis proceeds as above, and we find that B's optimal strategy is of the form

$$\psi_i^*(\eta) = \begin{cases} 0 & \eta < b_i, \\ \alpha_i & \eta = b_i, \\ 1 & \eta > b_i, \end{cases}$$

where $0 \leq \alpha_i < 1$. The quantity α_i indicates that there is a possibility that B will randomize for a given hand between seeing or passing a bet. However, it can be shown that B will randomize for at most one hand if A bets a_i. Similarly,

$$\phi_i^*(\xi) = \begin{cases} 0 & b_1 < \xi < c_i, \\ 1 & c_i < \xi < c_{i+1}, \\ 0 & c_{i+1} < \xi < 1, \end{cases}$$

where c_i are critical numbers similar to those obtained above and $c_1 = b_1$, $c_{n+1} = 1$. There is also a possibility that A will randomize for $\xi = c_i$ and $\xi = c_{i+1}$. The details are omitted.

The above analysis extends to the case of an arbitrary c.d.f.; usually the optimal strategies will involve randomization at a few isolated points. In this aspect—i.e., optimal strategies involving randomization for the case of discrete distributions as distinguished from nonrandomizing optimal strategies for continuous distributions—the model discussed in this section is very similar to statistical testing models.

9.4 Poker model with two rounds of betting. In this section we generalize the poker model of the preceding section to include two rounds of betting, but at the same time we restrict it by permitting only one size of bet. We assume again, for convenience, that hands are dealt at random to each player from the unit interval according to the uniform distribution.

Pay-off and strategies. After making the initial bet of one unit, A acts first and has two choices: he may fold or bet a units. B acts next and has three choices: he may fold, he may see, or he may raise by betting $a + b$ units. If B has raised, A must either fold or see.

If A and B draw cards ξ and η, respectively, their strategies may be described as follows:

$$\begin{aligned}
\phi_1(\xi) &= \text{probability that } A \text{ bets } a \text{ and folds later if } B \text{ raises.}\\
\phi_2(\xi) &= \text{probability that } A \text{ bets } a \text{ and sees if } B \text{ raises.}\\
1 - \phi_1(\xi) - \phi_2(\xi) &= \text{probability that } A \text{ folds initially.}\\
\psi_1(\eta) &= \text{probability that } B \text{ sees the initial bet.}\\
\psi_2(\eta) &= \text{probability that } B \text{ raises.}\\
1 - \psi_1(\eta) - \psi_2(\eta) &= \text{probability that } B \text{ folds.}
\end{aligned}$$

The expected return is

$$\begin{aligned}
K(\phi, \psi) = &-\int [1 - \phi_1(\xi) - \phi_2(\xi)]\, d\xi \\
&+ \iint [\phi_1(\xi) + \phi_2(\xi)][1 - \psi_1(\eta) - \psi_2(\eta)]\, d\xi\, d\eta \\
&+ (a + 1) \iint \phi_1(\xi)\psi_1(\eta) L(\xi, \eta)\, d\xi\, d\eta \\
&- (a + 1) \iint \phi_1(\xi)\psi_2(\eta)\, d\xi\, d\eta \\
&+ (a + 1) \iint \phi_2(\xi)\psi_1(\eta) L(\xi, \eta)\, d\xi\, d\eta \\
&+ (1 + a + b) \iint \phi_2(\xi)\psi_2(\eta) L(\xi, \eta)\, d\xi\, d\eta,
\end{aligned}$$

where $L(\xi, \eta) = \text{sign}(\xi - \eta)$. (This expected yield is derived by considering mutually exclusive possibilities: A folds; A bets and B folds; A acts according to ϕ_1 and B sees or raises; A acts according to ϕ_2 and B sees or raises.)

If we denote the optimal strategies by

$$\boldsymbol{\phi}^*(\xi) = \langle \phi_1^*(\xi), \phi_2^*(\xi)\rangle, \qquad \boldsymbol{\psi}^*(\eta) = \langle \psi_1^*(\eta), \psi_2^*(\eta)\rangle$$

and rearrange the terms as we have done in the previous examples, we obtain

$$\begin{aligned} K(\boldsymbol{\phi}, \boldsymbol{\psi}^*) = -1 &+ \int_0^1 \phi_1(\xi)\Big[2 + a\int_0^\xi \psi_1^*(\eta)\,d\eta \\ &- (a+2)\int_\xi^1 \psi_1^*(\eta)\,d\eta - (a+2)\int_0^1 \psi_2^*(\eta)\,d\eta\Big]d\xi \\ &+ \int_0^1 \phi_2(\xi)\Big[2 + a\int_0^\xi \psi_1^*(\eta)\,d\eta - (a+2)\int_\xi^1 \psi_1^*(\eta)\,d\eta \\ &+ (a+b)\int_0^\xi \psi_2^*(\eta)\,d\eta - (a+b+2)\int_\xi^1 \psi_2^*(\eta)\,d\eta\Big]d\xi \end{aligned} \tag{9.4.1}$$

and

$$\begin{aligned} K(\boldsymbol{\phi}^*, \boldsymbol{\psi}) = \int_0^1 &[-1 + 2\phi_1^*(\xi) + 2\phi_2^*(\xi)]\,d\xi \\ &+ \int_0^1 \psi_1(\eta)\Big[-(a+2)\int_0^\eta [\phi_1^*(\xi) + \phi_2^*(\xi)]\,d\xi \\ &+ a\int_\eta^1 [\phi_1^*(\xi) + \phi_2^*(\xi)]\,d\xi\Big]d\eta \\ &+ \int_0^1 \psi_2(\eta)\Big[-(a+2)\int_0^1 \phi_1^*(\xi)\,d\xi \\ &- (a+b+2)\int_0^\eta \phi_2^*(\xi)\,d\xi + (a+b)\int_\eta^1 \phi_2^*(\xi)\,d\xi\Big]d\eta. \end{aligned} \tag{9.4.2}$$

Search for the optimal strategies. We shall search again for the functions $\boldsymbol{\phi}^*(\xi)$ that maximize (9.4.1) and the functions $\boldsymbol{\psi}^*(\eta)$ that minimize (9.4.2). Intuitive considerations suggest that A may do a certain amount of bluffing with low cards with the intention of folding if raised; he will also choose $\phi_1^*(\xi) = 1$ in an intermediate range, say $c < \xi < e$, and he will select the strategy $\phi_2^*(\xi) = 1$ for $e \leq \xi \leq 1$. B should sometimes bluff by raising hands in the interval $0 \leq \eta < c$; he should choose $\psi_1^* = 1$ in an interval $c \leq \eta < d$, and $\psi_2^* = 1$ in $d \leq \eta \leq 1$.

This is not to imply that this is the only possible form of the optimal strategies. In fact, we shall see later that optimal strategies of different character exist. However, once one pair of optimal strategies is determined, it is then relatively easy to calculate all solutions; hence we first concentrate on determining optimal strategies of the form indicated. To this end we shall attempt to find values of c, d, and e that produce a solution of the given type. In view of the construction of ψ_1^* we see immediately that the coefficient of $\phi_1(\xi)$ in (9.4.1) is constant for $\xi \leq c$; evaluating this constant and setting it equal to zero, we obtain

$$2 - (a+2)\int_0^1 \psi_2^*(\eta)\, d\eta - (a+2)(d-c) = 0. \tag{9.4.3}$$

The coefficients of ψ_1^* and ψ_2^* should be equal at the point d, where B changes the character of his action; this requires

$$(2a+2)\int_d^1 \phi_1^*(\xi)\, d\xi = -b\int_0^d \phi_2^*(\xi)\, d\xi + b\int_d^1 \phi_2^*(\xi)\, d\xi. \tag{9.4.4}$$

A similar condition at $\xi = e$ requires

$$(2a+b+2)\int_0^e \psi_2^*(\eta)\, d\eta = b\int_e^1 \psi_2^*(\eta)\, d\eta. \tag{9.4.5}$$

At the point $\eta = c$, where B begins to play ψ_1^* and ψ_2^* without bluffing, the corresponding coefficients (which are decreasing functions) should change from nonnegative to nonpositive, i.e., they should be zero at $\eta = c$. Hence

$$\begin{aligned} -(a+2)\int_0^c [\phi_1^*(\xi) + \phi_2^*(\xi)]\, d\xi + a\int_c^1 [\phi_1^*(\xi) + \phi_2^*(\xi)]\, d\xi &= 0, \\ -(a+2)\int_0^1 \phi_1^*(\xi)\, d\xi + (a+b)\int_c^1 \phi_2^*(\xi)\, d\xi &= 0. \end{aligned} \tag{9.4.6}$$

(In writing (9.4.6) we postulate that $\phi_2^*(\xi) = 0$ for $0 \leq \xi \leq c$; this is intuitively clear.) At this point we introduce the notation

$$m_1 = \int_0^c \phi_1^*(\xi)\, d\xi, \qquad m_2 = \int_0^c \psi_2^*(\eta)\, d\eta.$$

Recalling the assumptions made on the form of the solution and assuming that $c < e < d$, equations (9.4.3–6) may be written as follows:

$$2 = (a+2)(m_2 + 1 - c), \tag{9.4.7}$$

$$1 - d = d - e \qquad \text{or} \qquad 2(1-d) = (1-e), \tag{9.4.8}$$

$$(2a + b + 2)m_2 = b(1 - d), \tag{9.4.9}$$

$$(a + 2)m_1 = a(1 - c), \tag{9.4.10}$$

$$(a + 2)(m_1 + e - c) = (a + b)(1 - e). \tag{9.4.11}$$

We have obtained a system of five equations in the five unknowns m_1, m_2, c, d, e; we now prove that this system of equations has a solution which is consistent with the assumptions made previously, namely that $0 < c < e < d < 1; 0 < m_1 < c; 0 < m_2 < c$.

Solution of equations (9.4.7–11). The system of equations may be solved explicitly as follows. We write the last equation as:

$$(a + 2)(m_1 + 1 - c) = (2a + b + 2)(1 - e).$$

Eliminating m_1 and $1 - e$ by means of (9.4.10) and (9.4.8), we obtain

$$(a + 1)(1 - c) = (2a + b + 2)(1 - d). \tag{9.4.12}$$

From the remaining equations we eliminate m_2; then

$$\frac{2}{a + 2} - (1 - c) = \frac{b}{2a + b + 2}(1 - d). \tag{9.4.13}$$

Therefore

$$(1 - d)\left(\frac{b}{2a + b + 2} + \frac{2a + b + 2}{a + 1}\right) = \frac{2}{a + 2}\cdot \tag{9.4.14}$$

Having obtained $1 - d$, we can solve (9.4.13) for $1 - c$, and the remaining unknowns are then calculated from the original equations.

In order to show that the solution is consistent, we first note that (9.4.14) implies that $1 - d > 0$. Equation (9.4.12) shows that $1 - c > 1 - d$, and therefore $c < d$. Also, from (9.4.13),

$$c = \left(\frac{b}{2a + b + 2}\right)(1 - d) + \left(1 - \frac{2}{a + 2}\right) > 0. \tag{9.4.15}$$

Since $2(a + 1)(1 - c) = (2a + b + 2)(1 - e)$, we infer that $1 - e < 1 - c$, or $c < e$; and since $2d = 1 + e$, we must have $e < d$. Summing up, we have shown that $0 < c < e < d < 1$. For the two remaining conditions we note that (9.4.7) implies that

$$m_2 = c - \left(1 - \frac{2}{a + 2}\right),$$

so that $m_2 < c$, and by (9.4.15), $m_2 > 0$. Finally, using (9.4.10) and (9.4.7), we conclude that

$$m_1 = \frac{a}{a+2}(1-c) = (1-c) - \frac{2}{a+2}(1-c)$$

$$= (1-c)[1-(m_2+1-c)] = (1-c)(c-m_2),$$

so that $0 < m_1 < c$.

Optimality of the strategies ϕ^* *and* ψ^*. We summarize the representation of ϕ^* and ψ^* in terms of the values of c, e, d, m_1, and m_2 as computed above:

$$\phi_1^*(\xi) = \begin{cases} 1 & c \le \xi < e, \\ 0 & e \le \xi \le 1; \end{cases} \qquad \phi_2^*(\xi) = \begin{cases} 0 & 0 \le \xi < e, \\ 1 & e \le \xi \le 1; \end{cases}$$

$$\psi_1^*(\eta) = \begin{cases} 0 & 0 \le \eta < c, \\ 1 & c \le \eta < d, \\ 0 & d \le \eta \le 1; \end{cases} \qquad \psi_2^*(\eta) = \begin{cases} 0 & c \le \eta < d, \\ 1 & d \le \eta \le 1. \end{cases} \tag{9.4.16}$$

In the remaining interval $0 \le \eta < c$, the functions $\phi_1^*(\xi)$ and $\psi_2^*(\eta)$ are chosen arbitrarily but bounded between 0 and 1, satisfying

$$\int_0^c \phi_1^*(\xi)\,d\xi = m_1 \qquad \text{and} \qquad \int_0^c \psi_2^*(\eta)\,d\eta = m_2,$$

respectively.

It remains to verify that the strategies ϕ^* and ψ^* prescribed above maximize $K(\phi, \psi^*)$ of (9.4.1) and minimize $K(\phi^*, \psi)$ of (9.4.2), respectively. To do this, we first examine the coefficients $M_1(\xi)$ and $M_2(\xi)$ of ϕ_1 and ϕ_2 in $K(\phi, \psi^*)$. By construction the coefficient $M_1(\xi)$ of ϕ_1 is identically zero on $[0, c)$, increases linearly on $[c, d)$, and afterwards remains constant. Also, $M_1(\xi)$ is continuous throughout $[0, 1]$. Next we notice that $M_2(\xi)$ is linear on $[c, d]$ with the same slope as $M_1(\xi)$. Furthermore, they agree at $\xi = e$ in $[c, d]$ by (9.4.5), and hence $M_1 = M_2$ for ξ in $[c, d]$.

We may also deduce immediately from the definition of ψ^* that M_2 increases strictly throughout $[0, 1]$ (see Fig. 9.1). With these facts the maximization of $K(\phi, \psi^*)$ is now easy to perform. Clearly, the maximum is achieved for any ϕ with the properties (a) $\phi_2 = 0$ and ϕ_1 arbitrary $(0 \le \phi_1 \le 1)$ for ξ in $[0, c)$; (b) $\phi_1 + \phi_2 = 1$ and otherwise arbitrary for ξ in $[c, d)$; (c) $\phi_2 = 1$ for ξ in $[d, 1]$. It is clear that ϕ^* as specified above fulfills these conditions.

A study of the coefficients $N_1(\eta)$ and $N_2(\eta)$ of $\psi_1(\eta)$ and $\psi_2(\eta)$ in $K(\phi^*, \psi)$ shows that they are as indicated in Fig. 9.2. We observe that

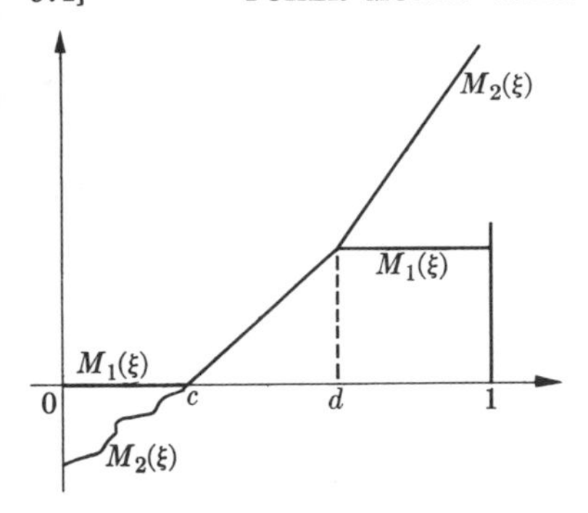

FIGURE 9.1

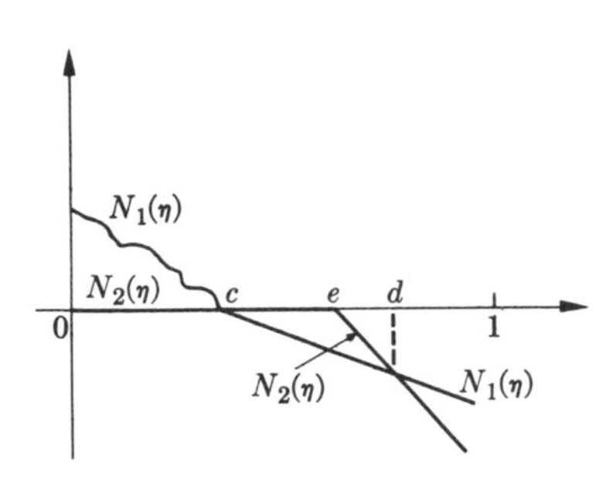

FIGURE 9.2

$K(\phi^*, \psi)$ is minimized by any ψ with the properties (a′) $\psi_1 = 0$ and ψ_2 arbitrary $(0 \leq \psi_2 \leq 1)$ for η in $[0, c)$; (b′) $\psi_1 = 1$ for η in $[c, d)$; (c′) $\psi_2 = 1$ for η in $[d, 1]$. Clearly, ψ^* as specified above obeys these requirements. The proof of the optimality of ϕ^* and ψ^* is now complete.

For illustrative purposes we append the following example. Let $a = b = 2$; then

$$\phi_1^*(\xi) = \begin{cases} \int_0^{19/35} \phi_1^* = \frac{8}{35} & \\ 1 & \frac{19}{35} \leq \xi < \frac{23}{35}, \\ 0 & \frac{23}{35} \leq \xi \leq 1; \end{cases}$$

$$\phi_2^*(\xi) = \begin{cases} 0 & 0 \leq \xi < \frac{23}{35}, \\ 1 & \frac{23}{35} \leq \xi \leq 1; \end{cases}$$

$$\psi_1^*(\eta) = \begin{cases} 0 & 0 \leq \eta < \frac{19}{35}, \\ 1 & \frac{19}{35} \leq \eta < \frac{29}{35}, \\ 0 & \frac{29}{35} < \eta \leq 1; \end{cases}$$

$$\psi_2^*(\eta) = \begin{cases} \int_0^{19/35} \psi_2^* = \frac{3}{70} & \\ 0 & \frac{19}{35} \leq \eta < \frac{29}{35}, \\ 1 & \frac{29}{35} \leq \eta \leq 1. \end{cases}$$

The value is —11/35.

The general solution. It follows readily from Fig. 9.2 that the minimizing solution cannot have any other form than that given in (9.4.16). However, in maximizing $K(\phi, \psi^*)$ we found that on the interval $[c, d]$ the only necessary condition for ϕ to qualify as a maximum is that $\phi_1 + \phi_2 = 1$. Altering ϕ_1 and ϕ_2 in this interval subject to this condition, we can determine all possible optimal strategies for A. To this end, we specify $\tilde{\phi}_1(\xi)$ on $[0, c]$ such that

$$\int_0^c \tilde{\phi}_1 = m_1$$

(calculated above) and $\tilde{\phi}_2(\xi) = 1$ on $[d, 1]$. For ξ in $[c, d]$ we require only that $\tilde{\phi}_1 + \tilde{\phi}_2 = 1$. Writing out the conditions under which $K(\tilde{\phi}, \psi)$ is minimized for ψ^*, we obtain

$$1 - d = \int_c^d \tilde{\phi}_2(\xi)\, d\xi \quad \text{and} \quad \int_\eta^d \tilde{\phi}_1(\xi)\, d\xi \leq \frac{b(d - \eta)}{a + b + 1} \quad \text{for} \quad \eta \in [c, d], \tag{9.4.17}$$

where c and d are the same as before. We obtain these constraints by equating the coefficients of $\psi_1(\eta)$ and $\psi_2(\eta)$ in (9.4.2) at $\eta = d$ and by requiring that $N_1(\eta) \leq N_2(\eta)$ for η in $[c, d]$.

This relation is easily seen to be necessary and sufficient for $\tilde{\phi}$ to be optimal (see Problem 8).

***9.5 Poker model with k raises.** In this section we indicate the form of the optimal strategies of a poker model with several rounds of betting. The methods of analysis are in principle extensions of those employed in the preceding section, but far more complicated in detail. We omit the proofs, referring the reader to the references at the close of this chapter.

Rules, strategies, and pay-off. The two players ante one unit each and receive independently hands ξ and η (which are identified with points of the unit interval) according to the uniform distribution. There are $k + 1$ rounds of betting ("round" in this section means one action by one player). In the first round A may either fold (and lose his ante) or bet a units. A and B act alternately. In each subsequent round a player may either fold or see (whereupon the game ends), or raise the bet by a units. In the last round the player can only fold or see. If k is even, the last possible round ends with A; if k is odd, the last possible round ends with B.

A strategy for A can be described by a k-tuple of functions $\phi = \langle \phi_1(\xi), \phi_2(\xi), \ldots, \phi_k(\xi) \rangle$. These functions indicate A's course of action when he receives the hand ξ. Explicitly,

$$1 - \sum_{i=1}^{k} \phi_i(\xi)$$

is the probability that A will fold immediately, and

$$\sum_{i=1}^{k} \phi_i(\xi)$$

---

* Starred sections may be omitted at first reading without loss of continuity.

is the probability that A will bet at his first opportunity. Further,

$$\phi_1(\xi) = \text{probability that } A \text{ will fold in his second round,}$$

$$\phi_2(\xi) = \text{probability that } A \text{ will see in his second round,}$$

$$\textstyle\sum_{i=3}^k \phi_i(\xi) = \text{probability that } A \text{ will raise in his second round}$$

if the occasion arises, i.e., if B has raised in his first round and kept the game going. Similarly, if the game continues until A's rth round, then

$$\phi_{2r-3}(\xi) = \text{probability that } A \text{ will fold in his } r\text{th round,}$$

$$\phi_{2r-2}(\xi) = \text{probability that } A \text{ will see in his } r\text{th round,}$$

$$\textstyle\sum_{i=2r-1}^k \phi_i(\xi) = \text{probability that } A \text{ will raise in his } r\text{th round.}$$

Analogously, a strategy for B is expressible as a k-tuple

$$\psi = \langle \psi_1(\eta), \ldots, \psi_k(\eta) \rangle$$

which indicates B's course of action when he receives the hand η. The probability that B will fold at his first opportunity is

$$\psi_0(\eta) = 1 - \sum_{j=1}^k \psi_j(\eta).$$

If the game continues until B's rth round, then

$$\psi_{2r-2}(\eta) = \text{probability that } B \text{ will fold in his } r\text{th round,}$$

$$\psi_{2r-1}(\eta) = \text{probability that } B \text{ will see in his } r\text{th round,}$$

$$\textstyle\sum_{j=2r}^k \psi_j(\eta) = \text{probability that } B \text{ will raise in his } r\text{th round.}$$

If the two players receive hands ξ and η and choose the strategies ϕ and ψ, respectively, then the pay-off to A can be computed as in the previous examples by considering the mutually exclusive ways in which the betting may terminate. The pay-off to A is as follows:

$$P[\phi(\xi), \psi(\eta)] = (-1)\left(1 - \sum_{i=1}^k \phi_i(\xi)\right)$$

$$+ \sum_{i=1}^k \phi_i(\xi)\left[1 - \sum_{j=1}^k \psi_j(\eta) + (a+1)\psi_1(\eta)L(\xi, \eta)\right]$$

$$+ \sum_{j=2}^k \psi_j(\eta)\{-(a+1)\phi_1(\xi) + (2a+1)\phi_2(\xi)L(\xi, \eta)\}$$

$$\vdots$$

$$+ \sum_{j=2r-2}^{k} \psi_j(\eta)\{-[(2r-3)a+1]\phi_{2r-3}(\xi)$$

$$+ [(2r-2)a+1]\phi_{2r-2}(\xi)L(\xi,\eta)\}$$

$$+ \sum_{i=2r-1}^{k} \phi_i(\xi)\{[(2r-2)a+1]\psi_{2r-2}(\eta)$$

$$+ [(2r-1)a+1]\psi_{2r-1}(\eta)L(\xi,\eta)\}$$

$$+ \sum_{j=2r}^{k} \psi_j(\eta)\{-[(2r-1)a+1]\phi_{2r-1}(\xi) + (2ra+1)\phi_{2r}(\xi)L(\xi,\eta)\}$$

$$+ \sum_{i=2r+1}^{k} \phi_i(\xi)\{(2ra+1)\psi_{2r}(\eta) + [(2r+1)a+1]\psi_{2r+1}(\eta)L(\xi,\eta)\}$$

$$\vdots$$

where $L(\xi, \eta)$ is the function defined on p. 245. The expected pay-off is

$$K(\phi, \psi) = \int_0^1 \int_0^1 P[\phi(\xi), \psi(\eta)]\, d\xi\, d\eta. \tag{9.5.1}$$

Description of the optimal strategies. There exist optimal strategies ϕ^* and ψ^* characterized by $2k+1$ numbers $b, c_1, \ldots, c_k, d_1, \ldots, d_k$. When a player gets a hand ξ in $(0, b)$ he will bluff part of the time and fold part of the time. We shall write

$$m_i = \int_0^b \phi_i^*(\xi)\, d\xi \qquad (i = 1, 3, 5, \ldots) \tag{9.5.2}$$

and

$$n_j = \int_0^b \psi_j^*(\eta)\, d\eta \qquad (j = 2, 4, 6, \ldots) \tag{9.5.3}$$

to represent the probabilities of bluffing in the various rounds of betting. If A receives a hand ξ in (c_{i-1}, c_i), where $c_0 = b$, he will choose $\phi_i^*(\xi) = 1$ and $\phi_l^*(\xi) = 0$ for $l \neq i$. Similarly, if B gets a hand η in (d_{j-1}, d_j), where $d_0 = b$, he will choose $\psi_j^*(\eta) = 1$ and $\psi_l^*(\eta) = 0$ for $l \neq j$. The solution is represented by Fig. 9.3. The fact that

$$c_{2r-1} < d_{2r-1} < d_{2r} < c_{2r} \qquad (r = 1, 2, \ldots)$$

is important.

The constants c_i, d_j, m_i, and n_j are determined by solving an elaborate system of equations analogous to (9.4.7–11). Explicitly, if k is even, b, c_i, d_j, m_i, and n_j are evaluated as solutions of the following equations:

$$[(4r-1)a+2]\sum_{j=2r}^{k} n_j = a(1-d_{2r-1}) \qquad (2r = 2, 4, \ldots, k),$$

$$a(c_{2r-2}-d_{2r-2}) = a(1-c_{2r-2}) + [(4r-3)a+2]\sum_{j=2r}^{k} n_j \qquad (2r = 4, 6, \ldots, k),$$

$$[(4r-3)a+2]\sum_{i=2r-1}^{k} m_i = a(1-c_{2r-2}) \qquad (2r = 2, 4, \ldots, k),$$

$$a(d_{2r-1}-c_{2r-1}) = a(1-d_{2r-1}) + [(4r-1)a+2]\sum_{i=2r+1}^{k} m_i \qquad (2r = 2, 4, \ldots, k),$$

$$(4ra+2)(c_{2r}-d_{2r-1}) = [(4r+2)a+2](c_{2r}-d_{2r}) \qquad (2r = 2, 4, \ldots, k-2),$$

$$[(4r-2)a+2](d_{2r-1}-c_{2r-2}) = (4ra+2)(d_{2r-1}-c_{2r-1}) \qquad (2r = 2, 4, \ldots, k),$$

$$2 = (a+2)\int_0^1 \sum_{j=1}^{k} \psi_j^*(\eta)\, d\eta.$$

An analogous system applies for k odd. The solutions obtained are consistent with the requirements of Fig. 9.3.

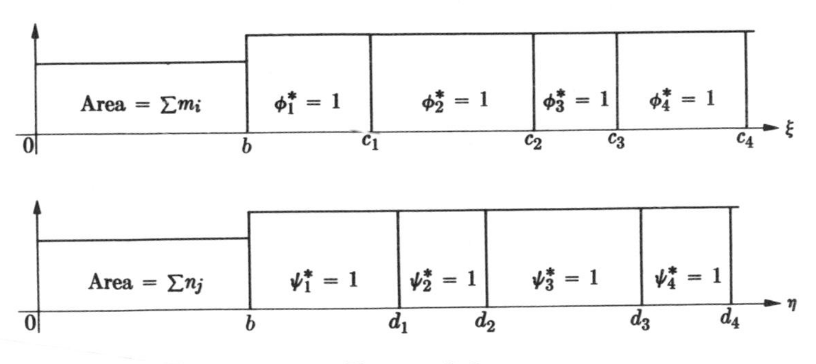

FIGURE 9.3

9.6 Poker with simultaneous moves. Two players, A and B, make simultaneous bets after drawing hands according to the uniform distribution. The initial bet can be either b (the low bet) or a (the high bet). If both bets are equal, the player with the higher hand wins. If one player bets high and the other bets low, the low bettor has a choice: he may either fold (losing the low bet) or see by making an additional bet of $a - b$. If the low bettor sees, the player with the higher hand wins the pot.

Since the game is symmetric, we need only describe the strategies of one player. If A draws ξ, we shall write

$\phi_1(\xi)$ = probability that A will bet low and fold if B bets high,

$\phi_2(\xi)$ = probability that A will bet low and subsequently see,

$\phi_3(\xi)$ = probability that A will bet high.

These functions are of course subject to the constraints

$$\phi_i(\xi) \geq 0, \qquad \sum_{i=1}^{3} \phi_i(\xi) = 1.$$

The expected yield to A if he uses strategy ϕ while B employs strategy ψ reduces to

$$\begin{aligned} K(\phi, \psi) = {} & b\int_0^1\int_0^1 [\phi_1(\xi) + \phi_2(\xi)][\psi_1(\eta) + \psi_2(\eta)]L(\xi, \eta)\, d\xi\, d\eta \\ & - b\int_0^1\int_0^1 \phi_1(\xi)\psi_3(\eta)\, d\xi\, d\eta + b\int_0^1\int_0^1 \phi_3(\xi)\psi_1(\eta)\, d\xi\, d\eta \\ & + a\int_0^1\int_0^1 \phi_2(\xi)\psi_3(\eta)L(\xi, \eta)\, d\xi\, d\eta \\ & + a\int_0^1\int_0^1 \phi_3(\xi)\psi_2(\eta)L(\xi, \eta)\, d\xi\, d\eta \\ & + a\int_0^1\int_0^1 \phi_3(\xi)\psi_3(\eta)L(\xi, \eta)\, d\xi\, d\eta. \end{aligned}$$

Because of the symmetry of the game, we may replace B's strategy $\psi(\eta)$ in this expression by an optimal strategy $\phi^*(\eta)$. We make the plausible assumption that in this strategy $\phi_2^*(\eta) = 0$, since there would appear to be no clear justification for making a low bet initially and then seeing. The consistency of this assumption will be established later. With $\phi_2^*(\eta) = 0$ we may write $K(\phi, \phi^*)$ as

$$\begin{aligned} K(\phi, \phi^*) = {} & b\int_0^1\int_0^1 [\phi_1(\xi) + \phi_2(\xi)]\phi_1^*(\eta)L(\xi, \eta)\, d\xi\, d\eta \\ & - b\int_0^1\int_0^1 \phi_1(\xi)\phi_3^*(\eta)\, d\xi\, d\eta + b\int_0^1\int_0^1 \phi_3(\xi)\phi_1^*(\eta)\, d\xi\, d\eta \\ & + a\int_0^1\int_0^1 [1 - \phi_1(\xi) - \phi_3(\xi)]\phi_3^*(\eta)L(\xi, \eta)\, d\xi\, d\eta \end{aligned}$$

$$+ a\int_0^1\int_0^1 \phi_3(\xi)\phi_3^*(\eta)L(\xi, \eta)\,d\xi\,d\eta$$

$$= \int_0^1 \phi_1(\xi)\left[b\int_0^\xi \phi_1^*(\eta)\,d\eta - b\int_\xi^1 \phi_1^*(\eta)\,d\eta - b\int_0^1 \phi_3^*(\eta)\,d\eta - a\int_0^\xi \phi_3^*(\eta)\,d\eta + a\int_\xi^1 \phi_3^*(\eta)\,d\eta\right]d\xi$$

$$+ \int_0^1 \phi_2(\xi)\left[b\int_0^\xi \phi_1^*(\eta)\,d\eta - b\int_\xi^1 \phi_1^*(\eta)\,d\eta\right]d\xi$$

$$+ \int_0^1 \phi_3(\xi)\left[b\int_0^1 \phi_1^*(\eta)\,d\eta\right]d\xi + \int_0^1\int_0^1 \phi_3^*(\eta)L(\xi, \eta)\,d\xi\,d\eta, \tag{9.6.1}$$

or

$$K(\boldsymbol{\phi}, \boldsymbol{\phi}^*) = \int_0^1 \sum \phi_i(\xi)T_i(\xi)\,d\xi + Z,$$

where Z is a term independent of the ϕ_i. The $\boldsymbol{\phi}$ maximizing $K(\boldsymbol{\phi}, \boldsymbol{\phi}^*)$ is evaluated by choosing the component ϕ_i as large as possible whenever $T_i(\xi) = \max_j T_j(\xi)$. If the maximum of $T_j(\xi)$ is attained simultaneously by two of the T_i, then the corresponding ϕ_i may share any positive values provided their sum is 1.

Some bluffing is anticipated on low hands. This suggests that for $\xi < \xi_0$ we should have $T_1(\xi) = T_3(\xi) > T_2(\xi)$. Differentiating the identity $T_1(\xi) = T_3(\xi)$, and remembering that $\phi_1^*(\xi) + \phi_3^*(\xi) = 1$ on the interval $[0, \xi_0]$, we deduce that

$$\phi_1^*(\xi) = \frac{a}{a+b} \text{ a.e.} \qquad \text{for} \quad \xi < \xi_0.$$

With this choice of ϕ_1^*, and where $\phi_3^*(\xi) = 1$ for $\xi > \xi_0$, we obtain that $T_1(\xi) = T_3(\xi)$ is possible only if

$$\xi_0 = \frac{a-b}{a}.$$

The proposed solution is as follows:

$$\boldsymbol{\phi}^* = \begin{cases} \phi_1^*(\xi) = \dfrac{a}{a+b}, \quad \phi_3^*(\xi) = \dfrac{b}{a+b} & \xi < \xi_0, \\ \phi_3^*(\xi) = 1 & \xi > \xi_0. \end{cases} \tag{9.6.2}$$

It is now routine to verify that $\boldsymbol{\phi}^*$ as exhibited is indeed optimal. It is clear that $T_1(\xi) = T_3(\xi) > T_2(\xi)$ for $\xi < \xi_0$, and hence the maximum is achieved provided only that $\phi_1 + \phi_3 = 1$, which is certainly satisfied for $\boldsymbol{\phi}^*$ of (9.6.2). Moreover, we have seen that $T_1(\xi) = T_3(\xi)$ uniquely determines $\boldsymbol{\phi}^*$ to be as in (9.6.2) for $\xi < \xi_0$.

For $\xi > \xi_0$, by examining (9.6.1) we find that $T_2(\xi) = T_3(\xi) > T_1(\xi)$. Hence the maximization of $K(\phi, \phi^*)$ requires ϕ to be such that $\phi_2 + \phi_3 = 1$. But, if $\phi_2^* > 0$ in this interval, a simple calculation shows that $T_1(\xi) < T_2(\xi)$ for $\xi \geq \xi_1$ where $\xi_1 < \xi_0$. All these inferences in conjunction prove that the ϕ^* of (9.6.2) is the unique optimal strategy of the game.

9.7 The Le Her game. The Le Her game is described in Todhunter's *History of the Mathematical Theory of Probability* as follows:

"Peter holds a common pack of cards; he gives a card at random to Paul and takes one himself; the main object is for each to obtain a higher card than his adversary. The order of values is *Ace, two, three, . . . , ten, Knave, Queen, King.*

Now if Paul is not content with his card, he may compel Peter to change with him; but if Peter has a *King*, he is allowed to retain it. If Peter is not content with the card which he at first obtained, or which he has been compelled to receive from Paul, he is allowed to change it for another taken out of the pack at random; but if the card he then draws is a *King*, he is not allowed to have it, but must retain the card with which he was dissatisfied. If Paul and Peter finally have cards of the same value, Paul is considered to lose."

We shall not give a solution of this game here. Instead we shall consider a slightly modified continuous form of this game which yields a solution more easily.

Players A and B are assigned numbers ξ and η, respectively, equally likely in the unit interval. A moves first and can either keep his number ξ or exchange it for B's η. B then moves and can either keep his number, whether it be his original η or the ξ that A gave him, or exchange it for a new number ζ chosen at random from the unit interval. The pay-off to A is 1 if his final number is greater than B's final number and -1 if his number is smaller than B's. We need not worry about the case of the two numbers being equal, since this occurs with probability 0.

We proceed to the analysis of this game, using methods similar to those of the preceding models. Again we allow intuitive considerations to guide us.

Let $u(\xi)$ denote the probability that if A receives ξ he will keep it.

Let $\psi(\eta)$ denote the probability that B will ask for a new number ζ if he receives η and A has kept ξ.

Let $\phi(\eta, \xi)$ denote the probability that B will ask for a new number ζ if A has traded him ξ for η.

The conditions that $u(\xi)$, $\psi(\eta)$, and $\phi(\eta, \xi)$ must satisfy are $0 \leq u(\xi)$, $\psi(\eta)$, $\phi(\eta, \xi) \leq 1$. By enumerating all the possibilities and defining sign $(\xi - \eta) = L(\xi, \eta)$, we find that the expected pay-off $K(u; \psi, \phi)$ to A is

$$\int_0^1\int_0^1 u(\xi)[1 - \psi(\eta)]L(\xi, \eta)\,d\xi\,d\eta + \int_0^1\int_0^1\int_0^1 u(\xi)\psi(\eta)L(\xi, \zeta)\,d\xi\,d\eta\,d\zeta$$
$$+ \int_0^1\int_0^1 [1 - u(\xi)][1 - \phi(\eta, \xi)]L(\eta, \xi)\,d\xi\,d\eta$$
$$+ \int_0^1\int_0^1\int_0^1 [1 - u(\xi)]\phi(\eta, \xi)L(\eta, \zeta)\,d\xi\,d\eta\,d\zeta. \quad (9.7.1)$$

Common sense tells us to guess that the optimal $\phi^0(\eta, \xi)$ is of the form

$$\phi^0(\eta, \xi) = \begin{cases} 0 & \xi > \eta, \\ 1 & \xi < \eta, \end{cases}$$

since if B holds ξ and $\xi > \eta$ he will surely win if he holds on to ξ, and he will surely lose if $\xi < \eta$ and he does not draw another card. Hence the part of K which involves u can be reduced to

$$\int_0^1 u(\xi)\Big[(\xi^2 + 2\xi - 1) + (2\xi - 1)\int_0^1 \psi(\eta)\,d\eta$$
$$+ \int_\xi^1 \psi(\eta)\,d\eta - \int_0^\xi \psi(\eta)\,d\eta\Big]d\xi. \quad (9.7.2)$$

When the expression in the brackets is positive A wants $u(\xi) = 1$, and when it is negative he wants $u(\xi) = 0$, since he is maximizing.

Now let us look at the part of K which involves ψ. This is expressible as

$$\int_0^1 \psi(\eta)\Big[\int_0^1 u(\xi)(2\xi - 1)\,d\xi + \int_0^\eta u(\xi)\,d\xi - \int_\eta^1 u(\xi)\,d\xi\Big]d\eta. \quad (9.7.3)$$

When the expression in the brackets is negative, B wants $\psi(\eta) = 1$ because he wants to minimize. When it is positive, he wants $\psi(\eta) = 0$. It is clear that since

$$\int_0^\eta u(\xi)\,d\xi \qquad \text{and} \qquad \int_\eta^1 u(\xi)\,d\xi$$

are monotone-increasing and monotone-decreasing functions of η, respectively, there exists a value η_0 such that $\psi^0(\eta) = 1$ for $\eta < \eta_0$ and $\psi^0(\eta) = 0$ for $\eta > \eta_0$.

Now that we know the form of $\psi^0(\eta)$, let us see what we can say about the form of $u^0(\xi)$. When ψ is replaced by ψ^0 in the bracketed expression in (9.7.2), it becomes for $\xi < \eta_0$

$$[\xi^2 + 2\xi\eta_0 - 1] \quad (9.7.4)$$

and for $\xi \geq \eta_0$

$$[\xi^2 + 2\xi + 2\xi\eta_0 - 2\eta_0 - 1]. \quad (9.7.5)$$

Since this expression is a monotone-increasing function of ξ, there exists a value ξ_0 such that $u^0(\xi) = 0$ for $\xi < \xi_0$ and $u^0(\xi) = 1$ for $\xi > \xi_0$.

We can now substitute u^0 in (9.7.3) and determine the relationship between ξ_0 and η_0. For $\eta \leq \xi_0$ the bracketed expression in (9.7.3) becomes

$$\left[2\int_{\xi_0}^{1} \xi\, d\xi - 2\int_{\xi_0}^{1} d\xi\right]. \tag{9.7.6}$$

Since this expression is nonpositive, $\psi^0(\eta) = 1$ for $\eta \leq \xi_0$. For $\eta \geq \xi_0$ the bracketed expression becomes

$$[-1 - \xi_0^2 + 2\eta]; \tag{9.7.7}$$

hence $\psi^0(\eta) = 1$ for $\eta < (\xi_0^2 + 1)/2$ and $\psi^0(\eta) = 0$ for $\eta > (\xi_0^2 + 1)/2$.

Clearly, $\xi_0 < (\xi_0^2 + 1)/2 = \eta_0$; hence setting (9.7.4) equal to zero and solving for ξ gives the value of ξ_0. The value ξ_0 is the unique root between 0 and 1 of the equation $\xi^3 + \xi^2 + \xi - 1 = 0$. Solving this equation, we obtain

$$\xi_0 \sim 0.5437.$$

From the derivations of u^0 and ψ^0 we know that u^0 maximizes $K(u; \psi^0, \phi^0)$ and that ψ^0 minimizes $K(u^0; \psi, \phi^0)$, but we must also check that ϕ^0 simultaneously minimizes $K(u^0; \psi, \phi)$. By integrating out ζ, the part of K involving ϕ can be reduced to

$$\int_0^1\int_0^1 \phi(\eta, \xi)[1 - u^0(\xi)][(2\eta - 1) - L(\eta, \xi)]\, d\xi\, d\eta,$$

or

$$\int_0^1\int_0^{\xi_0} \phi(\eta, \xi)[2\eta - 1 - L(\eta, \xi)]\, d\xi\, d\eta.$$

Hence it is clear that our assumed ϕ^0 actually minimizes $K(u^0; \psi, \phi)$. Thus u^0, ψ^0, ϕ^0 as constructed are optimal strategies. The value of the game is

$$-\frac{\xi_0^4}{4} - \frac{\xi_0^3}{3} - \frac{\xi_0^2}{2} - \frac{1}{4} + \xi_0 \sim 0.0705.$$

In final form, the optimal strategies are

$$u^0(\xi) = \begin{cases} 1 & \xi > \xi_0, \\ 0 & \xi < \xi_0; \end{cases}$$

$$\psi^0(\eta) = \begin{cases} 1 & \eta < \eta_0 = \dfrac{1 + \xi_0^2}{2} \sim 0.6478, \\ 0 & \eta > \eta_0; \end{cases} \qquad \phi^0(\eta, \xi) = \begin{cases} 1 & \xi < \eta, \\ 0 & \xi > \eta; \end{cases}$$

where ξ_0 is the unique root lying in the unit interval of the equation $\xi^3 + \xi^2 + \xi - 1 = 0$.

When the uniform distribution is replaced by an arbitrary continuous distribution F the form of the optimal strategies is unchanged. However, if F is taken to be a discrete distribution, $u^0(\xi)$ and $\psi^0(\eta)$ may involve randomization at the critical values ξ_0 and η_0, respectively.

***9.8 "High hand wins."** Each of two players, A and B, receives a hand in (0, 1) according to the uniform distribution and bets one of the amounts $a_1, a_2, \ldots, a_n$ (we assume that $a_i < a_{i+1}$). The pay-off is as follows: If one of the bets is higher than the other, the player who made the high bet wins the pot; if the two bets are equal, the players compare hands and the player with the better hand wins the pot. Since the game is symmetric, the value is zero and the optimal strategies are the same for both players. We shall denote by $\phi_i(\xi)$ the probability that A bets the amount a_i on the hand ξ. These functions are subject to the restrictions

$$\phi_i(\xi) \geq 0, \qquad \sum_{i=1}^{n} \phi_i(\xi) = 1.$$

If the two players choose the strategies $\phi_i(\xi)$, $\psi_i(\eta)$, the expected pay-off to A is

$$K(\phi, \psi) = \sum_{i=1}^{n} \int_0^1 \phi_i(\xi) \left[\sum_{k=1}^{i-1} a_k \int_0^1 \psi_k(\eta)\, d\eta - a_i \sum_{k=i+1}^{n} \int_0^1 \psi_k(\eta)\, d\eta \right.$$
$$\left. + a_i \int_0^1 \psi_i(\eta) L(\xi, \eta)\, d\eta \right] d\xi,$$

where $L(\xi, \eta) = \text{sign}\,(\xi - \eta)$. Recalling that $\sum \psi_i(\eta) = 1$, we may write the pay-off as

$$K(\phi, \psi) = \sum_{i=1}^{n} \int_0^1 \phi_i(\xi)$$
$$\times \left[-a_i + \sum_{k=1}^{i-1} (a_i + a_k) \int_0^1 \psi_k(\eta)\, d\eta + 2a_i \int_0^{\xi} \psi_i(\eta)\, d\eta \right] d\xi.$$

Since the game is symmetric, the optimal strategies for both players are the same. Following the method used in the previous examples, we shall denote an optimal strategy by ϕ^*. The problem is then to find ϕ^* such that the maximum of $K(\phi, \phi^*)$ occurs when $\phi = \phi^*$. In order to simplify the notation, we shall denote the bracketed expression in the last equation by $L_i(\xi)$; thus

$$K(\phi, \phi^*) = \int_0^1 \sum_{i=1}^{n} \phi_i(\xi) L_i(\xi)\, d\xi. \tag{9.8.1}$$

We shall assume that the solution is of the following form: There exist numbers

$$0 < \xi_1 \leq \xi_2 \leq \xi_3 \cdots \leq \xi_{n-1} \leq \xi_n \qquad (\xi_n = 1)$$

such that in the interval $(0, \xi_1)$ all $\phi_i^*(\xi)$ are used with some probabilities $\alpha_i^1 > 0$; in the second interval (ξ_1, ξ_2) only the bets $a_2, \ldots, a_n$ are made with probabilities α_i^2; and so forth for the other intervals.

This proposed strategy was arrived at after much trial and manipulation of $L_i(\xi)$. The reasoning behind it is clear. Clearly, $\phi_i(\xi) = 1$ wherever $L_i(\xi) > L_j(\xi)$ for $j \neq i$. The only flexibility available occurs if $L_i(\xi) = L_{i_0}(\xi) > L_j(\xi)$ for $j \neq i, i_0$. In this case we may determine the values of $\phi_i(\xi)$ and $\phi_{i_0}(\xi)$ arbitrarily, subject only to the condition $\phi_i(\xi) + \phi_{i_0}(\xi) = 1$. Further examination of the coefficients $L_i(\xi)$ ultimately suggests the form of the solution as stated above.

It is necessary that if in some interval two of the functions, say $\phi_i(\xi)$ and $\phi_j(\xi)$, are nonzero, their corresponding coefficients in (9.8.1) are equal, that is, $L_i(\xi) = L_j(\xi)$. We note that the function L_i consists of a term

$$c_i = -a_i + \sum_{k=1}^{i-1} (a_i + a_k) \int_0^1 \phi_k^*(\eta)\, d\eta \qquad (9.8.2)$$

independent of ξ, and a term

$$2a_i \int_0^\xi \phi_i^*(\xi)\, d\xi.$$

We shall determine the unknown constants α_i^r so that the last term of L_i equals the corresponding term of L_j. Then we shall determine ξ_r so that $c_i = c_j$. Since in the interval $(0, \xi_1)$ we assumed that $\phi_i^*(\xi) = \alpha_i^1$, the equation

$$2a_i \int_0^\xi \phi_i^*(\xi)\, d\xi = 2a_j \int_0^\xi \phi_j^*(\xi)\, d\xi$$

implies that $2a_i\alpha_i^1 = 2a_j\alpha_j^1$; in particular, if $j = 1$,

$$\alpha_i^1 = \frac{a_1}{a_i}\, \alpha_1^1.$$

Since $\sum \alpha_i^1 = \sum \phi_i^*(\xi) = 1$, we finally obtain $\alpha_i^1 = 1/a_i b_1$ $(i = 1, \ldots, n)$, where

$$b_1 = \sum_{k=1}^{n} \frac{1}{a_k}\cdot$$

If we continue the argument, we deduce that in the interval (ξ_{i-1}, ξ_i) we have $\alpha_i^r = 1/a_i b_r$ $(i = r, r+1, \ldots, n)$, where

$$b_r = \sum_{k=r}^{n} \frac{1}{a_k}\cdot$$

With these values of α_i^r the numbers c_i defined in (9.8.2) may be written as follows:

$$c_1 = -a_1,$$

$$c_2 = -a_2 + (a_2 + a_1)\frac{\xi_1}{a_1 b_1},$$

$$c_3 = -a_3 + (a_3 + a_1)\frac{\xi_1}{a_1 b_1} + (a_3 + a_2)\frac{\xi_1}{a_2 b_1} + (a_3 + a_2)\frac{(\xi_2 - \xi_1)}{a_2 b_2},$$

and so on. If all these constants are to be equal, $c_k = -a_1$ for all k. The numbers $\xi_1, \ldots, \xi_n$ can now be determined uniquely from these equations. We must point out that the values so obtained produce a solution if and only if $0 \leq \xi_1 \leq \xi_2 \leq \xi_3 \cdots \leq \xi_n = 1$. This condition is not always satisfied; whether it is satisfied or not depends on the values of $a_1, \ldots, a_n$.

For $n = 2$ we always obtain a solution, with

$$\xi_1 = \frac{a_2 - a_1}{a_2 + a_1} a_1 b_1 = \frac{a_2 - a_1}{a_2}.$$

For $n = 3$ the equations for ξ_1 and ξ_2 become

$$\xi_1 = \frac{a_2 - a_1}{a_2 + a_1} a_1 b_1,$$

$$\xi_2 - \xi_1 = \frac{a_1 b_2(a_3 a_1 + a_3 a_2 + a_2 a_1 - 3a_2^2)}{(a_2 + a_1)(a_3 + a_2)}.$$

If $a_1 = 1$, $a_2 = 2$, $a_3 = 4$, then $\xi_1 = 7/12$, $\xi_2 = 8/12$.

The modifications of the solution, in case not all ξ_i lie in the unit interval or in case $\xi_j < \xi_i$ for some $i < j$, will not be given (see Problem 10).

Continuous betting range. The only difference between the following model and the previous one is that instead of having a discrete set of permissible bets $a_1, \ldots, a_n$, we now allow any bet α in the interval $0 < b \leq \alpha \leq a$. The pay-off is the same as in the previous model; the game is clearly symmetric and its value must be zero.

Suppose that the minimizing player (B) chooses the following strategy: If his hand ξ is better than a fixed value ξ_0 (which will be determined later), he will make the highest possible bet a; if his hand is smaller than ξ_0, he will follow a distribution with a continuous density $\phi(\alpha)$ irrespective of which hand $\xi < \xi_0$ he has drawn. In order to compute the expected pay-off to the maximizing player (A), we consider three cases:

Case 1. A draws a hand ξ and chooses the bet $\beta < a$. Since B's distribution is continuous, the probability that the cards will be compared is zero. The expected pay-off depends only on β, since it depends on which is the higher bet. Denoting this expected pay-off to A by $R(\beta)$, we have

$$R(\beta) = \int_b^\beta \alpha\phi(\alpha)\,d\alpha - \beta\int_\beta^a \phi(\alpha)\,d\alpha - \beta[1 - \xi_0], \tag{9.8.3}$$

where we have normalized ϕ so that $\int \phi = \xi_0$.

Case 2. If A draws a hand $\xi \leq \xi_0$ and makes the highest possible bet a, the pay-off is still given by (9.8.3), with $\beta = a$.

Case 3. If A draws a hand $\xi > \xi_0$ and chooses the high bet a, the expected pay-off is

$$r(\xi) = \int_b^a \alpha\phi(\alpha)\,d\alpha + a(\xi - \xi_0) - a(1 - \xi), \tag{9.8.4}$$

since he wins if B bets either $\alpha < a$ or a with a hand smaller than ξ; but he loses a if B bets a with a hand between ξ and 1.

We now choose the function $\phi(\alpha)$ such that $R(\beta)$ will be a constant. If we differentiate $R(\beta)$ with respect to β, we obtain

$$2\beta\phi(\beta) - \int_\beta^a \phi(\alpha)\,d\alpha - (1 - \xi_0) = 0. \tag{9.8.5}$$

Another differentiation yields $2\beta\phi'(\beta) + 3\phi(\beta) = 0$, or

$$\phi(\beta) = k\beta^{-3/2}. \tag{9.8.6}$$

If we substitute this value for $\phi(\beta)$ in (9.8.5) and integrate, we obtain

$$2ka^{-1/2} = 1 - \xi_0. \tag{9.8.7}$$

Also, since the strategy that B is following is a probability distribution, we must have

$$\int_b^a \phi(\alpha)\,d\alpha + 1 - \xi_0 = 1, \tag{9.8.8}$$

which implies that

$$2kb^{-1/2} - 2ka^{-1/2} = \xi_0. \tag{9.8.9}$$

If we add (9.8.7) and (9.8.9), we obtain

$$k = \frac{\sqrt{b}}{2}, \qquad \phi(\alpha) = \frac{\sqrt{b}}{2}\,\alpha^{-3/2},$$

and from (9.8.7)

$$\xi_0 = 1 - \left(\frac{b}{a}\right)^{1/2}.$$

With this choice of $\phi(\alpha)$ and ξ_0 the value of $R(\beta)$ is independent of β and can be evaluated for $\beta = b$; then

$$R(b) = -b\left\{\int_b^a \phi(\alpha)\,d\alpha + 1 - \xi_0\right\} = -b$$

by (9.8.8).

We now consider the expression (9.8.4) for the function $r(\xi)$ that arises in the remaining case. Clearly, $r'(\xi) = 2a$, and for $\xi = \xi_0$

$$r(\xi_0) = \int_b^a \alpha\phi(\alpha)\,d\alpha - a(1 - \xi_0) = R(a).$$

Since we know that $R(a) = -b$, we conclude that

$$r(\xi) = -b + 2a(\xi - \xi_0).$$

The results may be summarized as follows: If the minimizing player follows the strategy described above, the expected value to A is $-b$ for all his hands and bets except when $\xi > \xi_0$ and the bet is a; in this case the expected value is $-b + 2a(\xi - \xi_0)$.

It is clear that if A wishes to maximize his return against the given strategy, he may allow any set of bets for hands $\xi \leq \xi_0$. These do not affect the return. But for hands greater than ξ_0 he must make the highest possible bet, so that he obtains $-b + 2a(\xi - \xi_0) \geq -b$. Clearly, the strategy described above for the minimizing player fulfills this description and is therefore optimal against itself. Since the game is symmetric, this assertion shows that the given strategy is an optimal strategy. The expected return is

$$-b + 2a\int_{\xi_0}^1 (\xi - \xi_0)\,d\xi = -b + a(1 - \xi_0)^2 = 0,$$

as it should be.

9.9 Problems.

1. Consider the first model of Section 9.1, with the same rules governing the play of the game. However, allow the bettor the option of either folding, thus forfeiting his initial bet, or making a bet of size a_i ($i = 1, \ldots, N$), where $1 \leq a_1 < a_2 < \cdots < a_N$. The dealer must always cover the bet. Find the strategy which maximizes the expected value to the bettor.

2. Solve the first model of Section 9.1 when G is a finite discrete distribution (i.e., describe how the solution differs from the solution when G is continuous).

3. Solve the poker model of Section 9.2 where A's hands are dealt according to a continuous distribution $F(\xi)$ and B's hands are dealt according to a continuous distribution $G(\eta)$.

4. Calculate the value of the game for the poker model of Section 9.3.

5. Show that any optimal strategy for Player I in the game of Section 9.3 is necessarily of the form given in (9.3.16).

6. Calculate the value of the game of Section 9.4.

7. Determine the explicit solution of the poker model of Section 9.4 for the bets $a = 2$, $b = 4$.

8. Prove that all solutions of the poker model of Section 9.4 are characterized by the parameters c, d, m_1, m_2, and $\phi_2(\xi)$, where the first four are determined by equations (9.4.7–11) and the last satisfies (9.4.17) and the condition that $\phi_1(\xi) + \phi_2(\xi) = 1$ for ξ in $[c, d]$.

9. Consider the poker model of Section 9.6, but with each player given the added option of folding initially, thus giving up an ante of one unit. Suppose also that $a > b > 1$. If both drop initially, the pay-off is zero to each player; in all other respects the play of the game is the same. Show that the optimal strategy (remember the game is symmetric) has the following form: On an interval $[0, c]$ fold immediately; on $[c, d]$ randomize between a low and high bet; on $[d, 1]$ make a high bet. Determine the parameters of the solution.

10. Determine the complete form of the solution of the game of Section 9.8 for the case of three bet levels ($n = 3$).

11. Consider the poker model of Section 9.6 but with three bet levels: c (low bet), b (medium bet), and a (high bet). The rules of play are the same: i.e., both players make simultaneous bets; if the bets are of unequal size, they must be made equal or else the player making the lower bet forfeits his bet to his opponent; if the bets are equal, or are made equal, the hands are compared and the higher hand wins the pot. Solve the game.

12. You are dealt a real random variable ξ according to the distribution $F(\xi)$. You may either keep this hand, thus ending the game, or request another. You have this same option a total of N times; if you exercise it that many times, you are obliged to keep the Nth hand you are dealt. Determine the strategy that maximizes your expected value. Give an explicit answer if F is the uniform distribution on the unit interval for $N = 2$ and $N = 3$. (*Hint:* Solve recursively.)

13. Consider the same game as in Problem 12, except that you must keep two of the N hands, so that if you have rejected the first $N - 2$ hands offered you, you must accept the last two, and if you have accepted only one of the first $N - 1$ hands, you must accept the Nth hand. Find the procedure which maximizes the expected sum of these two values. Give an explicit answer for $N = 3$ for the case where F is the uniform distribution on the unit interval.

Problems 14–18 are based on the following model. Strategies for Players I and II are n-tuples

$$\boldsymbol{\phi}(\xi) = \langle \phi_1(\xi), \ldots, \phi_n(\xi) \rangle \qquad \text{and} \qquad \boldsymbol{\psi}(\eta) = \langle \psi_1(\eta), \ldots, \psi_n(\eta) \rangle,$$

respectively. The pay-off is

$$K(\boldsymbol{\phi}, \boldsymbol{\psi}) = \int_0^1 \int_0^1 R(\xi, \eta; \boldsymbol{\phi}, \boldsymbol{\psi})\, d\xi\, d\eta,$$

where

$$R(\xi, \eta; \boldsymbol{\phi}, \boldsymbol{\psi}) = \sum_{i=1}^{n}$$

$$\times \{a_i\phi_i(\xi) - b_i\psi_i(\eta) + [c_i + d_i \operatorname{sign}(\xi - \eta)]\phi_i(\xi)\psi_i(\eta)\}.$$

We assume for definiteness that $a_i \geq 0$, $b_i \geq 0$, $d_i > 0$, and we assume also that $0 \leq \phi_i(\xi) \leq 1$ and $0 \leq \psi_i(\eta) \leq 1$.

If no other constraints are imposed, then $K(\boldsymbol{\phi}, \boldsymbol{\psi})$ is a sum of n independent games which may be solved separately. In Problems 17 and 18 we add a further constraint: $\sum \phi_i(\xi) \leq 1$. In this case the kernel K must be dealt with as a whole. Use the techniques of this chapter to solve the following problems.

14. If $|c_i| \geq d_i$ and $-b_i + c_i + d_i \leq 0$, show that the strategies $\psi_i(\eta) \equiv 1$ for all i are uniformly optimal against any $\boldsymbol{\phi}$. Construct an optimal $\boldsymbol{\phi}^*$ in the game $K(\boldsymbol{\phi}, \boldsymbol{\psi})$.

15. Solve the game in which $|c_i| \geq d_i$ and $a_i + c_i - d_i \geq 0$.

16. Consider the case in which $|c_i| < d_i$, $a_i + c_i - d_i < 0$, and $-b_i + c_i + d_i > 0$. Define

$$\alpha_i = 1 - \frac{a_i}{d_i - c_i}, \qquad \beta_i = 1 - \frac{b_i}{d_i + c_i},$$

$$\gamma_i = \frac{d_i + c_i}{2d_i}, \qquad \delta_i = \frac{d_i - c_i}{2d_i}.$$

Show that (a) if $\gamma_i \cdot \beta_i > \alpha_i$, optimal strategies are

$$\phi_i^*(\xi) = \begin{cases} 0 & \xi < \alpha_i, \\ 1 & \xi > \alpha_i, \end{cases} \qquad \psi_i^*(\eta) = \begin{cases} 0 & \eta < \gamma_i \cdot \beta_i, \\ 1 & \eta > \gamma_i \cdot \beta_i; \end{cases}$$

(b) if $\beta_i \geq \alpha_i \geq \gamma_i \cdot \beta_i$, optimal strategies are

$$\phi_i^*(\xi) = \begin{cases} m_i & \xi < \alpha_i, \\ 1 & \xi \geq \alpha_i, \end{cases} \qquad \psi_i^*(\eta) = \begin{cases} 0 & \eta < \alpha_i, \\ 1 & \eta > \alpha_i, \end{cases}$$

where m_i satisfies

$$\alpha_i = \frac{d_i + c_i - b_i}{d_i + c_i + m_i(d_i - c_i)};$$

(c) if $\alpha_i \geq \beta_i \geq \delta_i \cdot \alpha_i$, optimal strategies are

$$\phi_i^*(\xi) = \begin{cases} 0 & \xi < \beta_i, \\ 1 & \xi > \beta_i, \end{cases} \qquad \psi_i^*(\eta) = \begin{cases} n_i & \eta < \beta_i, \\ 1 & \eta \geq \beta_i, \end{cases}$$

where n_i satisfies

$$\beta_i = \frac{d_i - c_i - a_i}{d_i - c_i + n_i(d_i + c_i)};$$

(d) if $\delta_i \cdot \alpha_i > \beta_i$, optimal strategies are

$$\phi_i^*(\xi) = \begin{cases} 0 & \xi < \delta_i \cdot \alpha_i, \\ 1 & \xi > \delta_i \cdot \alpha_i, \end{cases} \qquad \psi_i^*(\eta) = \begin{cases} 0 & \eta < \delta_i \cdot \alpha_i, \\ 1 & \eta > \delta_i \cdot \alpha_i. \end{cases}$$

17. Suppose that we require the conditions listed in the description of the model plus the further condition that $\sum \phi_i(\xi) = 1$. Show that an optimal strategy ψ^* for Player II has the form

$$\psi_i^*(\eta) = \begin{cases} 0 & \eta < \eta_i, \\ 1 & \eta > \eta_i. \end{cases}$$

18. For the same situation as in Problem 17, except that $|c_i| < d_i$ and we now assume that $b_i \leq 0$, show that an optimal strategy ψ^* for Player II is characterized by the critical numbers η_i as above, with $\eta_i = 1$ or

$$\eta_i = 1 - \frac{a_i}{d_i - c_i} + \frac{\lambda^*}{d_i - c_i},$$

where λ^* is independent of i.

19. (Liar's dice.) Player I receives a random hand from the unit interval. He announces one of the values k/n ($k = 0, 1, \ldots, n$) from the unit interval, stating that his hand is superior or equal to this value. Player II says whether he believes or disbelieves Player I's call. If he believes and the call is true, or if he disbelieves and the call is false, he wins one unit. In all other circumstances Player I wins one unit. Find a pair of optimal solutions for the game.

Notes and References to Chapter 9

In many respects infinite game theory may be said to have originated in the study of poker models. The first mathematical analysis of infinite games was carried out by von Neumann in 1926–28 [238]. (The full results of this analysis, which was concerned with parlor games, were not published until 1944.)

General poker studies, approached from a purely probabilistic point of view, have received much attention through the years. Numerous authors have tabulated the various probabilities of acquiring a specified hand in a shuffled deal, of drawing successfully to an inside straight, etc. Borel [38] devoted an extensive treatise to an enumeration of the percentages associated with bridge and poker. Ville was attracted to the analysis of poker as a mathematical game in 1938 [237], and as a corollary to his investigation of poker games he presented the first proof of the min-max theorem for games with continuous pay-off kernels.

Poker is a rich source of challenging game models. Throughout our discussion of the many examples of this chapter, we have tried to stress the common pattern to be found in their solutions, in addition to exhibiting a more or less standard methodology for solving them.

The reader should be reminded of the poker models dealt with earlier (see Section 4.3 and the problems of Chapter 4 in Vol. I). In this connection, we also call attention to Kuhn [156] and Bohnenblust [34].

Three-person poker has been studied by Nash and Shapley [189]. They calculate all possible equilibrium points associated with a version of a three-person game.

§ 9.1. The idealized model of blackjack and Red Dog discussed in this section is due to Bellman. Of some related interest is the semirigorous empirical solution of the real game of blackjack presented by Baldwin, Contey, Maisel, and McDermott [17].

§ 9.2. The simple poker model analyzed here, involving one size of bet and one round of betting, was formulated originally by Bellman [20]. The method of this section is new and typifies a special instance of the general fixed-point technique.

§ 9.3. The model discussed in this section was proposed by Karlin and Restrepo [138]. The relevant features of the pay-off kernel were abstracted in 1958 by Restrepo (unpublished). By the same techniques, employed in a more complicated way, he was able to solve a considerably larger class of analogous games. His findings should ultimately be useful in solving new types of poker models and decision problems. (See Problems 14–18 of this chapter.)

§ 9.4. The poker model of this section was proposed by Bellman [20], who partially solved it. A complete solution, based on the fixed-point method of solving general games, is presented in Karlin and Restrepo [138]. Independently, Goldman and Stone (unpublished) fully treated the same model.

§ 9.5. The first elaborate poker model of k rounds of betting was conceived and solved by Karlin and Restrepo [138].

§ 9.6. A poker model with simultaneous moves is treated in the classic von Neumann and Morgenstern book on game theory ([242], Chap. 5). Von Neumann's analysis is based on a discrete version of the poker model, from which he solves a continuous form of the problem by proceeding to a limit. A slight generalization of this model is discussed by Alan Goldman and Jeremy Stone (see Problem 9). The techniques of this section conform to the fixed-point methodology.

§ **9.7.** The formulation of the Le Her game as an example of game theory is due to Dresher [69], who analyzed the original discrete version of the problem. The continuous Le Her game was investigated in [138].

§ **9.8.** The two models of this section are due to Gillies, Mayberry, and von Neumann [105].

§ **9.9.** Problem 9 is due to Goldman and Stone. The assertions of Problems 14–18 were first established by Restrepo [206].

SOLUTIONS TO PROBLEMS

CHAPTER 1

Problem 2. A strategy for Player I consists of two components, the first giving the probability of betting if he draws Lo, and the second the probability of betting if he draws Hi. A strategy for Player II is given by a single component, the probability of calling if Player I bets. The optimal strategies are

$$\left\{\frac{b-a}{b+a}, 1\right\} \quad \text{and} \quad \left\{\frac{2a}{b+a}\right\},$$

respectively. The value is $a(b-a)/(b+a)$.

Problem 4. Let $\mathbf{A} = \|a_{ij}\|$ $(i, j = 1, 2, \ldots, 9)$ denote the pay-off matrix. Then

$$a_{ij} = \begin{cases} -1 & \text{if } j - i \geq 2, \\ +1 & \text{if } i - j \geq 2, \\ +2 & \text{if } j - i = 1, \\ -2 & \text{if } i - j = 1, \\ 0 & \text{if } i - j = 0. \end{cases}$$

Problem 6. The optimal strategies are $\langle \frac{5}{7}, \frac{2}{7} \rangle$ and $\langle \frac{5}{7}, \frac{2}{7}, 0, 0 \rangle$; the value is $\frac{13}{7}$.

Problem 8. In both $\mathbf{A}_1$ and $\mathbf{A}_2$ the element 1 is a saddle point. Hence $v(\mathbf{A}_1) = v(\mathbf{A}_2) = 1$ and $v(\mathbf{A}_1) + v(\mathbf{A}_2) = 2$. Setting $x = 3$ yields $v(\mathbf{A}_1 + \mathbf{A}_2) = 3 > 2$, and setting $x = -1$ yields $v(\mathbf{A}_1 + \mathbf{A}_2) = 1 < 2$.

Problem 9. Consider the game

$$\begin{Vmatrix} 2 & 0 \\ 0 & 2 \end{Vmatrix}.$$

The value is 1. However, note that $\mathbf{x}^* = \langle \frac{1}{2}, \frac{1}{2} \rangle$, $\mathbf{y}^* = \langle 1, 0 \rangle$ yields $(\mathbf{x}^*, \mathbf{A}\mathbf{y}^*) = 1$ and yet $\mathbf{y}^*$ is not optimal.

Problem 10. Let $\mathbf{u} = \langle u_1, \ldots, u_n \rangle$ be an optimal strategy for Player I and $\mathbf{w} = \langle w_1, \ldots, w_m \rangle$ be an optimal strategy for Player II. Then

$$\sum_{i=1}^{n} u_i a_{ij} \geq v \geq \sum_{j=1}^{m} a_{ij} w_j$$

for all i and j. It follows that if $v \geq 0$ the first alternative holds, while if $v < 0$ the second alternative holds.

Problem 13. In the game with matrix

$$\mathbf{C} = \begin{Vmatrix} a & b \\ 0 & d \end{Vmatrix},$$

the element 0 is a saddle point and $\xi = 0$, $\eta = 1$ are unique optimal strategies. Thus

$$\mathbf{x} = \langle 0, 0, \ldots, 0, \overset{*}{x}_{n_1+1}, \ldots, \overset{*}{x}_{n_1+n_2} \rangle,$$

$$\mathbf{y} = \langle \overset{0}{y}_1, \ldots, \overset{0}{y}_{m_1}, 0, \ldots, 0 \rangle.$$

If Player II uses any of the first m_1 columns, Player I obtains zero; if Player II uses any of the last m_2 columns, Player I obtains at least $d > 0$. Thus, by taking combinations, $C(\mathbf{x}, \tilde{\mathbf{y}}) \geq 0$ for any $\tilde{\mathbf{y}}$. Similarly $C(\tilde{\mathbf{x}}, \mathbf{y}) \leq 0$ for any $\tilde{\mathbf{x}}$, and $\{\mathbf{x}, \mathbf{y}\}$ is optimal.

Problem 15. Consider the game with matrix $a_{ij} = f(\mathbf{x}_i, \mathbf{y}_j)$. If $\boldsymbol{\xi} = \langle \xi_1, \ldots, \xi_n \rangle$ and $\boldsymbol{\eta} = \langle \eta_1, \ldots, \eta_m \rangle$ are optimal strategies, then

$$\mathbf{x}^0 = \sum_{i=1}^{n} \xi_i \mathbf{x}_i \quad \text{and} \quad \mathbf{y}^0 = \sum_{j=1}^{m} \eta_j \mathbf{y}_j$$

satisfy the inequalities.

Problem 16. If there exists no $\mathbf{x}$ in X for which $f(\mathbf{x}, \mathbf{y}) > \alpha$ for all $\mathbf{y}$ in Y, then the set $B_x = \{\mathbf{y} | f(\mathbf{x}, \mathbf{y}) \leq \alpha\}$ is nonvoid, closed because f is lower-semicontinuous, and convex because f is convex in $\mathbf{y}$. For any finite selection $\mathbf{x}_1, \mathbf{x}_2, \ldots, \mathbf{x}_n$ the set $B_{x_1, x_2, \ldots, x_n} = \{\mathbf{y} | f(\mathbf{x}_i, \mathbf{y}) \leq \alpha \text{ all } i\}$ is also nonvoid, closed, and convex. Otherwise the image of Y in E^n defined by the mapping $\mathbf{y} \to \{f(\mathbf{x}_i, \mathbf{y}) - \alpha\}$ has a convex closure which never touches the negative orthant. A strict separating plane exists with components of the normal ξ_i ($i = 1, 2, \ldots, n$) satisfying $\xi_i \geq 0$ and not all zero. It follows that B_{x_0} is empty where

$$\mathbf{x}_0 = \sum_{i=1}^{n} \xi_i \mathbf{x}_i,$$

a contradiction. Hence the intersection $\bigcap_\gamma B_{x_1, x_2, \ldots, x_n}$ taken over all finite collections of $\mathbf{x}_i$ contains an element $\mathbf{y}_0$, satisfying $f(\mathbf{x}, \mathbf{y}_0) \leq \alpha$ for all $\mathbf{x}$ in X.

Problem 17. Let $\mathbf{A} = \|a_{ij}\|_{n \times m}$, $b_j = \max_{1 \leq i \leq n} a_{ij}$, and i_j be a value of i such that $a_{i_j j} = b_j$. We proceed by induction on m. If $m = 2$, consider the 2×2 submatrix

$$\begin{Vmatrix} a_{i_1 1} & a_{i_1 2} \\ a_{i_2 1} & a_{i_2 2} \end{Vmatrix}.$$

(If $i_1 = i_2$, use this row and any other; if $n = 1$, the theorem is trivial.) If $a_{i_1 1}$

is the saddle point of this 2×2 matrix, then $a_{i_11} \leq a_{i_12}$ and $a_{i_11} = b_1 \geq a_{i1}$ for all i, and hence a_{i_11} is a saddle point of the original $n \times 2$ matrix. If a_{i_21} is the saddle point of the 2×2 matrix, then $a_{i_21} \leq a_{i_22}$ and $a_{i_21} \geq a_{i_11} = b_1 \geq a_{i1}$ for all i, and hence a_{i_21} is a saddle point of the original $n \times 2$ matrix. Similar arguments hold for a_{i_12} and a_{i_22}.

Suppose now that the theorem is true for $m - 1$, and let $a_{i_0j_0}$ be a saddle point for the $n \times (m - 1)$ submatrix consisting of the first $m - 1$ columns of our $n \times m$ matrix. If $a_{i_0j_0} \leq a_{i_0m}$, then $a_{i_0j_0}$ will be a saddle point of the full $n \times m$ matrix. Hence we consider $a_{i_0j_0} > a_{i_0m}$. If $a_{i_0m} = a_{i_mm}$, then a_{i_0m} is a saddle point of the $n \times m$ matrix. Hence we may also consider only the case $a_{i_0m} < a_{i_mm}$.

Case I: $a_{i_mm} \leq a_{i_mj_0}$. Let k be an arbitrary integer between 1 and $n - 1$. Consider the 2×2 submatrix

$$\begin{Vmatrix} a_{i_mk} & a_{i_mm} \\ a_{i_0k} & a_{i_0m} \end{Vmatrix}.$$

Now $a_{i_0m} < a_{i_0j_0} \leq a_{i_0k}$, and thus a_{i_0k} cannot be a saddle point of this 2×2 submatrix. Also since $a_{i_0m} < a_{i_mm}$, it follows that a_{i_0m} cannot be a saddle point. If a_{i_mk} is a saddle point, then $a_{i_mk} \geq a_{i_0k} \geq a_{i_0j_0} \geq a_{i_mj_0} \geq a_{i_mm}$. On the other hand, if a_{i_mm} is a saddle point, we also have $a_{i_mk} \geq a_{i_mm}$. Thus a_{i_mm} is a saddle point of the $n \times m$ matrix.

Case II: $a_{i_mm} > a_{i_mj_0}$. We proceed as above; in case a_{i_mk} is the saddle point of the 2×2 submatrix, we have $a_{i_mk} \geq a_{i_mj_0}$ as above. If a_{i_mm} is the saddle point, then $a_{i_mk} \geq a_{i_mm} > a_{i_mj_0}$. Thus $a_{i_mj_0} \leq a_{i_mk}$ $(k = 1, 2, \ldots, m)$. Now consider the 2×2 submatrix

$$\begin{Vmatrix} a_{i_mj_0} & a_{i_mm} \\ a_{i_0j_0} & a_{i_0m} \end{Vmatrix}.$$

Clearly $a_{i_0m} < a_{i_0j_0}$, hence $a_{i_0j_0}$ is not a saddle point of this 2×2; $a_{i_0m} < a_{i_mm}$, hence a_{i_0m} is not a saddle point; and $a_{i_mm} > a_{i_mj_0}$, hence a_{i_mm} is not a saddle point. Thus $a_{i_mj_0}$ must be the saddle point of this 2×2 submatrix, and hence $a_{i_mj_0} \geq a_{i_0j_0} \geq a_{ij_0}$ $(i = 1, 2, \ldots, n)$. Thus $a_{i_mj_0}$ is a saddle point of the $n \times m$ matrix.

Problem 18. The example

$$\begin{Vmatrix} 3 & 0 & 4 \\ 2 & 1 & 0 \\ 0 & 1 & 3 \end{Vmatrix}$$

shows that the result is not correct if the strictness assumption is removed.

Problem 19. See Wolfe [255].

CHAPTER 2

Problem 1. Let

$$m_n = \min_{0 \le \eta \le 1} \int K(\xi, \eta) d\left(\sum_{i=0}^{n} \lambda_i^n x_{i/n}(\xi)\right),$$

and define $\epsilon_n = v_n - m_n$. Given $\epsilon > 0$, let δ be such that

$$(\xi - \xi')^2 + (\eta - \eta')^2 < \delta^2 \qquad \text{implies} \qquad |K(\xi, \eta) - K(\xi', \eta')| < \epsilon,$$

which is possible owing to the uniform continuity of K. Then for any

$$n > \frac{1}{\delta} \qquad \text{and} \qquad \frac{j-1}{n} \le \eta \le \frac{j+1}{n},$$

$|K(\xi, \eta) - K(\xi, j/n)| < \epsilon$ implies

$$\left| \int [K(\xi, \eta) - K(\xi, j/n)] d\left(\sum_{i=0}^{n} \lambda_i^n x_{i/n}(\xi)\right) \right| < \epsilon.$$

Hence

$$m_n > \min_j \int K(\xi, j/n) d\left(\sum_{i=0}^{n} \lambda_i^n x_{i/n}(\xi)\right) - \epsilon = v_n - \epsilon,$$

and thus $\epsilon_n < \epsilon$.

Problem 2. First, sup inf $K \le$ inf sup K by the usual argument. Next, observe that

$$x^n = \sum_{i=0}^{n} \lambda_i^n x_{i/n}(\xi)$$

and that

$$\int K(\xi, \eta)\, dx^n(\xi) \ge v_n - \epsilon_n$$

for all η implies sup inf $K \ge v_n - \epsilon_n$. Similarly, inf sup $K \le v_n + \epsilon_n$. These inequalities combined yield inf sup $K \le$ sup inf $K + 2\epsilon$, with ϵ arbitrary.

Problem 4. By direct calculation,

$$\int_0^1 K(\xi, \eta)\, d\xi = +1 \qquad \text{and} \qquad \int_0^1 K(\xi, \eta)\, d\eta = -1,$$

from which the result follows.

Problem 5. Given x, choose ξ so that $x(1-) - x(\xi-) < \epsilon$; let y^x depending on x be determined as y_ξ. Then

$$\begin{aligned} K(x, y^x) &= (-1)x(\xi-) + x(1-) - x(\xi) - [1 - x(1-)] \\ &= -1 + x(1-) - x(\xi) + x(1-) - x(\xi-) \\ &\le -1 + 2\epsilon. \end{aligned}$$

Thus $\sup_x \inf_y K(x, y) \leq -1$, and $K \geq -1$ everywhere implies $\sup \inf K = -1$. By symmetry, $\inf \sup K = +1$.

Problem 6. See [223].

Problem 8. Consider $\int K^*(\xi, \eta)\, dx^0(\xi)$; since $L_{ij}(i/n, j/n) = 0$, we have

$$\int K^*(\xi, \eta)\, dx^0(\xi) = \int K(\xi, \eta)\, dx^0(\xi) \geq v$$

for $\eta = j/n$ $(j = 0, 1, \ldots, n)$. On the other hand, $L_{ij}(i/n, \eta) > 0$ for $\eta \neq j/n$ leads to

$$\int K^*(\xi, \eta)\, dx^0(\xi) > \int K(\xi, \eta)\, dx^0(\xi) \geq v$$

for $\eta \neq j/n$. Hence no optimal strategy for Player II can have support outside of the points j/n $(j = 0, 1, \ldots, n)$ by Lemma 2.2.1. It remains only to apply the uniqueness hypothesis concerning the optimal strategy μ^0 of the matrix game.

Problem 10. The hypothesis means that there exists a constant k such that

$$\int_0^1 \frac{(1+\xi)^a(1+\eta)^b}{(1+\xi\eta)^c} K \frac{\xi^m(1-\xi)^n}{(1+\xi)^p}\, d\xi = k,$$

or

$$K\int_0^1 (1+\xi)^{a-p}\xi^m(1-\xi)^n(1+\xi\eta)^{-c}\, d\xi = k(1+\eta)^{-b},$$

$$K\int_0^1 (1+\xi)^{a-p}\xi^m(1-\xi)^n \sum_{\alpha=0}^{\infty} (-1)^\alpha \binom{\alpha+c-1}{\alpha} \xi^\alpha\eta^\alpha\, d\xi$$

$$= k\sum_{\alpha=0}^{\infty} (-1)^\alpha \binom{\alpha+b-1}{\alpha} \eta^\alpha.$$

Equating coefficients of η^α yields

$$K\int_0^1 (1+\xi)^{a-p}\xi^{m+\alpha}(1-\xi)^n \binom{\alpha+c-1}{\alpha}\, d\xi$$

$$= k\binom{\alpha+b-1}{\alpha} \qquad (\alpha = 0, 1, 2, \ldots).$$

Letting $p = a$ and evaluating the integral, we have

$$K\frac{\Gamma(m+\alpha+1)\Gamma(n+1)}{\Gamma(m+n+\alpha+2)}\frac{\Gamma(\alpha+c)}{\Gamma(c)} = k\frac{\Gamma(\alpha+b)}{\Gamma(b)} \qquad (\alpha = 1, 2, \ldots). \quad (*)$$

If we determine $m = b - 1$ and $m + n + 2 = c$ with proper choice of the constant k, then the relation (*) is correct for every integer α. Thus $m = b - 1$ and $n = c - b - 1$.

Now K is determined by

$$K\int \frac{\xi^m(1-\xi)^n}{(1+\xi)^p}\,d\xi = 1,$$

which is possible since $c > b > 0$. A similar computation shows that for Player II, $m = a - 1$, $n = c - a - 1$, and $p = b$.

Problem 11.

(a) $x^0 = y^0 = \frac{1}{3}I_0 + \frac{1}{3}I_{1/2} + \frac{1}{3}I_1; \qquad v = \dfrac{2+\lambda}{3}.$

(b) $x^0 = \frac{1}{2}I_0 + \frac{1}{2}I_1; y^0 = \frac{1}{2}I_{1/2} + \frac{1}{2}I_1; \qquad v = \dfrac{\lambda+1}{2} \quad \text{if } \lambda < 1;$

$x^0 = \frac{1}{2}I_{1/2} + \frac{1}{2}I_1; y^0 = \frac{1}{2}I_0 + \frac{1}{2}I_1; \qquad v = \dfrac{\lambda+1}{2} \quad \text{if } \lambda > 1.$

Problem 12. Consider the relation

$$\begin{aligned}\int_0^{B+1} K(\xi,\eta)\,dx(\xi) &= (1-\theta)x(\eta) + x(\eta+1) - x(\eta)\\ &= x(\eta+1) - \theta x(\eta) = c \qquad (0 \leq \eta \leq B),\end{aligned}$$

where c is a constant to be determined later. Then

$$\begin{aligned}x(1) = c + \theta x(0);\ x(2) = c\cdot(1+\theta) + \theta^2 x(0);\ \ldots;\\ x(n+1) = c\cdot(1+\theta+\theta^2+\cdots+\theta^n) + \theta^{n+1}x(0).\end{aligned}$$

Since Player I desires c to be a maximum, he should take

$$x(n+1) = x(B+1) = 1 \qquad \text{and} \qquad x(0) = 0.$$

Thus

$$x(1) = c;\ x(2) = c\cdot(1+\theta);\ \ldots;\ x(n+1) = c\cdot(1+\theta+\cdots+\theta^n) = 1.$$

But x is constant on $[n+1, B+1]$, and therefore it must be constant on $[n-i+1, B-i+1]$, where $i = 2, \ldots, n+1$. No conditions other than these are implied by the functional equation; i.e., we may take any form for x on $[B-n, 1]$ so long as it is a distribution function and $x(1) = c$. Consequently we take the simplest form, which consists of a jump of c at 1. Then

$$x(\xi) = \begin{cases} 0 & \text{if } \xi < 1,\\ c(1+\theta+\cdots+\theta^{i-1}) & \text{if } i \leq \xi < i+1 \quad (i = 1, 2, \ldots, n+1),\end{cases}$$

where

$$c = \frac{1}{1+\theta+\cdots+\theta^n} = \frac{1-\theta}{1-\theta^{n+1}}\,.$$

Similarly, it is easy to show that

$$y(\eta) = \begin{cases} 0 & \text{for} \quad \eta < 0, \\ c(\theta^n + \cdots + \theta^{n-i}) & \text{for} \quad i \leq \eta < i + 1 \quad (i = 0, 1, \ldots, n) \end{cases}$$

is optimal for Player II.

CHAPTER 3

Problem 1. $\xi^0 = \langle \frac{1}{7}, 0, \frac{6}{7} \rangle; \quad \eta^0 = \langle 0, \frac{5}{7}, \frac{2}{7} \rangle; \quad v = \frac{2}{7}.$

Problem 3. Anything is optimal for Player I; $y^0 = I_0$; $v = 0$.

Problem 5. (a) $x^0 = y^0 = I_0$; $v = 0$.
(b) $x^0 = I_1$; $y^0 = \alpha I_0 + (1 - \alpha) I_1$, where $0 \leq \alpha \leq 1$; $v = 0$.

Problem 6. Let $y^0 = I_1$. Then the fact that $\int K(\xi, \eta)\, dy^0(\eta) = \xi^3 - 1 < 1$ except for $\xi = 1$ means that 1 is the only point in the spectrum of Player I's optimal strategies. Hence $x^0 = I_1$. On the other hand, the fact that

$$\int K(\xi, \eta)\, dx^0(\xi) = \eta - \eta^2 > 0$$

except for $\eta = 0, 1$ implies that 0 and 1 are the only possible points in the spectrum of Player II's optimal strategies.

Problem 8. Any strategy for Player I whose first two moments r_1 and r_2 satisfy $3r_2 + 2r_1 = 2$ is optimal. Similarly, any strategy for Player II whose first two moments s_1 and s_2 satisfy $4s_2 + 3s_1 = 2$ is optimal. In particular, $x^0 = \frac{3}{5} I_0 + \frac{2}{5} I_1$ and $y^0 = \frac{5}{7} I_0 + \frac{2}{7} I_1$ are optimal.

Problem 9. See [70].

Problem 11. In the perturbed game $x^0 = x_0 = I_0$ is the unique optimal strategy for Player I and $v = 0$, but $y^0 = y_0$ and $y^0 = y_\epsilon$ are both optimal for Player II.

Problem 12. By the hypothesis, $\mathbf{r}^0$ must be the unique intersection of the sets defined by the equations $f_j(\mathbf{r}) = 0$ and $\mathbf{s}^0$ the unique intersection of the sets defined by the equations $g_i(\mathbf{s}) = 0$. Moreover, sufficiently small perturbations preserve the unique intersection of $f_j(\mathbf{r})$ and $g_i(\mathbf{s})$ on the interior of R and S, respectively, and hence give unique equalizing optimal strategies. But since both strategies are interior, all optimal strategies for both players must be equalizers.

Problem 14. $P^* = P$.

Problem 15. See [70], p. 80.

Problem 16. See [70], p. 81.

Problem 17. Consider

$$A = \begin{vmatrix} v+1 & -\dfrac{1}{ns_1^0} & \cdots & -\dfrac{1}{ns_n^0} \\ -\dfrac{1}{nr_1^0} & \dfrac{1}{nr_1^0 s_1^0} & & 0 \\ \vdots & & \ddots & \\ -\dfrac{1}{nr_n^0} & 0 & & \dfrac{1}{nr_n^0 s_n^0} \end{vmatrix}.$$

Clearly, $\mathbf{As}^0 = \langle v, 0, \ldots, 0\rangle$ and $\mathbf{r}^0\mathbf{A} = \langle v, 0, \ldots, 0\rangle$, and the rank of $\mathbf{A}$ is n. The fact that $\mathbf{r}^0$ is interior implies that any optimal strategy for Player II must be an equalizer. Also, if $\mathbf{s} = \langle 1, s_1, \ldots, s_n\rangle$ is optimal for Player II, then for $1 \leq i \leq n$,

$$-\frac{1}{nr_i^0} + \frac{1}{nr_i^0 s_i^0} s_i = 0.$$

It follows that $-1 + (s_i/s_i^0) = 0$, hence $s_i = s_i^0$ and Player II's optimal strategy is unique. An analogous argument applies for Player I.

Problem 18. Analogous to the case of powers, i.e., when $h_r(t) = t^r$.

CHAPTER 4

Problem 1. $x^0 = I_1$; $y^0 = I_0$; $v = K(1, 1)$.

Problem 3. $x^0 = \frac{1}{2}I_0 + \frac{1}{2}I_1$; $y^0 = I_{1/2}$; $v = \frac{1}{4}$.

Problem 5. (a)

$$K(\xi, \eta_0) = K(\xi \cdot 1 + (1 - \xi) \cdot 0, \eta_0) \leq \xi K(1, \eta_0) + (1 - \xi)K(0, \eta_0) \leq K(0, \eta_0) \leq K(0, \eta).$$

Hence $\{0, \eta_0\}$ is a saddle point.

(b) Similar to (a)

(c) Let $L_0(\eta)$ be a supporting line to $K(0, \eta)$ at η^* and $L_1(\eta)$ a supporting line to $K(1, \eta)$ at η^*. Then since one of these is increasing and the other decreasing, there exists a value λ $(0 < \lambda < 1)$ such that

$$\lambda L_0(\eta) + (1 - \lambda)L_1(\eta) = K(0, \eta^*) \qquad \text{for all} \quad \eta.$$

Set $x^0 = \lambda I_0 + (1 - \lambda)I_1$; then we have

$$K(x^0, \eta) = \lambda K(0, \eta) + (1 - \lambda)K(1, \eta) \geq \lambda L_0(\eta) + (1 - \lambda)L_1(\eta) = K(0, \eta^*).$$

Also, $K(\xi, \eta^*) \leq \xi K(1, \eta^*) + (1 - \xi)K(0, \eta^*) = K(0, \eta^*)$. By integration, we obtain $K(x^0, y) \geq K(0, \eta^*) \geq K(x, \eta^*)$, and hence x^0 and y_{η^*} are optimal with $v = K(0, \eta^*)$.

Problem 7. $x^0 = I_0$ or I_1 (or a combination of these); $y^0 = I_0$ or I_1; $v = 0$.

Problem 10. Apply Theorem 4.3.1 for both players.

Problem 11. Suppose the contrary. Let $h(\eta) = \int K(\xi, \eta)\, dx^*(\xi)$, where x^* is optimal for Player I. Then $h^{(3)}(\eta) > 0$, and $h(\eta) \geq v$ with equality at η_0 and 1. Thus $h^{(2)}(\eta_0) > 0$, and h attains a relative maximum somewhere in $(\eta_0, 1)$; hence $h^{(2)}(\eta_1) < 0$. The mean value theorem implies that $h^{(3)}(\eta) < 0$ for some η ($\eta_0 \leq \eta \leq \eta_1$), a contradiction.

Problem 12. In the notation of Section 4.3, $\eta_1^0 = 5$, $\eta_2^0 = 3\frac{3}{4}$, $\eta_3^0 = 1\frac{1}{4}$; $\lambda_1^0 = \frac{1}{4}$, $\lambda_2^0 = \frac{5}{16}$, $\lambda_3^0 = \frac{7}{16}$.

Problem 13. The pay-off is convex in $\boldsymbol{\eta}$ for every $\boldsymbol{\xi}$. Therefore the theory of Section 4.4 applies: Player II has an optimal pure strategy, and Player I has an optimal strategy with $n + 1$ points in its spectrum. In fact, the origin is optimal for Player II, and any symmetric distribution on the boundary is optimal for Player I—e.g., two points suffice.

Problem 14. Similar to Problem 13.

Problem 15. If we assume

$$\sum_{i=1}^{m} \lambda_i^0 \mathbf{a}_i = 0,$$

then

$$1 = \sum_{i=1}^{m} \lambda_i^0(\mathbf{a}_i, \mathbf{a}_i - \boldsymbol{\eta}) \leq \sum_{i=1}^{m} \lambda_i^0 |\mathbf{a}_i|\, |\mathbf{a}_i - \boldsymbol{\eta}| = \sum \lambda_i^0 |\mathbf{a}_i - \boldsymbol{\eta}|.$$

Hence $K(\boldsymbol{\lambda}^0, \boldsymbol{\eta}) \geq 1 = K(\boldsymbol{\lambda}^0, \mathbf{0}) = K(\boldsymbol{\lambda}, \mathbf{0})$ for all $\boldsymbol{\eta}$ and $\boldsymbol{\lambda}$.

On the other hand, $\boldsymbol{\lambda}^*$ optimal implies $K(\boldsymbol{\lambda}^*, \boldsymbol{\eta}) = \sum \lambda_i^* |\mathbf{a}_i - \boldsymbol{\eta}| \geq 1$ for all $\boldsymbol{\eta}$. Thus

$$1 \leq \left(\sum_{i=1}^{m} \sqrt{\lambda_i^*} \sqrt{\lambda_i^*}\, |\mathbf{a}_i - \boldsymbol{\eta}| \right)^2 \leq \sum_{i=1}^{m} \lambda_i^* \sum_{i=1}^{m} \lambda_i^* |\mathbf{a}_i - \boldsymbol{\eta}|^2$$
$$= 1 - 2(\mathbf{z}, \boldsymbol{\eta}) + |\boldsymbol{\eta}|^2,$$

where

$$\mathbf{z} = \sum_{i=1}^{m} \lambda_i^* \mathbf{a}_i.$$

Now consider $\boldsymbol{\eta} = \mathbf{z}/2$. Then $-|\mathbf{z}|^2 + \frac{1}{4}|\mathbf{z}|^2 \geq 0$, requires $\mathbf{z} = \mathbf{0}$.

Problem 16. By Theorem 4.4.1 there exist numbers ξ_i ($i = 1, \ldots, n + 1$) and supporting planes $L_i(\boldsymbol{\eta})$ to $K(\xi_i, \boldsymbol{\eta})$ at the point $\boldsymbol{\eta}_0$ ($\boldsymbol{\eta}_0$ = the pure optimal strategy for Player II) such that

$$\sum_{i=1}^{n+1} \lambda_i K(\xi_i, \boldsymbol{\eta}) \geq \sum_{i=1}^{n+1} \lambda_i L_i(\boldsymbol{\eta}) \geq v \qquad (*)$$

for all $\boldsymbol{\eta}$ in B, with $\lambda_i \geq 0$ and

$$\sum_{i=1}^{n+1} \lambda_i = 1.$$

We now prove that a convex combination having this property exists which uses at most n distinct ξ_i's. To this end, we enlarge B as follows. For each extreme point of the set of supporting planes to B at $\boldsymbol{\eta}^0$ we choose the closed half-space that contains B. There are at most n extreme points, and the intersection of the corresponding half-spaces will be a closed convex set C containing B. The set C has the property that $\boldsymbol{\eta}$ in C implies $f(\boldsymbol{\eta}) = \sum \lambda_i L_i(\boldsymbol{\eta}) \geq v$. Indeed, if $\boldsymbol{\eta}$ is an interior point of C, the ray joining $\boldsymbol{\eta}$ to $\boldsymbol{\eta}_0$ must contain some $\boldsymbol{\eta}_1 \neq \boldsymbol{\eta}_0$, where $\boldsymbol{\eta}_1 \in B$. Then $\boldsymbol{\eta}_1 = \alpha\boldsymbol{\eta}_0 + (1 - \alpha)\boldsymbol{\eta}$ and $f(\boldsymbol{\eta}_1) = \alpha f(\boldsymbol{\eta}_0) + (1 - \alpha)f(\boldsymbol{\eta})$. Since $f(\boldsymbol{\eta}_0) = v$ and $f(\boldsymbol{\eta}_1) \geq v$, we obtain $f(\boldsymbol{\eta}) \geq v$. If $\boldsymbol{\eta}$ is on the boundary of C, then the same holds by the continuity of f. Finally, since the number k of extreme planes is at most n, we can find a simplex S with exactly $n + 1$ corners $\mathbf{d}_1, \ldots, \mathbf{d}_{n+1}$ such that $C \supset S \supset B$. The point $\boldsymbol{\eta}_0$ is still a boundary point of S.

Next, we define an auxiliary matrix game by setting $a_{ij} = L_i(\mathbf{d}_j)$. The strategy spaces will be the usual simplices of vectors in E^{n+1}. If the maximizing player chooses the $\boldsymbol{\lambda}$ vector of (*), he obtains $\sum \lambda_i L_i(\mathbf{d}_j) \geq v$. On the other hand, the original strategy $\boldsymbol{\eta}_0$ can be expressed as a convex representation $\alpha_1 \mathbf{d}_1 + \cdots + \alpha_{n+1} \mathbf{d}_{n+1}$. If the minimizing player in the matrix game uses $\boldsymbol{\alpha} = \langle \alpha_1, \ldots, \alpha_{n+1} \rangle$, the return to Player I is $L_i(\boldsymbol{\eta}_0) = v$ for each i. Therefore $\boldsymbol{\lambda}$ and $\boldsymbol{\alpha}$ are optimal strategies and v is the value. Since $\boldsymbol{\eta}_0$ is on the boundary of S, the optimal $\boldsymbol{\alpha}$ involves at most n nonzero components. By Problem 20 of Chapter 3 in Vol I, the matrix game has an optimal strategy $\boldsymbol{\lambda}^*$ with at most n nonzero components. For simplicity of notation we assume that these correspond to the first n components. It is asserted that

$$x^* = \sum_{i=1}^{n} \lambda_i^* I_{\xi_i}$$

is optimal for Player I in the convex game. Explicitly,

$$\int K(\xi, \boldsymbol{\eta})\, dx^*(\xi) = \sum_{i=1}^{n} \lambda_i^* K(\xi_i, \boldsymbol{\eta}) \geq \sum_{i=1}^{n} \lambda_i^* L_i(\boldsymbol{\eta})$$

$$= \sum_{i=1}^{n} \sum_{j=1}^{n+1} \lambda_i^* \beta_j L_i(\mathbf{d}_j) \geq v,$$

where

$$\boldsymbol{\eta} = \sum_{j=1}^{n+1} \beta_j \mathbf{d}_j, \qquad \beta_j \geq 0, \qquad \text{and} \qquad \sum_{j=1}^{n+1} \beta_j = 1.$$

Problem 17. There can be at most a one-dimensional family of supporting planes at any point, and hence at most two extreme points.

Problem 18. See [162 Chap. 7].

CHAPTER 5

Problem 2. $n = 3, a = 0.568, \beta = 0.124$; $n = 4, a = 0.626, \beta = 0.160$.

Problem 4. $x^0 = y^0$ has density

$$f(\xi) = \begin{cases} 0 & 0 \leq \xi < \frac{1}{4}, \\ \frac{1}{2}\xi^{-3/2} & \frac{1}{4} \leq \xi \leq 1; \end{cases}$$

the value is zero.

Problem 5. The pay-off function has the form

$$K(\xi, \eta) = \begin{cases} L(\xi, \eta) = p_1p_2\xi\eta - p_2\eta + p_1\xi & \xi < \eta, \\ \phi(\xi) \quad = (p_1 - p_2)\xi & \xi = \eta, \\ M(\xi, \eta) = -p_1p_2\xi\eta - p_2\eta + p_1\xi & \xi > \eta. \end{cases}$$

Since $L(0, 0) = M(0, 0)$, this kernel falls into Group 2. Here

$$T(\xi, t) = \begin{cases} \dfrac{1 - p_1\xi}{2p_1t^2} & a \leq \xi < t \leq 1, \\ \dfrac{1 + p_1\xi}{2p_1t^2} & a \leq t \leq \xi \leq 1. \end{cases}$$

Then $(I - T_a)f = 0$ becomes

$$2p_1t^2f(t) - 1 + p_1\int_a^t \xi f(\xi)\, d\xi - p_1\int_t^1 \xi f(\xi)\, d\xi = 0,$$

which implies $f(t) = kt^{-3}$, where k and a must satisfy

$$\frac{k}{2}\left(\frac{1}{a} - 1\right)\left(\frac{1}{a} + 1\right) = 1 \qquad \text{and} \qquad p_1k\left(1 + \frac{1}{a}\right) = 1.$$

Hence $a = 1/(1 + 2p_1)$; similarly $b = 1/(1 + 2p_2)$. Without loss of generality, we consider only the case $p_2 > p_1$, that is, $a > b$. We must consider

$$(I - U_a)g_a = \delta g_1.$$

Since $L_\xi(u, \eta)$ is not a function of u, this equation must have the same form of solution as before, that is, $g_a(u) = k_2u^{-3}$, where k_2 and δ satisfy

$$p_2k_2\left(1 + \frac{1}{a}\right) = 1 + \delta p_2 \qquad \text{and} \qquad \frac{k_2}{2}\left(\frac{1}{a} + 1\right)\left(\frac{1}{a} - 1\right) = 1 - \delta.$$

Next, solving for k, k_2, and δ, we find that the density for x^0 is

$$f(\xi) = \begin{cases} 0 & 0 \leq \xi < \dfrac{1}{1 + 2p_1}, \\ \dfrac{1}{2p_1(1 + p_1)}\xi^{-3} & \dfrac{1}{1 + 2p_1} \leq \xi \leq 1, \end{cases}$$

and the density for y^0 is

$$g(\eta) = \begin{cases} 0 & 0 \le \eta < \dfrac{1}{1 + 2p_1}, \\ \dfrac{1 + p_2}{2p_2(1 + p_1)^2}\eta^{-3} & \dfrac{1}{1 + 2p_1} \le \eta \le 1; \end{cases}$$

and y^0 has a jump of magnitude $(p_2 - p_1)/p_2(1 + p_1)$ at 1. The value is $-(p_2 - p_1)/(1 + p_1)$.

Problem 6. The pay-off function has the form

$$K(\xi, \eta) = \begin{cases} L(\xi, \eta) = p_1\xi\alpha + p_2\eta\beta - p_1p_2\xi\eta\beta & \xi < \eta, \\ \phi(\xi) = (p_1\alpha + p_2\beta)\xi & \xi = \eta, \\ M(\xi, \eta) = p_1\xi\alpha + p_2\eta\beta - p_1p_2\xi\eta\alpha & \xi > \eta. \end{cases}$$

Since $L(0, 0) = M(0, 0)$, this kernel falls into Group 2, and

$$T(\xi, t) = \begin{cases} \dfrac{-\beta + p_1\beta\xi}{p_1t^2(\alpha - \beta)} & a \le \xi < t \le 1, \\ \dfrac{-\beta + p_1\alpha\xi}{p_1t^2(\alpha - \beta)} & a \le t \le \xi \le 1. \end{cases}$$

Now $(I - T_a)f = 0$ reduces to

$$p_1t^2(\alpha - \beta)f(t) + \beta - p_1\beta\int_a^t \xi f(\xi)\, d\xi - p_1\alpha\int_t^1 \xi f(\xi)\, d\xi = 0,$$

which implies $f(t) = kt^{-3}$, where k and a must satisfy

$$\frac{k}{2}\left(\frac{1}{a} - 1\right)\left(\frac{1}{a} + 1\right) = 1 \quad \text{and} \quad p_1k\left(\alpha - \frac{\beta}{a}\right) = -\beta.$$

Hence

$$a = \frac{\sqrt{p_1^2 + 1 - 2(\alpha/\beta)p_1} - p_1}{1 - 2(\alpha/\beta)p_1}.$$

Similarly,

$$b = \frac{\sqrt{p_2^2 + 1 - 2(\beta/\alpha)p_2} - p_2}{1 - 2(\beta/\alpha)p_2}.$$

(Notice that

$$\frac{1}{a} = p_1 + \sqrt{p_1^2 + 1 - 2p_1(\alpha/\beta)};$$

this is easier to work with.) Now $a > b$ is equivalent to

$$p_2 > p_1 \cdot \frac{p_1 + \sqrt{p_1^2 + 1 - 2(\alpha/\beta)p_1}}{2p_1 - (\beta/\alpha)},$$

and we consider only this case as before. (Note that the factor multiplying p_1 is greater than 1 if $-\alpha/\beta > 1$ and less than 1 if $-\alpha/\beta < 1$.) As in the previous problem, we see that $g_a(u) = k_2 u^{-3}$, where k_2 and δ satisfy

$$-\alpha + \frac{p_2 \alpha k_2}{a} - p_2 \beta k_2 + p_2 \delta \beta = 0 \qquad \text{and} \qquad \frac{k_2}{2}\left(\frac{1}{a^2} - 1\right) = 1 - \delta.$$

Solving for k, k_2, and δ yields

$$k = \frac{1}{p_1[(1/a) - (\alpha/\beta)]},$$

$$k_2 = \frac{1 - p_2(\beta/\alpha)}{p_2[(1/a) - (\beta/\alpha)] - p_1 p_2 (\beta/\alpha)[(1/a) - (\alpha/\beta)]},$$

$$\delta = \frac{p_2[(1/a) - (\beta/\alpha)] - p_1[(1/a) - (\alpha/\beta)]}{p_2[(1/a) - (\beta/\alpha)] - p_1 p_2 (\beta/\alpha)[(1/a) - (\alpha/\beta)]}.$$

It is interesting that the strategies depend only on the ratio of α to β and hence reduce to the previous case if $\alpha/\beta = -1$. The value is

$$\alpha\left[\frac{2p_1}{(1/a) + 1} - \frac{p_2[1 - (\beta/\alpha)]}{[(1/a) - (\alpha/\beta)]}\right].$$

Problem 7. Employing the method, one obtains that x^0 has density

$$f(\xi) = \begin{cases} 0 & 0 \le \xi < e^{-2}, \\ \dfrac{e^{-2}}{2\xi^2} & e^{-2} \le \xi \le 1 \end{cases}$$

and has a jump of magnitude $(1 + e^{-2})/2$ at 1; y^0 has density

$$g(\eta) = \begin{cases} 0 & 0 \le \eta < e^{-2}, \\ \dfrac{1}{2\eta} & e^{-2} \le \eta \le 1, \end{cases}$$

and $v = (1 + e^{-2})/2$.

Problem 8. The pay-off function is

$$K(\xi, \eta) = \begin{cases} lp(\xi) - sp(\eta) + lsp(\xi)p(\eta) & \xi > \eta, \\ (l - s)p(\xi) & \xi = \eta, \\ lp(\xi) - sp(\eta) - lsp(\xi)p(\eta) & \xi < \eta. \end{cases}$$

The relation $(I - T_a)f = 0$ yields $f(t) = kp'(t)/p^3(t)$, where k and a must satisfy

$$lk\left(\frac{1}{p(a)} + \frac{1}{p(1)}\right) = 1 \qquad \text{and} \qquad \frac{k}{2}\left(\frac{1}{p^2(a)} - \frac{1}{p^2(1)}\right) = 1,$$

and thus a is the unique number in (0, 1) satisfying $p(a) = p(1)/[1 + 2lp(1)]$. Similarly, b is the unique number in (0, 1) satisfying $p(b) = p(1)/[1 + 2sp(1)]$. Since $l > s$, it follows that $p(a) < p(b)$, and hence $a < b$. Again $f_a(t) = k_1 p'(t)/p^3(t)$, where

$$k_1 l\left(\frac{1}{p(b)} + \frac{1}{p(1)}\right) = 1 + \beta l p(1) \qquad \text{and} \qquad \frac{k_1}{2}\left(\frac{1}{p^2(b)} - \frac{1}{p^2(1)}\right) = 1 - \beta.$$

Solving, we find that the optimal strategy for the large company has density

$$f(t) = \begin{cases} 0 & 0 \le t \le p^{-1}\left(\dfrac{p(1)}{1 + 2sp(1)}\right), \\[2ex] \dfrac{p(1)[1 + lp(1)]}{2l[1 + sp(1)]^2} \dfrac{p'(t)}{p^3(t)} & p^{-1}\left(\dfrac{p(1)}{1 + 2sp(1)}\right) \le t \le 1 \end{cases}$$

and a jump of magnitude $(l - s)/[1 + sp(1)]l$ at 1. The optimal strategy for the small company has density

$$g(u) = \begin{cases} 0 & 0 \le u \le p^{-1}\left(\dfrac{p(1)}{1 + 2sp(1)}\right), \\[2ex] \dfrac{p(1)}{2s[1 + sp(1)]} \dfrac{p'(u)}{p^3(u)} & p^{-1}\left(\dfrac{p(1)}{1 + 2sp(1)}\right) \le u \le 1. \end{cases}$$

The value is $(l - s)p(1)/[1 + sp(1)]$.

Problem 10. If either a or b [as given by (5.5.4) and (5.5.6)] is nonnegative, the kernel is Group 2 and the solution is as given. If both are negative, the kernel is either Group 3 or 4; an easy computation shows that it is Group 3 if $P_1(0) + P_2(0) < 1$ and Group 4 if $P_1(0) + P_2(0) \ge 1$. Also it is clear that cases E, I, L, and N cannot arise. However, each of the other three cases in each group can clearly arise.

Problem 12. Suppose both players have optimal strategies possessing a jump at 1 and $M(1, 1) < \Phi(1)$. Since Player II has a jump at 1,

$$\int_0^1 K(\xi, 1)\, dx^0(\xi) = v.$$

We choose η sufficiently close to one and such that:

(1) $L(\xi, \eta) - L(\xi, 1) < \epsilon$ for $0 \le \xi \le \eta$, and $M(1, \eta) - M(1, 1) < \epsilon$ by the uniform continuity of L and M;

(2) $1 - \alpha - x^0(\eta) < \epsilon$, where α is the jump of x^0 at 1;

(3) η is a point of continuity of x^0.

Then, letting C be a bound for $L(\xi, \eta)$and $M(\xi, \eta)$, we have

$$\begin{aligned}\int K(\xi, \eta)\, dx^0(\xi) &= \int_0^\eta L(\xi, \eta)\, dx^0(\xi) + \int_\eta^1 M(\xi, \eta)\, dx^0(\xi) \\ &\leq \int_0^\eta [L(\xi, 1) + \epsilon]\, dx^0(\xi) + [M(1, 1) + \epsilon][1 - x^0(\eta)] \\ &\leq \int_0^{1-} L(\xi, 1)\, dx^0(\xi) + [1 - \alpha - x^0(\eta)][C + M(1, 1)] \\ &\qquad + \alpha M(1, 1) + \epsilon \\ &\leq \int_0^1 K(\xi, 1)\, dx^0(\xi) + \epsilon(2C + 1) + \alpha[M(1, 1) - \Phi(1)] \\ &= v + \epsilon(2C + 1) + \alpha[M(1, 1) - \Phi(1)] < v\end{aligned}$$

for ϵ sufficiently small, a contradiction. Thus $\Phi(1) \leq M(1, 1)$. A similar absurdity results under the condition $\Phi(1) < L(1, 1)$. Thus, $\Phi(1) \geq L(1, 1)$ and $\{1, 1\}$ is clearly a saddle point.

Problem 15. The integral equation (5.2.2) reduces to

$$f(t) = \frac{1}{2\sum_{i=1}^n p_i(t)q_i(t)}\left[-\int_a^t \sum_{i=1}^n p_i(\xi)q_i'(t)f(\xi)\, d\xi + \int_t^1 \sum_{i=1}^n p_i'(t)q_i(\xi)f(\xi)\, d\xi \right.$$
$$\left. -\alpha \sum_{i=1}^n p_i(0)q_i'(t) + \beta \sum_{i=1}^n p_i'(t)q_i(1)\right].$$

Define

$$P_i(t) = \int_t^1 p_i(\xi)f(\xi)\, d\xi \qquad \text{and} \qquad Q_i(t) = \int_t^1 q_i(\xi)f(\xi)\, d\xi.$$

Thus $P_i'(t) = -p_i(t)f(t)$, $Q_i'(t) = Q_i(t)f(t)$; $P_i(1) = 0$ and $Q_i(1) = 0$. Hence

$$P_j'(t) = \frac{p_j(t)}{2\sum_{i=1}^n p_i(t)q_i(t)}$$
$$\times \left\{\sum_{i=1}^n q_i'(t)[P_i(a) - P_i(t) + \alpha p_i(0)] - \sum_{i=1}^n p_i'(t)[Q_i(t) + \beta q_i(1)]\right\}$$

and a similar expression holds for $Q_j'(t)$ $(j = 1, \ldots, n)$. This system is then solved with the appropriate boundary conditions, and $f(t)$ is determined from, say, $P_1(t)$.

Problem 16. We already have the existence of an optimal strategy for Player II such that

$$\int_0^1 K(\xi, \eta)\, dy^0(\eta) < v \qquad (0 < \xi < a).$$

(If $a = 0$, this is irrelevant.) Thus the spectrum for any optimal strategy for Player I must be restricted to $\{0\}$ and the interval $[a, 1]$. Let η_0 be in $(a, 1)$, and suppose that $x(\xi)$ is optimal for Player I and has a discontinuity at η_0. Then

$$\int_0^\eta L(\xi, \eta)\, dx(\xi) + \int_\eta^1 M(\xi, \eta)\, dx(\xi) = v \qquad (a < \eta < 1)$$

implies

$$\int_0^{\eta_0-} L(\xi, \eta_0)\, dx(\xi) + \int_{\eta_0-}^1 M(\xi, \eta_0)\, dx(\xi) = v$$

and

$$\int_0^{\eta_0+} L(\xi, \eta_0)\, dx(\xi) + \int_{\eta_0+}^1 M(\xi, \eta_0)\, dx(\xi) = v$$

by the Lebesgue dominated-convergence theorem. Hence, if ϵ_0 is the magnitude of the jump at η_0, we have $\epsilon_0[L(\eta_0, \eta_0) - M(\eta_0, \eta_0)] = 0$; it follows that $\epsilon_0 = 0$, since $L(\eta, \eta) > M(\eta, \eta)$ on $[a, 1]$. Considering the point a, we still have

$$\int_0^{a+} L(\xi, a)\, dx(\xi) + \int_{a+}^1 M(\xi, a)\, dx(\xi) = v$$

and

$$\int_0^{a-} L(\xi, a)\, dx(\xi) + \int_{a-}^1 M(\xi, a)\, dx(\xi) \leq v.$$

Hence $\epsilon_0[M(a, a) - L(a, a)] \geq 0$. As before, $L(a, a) > M(a, a)$ implies $\epsilon_0 \leq 0$ and consequently $\epsilon_0 = 0$.

Problem 17. Since the known optimal strategy for Player II has support including $[a, 1]$, we know that x optimal implies

$$\int K(\xi, \eta)\, dx(\xi) = v \qquad \text{for} \quad a < \eta < 1.$$

Thus letting α be the possible jump of x at 0 and β the possible jump at 1, we know that

$$\begin{aligned} v &= \alpha L(0, \eta) + \int_a^\eta L(\xi, \eta)\, dx(\xi) + \int_\eta^{1-} M(\xi, \eta)\, dx(\xi) + \beta M(1, \eta) \\ &= \alpha L(0, \eta) + L(\eta, \eta)x(\eta) - L(a, \eta)x(a) - \int_a^\eta x(\xi) L_\xi(\xi, \eta)\, d\xi \\ &\qquad + M(1, \eta)x(1-) - M(\eta, \eta)x(\eta) - \int_\eta^1 x(\xi) M_\xi(\xi, \eta)\, d\xi \\ &\qquad + \beta M(1, \eta) \qquad (a \leq \eta < 1), \end{aligned}$$

the second equality following by an integration by parts. Now all the functions involved are absolutely continuous, having bounded derivatives. Since $L(\eta, \eta) - M(\eta, \eta) > 0$ in this interval, division by this quantity yields the absolute continuity of $x(\xi)$. Differentiation of the last equation above and division by $L(\eta, \eta) - M(\eta, \eta)$ yields the continuity of $x'(\xi)$.

Problem 18. This final step depends on the individual case. Case F is chosen as being representative. All cases are similar. We know from the preceding two problems that $x = (\alpha_1 I_0, f, \beta_1 I_1)$. Suppose $\beta_1 > 0$. Then

$$v = \int_0^1 K(1, \eta)\, dy^0(\eta) = \gamma M(1, 0) + \int_a^1 M(1, \eta) g_a(\eta)\, d\eta + \delta\Phi(1),$$

and since $\delta > 0$,

$$v = \int_0^1 K(\xi, 1)\, dx(\xi) = \alpha_1 L(0, 1) + \int_a^1 L(\xi, 1) f(\xi)\, d\xi + \beta_1 \Phi(1).$$

Also, by taking limits as we did in Problem 16,

$$v = \gamma M(1, 0) + \int_a^1 M(1, \eta) g_a(\eta)\, d\eta + \delta L(1, 1),$$

$$v = \alpha_1 L(0, 1) + \int_a^1 L(\xi, 1) f(\xi)\, d\xi + \beta_1 M(1, 1).$$

This implies that

$$0 = \beta_1[\Phi(1) - M(1, 1)] = \delta[\Phi(1) - L(1, 1)],$$

and it follows that $M(1, 1) = L(1, 1)$, a contradiction. Hence $\beta_1 = 0$.
Now differentiation of

$$v = \alpha_1 L(0, \eta) + \int_a^\eta L(\xi, \eta) f(\xi)\, d\xi + \int_\eta^1 M(\xi, \eta) f(\xi)\, d\xi$$

shows that f satisfies the original integral equation with $\beta = 0$. This cannot have a solution with $\alpha_1 = 0$, since $\lambda(a) < 1$. Thus the solution of the integral equation is unique for fixed α_1, and α_1 is uniquely determined by

$$\int_a^1 f(\xi)\, d\xi = 1 - \alpha_1.$$

Thus this solution must be the same as the original solution.

Problem 19. Requires analysis similar to Problems 16–18.

Problem 20. First, consider the case in which

$$L(1, 1) \leq M(1, 1) \qquad \text{but} \qquad \Phi(1) > M(1, 1).$$

(Here $\Phi(0)$ is immaterial.) Then Player I still plays 1, but Player II must play a uniform density on $(1 - \epsilon, 1)$ to "rub out" the $\Phi(1)$. The value is $M(1, 1)$, since we have demonstrated ϵ-optimal strategies. If $\Phi(1) < L(1, 1)$, a similar proof shows that $L(1, 1)$ is the value. Now we need consider only cases in which $L(1, 1) > M(1, 1)$. Suppose that L and M are Group 2 kernels, but that $\Phi(1) > L(1, 1)$; $\Phi(0)$ is still immaterial. Consider the new kernel $K^*(\xi, \eta)$, with

$\Phi(1)$ replaced by $L(1, 1)$ but otherwise unchanged. The optimal strategies for this kernel are $(f_a, \beta I_1)$, $(g_a, \delta I_1)$, where $\beta, \delta \geq 0$. We shall show that v^*, the value of this game, is also the value of the original game. For $0 \leq \eta < 1$,

$$\int_0^1 K(\xi, \eta)\, dx^0(\xi) = \int_0^1 K^*(\xi, \eta)\, dx^0(\xi) \geq v^*.$$

Also

$$\begin{aligned}\int_0^1 K(\xi, 1)\, dx^0(\xi) &= \int_a^1 L(\xi, 1) f_a(\xi)\, d\xi + \beta\Phi(1)\\ &\geq \int_a^1 L(\xi, 1) f_a(\xi)\, d\xi + \beta L(1, 1)\\ &= \int_0^1 K^*(\xi, 1)\, dx^0(\xi) \geq v^*.\end{aligned}$$

As in the previous case, we find only an ϵ-optimal strategy for Player II. This is done by using the density $g_a(\eta)$ on $[a, 1 - \epsilon]$ and $g_a(\eta) + \delta/\epsilon$ on $[1 - \epsilon, 1]$. As before, this removes the effect of the value of $\Phi(1)$, for

$$\int K(\xi, \eta)\, dy^0(\eta) = \int_a^1 K(\xi, \eta) g_a(\eta)\, d\eta + \int_{1-\epsilon}^1 \frac{\delta}{\epsilon} K(\xi, \eta)\, d\eta.$$

For $\xi = 1$ the last term is approximately $\delta M(1, 1)$, and for $\xi < 1$ the last term is approximately $\delta L(1, 1)$; hence $K^*(\xi, y^0) \leq v^* + \epsilon'$ for all $0 \leq \xi \leq 1$, where ϵ' tends to zero with ϵ. If $\Phi(1) < M(1, 1)$, a similar proof works.

If Group 3 or Group 4 is involved, the method is similar; there are now more cases, however, since $\Phi(0)$ becomes relevant.

Problem 21. We prove that the linear operator U_a transforms positive continuous functions into continuous monotone-decreasing functions. To this end, we differentiate

$$U_a g = \int_a^\xi \frac{M_\xi(\xi, \eta) g(\eta)}{L(\xi, \xi) - M(\xi, \xi)}\, d\eta + \int_\xi^1 \frac{L_\xi(\xi, \eta) g(\eta)}{L(\xi, \xi) - M(\xi, \xi)}\, d\eta,$$

getting

$$\begin{aligned}\frac{d(U_a g)}{d\xi} &= \frac{1}{[L(\xi, \xi) - M(\xi, \xi)]}\\ &\quad\times \left\{\int_a^\xi M_{\xi\xi}(\xi, \eta) g(\eta)\, d\eta + \int_\xi^1 L_{\xi\xi}(\xi, \eta) g(\eta)\, d\eta + [M_\xi(\xi, \xi) - L_\xi(\xi, \xi)] g(\xi)\right\}\\ &\quad - \frac{\left(\int_a^\xi M_\xi g\, d\eta + \int_\xi^1 K_\xi g\, d\eta\right)[L'(\xi, \xi) - M'(\xi, \xi)]}{[L(\xi, \xi) - M(\xi, \xi)]^2}.\end{aligned}$$

The hypothesis clearly implies that $d(U_a g)/d\xi < 0$. Also,

$$\psi_a(\eta) = \sum_{n=0}^{\infty} (U_a^n q)(\eta)$$

is positive continuous and decreasing, since each term of this series has these properties.

CHAPTER 6

Problem 1. In this example

$$M_1(\eta) = \max_{\theta \geq \eta} P_1(\theta) = \tfrac{2}{3}, \qquad M_2(\xi) = \max_{\theta \geq \xi} P_2(\theta) = \begin{cases} \tfrac{1}{2} & \text{on } \xi \leq \tfrac{1}{2}, \\ 1 - \xi & \text{on } \xi > \tfrac{1}{2}. \end{cases}$$

Hence

$$f(\xi) = \begin{cases} \tfrac{3}{2}\left|\tfrac{1}{3} - \xi\right| - \tfrac{1}{2} & \text{on } 0 \leq \xi \leq \tfrac{1}{2}, \\ -\xi^2 + \tfrac{10}{3}\xi - \tfrac{5}{3} & \text{on } \tfrac{1}{2} < \xi \leq 1, \end{cases} \qquad \text{and} \qquad g(\eta) = -\tfrac{1}{6} + \tfrac{5}{3}\left|\tfrac{1}{2} - \eta\right|.$$

Examination of the graphs of $G(\eta)$ and $F(\xi)$ shows that case 1 occurs and $v = 0, b = 0, c = 2/5$. From the definition, $\phi(\xi) = |\tfrac{1}{3} - \xi| - \tfrac{1}{2} + |\tfrac{1}{2} - \xi|$, and hence $\phi(b) = \tfrac{1}{3} \geq 0, \phi(c) = -\tfrac{1}{3} \leq 0$. Thus $\{I_0, I_{2/5}\}$ is a pair of optimal strategies for Players I and II, respectively. In fact, I_0 is unique for Player I, but Player II may use any pure strategy between 2/5 and 3/5 (or, of course, a mixture of these).

Problem 2. The pay-off kernel reduces to

$$K(\xi, \eta) = \begin{cases} L(\xi, \eta) = p_1\xi\alpha + p_2\eta\beta + p_1p_2\beta(1 - \xi - \eta) & \xi > \eta, \\ \phi(\xi) = p_1\xi\alpha + p_2\xi\beta & \xi = \eta, \\ M(\xi, \eta) = p_1\xi\alpha + p_2\eta\beta + p_1p_2\alpha(1 - \xi - \eta) & \xi < \eta. \end{cases}$$

The integral equation $(I - T_a)f = 0$ becomes

$$f(t) - \int_a^t \frac{p_2\beta(p_1 - 1)}{p_1p_2(\beta - \alpha)(1 - 2t)} f(\xi)\, d\xi - \int_t^1 \frac{p_2(\alpha p_1 - \beta)}{p_1p_2(\beta - \alpha)(1 - 2t)} f(\xi)\, d\xi = 0$$

or

$$p_1(\beta - \alpha)(1 - 2t)f(t) - \beta(p_1 - 1)\int_a^t f(\xi)\, d\xi - (\alpha p_1 - \beta)\int_t^1 f(\xi)\, d\xi = 0.$$

Hence

$$-2p_1(\beta - \alpha)f(t) + p_1(\beta - \alpha)\,(1 - 2t)f'(t) - \beta(p_1 - 1)f(t) + (\alpha p_1 - \beta)f(t) = 0,$$

which simplifies to $-3f(t) + (1 - 2t)f'(t) = 0$. Its solution is the function

$f(t) = k_1(2t - 1)^{-3/2}$, where k_1 and a necessarily satisfy

$$(\alpha p_1 - \beta) = \beta(p_1 - 1)(2a - 1)^{-1/2} \quad \text{and} \quad k_1\left(-1 + \frac{1}{\sqrt{2a-1}}\right) = 1.$$

Thus

$$a = \tfrac{1}{2} + \tfrac{1}{2}\frac{(1 - p_1)^2}{[1 - (\alpha/\beta)p_1]^2}.$$

Similarly,

$$b = \tfrac{1}{2} + \tfrac{1}{2}\frac{(1 - p_2)^2}{[1 - (\beta/\alpha)p_2]^2}.$$

Now $a > b$ is equivalent to

$$p_2 > p_1 \frac{1}{[1 + (\beta/\alpha)]p_1 - (\beta/\alpha)},$$

and we consider this case as before. We find that $g_a(u) = k_2(2u - 1)^{-3/2}$, where k_2 and δ must satisfy

$$k_2\left(-1 + \frac{1}{\sqrt{2a-1}}\right) = 1 - \delta$$

and

$$k_2\left(-1 + \frac{\alpha(1 - p_2)}{\alpha - p_2\beta}\frac{1}{\sqrt{2a-1}}\right) = -\delta.$$

Solving for k_1, k_2, and δ yields

$$k_1 = \frac{1 - p_1}{p_1[1 - (\alpha/\beta)]}, \qquad k_2 = \frac{[p_2 - (\alpha/\beta)](1 - p_1)}{p_2[1 - (\alpha/\beta)p_1][1 - (\alpha/\beta)]},$$

$$\delta = 1 - \frac{p_1[p_2 - (\alpha/\beta)]}{p_2[1 - (\alpha/\beta)p_1]}.$$

The value is

$$\frac{\alpha}{\beta}\frac{[1 - (\alpha/\beta)p_1][p_2 - (\alpha/\beta)]}{1 - (\alpha/\beta)} - \alpha p_2 \frac{1 - p_1}{1 - (\alpha/\beta)} + \alpha\frac{1 - (\alpha/\beta)p_1}{1 - (\alpha/\beta)}.$$

It is interesting to note that the value of this game is the same as that of Problem 2 of Chapter 5 when $\alpha = 1, \beta = -1$.

Problem 3. The interpretation of this game is as follows: Player I has two guns available, Player II has only one. If Player I uses the first gun and fires first, he wins one unit; if both players fire at the same time, Player I loses one unit. If he uses the second gun and fires first, he loses one unit; if both players fire at the same time, Player I gains one unit. In all circumstances, the pay-off is zero if Player II fires first.

We first show that $\sup_x \inf_y K(x, y) = 0$. If x assigns a probability to A_1 at 0 that is at least as large as the probability assigned to A_2 at 0, then choose a strategy y depending on x; specifically, $y_x = y_0$. If the probability of x assigned to A_1 at 0 is smaller than the probability assigned to A_2 at 0, then let y_x spread uniformly over $[0, \epsilon]$ for ϵ small. In either case we have $\inf_y K(x, y) \leq K(x, y_x) \leq$ the probability that x assigns to $(0, \epsilon)$ in A_1, which implies $\inf_y K(x, y) \leq 0$ for all x. Then consider x^0 assigning probability $\frac{1}{2}$ to A_1 at 0 and $\frac{1}{2}$ to A_2 at 0. It follows that the yield to Player II is identically 0; hence $\sup_x \inf_y K(x, y) = 0$.

Next we show that $\inf_y \sup_x K(x, y) = \frac{1}{3}$. If the probability that the strategy y assigns at 0 is at least $\frac{2}{3}$, let x_y assign probability 1 to A_2 at 0. If the probability that y assigns at 0 is smaller than $\frac{2}{3}$, choose x_y spread uniformly over $[0, \epsilon]$ in A_1. In either case we obtain that $\sup_x K(x, y) \geq \frac{1}{3}$ for all y. Then consider $y^0 = \frac{2}{3}y_0 + \frac{1}{3}y_1$. We have $K(x, y^0) \leq \frac{1}{3}$ for all x, and hence $\inf_y \sup_x K(x, y) = \frac{1}{3}$.

Problem 4. See Blackwell and Girschick [30].

Problem 5. As in the first problem in Chapter 5, this involves only a change in time scale $\tilde{\xi} = P(\xi)$. The value is still $(m - n)/(m + n)$, and the first critical time becomes $P^{-1}[1/(m + n)]$.

Problem 6. If $1/n \leq \lambda < 1/(n - 1)$ $(n = 2, 3, 4, \ldots)$, then an optimal strategy for Player I is

$$\frac{1}{n} \sum_{i=0}^{n-1} I_{i/(n-1)};$$

an optimal strategy for Player II is

$$\frac{1}{n} \sum_{i=1}^{n} I_{i/n},$$

and $v = (n - 1 + \lambda)/n$.

Problem 7. It follows from Theorem 6.5.1 that all optimal strategies for both players are of similar form on an interval $[a, b]$. Suppose $a > 0$. Now

$$h(\xi) = \int K(\xi, \eta)\, dy^0(\eta) = \gamma M(\xi, 0) + \int_a^1 L(\xi, \eta)\, dy^0(\eta) \qquad \text{for} \quad 0 < \xi \leq a,$$

but $h'(a)$ exists and hence must equal 0, from which it follows that

$$\gamma M_\xi(a, 0) + \int_a^1 L_\xi(a, \eta)\, dy^0(\eta) = 0.$$

But $L_\xi < 0$ requires $\gamma > 0$. It now follows that

$$v = \int K(\xi, a)\, dx^0(\xi) = \int_a^1 M(\xi, a)\, dx^0(\xi),$$

since a belongs to the spectrum of Player II's optimal strategies; and

$$v = \int K(\xi, 0)\, dx^0(\xi) = \int M(\xi, 0)\, dx^0(\xi),$$

since $\gamma > 0$. This is a contradiction, since M is strictly decreasing in η. Thus $a = 0$, and similarly $b = 1$.

Suppose $\alpha = 0$. Then

$$m(\eta) = \int K(\xi, \eta)\, dx(\xi) = \int_0^\eta L(\xi, \eta) f(\xi)\, d\xi + \int_\eta^1 M(\xi, \eta) f(\xi)\, d\xi + \beta M(1, \eta),$$

$$m'(\eta) = \int_0^\eta L_\eta(\xi, \eta) f(\xi)\, d\xi + \int_\eta^1 M_\eta(\xi, \eta) f(\xi)\, d\xi + \beta M_\eta(1, \eta).$$

and thus

$$m'(0) = \int_0^1 M_\eta(\xi, 0) f(\xi)\, d\xi + \beta M_\eta(1, 0) < 0,$$

a contradiction. Hence $\alpha > 0$. Similarly, $\gamma > 0$. This implies that $\lambda(T) < 1$ and $\mu(U) < 1$, from the proof of Theorem 6.5.1. Thus f_0 and g_0 may be expressed as given and are uniquely determined if α, β, γ, and δ are uniquely determined. These constants are determined from the four given equations. If the two equations for α and β had more than one solution, then there would be at least a one-parameter family; thus α could be zero, which we have already seen to contradict the second equation. Thus α and β are uniquely determined, and similarly so are γ and δ.

Problem 8. The expected pay-off functions are clearly

$$K_1(\xi, \eta) = \begin{cases} \xi - c_1 & \text{if } \xi < \eta, \\ \dfrac{\xi - c_1}{2} & \text{if } \xi = \eta, \\ 0 & \text{if } \xi > \eta; \end{cases} \qquad K_2(\xi, \eta) = \begin{cases} 0 & \text{if } \xi < \eta, \\ \dfrac{\eta - c_2}{2} & \text{if } \xi = \eta, \\ \eta - c_2 & \text{if } \xi > \eta. \end{cases}$$

For fixed ξ_0, the value of $K_2(\xi_0, \eta)$ is maximized by any $\eta > \xi_0$ if $\xi_0 < c_2$ (maximum is 0), by any $\eta \geq \xi_0$ if $\xi_0 = c_2$ (maximum is 0), and nowhere if $\xi_0 > c_2$ (the supremum is $\xi_0 - c_2$). Similar relations hold for K_1 with η_0 fixed. A pure equilibrium point $\{\xi_0, \eta_0\}$ can therefore have neither $\xi_0 > \eta_0$ nor $\xi_0 < \eta_0$, and $\xi_0 = \eta_0$ works only if $\xi_0 = \eta_0 = c_1 = c_2$.

Problem 9. Assume that Player I has an absolutely continuous strategy which makes Player II's expectation independent of η in an interval $[a, \infty)$. This leads to the integral equation

$$v_2 = (\eta - c_2) \int_\eta^\infty f(\xi)\, d\xi \qquad (a \leq \eta).$$

From this relation and the normalization condition

$$\int_a^\infty f(\xi)\, d\xi = 1,$$

we obtain

$$f(\xi) = \frac{a - c_2}{(\xi - c_2)^2}$$

in the interval $a \leq \xi < \infty$ and $v_2 = a - c_2$. Similar considerations show that

$$g(\eta) = \frac{a - c_1}{(\eta - c_1)^2}$$

in the interval $[a, \infty)$, and hence the densities f and g so constructed constitute an equilibrium strategy.

CHAPTER 7

Problem 1. $x^0(\xi) = y^0(\xi) = (\xi + \xi^2)/2$.

Problem 2. $v = 1/\ln 2$. By Theorem 7.1.1 every optimal strategy is an equalizer and hence cannot have a finite support.

Problem 3. If the conclusion is false, then for any optimal y

$$v \equiv \int \frac{P(\xi, \eta)}{Q(\xi, \eta)} \, dy(\eta),$$

and letting $\xi \to \infty$ along C, we obtain $v = 0$, which is impossible.

Problem 5. (a)

$$\frac{1}{\cosh (u - v)} = 2e^v e^u \frac{1}{e^{2v} + e^{2u}}.$$

Hence, for u's and v's arranged in increasing order, we have

$$\text{sign det} \frac{1}{\cosh (u_i - v_j)} = \text{sign det} \frac{1}{z_i + w_j},$$

where $z_i = e^{2u_i}$, $w_j = e^{2v_j}$ and $0 < z_1 < z_2 < \cdots < z_n$, $0 < w_1 < w_2 < \cdots < w_n$. The last determinant is the classical Cauchy determinant, which may be easily evaluated and shown to be positive.

(b) The determinant (ii) of p. 181 applied to e^{u-e^u}, apart from a constant factor, reduces to a Vandermonde-type determinant of the form det $(e^{z_i w_j})$, with z_i and w_j strictly increasing. Its value is trivially positive.

(c) The case $n = 0$ is trivial. The case of $n > 0$ is obtained by successive convolutions of the case of $n = 0$.

Problem 6. $y^0 = \frac{1}{2}I_0 + \frac{1}{2}I_1$; $x^0 = \frac{1}{2}I_a + \frac{1}{2}I_{1-a}$, where a is approximately 0.072.

Problem 7. It is the value λ for which the derivative of

$$\varphi(u) = \frac{1}{1 + \lambda u^2} + \frac{1}{1 + \lambda(1 - u)^2}$$

has a multiple zero at $u = \frac{1}{2}$; that is, $\lambda = \frac{4}{3}$.

Problem 8. $\lambda = 2$;

$$1 - \sqrt{1 - e^{-\lambda/2}} = [\exp(-\lambda\sqrt{1 - e^{-\lambda/2}})]\,(1 + \sqrt{1 + e^{-\lambda/2}}).$$

Problem 9. Similar to Lemma 7.2.3.

Problem 11. (a) Use the characterization of the Cantor distribution given in Example 2, p. 178.

Problem 13. The kernel is clearly invariant under the compact group of all rotations of the unit circle. Hence there is a rotation-invariant optimal strategy.

Problem 14. Let $K_n(\xi, \eta)$ be the kernel obtained from $K(\xi, \eta)$ by deleting the nth term. Clearly, K_n converges uniformly to K as $n \to \infty$, but the games with kernels K_n do not have unique optimal strategies (Theorem 7.5.3).

Problem 15. The strategies are equalizers and yield the value

$$\sum_{n=1}^{\infty} \frac{1}{2^n - 1}.$$

Problem 16. By direct verification it follows that x and y are optimal and the value of the game is zero. Also, for the specified x and any optimal y^*,

$$0 \le \iint K(\xi, \eta)\, dx(\xi)\, dy^*(\eta) = \int e^{-\eta^2}\, dy^*(\eta),$$

implying that $y^* = I_0$. Finally, for any optimal x^* with moments μ_n^*, we obtain

$$0 \le \int K(\xi, \eta)\, dx^*(\eta) = \sum_{n=1}^{\infty} \frac{1}{2^n} (\mu_n^* - \mu_n)\eta^n \sin\frac{1}{\eta} + e^{-\eta^{-2}}.$$

Clearly, for η small, the first nonzero term in the series dominates; and since $\sin(1/\eta)$ changes sign, we must have $\mu_n^* = \mu_n$ for all n. Hence $x^* = x$.

Problem 19. Let x^* be any other optimal strategy, and let

$$dx^0(\xi) = \frac{dx^*(\xi)}{f(\xi)},$$

where

$$f(\xi) = \int_0^1 h(\xi t)\, dt.$$

By substitution, $\int k(\xi, \eta)\, dx^*(\xi) \equiv v$ reduces to

$$\int_0^1 h(\xi\eta)\, dx^0(\xi) = v \int_0^1 h(\eta t)\, dt.$$

Expanding both sides in power series, one notes that the n_i moment of x^0 is determined whenever $a_{n_i} \ne 0$. By virtue of the classical Müntz theorem, x^0 is unique.

Problem 20. Similar to Problems 17 and 19.

Problem 21. An application of Problems 18 and 19 with $h(\xi\eta) = (1 + \xi\eta)^{-2}$.

Problem 22. Clearly, there exists some α_i ($i \leq \alpha_i \leq i + 1$) such that

$$\int_0^1 \xi^{\alpha_i}\, d_\xi(x - x^*) = 0.$$

Hence, by the Müntz theorem, $x^* = x$.

Problem 23. The value of the game is zero and x^0 is optimal. Also using any optimal x^* and $y^{(n)} = I_{1/n}$ implies that $\mu_{2n}^* \geq \mu_{2n}$ and $\mu_{2n+1}^* \leq \mu_{2n+1}$. Thus $x = x^*$.

Problem 24. See [137], p. 382.

Problem 25. See [137], p. 384.

Problem 26. See [110].

Problem 27. $K(\xi, \eta) = -\beta\xi + (2 - \alpha)\eta + \xi^2 + (1 - \alpha - \beta)\xi\eta - \eta^2$. Observe that this kernel is strictly convex in ξ and strictly concave in η.

Problem 28. This is the assertion that a convex set is determined by a denumerable number of supporting hyperplanes.

Problem 29. Use Problem 28.

CHAPTER 8

Problem 2. Let $L_0(t)$ represent the solution (it is unique since the functional being minimized is strictly convex). By the methods of p. 211, $L_0(t)$ is the solution if and only if

$$\min_L \int_0^1 [L_0(t) - t]L(t)\, dt = \int [L_0(t) - t]L_0(t),$$

where the minimum is extended over all functions $L(t)$ satisfying the constraints. By the Neyman-Pearson lemma there exists a constant k such that

$$L_0(t) = \begin{cases} 1 & \text{if } t - L_0(t) > k, \\ 0 & \text{if } t - L_0(t) < k, \end{cases} \qquad (*)$$

and $L_0(t) - t = -k$ elsewhere. We want $L_0(t)$ to resemble t as much as possible, so a solution of the form

$$L_0(t) = \begin{cases} 0 & \text{on } t < c_1, \\ t - k & \text{on } c_1 < t < c_2, \\ 1 & \text{on } c_2 < t < 1 \end{cases}$$

seems reasonable. However, if this is the case, we must have $c_2 - L_0(c_2) = c_2 - 1 \geq k$ and $c_1 - L_0(c_1) = c_1 \leq k$, which implies $c_2 - 1 \geq c_1$ or $c_2 - c_1 \geq 1$. This type of solution is impossible unless $p = \frac{1}{2}$, in which case $k = 0$. Thus we try a solution of the form

$$L_0(t) = \begin{cases} 0 & \text{on} \quad 0 \leq t < c_1, \\ t - k & \text{on} \quad c_1 < t \leq 1, \end{cases} \qquad \text{or} \qquad L_0(t) = \begin{cases} t - k & \text{on} \quad 0 \leq y < c_2, \\ 1 & \text{on} \quad c_2 < t \leq 1. \end{cases}$$

In the first case $c_1 - L_0(c_1) = c_1 \leq k$ and $c_1 - k \geq 0$, which gives $c_1 = k$. Thus we obtain for a "solution" max $(0, t - c_1)$, whose integral is

$$\frac{(1 - c_1)^2}{2} \leq \frac{1}{2}.$$

If $p \leq \frac{1}{2}$, we see that $L_0(t) = \max(0, t - c_1)$, where $c_1 = 1 - \sqrt{2p}$ satisfies the consistency requirements (*) and is thus indeed the solution. If $p \geq \frac{1}{2}$, the second case applies and we obtain $L_0(t) = \min(1, t + 1 - c_2)$, where $c_2 = \sqrt{2(1 - p)}$.

Problem 3. First, restrict consideration to the set

$$\Omega = \{\phi | 0 \leq \phi(t) \leq M, \phi \text{ vanishes outside } [-T, T] \text{ and } \int \phi(t)\, dt = \alpha\}.$$

We obtain, according to the method of p. 211, that $\phi_0(t)$ is the minimum if and only if

$$\max_{\phi \in \Omega} \int_{-T}^{T} p(t) e^{-\phi_0(t)} \phi(t)\, dt = \int_{-T}^{T} p(t) e^{-\phi_0(t)} \phi_0(t)\, dt.$$

With the help of the Neyman-Pearson lemma, we obtain for an appropriate constant k

$$\phi_0(t) = \begin{cases} M & \text{if} \quad \phi_0(t) < \log \dfrac{p(t)}{k}, \\ 0 & \text{if} \quad \phi_0(t) > \log \dfrac{p(t)}{k}, \end{cases} \tag{**}$$

and $\phi_0(t) = \log (p(t)/k)$ elsewhere.

We may try a solution of the form

$$\phi_0(t) = \begin{cases} 0 & \text{if} \quad \log \dfrac{p(t)}{k} < 0, \\ \dfrac{\log p(t)}{k} & \text{if} \quad 0 \leq \log \dfrac{p(t)}{k} \leq M, \\ M & \text{if} \quad \log \dfrac{p(t)}{k} > M, \end{cases}$$

which satisfies the consistency relations (**).

It is readily verified by use of the Lebesgue dominated-convergence theorem that

$$\int_{-T}^{T} \phi_0(t)\, dt$$

is a continuous function of k, and that the integral tends to zero as k goes to infinity. Also the integral converges to $2TM$ when k tends to zero. Thus for T and M sufficiently large we can satisfy

$$\int_{-T}^{T} \phi_0(t)\, dt = \alpha.$$

Now, if we let T and M increase to ∞, $\phi_0(t)$ converges to $\max \{0, \log p(t)/k\}$, where k is determined by the condition

$$\int_{-\infty}^{\infty} \max \left\{0, \log \frac{p(t)}{k}\right\} dt = \alpha$$

(again by the dominated-convergence theorem). If we truncate the values of an arbitrary ϕ at a sufficiently large value of M, and if we further limit the argument to an interval $2T$, we can compare ϕ and ϕ_0. It follows from this comparison that

$$\phi_0(t) = \max \left\{0, \log \frac{p(t)}{k}\right\}$$

is the solution.

Problem 4. The solution is

$$\phi_0(t) = \begin{cases} 0 & \text{if } \dfrac{p(t)g(t)}{h(t)} < k, \\[2ex] \dfrac{1}{g(t)} \log \dfrac{p(t)g(t)}{kh(t)} & \text{if } \dfrac{p(t)g(t)}{h(t)} \geq k, \end{cases}$$

where k is determined so that

$$\int_{-\infty}^{\infty} h(t)\phi_0(t)\, dt = \alpha.$$

The procedure of verifying this solution is similar to that of Problem 3.

Problem 5. Suppose we decide to look in box i a total of n times, at times $t_1, t_1 + t_2, \ldots, t_1 + t_2 + \ldots + t_n$. The expected gain from these "looks" is

$$\frac{p_i}{N} [G(t_1) + G(t_2) + \cdots + G(t_n)].$$

Then we want to maximize

$$\sum_{j=1}^{n} G(t_j)$$

subject to

$$\sum_1^n t_j \leq N, \qquad 0 \leq t_j \leq N.$$

Let t_i^0 $(i = 1, \ldots, n)$ denote a solution to the problem. We define

$$I(\lambda) = \sum_{j=1}^n G(\lambda t_j^0 + (1 - \lambda)t_j).$$

Then $I(\lambda)$ is strictly concave as λ traverses the unit interval and achieves its maximum at $\lambda = 1$. Equivalently, $I'(1) \geq 0$ or

$$\sum_{j=1}^n g(t_j^0)t_j^0 \geq \sum_{j=1}^n g(t_j^0)t_j.$$

By the Neyman-Pearson lemma there exists a constant h such that

$$t_j^0 \begin{cases} = N & \text{if } g(t_j^0) > h, \\ = 0 & \text{if } g(t_j^0) < h, \end{cases}$$

and $g(t_j^0) = h$ otherwise. A consistent choice for t_j^0 satisfying these relations is $t_j^0 = N/n$ for all j. Next set $h = g(t_j^0)$. The expected gain from this policy is

$$\sum_{i=1}^k \frac{p_i}{N} n_i G(m_i) = \sum_1^k \frac{p_i}{m_i} G(m_i).$$

Of course,

$$\sum_{i=1}^k \frac{1}{m_i} = \sum_{i=1}^k \frac{n_i}{N} = 1.$$

Problem 6. Let $x_i = 1/m_i$ represent a search procedure, and let x_i^0 denote the optimum. A direct calculation shows that

$$I(\lambda) = \sum_{i=1}^k p_i[\lambda x_i^0 + (1 - \lambda)x_i]G\left(\frac{1}{\lambda x_i^0 + (1 - \lambda)x_i}\right)$$

is strictly concave. Thus, by the method of Section 8.2, we consider $I'(1) \geq 0$, that is,

$$\sum_{i=1}^k p_i \left\{G\left(\frac{1}{x_i^0}\right) - \frac{1}{x_i^0} g\left(\frac{1}{x_i^0}\right)\right\} (x_i^0 - x_i) \geq 0.$$

By the Neyman-Pearson lemma there exists a constant h such that

$$x_i^0 = \begin{cases} 1 \\ \dfrac{1}{N} \end{cases} \quad \text{if} \quad G(m_i^0) - m_i^0 g(m_i^0) \begin{cases} > \dfrac{h}{p_i}, \\ < \dfrac{h}{p_i}, \end{cases}$$

and x_i^0 is arbitrary otherwise. Let $f(y) = G(y) - yg(y)$. Note that $f'(y) = -yg'(y) > 0$, which means that f is increasing. Now we define

$$x_i^0 = \frac{1}{f^{-1}(h/p_i)} \quad \text{if } \frac{h}{p_i} \le f(N) \quad \text{and} \quad x_i^0 = \frac{1}{N} \quad \text{if } \frac{h}{p_i} > f(N).$$

If h is chosen very large, then $\sum x_i^0 = kN^{-1} < 1$; if h is chosen sufficiently small, then $x_i^0 \ge 1$ for all i implies $\sum x_i^0 > 1$, since $f(0) = 0$. Since $\sum x_i^0$ is a continuous function of k, we obtain a solution by continuity. It is easy to show that the solution constructed obeys the consistency requirements stated above.

Problem 8. See [54].

Problem 9. By virtue of the Neyman-Pearson lemma, since $\max_\phi K(\phi, \psi^0)$ and $\min_\psi K(\phi^0, \psi)$ equal $K(\phi^0, \psi^0)$, we deduce that

$$\phi^0 = \begin{cases} 1 & \text{if } f(t)\psi^0(t) > h, \\ \text{arbitrary} & \text{if } f(t)\psi^0(t) = h, \\ 0 & \text{if } f(t)\psi^0(t) < h, \end{cases} \qquad \psi^0 = \begin{cases} 1 & \text{if } f(t)\phi^0(t) < k, \\ \text{arbitrary} & \text{if } f(t)\phi^0(t) = k, \\ 0 & \text{if } f(t)\phi^0(t) > k. \end{cases} \tag{***}$$

(We consider only $0 < A, B < 1$; if either A or B equals 0 or 1, the hypothesis of the problem is violated. Clearly, $k \ge 0$, since $k < 0$ implies $\psi^0 \equiv 0$; and $h \ge 0$, since $h < 0$ implies $\phi^0 \equiv 1$. We prove that $k = 0$ is impossible.

Suppose the contrary. Then $A > 0$ means that $\psi^0(t_0) > 0$ for some t_0, which implies $\phi^0(t_0) = 0$. Hence $f(t_0)\psi^0(t_0) \le h$, and thus $h > 0$. Moreover, for any t, if $\psi^0(t) > 0$, then $\phi^0(t) = 0$ as above; and if $\psi^0(t) = 0$, then $f(t)\psi^0(t) = 0 < h$, implying $\phi^0(t) = 0$. Hence $\phi^0 \equiv 0$, a contradiction. It follows that $k > 0$. However, in general h may be zero.

Let h and k be defined by the equations $h/f(P) = 1$ and $k/f(Q) = 1$, respectively, where P and Q satisfy (+) and the inequalities $0 < P < Q < 1$. On $[0, P)$, we have $h/f(t) > 1 \ge \psi^0(t)$, which implies $\phi^0(t) = 0$, and $k/f(t) > 1 \ge \phi^0(t)$, which implies $\psi^0(t) = 1$. On (P, Q), we have $k/f(t) > 1$, which implies $\psi^0(t) = 1$; hence $\psi^0(t) > h/f(t)$, and thus $\phi^0(t) = 1$. On $(Q, 1)$ we must have $\phi^0(t) = k/f(t)$ and $\psi^0(t) = h/f(t)$. By construction, $k = f(Q)$ and $h = f(P)$, and ϕ^0 and ψ^0 are now completely specified. These "solutions" clearly satisfy the consistency requirements (***). Note that

$$\int_0^1 \phi^0(t)\,dt = Q - P + f(Q)\int_Q^1 dt/f(t) = B$$

and

$$\int_0^1 \psi^0(t)\,dt = Q + f(P)\int_Q^1 dt/f(t) = A,$$

as required.

Problem 10. Similar to Problem 9.

Problem 11. If the cases are not mutually exclusive, then there is a pair A, B having a solution P_1, Q_1 of (+) satisfying $0 < P_1 < Q_1 < 1$, and a solution P_2, Q_2 of (++) satisfying $0 \leq Q_2 \leq P_2 \leq 1$.

$$Q_1 + f(Q_1) \int_{Q_1}^1 dt/f(t) > Q_1 + f(P_1) \int_{Q_1}^1 dt/f(t) = A = P_2 + f(P_2) \int_{P_2}^1 dt/f(t);$$

and since

$$x + f(x) \int_x^1 dt/f(t)$$

is strictly increasing, we have $Q_1 > P_2$. Similarly,

$$P_1 + f(P_1) \int_{P_1}^1 dt/f(t) < Q_1 + f(P_1) \int_{Q_1}^1 dt/f(t) = P_2 + f(P_2) \int_{P_2}^1 dt/f(t)$$

implies $P_1 < P_2$. But

$$\begin{aligned} Q_1 - P_1 + f(Q_1) \int_{Q_1}^1 dt/f(t) = B &= f(Q_2) \int_{P_2}^1 dt/f(t) \\ &\leq f(P_2) \int_{P_2}^1 dt/f(t) \\ &< P_2 - P_1 + f(P_2) \int_{P_2}^1 dt/f(t) \end{aligned}$$

reduces to

$$Q_1 + f(Q_1) \int_{Q_1}^1 dt/f(t) < P_2 + f(P_2) \int_{P_2}^1 dt/f(t),$$

which yields $Q_1 < P_2$, a contradiction.

Problem 12. Consider the pay-off function

$$\begin{aligned} V(t, p) &= A(t) \exp\left[- \int_0^t r(u)p(u)\, du\right] \\ &= \exp\left[- \left(\int_0^t r(u)p(u)\, du - \log A(t)\right)\right]. \end{aligned}$$

By Jensen's inequality,

$$\begin{aligned} \int_0^1 V(t, p)\, dx_0(t) &\geq \exp\left[- \int_0^1 \left(\int_0^t r(u)p(u)\, du - \log A(t)\right) dx_0(t)\right] \\ &= \exp\left[- \int_0^1 r(t)p(t)\, dt + \log A(1) + \int_0^1 r(t)p(t)x_0(t)\, dt \right. \\ &\qquad \left. - \int_0^1 \frac{A'(t)}{A(t)} x_0(t)\, dt\right] \\ &= \exp\left[- \int_0^1 r(t)p(t)(1 - x_0(t))\, dt + \log A(1) \right. \\ &\qquad \left. - \int_0^1 \frac{A'(t)}{A(t)} x_0(t)\, dt\right]. \end{aligned}$$

For Case I, this yields

$$\int_0^1 V(t, p)\, dx_0(t) \geq \exp\left[-\int_0^d r(t)p(t)\, dt - \int_d^1 r(c)p(t)\, dt + \log A(d) + r(c)\int_d^1 \frac{A'(t)}{A(t)r(t)}\, dt\right]$$

$$= A(d)\exp\left[\int_0^d (r(c) - r(t))p(t)\, dt - r(c)\,\delta - r(c)\int_c^d dt + r(c)\,\delta\right]$$

$$= A(d)\exp\left[\int_0^c (r(c) - r(t))p(t)\, dt + \int_c^d (r(t) - r(c))(1 - p(t))\, dt - \int_c^d r(t)\, dt\right]$$

$$\geq A(d)\exp\left[-\int_c^d r(t)\, dt\right] = M_1 = V.$$

For Case II and Case III, the analysis is similar and simpler

CHAPTER 9

Problem 1. The expected yield to the bettor for a strategy ϕ is given by the expression

$$K(\phi) = -1\int[1 - \phi_1(\xi) - \cdots - \phi_N(\xi)]\, dF(\xi) + (a_1 + 1)\iint \phi_1(\xi)L(\xi, \eta)\, dF(\xi)\, dG(\eta) + \cdots + (a_N + 1)\iint \phi_N(\xi)L(\xi, \eta)\, dF(\xi)\, dG(\eta)$$

$$= -1 + \int \phi_1(\xi)[1 + (a_1 + 1)(2G(\xi) - 1)]\, dF(\xi) + \cdots + \int \phi_N(\xi)[1 + (a_N + 1)(2G(\xi) - 1)]\, dF(\xi).$$

If ξ_0 is defined as previously (p. 242), it is clear that the coefficient of $\phi_N(\xi)$ is largest and positive for $\xi > \xi_0$, which implies that $\phi_N^0(\xi) = 1$ for $\xi > \xi_0$, where ϕ^0 denotes the optimal strategy. Let ζ_0 be the root of

$$1 + (a_1 + 1)(2G(\xi) - 1) = 0.$$

Then $\zeta_0 < \xi_0$ and the coefficient of $\phi_1(\xi)$ is largest and positive on $\zeta_0 < \xi < \xi_0$, which means that $\phi_1^0(\xi) = 1$ on this interval. On $[0, \zeta_0)$ he should fold as before, since all coefficients of the ϕ_i are negative.

Problem 2. The only change is that now

$$\int L(\xi, \eta)\, dG(\eta) = G(\xi-) + G(\xi+) - 1.$$

Wherever this is positive, $\phi_2^0(\xi) = 1$. Now the end points will usually be relevant, whereas they were not in the continuous case. On the other hand, there may be an entire interval that is not relevant. If $G(\xi-) + G(\xi+) - 1 < 0$ and $1 + 2(G(\xi-) + G(\xi+) - 1) > 0$, then $\phi_1^0 = 1$. If the latter quantity is negative, the player should fold.

Problem 3. Define c_1 by $G(c_1) = a/(a+2)$ (not necessarily unique). If $F(c_1) \geq a/[2(a+1)]$, the solution has the following form:

$$\psi^*(\eta) = \begin{cases} 0 & 0 \leq \eta \leq c_1, \\ 1 & \eta > c_1; \end{cases}$$

$$\phi^*(\xi) = \begin{cases} \text{arbitrary but satisfying} & \\ \displaystyle\int_0^{c_1} \phi^*(\xi)\, dF(\xi) = \frac{a[1 - F(c_1)]}{a+2} & 0 \leq \xi \leq c_1, \\ 1 & \xi > c_1. \end{cases}$$

The consistency requirements in this case are checked exactly as in the given model. The requirement $F(c_1) \geq a/[2(a+1)]$ is needed to ensure that

$$\int_0^{c_1} \phi^*(\xi)\, dF(\xi) = \frac{a[1 - F(c_1)]}{a+2}$$

is possible. If $F(c_1) < a/[2(a+1)]$, define c_2 by $F(c_2) = a/[2(a+1)]$ and

$$\psi^*(\eta) = \begin{cases} 0 & \eta \leq c_2, \\ 1 & \eta < c_2, \end{cases} \qquad \phi^*(\xi) = 1.$$

Then the coefficient of ϕ at c_2 is given by

$$-(a+2)F(c_2) + a[1 - F(c_2)] = a - 2(a+1)F(c_2) = 0,$$

and thus ϕ^* minimizes, since the coefficient is monotonic. Also the coefficient of ϕ at 0 is given by $2 - (a+2)(1 - G(c_2)) = -a + (a+2)G(c_2)$. But $F(c_1) < F(c_2)$ implies $c_1 < c_2$, and thus $G(c_2) \geq G(c_1) = a/(a+2)$. Thus $-a + (a+2)G(c_2) \geq -a + a = 0$, and since the coefficient of ϕ is monotonic ϕ^* maximizes.

Problem 4. The value of the game is

$$v = K(\boldsymbol{\phi}^*, \boldsymbol{\phi}^*) = -1$$

$$+ \sum_{i=1}^{n} \int_0^1 \phi_i^*(\xi) \left(2 + a_i \int_0^{\xi} \psi_i^*(\eta)\, d\eta - (a_i + 2) \int_{\xi}^1 \psi_i^*(\eta)\, d\eta\right) d\xi.$$

We know that the coefficient of ϕ_i^* is zero for $\xi < b_i$, and that ϕ_i^* is zero elsewhere except for the interval (c_i, c_{i+1}); thus we need compute the coefficient of ϕ_i^* only for this interval. Since $c_i \geq b_i$, the coefficient is

$$\begin{aligned} 2 + a_i\int_{b_i}^{\xi} d\eta - (a_i + 2)\int_{\xi}^{1} d\eta &= 2 + a_i(\xi - b_i) - (a_i + 2)(1 - \xi) \\ &= 2 + a_i(\xi - b_i) + (a_i + 2)(\xi - b_i) \\ &\qquad - (a_i + 2)(1 - b_i) \\ &= 2(a_i + 1)(\xi - b_i). \end{aligned}$$

Hence

$$\begin{aligned} v &= -1 + \sum_{i=1}^{n} \int_{c_i}^{c_{i+1}} 2(a_i + 1)(\xi - b_i)\, d\xi \\ &= -1 + \sum_{i=1}^{n} (a_i + 1)[(c_{i+1} - b_i)^2 - (c_i - b_i)^2] \\ &= -1 + \sum_{i=1}^{n} (a_i + 1)(c_{i+1} - c_i)(c_{i+1} + c_i - 2b_i). \end{aligned}$$

Problem 6. The value is

$$\begin{aligned} v = K(\phi^*, \psi^*) &= -1 + 2\int_0^1 [\phi_1^*(\xi) + \phi_2^*(\xi)]\, d\xi \\ &\quad + \int_0^1 \psi_1^*(\eta)\Big[-(a + 2)\int_0^{\eta} [\phi_1^*(\xi) + \phi_2^*(\xi)]\, d\xi \\ &\quad + a\int_{\eta}^1 [\phi_1^*(\xi) + \phi_2^*(\xi)]\, d\xi\Big]\, d\eta \\ &\quad + \int_0^1 \psi_2^*(\eta)\Big[-(a + 2) \int_0^1 \phi_1^*(\xi)\, d\xi \\ &\quad - (a + b + 2)\int_0^{\eta} \phi_2^*(\xi)\, d\xi + (a + b)\int_{\eta}^1 \phi_2^*(\xi)\, d\xi\Big]\, d\eta \\ &= -1 + 2(m_1 + 1 - c) + \int_c^d [-(a + 2)(m_1 + \eta - c) \\ &\quad + a(1 - \eta)]\, d\eta + \int_d^1 [-(a + 2)(m_1 + e - c) \\ &\quad - (a + b + 2)(\eta - e) + (a + b)(1 - \eta)]\, d\eta \\ &= 1 + 2m_i - 2c - (a + 2)(m_1 - c)(d - c) + a(d - c) \\ &\quad - (a + 1)(d^2 - c^2) - (a + 2)(m_1 + e - c)(1 - d) \\ &\quad + (a + b)(1 + e)(1 - d) + 2e(1 - d) \\ &\quad - (a + b + 1)(1 - d^2). \end{aligned}$$

Using $(a+2)m_1 = a(1-c)$ [equation (9.4.10)] in the second term, we obtain

$$\begin{aligned} v = 1 + 2m_i - 2c + 2c(a+1)(d-c) + (a+1)c^2 \\ + bd^2 - (a+b+1) - (a+2)(m_1+e-c)(1-d) \\ + (a+b)(1+e)(1-d) + 2e(1-d). \end{aligned}$$

Now using $(a+2)(m_1+e-c) = (a+b)(1-e)$ [equation (9.4.11)] in the third term from the end, we obtain

$$v = 2m_1 - 2c - (a+b) + c(a+1)(2d-c) + bd^2 + 2e(1-d)(a+b+1).$$

Problem 7. $c = \frac{31}{56}$; $d = \frac{97}{112}$; $e = \frac{41}{56}$; $m_1 = \frac{25}{112}$; $m_2 = \frac{3}{56}$. $v = -\frac{37}{112}$.

Problem 8. We see from Fig. 9.2 that we must have $\psi_1^*(\eta) = 1$ for $c < \eta < d$ and $\psi_2^*(\eta) = 1$ for $d < \eta < 1$. Also $\psi_1^*(\eta) = 0$ for $0 < \eta < c$, and ψ_2^* is arbitrary on that interval. However, we must have $M_1(c) = 0$ for the given $\langle\phi_1^*, \phi_2^*\rangle$ to be optimal, and this implies

$$\int_0^c \psi_2^*(\eta)\, d\eta = m_2.$$

Again from the figure we see that $\phi_2^*(\xi) = 1$ on $d < \xi < 1$ and $\phi_2^*(\xi) = 0$ on $0 < \xi < c$. However, $\phi_1^*(\xi)$ is arbitrary for $0 < \xi < c$, and ϕ_1^* and ϕ_2^* are arbitrary for $c < \xi < d$, except that $\phi_1^* + \phi_2^* = 1$. The requirement $N_1(d) = N_2(d)$ yields

$$\int_0^d \tilde{\phi}_2(\xi)\, d\xi = 1 - d.$$

Using this result and the condition $N_2(c) = 0$ yields

$$\int_0^c \tilde{\phi}_1(\xi)\, d\xi = m_1.$$

The last necessary requirement is that $N_2(\eta) \geq N_1(\eta)$ for $c < \eta < d$, and this yields the second part of (9.4.17).

It is clear that any strategy satisfying these conditions satisfies the consistency relations.

Problem 9. Let ϕ_1, ϕ_2 and ϕ_3 have the same meaning as before, but relax the sum constraint to the form $\phi_1 + \phi_2 + \phi_3 \leq 1$, so that $1 - \phi_1 - \phi_2 - \phi_3$ now represents the probability of folding initially. The pay-off is

$$\begin{aligned} K^*(\phi, \psi) &= K(\phi, \psi) \\ &+ (-1)\int_0^1\int_0^1 [1 - \phi_1(\xi) - \phi_2(\xi) - \phi_3(\xi)][\psi_1(\eta) + \psi_2(\eta) + \psi_3(\eta)]\, d\xi\, d\eta \\ &+ \int_0^1\int_0^1 [\phi_1(\xi) + \phi_2(\xi) + \phi_3(\xi)][1 - \psi_1(\eta) - \psi_2(\eta) - \psi_3(\eta)]\, d\xi\, d\eta, \end{aligned}$$

where $K(\phi, \psi)$ is as given.

Thus

$$K^*(\phi, \phi^*) = \text{constant} + \int \phi_1(\xi)\Big[1 + b\int_0^\xi (\phi_1^* + \phi_2^*)\, d\eta - b\int_\xi^1 (\phi_1^* + \phi_2^*)\, d\eta - b\int_0^1 \phi_3^*\, d\eta\Big] d\xi$$

$$+ \int \phi_2(\xi)\Big[1 + b\int_0^\xi (\phi_1^* + \phi_2^*)\, d\eta - b\int_\xi^1 (\phi_1^* + \phi_2^*)\, d\eta + a\int_0^\xi \phi_3^*\, d\eta - a\int_\xi^1 \phi_3^*\, d\eta\Big] d\xi$$

$$+ \int \phi_3(\xi)\Big[1 + b\int_0^1 \phi_1^*\, d\eta + a\int_0^\xi (\phi_2^* + \phi_3^*)\, d\eta - a\int_\xi^1 (\phi_2^* + \phi_3^*)\, d\eta\Big] d\xi.$$

The solution is given by $\phi_1^* = a/(a + b)$, $\phi_3^* = b/(a + b)$ on the interval (c, d), and $\phi_3^* = 1$ on the interval $(d, 1)$, where $c = 1 - 1/b$ and $d = 1 - 1/a$. The constants c and d are determined by setting the coefficients of both $\phi_1(\xi)$ and $\phi_3(\xi)$ equal to 0 at c.

Problem 10. Two cases arise. The indicated solution for $n = 3$ is valid so long as $a_3a_1 + a_3a_2 + a_2a_1 - 3a_1^2 \geq 0$. If the opposite inequality holds, the solution has the same general form except that the middle bet is the one that is not relevant on the middle interval. Specifically,

$$\phi_1(\xi) = \begin{cases} \dfrac{1}{a_1b_1}, \\ \dfrac{a_3}{a_1 + a_3}, \\ 0, \end{cases} \qquad \phi_2(\xi) = \begin{cases} \dfrac{1}{a_2b_1}, \\ 0, \\ 0, \end{cases}$$

$$\phi_3(\xi) = \begin{cases} \dfrac{1}{a_3b_1} & \text{for} \quad 0 \leq \xi \leq \xi_1, \\ \dfrac{a_1}{a_1 + a_3} & \text{for} \quad \xi_1 \leq \xi < \xi_2, \\ 1 & \text{for} \quad \xi_2 \leq \xi \leq 1, \end{cases}$$

where

$$\xi_1 = \frac{2a_1(a_3 - a_2)}{(a_1 + a_2)(a_2 + a_3)} a_2b_1$$

and

$$\xi_2 - \xi_1 = \frac{(a_1 + a_3)(3a_2^2 - a_1a_2 - a_1a_3 - a_2a_3)}{a_3(a_2 + a_1)(a_3 + a_2)}.$$

Problem 11. As in Section 9.6, it suffices to describe the strategies of one player, since the game is symmetric. If A draws ξ, we write

$\phi_1(\xi)$ = probability A will bet c and fold if B bets b or a,
$\phi_2(\xi)$ = probability A will bet c and see if B bets b but fold if B bets a,
$\phi_3(\xi)$ = probability A will bet c and fold if B bets b but see if B bets a,
$\phi_4(\xi)$ = probability A will bet c and see if B bets b or a,
$\phi_5(\xi)$ = probability A will bet b and fold if B bets a,
$\phi_6(\xi)$ = probability A will bet b and see if B bets a,
$\phi_7(\xi)$ = probability A will bet a.

These functions are subject to the constraints $\phi_i(\xi) \geq 0, \sum_{i=1}^{7} \phi_i(\xi) = 1$. The form of the optimal strategy in this game depends on the ratios a/c and b/c. The various forms are fairly complex and are discovered by using a combination of intuitive considerations and formal manipulations. Once the correct form of optimal strategy has been discovered, the conditions under which it holds and the critical numbers involved may be determined as in Section 9.6.

We describe the solution for two of the five cases that arise.

Case I. $a^2b + a^2c - 2ac^2 - 4abc - 2ab^2 - b^3 + 7b^2c \leq 0$; $2c - b \geq 0$.

For $0 < \xi < \xi_1$:

$$\phi_1^*(\xi) = \frac{ab}{ab + ac + bc}, \quad \phi_5^*(\xi) = \frac{ac}{ab + ac + bc}, \quad \phi_7^*(\xi) = \frac{bc}{ab + ac + bc}.$$

For $\xi_1 < \xi < \xi_2$: $\phi_1^*(\xi) = \dfrac{a}{a + c}, \quad \phi_7^*(\xi) = \dfrac{c}{a + c}.$

For $\xi_2 < \xi < \xi_3$: $\phi_2^*(\xi) = \dfrac{a}{a + c}, \quad \phi_7^*(\xi) = \dfrac{c}{a + c}.$

For $\xi_3 < \xi < \xi_4$: $\phi_2^*(\xi) = 1.$

For $\xi_4 < \xi < \xi_5$: $\phi_6^*(\xi) = 1.$

For $\xi_5 < \xi < 1$: $\phi_7^*(\xi) = 1.$

Here, $$\eta_1 = \xi_1 = \frac{2(a - b)(b - c)^2(ab + ac + bc)}{a(b + c)(a^2b + a^2c + 2ac^2 + b^3 - 2abc - 3b^2c)},$$

$$\eta_2 = \xi_2 - \xi_1 = \frac{(a + c)(b - c)(a^2b + a^2c + 2ac^2 + 3b^3 - 5b^2c - 2ab^2)}{a(b + c)(a^2b + a^2c + 2ac^2 + b^3 - 2abc - 3b^2c)},$$

$$\eta_3 = \xi_3 - \xi_2 = \frac{2(a + c)(a - b)^2(2c - b)}{a(a^2b + a^2c + 2ac^2 + b^3 - 2abc - 3b^2c)},$$

$$n_4 = \xi_4 - \xi_3 = \frac{2(a - b)^2(b - c)}{a^2b + a^2c + 2ac^2 + b^3 - 2abc - 3b^2c},$$

$$\eta_5 = \xi_5 - \xi_4 = \frac{2c(a - b)(b - c)}{a^2b + a^2c + 2ac^2 + b^3 - 2abc - 3b^2c},$$

$$\eta_6 = 1 - \xi_5 = \frac{c(a^2b^2 + a^2c^2 - 6ab^2c + 2a^2bc + 2b^3c + b^4 + 2ac^3 - 7b^2c^2 + 4abc^2)}{a(b + c)(a^2b + a^2c + 2ac^2 + b^3 - 2abc - 3b^2c)}.$$

Case II. $a^2b + a^2c - 2ac^2 - 4abc - 2ab^2 - b^3 + 7b^2c \leq 0; 2c - b \leq 0.$

For $0 < \xi < \xi_1$:

$$\phi_1^*(\xi) = \frac{ab}{ab + ac + bc}, \quad \phi_5^*(\xi) = \frac{ac}{ab + ac + bc}, \quad \phi_7^*(\xi) = \frac{bc}{ab + ac + bc}.$$

For $\xi_1 < \xi < \xi_2$: $\phi_1^*(\xi) = \dfrac{a}{a + c}$, $\phi_7^*(\xi) = \dfrac{c}{a + c}$.

For $\xi_2 < \xi < \xi_3$: $\phi_1^*(\xi) = \dfrac{b - 2c}{b - c}$, $\phi_2^*(\xi) = \dfrac{c}{b - c}$.

For $\xi_3 < \xi < \xi_4$: $\phi_6^*(\xi) = 1$.

For $\xi_4 < \xi < 1$: $\phi_7^*(\xi) = 1$.

Here, $\eta_1 = \xi_1 = \dfrac{2(a - b)(b - c)^2(ab + ac + bc)}{a(b + c)(a^2b + a^2c + 2ac^2 + b^3 - 2abc - 3b^2c)}$,

$$\eta_2 = \xi_2 - \xi_1 = \frac{(a + c)(3a^2c^2 - a^2b^2 - 2ab^2c + 2a^2bc - 6b^3c + b^4 + 2ab^3 - 2ac^3 + 9b^2c^2 - 6abc^2)}{a(b + c)(a^2b + a^2c + 2ac^2 + b^3 - 2abc - 3b^2c)},$$

$$\eta_3 = \xi_3 - \xi_2 = \frac{2(a - b)^2(b - c)}{a^2b + a^2c + 2ac^2 + b^3 - 2abc - 3b^2c},$$

$$\eta_4 = \xi_4 - \xi_3 = \frac{2c(a - b)(b - c)}{a^2b + a^2c + 2ac^2 + b^3 - 2abc - 3b^2c},$$

$$\eta_5 = 1 - \xi_4 = \frac{c(a^2b^2 + a^2c^2 - 6ab^2c + 2a^2bc + 2b^3c + b^4 + 2ac^3 - 7b^2c^2 + 4abc^2)}{a(b + c)(a^2b + a^2c + 2ac^2 + b^3 - 2abc - 3b^2c)}.$$

Problem 12. Let I_n represent the maximum expected yield if an optimal procedure is employed when a total of N opportunities are available. Let $\phi(\xi)$ denote the probability of keeping ξ when it is offered at the first trial. Then clearly

$$I_n = \max_{0 \leq \phi \leq 1} \left\{ \int \xi\phi(\xi)\, dF(\xi) + I_{n-1} \int [1 - \phi(\xi)]\, dF(\xi) \right\}$$

$$= \max_{0 \leq \phi \leq 1} \left\{ \int \phi(\xi)(\xi - I_{n-1})\, dF(\xi) + I_{n-1} \right\}.$$

The maximum is achieved when $\phi(\xi) = 1$ for $\xi \geq I_{n-1}$ and $\phi(\xi) = 0$ for $\xi < I_{n-1}$. Thus I_n satisfies the recursion formula

$$I_n = I_{n-1} + \int_{I_{n-1}}^{1} (\xi - I_{n-1})\, dF(\xi).$$

In the case of the uniform distribution $I_n = (1 + I_{n-1}^2)/2$. For $N = 2$, the critical value is 5/8. For $N = 3$, the critical value is 89/128.

Problem 13. Let S_n represent the maximum expected yield if an optimal procedure is employed when a total of N opportunities are available. Let $\phi(\xi)$ denote the probability of selecting the offer ξ at the first trial. Then clearly

$$S_n = \max_{0 \leq \phi \leq 1} \left\{ \int (\xi + I_{n-1})\phi(\xi)\, dF(\xi) + S_{n-1} \int [1 - \phi(\xi)]\, dF(\xi) \right\},$$

where I_n has the same meaning as in Problem 12. We deduce as above that the maximum is achieved when

$$\phi(\xi) = \begin{cases} 1 & \text{for} \quad \xi > S_{n-1} - I_{n-1}, \\ 0 & \text{for} \quad \xi < S_{n-1} - I_{n-1}, \end{cases}$$

and

$$S_n = \int_{S_{n-1} - I_{n-1}}^{1} (\xi + I_{n-1} - S_{n-1})\, dF(\xi) + S_{n-1}.$$

In the case of the uniform distribution the critical number with respect to the first selection is 39, 138. The second choice is made by applying the criteria determined in Problem 12.

In Problems 14–18, let $C_i(\xi, \phi)$ denote the coefficient of Q_i when $K(\phi, \psi)$ has been rearranged in a form analogous to that of equation (9.2.1). Also let $D_i(\eta, \phi)$ denote the coefficient for ϕ_i in the other rearrangement of $K(\phi, \psi)$.

Problem 14. $\psi_i^*(\eta) = 1$ is optimal, since $D_i(\eta, \phi) \leq D_i(0, \phi) \leq 0$ for all ϕ. ϕ_i^* is the characteristic function for the interval

$$\left[\frac{a_i + c_i - d_i}{2d_i}, 1\right] \cap [0, 1].$$

Problem 15. This problem can be reduced to Problem 14 with the roles of the players interchanged.

Problem 16. Using the rearrangements suggested above, we can verify directly that $K(\phi^*, \psi^*) = \max_\phi K(\phi, \psi^*)$ and $K(\phi^*, \psi^*) = \min_\psi K(\phi^*, \psi)$.

Problem 17. Let ϕ^*, ψ^* be any pair of optimal strategies, and let ψ^0 be obtained from ψ^* by replacing a fixed component ψ_i^* by the characteristic function of the interval $[\eta_i, 1]$, where

$$1 - \eta_i = \int_0^1 \psi_i^*\, d\eta.$$

The pair of strategies $\{\phi^*, \psi^0\}$ is still optimal since $K(\phi^*, \psi^0) \leq K(\phi^*, \psi) = \min_\psi K(\phi^*, \psi)$, and $K(\phi^*, \psi^0) = \max_\phi K(\phi, \psi^*)$.

Problem 18. Let $\{\phi^*, \psi^*\}$ be a pair of optimal strategies, with ψ^* of the form given in the previous problem. Then $C_i(\xi, \psi^*)$ has a constant value λ_i in the interval $0 \leq \xi \leq \eta_i$. If $\lambda_i < \lambda_j$ and $0 < \eta_i < 1$, then the value of η_i can be altered—increasing the value of λ_i until $\lambda_i = \lambda_j$—without destroying the optimal character of the pair $\{\phi^*, \psi^*\}$. Thus there is an optimal strategy with $\lambda_i = \lambda^*$ for all indices for which $\eta_i \neq 1$. For these indices, η_i is as described in the problem.

Problem 19. Let $\phi_i(\xi)$ denote the probability of calling i/n, where

$$i = 0, 1, \ldots, n \qquad \text{and} \qquad \sum \phi_i(\xi) = 1.$$

Let a_i denote the probability of Player II's believing Player I if he calls i/n. Corresponding to these strategies, we have

$$\begin{aligned} K(\phi, a) &= -\int_0^1 \phi_0(\xi) a_0 \, d\xi + \int_0^1 \phi_0(\xi)(1 - a_0) \, d\xi + \int_0^1 \phi_n(\xi) a_n \, d\xi \\ &\quad - \int_0^1 \phi_n(\xi)(1 - a_n) \, d\xi \\ &\quad + \sum_{i=1}^{n-1} \left[\int_0^{i/n} \phi_i(\xi) a_i \, d\xi - \int_0^{i/n} \phi_i(\xi)(1 - a_i) \, d\xi - \int_{i/n}^1 \phi_i(\xi) a_i \, d\xi \right. \\ &\quad \left. + \int_{i/n}^1 \phi_i(\xi)(1 - a_i) \, d\xi \right] \\ &= (1 - 2a_0) \int_0^1 \phi_0(\xi) \, d\xi + (2a_n - 1) \int_0^1 \phi_n(\xi) \, d\xi \\ &\quad + \sum_{i=1}^{n} \left[(2a_i - 1) \int_0^{i/n} \phi_i(\xi) \, d\xi - (2a_i - 1) \int_{i/n}^1 \phi_i(\xi) \, d\xi \right]. \end{aligned}$$

Clearly, $\phi_0^0(\xi) = 0$ and $\phi_n^0(\xi) = 0$. Since $a_i = \frac{1}{2}$ for all i is an equalizer for Player II, we look for an equalizer for Player I. This would require

$$\int_0^{i/n} \phi_i(\xi) \, d\xi = \int_{i/n}^1 \phi_i(\xi) \, d\xi \qquad (i = 1, \ldots, n - 1).$$

Continued

One way of doing this is to have

ϕ_1^0	ϕ_2^0	ϕ_3^0	$\cdots$	ϕ_{n-2}^0	ϕ_{n-1}^0	Interval of definition
$\frac{1}{2}+\frac{1}{2(n-1)}$	$\frac{1}{2(n-1)}$	$\frac{1}{2(n-1)}$	$\cdots$	$\frac{1}{2(n-1)}$	$\frac{1}{2(n-1)}$	$\left(0, \frac{1}{n}\right)$
$\frac{1}{2}$	$\frac{1}{2}$	0	$\cdots$	0	0	$\left(\frac{1}{n}, \frac{2}{n}\right)$
0	$\frac{1}{2}$	$\frac{1}{2}$	$\cdots$	0	0	$\left(\frac{2}{n}, \frac{3}{n}\right)$
$\vdots$						
0	0	0	$\cdots$	$\frac{1}{2}$	$\frac{1}{2}$	$\left(\frac{n-2}{n}, \frac{n-1}{n}\right)$
$\frac{1}{2(n-1)}$	$\frac{1}{2(n-1)}$	$\frac{1}{2(n-1)}$	$\cdots$	$\frac{1}{2(n-1)}$	$\frac{1}{2}+\frac{1}{2(n-1)}$	$\left(\frac{n-1}{n}, 1\right)$

APPENDIXES

In order to make this volume as complete in itself as possible and accessible to the mature student who has only a modicum of mathematical training beyond the calculus, we devote these appendixes to a brief review of those topics of upper-class undergraduate mathematics which play a fundamental role in the studies of this book. Where an analysis requires more advanced mathematics, the essentials are usually directly incorporated in the text.

In Appendix A we summarize the salient features of the theory of vector spaces and matrices. Our approach to this theory, unlike the usual approach, is primarily geometric rather than algebraic. We do not develop matrix theory abstractly in terms of general rings and fields, but restrict the scalar field of our vector spaces to the real or complex numbers. Proofs that are not easily accessible in the literature are indicated in detail, and in all circumstances we have at least sketched the arguments verifying the main statements of the theory.

Appendix B treats some aspects of the theory of convex sets and convex functions. Much of the theory of finite games, linear programming, and mathematical economics is based on operations with sets and functions derived from linear and convex inequalities. We summarize here the appropriate versions of the separation theorems for convex sets and discuss simple forms of the duality theory of convex cones. Some advanced topics in the theory of convex sets—for example, the development of the theory of conjugate convex functions and certain convexity results connected with moment spaces—are presented in the text.

Appendix C reviews several miscellaneous facts regarding functions, integrals, and fixed-point theorems which are explicitly called upon in making some of the analyses in this volume. The concepts of semicontinuity and equi-continuity are defined precisely, and some of the properties of these functions are listed. The important Kakutani fixed-point theorem is noted in Section C.2. The relevant properties of Lebesgue-Stieltjes integrals are recorded for reference in Section C.3.

The following books and articles may profitably be consulted in connection with the topics covered in these appendixes:

Appendix A: Perlis [195], Birkoff and McLane [26], Gantmacher [101].
Appendix B: Fenchel [81], Hardy, Littlewood, and Pólya [118, Chap. 2].
Appendix C: Graves [112], Rudin [209].

APPENDIX A

VECTOR SPACES AND MATRICES

A.1 Euclidean and unitary spaces. A real vector space may be defined as follows: We have a set V; an associative and commutative addition, $+$, of pairs of its elements; a unique zero element $\boldsymbol{\theta}$; and a multiplication of elements of V by real numbers, for which, with α, β real and $\mathbf{x}$, $\mathbf{y}$ in V

$$\mathbf{x} + \boldsymbol{\theta} = \mathbf{x} \qquad 1 \cdot \mathbf{x} = \mathbf{x}, \qquad 0 \cdot \mathbf{x} = \boldsymbol{\theta},$$

$$\alpha(\mathbf{x} + \mathbf{y}) = \alpha\mathbf{x} + \alpha\mathbf{y}, \qquad (\alpha + \beta)\mathbf{x} = \alpha\mathbf{x} + \beta\mathbf{x}, \qquad \alpha(\beta\mathbf{x}) = (\alpha\beta)\mathbf{x}.$$

As a prime example we have the Euclidean n-dimensional space E^n, which consists of the set of all ordered n-tuples $\mathbf{x} = \langle x_1, x_2, \ldots, x_n \rangle$ of real numbers $x_1, x_2, \ldots, x_n$, with addition and multiplication by a real number γ defined by

$$\langle x_1, x_2, \ldots, x_n \rangle + \langle y_1, y_2, \ldots, y_n \rangle = \langle x_1 + y_1, x_2 + y_2, \ldots, x_n + y_n \rangle,$$

$$\gamma\langle x_1, x_2, \ldots, x_n \rangle = \langle \gamma x_1, \gamma x_2, \ldots, \gamma x_n \rangle.$$

Here $\boldsymbol{\theta}$ (which we shall henceforth call $\mathbf{0}$) is $\langle 0, 0, \ldots, 0 \rangle$; each $\mathbf{x} = \langle x_1, x_2, \ldots, x_n \rangle$ has a negative $-\mathbf{x} = \langle -x_1, -x_2, \ldots, -x_n \rangle$ for which $\mathbf{x} + (-\mathbf{x}) = \mathbf{0}$ (the same is true abstractly: $\mathbf{x} + (-1) \cdot \mathbf{x} = 1 \cdot \mathbf{x} + (-1) \cdot \mathbf{x} = [1 + (-1)] \cdot \mathbf{x} = 0 \cdot \mathbf{x} = \boldsymbol{\theta}$); and we may define subtraction by $\mathbf{y} - \mathbf{x} = \mathbf{y} + (-\mathbf{x})$. It should be noted that in this example the real numbers may be replaced by the complex numbers; in that case the corresponding vector space U^n is referred to as the unitary n-dimensional space.

Actually E^n (and U^n) are not merely vector spaces; we can perform the additional operation of forming the inner product $(\mathbf{x}, \mathbf{y})$ of two elements $\mathbf{x}$, $\mathbf{y}$, which has a value

$$(\mathbf{x}, \mathbf{y}) = \sum_{i=1}^{n} x_i \bar{y}_i,$$

where $\bar{y}_i$ is the complex conjugate of y_i. (Two elements $\mathbf{x}$, $\mathbf{y}$ are said to be *orthogonal* if $(\mathbf{x}, \mathbf{y}) = 0$.) By means of this operation we define the usual distance function

$$d(\mathbf{x}, \mathbf{y}) = (\mathbf{x} - \mathbf{y}, \mathbf{x} - \mathbf{y})^{\frac{1}{2}} = \left(\sum_{i=1}^{n} |x_i - y_i|^2 \right)^{\frac{1}{2}},$$

where $d(\mathbf{x}, \mathbf{y}) = d(\mathbf{y}, \mathbf{x}) \geq 0$ (with equality holding if and only if $\mathbf{x} = \mathbf{y}$), and the triangle inequality

$$d(\mathbf{x}, \mathbf{y}) \leq d(\mathbf{x}, \mathbf{z}) + d(\mathbf{z}, \mathbf{y}).$$

Conventionally, x_i is called the ith *component* of $\mathbf{x} = \langle x_1, x_2, \ldots, x_n \rangle$, and $d(\mathbf{0}, \mathbf{x})$ is called the *length* of $\mathbf{x}$. In all that follows we shall be considering E^n and U^n as the vector spaces involved, and any statements made concerning E^n may be applied with equal force to U^n (but not conversely). To eliminate reference to real and complex numbers we shall refer to the numbers involved as *scalars* (α, β, γ, etc.), and to the elements of a vector space as *vectors* ($\mathbf{x}$, $\mathbf{y}$, $\mathbf{z}$, etc.).

Exercises

1. Let $\mathfrak{P}^n$ be the set of all polynomials $p(t)$ defined on $-\infty < a \leq t \leq b < \infty$, with real (complex) coefficients such that the degree of $p(t)$ is at most $n - 1$. Verify that under the ordinary notions of addition of polynomials and multiplication by a real (complex) scalar, $\mathfrak{P}^n$ is a real (complex) vector space. Show that an inner product for p_1 and p_2 in $\mathfrak{P}^n$ may be defined by

$$(p_1, p_2) = \int_a^b p_1(t)\bar{p}_2(t)\, dt.$$

2. Let C be the set of complex numbers, addition and multiplication by *real* scalars being defined as usual. Show that C is a real vector space, which possesses an inner product for $x = a + bi$ and $y = c + di$ given by $(x, y) = ac + bd$.

3. In E^3, for instance, we can associate each vector $\langle x, y, z \rangle$ with the point in three-dimensional space having Cartesian coordinates $\{x, y, z\}$. Thus in referring to sets of vectors we may speak of points, lines, and planes. Interpret this statement.

4. For E^n let

$$d(\mathbf{x}, \mathbf{y}) = \max_{1 \leq i \leq n} |x_i - y_i|.$$

Show that $d(\mathbf{x}, \mathbf{y})$ satisfies the usual requirements for a distance function; namely,

$$d(\mathbf{x}, \mathbf{y}) = d(\mathbf{y}, \mathbf{x}) \geq 0 \qquad \text{(with equality if and only if } \mathbf{x} = \mathbf{y}\text{)},$$
$$d(\mathbf{x}, \mathbf{y}) \leq d(\mathbf{x}, \mathbf{z}) + d(\mathbf{z}, \mathbf{y}).$$

5. Using the usual distance function for E^n and noticing the relationships between "orthogonal" and the usual definition of "perpendicular," state and prove a general Pythagorean theorem for E^n.

A.2 Subspaces, linear independence, basis, direct sums, orthogonal complements. A subspace V_1 of a vector space V is a subset which is closed under addition and scalar multiplication: if $\mathbf{x}$ and $\mathbf{y}$ are in V_1, and α is a scalar, then $\mathbf{x} + \mathbf{y}$ and $\alpha\mathbf{y}$ are in V_1. Under these operations V_1 is a vector space in its own right. For example, the set of all linear combinations

$$\sum_{i=1}^{k} \alpha_i \mathbf{x}^i = \alpha_i \mathbf{x}^1 + \cdots + \alpha_k \mathbf{x}^k$$

of a fixed set $\{\mathbf{x}^1, \mathbf{x}^2, \ldots, \mathbf{x}^k\}$ of *vectors* forms a subspace called the span of $\{\mathbf{x}^1, \mathbf{x}^2, \ldots, \mathbf{x}^k\}$. Alternatively, if V_1 is the span of $\{\mathbf{x}^1, \ldots, \mathbf{x}^k\}$, then $\{\mathbf{x}^1, \ldots, \mathbf{x}^k\}$ *spans* V_1.

A subsct $\{\mathbf{x}^1, \ldots, \mathbf{x}^k\}$ is *linearly independent* if

$$\sum_{i=1}^{k} \alpha_i \mathbf{x}^i = \mathbf{0}$$

implies $\alpha_i = 0$ $(i = 1, \ldots, k)$. A *linearly independent* subset $\{\mathbf{x}^1, \ldots, \mathbf{x}^k\}$ which *spans* V is a *basis* for V. In E^n,

$$\mathbf{e}^1 = \langle 1, 0, 0, \ldots, 0\rangle, \qquad \mathbf{e}^2 = \langle 0, 1, 0, \ldots, 0\rangle, \qquad \ldots, \qquad \mathbf{e}^n = \langle 0, 0, \ldots, 0, 1\rangle$$

clearly forms a basis. (Not all vector spaces have finite bases; a trivial example is the case in which V has exactly one element, $\mathbf{0}$, where no subset is independent. In this case we shall not usually regard V as a vector space.)

If $\{\mathbf{x}^1, \ldots, \mathbf{x}^k\}$ spans V, then a subset forms a basis for V. This is clear if the set is linearly independent; otherwise we have

$$\sum_{i=1}^{k} \alpha_i \mathbf{x}^i = \mathbf{0},$$

with, say, $\alpha_1 \neq 0$. It follows that

$$\mathbf{x}^1 = \sum_{i=2}^{k} (-\alpha_i \cdot \alpha_1^{-1})\mathbf{x}^i,$$

and clearly $\{\mathbf{x}^2, \ldots, \mathbf{x}^k\}$ spans V. If this subset is not independent, we repeat the argument until we arrive at one that is. The advantage of recognizing a basis $\{\mathbf{y}^1, \ldots, \mathbf{y}^l\}$ for V is that each $\mathbf{x}$ in V is *uniquely* expressible in the form

$$\sum_{i=1}^{l} \alpha_i \mathbf{y}^i,$$

where we refer to the unique scalars α_i as the coordinates of $\mathbf{x}$ relative to the basis $\{\mathbf{y}_1, \ldots, \mathbf{y}_l\}$.

Some important facts about bases are the following, which we state without proof. Each nontrivial subspace V_1 of E^n has a basis, and each basis for V_1 has the same number of elements, the *dimension* (abbreviated dim) of V_1. If $V_1 = \{\mathbf{0}\}$, then dim $V_1 = 0$. If $V_1 \neq E^n$ (i.e., V_1 is a *proper* subspace), then each basis $\{\mathbf{x}^1, \mathbf{x}^2, \ldots, \mathbf{x}^k\}$ of V_1 can be enlarged to a basis $\{\mathbf{x}^1, \mathbf{x}^2, \ldots, \mathbf{x}^k, \mathbf{y}^1, \ldots, \mathbf{y}^l\}$ of E^n (thus V_1 has dimension $k < n =$ dimension E^n), and indeed any linearly independent subset can be so enlarged, being a basis of its span. Consequently, any linearly independent subset of E^n with n elements is already a basis for E^n. It should be noted that V_1 as a vector space is essentially E^k imbedded in E^n, since the one-to-one (1-1) mapping

$$\langle \alpha_1, \alpha_2, \ldots, \alpha_k \rangle \to \sum_{i=1}^{k} \alpha_i \mathbf{x}^i$$

(α_i scalars) of E^k onto V_1 preserves all sums and scalar multiples, as does its inverse.

In E^n or U^n a basis is *orthonormal* if

$$(\mathbf{y}^i, \mathbf{y}^j) = \begin{cases} 0 & i \neq j, \\ 1 & i = j, \end{cases}$$

i.e., if the distinct elements of the basis are orthogonal and each is of length 1. The components of a vector relative to an orthonormal basis are computed from the inner product:

$$\mathbf{x} = \sum_{i=1}^{k} (\mathbf{x}, \mathbf{y}^i)\mathbf{y}^i$$

for $\mathbf{x}$ in the span of $\{\mathbf{y}^1, \ldots, \mathbf{y}^k\}$, since the difference of the two sides is orthogonal to each $\mathbf{y}^i$, and hence to itself. *Every subspace of E^n (or U^n) has an orthonormal basis.*

If V_1 and V_2 are subspaces of V and

$$V = V_1 + V_2 = \{\mathbf{x} + \mathbf{y} \mid \mathbf{x} \in V_1, \mathbf{y} \in V_2\}$$

while $V_1 \cap V_2 = \{\mathbf{0}\}$, then V is the *direct sum* of V_1 and V_2; symbolically $V = V_1 \oplus V_2$. Each $\mathbf{z}$ in V is then uniquely expressible as $\mathbf{x} + \mathbf{y}$, where $\mathbf{x} \in V_1, \mathbf{y} \in V_2$; for $\mathbf{x} + \mathbf{y} = \mathbf{x}' + \mathbf{y}'$ implies that $\mathbf{x} - \mathbf{x}' = \mathbf{y}' - \mathbf{y}$ is in both V_1 and V_2, and hence equals zero. Consequently, if $\{\mathbf{x}^1, \ldots, \mathbf{x}^k\}$ is a basis for V_1 and $\{\mathbf{y}^1, \ldots, \mathbf{y}^m\}$ is a basis for V_2, then

$$\{\mathbf{x}^1, \ldots, \mathbf{x}^k, \mathbf{y}^1, \ldots, \mathbf{y}^m\}$$

is a basis for V; for it clearly spans V, while $\sum \gamma_i \mathbf{x}^i + \sum \delta_1 \mathbf{y}^j = \mathbf{0}$ or $\sum \gamma_i \mathbf{x}^i = -\sum \delta_1 \mathbf{y}^j$ implies that each is zero, and hence that all γ_i and δ_j are 0.

For any subspace V of E^n (or U^n) the set

$$V^\perp = \{\mathbf{x} \mid (\mathbf{x}, \mathbf{y}) = 0 \quad \text{for } \mathbf{y} \text{ in } V\}$$

is called the *orthogonal complement* of V. It is clearly a subspace, and $E^n = V \oplus V^\perp$; also $V^{\perp\perp} = (V^\perp)^\perp = V$. Since clearly $\dim (V_1 \oplus V_2) = \dim V_1 + \dim V_2$, it follows that $\dim V^\perp = n - \dim V$.

Exercises

1. In $\mathcal{P}^n$ (see Problem 1 of Section A.1) show that the polynomials $1, t, \ldots, t^{n-1}$ comprise a basis, and that $\mathcal{P}^n$ has dimension n.

2. With C as specified in Problem 2 of Section A.1, prove that an orthogonal basis is formed from the elements 1 and i, and thus that C is two-dimensional.

3. Prove that any nonvoid proper subspace of E^3 corresponds to a line or plane passing through the origin, and that its orthogonal complement corresponds to the perpendicular plane or line also passing through the origin.

4. Let $\mathbf{a}$ be a nonzero vector in E^n. Let $H = \{\mathbf{x} | \mathbf{x} \in E^n, (\mathbf{x}, \mathbf{a}) = 0\}$. Show that H, called a hyperplane, is a subspace of E^n and find its dimension.

5. Find an orthogonal basis for $\mathcal{P}^2$ and express the polynomials 1 and t in terms of this basis.

A.3 Linear transformations, matrices, and linear equations. A linear transformation $\mathbf{T}$ mapping E^n into E^m is a function with domain E^n and values in E^m which preserves sums and scalar multiples: $\mathbf{T}(\mathbf{x} + \mathbf{y}) = \mathbf{Tx} + \mathbf{Ty}$, and $\mathbf{T}(\alpha\mathbf{x}) = \alpha\mathbf{Tx}$. Clearly, $\mathbf{T0} = \mathbf{0}$, since $\mathbf{T0} = \mathbf{T}(\mathbf{0} + \mathbf{0}) = \mathbf{T0} + \mathbf{T0}$; similarly, $\mathbf{T}(\mathbf{x} - \mathbf{y}) = \mathbf{Tx} - \mathbf{Ty}$, since

$$\mathbf{T}(\mathbf{x} - \mathbf{y}) + \mathbf{Ty} = \mathbf{T}(\mathbf{x} - \mathbf{y} + \mathbf{y}) = \mathbf{Tx}.$$

The null space of $\mathbf{T}$, $\mathfrak{N}(\mathbf{T}) = \{\mathbf{x} | \mathbf{Tx} = \mathbf{0}\}$, is a subspace of the domain E^n (that is, $\mathbf{x}$, $\mathbf{y}$ in $\mathfrak{N}(\mathbf{T})$ yields $\alpha\mathbf{x} + \beta\mathbf{y}$ in $\mathfrak{N}(\mathbf{T})$ for any scalars α and β); however, $\mathfrak{N}(\mathbf{T})$ can reduce to the trivial subspace $\{\mathbf{0}\}$. This occurs exactly when $\mathbf{T}$ is one-to-one, and $\mathbf{T}$ being one-to-one is equivalent to $\mathfrak{N}(\mathbf{T}) = \{\mathbf{0}\}$, or $\dim \mathfrak{N}(\mathbf{T}) = 0$.

Similarly, the *range* of $\mathbf{T}$, $\mathcal{R}(\mathbf{T}) = \{\mathbf{Tx} | \mathbf{x} \in E^n\}$, is a subspace of E^m, and the following dimensional relationship holds:

$$\dim \mathfrak{N}(\mathbf{T}) + \dim \mathcal{R}(\mathbf{T}) = n. \tag{A.3.1}$$

We demonstrate this relationship as follows. If we select a basis

$$\{\mathbf{x}^1, \ldots, \mathbf{x}^k\}$$

of $\mathfrak{N}(\mathbf{T})$ and enlarge it to a basis $\{\mathbf{x}^1, \ldots, \mathbf{x}^k, \mathbf{y}^{k+1}, \ldots, \mathbf{y}^n\}$ of E^n, then, with $V = \text{span}\,\{\mathbf{y}^{k+1}, \ldots, \mathbf{y}^n\}$, we have $E^n = \mathfrak{N}(\mathbf{T}) \oplus V$. We have only to check $\mathfrak{N}(\mathbf{T}) \cap V = \{\mathbf{0}\}$. But $\mathbf{z}$ in $\mathfrak{N}(\mathbf{T}) \cap V$ implies

$$\mathbf{z} = \sum_{i=1}^{k} \alpha_i \mathbf{x}^i = \sum_{j=k+1}^{n} \beta_j \mathbf{y}^j;$$

hence all $\alpha_i, \beta_j = 0$, since $\{\mathbf{x}^1, \ldots, \mathbf{x}^k, \mathbf{y}^{k+1}, \ldots, \mathbf{y}^n\}$ is linearly independent, and thus $\mathbf{z} = \mathbf{0}$. But the image of the general element $\sum \alpha_i \mathbf{x}^i + \sum \beta_j \mathbf{y}^j$ of E^n is $\mathbf{T}(\sum \beta_j \mathbf{y}^j) = \sum \beta_j \mathbf{T}\mathbf{y}^j$, and hence $\{\mathbf{T}\mathbf{y}^{k+1}, \ldots, \mathbf{T}\mathbf{y}^n\}$ spans

$\mathcal{R}(\mathbf{T})$. On the other hand, the set is linearly independent, since $\sum \beta_j \mathbf{T}\mathbf{y}^j = \mathbf{0}$ implies that $\mathbf{T}(\sum \beta_j \mathbf{y}^j) = \mathbf{0}$, which means that $\sum \beta_j \mathbf{y}^j \in \mathfrak{N}(\mathbf{T}) \cap V = \{\mathbf{0}\}$, and hence all $\beta_j = 0$. Thus dim $\mathcal{R}(\mathbf{T}) = n - k$, and our relation holds. It should be no[illegible]d that V is in no way unique.

As consequen[illegible]s, note that dim $\mathcal{R}(\mathbf{T}) \leq n$, with equality holding if and only if $\mathbf{T}$ is 1-1. Thus when $m = n$, the transformation $\mathbf{T}$ maps E^n onto itself if and only if it is 1-1.

Matrices. A rectangular array of numbers

$$\mathbf{A} = \|a_{ij}\| = \left\| \begin{matrix} a_{11} & a_{12} & \cdots & a_{1n} \\ a_{21} & a_{22} & \cdots & a_{2n} \\ \vdots & \vdots & & \vdots \\ a_{m1} & a_{m2} & \cdots & a_{mn} \end{matrix} \right\|$$

is called a *matrix*—in this case an $m \times n$ (m-row by n-column) matrix—with *entries* a_{ij}. If $m = n$, we call the matrix *square*.

$\mathbf{A}$ determines a linear transformation $\mathbf{T}$ of E^n into E^m defined by setting

$$\mathbf{T}\langle x_1, x_2, \ldots, x_n \rangle = \left\langle \sum_{j=1}^{n} a_{1j}x_j, \quad \sum_{j=1}^{n} a_{2j}x_j, \ldots, \quad \sum_{j=1}^{n} a_{mj}x_j \right\rangle.$$

Conversely, if we denote the usual orthonormal basis of E^n by $\{\mathbf{e}^1, \ldots, \mathbf{e}^n\}$ and that of E^m by $\{\mathbf{f}^1, \ldots, \mathbf{f}^m\}$, each linear transformation $\mathbf{T}$ is so determined. For $\mathbf{T}$ is determined by its values at the $\mathbf{e}^j$, since

$$\mathbf{T}\left(\sum_{j=1}^{n} x_j \mathbf{e}^j\right) = \sum_{j=1}^{n} x_j \mathbf{T}\mathbf{e}^j;$$

and since each $\mathbf{T}\mathbf{e}^j$ is expressible as

$$\sum_{i=1}^{m} a_{ij}\mathbf{f}^i$$

for some unique $a_{1j}, a_{2j}, \ldots, a_{mj}$, we obtain

$$\mathbf{T}\mathbf{x} = \sum_{j=1}^{n} x_j \left(\sum_{i=1}^{m} a_{ij}\mathbf{f}^i\right) = \sum_{j=1}^{n}\sum_{i=1}^{m} x_j a_{ij}\mathbf{f}^i = \sum_{i=1}^{m}\left(\sum_{j=1}^{n} a_{ij}x_j\right)\mathbf{f}^i,$$

and the components of $\mathbf{T}\mathbf{x}$ are obtained as before from $\mathbf{A}$. Moreover, it is clear that we may replace $\{\mathbf{e}^1, \ldots, \mathbf{e}^n\}$ and $\{\mathbf{f}^1, \ldots, \mathbf{f}^m\}$ by any other fixed ordered bases in E^n and E^m and set up a similar one-to-one correspondence between linear transformations mapping E^n into E^m and $m \times n$ matrices. The exact correspondence will of course depend greatly on the bases chosen (and on their indexing). However, when we consider a matrix as a transformation we shall have in mind our original correspondence.

Corresponding to the linear transformations **T** and **S** mapping E^n into E^m there is a natural sum $\mathbf{T} + \mathbf{S}$, again a linear transformation, such that $(\mathbf{T} + \mathbf{S})(\mathbf{x}) = \mathbf{Tx} + \mathbf{Sx}$. Similarly, $c\mathbf{T}$ at $\mathbf{x}$ is defined to be $c \cdot \mathbf{Tx}$ (c a scalar). The corresponding operations on matrices are clearly the addition of corresponding entries:

$$\mathbf{A} + \mathbf{B} = \begin{Vmatrix} a_{11} + b_{11} & a_{12} + b_{12} & \cdots \\ a_{21} + b_{21} & \cdots\cdots\cdots & \cdots \\ \vdots & & \\ a_{m1} + b_{m1} & \cdots\cdots\cdots & \cdots \end{Vmatrix}$$

and

$$c\mathbf{A} = \begin{Vmatrix} ca_{11} & ca_{12} & \cdots \\ \vdots & \vdots & \\ ca_{m1} & ca_{m2} & \cdots \end{Vmatrix}$$

(each entry multiplied by c). Clearly the $m \times n$ matrices thus form a vector space with an obvious basis $\mathbf{e}_{ij}$ ($i = 1, \ldots, m; j = 1, \ldots, n$), where $\mathbf{e}_{ij}$ has all entries 0 except for a single 1 in the ijth place. Thus the space is mn-dimensional.

If **T** maps E^n into E^m and **S** maps E^m into E^l, we can compose the mappings in such a way that **ST** maps E^n into E^l. Explicitly, we set $\mathbf{ST}(\mathbf{x}) = \mathbf{S}(\mathbf{Tx})$. The linear transformation **ST** is called the product of **S** and **T**; the corresponding matrix product is easily computed: with **S** corresponding to $\mathbf{A}(l \times m)$ and **T** corresponding to $\mathbf{B}(m \times n)$, **AB** is an $l \times n$ matrix with ijth entry

$$\sum_{k=1}^{m} a_{ik}b_{kj} \qquad (i = 1, \ldots, l; \quad j = 1, \ldots, n).$$

If we view the elements of E^n as $n \times 1$ matrices

$$\begin{Vmatrix} x_1 \\ x_2 \\ \vdots \\ x_n \end{Vmatrix}$$

(a *column* vector), our computation of the components of **Tx** (where **T** corresponds to $\|b_{ij}\|$) is simply the formation of the matrix product

$$\begin{Vmatrix} b_{11} & b_{12} & \cdots & b_{1n} \\ b_{21} & b_{22} & \cdots & b_{2n} \\ \vdots & \vdots & & \vdots \\ b_{m1} & b_{m2} & \cdots & b_{mn} \end{Vmatrix} \begin{Vmatrix} x_1 \\ x_2 \\ \vdots \\ x_n \end{Vmatrix} = \begin{Vmatrix} \sum_{j=1}^{k} b_{1j}x_j \\ \vdots \\ \sum_{j=1}^{n} b_{mj}x_j \end{Vmatrix}$$

giving the image elements of E^m as column vectors ($m \times 1$). The resulting column vector can also be expressed as

$$x_1 \begin{Vmatrix} b_{11} \\ b_{21} \\ \vdots \\ b_{m1} \end{Vmatrix} + x_2 \begin{Vmatrix} b_{12} \\ b_{22} \\ \vdots \\ b_{m2} \end{Vmatrix} + \cdots + x_n \begin{Vmatrix} b_{1n} \\ b_{2n} \\ \vdots \\ b_{mn} \end{Vmatrix}$$

so that the *dimension of* $\mathcal{R}(\mathbf{T})$, *called the rank of* $\mathbf{T}$ *or* $\mathbf{B}$, *is the number of linearly independent columns of* $\mathbf{B}$ (viewed as vector elements of E^m). One can obtain the rank by eliminating dependent columns, or by determinants; the rank is k if after deleting rows and columns of $\mathbf{B}$, the largest square submatrices with nonzero determinant are $k \times k$.

Finally, a linear transformation from E^n into E^1, which we may regard as the real line, is called a *linear functional*. As we have seen, it is given (relative to the bases $\{\mathbf{e}^1, \ldots, \mathbf{e}^n\}$ and $\{\mathbf{f}^1\}$, where $\mathbf{f}^1 = \langle 1 \rangle$) by a $1 \times n$ matrix, and its value at $\langle x_1, x_2, \ldots, x_n \rangle$ is of the form $c_1x_1 + c_2x_2 + \cdots + c_nx_n$, with c_i fixed. Thus each linear functional is obtained from the inner product and maps $\mathbf{x}$ into $(\mathbf{x}, \mathbf{y})$ for some fixed $\mathbf{y}$ in E^n.

Linear equations. The correspondence between matrices and linear transformations can be applied to systems of linear equations. Consider the following system of m equations in n unknown (real or complex) numbers $x_1, x_2, \ldots, x_n$:

$$\begin{aligned} a_{11}x_1 + a_{12}x_2 + \cdots + a_{1n}x_n &= c_1 \\ a_{21}x_1 + a_{22}x_2 + \cdots + a_{2n}x_n &= c_2 \\ &\vdots \\ a_{m1}x_1 + a_{m2}x_2 + \cdots + a_{mn}x_n &= c_m. \end{aligned} \tag{A.3.2}$$

Solving the system is equivalent to finding an $\mathbf{x} = \langle x_1, x_2, \ldots, x_n \rangle$ in E^n mapping onto $\mathbf{c} = \langle c_1, c_2, \ldots, c_m \rangle$. A solution can be obtained for every c if and only if the rank of $\mathbf{A} = \|a_{ij}\|$ is m.

If each $c_i = 0$, the system is called *homogeneous* (otherwise inhomogeneous) and has the trivial solution $\langle 0, 0, \ldots, 0 \rangle$; nontrivial solutions exist whenever $\mathfrak{N}(\mathbf{A}) \neq \{\mathbf{0}\}$. If $n > m$ (more unknowns than equations), this must be the case, for otherwise $\dim \mathcal{R}(\mathbf{A}) = n$, which is greater than the dimension of E^m, while $\mathcal{R}(\mathbf{A}) \subset E^m$. If $n = m$, (A.3.2) is solvable for each $\mathbf{c}$ if and only if each solution is unique, or indeed if the solution for $\mathbf{c} = \mathbf{0}$ is unique; this amounts to saying that $\mathbf{A}$ maps E^n onto itself if and only if $\mathfrak{N}(\mathbf{A}) = \{\mathbf{0}\}$.

Transformations from E^n *into* E^n. Let $L(E^n)$ be the set of all linear transformations $\mathbf{T}$ mapping E^n into E^n. If $\mathbf{S}$ and $\mathbf{T}$ are in $L(E^n)$, then $\mathbf{S} + \mathbf{T}$, $\mathbf{ST}$, and $c\mathbf{T}$ are all elements of $L(E^n)$, and $L(E^n)$ forms a vector space with a (noncommutative) associative and distributive multiplication. The zero element maps each $\mathbf{x}$ into $\mathbf{0}$, while the identity transformation $\mathbf{I}$ mapping $\mathbf{x}$ into $\mathbf{x}$ yields $\mathbf{TI} = \mathbf{IT} = \mathbf{T}$. In terms of matrices, the zero transformation corresponds to the matrix with all entries 0, and

$$\mathbf{I} = \begin{Vmatrix} 1 & 0 & \cdots & 0 \\ 0 & 1 & \cdots & 0 \\ \vdots & \vdots & \ddots & \vdots \\ 0 & 0 & \cdots & 1 \end{Vmatrix}$$

(with 1's only on the main diagonal). We know that $\mathbf{T}$ is 1-1 if and only if $\mathbf{T}$ is onto; in this case the inverse transformation $\mathbf{T}^{-1}$ (which has as its domain all of E^n) is also linear and thus in $L(E^n)$; $\mathbf{T}^{-1}\mathbf{y} = \mathbf{x}$ is equivalent by definition to $\mathbf{y} = \mathbf{Tx}$. The 1-1 transformations in $L(E^n)$ are just those with a multiplicative inverse $\mathbf{T}^{-1}$: $\mathbf{TT}^{-1} = \mathbf{T}^{-1}\mathbf{T} = \mathbf{I}$. They are called *nonsingular* or invertible; all others are *singular*. The same terminology is applied to matrices. In particular, $\mathbf{A}$ is nonsingular if $\mathbf{AB} = \mathbf{I}$ (or $\mathbf{BA} = \mathbf{I}$) for some matrix $\mathbf{B}$ (for $\mathbf{A}$, considered as a linear transformation, cannot lower dimension, and therefore is onto and 1-1). $\mathbf{B}$ of course must be the (unique) inverse matrix, denoted by $\mathbf{A}^{-1}$, and

$$\sum_{j=1}^{n} a_{ij}b_{jk} = 0$$

if $i \neq k$. If $\mathbf{A}$ is singular, we have numbers $x_1, x_2, \ldots, x_n$, not all zero, with $\mathbf{A}\langle x_1, \ldots, x_n\rangle = \langle 0, \ldots, 0\rangle$.

Finally note that a product of nonsingular matrices or transformation is nonsingular: $(\mathbf{ST}) \cdot (\mathbf{T}^{-1}\mathbf{S}^{-1}) = \mathbf{T}^{-1}\mathbf{S}^{-1} \cdot \mathbf{ST} = \mathbf{I}$.

Exercises

1. In E^2 with basis $\{\mathbf{e}^1, \mathbf{e}^2\}$ let the transformation $\mathbf{A}_\theta$ describe the rotation of all points θ degrees counterclockwise about the origin. Show that the matrix of $\mathbf{A}_\theta$ is given by

$$\mathbf{A}_\theta = \begin{pmatrix} \cos\theta & \sin\theta \\ -\sin\theta & \cos\theta \end{pmatrix}.$$

Show further that $\mathbf{A}_{\theta_1+\theta_2} = \mathbf{A}_{\theta_1}\mathbf{A}_{\theta_2}$ and $\mathbf{A}_\theta^{-1} = \mathbf{A}_{-\theta}$.

2. In the space $\mathcal{P}^n$ show that

$$\int_a^b p(t)\,dt$$

is a linear functional and express it in the form $(\mathbf{q}, \mathbf{p})$, where $\mathbf{q}$ is some element in $\mathcal{P}^n$ and $(,)$ is the inner product specified in Problem 1 of Section A.1.

3. On the space $\mathcal{P}^n$ with the basis $\{1, t, t^2, \ldots, t^{n-1}\}$, define the linear transformation $\mathbf{A}_k$ $(k \geq 1)$ as follows:

$$\mathbf{A}_k \mathbf{p} = \frac{d^k}{dt^k} p(t).$$

Verify that $\mathbf{A}_k$ is a singular transformation whose matrix representation in terms of this basis is

$$\left\| \begin{matrix} 0 & \cdots & 0 & \frac{k!}{0!} & 0 & \cdots & \cdots & 0 \\ 0 & \cdots & \cdots & 0 & \frac{(k+1)!}{1!} & 0 & \cdots & 0 \\ \vdots & & & & & & & \\ 0 & \cdots & \cdots & \cdots & \cdots & \cdots & \cdots & \frac{(n-1)!}{(n-k-1)!} \\ \vdots & & & & & & & \vdots \\ 0 & \cdots & \cdots & \cdots & \cdots & \cdots & \cdots & 0 \end{matrix} \right\|$$

Show also that $\mathcal{R}(\mathbf{A}_k)$ is the span of $\{t^{n-k}, t^{n-k+1}, \ldots, t^{n-1}\}$, and that $\mathfrak{N}(\mathbf{A}_k)$ is the span of $\{1, t, \ldots, t^{n-k-1}\}$.

A.4 Eigenvalues, eigenvectors, and the Jordan canonical form. If $\mathbf{T} - \lambda\mathbf{I}$ is singular for some scalar λ, then λ is called an *eigenvalue* of $\mathbf{T}$. Associated with λ we have some $\mathbf{x} \neq \mathbf{0}$ such that $(\mathbf{T} - \lambda\mathbf{I})\mathbf{x} = \mathbf{0}$ or $\mathbf{T}\mathbf{x} = \lambda\mathbf{x}$; this vector $\mathbf{x}$ is called an *eigenvector*, corresponding to the eigenvalue λ.

Each $\mathbf{T}$ in $L(U^n)$ has eigenvalues, and hence eigenvectors. The corresponding statement for $L(E^n)$, where λ is taken to be real, is false. However, any real matrix corresponds to an element of $L(U^n)$ and thus has complex eigenvalues and eigenvectors.

The existence of eigenvalues is established as follows. Note that since $L(U^n)$ is n^2-dimensional, the $n^2 + 1$ elements $\mathbf{I}, \mathbf{T}, \mathbf{T}^2, \ldots, \mathbf{T}^{n^2}$ are linearly dependent; we have $a_0, a_1, \ldots, a_{n^2}$ not all zero, with $a_0\mathbf{I} + a_1\mathbf{T} + \cdots + a_{n^2}\mathbf{T}^{n^2} = \mathbf{0}$. The corresponding polynomial $p(z) = a_0 + a_1 z + \cdots + a_{n^2} z^{n^2}$ can of course be factored by the fundamental theorem of algebra:

$$p(z) = \prod_{i=1}^{k} (z - \lambda_i).$$

Since the identity of the polynomials says exactly that we obtain the coefficients of p by multiplying out the right side, and the same coefficients are obtained in multiplying out

$$\prod_{i=1}^{k} (\mathbf{T} - \lambda_i \mathbf{I}),$$

we have

$$\prod_{i=1}^{k} (\mathbf{T} - \lambda_i \mathbf{I}) = a_0\mathbf{I} + a_1\mathbf{T} + \cdots + a_{n^2}\mathbf{T}^{n^2} = p(\mathbf{T}) = 0;$$

consequently one of the $\mathbf{T} - \lambda_i\mathbf{I}$ is singular.

For any basis $\{\mathbf{y}^1, \ldots, \mathbf{y}^n\}$ of U^n, by writing

$$\mathbf{x} = \sum_{i=1}^{n} x_i \mathbf{y}^i$$

and similarly expressing $\mathbf{Tx}$ in terms of its components relative to $\{\mathbf{y}^1, \ldots, \mathbf{y}^n\}$, we can represent $\mathbf{T}$ by a matrix $\mathbf{A}$ mapping $\langle x_1, \ldots, x_n\rangle$ into the vector of components of $\mathbf{Tx}$ relative to the basis. For an appropriate choice of basis, $\mathbf{A}$ has the Jordan *canonical form*

$$\mathbf{A} = \begin{Vmatrix} \mathbf{A}_1 & \mathbf{0} & \cdots & \mathbf{0} \\ \mathbf{0} & \mathbf{A}_2 & \cdots & \mathbf{0} \\ \vdots & \vdots & \ddots & \vdots \\ \mathbf{0} & \mathbf{0} & \cdots & \mathbf{A}_k \end{Vmatrix}$$

where $\mathbf{0}$ represents appropriate rectangular matrices with all entries zero, and each $\mathbf{A}_i$ is a square matrix which can similarly be written (in so-called block-diagonal form) with diagonal blocks of the form

$$\mathbf{B} = \begin{Vmatrix} \lambda_i & 0 & 0 & \cdots & 0 \\ 1 & \lambda_i & 0 & \cdots & 0 \\ 0 & 1 & \lambda_i & \cdots & 0 \\ \vdots & \vdots & \vdots & & \vdots \\ 0 & 0 & 0 & \cdots & \lambda_i \end{Vmatrix}$$

(λ_i on the main diagonal, 1's below, 0's elsewhere). The same λ_i appears in all sub-blocks of $\mathbf{A}_i$, but $\lambda_i \neq \lambda_j$, $i \neq j$, and $\lambda_1, \ldots, \lambda_k$ comprise the set of all eigenvalues of $\mathbf{T}$. The effect of this decomposition of $\mathbf{A}$ into blocks is the following: Any vector in the span of the basis elements

$$\{\mathbf{y}^j, \mathbf{y}^{j+1}, \ldots, \mathbf{y}^k\}$$

corresponding to the block $\mathbf{B}$ maps into a vector in this same subspace; moreover, on this subspace $\mathbf{T}$ acts as the sum of two simple transformations: the transformation that maps $\mathbf{x}$ into $\lambda_i\mathbf{x}$ and the transformation that maps $\mathbf{y}^j$ into $\mathbf{y}^{j+1}$, $\mathbf{y}^{j+1}$ into $\mathbf{y}^{j+2}, \ldots, \mathbf{y}^{k-1}$ into $\mathbf{y}^k$, $\mathbf{y}^k$ into $\mathbf{0}$ (thus $\mathbf{T} - \lambda_i\mathbf{I}$ reduces to the latter transformation on the subspace). In effect, we have a decomposition of E^n into (a direct sum of) pieces, and $\mathbf{T}$ has a simple form on each piece. In many important cases each block $\mathbf{B}$ is 1×1; the matrix $\mathbf{A}$ assumes the pure diagonal form

$$\begin{Vmatrix} \lambda_1 & 0 & \cdots & 0 \\ 0 & \lambda_2 & \cdots & 0 \\ \vdots & \vdots & \ddots & \vdots \\ 0 & 0 & \cdots & \lambda_n \end{Vmatrix},$$

where now the λ_i need not all be distinct; and the eigenvectors of $\mathbf{T}$ span U^n. Moreover, if $\mathbf{T}$ *has n distinct eigenvalues,* it is clear that each block $\mathbf{B}$ is 1×1 and $\mathbf{A}$ has the pure diagonal form.

We now indicate how the basis $\{\mathbf{y}^1, \ldots, \mathbf{y}^n\}$ is constructed. For a singular $\mathbf{S}$ in $L(U^n)$, or indeed $L(E^n)$, consider $\mathfrak{N}(\mathbf{S})$, $\mathfrak{N}(\mathbf{S}^2)$, ... Clearly, $\mathfrak{N}(\mathbf{S}) \subset \mathfrak{N}(\mathbf{S}^2) \subset \cdots$, and since proper subspaces have smaller dimensions, there is some integer k $(\leq n)$ for which

$$\mathfrak{N}(\mathbf{S}) \subsetneqq \mathfrak{N}(\mathbf{S}^2) \subsetneqq \cdots \subsetneqq \mathfrak{N}(\mathbf{S}^k) = \mathfrak{N}(\mathbf{S}^{k+1}).$$

Then $\mathfrak{N}(\mathbf{S}^k) = \mathfrak{N}(\mathbf{S}^{k+j})$ for all $j \geq 0$. For since $\mathfrak{N}(\mathbf{S}^k) = \mathfrak{N}(\mathbf{S}^{k+1})$, if $\mathbf{S}^{k+1}\mathbf{x} = \mathbf{0}$ it follows that $\mathbf{S}^k\mathbf{x} = \mathbf{0}$; and thus $\mathbf{S}^{k+j}\mathbf{x} = \mathbf{0} = \mathbf{S}^{k+1}(\mathbf{S}^{j-1}\mathbf{x})$ implies $\mathbf{S}^k(\mathbf{S}^{j-1}\mathbf{x}) = \mathbf{S}^{k+j-1}\mathbf{x} = \mathbf{0}$, from which it follows that

$$\mathbf{S}^{k+j-1}\mathbf{x} = \mathbf{S}^{k+j-2}\mathbf{x} = \cdots = \mathbf{S}^k\mathbf{x} = 0.$$

By the dimension relation (A.3.1),

$$\mathfrak{R}(\mathbf{S}) \supsetneqq \mathfrak{R}(\mathbf{S}^2) \supsetneqq \cdots \supsetneqq \mathfrak{R}(\mathbf{S}^k) = \mathfrak{R}(\mathbf{S}^{k+1}) = \cdots$$

Thus $\mathbf{S}$ restricted to $\mathfrak{R}(\mathbf{S}^k)$ is 1-1 because it is onto. But, since $\mathbf{S}^k$ is 1-1, we also have $\mathfrak{N}(\mathbf{S}^k) \cap \mathfrak{R}(\mathbf{S}^k) = \{\mathbf{0}\}$. If $\{\mathbf{x}^1, \ldots, \mathbf{x}^l\}$ is a basis for $\mathfrak{N}(\mathbf{S}^k)$ and $\{\mathbf{y}^1, \ldots, \mathbf{y}^m\}$ is a basis for $\mathfrak{R}(\mathbf{S}^k)$, it follows by (A.3.1) that $l + m = n$. The conjunction $\{\mathbf{x}^1, \ldots \mathbf{x}^l; \mathbf{y}^1, \ldots, \mathbf{y}^m\}$ forms a basis for U^n, and $U^n = \mathfrak{N}(\mathbf{S}^k) \oplus \mathfrak{R}(\mathbf{S}^k)$. Corresponding to our basis, $\mathbf{S}$ would have the matricial structure

$$\begin{pmatrix} \mathbf{A} & \mathbf{0} \\ \mathbf{0} & \mathbf{B} \end{pmatrix},$$

where $\mathbf{A}$ is $l \times l$ and $\mathbf{B}$ is $(n - l) \times (n - l)$.

Now suppose that $\mathbf{T}$ is any element of $L(U^n)$. We know that there is an eigenvalue λ_1; hence we may let $\mathbf{S} = \mathbf{T} - \lambda_1\mathbf{I}$ and let

$$\mathfrak{N}_1 = \mathfrak{N}[(\mathbf{T} - \lambda_1\mathbf{I})^{k_1}],$$

with k_1 the k obtained for $\mathbf{S}$ above. Accordingly, $\mathfrak{R}_1 = \mathfrak{R}[(\mathbf{T} - \lambda_1\mathbf{I})^{k_1}]$. We have $U^n = \mathfrak{N}_1 \oplus \mathfrak{R}_1$, with $(\mathbf{T} - \lambda_1\mathbf{I})\mathbf{x}$ in $\mathfrak{N}_1$ or $\mathfrak{R}_1$ according as $\mathbf{x}$ is in $\mathfrak{N}_1$ or $\mathfrak{R}_1$. Since $\lambda_1\mathbf{x}$ is in a given subspace whenever $\mathbf{x}$ is, $\mathbf{T}\mathbf{x}$ is in $\mathfrak{N}_1$ if $\mathbf{x}$ is in $\mathfrak{N}_1$, and $\mathbf{T}\mathbf{x}$ is in $\mathfrak{R}_1$ if $\mathbf{x}$ is in $\mathfrak{R}_1$. Let $\mathbf{T}_1$ be the restriction of $\mathbf{T}$ to

$\mathcal{R}_1$, mapping $\mathcal{R}_1$ into itself. Since we can identify $\mathcal{R}_1$ with some U^m, it follows that $\mathbf{T}_1$ has an eigenvalue λ_2 such that $\mathbf{Tx} = \mathbf{T}_1\mathbf{x} = \lambda_2\mathbf{x}$ for $\mathbf{x} \neq \mathbf{0}$ in $\mathcal{R}_1$. Since $(\mathbf{T} - \lambda_1\mathbf{I})\mathbf{x} \neq \mathbf{0}$ ($\mathbf{T} - \lambda_1\mathbf{I}$ being 1-1 on $\mathcal{R}_1$), we deduce that $\lambda_2 \neq \lambda_1$. Defining $\mathfrak{N}_2$ and $\mathcal{R}_2$ in the obvious fashion, we obtain $\mathcal{R}_1 = \mathfrak{N}_2 \oplus \mathcal{R}_2$ (hence $U^n = \mathfrak{N}_1 \oplus \mathfrak{N}_2 \oplus \mathcal{R}_2$), with

$$\mathbf{T}\mathfrak{N}_2 = \mathbf{T}_1\mathfrak{N}_2 = \{\mathbf{T}_1\mathbf{x} | \mathbf{x} \in \mathfrak{N}_2\} \subset \mathfrak{N}_2, \qquad \mathbf{T}\mathcal{R}_2 \subset \mathcal{R}_2.$$

We continue and obtain $U^n = \mathfrak{N}_1 \oplus \cdots \oplus \mathfrak{N}_r$, with associated distinct eigenvalues $\lambda_1, \ldots, \lambda_r$ such that $(\mathbf{T} - \lambda_i\mathbf{I})^{k_i}\mathbf{x} = 0$ for all $\mathbf{x}$ in $\mathfrak{N}_i$ and some appropriate $k_i \geq 1$. Choosing any basis for U^n which is formed by combining bases of the $\mathfrak{N}_i$ now yields a matrix $\mathbf{A}$ in block-diagonal form, since this amounts to saying $\mathbf{T}\mathfrak{N}_i \subset \mathfrak{N}_i$.

The further decomposition into sub-blocks is delicate: we shall only note that if we choose an $\mathbf{x}_1$ in $\mathfrak{N}_1$ for which $(\mathbf{T} - \lambda_1\mathbf{I})^{k_1-1}\mathbf{x}_1 \neq \mathbf{0}$ (i.e., an element of $\mathfrak{N}[(\mathbf{T} - \lambda_1\mathbf{I})^{k_1}]$ not in $\mathfrak{N}[(\mathbf{T} - \lambda_1\mathbf{I})^{k_1-1}]$, a choice that is made possible by the meaning of k_1), then for $\mathbf{S} = \mathbf{T} - \lambda_1\mathbf{I}$ the elements $\mathbf{x}_1$, $\mathbf{Sx}_1, \ldots, \mathbf{S}^{k-1}\mathbf{x}_1$ are independent: for if

$$\gamma_1\mathbf{x}_1 + \gamma_2\mathbf{Sx}_1 + \cdots + \gamma_{k_1-1}\mathbf{S}^{k_1-1}\mathbf{x}_1 = \mathbf{0},$$

then by applying $\mathbf{S}^{k_1-1}$ we obtain $\gamma_1\mathbf{S}^{k_1-1}\mathbf{x}_1 = \mathbf{0}$ and hence $\gamma_1 = 0$; we then apply $\mathbf{S}^{k_1-2}, \mathbf{S}^{k_1-3}, \ldots$ successively to deduce $\gamma_2 = \gamma_3 = \cdots = 0$. These vectors need not constitute a basis for $\mathfrak{N}_1$, but it can be shown that there exists a finite set $\mathbf{x}_1, \mathbf{x}_2, \ldots$ such that $\mathbf{x}_1, \mathbf{Sx}_1, \ldots, \mathbf{S}^{k_1-1}\mathbf{x}_1$; $\mathbf{x}_2, \mathbf{Sx}_2, \ldots, \mathbf{S}^{j_2-1}\mathbf{x}_2$; $\mathbf{x}_3, \mathbf{Sx}_3, \ldots, \mathbf{S}^{j_3-1}\mathbf{x}_3$; ... form a basis for $\mathfrak{N}_1$ where $\mathbf{S}^{j_2}\mathbf{x}_2 = \mathbf{0}$, $\mathbf{S}^{j_3}\mathbf{x}_3 = \mathbf{0}, \ldots j_2 \leq k_1, j_3 \leq j_2$, etc.* It is easily seen that on the span V_0 of $\mathbf{x}_1, \mathbf{Sx}_1, \ldots, \mathbf{S}^{k_1-1}\mathbf{x}_1$ we can write $\mathbf{S}$ as a matrix of the form

$$\begin{Vmatrix} 0 & 0 & \ldots & 0 & 0 \\ 1 & 0 & \ldots & 0 & 0 \\ 0 & 1 & \ldots & 0 & 0 \\ \vdots & \vdots & \ddots & \vdots & \vdots \\ 0 & 0 & \ldots & 1 & 0 \end{Vmatrix};$$

hence on V_0

$$\mathbf{T} = \lambda_1\mathbf{I} + \mathbf{S} = \begin{Vmatrix} \lambda_1 & 0 & \ldots & 0 \\ 1 & \lambda_1 & \ldots & 0 \\ 0 & 1 & \ldots & 0 \\ \vdots & \vdots & & \vdots \\ 0 & 0 & \ldots & \lambda_1 \end{Vmatrix}.$$

* For the sake of preserving the continuity here, the proof of this assertion has been placed at the end of the argument (see p. 334).

Clearly, each such sub-block of our final matrix $\mathbf{A}$ yields exactly one eigenvector (here $\mathbf{S}^{k_1-1}\mathbf{x}_1$). Similar statements apply to the span of

$$\mathbf{x}_2, \mathbf{S}\mathbf{x}_2, \ldots, \mathbf{S}^{j_2-1}\mathbf{x}_2, \text{ etc.}$$

The number of eigenvectors we thus obtain corresponding to λ_1 is called the *geometric multiplicity* of λ_1; it is precisely the dimension of the subspace spanned by the eigenvectors corresponding to λ_1. Since $\mathbf{T}$ maps

$$\begin{aligned}
&\mathbf{x}_1 \to \lambda_1\mathbf{x}_1 + \mathbf{S}\mathbf{x}_1,\\
&\mathbf{S}\mathbf{x}_1 \to \lambda_1\mathbf{S}\mathbf{x}_1 + \mathbf{S}^2\mathbf{x}_1,\\
&\vdots\\
&\mathbf{S}^{k_1-2}\mathbf{x}_1 \to \lambda_1\mathbf{S}^{k_1-2}\mathbf{x}_1 + \mathbf{S}^{k_1-1}\mathbf{x}_1,\\
&\mathbf{S}^{k_1-1}\mathbf{x}_1 \to \lambda_1\mathbf{S}^{k_1-1}\mathbf{x}_1,
\end{aligned}$$

$\mathbf{T} - \lambda\mathbf{I}$ maps

$$\begin{aligned}
&\mathbf{x}_1 \to (\lambda_1 - \lambda)\mathbf{x}_1 + \mathbf{S}\mathbf{x}_1,\\
&\mathbf{S}\mathbf{x}_1 \to (\lambda_1 - \lambda)\mathbf{S}\mathbf{x}_1 + \mathbf{S}^2\mathbf{x}_1,\\
&\vdots\\
&\mathbf{S}^{k_1-2}\mathbf{x}_1 \to (\lambda_1 - \lambda)\mathbf{S}^{k_1-2}\mathbf{x}_1 + \mathbf{S}^{k_1-1}\mathbf{x}_1,\\
&\mathbf{S}^{k_1-1}\mathbf{x}_1 \to (\lambda_1 - \lambda)\mathbf{S}^{k_1-1}\mathbf{x}_1.
\end{aligned}$$

Consequently, if $\lambda \neq \lambda_1$, it follows that $\mathcal{R}(\mathbf{T} - \lambda\mathbf{I})$ contains $\mathbf{S}^{k_1-1}\mathbf{x}_1$, hence $\mathbf{S}^{k_1-2}\mathbf{x}_1$, and thus all the elements spanning V_0. Therefore $\mathbf{T} - \lambda\mathbf{I}$ maps V_0 onto itself for any $\lambda \neq \lambda_i$ and all i, yielding $\mathfrak{N}_i \subset \mathcal{R}(\mathbf{T} - \lambda\mathbf{I})$. Hence $\mathbf{T} - \lambda\mathbf{I}$ is onto and $(\mathbf{T} - \lambda\mathbf{I})^{-1}$ exists, and thus $\lambda_1, \ldots, \lambda_k$ is the full set of eigenvalues.

Finally, note that if $\mathbf{S}_0$ is the unique linear transformation mapping $\mathbf{e}^i$ onto $\mathbf{y}^i$—that is, if

$$\mathbf{S}_0 \sum_{i=1}^{n} \gamma_i\mathbf{e}^i = \sum_{i=1}^{n} \gamma_i\mathbf{y}^i$$

—then for $\mathbf{z} = \langle \gamma_1, \gamma_2, \ldots, \gamma_n \rangle$ we have

$$\mathbf{T}\mathbf{S}_0\mathbf{z} = \sum_{i=1}^{n} \left(\sum_{j=1}^{n} a_{ij}\gamma_j \right) \mathbf{y}^i;$$

and therefore (since $\mathbf{S}_0$, being onto, is nonsingular)

$$\mathbf{S}_0^{-1}\mathbf{T}\mathbf{S}_0\mathbf{z} = \sum_{i=1}^{n} \left(\sum_{j=1}^{n} a_{ij}\gamma_j \right) \mathbf{e}^i \qquad (\mathbf{S}_0^{-1}\mathbf{y}^i = \mathbf{e}^i).$$

Thus if $\mathbf{T}$ corresponds to the matrix $\mathbf{B}$ relative to the basis $\{\mathbf{e}^1, \ldots, \mathbf{e}^n\}$, and $\mathbf{S}_0$, $\mathbf{S}_0^{-1}$ to $\mathbf{C}$, $\mathbf{C}^{-1}$, respectively, we have $\mathbf{C}^{-1}\mathbf{BC} = \mathbf{A}$, or equivalently $\mathbf{B} = \mathbf{C}\,(\mathbf{C}^{-1}\mathbf{BC})\mathbf{C}^{-1} = \mathbf{CAC}^{-1}$. Two matrices $\mathbf{A}$ and $\mathbf{B}$ for which $\mathbf{B} = \mathbf{CAC}^{-1}$ for some $\mathbf{C}$ are called *similar*, and any matrix is therefore similar to one of the (block-diagonal) Jordan canonical form. Any two matrices $\mathbf{A}$ and $\mathbf{B}$ are similar if and only if their Jordan canonical forms differ only in the order in which the blocks and sub-blocks appear.

We now present the formal proof referred to in the footnote on p. 332.

(i) Suppose that we have a linearly independent set of vectors of $\mathfrak{N}_1$ of the form

$$\begin{array}{llll} \mathbf{x}_1, & \mathbf{Sx}_1, & \ldots, & \mathbf{S}^{j_1-1}\mathbf{x}_1; \\ \mathbf{x}_2, & \mathbf{Sx}_2, & \ldots, & \mathbf{S}^{j_2-1}\mathbf{x}_2; \\ \vdots & & & \\ \mathbf{x}_r, & \mathbf{Sx}_r, & \ldots, & \mathbf{S}^{j_r-1}\mathbf{x}_r, \end{array}$$

where $k_1 = j_1 \geq j_2 \geq \cdots \geq j_r$ and $\mathbf{S}^{j_i}\mathbf{x}_i = 0$ $(i = 1, 2, \ldots, r)$.

(ii) Suppose also that for each $l = 1, 2, \ldots, r$ there exists no $\mathbf{x}$ satisfying $\mathbf{S}^{k_1}\mathbf{x} = \mathbf{0}$ and such that $\mathbf{x}, \mathbf{Sx}, \ldots, \mathbf{S}^{j_l}\mathbf{x}$ together with the first $l - 1$ rows of vectors in (i) are linearly independent.

Then either the vectors in (i) form a basis for $\mathfrak{N}_1$ or they can be extended so as to satisfy both (i) and (ii).

Proof. Let j_{r+1} be the largest integer satisfying the properties that for some $\mathbf{x}_{r+1}$, $\mathbf{S}^{k_1}\mathbf{x}_{r+1} = \mathbf{0}$ and $\mathbf{x}_{r+1}, \mathbf{Sx}_{r+1}, \ldots, \mathbf{S}^{j_{r+1}-1}\mathbf{x}_{r+1}$ together with the vectors in (i) are linearly independent. Then, since $j_{r+1} \leq j_r$ by definition, if $\mathbf{S}^{j_{r+1}}\mathbf{x}_{r+1} = \mathbf{0}$ we are through.

Suppose, then, that $\mathbf{0} \neq \mathbf{S}^{j_{r+1}}\mathbf{x}_{r+1} = \sum \alpha_i \mathbf{S}^{v_i}\mathbf{x}_{\mu_i}$, with all $\alpha_i \neq 0$ and $v_1 \leq v_2 \leq \ldots$ Clearly, we must have all $\mu_i \leq r$. For otherwise we could repeatedly multiply both sides by $\mathbf{S}$, substitute the above equality for $\mathbf{S}^{j_{r+1}}\mathbf{x}_{r+1}$ when it appears on the right, and obtain all $\mathbf{S}^{j_{k+1}+r}\mathbf{x}_{r+1} \neq \mathbf{0}$ for $r = 0, 1, 2, \ldots$ (since $\alpha_i \neq 0$), contradicting $\mathbf{S}^{k_1}\mathbf{x}_{r+1} = \mathbf{0}$. If $v_1 \leq j_{r+1} - 1$, then $j_{r+1} + \mu_1 - v_1 - 1 \geq \mu_1$. But

$$\mathbf{S}^{j_{r+1}+\mu_1-v_1-1}\mathbf{x}_{r+1} = \alpha_1\mathbf{S}^{\mu_1-1}\mathbf{x}_{\mu_1} + \cdots$$

together with $\mathbf{x}_{r+1}, \mathbf{Sx}_{r+1}, \ldots, \mathbf{S}^{j_{r+1}+\mu_1-v_1-2}\mathbf{x}_{r+1}$, and the first $\mu_1 - 1$ rows of vectors in (i) are independent, contradicting (ii). Hence $\mathbf{S}^{j_{r+1}}\mathbf{x}_{r+1}$ has the representation $\mathbf{S}^{j_{r+1}}\mathbf{x}_{r+1} = \mathbf{S}^{j_{r+1}}\,(\sum \alpha_i \mathbf{S}^{v'_i}\mathbf{x}_{\mu_i})$, with all $\mu_i \leq r$. Hence the vector $\mathbf{x}_{r+1} - \sum \alpha_i \mathbf{S}^{v'_i}\mathbf{x}_{\mu_i}$ satisfies all the necessary requirements.

The case of distinct eigenvalues. If a matrix $\mathbf{B}$ has n distinct eigenvalues, it is similar to a diagonal matrix $\mathbf{A}$ given by its canonical form $\mathbf{C}^{-1}\mathbf{BC} = \mathbf{A}$. In this case $\mathbf{C}$ can be constructed directly from the eigenvectors of $\mathbf{B}$ (which form the basis $\{\mathbf{y}^1, \ldots, \mathbf{y}^n\}$ of the preceding section). For $\mathbf{B}\mathbf{y}^i = \lambda_i \mathbf{y}^i$, and if $\mathbf{y}^i = \langle y_1^i, y_2^i, \ldots, y_n^i \rangle$, then, writing $\mathbf{y}^i$ as a column vector, we have

$$\left\| \begin{matrix} b_{11} & b_{12} & \cdots & b_{1n} \\ b_{21} & b_{22} & \cdots & b_{2n} \\ \vdots & \vdots & & \vdots \\ b_{n1} & b_{n2} & \cdots & b_{nn} \end{matrix} \right\| \left\| \begin{matrix} y_1^i \\ y_2^i \\ \vdots \\ y_n^i \end{matrix} \right\| = \lambda_i \left\| \begin{matrix} y_1^i \\ y_2^i \\ \vdots \\ y_n^i \end{matrix} \right\|,$$

whence, by quick computation

$$\left\| \begin{matrix} b_{11} & b_{12} & \cdots & b_{1n} \\ b_{21} & b_{22} & \cdots & b_{2n} \\ \vdots & \vdots & & \vdots \\ b_{n1} & b_{n2} & \cdots & b_{nn} \end{matrix} \right\| \left\| \begin{matrix} y_1^1 & y_1^2 & \cdots & y_1^n \\ y_2^1 & y_2^2 & \cdots & y_2^n \\ \vdots & \vdots & & \vdots \\ y_n^1 & y_n^2 & \cdots & y_n^n \end{matrix} \right\|$$

$$= \left\| \begin{matrix} y_1^1 & y_1^2 & \cdots & y_1^n \\ y_2^1 & y_2^2 & \cdots & y_2^n \\ \vdots & \vdots & & \vdots \\ y_n^1 & y_n^2 & \cdots & y_n^n \end{matrix} \right\| \left\| \begin{matrix} \lambda_1 & 0 & \cdots & 0 \\ 0 & \lambda_2 & \cdots & 0 \\ \vdots & \vdots & & \vdots \\ 0 & 0 & \cdots & \lambda_n \end{matrix} \right\|.$$

Since the matrix $\mathbf{Y} = \|y_j^i\|$ maps the basic column vectors $\langle 1, 0, \ldots, 0 \rangle$, $\langle 0, 1, 0, \ldots, 0 \rangle, \ldots$ onto $\mathbf{y}^1, \mathbf{y}^2, \ldots$, respectively, it maps U^n onto itself; hence $\mathbf{Y}^{-1}$ exists and $\mathbf{Y}^{-1}\mathbf{B}\mathbf{Y} = \mathbf{\Lambda}$, where $\mathbf{\Lambda}$ denotes the above diagonal matrix of the eigenvalues.

Exercises

1. Find the Jordan forms for the matrices

$$\left\| \begin{matrix} 1 & 0 & 2 \\ 0 & 3 & 1 \\ 2 & -1 & 0 \end{matrix} \right\| \quad \text{and} \quad \left\| \begin{matrix} 1 & -1 \\ 1 & 1 \end{matrix} \right\|.$$

2. A common type of square matrix is one in which all elements are nonnegative and the sum of the elements in each row is unity. For such a matrix $\mathbf{A}$ prove that 1 is an eigenvalue, that all eigenvalues are less than or equal to 1 in absolute value, and that the index of the eigenvalue 1 is 1; that is, the null space of $\mathbf{A} - \mathbf{I}$ and $(\mathbf{A} - \mathbf{I})^2$ coincide.

3. Construct an $n \times n$ matrix $\mathbf{A}$ with distinct eigenvalues such that $\mathbf{A}^n = \mathbf{I}$.

4. An equivalence relation on transformations is one which satisfies the following three postulates:

(1) Any transformation is equivalent to itself.
(2) If **A** is equivalent to **B**, then **B** is equivalent to **A**.
(3) If **A** is equivalent to **B** and **B** is equivalent to **C**, then **A** is equivalent to **C**.

Prove that the notion of "similarity" is an equivalence relation.

5. In E^3 with basis $\mathbf{e}^1$, $\mathbf{e}^2$, $\mathbf{e}^3$ let $\mathbf{a}$ be a fixed vector with $(\mathbf{a}, \mathbf{a}) = 1$. Consider the hyperplane $H = \{\mathbf{x} | (\mathbf{a}, \mathbf{x}) = 0\}$. By the orthogonal projection $\mathbf{y}'$ of a vector onto H we mean the unique vector $\mathbf{y}'$ in H such that $\mathbf{y}' = \mathbf{y} - \alpha\mathbf{a}$ for some scalar α. Define $\mathbf{P}\mathbf{y} = \mathbf{y}'$. Then $\mathbf{P}\mathbf{y} = \mathbf{y} - (\mathbf{y}, \mathbf{a})\mathbf{a} \cdot \mathcal{R}(P) = H$. The null space of P is the space spanned by the vector $\mathbf{a}$. The eigenvalues of $\mathbf{P}$ are 0 and 1, and the Jordan form of $\mathbf{P}$ is given by

$$\begin{Vmatrix} 0 & 0 & 0 \\ 0 & 1 & 0 \\ 0 & 0 & 1 \end{Vmatrix}.$$

Verify all these statements.

A.5 Transposed, normal, and hermitian matrices; orthogonal complement.

If $\mathbf{A}$ is any complex $m \times n$ matrix, the *conjugate transpose* $\mathbf{A}^*$ of $\mathbf{A}$ is the $n \times m$ matrix with ijth entry

$$a_{ij}^* = \bar{a}_{ji} \qquad (i = 1, \ldots, n; j = 1, \ldots, m).$$

Clearly $(\mathbf{A} + \gamma\mathbf{B})^* = \mathbf{A}^* + \bar{\gamma}\mathbf{B}^*$, and $(\mathbf{AB}^*) = \mathbf{B}^*\mathbf{A}^*$ is easily verified when $\mathbf{AB}$ makes sense. For $\mathbf{x}$ in U^n and $\mathbf{y}$ in U^m

$$(\mathbf{Ax}, \mathbf{y}) = \sum_{i=1}^{m} \sum_{j=1}^{n} a_{ij}x_j\bar{y}_i = \sum_{j=1}^{n} x_j \sum_{i=1}^{m} \bar{a}_{ij}y_i = (\mathbf{x}, \mathbf{A}^*\mathbf{y}).$$

As a consequence, $\mathcal{R}(\mathbf{A})$ is orthogonal to $\mathfrak{N}(\mathbf{A}^*)$; that is, $(\mathbf{Ax}, \mathbf{y}) = 0$ if $\mathbf{A}^*\mathbf{y} = \mathbf{0}$. Indeed, $(\mathbf{Ax}, \mathbf{y}) = 0$ for all $\mathbf{x}$ if and only if $\mathbf{A}^*\mathbf{y} = \mathbf{0}$. Hence, if we denote by $\mathcal{R}(\mathbf{A})^\perp$ the set of all elements of U^m orthogonal to the subspace $\mathcal{R}(\mathbf{A})$, we may write $\mathcal{R}(\mathbf{A})^\perp = \mathfrak{N}(\mathbf{A}^*)$; similarly $\mathcal{R}(\mathbf{A}^*)^\perp = \mathfrak{N}(\mathbf{A})$. $\mathcal{R}(\mathbf{A})^\perp$ is known as the *orthogonal complement of* $\mathcal{R}(\mathbf{A})$. (The orthogonal complement of any subset of U^n is always a subspace; if V is a subspace, it is easily shown that $V^{\perp\perp} = (V^\perp)^\perp = V$.) When we are dealing with the real vector space E^n, the conjugate transpose matrix $\mathbf{A}^*$ reduces to the ordinary transpose matrix $\mathbf{A}'$(obtained from $\mathbf{A}$ by interchanging corresponding rows and columns), and all the corresponding operations and relations of $\mathbf{A}'$, $\mathcal{R}(\mathbf{A}')^\perp$, and $\mathfrak{N}(\mathbf{A}')$ remain valid.

When $\mathbf{A}$ is $n \times n$, $\mathbf{A}$ is called *normal* if $\mathbf{A}^*\mathbf{A} = \mathbf{AA}^*$ (in general, $\mathbf{A}^*\mathbf{A}$ is $n \times n$ and $\mathbf{AA}^*$ is $m \times m$). One form of normal matrix is the hermitian matrix: $\mathbf{A}$ is *hermitian* if $\mathbf{A} = \mathbf{A}^*$. (A real hermitian matrix is called symmetric; here $\mathbf{A} = \mathbf{A}'$.) Since $(\mathbf{AB})^* = \mathbf{B}^*\mathbf{A}^*$, and clearly $\mathbf{A}^{**} = \mathbf{A}$, each $n \times n$ matrix $\mathbf{A}$ has $\mathbf{A}^*\mathbf{A}$ (or $\mathbf{AA}^*$) hermitian.

Normal matrices have elegant properties compared with general matrices. As we shall see, *the eigenvectors of a normal matrix* $\mathbf{A}$ *span* U^n *and those corresponding to distinct eigenvalues are orthogonal, so that we may obtain an orthonormal basis of eigenvectors for* U^n; moreover, $\mathbf{Ax} = \lambda\mathbf{x}$ ($\mathbf{x} \neq \mathbf{0}$) *is equivalent to* $\mathbf{A}^*\mathbf{x} = \bar{\lambda}\mathbf{x}$, that is, $\mathbf{A}$ *and* $\mathbf{A}^*$ *share the same eigenvectors with conjugate eigenvalues.*

To obtain these facts, we begin by noting that when $\mathbf{A}$ is normal, $\mathbf{B} = \mathbf{A} - \lambda\mathbf{I}$ is also normal. Thus if $\mathbf{B}^k\mathbf{x} = \mathbf{0}$, we have $\mathbf{B}^{*k}\mathbf{B}^k = \mathbf{0}$; and since $\mathbf{B}^*$ and $\mathbf{B}$ commute, $(\mathbf{B}^*\mathbf{B})^k\mathbf{x} = \mathbf{0}$, and surely $(\mathbf{B}^*\mathbf{B})^{2^k}\mathbf{x} = \mathbf{0}$. Let $\mathbf{C} = \mathbf{B}^*\mathbf{B} = \mathbf{C}^*$. Then

$$(\mathbf{C}^{2^k}\mathbf{x}, \mathbf{x}) = 0 = (\mathbf{C}^{2^{k-1}}\mathbf{x}, \mathbf{C}^{2^{k-1}}\mathbf{x}),$$

and thus $\mathbf{C}^{2^{k-1}}\mathbf{x} = \mathbf{0}$; similarly $\mathbf{C}^{2^{k-2}}\mathbf{x} = \mathbf{0}$, and continuing, we find ultimately that $\mathbf{Cx} = \mathbf{0}$, or $\mathbf{B}^*\mathbf{Bx} = \mathbf{0}$. But again $(\mathbf{B}^*\mathbf{Bx}, \mathbf{x}) = 0 = (\mathbf{Bx}, \mathbf{Bx})$, and thus $\mathbf{Bx} = \mathbf{0}$. Consequently each element of $\mathbf{x}_i$ obtained in our Jordan decomposition is an eigenvector of $\mathbf{A}$, and U^n is spanned by eigenvectors. Further, if $\mathbf{Bx} = \mathbf{0}$, then

$$(\mathbf{Bx}, \mathbf{Bx}) = 0 = (\mathbf{B}^*\mathbf{Bx}, \mathbf{x}) = (\mathbf{BB}^*\mathbf{x}, \mathbf{x}) = (\mathbf{B}^*\mathbf{x}, \mathbf{B}^*\mathbf{x}),$$

and thus $\mathbf{B}^*\mathbf{x} = \mathbf{0}$. In terms of $\mathbf{A} - \lambda\mathbf{I}$, if $\mathbf{Ax} = \lambda\mathbf{x}$, then $\mathbf{A}^*\mathbf{x} = \bar{\lambda}\mathbf{x}$; applying this to $\{\mathbf{A}^*, \bar{\lambda}\}$ in place of $\{\mathbf{A}, \lambda\}$ yields the converse.

Consequently, if $\mathbf{Ax} = \lambda\mathbf{x}$ and $\mathbf{Ay} = \mu\mathbf{y}$, with $\lambda \neq \mu$ and $\mathbf{x}, \mathbf{y} \neq \mathbf{0}$, we have

$$(\mathbf{Ax}, \mathbf{y}) = (\lambda\mathbf{x}, \mathbf{y}) = \lambda(\mathbf{x}, \mathbf{y}) = (\mathbf{x}, \mathbf{A}^*\mathbf{y}) = (\mathbf{x}, \bar{\mu}\mathbf{y}) = \mu(\mathbf{x}, \mathbf{y}).$$

It follows that $(\lambda - \mu)(\mathbf{x}, \mathbf{y}) = 0$; thus $(\mathbf{x}, \mathbf{y}) = 0$, and distinct eigenvalues have orthogonal eigenvectors. Since the set of eigenvectors corresponding to λ_i forms the subspace Y_i when $\mathbf{0}$ is adjoined, we can find an orthonormal basis for each Y_i; obviously the union of these bases is the required orthonormal basis for U^n.

Applying the construction of the last section, we find that $\mathbf{Y}^{-1}\mathbf{AY} = \mathbf{\Lambda}$, a diagonal matrix, where the column vectors of $\mathbf{Y}$ are orthonormal (being our basis of orthonormal eigenvectors). Since this amounts to saying

$$\sum_{j=1}^{n} y_j^i \bar{y}_j^k = \delta_{ik},$$

we have $\mathbf{YY}^* = \mathbf{I}$, and thus $\mathbf{Y}^{-1} = \mathbf{Y}^*$. Such a matrix is called *unitary,* and $\mathbf{A}$ is said to be unitarily equivalent (rather than unitarily similar) to $\mathbf{\Lambda}$. A unitary matrix preserves inner products (hence length and orthogonality), for $(\mathbf{Yx}, \mathbf{Yy}) = (\mathbf{Y^*Yx}, \mathbf{y}) = (\mathbf{x}, \mathbf{y})$.

Hermitian matrices have the further property that each λ_i is real, for the equal matrices $\mathbf{A}$ and $\mathbf{A}^*$ have conjugate eigenvalues; moreover, any normal matrix with this property is hermitian, for then

$$\mathbf{A} = \mathbf{Y\Lambda Y^*} = (\mathbf{Y\Lambda^* Y^*})^* = (\mathbf{Y\Lambda Y^*})^* = \mathbf{A}^*.$$

A unitary matrix is of course normal, and each eigenvalue is of unit modulus ($|\lambda| = 1$; for $\mathbf{Ux} = \lambda\mathbf{x}$ has the same length as $\mathbf{x}$). Conversely, any normal matrix with unimodular eigenvalues is unitary. A real unitary matrix is called *orthogonal*; any real eigenvalue is ± 1 and has corresponding real eigenvectors, which either are fixed points of the corresponding transformation or are reflected into their negatives. (No real eigenvalues need exist if n is even.) An orthogonal matrix also preserves angles.

Projections. If $U^n = V \oplus W$, then the unique expression of $\mathbf{x}$ as $\mathbf{v} + \mathbf{w}$ ($\mathbf{v} \in V$; $\mathbf{w} \in W$) leads to a linear transformation $\mathbf{P}$ mapping $\mathbf{x}$ into $\mathbf{v}$, and $\mathbf{P}^2\mathbf{x} = \mathbf{Pv} = \mathbf{v} = \mathbf{Px}$, that is, $\mathbf{P}^2 = \mathbf{P}$. If V and W are orthogonal subspaces ($W = V^\perp$), then $\mathbf{P} = \mathbf{P}^2 = \mathbf{P}^*$. Conversely, any transformation $\mathbf{P}$ such that $\mathbf{P} = \mathbf{P}^2 = \mathbf{P}^*$ (called an *orthogonal projection*, or *projection*) yields a direct-sum decomposition $U^n = \mathcal{R}(\mathbf{P}) + \mathfrak{N}(\mathbf{P})$. As a normal matrix, $\mathbf{P}$ can be represented (by an appropriate choice of orthonormal basis) as a diagonal matrix with 0's and 1's on the main diagonal and 0's elsewhere.

Exercises

1. Interpret and prove the statement that any orthogonal transformation may be regarded as a rotation of axes about a fixed origin.

2. We have defined $\mathbf{A}^*$ with respect to a particular basis of U^n, but it may also be defined directly from the inner product. Prove that for each $\mathbf{y} \in U^n$, $\mathbf{A}^*\mathbf{y}$ is the unique vector $\mathbf{y}'$ such that $(\mathbf{Ax}, \mathbf{y}) = (\mathbf{x}, \mathbf{y}')$ for all $\mathbf{x} \in U^n$.

3. On the space $\mathcal{P}^n$ of polynomials with real coefficients over the interval $[a, b]$, let $f(t)$ be a specific even polynomial. Define the linear transformation $\mathbf{A}$ on $\mathcal{P}^n$ as follows:

$$(\mathbf{A}p)(t) = \int_a^b f(t - y)p(y)\, dy.$$

Show that $\mathbf{A}$ is a symmetric transformation.

4. Prove that a matrix is unitary if and only if it transforms orthonormal vectors into orthonormal vectors.

5. Let $\mathbf{A}$ be a normal transformation with eigenvalues $\lambda_1, \lambda_2, \ldots, \lambda_r$ (eigenvalues of multiplicity higher than one appearing only once). Let $\mathbf{E}_k$ be the per-

pendicular projection on the subspace of eigenvectors of λ_k. Show that if $i \neq j$, then $\mathbf{E}_i\mathbf{E}_j = \mathbf{0}$. Show also that $\mathbf{E}_1 + \mathbf{E}_2 + \cdots + \mathbf{E}_k$ is a perpendicular projection for all $k \leq r$, and that $\mathbf{E}_1 + \mathbf{E}_2 + \cdots + \mathbf{E}_r = \mathbf{I}$.

6. For the same setup as in Problem 5, prove the matrix relation

$$\mathbf{A} = \lambda_1\mathbf{E}_1 + \lambda_2\mathbf{E}_2 + \cdots + \lambda_r\mathbf{E}_r.$$

A.6 Quadratic form. Any expression

$$\sum_{i,j=1}^{n} a_{ij}x_ix_j$$

in the n real variables $x_1, \ldots, x_n$ (where a_{ij} is a real constant) is called a quadratic form. Since

$$Q(\mathbf{x}) = \sum_{i,j=1}^{n} a_{ij}x_ix_j = \sum \frac{1}{2}(a_{ij} + a_{ji})x_ix_j,$$

one can replace the original a_{ij} by a symmetric array with $a_{ij} = a_{ji}$. The corresponding matrix $\mathbf{A}$ is then a hermitian transformation of U^n into itself and has only real eigenvalues. Let λ denote one of these eigenvalues. Then since $\mathbf{A}(\mathbf{x} + i\mathbf{y}) = \lambda(\mathbf{x} + i\mathbf{y})$ implies $\mathbf{Ax} = \lambda\mathbf{x}$ and $\mathbf{Ay} = \lambda\mathbf{y}$, and one of these real vectors is nonzero, λ has a corresponding eigenvector in E^n. Repeating our former arguments (applying them now to E^n instead of U^n), we obtain a real orthonormal basis $\{\mathbf{y}^1, \ldots, \mathbf{y}^n\}$ of eigenvectors corresponding to the eigenvalues $\lambda_1, \ldots, \lambda_n$.

Further, the matrix $\mathbf{Y}$ we obtain from $\mathbf{y}^1, \ldots, \mathbf{y}^n$ is now a real unitary (i.e., orthogonal) matrix, which yields $\mathbf{Y}^*\mathbf{A}\mathbf{Y} = \mathbf{\Lambda}$ (diagonal), or $\mathbf{A} = \mathbf{Y}\mathbf{\Lambda}\mathbf{Y}^*$. Consequently, if we write

$$\mathbf{x} = \sum_{i=1}^{n} x'_i\mathbf{y}^i \qquad (\mathbf{x} \in E^n),$$

from which it follows that

$$\mathbf{Y}\mathbf{x}' = \sum_{i=1}^{n} x'_i\mathbf{y}^i = \mathbf{x},$$

then $\mathbf{Y}\overset{*}{\mathbf{x}} = \mathbf{Y}^{-1}\mathbf{x} = \mathbf{x}'$ and we have

$$Q(\mathbf{x}) = (\mathbf{Ax}, \mathbf{x}) = (\mathbf{Y}\mathbf{\Lambda}\mathbf{Y}^*\mathbf{x}, \mathbf{x}) = (\mathbf{\Lambda}\mathbf{Y}^*\mathbf{x}, \mathbf{Y}^*\mathbf{x}) = \sum_{i=1}^{n} \lambda_i(x'_i)^2.$$

Thus, by selecting an appropriate orthonormal basis we may express Q in diagonal form, with no mixed products of components appearing. Geometrically, then, the locus of all $\mathbf{x}$ satisfying $Q(\mathbf{x}) = 1$ is an ellipsoid if all

$\lambda_i > 0$. This occurs if and only if $Q(\mathbf{x})$ is nonnegative and $Q(\mathbf{x}) = 0$ implies $\mathbf{x} = \mathbf{0}$, i.e., when Q is *positive definite*.

Two quadratic forms Q and Q_1 are said to be unitarily equivalent if one arises from the other via a unitary transformation of E^n into itself, i.e., if $Q(\mathbf{x}) = Q_1(\mathbf{Ux})$. (In our discussion $\mathbf{U}$ is of course an orthogonal matrix, but similar results are available for complex quadratic forms.) This occurs if and only if $\mathbf{A}$ and $\mathbf{A}_1$ are unitarily equivalent, for then $(\mathbf{Ax}, \mathbf{x}) = (\mathbf{A}_1\mathbf{Ux}, \mathbf{Ux}) = (\mathbf{U}^*\mathbf{A}_1\mathbf{Ux}, \mathbf{x})$ and it follows that $\mathbf{A} = \mathbf{U}^*\mathbf{A}_1\mathbf{U}$. Thus Q and Q_1 are unitarily equivalent if and only if their diagonal forms feature the same eigenvalues (of identical multiplicities, but possibly in a different arrangement).

If Q is a positive definite quadratic form, then one can replace the usual inner product by

$$[\mathbf{x}, \mathbf{y}] = (\mathbf{Ax}, \mathbf{y}),$$

where $\mathbf{A}$ is the corresponding hermitian matrix; for $[\mathbf{x}, \mathbf{y}]$ will have all the required properties. Geometrically, this amounts to replacing the unit sphere $(\mathbf{x}, \mathbf{x}) = 1$ by an ellipsoid. In terms of the components $x_1', x_2', \ldots, x_n'$ of $\mathbf{x}$ relative to this basis, the new length of $\mathbf{x}$ is computed as the ordinary length of $(\sqrt{\lambda_1}\, x_1', \ldots, \sqrt{\lambda_n}\, x_n')$.

The new inner product will of course give rise to a new version of the conjugate transpose of a matrix $\mathbf{A}$ if one insists that $[\mathbf{Ax}, \mathbf{y}] = [\mathbf{x}, \mathbf{A}^*\mathbf{y}]$. Thus one obtains an analogous concept of normal and hermitian linear transformations relative to the new inner product; these are of course related to the old by virtue of the obvious change of scale (in each x_i') which transforms our new inner product into the old.

Exercises

1. Show that for any real matrix transformation $\mathbf{A}$ on E^n,

$$2(\mathbf{Ax}, \mathbf{y}) = (\mathbf{A}(\mathbf{x} + \mathbf{y}), \mathbf{x} + \mathbf{y}) - (\mathbf{Ax}, \mathbf{x}) - (\mathbf{Ay}, \mathbf{y}).$$

With the aid of this identity prove that if $(\mathbf{Ax}, \mathbf{x}) = (\mathbf{Bx}, \mathbf{x})$ for all $\mathbf{x}$ in E^n, then $\mathbf{A} = \mathbf{B}$.

2. Prove that $[\mathbf{x}, \mathbf{y}]$ as defined in this section satisfies the requirements for an inner product. Is it necessary that $\mathbf{A}$ be hermitian?

3. Prove that if

$$\sum_{i=1}^{n} x_i^2 = Q_1 + Q_2,$$

where Q_1 and Q_2 are quadratic forms of $\mathbf{x} = \langle x_1, x_2, \ldots, x_n\rangle$ of ranks r and $n - r$, respectively, then there is an orthogonal transformation $\mathbf{C}$ such that

$$\mathbf{x} = \mathbf{Cy}, \qquad Q_1 = \sum_{i=1}^{r} y_i^2, \qquad Q_2 = \sum_{i=r+1}^{n} y_i^2.$$

4. Prove that if Q is positive definite, then the locus of all $\mathbf{x}$ such that $Q(\mathbf{x})$ = constant is an ellipsoid with semi-axes proportional to $1/\sqrt{\lambda_r}$, where the λ_r's are the eigenvalues of A. Any direction corresponding to an eigenvector of λ_r is a semi-axis of the ellipsoid.

A.7 Matrix-valued functions. For any $n \times n$ matrix $\mathbf{A}$ and polynomial $p(t) = a_0 + a_1 t + \cdots + a_k t^k$ one may clearly form $a_0\mathbf{I} + a_1\mathbf{A} + \cdots + a_k\mathbf{A}^k = p(\mathbf{A})$; thus we obtain a matrix-valued function of $\mathbf{A}$; that is, $\mathbf{A}$ maps into $p(\mathbf{A})$. Similarly, for appropriate power series

$$f(t) = \sum_{i=0}^{\infty} a_i t^i$$

the corresponding matrix series

$$f(\mathbf{A}) = \sum_{i=0}^{\infty} a_i \mathbf{A}^i$$

(where $\mathbf{A}^0 = \mathbf{I}$) will be convergent [in the sense that each entry of the partial sums converges, with its limit the value of the corresponding entry in $f(\mathbf{A})$]. For example, if f is an entire function, $f(\mathbf{A})$ will be defined for every square matrix; in particular,

$$e^{\mathbf{A}} = \mathbf{I} + \mathbf{A} + \frac{1}{2!}\mathbf{A}^2 + \cdots$$

is well defined. *If $\lambda_1, \ldots, \lambda_k$ are the eigenvalues of $\mathbf{A}$, then $f(\lambda_1), \ldots, f(\lambda_k)$ are the eigenvalues of $f(\mathbf{A})$.* For $\mathbf{A}\mathbf{x} = \lambda\mathbf{x}$ implies $\mathbf{A}^2\mathbf{x} = \lambda\mathbf{A}\mathbf{x} = \lambda^2\mathbf{x}$, $\mathbf{A}^3\mathbf{x} = \lambda^3\mathbf{x}, \ldots$; hence $f(\mathbf{A})\mathbf{x} = f(\lambda)\mathbf{x}$, so that each $f(\lambda_i)$ is an eigenvalue of $f(\mathbf{A})$ whose associated eigenvectors are those belonging to the eigenvalue λ_i of $\mathbf{A}$. To see that all eigenvalues of $f(\mathbf{A})$ are so obtained, consider the Jordan canonical form $\mathbf{B}^{-1}\mathbf{A}\mathbf{B}$ of $\mathbf{A}$, noting that $\mathbf{B}^{-1}\mathbf{A}^k\mathbf{B} = (\mathbf{B}^{-1}\mathbf{A}\mathbf{B})^k$ and thus $\mathbf{B}^{-1}f(\mathbf{A})\mathbf{B} = f(\mathbf{B}^{-1}\mathbf{A}\mathbf{B})$; the computation of $f(\mathbf{B}^{-1}\mathbf{A}\mathbf{B})$ can be made sub-block by sub-block, and its eigenvalues become apparent.

A particularly important example of a matrix-valued function is furnished by $(\lambda\mathbf{I} - \mathbf{A})^{-1}$. When $|\lambda|$ exceeds the modulus of each eigenvalue, the formal power series expansion

$$\begin{aligned}(\lambda\mathbf{I} - \mathbf{A})^{-1} &= \lambda^{-1}\left(\mathbf{I} - \frac{1}{\lambda}\mathbf{A}\right)^{-1} \\ &= \lambda^{-1}\left(\mathbf{I} + \frac{1}{\lambda}\mathbf{A} + \frac{1}{\lambda^2}\mathbf{A}^2 + \cdots\right) = \sum_{i=0}^{\infty} \frac{\mathbf{A}^i}{\lambda^{i+1}}\end{aligned}$$

is valid. Moreover, by fixing $\mathbf{A}$ we obtain an analytic function of the complex variable λ; that is, λ maps into $(\lambda\mathbf{I} - \mathbf{A})^{-1}$, defined for all $\lambda \neq \lambda_1, \ldots, \lambda_k$. The function is analytic in the sense that

$$\frac{d}{d\lambda}(\lambda\mathbf{I} - \mathbf{A})^{-1} = \lim_{\lambda' \to \lambda} \frac{1}{\lambda' - \lambda}[(\lambda'\mathbf{I} - \mathbf{A})^{-1} - (\lambda\mathbf{I} - \mathbf{A})^{-1}]$$

$$= -(\lambda\mathbf{I} - \mathbf{A})^{-2}$$

exists where $\lambda \neq \lambda_1, \ldots, \lambda_k$. One may also apply the Cauchy integral theorem, its validity being obtained in the usual fashion; here, however, we can interpret the function's analyticity as meaning that it has a power series expansion about each $\lambda \neq \lambda_1, \ldots, \lambda_k$, with matrix coefficients. Each λ_i is a pole of the function; for if $\mathbf{C} = \mathbf{B}^{-1}\mathbf{A}\mathbf{B}$ is again the Jordan canonical form, we have $\mathbf{A} = \mathbf{B}\mathbf{C}\mathbf{B}^{-1}$ and

$$\lambda\mathbf{I} - \mathbf{A} = \lambda\mathbf{I} - \mathbf{B}\mathbf{C}\mathbf{B}^{-1} = \mathbf{B}(\lambda\mathbf{I} - \mathbf{C})\mathbf{B}^{-1};$$

hence $(\lambda\mathbf{I} - \mathbf{A})^{-1} = \mathbf{B}(\lambda\mathbf{I} - \mathbf{C})^{-1}\mathbf{B}^{-1}$, and it suffices to consider $(\lambda\mathbf{I} - \mathbf{C})^{-1}$, or indeed $(\mathbf{C} - \lambda\mathbf{I})^{-1}$. Since $\mathbf{C} - \lambda\mathbf{I}$ has the same block-diagonal form as $\mathbf{C}$, and since block-diagonal matrices (with blocks of similar size) multiply block by block—i.e., since

$$\begin{pmatrix} \mathbf{A} & \mathbf{0} \\ \mathbf{0} & \mathbf{B} \end{pmatrix}\begin{pmatrix} \mathbf{C} & \mathbf{0} \\ \mathbf{0} & \mathbf{D} \end{pmatrix} = \begin{pmatrix} \mathbf{AC} & \mathbf{0} \\ \mathbf{0} & \mathbf{BD} \end{pmatrix}$$

if $\mathbf{A}$ and $\mathbf{C}$ are $k \times k$, and $\mathbf{B}$ and $\mathbf{D}$ are $l \times l$—it is sufficient to compute the inverse of each sub-block

$$\left\| \begin{matrix} \lambda_i - \lambda & 0 & \cdots & \cdots & 0 \\ 1 & \lambda_i - \lambda & \cdots & \cdots & 0 \\ 0 & 1 & \cdots & \cdots & 0 \\ \vdots & \vdots & & & \vdots \\ 0 & 0 & \cdots & \cdots & \lambda_i - \lambda \end{matrix} \right\|$$

of $\mathbf{C} - \lambda\mathbf{I}$. Writing

$$\mathbf{D} = (\lambda_i - \lambda)\mathbf{I} + \mathbf{E} = (\lambda_i - \lambda)\left(\mathbf{I} + \frac{1}{\lambda_i - \lambda}\mathbf{E}\right),$$

where $\mathbf{E}$ is the $k \times k$ matrix with 1's below the main diagonal and 0's elsewhere, and formally expanding in analogy with $(1 + x)^{-1} = 1 - x + x^2 - x^3 + \cdots$, we obtain

$$\mathbf{D}^{-1} = (\lambda_i - \lambda)^{-1}\left(\mathbf{I} + \frac{1}{\lambda_i - \lambda}\mathbf{E}\right)^{-1}$$

$$= (\lambda_i - \lambda)^{-1}\left(\mathbf{I} - \frac{1}{\lambda_i - \lambda}\mathbf{E} + \frac{1}{(\lambda_i - \lambda)^2}\mathbf{E}^2 - \cdots\right),$$

which converges, since $\mathbf{E}^k = \mathbf{0}$. Moreover, for the same reason we may check that the value is the appropriate inverse by multiplying by

$$\mathbf{I} + \frac{1}{\lambda_i - \lambda}\mathbf{E}.$$

Thus the highest power of $(\lambda_i - \lambda)^{-1}$ which appears is the kth, and $\mathbf{D}^{-1}$ appears as a Laurent expansion (with constant matrix coefficients) corresponding to a pole of order k at λ_i. Consequently the order of the pole we obtain at λ_i for $(\lambda\mathbf{I} - \mathbf{C})^{-1}$ [or $(\lambda\mathbf{I} - \mathbf{A})^{-1}$] is the size of the largest sub-block featuring λ_i; recall that this is the least k for which

$$\mathfrak{N}[(\lambda\mathbf{I} - \mathbf{A})^k] = \mathfrak{N}[(\lambda\mathbf{I} - \mathbf{A})^{k+1}].$$

Exercises

1. Show that if $\mathbf{AB} = \mathbf{BA}$, then $e^{A+B} = e^A e^B$.

2. Find a Laurent series expansion for $(\lambda\mathbf{I} - \mathbf{A})^{-n}$ and show that it converges when λ exceeds the modulus of each eigenvalue. Under these conditions, show that

$$\frac{d}{d\lambda}(\lambda\mathbf{I} - \mathbf{A})^{-n} = -n(\lambda\mathbf{I} - \mathbf{A})^{-n-1}.$$

3. Let $\lambda_1, \lambda_2, \ldots, \lambda_r$ and $\mathbf{E}_1, \mathbf{E}_2, \ldots, \mathbf{E}_r$ (as defined in Problem 5 of Section A.5) be the eigenvalues and perpendicular projections corresponding to a normal matrix $\mathbf{A}$. Let $f(\lambda)$ be a convergent power series. Show that

$$f(\mathbf{A}) = \sum_{j=1}^{r} f(\lambda_j)\mathbf{E}_j.$$

4. Show that if $f(d) = \mathbf{A}_0 + \mathbf{A}_1 d + \mathbf{A}_2 d^2 + \cdots$ converges absolutely in some interval about d_0, then $f'(d_0) = \mathbf{A}_1 + 2\mathbf{A}_2 d_0 + 3\mathbf{A}_3 d_0^2 + \cdots$.

5. If $\mathbf{A}$ is a normal matrix with nonzero eigenvalues, how can you define $\log \mathbf{A}$?

A.8 Determinants; minors, cofactors. The determinant $|\mathbf{A}|$ (sometimes written $\det \mathbf{A}$) of an $n \times n$ matrix $\mathbf{A}$ is given by the sum

$$\sum_{\pi} \delta(\pi) a_{1,\pi(1)} a_{2,\pi(2)} \cdots a_{n,\pi(n)} = \sum_{\pi} \delta(\pi) a_{\pi(1),1} a_{\pi(2),2} \cdots a_{\pi(n),n}, \tag{A.8.1}$$

where π runs over all permutations of $\{1, 2, \ldots, n\}$, and $\delta(\pi)$ is 1 if π is an even permutation and -1 if π is an odd permutation. Interchange of any two rows (or columns) of $\mathbf{A}$ produces a matrix $\mathbf{B}$ such that $|\mathbf{B}| = -|\mathbf{A}|$, since the corresponding terms in (A.8.1) occur with permutations of opposite parity; thus if $\mathbf{A}$ has two identical rows (or columns), $|\mathbf{A}| = 0$. The determinant of the matrix obtained by deleting the ith row and jth column of $\mathbf{A}$ is the *minor* m_{ij} of a_{ij}; the *cofactor* c_{ij} of a_{ij} is then defined by $c_{ij} = (-1)^{i+j}m_{ij}$.

It follows directly from (A.8.1) that

$$|\mathbf{A}| = \sum_{k=1}^{n} a_{ik}c_{ik} = \sum_{k=1}^{n} a_{kj}c_{kj} \tag{A.8.2}$$

for all i, j; these are the so-called row and column expansions of $|\mathbf{A}|$. Further, if $i \neq j$,

$$\sum_{k=1}^{n} a_{ik}c_{jk} = 0 = \sum_{k=1}^{n} a_{ki}c_{kj},$$

since these expressions are the row and column expansions of $|\mathbf{B}|$ where $\mathbf{B}$ is a matrix with two identical rows or columns. Consequently

$$|\mathbf{A}| = \sum_{k=1}^{n} a_{ik}c_{ik} - \lambda \sum_{k=1}^{n} a_{jk}c_{ik} = \sum_{k=1}^{n} (a_{ik} - \lambda a_{jk})c_{ik}$$

if $i \neq j$, so that $|\mathbf{A}|$ coincides with the value of the determinant of the matrix obtained from $\mathbf{A}$ by subtracting λ times the jth row from the ith row. Writing $|\mathbf{A}|$ in the descriptive form

$$\begin{vmatrix} a_{11} & a_{12} & \cdots & a_{1n} \\ a_{21} & a_{22} & \cdots & a_{2n} \\ \vdots & \vdots & & \vdots \\ a_{n1} & a_{n2} & \cdots & a_{nn} \end{vmatrix},$$

we may thus subtract any multiple of a row (or column) of the determinant from any other without changing its value.

Let

$$\mathbf{A}\begin{pmatrix} i_1, i_2, \ldots, i_r \\ j_1, j_2, \ldots, j_r \end{pmatrix}$$

denote the determinant of the matrix obtained from $\mathbf{A}$ by deleting all rows but $i_1, i_2, \ldots, i_r$ and all columns but $j_1, \ldots, j_r$, where $i_1 < i_2 < \cdots < i_r, j_1 < j_2 < \cdots < j_r$; that is,

$$\mathbf{A}\begin{pmatrix} i_1, i_2, \ldots, i_r \\ j_1, j_2, \ldots, j_r \end{pmatrix} = \begin{vmatrix} a_{i_1j_1} & a_{i_1j_2} & \cdots & a_{i_1j_r} \\ a_{i_2j_1} & \cdots & \cdots & \cdots \\ \vdots & & & \\ a_{i_rj_1} & \cdots & \cdots & a_{i_rj_r} \end{vmatrix}.$$

Such subdeterminants of a product **AB** are related to those of **A** and **B** by the expansion

$$\mathbf{AB}\begin{pmatrix} i_1, i_2, \ldots, i_r \\ j_1, j_2, \ldots, j_r \end{pmatrix} = \sum_{k_1 < k_2 < \cdots < k_r} \mathbf{A}\begin{pmatrix} i_1, i_2, \ldots, i_r \\ k_1, k_2, \ldots, k_r \end{pmatrix} \mathbf{B}\begin{pmatrix} k_1, k_2, \ldots, k_r \\ j_1, j_2, \ldots, j_r \end{pmatrix}, \quad \text{(A.8.3)}$$

where the sum extends over all $\{k_1, k_2, \ldots, k_r\}$ satisfying the indicated relation.

From (A.8.2) and our succeeding remarks, if adj (**A**) denotes the matrix with ijth entry

$$b_{ij} = c_{ji},$$

the *adjoint matrix* of **A**, then $\mathbf{A} \cdot \text{adj}(\mathbf{A}) = |\mathbf{A}|\mathbf{I} = \text{adj}(\mathbf{A}) \cdot \mathbf{A}$. Thus if $|\mathbf{A}| \neq 0$,

$$\frac{1}{|\mathbf{A}|} \text{adj}(\mathbf{A})$$

is the inverse of **A**. But $|\mathbf{AB}| = |\mathbf{A}||\mathbf{B}|$ by (A.8.3); thus if $\mathbf{A}^{-1}$ exists, $|\mathbf{A}||\mathbf{A}^{-1}| = |\mathbf{I}| = 1$ and $|\mathbf{A}| \neq 0$. *Hence the nonsingular matrices are those with nonzero determinants.*

If $\mathbf{B}^{-1}\mathbf{AB}$ is the Jordan canonical form of **A**, then, since $|\mathbf{B}^{-1}\mathbf{AB}| = |\mathbf{B}^{-1}||\mathbf{A}||\mathbf{B}| = |\mathbf{A}|$, it follows that $|\mathbf{A}|$ is the product of the powers of its eigenvalues. Consequently, if 0 appears among the eigenvalues (say $\lambda_i = 0$), we may slightly change λ_i so as to have nonzero eigenvalues and a nonsingular matrix; the corresponding changes in the entries of **A** are small, and thus *each singular* **A** *is a limit of nonsingular matrices.*

Further, since

$$|\mathbf{B}^{-1}\mathbf{AB} - \lambda\mathbf{I}| = |\mathbf{B}^{-1}(\mathbf{A} - \lambda\mathbf{I})\mathbf{B}| = |\mathbf{A} - \lambda\mathbf{I}|,$$

with **B** as above, it follows that

$$|\mathbf{B}^{-1}\mathbf{AB} - \lambda\mathbf{I}| = \prod_{i=1}^{k} (\lambda_i - \lambda)^{n_i}.$$

Thus the λ_i are exactly the roots of the polynomial equation

$$|\mathbf{A} - \lambda\mathbf{I}| = 0$$

in λ, the *characteristic equation* of **A**. The multiplicity of λ_i as a root is called the *algebraic multiplicity* of λ_i (the algebraic multiplicity $\geq$ the geometric multiplicity = number of linearly independent eigenvectors corresponding to λ_i); it is of course the number of times λ_i appears on the main diagonal of the Jordan form.

Exercises

1. Show that if $\langle x_1, y_1\rangle$ and $\langle x_2, y_2\rangle$ are distinct points on the real plane, the equation

$$\begin{vmatrix} x & y & 1 \\ x_1 & y_1 & 1 \\ x_2 & y_2 & 1 \end{vmatrix} = 0$$

represents a straight line through the points.

2. Prove the following statement:

Given a nonsingular $n \times n$ matrix $\mathbf{A}$ and a vector $\mathbf{y}$ in U^n, let $\mathbf{A}_{iy}$ be the matrix in which column i has been replaced by y and all other columns remain unchanged. Then the solution of the system of equations $\mathbf{Ax} = \mathbf{y}$, where $\mathbf{x} = \langle x_1, x_2, \ldots, x_n\rangle$, is given by

$$x_i = \frac{|\mathbf{A}_{iy}|}{|\mathbf{A}|} \qquad (i = 1, \ldots, n).$$

3. Show that if all the elements in a matrix above the main diagonal are zero, the diagonal elements are the eigenvalues.

4. Let $\mathbf{A}$ have eigenvalues $\lambda_1, \lambda_2, \ldots, \lambda_n$, with repetitions if necessary. Find $|(I - \lambda\mathbf{A})^{-1}|$.

5. By $|\mathbf{x}|$ we mean $(\mathbf{x}, \mathbf{x})^{\frac{1}{2}}$. We say that a matrix or transformation $\mathbf{A}$ is bounded on U if there is a constant K such that $\|\mathbf{Ax}\| \leq K\|\mathbf{x}\|$ for all $\mathbf{x}$ in U. The greatest lower bound, K_0, of all such K's is the bound of $\mathbf{A}$. Prove that all transformations on U^n are bounded. Show that if $\mathbf{A}$ is the limit of nonsingular matrices, then $\mathbf{A}$ is singular if and only if the bounds of the inverses of these matrices approach infinity.

A.9 Some identities. Let $\mathbf{U}$ be the $n \times n$ matrix with all entries 1, and let $\mathbf{u} = \langle 1, 1, \ldots, 1\rangle$ be an n-component vector of $\mathbf{U}$ with all entries 1. The following identities relating the determinants and adjoints of $\mathbf{A}$ and $\mathbf{A} + a\mathbf{U}$ are used in Section 2.4:

$$\begin{aligned} |\mathbf{A} + a\mathbf{U}| &= |\mathbf{A}| + a\,(\mathbf{u}, \text{adj}\,(\mathbf{A})\,\mathbf{u}), \\ \mathbf{u}\,\text{adj}\,(\mathbf{A} + a\mathbf{U}) &= \mathbf{u}\,\text{adj}\,(\mathbf{A}), \\ \text{adj}\,(\mathbf{A} + a\mathbf{U})\mathbf{u} &= \text{adj}\,(\mathbf{A})\mathbf{u}. \end{aligned}$$

Here $(\mathbf{u}, \text{adj}\,(\mathbf{A})\mathbf{u})$ when calculated yields

$$\sum_{i,j=1}^{n} A_{ij},$$

where A_{ij} is the cofactor of a_{ij}. If we regard $|\mathbf{A}|$ as a function of the n^2 variables a_{ij}, the row expansion

$$|\mathbf{A}| = \sum_{k=1}^{n} a_{ik}A_{ik}$$

implies $\partial|\mathbf{A}|/\partial a_{ij} = A_{ij}$. Thus, regarding $|\mathbf{A} + a\mathbf{U}|$ as a function of a, setting $a'_{ij} = a_{ij} + a$, and denoting by A'_{ij} the cofactor of a'_{ij} in $\mathbf{A} + a\mathbf{U}$, we have

$$\frac{d}{da}|\mathbf{A} + a\mathbf{U}| = \sum_{i,j=1}^{n} \frac{\partial}{\partial a'_{ij}}|\mathbf{A} + a\mathbf{U}| \frac{d}{da}(a'_{ij}) = \sum_{i,j=1}^{n} A'_{ij}.$$

But since $\sum_{j=1}^{n} A'_{ij} \cdot 1$ is evidently the expansion about the ith row of a matrix obtained from $\mathbf{A} + a\mathbf{U}$ by replacing the ith row by $1, 1, \ldots, 1$, that is, the matrix

$$\begin{vmatrix} a_{11}+a & a_{12}+a & \cdots & a_{1n}+a \\ \vdots & & & \\ a_{i-1,1}+a & a_{i-1,2}+a & \cdots & a_{i-1,n}+a \\ 1 & 1 & \cdots & 1 \\ a_{i+1,1}+a & a_{i+1,2}+a & \cdots & a_{i+1,n}+a \\ \vdots & & & \\ a_{n1}+a & a_{n2}+a & \cdots & a_{nn}+a \end{vmatrix},$$

we can subtract a times the ith row from all others to obtain

$$\sum_{j=1}^{n} A'_{ij} = \sum_{j=1}^{n} A_{ij}. \tag{A.9.1}$$

Thus

$$\frac{d}{da}|\mathbf{A} + a\mathbf{U}| = \sum_{i,j=1}^{n} A_{ij},$$

which is independent of a. It follows that $|\mathbf{A} + a\mathbf{U}|$ is the linear function

$$|\mathbf{A}| + a \cdot \sum_{i,j=1}^{n} A_{ij},$$

which is our first identity. Further, (A.9.1) states precisely that

$$\left\|\begin{matrix} A'_{11} & A'_{12} & \cdots & A'_{1n} \\ A'_{21} & A'_{22} & \cdots & A'_{2n} \\ \vdots & \vdots & & \vdots \\ A'_{n1} & A'_{n2} & \cdots & A'_{nn} \end{matrix}\right\| \left\|\begin{matrix} 1 \\ 1 \\ \vdots \\ 1 \end{matrix}\right\| = \left\|\begin{matrix} \sum A'_{1j} \\ \sum A'_{2j} \\ \vdots \\ \sum A'_{nj} \end{matrix}\right\| = \left\|\begin{matrix} A_{11} & A_{12} & \cdots & A_{1n} \\ A_{21} & A_{22} & \cdots & A_{2n} \\ \vdots & \vdots & & \vdots \\ A_{n1} & A_{n2} & \cdots & A_{nn} \end{matrix}\right\| \left\|\begin{matrix} 1 \\ 1 \\ \vdots \\ 1 \end{matrix}\right\|,$$

that is, our third identity. The second follows by the original argument concerning

$$\sum_{j=1}^{n} A'_{ij} \qquad \text{applied to} \quad \sum_{i=1}^{n} A'_{ij}$$

as a column expansion.

It follows directly from the definition that $|\lambda \mathbf{A}| = \lambda^n |\mathbf{A}|$ for any $n \times n$ matrix $\mathbf{A}$. Consequently, since

$$\mathbf{A}^{-1} = \frac{1}{|\mathbf{A}|} \operatorname{adj}(\mathbf{A})$$

when $\mathbf{A}$ is nonsingular, we have

$$|\mathbf{A}^{-1}| = \frac{1}{|\mathbf{A}|} = \frac{1}{|\mathbf{A}|^n} |\operatorname{adj}(\mathbf{A})|,$$

and thus $|\operatorname{adj}(\mathbf{A})| = |\mathbf{A}|^{n-1}$. Since this last relation holds for any limit of nonsingular matrices, it holds for all $\mathbf{A}$.

For any nonsingular matrix $\mathbf{A}$,

$$\mathbf{A}^{-1}\begin{pmatrix} i_1, i_2, \ldots, i_p \\ j_1, j_2, \ldots, j_p \end{pmatrix} = \frac{\mathbf{A}\begin{pmatrix} j'_1, j'_2, \ldots, j'_{n-p} \\ i'_1, i'_2, \ldots, i'_{n-p} \end{pmatrix}}{|\mathbf{A}|} (-1)^{\Sigma_{\nu=1}^{p}(i_\nu + j_\nu)} \tag{A.9.2}$$

where $\{j'_1, j'_2, \ldots, j'_{n-p}\}$ forms the complementary set of column indices to $\{j_1, \ldots, j_p\}$, and similarly for $\{i'_1, \ldots, i'_{n-p}\}$. Equation (A.9.2) is known as Sylvester's identity.

Consider the case in which $i_1 = j_1 = 1$, $i_2 = j_2 = 2$, $\ldots$, $i_p = j_p = p$, so that the identity becomes

$$\mathbf{A}^{-1}\begin{pmatrix} 1, 2, \ldots, p \\ 1, 2, \ldots, p \end{pmatrix} = \frac{1}{|\mathbf{A}|} \mathbf{A}\begin{pmatrix} p+1, p+2, \ldots, n \\ p+1, p+2, \ldots, n \end{pmatrix}.$$

Noting that the transpose of a matrix has the same determinant as the matrix and that $|\mathbf{A}||\mathbf{B}| = |\mathbf{AB}|$, and denoting by A_{ij} the cofactor of a_{ij}, we have the product

$$\begin{vmatrix} A_{11} & A_{12} & \cdots & A_{1p} & A_{1,p+1} & \cdots & \cdots & A_{1n} \\ A_{21} & A_{22} & \cdots & A_{2p} & A_{2,p+1} & \cdots & \cdots & A_{2n} \\ \vdots & & & & & & & \\ A_{p1} & A_{p2} & \cdots & A_{pp} & A_{p,p+1} & \cdots & \cdots & A_{pn} \\ 0 & 0 & \cdots & 0 & 1 & 0 & \cdots & 0 \\ 0 & 0 & \cdots & 0 & 0 & 1 & \cdots & 0 \\ \vdots & \vdots & & \vdots & \vdots & & & \vdots \\ 0 & 0 & \cdots & 0 & 0 & \cdots & \cdots & 1 \end{vmatrix}$$

$$\times \begin{vmatrix} a_{11} & a_{21} & \cdots & a_{n1} \\ a_{12} & a_{22} & \cdots & a_{n2} \\ \vdots & & & \\ a_{1p} & a_{2p} & \cdots & a_{n,p} \\ a_{1,p+1} & a_{2,p+1} & \cdots & a_{n,p+1} \\ \vdots & \vdots & & \vdots \\ a_{1n} & a_{2n} & \cdots & a_{nn} \end{vmatrix}$$

(where the first p rows of the first determinant coincide with those of adj (**A**), while the last $n - p$ feature only 1's on the main diagonal as non-zero entries), equal to

$$\begin{vmatrix} |\mathbf{A}| & 0 & \cdots & \cdots & \cdots & 0 & 0 & \cdots & 0 \\ 0 & |\mathbf{A}| & 0 & \cdots & \cdots & \cdots & \cdots & \cdots & \cdots \\ \vdots & & & & & & & & \\ 0 & \cdots & \cdots & \cdots & 0 & |\mathbf{A}| & 0 & \cdots & 0 \\ a_{1,p+1} & \cdots & \cdots & \cdots & \cdots & a_{p,p+1} & a_{p+1,p+1} & \cdots & a_{n,p+1} \\ a_{1,p+2} & \cdots & \cdots & \cdots & \cdots & a_{p,p+2} & a_{p+1,p+2} & \cdots & a_{n,p+2} \\ \vdots & & & & & & & & \\ a_{1,n} & \cdots & \cdots & \cdots & \cdots & a_{p,n} & a_{p+1,n} & \cdots & a_{n,n} \end{vmatrix}$$

$$= |\mathbf{A}|^p \mathbf{A}\begin{pmatrix} p+1, \ldots, n \\ p+1, \ldots, n \end{pmatrix}$$

(expanding by rows, using the first p rows). On the other hand, if we expand the first factor of our product by rows (using the last $n - p$ rows), it becomes, with $\mathbf{C} = \text{adj}(\mathbf{A})$,

$$\mathbf{C}\begin{pmatrix} 1, 2, \ldots, p \\ 1, 2, \ldots, p \end{pmatrix} |\mathbf{A}|;$$

hence

$$\mathbf{C}\begin{pmatrix} 1, 2, \ldots, p \\ 1, 2, \ldots, p \end{pmatrix} = |\mathbf{A}|^{p-1} \mathbf{A}\begin{pmatrix} p+1, \ldots, n \\ p+1, \ldots, n \end{pmatrix}.$$

But

$$\mathbf{A}^{-1} = \frac{1}{|\mathbf{A}|}\mathbf{C},$$

and thus

$$\mathbf{A}^{-1}\begin{pmatrix} 1, \ldots, p \\ 1, \ldots, p \end{pmatrix} = \frac{1}{|\mathbf{A}|^p}\mathbf{C}\begin{pmatrix} 1, \ldots, p \\ 1, \ldots, p \end{pmatrix} = \frac{1}{|\mathbf{A}|}\mathbf{A}\begin{pmatrix} p+1, \ldots, n \\ p+1, \ldots, n \end{pmatrix},$$

which is exactly the identity in our special case.

By further interchanging appropriate rows and columns we may achieve (A.9.2). In terms of $\mathbf{C} = \operatorname{adj}(\mathbf{A})$ the identity (A.9.2) becomes

$$\frac{1}{|\mathbf{A}|^p}\mathbf{C}\begin{pmatrix} i_1, \ldots, i_p \\ j_1, \ldots, j_p \end{pmatrix} = (-1)^{\sum_{\nu=1}^{p}(i_\nu + j_\nu)} \frac{1}{|\mathbf{A}|}\mathbf{A}\begin{pmatrix} j'_1, \ldots, j'_{n-p} \\ i'_1, \ldots, i'_{n-p} \end{pmatrix},$$

or

$$\mathbf{C}\begin{pmatrix} i_1, \ldots, i_p \\ j_1, \ldots, j_p \end{pmatrix} = (-1)^{\sum_{\nu=1}^{p}(i_\nu + j_\nu)} |\mathbf{A}|^{p-1}\mathbf{A}\begin{pmatrix} j'_1, \ldots, j'_{n-p} \\ i'_1, \ldots, i'_{n-p} \end{pmatrix},$$

since

$$\frac{1}{|\mathbf{A}|}\mathbf{C} = \mathbf{A}^{-1}.$$

The determinants

$$\mathbf{A}\begin{pmatrix} i_1, \ldots, i_p \\ j_1, \ldots, j_p \end{pmatrix}$$

are clearly generalized minors of $\mathbf{A}$; but in particular the ijth minor of $\mathbf{A}$ is

$$\mathbf{A}\begin{pmatrix} 1, 2, \ldots, i-1, i+1, \ldots, n \\ 1, 2, \ldots, j-1, j+1, \ldots, n \end{pmatrix},$$

while of course

$$\mathbf{A}\begin{pmatrix} 1, \ldots, n \\ 1, \ldots, n \end{pmatrix} = |\mathbf{A}|.$$

Let $p < n$ be fixed, and set

$$d_{ij} = \mathbf{A}\begin{pmatrix} 1, \ldots, p, i \\ 1, \ldots, p, j \end{pmatrix} \qquad (i, j \geq p+1),$$

so that the $(n - p) \times (n - p)$ array $\mathbf{D} = \|d_{ij}\|$ forms a matrix. Then (*Sylvester's theorem*)

$$\mathbf{D}\begin{pmatrix} i_1, \ldots, i_q \\ j_1, \ldots, j_q \end{pmatrix} = \mathbf{A}\begin{pmatrix} 1, \ldots, p \\ 1, \ldots, p \end{pmatrix}^{q-1} \mathbf{A}\begin{pmatrix} 1, \ldots, p, i_1, \ldots, i_q \\ 1, \ldots, p, j_1, \ldots, j_q \end{pmatrix}. \quad \text{(A.9.3)}$$

Since we shall be concerned only with the matrix obtained from $\mathbf{A}$ by deleting all rows but $1, 2, \ldots, p, i_1, \ldots, i_q$ and all columns but $1, 2, \ldots, p, j_1, \ldots, j_q$, we can assume without loss of generality that this matrix forms our original $\mathbf{A}$ and thus prove the identity in the one case in which $q = n - p$:

$$\mathbf{D}\begin{pmatrix} p+1, \ldots, n \\ p+1, \ldots, n \end{pmatrix} = \mathbf{A}\begin{pmatrix} 1, \ldots, p \\ 1, \ldots, p \end{pmatrix}^{n-p-1} \mathbf{A}\begin{pmatrix} 1, \ldots, n \\ 1, \ldots, n \end{pmatrix}$$

$$= \mathbf{A}\begin{pmatrix} 1, \ldots, p \\ 1, \ldots, p \end{pmatrix}^{n-p-1} \cdot |\mathbf{A}|.$$

Let $\mathbf{C} = \operatorname{adj}(\mathbf{A})$. Then

$$c_{ij} = (-1)^{i+j}\mathbf{A}\begin{pmatrix} 1, \ldots, j-1, j+1, \ldots, n \\ 1, \ldots, i-1, i+1, \ldots, n \end{pmatrix}.$$

In view of our first identity (written in terms of the adjoint), we have

$$\begin{aligned} \mathbf{C}&\begin{pmatrix} p+1, \ldots, r-1, r+1, \ldots, n \\ p+1, \ldots, s-1, s+1, \ldots, n \end{pmatrix} \\ &= (-1)^{r+s}|\mathbf{A}|^{n-p-2}\mathbf{A}\begin{pmatrix} 1, \ldots, p, s \\ 1, \ldots, p, r \end{pmatrix} \qquad \text{(A.9.4)} \\ &= (-1)^{r+s}|\mathbf{A}|^{n-p-2}d_{sr}, \end{aligned}$$

and

$$\mathbf{C}\begin{pmatrix} p+1, \ldots, n \\ p+1, \ldots, n \end{pmatrix} = |\mathbf{A}|^{n-p-1}\mathbf{A}\begin{pmatrix} 1, \ldots, p \\ 1, \ldots, p \end{pmatrix}. \qquad \text{(A.9.5)}$$

But

$$b_{sr} = (-1)^{s+r}\mathbf{C}\begin{pmatrix} p+1, \ldots, r-1, r+1, \ldots, n \\ p+1, \ldots, s-1, s+1, \ldots, n \end{pmatrix} \qquad (r, s \geq p+1)$$

defines the elements of the adjoint $\mathbf{B}$ of the $(n - p) \times (n - p)$ matrix

$$\left\| \begin{matrix} c_{p+1,p+1} & c_{p+1,p+2} & \cdots & c_{p+1,n} \\ c_{p+2,p+2} & \cdots & \cdots & c_{p+2,n} \\ \vdots & & & \vdots \\ c_{n,p+1} & \cdots & \cdots & c_{n,n} \end{matrix} \right\|,$$

and thus

$$\mathbf{B}\begin{pmatrix} p+1, \ldots, n \\ p+1, \ldots, n \end{pmatrix} = |\mathbf{B}| = \mathbf{C}\begin{pmatrix} p+1, \ldots, n \\ p+1, \ldots, n \end{pmatrix}^{n-p-1}$$

since

$$\mathbf{C}\begin{pmatrix} p+1, \ldots, n \\ p+1, \ldots, n \end{pmatrix}$$

is just the determinant of this matrix and $|\text{adj}\,(\mathbf{A}_0)| = |\mathbf{A}_0|^{k-1}$ for any $k \times k$ matrix $\mathbf{A}_0$. But clearly $b_{sr} = |\mathbf{A}|^{n-p-2} d_{sr}$; hence

$$|\mathbf{B}| = |\mathbf{A}|^{(n-p-2)(n-p)} \mathbf{D}\begin{pmatrix} p+1, \ldots, n \\ p+1, \ldots, n \end{pmatrix}$$

and

$$\mathbf{C}\begin{pmatrix} p+1, \ldots, n \\ p+1, \ldots, n \end{pmatrix}^{n-p-1} = |\mathbf{A}|^{(n-p)(n-p-2)} \mathbf{D}\begin{pmatrix} p+1, \ldots, n \\ p+1, \ldots, n \end{pmatrix}.$$

Thus by (A.9.5)

$$|\mathbf{A}|^{(n-p-1)^2} \mathbf{A}\begin{pmatrix} 1, \ldots, p \\ 1, \ldots, p \end{pmatrix}^{n-p-1} = |\mathbf{A}|^{(n-p)(n-p-2)} \mathbf{D}\begin{pmatrix} p+1, \ldots, n \\ p+1, \ldots, n \end{pmatrix},$$

or

$$\mathbf{D}\begin{pmatrix} p+1, \ldots, n \\ p+1, \ldots, n \end{pmatrix} = |\mathbf{A}|\mathbf{A}\begin{pmatrix} 1, \ldots, p \\ 1, \ldots, p \end{pmatrix}^{n-p-1},$$

completing the proof.

A.10 Compound matrices. From an $n \times n$ matrix $\mathbf{A}$ we form the pth associated (compound) matrix $\mathfrak{A}_p$, whose elements comprise all $p \times p$ subdeterminants of $\mathbf{A}$, arranged in alphabetical order (viz., $\{1, 3\}$ precedes $\{1, 8\}$; $\{2, 6\}$ precedes $\{3, 1\}$). For example, the following matrix $\mathfrak{A}_2$ is a square matrix of size $\binom{n}{2} \times \binom{n}{2}$:

$$\left\|\begin{matrix} A\begin{pmatrix}1,2\\1,2\end{pmatrix} & A\begin{pmatrix}1,2\\1,3\end{pmatrix} & \cdots & A\begin{pmatrix}1,2\\1,n\end{pmatrix} & A\begin{pmatrix}1,2\\2,3\end{pmatrix} & \cdots & A\begin{pmatrix}1, & 2\\n-1, & n\end{pmatrix} \\ A\begin{pmatrix}1,3\\1,2\end{pmatrix} & A\begin{pmatrix}1,3\\1,3\end{pmatrix} & \cdots & A\begin{pmatrix}1,3\\1,n\end{pmatrix} & A\begin{pmatrix}1,3\\2,3\end{pmatrix} & \cdots & A\begin{pmatrix}1, & 3\\n-1, & n\end{pmatrix} \\ \vdots & & & & & & \vdots \\ A\begin{pmatrix}n-1, & n\\1, & 2\end{pmatrix} & \cdots & \ddots & \cdots & \cdots & \cdots & A\begin{pmatrix}n-1, n\\n-1, n\end{pmatrix} \end{matrix}\right\|$$

Similarly, $\mathcal{A}_p$ is a matrix of size $\binom{n}{p} \times \binom{n}{p}$.

It follows immediately from formula (A.8.3) that if $\mathbf{A} \cdot \mathbf{B} = \mathbf{C}$, then $\mathcal{A}_p \cdot \mathcal{B}_p = \mathcal{C}_p$, where $\mathcal{A}_p$, $\mathcal{B}_p$, and $\mathcal{C}_p$ are the pth compound matrices of $\mathbf{A}$, $\mathbf{B}$, and $\mathbf{C}$, respectively. In particular, if $\mathbf{B} = \mathbf{A}^{-1}$, then $\mathcal{B}_p$ is the inverse matrix of $\mathcal{A}_p$.

Let $\lambda_1, \lambda_2, \ldots, \lambda_n$ denote a complete system of eigenvalues (allowing repetitions when required) of the matrix $\mathbf{A}$, and suppose that $\mathbf{A} = \mathbf{PTP}^{-1}$, where

$$\mathbf{T} = \left\|\begin{matrix} \lambda_1 & & & \bigcirc \\ \cdot\ \cdot & \lambda_2 & & \\ & & \ddots & \\ \cdot & \cdot & & \lambda_n \end{matrix}\right\|$$

represents the Jordan canonical form of $\mathbf{A}$. Then manifestly

$$\mathcal{A}_p = \mathcal{P}_p \mathfrak{T}_p \mathcal{P}_p^{-1}, \tag{A.10.1}$$

where $\mathfrak{T}_p$ is a triangular matrix whose diagonal elements are

$$\mathbf{T}\begin{pmatrix} i_1, \ldots, i_p \\ i_1, \ldots, i_p \end{pmatrix} = \lambda_{i_1}\lambda_{i_2} \cdots \lambda_{ip}.$$

Thus the eigenvectors of $\mathcal{A}_p$ are apparent, being the diagonal entries of $\mathfrak{T}_p$.

In the important special case in which the eigenvalues are distinct, we have verified (p. 334) that the similarity transformation $\mathbf{P}$ can be formed as a matrix whose column vectors are the eigenvectors of $\lambda_1, \lambda_2, \ldots, \lambda_n$, respectively. But we see by inspection of (A.10.1) that since $\mathfrak{T}_p$ is diagonal, the eigenvector corresponding to the eigenvalue $\lambda_{k_1}\lambda_{k_2} \ldots \lambda_{k_p}$ is the vector

$$\mathbf{P}\begin{pmatrix} \alpha_1, \alpha_2, \ldots, \alpha_p \\ \\ k_1, k_r, \ldots, k_p \end{pmatrix},$$

where $(\alpha_1, \ldots, \alpha_p)$ designates the variable component index.

A matrix $\mathbf{A}$ is said to be totally positive (T.P.) of order p if

$$\mathbf{A}\begin{pmatrix} i_1, i_2, \ldots, i_p \\ \\ j_1, j_2, \ldots, j_p \end{pmatrix} \geq 0$$

for every pair of index sets $1 \leq i_1 < i_2 < \cdots < i_p \leq n$ and $1 \leq j_1 < j_2 < \cdots < j_p \leq n$. Because of (A.8.3) the product of two totally positive matrices of order p is again totally positive of order p. A matrix is said to be weakly totally positive if

$$(-1)^{\Sigma_{\nu=1}^{p}(i_\nu + j_\nu)}\, \mathbf{A}\begin{pmatrix} i_1, \ldots, i_p \\ \\ j_1, \ldots, j_p \end{pmatrix} \geq 0 \qquad \text{for}$$

$$1 \leq \begin{matrix} i_1 < i_2 < \cdots < i_p \\ j_1 < j_2 < \cdots < j_p \end{matrix} \leq n.$$

We may immediately deduce from Sylvester's identity (A.9.2) that if $\mathbf{A}$ is T.P. of order p, then $\mathbf{B} = \mathbf{A}^{-1}$ (provided the inverse exists) is weakly T.P. of order p.

By way of illustration, we suppose that $\mathbf{A} = \|a_{ij}\|$, where

$$a_{ij} = \begin{cases} b_i c_j & i \leq j, \\ b_j c_i & i \geq j, \end{cases}$$

and all b_i and c_i have the same sign and are nonzero. A calculation will show that A is T.P. of all orders if and only if

$$\frac{b_1}{c_1} \leq \frac{b_2}{c_2} \leq \cdots \leq \frac{b_n}{c_n}.$$

Exercises

1. If $\mathbf{A}$ is normal or unitary, show that the same is true for $\mathcal{A}_p$.
2. Prove that if $\mathbf{A}$ is symmetric, $\mathcal{A}_p$ is symmetric; and that if the quadratic form for $\mathbf{A}$ is positive definite, the same holds for $\mathcal{A}_p$.
3. Prove that if $|\mathbf{A}| > 0$ and $\mathbf{A}$ is weakly T.P. of order p, then $\mathbf{A}^{-1}$ is T.P. of order p.
4. Find $|\mathcal{A}_p|$ in terms of $|\mathbf{A}|$.

APPENDIX B

CONVEX SETS AND CONVEX FUNCTIONS

B.1 Convex sets in $\mathbf{E}^n$. A set X of vectors in E^n is called a convex set if, for any two vectors $\mathbf{x}$ and $\mathbf{y}$ in X, the following relations hold: $\alpha\mathbf{x}+\beta\mathbf{y} \in X$; $\alpha, \beta \geq 0$; $\alpha + \beta = 1$. Among the simpler convex sets are linear subspaces, spheres, and triangles.

For a finite number of vectors $\mathbf{a}^1, \ldots, \mathbf{a}^r$ and real numbers $\alpha_1, \ldots, \alpha_r$, the set H defined by

$$H = \{\mathbf{x} | (\mathbf{a}^i, \mathbf{x}) = \alpha_i, \qquad i = 1, \ldots, r\}$$

is a closed convex set. In particular, for a nonzero vector $\mathbf{a}$ and a real number α, the set $H = \{\mathbf{x} | (\mathbf{a}, \mathbf{x}) = \alpha\}$ is called a hyperplane [or an $(n - 1)$-dimensional linear variety].

A hyperplane H determines two *closed half-spaces*

$$H_1 = \{\mathbf{x} | (\mathbf{a}, \mathbf{x}) \geq \alpha\}, \qquad H_2 = \{\mathbf{x} | (\mathbf{a}, \mathbf{x}) \leq \alpha\},$$

both of which are closed, convex sets.

A hyperplane H is said to be a *supporting hyperplane* to a convex set X if X is contained in one of the half-spaces of H and the boundary of X has a point in common with H. More precisely, a hyperplane $\{\mathbf{a}, \alpha\}$ is a supporting hyperplane to X if

$$\inf_{x \in X} (\mathbf{a}, \mathbf{x}) = \alpha.$$

(Here the appropriate infinite point must be admitted as a possible point.)

▶LEMMA B.1.1. Let X be a convex set and $\mathbf{y}$ be a point exterior to the closure of X. Then there exists a vector $\mathbf{a}$ such that

$$\inf_{x \in X} (\mathbf{a}, \mathbf{x}) > (\mathbf{a}, \mathbf{y}).$$

Proof. Let $\mathbf{x}^0$ be a boundary point of X such that

$$\sqrt{(\mathbf{y} - \mathbf{x}^0, \mathbf{y} - \mathbf{x}^0)} = |\mathbf{y} - \mathbf{x}^0| = \inf_{x \in X} |\mathbf{y} - \mathbf{x}|. \tag{B.1.1}$$

If $\mathbf{x} \in X$, then $\mathbf{x}^0 + t(\mathbf{x} - \mathbf{x}^0) \in \overline{X}$ $(0 \leq t \leq 1)$, where $\overline{X}$ denotes the closure of X.

By (B.1.1)

$$|\mathbf{x}^0 + t(\mathbf{x} - \mathbf{x}^0) - \mathbf{y}|^2 \geq |\mathbf{x}^0 - \mathbf{y}|^2. \tag{B.1.2}$$

Expanding (B.1.2), we obtain

$$2t(\mathbf{x} - \mathbf{x}^0, \mathbf{x}^0 - \mathbf{y}) + t^2(\mathbf{x} - \mathbf{x}^0, \mathbf{x} - \mathbf{x}^0) \geq 0 \qquad (0 \leq t \leq 1).$$

Hence $(\mathbf{x} - \mathbf{x}^0, \mathbf{x}^0 - \mathbf{y}) \geq 0$, or

$$(\mathbf{x}, \mathbf{x}^0 - \mathbf{y}) \geq (\mathbf{x}^0, \mathbf{x}^0 - \mathbf{y}) > (\mathbf{y}, \mathbf{x}^0 - \mathbf{y}).$$

Let $\mathbf{a} = \mathbf{x}^0 - \mathbf{y}$. Then $\mathbf{a} \neq \mathbf{0}$ and

$$\inf_{x \in X} (\mathbf{x}, \mathbf{a}) \geq (\mathbf{x}^0, \mathbf{a}) > (\mathbf{y}, \mathbf{a}),$$

as asserted.

▶LEMMA B.1.2. Let X be a convex set and $\mathbf{y}$ be on the boundary of X. Then there exists a supporting plane through $\mathbf{y}$; i.e., there exists a nonzero vector $\mathbf{a}$ such that

$$\inf_{x \in X} (\mathbf{a}, \mathbf{x}) = (\mathbf{a}, \mathbf{y}).$$

Proof. Consider a sequence $\{\mathbf{y}^\nu\}$, with $\mathbf{y}^\nu$ exterior to the closure of X and such that $\lim_\nu \mathbf{y}^\nu = \mathbf{y}$.

By Lemma B.1.1, there is a sequence $\{\mathbf{a}^\nu\}$ which may be normalized such that

$$(\mathbf{a}^\nu, \mathbf{a}^\nu) = 1, \qquad \inf_{x \in X} (\mathbf{x}, \mathbf{a}^\nu) > (\mathbf{y}^\nu, \mathbf{a}^\nu) \qquad (\nu = 1, 2, \ldots).$$

Taking a limiting point $\mathbf{a}$ of $\{\mathbf{a}^\nu\}$, we have, for any $\mathbf{x} \in X$,

$$(\mathbf{x}, \mathbf{a}) = \lim_\nu (\mathbf{x}, \mathbf{a}^\nu) \geq \lim_\nu (\mathbf{y}^\nu, \mathbf{a}^\nu) = (\mathbf{y}, \mathbf{a}),$$

as asserted.

The preceding two lemmas in conjunction yield the following theorem.

▶THEOREM B.1.1. A closed convex set is the intersection of all its supporting half-spaces, and every boundary point of the set lies on a supporting hyperplane.

▶THEOREM B.1.2. If X and Y are two convex sets with no interior point in common, then there is a hyperplane H that separates X and Y; that is, there exist a nonzero vector $\mathbf{a}$ and a scalar α such that $(\mathbf{a}, \mathbf{x}) \geq \alpha$ for all $\mathbf{x} \in X$ and $(\mathbf{a}, \mathbf{y}) \leq \alpha$ for all $\mathbf{y} \in Y$.

Proof. Consider the set

$$X - Y = \{\mathbf{x} - \mathbf{y} | \mathbf{x} \in X, \mathbf{y} \in Y\},$$

which is convex and does not contain $\mathbf{0}$ in the interior. By Lemmas B.1.1 and B.1.2 such a set has a vector $\mathbf{a}$ ($\mathbf{a} \neq \mathbf{0}$) such that

$$(\mathbf{a}, \mathbf{x} - \mathbf{y}) \geq (\mathbf{a}, \mathbf{0}) = 0 \qquad (\mathbf{x} \in X, \mathbf{y} \in Y).$$

The assertion of the theorem follows immediately.

▶ THEOREM B.1.3. If X and Y are closed convex sets having no point in common and at least one of them is bounded, then there exists a hyperplane H determined by the parameters $\{\mathbf{a}, \alpha\}$ that strictly separates X and Y: that is, $(\mathbf{a}, \mathbf{x}) > \alpha$ for all $\mathbf{x} \in X$ and $(\mathbf{a}, \mathbf{y}) < \alpha$ for all $\mathbf{y} \in Y$.

Proof. In this case $X - Y$ is closed and does not contain $\mathbf{0}$; hence the theorem follows immediately from Lemma B.1.1.

The following result is frequently used in the text.

▶ LEMMA B.1.3. Let X be a convex set which has no point in common with the nonnegative orthant. Then there exists a vector $\mathbf{a} > \mathbf{0}$ (every component of $\mathbf{a}$ is nonnegative and at least one component is positive) such that $(\mathbf{a}, \mathbf{x}) \leq 0$ for all $\mathbf{x} \in X$.

Proof. Apply Theorem B.1.2 to X and Y, where Y denotes the nonnegative orthant. Then there exists a vector $\mathbf{a} \neq \mathbf{0}$ such that $(\mathbf{a}, \mathbf{x}) \leq 0$ for all $\mathbf{x} \in X$ and $(\mathbf{a}, \mathbf{y}) \geq 0$ for all $\mathbf{y} \in Y$. It is easy to show that $\mathbf{a} > \mathbf{0}$.

EXERCISES

1. Theorem B.1.3 is not necessarily true if both X and Y are unbounded. Show that a counterexample can be constructed for $E^2 = \{\langle x_1, x_2\rangle\}$ by taking $X = \{\langle x_1, x_2\rangle | x_1 \leq 0\}$ $Y = \{\langle x_1, x_2\rangle | x_2 \geq e^{-x_1}\}$.

2. Prove that every set and its closure have the same exterior points. Show also that for convex sets it is also true that every convex set and its closure have the same inner points, hence also the same boundary points.

3. Prove that in E^2 every polygon is convex if and only if its interior angles are less than 180 degrees.

4. Prove that the intersection of any number of convex sets is convex. Prove that the closure of a convex set is convex.

5. Prove that every set in E^n can be embedded uniquely in a subspace of lowest possible dimension.

6. To obtain the convex hull of a closed bounded set in E^2 we can "wrap a string around the set and pull it tight." The area determined by the string and its interior is the convex hull. Show that this method works for finding the convex hull of two triangles, or two circles. What does Lemma B.2.2 say about the convex hull of two triangles (or circles)?

B.2 Convex hulls of sets and extreme points of convex sets. The smallest convex set containing a set X is called the *convex hull* of X. It is usually denoted by $[X]$, sometimes by Co (X). The set $[X]$ may be constructively formed as follows:

$$[X] = \left\{\sum_{k=1}^{r} \alpha_k \mathbf{x}^k | \mathbf{x}^k \in X, \sum_{k=1}^{r} \alpha_k = 1, \alpha_k \geq 0, k = 1, \ldots, r\right\},$$

where r is an arbitrary positive integer. If in particular $X = \{\mathbf{a}^1, \ldots, \mathbf{a}^r\}$, then $[\mathbf{a}^1, \ldots, \mathbf{a}^r]$ is referred to as the polyhedral convex set spanned by $\mathbf{a}^i$. It is easily shown that if X is bounded, $[X]$ is bounded, and that if X is compact, $[X]$ is compact. The set $[\mathbf{a}^1, \ldots, \mathbf{a}^r]$ is clearly closed and bounded.

The following well-known representation is of fundamental value.

▶ LEMMA B.2.1. Let X be a subset of E^n, and let $[X]$ be the convex hull of X. Then every point of $[X]$ can be represented as a convex combination of at most $n + 1$ points of X.

Proof. It suffices to show that if

$$\mathbf{x} = \sum_{k=1}^{r} \alpha_k \mathbf{x}^k \ (\mathbf{x}^k \in X, \alpha_k > 0), \quad \text{and} \quad \sum_{k=1}^{r} \alpha_k = 1 \quad (r > n + 1),$$

then the number of vectors $\mathbf{x}^1, \ldots, \mathbf{x}^r$ of X used in the representation of $\mathbf{x}$ can be reduced. Since $\mathbf{x}^1, \ldots, \mathbf{x}^r$ $(r > n + 1)$ are linearly dependent, there exist real numbers $\beta_1, \ldots, \beta_r$ not all zero such that

$$\sum_{k=1}^{r} \beta_k \mathbf{x}^k = \mathbf{0}, \sum_{k=1}^{r} \beta_k = 0.$$

Let ϵ be a real number such that $\alpha_k + \epsilon\beta_k \geq 0$ $(k = 1, \ldots, r)$ and $\alpha_k + \epsilon\beta_k = 0$ for some k, say k_0. Then

$$\mathbf{x} = \mathbf{x} + \sum_{k=1}^{r} \epsilon\beta_k \mathbf{x}^k = \sum_{k=1}^{r} (\alpha_k + \epsilon\beta_k)\mathbf{x}^k,$$

where $\alpha_k + \epsilon\beta_k \geq 0$ $(k = 1, \ldots, r)$, $\alpha_{k_0} + \epsilon\beta_{k_0} = 0$; and

$$\sum_{k=1}^{r} (\alpha_k + \epsilon\beta_k) = 1,$$

as was to be proved.

By imposing further restrictions on X we may sharpen the preceding result as follows.

▶ LEMMA B.2.2. If X has at most n connected components, then the number $n + 1$ in Lemma B.2.1 may be decreased to n.

Proof. Consider the set S spanned by a fixed set of $n + 1$ points of X, and assume that S has an interior point $\mathbf{x}^0$ that cannot be represented by a combination of n or fewer points of X. Corresponding to each $(n-2)$-dimensional face L of S, let

$$T_L = \{\mathbf{x} = (1-\lambda)\mathbf{x}^0 + \lambda\mathbf{y} \mid \lambda \leq 0, \mathbf{y} \in L\}.$$

The set T_L cannot intersect X, because if $\mathbf{c} = (1-\lambda)\mathbf{x}^0 + \lambda\mathbf{y}$ were in X, then $\lambda \leq 0$ and

$$\mathbf{x}^0 = \frac{1}{1-\lambda}\mathbf{c} + \frac{-\lambda}{1-\lambda}\mathbf{y}$$

would represent $\mathbf{x}^0$ as a combination of n or fewer points of X. The sets T_L obtained by traversing the $(n-2)$-dimensional faces of S form the boundaries of $n+1$ nonoverlapping cones whose vertices are at $\mathbf{x}^0$ and whose union is all of E^n. Furthermore, each vertex $\mathbf{x}^i$ of S belongs to a different cone. Since X cannot intersect the boundaries of these cones, X must have at least $n+1$ connected components, thus contradicting the hypothesis.

▶ LEMMA B.2.3. If X is a bounded set in E^n and

$$\mathbf{x} = \sum_{i=1}^{\infty} \alpha_i \mathbf{x}^i,$$

where

$$\mathbf{x}^i \in X, \ \alpha_i > 0, \ \sum_{i=1}^{\infty} \alpha_i = 1,$$

then $\mathbf{x}$ belongs to the convex hull of X.

This is easily proved by employing Theorem B.1.1 plus an induction argument based on the dimension of the smallest linear space containing the point $\mathbf{x}$. Suppose that the result has been proved for sets in E^n. Then let X be a bounded set in E^{n+1}. If $\mathbf{x}$ is a boundary point of the closure of the convex span Γ of $\{\mathbf{x}^i\}$, then any supporting plane p to Γ at $\mathbf{x}$ necessarily contains all $\mathbf{x}^i$. But the part common to p and Γ lies in a Euclidian space of dimension at most n, and therefore the induction hypothesis applies. Finally, if $\mathbf{x}$ is interior to Γ, the result is immediate.

A point $\mathbf{x}$ in X is called an *extreme point* of X if there are no points $\mathbf{x}^1$ and $\mathbf{x}^2$ in X such that $\mathbf{x} = \lambda\mathbf{x}^1 + (1-\lambda)\mathbf{x}^2$ for some λ $(0 < \lambda < 1)$, where $\mathbf{x}^1 \neq \mathbf{x}^2$.

▶ LEMMA B.2.4. A closed, bounded, convex set X is spanned by its extreme points. That is, every $\mathbf{x}$ in X can be represented in the form

$$\mathbf{x} = \sum_{k=1}^{r} \lambda_k \mathbf{x}^k \ \left(\lambda_k \geq 0; k = 1, \ldots, r; \sum_k \lambda_k = 1\right),$$

where $\mathbf{x}^1, \ldots, \mathbf{x}^r$ *are extreme points* of X.

The proof again follows by an induction argument on the dimension of the smallest linear space containing X. We omit the formal details.

Exercises

1. Let S be a convex subset of E^n containing the origin and n linearly independent vectors. Prove that S has inner points relative to E^n.
2. Apply Lemma B.2.1 to E^2 to show that the convex hull of a set S is obtained by taking the union of all possible triangles in E^2 with vertices in S.
3. Let $\mathbf{e}$ be the point in the closed bounded convex set S which is farthest from the origin. Prove that $\mathbf{e}$ is an extreme point of S.
4. Show that any closed convex set in E^n is the intersection of a denumerable number of closed half-spaces.
5. Prove that a closed convex set in E^n has a finite number of extreme points if and only if it is the intersection of a finite number of closed half-spaces.
6. Establish that a closed, bounded, convex set has extreme points in every supporting hyperplane.

B.3 Convex cones. A set X in E^n is called a *convex cone* if, for any $\mathbf{x}$ and $\mathbf{y}$ in X,

$$\alpha\mathbf{x} + \beta\mathbf{y} \in X \qquad (\alpha, \beta \geq 0).$$

A convex cone, of course, is a *convex set.* Linear subspaces and half-spaces are examples of convex cones.

The smallest convex cone that contains a set X is called the *convex cone spanned by* X and is denoted by $\mathcal{P}(X)$. Equivalently,

$$\mathcal{P}(X) = \left\{\sum_{k=1}^{r} \alpha_k \mathbf{x}^k \middle| \alpha_k \geq 0, \mathbf{x}^k \in X, k = 1, \ldots, r; r \text{ arbitrary}\right\}.$$

The polar cone X^+ is defined as the set of all directed separating hyperplanes to a set X which contain the origin. In symbols,

$$X^+ = \{\mathbf{a} | (\mathbf{a}, \mathbf{x}) \geq 0 \quad \text{for all} \quad \mathbf{x} \in X\}. \tag{B.3.1}$$

Clearly X^+ is a closed convex cone.

For any sets X and Y, the following four properties are simple consequences of the definition (B.3.1) of polar cones.

(i) $X \subset Y$ implies $X^+ \supset Y^+$;

(ii) $X \subset X^{++}$;

(iii) $X^+ = X^{+++}$;

(iv) $(X + Y)^+ = X^+ \cap Y^+ \quad$ if $\mathbf{0} \in X, Y$.

▶ THEOREM B.3.1. For any closed convex cones X and Y

$$\text{(I)}\ X = X^{++} \qquad \text{(duality)},$$

$$\left.\begin{aligned}\text{(II)}\ (X + Y)^+ &= X^+ \cap Y^+,\\ \text{(III)}\ (X \cap Y)^+ &= \overline{X^+ + Y^+},\end{aligned}\right\} \qquad \text{(modularity)}.$$

(Recall that a bar symbol over a set denotes closure.)

Proof. (I) Let $\mathbf{y} \notin X$. Then by Theorem B.1.3 there exists a vector $\mathbf{a}$ such that $(\mathbf{a}, \mathbf{x}) > (\mathbf{a}, \mathbf{y})$ for all $\mathbf{x} \in X$. Since X is a cone, $(\mathbf{a}, \mathbf{x}) \geq 0 > (\mathbf{x}, \mathbf{y})$ for all $\mathbf{x} \in X$. Hence $\mathbf{a} \in X^+$, $\mathbf{y} \notin X^{++}$, as was to be shown.

(II) This assertion was noted in property (iv) above.

(III) $(X \cap Y)^+ = (X^{++} \cap Y^{++})^+ = (X^+ + Y^+)^{++} = \overline{X^+ + Y^+}$.

The following theorem can be proved in a similar manner.

▶ THEOREM B.3.2. For any set X,

$$\overline{P(X)} = X^{++}.$$

Convex polyhedral cones. The convex cone spanned by a finite set of vectors $\mathbf{a}^1, \ldots, \mathbf{a}^r$ is called a *convex polyhedral cone* and denoted by

$$\mathcal{P}(\mathbf{a}^1, \ldots, \mathbf{a}^r) = \left\{\sum_{k=1}^{r} \lambda_k \mathbf{a}^k \middle| \lambda_k \geq 0,\ k = 1, \ldots, r\right\}.$$

Using matrix notation, we may write

$$\mathcal{P}(\mathbf{a}^1, \ldots, \mathbf{a}^r) = \mathcal{P}(\mathbf{A}),$$

where $\mathbf{A}$ is the $n \times r$ matrix whose column vectors are $\mathbf{a}^1, \ldots, \mathbf{a}^r$. In particular, for a nonzero vector $\mathbf{a}$, we have $\mathcal{P}(\mathbf{a}) = \{\lambda \mathbf{a} | \lambda \geq 0\}$. The set $\mathcal{P}(\mathbf{a})$ is referred to as the *ray* spanned by $\mathbf{a}$.

A convex polyhedral cone is closed. This is a consequence of the following more general theorem.

▶ THEOREM B.3.3. A convex polyhedral cone is represented as the intersection of a finite set of supporting half-spaces. Equivalently, for any finite set of vectors $\mathbf{a}^1, \ldots, \mathbf{a}^r$, there exist vectors $\mathbf{b}^1, \ldots, \mathbf{b}^s$ such that

$$\mathcal{P}(\mathbf{a}^1, \ldots, \mathbf{a}^r) = (\mathbf{b}^1)^+ \cap \ldots \cap (\mathbf{b}^s)^+,$$

and conversely.

The vectors $\mathbf{b}^j$ may be found by taking all supporting planes passing through the origin to the convex space spanned by the vectors $\mathbf{a}^i$.

The duality theorem for convex polyhedral cones may be expressed in matrix notation as follows.

▶ THEOREM B.3.4. The inner product $(\mathbf{x}, \mathbf{y})$ is nonnegative for all $\mathbf{yA} \geq \mathbf{0}$ if and only if $\mathbf{x} = \mathbf{Au}$ for some $\mathbf{u} \geq \mathbf{0}$.

The following theorem is useful in our study of linear programming.

▶ THEOREM B.3.5. Let X be a bounded convex polyhedral set or a convex closed cone having no point but $\mathbf{0}$ in common with the nonnegative orthant of E^n. Then there exists a vector $\mathbf{a}$ with strictly positive components ($\mathbf{a} \gg \mathbf{0}$) such that $(\mathbf{a}, \mathbf{x}) \leq 0$ for all $\mathbf{x} \in X$.

Proof. It is enough to prove the case when X is a cone. Let us suppose that $\mathbf{a} \geq \mathbf{0}$ and $(\mathbf{a}, \mathbf{x}) \leq 0$ for all $\mathbf{x} \in X$ imply that $a_i = 0$ for some component index i, say $i = 1$. Then

$$(-X)^+ \cap (I)^+ \subset H = \{\langle u_1, \ldots, u_n\rangle | u_1 = 0\},$$

where I represents the nonnegative orthant. By the duality theorem (Theorem B.3.1),

$$\overline{-X + I} \supset H^+,$$

and it is readily shown that $-X + I$ is already closed.

Since $\langle \alpha, 0, 0, \ldots, 0\rangle \in H^+$ for all α, there exist vectors $\mathbf{x} \in X$ and $\mathbf{b} \geq \mathbf{0}$ such that

$$-\mathbf{x} + \mathbf{b} = \langle \alpha, 0, \ldots, 0\rangle = \mathbf{a}.$$

If we take $\alpha < 0$, it is clear that $\mathbf{x} = \mathbf{b} - \mathbf{a}$ is a nonnegative vector belonging to X and with a positive first component, in contradiction to the hypothesis of the theorem.

EXERCISES

1. In E^3 let $X = \{(x, y, 1) | x^2 + y^2 = 1\}$. Find $\mathcal{P}(x)$.
2. In E^2 let $X = \{(x_1, x_2) | x_1 = \sqrt{x_2^2 + 1}\}$. Then $\mathcal{P}(X) = \{(0, 0)\} + \{(x_1, x_2) | 0 < |x_2| < x_1\}$. Thus X and its convex hull may be closed without the same being true of $\mathcal{P}(X)$. Verify this statement.
3. Prove that $\overline{\mathcal{P}(X)} \supset \mathcal{P}(\overline{X})$.
4. Let C be a convex set. Show that $\mathcal{P}(C)$ consists of the union of all rays spanned by points of C.
5. Prove that if C is a closed bounded convex set, $\mathcal{P}(C)$ is the convex hull of the union of all rays spanned by extreme points of C.
6. Let $\mathbf{x}_0$ be a vector in X, a bounded convex polyhedral set. Prove that there exists a vector $\mathbf{a} \gg \mathbf{0}$ such that $(\mathbf{a}, \mathbf{x}) \leq (\mathbf{a}, \mathbf{x}_0)$ for all $\mathbf{x}$ in X, if and only if there is no vector $\mathbf{x}_1$ in X such that $\mathbf{x}_1 - \mathbf{x}_0 > \mathbf{0}$.

B.4 Convex and concave functions. A function $f(\mathbf{x})$ defined on a convex set X of E^n is called a *convex* function if

$$f[t\mathbf{x} + (1 - t)\mathbf{y}] \leq tf(\mathbf{x}) + (1 - t)f(\mathbf{y}) \qquad (\mathbf{x}, \mathbf{y} \in X; 0 \leq t \leq 1). \tag{B.4.1}$$

If (B.4.1) holds with strict inequality whenever $\mathbf{x} \neq \mathbf{y}$ and $0 < t < 1$, then $f(\mathbf{x})$ is called *strictly convex*. A function $f(\mathbf{x})$ is called *concave* (*strictly concave*) if $-f(\mathbf{x})$ is convex (strictly convex).

Examples of convex functions are such functions as e^x and $x_1^2 + x_2^2$. The condition (B.4.1) implies that

$$f\left(\sum_{k=1}^{r} t_k \mathbf{x}^k\right) \leq \sum_{k=1}^{r} t_k f(\mathbf{x}^k), \tag{B.4.2}$$

where $\mathbf{x}^k \in X$, $\sum t_k = 1$, $t_k \geq 0$, $k = 1, \ldots, r$.

We list the following important facts concerning convex functions, each of which is proved in a straightforward fashion.

(i) Let

$$[X, f] = \left\{ \begin{pmatrix} \mathbf{x} \\ \xi \end{pmatrix} \middle| \mathbf{x} \in X,\ \xi \geq f(\mathbf{x}) \right\}.$$

Then $f(\mathbf{x})$ is a convex function defined on X if and only if $[X, f]$ is a convex set in E^{n+1}.

If f is strictly convex, then $[X^0, f]$ is a strictly convex set S; that is, the midpoint of any two points of the set S lies in the relative interior of S. Here X^0 denotes the interior of X.

(ii) If $f_k(\mathbf{x})$ $(k = 1, \ldots, r)$ are convex functions defined on X, and $\lambda_k \geq 0$ $(k = 1, \ldots, r)$, then $\sum_k \lambda_k f_k(\mathbf{x})$ is also a convex function on X.

(iii) Let $f_v(\mathbf{x})$ be a set of convex functions on X; then

$$X_0 = \{\mathbf{x} | \mathbf{x} \in X,\ \sup_v f_v(\mathbf{x}) < \infty\}$$

is a convex set, and $\sup_v f_v(\mathbf{x})$ is convex on X_0.

(iv) If $f(\mathbf{x})$ is strictly convex on X, then $f(\mathbf{x})$ has at most one local minimum. Any local minimum is an absolute minimum.

(v) If $f(\mathbf{x})$ is convex on X, it is continuous in the relative interior of X.

(vi) For a boundary point $\mathbf{y}$ of X,

$$\underline{\lim}_{x \to y} f(\mathbf{x}) \leq f(\mathbf{y}).$$

(vii) If X is an open convex set and $f(\mathbf{x})$ is differentiable in X, we have another characterization of the convexity of $f(\mathbf{x})$. Specifically, $f(\mathbf{x})$ is convex if and only if, for $\mathbf{x}^0$ and $\mathbf{x}$ in X,

$$f(\mathbf{x}) - f(\mathbf{x}^0) \geq \sum_i \left(\frac{\partial f}{\partial x_i}\right)^0 (x_i - x_i^0).$$

(viii) If $f(\mathbf{x})$ is convex on an open convex set X, then $f(\mathbf{x})$ has second-order partial derivatives existing *almost everywhere* (in the sense of the Lebesgue measure).

(ix) If X is one-dimensional and f is convex on X, then f has a derivative except for at most a countable set of points. Moreover, the right-hand and left-hand derivatives exist everywhere and are right and left continuous, respectively.

(x) If $f(\mathbf{x})$ is a twice-differentiable function, and X is a general convex set in E^n, then $f(\mathbf{x})$ is convex if and only if the matrix of second derivatives

$$\left(\frac{\partial^2 f}{\partial x_i \partial x_j}\right)_{i,j}$$

is positive semidefinite for every x in X. If

$$\left(\frac{\partial f}{\partial x_i \partial x_j}\right)_{i,j}$$

is positive definite, then $f(\mathbf{x})$ is strictly convex.

Exercises

1. Consider the following examples: Prove that $(\mathbf{a}, \mathbf{x})$ is a convex function on E^n, but never strictly convex. For $\mathbf{x} = \langle x_1, x_2, \ldots, x_n \rangle$ in E^n, $f(\mathbf{x}) = \sum_1^r \lambda_i |x_i|^{\alpha_i}$ $(\alpha_i \geq 1)$ is convex on E^n if all $\lambda_i \geq 0$.

2. Let $f(\mathbf{x})$ be a convex function on E^n. Show by applying the theorem on supporting hyperplanes that we obtain the result that for any point $\mathbf{x}_0$ in E^n there is a linear function $l(\mathbf{x})$ (that is, $(\mathbf{l}, \mathbf{x})$) such that $l(\mathbf{x}) \leq f(\mathbf{x})$ and $l(\mathbf{x}_0) = f(\mathbf{x}_0)$. Compare this with (vii).

3. If $f_k(x)$, $k = 1, 2, \ldots$ are nonnegative monotonically increasing convex functions on E^1, so is $\Pi_{k=1}^n f_k(x)$.

4. If in E^n, $f(\mathbf{x})$ is convex (strictly convex) and $\mathbf{A} \not\equiv \mathbf{0}$, then so is $f(\mathbf{Ax} + \mathbf{b})$, where $\mathbf{A}$ and $\mathbf{b}$ are a matrix and a vector of appropriate dimensions.

5. Let $f(\mathbf{x})$ be a continuous function on a convex set X of E^n. Then f is convex if and only if

$$f\left(\frac{\mathbf{x} + \mathbf{y}}{2}\right) \leq \frac{f(\mathbf{x}) + f(\mathbf{y})}{2} \qquad \mathbf{x}, \mathbf{y} \in X.$$

6. Construct a convex function on $[0, 1]$ whose derivative fails to exist for an infinity of points.

APPENDIX C

MISCELLANEOUS TOPICS

C.1 Semicontinuous and equicontinuous functions. *Semicontinuous functions.* A real-valued function $f(\mathbf{x})$ defined on a set X in E^n is said to be continuous at $\mathbf{x}_0$ in X if whenever $\mathbf{x}_n \to \mathbf{x}_0$ (i.e., whenever the Euclidian distance $d(\mathbf{x}_n, \mathbf{x}_0)$ tends to zero), $f(\mathbf{x}_n) \to f(\mathbf{x}_0)$. An equivalent definition expressed in terms of neighborhoods runs as follows: For any $\epsilon > 0$ there exists some $\delta(\epsilon, \mathbf{x}_0) > 0$ such that if $d(\mathbf{x}, \mathbf{x}_0) < \delta$, then

$$-\epsilon \leq f(\mathbf{x}) - f(\mathbf{x}_0) \leq \epsilon. \tag{C.1.1}$$

A function is continuous in X if it is continuous at every point of X.

If δ may be chosen independent of $\mathbf{x}_0$ in X, then f is said to be uniformly continuous in X.

A function f is said to be upper semicontinuous at $\mathbf{x}_0$ if the right-hand inequality of (C.1.1) holds for any arbitrary prescribed ϵ ($\epsilon > 0$) and all values $\mathbf{x}$ satisfying $d(\mathbf{x}, \mathbf{x}_0) \leq \delta$ for a suitable positive δ. Similarly, f is lower semicontinuous at $\mathbf{x}_0$ if the left-hand inequality of (C.1.1) holds for any prescribed $\epsilon > 0$ and the corresponding δ. Equivalently, f is upper (lower) semicontinuous at $\mathbf{x}_0$ if and only if

$$\overline{\lim_{x \to x_0}}\, f(\mathbf{x}) \leq f(\mathbf{x}_0)\Big(\underline{\lim_{x \to x_0}}\, f(\mathbf{x}) \geq f(\mathbf{x}_0)\Big).$$

If f is upper semicontinuous on X (i.e., at every point of X), then the set $L = \{\mathbf{x} | f(\mathbf{x}) < \alpha\}$ for each real α is relatively open in X, and conversely. For if $\mathbf{x}_0$ belongs to L, then there exists an ϵ such that $f(\mathbf{x}_0) + \epsilon < \alpha$. By the definition of an upper semicontinuous function, there exists about $\mathbf{x}_0$ an open sphere of radius δ all of whose members, also in X, belong to L; hence L is relatively open. The converse is proved similarly. Analogously, a function f is lower semicontinuous on X if $\{\mathbf{x} | f(\mathbf{x}) > \alpha\}$ is relatively open for each real α.

With the aid of these characterizations we may readily establish the following properties:

(i) The set of all upper (lower) semicontinuous functions is closed under addition. The negative of an upper semicontinuous function is lower semicontinuous, and conversely.

(ii) If f_α are upper (lower) semicontinuous on X and bounded below (bounded above), then $\inf_\alpha f_\alpha$ ($\sup_\alpha f_\alpha$) is upper (lower) semicontinuous. (Here the last equivalent definition of upper semicontinuity applies most easily.)

(iii) An upper (lower) semicontinuous function defined on a compact set achieves its maximum (minimum).

(iv) An upper (lower) semicontinuous function defined on a compact set may be approached by a decreasing (increasing) sequence of continuous functions, and conversely.

Equicontinuous functions. A family of functions f_α defined on X is said to be equicontinuous at $\mathbf{x}_0$ if for any $\epsilon > 0$ there exists some $\delta(\epsilon, \mathbf{x}_0) > 0$ (independent of α) such that $|f_\alpha(\mathbf{x}) - f_\alpha(\mathbf{x}_0)| < \epsilon$ for $d(\mathbf{x}, \mathbf{x}_0) \leq \delta$ and all α. The functions f_α are said to be equicontinuous on X if they are equicontinuous at each point of X. When δ may be chosen independent of $\mathbf{x}_0$ in X, we speak of uniform equicontinuity for f_α. To illustrate this concept, we exhibit two families of equicontinuous functions.

(1) Let $K(s, t)$ be a uniformly continuous function defined on the unit square, and let $f_\alpha(t)$ constitute a family of uniformly bounded functions $(0 \leq t \leq 1)$. Then

$$g_\alpha(s) = \int_0^1 K(s, t) f_\alpha(t)\, dt$$

represents a uniform equicontinuous family of functions. The estimate

$$|g_\alpha(s_1) - g_\alpha(s_2)| \leq M \max_{0 \leq t \leq 1} |K(s_1, t) - K(s_2, t)|,$$

where M is the bound of f_α, clearly implies the assertion.

(2) Let $f_\alpha(z)$ constitute a family of uniformly bounded functions that are analytic in the unit circle C of the complex plane. Then on X (any interior circle of C) the functions f_α are uniformly equicontinuous. We prove this statement by the same reasoning as in Example (1), using the Cauchy integral formula to represent f_α for points of X.

For our purposes the most important property of equicontinuous functions is given by the classical Ascoli theorem, which may be stated as follows: *Let f_α be a uniformly bounded equicontinuous family of real functions defined on a compact separable set X; then we can select a subsequence f_{α_i} which converges uniformly.*

We sketch the idea of the proof. We specify a countable dense subset $\{\mathbf{x}_i\}$ of X. By the diagonal procedure we may select a subsequence f_{α_i} which converges for all $\mathbf{x}_i$. Owing to the equicontinuity, it follows that f_{α_i} converges uniformly as desired.

C.2 Fixed-point theorems. The following theorems are reviewed with no proofs included. For detailed discussion of these results, see [130].

Brouwer's fixed-point theorem. Let $\phi(\mathbf{x})$ be a continuous point-to-point mapping of a closed simplex X into itself. Then there exists a point $\mathbf{x}_0$ in

X such that $\mathbf{x}_0 = \phi(\mathbf{x}_0)$. (For our purposes, we define a simplex in E^n as a convex set spanned by the origin and n linearly independent points.)

The theorem is actually correct for any set which is homeomorphic to an n-dimensional simplex.

The fixed-point theorem was generalized by Kakutani to cover point-to-set mappings. Prior to stating this result, we clarify a few terms. A mapping ϕ which transforms every point $\mathbf{x}$ in X into a subset of X is called a *point-to-set* mapping. A point-to-set mapping ϕ is called *upper semicontinuous* if $\mathbf{x}_n \to \mathbf{x}_0$, $\mathbf{y}_n \to \mathbf{y}_0$, $\mathbf{x}_n \in X$, $\mathbf{y}_n \in \phi(\mathbf{x}_n)$ imply $\mathbf{y}_0 \in \phi(\mathbf{x}_0)$.

It is easily verified that a mapping ϕ is upper semicontinuous if and only if the graph $\{\langle \mathbf{x}, \mathbf{y} \rangle | \mathbf{x} \in X, \mathbf{y} \in \phi(\mathbf{x})\}$ is closed in E^{2n}.

A point-to-set mapping $\phi(\mathbf{x})$ is said to be lower semicontinuous if for every $\mathbf{y} \in \phi(\mathbf{x}_0)$ whenever $\mathbf{x}_n \to \mathbf{x}_0$ there exists $\mathbf{y}_n \in \phi(\mathbf{x}_n)$ such that $\mathbf{y}_n \to \mathbf{y}$.

A point-to-set mapping is continuous if it is both lower and upper semicontinuous. The definition of upper semicontinuous for a point-to-set mapping ϕ reduces to that of Section C.1 when ϕ is a point function, provided we define the containing relation $\mathbf{y} \in \phi(\mathbf{x})$ as $\mathbf{y} \leq \phi(\mathbf{x})$. It is also possible to express the properties of point-to-set mappings in terms of neighborhood concepts. For example, $\phi(\mathbf{x})$ is upper semicontinuous at $\mathbf{x}_0$ if corresponding to any open set U containing $\phi(\mathbf{x}_0)$ there exists some $\delta(U) > 0$ such that $d(\mathbf{x}, \mathbf{x}_0) < \delta$ implies $\phi(\mathbf{x}) \subset U$.

Kakutani's fixed-point theorem. Let X be a closed simplex, and let ϕ represent an upper semicontinuous mapping which maps each point of X into a closed convex subset of X. Then there exists a point $\mathbf{x}_0 \in X$ such that $\mathbf{x}_0 \in \phi(\mathbf{x}_0)$.

This fixed-point theorem has been further generalized by Eilenberg and Montgomery [73] and Begle [18].

C.3 Set functions and probability distributions. We review very briefly, without proofs, some of the theory behind the Lebesgue-Stieltjes integral. For full discussions, see Graves [112].

The class of Borel sets. The set of all points that belong to some member of a sequence $\{S_n\}$ is the union of the sequence, and the set of all points that belong to every member of the sequence is called the intersection. These two sets are denoted by

$$\bigcup_{n=1}^{\infty} S_n \qquad \text{and} \qquad \bigcap_{n=1}^{\infty} S_n,$$

respectively. The set of points that belong to S_1 but do not belong to S_2 is called the difference, and it is denoted by $S_1 - S_2$.

A class $\mathcal{S}$ of subsets of E^k is said to be an additive class of sets if it satisfies the following conditions:

(1) E^k belongs to $\mathcal{S}$.

(2) If $S_1, \ldots, S_n, \ldots$ all belong to $\mathcal{S}$, then

$$\bigcup_{n=1}^{\infty} S_n$$

belongs to $\mathcal{S}$.

(3) If S_1 and S_2 belong to $\mathcal{S}$, then $S_1 - S_2$ belongs to $\mathcal{S}$.

The smallest additive class of sets that contains all the "rectangles" in E^k (or all the intervals if $k = 1$) is called the class of Borel sets of E^k.

Nonnegative set functions. A nonnegative set function P is a real-valued function that is defined on a class of Borel sets $\mathcal{S}$ and satisfies the following conditions:

(1) $P(S) \geq 0 \qquad$ for all $\quad S \in \mathcal{S}$.

(2) $P\left(\bigcup_{n=1}^{\infty} S_n\right) = \sum_{n=1}^{\infty} P(S_n)$ if $S_n \cap S_m = \phi$ for $n \neq m$.

(3) $P(S) < \infty \qquad$ if S is bounded.

The class of all nonnegative set functions defined on the class of Borel sets of real numbers is equivalent to the class of all monotone-increasing functions (defined on the real line) that are continuous from the right; this equivalence is unique if we can identify two increasing functions that differ everywhere by a fixed constant. The equivalence is constructed as follows: Given P and any real number α, define

$$F(x, \alpha) = \begin{cases} P\{\xi | \alpha < \xi \leq x\} & x > \alpha, \\ 0 & x = \alpha, \\ -P\{\xi | x < \xi \leq \alpha\} & x < \alpha. \end{cases}$$

$F(x, \alpha)$ is an increasing function for each α, and it is continuous from the right; furthermore, if $\alpha_1 < \alpha_2$, then

$$F(x, \alpha_1) - F(x, \alpha_2) = P\{\xi | \alpha_1 < \xi \leq \alpha_2\}.$$

That is, $F(x, \alpha_1)$ and $F(x, \alpha_2)$ differ by a constant, independent of x. On the other hand, if $F(x)$ is any monotone-increasing function that is also continuous from the right, one may define

$$P\{\xi | a < \xi \leq b\} = F(b) - F(a).$$

The resulting set function defined on intervals can be extended to a nonnegative set function defined on all Borel sets.

If $P\{\xi | -\infty < \xi < \infty\} = 1$, the set function is called a probability measure, and the corresponding point function F is commonly normalized so that $F(-\infty) = 0$ and $F(\infty) = 1$. The function F, when normalized, is referred to as the (cumulative) distribution function. If $F'(x) = f(x)$

exists, then $f(x)$ is called the probability density function. Clearly,

$$f(x) \geq 0, \qquad \int_{-\infty}^{\infty} f(x)\, dx = 1.$$

Any function f that satisfies the last two conditions is the density function of some probability distribution.

These concepts can be extended to E^n. The probability distributions are the nonnegative set functions P with the property that $P(E^n) = 1$. To each P there corresponds a cumulative distribution function F defined by

$$F(x_1, \ldots, x_n) = P\{\xi \in E^n | \xi_1 \leq x_1, \ldots, \xi_n \leq x_n\},$$

etc. This function F is monotone-increasing in each variable;

$$F(-\infty, x_2, \ldots, x_n) = \cdots = F(x_1, x_2, \ldots, x_{n-1}, -\infty) = 0$$

and

$$F(\infty, \infty, \ldots, \infty) = 1.$$

Furthermore, F is continuous from the right in each variable; and the difference $\Delta_1 \Delta_2 \cdots \Delta_n F$, where by definition

$$\begin{aligned} \Delta_i F = {} & F(a_1, \ldots, a_{i-1}, a_i + \delta_i, a_{i+1}, \ldots, a_n) \\ & - F(a_1, \ldots, a_{i-1}, a_i, a_{i+1}, \ldots, a_n), \end{aligned}$$

is nonnegative since this difference is

$$P\{\xi | a_1 < \xi_1 \leq a_1 + \delta_1, \ldots, a_n < \xi_n \leq a_n + \delta_n\}.$$

Conversely, any function with the last four properties is the cumulative distribution function of some probability distribution.

Marginal distributions. If $F(x_1, \ldots, x_n)$ is a cumulative distribution function defined on E^n, then $F(x_1, \ldots, x_{n-1}, \infty)$ is also a distribution function defined on E^{n-1}. In general, if k of the variables are assigned the value $+\infty$, then F induces a distribution function of the remaining $n - k$ variables, which is called the marginal distribution.

Lebesgue-Stieltjes integral. Let g be a bounded function defined on E^n, and let P be a probability distribution defined on the class of Borel sets of E^n. In order to define the upper and lower integrals of g with respect to P, we consider first any subdivision of E^n into disjoint Borel sets $E_1, \ldots, E_k$ and define

$$m_i = \inf_{x \in E_i} g(x), \qquad M_i = \sup_{x \in E_i} g(x),$$

$$s = \sum_{i=1}^{k} m_i P(E_i), \qquad S = \sum_{i=1}^{k} M_i P(E_i).$$

The numbers s and S are called the lower and upper Darboux sums of g for the given subdivision. By varying the subdivision one obtains different values for the Darboux sums; if a subdivision is a refinement of another, the lower Darboux sum increases and the upper Darboux sum decreases. The supremum of the lower Darboux sums is called the lower integral of g with respect to p; the infimum of the upper Darboux sums is called the upper integral. If both integrals are equal, then the function g is said to be integrable with respect to P, and its integral is the common value of the upper and lower integrals; this value is usually denoted by

$$\int g(x)\, dP(x).$$

If F is the point function associated with the set function P, the previous notation is commonly replaced by

$$\int g(x)\, dF(x).$$

Besides the obvious additive properties of integrals, we record for reference the Lebesgue criterion of dominated convergence. Let $|f_n| \leq g$ and let g possess an integral with respect to P. If f_n converges everywhere to f (actually a.e. with respect to P would suffice), then

$$\lim_{n\to\infty} \int f_n\, dP = \int f\, dP. \tag{C.3.1}$$

If f_n forms a monotone-increasing sequence of integrable functions with no other restrictions and P is a positive measure, then the same conclusion holds, provided that the value infinity is allowed.

Sequences of distributions. Weak convergence.* A sequence $\{F_n\}$ of distribution functions converges weak* to a distribution function F if $F_n(x) \to F_0(x)$ at every point of continuity of F_0. Every sequence of distribution functions contains a subsequence F_{n_k} that converges weak* to a function F, but F need not be a distribution function. However, if g is continuous $g(\pm\infty) = 0$, and $F_{n_k} \to F$ (weak*), then

$$\lim_{k\to\infty} \int g(x)\, dF_{n_k}(x) = \int g(x)\, dF(x). \tag{C.3.2}$$

If, in addition, F is a distribution function, the convergence of (C.3.2) is valid for any g which is bounded and continuous.

These results are known as *Helly's convergence theorem.*

BIBLIOGRAPHY

Each reference work is listed once under the name of the first author as the article or book appears in the literature. At the end of the bibliography, each co-author is cited for the articles or books not previously listed directly under his name. In this case, the relevant reference is indicated by the appropriate numbered article in the bibliography.

[1] ALLAIS, M., *Traité d'economie pure,* Imprimerie Nationale, Paris, (1952), 852 pp.

[2] ALLEN, R. G. D., *Mathematical Economics,* London: Macmillan, 1956.

[3] ARROW, K. J., "Alternative proof of the substitution theory for Leontief models in the general case," Chap. 9 of [146].

[4] ARROW, K. J., "An extension of the basic theorems of classical welfare economics," in [200], pp. 507–32.

[5] ARROW, K. J., E. W. BARANKIN, and D. BLACKWELL, "Admissible points of convex sets," in [162], pp. 87–91.

[6] ARROW, K. J., D. BLOCK, and L. HURWICZ, "On the stability of the competitive equilibrium, II," *Econometrica,* **27** (1959), 82–109.

[7] ARROW, K. J., and G. DEBREU, "Existence of an equilibrium for a competitive economy," *Econometrica,* **22** (1954), 265–90.

[8] ARROW, K. J., T. E. HARRIS, and J. MARSCHAK, "Optimal inventory policy," *Econometrica,* **19** (1951), 250–72.

[9] ARROW, K. J., and L. HURWICZ, "Reduction of constrained maxima to saddle-point problems," in [201], V, 1–20.

[10] ARROW, K. J., and L. HURWICZ, "The gradient method for concave programming," Chap. 6 of [12].

[11] ARROW, K. J., and L. HURWICZ, "On the stability of the competitive equilibrium, I," *Econometrica,* **26** (1958), 522–552.

[12] ARROW, K. J., L. HURWICZ, and H. UZAWA (eds.), *Studies in Linear and Non-Linear Programming,* Stanford, Calif.: Stanford University Press, 1958.

[13] ARROW, K. J., and M. MCMANUS, "A note on dynamic stability," *Econometrica,* **26** (1958), 448–54.

[14] ARROW, K. J., and M. NERLOVE, "A note on expectation and stability," *Econometrica,* **26** (1958), 297–305.

[15] ARROW, K. J., and S. KARLIN, "Price speculation under certainty," Chap. 13 of [12].

[16] ARROW, K. J., S. KARLIN, and H. SCARF, *Studies in the Mathematical Theory of Inventory and Production,* Stanford, Calif.: Stanford University Press, 1958.

[17] BALDWIN, R. R., W. E. CANTEY, H. MAISEL, and J. P. MCDERMOTT, "The Optimum Strategy in Blackjack," *J. Amer. Stat. Assoc.,* **51** (1956), 429–39.

[18] BEGLE, E. G., "A fixed point theorem," *Ann. Math.,* **51** (1950), 544–50.

[19] BELLMAN, R., *Dynamic Programming,* Princeton, N. J.: Princeton University Press, 1957.

[20] BELLMAN, R., "On games involving bluffing," *Rendiconti del Circolo Mathematico di Palermo*, Series 2, Vol. 1 (1952), 139–56.

[21] BELLMAN, R., and M. A. GIRSHICK, "An extension of results on duels with two opponents, one bullet each, silent guns, equal accuracy," The RAND Corporation, D-403, 1949.

[22] BELLMAN, R., I. GLICKSBERG, and O. GROSS, "On some variational problems occurring in the theory of dynamic programming," *Rendiconti del Circolo Mathematico di Palermo*, Series 2, Vol. 3 (1954), 1–35.

[23] BELLMAN, R., and M. SHIFFMAN, "On the Min-Max of

$$\int_0^1 f(x)a(x)\, d(x)\, dt(x),"$$

The RAND Corporation, RM-308-1, 1949.

[24] BELZER, R. L., "Silent duels, specified accuracies, one bullet each," The RAND Corporation, RAD(L)-301, 1948.

[25] BERGE, C., "Sur une théorie ensembliste des jeux alternatifs," *Journal de mathématiques pures et appliquées*, **32** (1953), 129–84.

[26] BIRKOFF, G., and S. MACLANE, *A Survey of Modern Algebra*, New York: Macmillan, 1941.

[27] BLACKETT, D. W., "Some blotto games," *Naval Research Logistics Quarterly*, **1** (1954), 55–60.

[28] BLACKWELL, D., "The silent duel, one bullet each, arbitrary accuracy," The RAND Corporation, RM-302, 1948.

[29] BLACKWELL, D., "The noisy duel, one bullet each, arbitrary, non-monotone accuracy," The RAND Corporation, D-442, 1949.

[30] BLACKWELL, D., and M. A. GIRSHICK, *Theory of Games and Statistical Decisions*, New York: Wiley, 1954.

[31] BLACKWELL, D., and M. A. GIRSHICK, "A loud duel with equal accuracy where each duelist has only a probability of possessing a bullet," The RAND Corporation, RM-219, 1949.

[32] BLACKWELL, D., and M. SHIFFMAN, "A bomber-fighter duel," The RAND Corporation, D-509, 1949.

[33] BLACKWELL, D., and M. SHIFFMAN, "A bomber-fighter duel," The RAND Corporation, RM-193, 1949.

[34] BOHNENBLUST, H. F., "The Theory of Games," E. F. BECKENBACH (ed.), *Modern Mathematics for the Engineer*, New York: McGraw-Hill, 1956.

[35] BOHNENBLUST, H. F., and S. KARLIN, "On a theorem of Ville," in [161], pp. 155–61.

[36] BOHNENBLUST, H. F., S. KARLIN, and L. S. SHAPLEY, "Solutions of discrete two-person games," in [161], pp. 51–73.

[37] BOHNENBLUST, H. F., S. KARLIN, and L. S. SHAPLEY, "Games with continuous, convex pay-off," in [161], pp. 181–92.

[38] BOREL, E., "Applications aux jeux de hasard," *Traité du calcul des probabilités et de ses applications*, Paris: Gauthier-Villars, 1938.

[39] BOREL, E., "The theory of play and integral equations with skew symmetrical kernels," "On games that involve chance and the skill of the players,"

and "On systems of linear forms of skew symmetric determinants and the general theory of play," trans. L. J. Savage, *Econometrica*, **21** (1953), 97–117.

[40] Brown, G. W., "Some notes on computation of games solutions," The RAND Corporation, D-436, 1949.

[41] Brown, G. W., "Iterative solution of games by fictitious play," Chap. 24 of [146].

[42] Brown, G. W., and J. von Neumann, "Solutions of games by differential equations," in [161], pp. 73–79.

[43] Brown, R. H., "The solution of a certain two-person zero-sum game," *Operations Research*, **5** (1957), 63–67.

[44] Charnes, A., "Optimality and degeneracy in linear programming," *Econometrica*, **20** (1952), 160–70.

[45] Charnes, A., and W. W. Cooper, "Generalizations of the warehousing model," *Operations Research Quarterly*, **6** (1955), 131–72.

[46] Charnes, A., W. W. Cooper, and A. Henderson, *Introduction to Linear Programming*, New York: Wiley, 1953.

[47] Chenery, H. B., "The role of industrialization in development programs," *Amer. Econ. Rev.*, **45** (1955), 40–56.

[48] Chenery, H. B., and K. S. Kretchmer, "Resource allocation for economic development," *Econometrica*, **24** (1956), 365–99.

[49] Chenery, H. B., and H. Uzawa, "Non-linear programming in economic development," Chap. 15 of [12].

[50] Churchman, C. W., R. L. Ackoff, and E. L. Arnoff, *Introduction to Operations Research*, New York: Wiley, 1957.

[51] Danskin, J. M., "Fictitious play for continuous games," *Naval Research Logistics Quarterly*, **1** (1954), 313–20.

[52] Danskin, J. M., "Mathematical treatment of a stockpiling problem," *Naval Research Logistics Quarterly*, **2** (1955), 99–109.

[53] Danskin, J. M., and L. Gillman, "Explicit solution of a game over function space," The RAND Corporation, P-235, 1951.

[54] Danskin, J. M., and L. Gillman, "A game over function space," *Rivista Mat. Univ. Parma*, **4** (1953), 83–94.

[55] Dantzig, G. B., "A proof of equivalence of the programming problem and the game problem," Chap. 20 of [146].

[56] Dantzig, G. B., "Maximization of a linear function of variables subject to linear inequalities," Chap. 21 of [146].

[57] Dantzig, G. B., *The Theory of Mathematical Programming*, notes, The RAND Corporation.

[58] Dantzig, G. B., and A. J. Hoffman, "Dilworth's theorem on partially ordered sets," in [163], pp. 207–14.

[59] Dantzig, G. B., A. Orden, and P. Wolfe, "The generalized simplex method for minimizing a linear form under linear inequality restraints," *Pac. J. Math.*, **5** (1955), 183–95.

[60] Dantzig, G. B., and A. Wald, "On the fundamental lemma of Neyman and Pearson," *Ann. Math. Stat.*, **22** (1951), 87–93.

[61] Debreu, G., "A social equilibrium existence theorem," *Proc. Nat. Acad. Sci.*, **38** (1952), 886–93.

[62] Debreu, G., "Valuation equilibrium and Pareto optimum," *Proc. Nat. Acad. Sci.*, **40** (1954), 588–92.

[63] Debreu, G., and I. N. Herstein, "Non-negative square matrices," *Econometrica*, **21** (1953), 597–607.

[64] Dorfman, R., *Application of Linear Programming to the Theory of the Firm*, Berkeley and Los Angeles: University of California Press, 1951.

[65] Dorfman, R., P. A. Samuelson, and R. M. Solow, *Linear Programming and Economic Analysis*, New York: McGraw-Hill, 1958.

[66] Dresher, M., *Theory and Applications of Games of Strategy*, The RAND Corporation, R-216 (1951).

[67] Dresher, M., "Methods of solution in game theory," *Econometrica*, **18** (1950), 179–81.

[68] Dresher, M., "Games of strategy," *Mathematics Magazine*, **25** (1951), 93–99.

[69] Dresher, M., "Le Her," RAND report (unpublished).

[70] Dresher, M., and S. Karlin, "Solutions of convex games as fixed points," in [162], pp. 75–86.

[71] Dresher, M., S. Karlin, and L. S. Shapley, "Polynomial games," in [161], pp. 161–180.

[72] Dresher, M., A. W. Tucker, and P. Wolfe (eds.), *Contributions to the theory of Games, III*, (*Ann. Math. Studies*, Vol. 39), Princeton, N. J.: Princeton University Press, 1957.

[73] Eilenberg, S., and D. Montgomery, "Fixed point theorems for multi-valued transformations," *Amer. J. Math.*, **68** (1946), 214–22.

[74] Eisemann, K., "Linear programming," *Quart. Appl. Math.*, **13** (1955), 209–32.

[75] Enthoven, A. C., and K. J. Arrow, "A theorem on expectations and the stability of equilibrium," *Econometrica*, **24** (1956), 288–93.

[76] Fan, K., "Fixed point and minimax theorems in locally convex topological linear spaces," *Proc. Nat. Acad. Sci. U. S. A.*, **38** (1952), 121–26.

[77] Fan, K., "Minimax theorems," *Proc. Nat. Acad. Sci. U. S. A.*, **39** (1953), 42–47.

[78] Fan, K., I. Glicksberg, and A. J. Hoffman, "Systems of inequalities involving convex functions," *Proc. Amer. Math. Soc.*, **8** (1957), 617–22.

[79] Farkas, J., "Theorie der einfachen Ungleichungen," *Journal für reine und angewandte Mathematik*, **124** (1901), 1–27.

[80] Feller, W., *An Introduction to Probability Theory and its Applications, I*, New York: Wiley, 1950.

[81] Fenchel, W., "Convex Cones, Sets, and Functions," Lecture notes, Department of Mathematics, Princeton University, 1953.

[82] Ferguson, R. O., and L. F. Sargent, *Linear Programming*, New York: McGraw-Hill, 1958.

[83] Fleming, W. H., "On a class of games over function space and related variation problems," *Ann. Math.*, **60** (1954), 578–94.

[84] Flood, M. M., "On the Hitchcock distribution problem," *Pac. J. Math.*, **3** (1953), 369–86.

[85] FORD, L. R., JR., and D. R. FULKERSON, "Maximal flow through a network," *Canadian J. Math.*, **8** (1956), 399–404.

[86] FORD, L. R., JR., and D. R. FULKERSON, "A simple algorithm for finding maximal network flows and an application to the Hitchcock Problem," *Canadian J. Math.*, **9** (1957), 210–18.

[87] FORD, L. R., JR., and D. R. FULKERSON, "A primal dual algorithm for the capacitated Hitchcock problem," *Naval Research Logistics Quarterly*, **4** (1957), 47–54.

[88] FRANK, M., and P. WOLFE, "An algorithm for quadratic programming," *Naval Research Logistics Quarterly*, **3** (1956), 95–110.

(89) FROBENIUS, G., "Über Matrizen aus positiven Elementen," *Sitzungsberichte*, **8** (1908), 471–76.

[90] FULKERSON, D. R., "Hitchcock transportation problem," The RAND Corporation, P-890, 1956.

[91] FULKERSON, D. R., "Notes on linear programming. Part XLV. A network-flow feasibility theorem and combinatorial applications," The RAND Corporation, RM-2159, 1958.

[92] GADDUM, J. W., A. J. HOFFMAN, and D. SOKOLOWSKY, "On the solution of the caterer problem," *Naval Research Logistics Quarterly*, **1** (1954), 223–29.

[93] GALE, D., "The law of supply and demand," *Mathematica Scandinavica*, **3** (1955), 155–69.

[94] GALE, D., "The closed linear model of production," in [163], pp. 285–303.

[95] GALE, D., "Information in games with finite resources," in [72], pp. 141–45.

[96] GALE, D., Lecture notes, Brown University (1957).

[97] GALE, D., and O. GROSS, "A note on polynomial and separable games," The RAND Corporation, P-1216, 1957.

[98] GALE, D., H. W. KUHN, and A. W. TUCKER, "On symmetric games," in [161], pp. 81–87.

[99] GALE, D., H. W. KUHN, and A. W. TUCKER, "Linear programming and the theory of games," Chap. 19 of [146].

[100] GALE, D., and S. SHERMAN, "Solutions of finite two-person games," in [161], pp. 37–49.

[101] GANTMACHER, F., *Theory of Matrices* (in Russian), Moscow, 1954.

[102] GANTMACHER, F., and M. KREIN, *Oscillatory Matrices and Kernels and Small Vibrations of Mechanical Systems*, 2d ed. (in Russian), Moscow, 1950.

[103] GASS, S. I., *Linear Programming, Methods and Applications*, New York: McGraw-Hill, 1958.

[104] GEORGESCU-ROEGEN, N., "The aggregate linear production function and its application to von Neumann's economic model," Chap. 4 of [146].

[105] GILLIES, D. B., J. P. MAYBERRY, and J. VON NEUMANN, "Two variants of poker," in [162], pp. 13–50.

[106] GILLMAN, L., "Operations analysis and the theory of games: an advertising example," *J. Amer. Stat. Assoc.*, **45** (1950), 541–46.

[107] GLICKSBERG, I., "Noisy duel, one bullet each, with simultaneous fire and unequal worths," The RAND Corporation, RM-474, 1950.

[108] GLICKSBERG, I., "A further generalization of the Kakutani fixed point theorem, with application to Nash equilibrium points," *Proc. Amer. Math. Soc.*, **3** (1952), 170–74.

[109] GLICKSBERG, I., and O. GROSS, "Butterfly games," The RAND Corporation, RM-655, 1951.

[110] GLICKSBERG, I., and O. GROSS, "Notes on games over the square," in [162], pp. 173–84.

[111] GOLDMAN, A. J., and A. W. TUCKER, "Theory of linear programming," in [163], pp. 53–97.

[112] GRAVES, L. M., *The Theory of Functions of Real Variables*, New York: McGraw-Hill, 1946.

[113] GROSS, O., "A rational payoff characterization of the Cantor distribution," The RAND Corporation, D-1349, 1952.

[114] GROSS, O., "The derivatives of the value of a game," The RAND Corporation, RM-1286, 1954.

[115] GROSS, O., "A simple linear programming problem explicitly solvable in integers," The RAND Corporation, RM-1560, 1955.

[116] GROSS, O., "A rational game on the square," in [72], pp. 307–11.

[117] HAHN, F. H., "Gross substitutes and the dynamic stability of general equilibrium," *Econometrica*, **26** (1958), 169–70.

[118] HARDY, G. H., J. E. LITTLEWOOD, and G. PÓLYA, *Inequalities*, Cambridge, England: Cambridge University Press, 1952.

[119] HAWKINS, D., and H. A. SIMON, "Note: Some conditions of macroeconomic stability," *Econometrica*, **17** (1949), 245–48.

[120] HELMER, O., "Open problems in game theory," *Econometrica*, **20** (1952), 90 abstract.

[121] HICKS, J. R., *Value and Capital*, 2d ed., Oxford: Oxford University Press, 1953.

[122] HILDRETH, C., "A quadratic programming procedure," *Naval Research Logistics Quarterly*, **4** (1957), 79–85.

[123] HIRSCH, W., and G. B. DANTZIG, "The fixed charge problem," The RAND Corporation, RM-1383, 1954.

[124] HITCHCOCK, F. L., "Distribution of a product from several sources to numerous localities," *J. Math. Phys.*, **20** (1941), 224–30.

[125] HOHN, F. E., and F. MODIGLIANI, "Production planning over time and the nature of the expectation and planning horizon," *Econometrica*, **23** (1955), 46–66.

[126] HOUTHAKKER, H. S., "Revealed preference and the utility function," *Economica*, **17** (1950), 159–74.

[127] HOUTHAKKER, H. S., "La forme des courbes d'Engel," *Cahiers du Séminaire d'Économétrie*, (1953), 59–66.

[128] HURWICZ, L., "Programming in linear spaces," Chap. 4 of [12].

[129] JACOBS, W. W., "The caterer problem," *Naval Research Logistics Quarterly*, **1** (1954), 154–65.

[130] KAKUTANI, S., "A generalization of Brouwer's fixed point theorem," *Duke Math. Journal*, **8** (1941), 457–58.

[131] KANTOROVITCH, L., "On the transformation of masses," *Dokl. Akad. Nauk USSR*, **37** (1942), 199–201.

[132] KAPLANSKY, I., "A contribution to von Neumann's theory of games," *Ann. Math.*, **46** (1945), 474–79.

[133] KARLIN, S., "Operator treatment of minmax principle," in [161], pp. 133–54.

[134] KARLIN, S., "Reduction of certain classes of games to integral equations," in [162], pp. 125–58.

[135] KARLIN, S., "The theory of infinite games," *Ann. Math.*, **58** (1953), 371–401.

[136] KARLIN, S., "Polya type distributions, II," *Ann. Math. Stat.*, **28** (1957), 281–308.

[137] KARLIN, S., "On games described by bell-shaped kernels," in [72], pp. 365–91.

[138] KARLIN, S., and R. RESTREPO, "Multistage poker models," in [72], pp. 337–63.

[139] KARLIN, S., "Positive Operators," *Journal of Math. and Mech.*, (1960).

[140] KARLIN, S., and L. S. SHAPLEY, "Some applications of a theorem on convex functions," *Ann. Math.*, **52** (1950), 148–53.

[141] KARLIN, S., and L. S. SHAPLEY, "Geometry of Moment Spaces," *Memoirs of the American Mathematics Society*, **12** (1953), 1–93.

[142] KEMENY, J. G., O. MORGENSTERN, and G. L. THOMPSON, "A generalization of the von Neumann model of an expanding economy," *Econometrica*, **24** (1956), 115–35.

[143] KIEFER, J., "Invariance, minimax sequential estimation and continuous time processes," *Ann. Math. Stat.*, **28** (1957), 573–601.

[144] KNIGHT, F. H., "A note on Professor Clark's illustration of marginal productivity," *Journal of Political Economics*, **33** (1925), 550–53.

[145] KOOPMAN, B. O., "The theory of search, I: kinematic bases," *Operations Research*, **4** (1956), 324–46.

[146] KOOPMANS, T. C. (ed.), *Activity Analysis of Production and Allocation*, Cowles Commission Monograph No. 13, New York: Wiley, 1951.

[147] KOOPMANS, T. C., *Three Essays on the State of Economic Analysis*, New York: McGraw-Hill, 1957.

[148] KOOPMANS, T. C., "Alternative proof of the substitution theorem for Leontief models in the case of three industries," Chap. 8 of [146].

[149] KOOPMANS, T. C., "Water storage policy in a simplified hydroelectric system," in *Proc. International Conference on Operations Research*, Bristol, England: John Wright and Sons, 1958.

[150] KOOPMANS, T. C., and S. REITER, "A model of transportation," Chap. 14 of [146].

[151] KOTELYANSKII, D. M., "On some properties of matrices with positive elements," (in Russian), *Mat. Sbornik* (*N. S.*), **31** (1952), 497–506.

[152] KOTELYANSKII, D. M., "On a property of sign-symmetric matrices," *Uspehi Matem. Nauk* (*N. S.*), **8** (1953), 163–67.

[153] KREIN, M. G., and M. A. RUTMAN, "Linear operators leaving invariant a

cone in a Banach space," *Uspehi Matem. Nauk* (*N. S.*), **3** (1948), 3–95 (Amer. Math. Soc. Translation No. 26, New York, 1950).

[154] KRETCHMER, K. S., "Linear programming in locally convex spaces and its use in analysis," Ph.D. thesis, Carnegie Inst. of Tech.

[155] KUDO, H., "On minimax invariant estimates of the transformation parameter," *National Science Report of the Ochanomizu University*, **6** (1955), 31–73.

[156] KUHN, H. W., "A simplified two-person poker," in [161], pp. 97–103.

[157] KUHN, H. W., "A combinatorial algorithm for the assignment problem," Issue 11, Logistics Papers, George Washington University Logistics Research Project, 1954.

[158] KUHN, H. W., "The Hungarian method for the assignment problem," *Naval Research Logistics Quarterly*, **2** (1955), 83–97.

[159] KUHN, H. W., "On a theorem of Wald," in [163], pp. 265–73.

[160] KUHN, H. W., "Variants of the Hungarian method for assignment problems," *Naval Research Logistics Quarterly*, **3** (1956), 253–58.

[161] KUHN, H. W., and A. W. TUCKER (eds.), *Contributions to the Theory of Games, I*, (*Ann. Math. Studies*, Vol. 24), Princeton, N. J.: Princeton University Press, 1950.

[162] KUHN, H. W., and A. W. TUCKER (eds.), *Contributions to the Theory of Games, II*, (*Ann. Math. Studies*, Vol. 28), Princeton, N. J.: Princeton University Press, 1953.

[163] KUHN, H. W., and A. W. TUCKER (eds.), *Linear Inequalities and Related Systems*, (*Ann. Math. Studies*, Vol. 38), Princeton, N. J.: Princeton University Press, 1956.

[164] KUHN, H. W., and A. W. TUCKER, "Non-linear programming," in [200], pp. 481–92.

[165] LEMKE, C. E., "The dual method of solving the linear programming problem," *Naval Research Logistics Quarterly*, **1** (1954), 36–47.

[166] LEONTIEF, W. W., *The Structure of the American Economy*, Cambridge, Mass: Harvard University Press, 1941.

[167] *Linear Programming*, Vols. I and II, Second Symposium, National Bureau of Standards, U. S. Department of Commerce, January 27–29, 1955.

[168] LOOMIS, L. H., "On a theorem of von Neumann," *Proc. Nat. Acad. Sci. U. S. A.*, **32** (1946), 213–15.

[169] LUCE, R. D., and H. Raiffa, *Games and Decisions*, New York: Wiley, 1957.

[170] LUCE, R. D., and A. W. TUCKER (eds.), *Contributions to the Theory of Games, IV* (*Ann. Math. Studies*, Vol. 40), Princeton, N. J.: Princeton University Press, 1959.

[171] MANNE, A. S., *Scheduling of Petroleum Refinery Operations*, Cambridge, Mass.: Harvard University Press, 1956.

[172] MARKOWITZ, H. M., "Portfolio selection," *Journal of Finance*, **7** (1952), 77–91.

[173] MARKOWITZ, H. M., "The optimization of a quadratic function subject to linear constraints," *Naval Research Logistics Quarterly*, **3** (1956), 111–13.

[174] MARKOWITZ, H. M., and A. S. MANNE, "On the solution of discrete programming problems," *Econometrica*, **25** (1957), 84–110.

[175] McCLOSKEY, J. F., and F. N. TREFETHEN, *Operations Research for Management*, Baltimore: Johns Hopkins University Press, 1954.

[176] McDONALD, J., *Strategy in Poker, Business and War*, New York: Norton, 1950.

[177] McKENZIE, L. W., "Competitive equilibrium with dependent consumer preferences," in [167], 277–94.

[178] McKENZIE, L. W., "Demand theory without a utility index," *Review of Economic Studies*, **24** (1956–57), 185–89.

[179] McKINSEY, J. C. C., *Introduction to the Theory of Games*, New York: McGraw-Hill, 1952.

[180] METZLER, L. A., "Stability of multiple markets: the Hicks conditions," *Econometrica*, **13** (1945), 277–92.

[181] MILLS, H. D., "Marginal values of matrix games and linear programs," in [163], pp. 183–93.

[182] MORGENSTERN, O. (ed.), *Economic Activity Analysis*, New York: Wiley, 1954.

[183] MORISHIMA, M., "On the laws of change of the price system in an economy which contains complementary commodities," *Osaka Economics Papers*, **1** (1952), 101–13.

[184] MORISHIMA, M., "Prices, interest and profits in a dynamic Leontief system," *Econometrica*, **26** (1958), 358–80.

[185] MOSAK, J. L., *General Equilibrium Theory in International Trade*, Bloomington, Indiana: The Principia Press, 1944.

[186] MOTZKIN, T. S., *Beiträge zur Theorie der linearen Ungleichungen*, [Inaugural Dissertation] Jerusalem, 1936.

[187] MOTZKIN, T. S., H. RAIFFA, G. L. THOMPSON, and R. M. THRALL, "The double description method," in [162], pp. 51–73.

[188] NASH, J. F., "Equilibrium points in N-person games," *Proc. Nat. Acad. Sci., U. S. A.*, **36** (1950), 48–49.

[189] NASH, J. F., and L. S. SHAPLEY, "A simple three-person poker game," in [161], pp. 105–16.

[190] NEYMAN, J., and E. PEARSON, "On the problem of the most efficient tests of statistical hypotheses," *Philosophical Transactions of the Royal Society*, **A-231** (1933), 289–337.

[191] NIKAIDO, H., "On a minimax theorem and its applications to functional analysis," *Journal of the Mathematical Society of Japan*, **5** (1953), 86–94.

[192] NIKAIDO, H., "Note on the general economic equilibrium for non-linear production functions," *Econometrica*, **22** (1954), 49–53.

[193] NIKAIDO, H., "On von Neumann's minimax theorem," *Pac. J. Math.*, **4** (1954), 65–72.

[194] PEISAKOFF, M. P., "Transformation parameters," Ph.D. thesis, Princeton University, 1950.

[195] PERLIS, S., *Theory of Matrices*, Cambridge, Mass.: Addison-Wesley, 1952.

[196] PITMAN, E. J. G., "The estimation of location and scale parameters of a continuous population of any given form," *Biometrika*, **30** (1939), 391–421.

[197] PITMAN, E. J. G., "Tests of hypotheses concerning location and scale parameters," *Biometrika*, **31** (1939), 200–215.

[198] Prager, W., "On the role of congestion in transportation problems," *Zeit. für angewandte Math. Mech.*, **35** (1955), 381–96.

[199] Prager, W., "On the caterer problem," *Management Science*, **3** (1956), 15–23.

[200] *Proceedings of the Second Berkeley Symposium on Mathematical Statistics and Probability*, Berkeley and Los Angeles: University of California Press, 1951.

[201] *Proceedings of the Third Berkeley Symposium on Mathematical Statistics and Probability*, Berkeley and Los Angeles: University of California Press, 1955. 5 vols.

[202] Rademacher, H., and I. J. Schoenberg, "Helly's theorems on convex domains and Tchebycheff's approximation problem," *Canadian Math.*, **2** (1950), 245–56.

[203] Reinfeld, N. V., and W. R. Vogel, *Mathematical Programming*, Englewood Cliffs, N. J.: Prentice-Hall, 1958.

[204] Reiter, S., "Efficiency and prices in the theory of an international economy," Technical Report No. 13, Stanford University, 1954.

[205] Restrepo, R., "Tactical problems involving several actions," in [72], pp. 313–35.

[206] Restrepo, R., "Games with a random move," unpublished.

[207] Riley, V., and S. I. Gass, *A Bibliography of Linear Programming and Associated Techniques*, Baltimore: Johns Hopkins University Press, 1958.

[208] Robinson, J., "An iterative method of solving a game," *Ann. Math.*, **54** (1951), 296–301.

[209] Rudin, W., *Principles of Mathematical Analysis*, New York: McGraw-Hill, 1953.

[210] Samuelson, P. A., *Foundations of Economic Analysis*, Cambridge, Mass.: Harvard University Press, 1955.

[211] Samuelson, P. A., "A note on the pure theory of consumers' behavior," *Economica*, **18** (1938), 61–71, 353–54.

[212] Samuelson, P. A., "Frank Knight's theorem in linear programming," The RAND Corporation, D-782, 1950 (hectographed).

[213] Samuelson, P. A., "Abstract of a theorem concerning substitutability in open Leontief models," Chap. 7 of [146].

[214] Savage, I. R., "Cycling," *Naval Research Logistics Quarterly*, **3** (1956), 163–75.

[215] Schoenberg, I. J., "On smoothing operations and their generating functions," *Bull. Amer. Math. Soc.*, **59** (1953), 199–230.

[216] Shapley, L. S., *The Theory of n-Person Games*, unpublished notes.

[217] Shapley, L. S., "The silent duel, one bullet *vs.* two, equal accuracy," The RAND Corporation, RM-445, 1950.

[218] Shapley, L. S., "An example of an infinite non-constant-sum game," The RAND Corporation, RM-898, 1952.

[219] Shapley, L. S., and R. N. Snow, "Basic solutions of discrete games," in [161], pp. 27–37.

[220] Sherman, S., "Games and subgames," *Proc. Amer. Math. Soc.*, **2** (1951), 186–87.

[221] SHIFFMAN, M., "On the equality of min max = max min, and the theory of games," The RAND Corporation, RM-243, 1941.

[222] SHIFFMAN, M., "Games of timing," in [162], pp. 97–123.

[223] SION, M., and P. WOLFE, "On a game without a value," in [72], pp. 299–305.

[224] SION, M., "On general minimax theorems," *Pac. J. Math.*, **8** (1958), 171–76.

[225] SLUTSKY, E., "Sulla teoria del bilancio del consumatore," *Giornale degli Economisti*, **51** (1915), 19–23.

[226] SMITH, W. E., "Various optimizers for single-stage production," *Naval Research Logistics Quarterly*, **3** (1956), 59–66.

[227] SOLOW, R. M., "On the structure of linear models," *Econometrica*, **20** (1952), 29–46.

[228] SOLOW, R. M., and P. A. SAMUELSON, "Balanced growth under constant returns to scales," *Econometrica*, **21** (1953), 412–24.

[229] STIGLER, G. J., "The cost of substance," *Journal of Farm Economics*, **27** (1945), 303–14.

[230] SUITS, D. B., "Dynamic growth under diminishing returns to scales," *Econometrica*, **22** (1954), 496–501.

[231] THOMPSON, G. L., "On the solution of a game-theoretic problem," in [163], pp. 275–84.

[232] UZAWA, H., "On the logical relation between preference and revealed preference," Technical Report No. 38, Department of Economics, Stanford University, 1956.

[233] UZAWA, H., "A note on the stability of equilibrium," Technical Report No. 44, Department of Economics, Stanford University, 1957.

[234] UZAWA, H., "A note on the Menger-Wieser theory of imputation," Technical Report No. 45, Department of Economics, Stanford University, 1957.

[235] UZAWA, H., "The gradient method for concave programming, II. Global results," Chap. 7 of [12].

[236] VAJDA, S., *Theory of Games and Linear Programming*, New York: Wiley, 1956.

[237] VILLE, J., "Note sur la théorie générale des jeux où intervient l'habilité des jouers," in "Applications aux jeux de hasard," by E. Borel and J. Ville, Tome IV, Fascicule II, in Borel [1938].

[238] VON NEUMANN, J., "Zur Theorie der Gesellschaftsspiele," *Mathematische Annalen*, **100** (1928), 295–320.

[239] VON NEUMANN, J., "Über ein ökonomisches Gleichungssystem und eine Verallgemeinerung des Brouwerschen Fixpunktsatzes," *Ergebnisse eines Mathematischen Kolloquiums*, **8** (1937), 73–83.

[240] VON NEUMANN, J., "A certain zero-sum two-person game equivalent to the optimum assignment problem," in [162], pp. 5–12.

[241] VON NEUMANN, J., "A numerical method to determine optimum strategy," *Naval Research Logistics Quarterly*, **1** (1954), 109–15.

[242] VON NEUMANN, J., and O. MORGENSTERN, *Theory of Games and Economic Behavior*, 2d ed., Princeton, N. J.: Princeton University Press, 1947 (1st ed., 1944).

[243] WAGNER, H., "A practical guide to the dual theorem," *Operations Research*, **6** (1958), 364–84.

[244] WALD, A., *Sequential Analysis*, New York: Wiley, 1947.

[245] WALD, A., *Statistical Decision Functions*, New York: Wiley, 1950.

[246] WALD, A., "Über die eindeutige positive Lösbarkeit der neuen Produktionsgleichungen," *Ergebnisse eines mathematischen Kolloquiums*, **6** (1935), 12–20.

[247] WALD, A., "Über die Produktionsgleichungen der ökonomischen Wertlehre, II," *Ergebnisse eines mathematischen Kolloquiums*, **7** (1936), 1–6.

[248] WALD, A., "Statistical decision functions which minimize the maximum risk," *Ann. Math.*, **46** (1945), 265–80.

[249] WALD, A., "Statistical decision theory," *Ann. Math. Stat.*, **20** (1949), 165–205.

[250] WALD, A., "Note on zero-sum two-person games," *Ann. Math.*, **52** (1950), 739–42.

[251] WALD, A., "On some systems of equations of mathematical economics," *Econometrica*, **19** (1951), 368–403.

[252] WALRAS, L., *Elements of Pure Economics, or the Theory of Social Wealth*, trans. W. Jaffe, Homewood, Ill.: Irwin, 1954.

[253] WEYL, H., "The elementary theory of convex polyhedra," in [161], pp. 3–18.

[254] WILLIAMS, J. D., *The Compleat Strategyst*, New York: McGraw-Hill, 1954.

[255] WOLFE, P., "Determinateness of polyhedral games," in [163], pp. 195–98.

[256] WOLFE, P., "The simplex method for quadratic programming," The RAND Corporation, P-1295, 1957.

[257] WOOD, M. K., and G. B. DANTZIG, "Programming of interdependent activities, I. General discussion," *Econometrica*, **17** (1949), 193–99. (Reprinted in revised form as Chap. 1 of [146].)

INDEX

A CATALOG OF SELECTED

DOVER BOOKS

IN SCIENCE AND MATHEMATICS

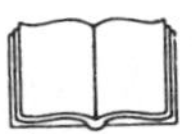